MW00844788

HANDBOOK OF LC-MS BIOANALYSIS

HANDBOOK OF LC-MS BIOANALYSIS

Best Practices, Experimental Protocols, and Regulations

Edited by

WENKUI LI, PhD
Novartis Institutes for BioMedical Research,
East Hanover, NJ, USA

JIE ZHANG, PhD
Novartis Institutes for BioMedical Research,
East Hanover, NJ, USA

FRANCIS L.S. TSE, PhD
Novartis Institutes for BioMedical Research,
East Hanover, NJ, USA

Published by John Wiley & Sons, Inc., Hoboken, New Jersey.
Published simultaneously in Canada.

Library of Congress Cataloging-in-Publication Data:

Handbook of LC-MS bioanalysis : best practices, experimental protocols, and regulations / edited by Wenkui Li, Ph.D., Novartis Institutes for BioMedical Research, Jie Zhang, Ph.D., Novartis Institutes for BioMedical Research, Francis L.S. Tse, Ph.D., Novartis Institutes for BioMedical Research.

pages cm

Includes bibliographical references and index.

ISBN 978-1-118-15924-8 (cloth)

1. Drugs–Spectra–Handbooks, manuals, etc. 2. Drugs–Analysis–Handbooks, manuals, etc. 3. Mass spectrometry–Handbooks, manuals, etc. 4. Liquid chromatography–Handbooks, manuals, etc. I. Li, Wenkui, 1964- editor of compilation. II. Zhang, Jie, 1962- editor of compilation. III. Tse, Francis L. S., editor of compilation. IV. Title: Handbook of liquid chromatography-mass spectrometry bioanalysis.

RS189.5.S65H36 2013

615.1′901–dc23

2013004581

ISBN: 9781118159248

10 9 8 7 6 5 4 3 2

CONTENTS

PREFACE

Bioanalysis is the most heavily regulated area within the discipline of drug metabolism and pharmacokinetics, which supports a large sector of drug development. The health authorities have very specific requirements regarding quality and integrity of bioanalytical results, and different customers usually have additional expectations on the performance of bioanalytical assays.

Much has happened in recent years toward faster, cheaper, and better ways of providing quality bioanalytical results. Both the European Medicines Agency (EMA) and the US Food and Drug Administration (FDA) have renewed or are renewing their guidances on bioanalytical method validation (e.g., 21July2011/EMEA/CHMP/EWP/192217/2009) with the ultimate goal of improving the quality of bioanalytical results. Novel approaches to bioanalytical method development as well as advents of new liquid chromatographic (LC) techniques and mass spectrometric (MS) instruments have been reported. Various automatic laboratory procedures, electronic laboratory notebooks, and data management systems are now available. All of these culminate in a remarkable improvement in the quality, speed, and cost-effectiveness of bioanalytical work and contribute to the value we deliver to the patient.

Given the rapid changes within the field of bioanalysis and also in the larger area of drug development in which we operate, it is timely to conduct a broad overview of our discipline. This book is the first comprehensive handbook for LC-MS bioanalysis and provides an update on all important aspects of LC-MS bioanalysis of both small molecules and macromolecules. It not only addresses the needs of the bioanalytical scientists on the pivotal projects but also features perspectives on some advanced and emerging technologies including high-resolution mass spectrometry and dried blood spot (DBS) microsampling.

The 51 chapters of the book are divided into four parts. Part I provides a comprehensive overview on the role of LC-MS bioanalysis in drug discovery and development and therapeutic drug monitoring (Chapter 1), the key elements of a regulated bioanalytical laboratory (Chapter 2), and the current international regulations and quality standards of bioanalysis (Chapter 3).

In Part II, the global regulations and quality standards related to LC-MS bioanalysis are reviewed and compared. Chapter 4 highlights the current regulations governing bioanalytical method validations from a number of countries and regions including Brazil (ANVISA), Canada, China, the European Union (EMA), India, Japan, and the United States (FDA). This is followed by two in-depth reviews on the topics of assay reproducibility (Chapter 5) and method transfer (Chapter 6). Chapter 7 presents the current practices and regulatory requirements on Metabolites in Safety Testing (MIST). The guidances of regulatory bodies worldwide on bioanalysis for bioequivalence (BE)/bioavailability (BA) studies are compared in Chapter 8, whereas Chapter 9 concerns the specific topic of good laboratory practice (GLP) and its interpretation and application by different agencies, countries, and regions. Of special interest is the rapid evolvement of regulations on bioanalytical data management, which is discussed in Chapter 10. Chapter 11 concludes Part II by giving a detailed analysis of regulatory inspections including health authority expectation, reported inspectional trends, citations, and regulatory followup letters. Recent FDA 483 observations as well as other "hot topics" in bioanalysis compliance that have raised concerns about data integrity are reviewed. Applicable best practices in LC-MS bioanalysis are portrayed in Part III. From this section of the book, the reader will find sound scientific rationale and helpful practical instructions on the assessment of whole blood stability and blood/plasma distribution (Chapter 12), and on biological sample collection, processing and storage (Chapter 13). Chapter 14 introduces various sample preparation techniques for LC-MS bioanalysis, while the best practices in LC separation and MS detection are discussed in Chapters 15 and 16, respectively. A good bioanalytical method must be sensitive, specific, selective, reproducible, high-throughput, and fundamentally robust. Many factors that can contribute to the success of an assay are reviewed including the choice of internal standard (Chapter 17), evaluation of system suitability (Chapter 18), sensitivity enhancement via derivatization of analyte(s) of interest (Chapter 19), evaluation and elimination of matrix effect (Chapter 20), evaluation and elimination of carryover and/or contamination (Chapter 21), and robotic automation (Chapter 22). Chapters 23–29 describe the bioanalysis of drugs, biomarkers, and other analytes of interest in various body fluids and tissues, and Chapter 30 is devoted to DBS sampling and related bioanalytical issues. Chapter 31 offers some useful strategies for enhancing MS detection, and Chapter 32 shows the proper use of statistics as a tool for ensuring adequate method performance in LC-MS bioanalysis. The simultaneous quantitative and qualitative LC-MS bioanalysis of drugs and metabolites are discussed in Chapter 33.

Part IV aims to provide detailed instructions with representative experimental protocols for the LC-MS bioanalysis of various types of drug molecules commonly encountered in the bioanalytical laboratory today (Chapters 34–49). Chapter 50 describes a typical procedure using microflow LC-MS for the quantitative analysis of drugs in support of microsampling. Finally, a protocol on the quantification of endogenous analytes in biofluids without a true blank matrix is given in Chapter 51.

Our purpose in committing to this project was to provide scientists in industry, academia, and regulatory agencies with not only all the "important points to consider" but also all "practical tricks to implement" in LC-MS bioanalysis of various molecules according to current health authority regulations and industry practices. In this book, we are confident that we have accomplished our goal. The book represents a major undertaking, which would not have been possible without the contributions of all the authors and the patience of their families. We also thank the terrific editorial staff at John Wiley & Sons and give a special acknowledgment to Michael Leventhal, Associate Editor, and Robert Esposito, Associate Publisher, at John Wiley & Sons for their premier support of this project.

WENKUI LI, PhD
JIE ZHANG, PhD
FRANCIS L.S. TSE, PhD

CONTRIBUTORS

Arnold, Mark E., Bioanalytical Sciences, Bristol-Myers Squibb Co., Princeton, NJ, USA

Aubry, Anne-Françoise, Bioanalytical Sciences, Bristol-Myers Squibb Co., Princeton, NJ, USA

Awaiye, Kayode, Bioanalytical, BioPharma Services Inc., Toronto, ON, Canada

Bansal, Surendra K., Bioanalytical Research & Development, Non-Clinical Safety, Hoffmann-La Roche Inc., Nutley, NJ, USA

Barrientos-Astigarraga, Rafael E., Magabi Pesquisas Clínicas e Farmacêuticas Ltda, Itaqui Itapevi, Sao Paulo, Brazil

Bartels, Michael J., Toxicology and Environmental Research & Consulting, The Dow Chemical Company, Midland, MI, USA

Bartlett, Michael G., Department of Pharmaceutical & Biomedical Sciences, College of Pharmacy, University of Georgia, Athens, GA, USA

Bennett, Patrick, Thermo Fisher Scientific, San Jose, CA, USA

Briscoe, Chad, US Bioanalysis, PRA International, Lenexa, KS, USA

Bruenner, Bernd A. Bioanalytical Sciences, Pharmacokinetics & Drug Metabolism, Amgen Inc., Thousand Oaks, CA, USA

Carter, Spencer, Tandem Labs, a Labcorp Company, Salt Lake City, UT, USA

Chen, Buyun, Department of Pharmaceutical & Biomedical Sciences, College of Pharmacy, University of Georgia, Athens, GA, USA

Chow, Frank, Lachman Consultant Services Inc., Westbury, NY, USA

Cohen, Lucinda, NJ Discovery Bioanalytical Group, Merck, Rahway, NJ, USA

Cohen, Sabine, Hospices Civils de Lyon, Laboratoire de biochimie-toxicologie, Centre Hospitalier Lyon-Sud, Pierre Bénite, Cedex, France

de Boer, Theo, Analytical Biochemical Laboratory (ABL) B.V., W.A. Scholtenstraat 7, Assen, The Netherlands

Demers, Roger, Tandem Labs, a Labcorp Company, West Trenton, NJ, USA

Duggan, Jeffrey X., Bioanalysis & Metabolic Profiling, DMPK, Boehringer-Ingelheim Pharmaceuticals Inc., Ridgefield, CT, USA

Edom, Richard, Bioanalysis, Janssen Research & Development, Global Development Operations, Raritan, NJ, USA

Evans, Christopher A., PTS-DMPK Bioanalytical Science & Toxicokinetics (BST), GlaxoSmithKline, King of Prussia, PA, USA

Flarakos, Jimmy, Drug Metabolism & Pharmacokinetics, Novartis Institutes for BioMedical Research, East Hanover, NJ, USA

Fu, Yunlin, Drug Metabolism & Pharmacokinetics, Novartis Institutes for BioMedical Research, East Hanover, NJ, USA

Gagnieu, Marie-Claude, Hospices Civils de Lyon, Laboratoire de pharmacologie, Hôpital E. Herriot, Lyon, France

Gao, Hong, Drug Metabolism & Pharmacokinetics, Vertex Pharmaceuticals Inc., Cambridge, MA, USA

Garofolo, Fabio, Bioanalytical, Algorithme Pharma Inc., Laval (Montreal), Quebec, Canada

Guitton, Jérôme, Hospices Civils de Lyon, Laboratoire de ciblage thérapeutique en cancérologie, Centre Hospitalier Lyon-Sud, Pierre Bénite, Cedex, France; Université Claude Bernard Lyon I, Laboratoire de Toxicologie, ISPBL, Lyon, France

Hawthorne, Glen, Department of Bioanalysis, Huntingdon Life Sciences, Alconbury, Cambridgeshire, UK

Hayes, Michael, Drug Metabolism & Pharmacokinetics, Novartis Institutes for BioMedical Research, East Hanover, NJ, USA

Hill, Howard M., Pharmaceutical Development, Huntingdon Life Sciences, Alconbury, Cambridgeshire, UK

Ho, Stacy, Drug Metabolism & Pharmacokinetics, Preclinical Biosciences, Sanofi, Waltham, MA, USA

Hoffman, David, Early Development Biostatistics, Sanofi, Bridgewater, NJ, USA

Huang, Qingtao (Mike), Bioanalysis, Janssen Research & Development, Global Development Operations, Raritan, NJ, USA

James, Christopher A., Bioanalytical Sciences, Pharmacokinetics & Drug Metabolism, Amgen Inc., Thousand Oaks, CA, USA

Ji, Allena J., Clinical Specialty Lab, Genzyme, A Sanofi Company, Framingham, MA, USA

Jian, Wenying, Bioanalysis, Janssen Research & Development, Global Development Operations, Raritan, NJ, USA

Kindt, Erick, Pharmacokinetics, Dynamics & Metabolism, Pfizer Inc., San Diego, CA, USA

Kudoh, Shinobu, Pharmaceuticals & Life-sciences Division, Shimadzu Techno-Research, Inc., Nakagyo-ku, Kyoto-shi, Kyoto-fu, Japan

Lachman, Leon, Lachman Consultant Services, Inc., Westbury, NY, USA

Lefebvre, Isabelle, Institut des Biomolécules Max Mousseron, UMR 5247 CNRS-UM1-UM2, Université Montpellier 2, Montpellier, Cedex, France

Li, Feng (Frank), Alliance Pharma Inc., Malvern, PA, USA

Li, Hongyan, Bioanalytical Sciences, Pharmacokinetics & Drug Metabolism, Amgen Inc, Thousand Oaks, CA, USA

Li, Wenkui, Drug Metabolism & Pharmacokinetics, Novartis Institutes for BioMedical Research, East Hanover, NJ, USA

Li, Wenlin (Wendy), Pharmacokinetics, Dynamics & Metabolism, Pfizer Inc., San Diego, CA, USA

Licea-Perez, Hermes, PTS-DMPK Bioanalytical Science & Toxicokinetics (BST), GlaxoSmithKline, King of Prussia, PA, USA

Lin, Zhongping (John), Bioanalytical Services, Frontage Laboratories Inc., Malvern, PA, USA

Liu, Guowen, Bioanalytical Sciences, Bristol-Myers Squibb Co., Princeton, NJ, USA

Love, Iain, Department of Bioanalysis, Huntingdon Life Sciences, Alconbury, Cambridgeshire, UK

Majumdar, Tapan K., Drug Metabolism & Pharmacokinetics, Novartis Institutes for BioMedical Research, East Hanover, NJ, USA

Martinez, Elizabeth M., Department of Medicinal Chemistry & Pharmacognosy, University of Illinois College of Pharmacy, Chicago, IL, USA

McGinnis, A. Cary, Department of Pharmaceutical & Biomedical Sciences, College of Pharmacy, University of Georgia, Athens, GA, USA

Meng, Min, Tandem Labs, a Labcorp Company, Salt Lake City, UT, USA

Miller, Jeffrey D., Product Applications, AB Sciex, Framingham, MA, USA

Moyer, Michael, Bioanalytical Services, Frontage Laboratories Inc., Malvern, PA, USA

Nash, Bradley, Global Quality & Compliance, PPD, Richmond, VA, USA

Ohnmacht, Corey M., US Bioanalysis, PRA International, Lenexa, KS, USA

Patel, Shefali, Bioanalysis, Janssen Research & Development, Global Development Operations, Raritan, NJ, USA

Pawula, Maria, Department of Bioanalysis, Huntingdon Life Sciences, Alconbury, Cambridgeshire, UK

Rahavendran, Sadayappan V., Pharmacokinetics, Dynamics & Metabolism, Pfizer Inc., San Diego, CA, USA

Rajarao, Joe, Global Professional Services, IDBS, Bridgewater, NJ, USA

Ramanathan, Dil, Kean University, New Jersey Center for Science, Technology & Mathematics, Union, NJ, USA

Ramanathan, Ragu, Drug Metabolism and Pharmacokinectics, QPS LLC, Newark, DE, USA

Reuschel, Scott, Tandem Labs, a Labcorp Company, Salt Lake City, UT, USA

Rudewicz, Patrick J., Metabolism & Pharmacokinectics, Novartis Institutes for BioMedical Research, Emeryville, CA, USA

Santa, Tomofumi, Graduate School of Pharmaceutical Sciences, The University of Tokyo, Bunkyo-ku, Tokyo, Japan

Savale, Shrinivas S., Bioevaluation Centre, Torrent Pharmaceuticals Limited, Gandhinagar, Gujarat, India

Shrivastav, Pranav S., Department of Chemistry, School of Sciences, Gujarat University School of Science, Navrangpura, Ahmedabad, Gujarat, India

Singhal, Puran, Bioanalytical, Alkem Laboratories Ltd, Navi Mumbai, Maharashtra, India

Skor, Heather, Applied Proteomics, Inc., San Diego, CA, USA

Smith, Graeme T., Department of Bioanalysis, Huntingdon Life Sciences, Alconbury, Cambridgeshire, UK

Smith, Harold T., Drug Metabolism & Pharmacokinetics, Novartis Institutes for BioMedical Research, East Hanover, NJ, USA

Smith, J. Kirk, Regulatory Affairs & Quality Systems, Smithers, Wareham, MA, USA

Stanczyk, Frank, Department of Obstetrics and Gynecology, Department of Preventive Medicine, University of Southern California Keck School of Medicine, Los Angeles, CA, USA

Tan, Aimin, Bioanalytical, BioPharma Services Inc., Toronto, ON, Canada

Tang, Daniel, ICON Development Solutions APAC, Pudong, Shanghai, China

Timmerman, Philip, Bioanalysis, Janssen Research & Development, Division of Janssen Pharmaceutica N.V., Beerse, Belgium

Tse, Francis L.S., Drug Metabolism & Pharmacokinetics, Novartis Institutes for BioMedical Research, East Hanover, NJ, USA

Tweed, Joseph A., Regulated Bioanalytical, Pharmacokinetics, Dynamics & Metabolism, Pfizer Inc., Groton, CT, USA

Unger, Steve, Bioanalytical Sciences, Worldwide Clinical Trials, Austin, TX, USA

van Amsterdam, Peter, Abbott Healthcare Products BV, 1380 DA WeespC. J. van Houtenlaan 36, The Netherlands.

van Breemen, Richard B., Department of Medicinal Chemistry & Pharmacognosy, University of Illinois College of Pharmacy, Chicago, IL, USA

van de Merbel, Nico, Bioanalytical Laboratories, PRA International, Assen, The Netherlands

Voelker, Troy, Tandem Labs, a Labcorp Company, Salt Lake City, UT, USA

Wang, Laixin, Tandem Labs, a Labcorp Company, Salt Lake City, UT, USA

Weng, Naidong, Bioanalysis, Janssen Research & Development, Global Development Operations, Raritan, NJ, USA

Wieling, Jaap, QPS Netherlands BV, Groningen, The Netherlands

Williams, John, Vertex Pharmaceuticals Inc., Cambridge, MA, USA

Xia, Yuan-Qing, Product Applications, AB Sciex, Framingham, MA, USA

Yadav, Manish S., Clinical Research & Bioanalysis, Alkem Laboratories Ltd, Navi Mumbai, Maharashtra, India

Yang, Yi (Eric), PTS-DMPK Bioanalytical Science & Toxicokinetics, GlaxoSmithKline, King of Prussia, PA, USA

Yang, Ziping, Drug Metabolism & Pharmacokinetics, Novartis Institutes for BioMedical Research, East Hanover, NJ, USA

Yau, Martin, Lachman Consultant Services, Inc., Westbury, NY, USA

Yuan, Weiwei, QPS LLC, Newark, DE, USA

Zeng, Jianing, Bioanalytical Sciences, Bristol-Myers Squibb Co., Princeton, NJ, USA

Zhang, Duxi, Department of Bioanalysis, WuXi AppTec (Suzhou) Co., Ltd., Suzhou, Jiangsu, China

Zhang, Fagen, Toxicology & Environmental Research & Consulting, The Dow Chemical Company, Midland, MI, USA

Zhang, Jie, Drug Metabolism & Pharmacokinetics, Novartis Institutes for BioMedical Research, East Hanover, NJ, USA

Zhou, Jin, Drug Metabolism & Pharmacokinetics, Boehringer-Ingelheim Pharmaceuticals Inc., Ridgefield, CT, USA

ABBREVIATIONS

AAPS	American association of pharmaceutical scientists
AAS	Atomic absorption spectroscopy
AC	Absolute carryover
Ach	Acetylcholine
ACUP	Animal care and use protocol
ADC	Antibody–drug conjugate
ADME	Absorption, distribution, metabolism, and excretion
AFA	Adaptive focused acoustics
ALQ/AULOQ	Above the upper limit of quantification
ANDA	Abbreviated new drug application
ANVISA	National health surveillance agency (in Portuguese, Agência Nacional de Vigilância Sanitária)
AP	Analytical procedure
APCI	Atmospheric pressure chemical ionization
API	Atmospheric pressure ionization
APPI	Atmospheric pressure photoionization
ASE	Accelerated solvent extraction
ASEAN	Association of southeast asian nations
AUC	Area under the curve
BA	Bioavailability
BDMA	Butyldimethylamine
BE	Bioequivalence
BIMO	Bioresearch monitoring
BLQ/BLLOQ	Below the lower limit of quantification
BMAB	Butylmethylamine bicarbonate
BNPP	Bis(4-nitrophenyl)-phosphate
BSA	Bovine serum albumin
CAD	Charged aerosol detection
CAD	Collision-associated disassociation
CAPA	Corrective and preventive action
CD	Compact disc
CDC	Centers for disease control and prevention
CDSCO	Central drugs standard control organization
CE	Collision energy
CFR	Code of federal regulations
CID	Collision-induced dissociation
CL	Clearance
CNS	Central nervous system
CoA	Certificate of analysis
COV	Compensation voltage
CPGM	Compliance program guidance manual
CR	Concentration ratio
CRO	Contract research organization
Cs	Calibration standard
CSF	Cerebrospinal fluid
CSI	Captive spray ionization
CV	Coefficient of variation
CXP	Collision exit potential
CZE	Capillary zone electrophoresis
DBS	Dried blood spot
DDI	Drug–drug interaction
DDTC	Diethyldithiocarbamate
DDVP	2,2-Dichlorovinyl dimethyl phosphate
DHEA	Dehydroepiandrosterone
DHT	Dihydrotestosterone
DIFP/DFP	Diisopropyl fluorophosphate
DM	Drug metabolism
DMS	Differential mobility spectrometry
DMS	Dried matrix spot
DNA	Deoxyribonucleic acid
Dns-Cl	Dansyl chloride
Dns-Hz	Dansyl hydrazine
DP	Declustering potential
DPS	Dried plasma spot
DPX	Disposable pipette extraction

DQ	Design qualification
DTNB	5, 5'-Dithiobis-(2-nitrobenzoic acid)
DTT	Dithiothreitol
EB	Endogenous baseline
EBF	European bioanalysis forum
EDMS	Electronic data management system
EDTA	Ethylenediaminetetraacetic acid
EFPIA	European federation of pharmaceutical industries associations
EHNA	Erythro-9-(2-hydroxy-3-nonyl) adenine
ELN	Electronic laboratory notebook
ELSD	Evaporative light scattering detection
EMA	European medicines agency
EP	Entrance potential
EPA	Environmental protection agency
ESI	Electrospray ionization
FAIMS	Field-asymmetric waveform ion mobility spectrometry
FDA	Food and drug administration
FIH	First-in-human
FOIA	Freedom of information act
FP	Focusing potential
FTICR	Fourier transform ion cyclotron resonance
FWHM	Full width at half maximum
GAMP	Good automated manufacturing practice
GBC	Global bioanalytical consortium
GC-MS	Gas chromatography-mass spectrometry
GCP	Good clinical practice
GLP	Good laboratory practice
GMP	Good manufacturing practice
GPhA	Generic pharmaceutical association
HAA	Hexylammonium acetate
HCT	Hematocrit
HETP	Height equivalent of a theoretical plate
HFIP	Hexafluoroisopropanol
HILIC	Hydrophilic interaction liquid chromatography
HMP	2-Hydrazino-1-methyl-pyridine
HP	2-Hydrazinopyridine
HPFB	Health products and food branch
HPLC	High pressure liquid chromatography or high performance liquid chromatography
HRMS	High resolution mass spectrometry
HSA	Human serum albumin
HTLC	High-turbulence liquid chromatography
IA	Immunoaffinity
IACUC	Institutional animal care and use committee
ICH	International conference on harmonization
ICP-MS	Inductively coupled plasma–mass spectrometry
ID	Inner diameter
IDMS	Isotope dilution mass spectrometry
IEC	Ion-exchange chromatography
IMS	Ion mobility spectrometry
IND	Investigational new drug
IP	Ion-pairing
IQ	Installation qualification
IS/ISTD	Internal standard
ISA	Incurred sample accuracy
ISR	Incurred sample reanalysis or incurred sample reproducibility
ISS	Incurred sample stability
IV	Intravenous
KFDA	Korea food and drug administration
LC-MS	Liquid chromatography–mass spectrometry
LC-MS/MS	Liquid chromatography-tandem mass spectrometry
LIMS	Laboratory information management system
LLE	Liquid–liquid extraction
LLOQ	Lower limit of quantification
LOD	Limit of detection
LUV	Large unilamellar vesicle
MAD	Multiple ascending dose
MAX	Mixed mode anion exchange
MCD	Maximum concentration difference
MCX	Mixed mode cation exchange
MD	Method development
MDF	Mass defect filter
MEPS	Microextraction by packed sorbent
MF	Matrix factor
MFC	Microfluidic flow control
MHFW	Ministry of health and family welfare
MHLW	Ministry of health, labour and welfare
MHRA	Medicines and healthcare products regulatory agency
MIP	Molecularly imprinted polymers
MIST	Metabolites in safety testing
MLV	Multilamellar vesicle
MRM	Multiple reaction monitoring
MS	Mass spectrometry
MTBE	Methyl tert-butyl ether
MV	Method validation
MVS	Multichannel verification system
MWCO	Molecular weight cutoff
NCCLS	National committee for clinical laboratory standards
NCE	New chemical entity
NDA	New drug application
NHS	National health service
NIH	National institutes of health
NL	Neutral loss
NME	New molecular entity
NMR	Nuclear magnetic resonance
NPLC	Normal phase liquid chromatography

NRTI	Nucleoside reverse transcriptase inhibitor
NSB	Nonspecific binding
NSI	Nanospray ionization
OC	Oral contraceptive
OECD	Organization for economic cooperation and development
OEM	Original equipment manufacturer
OOS	Out-of-specification
OQ	Operational qualification
ORA	Office of regulatory affairs
OSI	Office of scientific investigations
PB	Protein binding
PBMC	Peripheral blood mononuclear cell
PBS	Phosphate buffered saline
PCT	Pressure cycling technology
PCV	Packed cell volume
PD	Pharmacodynamics
PDA	Photodiode array
PDF	Portable document format
PEEK	Polyether ether ketone
PEG	Polyethylene glycol
PGC	Porous graphitic carbon
PI	Principal investigator
PK	Pharmacokinetics
PM	Preventive maintenance
PMP	Pressure monitoring pipetting
PMSF	Phenylmethylsulfonyl fluoride
PNPA	p-Nitrophenyl acetate
PoC	Proof of concept
PPE	Protein precipitation extraction
PPT	Protein precipitation
PQ	Performance qualification
PTM	Posttranslational modification
QA	Quality assurance
QAS	Quality assurance statement
QAU	Quality assurance unit
QC	Quality control
QqQ	Triple quadrupole
QqQ_{LIT}	Hybrid triple quadrupole-linear ion trap
QqTOF	Hybrid quadrupole time-of-flight
Q-TOF	Quadrupole time-of-flight
RAD	Radioactivity detection
RAM	Restricted access material
RBC	Red blood cell
RC	Relative carryover
RE	Recovery
RED	Rapid equilibrium dialysis
RF	Response factor
RFID	Radiofrequency identifier
RIA	Radioimmunoassay
RNA	Ribonucleic acid
RT	Retention time
RT	Room temperature
RT-qPCR	Real-time reverse transcription polymerase chain reaction
SAD	Single ascending dose
SALLE	Salting-out assisted LLE
SAX	Strong anion ion exchange
SBSE	Stir bar sorptive extraction
SCX	Strong cation ion exchange
SD	Standard deviation
SDMS	Scientific data management system
SE	Standard error
SFC	Supercritical fluid chromatography
SFDA	State food and drug administration
SIL-IS	Stable isotope labeled internal standard
SIM	Selected ion monitoring
siRNA	Small interfering RNA
SLE	Supported liquid extraction
S/N	Signal-to-noise
SOP	Standard operating procedure
SPE	Solid phase extraction
SPME	Solid phase microextraction
SRM	Selected reaction monitoring
SSBG	Sex steroid binding globulin
STD	Standard
SUV	Small unilamellar vesicle
SV	Separation voltage
TADM	Total aspirate and dispense monitoring
TDM	Therapeutic drug monitoring
TEA	Triethylamine
TEAA	Triethylammonium acetate
TEAB	Triethylammonium bicarbonate
TFA	Trifluoroacetic acid
TGA	Therapeutic goods administration
THU	Tetrahydrouridine
TIC	Total ion chromatogram
TK	Toxicokinetics
TMA	Trimethyl ammonium
TPD	Therapeutic products directorate
TSCA	Toxic substance control act
TTFA	Thenoyltrifluoroacetone
UHPLC	Ultra high performance liquid chromatography
ULOQ	Upper limit of quantification
UPLC	Ultra Performance liquid chromatography
URS	User requirement specification
UV	Ultraviolet
WAX	Weak anion exchange
WBC	White blood cell
WHO	World health organization

PART I

OVERVIEW OF LC-MS BIOANALYSIS

1

ROLES OF LC-MS BIOANALYSIS IN DRUG DISCOVERY, DEVELOPMENT, AND THERAPEUTIC DRUG MONITORING

STEVE UNGER, WENKUI LI, JIMMY FLARAKOS, AND FRANCIS L.S. TSE

1.1 INTRODUCTION

Bioanalysis is a subdiscipline of analytical chemistry for the quantitative measurement of xenobiotics (chemically synthesized or naturally extracted drug candidates and genetically produced biological molecules and their metabolites or post-translationally modified products) and biotics (macromolecules, proteins, DNA, large molecule drugs, metabolites) in biological systems. Many scientific decisions regarding drug development are dependent upon the accurate quantification of drugs and endogenous components in biological samples. Unlike its sister subdisciplines of analytical chemistry such as drug substance and drug product analysis, one very unique feature of contemporary bioanalysis is that its measurement target is always at very low concentration levels, typically at low ng/ml concentration range and even at pg/ml for highly potent medicines. It is this very low concentration, compounded by coexisting endogenous or exogenous compounds with similar chemical structures to the target analytes at a much higher concentration (typically at μg/ml to mg/ml range), that challenges bioanalytical scientists to accurately and definitively measure the analytes of interest.

Since its commercial introduction in the 1980s, liquid chromatography–mass spectrometry (LC-MS), or much more predominantly, tandem mass spectrometry (LC-MS/MS) has rapidly become standard instrumentation in any well-equipped bioanalytical laboratory. LC-MS is a combination of the physicochemical separation capabilities of liquid chromatography (LC) and the mass (MS or MS/MS) separation/detection capabilities of mass spectrometry. In LC-MS bioanalysis, assay selectivity can be readily achieved by three stages of separation of the analyte(s) of interest from unwanted components in the biological matrix: (1) sample extraction (protein precipitation, liquid–liquid extraction, solid-phase extraction, etc.), (2) column chromatography, and (3) tandem mass spectrometric detection in selected reaction monitoring (SRM) or multiple reaction monitoring (MRM) mode. Nevertheless, many factors, including matrix effect, ion suppression, and in-source breakdown of labile metabolites, can compromise the reliability of a LC-MS bioanalytical assay. These factors should be carefully evaluated during method development.

The focus of LC-MS bioanalysis in the pharmaceutical industry is to provide a quantitative measurement of the active drug and/or its metabolite(s) for the accurate assessment of pharmacokinetics, toxicokinetics, bioequivalence (BE), and exposure–response (pharmacokinetics/ pharmacodynamics) relationships (Figure 1.1). The quality of these studies, which are often used to support regulatory filings and other evaluations, is directly related to the conduct of the underlying bioanalysis. Therefore, the application of best practices in bioanalytical method development, validation, and associated sample analysis is key to an effective discovery and development program leading to

Handbook of LC-MS Bioanalysis: Best Practices, Experimental Protocols, and Regulations, First Edition. Edited by Wenkui Li, Jie Zhang, and Francis L.S. Tse.

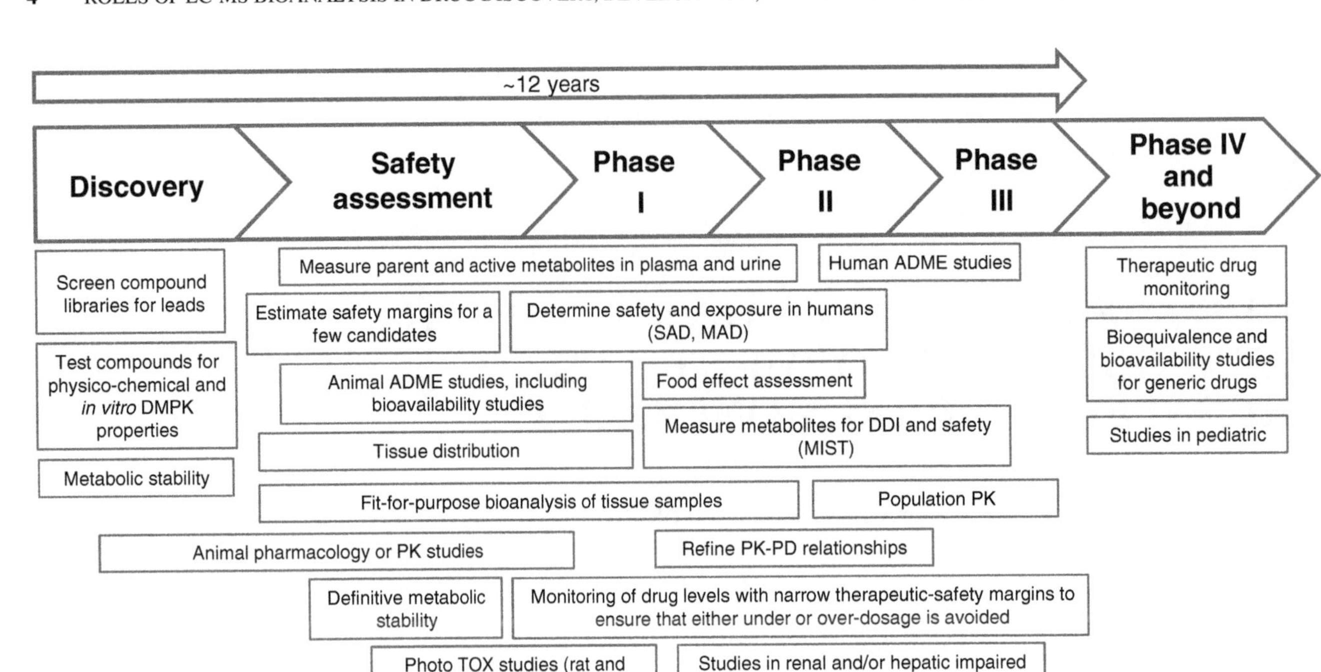

FIGURE 1.1 A flowchart of drug discovery and development, and postapproval studies of drugs where LC-MS bioanalysis plays important roles.

the successful registration and commercialization of a drug product.

1.2 LC-MS BIOANALYSIS IN DRUG DISCOVERY

Before the introduction of combinatorial chemistry, many drug candidates came from natural products where an active compound was isolated and its chemical structure was characterized using NMR, MS, IR, and derivatization or selective degradation chemistry. Screening entailed an assessment of bioactivity and physicochemical data compared to known databases. High-resolution mass spectrometry played a critical role allowing molecular formula searches from accurate mass data. Similarly, spectral databases allowed positive confirmation or class assessments. This process helped to ensure that novel compounds were selected. Since the introduction of combinatorial chemistry 20 years ago, the analyst's role in early drug discovery has shifted to the development of highly efficient LC-MS analytical methods to support quantitative analysis. The drug discovery process begins with compound library development and ends with the selection of preclinical drug candidates for preclinical safety assessment. LC-MS bioanalysis plays an important role throughout this process.

1.2.1 Structure-Activity Relationships from High-Throughput Screening

High-throughput LC-MS assays can be employed for the determination of solubility, membrane permeability or transport, protein binding, and chemical and metabolic stability for a large number of compounds that have been identified as "hits" (Janiszewski et al., 2008). Thousands of compounds per year go through some or all of these screening procedures. The *in vitro* studies validate *in silico* assessments performed prior to synthesis and select compounds for moving forward in development.

1.2.2 Structure—PK-PD Relationships

Selected compounds from high-throughput screening are subsequently evaluated in pharmacology models for efficacy. Provided the targeted biochemistry is applicable to LC-MS analysis, high-throughput screening of potential biomarkers can be performed in pharmacology studies via either a targeted pathway or a metabolomics approach. If successful, discovery biomarkers may be useful in preclinical and clinical studies. Simple examples include steroid biomarkers such as testosterone or dihydrotestosterone for

5-α-reductase inhibitors or estrogen for selective estrogen receptor modulators.

Integration of drug metabolism and pharmacokinetics (DMPK), pharmacology, and biology studies in drug discovery can greatly accelerate an understanding of the pharmacokinetic–pharmacodynamic (PK-PD) relationships of lead compounds. The minimum effective dose observed in the pharmacology model is validated from knowledge of drug and active metabolite levels at the target site and compared with *in vitro* efficacy. Compounds known to have *in vitro* potency but are devoid of *in vivo* activity are suspected of having poor bioavailability (BA) or other DMPK properties (transport to target site, rapid clearance, etc.). Alternatively, compounds with an unanticipated high *in vivo* activity may have superior access to the site of action or form active metabolites.

LC-MS has a fundamental role in the success of many of these discovery studies. An appropriately designed, early *in vitro* study can determine intrinsic clearance in multiple species. *In vitro* assessments have improved our ability to predict systemic clearance using intrinsic clearance. However, predicting volume of distribution and tissue concentrations is far more difficult. Combinatorial approaches such as cassette dosing or coadministration of many compounds is one means of quickly assessing penetration into target sites. Typically, ~20 compounds are coadministered, but as many as 100 have been attempted (Berman et al., 1997). The specificity of MS detection allows one to simultaneously measure numerous compounds in biofluids and tissues and rapidly screen drug candidates for their ability to penetrate into the site of action (Wu et al., 2000).

1.2.3 Candidate Selection

Within a therapeutic program, a limited number of compounds may be investigated in greater detail as possible preclinical drug candidates. These include assessments at various doses in the rodent and nonrodent toxicology species. Defining the systemic and local exposures, refining PK-PD models and exploring dose-proportionality are among the objectives of this phase. Studies with both single and multiple ascending doses may be undertaken in an effort to assess accumulation, induction and toxicity. Whereas a "generic" LC-MS assay may suffice in supporting these non-GLP assessments of drug properties, one needs to be aware of the potential pitfalls, including stability of parent and metabolites and matrix effects from unknown metabolites, endogenous components, and dosing vehicles such as polyethylene glycols, a frequently used formulation for IV dosage.

As a drug candidate progresses further, translational medicine often will define biomarkers from pharmacology or metabolomics studies that can be used in clinical trials. Over the past 15 years, there has been considerable progress in the use of LC-MS to measure small biochemicals and peptides. The ability to use biomarkers as a surrogate endpoint and to ensure a reliable PK-PD relationship is a common strategy for most drug development programs.

1.3 LC-MS BIOANALYSIS IN PRECLINICAL DEVELOPMENT OF DRUGS

1.3.1 Toxicokinetics

Drug safety assessment studies regulated under good laboratory practice (GLP) are an important part of the preclinical development activities. In a typical toxicology study, toxicokinetic evaluation is performed in order to ascertain adequate drug exposure in the study animals. To support bioanalysis of toxicokinetic samples from the GLP studies, generic LC-MS methods used during drug discovery may no longer be suitable. Modification of the generic method or redevelopment of the respective method is often needed, followed by full assay validation according to the current regulatory guidance and industrial practices (EMA, 2011; FDA 2001; Viswanathan et al., 2007). These requirements are implemented to ensure adequate sensitivity, selectivity, accuracy, precision, reproducibility, and a number of other performance related criteria for a given method.

Preclinical toxicity studies typically employ a broad dose range that could result in a wide range of circulating concentrations of the test compound. Test samples containing analyte levels exceeding the upper limit of quantification (ULOQ) need to be diluted, a step that can sometimes introduce errors. On the other hand, the lower limit of quantification (LLOQ) must be established so that the assay is sensitive enough to measure trough levels from the lowest dose, yet not too sensitive that background noise (false positives) in specimens collected from control animals is detected. A useful rule-of-thumb is to set the LLOQ at ca. 5% of the anticipated peak concentration following the low dose, which should allow accurate analyte measurement for approximately four half-lives.

Different strains of rats such as Sprague Dawley, Wistar Hannover, and Fischer are used in toxicology studies. The LC-MS assay method should be validated using the matrix from the same strain. The beagle dog is generally the default nonrodent species. Nonhuman primates, such as cynomolgus, rhesus, or marmoset monkeys, are occasionally used. The most common use of nonhuman primates is when assessing immunogenicity of large molecule drugs or when the metabolic profiles of dogs differ significantly from human. Drug metabolizing enzymes, such as aldehyde oxidase, can have pronounced differences across species. Matching metabolic profiles to human assures good safety coverage for all metabolites. When metabolism differs across species, metabolism-mediated toxicity can result in sensitivity within one species relative to others. For this reason, there may be a need to measure metabolites in GLP preclinical studies. Although metabolite measurement in those toxicokinetic (TK) samples might be exempt from full GLP compliance due to various reasons, for example, absence of purity certification of reference metabolites and lack of full validation of the intended LC-MS assays, care must be taken to ensure the integrity of the results generated. Often, an assay separate from the parent measurement may be set up for the occasional metabolite quantification. New guidance requires that steady-state exposures of significant metabolites in all species are obtained (Anderson et al., 2010). Non-GLP or tiered assays allow these decisions to be made without extensive validation of multiple assays (Viswanathan et al., 2007).

In parallel with clinical drug development is the continued testing of the compound in animal toxicology studies. This includes extending the safety in primary toxicology animals with longer study durations. Dose range-finding studies are conducted in preparation for the 2-year carcinogenicity studies in mouse and rat. Phototoxicity studies are performed in mice. Reproductive toxicology is performed in rats and rabbits. Bioanalytical assays need to be validated in these additional species. Again, metabolites unique to these species need to be considered.

The bioanalyst should be prepared to support LC-MS bioanalysis of tissue samples for certain programs. Extensive validation and stability determinations might be needed, sometimes for both parent drug and metabolites. Having a stable isotope labelled internal standard can help avoid problems such as differences in extraction recovery and compensate for variability due to sample processing, transfer and analysis of study tissue samples. Homogenization prior to freezing is also preferred. Nevertheless, one can never fully ensure consistent analysis from tissue samples since the spiked quality control (QC) samples cannot fully mimic the incurred tissues. The most definitive approach would be to compare tissue results obtained using LC-MS to those from LC analysis in a radiolabeled study.

1.3.2 Preclinical ADME and Tissue Distribution Studies in Animals

Preclinical studies to elucidate the absorption, distribution, metabolism, and excretion (ADME) of drug candidates are usually conducted before and during the clinical phase. Radiolabeled drug is often needed for the animal ADME or tissue distribution (quantitative whole body autoradiography) studies, although with today's LC-MS instrumentation, much information can be gathered without the use of radiolabeled isotopes. Parent drug absorption and elimination can be readily assessed using LC-MS assays. Metabolites can be determined using LC-MS under unit or high resolution conditions. Blood-to-plasma partitioning and protein binding, once done exclusively using radiolabeled drug can now be performed using highly sensitive LC-MS assays. The question of whether radiolabeled mass-balance studies in laboratory animals are still needed today has generated much discussion (Obach et al., 2012; White et al., 2013). The advance in LC-MS technology was the catalyst for this change.

1.4 LC-MS BIOANALYSIS IN CLINICAL DEVELOPMENT OF DRUGS

1.4.1 First-in-Human Studies

Upon successful completion of the preclinical safety assessment of drug candidates, the investigational new drug (IND) submission is prepared. Traditionally, first-in-human (FIH) studies have included separate single and multiple ascending dose (SAD and MAD) studies. Today, adaptive studies can include a combination of SAD and MAD. To ensure safety, a sufficiently low starting dose is selected, and the supporting bioanalytical assay usually requires an LLOQ much lower than that used in toxicology studies. For a drug candidate with a wide safety margin, a bioanalytical method with a similar dynamic range will be needed. While it might be difficult to obtain a full PK profile on the earliest doses of an ascending dose study, a full PK profile will be required when an efficacious dose is reached. In addition to defining the maximum tolerable dose and possibly biological effect, the DMPK objectives in FIH studies include defining drug absorption, dose proportionality, and systemic clearance. Metabolite profiling and measurement will also be conducted to make sure unique human metabolites do not exist and major circulating metabolites at or above 10% of total drug-related exposure at steady state are also present at comparable or greater exposure levels in at least one of the main preclinical toxicology species (FDA, 2008).

A bioanalytical LC-MS method should be developed and validated prior to completion of the study protocol. Important information such as conditions for blood sample collection, plasma harvest, sample storage, and transfer must also be verified. If samples need to be stabilized because of the presence of labile parent or metabolites, the information should be provided well in advance so that the clinical staff can be properly trained in the required sample handling procedures.

The SAD/MAD study may also include an arm to study the food effect (fasted vs. fed) on the BA of the drug. Some drugs bind to food resulting in decreased absorption. In contrast, food can stimulate bile acid secretion that helps to dissolve less soluble drugs, making them more bioavailable. A bioanalytical LC-MS method should, therefore, be evaluated in both normal and lipemic plasma. The assay should be insensitive to changes in phospholipid concentration, a common issue in electrospray ionization that requires attention during method development and validation.

Drug concentrations in urine are also typically measured to assess renal clearance. Unlike plasma, blood or serum, urine does not normally contain significant amounts of proteins and lipids. The lack of proteins and lipids in urine samples can be associated with the issue of nonspecific binding or container surface adsorption of drug molecules, especially those lipophilic and highly protein bound, in quantitative analysis of urine samples. The issue is often evidenced by the unusually low extraction recovery of the analytes of interest and/or nonlinearity of the calibration curves or highly variable QC sample results. Quick identification and effective prevention of analyte loss due to nonspecific binding or container surface adsorption must be conducted by bioanalytical scientists prior to the study so that the correct collection and storage condition can be provided (Li et al., 2010).

1.4.2 Human ADME Studies

Comprehensive information on the ADME of a drug in humans can be obtained from mass-balance studies using a radiolabeled compound, and this should be an early objective in clinical drug development (Pellegatti, 2012). Information on drug tissue distribution in rodents (e.g., rat) and the anticipated therapeutic dose are needed for planning a human ADME study. Some knowledge of the drug metabolism *in vitro* and in animals can help to select the position and desired specific activity of the radiolabel. Quantitative whole body autoradiography is a common tool for tissue distribution studies. Disposition of radioactivity into specific organs is quantified and scaled to human. Dosimetry calculations are performed to ensure safe radioactivity exposure limits in dosing of humans. Typically the maximum exposure limit is 1 mSv (ICRP 103, 2007). Traditional ADME studies generally use liquid scintillation counting and doses of $\sim$100 μCi of ^{14}C labeled drug mixed with unlabeled drug. LC-MS for measuring unlabeled drug is often used in human ADME studies to differentiate the parent compound from its metabolite(s). For studies employing microdoses ($<$100 μg) or doses of low radioactivity ($<$1 μCi), accelerator mass spectrometry may be needed to measure the ^{14}C labeled drug (Garner, 2005), whereas high sensitivity LC-MS methods have been used to determine unlabeled drug concentrations (Balani et al., 2005).

ADME studies, though limited by their single dose nature, do illuminate what is important to measure in toxicology and clinical studies to satisfy Metabolites in Safety Testing requirements (Anderson et al., 2010). Obach et al. (2012) have advocated deferring the cost of this study until after proof of concept (POC) and relying on pharmacokinetic information derived from nonradiolabeled studies, namely SAD and MAD. The risk of delaying the human ADME study is that unique human metabolites may be uncovered after POC. The surprise of having significant metabolites found late in drug development can expose a lack of safety coverage or protection of intellectual property if the metabolite is active. The advancement of more powerful high resolution mass spectrometry for metabolite identification in LC-MS bioanalysis of early stage study samples helps to mitigate the risk.

1.4.3 Human Drug–Drug Interaction Studies

A drug–drug interaction (DDI) is a situation where a drug affects the activity or toxicity of another drug when both are coadministered. Interactions can be found where saturable or inducible enzymes or transporters are expressed and play a role in the absorption and disposition of the drug. DDI can increase or decrease the activity of the drug or a new effect can be produced that neither produces on its own. This interaction can occur between the drug to be developed and other concomitantly administered drugs, foods, or medicinal plants or herbs. During clinical development, DDI studies are normally conducted for the drug candidate in healthy volunteers or patients to confirm any significant observations seen during *in vitro* DDI studies.

From the perspective of LC-MS bioanalysis, assay specificity against the coadministered medicines and their significant metabolites needs to be demonstrated. In the case of metabolites that are difficult to obtain, interference could be discounted based upon MS detection (e.g., differing MW or MRM). On the other hand, possible interference due to drug candidates and/or their major metabolites on the accuracy of determination of DDI compounds and their significant metabolites must be checked to ensure the quality of LC-MS bioanalytical results for the DDI assessment.

1.4.4 Renal Impaired and Hepatic Impaired Studies in Human

Kidney (or renal) failure is a medical condition in which the kidneys fail to adequately filter toxins and waste products from the blood. Similarly, liver (or hepatic) failure is the inability of the liver to perform its synthetic and metabolic function as part of normal physiology. Either can be acute or chronic. Drug elimination may occur by filtration in the kidney or metabolism in the liver. When impacted by disease, drug accumulation can result in toxicity. Depending on the properties of metabolism and excretion of a drug candidate, clinical studies in renal impaired or hepatic impaired patients need to be conducted. In addition to conventional plasma samples, urine samples may be collected and analyzed. Some drugs may be metabolically activated, resulting in idiosyncratic liver toxicities. Therefore, it is important to understand both the impact of an impaired liver on the normal pharmacokinetic properties of a drug as well as the potential of a drug to impact liver function.

From the perspective of LC-MS bioanalysis, assay dynamic range must be suitable to measure exposures from any given dose, or assay integrity of sample dilution must be checked to ensure data integrity for samples with unexpected high analyte concentrations due to the impaired liver or kidney function.

1.4.5 Phase II and Phase III Studies

Moving beyond preliminary safety studies to POC studies is a milestone goal for clinical drug development. A successful program will demonstrate POC before the end of phase II studies. Therefore, moving from healthy subjects to the intended patient population is an important transition. However, patients might take more medications or are under treatment with drug combinations. With this regard, the robustness of the intended LC-MS assay should be validated free from possible interference of combination drugs and their metabolites.

Phase II and III studies are larger and more expensive. In order to support the bioanalysis of a large number of samples from these large multicentered trials, automation is an important consideration. For long-term, multicentered studies, the assay must be rugged enough to ensure storage stability. A well-planned stability assessment of drug candidate and its metabolite(s) of interest is critical as stability must cover all reported results. Any significant assay bias must also be well characterized. The entire bioanalytical work is represented in the new drug application (NDA) submission. This includes tabular and written summaries of assay validation performance of both nonclinical and clinical assays. Given that the development process of a drug may last more than a decade, it is important to maintain institutional knowledge to avoid gaps at filing.

1.4.6 "Fit-for-Purpose" Biomarker Measurement Using LC-MS in Clinical Samples

As drug candidates progress through POC studies, there is great need for LC-MS assays to measure biomarkers in clinical studies. There are numerous examples, including steroids, lipids, nucleotides, and peptides, which are directly amenable to LC-MS. Due to the endogenous nature of biomarkers, bioanalysis of those compounds usually encounters a series of challenges in maintaining analyte integrity from collection to analysis, achieving specificity, and obtaining sufficient sensitivity, especially when endogenous concentrations are downregulated. Those challenges entail special consideration and meticulous experimental design in method development, validation, and study conduct. Among the four common approaches to the preparation of standards, i.e. (1) authentic analyte in authentic matrix, (2) authentic analyte in surrogate matrix, (3) surrogate analyte in authentic matrix, and (4) charcoal or chemical stripping and immunodepletion, the last three are the ones most often applied.

Currently, there is no regulatory guidance specifically for biomarker bioanalysis. Therefore, whatever needs to be done in LC-MS bioanalysis of biomarkers should fit for the purpose of the intended use of the data. This approach has gained consent within the bioanalysis community. The term "fit for purpose" reflects flexible inclusion/exclusion of validation experiments, experiment design, and acceptance criteria. In general, assessment of accuracy, precision, and stability is considered the essential part of assay validation, while others, for example, matrix effect and recovery, are considered optional, especially when a stable isotope labeled IS is used.

Another emerging trend in biomarker quantitation is the LC-MS bioanalysis of peptide or protein biomarkers although ligand-binding assays, for example, enzyme-linked immuno sandwich assay (ELISA), still play an important role. Compared to ELISA, LC-MS assay development is relatively fast with no need to raise antibodies. More importantly, LC-MS assays can measure proteins as peptide surrogate with similar sensitivity and specificity to many immunoassays. However, introduction of stable label protein internal standard can be very challenging and costly.

1.4.7 Other LC-MS Assays Needed for Clinical Development of Drugs

As clinical drug development progresses, there can be other needs for LC-MS assays. Metabolism-mediated toxicity or adverse events often trigger these requests. In toxicology studies, this can include an assessment of parent and metabolite concentrations in various tissues from the most sensitive species. In man, penetration or distribution questions may be difficult to answer. For blood–brain barrier penetration, only cerebrospinal fluid surrogate sampling may be possible.

For antiinfective drugs, penetration studies are critical to ensuring that trough concentrations greater than IC_{50} levels are maintained where needed. When this is not achieved, resistance can develop. A similar objective to define cellular penetration can be achieved by analyzing peripheral blood mononuclear cells (PBMCs) after dosing with virology drug candidates. Plasma concentration is a poor indicator of drug activities in the cell since the activation of the drug (nucleoside) to its triphosphate involves multiple enzymatic processes that may vary by individual (Rodman et al., 1996). The pharmacokinetics of the intracellular triphosphate is also very different from that of the nucleoside. For example, the intracellular triphosphate form of emtricitabine has a much longer half-life than the plasma half-life of emtracitabine (Wang et al., 2004). Analysis of drug concentrations at the target site is often fundamental to prove target engagement and can serve to build the clinical PK-PD model. For instance, both intracellular penetration and phosphorylation is needed to activate nucleoside reverse transcriptase inhibitors. Therefore, determination of the active phosphorylated drug in PBMCs can illustrate a drug candidate's limitation as a substrate for kinase activation (Shi et al., 2002).

Drugs that have synergistic effect, such as oncology and virology agents, are generally dosed as a combination therapy. Developing an efficient LC-MS assay to simultaneously measure dose combinations is a worthwhile investment for both clinical drug development and later therapeutic drug monitoring (TDM) (Taylor et al., 2011). As many as 17 antiretroviral agents in human plasma have been measured in highly active antiretroviral therapy (HAART) (Jung et al., 2007).

Performing cost-effective LC-MS monitoring of combination therapeutics helps to ensure that safety margins of each therapeutic agent are maintained. Assays can be challenging due to the potential effect of one drug on the other. Inhibition of clearance may elevate parent concentrations while reducing those of the metabolites. Likewise, induction may reduce parent drug concentrations while elevating systemic metabolite levels. In either case, a wider range of assay or higher dilutions would be needed. The assay must also be demonstrated not to interfere with the dosed combination as well as other common medications or their metabolites. Testing of interfering drugs is readily accomplished in validation; however, testing metabolites from such complex mixtures is difficult. Using plasma from subjects receiving the concomitant drug therapies may be the best approach in method development.

1.4.8 LC-MS Bioanalysis in Postapproval Studies (Phase IV) of Drugs

Following drug approval, additional studies are often undertaken at the request of the regulatory agency. Pediatric studies can require a reduction of the sample volume or additional work to obtain a lower LLOQ. Antiinfective drugs may require an assessment in special fluids such as otitis media before being approved for ear infections. Furthermore, a regulatory agency may require the routine monitoring of drugs with a narrow therapeutic index such as many oncology or immunology drugs in larger patient populations.

1.4.9 LC-MS Bioanalysis in BE and BA Studies for Generic Drugs

At patent expiration of a brand drug, generic versions that demonstrate BE to the innovator's product may be marketed via the Abbreviated New Drug Application (ANDA) process. In the United States, a first-to-file company enjoys 6-month exclusivity for its generic product. In order to demonstrate the BE of two proprietary preparations of the same drug molecule, studies must be conducted to show an equivalent rate and extent of BA of the two products. The rate is usually measured in terms of the peak plasma concentration (C_{max}), whereas the extent is represented by the area under the concentration–time curve. Having a robust assay that can provide precise and accurate results is important in supporting BE trials (Shah and Bansal, 2011), and LC-MS assays have eclipsed GC-MS or HPLC methods for this purpose (Marzo and Dal Bo, 2007).

BE studies may also be required for a change in drug substance, drug product, or manufacturing site. When comparing two treatments the study design must include a sufficient washout between treatments. To eliminate inter-run bias, all treatments for a given subject must be analyzed within the same run. Dilutions are to be avoided, thus the dynamic range may need to be adjusted to accommodate the anticipated concentrations. The number and placement of QC samples in BE studies must mimic study sample concentrations. Due to the need for high accuracy and precision to assess formulations, many countries have instituted special guidances for BA and BE studies. These rules are in addition to normal bioanalytical requirements that have been established over the past two decades (FDA, 2001; Viswanathan et al., 2007; EMA, 2011).

1.4.10 LC-MS Bioanalysis in Therapeutic Drug Monitoring

TDM is needed for drugs with a narrow therapeutic index or when used in impaired populations. While TDM is mostly used to avoid overdosing drugs, it may also help to avoid underdosing or to monitor compliance. TDM is useful in many therapeutic areas, from antiinfectives to immunosupressants (Adaway and Keevil, 2012). For many antibiotics with a good safety margin, underdosing is of primary concern. On the other hand, antibiotics including the aminoglycosides, vancomycin, and colistins are associated with nephro- or neurotoxicity, thus overdosing is a primary concern.

Antiinfective medications can rapidly develop drug resistance so mono-therapy of HIV medicines is not likely. Comedication or HAART is the norm. There can be significant DDI in combination therapy. Ritonavir is both a HIV protease inhibitor as well as a potent cytochrome P450 CYP3A4 inhibitor used to intentionally boost the exposure to other drugs. Knowing individual drug concentrations in plasma will, therefore, aid therapy. A TDM LC-MS method measured 21 antiviral drugs from four drug classes in human plasma (Gehrig et al., 2007). Protein precipitation was needed to achieve high recovery of all diverse drugs.

Maintaining an effective level of exposure can be improved by determining drug concentrations at the site of action. As many as 10 antiretroviral drugs were determined in human PBMCs (Elens et al., 2009). Intracellular activation of reverse transcriptase inhibitors by phosphatases to their mono-, di-, and active triphosphate plays a critical role in the effectiveness of the administered prodrug.

Free drug concentrations can often provide a better correlation to drug effect than total (free and protein bound) drug levels. While ultrafiltrate or dialysate may be prepared from plasma and assayed, a simpler solution is to assay saliva. Saliva can be obtained in a noninvasive manner and saliva concentrations represent the free drug since protein-bound drugs cannot enter saliva. Other noninvasive matrices such as hair may be useful to establish patient compliance. Slightly more invasive is finger-prick sampling. Assay sensitivity is always critical when working with these low sample volumes. However, a finger-prick sample may be useful in pediatrics and subjects with compromised vascularization, including the elderly and drug addicts.

Dried blood spots (DBS) with LC-MS bioanalysis have gained popularity in TDM for a wide spectrum of drug molecules. Because DBS samples can be collected by patients themselves or their guardians with very minimum training, it opens up the possibility of collecting clinical pharmacokinetic samples not only from various in-patients but also from out-patients, especially those from remote areas (Burhenne et al., 2008). This is especially important when there is a need to monitor drugs with a narrow therapeutic index, for example, tacrolimus and cyclosporine A, with a wide variation in intra- and interpatient pharmacokinetics. DBS samples can be promptly taken whenever a concentration-related side effect appears (Li and Tse, 2010).

TDM is often limited by the difficulty in securing a representative sample and the cost of its analysis. Immunoassays, while highly cost-effective in multiplexed formats, may not provide the specificity required for combination therapies in diverse patient populations. Assay bias may result from not being able to distinguish parent drug from inactive metabolites and endogenous components. Due to its high specificity and sensitivity, LC-MS is evolving as a key player in TDM for various drugs, including anticancer and antiinfective drugs and immunosuppressants (Saint-Marcoux et al., 2007). Since LC-MS has the potential to simultaneously determine multiple analytes, its cost-effectiveness is beginning to rival immunoassays.

Within the field of immunosuppressants, TDM using LC-MS is well established to measure drugs targeted to the mammalian target of rapamycin (mTOR) such as temsirolimus or everolimus. Assays often measure combination therapies that include cyclosporine A, mycophenolic acid, everolimus, sirolimus, tacrolimus, and their metabolites (Saint-Marcoux et al., 2007). LC-UV assays can be used for some of these drugs, but the specificity, speed, and ability to measure all within a single assay makes LC-MS the preferred approach. Due to its preferential partitioning into red blood cells, cyclosporine A and the mTOR inhibitors sirolimus and everolimus are measured in blood. Mycophenolic acid is highly bound to albumin and therefore analyzed from plasma. Monitoring of both the ether and acyl glucuruonides of mycophenolic acid may also be needed (Yang and Wang, 2008). While individual extracts from blood and plasma may be needed, a common LC-MS method can be used. Sampling a few hours after the last dose and at trough is common, requiring an assay with a wide dynamic range (Sallustio, 2010).

For mTOR inhibitors, fragmentation may be limited. When monitoring the deprotonated molecule in negative ion mode, methanol loss is a commonly used transition. In positive ions, cationization with Na^+, K^+, or NH_4^+ is preferred over the MH^+ ion. Less specific transitions such as ammonia loss are compensated by the greater specificity of a high mass parent. Stable label internal standards improve assay performance, particularly for blood assays or when cationization adducts are monitored.

The high risk of organ transplant rejection warrants TDM. Clinical laboratories are becoming increasingly populated with LC-MS instrumentation and moving away from immunoassays. In addition to validating assays, proficiency tests from the College of American Pathologists or UK National External Quality Assessment Service require pooled samples from patients to be analyzed at regular intervals. Results from many hundreds of laboratories are compared to ensure standardized results are achieved.

1.5 LC-MS BIOANALYSIS OF LARGE MOLECULE DRUGS AND BIOPHARMACEUTICALS

Peptide and oligonucleotide drugs have special requirements for their bioanalysis (Nowatzke et al., 2011). Neither is generally a substrate for CYP enzymes but rather cleaved by normal proteases or nucleases. Stabilization of peptides, thus extending the half-life *in vivo*, can be accomplished using non-native enantiomers and sterically hindered amino acids or by attaching fatty acids and polyethylene glycols. The latter also serves to reduce immunogenicity. Bioanalysis using LC-MS can be difficult due to the adsorptive nature and instability of the analytes. Sensitivity loss due to lack of distinguished fragmentation is another issue. Pegylated peptides represent particular challenges due to their heterogenous nature and large molecular weight. Larger peptides or protein therapeutics may need to be selectively reduced in size to make them tractable to LC-MS.

Due to their size, proteins (>10 kDa) are less amenable to low level, direct quantification. The broad isotope distribution and presence of multiply charged species distributes their ion abundance among several response peaks thereby "diluting" the overall characterizing signal. Direct measurement of intact proteins may be made more specific using high resolution instrumentation such as time-of-flight or Orbitraps. Data processing algorithms can plot a summation of ions resulting from individual isotopes and charged forms. However, the individual ions need to be free of interference to maintain specificity. Furthermore, many proteins have abundant heterogeneity due to post-translational modifications (PTMs). LC-MS of enzyme digests have been increasingly used for quantification of proteins using a stable label or surrogate internal standard. However, one needs to be aware of the potential loss of selectivity of this approach due to the similarity between intact and post-translational modified proteins. When the protein therapeutic has extensive PTMs, an immunoassay which responds to all forms may be better at assessing its overall exposure.

Toxins attached to antibody–drug conjugates (ADC) are tractable to LC-MS. The antibody is immunocaptured and the toxin released by hydrolysis. In the case of a peptide toxin, enzymatic hydrolysis is used and the peptide measured using LC-MS. A PK profile of the toxin load versus time is established. Immunoassays of differing specificity can be used to assess antibody concentrations (Xu et al., 2011). ADC development is now showing the long sought promise of antibodies for drug targeting. LC-MS has a critical role in characterizing the "payload" of these drug delivery agents.

1.6 GUIDANCE AND REGULATIONS FOR LC-MS BIOANALYSIS

There is no doubt that LC-MS bioanalysis in support of regulated preclinical safety assessment of drug candidates submitted to the FDA must be conducted in compliance with the Good Laboratory Practice Regulations (Fed. Reg., Vol.43, 21CFR Part 58, 22-Dec-1978) and all subsequent amendments to these regulations. Submissions to European Medicines Agency (EMA) require compliance with the Organization for Economic Cooperation and Development (OECD) Principles on Good Laboratory Practice (ENV/MC/CHEM(98)17) and all subsequent OECD consensus documents. Other regulations include the Directive 2004/10/EC of the European Parliament and of the Council of Feb 11, 2004 on the harmonization of laws, regulations, and administrative provisions relating to the application of the principles of Good Laboratory Practice and the verification of their applications for tests on chemical substances (OJ No. L 50 of 20.2.2004), and other country specific regulations, for example, State Food and Drug Administration of China (SFDA).

The majority of bioanalysts around the world have become familiar with the procedures and requirements of bioanalysis as provided by the FDA guidance (FDA, 2001) and EMA guidance (EMA, 2011), and a series of Crystal City whitepapers (Viswanathan et al., 2007). Both the FDA and EMA guidances cover all aspects of bioanalysis from method validation to samples analysis from all nonclinical and clinical studies.

1.7 GENERAL CONSIDERATIONS OF A ROBUST LC-MS BIOANALYTICAL METHOD

All requirements on what should be performed in validating a LC-MS assay method in support of preclinical or clinical studies have been highlighted in the major health authority guidances (EMA, 2011; FDA, 2001). The items that need to be validated include but are not limited to: (1) selectivity and specificity, (2) sensitivity, (3) linearity, (4) intra- and interday precision and accuracy, (5) stability (stock/spiking solution stability, stability in QC samples that undergo freeze–thaw cycles and at room temperature, stability in the reconstituted sample extract stored under autosampler condition, long term stability under intended storage condition, stability in blood), (6) dilution integrity, (7) carryover, and (8) assay batch size.

The initial assessment of assay calibration regression model will include the slope or sensitivity of the calibration curve and weighting. Once established, all future analysis must be performed according to this model. Whenever possible, standards and QC samples must be prepared in the identical matrix to study samples. Acceptance criteria is three-fourth of all standards and two-thirds of all QC samples are within 15% (20% at LLOQ) of their nominal value and that 50% of QC samples at each concentration are acceptable. Tests for detection specificity and assay selectivity will employ either blank matrix or samples spiked at the LLOQ, respectively, in matrix from numerous individuals. Lipemic and hemolyzed plasma should be tested to ensure that the assay is insensitive to sample condition. Establishing the assay range by testing at the LLOQ and analytical QC levels (low, mid, and high) is required in replicates ≥ 5 for at least three batches. These accuracy and precision tests are critical to a validation.

The numerous stability campaigns needed to support sample analysis are performed using either a bracketed range of stock solutions or QC samples, minimally tested at low and high QC concentrations. Each part of the sample collection and storage process must be tested. Plasma assays can be used to assess blood stability by harvesting and analyzing plasma at different time points (e.g., 0, 60, and 120 min). Multiple dilutions are generally tested during validation to ensure that the most diluted study sample can be reproducibly analyzed. While tested in validation, it is important to also demonstrate performance during study sample analysis by including dilution QC samples within runs. Carryover must be minimized in method development and controlled in validation or sample analysis to less than 20% LLOQ. Tests must include an assessment of ruggedness by ensuring that the maximum batch used in sample analysis is tested during validation.

During study sample analysis, incurred sample reproducibility must also be demonstrated (Fast et al., 2009). One should recognize that incurred samples are much more complex in their composition than the QC samples. QC samples do not contain the various drug metabolites, drug isomers/epimers, coadministered drug(s) and their metabolites, and/or dosing vehicles as often seen in the incurred samples. In some cases, the plasma concentration of metabolites can reach an order of magnitude greater than that of the parent drug. Heightened awareness among bioanalytical scientists to the differences between QC samples and incurred samples will reduce the risk of subsequent bioanalytical failure (Li et al., 2011). The integrity of individual samples can also impact this assessment, as seen for hemolyzed plasma or inhomogeneous urine samples. The requirements of conducting incurred sample reanalysis (ISR) have been established over the past decade (Fast et al., 2009; EMA, 2011). The concentration obtained for the initial analysis and the concentration obtained by reanalysis should be within 20% of their mean for at least two-thirds of the repeats. ISR can be performed during next run (rolling ISR) or analyzed in separate batch at the end of the study. Testing at the end of the study examines both stability and imprecision in the assessment. Testing at the end of the study is also more reflective of any repeat analysis and can better understand issues related to interference from unstable concomitant medications or its metabolites. ISR on the following day's run not only examines assay reproducibility but also will give an earlier diagnosis of any problems. For this reason, some laboratories perform a mixture of both early rolling and later batch ISR. The assay should be tested for ISR in each clinical trial, as new populations or new clinics can spawn new problems.

Prior to transferring bioanalytical work for an ongoing study from one laboratory to another, for example, from the sponsor's laboratory to a contract research organization, cross-validation or conformance testing using QC and incurred samples must be conducted. The success of an LC-MS assay method transfer is critical to ensure the quality and integrity of subsequent PK-PD assessments.

1.8 CONCLUSIONS

LC-MS bioanalysis has become a primary tool in every stage of drug discovery, drug development, and postapproval TDM. It helps in not only selecting better drug candidates but also improving our understanding of safety and pharmacokinetics of the drugs. Furthermore, LC-MS bioanalysis of toxicology biomarkers will no doubt help to avoid the selection of poor drug candidates that would fail in longer term toxicology studies. LC-MS is increasingly being used to assay clinical biomarkers and ensure that the drug effects observed in animal models translate into success in human proof-of-concept studies.

REFERENCES

Adaway JE, Keevil BG. Therapeutic drug monitoring and LC-MS/MS. J Chromatogr B 2012;883–884:33–49.

Anderson S, Kanadler MP, Luffer-Atlas D. Overview of metabolite safety testing from an industry perspective. Bioanalysis 2010;2(7):1249–1261.

Balani SK, Nagaraja NV, Qian MG, et al. Evaluation of microdosing to assess pharmacokinetic linearity in rats using liquid chromatography-tandem mass spectrometry. Drug Metab Dispos 2005;34:384–388.

Berman J, Halm K, Adkison K, Shaffer J. Simultaneous pharmacokinetic screening of a mixture of compounds in the dog using API LC/MS/MS analysis for increased throughput. J Med Chem 1997;40: 827–829.

Burhenne J, Riedel K-D, Rengelshausen J, et al. Quantification of cationic anti-malaria agent methylene blue in different human biological matrices using cation exchange chromatography coupled to tandem mass spectrometry. J Chromatogr B 2008;863(2):273–282.

Elens L, Veriter S, Yombi JC, et al. Validation and clinical application of a high performance liquid chromatography tandem mass spectrometry (LC-MS/MS) method for the quantitative determination of 10 anti-retrovirals in human peripheral blood mononuclear cells. J Chromatogr B 2009;877:1805–1814.

European Medicines Agency. Guideline on bioanalytical method validation. Jul 2011. CHMP/EWP/192217/2009. Available at http://www.ema.europa.eu/docs/en_GB/document_library/Scientific_guideline/2011/08/WC500109686.pdf. Accessed Mar 1, 2013.

Fast, DM, Kelly M, Viswanathan CT, et al. Workshop report and follow-up–AAPS workshop on current topics in GLP bioanalysis: assay reproducibility for incurred samples–Implications of Crystal City recommendations. AAPS J 2009;11:238–241.

FDA. Guidance for industry: safety testing of drug metabolites. Feb 2008. Available at http://www.fda.gov/OHRMS/DOCKETS/98fr/FDA-2008-D-0065-GDL.pdf. Accessed Mar 1, 2013.

FDA. Guidance for industry: bioanalytical method validation. May 2001. Available at http://www.fda.gov/downloads/Drugs/GuidanceComplianceRegulatoryInformation/Guidances/ucm070107.pdf. Accessed Mar 1, 2013.

Garner RC. Less is more: the human microdosing concept. Drug Discovery Today 2005;10:449–451.

Gehrig A, Mikus G, Haefeli W, Burhenne J. Electrospray tandem mass spectroscopic characterisation of 18 antiretroviral drugs and simultaneous quantification of 12 antiretrovirals in plasma. Rapid Commun Mass Spectrom 2007;21:2704–2716.

ICH Guidance M3(R2) Nonclinical safety studies for the conduct of human clinical trials and marketing authorization for pharmaceuticals. Jun 2009. CPMP/ICH/286/95. Available at http://www.emea.europa.eu/docs/en_GB/document_library/Scientific_guideline/2009/09/WC500002720.pdf. Accessed Mar 1, 2013.

International Commission on Radiological Protection (ICRP). The 2007 Recommendations of the International Commission on Radiological Protection 2007, volume 103, Elsevier.

Janiszewski JS, Liston TE, Cole MJ. Perspectives in bioanalytical mass spectrometry and automation in drug discovery. Curr Drug Metabol 2008;9:986–994.

Jung BH, Rezk NL, Bridges AS, Corbett AH, Kashuba AD. Simultaneous determination of 17 antiretroviral drugs in human plasma for quantitative analysis with liquid chromatography-tandem mass spectrometry. Biomed Chromatogr 2007;21:1095–1104.

Li W, Luo S, Smith HT, Tse FL. Quantitative determination of BAF312, a S1P-R modulator, in human urine by LC-MS/MS: prevention and recovery of lost analyte due to container surface adsorption. J Chromatogr B Analyt Technol Biomed Life Sci 2010;878(5–6):583–589.

Li W, Tse FL. Dried blood spot sampling in combination with LC-MS/MS for quantitative analysis of small molecules. Biomed Chromatogr 2010;24(1):49–65.

Li W, Zhang J, Tse FL. Strategies in quantitative LC-MS/MS analysis of unstable small molecules in biological matrices. Biomed Chromatogr 2011;25(1–2):258–277.

Marzo A, Dal Bo L. Tandem mass spectrometry (LC-MS-MS): a predominant role in bioassays for pharmacokinetic studies. Arzneim.-Forsch 2007;57(2):122–128.

Nowatzke W, Rogers K, Wells E, Bowsher R, Ray C, Unger S. Unique challenges of providing bioanalytical support for biological therapeutic pharmacokinetic programs. Bioanalysis 2011;3:509–521.

Obach RS, Nedderman AN, Smith DA. Radiolabelled mass-balance excretion and metabolism studies in laboratory animals: are they still necessary? Xenobiotica 2012;42(1):46–56.

Pellegatti M. Preclinical *in vivo* ADME studies in drug development: a critical review. Expert Opin Drug Metab Toxicol 2012; 8(2):161–172.

Rodman JH, Robbins B, Flynn PM, Fridland A. A systemic and cellular model for zidovudine plasma concentrations and intracellular phosphorylation in patients. J Infect Dis 1996;174:490–499.

Saint-Marcoux F, Sauvage F-L, Marquet P. Current role of LC-MS in therapeutic drug monitoring. Anal Bioanal Chem 2007;388:1327–1349.

Sallustio BC. LC-MS/MS for immunosuppressant therapeutic drug monitoring. Bioanalysis 2010;2(6):1141–1153.

Shah VP, Bansal S. Historical perspective on the development and evolution of bioanalytical guidance and technology. Bioanalysis 2011;3(8):823–827.

Shi G, Wu JT, Li Y, et al. Novel direct detection method for quantitative determination of intracellular nucleoside triphosphates using weak anion exchange liquid chromatography/tandem mass spectrometry. Rapid Commun Mass Spectrom 2002;16:1092–1099.

Taylor PJ, Tai C-H, Franklin ME, Pillans PI. The current role of liquid chromatography-tandem mass spectrometry in therapeutic drug monitoring of immunosuppressant and antiretroviral drugs. Clin Biochem 2011;44:14–20.

Viswanathan CT, Bansal S, Booth B, et al. Workshop/Conference Report – Quantitative Bioanalytical Methods Validation and Implementation: Best Practices for Chromatographic and Ligand Binding Assays. AAPS J 2007;9:E30–E42.

Wang LH, Begley J, St Claire III RLS, Harris J, Wakeford C, Rousseau FS. Pharmacokinetic and pharmacodynamic characteristics of emtricitabine support its once daily dosing for the treatment of HIV infection. AIDS Res Hum Retroviruses 2004;20:1173–1182.

White RE, Evans DC, Hop CE, Moore DJ, Prakash C, Surapaneni S, Tse FLS. Radiolabelled mass-balance excretion and metabolism studies in laboratory animals: a commentary on why they are still necessary. Xenobiotica 2013;43: 219–225.

Wu JT, Zeng H, Qian M, Brogdon BL, Unger SE. Direct plasma sample injection in multiple-component LC-MS-MS assays for high-throughput pharmacokinetic screening. Anal Chem 2000;72:61–67.

Xu K, Liu L, Saad OM, et al. Characterization of intact antibody–drug conjugates from plasma/serum in vivo by affinity capture capillary liquid chromatography–mass spectrometry. Anal Biochem 2011;412(1):56–66.

Yang Z, Wang S. Recent development in application of high performance liquid chromatography-tandem mass spectrometry in therapeutic drug monitoring of immunosuppressants. J Immunological Methods 2008;336:98–103.

2

OVERVIEW: FUNDAMENTALS OF A BIOANALYTICAL LABORATORY

SHEFALI PATEL, QIANGTAO (MIKE) HUANG, WENYING JIAN, RICHARD EDOM, AND NAIDONG WENG

2.1 INTRODUCTION

Progression from a new drug candidate to a marketed pharmaceutical takes significant scientific resources and financial investment. While in the last decades the pharmaceutical industry has steadily increased its investment in drug discovery and development, fewer drugs have been approved by the Food and Drug Administration (FDA). The risk in drug development is still tremendously high, maybe higher than ever, due to the raised barrier of entry. "Me-too" drugs are less likely to be approved and only the best of the best with superior efficacy and safety will be able to enter the market. To select the optimal drug candidates and eliminate weak ones for efficient utilization of resources, questions concerning toxicity and efficacy need to be addressed during preclinical and clinical studies. Figure 2.1 summarizes some of the major types of studies to answer these questions.

Timely bioanalytical (BA) support during the entire drug discovery and development process is essential to allow rapid decision making. Guided by data, drug candidates may be modified to improve their properties, or in some cases, programs may be terminated entirely. Significant cost savings may incur if reliable decisions can be made quickly (Lee and Kerns, 1999). Since BA is one of the few disciplines to cover the entire drug discovery and development process, there are a number of unique requirements at each of the various stages. Discovery BA support is operated under non-GLP (good laboratory practices) conditions. Here, generic methods and calculations with Excel spreadsheets are often used, which make it easier to adopt a universal strategy to streamline and automate the process. The ultimate goal is to generate data as quickly as possible to enable the elimination of less optimal drug candidates. The BA mindset certainly shifts from more flexible discovery BA support to more compliance-oriented activities in the drug development stage, where many preclinical BA activities are conducted under strict GLP regulations and the data routinely undergoes rigorous regulatory scrutiny. New technologies are used cautiously, and only after they meet the challenges of vigorous compliance validation will they be widely accepted as a primary tool. Quick turnaround time for method development, validation and sample analysis for multiple animal species (rat, dog, monkey, rabbit, mouse, etc.) and oftentimes for multiple matrices (plasma, tissues, etc.) is also a prerequisite for preclinical bioanalysis. Clinical bioanalysis has its own unique challenges that can vary from extremely quick turnaround time for first-in-human (FIH) studies to extremely complicated sample logistics for multisite, long-term phase III studies. Because of the diversified BA support from discovery to late phase clinical trials, the process and personnel in a typical BA laboratory should be flexible. For preclinical BA support, there is a need for a very quick turnaround of sample analysis in addition to the heavy workload on method development and validation. Therefore, transitioning of the method from one group to another occurs less in preclinical support. For clinical BA support, it is common to have a group of scientists dedicated to method development and validation, and then the methods are transferred to other chemists for large sets of routine sample analysis that could last for several months to several years. In general, more method development and validation expertise is needed in GLP bioanalysis than in non-GLP one. On the other hand, in clinical BA support, communicating

Handbook of LC-MS Bioanalysis: Best Practices, Experimental Protocols, and Regulations, First Edition. Edited by Wenkui Li, Jie Zhang, and Francis L.S. Tse.

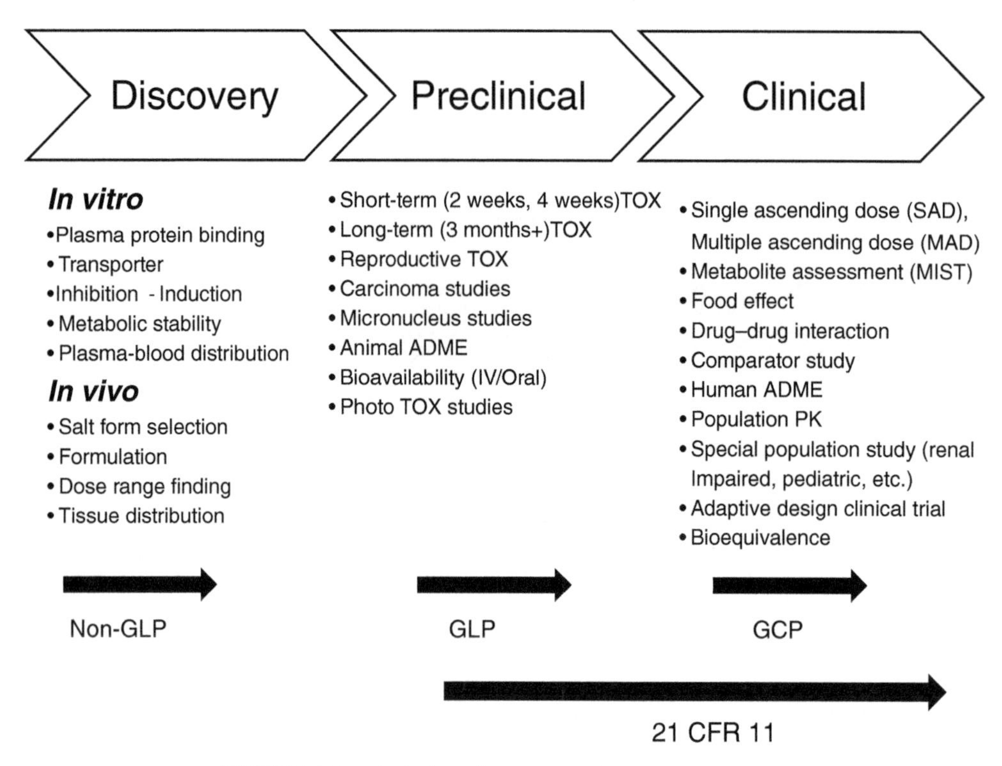

FIGURE 2.1 Typical studies supported in a BA laboratory.

with clinical teams and coordinating with multiple sites for sample logistics becomes a much more important task. The project team also relies heavily on the BA team for the preparation of a New Drug Application (NDA) submission.

The BA laboratory is facing constant pressure to reduce costs while optimizing efficiency to be competitive in the marketplace with other BA laboratories in the industry and with contract research organizations (CROs). While reducing operational costs is important, proper emphasis should be placed on maintaining high-quality data. The purpose should be to create and maintain laboratories that (1) are responsive to current and future needs, (2) recruit and retain qualified scientists, (3) encourage interaction among scientists from various disciplines, and (4) facilitate cooperation among all of the drug development partners.

2.2 KEY ELEMENTS OF A BA LABORATORY

Bioanalysis is complex with many personnel, departments, instruments and processes involved. BA laboratories use a range of technologies to ensure that samples are processed efficiently and accurately and quality results are delivered in a timely manner. The technologies involve advanced instrumentation, intelligent robotics, and powerful computers and software. During the BA process, efficient assay development and validation, rapid analysis of samples, seamless instrument interfacing, detailed sample tracking, and prompt reporting of results are just a few of the key steps. Figure 2.2 is a representative process map of the typical operation in a BA laboratory. The four major components for any GLP BA laboratory to function successfully are:

1. Facility
2. Infrastructure
3. Compliance
4. Documentation

In this chapter, we focus on the vital role of each of the above four components. In addition, quality assurance (QA) and other supporting functions such as quality control (QC), sample management, archivist, technical support, planning, and reporting also play pivotal roles. Monitoring BA activities in CROs has also become more important in recent years. These functions will also be discussed in this chapter.

Coordinating and managing all these functions into one organization makes a bioanalysis operation quite complicated. It is imperative for every function to work together efficiently as an organization and to effectively communicate with each other.

2.2.1 Facility

2.2.1.1 Space For a well-organized bioanalysis facility, it is important that sufficient, clean space be allocated. In

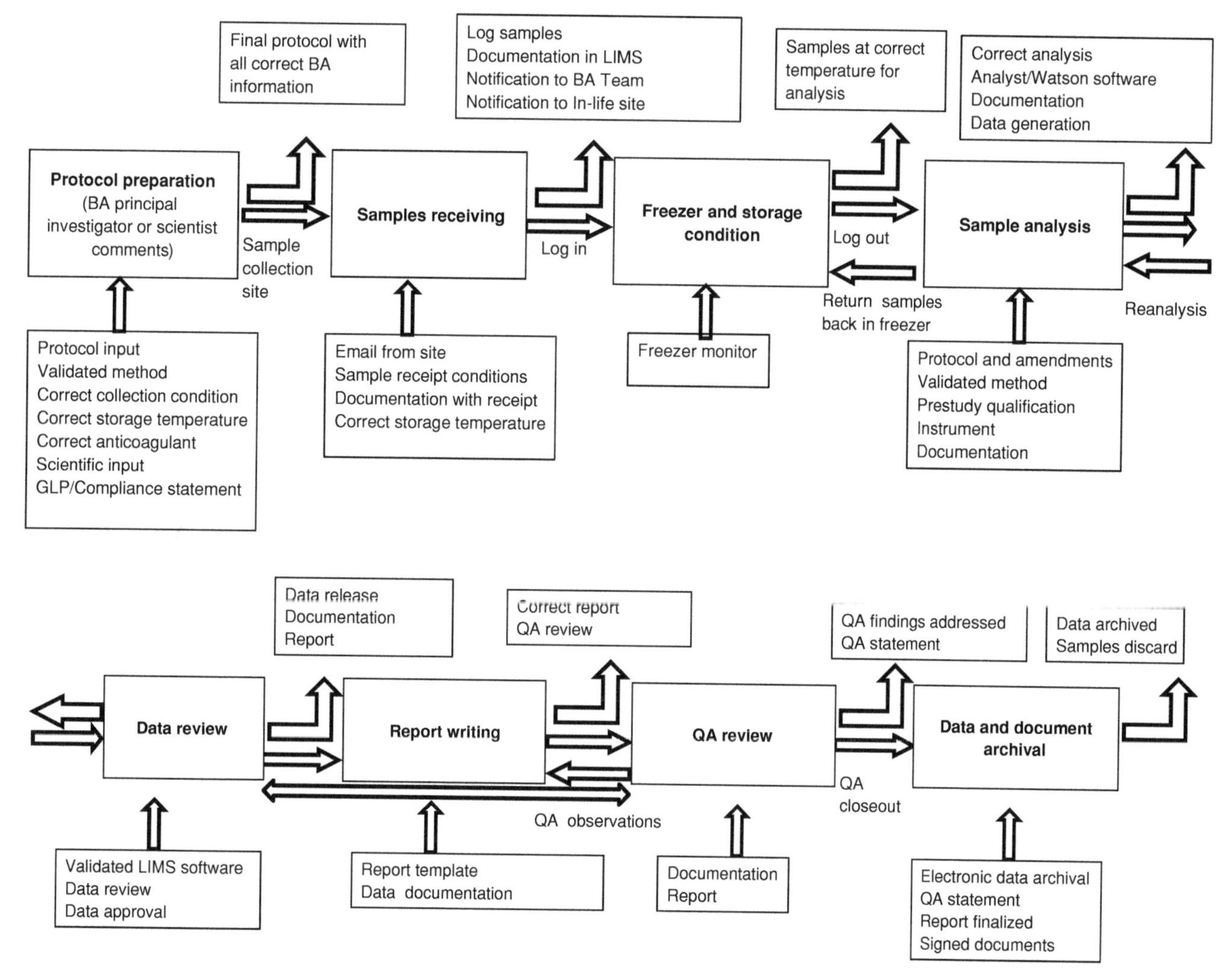

FIGURE 2.2 A representative bioanalysis process map.

the process workflow of any laboratory, the following considerations should be given to ensure efficient operation and effective communication:

- Efficient arrangement of instrumentation.
- Convenient location for the sample preparation.
- Smooth flow of analytical work between various tasks.
- Logical arrangement of paperwork and nonlaboratory areas.
- Proximal location of communication equipments (voice, data, and Internet).
- Effective personnel communication.

For instruments such as mass spectrometers to function properly, they should be placed in a well-ventilated area. Noise, temperature, and humidity should be given due consideration and must be controlled when designing the BA laboratory. Certain instruments such as 96-well liquid handling robotics and weighing balances should have proper ventilation so that the solvent fumes or chemical dust does not circulate within the laboratory area. The instruments should be properly spaced so that they do not generate excessive heat and noise. Careful considerations should be given to the electrical voltage and amperage needed for various instruments. Waste gas connections must be supplied, and proper heating, ventilation, and air conditioning must be maintained year-around. Fume hood placement should also be considered for the proper air movement in the building. The instruments should be connected to a backup electrical system such as an uninterruptible power supply and a backup generator in case of an emergency loss of power. Ideally, an isolated area should be used to weigh compounds to avoid accidental cross-contamination. It is preferable to have an individual weighing room with controlled access. The room should be quiet, and the balance should be located within a ventilated hood for the proper protection of the operators. If controlled substances are used, they must be properly secured and tracked. A BA laboratory should be solely used for BA activities, and not for conducting other tasks such as dosing formulation preparation to avoid any potential contamination.

All refrigerators and freezers in the laboratory should be monitored via an electronic system to record the temperature and/or humidity environment. A contingency plan should be in place, such as a backup freezer, to rescue samples in case of failure. Freezers and refrigerators should initiate an alarm if the temperature goes out of the designated ranges. Outside of normal working hours, an emergency system will notify personnel to rush to the site and transfer the samples to another functioning unit.

The sample preparation area should preferably be near the instrumentation. Too often, equipment ends up in whatever available space, which may not be the most convenient for efficient workflow. Personnel movement should also be considered while designing a laboratory so that minimal time is required when traveling from one bench to another. For example, sample extraction plates should not be carried from one laboratory to another via a long hallway.

The day-to-day data handling and processing also plays an important role in workflow design. Scientists' workstations should be placed in such a way that data is easily processed as soon as it is generated. Paperwork and raw data should be placed in a secure location before the end of the working day to prevent accidental loss. Several communication tools, such as phones, printers and computers, should be readily available and accessible to minimize back and forth movement of personnel. In addition to the laboratory area, the building should have sufficient supporting facilities such as offices, work areas, lunchrooms, washrooms, and parking.

2.2.1.2 ***Security*** Protection of the BA facility from unauthorized access is also of critical importance. The building in which regulated activities are conducted must be a secured environment with restricted access. Only authorized personnel should be allowed to enter nonpublic areas such as laboratories. Even more rigorous security is required for areas such as archive or computer server facilities. Only a very few staff members should have routine access, and the movements of personnel and documents should be carefully tracked using devices such as access card readers. These records should be maintained and must be readily available for inspection.

BA facilities should have alarms and fire detection systems for off-hours protection. Police and fire authorities, as well as facility senior management, should be automatically alerted in case of intrusion or catastrophe. Needless to say, proper insurance policies should be carried.

2.2.1.3 ***Instrument Qualification and Software Validation*** BA laboratories need a variety of instruments. Table 2.1 lists instruments typically present in a BA laboratory along with their classification based on different requirements for performance verification. Analytical instruments must be calibrated and verified at intervals according to the standard operating procedures (SOPs) of the department. For instrumentation involved in regulated studies, proper installation qualification (IQ), operational qualification (OQ), performance qualification (PQ), and appropriate software validation should be performed. The IQ verifies that the instrument is correctly installed and in good working order. The OQ is used to demonstrate and provide documented evidence that, after installation, the instrument performs according to the manufacturer's specifications. The PQ demonstrates that the instrument is functioning properly for the intended use in the individual laboratory. A variety of different systems should undergo IQ/OQ/PQ testing before starting any GLP or regulated activities in the laboratory. Some of the systems that need qualification and software validation are:

TABLE 2.1 List of Common BA Instrumentation by Class

Instrument	Class of instrument
Balance	B
Centrifuge	B
LC (liquid chromatography)	C
MS (mass spectrometer)	C
Liquid handler	C
Pipette	B
Column heater	A
Tissue homogenizer	B
Vortex mixers	A
pH meter	B
Evaporator	A
Incubator	A
Refrigerators and freezers	B
Water purification system	B

Group A, nonqualified instruments. Visual verification of performance.
Group B, qualified instruments. Controlled by firmware that is nonuser configurable.
Group C, qualified instruments. Controlled by software that is user configurable.

- Chromatographic data systems—qualification and software validation.
- Detectors such as mass spectrometers—qualification and software validation.
- LIMS (laboratory information management system)—software validation.
- Electronic laboratory notebooks—software validation.
- Data archival systems—software validation.

At a minimum, the IQ/OQ/PQ must be performed at the following times:

- At the time of instrument installation (IQ/OQ/PQ).
- When an instrument is moved from one location to another (IQ/OQ/PQ).
- Major instrument repair (OQ/PQ).

In addition to qualification of instruments and software, change control documentation should be in place to demonstrate that any change of an existing system is introduced in a controlled and coordinated manner. Examples of changes are:

- Any software or hardware upgrade.
- Any new components, such as an additional HPLC pump, are added to a system.
- Any components are removed due to a repair or failure.

A preventive maintenance (PM) plan should be in place for every 6–12 months as per the use of the instrument or SOP specifications. A PM is done to maintain the functional performance of the instrument. In addition to regular PMs, instruments should be routinely tested to confirm performance and to identify potential problems from normal wear or inadequate user maintenance. All IQ/OQ/PQ activities, as well as PMs, should be recorded in the logbook associated with each instrument, and the related paper documents should be periodically archived. Any major repairs and routine or nonroutine maintenance should also be documented in the logbook along with a brief description of when the defect was found, how it was fixed and who conducted the repair. In all cases, the signature and date of the person performing the work should be included in the logbook. Any software updates, in particular patches that can fix existing bugs, should be timely evaluated and implemented.

*2.2.1.4 **Archival Facility*** Protection of documents is a critical component of any organization. For temporary storage of paper raw data, it is recommended that fireproof cabinets be placed in the office area, and these data should be secured in them at the end of each working day. Longer term archive requires a dedicated, fire-protected, waterproof, humidity-controlled room. The facility must be secured to prevent unauthorized access. Since all raw paper and electronic data should be kept for a certain period of years, ideally they should be archived at an off-site facility for long-term storage.

Electronic data should be maintained, protected, and archived similarly to paper data, but there are some additional considerations. The facility should have a secure computer server room for archival of electronic data. The data should be backed up daily, weekly, or monthly in accordance with its criticality. This policy should be outlined in an SOP. In addition to routine on-site backup, it is recommended to have an off-site backup system in case of catastrophic event. It is also important that the records should be readable in the future. With new instrumentation emerging each year and software changing rapidly, this creates an added burden on electronic archive facilities. Forward file compatibility or the continued availability of legacy versions of software from each instrument manufacturer is challenging. Legacy computer equipment capable of running outdated versions of software is also recommended, along with the staff who have appropriate training to use these software.

Vital records that are needed to reconstruct studies, such as communications with study directors, protocols and amendments, legacy reports, QA study records, and certificates of analysis, should all be archived. Routine QA inspections of the archive facility should be conducted. In addition to study data, facility records (e.g., training records, equipment records, computer system validations, SOPs, temperature records, and master schedules) should also be archived.

*2.2.1.5 **Safety*** Protecting human health and life is paramount, and safety must always be the first priority in the laboratory. Proper personal protective equipment (PPE) such as safety glasses, gloves, laboratory coats, and bench-top shields should be used all the times. The laboratories must include safety showers and eye wash stations. Staff must be properly trained to use them in an emergency, and they should also be familiar with the location of fire extinguishers, spill containment equipment, and most importantly emergency exit. Devices such as bio-safety cabinets and fume hoods should be strategically placed and well ventilated. Hazardous experiments must be done in the fume hoods. Procedures for safe and environmentally acceptable disposal of laboratory wastes like paper, glass, sharps, chemicals, radioactives, and biologics should be developed and must be followed. All chemicals should be properly labeled to show the nature, degree of hazard, and the expiration date. All flammable chemicals should be stored in fireproof cabinets or explosion proof refrigerators. Biohazardous and chemical wastes should be picked up on a regular basis to prevent accumulation, and the disposal must follow federal, state and local regulations. This also applies to liquid and radioactive wastes (if any). If radioactivity is used, the areas should be properly labeled and segregated, and appropriate training in radiation safety should be given to the staff. All laboratory areas, workstations, and fume hoods should be kept clutter free, and access to fire extinguishers and escape routes must not be impeded. Laboratories must be designed to have two routes of egress from each area in case of emergency. It is advisable to develop an emergency evacuation plan in case of fire, medical emergency, chemical release, radioactive spill, and dangerous weather situations. Regular laboratory inspections should be conducted on a bi-weekly or monthly basis to keep the laboratory safe, to identify and correct hazardous conditions and unsafe practices, and to keep the laboratory GLP compliant.

2.2.2 Infrastructure

For any BA laboratory, personnel must have appropriate training and experience. In addition, management oversight is a requirement in a GLP environment, so a proper management structure must be set up to enable adequate supervision

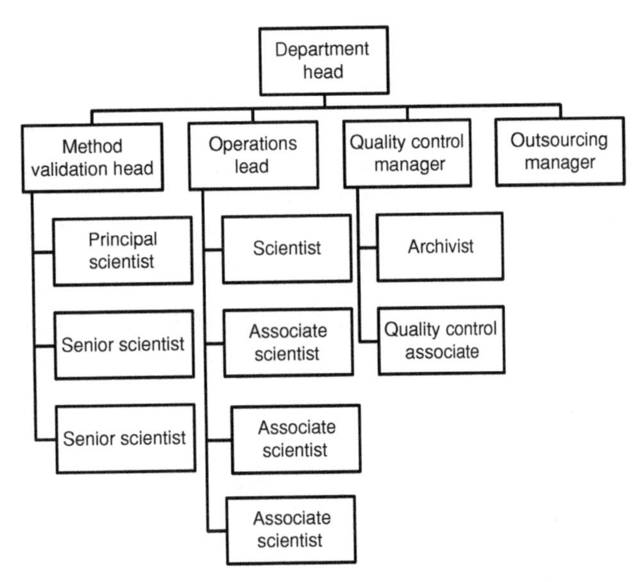

FIGURE 2.3 An example of a BA organization chart.

and review. Management should oversee the daily operation and execution of all activities.

A BA laboratory should have an organizational chart reflecting current personnel and reporting relationships. Inspectors from regulatory agencies will often request the organizational chart as a first step in an audit. Typically, the chart is depicted in a hierarchical format (top-down), with staff clustered into logical groups by function or department. The organization chart should show each employee's name and title (or function). Direct reporting relationships are typically shown with solid lines connecting staff members. Indirect relationships are shown with dotted lines. The chart must be kept current, be reviewed periodically, and must be available for review immediately upon request. Figure 2.3 depicts an example of a BA organizational chart. A typical organization may consist of different roles like method validation scientist, sample analysis scientist, sample management, contributing scientist, and principal investigator (PI). These roles should be at different levels of authority, as per the experience and expertise of the personnel. For a larger organization, each role may be more specialized and disseminated due to the opportunity presented by staff size. For a smaller organization, multiple roles might be combined into one. Support functions should also be set up for several specialized laboratory activities such as management of sample inventory, QC, resource planning, and writing of reports. Serious consideration should be given to laboratory staffing decisions, not just staff numbers, but also their ability and experience.

Staff must be well trained to function in the laboratory. Proper training should be designed for new and existing personnel in each aspect of their job. Training refreshers should also be given to all staff on a regular basis. All training should be documented in order to demonstrate that personnel are fully qualified for doing the assigned job.

Each position or function in a BA laboratory should have a written job description. The descriptions should be narrative, and they should include an outline of the key activities and requirements for each major type of job in the laboratory. Relationships with other people or functions in the organization, such as their supervisory level and managerial requirements, should also be disclosed. The job descriptions should be reviewed periodically to ensure that they are an accurate reflection of the tasks. They should be approved by laboratory management and should be readily available for inspection.

Each employee of a BA laboratory should have an up-to-date record of their training and qualifications. The records should include a curriculum vitae (CV) that summarizes their educational background, prior experience, and scholarly activities such as scientific publications and presentations. In addition to the CV, each employee should have a training file that outlines their specific training accomplishments and "on-the-job" experience. Training records should be reviewed periodically by laboratory management to ensure that they are complete. They should also be adequately protected (e.g., stored in the archive facility or a suitably protected cabinet). These records are used during regulatory inspections to verify whether an employee has adequate training and experience to perform their job function. This tends to be an important part of all inspections by regulatory agencies.

2.2.3 Compliance

2.2.3.1 Industry Regulations GLP regulations issued by the USFDA deal with the organization, process, and conditions under which preclinical laboratory studies are planned, performed, monitored, recorded, and reported. Published GLP regulations and guidelines have a significant impact on the daily operation of a BA laboratory. Evaluation of GLP compliance is an important tool for measuring a facility's fitness for operation and for assuring data integrity. GLP regulations are mainly driven by FDA 21CFR58 (USFDA, 2011) and Organization for Economic Cooperation and Development (OECD) principles on GLP and consensus documents (OECD, 1998). For conduct of activities in support of clinical trials, good clinical practice required by FDA and international conference on harmonization (ICH) must be followed (ICH, 1996; USFDA, website). Another important regulation is FDA's 21CFR11, which covers e-records and e-signatures (USFDA, 2011). In today's modern laboratory, more and more information and data are generated and stored electronically, which means 21CFR11 can have a significant impact to a laboratory's compliance. Especially when electronic signatures are implemented, thorough validation should be in place to demonstrate that they are 21CFR11 compliant.

Other guidance and/or best practices for bioanalysis include Guidance for Industry Bioanalytical Method Validation (USFDA, 2001); European Medicines Agency (EMA)

Guideline on Validation of Bioanalytical Methods (European Medicines Agency, 2011); Quantitative Bioanalytical Methods Validation and Implementation: Best Practices for Chromatographic and Ligand Binding Assays (Viswanathan et al., 2007); Workshop Report and Follow-up–AAPS Workshop on Current Topics in GLP Bioanalysis: Assay Reproducibility for Incurred Samples–Implications of Crystal City Recommendations (Fast et al., 2009). Figure 2.1 shows what kinds of regulations are applied during different phases of drug development.

2.2.3.2 Standard Operating Procedures A BA laboratory should have a complete set of SOPs that dictate consistent practices for each task. SOPs should be written in a clear, concise manner for the reader(s) to easily follow. Overly detailed SOPs should be avoided since they can create a compliance problem—too many details or steps will make the SOP difficult to follow. A balance must be struck between clarity and brevity. Table 2.2 lists typical SOPs in a BA laboratory. In general, the first SOP (e.g., SOP 001) outlines the process for creating, reviewing, issuing, and maintaining SOPs. It grants authority for specific people to issue SOPs, requires training for in-scope staff members, and also sets the process to update, replace or to make obsolete an SOP. Each SOP should be organized into sections such as Purpose, Scope, Applicable Groups, Procedures, and so on. For consistency, all SOPs within a laboratory or organization should follow the same format. Each SOP should be approved by individuals qualified in the subject matter of the SOP. Typically, SOPs are required to have a periodic review. This helps to ensure that SOPs stay current with the everyday practices of the laboratory, the current thinking in the industry, and the continually changing regulatory requirements. A record of training on SOPs should be kept for all staff members, and it must be updated when each new SOP is issued or each existing SOP is revised and versioned.

TABLE 2.2 Typical SOPs in a Bioanalytical Laboratory

SOP No.	SOPs
001	SOP on SOPs
002	Raw data
003	Project and laboratory notebooks
004	Sample handling and tacking
005	Method validation for chromatographic assays
006	Sample analysis for chromatographic assays
007	Method qualification for chromatographic assays
008	Chromatographic instruments
009	Watson user and support procedures
010	Sciex user and support procedures
011	Handling of bioanalytical data
012	Study sample repeat analysis
013	Analytical reference standards
014	Reagent and solution labeling
015	Pipettes
016	Reports
017	Balance and reference weights
018	pH meters
019	Refrigerators and freezers
020	Instruments logbooks
021	Liquid handling System
022	Centrifuges
023	Organization charts
024	Multisite studies
025	Laboratory instrument qualification
026	Change control of qualified instruments
027	Computerized system validation
028	Change control of computerized system
029	Quality control review of bioanalytical activity
030	Outsourcing study
031	Master schedule

Equally important to SOPs, a parallel system must be put in place to document that SOPs have been followed. For example, if an SOP requires that an instrument be calibrated yearly, then records must be kept showing that it was, in fact, calibrated each year. Any deviation from SOPs must be documented and approved. However, frequent deviations from an SOP may indicate a lack of system control or training. These records, which document compliance with SOPs, must be maintained for a significant period of time. This period is frequently specified in one of the SOPs, but it can also be dictated by company policy. In some circumstances, it may be appropriate to include the records into study raw data.

Although SOPs are considered authoritative instructions for a task or activity, it must be remembered that specific instructions in a protocol can override SOP requirements. In the hierarchy of compliance, study-specific protocol requirements carry the highest order. SOPs should also address the processes to be followed for "multisite studies" where different portions of the studies are done at different places. In addition, for a GLP facility, a master schedule should be in place to list which studies are conducted in the organization and their status.

2.2.4 Documentation

2.2.4.1 How to Document All data should be documented properly so that it can be fully reconstructed (i.e., what was done, how it was done, where it was done, who did it, and when). For compliance in regulated studies, all documentation must be done in real time, as the data is generated. Data can be documented in different ways, across different companies. For example, paper laboratory notebooks can be bound books, or they can be prepaginated, loose leaf binders, or they can be nonpaginated loose leaf binders that are paginated after the notebook is completed. If an electronic laboratory notebook is used, it should capture full audit trail details, as well as be thoroughly validated and 21 CFR 11 compliant (USFDA, 2011).

Data recording must be direct, prompt, legible, accurate, and if paper records are used, it must be done in permanent ink. The date and the signature or initials of the recorder should be included in each entry. Backdating is unacceptable. When making corrections or changes, the original entry should not be obscured, and the reason for the change must be given for each correction.

Even for nonregulated studies, proper documentation and compliance practices should be in place to ensure the data is legible, accurate and can be reconstructed.

2.2.4.2 Notebook Documentation Documentation of each day's work should be recorded in the laboratory notebook. The laboratory should have an SOP on how to document and what should be included in the notebook. Predesigned forms such as "daily run packs" can be a handy tool to ensure completeness and consistency of documentation. Every entry in the notebook should be clear and logical, and it should be signed/dated by the author and reviewed by another scientist who also signs/dates. A single notebook per study is the preferred practice, rather than entries in different notebooks for separate aspects of a study. All materials used in each experiment, such as compound certificate of analysis, reagents, solution preparations, and so on should be recorded in the notebook. Certain critical steps such as weighing of materials or preparation of calibration standards and QC samples should be reviewed and signed by another scientist as soon as possible so any mistakes could be corrected promptly and would not carry through the whole study.

2.2.4.3 Electronic Data All computerized systems in the laboratory (including data acquisition and processing) should have full security privileges enabled, including unique user identifications with encrypted passwords. A system administrator should be named who conducts account management, and a strict change control process must be in place.

Electronic audit trails (on software programs, instruments, LIMS, etc.) are essential for reconstruction of electronic data because they leave a record of who was operating the computer and what was done. Audit trails must be enabled at all times with only the administrator able to turn them off. The audit trails should be reviewed as part of the QC and QA process to confirm the integrity of the data.

In the typical laboratory process, data must be transferred from acquisition computers to processing computers and LIMS systems. In some circumstances, this involves the generation of intermediary files, and therefore, this transfer must be secure. An additional review process should be in place to ensure data integrity during the transfer process.

All electronic data, including the audit trails, should be maintained and archived at the end of the study.

2.2.4.4 Deviations Deviation from a SOP or from a study protocol must be documented and assessed for impact by the study director (for GLP studies) or BA management. An SOP should be in place on how to document a deviation. All deviations should be signed by the responsible scientist/PI, BA management, and the respective study director. Any remedial actions needed to rectify the deviation must be completed and documented. A master log file of deviations is often a useful tool to guide laboratory management. For example, if deviations to a particular SOP occur often, it may point to ambiguous wording or to a lack of training. Appropriate actions can then be taken.

2.2.4.5 Failure Investigations During method validation or routine sample analysis, any failed run or problematic run should be documented in the raw data. If study quality may potentially be impacted, or if it is deemed to be an unexpected event, a thorough investigation should be conducted. These investigations, conclusions and any corrective actions taken should all be documented in the study raw data and report so the events can be reconstructed during a review. It is recommended that a master log of investigations be kept for the entire laboratory that can be searchable and traceable. Periodic review of all investigations can reveal trends, which can be a useful tool for management to determine, for example, if certain instrumentation or assays are more unreliable than others.

2.2.4.6 Communications All relevant, study-related communications should be kept and archived. Communication with the study director (GLP) or management includes items such as receipt of the protocol and amendments, PI statement, receipt of samples, release of data, deviations, and so on. Another important communication to be kept in study records are any requests for sample reanalysis.

2.3 QUALITY ASSURANCE

A QA program is necessary for all laboratories conducting GLP activities to monitor the quality of the data generated by the laboratory and to promote confidence in BA results. QA audits ensure that the correct procedures were followed by qualified personnel in compliance with laboratory SOPs and the study protocol. In this section, we discuss the role of QA in the bioanalysis laboratory, even though QA must be an independent unit outside the jurisdiction of laboratory management.

QA for the BA laboratory encompasses all quality-related activities to ensure the validity of laboratory testing and analysis. A successful QA unit will not only provide an improved ability to pinpoint problems early but also act as a partner, resulting in the generation of reliable data. For any BA laboratory, specific QA program guidelines should be determined with input from laboratory personnel, management, and clients.

Ideally, QA personnel should be subject matter experts in current regulations and compliance, as well as be very familiar with the science they audit. QA audits for BA activities may cover items such as computer systems, LIMS, instrument qualifications, raw data audits, instrument audit trail reviews, reagent preparations, personnel training records, in-life inspections, and so on. In addition to study-directed audits, QA also conducts system audits to periodically review the overall compliance of the entire BA laboratory. Once an audit is completed, and corrective and preventive actions are taken to address any findings, a quality assurance statement (QAS) is issued stating that the quality of the data is acceptable, and the study is released.

In addition to GLP studies, clinical studies are also considered "regulated" and may be audited by QA in some companies. However, in general, a QAS is not issued for clinical studies. In our view, QA should also review BA method validations for both GLP and clinical applications since the reliability of any data depends on the robustness and quality of method validation.

2.4 SUPPORTING FUNCTIONS

For typical BA laboratory to operate effectively, several additional supporting functions are needed. Supporting functions include QC, sample management, archivist, technical support, planning, and reporting. Different companies have one or multiple personnel performing each of these tasks depending on their business needs. The supporting functions help BA to become a cohesive and successful BA operation.

2.4.1 Quality Control

The function of QC is to review laboratory procedures and data generated from assay validations and study sample analysis to ensure that the results are accurate and complete. The goal is to maintain the highest quality and dependability of the data. Unlike QA that is independent from the laboratory management structure, the QC group functions much like an internal business partner. However, it is important that the QC group have its own reporting structure. For example, they should not report directly to the manager responsible to generate the data to avoid potential conflicts of interest.

The QC process should be described in specific SOPs, which should define how to maintain the QC records for each particular study. The QC process also includes a detailed examination of the data, instrumentation, and laboratory processes. The QC records should preferably be archived along with the study results. In addition to SOPs, a very helpful tool is a QC checklist (see, e.g., the following list) that can serve as a record of what was reviewed:

- Review study protocols/amendments.
- Verify instruments/logbooks.
- Verify certificate of analysis for reference standard.
- Verify correct method version and acquisition method.
- Verify calculations and labels for stock solutions, sub stocks, calibration standards, and QCs.
- Verify adequate stability in solutions and biological samples.
- Review chromatograms for proper integrations and lack of carryover or interference.
- Verify run meets acceptance criteria.
- Review audit trails and timestamps of injection sequence.
- Verify documentation of failed run investigations and method/protocol/SOP deviations.
- Verify instrument raw data matches LIMS data input.
- Review project notebook for general documentation completeness.

2.4.2 Sample Management

The sample management group is one of the most indispensible supporting functions for a BA laboratory. Their core activity is the receipt, organization, and management of the biological samples, which is typically done with the assistance of a validated LIMS system. However, the sample management group may be involved in many other highly valuable activities such as the following:

- Creation of the study in the LIMS system.
- Generation of bar-code labels for the sample containers and shipping the labels to the collection site. In some cases, they may prelabel the containers to be shipped to the site.
- Maintain the freezer inventory to track sample location.
- Keep freezer temperature logs to demonstrate storage conditions for samples. This includes monitoring the freezers for proper function and initiating repairs when needed.
- Check that proper temperature was maintained during sample shipping and communicate with the collection site that the samples were received in good condition.
- Disposal of samples per SOPs when studies are completed. The major reasons for timely disposal of samples are twofold: (1) the stability of the analyte(s) of interest in a certain matrix may be limited, and the established freezer storage stability data may not cover an extended storage period; (2) sample storage space is limited at any company. Sample management groups normally work with the study PI, study director and/or the project team to dispose of study samples when study report is signed off. The same is true for outsourced studies.

- Maintain the compound material inventory, including tracking of storage conditions and chain of custody. Certificates of analysis and expiration dates for reference standards are also usually tracked by the sample management group. They also dispose of expired compounds.
- In some laboratories, the sample management group also maintains the QC samples and tracks their use for each study.
- If studies are contracted to an outside laboratory, the sample management group is usually involved in the shipment of the samples to the CRO.
- In some cases, internal freezer capacity becomes exhausted so the sample management group will arrange for storage at a contract facility and will manage the process, payments, and records.

2.4.3 Archivist

The archiving of study materials and test facility records is a crucial part of compliance with the principles of GLP. Well-organized and maintained raw data is the only means that can be used to reconstruct the study and enable the information contained in the final report to be verified. For laboratories conducting regulated studies, there should be a designated archivist on site. The archivist should be properly trained and is responsible for the day-to-day management of the archive facilities in accordance with established company policies, archive SOPs, and the principles of GLP. The archivist should also be able to efficiently retrieve the data in case of a regulatory agency audit. The responsibility of the archivist should include activities such as the following:

- Ensure that proper storage and retrieval of materials is maintained.
- Control and document movement of material in and out of the archives.
- Maintain chain of custody documentation for materials submitted to the archive.

Access to the archival facilities should be restricted and controlled by the archivist or their designee. If, for any reason, personnel other than archive staff need to access the archive room, they should be accompanied by a member of the archive staff, and the visits should be recorded with the identity of the visitor, the reason for the visit, the duration of the visit, and the materials reviewed during the visit.

2.4.4 Technical Support

The technical support team is responsible for assisting with many of the systems in the complex environment of a BA laboratory. The team will provide support to diverse technologies such as laboratory and scientific software applications, laboratory instrumentation, scientific instrument integration, computer and server networking, and numerous other IT applications. Some of the functions in which this team is typically involved are:

- LIMS support (e.g., Watson) such as acting as the administrator, creating studies, electronic archiving, troubleshooting.
- Software validations and instrument IQ/OQ/PQ qualifications.
- Software training and troubleshooting.
- Maintaining monitoring systems (e.g., freezer monitoring).
- Computer support for personal and equipment computers.

2.4.5 Planning

Recently, there is a growing trend toward dedicated planning groups within the laboratory organization. Forecasting future workloads, and planning resource utilization, are essential activities for an efficient operation. Planning groups can help management areas such as the following:

- Benchmarking historical performance to improve efficiency.
- Reviewing available resources for incoming projects.
- Tracking research expenditures and overhead.
- Assisting management to investigate business development prospects.

2.4.6 Report Generation

Driven by business need, a dedicated report writing function has become an integral part of many BA laboratories. Although scientists are ultimately responsible for accuracy and completion of reports, dedicated personnel may be hired and trained for the writing to improve the productivity of the BA laboratory.

For laboratories conducting regulated studies, the current FDA guidance (USFDA, 2001), EMA guidelines (European Medicines Agency, 2011), and FDA-Industry whitepaper (Viswanathan et al., 2007) contain specific recommendations on the content of BA reports. Since these documents are an important part of many regulatory submissions, the report format should be designed to include all of the recommendations.

For a BA report that supports a GLP study, the final report should contain the essential information required by the principles of GLP. The report should be audited by QA and it should include a QAS, According to GLP regulations, the BA PI is responsible for producing a BA report detailing the work performed under his or her supervision and for sending the report to the study director. There should be a PI statement certifying that the report accurately reflects the work performed and the results obtained, and that the work was conducted in compliance with the principles of GLP. If the work was done by a CRO, there should also be a QAS signed by the CRO's QA unit.

2.5 CRO MONITORING

Over the past decade, there has been a significant increase in the amount of outsourcing by pharmaceutical and biotechnology companies. Particularly in the past few years, along with the many challenges and changes in the pharma/biotech industries, this trend has been further accelerated, with emphasis shifting to the use of CROs in emerging market countries. During this time of increasing pressure to keep research and development (R&D) costs low, and to deliver more drugs to the market faster, there is a greater reliance on CROs to provide critical, knowledge-based services. CROs understand the merit of being a research partner to the pharmaceutical and biotechnology industries, and they have successfully been a valuable collaborator throughout all phases of drug development. It is essential for the sponsor and CRO to form a close partnership to ensure success of outsourced projects. In the following sections, we will focus on the roles and responsibilities of CRO monitoring, and on how CRO monitoring functions are a key partner in the bioanalysis operation.

2.5.1 Personnel

The personnel involved in CRO monitoring normally oversee the following activities: contract/purchase order (PO) setup and invoice approval, assay development/validation (including assay transfers), sample analysis, data transfer, and data/report review, and so on. These functions can be covered by a single individual or by different persons depending on the structure of the BA group and the available resources. For some companies, the contract/PO setup and invoice approval function may be done by a group of people who are outside of the BA group (e.g., the planning or financial group of the company). Proper experience and training are needed for people covering each of the outsourcing functional areas.

2.5.2 Outsource Process

The basic outsource process includes, but is not limited to, the following steps: (1) decide what needs to be outsourced; (2) select a qualified CRO; (3) set up the contract and PO; (4) approve invoices according an agreed schedule or milestones; (5) monitor assay development/validation and sample analysis activities; (6) work with the CRO to resolve issues (if any); (7) transfer data; and (8) review data and reports. The details of each of the above steps will be elaborated further in the following sections.

2.5.2.1 CRO Selection The scope of projects that need to be outsourced depends on the sponsor's outsourcing strategy, and their current workload, and available resources. Some sponsors may only outsource clinical studies after FIH activities are completed, while others may outsource all stages (discovery, preclinical, both non-GLP and GLP, and clinical studies), and some sponsors may even outsource the entire development program. After a sponsor decides what needs to be outsourced, the CRO will be chosen to conduct the activities. Different companies may have varying CRO selection processes, but the key factors to be considered during CRO selection are similar. These key factors may include qualifications in terms of both scientific expertise and regulatory compliance, ability to deliver quality data on time, and competitive pricing. The CRO selection team usually consists of people from the BA group, the QA unit and the procurement department (or other financial/outsource related departments). The CRO selection team will visit the CRO facilities and evaluate them on the above competencies. After the thorough evaluation process, a single CRO or several CROs are selected (the selected CROs may be referred to as "preferred vendors" by the sponsor). Before any actual outsourcing work is placed at a CRO, a confidentiality agreement and a master services agreement are established to make sure the sponsor's intellectual property is kept confidential and proper charges will be applied for the outsourcing activities. The outsourcing costs can be charged either by the study or by time equivalents.

2.5.2.2 CRO Site Audits and Visits Regular CRO site audits and visits are needed to ensure the selected CROs are in line with the current regulatory requirements and the sponsor's expectations. The frequency of CRO site audits and/or visits may vary depending on the workload outsourced to the CRO(s) and any "ad hoc" needs for certain studies. Normally, the sponsor's QA unit conducts a yearly audit to make sure the findings from previous audits have been properly addressed and the CRO's SOPs, data, facilities, staff, daily operations and practices are in compliance with current regulatory guidelines and standards. The sponsor's BA study monitor may also have regular visits (or "ad hoc" visits as needed) to review some specific areas or validations/sample analysis to make sure the CRO delivers high-quality data that meets both scientific and regulatory expectations.

2.5.2.3 SOPs There is no clear regulatory requirement whether a CRO must follow its own SOPs or the sponsor's

SOPs when performing each sponsor's projects. Before starting any actual outsourcing work, it is always a good idea to clarify which SOPs need to be followed. If the sponsor's SOPs will be used, they will be made available to the selected CRO, and the CRO's staff will be required to be trained on them before starting any projects. If SOPs from the CRO will be used, the BA study monitor from the sponsor will normally request the related SOPs and use them to guide their review of the CRO's data and reports.

2.5.2.4 Communications Effective communication between the selected CRO and the sponsor is key for the success of CRO monitoring. In addition to e-mail and phone communications, depending on the workload and project needs, a regular weekly, bi-weekly or monthly teleconference may be arranged between the CRO and sponsor to discuss project progress and possible options for resolving issues (if any). "Ad hoc" meetings may also be held to address some special or urgent needs and issues. Meeting minutes should be captured and documented with action items requiring follow-up.

2.5.2.5 Assay Development/Validation and Sample Analysis The CRO may start assay development from "scratch" if no previous activities have been conducted by the sponsor for a given project. On the other hand, the CRO may only need to do an assay transfer if a validated assay is already available from the sponsor's laboratory. Before the beginning of assay development/validation, or assay transfer activities, a validation plan/protocol (describing experiments to be done and the acceptance criteria, etc.) is normally drafted, reviewed, and approved by both the CRO and the sponsor. However, not all CROs and sponsors use a formal validation plan/protocol. The work scope of the validation and the acceptance criteria may be communicated and established in different ways (e.g., included in the contract, or referred to SOPs).

Delivering quality data on time for sample analysis is the main goal for BA outsourcing work. The sponsor's study monitor needs to make sure assays are properly established (i.e., validated assays for GLP and clinical studies, and qualified assays for non-GLP studies or for biomarkers, metabolites and urine samples in clinical studies) at the CRO before the study starts. Timelines for delivering results should be clearly discussed and established according to sponsor's requirements prior to beginning the study. The sponsor's study monitor needs to make sure that study protocols, including randomization codes (for blinded clinical studies), are provided to the CRO before start of sample analysis. Any protocol amendments also need to be provided in a timely manner. The staff at the CRO who are responsible to conduct the sample analysis and data review need to read, understand, and follow the study protocol and amendments (if any) and related SOPs as well as regulatory guidelines during sample analysis.

The sponsor's study monitor needs to work closely with the CRO to make sure the assay validations and sample analysis are done timely and properly from both scientific and regulatory points of view. The sponsor's study monitor and CRO need to work together to resolve any issues that may come up. Particularly for assay transfer issues, since the sponsor already knows the assay well, the sponsor will share the assay and the compound or metabolite related information with the CRO, and may also send the scientist who developed and validated the assay to the CRO's laboratory to help resolve any issues if deemed necessary.

2.5.2.6 Data/Report Review and Transfer The data/report review and transfer procedures for both validation and sample analysis vary among different sponsors and CROs. After the preliminary data or draft report is generated, the CRO will normally review the data (including audit trails) and report, and then deliver them as drafts to the sponsor's BA study monitor for their initial review. Depending on the nature of the studies (e.g., nonclinical (GLP vs. non-GLP) and clinical studies), the data review process and report format may be different. In general, for GLP and clinical studies, the final data and report must be reviewed by the CRO's QA unit before they can be released. For non-GLP (nonclinical) studies, QA review is usually not required. The sponsor's study monitor also needs to review the data and report (both the draft and final versions) carefully to further ensure quality data and report are delivered.

The data transfer process is also different among sponsors and CROs. In general, the data needs to be delivered to the sponsor in a secured way (e.g., encrypted files through e-mail or secured data exchange via online services). Normally, the raw data (both paper and electronic) is archived at the CRO according to their SOPs. The CRO's archival SOP should clearly indicate the process including how long the raw data will be archived at the CRO-specified archival site, and what will happen with raw data that exceeds the specified archival period (e.g., return the raw data to sponsor).

2.5.2.7 Pharma BA Laboratory versus CRO BA Laboratory BA laboratories in Pharma companies and CROs have many similarities and some differences. The major similarities are: (1) following the same guidance from FDA, EMA and other regulatory agencies for BA assay validation, sample analysis and all related documentations; (2) applying sound scientific judgment to guide daily work and solve problems; (3) utilizing similar instrumentation (e.g., liquid handler and LC-MS/MS system) for sample preparation and analysis; (4) implementing good quality system to ensure the data integrity and reconstructability; and (5) delivering quality data within established timelines.

The major differences are: (1) Pharma BA laboratory has many interactions with both CRO and sponsor's project teams while CRO laboratory normally interacts only with sponsor's study monitor, study director and QA unit. CROs do not usually interact with sponsor's project teams directly; (2) Pharma BA laboratory normally has a more comprehensive view of the entire process of the compound development (from discovery to development, and all the way to the final filing of the compound) while CRO BA laboratory in general is limited to the knowledge for a particular stage of the compound development; and (3) scientists working at a CRO may see many different types of molecules in different therapeutical areas, while the Pharma BA scientist may have more thorough knowledge about other important aspects such as metabolites and biomarkers for the projects.

There should be a complimentary rather than competitive collaboration between the scientists from both sides. Pharma BA laboratory tends to outsource established methods, as well as nonproprietary assays used for supporting drug–drug interaction and comparator studies. Currently, there is a trend of decreasing the internal resource at Pharma BA, while the CROs are expanding steadily and are able to attract and maintain talents. Many of the first class BA scientists are now working at the CROs on many cutting-edge technology applications. The relationship between Pharma and CROs are evolving from pure commodity of cost saving to business partnership of risk sharing.

2.6 CONCLUSIONS

BA is one of the few disciplines to cover all stages of product development, from early discovery to preclinical studies and all phases of clinical development. BA plays a crucial role in target identification, lead optimization, bioavailability estimation, patient stratification, and therapeutic drug monitoring. Because of the diversified BA support from discovery to late phase clinical trials, the process and personnel in any BA laboratory need both scientific and compliance rigor. Optimal design of laboratory and workflow, strategy for meeting today's regulatory requirements and scientific advances, as well as implementation of best practices are absolutely required so that resources are optimized, project timelines are met, and compliances are followed.

The demand is greater than ever to improve efficiency and reduce costs, but maintaining compliance is also a must in our complex regulatory environment. Compliance can never be sacrificed for the purpose of reducing costs. The BA laboratory also needs to keep up with the current regulations and industrial standards so that appropriate measurements are in place to de-risk any potential compliance issue. The BA laboratory should have a well-established procedure to properly document paper and electronic data to ensure data integrity. Proper documentation from deviations and timely investigations are required.

The laboratory should continually improve their SOPs and utilize the best practices for optimal performance. The BA laboratory also wants to benchmark against current industrial practices. Proper design of laboratory space and instrument location is needed for efficient workflow. Security and safety of the facility and staff are also required. Timely archival and retrieval of paper and electronic raw data is needed to be in GLP compliance. Sufficient investment and maintenance on the instruments and software are important. Both instruments and software should be vigorously tested for compliance before being used in regulated studies.

With a significant increase in the amount of outsourcing by pharmaceutical and biotechnology companies, CRO partnership has been essential in BA operation. From the outsourcing process to CRO selection along with CRO site audits, visits, and communication, all activities requires keen collaboration and partnership to meet scientific and regulatory compliance.

Investment in personnel is the most important task for the BA management. High priority should be given to hiring, training, and retaining every function of the BA laboratory. Scientific, technical, and managerial staff should be given an opportunity to develop their careers and fulfill their personal aspirations as well as meet business needs. Staff should be encouraged to make scientific presentations and publications. Exploration of emerging technologies and adaptation of industry best practices enhances the resilience of a BA organization in the current competitive environment.

REFERENCES

European Medicines Agency, Guideline on bioanalytical method validation, EMEA/CHMP/EWP/192217/2009. 2011.

Fast DM, Kelley M, Viswanathan CT, O'Shaughnessy J, King SP, Chaudhary A, Weiner R, DeStefano AJ, Tang D. Workshop report and follow-up—AAPS Workshop on current topics in GLP Bioanalysis: assay reproducibility for incurred samples—implications of Crystal City recommendations. AAPS J 2009;11(2):238–241.

ICH. Guidance for Industry E6 Good Clinical Practice: Consolidated Guidance. 1996.

Lee MS, Kerns EH. LC/MS applications in drug development. Mass Spectrom Rev 1999;18 (3–4):187–279.

OECD. OECD Principles on Good Laboratory Practice. 1998.

USFDA. Available at http://www.fda.gov/ScienceResearch/SpecialTopics/RunningClinicalTrials/default.htm. Accessed Feb 25, 2013.

USFDA. CFR—Code of Federal Regulations Title 21, Part 11 Electronic Records; Electronic Signatures. 2011.

USFDA. CFR—Code of Federal Regulations Title 21, Part 58 Good Laboratory Practice for Nonclinical Laboratory Studies. Last Updated 2011.

USFDA. Guidance for Drug Evaluation and Research. Guidance for Industry: Bioanalytical Method Validation. May, 2001.

Viswanathan CT, Bansal S, Booth B, et al. Quantitative bioanalytical methods validation and implementation: best practices for chromatographic and ligand binding assays. Pharm Res 2007;24(10):1962–1973.

3

INTERNATIONAL REGULATIONS AND QUALITY STANDARDS OF BIOANALYSIS

SURENDRA K. BANSAL

3.1 INTRODUCTION

Year 1986 – while working in a leading contract research organization (CRO), I went to a Pharma company to negotiate a bioanalytical project, and quoted them for method validation and samples analysis. The client was baffled —"why the validation cost? Just analyze the samples." I politely declined to perform just the samples analysis, saying that the validation was required to establish confidence in the assay before analyzing samples. I did not get the project, but got a good lesson on the state of bioanalytical regulation existing in those days. There were no standard rules for bioanalysis before 1990s. Sponsors would provide their own criteria for assessing the sensitivity, accuracy, and precision of analysis—the parameters required for example, by US regulations: 21 CFR: Sec. 320.29 (US 21 CFR 320.29). Since 1980s, the formal regulations for bioanalysis may have changed only a little, but the regulatory bioanalytical guidance has expanded considerably. The era of bioanalytical guidance started in 1990, when pioneers in regulated bioanalysis met in a global workshop on bioanalysis in Crystal City, Arlington, VA. This workshop started a series that later became popularly known as "Crystal City Bioanalytical Workshops." See Shah and Bansal, 2011 for details on historical perspective. This workshop provided the first whitepaper on the procedures and requirements for bioanalytical method validation and sample analysis (Shah et al., 1992). These procedures and requirements were acceptable to both, the practitioners and regulators (mostly US) of bioanalysis. The momentum generated by Crystal City Workshops led to the release of the first regulatory guidance in the world by US Food and Drug Administration (FDA) (US FDA guidance, 2001). Although FDA provided the first formal regulatory guidance (US FDA guidance, 2001) to the industry, it was not very different from the first Crystal City whitepaper (Shah et al., 1992), and therefore, the transition to adopt the guidance was not difficult for the bioanalysts.

3.2 GLOBAL BIOANALYTICAL GUIDANCE

Now, more than two decades later, majority of the bioanalysts around the world have become familiar with the procedures and requirements of bioanalysis as provided by the FDA guidance (US FDA guidance, 2001) and Crystal City whitepapers (Shah et al., 1992, 2000; Viswanathan et al., 2007). Also, during this period, there has been a proliferation of guidance for the analysis of samples from bioavailability (BA)/bioequivalence (BE) studies issued by various regulatory agencies around the world. The multitude of bioanalytical guidance available for BA/BE studies are not materially different from the FDA guidance, and provide only limited procedures and requirements for the bioanalysis of samples from BA/BE studies as applicable in different parts of the world (Table 3.1). Detailed guidance on bioanalytical validation and analysis was drafted by one other major regulatory agency, European Medicines Agency (EMA draft 2009), and its final version was issued in 2011 (EMA guidance, 2011). The EMA authorities respected the existing bioanalytical experience in the world, and their final released guidance (EMA guidance, 2011) does not conflict with the existing bioanalytical guidance and experience but supplements it. The FDA has already announced that they will be releasing a revision of their 2001 guidance in around 2013. It is hoped

Handbook of LC-MS Bioanalysis: Best Practices, Experimental Protocols, and Regulations, First Edition. Edited by Wenkui Li, Jie Zhang, and Francis L.S. Tse.

TABLE 3.1 Listing of the Comprehensive and a Sample of the Global Bioequivalence/Bioavailability (BA/BE) Studies-Specific Bioanalytical Guidance

Regulatory agency	Document	Year	Comments
Comprehensive guidance			
Food and Drug Administration (FDA), USA	Guidance for industry: bioanalytical method validation	2001	A comprehensive guidance
European Medicines Agency (EMA), European Union (EU)	Guideline on validation of bioanalytical methods	2011	A comprehensive guidance
Limited, specific guidance for BA/BE studies			
Agencia Nacional de Vigilancia Sanitaria (ANVISA), Brazil	Resolution RE No. 899, May 23, 2003: Guide for validation of analytical and bioanalytical methods	2003	Validation procedures used for chromatographic methods employed in BA/BE studies
	Manual for good bioavailability bioequivalence practices	2002	Detailed instruction for conducting bioanalysis specifically for BA/BE studies
	Resolution RDC No. 27, May 17, 2012	2012	Guidance for bioanalytical method validation to be used for studies in registration of drugs in Brazil
FDA, USA	Bioavailability and bioequivalence studies for orally administered drug products—general considerations	2003	Refers to FDA Bioanalytical Guidance of 2001
EMA, EU	Guideline on the investigation of bioequivalence	2002 (New), 2010 (rev 1)	Brief directive to use quality principles for bioanalysis
Therapeutic Products Directorate (TPD), Health Products and Food Branch (HPFB), Canada	Conduct and analysis of bioavailability and bioequivalence studies	2002	Refers to Crystal City 1 (Shah et al., 1992) report and provides brief description of bioanalytical requirements
	Conduct and analysis of comparative bioavailability studies (draft)	2009	Refers to Crystal City 1 (Shah et al., 1992) and 3 (Viswanathan et al., 2007) reports and provides brief description of bioanalytical requirements. It is being revised to adopt EMA's guideline on bioanalytical method validation (Ormsby, 2012)
Therapeutic Goods Administration (TGA), Australia	Note for guidance on investigation of bioavailability and bioequivalence	2002	Adopted EMA guidance
Medsafe, New Zealand	New Zealand regulatory guidelines	2001	ICH, EMA, or FDA guidelines are recognized
State Food and Drug Administration (SFDA), China	Technique guideline for human bioavailability and bioequivalence studies on chemical drug products	2004	Refers to FDA Bioanalytical guidance (US FDA guidance, 2001) and provides brief description of bioanalytical requirements
Ministry of Health and Labor Welfare (MHLW), Japan	Clinical pharmacokinetic studies of pharmaceuticals	2001	Brief description of bioanalytical requirements
Ministry of Health and Family Welfare (MHFW), Central Drugs Standard Control Organization (CDSCO), India	Guidelines for bioavailability and bioequivalence studies	2005	Brief description of bioanalytical requirements
ASEAN: Assoc. of Southeast Asian Nations (10 nations)	ASEAN guidelines for the conduct of bioavailability and bioequivalence studies	2001	Adapted from CPMP (EMA) guideline for BA/BE studies
Saudi FDA, Saudi Arabia	Bioeqivalence requirements guidelines	2005	Refers to FDA Bioanalytical Guidance (US FDA guidance, 2001)
Egyptian Drug Authority, Egypt	Guidelines for bioequivalence Studies for marketing authorization of generic products	2010	Brief description of bioanalytical requirements
Korea Food and Drug Administration (KFDA), S. Korea	Guidance document for bioeqivalence study	2008	Brief description of bioanalytical requirements

that FDA's revised guidance will also not change the bioanalytical guidance in principle, but would only clarify and extend to areas where guidance may be lacking.

The existing bioanalytical guidance documents from various national regulatory agencies can be categorized into "comprehensive" or "limited" guidance (Table 3.1). Comprehensive bioanalytical guidance, released by FDA (US FDA guidance, 2001) and EMA (EMA guidance, 2011) covers all aspects of bioanalysis from method validation to samples analysis from all nonclinical and clinical studies. Limited bioanalytical guidance documents, released by all other global regulatory agencies, provide various levels of details for bioanalysis—from a reference to comprehensive guidance to almost full guidance for analyzing samples from BA/BE studies. Table 3.1 provides a listing of the comprehensive bioanalytical guidance and a sample of BA/BE study specific guidance documents available globally. The limited local guidance available for BA/BE studies (Table 3.1) is not sufficient to perform the bioanalytical work entirely. Therefore, the comprehensive guidance documents from US FDA and EMA must be used to supplement the local guidance available from various countries in the world.

In the current economy, the global boundaries for performing and submitting bioanalytical work are shrinking. Therefore, having multiple guidance and regulations for bioanalysis (Table 3.1) only create additional burden to performing quality bioanalysis (Bansal et al., 2010; Timmerman et al., 2010). While globalization of bioanalytical work was already happening, the appearance of the second comprehensive bioanalytical draft guidance from EMA in 2009 (EMA draft, 2009) led to several discussions within the global bioanalytical community (Bansal et al., 2010; Timmerman et al., 2010; van Amsterdam et al., 2010) on the need for global harmonization of bioanalytical guidance. The discussions led to the creation of "global bioanalysis consortium" (GBC), which is organized by bioanalysts primarily from the global Pharma industry and CROs. GBC's mission is to generate a globally harmonized document on bioanalytical procedures and requirements that can be presented to the regulators of various countries or regions of the world. The intention and expectation is that such a document may influence global regulatory agencies to adopt the recommendations in their bioanalytical guidance (van Amsterdam et al., 2010). This would eventually lead to a globally harmonized bioanalytical guidance, which is the need of the day and is highly sought after by the industry. The harmonized guidance would benefit both the practitioners and regulators of bioanalysis and would improve scientific quality and compliance.

3.3 BIOANALYTICAL QUALITY

The available regulatory guidance applies to the validation and analysis of bioanalytical samples, and covers both chromatographic and ligand binding assays. Methods used for determining concentrations of biomarkers for assessing pharmacodynamic endpoints are out of scope for the available guidance. In future, the regulatory guidance may also cover some aspects of biomarker quantification. Adhering to bioanalytical guidance and applicable regulations should ensure high quality of the submitted data. But what is high quality in bioanalytical data? Is the "quality" understood and achieved in a uniform manner globally by the practitioners and regulators of bioanalysis? Strategically, the bioanalytical quality is further explored further in the following sections.

3.3.1 Bioanalytical Quality Strategies

Bioanalysis is a systematic comparative technology. Concentrations in bioanalytical test samples are measured by comparing them to reference standards. Concentrations in bioanalytical samples are generally unknown and are within a wide range. At most, the bioanalysts would have rough estimates of the concentrations. Despite these challenges, the bioanalysis produces quite accurate, precise, and reproducible results when the bioanalytical work is performed with proactive high-quality strategies and principles. The bioanalytical quality should be assessed both quantitatively and qualitatively.

3.3.2 Quantitative Assessment

Quantitatively, the quality is assessed by determining accuracy, precision, and reproducibility of the measured concentrations (Figure 3.1). True concentration in test sample is the target for quantification, but it is unknown. Therefore, surrogate quality control (QC) samples are utilized to assess the accuracy and precision of bioanalysis. Regulatory guidance defines the allowed precision and accuracy for the QC samples, and it is depicted in Figure 3.1 for

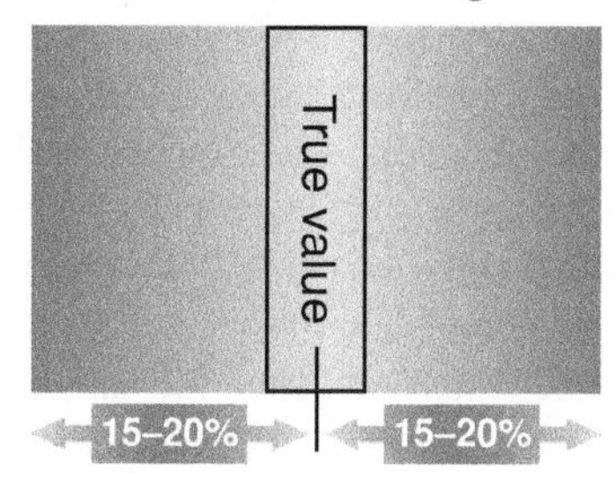

FIGURE 3.1 Quantitative measures of quality.

chromatographic methods. Reproducibility of measured concentrations is assessed quantitatively by reanalyzing a fraction of already analyzed test samples through incurred sample reanalysis (ISR).

3.3.3 Qualitative Assessment

Qualitatively, high quality in bioanalysis should be built proactively through good documentation and compliance with the regulatory guidance, standard operating procedures (SOPs), and the bioanalytical method. To ensure high quality, appropriate quality checks and controls should be built in and monitored throughout the bioanalytical process. For this discussion, the bioanalytical process has been divided into four stages. The checks and controls required in each stage are described below and listed in Table 3.2. The checks and controls are built in all stages of the bioanalytical process—from writing of the bioanalytical method to writing of the bioanalytical report. If errors or anomalies are observed, the built-in quality checks and controls should allow an unbiased investigation and diagnosis of the problem. The check and controls should be part of the routine process for performing the bioanalytical work in compliance with the global bioanalytical regulations and guidance. The descriptions provided in the following sections highlight the importance of proactive checks and control in achieving high quality in bioanalysis. The following sections highlight the quality principles, and are not meant to provide a full process for conducting bioanalysis. The users should follow their SOPs and global regulatory guidance to perform the analysis. Appropriate quality checks and controls inserted within the bioanalytical process would not only enhance the quality, but would also prove valuable in diagnosing and solving any unexpected problems observed during the analysis.

3.3.3.1 Written Instructions

a. *Standard operating procedures*: The bioanalytical laboratory should have a set of strategically written SOPs, complying with global regulations and guidance. The SOPs should provide strategies and procedures to perform high-quality work. Besides providing the operating procedures for bioanalysis, the SOPs also define the culture in the laboratory. Therefore, each laboratory site should have its own set of SOPs, and all analysts should be well trained on working with this set of SOPs and other relevant procedures required in the laboratory.
b. *Bioanalytical method*: The bioanalytical method should be written as a method document (preferred) or SOP before analyzing samples. The method should provide all operating instructions for the analyst to follow so that the method can be used reproducibly throughout the samples analysis.

3.3.3.2 Written method → Analytical Run

a. *Reference standards*: Only purity certified reference standards within their stability period should be used for regulated bioanalysis, and they should be stored according to the specified instructions.
b. *Stock solutions*: Calculations for making stock solutions must be accurate and should be carefully verified to avoid propagating any errors. Purity and molecular correction factors for the salt form of the reference standard should be taken into consideration when determining the correct weight of the free form of the reference standard. The homogeneity and accuracy of the stock solutions preparation from solid reference standards must be ensured by making stock solutions from two independent weighings and comparing them. The response factor for the two stock solutions should match within a narrow window (e.g., 5% using LC-MS measurement) before utilizing them for the preparation of spiked calibration standards and QC samples. Careful consideration should be given to the storage of stock solutions and when taking out aliquots from stock solutions if they are stored over a long period of time. Concentrations of stock solutions can change due to evaporation of volatile solvents during storage and handling even when the reference standard may be stable in the stock solution.
c. *Calibration standards and QC samples*: Stock solutions are spiked to biological matrices to prepare calibration standard and QC samples. Extra care must be taken to calculate and prepare these samples. The calibration standards and QC samples are the backbone for achieving quality results in bioanalysis, and must therefore be prepared without any error. To avoid mistakes in calculations, it is recommended that the calculations are performed using a verified spreadsheet or software. If the calculations are performed manually, they should be carefully checked by a second person to avoid any possibility of errors.
d. *Sample processing*: For reproducible incurred samples reanalysis, it is important that homogenous and accurate aliquots are taken for analysis. To ensure homogeneity of aliquots, the frozen test sample should be fully thawed and thoroughly mixed prior to taking the aliquots. The internal standard must be added precisely using an accurate and precise pipette. The analysts should take extra precautions to avoid sample mix-up or contamination during samples processing.
e. *Instrument settings*: Instruments should be set within the tolerance as described in the method. If changes in settings are made, full documentation of the revised settings should be kept.

TABLE 3.2 Proactive Quality Checks and Controls for Achieving High Quality in Bioanalysis

Table 3.2a: Written instructions

Quality Check or Control	Comment
SOPs (standard operating procedures) • Provide operating procedures • Define culture in the lab	Strategic SOPs complying with regulatory guidance are preferred over too prescriptive SOPs.
Bioanalytical Method • Description of reference standards and reagents • Instrument settings • Preparation of calibration standards and QC samples • Sample extraction procedure • Calculation procedures • Analytical notes and precautions	Written method should be available as independent method report (preferred), or SOP.

Table 3.2b: Written method → analytical run

Quality Check or Control	Review or Perform
Reference standards	• Purity certification. • Stability.
Stock solution	• Accurate Calculation. • Ensure homogeneity and accurate preparation by duplicate weighing and comparison of stock solutions.
Calibration standards and QCs	• Accurate preparation scheme. • Accurate calculations and preparation. Calibration standards and QC samples are the backbone for quality bioanalysis
Sample processing	• Accurate and homogeneous aliquot taking. • Precise addition of internal standard. • Prevent mix-up and contamination.
Instrument settings	• Instrument settings are as described in the written method. • Changes in settings from written instructions permitted only as specified in the method to obtain optimal performance ○ Keep adequate documentation and/or audit trail for changes in settings.
System suitability	• Performed and checked prior to starting the analytical run.

Table 3.2c: Analytical run → reportable data

Quality Check or Control	Review or Perform
Chromatographic integration and review	• Integrate all chromatograms (standards, QCs, and samples) uniformly. Review chromatography for proper integration.
Calculate data	• Calculate data as specified in the method.
Incurred samples reanalysis	• Perform and review throughout analysis to catch ○ Inadvertent mistakes. ○ Inhomogeneous sample aliquots. ○ Method problems (new methods).
Unexpected problem resolution	• Review independently without bias. • Instruments, method, analysts.

(*continued*)

TABLE 3.2 ***(Continued)***

Table 3.2d: Reportable data → final report

Quality check or control	Review or perform
Reports	• Create accurate and easy to follow tables and transfer tables to report. ○ Ensure right tables are transferred. ○ Automation in report writing is beginning but is not yet widely used.
Report review	• Review report for errors. • Independent audit of report by quality assurance.
Archive data	• Archive electronic data ○ May be used for reprocessing if needed later.

f. *System suitability*: System suitability for chromatographic assay is very important and should be performed prior to starting the analytical run. The chromatographic system should be checked for required sensitivity and specificity for the assay. Reproducibility should also be checked during system suitability if the system is not inherently reproducible, for example, in chromatographic assays with no internal standard. If the system is not suitable, the analytical run should not be started. After starting the analytical run, the acceptance of the analytical run should be assessed by the calibration standards and QC samples and not from the system suitability samples.

3.3.3.3 *Analytical Run → Reportable Data*

a. *Chromatographic integration and review*: All chromatograms from calibration standards, QC, and test samples should be integrated uniformly, preferably without any manual integration. The chromatograms should be reviewed for proper peak integration prior to using the peak response for concentration calculation to avoid any bias.
b. *Concentration calculations*: The data must be calculated using the same standard curve fit and weighting as specified in the method. Changing of calculation method should not be allowed during samples analysis.
c. *Incurred samples reanalysis*: To get the most value from the incurred samples reanalysis, the ISR must be performed throughout samples analysis to catch any inadvertent mistakes, inhomogeneous sample aliquot taking, or other method problems, especially for the newly adopted methods. Performing ISR at the end of the study will not catch the errors in time to correct the problems when observed.
d. *Unexpected problem resolution*: When unexpected problems or anomalous results are observed, the checks and controls built within the bioanalytical process should be used to review the problem independently without any bias. The independent unbiased review should include the review of instruments, method, and the analyst's performance. These checks and controls also allow early observation of the problem. It is easier to resolve the problem if it is discovered early.

3.3.3.4 *Reportable Data → Final Report*

a. *Reports*: Accurate and easy to follow bioanalytical reports should be created. Lengthy bioanalytical reports are generally not helpful. Inclusion of extra raw data only makes the report difficult to understand. Also, the unnecessary information in the report tends to hide the real problems in the bioanalytical work. The reports must be concise and transparent. The reports should contain brief text to explain what was done, and provide results in clear, easy to follow tables. When the work is performed using a laboratory information management system, it is easy to extract all tables from the database. But, care must be taken to select the right tables for the report under preparation. Automation in creating bioanalytical reports is beginning, but it is not prevalent yet.
b. *Report review*: The reports should be reviewed for errors by an independent reviewer. The regulatory reports should also be audited by an independent quality assurance.
c. *Archive data*: Both the electronic data and paper documents should be archived. The electronic data would be required, should there be a need to reprocess bioanalytical results at a later time. With the continued increase in the use of electronic notebook and archiving systems, it will not be long when the paper archives will become a dinosaur in the bioanalytical world.

3.4 SCIENCE, QUALITY, AND REGULATION

Modern bioanalytical science is not the same as it was in pre-regulatory guidance days prior to 1990. Excellent progress has been made in multiple areas of bioanalysis, e.g. in speed, specificity, and sensitivity. It is now a routine to develop highly sensitive and specific bioanalytical methods for multiple analytes in complex matrices within a short time. Sample volume requirements have also been reduced to only a few micro liters of biological fluid. This continuous progress in science requires additional or revised regulatory guidance and quality measures. To ensure continued scientific progress, the regulatory guidance and quality measures cannot remain very rigid. A process should be available to revise them as needed. It is very reassuring to see that the regulators are willing to work with the practitioners of bioanalysis. As the science evolves, the regulators are willing to revise their guidance. With this encouragement, hundreds of bioanalysts are working under the umbrella of GBC to review and harmonize the best procedures and requirements for bioanalysis, which may become the basis for future guidance and regulations. Working together, the regulators and bioanalysts can enhance the quality and global consistency of the bioanalytical data. The bioanalytical science, quality, and regulations are truly inter-related within a logical and harmonious cycle. No part of this cycle is above the other, and any part should be able to initiate the cycle of logic and harmony (Figure 3.2). In the beginning, Science may have initiated the cycle, but in a mature bioanalytical field where high quality is in the minds of both the practitioners and regulators, it is not uncommon for quality or regulation to start the bioanalytical cycle of logic and harmony. It is the responsibility of everyone working in the bioanalytical field to keep bioanalytical science logical and of highest quality. Then, everyone will benefit from the quality in bioanalysis. The practitioners will receive clear harmonious guidance for performing their bioanalytical work and the regulators will receive high-quality filings to review. Ultimately, the patients using the pharmaceutical products will benefit from this global bioanalytical harmony, through rapid and consistent reviews of the studies by the regulators.

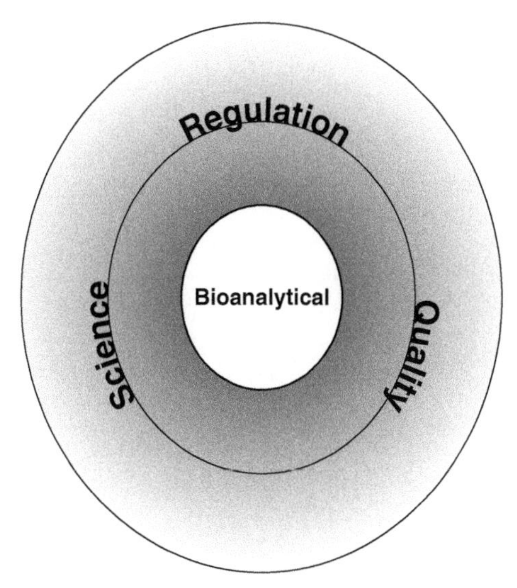

FIGURE 3.2 Bioanalytical cycle of logic and harmony.

ACKNOWLEDGMENTS

The author thanks Lisa Benincosa, Faye Vazvaei, and Sarika Bansal for their valuable edits and suggestions, and Bhavna Malholtra for the help in creating figures.

REFERENCES

Bansal SK, Arnold M, Garofolo F. International harmonization of bioanalytical guidance. *Bioanalysis* 2010; 2(4): 685–687.

EMA: European Medicines Agency, Committee for Medicinal Products for Human Use. Guideline on Validation of Bioanalytical Methods (Draft) [EMEA/CHMP/EWP/192217/2009]. 2009. Available at http://www.ema.europa.eu/pdfs/human/ewp/19221709en.pdf.

EMA: European Medicines Agency, Committee for Medicinal Products for Human Use. Guideline on Validation of Bioanalytical Methods [EMEA/CHMP/EWP/192217/2009]. 2011. Available at http://www.ema.europa.eu/docs/en_GB/document_library/Scientific_guideline/2011/08/WC500109686.pdf.

Ormsby, E. Canada's New Bioequivalence Guidances, presented at 6th Workshop on Recent Issues in Bioanalysis, San Antonio, TX, 2012.

Shah VP, Bansal, S. Historical perspective on the development and evolution of bioanalytical guidance and technology. *Bioanalysis* 2011;3(8):823–827.

Shah VP, Midha KK, Dighe SV, et al. Analytical methods validation: bioavailability, bioequivalence and pharmacokinetic studies. *Pharm Res* 1992;9(4):588–592.

Shah VP, Midha KK, Findlay JW, et al. Bioanalytical method validation—a revisit with a decade of progress. *Pharm Res* 2000; 17(12):1551–1557.

Timmerman P, Lowes S, Fast DM, Garofolo F. Request for global harmonization of the guidance for bioanalytical method validation and sample analysis. *Bioanalysis* 2010;2(4): 683.

US 21 CFR 320.29 (Code of Federal Regulations), Food and Drug Administration. Analytical methods for an in vivo bioavailability study. 42 FR 1648, January 7, 1977, as amended at 67 FR 77674, December 19, 2002. Available at http://www.accessdata.fda.gov/scripts/cdrh/cfdocs/cfcfr/CFRSearch.cfm?fr=320.29.

US FDA: US Department of Health and Human Services, Center for Drug Evaluation and Research. Guidance for Industry: Bioanalytical Method Validation. 2001. Available at www.fda.gov/downloads/Drugs/GuidanceComplianceRegulatoryInformation/Guidances/UCM070107.pdf.

van Amsterdam P, Arnold M, Bansal S, et al. Building the Global Bioanalysis Consortium – working towards a functional globally acceptable and harmonized guideline on bioanalytical method validation. *Bioanalysis* 2010;2(11):1801–1803.

Viswanathan CT, Bansal S, Booth B, et al. Quantitative bioanalytical methods validation and implementation: best practices for chromatographic and ligand binding assays. *Pharm Res* 2007;24(10):1962–1973.

PART II

CURRENT UNDERSTANDING OF LC-MS BIOANALYSIS-RELATED REGULATIONS

4

CURRENT REGULATIONS FOR BIOANALYTICAL METHOD VALIDATIONS

MARK E. ARNOLD, RAFAEL E. BARRIENTOS-ASTIGARRAGA, FABIO GAROFOLO, SHINOBU KUDOH, SHRINIVAS S. SAVALE, DANIEL TANG, PHILIP TIMMERMAN, AND PETER VAN AMSTERDAM

4.1 INTRODUCTION

Ensuring patient health and safety is a common objective for the pharmaceutical industry, their contract research partners and most certainly that of health authorities that regulate both. Over the past 30 years, drugs have become more and more potent and only advances in technology have allowed the measurement of drugs and their metabolites at the lower circulating concentrations. In parallel, the users of the drug concentration data, the pharmacokineticists, have asked for data with improved accuracy and precision to enable the development of better models that can support drug registrations. The response of health authorities to the increased significance of the pharmacokinetic data within a filing has been to increase their levels of scrutiny and implement new regulations and guidance designed to ensure sound science and data quality in bioanalytical validations and the analysis of nonclinical and clinical study samples for drugs and metabolites. Bioanalytical scientists operate in this milieu of sound science, data quality, regulatory compliance and advancing technology. Over the past decade, additional pressures have been felt as the pharmaceutical industry is facing pressures to reduce its operating costs while at the same time seeks to file its drugs across the world. Reducing costs for the bioanalysis has been achieved in many laboratories through implementing appropriate technology, but it is the globalization of the filings that has caused the bioanalyst to move from inwardly looking at their laboratory operations in relation to their own country's regulations, to outwardly looking at the regulations that govern bioanalysis from multiple countries and regions. While a core of commonality exists among the rules and recommendations promulgated by various health authorities, differences do exist and sometimes conflict. To aid the bioanalytical scientist in dealing with the variety of international regulations, this chapter is intended to provide insight into the current regulations applicable to the use of liquid chromatography–tandem mass spectrometry (LC-MS/MS) in bioanalysis from a number of countries and regions including Brazil, Canada, China, the European Union, India, Japan, and the United States.

4.2 CONTEXT OF THE REGULATORY ENVIRONMENT

Regulation of bioanalysis has had many roots. At their core, all are designed to show that the method, and, thereby the data being generated, is sound and reliable for its intended purpose. There can be many ways to achieve this and for more than two decades, regulators have been responding to observations seen in the field and within pharmaceutical filings, while the scientific community has implemented new technologies and sought ways to demonstrate their validity.

A challenge to LC-MS/MS bioanalytical scientists in working through the regulations from various countries is the understanding of the terminology. Like any area of new science, scientists involved in validations and regulated bioanalysis started without consistent and clear definitions; each having their own perspectives on what was needed to deliver

Handbook of LC-MS Bioanalysis: Best Practices, Experimental Protocols, and Regulations, First Edition. Edited by Wenkui Li, Jie Zhang, and Francis L.S. Tse.

scientifically sound, quality data. But as the scientific community has published methodology, met in conferences, created white papers and later, interacted with the regulatory community, they have accepted and used single definitions to enable clear and effective communications.

An example of such evolution can be seen in the definitions of validation, partial validation, and cross-validation that occurred in 16 years from the first to the third AAPS/FDA Crystal City conferences on bioanalytical method validation. The first meeting's discussion put forth basic principles for the experiments and practices required to demonstrate and then use reliable, sensitive, specific bioanalytical methods with known accuracy and precision (Shah, 1992) to measure stable drugs and metabolites in nonclinical and clinical samples. These values of these validation principles were immediately realized by Health Canada and rapidly introduced into regulatory guidance in 1992 (Health Canada, 1992). However, the terminology related to partial validation and cross-validation was somewhat inconsistently applied within the conference report. That confusion was clarified within the second Crystal City meeting's report (Shah, 2000) and then codified in the FDA's 2001 guidance on bioanalytical method validation (FDA, 2001). The definitions described there quickly became the *de facto* standard for many countries, and bioanalysts at the third Crystal City meeting were seeing requests from a variety of counties for cross-validation data. Similar language and concepts on validation and partial validation were placed within Brazil's 2003 ANVISA Resolution RE No. 899 (ANVISA 2003), which has been substantially updated in RDC 27 (ANVISA 2012). The European Medicines Agency (EMA) released their guideline on bioanalytical method validation in 2011 (EMA, 2011) that is consistent with the basic concepts previously noted but goes on to provide some further refinements. To avoid additional guidances, Canada has decided to adopt the EMA guideline (Health Canada (2012)). In contrast, while this chapter was being finalized, Japan has chosen to issue its own guideline as a draft for comment.

Technical advances have preceded the evolution in regulations, and in fact driven the need for some specifics found in today's regulatory documents. In 1990, most drug concentration measurements were performed by high-pressure liquid chromatography (HPLC), gas chromatography mass spectrometry (GC-MS), or enzyme-linked immunosorbent assay (ELISA). Since then, small molecule drug analysis has shifted to LC-MS/MS analysis with its increased sensitivity and improved specificity that also provide for shorter run times. ELISA assays for small molecules are rare today, and most HPLC assays cannot achieve the needed sensitivity of today's more potent drug molecules. This shift to LC-MS/MS, driven primarily by the simplified ease of use and improved ruggedness, has not been without problem. In-source degradation, ionization issues, suppression or enhancement, and adduct formation were found through advanced use of the technology as users tried to get the most from the instruments to offset their higher cost. Stable isotope labeled internal standards help to a large extent to recognize and reduce the effect of these kinds of problems and are thus, nowadays commonly used within regulated bioanalysis. What was thought to be the ultimate panacea for the bioanalyst has been shown to be of extremely high value, but one that still requires that the bioanalyst be a sound scientist and chromatographer. Bioanalysts are once again being challenged today as LC-MS/MS is applied to quantify large biologics from a specific enzymatic-digestion fragment; having to develop new experiments for the validation to deal with the potential for many fragments of similar chromatographic characteristics and molecular weights to exist within the enzymatic digest.

4.3 VALIDATIONS OF METHODS

A bioanalytical method validation includes all of the procedures that demonstrate that a particular method used for quantitative measurement of analytes in a given biological matrix, such as blood, plasma, serum, or urine, is reliable and reproducible for its intended use. The fundamental parameters for a validation include (1) accuracy, (2) precision, (3) selectivity, (4) sensitivity, (5) reproducibility, and (6) stability. To achieve this, a number of experiments must be performed to fully characterize the method specific to those parameters. While not previously specified, additional characteristics, for example, linearity, recovery, and IS normalized matrix factor, are now routinely evaluated to complete the validation's characterization of the method in compliance with the EMA's 2011 guideline. The literature is replete with articles describing multiple approaches to complete each of the experiments and the authors will not repeat them here. One should be aware, however, that the experimental designs that, while scientifically sound, are unique or nonstandard may result in questions being asked by regulatory authorities who may be less familiar with the science or approach. A best practice is to define the experiments needed to make the method fit-for-purpose in a validation plan. While this is not required under FDA good laboratory practices (GLPs), most organizations operating under Organization for Economic Cooperation and Development (OECD) GLPs (OECD, 1998) do implement validation plans with study director oversight. Changes to a validated method, when major, may require a complete revalidation of many of the measures of performance, while minor changes may only require a partial validation.

Partial validations are needed when a method has been changed slightly or when a method is transferred to another laboratory. Similar to the approach used for deciding the extent of a validation needed for a method to be fit-for-purpose, determining what experiments are needed in a partial validation is best based on a scientific understanding of the method and the implications of the changes to a method. For instance, changes in extraction solvent may result in a change in the background interferences present in the column and during ionization in the mass spectrometer source, while a change in solvent manufacturer where the purity is identical is likely to have little change in the chromatography; thus, the experiments to prove the lack of impact should be designed appropriately. More experiments are needed for the change in extraction solvent; whereas an assessment may determine that experiments are not needed for the change in solvent manufacturer. The extent of the testing needed is best expressed in the EMA Guideline: "Partial validation can range from as little as the determination of the within-run precision and accuracy, to an almost full validation." Most countries agree that at a minimum, a single accuracy and precision run is needed.

Cross-validations are needed under two scenarios: (1) when two or more unique methods are used in the same study and (2) two or more methods are used in a filing. In both cases, the intent is to demonstrate to the regulatory authority reviewing the data that it is comparable irrespective of the method and that it may be fully relied on for decision making. Experience has shown that even the execution of the same method in two laboratories does not ensure identical results. Owing mainly to the interpretation of the written procedure, differences in equipment and personnel techniques, differing results can be readily detected during cross-validations and corrected prior to putting study samples at risk. Cross-validations may be conducted in a variety of manners using quality control (QC) samples, incurred samples or incurred sample pools. Where identical methods are used in two different laboratories, QC samples may be sufficient. When two different methods are being compared, QC samples and incurred samples or incurred sample pools are typically used. Consideration to the use of individual incurred samples or pooled samples must be made. The former may not have sufficient volume to be tested in replicates in each laboratory, while the latter allow replicates to gain accuracy and precision information, however they have the potential to hide differences in the methods by diluting interferences present at high levels in some samples. In the end, good scientific rationale should be used in the decision making of the design of the cross-validation. An SOP or validation plan, developed *a priori*, are also best practice in ensuring that the acceptability of the cross-validation is not influenced by the interpretation of the results. Lastly, a consideration must be given to the use of incurred clinical samples, where their use must be covered within the patient's or subject's informed consent. Only the EMA has issued acceptance criteria for the cross-validation: the mean accuracy of QCs should not differ by more than 15% from their nominal concentration, and when incurred samples are used, two-thirds of the samples tested may not differ by more than 20% from the mean.

4.3.1 Matrix Requirements

Most countries agree that the matrix used in validation should be the same as that used for the study samples. Brazil and Japan agree in practice that alternate matrices can be used for endogenous analytes. The ANVISA, FDA and EMA have flexibility around the matrix requirements when working in rare or hard to obtain matrices (e.g., vitreal fluid). It is generally agreed that the anticoagulant used in the validation and for standards and QCs during sample analysis must match that of the study samples. Much discussed but not defined within any of the regulations is the need to use the same salt form of the anticoagulant as used in study samples. Partial validations are required when changing anticoagulants, and one should always consider room temperature and frozen stability within those validations, as anticoagulants have different abilities to inhibit enzymes capable of metabolizing drugs and metabolites.

4.3.2 Reference Standards

Most countries are in agreement that for the drug of interest, a certificate of analysis (CoA) is needed. What is not specified or specified as having lesser requirements is the CoA for metabolites or internal standards. In general, the CoA for the drug should contain the identity of the material, lot or batch number, expiration date, purity, storage condition and the manufacturer. In Brazil, only nonpharmacopeic reference standards must present a detailed CoA prepared by the manufacturer. EMA and FDA specify that no CoA is needed for internal standards, as long as they are demonstrated to be suitable for their intended purpose. In contrast, Brazil requires a CoA for the internal standard. When considering metabolites, which are frequently made or isolated in limited quantities, characterization of the reference material should be performed to the extent possible; which will frequently require the analytical facility to implement procedures that are material sparing.

4.3.3 Robustness Testing

Previously, in RE 899, ANVISA was somewhat unique in requiring robustness testing, however, this requirement has been recently removed (ANVISA 2012). More recently, robustness testing has been accomplished through the growing use of design-of-experiment procedures during method development. However approached, understanding how the subtle shifts in pH of extraction buffers or mobile phases, or organic contents, mixing times, etc., influence a method are important to ensure robustness and the ability of multiple people to successfully use the method.

4.3.4 Sensitivity

Practically, bioanalytical methods must be sufficiently sensitive to measure the analytes to enable proper characterization of the pharmacokinetics to answer the questions posed by the study design. EMA provides an example for bioequivalence (BE) studies that the lower limit of quantitation (LLOQ) should be no higher than 5% of the expected C_{max}. In implementing, regulators are in agreement and require that the accuracy and precision of the method must be within 20% at the LLOQ. EMA, and FDA require that the signal-to-noise at the LLOQ be $\geq$5:1. The Crystal City III conference report also provides some guidance for methods on endogenous analytes and the need for bioanalytical scientists to establish adequate procedures for dealing with concentrations of endogenous analytes in blank matrix that exceed 20% of the LLOQ.

4.3.5 Selectivity

The selectivity of a method in its simplest form is the ability of a method to "measure unequivocally and differentiate the analyte(s) in the presence of components, which may be expected to be present" (FDA, 2001) in the sample. This means that the method must be able to differentiate the analyte (drug or metabolite) from endogenous components, comedications and other drug related compounds (i.e., other metabolites and stereoisomers).

While more numerous, the endogenous components are most easily addressed during method development. All regions have an expectation for demonstrating the difference from endogenous components. Japan prefers to have matrix from three male and three female donors tested, while both the EMA and Brazil's ANVISA require testing plasma methods during validation with hemolyzed and lipemic matrices in addition to at least four lots of normal or preferably, disease state matrix. These tests compare the response at the retention time of the analyte in blank matrix against the analyte spiked into the same lots. When looking at lipemic and hemolyzed plasma, one can also detect the influence of these conditions on the robustness of the method. ANVISA, FDA, and EMA specify that endogenous components should not have responses greater than 20% of the same lot of matrix spiked to the LLOQ of the method.

With regard to comedications, many laboratories have continued their practice from the use of HPLC-UV methodology of using a cocktail of comedications spiked into QCs to demonstrate the lack of impact on the analytes. With the expansion of the formulary seen in population PK studies, it is impractical and may not be necessary for most drugs when using LC-MS/MS due to its superior selectivity. However, one must understand the drug's expected *in vivo* characteristics to be able to scientifically state that comedication method challenges are not needed. Those *in vivo* characteristics include protein binding and the capacity of those binding proteins, since analytes highly bound to low-capacity proteins may be displaced, and therefore, behave differently during the extraction procedures.

Specificity from drug-related compounds during early drug development is one of the more challenging characteristics of a method to prove. It is not unknown for some metabolites or chemical degradants to be isobaric with similar fragmentation patterns to the drug or for metabolic stereoisomers of the drug to be formed. Additionally, *N*-glucuronides are known to degrade in the source back to the aglycone and, if not chromatographically resolved, to interfere with the analysis. In the absence of *ex vivo* samples to challenge a method during its development, it is recommended that once they are available, alternate chromatographies be employed on sample extracts to demonstrate the specificity of the method from these potential components. The results of such testing should be documented through addenda to the method validation reports.

4.3.6 Recovery

Recovery is required by most health authorities as a key characteristic of the method. The approaches to generating a number vary between countries as do the interpretation of the results. India required recovery testing at low, medium, and high concentrations with an expectation of methods with low recovery being redeveloped to improve recovery. Brazil has no absolute requirement, but does require that the %CV (coefficient of variation) of the value be <20%; demonstrating the reproducibility of the method. The FDA only requires that the recovery is consistent, precise, and reproducible. EMA and ANVISA do not address recovery, preferring to focus on the matrix effect as the more influential factor affecting the reliability of the data.

4.3.7 Matrix Effects

At the time of writing, only the 2011 EMA and 2012 ANVISA Guidelines have a health authority requirement for testing matrix effects. Its first recommended use, however, appears in the 2007 AAPS-FDA Crystal City III Conference report, which due to the involvement of the FDA and industry is considered to be an agreed upon position and extension of practices both parties agree to operate by, and therefore, the *de facto* standard from then to the release of the EMA guideline. In both writings, the matrix factor is defined as the IS normalized response for the analyte in the presence of matrix divided by the IS normalized response in the absence of matrix (i.e., in buffer, water, or solvent). It is typically measured at low and high concentrations in at least six lots of matrix with the %CV of the results to be $\leq$15%. Interestingly, the AAPS-FDA report recognized that using stable labeled internal standards obviated the need for this test, while the

EMA requires it in all cases and adds that the impact of excipients (i.e., polyethylene glycol or polysorbate) that may be in the circulation should be considered as part of the testing. The EMA guideline also calls for testing additional matrix conditions that may impact the results, including hemolyzed, lipemic and special populations, hepatic or renally impaired patients. ANVISA regulations require testing eight matrix lots: four normal, two lipemic, and two hemolyzed. However tested, the key success factor is for the method to have a consistent matrix factor in all samples.

4.3.8 Calibration Curves

The fundamental requirements have not changed since the AAPS-FDA Crystal City I conference and have been included in most countries regulations: a calibration curve shall consist of at least six nonzero concentration samples covering the expected concentration range of study samples or the extent of the instrument's achievable range. The number of concentrations used has drifted downward to five for many countries, but was reinforced at six by the EMA in 2011. Other augmentation of this basic requirement has occurred over the past two decades by various agencies. The FDA has required the curve range to be reduced, or additional QC samples added when the range of concentrations observed in study samples is significantly smaller than the range of concentrations validated and that more standards be used for large ranges or nonlinear ranges (i.e., quadratic regression models) to ensure accurate characterization of the concentration-response relationship. The EMA has provided similar guidance. Broad latitude is given by all countries on the selection of the concentrations used within the calibration curves. Recognizing that the instrument responses of early generations of LC-MS/MSs tended to drift over the course of long runs, the industry moved to placing one set of calibrators at the beginning and one at the end of the batch, thus averaging the drift. Proponents and opponents have argued about this, but as yet no region has codified bracketing as a requirement. All regions include the need for at least one zero concentration sample (blank) to be included in each batch, but to not be included the regression analysis.

All countries agree that the simplest regression model and weighting scheme that fits the calibration data should be used; however, the interpretation of what simplest means does vary between countries. At the Crystal City II conference, considerable dialog occurred about the use of models beyond linear or quadratic. The FDA Guidance permits more complex models, but with the proviso that their use be justified. This definition of complexity evolved over time. Brazil restricted this further and required that for any model other than straight-line linear, $y = mx + b$ (note that a quadratic model is considered linear) (Massart, 1988), the justification must include statistical demonstration of the appropriateness of that model. Similarly, Japan prefers a pure linear model for small molecule analysis, but has provided flexibility within its draft regulation. When considering the nature of many LC-MS/MS systems and the use of multifold calibration ranges that demonstrate a quadratic nature, this has required companies to develop and implement appropriate statistical tests as part of their validation to support the selection of the regression model. The EMA has aligned its regulations with those of the FDA.

All countries agree that when considering the regression, blanks should not be included but beyond that, the rules are complex. The United States, Japan, Brazil, China, and India agree that at least 75% of the calibration standards should be within 15% of their nominal, 20% at the LLOQ. Brazil requires that at least six of the standards meet these acceptance limits. The EMA provided some additional guidance on how to handle the regression process by clearly defining how standards are to be rejected to obtain the best model for each batch:

> In case replicates are used, the criteria, within $\pm 15\%$ (or $\pm 20\%$ for LLOQ) should also be fulfilled for at least 50% of the calibration standards tested per concentration level. In case a calibration standard does not comply with these criteria, this calibration standard sample should be rejected, and the calibration curve without this calibration standard should be re-evaluated, including regression analysis.

4.3.9 QC Samples

All regulatory authorities are in agreement that QC samples at concentrations representative of study samples are needed with each run. However, the specific concentration requirements for QCs do differ. Most countries agree that a low QC sample at $\leq$3 times the LLOQ is needed and that the high QC should be in the upper quartile of the curve. The placement of the mid or medium QC is frequently debated as its placement must meet multiple requirements; it should be near the middle ($\sim$50%) of the calibration range, be representative of the expected concentrations in study samples and not be at concentrations used for the calibration standards. The latter is an expectation of India's CDSCO (CDSCO, 2005). When considering what the midpoint means is also debated—is it placed near the middle of a linear or logarithmic assessment of the calibration curve? Scientific judgment should be used to ensure that the QC concentrations are representative of the concentrations expected for study samples. In preparing the QC samples, the FDA and EMA agree on requiring two different stock solutions be used—one for the calibration curve and one for QCs—unless a single stock solution has been verified against another stock, and then it may be used for both standards and QCs.

In addition to these QCs, additional QCs may be required during validation. Most countries permit or require the use of QCs at additional concentrations for calibration curves with large ranges or during sample analysis when the normal QCs do not adequately reflect the concentrations of the study samples. It is generally accepted that a QC sample at the LLOQ) concentration is needed in validation to adequately describe the accuracy and precision of the method at the LLOQ. This QC is not used during sample analysis. The use of one or more dilution QCs, with concentrations above the range of the calibration curve and diluted in the same manner as projected for the study samples, is also expected. Lesser agreement exists on the need for the dilution QC concentrations to be reflective of the expected concentrations of the study samples and the need to use dilution QCs during study sample analysis. However, both are best practices and are able to provide evidence of potential problems: solubility problems for the former and errors in preparation for the latter.

Accuracy acceptance criteria are generally agreed to as two-thirds of all QCs must be within ± 15% of their nominal concentration; ± 20% for QCs at the LLOQ, with at least 50% of the QCs meeting this criterion at each concentration. Precision acceptance criteria are similar where the %CV at each concentration is to be ≤15% (≤20% at the LLOQ), with most regulatory agencies also looking for these values for the intra- (within) and inter- (between) batch precision.

In an interesting juxtaposition, the FDA requires the overall %CV during validation be ≤15% (≤20% at the LLOQ) but not for sample analysis, while the EMA requires, during sample analysis, the overall mean accuracy and precision of the QC samples in accepted study sample runs be ≤15% but not during validation. Typically achieving both recommendations is not a problem, but both require awareness of the requirements and a check prior to validation or study wrap up. The FDA also provides an option for identifying statistical outliers and reporting the QC accuracy and precision data with and without the outliers.

4.3.10 Stability in Matrix

The use of QC samples as surrogates for analyte stability in study samples is a well-established and expected practice. The testing of stability should mimic the expected conditions for the study samples from anticoagulant, container material, presence of stabilizers, storage conditions, etc. Regulators routinely look for a variety of stability tests and will ask for additional documentation of stability if something unexpected happens with study samples (i.e., thawing during shipment). Routine experiments for stability performed within a validation include room temperature stability for 4–24 h, freeze–thaw stability for a minimum of three cycles, initially frozen for 24 h with 12 h frozen between thaws, autosampler stability for a minimum of 24 h with reanalysis against both the original calibration curve and a freshly prepared set of calibration standards and QCs, and long-term stability at one or more temperatures (e.g., −20°C and −70°C). It is unclear why regulatory authorities are not in agreement with applying the Arrhenius equation for chemical stability to analyte stability in matrix, but most are of the opinion that stability at −20°C is insufficient to demonstrate it at −70°C, and vice versa, but do allow a bracketing approach for temperatures in between when stability is demonstrated at −20°C and −70°C.

To test stability, most regulatory authorities require testing of low and high concentrations, the low and high QCs in practice, with a minimum of three replicates. Not addressed by most regulations, testing of dilution QCs in the same manner is a best practice. Until 2012, Brazil required stability calculations to compare to the Day 0 measurement. However, in 2012, its RDC 27 aligned its practices with other countries that required stability to be calculated against nominal concentrations. Most countries recognize the need for stability tests to be performed against freshly spiked calibration curves, not permitting the use of frozen calibration standards. Analytical QCs to demonstrate the performance of the run on the day of the stability experiment are required, with most countries accepting QCs within their period of demonstrated stability. The Crystal City III meeting conference report recommends that stability samples be measured on day 0 or day 1 to ensure that the preparation of the stability sample is correct.

Only addressed within the FDA and EMA regulations is the need to assess stability during the sample collection and handling period. In both cases, no detailed experiments or expectations are defined, so the bioanalyst is left to determine appropriate procedures and criteria. It is suggested that the criteria not be more lenient than ± 15%, so as to minimize variability within the data.

4.3.11 Stock Solution Stability

All countries recognize the need to perform stability tests on stock and working solutions of the analytes, but some differences exist in the type of testing and if it should also be applied to the internal standard. Typically, 6–8 h room temperature testing and testing under the conditions of storage (e.g., refrigerator, −20°C or −70°C) for the expected duration of use. Agreement exists that if different solvent systems are used between the stock and working solutions, they each should be tested. Also, comparisons should be made against fresh stock solutions; the FDA agreeing to a broader definition of what fresh means in the Crystal City III report. The EMA specifies that a bracketing approach can be used where the highest and lowest concentrations in a particular solvent system can be used to demonstrate stability at all intermediate dilutions. Over time, the period of documented stability will extend and addenda to the validation report should be made with the new information.

Brazil looks to have stability demonstrated for analog but not stable labeled isotope internal standard aligning with the EMA, that notes stability of stable isotope labeled internal standards are not needed if it is demonstrated that no isotope exchange reactions occurs, this leaves open the need to test stability for analog internal standards. The FDA is less clear on the need to test the stability of the internal standard, but notes the need for the internal standards or degradants to not interfere in the analysis of the analytes.

4.3.12 System Suitability

System suitability is not mentioned within most counties regulations, but the FDA recommends that the optimum performance of the instrument be demonstrated. This is typically conducted through system suitability samples used to demonstrate different characteristics of the LC-MS/MS in different labs, including but not limited to sensitivity, peak shape, and retention time.

4.3.13 Carryover

China and India do not address carryover within their regulations. The EMA limits it to that derived from robotic liquid handlers. Brazil discusses it as residual effect. The FDA did not address carryover in its 2001 Guidance, but it was addressed in the Crystal City III conference report as multifaceted problem; potentially derived from one or more of the following: liquid handler carryover, retention of sample components within the chromatographic system eluted in the next or later samples, etc. Japan has an expectation to analyze three blank samples following the ULOQ standard with an expectation that the signal response at the retention time of the analyte for all three samples is $\leq 20\%$ of the LLOQ sample. Thus, Japan addresses the potential for late eluting interferences. The FDA and EMA criteria are the same as those for Japan. Brazil has a similar criteria but also includes demonstrating that the response at the retention time of the internal standard in these three samples must be less than or equal to 5% of the typical response.

4.3.14 Determination of Metabolites

Both EMA (EMA, 2011) and FDA (FDA, 2001) have supported the concept of a tiered approach to the validation of methods to measure metabolites during the drug development process, starting with scientifically sound methods and proceeding to methods as fully validated as those for the drug. In discussing metabolites with colleagues in drug safety (animal toxicologists) and clinical pharmacologists, different import is made on the need for reliable metabolite data. Key is the definition of what constitutes a significant human metabolite (Baillie, 2002; Smith, 2005; FDA, 2008; ICH, 2009). Both toxicologists and clinical pharmacologists agree that for significant human metabolites, the methods need to be very reliable (e.g., more highly validated) to establish the safety margins. This is very important where the metabolite has activity or toxicity and the safety multiple is small. The toxicologist is also frequently interested in having reliable data on many more metabolites than just the significant ones in humans, as they would like to correlate with, if not fully assign, the toxicity in a species to a particular metabolite. If they can show that the metabolite is species specific and the direct cause of a toxicity, they can remove that toxicity as one that may be of concern for humans, resulting in the drug having fewer potential toxicities and improving the chance for drug approval.

In discovery, companies vary in the level of method characterization performed for drugs and metabolites; this variability carries over into development as companies apply risk-based assessments to the issue of metabolite analysis. Some companies prefer to have robust fully validated methods in Investigational New Drug toxicity studies and first-in-human studies for major animal metabolites identified in discovery. Others prefer to delay metabolite characterization until after the definitive ADME (absorption, distribution, metabolism and excretion) studies are performed, thus deferring and limiting the use of resources only to the identified significant human metabolites. In this model, fully validated methods for the metabolite(s) would be used. Due to the number of variations on this theme that are possible, and growing with the advent of newer approaches to metabolite identification (e.g., the mass defect filter) (Zhu, 2006), a simplified and tiered approach can be taken starting in discovery.

4.3.15 Incurred Sample Reanalysis

For over 6 years, incurred sample reanalysis (ISR) has enjoyed an inordinate amount of attention from the scientific, quality assurance, and regulatory world. It will be addressed in detail elsewhere within this book, but suffice it to say that like many concepts, it has morphed from its initial discussions and is now supported by the FDA and EMA as an important in-study demonstration of the quality of the method and its practical in-laboratory execution. The latter being perhaps the more oft-seen detector of problems within the study data. Canada previously had a requirement for ISR, but dropped it in 2003 (Health Canada, 2003). Brazil, India, China, and Japan do not have regulations calling for ISR, but based on the authors' experiences, many countries are now routinely looking for or requesting it.

4.3.16 Run Size

The FDA and EMA agree that during validation, a run be created that contains the same number of samples as expected to be analyzed during study sample analysis. This is typically performed to demonstrate that the chromatography is sustained, as well as the signal within the mass spectrometer.

4.3.17 Reporting

Regulatory authorities have yet to provide specific guidance on the format of the bioanalytical method validation report.

Guidance on the content of the report has been provided in most of the regulations and Crystal City conference reports, with the report on the Crystal City III conference providing a table on specifics for inclusion in the report. The EMA Guideline has provided the most recent standard, and it includes the requirement to present the individual data values and concentrations for each experiment performed. This level of details simplifies the review process, as the reviewer is able to move from the summary results into the details to probe and assess the fundamental quality of the method. It does, however, impose more extensive requirements on the scientific community to publish detailed experimental results in their reports.

4.3.18 Other Topics

Over the years, various topics have arisen from experimental or audit observations that have generated concern or dialog within the scientific and regulatory communities. Stability relationships between −20°C and −70°C were hotly debated but closed through regulatory guidance on the subject. Topics such as the need to validate matrices using different counterions of the same anticoagulant, and the need to demonstrate stability in the presence of comedications have been discussed at conferences and in the literature but have yet to be resolved.

4.4 CONCLUSION

The generation of reliable bioanalytical drug and metabolite concentrations starts with sound science-based method development, is demonstrated through method validations, and finally, proven through the use of the method on nonclinical and clinical study samples. Validations characterize the method, testing its key parameters to ensure that, when used, the inherent variability between animals or humans does not influence the result. For most drugs, several methods may be used during the drug's development as information is gathered on the drug, metabolites are added and dropped from the method, doses are refined, combination drug products are developed or the method is transferred to other laboratories including contract research laboratories. Through these evolutions, the validations and cross-validations provide the evidence regulatory authorities need to accept the data from various studies as reliable and comparable.

REFERENCES

ANVISA. *Resolution 895, Guidelines for elaborating a relative bioavailability/ bioequivalence study's technical report.* Brazil; 2003.

ANVISA. Resolution RDC 27, Minimum requirements for Bioanalytical Method Validation used in studies with the purpose of registration and post-registration of medicines. Brazil; 2012.

Baillie TA, Cayen MN, Fouda H, et al. Drug metabolites in safety testing. Toxicol Appl Pharmacol 2002;182(3):188–196.

CDSCO, Guidelines for Bioavailability and Bioequivalence Studies. Directorate General of Health Services, Ministry of Health and Family Welfare, Government of India, New Delhi, India; 2005.

EMA. Guideline on Bioanalytical Method Validation. Available at http://www.ema.europa.eu/docs/en_GB/document_library/Scientific_guideline/2011/08/WC500109686.pdf. Accessed Feb 25, 2013. London: European Medicines Agency; 2011.

FDA. Guidance for Industry: Bioanalytical Method Validation. Food and Drug Administration. Rockville, MD, USA; 2001.

FDA. Guidance for Industry—Safety Testing of Drug Metabolites. Food and Drug Administration. Rockville, MD, USA; 2008.

Health Canada. *Conduct and Analysis of Bioavailability and Bioequivalence Studies – Part A: Oral Dosage Formulations.* Ministry of Health, Heath Products and Food Branch. Canada; 1992.

Health Canada Guidance Document: Conduct and Analysis of Comparative Bioavailability Studies, Health Products and Food Branch; 2012.

Health Canada. Notice to Industry: Removal of Requirement for 15% Random Replicate Samples. Canada: Ministry of Health, Heath Products and Food Branch; 2003.

ICH. M3(R2): Guidance on Non-Clinical Safety Studies for the Conduct of Human Clinical Trials and Marketing Authorization for Pharmaceuticals. International Committee on Harmonization; 2009.

Japan Ministry of Health, Labour and Welfare. Draft Guideline on Bioanalytical Method Validation in Pharmaceutical Development, Japan; 2013.

Massart DL. *Chemometrics a Textbook.* Amsterdam: Elsevier; 1988.

OECD. Principles for Good Laboratory Practice (GLP). Available at http://www.oecd.org/document/63/0,3746,en_2649_37465_2346175_1_1_1_37465,00.html. Accessed Feb 25, 2013. A Paris, France: Organisation for Economic Co-operation and Development, 1998.

Shah VP, Midha KK, Dighe S, et al. Analytical methods validation: bioavailability, bioequivalence and pharmacokinetic studies. Pharm Res 1992;9:588–592.

Shah VP, Midha KK, Findlay JW, et al. Bioanalytical method validation—a revisit with a decade of progress. Pharm Res 2000;17(12):1551–1557.

Smith DA, Obach RS. Seeing through the mist: abundance versus percentage. Commentary on metabolites in safety testing. Drug Metab Dispos 2005;33(10):1409–1417.

Zhu M, Ma L, Zhang D, et al. Detection and characterization of metabolites in biological matrices using mass defect filtering of liquid chromatography/high resolution mass spectrometry data. Drug Metab Dispos 2006;34(10):1722–1733.

Viswanathan CT, Bansal S, Booth B, et al. Quantitative bioanalytical methods validation and implementation: best practices for chromatographic and ligand binding assays. Workshop/Conference Report. AAPS J 2007;9(1): E30–E42.

5

CURRENT UNDERSTANDING OF BIOANALYTICAL ASSAY REPRODUCIBILITY: INCURRED SAMPLE REANALYSIS, INCURRED SAMPLE STABILITY, AND INCURRED SAMPLE ACCURACY

MANISH S. YADAV, PRANAV S. SHRIVASTAV, THEO DE BOER, JAAP WIELING, AND PURAN SINGHAL

5.1 INTRODUCTION

The need to demonstrate assay reproducibility was envisioned way back in 1990 when Health Canada guidance on bioanalytical method validation recommended reanalysis of incurred samples to ensure the reliability of a validated bioanalytical method. This was essential largely due to differences found in the results for spiked standard samples from real time study samples, which are exposed not only to systemic circulation but also to different systemic metabolic pathways and transformations. The major reason for reanalysis of study samples is attributed to the instability of analyte and/or its labile metabolite and associated matrix components in a biological sample. Incurred sample reanalysis (ISR) inherently displays confidence in regulated sample analysis by reanalysis of a selected portion of the incurred samples to determine the reproducibility of original analytical results. Incurred sample stability (ISS) is another instrument to monitor the stability of drugs and their metabolites in a certain time frame. It is closely related to ISR and both have now become an integral part of bioanalytical testing. Incurred sample accuracy (ISA) is a relatively new approach to assess method reproducibility by scheming the systematic errors during ISR or ISS study. Thus, ISA can be considered as a supplemental theme to review the method reproducibility.

This chapter aims to summarize the rationale for reanalysis based on these three important indicators of method reproducibility and robustness. The underlining principles of these indicators and some selected case studies are outlined to understand their significance and relevance in bioanalytical research.

5.2 INCURRED SAMPLE REANALYSIS

ISR can be effectively summarized as a fundamental concept in bioanalysis to assess the reproducibility and accuracy of bioanalytical methods in the conduct of pharmacokinetic, bioequivalence (BE), and preclinical safety studies. ISR can work as a tool to assess whether the validated method developed with spiked samples is in fact robust for study samples. The importance of ISR study can be envisaged from its role in clinical as well as in preclinical studies. Incurred samples can differ significantly in their composition when compared with the calibration standards (CS) and quality control (QC) samples that are used to validate the method. ISR study data reaffirms the reproducibility and reliability of a validated bioanalytical method. Inherent complexities in working with different biological matrices and a variety of drug formulations has made it imperative to test incurred (study) samples, which is key in assessing the performance of any self-sustaining/an error-free bioanalytical method. The necessity of this practice has now been widely accepted in regulated bioanalytical laboratories of pharmaceutical companies and contract research organizations (CRO) worldwide.

Handbook of LC-MS Bioanalysis: Best Practices, Experimental Protocols, and Regulations, First Edition. Edited by Wenkui Li, Jie Zhang, and Francis L.S. Tse.

Canadian authorities (Health Canada, 1992) were the first in suggesting reassay of randomly selected 15% of the incurred samples. Nevertheless, the outcome of the first bioanalytical workshop in 1990 (Shah et al., 1992) set the benchmark for the standard procedures to be adopted for bioanalysis. This workshop was the joint venture of the American Association of Pharmaceutical Scientists (AAPS), the United States Food and Drug Administration (US FDA), the International Pharmaceutical Federation, the Health Protection Branch, and the Association of Analytical Chemists. It was the first major workshop dedicated to investigating and harmonizing procedures required in bioanalytical method validations. The focuses of the workshop were on the requirements for bioanalytical method validation, procedures to establish reliability of a bioanalytical method, parameters to ensure acceptability of analytical method performance in prestudy validation, and within-study validation (Shah et al., 2000; Shah, 2007). The outstanding features of this conference laid the foundation for the formulation of US FDA guidelines (US FDA, 2001).

In the past years, ISR has been a subject of discussion and debate at several forums, meetings/conferences/workshops (Health Canada, 2003; Bansal, 2006a, 2006b, 2008, 2010; European Medical Agency, 2012). Through those discussions, some critical observations/opinions (Bryan, 2008; Smith, 2010; Garofolo, 2011; Kelley, 2011; Viswanathan, 2011) and White papers (Timmerman et al., 2009; Savoie et al., 2010; Lowes et al., 2011) have been published to strength the fundamental concept of ISR. The acceptance of this concept can be envisaged from several published reports (Gupta et al., 2011; Parekh et al., 2010, Patel et al., 2011a, 2011b; Yadav et al., 2009, 2010a, 2010b, 2010c, 2010d, 2012), where ISR has been considered as a part of standard laboratory practice in development of new bioanalytical assays.

The urgency in introducing ISR in regulated bioanalysis has mainly originated from the audits and reviews conducted by the regulatory authorities, who observed significant difference in results between initial analysis and repeat analysis in the reports from pharmaceutical companies and CROs. Because of this reason, the authorities in the past two decades have suggested reanalysis of study samples to demonstrate assay reproducibility. However, it was put on hold till 2006, presumably due to limited consensus and its implications among various partners associated with bioanalytical work. The third AAPS/FDA Bioanalytical Workshop at Crystal City III meeting held in May 2006 emphasized the need for ISR (AAPS, 2006). The workshop recommended that ISR should be evaluated in addition to the conventional prestudy validation. This workshop provided a general framework and rationale for ISR, although questions remained concerning its demeanor and the acceptable outcomes. The following year witnessed several articles on best practice for bioanalytical method validation and study sample analysis (Bansal and DeStefano, 2007; James and Hill, 2007; Kelley and DeSilva, 2007; Nowatzke and Woolf, 2007). Since then it has been suggested that ISR should be performed for each species used in regulatory toxicology assessments, as well as in clinical studies when appropriate (Viswanathan et al., 2007). Further, in a detailed report (Rocci et al., 2007) several recommendations were provided concerning the number and types of samples that should be analyzed as well as the manner in which the resultant data should be examined. These recommendations broadly summarized some practical and scientific questions that need to be considered to check for reproducibility of incurred samples. These include type of studies where samples should be reanalyzed, how many and which samples are to be reanalyzed, statistical acceptance criteria for assay reproducibility, the approach for deriving valid conclusions, and corrective and preventive actions to be taken after completion of the entire study. The 2008 AAPS Workshop on "Current Topics in GLP Bioanalysis: Assay Reproducibility for Incurred Samples" was the cornerstone in establishing ISR as an integral part of an intended bioanalytical assay in demonstrating assay reproducibility using incurred (study) samples (AAPS, 2008). The workshop was convened to provide a forum for discussion on regulatory perspectives and implications of ISR for small and large molecules and consensus building regarding ISR for both preclinical and clinical studies. Following this workshop, a report with specific recommendations on the conduct and implementation of ISR were outlined (Fast et al., 2009). In particular, the report outlines the basis for ISR, general operational principles, timing of assessment, sample selection and acceptance criteria as well.

5.2.1 Principles and Practice of ISR

The purpose of ISR is to evaluate the reproducibility of a validated method by reanalysis of selected *in vivo* samples after initial analysis. A well-designed bioanalytical method validation protocol with an integrated ISR plan can lead to continuous review of assay performance and improvement if needed for the intended bioanalytical assays. A good scientific practice is instrumental in developing robust analytical methods with accurate bioanalytical results. Basically, ISR reinforces the confidence in a bioanalytical method if reproducibility in the study sample analysis is demonstrated. A conventional ISR involves analysis of at least 20 samples taken from an *in vivo* study a second time using the method that was described in prestudy validation and employed in generating the initial study sample results.

The possible causes of irreproducibility in sample reanalysis can be due to one or more reasons, including but not limited to:

1. instability of parent drug or metabolites in collected or stored samples, in-source conversion;

2. sample inhomogeneity due to variability in collection, handling, and storage;
3. execution error, that is, wrong labeling, insufficient vortexing or thawing of samples prior to sampling or contamination during collection, storage or sample processing, and analysis;
4. drug–protein binding differences in subject samples;
5. matrix interference, in particular late eluting phospholipids in reversed phase conditions;
6. variations in extraction efficiency;
7. presence of typical concomitant medications;
8. impact of dosing vehicle(s);
9. nonruggedness of the intended bioanalytical method;
10. nonoptimal tracking of internal standard (IS) for the variations in the liquid chromatography–tandem mass spectrometry (LC-MS/MS) responses of the analyte(s);
11. purity of materials, reagents, and solutions used in the study;
12. interference due to the presence of isobaric compounds.

The salient features of AAPS workshop (AAPS, 2008) on the conduct, utility, and acceptance criteria of ISR can be summarized as follows:

1. It is essential to have a standard operating procedure (SOP) for ISR.
2. ISR should be conducted early in the study and not at the end except for small toxico-kinetics studies.
3. For preclinical studies, once per species, per method and per laboratory for ISR is adequate as animals are considered to be more homogeneous in genetics, diet, and housing than humans. Samples for analysis should be from first subchronic toxicology study.
4. For clinical studies, ISR is imperative for all BE studies with healthy volunteers or patients, and, additionally, for disease-state changes in patients population and drug–drug interaction studies. First-in-human oncology studies are areas where ISR assessment is imperative as it generally involves multiple medications, which may lead to changes in metabolism and endogenous compound formation.
5. ISR must be conducted on individual samples instead of pooled samples.
6. To identify a questionable sample or subject, it is recommended to analyze more subjects with fewer samples per subject.
7. Select samples for ISR for which there is already a valid analytical value.
8. The selected sample should include one near the C_{max} and one near the elimination phase.
9. When possible, selection of <3 x LLOQ samples should be avoided for single-analyte assays. For multi-analyte assays, select ISR samples based on the concentrations of the primary active entity.
10. For multi-analyte assays containing two or more primary active entities (i.e., comedication), consideration should be given to ensuring that samples are selected across concentration profile for all analytes. In case of markedly different concentration profiles, the selection of sample should be such that it adequately compensate for individual C_{max} values (e.g., the sample corresponding to the succeeding value after C_{max} from the profile with lower T_{max} that coincides with the preceding value from the profile having higher T_{max} value).
11. Reporting of data should be included in the validation report, in addition to the study report.
12. Failed ISR analysis must be investigated and appropriate follow-up action be taken (ISR does not accept or reject a study).
13. Over and above these points, good scientific judgment and experience may be exercised for the overall conduct of ISR.
14. The acceptance criteria of ISR for a small and large molecule permits two-thirds of repeats should be within 20% and 30% of the mean value, respectively.

The European Bioanalysis Forum (EBF) has provided recommendations on how to integrate incurred sample reproducibility (ISR) in the bioanalytical process (Timmerman et al., 2009). These recommendations aim to provide a comprehensive guidance to the analyst throughout the lifecycle of a validated method, including the application of the method for study samples. To make the validated method robust, it is essential to perform ISR in the early part of method's lifecycle with available study samples. Thus, including ISR as "part of method validation" will ensure the robustness of the intended methods that do not require frequent reinvestigation. ISR as "part of regular process check" can make sure whether the standard laboratory procedures are followed continuously. Based on these understandings, ISR can be set initially as first time in new matrix (preclinical species or human) or new target population. These include the first preclinical study in toxicity species, the first in humans (single vs. multiple dose), the first time into patients, the application of a method in special populations (pediatrics, renal, or hepatic impairment), and new disease indications. Additionally, ISR should be considered if a particular laboratory performs an assay for the first time and incurred samples are not reanalyzed as part of method transfer and cross-validation. Further, it was agreed upon that a regular process check is adequate for selectivity or potential matrix effects toward comedicated drugs (and their metabolites) in drug–drug interaction studies. The acceptance criteria as per EBF recommendation include (a) two-thirds of repeat values within 80–120% of the mean value for chromatographic assays and (b) two-thirds of repeat values within 70–130% of the mean value for ligand binding assays. The criterion for selection of number of samples is based on analysis of multiple subjects and limited samples per subject. This generally results in reanalysis of about 20–50 samples. Further, ISR should be performed on individual samples and not pooled samples. The dilution factors for overrange samples to generate the first value should be identical for ISR. The selection of sample for reanalysis should be such that the analyte concentration is within the established linear dynamic range. Finally, any failed ISR must be followed to find the root cause, the bioanalytical phase should be put on hold until the investigation is completed and appropriate follow-up procedures are implemented.

The acceptable practice for conducting ISR based on current understanding can be summarized through the following points (Viswanathan, 2011; EMA, 2012):

1. Adherence to SOP.
2. Must for all BE studies (healthy volunteer/patient population studies as appropriate).
3. Small molecule drug–drug interaction studies as appropriate.
4. For pharmacokinetic determinations, ISR experiments can be designed on a case-to-case basis.
5. For preclinical studies, ISR should be assessed once per method/species/laboratory.
6. For sample size consideration, the number of samples repeated should be 10% for sample size $\leq$1000, while for larger studies ($>$1000) an additional 5% for all samples above 1000, in addition to 10% for the first 1000 samples.
7. The acceptance criterion implies that 67% of repeats agree within 20% for small molecules and 30% for large molecules.
8. The ISR results should be expressed as percentage difference, where

$$\% \text{ Difference} = \frac{\text{Repeat value} - \text{original value}}{\text{Mean of original and repeat values}} \times 100.$$

9. For failed ISR the root cause must be thoroughly investigated, resolved; corrective and preventive action must be documented.
10. The results of ISR must be incorporated in the study report

Good ISR assessment bolsters the confidence in a validated bioanalytical method; however, ISR should not be the sole reason to accept or reject the results from a particular study. ISR is one facet of the overall performance of the assay and the final outcome should include other relevant aspects based on good scientific judgment and experience. An unsuccessful/failed ISR should be thoroughly reviewed with appropriate corrective/preventive actions.

5.2.2 Statistical Approaches for Assessment of ISR

ISR can be considered as a specific form of cross-validation used for bioanalytical methods to compare two methods or results obtained from two laboratories. To investigate reproducibility, the data-processing method must be capable to assess the accuracy and precision and at the same time capable to identify correctly the values that lie outside the defined limits. Different statistical approaches can be classified as follows:

a. The Bland–Altman plot gives a good assessment for visual examination of a set of data obtained from ISR. Inherent limitations in regression-based methods for cross-validation are evident when the two sets of data represent measurements on paired samples (Bland and Altman, 1995; Bland and Altman, 1999).
b. Plotting the difference of each data pair against the average of the two values and establishing limits of agreement against the average of the differences, which can include a specified percentage of the measured values.
c. Plotting of log differences may be more appropriate when the coefficient of variation is constant across the range of means. This approach has been utilized for potency estimates through replicate experiments and to obtain relevant statistical parameters (Eastwood et al., 2006).

The use of the Bland–Altman plot to treat incurred plasma samples in the analysis of a small molecule by LC-MS/MS and a macromolecule by enzyme-linked immunosorbent assay has been demonstrated previously (Rocci et al., 2007). It was suggested that this approach can help in the assessment of reproducibility, which includes testing of systematic differences between the results and in characterizing the degree of agreement between the results to detect random differences. Further, the estimation of 67% "limits of agreement" (range within which the ratio of sample results is expected to fall two-thirds of the time) between the results can also be verified. The specific application of content tolerance limits to Bland–Altman plots has also been suggested (Petersen et al., 1997). A combination of Bland–Altman plot and tolerance intervals has been suggested to evaluate method performance that can work as a tool in determining minimum sample size (Lytle et al., 2009). They worked out a data set thoroughly and gave an insight of its real importance that can help to analyze the data for ISR study. A detailed study has been demonstrated on different evaluation approaches for ISR data obtained for spironolactone and its active metabolite (Voicu et al., 2011). The reanalysis of samples following a BE study for spironolactone formulations was carried out twice—one immediately after the end of the study and the other after the long-term stability (9 months) achieved during method validation. The Bland–Altman approach was used to assess both the results. Although ISR was successful over the short analysis period for both the compounds, there was a systematic positive bias for the metabolite and a strong negative systematic bias for spironolactone and failed to support reproducibility. This was due to the continuous conversion of the parent drug to its active metabolite in plasma samples. Nevertheless, the reproducibility of the method was sustained by comparing the original and repeat values of concentration in samples by means of paired *t*-test, Wilcoxon sign rank-sum test and linear regression. Statistical considerations in ISR for macromolecules have also been described for five representative case studies, each associated with one of the following four modalities—humanized monoclonal antibody, peptibody, recombinant protein, or fully human antibody (Thway et al., 2010). These case studies included various stages of drug development from preclinical to clinical. The bioanalytical method reproducibility was examined using modified Bland–Altman approach and the 30% acceptance criteria. Simulation studies indicated high concordance between the ISR criteria and Bland–Altman approach.

5.2.3 Case Studies on ISR

Ever since the third AAPS/FDA Bioanalytical Workshop in May 2006, significant number of case studies on failed ISR tests have been documented in literature, with systematic investigation of their root cause and followup actions for small (Cote et al., 2011; Dicaire et al., 2011; Fu et al., 2011; Meng et al., 2011; Rocci et al., 2011; Yadav and Shrivastav, 2011) and macromolecules (Sailstad et al., 2011).

Although the ISR success rate in regulated bioanalytical laboratories is very high (~95%), the failed tests deserve a thorough scientific evaluation to unearth the root cause. EBF has reported an ISR failure rate of only 3.6% and 4.1% for preclinical and clinical studies, respectively, based on the data available from member companies (Timmerman et al., 2009). Similarly, statistics from some bioanalytical CROs have shown a failure rate of about 5% (Meng et al., 2011) and 4% (Tan et al., 2011), respectively. In the last 3 years, the outcome of our lab (MY) showed a success rate of 96.7% based on reanalysis of 10 209 incurred samples. In spite of this high success rate, the real challenge for the bioanalytical community is to explore the unmatched results in the remaining 33% (based on 67% acceptance criteria), which can be associated with very high percentage difference from the original analysis. This understanding has been prudently demonstrated through investigations beyond successful ISR for unmatched reassay results (Tan et al., 2011). Nevertheless, a failed ISR does not immediately invalidate the entire study, but it does call for suspension of the bioanalytical portion of the study until an investigation is completed, documented, and suitable follow-up actions are in place (Fast et al., 2009).

Based on reported case studies in the past 5–6 years, the major contributing factor in ISR failure investigations can be broadly classified into the following four categories:

a. Issues related to instability of drug or metabolite
b. Issues related to human error
c. Issues involving sample processing
d. Issues related to method error

Figure 5.1 gives a comparative status of different categories of root cause for ISR failure based on reported case studies. Some selected investigations have been comprehensively summarized and can work as a guide for developing more rugged, robust and reproducible bioanalytical assays.

5.2.3.1 Issues Related to Instability of Drug or Metabolite

Case 1 *Impact of methylation of acyl glucuronide metabolites* (Cote et al., 2011)

In-source conversion of acyl glucuronides into their parent compounds is a major concern in the development of

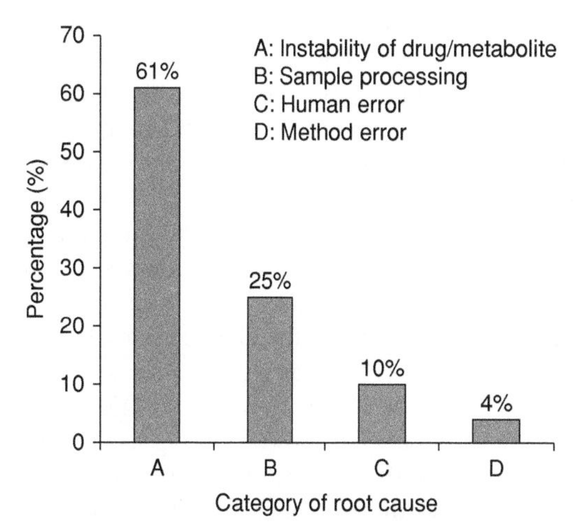

FIGURE 5.1 Contribution from different categories of root causes for ISR failure based on reported case studies.

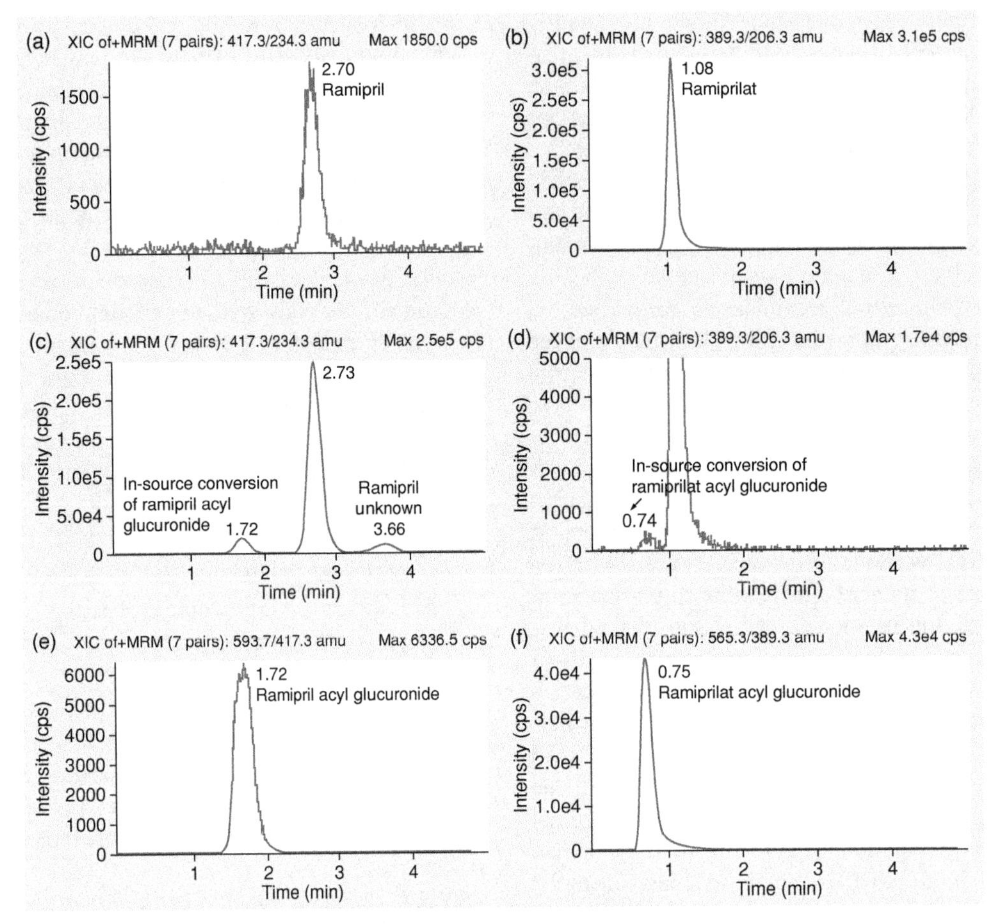

FIGURE 5.2 Chromatograms of (a) ramipril and (b) ramiprilat in a calibrant. Incurred sample chromatograms at (c) ramipril and (d) ramiprilat mass transitions. Incurred sample chromatograms at (e) ramipril acyl glucuronide and (f) ramiprilat acyl glucuronide mass transitions. All chromatograms were obtained under the initial chromatographic conditions (reproduced from De Boer et al. (2011) with permission of Future Science Ltd).

bioanalytical methods. A bioanalytical method was developed for the quantification of ramipril and its active metabolite ramiprilat in human plasma using deuterated ISs. During method development of ramipril/ramiprilat stage, due to unavailability of glucuronide metabolites and thus incurred samples were used to confirm the chromatographic separation of ramipril and ramiprilat glucuronides from ramipril and ramiprilat, respectively. The incurred samples showed a shouldering (coeluting) peak for both the analytes; thus, the samples were reinjected and the results showed a difference greater than 20% for ramiprilat. Thus, a thorough investigation was initiated to identify the root cause. The chromatograms of ramipril and ramiprilat calibrators are shown in Figures 5.2a and b. The incurred sample injected with the initial sample preparation showed three peaks at the multiple reaction monitoring (MRM) of ramipril (Figure 5.2c) at 2.70 min corresponding to ramipril and two unknown peaks at 1.72 and 3.66 min. However, only two peaks were observed for ramiprilat (1.08 min) and one unassigned at 0.74 min at the MRM of ramiprilat (Figure 5.2d). In-source conversion of the acyl glucuronide metabolites into the analytes of interest was assigned to the first eluting peak in the ramipril and ramiprilat chromatograms (Figures 5.2c and d). This was confirmed by monitoring the acyl glucuronide metabolite in incurred samples by the initial chromatography. Figures 5.2e and b present chromatograms obtained at ramipril and ramiprilat acyl glucuronide MRM transitions. Thus, it was concluded that in-source conversion of ramiprilat acyl glucuronide metabolite into ramiprilat was not the reason for unmatched results, as they were chromatographically separated. The main concern was the absence of a third peak in the ramiprilat MRM transition, which was clearly evident in ramipril at 3.66 min and was well separated from the ramipril peak at 2.73 min. Subsequently, extended chromatography was conducted by decreasing the organic solvent content compared to the initial chromatography to improve the separation of any potential interference from the analyte peak. Figure 5.3a and b show the chromatograms for ramipril and

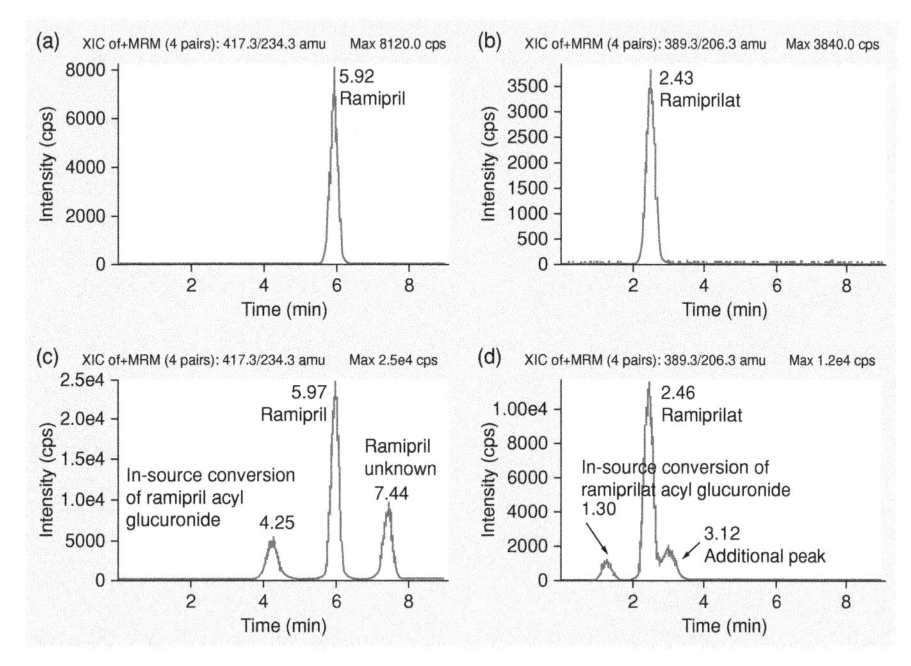

FIGURE 5.3 Chromatograms of (a) ramipril and (b) ramiprilat in a calibrant sample and an incurred sample at (c) ramipril and (d) ramiprilat mass transition injected with the extended chromatographic conditions (reproduced from De Boer et al. (2011) with permission of Future Science Ltd).

ramiprilat, respectively, from the extended chromatography and the corresponding incurred sample chromatograms in Figures 5.3c and d. Noticeably, a third peak was observed at the MRM transition of ramiprilat at 3.12 min, which was not observed with the initial chromatographic conditions. After extensive full scan MS study, it was found that the unknown peaks (for ramipril and ramiprilat) in the incurred sample was due to methylation of acyl glucuronide metabolite. This conversion was due to the interaction between the carboxylic acid of the acyl glucuronide and the solid phase extraction (SPE) sorbent to produce methyl ester acyl glucuronide. The methylation of the acyl glucuronide metabolite can interfere in the quantitation of the parent drug through in-source conversion in absence of chromatographic separation. Based on these observations, an accurate and reproducible method was developed using protein precipitation instead of SPE without evaporation step.

Case 2 *In-source fragmentation of oxcarbazepine sulphate metabolite* (Dicaire et al., 2011)

This report highlighted the impact of oxcarbazepine sulphate (OCN sulphate) metabolite on the quantitation of oxcarbazepine (OCN) for an ISR that met the acceptance criteria. The method was developed for the simultaneous determination of OCN and its 10-monohydroxy primary metabolite in human plasma. For OCN and its metabolite, nine and three samples, respectively, showed a bias above 20%, while three samples showed a similar bias for the parent drug and the metabolite, suggesting an execution error rather than a method error. During subsequent sample analysis, a shouldering peak was observed intermittently in oxcarbazepine chromatograms for specific subject samples, 3.5–12.0 h postdose. This extra peak was erratic, which sometimes interfered with the quantitation of oxcarbazepine or was well resolved with no interference. Thus, the metabolism of oxcarbazepine was studied thoroughly to find the root cause. In-source/interface conversion of 10,11-dihydrocarbamazepine metabolite to OCN was observed but could not induce the problematic peak, since the retention times of 10,11-dihydrocarbamazepine and the questionable peak were different. The possibility of in-source/interface conversion of OCN glucuronide was also ruled out as no visible peaks were observed for the reference MRM chromatogram of OCN glucuronide with the validated method conditions using atmospheric pressure chemical ionization (APCI) source. Further, precursor ion scans of the extracted problematic subject sample were done with the validated method parameters; however, it was nonconclusive. Finally, the same experiments were repeated in ESI mode to modify the ionization process. The use of ESI instead of APCI source allowed detection of an ion at *m/z* 333 as the precursor for *m/z* 253 (OCN) and *m/z* 208 (OCN product ion). The difference of 80 amu indicated the presence of OCN sulphate metabolite. The impact of OCN sulphate is minimal if it does not coelute with OCN. Use of multiple Zorbax SB-Phenyl columns showed that the retention time of OCN sulphate was column dependent, while the retention time for other analytes and IS were unaffected. This conclusively proved the erratic behavior of the peak and the problem encountered in the quantification of OCN due to in-source conversion.

Case 3 *With reference to freeze–thaw stability* (Rocci Jr., 2011)

An LC-MS/MS method was used to analyze a drug (~450 Da) and one of its metabolite in plasma. Plasma samples were prepared by SPE using deuterated ISs for both the analytes. ISR testing revealed negative bias for both the analytes, which suggested stability issue. There was greater and somewhat more severe failed ISR results toward a latter half of the run. Thus, preliminary investigations were conducted for standard and QC samples using different types of anticoagulants and storage temperature. However, none of these affected the reproducibility of QC samples. Further, liquid–liquid extraction was additionally validated and the results were compared with SPE using a subset of incurred samples. Regardless of the extraction procedure the results were lower compared to the original analysis and the issue remained unresolved. During the next clinical study, samples were processed as quickly as possible after collection and split into duplicates. Original results were obtained from the one set of duplicate samples and ISR testing was performed on the other duplicate set of samples. This resulted in only single thawing and the results from ISR evaluations for both the analytes were in excellent agreement with the original results. Thus, the negative bias in ISR results initially was primarily due to the instability of the parent drug and its metabolite during freeze–thaw process.

5.2.3.2 Issues Involving Sample Processing

Case 1 *In effective mixing of study samples in urine* (Fu et al., 2011)

This study described incurred urine sample reanalysis for a compound "A." The urine samples collected for compound "A" were once transferred from bulk containers to 15-ml polypropylene tubes before a small degree of nonspecific binding (~14%) was observed for the analyte. For accurate determination of urine concentration of compound "A," the CS and QC were prepared in a manner identical with the study samples. A surfactant, Tween-80 was added to CS, QC and study samples to prevent and/or retrieve the adsorbed analyte in the samples. After the completion of prestudy validation and study sample analysis, ISR was conducted for 21 randomly selected samples using the initial sample analysis. However, 14 samples failed to meet the acceptance criteria. A further reanalysis was conducted with the same samples but the results were outside ±20% bias. Thereafter, basic checks like chromatography (for peak shape and response) and stability were thoroughly examined. However, no anomaly was observed. Finally, sample mixing was reexamined and it was noticed that the urine samples tubes were almost full for most samples with very little headspace (<5%) after the addition of Tween-80 solution. This resulted in inhomogeneity due to inefficient vortex-mixing of the samples and retrieving the adsorbed analyte due to nonspecific binding. Thus, the sample mixing process was modified involving rotary mixing followed by vortexing. This ensured successful confirmatory test for 11 randomly selected samples.

Case 2 *Nonhomogeneity during sample processing* (Yadav and Shrivastav et al., 2011)

A sensitive bioanalytical method was designed for the determination of mesalamine in human plasma. The method was applied to a pivotal BE study that included 36 healthy human subjects with oral administration of 750 mg mesalamine (tablet, 3 times) under fasting conditions. The number of samples analyzed for ISR was 145 from 1440 total subject samples studied. During ISR, all values of ISR-1 displayed 9.2–96.8% variation from initial values. After preliminary investigation involving reevaluation of CC and QC samples, chromatography and sample processing steps, the samples were reanalyzed (ISR-2) but the results were outside the acceptance criteria compared to initial analysis. However, the results were comparable with ISR-1. Further, the interday variation was checked by performing one more ISR experiment (ISR-3). All results (expressed as percentage difference) of ISR were found comparable and within the acceptance criteria but different from the initial analysis (Table 5.1). Further, all processing steps were checked again, and upon careful and extended scrutiny, it was found that sample tubes (5 ml) were filled up to 75% of their

TABLE 5.1 Incurred Sample Reanalysis Data for Mesalamine

		Percentage difference		
Sample index	Sample ID	ISR-1 vs. Initial analysis	ISR-2 vs. ISR-1	ISR-3 vs. ISR-2
1	01106	50.0	0.0	1.2
2	01117	32.5	−7.8	3.2
3	02105	52.5	−6.2	−4.0
4	02118	90.7	0.9	−4.7
5	03106	15.9	−6.7	0.0
6	03118	96.8	−8.9	3.1
7	04107	77.0	−8.4	−0.3
8	04119	69.1	−10.2	3.3
9	05108	9.2	−12.6	−3.0
10	05118	66.3	−8.8	2.0
11	06107	84.3	−9.8	−3.9
12	06119	54.2	−8.5	1.6

Reproduced from Yadav and Shrivastav (2011) with permission of Future Science Ltd.

capacity prior to initial analysis. Thus, the mixing/vortexing was insufficient that led to inhomogeneity in the samples during initial analysis. Subsequently, when ISR was performed the volume had reduced that afforded proper homogenization. Thus, the consistency in results obtained for ISR 1–3 was essentially due to the additional volume created in the tubes, which facilitated proper vortexing after the initial analysis. This proves that the irreproducibility was due to poor execution of a reliable method.

5.2.3.3 Issues Related to Human Error

Case 1 *Random human error* (Tan et al., 2011)

i. In this typical case, ISR passed for a two analyte study with 94% and 97% success rate for the parent and its metabolite, respectively. However, the result of one sample obtained during ISR (389.63 pg/ml) was almost half of the original value (800.16 pg/ml) for the metabolite, while the results for the parent drug showed no such discrepancy. The most likely cause for such errors is due to double aliquoting or double addition of IS. In such an event, the impact should be similar for both the analytes; nevertheless, it was not the case. By evaluating the IS responses of the original and reassay runs, it was clear that an incorrect amount of IS was added in the original run since the IS response of this sample was almost half of the mean IS response of the accepted CSs and QCs, while those of other samples were quite consistent. Similar IS response was also observed for the parent drug; however, it was accepted due to inhouse acceptance criteria of ±50% of the IS mean. On the other hand, the IS response for the metabolite was very close to the limit of rejection. Based on this investigation, the original value was deemed incorrect and was changed to 416.71 pg/ml.
ii. In another study involving 106 subjects' only two subject samples failed to meet the acceptance criteria, which were analyzed by one particular analyst. The reassay results were higher than those of the original ones for this analyst. During examination of the analysts, it was revealed that the concerned analyst did not inverse sample tubes between vortexing, which resulted in inadequate mixing of samples prior to being aliquoted.

5.2.3.4 Issues Related to Method Error

Case 1 *Impact of buffer pH during sample preparation* (Yadav and Shrivastav et al., 2011)

A bioanalytical method was developed for the estimation of clopidogrel in human plasma by LC-MS/MS using clopidogrel-d_3 as IS. The plasma samples were prepared by liquid–liquid extraction with ethyl acetate after acidifying with 10% acetic acid in water (pH 3.0). The method was applied to a pivotal BE study in 80 healthy human subjects (4 dropouts during clinical phase) after oral administration of 75 mg clopidogrel (tablet) under fasting conditions. The number of samples analyzed for ISR was 320 from 3192 total subject samples studied. During the ISR-1 experiment, it was found that 4 values were outside the acceptance criteria. Three of these results had a high negative percentage difference (>51%) from the initial analysis. With no success after preliminary investigation involving reevaluation of CC and QC samples, chromatography and sample processing steps, the problematic incurred samples from the same batch were reprocessed and analyzed but similar results were observed. Further, these samples were reanalyzed along with their two preceding subject samples. The results obtained (ISR-2) were still outside the acceptance criteria compared to initial analysis results. The percentage difference was in the range of −32% to −55%. In the next step, the standby aliquot (sample(s) were analyzed in duplicate) was similarly analyzed (ISR-3), but the percentage difference from the initial values was even greater (>60%). Thus, the possibility of degradation of the analyte into its metabolites in the plasma matrix under the experimental conditions was suspected. Chromatographic (evaluation based on RTs) and mass spectrometric (through additional MRMs) studies were then initiated in the presence of acyl glucuronide (484.9/308.1) and its carboxylic acid metabolite (308.0/198.0), a major circulating metabolite. However, the peaks for both the metabolites and clopidogrel were well resolved, with no interconversion as evident in Figure 5.4. Thus, the stability of clopidogrel under bench top conditions (room temperature) was checked by taking three samples from the C_{max} time point (sample ID 04107, 04208, and 10207) along with fresh CS and QC samples. The samples were analyzed at 0.0, 2.0, and 6.0 h. There was no apparent difference in concentration values for clopidogrel at these time intervals. Further, the effect of buffer (pH, 3.0) on subject samples was checked at 0.0, 2.0, and 6.0 h along with CC and QC samples at room temperature and below 10°C. The results showed a significant change in response from 0.0 to 6.0 h. Thus, a study was designed to see the affect in the presence (pH, 3.0) and absence of acidic buffer on randomly selected subject samples (which were out of and within the acceptance criteria) along with preceding samples, CS and QC samples. The study was done at different time intervals and the results were compared with the initial analysis. The concentration values obtained were considerably different from the initial value (large percentage difference) for buffered samples (Table 5.2), while all results were found within acceptance criteria for unbuffered samples (−8.5% to 12.1%). Thus, it was conclusively proved that acidic buffer addition to samples during sample processing caused nonreproducibility in ISR results. Since the subject samples

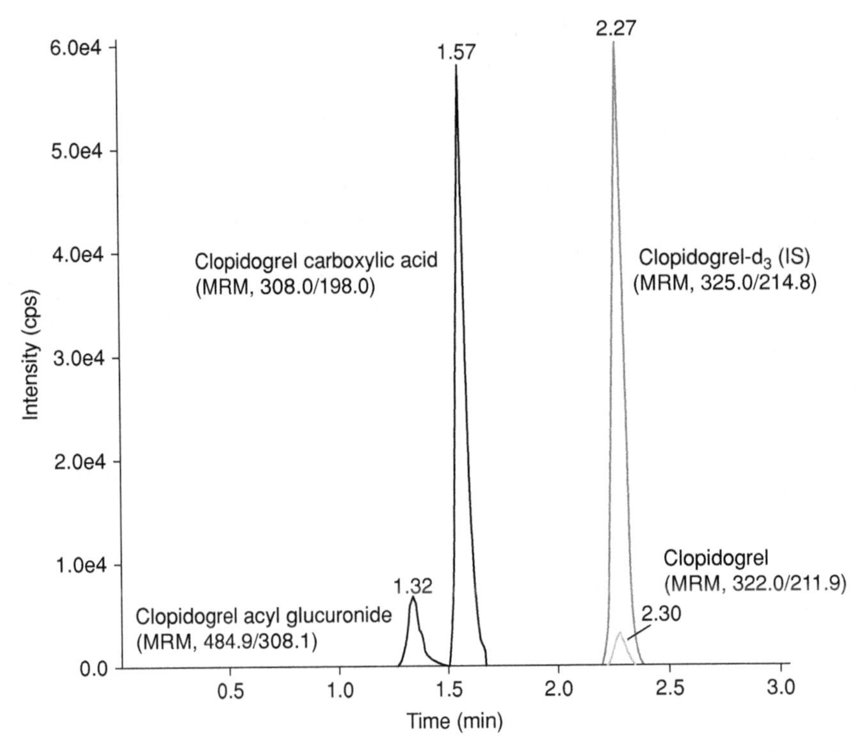

FIGURE 5.4 Representative chromatogram for an incurred sample of clopidogrel, clopidogrel carboxylic acid, clopidogrel acyl glucuronide and clopidogrel-d3 (IS) (reproduced from De Boer et al. (2011) with permission of Future Science Ltd).

TABLE 5.2 Incurred Sample Reanalysis Study for Clopidogrel in Presence and Absence of Buffer, pH 3.0 (6.0 h vs. 0.0 h)

Sample index	Sample ID	0.0 h (ng/ml)	6.0 h (ng/ml), with buffer	Percentage difference	6.0 h (ng/ml), without buffer	Percentage difference
1	15106	0.98	9.67	162.5	0.9	−8.5
2	15107	1.04	11.6	141.2	0.97	−7.0
3	15209	1.31	11.3	116.1	1.23	−6.3
4	15210	1.29	11.7	98.1	1.4	8.2
5	20107	0.293	2.89	−53.5	0.281	−4.2
6	20108	0.246	2.77	−73.7	0.27	9.3
7	20205	0.407	4.19	−50.2	0.41	0.7
8	20206	0.377	4.28	−60.6	0.42	10.8
9	21106	2.49	12.4	31.8	2.8	12.1
10	21107	2.72	13.0	26.1	2.9	5.0
11	21207	1.45	9.8	−11.2	1.6	9.8
12	21208	1.28	9.7	−21.2	1.5	12.5
13	44104	2.16	12.0	−8.0	2.0	−7.7
14	44105	1.76	14.6	4.2	1.9	8.7
15	45206	1.05	10.3	−37.2	1.11	5.6
16	45207	1.09	10.5	−41.5	1.14	4.5
17	52107	0.402	4.87	−110.9	0.44	9.0
18	52108	0.557	6.09	−98.9	0.61	9.1

Reproduced from Yadav and Shrivastav (2011) with permission of Future Science Ltd.

were significantly affected by the sample extraction in acidic conditions, the method was redeveloped by maintaining mild acidic conditions (pH 6.0) during sample processing/extraction without sacrificing recovery and sensitivity.

5.2.4 Dried Blood Spot ISR

Dried blood spot (DBS) sampling technique has created a paradigm change for bioanalytical analysis. DBS is now becoming increasingly popular due to manifold advantages over conventional wet sampling techniques. The renewed interest in this approach is based on animal and patient ethics, practical, scientific, and cost considerations. The ethical advantages are mainly related to refinement, reduction, or removal of laboratory animals used for preclinical animals (Timmerman et al., 2011). DBS involves collection, transport, storage, and analysis of blood samples on cellulose-based cards. It has shown considerable promise for toxicokinetic and pharmacokinetic analysis based on the following merits:

a. Considerable reduction in sample volume collection (~50 μl).
b. Less invasive (finger-prick) technique than conventional collection procedures, a boon for patients especially neonatal and children.
c. Requires minimum processing after collection.
d. Low biohazard risk.
e. Requires minimum facility for shipment and storage of DBS cards at ambient temperature, which eliminates the efforts and cost for the maintenance of cold chain (−20°C or −40°C) and hence the overall chain of custody is simplified.
f. Facilitates clinical studies in areas that have limited or no access to regulated bioanalytical laboratories.
g. All the above factors contribute toward considerable reduction in the overall cost of the study.

Based on discussions within the bioanalytical community and at scientific summits, EBF has come out with a set of recommendations that can serve as guidelines for the validation of bioanalytical methods for DBSs (Timmerman et al., 2011). EBF recognizes DBS as a case of microsampling with twofold advantage, one in lowering the sample volume requirement and the other of circumventing the problem of stability or facilitating phase III clinical studies in developing countries. Blood has markedly different physical behavior compared to liquid matrix, which can impact sample-to-sample variability and hence to the overall performance of the assay. In case of DBS, the important parameters that affect the bioanalytical behavior are related to hematocrit variability, hemolysis, and the impact of anticoagulants (Sennbro et al., 2011). Hemotocrit or the erythrocyte volume fraction is the percentage of red blood cells in blood and is one of the major factors that affect the assay reproducibility, as it impacts the spot formation, spot size, homogeneity of samples, and drying time. This is predominantly the case when a subsample punch is taken from the DBS instead of the entire punched sample for analysis. Although the principles of DBS ISR (spot and punch) are similar to that for liquid sample matrix ISR, certain generalizations can be drawn based on EBF recommendations, reported case studies and current practice. The scope of DBS ISR compared to liquid sample matrix ISR can be understood by first considering the basic difference between the two techniques (Barfield et al., 2011).

a. There are two approaches to sample collection for DBS compared to one for liquid matrix.
b. Homogeneity of liquid sample is more as compared to DBS that normally has three discrete samples (spots).
c. The accuracy of sample volume for liquid sample depends on a pipette, while the volume accuracy for DBS depends on the punch.
d. The possibility of contamination of DBS card is more likely than liquid matrix as the card is generally sealed 2 h after drying.
e. For repeat analysis, the volume of liquid sample may not be adequate especially for neonatal or transgenic studies while for DBS, three samples (15 μl for each spot) can be readily available.

DBS ISR can be done from the second spot for the same sample or from a second spot punch taken out of the same spot. GlaxoSmithKline (GSK) has been at the forefront in creating a large ISR database for DBS samples through different study protocols and for wide range of compounds (Barfield et al., 2011). Based on their report, the possible causes for unmatched results in DBS ISR can be understood through these case studies.

Case 1 *Difference in sampling techniques*

Typically, there are two ways to collect a sample, one that involves collecting blood into a tube, followed by vortexing and then spotting using a pipette. The other approach is to collect the blood sample into an EDTA (ethylenediaminetetraacetic acid)-coated capillary, followed by spotting using a capillary. During capillary sampling for each time point, three individual capillaries are used for spotting and each sample is separated by a small amount of time. This approach in some cases may results in irreproducibility that otherwise can be circumvented by using the first methodology.

Case 2 *Contamination of DBS cards*

In a representative study, during reviewing of data ISR analysis failed with a wide spread of data and the percentage differences were greater than 150 compared to initial analysis. Investigation with controls and physical examination revealed traces of fur on the cards. Samples taken from unspotted regions of the cards revealed contamination of the cards with the actual compound. Corrective actions were taken by using different laboratory areas to sample, spot, and dry the samples. Additionally, different analysts were employed for sampling and spotting.

In addition to these studies, intercard variability can also be considered an area for failed run investigations. Though it is still in the early phase of development and growth, DBS ISR shall gain in competence, as more case studies are documented by the bioanalytical community.

5.2.5 Approach for Efficient Conduct of ISR

Based on the knowledge gained through experience and published literature, an attempt is made to summarize some indispensable points to consider for minimization of ISR failures in regulated bioanalytical laboratories.

a. Judicious use of applicable guidance(s) including good laboratory practice (GLP), good clinical practice, and good documentation practice.
b. Know your analyte, a thorough study on the physicochemical and spectral properties, stability, metabolism, and biotransformations of analyte(s) before method development.
c. Knowledge of potential labile and isobaric metabolites and their expected concentrations.
d. Adequate information on sample handling and storage (in solution and biological matrix).
e. As the stability data generated from QC samples does exactly mimic the stability of analyte in incurred samples, it may be necessary in some specific cases to determine the stability of analyte of interest using incurred samples (Rocci et al., 2011).
f. Step-wise approach for blood stability testing when instability is likelihood.

5.2.6 Recommendations for ISR Investigation

a. The investigation must be initiated to examine the failed data and identify the root cause.
b. In case of an unassignable cause, one of these approaches may be adopted:
   - All ISR samples can be reanalyzed along with additional samples in 2 (or more) different batches based on sample size.
   - Results of all relevant batches can be evaluated to obtain interday ISR data.
c. The approach should clearly be defined: objective (hypothesis), experimental, results with valid conclusions.
d. The conclusions must include the cause of unmatched results, assessment of validity of the method, reliability of the original data and the need to reanalyze all the study samples using the original method or the improved method.
e. Conclusion of the investigation must be documented and approved by the head of the laboratory.
f. Regular training and discussions with the sample management group, relevant analysts, supervisors, management, and other concerned personnel.
g. Scientific insight obtained in the case should be shared with the bioanalytical lab, clinic, quality monitors, quality assurance auditors, and other concerned personnel.

5.3 INCURRED SAMPLE STABILITY

ISS is a critical element in regulated bioanalysis that requires detailed investigation. It is vital to understand the factors responsible for the instability of drugs and their metabolites in biological matrix. ISS is significantly different from ISR as it is solely associated with the stability of analyte during storage and handling. To avoid stability issues with incurred samples it is essential to ensure the stability of drug(s) and its active metabolites during the early part of method validation. Some of the factors responsible for the instability of drugs and their metabolites in biological matrices include oxidation and enzymatic degradation, temperature, light, and pH (Briscoe and Hage, 2009). Stability of chiral drugs is also an important area due to rapid intercoversion between *R* and *S* configurations under physiological pH conditions in some cases. Additionally, the effect of hemolysis on drug stability is another area that should be evaluated during development of bioanalytical assay (Bérubé et al., 2011). Recently, blood stability testing has gained momentum in the EBF discussions for its inclusion in method development and validation (Freisleben et al., 2011). The recommendations suggest analysis of the blood fraction using a qualified assay method in order to overcome experimental pitfalls.

Stability assessment in spiked matrix does not truly reflect ISS as the metabolites or other drug related compounds are likely to be present in the incurred sample but not in the spiked QC sample. As a typical case, an unstable metabolite could degrade or convert to the parent drug at any phase of bioanalytical process like sample collection, sample storage/restorage, sample processing, in autosampler, in analytical column or even in ion source of mass spectrometer. In majority of the cases, the profile of metabolite is not known till the bioanalytical assay is validated. Thus, it is difficult to estimate the role of ISS until an ISR analysis is carried out. Two regulatory documents (EMA, 2012; US FDA, 2001) and a white paper (AAPS, 2006) mentions the use of incurred samples in method validation, including some suggestions to perform stability experiments using incurred samples. This is mainly intended to improve the integrity and reliability of data generated using a validated method. However, there is no full description on the conduct ISS in these documents. For a drug with a rational stability profile, it is possible to initiate ISS for selected samples after a reasonable time interval (possibly around T_{max} and during elimination phase) following drug administration. In such a case, multiple dose studies would provide better assessment of the role of metabolites where it is reasonable to expect drug accumulation. Based on this criterion, it is not essential to conduct ISS immediately after sample collection, except in cases where the metabolites are highly unstable. As a starting point for the conduct of ISS, all three integrated parts of a bioanalytical assay can be investigated, namely sample preparation procedure, chromatography, and mass spectrometry, to verify the conversion of metabolite to its parent drug or prodrug to its active form. During sample preparation, the extraction procedure can be tested for robustness by some drastic changes in the extraction methodology like concentration of additives, pH, and temperature. Similarly, conditions can be altered to have an extended chromatography to check for any coelution of isobaric metabolites that may convert to the parent drug. Stability studies can be radically changed with respect to time to have better assessment. Selection of less intense reproducible peaks in MS and/or use of UV detection can also aid in ascertaining coeluting metabolites. Few case studies have been documented that highlight the importance of ISS in bioanalytical assays.

Case 1 *Instability due to change in laboratory environment* (Yadav and Shrivastav, 2011)

An LC-MS/MS assay method was developed to estimate rabeprazole in human plasma. The method employed a SPE technique, used omeprazole as IS and was successfully validated. The method was designed for a pivotal BE study with replicate design (four periods) in 50 healthy human subjects after oral administration of 20 mg rabeprazole (tablet) under fasting conditions. The number of samples analyzed for ISR (ISR-1) was 490 from 4800 total subject samples studied. The ISR-1 values displayed percentage differences from −69.8 to −20.7 compared to the initial results as shown in Table 5.3. The preliminary investigation was initiated for reevaluation of CC and QC samples, chromatography, and sample processing steps; however, no variation was found. Thus, ISR analysis was repeated (ISR-2) but the results did not match with the initial analysis. However, the results of ISR-1 and ISR-2 were comparable (percentage difference within −3.7% to 10.9%). Further, interday variation was also checked by performing one more ISR experiment (ISR-3); similar irregularity was observed with ISR-3 experimental results when compared against the initial analysis; however, when the results of ISR-3 were compared with ISR-2 they showed percentage differences well within acceptance limits (−2.3% to 11.8%). Thus, some issues with the initial sample were suspected. All processing steps were checked again, and upon careful and extended scrutiny, it was found that the initial analysis was performed during the summer season that is hot (40–45°C) with very little humidity (~50%). It was recognized that due to some internal delay the ISR-1 study was conducted after a gap of 3 months. By this time it was the monsoon season, which resulted in highly humid (~85%) conditions. Thus, all ISR studies (ISR1-3) were conducted during the monsoon season and the results obtained were consistent and comparable, nevertheless, they were different from the initial analysis results. Additionally, ISR-4 was conducted under controlled conditions of temperature (22°C ± 3) and humidity (50–70%), which supported the results obtained from all previous ISR (ISR1-3) studies. Based on the outcome, it was confirmed that change in environmental conditions resulted in degradation of rabeprazole, which lead to unmatched results with the initial analysis.

TABLE 5.3 Incurred Sample Reanalysis Data for Rabeprazole

		Percentage difference		
Sample index	Sample ID	ISR-1 vs. Initial analysis	ISR-2 vs. ISR-1	ISR-3 vs. ISR-2
1	02108	−31.4	−0.7	0.0
2	02218	−26.7	−3.7	4.9
3	03106	−27.4	1.2	9.7
4	03219	−69.8	−1.6	2.1
5	04105	−30.1	1.3	6.6
6	04220	−25.8	6.5	−0.5
7	07105	−25.6	7.1	11.8
8	07219	−27.6	4.5	−1.3
9	10106	−32.7	7.1	0.5
10	10221	−24.8	2.7	0.0
11	11104	−23.3	10.1	−2.3
12	11219	−44.1	4.1	5.6
13	12104	−35.8	4.7	8.6
14	12219	−20.7	3.7	5.4
15	14105	−27.6	10.9	0.4
16	14221	−26.8	5.2	3.8
17	15105	−29.4	2.6	11.3
18	15219	−21.3	4.7	3.4
19	16108	−21.4	5.0	−1.4
20	16221	−34.8	7.0	−0.5

Reproduced from Yadav and Shrivastav (2011) with permission of Future Science Ltd.

Case 2 *Instability during storage* (Meng et al., 2011)

A single-analyte study was reported using a stable-labeled IS by LC-MS/MS with atmospheric pressure chemical ionization source. The samples were prepared by protein precipitation from 50 μl monkey plasma sample. Out of 53 ISR samples (selected from three separate batches) that did not meet the acceptance criteria, 52 of the repeat values had a systematic negative bias (20–40%). Initially, the calibrator performance was checked and then the entire study was conducted within established long-term matrix storage stability, and all the study samples were analyzed without exceeding established freeze–thaw, bench-top, or extract stability. However, one observation was noticed regarding a small peak that eluted just before the analyte peak in study samples. This shoulder peak was noticeably absent in the standards, QCs, and control blank samples. As the size of the peak was very small and was adequately resolved, it was not given enough consideration. However, as part of investigation when the chromatogram from the original analysis was compared with ISR batch, it was apparent that the extra peak had increased substantially in size. The effect was more pronounced in individual samples that failed ISR compared to individual samples that passed ISR. Similar observation was also observed for a companion method used for analysis of rat plasma. However, the peak size was too small and thus had minimal impact on the ISR in rat plasma. Based on the evaluation of the data, it was hypothesized that the likely source for ISR failure was due to incurred sample instability caused by the formation of an undetermined, but potentially related compound in the matrix. Further a second ISR was performed with a view to investigate any correlation between the peak areas of the extra peak and the analyte peak in both the original and the ISR runs. Based on this action plan, 72 samples from the initial ISR run were analyzed for a third time, and all reportable results from the second ISR batch matched the results from the first ISR batch within 20% of the mean values. Finally, a few ISR samples were selected randomly to look for peak area of the analyte and the questionable peak. Interestingly, the percentage ISR bias after integration of both the peaks showed dramatic improvement relative to the original percentage ISR bias without inclusion of the extra peak. Thus, the ISR failure was essentially due to analyte instability in the incurred samples during storage.

5.4 INCURRED SAMPLE ACCURACY

It is commonly acknowledged that random and systematic analytical errors ("total error") contribute to poor data quality and, moreover, to imprecise and inaccurate toxicokinetic or pharmacokinetic parameters. Random errors (expressed as the standard deviation) can be caused by instrumental and sample instability, impure chemicals, environmental fluctuations, sample collection, and sample storage. Systematic errors (expressed as the bias) can be caused by insufficient selectivity, matrix effects, nonoptimal method calibration procedures, inadequate blank corrections, carryover issues, operator bias, and instrumental shifts. To investigate the random errors in GLP bioanalysis, it has found common ground in today's bioanalysis to assess the reproducibility of the method by reanalyzing part of the incurred samples (Fast et al., 2009; EMA, 2012). It was recently suggested that although assessment of ISR provides information about the precision (variability) of the assay it is only used as a surrogate for accuracy (Larsson and Han, 2007; Hill, 2009; De Boer and Wieling, 2011), that is, ISR provides information about the random errors within a predefined acceptance interval, but it will not give information about the systematic errors that may or may not have affected the accuracy of the obtained initial concentration. Therefore, in those cases where ISR does not detect any instability of the study samples, it is still possible that (relative) systematic errors have influenced the value of the initial detected concentration.

In order to obtain both precise and accurate data, the authors suggested conducting complementary experiments to calculate the (relative) systematic errors as an estimate for the ISA by applying standard addition.

5.4.1 Methodology for ISA Assessment

To demonstrate the applicability of ISA in GLP bioanalysis, De Boer and Wieling (2011) used samples from a BE study of alendronate in human urine. They reanalyzed 30 samples in duplicate without (common ISR) and after addition of a standard solution of alendronate at two concentration levels (ISA) utilizing a validated LC-MS/MS assay. Samples were selected as suggested previously (Fast et al., 2009; EMA, 2012). The samples were analyzed in duplicate to assess repeatability and were executed by two different technicians to include inter-analyst variation. The mean values of the repeated sample concentrations were used for calculation. Furthermore, to avoid additional dilution effects (parallelism), the total amount of artificial solution added to the incurred samples was equal for all samples, that is, for the ISR experiments as well as the initial analyses of the study samples, a fixed volume of organic solvent containing the IS was added, whereas for the ISA experiments the same volume was added containing both the IS as well as additional alendronate.

5.4.2 ISA Data Interpretation

De Boer and Wieling (2011) interpreted the obtained data as follows: the normally distributed data sets (ISR and ISA) were compared by calculating the correlation between the

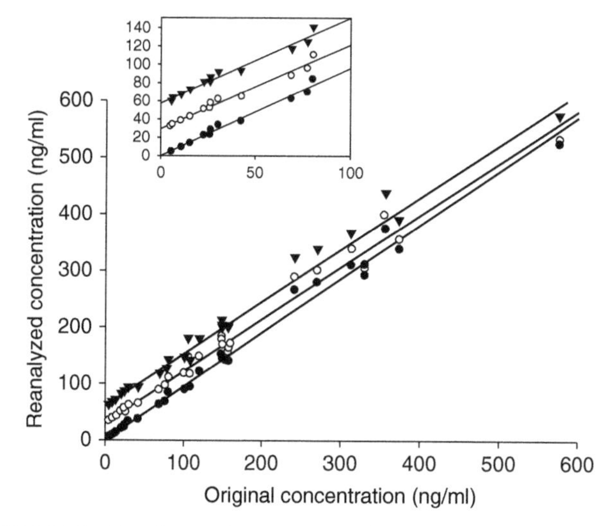

FIGURE 5.5 Correlations between the initial urine concentrations and the reanalyzed urine concentrations with and without addition of alendronate (• = ISR [$Y = 0.945X + 2.34$]; ○ = ISA [30.0 ng/ml; $Y = 0.915X + 32.4$]; ▼ = ISA [60.0 ng/ml; $Y = 0.923X + 60.2$]) (adapted from De Boer et al. (2011) with permission of Future Science Ltd).

data sets as a first indication of similarity (see Figure 5.5). By calculating the correlation between the data sets, information was obtained about the random errors (correlation coefficient), the relative systematic errors (slope), and the constant systematic errors (intercept) (Massart et al., 1988).

The calculated correlation coefficients (r) were 0.9947 (ISR), 0.9926 (ISA after standard addition of 30 ng/ml alendronate), and 0.9894 (ISA after standard addition of 60 ng/ml alendronate), respectively. The calculated slopes were 0.95 ± 0.02 [ISR], 0.92 ± 0.02 [ISA(30)] and 0.92 ± 0.03 [ISA(60)], respectively. Finally, the intercepts of the linear curves corresponded to the added amount of alendronate: 32 ± 4 and 60 ± 5 ng/ml, respectively. From these data the authors concluded that (1) based on the regression coefficient, there were no random errors affecting the data; (2) based on the obtained slope values, that were in the same order of magnitude and approaching 1.0, only minor relative systematic errors influenced the accuracy of the data; and (3) based on the intercepts, no constant systematic errors influence the accuracy of the data.

By calculating the correlation between the data sets, a proper estimation can be made for the accuracy of the initial reported concentrations. In Table 5.4, an overview is given of the ISR results and the ISA results. The ISA results were normalized by subtracting the added amount of alendronate (either 30 ng/ml or 60 ng/ml, respectively) from the measured concentration after standard addition. In the table the biases, which are a measure for the accuracy are shown. In comparison to ISR, that is calculated by dividing the difference between the initial value and the repeat value by the mean of the two values (EMA, 2012), it is suggested to calculate ISA by dividing the difference between the two values with the initial value, as for ISA it is of interest to see the actual deviation from initial value rather than the variability. From the table it can be readily seen that for the ISR experiments, no outliers were detected. For the 30 ng/ml addition, three outliers were detected, whereas for the 60 ng/ml addition, eight outliers were detected. These results would imply that the accuracy of the initial data was warranted, as these results adhere to the generally accepted criteria for ISR (i.e., two-thirds of the repeats should be within $\pm 20\%$ for small molecules (EMA, 2012). However, it can be clearly seen from the table that there is a strong relation between the observed high inaccuracies and the low concentration of the initial value. It was concluded by the authors that high spikes near low values result in statistical errors that influence the outcome of the ISA experiments. Therefore it was suggested by the authors to carefully select the concentration of the analyte that is used for the standard addition experiments to assess the ISA: e.g. ISA [low] ~3 × lowest observed study sample concentration (C_{min}) and ISA [high] ~0.5 × highest observed (undiluted) study sample concentration (C_{max}). To these samples, 50% of the observed initial concentration should be added as long as the final concentration does not exceed the upper limit of quantification.

5.4.3 Reflections on the Use of ISA in Bioanalytical Method Validation

At present, the regulatory agencies do not require ISA assessment because the ISR experiment is used as a surrogate for *accuracy*, whereas ISR is actually only a measure of method *precision* for incurred samples. The Global CRO Council recommended not performing ISA systematically unless there is a clear reason for performing it as part of an investigation (Lowes et al., 2011). Nevertheless, the assessment of ISA in combination with batch acceptance QC samples (which do not properly reflect the accuracy of the study sample concentrations) and assessment of ISR offers a complete overview of all analytical errors that affect the precision and accuracy of the reported data, information that will not be available when applying the current ISR approach.

5.5 SUMMARY

The overall method performance is the key to successful development and validation of bioanalytical assays for drug development programs. ISR, ISS, and ISA are the vital parameters to establish robust, scientifically sound, and high-quality bioassays. The recent EMA guidance has effectively established ISR as an integral concept in bioanalysis to ensure reproducibility of the bioanalytical method. ISR not only

TABLE 5.4 Overview of ISR and Normalized ISA Results

N	X	$\bar{X}_{R1,R2}$	ISR (%)	$\bar{X}_R[\text{xo}] - X_0$ where $X_0 = 30$ ng/ml	ISA (%)	$\bar{X}_R[\text{xo}] - X_0$ where $X_0 = 60$ ng/ml	ISA (%)
1	4.94	4.26	−14.8	2.95	**−39.3**	0.450	**−89.9**
2	5.83	5.14	−12.6	4.00	**−31.4**	3.50	**−40.0**
3	5.91	6.16	4.1	4.65	**−20.5**	2.90	**−50.9**
4	10.5	9.96	−5.3	8.55	−18.1	7.45	**−28.6**
5	15.2	14.2	−6.8	13.2	−13.2	12.5	−17.8
6	22.4	22.6	0.7	21.8	−2.7	21.1	−5.8
7	25.8	23.4	−10.0	22.5	−12.8	22.0	−14.7
8	26.1	28.9	10.2	27.9	6.9	26.1	0.0
9	30.1	34.0	12.0	32.6	8.3	32.2	7.0
10	42.2	37.7	−11.3	35.7	−15.4	33.6	**−20.4**
11	68.9	62.6	−9.7	58.4	−15.2	58.0	−15.8
12	77.3	69.1	−11.3	66.6	−13.8	65.0	−15.9
13	80.3	84.2	4.7	81.0	0.9	81.0	0.9
14	101	90.4	−11.1	88.5	−11.9	87.5	−12.9
15	106	120	12.4	118	11.3	118	11.3
16	108	95.3	−12.5	87.5	−18.5	81.0	**−25.0**
17	120	120	0.0	118	−1.7	119	−0.8
18	149	154	3.0	153	2.7	154	3.4
19	150	146	−2.7	144	−4.0	144	−4.0
20	151	142	−6.5	140	−7.3	138	−8.6
21	156	138	−12.2	130	−16.7	114	**−26.9**
22	157	140	−11.4	136	−13.4	138	−12.1
23	159	141	−12.0	143	−10.1	142	−10.7
24	242	266	9.3	258	6.6	263	8.7
25	270	279	3.3	271	0.4	276	2.2
26	313	310	−1.1	308	−1.6	304	−2.9
27	330	294	−11.5	276	−16.4	254	**−23.0**
28	356	373	4.7	368	3.4	375	5.3
29	373	338	−9.8	326	−12.6	329	−11.8
30	575	524	−9.3	502	−12.7	513	−10.8

N, sample number; X, initial concentration; $\bar{X}_{R1,R2}$, mean of reanalyzed concentration; $\bar{X}_R[\text{xo}] - X_0$, mean normalized concentration after standard addition of X_0 ng/ml alendronate. Adapted from De Boer et al. (2011) with permission of Future Science Ltd.

proves the reproducibility of the method but can also work as an investigation tool especially postanalysis, enabling bioscientists to authorize the integrity of bioanalytical methods. With diligent backup of available case studies together with good scientific practices and judgment, it is possible to ensure reliability of methods developed based on standard guidelines.

REFERENCES

AAPS. Third Bioanalytical Workshop: Quantitative Bioanalytical Methods Validation and Implementation. Crystal City, VA, 2006.

AAPS Workshop on Topics in GLP Bioanalysis: Assay Reproducibility for Incurred Samples. Crystal City, VA, 2008.

Bansal S. AAPS Bioanalytical Survey. Presented at: The AAPS Third Bioanalytical Workshop: Quantitative Bioanalytical Methods Validation and Implementation. Crystal City, VA, 2006a.

Bansal S. Repeat Bioanalysis selection and reporting. AAPS Workshop: quantitative bioanalytical methods validation and implementation: best practices for chromatographic and ligand binding assays, 2006b.

Bansal S, DeStefano A. Key Elements of bioanalytical method validation for small molecules. AAPS J 2007;9(2):E109–E114.

Bansal S. Did incurred sample reanalysis raise confidence in bioanalytical results? Presented at: The AAPS Workshop on Topics in GLP Bioanalysis: Assay Reproducibility for Incurred Samples. Crystal City, VA, 2008.

Bansal S, Arnold M, Garofolo F. International harmonization of bioanalytical guidance. Bioanalysis 2010;2(4):685–687.

Barfield M, Ahmed S, Busz M. GlaxoSmithKline's experience of incurred sample reanalysis for dried blood spot samples. Bioanalysis 2011;3(9):1025–1030.

Bérubé ER, Taillon MP, Milton Furtado M, Garofolo F. Impact of sample hemolysis on drug stability in regulated bioanalysis. Bioanalysis 2011;3(18):2097–2105.

Bland JM, Altman DG. Comparing methods of measurement: why plotting difference against standard method is misleading. Lancet 1995;346:1085–1087.

Bland JM, Altman DG. Measuring agreement in method comparison studies. Stat Methods Med Res 1999;8:135–160.

Briscoe CJ, Hage DS. Factors affecting the stability of drug and drug metabolites in biological matrices. Bioanalysis 2009;1(1): 205–220.

Bryan PD. What is incurred sample analysis and why is it important? AAPS Newsmagazine 2008;11(9):18–22.

Cote C, Lahaie M, Latour S, et al. Impact of methylation of acyl glucuronide metabolites on incurred sample reanalysis evaluation: ramiprilat case study. Bioanalysis 2011;3(9):951–965.

De Boer T, Wieling J. Incurred sample accuracy assessment: design of experiments based on standard addition. Bioanalysis 2011;3(9):983–992.

Dicaire C, Berube E-R, Dumont I, Furtado M, Garofolo F. Impact of oxcarbazepine sulphate metabolite on incurred sample reanalysis and quantification of oxcarbazepine. Bioanalysis 2011;3(9):973–982.

Eastwood BJ, Farmen MW, Iversen PW, et al. The minimum significant ratio: a statistical parameter to characterize the reproducibility of potency estimates from concentration-response assays and estimation by replicate-experiment studies. J Biomol Screen 2006;11(3):253–261.

European Medicines Agency (EMA). Committee for medicinal products for human use, guidelines on validation of bioanalytical methods (draft), EMA/CMP/EWP/192217/2011. Available at www.ema. europa.eu/ema. Accessed Apr 10, 2013.

Fast DM, Kelley M, Viswanathan CT, et al. AAPS workshop on current topics in GLP bioanalysis: assay reproducibility for incurred samples—implications of Crystal City recommendations. AAPS J 2009;11(2):238–241.

Freisleben A, Brudny-Koppel M, Mulder H, de Vries R, de Zwart M, Timmerman P. Blood stability testing: European Bioanalysis Forum view on current challenges for regulated bioanalysis. Bioanalysis 2011;3(12):1333–1336.

Fu Y, Li W, Smith HT, Tse FLS. An investigation of incurred human urine sample reanalysis failure. Bioanalysis 2011;3(9): 967–972.

Garofolo F. How to manage having no incurred sample reanalysis evaluation failures. Bioanalysis 2011;3(9):935–938.

Gupta A, Singhal P, Shrivastav PS, Sanyal M. Application of a validated ultra performance liquid chromatography—tandem mass spectrometry method for the quantification of darunavir in human plasma for a bioequivalence study in Indian subjects. J Chromatogr B 2011;879(24):2443–2453.

Health Canada. Guidance to Industry. Conduct and Analysis of Bioavailability and Bioequivalence Studies—Part A: Oral Dosage Formulations Used for Systemic Effects, 1992. Available at www.hc-sc.gc.ca/dhp-mps/prodpharma/applic-demande/guide-ld/bio/bio-a-eng.php. Accessed Mar 1, 2013.

Health Canada. Notice to industry-removal of requirement for 15% random replicate sample notice affecting guideline A and guideline B requirements, 2003. Available at www.hc-sc.gc.ca/dhp-mps/alt_formats/hpfb-dgpsa/pdf/prodpharma/15rep-eng.pdf. Accessed Mar 1, 2013.

Hill H. Developing trends in bioanalysis. Bioanalysis 2009; 1(8):1359–1364.

James CA, Hill HM. Procedural elements involved in maintaining bioanalytical data integrity for good laboratory practices studies and regulated clinical studies. AAPS J 2007;9(2): E123–E127.

Kelley M, DeSilva B. Key elements of bioanalytical method validation for macromolecules. AAPS J 2007;9(2): E156–E163.

Kelley M. Incurred sample reanalysis: it is just a matter of good scientific practice. Bioanalysis 2011;3(9):931–932.

Larsson M, Han F. Determination of rifalazil in dog plasma by liquid-liquid extraction and LC-MS/MS: Quality assessment by incurred samples. J Pharm Biomed Anal 2007;45(4): 616–624.

Lowes S, Jersey J, Shoup R, et al. Recommendations on: internal standard criteria, stability, incurred sample reanalysis and recent 483s by the Global CRO council for bioanalysis. Bioanalysis 2011;3(12):1323–1332.

Lytle FE, Julian RK, Tabert AM. Incurred sample reanalysis: enhancing the Bland-Altman approach with tolerance intervals. Bioanalysis 2009;1(4):705–714.

Massart DL, Vandeginste BGM, Deming SN, Michotte Y, Kaufman L. Chemometrics: A Textbook, 1st ed. Amsterdam, The Netherlands: Elsevier; 1988.

Meng M, Reuschel S, Bennett P. Identifying trends and developing solutions for incurred sample reanalysis failure investigations in a bioanalytical CRO. Bioanalysis 2011;3(4):449–465.

Nowatzke W, Woolf E. Best practices during bioanalytical method validation for the characterization of assay reagents and the evaluation of analyte stability in assay standards, quality controls and study samples. AAPS J 2007;9(2):E117–E122.

Parekh JM, Vaghela RN, Sutariya DK, Sanyal M, Yadav M, Shrivastav PS. Chromatographic separation and sensitive determination of teriflunomide, an active metabolite of leflunomide in human plasma by liquid chromatography tandem mass spectrometry. J Chromatogr B 2010;878(24):2217–2225.

Patel DS, Sharma N, Patel MC, Patel BN, Shrivastav PS, Sanyal M. Analysis of a second-generation tetracycline antibiotic minocycline in human plasma by LC–MS/MS. Bioanalysis 2011a;3(19):2177–2194.

Patel DS, Sharma N, Patel MC, Patel BN, Shrivastav PS, Sanyal M. Development and validation of a selective and sensitive LC–MS/MS method for determination of cycloserine in human plasma: Application to bioequivalence study. J Chromatogr B 2011b;879(23):2265–2273.

Petersen PH, Stöckl D, Blaabjerg O, et al. Graphical interpretation of statistical data from comparison of a field method with reference method by use of difference plots. Clin Chem 1997;43(11):2039–2046.

Rocci ML Jr., Collins E, Wagner-Caruso KE, Gibbs AD, Fellows DG. Investigation and resolution of incurred sample reanalysis failures: two case studies. Bioanalysis 2011;3(9):993–1000.

Rocci ML, Devanarayan V, Haughey DB, Jardieu P. Confirmatory reanalysis of incurred bioanalytical sample. AAPS J 2007;9(3):E336–E343.

Sailstad JM, Salfen BE, Bowsher RR. Incurred sample reanalysis: failures in macromolecules analysis-insight into possible causes. Bioanalysis 2011;3(9):1001–1006.

Savoie N, Garofolo F, van Amsterdam P, et al. White paper on recent issues in regulated bioanalysis & global harmonization of bioanalytical guidance. Bioanalysis 2010;2(12):1945–1960.

Sennbro CJ, Knutsson M, Timmerman P, van Amsterdam P. Anticoagulant counter ion impact on bioanalytical LC-MS/MS assay performance: additional validation required? Bioanalysis 2011;3(21):2389–2391.

Shah VP. The history of bioanalytical method validation and regulation: Evolution of a guidance document on bioanalytical methods validation. AAPS J 2007;9(1):E43–E47.

Shah VP, Midha KK, Dighe S, et al. Analytical methods validation: bioavailability, bioequivalence, and pharmacokinetic studies. Pharm Res 1992;9(4):588–592.

Shah VP, Midha KK, Findlay JW, et al. Bioanalytical method validation- a revisit with a decade of progress. Pharm Res 2000;17(12):1551–1557.

Smith G. Bioanalytical method validation: notable points in the 2009 draft EMA guideline and differences with the 2001 FDA guidance. Bioanalysis 2010;2(5):929–935.

Tan A, Gagnon-Carignan S, Lachance S, Boudreau N, Levesque A, Masse R. Beyond successful ISR: case-by-case investigations for unmatched reassay results when ISR passed. Bioanalysis 2011;3(9):1031–1038.

Thway TM, Macaraeg CR, Calamba D, et al. Bioanalytical method requirements and statistical considerations in incurred sample reanalysis for macromolecules. Bioanalysis 2010;2(9):1587–1596.

Timmerman P, Luedtke S, van Amsterdam P, Brudny-Kloeppel M, Lausecker B. Incurred sample reproducibility: views and recommendations by the European Bioanalysis Forum. Bioanalysis 2009;1(6):1049–1056.

Timmerman P, White S, Globig S, Ludtke S, Brunet L, Smeraglia J. EBF recommendation on the validation of bioanalytical methods for dried blood spots. Bioanalysis 2011;3(14):1567–1575.

US FDA. Guidance for industry bioanalytical method validation. US Department of Health and Human Services, FDA Centre for Drug Evaluation and Research, MD, 2001.

Viswanathan CT, Bansal S, Booth B, et al. Quantitative bioanalytical methods validation and implementation: best practices for chromatographic and ligand binding assays. AAPS J 2007;9(1):E30–E42.

Viswanathan CT. Incurred sample reanalysis: a global transformation. Bioanalysis 2011;3(23):2601–2602.

Voicu V, Gheorghe MC, Sora ID, Sârbu C, Medvedovici A. Incurred sample reanalysis: different evaluation approaches on data obtained for spironolactone and its active metabolite canrenone. Bioanalysis 2011;3(12):1343–1356.

Yadav M, Gupta A, Singhal P, Shrivastav PS. Development and validation of a selective and rapid liquid chromatography tandem mass spectrometry method for the quantification of abacavir in human plasma. J Chromatogr Sci 2010a;48(8):654–662.

Yadav M, Rao R, Kurani H, et al. Validated ultra high performance liquid chromatography-tandem mass spectrometry method for the determination of pramipexole in human plasma. J Chromatogr Sci 2010b;48(10):811–818.

Yadav M, Shrivastav PS. Incurred sample reanalysis (ISR): a decisive tool in bioanalytical research. Bioanalysis 2011;3(9):1007–1024.

Yadav M, Singhal P, Goswami S, Pande UC, Sanyal M, Shrivastav PS. Selective determination of antiretroviral agents tenofovir, emtricitabine and lamivudine in human plasma by a validated liquid chromatography tandem mass spectrometry method for bioequivalence study in healthy Indian subjects. J Chromatogr Sci 2010c;48(9):704–713.

Yadav M, Trivedi V, Upadhyay V, et al. Comparison of extraction procedures for assessment of matrix effect for selective and reliable determination of atazanavir in human plasma by LC-ESI-MS/MS. J Chromatogr B 2012. Forthcoming. DOI: 10.1016/j.jchromb.2011.12.031.

Yadav M, Upadhyay V, Chauhan V, et al. Chromatographic separation and simultaneous determination of tolterodine and its active metabolite, 5-hydroxymethyl tolterodine in human plasma by LC-ESI-MS/MS. Chromatographia 2010d;72(3–4):255–264.

Yadav M, Upadhyay V, Singhal P, Goswami S, Shrivastav PS. Stability evaluation and sensitive determination of antiviral drug, valacyclovir and its metabolite acyclovir in human plasma by a rapid liquid chromatography–tandem mass spectrometry method. J Chromatogr B 2009;877(8–9):680–688.

6

LC-MS BIOANALYTICAL METHOD TRANSFER

Zhongping (John) Lin, Wenkui Li, and Naidong Weng

6.1 INTRODUCTION

Bioanalytical methods are keys for accurate toxicokinetic and/or pharmacokinetic assessment of drug candidates in support of regional or worldwide regulatory submission. With globalization of drug development, bioanalytical methods are frequently transferred between sites within a company and/or from laboratories in pharmaceutical companies to contract research organizations (CROs), in particular, the CROs in the emerging market (e.g., China and India).

Transfer of bioanalytical method presents some challenges. The challenge can be specific to the method itself, process, communications, or cultural differences between the laboratories, and so on. If not well prepared, method transfer may end up with delay of the intended study. Then questions are: what should be prepared and what procedures should be in place to ensure a smooth method transfer? What are the acceptance criteria in line with the regulatory expectations and how to deal with failure in method transfer? Limited information on the above topics is available in the public domain. Rozet et al (2009) discussed the methodologies for analytical methods transfer with a focus on the study design, sample size, and statistical methodologies. Dewe (2009) reviewed the statistical methodologies to compare (bio) assays. These approaches were used for the evaluation of data sets of fortified quality control (QC) samples (Gansser, 2002; Rozet et al., 2008) and incurred samples (Gilbert et al., 1995). In addition, Shah and Karnes (2009) proposed a "fixed" range decision criterion for bioanalytical method transfer. In this chapter, based on our previous review article (Lin et al., 2011), we would like to summarize some common practices in transferring a LC-MS bioanalytical method from pharmaceutical companies to CROs in support of good laboratory practice (GLP) preclinical studies and regulated clinical studies. Precautions, procedures, acceptance criteria, root causes of method transfer failures and associated investigation are discussed here.

6.2 PREPARATION FOR METHOD TRANSFER

Bioanalytical method transfer should be considered as an integrated part of overall bioanalytical strategy in drug development program in pharmaceutical industry. Outsourcing some of the bioanalytical work to one or more CROs can certainly leverage limited internal resources to better support those high value-added and high priority project activities. As part of preparation for method transfer, as soon as a drug candidate progresses to a point where the intended bioanalytical method needs to be transferred to a CRO, the method should be thoroughly reviewed and/or retested, if needed, for robustness. Any new issues that were previously unnoticed should be addressed.

From an operational point of view, if the CRO (the receiving laboratory) to be considered for method transfer is new to the sponsor (the sending laboratory), precontract facility/site qualification(s) audits should be conducted well ahead along with other operational (including legal) processes. Both quality assurance (QA) unit and bioanalysis (BA) line function of the sponsor should be involved in the audit to determine the overall qualification, in particular, level of compliance of applicable GxP (GLP, GCP, etc.) regulations of the CRO as well as the scientific competencies. For CROs that have already established collaboration with the sponsor, periodical facility/site qualification(s) audits should still be conducted.

Handbook of LC-MS Bioanalysis: Best Practices, Experimental Protocols, and Regulations, First Edition. Edited by Wenkui Li, Jie Zhang, and Francis L.S. Tse.

Several key points need to be considered in those precontract and/or periodical facility/site qualification(s) audits:

- Quality of studies and services previously offered.
- Is the testing facility of suitable size and is building construction fitted with adequate security?
- *Regulatory compliance (e.g.,* Organization for Economic Cooperation and Development *and/or* Food and Drug Administration *(FDA) GLP compliance)*: Historical records of health authority inspection and the associated observation(s) (e.g., 483) and remedy, if any, should be reviewed.
- *Scientific capabilities and technical expertise*: This includes but are not limited to: bioanalytical method development, validation and study sample analysis, a sufficient numbers of laboratory state of art equipments/devices, computers and validated computerized devices, and data management system.
- Ability of achieving required timelines and results; ability of adhering to the study protocol/plan and standard operating procedures.
- Ability of archiving and retrieving data, including electronic data. Proper archiving procedure related to inspection and transfer of data as well as materials from one facility to another facility of the same CRO for long-term storage.
- Business continuity program, including written procedures for record and material retention in case of business discontinuity.

Other logistics also need to be considered in selection of a CRO:

- Is biological sample transportation to be a hurdle? This is especially important for clinical studies at multisites in multicountries.
- What kinds of documents are needed in order for the selected CRO to apply for sample importation permit(s) from the government? What is the estimated duration for the selected CRO to obtain a permit?

6.3 CURRENT UNDERSTANDING ON REGULATORY REQUIREMENTS FOR BIOANALYTICAL METHOD TRANSFER

By definition, a method transfer is to cross-validate an intended bioanalytical method in a different laboratory. In the 2001 FDA guidance, cross-validation is defined as "a comparison of validation parameters when two or more bioanalytical methods are used to generate data within the same study or across different studies (FDA, 2001)." More details on method cross-validation were discussed at the Third AAPS/FDA Bioanalytical Workshop of 2007 and some follow-up meetings (Viswanathan et al., 2007). In the 2011 European Medicines Agency (EMA) guidance on bioanalytical method validation, it is clearly stated that changes for which a partial validation may be needed include transfer of the bioanalytical method to another laboratory, change in equipment, calibration concentration range, limited sample volume, another matrix or species, change in anticoagulant, sample processing procedure, storage conditions, and so on. When data are to be obtained within a study from different laboratories, comparison of those data is needed and a cross-validation of the applied analytical methods should be carried out. Differences in sample preparation or the use of another analytical method (method modification) may result in different outcomes between the study sites. Cross-validation should be performed in advance of study samples being analyzed if possible. For the cross-validation, the same set of QC samples or incurred study samples should be analyzed using both methods. For QC samples, the obtained mean accuracy by the different methods should be within 15% and may be wider, if justified. For study samples, the difference between the two values obtained should be within 20% of the mean for at least 67% of the repeats (European Medicines Agency, 2011).

6.4 METHOD TRANSFER

A bioanalytical method transfer could be viewed as a cross-validation conducted by different analyst(s) on different instrumentation(s) using a different bioanalytical laboratory system (computer system, software, freezer, balance, pipette, solvent, automation, quality system, etc.). Procedurally, as soon as a laboratory is selected as the receiving laboratory, scientists from both the sending and receiving laboratories should discuss on all the details of the method, in particular method specificity and analyte stability. It should be emphasized that any trivial details that are not captured in the written method procedure may literally cause failure of method transfer; in particular, with those difficult ones. If possible, a face-to-face meeting and on-site training should be held on method transfer.

A successful transfer depends upon not only the scientific capabilities of both the sending and receiving laboratories but also their communication. Understanding the common goals of method transfer and appreciating the difference in culture will help build strong relationship between the two laboratories for a successful method transfer. The methods developed and validated at the sending laboratory need to be vigorously tested for its robustness prior to being transferred. The receiving laboratories also play a pivotal role. They should ensure a full understanding of the intended method and ask for clarification if there are any questions, including instruments, reference standards, storage stability, and so on.

Upon discussion with the sending laboratory, the receiving laboratory should draft a protocol for method transfer. As a bridge of the two laboratories, the protocol should be reviewed and agreed by both laboratories.

6.4.1 Partial/Cross-validation versus Full Validation

Depending on the nature of the program to be supported by the receiving laboratory using the method to be transferred, method transfer can be as simple as partial/cross-validation but can be as comprehensive as a full validation following the "fit for purpose" approach.

A partial/cross-validation can be as little as one intra-assay accuracy and precision determination along with necessary confirmation of matrix effect and recovery. Similar to regular method full validation, QC samples at lower limit of quantification (LLOQ), low, medium, and high levels should be analyzed in six replicates in the partial/cross-validation along with calibration standards. However, the current industrial trend on method transfer is to perform full validation at the receiving laboratory. In this case, method specificity/selectively, matrix effect and recovery, inter- and intrarun precision and accuracy of QC sample results for the LLOQ, low, medium and high QC samples, dilution integrity, batch size integrity and stabilities (stock/spiking solution stability, freeze/thaw stability, bench top stability, autosampler stability and long-term stability, etc.) must be fully validated according to the current guidance (FDA, 2001; European Medicines Agency, 2011).

Once the precision and accuracy of the method at the receiving laboratory has been established, QC samples (low, medium, and high) prepared and tested at the sending laboratory and/or incurred samples that have been analyzed in the sending laboratory are to be reanalyzed in the receiving laboratory. These QC and incurred samples and the associated analyte concentrations can be blinded or unblinded. The measured analyte concentrations in the QC and incurred samples are to be compared against those obtained at the receiving laboratory along with other validation data.

6.4.2 Method Modification

The practice in BA varies among the pharmaceutical companies and CROs. Although, by default, a method to be transferred should be fully validated and implemented to some extent at the sending laboratory, some companies may provide a discovery type of bioanalytical method for method transfer. In this case, method modification may be necessary. There are other reasons why a method needs to be modified. These reasons include but are not limited to (1) difference in instrument type, (2) difference in process, or (3) difference in laboratory settings between the sending and receiving laboratories. For example, the term of room temperature may be interpreted differently between laboratories in the United States and in China, and a high pressure liquid chromatography (HPLC) column purchased 2 years after the initial method validation may behave differently from the original one.

Upon review of assay documentation and all needed materials (reference compound and internal standard, etc.) from the sending laboratory, the receiving laboratory should test the assay method for transferability. Based on the evaluation outcomes together with internal experience, the receiving laboratory may request modification of the original method or method redevelopment. In both cases, a general recommendation is to ensure the basic feature (e.g., sample preparation) of the modified method be as close as possible to the original method. This is particularly important when the modified method is to be used in support of a study, for which the original method has already been employed.

6.4.3 Acceptance Criteria

Regardless of whether a single batch cross-validation or full validation is to be conducted for method transfer, the acceptance criteria should be consistent with the regulatory requirements (FDA, 2001; European Medicines Agency, 2011) and the expectations of the sending laboratory. The acceptance criteria should be discussed and agreed between the two laboratories and clearly stated in the method transfer protocol.

Accuracy is determined by replicate analysis of samples (QCs) containing known amounts of the analyte. Intra-run (cross/full validation) accuracy should be measured using six determinations at a minimum of four concentration levels (e.g., LLOQ, low, medium, and high). The mean measured value should be within $\pm 15\%$ (bias) of the nominal value for all QC concentration levels except for the LLOQ, where the bias (%) should be within $\pm 20\%$. For inter-run (full validation) accuracy, LLOQ, low, medium and high QC samples from at least three runs analyzed on at least two different days should be evaluated. The mean concentrations should be within $\pm 15\%$ of the nominal values for all QC samples, except for the LLOQ, for which the bias (%) value should be within $\pm 20\%$ of the nominal value.

Intra-run (cross/full validation) precision should be measured using six determinations at a minimum of four concentration levels (e.g., LLOQ, low, medium, and high). The precision determined at each concentration level should not exceed 15% of the coefficient of variation (CV) (%) except for the LLOQ, where the CV (%) should not exceed 20%. For inter-run (full validation) precision, LLOQ, low, medium, and high QC samples from at least three runs analyzed on at least two different days should be evaluated. The inter-run CV value should not exceed 15% for all QC samples, except for the LLOQ, for which the CV(%) value should not exceed 20%.

In addition to the above intra- (cross/full validation) and/or inter-run (full validation) assays at the receiving laboratory using QC samples prepared there, a set of QC samples (low, medium, and high) prepared and verified at the sending laboratory should be included at least in duplicate for each concentration level in the of the validation runs at the receiving laboratory. The difference in results of those QC samples obtained at the sending and receiving laboratory should be within ± 15% for all QC concentration levels tested.

As part of method transfer, a minimum of 20 incurred samples should be reanalyzed at the receiving laboratory. The concentrations obtained from the initial analysis at the sending laboratory and the concentrations obtained by reanalysis at the receiving laboratory should be within 20% of their mean values for at least 67% of the repeats. Large differences between results may indicate analytical issues and should be investigated.

6.5 COMMON CAUSES OF BIOANALYTICAL METHOD TRANSFER FAILURE

There is no doubt that a successful method transfer increases the confidence at both the sending and receiving laboratories. In contrast, a failed method transfer suggests deficiency in one or more areas. The common causes that may contribute to a failed method transfer are listed as follows:

- *Difference in laboratory settings between the sending and receiving laboratories*: A successful method transfer will not only depend on the experience and knowledge of both the sending and receiving laboratories in regulated BA but also on the communication on the characteristics of the method to be transferred. Seemingly irrelevant differences, such as HPLC system, MS, method of pipetting, automatic liquid handling system, reagents, storage of reagents, plate washers, method choice of siliconizing glass tubes, use of specialized equipment, type of vortex mixers, and even the fume hood air flow, and so on, between the two laboratories may cause a transfer validation failure. Therefore, it is essential for both the sending and receiving laboratories to ensure that details of the method are clearly communicated prior to the method transfer.
- *Inadequate stability*: Instability of the analyte or metabolites can also lead to poor assay reproducibility and accuracy. Many factors can cause the instability of a test compound and/or its metabolite(s). These include enzyme catalyzed degradation (e.g., ester-containing drugs or prodrugs), chemical reactivity (e.g., thiol compounds), catechol auto-oxidation, lactone and hydroxyl carboxylic acid interconversion (e.g., statin compounds), *N*-oxide decomposition and phase II metabolite deconjugation, chiral/epimeric/tautomeric/isomeric, and so on. Among the above causes, the most common one is the breakdown of phase II conjugates, in particular acyl glucuronide metabolite(s), to the parent analyte either during sample storage, processing or chromatography. There is no doubt that the breakdown of the unstable phase II conjugates result in over-estimation of the parent compound concentration and also significant positive bias values for the incurred samples upon reanalysis at the receiving laboratory. It is essential that stability information on the analyte and its metabolite(s) and the associated stabilization measures are reviewed. If the samples do contain labile metabolites, additional stabilization measures should be taken in place for sample collection and processing prior to LC-MS analysis. Any additional finding needs to be clearly communicated to the receiving laboratory as well.
- *Inadequate assay selectivity*: Assay selectivity should be assessed by fortifying at least six individual lots of matrix with the analyte at the LLOQ of the assay. The bias values of within ± 20% of the nominal concentration for five out of six individually spiked LLOQs are considered acceptable. For some methods with limited matrix availability, selectivity testing may have only been performed in a smaller number of lots or even in a single pool of matrix before the transfer. Testing in a single lot of matrix obscures lot-to-lot differences. This can result in method transfer failure if assay selectively is nonoptimal. In addition to certain common endogenous components, such as phospholipids, assay selectivity issues can also be caused by known/unknown metabolites, dosing vehicles, coadministered medicines.

  Chromatographic separation of analyte of interest is primarily based on the differences of physicochemical properties between the analyte and matrix components related to both mobile and stationary phases on the HPLC columns. When the mass spectrometer is operated in SRM or MRM mode, interference is expected to be rare. API techniques (e.g., ESI and APCI) are generally considered to be "soft" analytical techniques. However, any molecules with weak bonds can fragment during the ionization process (in-source fragmentation) before entering the Q1 chamber of the tandem mass spectrometer. The "in-source fragmentation" is particularly prominent if metabolites (e.g., *N*-oxides, *S*-oxides, glucuronide- or sulfate-conjugated metabolites) of the target analytes are present in sample extract. The in-source fragmentation of those metabolites can readily produce ions identical to the precursor ions of the parent compounds. In high-throughput LC-MS/MS analysis, chromatographic separation is often compromised by aggressively shortened LC run times. The in-source fragmentation of metabolite(s) can result in apparent difference in the measured incurred sample analyte concentrations between the sending and receiving laboratories if chromatographic condition differs from one laboratory to the other. In this case, chromatographic method often needs to be revamped prior to revalidation at one of the laboratories or both.

- *Common bench errors*: A failed incurred sample reanalysis in a method transfer may not be due to assay methodology (selectivity or stability issue), but due to bench errors, including but not limited to (1) inadequate sample thawing, (2) improper sample mixing prior to aliquotting, (3) improper sample dilution, and (4) lack of sample identification verification. As such, proper training of laboratory personnel is warranted.

6.6 INVESTIGATION OF A METHOD TRANSFER FAILURE

Whenever a failed method transfer occurs, necessary investigation should be conducted. A written procedure (e.g., method transfer protocol or protocol amendment) should be in place and be agreed upon by both the sending and receiving laboratories on the investigation. In general, the investigation includes two phases—phase I, review of the existing data and documentation; and phase II, laboratory investigation, which may occur at both the sending and receiving laboratory.

In phase I investigation, the documentation review should be focused on possible errors in sample identification, sample preparation and data processing/calculation, and possible wrong settings in the LC-MS systems, and so on. To conclude the phase I investigation, one of three determinations should be made: (1) assignable cause found, that is, root cause of method transfer failure is identified, (2) probable cause found, that is, root cause of a method transfer failure has been likely identified or has been narrow downed to a few items, and (3) no assignable cause found, that is, no any direct root cause is identified. If assignable cause is found, a corrective action should be immediately taken. A simple correction in data processing and calculation may conclude the investigation. Of course, necessary training should be followed to prevent reoccurrence. In the case where one or more possible bench or instrumentation errors is found, phase II laboratory investigation should be conducted to confirm the root cause(s). Based on the test results, determination can be made on whether (1) an assigned cause is identified or (2) no assigned cause is found, the former should trigger various immediate actions, including but are not limited to, reassay. For the later, further investigation should be followed and this investigation might also involve the sending laboratory. The method should be closely examined together with assessment of the possible impact due to inadequate sample preparation procedure, unknown unstable metabolite(s), poor chromatographic separation of analyte(s) from endogenous components, inadequate assay selectivity, and so on. One or more experiments need to be conducted to verify the possible root causes. The investigation can lead to a simple fix of the procedure followed by reassay or redevelopment and revalidation of the intended assay method.

6.7 SUMMARY

Depending on the purpose, a bioanalytical method transfer can be as simple as one-batch cross-validation, but can be as comprehensive as a full method validation. With the current trend in pharmaceutical industry of outsourcing more and more drug development work to CROs, in particular the CROs in emerging market, a full validation is likely necessary. Regardless of whether a simple cross-validation or a full validation is needed for a method transfer, the acceptance criteria as per the current regulatory bodies (FDA, EMA, etc.) should be met. Therefore, both the sending and receiving laboratories should work together with a proper planning. All technical challenges should be discussed via appropriate channels, including on-site visit or training. The receiving laboratory should not be afraid of challenging the robustness of the method for further improvement. By the end, a successful transfer can not only reduce drug development cost but also help build long-lasting relationship between the two parties, a win–win outcome.

REFERENCES

Dewe W. Review of statistical methodologies used to compare (bio) assays. J Chromatogr B 2009;877:2208–2213.

European Medicines Agency. Committee for Medicinal Products for Human Use Guideline on bioanalytical method validation. 21 July 2011 EMEA/CHMP/EWP/192217 /2009. Available at http://www.ema.europa.eu/docs/en_GB/document_library/Scientific_guideline/2011/08/WC500109686.pdf. Accessed Mar 1, 2013.

Gansser D. Chromatographia 2002;55:S–S71.

Gilbert MT, Barinov-Colligon I, Miksic JR. Cross-validation of bioanalytical methods between laboratories. J Pharm Biomed Anal 1995;13:385–394.

Lin ZJ, Li W, Weng N. Capsule review on bioanalytical method transfer: opportunities and challenges for chromatographic methods. Bioanalysis 2011;3(1):57–66.

Rozet E, Dewe W, Boukloze A, Boulanger B, Hubert Ph. Methodologies for the transfer of analytical methods: a review. J Chromatogr B 2009;877:2214–2223.

Rozet E, Dewe W, Morello R, et al. J Chromatogr A 2008;1189: 32–43.

Shah KA, Karnes HT. A proposed "fixed" range decision Criteria for transfer of bioanalytical methods. J Chromatogr B 2009;877:2270–2274.

US FDA. Guidance for Industry: Bioanalytical Method Validation. US Department of Health and Human Services, FDA, Center for Drug Evaluation and Research, Rockville, MD. 2001. Available at www.fda.gov/downloads/drugs/guidance compliance regulatory information/guidance/ucm070107.pdf. Accessed Mar 1, 2013.

Viswanathan CT, Bansal S, Booth B, et al. Quantitative bioanalytical methods validation and implementation: best practices for chromatographic and ligand binding assays. Workshop/Conference Report. AAPS J 2007;9:E30–E42.

7

METABOLITES IN SAFETY TESTING

Ragu Ramanathan and Dil M. Ramanathan

7.1 INTRODUCTION

Efficacy data and predictions used to move a new molecular entity (NME) into the clinical stage. Among these factors are cross-species differences in the absorption, distribution, metabolism, and excretion (ADME) properties of a NME. These factors can potentially lead to differential efficacy or toxicity between the preclinical species and humans (Lin et al., 2003; Bussiere, 2008; Gomase and Tagore, 2008). Since metabolic processes influence many of the ADME parameters, for example, bioavailability, systemic clearance, and toxicology of a drug candidate, results from drug metabolism studies are important in the selection of viable development candidates (Cheng et al., 2008). To improve patient safety and to limit late stage failures of NMEs, there has been a long-standing acceptance of the idea that drug metabolites present in humans should also be present in the species used for safety evaluation. There was no framework or criteria established for how to meet this goal until the release of the February 2008 US Food and Drug Administration (FDA) (FDA, 2008) and the July 2009 International Committee on Harmonization (ICH) (Harmonization, 2009) guidance for the industry on metabolites in safety testing (MIST). As shown in Figure 7.1, both guidance documents encourage sponsors to evaluate steady-state metabolite exposures in animals and humans as well as recommend doing this as early as feasible in the development process (Frederick and Obach et al., 2010). MIST recommendations posed some challenges to the pharmaceutical companies including the timing of ADME studies using radiolabeled materials, how best to perform metabolite profiling in early clinical studies and bioanalytical options needed to perform early metabolite profiling. This book chapter summarizes the current thinking and practices in the pharmaceutical industry in support of MIST related activities.

7.2 TIMING OF ADME STUDIES WITH RADIOLABELED MATERIALS

The benefits, difficulties, and timing of radiolabeled ADME studies were recently discussed in detail by several pharmaceutical scientists (Penner et al., 2009; Obach et al., 2012; Penner et al., 2012; White et al., 2013). Figure 7.2 compares two of several options available for conducting ADME studies using radiolabled materials. In the first option, radiolabeled studies are not scheduled until after the interpretation of human pharmacokinetics (PK) and metabolism in the first-in-human (FIH) studies, while in the second option, preclinical radiolabeled ADME studies are initiated before the completion of the FIH studies. Both options suggest completing preclinical radiolabeled ADME studies before the initiation of large-scale clinical studies. Similarly, in both options, before the start of FIH studies, one method for characterization of circulating metabolites would rely on *in vitro* metabolism systems for prediction of metabolites. While *in vitro* systems have made a huge impact on predicting overall metabolic clearance (Malinowski, 1997; O'Hara et al., 2001), these systems perform poorly in terms of predicting circulating metabolites. In retrospective studies, it was determined that *in vitro* systems predicted the primary and secondary human circulating metabolites between 41–70% and 12–56% of the time, respectively (Anderson et al., 2009; Dalvie et al., 2009). However, the characterization, synthesis, and quantification of *in vitro* system predicted metabolites would require extensive resources and may not capture metabolites relevant to humans.

Handbook of LC-MS Bioanalysis: Best Practices, Experimental Protocols, and Regulations, First Edition. Edited by Wenkui Li, Jie Zhang, and Francis L.S. Tse.

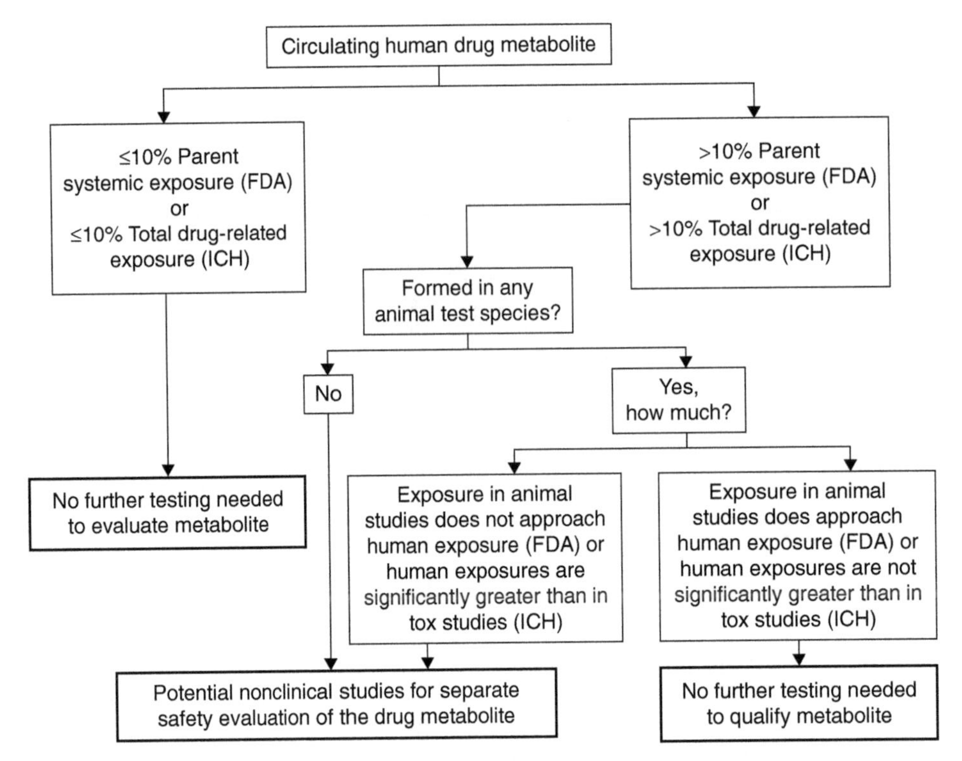

FIGURE 7.1 A metabolite in safety testing (MIST) decision tree incorporating suggestions from the 2008 FDA and 2009 ICH guidances. Generally, the ICH guidance supersedes the US FDA guidance (reprinted with permission from Frederick and Obach, 2010).

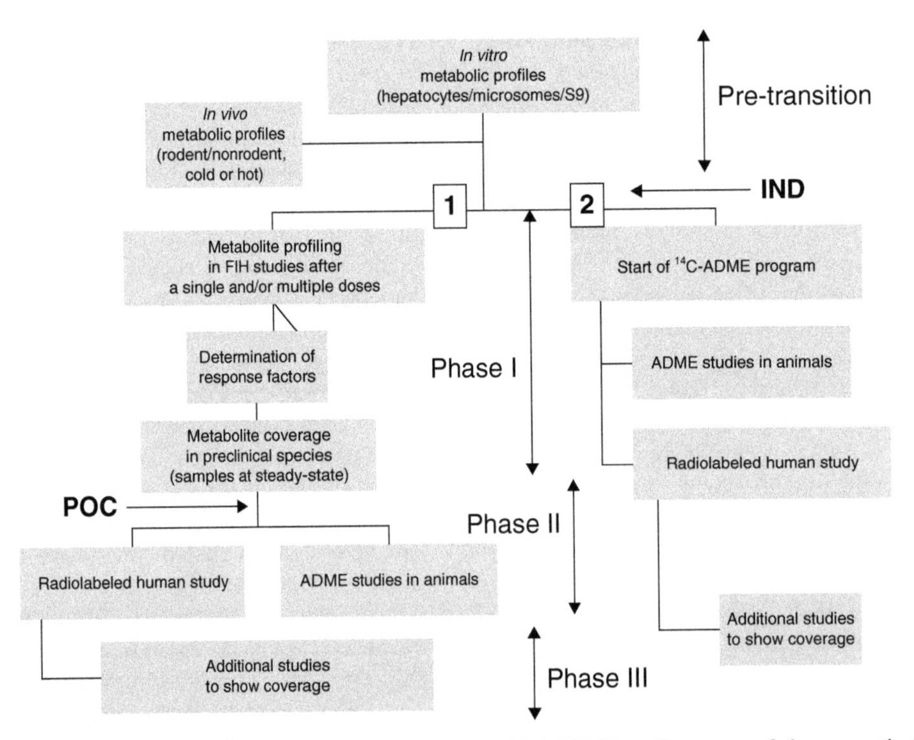

FIGURE 7.2 Options available for conducting radiolabled ADME studies as part of pharmaceutical research and development involved in bringing a new molecular entity to the market with changes in study timing due to MIST guidances (reprinted with permission from Penner et al., 2012).

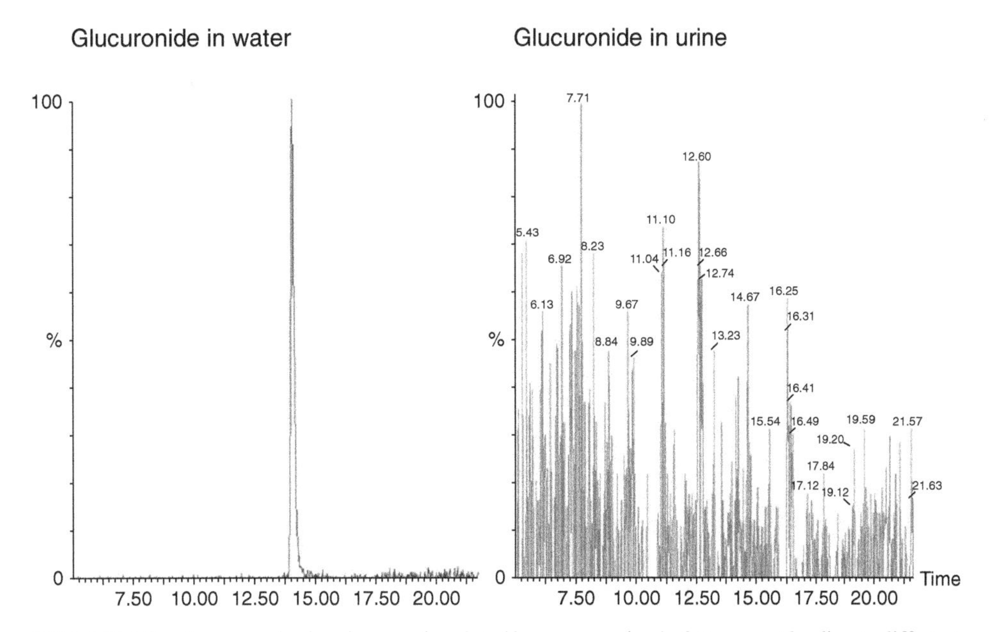

FIGURE 7.3 An example showing matrix related ion suppression/enhancement leading to different LC-MS responses for a glucuronide metabolite (reprinted with permission from Timmerman et al., 2010).

7.3 FIH STUDIES

The difficulties with prediction of circulating human metabolites have led many companies to invest resources on profiling of metabolites in FIH studies or option 1 shown in Figure 7.2. FIH studies, which are part of phase 1 clinical trials, start with single ascending dose (SAD) and advance into multiple ascending dose (MAD) administration. Goals of the SAD and MAD studies are to determine the safety and tolerability of a NME and a safe dosage range based on administration to a small number (20–100) of healthy adult volunteers (drugs with narrow therapeutic indices may be tested directly in patients) (Ramanathan et al., 2010). Determination of parent drug exposure to allow for full pharmacokinetic profiling is also a significant goal of these studies.

If another objective becomes popular in an effort to gain metabolite information from FIH studies, then mass spectrometry-based techniques are the methods of choice for detecting and characterizing metabolites. Several review papers and book chapters have discussed some of the liquid chromatography–mass spectrometry (LC-MS) methods available for detecting and characterizing metabolites (Ramanathan and LeLacheur et al., 2008, Wright, 2011; Zhu et al., 2011). LC-MS assays are further augmented by the availability of various high-resolution mass spectrometry (HRMS)-based post-data acquisition filtering techniques, including mass defect filtering (MDF), isotope pattern filtering (IPF), and background subtraction (BGS). These intelligent data mining tools, in the context of application to MIST workflow, were recently reviewed (Ma and Chowdhury, 2011, 2012; Zhu et al., 2011). Even with sophisticated software tools, without some form of quantification, there is still some difficulty in distinguishing major metabolites from those present at minor and trace levels.

7.4 STANDARD FREE QUANTIFICATION AND ITS LIMITATIONS

LC-MS responses, generated from electrospray ionization (ESI) techniques, are susceptible to variable ionization efficiencies due to interferences from endogenous matrix ions. Matrix-related ion suppression/enhancement, leading to different LC-MS responses for a glucuronide metabolite, is shown in Figure 7.3 (Timmerman et al., 2010). In this example, ESI-MS response for the glucuronide metabolite, present in urine, is completely suppressed possibly due to nonvolatile matrix materials and salts. Although matrix effects can be minimized by processing the urine or diluting the urine in solvents, the relative abundance of the glucuronide will not be reflected in the LC-MS data. This example underscores the need for having a reference standard for quantifying a metabolite. In other words, if reference standards of metabolite(s) of interest and stable isotope labeled internal standard(s) are available, LC-MS quantification of the metabolite(s) would have been straightforward. In reality, it is often that metabolite reference standards and internal standards are not readily available during early development stages.

An alternate approach to using reference standards is to use LC-radioactivity detection (RAD). RAD is nondiscriminating with respect to molecular weight and chemical structure, and is relatively immune to most matrix effects. In a recent study for vicriviroc (VCV) metabolism

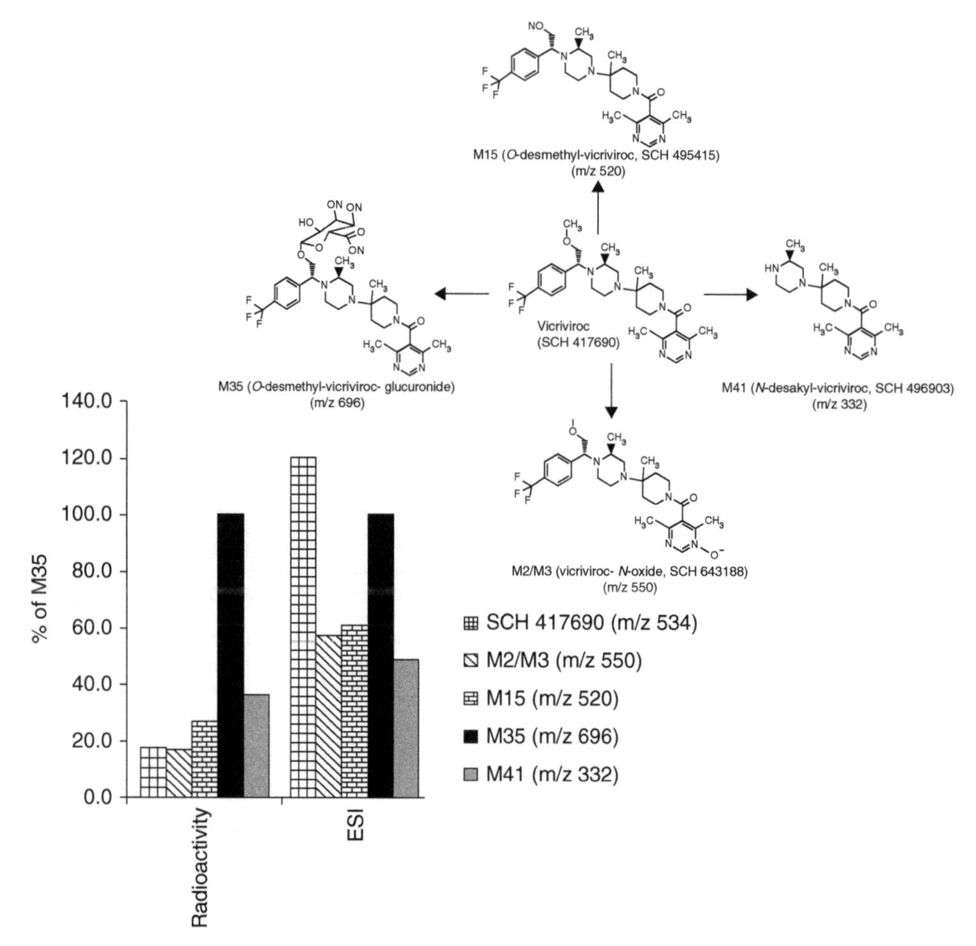

FIGURE 7.4 LC-radioactivity detector (LC-RAD) and LC–MS responses for vicriviroc and its phase I and phase II metabolites (reprinted with permission from Ramanathan et al., 2010).

(Ramanathan et al., 2007), LC-ESI-MS extracted ion chromatographic responses of the compound and its metabolites were compared to corresponding radiochromatographic profiles in Figure 7.4. Based on the LC-RAD radiochromatographic profile, M35 or VCV-*O*-desmethyl-glucuronide was determined to be the major human metabolite. In contrast, LC-ESI-MS detector signal showed greater variation for VCV and its metabolites in comparison to those obtained using LC-RAD. In the event that this study was conducted using nonradiolabeled drug and only the LC-ESI-MS had been used without suitable calibration curves, unchanged drug (VCV) would not have not been accurately measured in human urine or plasma samples. The major metabolite, M35, would have been underestimated, while all the other metabolites (M2/M3, M15, and M41) would have been overestimated. Examples shown in Figures 7.3 and 7.4 underscore the need to conduct radiolabeled ADME studies or quantification using reference standards in an effort to obtain true estimates of the metabolites. However, at the FIH trial stage, very often synthetic metabolite standards and a radiolabel form of the drug may not be available. Under such circumstances, metabolite profiling and analysis would have to rely solely on some of the new technological solutions (Wright et al., 2009; Ramanathan et al., 2010). Taking LC-MS response differences and availability of both reference standards and radiolabeled drug into considerations, exposure coverage, of drug metabolites, can be estimated via tiered assays. Tiered assays are categorized as (1) metabolite profiling, (2) standard free quantification/response factor determination, (3) qualified assay, and (4) validated assay. Overall, the tiered assay approach is accepted by regulatory agencies as well as the bioanalytical community as fit-for-purpose test, cost-disciplined, and scientifically appropriate for patient safety (Timmerman et al., 2010).

7.5 TIERED OPTIONS FOR DETERMINATION OF HUMAN METABOLITE EXPOSURES AND RESPECTIVE LIMITATIONS

All MIST-related activities can be performed in stages as shown in Figure 7.5 (Ramanathan et al., 2010). As a NME moves toward NDA, a more rigorous assay is utilized in

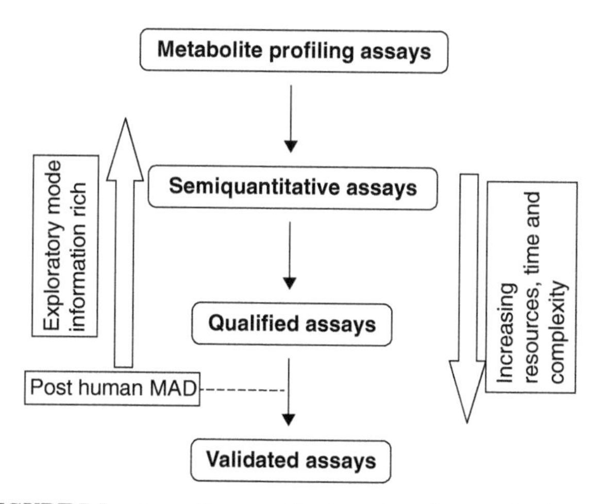

FIGURE 7.5 Assay tiers used for detection, characterization, and exposure estimation of metabolites in clinical studies, MAD = multiple ascending dose (reprinted with permission from Ramanathan et al., 2010).

the measurement of the metabolite concentrations. Additional details about the assays are provided in the following sections and summarized in Table 7.1. The bottom-line is that tiered options for determining human exposures to metabolites are scientifically sound, "fit-for-purpose" clearly defined, and qualified/validated as needed prior to each use.

7.5.1 Tier 1-Metabolite Profiling Assay

The most important step of metabolite profiling work is to include appropriate language and sample collection schemes within clinical and preclinical protocols. To a large degree, appropriate protocol language would eliminate the need for protocol amendments, resigning of subject consent forms, and additional institutional review board discussions. With appropriate protocol language present, upon completion of the bioanalysis of the parent drug, the left over PK/TK samples could be used for metabolite profiling assays. The only concern with regard to using left over PK/TK samples is sample integrity (freeze–thaw cycles, metabolite stability, etc.) and timing (period for bioanalysis of the parent to be completed). In SAD studies, as shown in Table 7.2, the scheduled PK blood samples could possibly be collected at six time points—post-dose from both placebo- and NME-dosed healthy volunteers from all the dosage groups. Metabolite profiling is conducted using samples from one or two of the high dose groups. Similarly, in monkey and dog TK studies, collecting separate samples (at all TK time points in metabolite profiling assays) does not provide a challenge. However, due to blood volume limitations, collecting additional samples from rat and mouse TK studies could be challenging. Therefore, the best option would be to resort to leftover TK samples.

Upon ensuring appropriate language in study protocols, the next step is to decide on the timing and which study samples are utilized for metabolite profiling assays. In most pharmaceutical companies, 1-month TK studies will precede the start of human SAD studies. Since the MIST guidance calls for evaluation of steady-state metabolite exposures in animals and humans, 1-month TK studies will give steady-state metabolite exposures for the preclinical species; however, the SAD will not provide the same exposures. Although human SAD studies provide the first opportunity to obtain a glimpse of circulating human metabolites, human SAD samples and one month TK study samples are used to initiate metabolite profiling related activities.

Since the volume of plasma samples from clinical and pre-clinical studies is limited, one option is to pool (separate pools for plasma from both drug dose versus placebo dose

TABLE 7.1 LC-MS Options Available for Determination of Metabolite Exposures in Clinical and Preclinical Studies

Assay Type	Metabolite Information	Metabolite Quantification	Sample Pooling to Increase Throughput	Metabolite Reference Standards	Time or cost
Metabolite Profiling	Full scan, all detectable metabolites; MDF, IPF, BGS	Not required or not achievable; head-to-head comparison of human and animal metabolite profiles	Sample pooling possible; only AUC information available	Not required	$
Standard-free quantification	Full scan, all detectable metabolites; MDF, IPF, BGS	UV, radioactivity, NMR, etc, to correct for LC-MS response differences	Sample pooling possible; only AUC information available	Not required	$$
Qualified	SRM; targeted	Standard Curve	Not preferred; other PK information	Required	$$$$
Validated	SRM; targeted	Standard Curve	Not preferred; other PK information	Required	$$$$$$

MDF, mass defect filter; IPF, isotope pattern filter; BGS, background subtraction; SRM, selected reaction monitoring (aka MRM, multiple reaction monitoring); UV, ultraviolet; NMR, nuclear magnetic resonance; AUC, area under the curve; $, US dollar.

TABLE 7.2 A Representative Pharmacokinetic and Metabolite Profiling Sampling Schedule that Shows the Options to Collect Separate Plasma and Urine Samples for Metabolite Profiling and Exposure Estimation

Pharmacokinetic and Metabolite Profiling Sampling Schedule					
Sample Collection Time		Time (Relative to Dosing) hour:min	PK Blood Collection	Metabolite Profiling Blood Collection	Urine Collection
Study Day	Time (Event)				
1	0 h (pre-dose)	00:00	x	x	x (pre-dose spot sample)
	0.25 h (post-dose)	00:15	x		x (0 h–12 h cumulative)
	0.5 h (post-dose)	00:30	x	x	
	1 h (post-dose)	01:00	x	x	
	1.5 h (post-dose)	01:30	x		
	2 h (post-dose)	02:00	x	x	
	3 h (post-dose)	03:00	x		
	4 h (post-dose)	04:00	x	x	
	6 h (post-dose)	06:00	x		
	8 h (post-dose)	08:00	x	x	
	10 h (post-dose)	10:00	x		
	12 h (post-dose)	12:00	x		
	18 h (post-dose)	18:00	x		x (12 h–24 h cumulative)
2	24 h (post-dose)	24:00	x	x	
	36 h (post-dose)	36:00	x		x (24 h–36 h cumulative)
3	48 h (post-dose)	48:00	x	x	x (36 h–48 h cumulative)
Number of samples			16	8	5

among human subjects/animals) the plasma samples across subjects and within subjects as shown in Figure 7.6. This type of pooling method, referred to as "area under the curve (AUC) pooling" or "time-proportional pooling," or "Hamilton pooling" has been described in detail by Hop et al. (1998) and Hamilton et al. (1981). AUC pooling scheme, as shown in Table 7.3, allows time-proportional integration of individual plasma time points from a subject into one sample rather than analyzing individual time points to generate a concentration–time curve. Once samples are generated for each subject, then equal volumes of plasma, from all NME dosed subjects, are pooled to generate a composite sample. Similarly, equal volumes of plasma, from all placebo subjects, are pooled to generate a composite placebo sample.

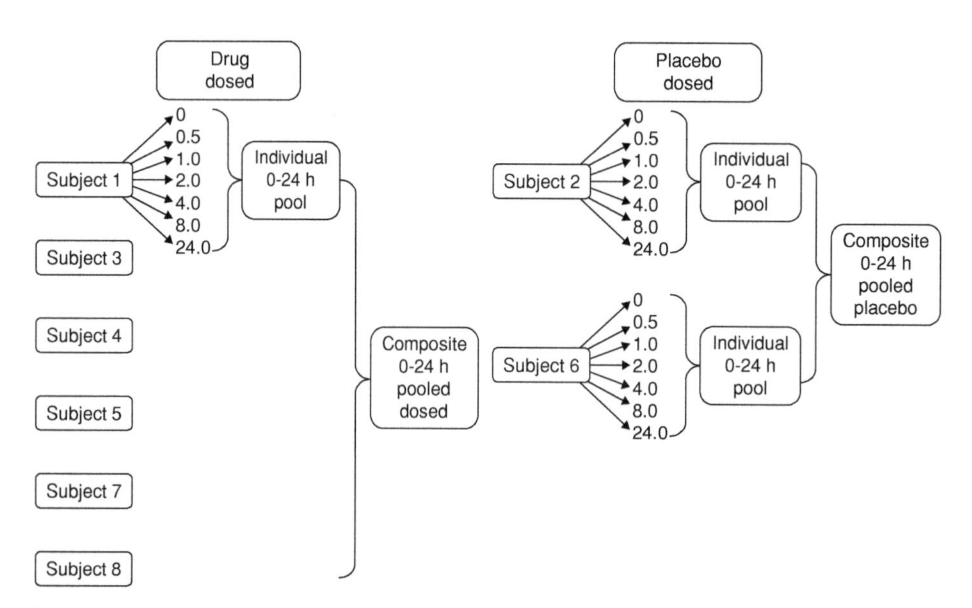

FIGURE 7.6 Sample pooling approach used to streamline metabolite profiling workflow established for analyzing plasma samples from human MAD and SAD studies (reprinted with permission from Ramanathan et al., 2010).

TABLE 7.3 A Representative Scheme Showing Plasma Volume and Corresponding 0–24 h Time Points Used for Generating a Time Proportionally Pooled Sample for Exposure Estimation

Time (h)	0	0.5	1	2	4	8	24
Δt_j	0.5	1	1.5	3	6	20	16
Vol= $\Delta t_j k$ (μl)	5	10	15	30	60	200	160

Where k is the proportionality constant, usually obtained after defining $t = 0$.
A value of 10 is used for pipetting convenience. Where $\Delta t_j = t_{j+1} - t_{j-1}$

The next step of metabolite profiling assays is to develop a suitable LC-MS method for head-to-head comparison of samples from preclinical and clinical studies. This step is achieved by using the approaches shown in Figures 7.7 and/or 7.8. The first option, described by Gao and Obach, involves targeted LC-MS/MS analysis of preclinical and clinical samples (Gao and Obach, 2011). A second option, described by Ma and Chowdhury (2011), involves LC-HRMS analysis of preclinical and clinical samples. In comparison to the targeted analysis, the LC-HRMS option provides the flexibility to acquire a combination of full scan and MS/MS data under LC time scale and preserves information regarding all metabolites. The combined full scan LC-HRMS and MS/MS workflow for drug metabolism applications were recently described (Fung et al., 2011; Ramanathan et al., 2011; Campbell and Le Blanc, et al., 2012; Ranasinghe et al., 2012). Another advantage to using HRMS full scan for head-to-head metabolite profiling is the option to apply post-acquisition data filtering techniques such as MDF, IPF, and BGS. In addition to the option to apply post-acquisition filters, other advantages to acquiring full scan HRMS data include preservation of information about additional analytes from an analysis and preservation of isotopic patterns of analytes of interest and background ions. Today, MDF is applied postacquisition on every HRMS platform and during real-time acquisition on selected platforms (Campbell and Le Blanc, 2012). For both options, discovery stage quantitative LC-MS/MS or metabolite profiling methods, used to generate data for candidate nomination activities, may serve as a

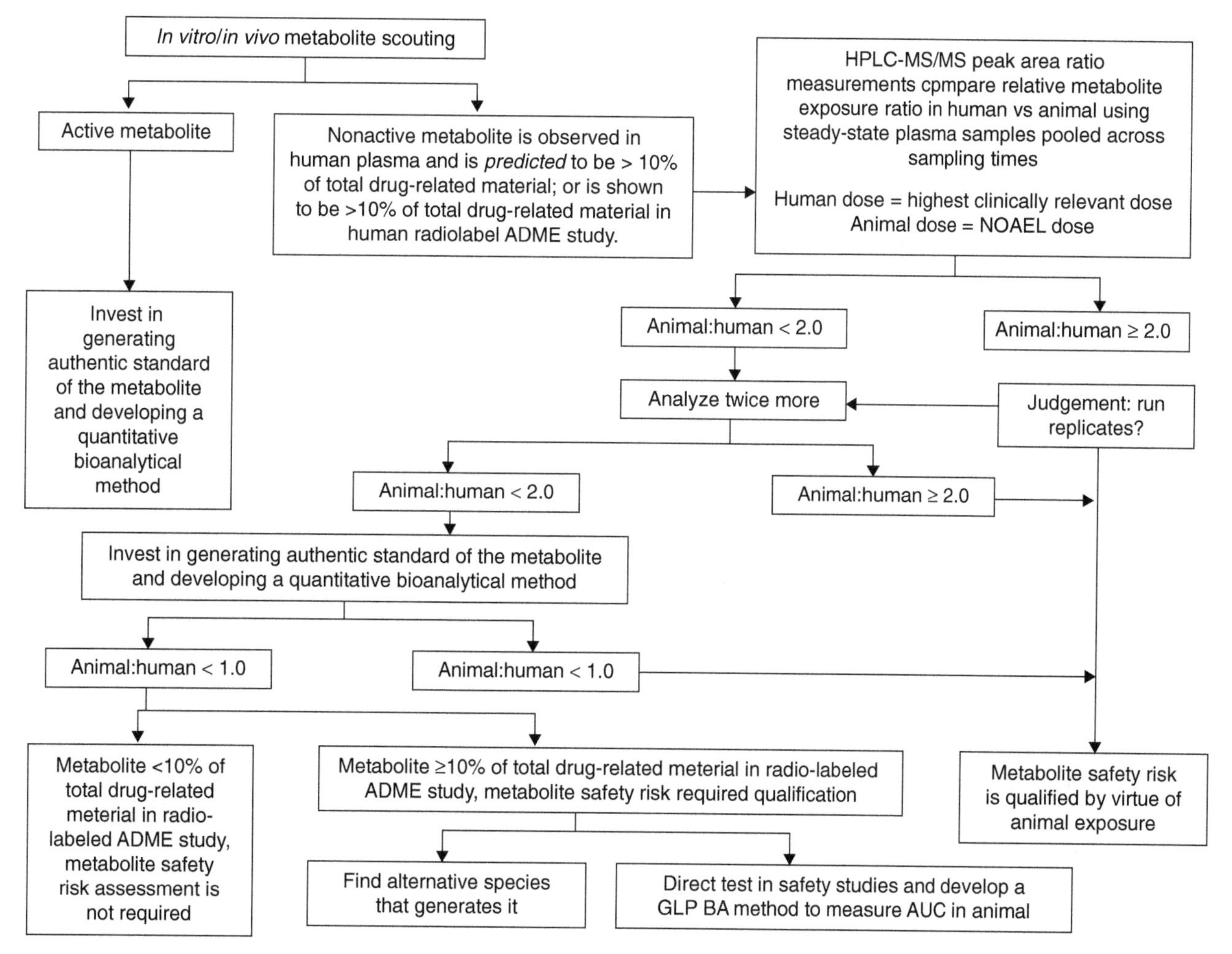

FIGURE 7.7 A LC-MS/MS based bioanalytical strategy for assessing human exposure coverage in preclinical species (reprinted with permission from Gao and Obach, 2011).

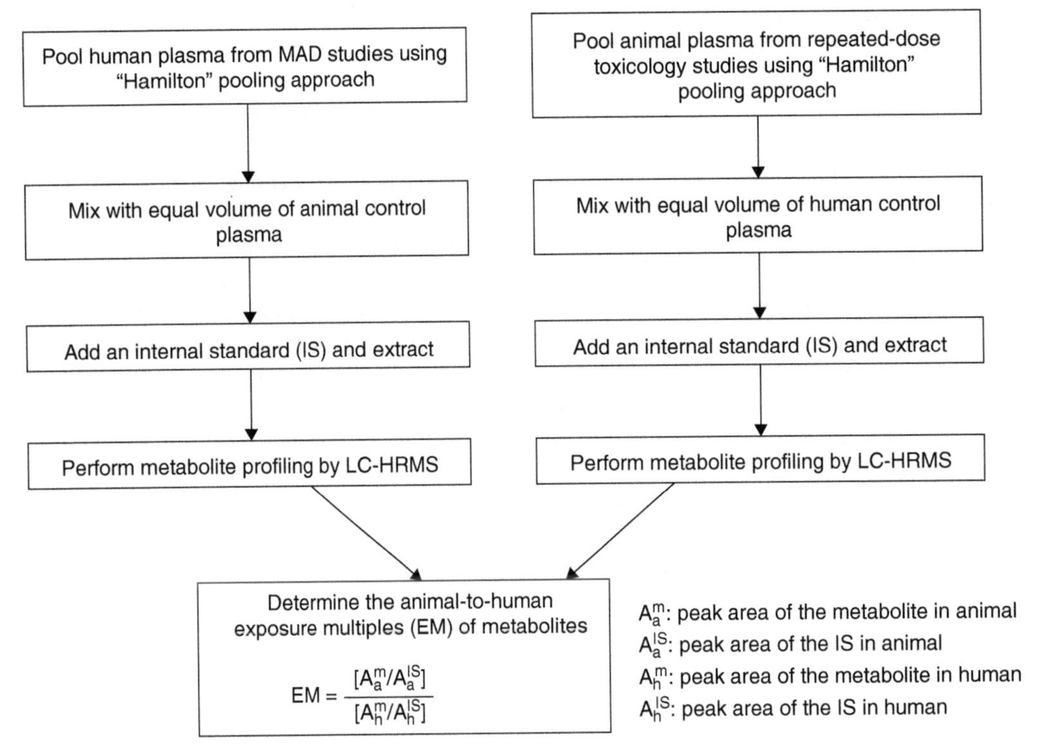

FIGURE 7.8 A LC-HRMS-based bioanalytical strategy for assessing human exposure coverage in preclinical species (reprinted with permission from Ma and Chowdhury, 2011).

starting point for assessing exposure multiples for metabolites.

For overall exposure estimation, sample extraction and reconstitution steps are also important (Chang et al., 2007). Some of the common goals regarding sample preparation are not only to remove the majority of matrix components but also to achieve good recovery for both the drug and metabolites for MS analysis with minimum manipulation and degradation (Ramanathan et al., 2010). In the absence of metabolite reference standards at the early stage of drug development, extraction recovery of metabolite(s) is typically assumed to be similar to the parent NME in the profiling work. This is an important yet unavoidable assumption at this stage. Definitive confirmation, of extraction recoveries for metabolites, is achieved upon completion of 14C-ADME studies or when synthetic reference standards become available for the metabolites.

In the example shown in Figure 7.9, extracted ion chromatograms, from low-resolution mass spectrometry (LRMS) full scan mass spectra, which was obtained following administration of placebo (bottom trace), drug for day 1 (second trace from the bottom), and drug for day 19 (top 2 traces) are compared (Ramanathan et al., 2010). Plasma profiles, from NME and placebo dosed subjects, were simultaneously investigated for all the metabolites reported in both *in vitro* and *in vivo* studies. Comparison of various spectra highlights the advantage of using HRMS for untargeted metabolite profiling. Overall, HRMS approach minimizes the time needed to investigate matrix-related peaks and shifts the focus to drug-related peaks. When using HRMS, as a second round of investigation, applying MDF, IPF, and BGS techniques can prove useful in the detection and characterization of additional metabolites, which are not readily picked-up by extracted ion chromatograms. Typically, these chromatograms are generated for common biotransformation-related metabolites. Once metabolite information from human MAD study plasma samples is available, the next step is to analyze plasma samples from preclinical toxicity study species. Plasma samples from animal species are carefully selected so that head-to-head comparisons of LC-MS profiles from multiple species are validated. In order to achieve a valid comparison, exposure to the parent NME should remain approximately the same between preclinical animal species and clinical human following multiple administration of the NME (Ramanathan et al., 2010). Plasma profiles, from humans and preclinical animal species, are then compared side-by-side to determine if there are disproportionate or human specific metabolites. Subsequent to this, experimentation, focusing on structural elucidation of metabolites, is initiated. Full characterization of metabolites would involve exploration of *in vitro* and/or *in vivo* methods, which are capable of producing larger quantities of the human metabolites for additional LC-MS, LC-MS/MS, and/or NMR characterization (Ramanathan et al., 2010).

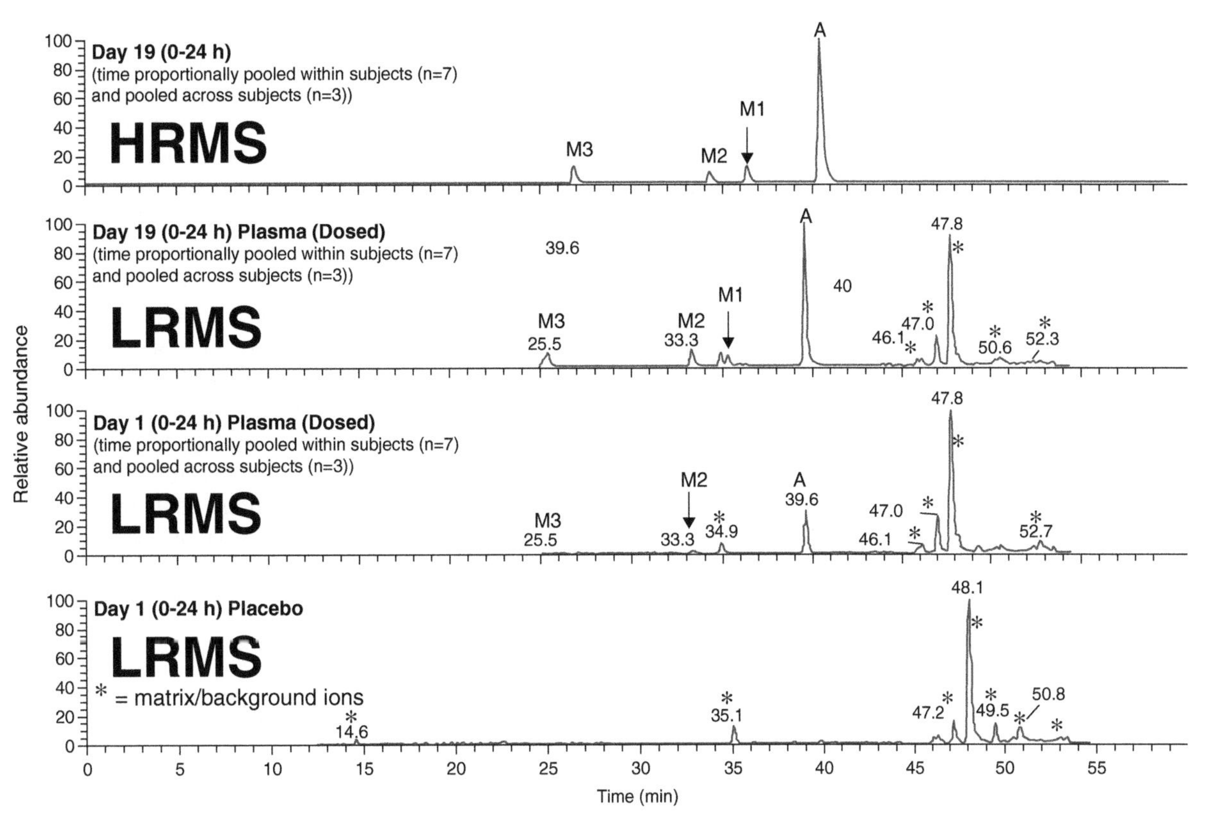

FIGURE 7.9 Plasma extracted ion chromatograms (unit mass resolution or low-resolution mass spectrometry; LRMS) for metabolites (M1, M2, and M3) and drug (A) following administration of drug (A) to human subjects (top panel depicts the high-resolution mass spectrometry (HRMS) metabolite profile under steady state conditions, the second panel shows the LRMS metabolite profile under steady state conditions, the third panel is obtained following administration of drug A for a day, and the bottom panel is from placebo dosed subjects which is crucial to weed out any nondrug derived components.

7.5.2 Tier 2-Standard Free Quantification/Response Factor Determination

Although head-to-head comparisons alleviate the need for response corrections and/or standard free quantification, chromatograms shown in Figures 7.3 and 7.4 underscore the importance of response factor correction for accurate determination of metabolite exposures. Figure 7.4 clearly shows that LC-MS response, observed for structurally related compounds, are considerably different. This limits the direct use of LC-MS responses for relative quantification of NMEs and associated metabolites in biological matrices. To overcome the difficulties in metabolite quantification associated with FIH metabolite profiling, several standard free quantification or response correction approaches have been developed. These approaches involve correlation of metabolite responses obtained via conventional LC-MS with those from non-LC-MS methods or LC-MS with nanospray techniques. The non-LC-MS methods include ultraviolet (UV), charged aerosol detection (CAD), nuclear magnetic resonance (NMR), and radioactivity (Yu et al., 2007; Zhang et al., 2007; Wright et al., 2009).

Among the non-LC-MS methods, UV-based techniques have been widely used to correct for LC-MS response differences at early stages of drug development. In order for UV-based techniques to be successful, the following four conditions must be met: (1) parent NME and its metabolites possess UV chromophores for the detection, (2) analysis of control (pre-dose or placebo) samples to eliminate endogenous UV peaks, (3) the drug-derived peaks should be chromatographically separated from matrix/endogenous peaks, and (4) sample extracts should be properly concentrated for suitable abundance of UV responses discernible from the matrix background. Some of these requirements limit the utility of UV-based methods to studies of high dose ranges. The technique becomes incapable of providing any useful information for studies involving highly potent low-dose drugs. Recently, detailed procedures, for using UV response for correcting for LC-MS response differences, were discussed (Vishwanathan et al., 2009; Yang et al., 2011). In the example shown in Figure 7.8 and summarized in Table 7.4, concentrations of all three metabolites (M1, M2, and M3) are reported following an appropriate correction for LC-MS response differences between parent NME and metabolites M1, M2,

TABLE 7.4 Exposure Estimates of Metabolites Following LC-MS Response Corrections Using LC-UV Responses for Compound A in a Multiple Ascending Dose (MAD) Study. Corresponding Metabolite Profiles Are Shown in Figure 7.9

Parent/ Metabolite	Rt (min)	LC-MS (% Parent after Correction)		AUC (0–24 h) ng*h/ml	
		Day 1	Day 19	Day 1	Day 19
Parent	40.2	100	100	8087[a]	43,642[a]
M1	36.1	7.18[b]	20.0[b]	581	8,728
M2	34.2	10.5[b]	36.7[b]	849	16,017
M3	25.1	4.84[b]	19.8[b]	391	8,641

[a]Obtained using a validated LC-MS/MS assay for the parent.
[b]LC-MS:LC-UV response correction factors used for normalizing equimolar responses for M3 = 1.6, M2 = 3.4, and M1 = 3.4.

and M3 (Ramanathan et al., 2010). LC-MS response correction factors of 1.6, 3.4, and 3.4, respectively, for M3, M2, and M1 were established following an analysis of a 0–24 h AUC pooled plasma samples from multiple subjects by LC-PDA-MS. LC-UV peaks were integrated at several wavelengths and response factors were derived using the procedure described by Vishwanathan et al. (2009). LC-UV peak areas, reported in Table 7.4, were integrated using 254 nm for affording comparison of human MAD data with discovery stage metabolite profiling data (Ramanathan et al., 2010). Since M1, M2, and M3 were all above the 10% of parent NME's AUC or above the 10% total drug-derived exposure (on Day 19, Standard free quantification only estimated exposure to total drug-derived material = 77,028 ng*h/ml and M1 = 11.3%, M2 = 20.8%, and M3 = 11.2% of the total drug-derived exposure), plasma samples, from preclinical species (rat and dog), were analyzed to assure coverage for these metabolites in species undergoing long-term toxicological testing (Ramanathan et al., 2010).

Another novel approach, to metabolite quantification in the absence of reference standards or radiolabeled materials, involves using low-flow techniques (Ramanathan et al., 2007, 2011; Nugent et al., 2009). In the approach described by Ramanathan et al. (2007), recently introduced nanospray ionization (NSI) source (Nanomate^{TM}), two HPLC systems, and a Q-TOF mass spectrometer were used to obtain normalized LC-MS response of drugs and associated metabolites. The data showed that the effects of the solvent environment on the ionization efficiency were minimized by maintaining the solvent composition unchanged throughout the HPLC run. When compared to responses obtained from radiochromatograms, conventional LC-ESI-MS overestimated the responses for the parent drug by 6- to 20-fold. The response normalization modification resulted in nearly uniform LC-NSI-MS response for all compounds evaluated. During the drug development process, if the AUC for a metabolite has to be determined as per the new FDA or ICH-M3 guidance, then a Hamilton pooled plasma sample can be analyzed with and without response normalization for estimating the response correction factor and the amount of the metabolite as well. Once the response correction factor is established, then the AUC of the parent drug obtained using a validated quantitative method can be used to roughly estimate the AUC of the metabolite (Ramanathan et al., 2010).

Another variant to NSI is the recently introduced captive spray ionization (CSI) source (Ramanathan et al., 2011). Overall, advantages of using a CSI source over a Nanomate or similar low-flow ion sources include no need for X, Y, Z optimization for optimum ion transfer and no need for a separate computer to control the source conditions either. Changing from ESI to CSI and back is relatively easy. Transmission of up to 100% of the ions through the ion transfer tube makes this technique more suitable for the intended work. Nevertheless, routine applications of these techniques, for metabolite exposure determination, are dependent on technological advances in the area of capillary flow LC, capillary column, ionization source designs, and fast scanning mass spectrometers.

7.5.3 Tier 3-Qualified Assays

As shown in Figure 7.5, qualified assay is also a fit-for-purpose assay and can be utilized as the third tier assay for quantifying metabolites in FIH and TK studies. Development of a qualified assay (Table 7.1) requires complete identification and synthesis of the metabolite of interest in order to generate a standard curve for quantification (Ramanathan et al., 2010). However, synthesis of the stable-labeled metabolite internal standard may not be required. The stable-labeled form of the parent NME could be used as internal standard (Penner et al., 2010), provided that structural differences between the parent NME and the metabolite are not significant (hydroxylation, demethylation, etc.) and both the parent and the metabolite show a similar extraction recovery and ionization efficiency. Generally, precision and accuracy acceptance criteria for all standards and quality control (QC) samples are set similar for non-GLP (good laboratory practice) assays at 20/25, where the acceptance criterion for lower limit of quantification (LLOQ) is at 25% deviation from the nominal and the acceptance criterion for other standards and QCs is at 20% deviation from nominal (Korfmacher, 2005).

Qualified assays are commonly employed in investigational new drug (IND)-enabling preclinical species toxicity studies to monitor metabolites, especially when a metabolite is discovered as a major active one and/or poses a safety concern. If the metabolite of interest is present at significant levels in FIH samples (especially under steady state conditions) as determined using the qualified assay method and/or the metabolite is pharmacologically active, then the metabolite should be analyzed using a validated assay in the follow-up long-term toxicity and clinical studies. On the other hand, if the metabolite of interest is minor or absent in FIH samples (especially under steady state conditions) as determined using qualified assay method, then the metabolite should be no longer monitored in the follow-up toxicity and clinical studies.

7.5.4 Tier 4-Validated Assays

Bioanalytical method validation involves performing a set of defined experiments to demonstrate that the intended method will satisfy or exceed the minimum acceptance criteria for linearity, LLOQ, matrix effect, recovery, accuracy, precision, selectivity, specificity, sensitivity, reproducibility, dilution integrity, carryover, and stability as highlighted in the health authority guidance for bioanalytical method validation (Ramanathan et al., 2010).

As outlined in Figure 7.5, a validated assay will be implemented as the final tier for monitoring a metabolite. Similar to a qualified assay, development of a validated assay requires identification and characterization of the metabolite and synthesis of the reference standard for the preparation of calibration standards. Stable isotope labeled internal standard for the metabolite is generally needed for the robustness of the LC-MS/MS assay. According to the FDA and EMEA guidance, the precision and accuracy, for both the standards and QC samples, are set at 15/20% for a validated assay; further guidance standards maintain that the observed QC concentrations should be within $\pm 15\%$ bias of the nominal concentration (accuracy) and $\leq 15\%$ CV (precision) for the intraday and interday assay at all levels, except at the LLOQ, where the intraday and interday assay accuracy and precision limits should be within $\pm 20\%$ bias and $\leq 20\%$ CV.

7.6 SUMMARY

Several factors contribute to the recent heightened interest in the early assessment of differences among steady-state metabolite exposure between animals and humans. One of the important factors is the need to assure patient safety. The recent regulatory guidance on MIST has provided the framework for early assessment of steady-state metabolite exposures in both humans and animals. With the availability of more efficient sample processing, separation techniques, and powerful computer systems than ever, LC in combination with various MS detection has been demonstrated as the most important tool for this task.

However, the noticeable differences, in MS response between a drug and its metabolites limits the use of conventional LC-MS or LC-HRMS as the choice of quantification for establishing the proportionality or disproportionality of metabolite exposure between the species. With this regard, UV, radioactivity, NMR, and micro-flow (<20 μl/min) LC in combination with nano- and micro-spray approaches provide better choices. As shown in recent publications (Wright et al., 2009; Ramanathan et al., 2010), upon correction of response differences, normalized responses in conventional LC-MS for NMEs and associated metabolites can be obtained. This in turn allows quantitative LC-MS analysis of metabolites without feasible reference standards. With this in mind, several fit-for-purpose and cost-disciplined bioanalytical methods have been introduced across the pharmaceutical industry to overcome the challenge around the unavailability of metabolite reference standards during early stages of drug development. Overall, bioanalytical assays, used for detection, characterization and exposure estimation of metabolites, are tiered into (1) metabolite profiling, (2) standard free quantification/response factor determination, (3) qualified assays, and (4) validated assays.

REFERENCES

Anderson S, Luffer-Atlas D, Knadler MP. Predicting circulating human metabolites: how good are we? Chem Res Toxicol 2009;22(2):243–256.

Bussiere JL. Species selection considerations for preclinical toxicology studies for biotherapeutics. Expert Opin Drug Metab Toxicol 2008;4(7):871–877.

Campbell JL, Le Blanc JC. Using high-resolution quadrupole TOF technology in DMPK analyses. Bioanalyis 2012;4(5):528–544.

Chang MS, Ji Q, Zhang J, El-Shourbagy TA. Historical review of sample preparation for chromatographic bioanalysis: pros and cons. Drug Dev Res 2007;68:107–133.

Cheng KC, Li C, Uss AS. Prediction of oral drug absorption in humans–from cultured cell lines and experimental animals. Expert Opin Drug Metab Toxicol 2008;4(5):581–590.

Dalvie D, Obach RS, Kang P, et al. Assessment of three human in vitro systems in the generation of major human excretory and circulating metabolites. Chem Res Toxicol 2009;22(2):357–368.

FDA, U. S. F. a. D. A. (2008). Guidance for industry: safety testing of drug metabolites. Available at http://www.fda.gov/cder/guidance/6897fnl.pdf, 1–25. Accessed March 4, 2013.

Frederick CB, Obach RS. Metabolites in safety testing: "MIST" for the clinical pharmacologist. Clin Pharmacol Ther 2010;87(3):345–350.

Fung EN, Xia YQ, Aubry AF, Zeng J, Olah T, Jemal M. Full-scan high resolution accurate mass spectrometry (HRMS) in regulated bioanalysis: LC-HRMS for the quantitation of prednisone and prednisolone in human plasma. J Chromatogr B Analyt Technol Biomed Life Sci 2011;879(27):2919–2927.

Gao H, Obach RS. Addressing MIST (metabolites in safety testing): bioanalytical approaches to address metabolite exposures in humans and animals. Curr Drug Metab 2011;12(6):578–586.

Gomase VS, Tagore S. Species scaling and extrapolation. Curr Drug Metab 2008;9(3):193–198.

Hamilton RA, Garnett WR, Kline BJ. Determination of mean valproic acid serum level by assay of a single pooled sample. Clin Pharmacol Ther 1981;29(3):408–413.

Harmonization ICO. (2009). Non-Clinical Safety Studies for the Conduct of Human Clinical Trials and Marketing Authorization for Pharmaceuticals. Available at http://www.fda.gov/downloads/Drugs/GuidanceComplianceRegulatoryInformation/Guidances/ucm073246.pdf. Accessed March 4, 2013.

Hop CE, Wang Z, Chen Q, Kwei G. Plasma-pooling methods to increase throughput for in vivo pharmacokinetic screening. J Pharm Sci 1998;87(7):901–903.

Korfmacher W. Bioanalytical assays in a drug discovery environment. In: W. Korfmacher, editor. *Using Mass Spectrometry for Drug Metabolism Studies*. Boca Raton, FL: CRC Press; 2005. p 1–34.

Lin J, Sahakian DC, de Morais SM, Xu JJ, Polzer RJ, Winter SM. The role of absorption, distribution, metabolism, excretion and toxicity in drug discovery. Curr Top Med Chem 2003; 3(10):1125–1154.

Ma S, Chowdhury SK. Analytical strategies for assessment of human metabolites in preclinical safety testing. Anal Chem 2011; 83(13):5028–5036.

Ma S, Chowdhury SK. Application of LC-high-resolution MS with 'intelligent' data mining tools for screening reactive drug metabolites. Bioanalysis 2012;4(5):501–510.

Ma S, Li Z, Lee KJ, Chowdhury SK. Determination of exposure multiples of human metabolites for MIST assessment in preclinical safety species without using reference standards or radiolabeled compounds. Chem Res Toxicol 2011;23(12):1871–1873.

Malinowski HJ (1997). The role of in vitro-in vivo correlations (IVIVC) to regulatory agencies. Adv Exp Med Biol 423:261–268.

Nugent K, Zhu Y, Kent P, Phinney B, Alvarado R. CaptiveSpray: A New Ionization Technique to Maximizing Speed, Sensitivity, Resolution and Robustness for LCMS Protein Biomarker Quantitation *Proceedings of the 57th ASMS Conference on Mass Spectrometry and Allied Topics*. Philadelphia, PA, ASMS, 2009.

O'Hara T, Hayes S, Davis J, Devane J, Smart T, Dunne A. In vivo-in vitro correlation (IVIVC) modeling incorporating a convolution step. J Pharmacokinet Pharmacodyn 2001;28(3):277–298.

Obach RS, Nedderman AN, Smith DA. Radiolabelled mass-balance excretion and metabolism studies in laboratory animals: are they still necessary? Xenobiotica 2012;42(1):46–56.

Penner N, Klunk LJ, Prakash C. Human radiolabeled mass balance studies: objectives, utilities and limitations. Biopharm Drug Dispos 2009;30(4):185–203.

Penner N, Ramanathan R, Zgoda-Pols J, Chowdhury S. Quantitative determination of hippuric and benzoic acids in urine by LC-MS/MS using surrogate standards. J Pharm Biomed Anal 2010;52(4):534–543.

Penner N, Xu L, Prakash C. Radiolabeled absorption, distribution, metabolism, and excretion studies in drug development: why, when, and how?. Chem Res Toxicol 2012;25(3):513–531.

Ramanathan, D, LeLacheur RM. Evolving role of mass spectrometry in drug discovery and development. In: R Ramanatha, editor. *Mass Spectrometry in Drug Metabolism and Pharmacokinetics*. Hoboken, NJ: John Wiley & Sons, Inc.; 2008. p 1–85.

Ramanathan R, Jemal M, Ramagiri S, et al. It is time for a paradigm shift in drug discovery bioanalysis: from SRM to HRMS. J Mass Spectrom 2011;46(6):595–601.

Ramanathan R, Josephs JL, Jemal M, Arnold M, Humphreys WG. Novel MS solutions inspired by MIST. Bioanalysis 2010;2(7):1291–1313.

Ramanathan R, Raghavan N, Comezoglu SN, Humphreys WG. A low flow ionization technique to integrate quantitative and qualitative small molecule bioanalysis. Int J Mass Spectrom 2011;301(7):127–135.

Ramanathan R, Zhong R, Blumenkrantz N, Chowdhury SK, and Alton KB. Response normalized liquid chromatography nanospray ionization mass spectrometry. J Am Soc Mass Spectrom 2007;18(10):1891–1899.

Ranasinghe A, Ramanathan R, Jemal M, D'Arienzo CJ, Humphreys WG, Olah T. Integrated quantitative and qualitative workflow for in-vivo bioanalytical support in drug discovery using hybrid Q-TOF MS. Bioanalyis 2012;4(5):511–528.

Timmerman P, Anders Kall M, Gordon B, Laakso S, Freisleben A, and Hucker R. Best practices in a tiered approach to metabolite quantification: views and recommendations of the European Bioanalysis Forum. Bioanalysis 2010;2(7):1185–1194.

Vishwanathan K, Babalola K, Wang J, et al. Obtaining exposures of metabolites in preclinical species through plasma pooling and quantitative NMR: addressing metabolites in safety testing (MIST) guidance without using radiolabeled compounds and chemically synthesized metabolite standards. Chem Res Toxicol 2009;22(2):311–322.

White RE, Evans DC, Hop CE, et al. Radiolabeled mass-balance excretion and metabolism studies in laboratory animals: a commentary on why they are still necessary. Xenobiotica 2013;43(2):219–225.

Wright P. Metabolite identification by mass spectrometry: forty years of evolution. Xenobiotica 2011;41(8):670–686.

Wright P, Miao Z, Shilliday B. Metabolite quantitation: detector technology and MIST implications. Bioanalyis 2009;1(4):831–845.

Yang Y, Grubb MF, Luk CE, Humphreys WG, Josephs JL. Quantitative estimation of circulating metabolites without synthetic standards by ultra-high-performance liquid chromatography/high resolution accurate mass spectrometry in combination with UV correction. Rapid Commun Mass Spectrom 2011;25(21):3245–3251.

Yu C, Chen CL, Gorycki FL., Neiss TG. A rapid method for quantitatively estimating metabolites in human plasma in the absence of synthetic standards using a combination of liquid chromatography/mass spectrometry and radiometric detection. Rapid Commun Mass Spectrom 2007;21(4):497–502.

Zhang D, Raghavan N, Chando T, et al. LC-MS/MS-based approach for obtaining exposure estimates of metabolites in early clinical trials using radioactive metabolites as reference standards. Drug Metab Lett 2007;1:293–298.

Zhu M, Zhang H, Humphreys WG. Drug metabolite profiling and identification by high-resolution mass spectrometry. J Biol Chem 2011;286(29):25419–25425.

8

A COMPARISON OF FDA, EMA, ANVISA, AND OTHERS ON BIOANALYSIS IN SUPPORT OF BIOEQUIVALENCE/ BIOAVAILABILITY STUDIES

BRADLEY NASH

8.1 INTRODUCTION TO BIOAVAILABILITY/ BIOEQUIVALENCY STUDIES

Before a discussion of the various international regulations and guidelines on bioanalysis in support of bioavailability (BA) and bioequivalency (BE) studies, it is necessary to briefly describe the purpose of both studies. BE is a pharmacokinetics term for assessing the expected *in vivo* biological equivalence of two proprietary preparations (formulations) of a drug. According to the US Food and Drug Administration (FDA), BE is

> the absence of a significant difference in the rate and extent to which the active ingredient or active moiety in pharmaceutical equivalents or pharmaceutical alternatives becomes available at the site of drug action when administered at the same molar dose under similar conditions in an appropriately designed study.
>
> —(Center for Drug Evaluation and Research, 2003)

BA is one of the essential tools in pharmacokinetics. By definition, BA is the fraction of an administered dose of unchanged drug that reaches the systemic circulation. When a drug is administered intravenously, its BA is 100%. However, when a drug is administered via nonintravenous routes (e.g., oral), in general, its BA decreases due to incomplete absorption and/or first-pass metabolism. A study of absolute BA is to compare the BA of the active drug in systemic circulation following nonintravenous administration (i.e., oral, rectal, transdermal, subcutaneous, or sublingual administration) with the BA of the same drug administered intravenously. In contrast, relative BA measures the BA (estimated as the AUC (area under the curve)) of a formulation (A) of a certain drug when compared with another formulation (B) of the same drug, usually a "brand name drug," or through administration via a different route. Relative BA is one of the measures for assessing BE between two drug products.

In order to demonstrate the BE of two proprietary preparations (formulations) of the same drug molecule, BE and/or BA studies must be conducted. For FDA approval, a generic drug manufacturer must demonstrate that the 90% confidence interval for the ratio of the mean responses (usually of AUC and the maximum concentration, C_{max}) of its product to that of the 'brand name drug' is within the limits of 80–125%. While AUC refers to the extent of BA, C_{max} refers to the rate of BA.

In recent years, both international regulatory bodies and industry groups have released new guidelines and/or regulations on the conduct of BE/BA studies. Liquid chromatography–mass spectrometry bioanalysis has and will continue to play an important role in support of these studies. This chapter is an attempt at providing an overview of the various international regulations and/or guidelines on the bioanalytical evaluations of BE studies.

8.2 REGULATIONS FROM THE US FDA

8.2.1 History and Validation

The US FDA has been the international leader in guiding BE studies since the early 1980s when the US Congress

Handbook of LC-MS Bioanalysis: Best Practices, Experimental Protocols, and Regulations, First Edition. Edited by Wenkui Li, Jie Zhang, and Francis L.S. Tse.

passed laws allowing for the approval of drugs based on BE. The conduct of BA/BE studies has since been regulated by 21CFR320 that goes into great detail on the overall conduct of the BA/BE study. However, only a brief mention of the bioanalysis of the study samples is made. It says little more than that the method used for bioanalysis shall be accurate, precise and suitably sensitive to measure the drug or metabolite concentrations achieved during the course of the study. Additional allowances are made for the overall conduct of the study if the bioanalytical method is not sufficiently sensitive; however, it has little detail on the conduct of the bioanalysis. While 21CFR320 does not direct the bioanalytical laboratory how to develop and validate an appropriate method, the guidances, 483's, and warning letters issued to the bioanalytical community have provided enlightenment to the development and validation of the bioanalytical method. In May 2001, the US FDA issued the first *Guidance for Industry on Bioanalytical Method Validation*. This guidance addressed both method validation requirements and conduct of sample analysis. The guidance for industry, *BA and BE Studies for Orally Administered Drug Products–General Considerations* issued by the agency in 2003 directs the BA/BE bioanalytical method to be validated according to the FDA guidance, *Bioanalytical Method Validation* of which a detailed discussion is made in Chapter 4 of this book and will not be repeated here.

8.2.2 Conduct of the Bioanalysis

The analysis of study samples for BA/BE studies should, in most aspects, mirror the principles followed during FDA good laboratory practices (GLP) studies that is discussed in depth in other chapters. In fact, it has been a general practice in the bioanalytical community to use a regulatory compliance statement of "this study was conducted according to the spirit of FDA GLPs," or something similar. However, the human BE study technically falls under the scope of good clinical practice (GCPs) and must follow all associated GCP regulations. Particular attention should be paid to the May 2001 guidance on *Bioanalytical Method Validation* and the March 2003 guidance regarding orally administered BA/BE studies. The latter states that the bioanalytical method must be reproducible. The current expectation is that incurred sample reproducibility (ISR) should be performed for all BE studies in order to demonstrate the reproducibility of the method. There have been many discussions in recent years regarding ISR analysis. From which a general rule has emerged indicating that approximately 10% of BA/BE study samples should be reanalyzed to demonstrate reproducibility. See the respective chapter of this book for a further discussion of ISR analysis.

In order to assure data integrity, it is advantageous to apply GLP principles to the bioanalysis of BA/BE study samples; however, it must be remembered that the human BE study is technically outside of the scope of GLP regulations and some GLP principles may not be applicable. In the high throughput environment of a modern chromatography laboratory, the qualification of instrumentation is often treated more as if it were in a good manufacturing practice (GMP) environment. This increase in instrument qualification requirements and routine maintenance is not surprising considering the large number of samples each instrument is capable of analyzing daily. This results in the potential for a small instrumentation malfunction to call into question the reliability of the sample results as well as cost the laboratory large sums of money to reanalyze. It is thus often the decision of individual laboratories to apply GMP and/or GLP principles to the bioanalysis of BA/BE study samples despite these clinical studies not being within the scope of either 21CFR58 or 21CFR211. Therefore, the bioanalytical laboratory must be vigilant in not inadvertently superseding requirements set forth in 21CFR320 with those voluntarily applied from 21CFR58 and 21CFR211.

8.3 REGULATIONS FROM THE EUROPEAN MEDICINES AGENCY

The guidelines released by the European Medicines Agency (EMA) on the investigation of BE are, in one very important way, identical to the recommendations released by the US FDA; the bioanalytical portion of the BE study should be conducted according to the principles of GLP. As was discussed in Section 8.2.2, it must again be noted that the BE study is technically outside of the scope of EMA GLP. So while an EMA regulated GLP study requires a laboratory to be monitored per a national GLP compliance program, a laboratory performing bioanalysis of samples in support of a clinical BE study is not required to be part of such a program. However, this overarching idea of applying GLP principles to a clinical study is critical to the proper conduct of the method validation and study sample analysis.

8.3.1 The Bioanalytical Method

Very similar to the FDA, the EMA also requires that the method be accurate, precise, selective, stable, and sufficiently sensitive. The EMA guidelines are more specific in defining the acceptable sensitivity of the method by indicating that the lower limit of quantitation should be at a minimum 1/20 of C_{max}. Additionally, while the *Guideline on the Investigation of BE* does not specifically address reproducibility or ISR analysis, the *Guideline on Bioanalytical Method Validation* issued in 2011 does indicate that ISR analysis should be performed on all pivotal BE studies. While the FDA has only indicated during inspections that lipemic and hemolyzed samples should be included during the selectivity portion of the method validation, the EMA has included these requirements in the recently released Bioanalytical Method Validation guidelines.

Due to the increased susceptibility for bias by the pharmacokineticist on the bioanalytical results during a BE study, the EMA strongly discourages reanalyzing samples based on the pharmacokinetic results. Therefore, all reasons for reanalysis should be clearly predefined in the method, protocol, or SOP. Similarly, the analysis of all study samples should be conducted blinded to avoid any potential bias on the part of the bioanalytical scientist.

8.3.2 Bioanalysis of Enantiomers

It is generally acceptable to use an achiral method for the bioanalysis of chiral compounds in the study samples; however, in certain instances, it is necessary to validate a method for a single enantiomer. If the enantiomers demonstrate different pharmacokinetics or pharmacodynamics, and if changes in the rate of absorption alter the exposure ratio of the enantiomers, then the enantiomers must be treated separately. Likewise, if it is known that only one of the enantiomers is pharmacologically active, then only the one enantiomer needs to be analyzed for in the study samples for BE.

8.4 REGULATIONS FROM THE BRAZILIAN SANITARY SURVEILLANCE AGENCY (ANVISA)

National Health Surveillance Agency (ANVISA), unlike the FDA and EMA, requires that all facilities that carry out the clinical, bioanalytical, or statistical portions of a BE study be certified by ANVISA. For bioanalytical laboratories, this certification process covers not only general facility appropriateness for conducting the study but also the processes the laboratory has in place for validating the method and performing the bioanalysis. In order to help both Brazilian and international facilities meet these requirements and to help standardize the growing pharmaceutical industry within Brazil, ANVISA issued a very detailed manual describing many aspects of the BE study. Module 2 of Volume I of this manual is dedicated to describing how the bioanalytical work in support of the study should be conducted (Brazilian Sanitary Surveillance Agency, 2002). It is far more detailed than any other regulations or guidelines available internationally on BE studies. However, the majority of these policies and processes are already reflected in the SOPs at most bioanalytical laboratories. The manual describes many common practices that all bioanalytical laboratories must adhere to; ranging from the proper choice of analytical standards to how glassware should be cleaned and maintained. It is critical that all of these recommendations be reviewed and closely adhered to during all human BE bioanalytical studies that are intended for submission to ANVISA. If the bioanalytical laboratory has already performed bioanalysis regulated by either the EMA or FDA, any additional requirements set forth by ANVISA should only require minimal changes to the GLP already being followed.

8.4.1 Method Validation

The parameters of the bioanalytical method that must be fully studied and validated include selectivity, recovery, limit of quantitation, accuracy, precision, linearity, and stability. Unlike the BE guidelines previously discussed that reference general method validation guidelines, ANIVSA thoroughly discusses each parameter of the validation within the BE guidelines. Those parameters, described in Table 8.1, should be evaluated for each analyte.

8.4.2 Conduct of Bioanalysis

When analyzing samples from any BE study, the fully validated method should be followed exactly as it was validated and all stability periods respected. Each batch or run should contain a set of calibration standards and quality controls at three concentrations spanning the calibration range to ensure that precision and accuracy determined during the validation is maintained. The run should be accepted based upon the performance of the quality controls. Sample reanalysis should be performed in triplicate, volume permitting, and all results documented.

8.5 OTHER INTERNATIONAL GUIDELINES

The guidelines issued by most national governments generally parallel the guidelines issued by the FDA and the EMA. While minor differences are present, this is often due to slight variations in how the science and current technology are interpreted. This section gives a brief overview of some of these regulations and guidelines present in emerging markets.

8.5.1 Indian Department of Health

The *Guidelines for BA and BE Studies* issued by the Central Drugs Standard Control Organization in 2005 very closely mirror those issued by the EMA discussed earlier. The guidelines expect a full validation to be performed in a similar manner as required by either FDA or EMA guidelines. The validation must demonstrate the stability of the drug and any metabolites in matrix, specificity and selectivity, sensitivity, recovery, range of quantitation, and precision and accuracy. In addition to those requirements, it is also desired that system stability be demonstrated during method validation by performing replicate analysis of the calibration curve both at the beginning and at the end of an analytical batch.

Again, the acceptance criteria for sample analysis are similar to the requirements of the EMA and quality controls

TABLE 8.1 Overview of ANVISA BA/BE Bioanalytical Method Validation Criteria

Parameter	Process	Criteria
Selectivity	Standard should be spiked into normal, lipemic, and hemolyzed plasmas and assayed in triplicate at low, medium, and high concentrations.	Ensure that the standard, internal standard, and possible interferents do not coelute.
Recovery	Repeated comparison of analytical results of low, medium, and high concentration samples with a standard solution.	100% recovery need not be attained; however, consistent and repeatable recovery must be demonstrated.
Limit of quantitation	Observed throughout the validation.	Generally is defined as the lowest calibration standard and should demonstrate a response at least 5 times greater than the blank and accuracy within 20% of the theoretical concentration.
Linearity (calibration curve)	A curve should be constructed with at least six-spiked samples as well as applicable blanks.	The simplest regression model capable of adequately describing the curve should be selected. The final curve should contain at least 5 points (consecutive points should not be excluded). The points should back calculate within 15% of the theoretical value (20% at the limit of quantitation).
Precision	A minimum of three concentration levels with at least 5 determinations should be tested for both inter- and intrarun precision.	Coefficient of variation should not exceed 15% (20% at the limit of quantitation).
Completeness (accuracy)	A minimum of 3 concentration levels with at least 5 determinations should be assayed.	Mean of the determinations should be within 15% of the nominal concentration (20% at the limit of quantitation).

should be analyzed to demonstrate run acceptance. Additionally, as discussed earlier, it is critical that reasons for reanalysis be clearly identified *a priori* and that all sample reanalysis be clearly identifiable and traceable.

8.5.2 Health Canada

The guidelines on BA/BE studies issued by the Canadian Minister of Health in 1992 and 1996, like the other major international guidelines, indicate that the bioanalytical method should be reproducible, specific, precise, accurate, and sufficiently sensitive to measure the entire expected concentration range of the study. In order to determine these parameters, the guidelines direct the bioanalytical method to be validated in accordance with Shah et al. (1992). As with other national guidelines the bioanalytical study should be documented according to GLP principles and likewise have *a priori* reassay criteria set.

8.5.3 China State FDA

The State Food and Drug Administration is actively trying to formulate and implement and standardize national regulations and guidelines on the development, analysis, and approval of pharmaceuticals in China. Until recently, China did not recognize the difference between a generic drug and a name-brand, patent-protected drug. Due to the large percentage of generic drug approvals in China (73% of all Chinese drug approvals in 2010), the recently adopted regulations and guidelines are critical to the safety and efficacy assessment of the products emerging from the rapidly increasing Chinese pharmaceutical industry. In 2009, the "Guideline for BA/BE Studies" was issued to define the parameters for the performance of the BE study in China. These regulations closely mirror those issued by both the FDA and EMA as the overall BE study must follow the "Guideline for Good Clinical Practice." In addition, it is stipulated that the method validation should include precision, accuracy, selectivity, matrix effect, sample stability, and determination of the lower limit of quantitation.

8.6 CONCLUSION

The reliable performance of the bioanalytical analysis of study samples during a BE or BA study is critical to the acceptance of the drug for use by national regulatory agencies. While all the national and international regulations and guidelines discussed in this chapter differ in some aspects, they all share two overall philosophies. First, a fully validated and reproducible bioanalytical method lays the groundwork for reliable and successful bioanalysis. Without a sound method validation, the entire bioanalytical study can be called into question. Second, in order to perform a successful bioanalytical study the philosophy surrounding GLPs should be respected during all aspects of the study, both validation and sample analysis.

With the constant advances in science and technology, the guidelines and regulations discussed here will be constantly updated. Due to the large number of different regulatory bodies, these constant updates may cause significant confusion for the industry. There have been efforts underway by both industry groups and government organizations to harmonize the regulations and guidelines internationally. For example, in 2005 the International Conference on Harmonization (ICH) put forth the ICH Q10 with a goal of providing a harmonized approach for clinical trials of BE studies. While such conferences and/or white papers in support of global harmonization are an appreciated movement, the ever-present differences between regulatory bodies will continue to present challenges to industry. Until a universally accepted guideline is created and implemented globally the pharmaceutical industry must continue to constantly adjust and navigate through an ever-fluid set of international expectations, guidelines, and regulations.

REFERENCES

Brazilian Sanitary Surveillance Agency. Manual for Good Bioavailability and Bioequivalence Practices—Volume 1. 2002.

Center for Drug Evaluation, State Food and Drug Administration, People's Republic of China. Guideline on BA/BE Studies—0980316268. 2009. Available at www.cde.org.tw/English/Regulations/SubLink/Document%2004.pdf. Accessed Mar 5, 2013.

Center for Drug Evaluation and Research. "Guidance for Industry: Bioavailability and Bioequivalence Studies for Orally Administered Drug Products—General Considerations." United States Food and Drug Administration. 2003. Available at http://www.fda.gov/downloads/Drugs/GuidanceCompliance RegulatoryInformation/Guidances/ucm070124.pdf. Accessed Mar 5, 2013.

Central Dugs Standard Control Organization, Directorate General of Health Services, Ministry of Health & Family Welfare, Government of India. Guidelines for Bioavailability & Bioequivalence Studies. 2005. Available at betest.kfda.go.kr/world/docu/%28etc%29india.pdf. Accessed Mar 5, 2013.

European Medicines Agency, Committee for Medicinal Products for Human Use (CHMP). Guideline on the Investigation of Bioequivalence. 2010. Available at www.emea.europa.eu/docs/en_GB/document_library/Scientific_guideline/2010/01/WC500070039.pdf. Accessed Mar 5, 2013.

European Medicines Agency, Committee for Medicinal Products for Human Use (CHMP). Guideline on Bioanalytical Method Validation. 2011. Available at www.ema.europa.eu/docs/en_GB/document_library/Scientific_guideline/2011/08/WC500109686.pdf. Accessed Mar 5, 2013.

Minister of Health, Health Canada, Health Products and Food Branch. Guidance for Industry: Conduct and Analysis of Bioavailability and Bioequivalence Studies—Part A: Oral Dosage Formulations Used for Systemic Effects. 1992. Available at www.hc-sc.gc.ca/dhp-mps/alt_formats/hpfb-dgpsa/pdf/prodpharma/bio-a-eng.pdf. Accessed Mar 5, 2013.

Minister of Health, Health Canada, Health Products and Food Branch. Guidance for Industry: Conduct and Analysis of Bioavailability and Bioequivalence Studies – Part B: Oral Modified Release Formulations. 1996. Available at www.hc-sc.gc.ca/dhp-mps/alt_formats/hpfb-dgpsa/pdf/prodpharma/bio-b-eng.pdf. Accessed Mar 5, 2013.

US Department of Health and Human Services, Food and Drug Administration. Center for Drug Evaluation and Research (CDER). Guidance for Industry: Bioanalytical Method Validation. 2001. Available at www.fda.gov/downloads/Drugs/GuidanceComplianceRegulatoryInformation/Guidances/ucm070107.pdf. Accessed Mar 5, 2013.

US Department of Health and Human Services, Food and Drug Administration. Center for Drug Evaluation and Research (CDER). Guidance for Industry: Bioavailability and Bioequivalence Studies for Orally Administered Drug Products—General Considerations. 2003. Available at www.fda.gov/downloads/Drugs/GuidanceComplianceRegulatoryInformation/Guidances/ucm070124.pdf. Accessed Mar 5, 2013.

US Department of Health and Human Services, Food and Drug Administration, Bioavailability and Bioequivalence Requirements. 2010. 21 C.F.R. § 320.

9

A COMPARISON OF THE GUIDANCE OF FDA, OECD, EPA, AND OTHERS ON GOOD LABORATORY PRACTICE

J. Kirk Smith

9.1 INTRODUCTION

Strictly, good laboratory practice (GLP) specifically refers to a quality system of management controls for research laboratories and organizations to ensure the uniformity, consistency, reliability, reproducibility, quality, and integrity of chemical (including pharmaceuticals) nonclinical safety tests, from physiochemical properties through acute to chronic toxicity tests. The original GLP regulatory mandate was promulgated in 1978 by the US FDA (US Food and Drug Administration—Federal Register). A similar mandate was followed by the US EPA (US Environmental Protection Agency) a few years later, and the Organization for Economic Cooperation and Development (OECD) in 1992. GLP applies to nonclinical studies conducted for the assessment of the safety or efficacy of chemicals (including pharmaceuticals) to man, animals, and the environment. An internationally recognized definition of GLP can be found on the website for the Medicines and Healthcare Products Regulatory Agency (MHRA)—UK, which defines GLP.

GLP embodies a set of principles that provides a framework within which laboratory studies are planned, performed, monitored, recorded, reported, and archived. These studies are undertaken to generate data by which the hazards and risks to users, consumers, and third parties, including the environment, can be assessed for pharmaceuticals (only preclinical studies), agrochemicals, cosmetics, food additives, feed additives and contaminants, novel foods, biocides, detergents, and so on. GLP helps assure regulatory authorities that the data submitted are a true reflection of the results obtained during the study and can, therefore, be relied upon when making risk/safety assessments. Although some of the liquid chromatography–mass spectrometry (LC-MS) bioanalysis work may not strictly fall within the scope of the GLP regulations (European Medicines Agency (EMA) good clinical practices (GCP)).

In general, LC-MS bioanalysis for investigational new drug (IND)-enabling toxicokinetic studies (preclinical) all fall within the scope of the GLP regulations. Analysis of the clinical trial samples (including bioavailability (BA) and bioequivalency (BE) study samples) fall under the scope of the GCP, but GLP principles should be applied (US FDA). While the FDA published guideline for Bioanalytical Method Validation do not explicitly state that validation for nonclinical studies is performed under GLP, the EMA guidelines Bioanalytical Method Validation clearly requires that validation for nonclinical studies is performed under GLP (EMA). Also discussed in earlier chapters, studies submitted as part of a marketing or registration application must be conducted according to the regional regulatory requirements, that is, if submitted in the United States and European Union, both regulatory requirements must be met. A clear understanding of the differences in the GLP requirements between the countries with the highest level of pharmaceutical research and development activity is essential to ensure effective program planning, development, and execution. This chapter provides a description of the GLP regulatory differences and how these apply to the bioanalytical laboratory.

At the time of this writing, the regions where significant pharmaceutical research and development activity is occurring are the United States, Japan, the European Union, and the BRIC countries (Brazil, Russia, India, and China) and the regulatory differences only for those countries listed will be considered; US GLP regulations (FDA and EPA) will be

Handbook of LC-MS Bioanalysis: Best Practices, Experimental Protocols, and Regulations, First Edition. Edited by Wenkui Li, Jie Zhang, and Francis L.S. Tse.

addressed first, followed by OECD and finally, the rest of the world.

9.2 FDA VERSUS EPA ON GLP

In my experiences, the regulated LC-MS bioanalytical studies conducted under the EPA GLP regulations are generally related to residue studies; that said, it is still beneficial to understand the differences between the EPA and FDA GLP regulations. There are actually two sets of regulations that were promulgated by the EPA, both found in the Code of Federal Regulations (CFR), under Chapter 40, part 160 (Federal Insecticide, Fungicide, and Rodenticide Act—FIFRA) and part 792 (Toxic Substance Control Act (TSCA)). These two regulations are nearly identical, the only notable difference in the two is the length of time required for record archival, section 160.195 being the most onerous by requiring the records be retained for "the period during which the sponsor holds any research or marketing permit to which the study is pertinent", generally interpreted as the life of the market registration. Due to the provisions in section 160.195, many organizations have retained records indefinitely since 1983, when the FIFRA regulation was first promulgated.

Differences between the EPA and the FDA GLP regulation are only slightly more significant; however, current industry and agency application of the regulations appear to be on an increasingly divergent path. While there may be many reasons for divergence, it is this author's opinion that one of the main contributors is the fact that the FDA has allowed many of the principles of Good Manufacturing Practices to infuse into the GLPs through the extensive use of guidance documents and current industry best practices.

An excellent document was published by the Office of Regulatory Affairs (ORA) in 2004 that provides a comparison of the FDA, EPA (TSCA), and OECD GLP regulations (US FDA—ORA). In the broadest sense, the primary differences between the FDA and the EPA GLPs are the provisions for the variety of test systems employed in the EPA studies (plants, invertebrates, aquatic species), field studies and the physical and chemical characterization studies, resulting from the broad scope covered by the EPA regulations. The FDA GLP is focused on those materials regulated by the FDA (drugs, food, food additives, biologics, and medical devices), the EPA is responsible for everything else. One of the most significant differences between the standards is the EPA compliance statement, failure to submit the completed compliance statement (signed by Study Director, the Sponsor, and the Submitter) may result in an immediate rejection of the study by the EPA. Details pertaining to the compliance statement have recently been revised and published (EPA). In a side-by-side comparison of the FDA (21CFR58) and EPA (40CFR160 or 40CFR792) GLP regulations, one should note the differences as described later.

The definitions of the study under the EPA GLP is significantly broader (see 40CFR160.3), while the FDA is limited to safety and does not include human, clinical or field trials in animals; the EPA definition of the study includes "the effects, metabolism, product performance (as required by 40 CFR 158.640), environmental and chemical fate, persistence and residue and other characteristics in humans, other living organisms or media," very broad indeed. EPA has a broader base for the terms applied to the materials associated with the study, it is helpful to understand the definitions for test article and substance, control article and substance, reference substance, carriers, and vehicles. Additional terminology specific to the EPA include the experimental start and end dates that are typically included on the EPA master schedule to indicate the current status of the study.

Roles and responsibilities as described in the two standards are identical with the exception to a single statement by the EPA directing the quality assurance unit (QAU) to conduct inspections and maintain records appropriate to the study.

One of the three most significant differences between the EPA and the FDA GLP regulations results from the broader scope of the EPA regulations, including the testing on plants, aquatic species (plants and animals), field trials (residue studies), therefore the sections on facilities are greatly expanded with the EPA GLP regulations (Subpart C, including sections 41, 43 and 45). The second most significant difference between the two US regulations is the archival period, with the FDA requirements extending to 5 years maximum after the data has been submitted in an application while the EPA archives may require the records be held for a "very long time," that is, the period during which the sponsor holds any research or marketing permit to which the study is pertinent (40CFR160.195). Finally, the third significant difference is found in Section 135 only in the EPA GLP regulations, physical and chemical characterization studies, the FDA regulations do not contain an equivalent section. The absence of this section has little impact on the vast majority of groups performing LC-MS bioanalysis, but interestingly, helps those groups performing studies in compliance with both OECD and EPA GLP requirements. This is discussed in more detail later.

Other, less significant differences in the US regulations pertain to the timing that the materials need to be characterized, EPA requires these are characterized prior to addition to the test systems, the FDA does not have a similar requirement. The characterization includes not only test materials, but solubility and stability of solutions as well.

The most striking difference between the US FDA and EPA regulations is not in the regulations themselves at all, but rather in the approach that the agencies take in interpretation and enforcement of the regulations. Key influences in regulatory interpretation and enforcement of FDA GLP include the 1987 preamble to 21CFR58, the regulations of

GCP and the regulations for BA and Bioequivalence Requirements (21CFR320 and more specifically 320.29(a)). In this author's opinion, these influences have resulted in the greatest divergence between the applications of the standards in the Bioanalytical Laboratory, based on the very simple requirement stated in section 320.29(a): "The analytical method used in an *in vivo* BA or bioequivalence study to measure the concentration of the active drug ingredient or therapeutic moiety, or its active metabolite(s), in body fluids or excretory products, or the method used to measure an acute pharmacological effect shall be demonstrated to be accurate and of sufficient sensitivity to measure, with appropriate precision, the actual concentration of the active drug ingredient or therapeutic moiety, or its active metabolite(s), achieved in the body." The interpretation of this section has led to the AAPS/FDA Crystal City workshops and resulting publications for the guidelines for Bioanalytical Method Validation including very prescriptive methods for conducting bioanalytical methods and studies employing bioanalytical methods (Viswanathan et al., 2007). The recent publication of the EU guidance includes the requirements that bioanalytical methods intended for use in GLP studies must be validated in accordance with the Principles of GLP (EMA). A thorough discussion of bioanalytical method development and validation is discussed elsewhere in this chapter. Because the GCP regulations do not specifically address the laboratory portion of the study, that is, the bioanalysis of the clinical samples, most work is executed according to the principles of GLP, and as a result, the general expectations of GLP are changing in the pharmaceutical industry—one might equate this to cGLP (current GLP) or sometimes referred to as cGMP (current good manufacturing practices) creep. Another area of divergence between FDA and EPA interpretation and compliance monitoring of the GLP regulations is specific to phase reports included in the GLP study, while both agencies require the phase reports to be signed by the responsible scientist (or Principal Investigator), expectations are different with respect to the timing of the signature on the reports.

In the preamble to the 1987 Final Rule, FDA has essentially defined the signed pathologist report as raw data, and as such, has developed an interpretation that the signed pathology report is required before draft reports can be issued to Sponsors to review and comment prior to finalization, using the rational that issuing a draft before all raw data is available is unacceptable (US FDA—Federal Register). While previously, it had been common practice to issue drafts including the draft pathologist report, the FDA has issued warning letters to CROs (contract research organizations) that included the citation that draft reports were issued to sponsors prior to having the signed pathologist report (US FDA—Warning Letters). Industry is still trying to address the concerns regarding performing and reporting the work in parallel. The EPA has no similar constraint on issuing the draft report for review prior to receiving the signed pathologist report.

9.3 FDA GLP VERSUS OECD GLP PRINCIPLES

Interestingly, the FDA issued a memorandum in December 2010 stating that they are considering updating the GLP regulations (US FDA, Federal Register). Public notice asked industry to consider 9 pertinent issues, including GLP Quality System, Multisite Studies, Electronic/Computerized Systems, Sponsor Responsibilities, Animal Welfare, Information on Quality Assurance Inspectional Findings, Process-Based Systems Inspections, Test and Control Article Information, and Sample Storage Container Retention. Each of the issues, if resolved as suggested by the FDA, will more closely align the FDA GLP regulations with the OECD Principles on GLP. At the present time, the EPA does not have a similar initiative; however, historically, the EPA has revised and reissued regulations to ensure consistency with the FDA rule. And while the proposed changes would all help the laboratory performing studies for multinational registration, rulemaking in the United States of America can take a very long time.

The most striking differences between the US and the OECD GLP are the breadth and prescriptive nature of the OECD GLPs. The OECD Series on Principles of Good Laboratory Practice and Compliance Monitoring is comprised of 15 monographs (see Table 9.1). These additional monographs help to provide guidance on topics and issues that were not conceivable when the US GLP regulations were promulgated; for example, Monograph 13 addresses the issues of multisite studies, very common for preclinical or clinical trials where the sponsor has subcontracted out various tasks/phases of the study, or monograph 11 that addresses the role and responsibilities of the sponsor in the GLP study. While there are no specific regulatory requirements, or guidance, that detail the expectations for bioanalysis of clinical trial samples (the US included), the EMA has perhaps made the strongest statements about employing GLP when analyzing clinical trial samples. For example, the recent reflection paper issued by the GCP Inspectors working group provides "To date no detailed guidance has been produced which outlines the expectations of national monitoring authorities with respect to the analysis or evaluation of samples collected as part of a clinical trial. In the absence of guidance, some laboratories apply the principles of GLP when conducting clinical analysis" (EMA-GCP Inspectors Working Group). The EMA regulations on bioequivalence, CPMP/EWP/QWP/1401/98 Rev. 1, provides "The bioanalytical part of bioequivalence trials should be performed in accordance with the principles of Good Laboratory Practice (GLP). However, as human bioanalytical studies fall outside the scope of GLP, the sites conducting the studies are not required to be monitored as part of a national GLP compliance programme (EMA Bioequivalence)."

TABLE 9.1 OECD Principles of GLP Monographs

Monograph number	Title of monograph	Applicable to LC/MS Bioanalytical Studies
No. 1	OECD Principles on Good Laboratory Practice	Yes
No. 2	Revised Guides for Compliance Monitoring Procedures for Good Laboratory Practice	Yes
No. 3	Revised Guidance for the Conduct of Laboratory Inspections and Study Audit	Yes
No. 4	Quality Assurance and GLP	Yes
No. 5	Compliance of Laboratory Suppliers with GLP Principles	Yes
No. 6	The Application of the GLP Principles to Field Studies	No
No. 7	The Application of the GLP Principles to Short Term Studies	No
No. 8	The Role and Responsibilities of the Study Director in GLP Studies	Yes
No. 9	Guidance for the Preparation of GLP Inspection Reports	Yes
No. 10	The Application of the Principles of GLP to Computerized Systems	Yes
No. 11	The Role and Responsibility of the Sponsor in the Application of the Principles of GLP	Yes
No. 12	Requesting and Carrying Out Inspections and Study Audits in Another Country	Yes
No. 13	The Application of the OECD Principles of GLP to the Organization and Management of Multi-Site Studies	Yes
No. 14	The Application of the Principles of GLP to in vitro Studies	Yes
No. 15	Establishment and Control of Archives that Operate in Compliance with the Principles of GLP	Yes

Some notable differences in the GLP regulations between the United States and OECD GLP regulations include full consideration to the multisite study and the use of the term Principal Investigator as the scientist responsible to the Study Director for conducting delegated phases of the multisite study (e.g., bioanalysis of preclinical samples). Other notable differences include consideration of the "short-term study," somewhat similar to the physical chemistry sections in the EPA GLP, and the use of process-based inspections in lieu of study specific inspections. OECD GLP requires that the role of the Test Facility Management is described, similar to the US GLP requirements for the QAU. Reflecting the more contemporaneous promulgation of the OECD GLP regulations is the requirement for validation of computerized systems, including aspects like audit trail, and the reason for changes to the inputted data (this is addressed in the US through 21CFR11, the regulations for electronic records and electronic signatures).

Minor differences are reflected in the definitions for terms like experimental starting date (the date study specific data is first collected, versus the date the test substance is first applied to the test system) or the role of the QAU with respect to the master schedule (QAU has access rather than maintaining a copy). One last notable difference is the length of time that archives need to be retained. In general, study specific records are held for at least one review cycle, and while industry standards may provide for periods up to 10 years (the Swiss Ordinance on Good Laboratory Practices requires that archives are held for a minimum of 10 years), most organizations charge for retention beyond 2–3 years (Federal Authorities of the Swiss Confederation).

As of January 2012, the OECD had 34 member countries (Table 9.2) that participate in MAD (mutual acceptance of data); that is, data generated according to the principles of OECD GLP submitted by one member country must be accepted by another member country, and an additional seven nonmember participant countries that were either working under full adherence or provisional adherence conditions. In most of the countries in emerging markets, the regulatory agencies have elected to adopt the OECD GLP Principles intact rather than developing unique or parallel regulatory requirements for nonclinical studies; hence, the

TABLE 9.2 34 Member Countries of the OECD as of January 2012

Australia	Austria	Belgium	Canada	Chile
Czech Republic	Denmark	Estonia	Finland	France
Germany	Greece	Hungary	Iceland	Ireland
Israel	Italy	Japan	Korea	Luxembourg
Mexico	Netherlands	New Zealand	Norway	Poland
Portugal	Slovak Republic	Slovenia	Spain	Sweden
Switzerland	Turkey	United Kingdom	United State	

TABLE 9.3 OECD MAD Nonmember Participants[a]

Full adherent	Provisional adherents
Argentina[b]	Malaysia
Brazil[b]	Thailand
India	
Singapore	
South Africa	

[a]As of January 2012.
[b]Limited to nonclinical environmental, health and safety data for pesticides, biocides and industrial chemicals.
Note: China and Russian Federation are not participants at this time; however, discussions with both countries are ongoing.

OECD GLP principles are practiced in at least 41 countries around the world today. The seven nonmember participants are provided in Table 9.3.

All countries that participate in the OECD GLP MAD program, with the exception of the United States, require that the testing laboratory undergo prospective approval to the GLP principles before they can legally undertake testing on studies in which the claim of GLP compliance is made. For example, in the United Kingdom, this is accomplished by submitting an application for Prospective Membership to the UK GLP Compliance Monitoring Program. The UK GLPMA (monitoring authority) then conducts an inspection of the test facility and issues an inspection report. Where issues need to be addressed (assuming no significant issues were identified), the UK GLPMA will review the remediation plan, and if acceptable, will issue a "Statement of GLP Compliance." Companies are then assessed on an ongoing basis according to a risk assessment for the organization, with contributing factors including the severity of the findings from recent inspections, the type of work being conducted (e.g., DMPK, tox or safety assessments), the type of the products being tested (sterile injectable or topical), significant changes at the facility (new management or new facility), and so on. Generally, facilities are inspected at least every 2 years, whereas in the US, the period between inspections can be considerably longer, although the US FDA and EPA try to maintain a 3-year cycle at the outside.

9.4 SOME COUNTRY SPECIFIC REQUIREMENTS ON GLP

While Japan is a full member of the OECD and a participant in MAD, they have developed their GLP regulations independent of the OECD GLP. The Japanese GLP regulations most recently revised and effective from August 2008, entitled "Ordinance on the GLP Standard for Conduct of Nonclinical Safety Studies of Drugs" is applicable only to drugs and not herbicides, pesticides, chemicals, or household products (similar to the United States, Japan has separate ministries for separate areas of commerce, the Ministry of Health and Welfare are responsible for drug approval, the Ministry of Agriculture, Forestry, and Fisheries is responsible for nondrug applications) (Ertz and Preu, 2008). Japanese GLP have gone far to harmonize with OECD, including aspects like Sponsor's Responsibilities (Article 4) and Multi-Site studies (Chapter 8). The few differences between the OECD GLP requirements and the Japan GLP Ordinance are summarized as follows:

- Specific requirement for an SOP on the health care of personnel engaged in the study;
- Study Director approval for unavoidable deviations to standard operating procedures;
- Protocol approval by both the Test Facility Manager (and the sponsor where applicable), although this provision is seen in various regional legislations for the adoption of the OECD GLP (Ertz and Preu, 2008).

Lastly, China, the largest nation in the world has promulgated its own GLP regulations, the most recently issued and effective from September 2003. Many aspects of the Chinese GLP regulations are consistent with the OECD GLP requirements. There are some similarities, as well as differences that are worth noting. Similar to the OECD requirements, the Chinese GLP requires a "complete management system," specific requirements for this is absent from the US GLP regulations. Interestingly, within the management system, the minimum educational requirements of the Organizational Director (as well as various staff roles) are stipulated. The director has more direct responsibility for each study than is required by OECD or the US GLP; the director has primary approval responsibility for the study protocol (approval date is considered the start date of the experiments or "commencement date") and for the final report (approval of the final report is the "completion date" of the study). The director is also required to approve all changes to the protocol as well. Other roles within the Chinese GLP are slightly different from the OECD requirements, including that of the quality unit, who review and sign standard operating procedures and review protocol prior to approval from the director. The quality unit also reviews changes to the protocol prior to director approval. In the US GLP, the quality unit must be entirely separate from the direction and conduct of that study; quality unit signature on protocols and standard operating procedures must be very clearly, and carefully, defined and limited only to ensuring regulatory consistency for the procedure. The Chinese GLP requires that an individual is identified as responsible for test and control substances, similar to the requirements of identifying the archivist. Chinese GLP regulations not only require standard operating procedures specific to the health examination system for the laboratory staff but also require that laboratory staff undergo periodic health examinations (more consistent with the Japanese GLP ordinance). Archive requirements are specified in the Chinese GLP, archives are required to be maintained for at least 5 years after the drug is marketed. Noteworthy absences from the Chinese GLP regulations are the topics addressing the sponsor's responsibilities (e.g., ensuring that the contracting laboratory is aware when studies must be conducted according to the GLP regulations and that the contracted facility can meet the GLP regulatory requirements, e.g., GLP certified laboratory) although sponsors are required to sign protocols for contracted studies and the multisite studies. One might anticipate that these elements will be included in future revision of the GLP regulations.

Chinese government requires laboratories conducting nonclinical laboratory studies for drug market approval in China are GLP certified. Since 2007, the certification is officially issued by the SFDA (State Food and Drug Administration). The certification inspection contains some 280 elements, and findings are classified by critical, major or minor (similar to OECD); inspections are carried out by an inspection team comprised of staff (Test Facility Management, QAU, Study Directors, etc.) from SFDA-GLP certified laboratories. To date, no foreign-owned laboratory has received SFDA-GLP certification. Foreign-owned laboratories in China that have been GLP certified have been certified to OECD standards by European certifying bodies. While this allows the laboratories to conduct GLP studies under OECD requirements (satisfying MAD requirements for application to all OECD member countries), this is not adequate for Chinese Drug application requirements. Laboratories must be SFDA-GLP certified to conduct nonclinical studies for submission packages for Chinese drug approval (China does not participate in MAD or OECD at this time). Over the past decades, there has been a significant migration of GLP work from the western countries to China solely for economic reasons. The above trade barriers clearly present additional burden on foreign regulatory inspection agencies.

9.5 GLP INSPECTION

Finally, the greatest difference between the US FDA, EPA, and OECD GLPs are the manner in which the regulations are enforced by the respective agencies. US EPA and OECD regulatory authorities take a somewhat similar approach to the inspection process, for example, the agencies will notify the test site in advance of the intent to inspect, the length of the advanced notification varies, but is typically in the order of weeks to months. The scope of the inspection is also generally indicated, including the specific studies that will be reviewed during the inspection. The EPA will notify the sponsor and the test facility simultaneously, if different. Inspections are generally relaxed, but under no circumstances are they less thorough or is their regulatory authority reduced in any manner, failure to adequately address findings in a timely manner can result in significant problems for the test facility, including having studies rejected or test facilities being removed from the compliance monitoring program altogether. Inspection findings from an OECD inspection are "graded," that is, categorized as critical, major or deficiencies, and often comments are provided by the auditor. Auditors may engage in significant discussion regarding aspects of the regulations, but do not provide instruction on "how" to comply with regulations. They encourage that the regulatory authorities are engaged whenever a company takes a nontraditional approach to compliance. While EPA does not grade the audit findings, the audit report will provide a statement indicating whether the findings present a risk to compliance. Responses to findings are due within 28 days of issuance of the audit report.

FDA inspections on the other hand are quite different from the EPA or OECD. Generally, the FDA will arrive at the test facility without prearrangement. Targeted studies will be identified during the opening sessions, or selected from the master index. Findings are listed on a form, # 483—notice of objectional observations," issued at the completion of the audit and all findings are weighted equally on the form. Response to the 483 observations must be addressed with 15 days of issuance or the test facility risks receiving a "warning letter." Industry has perhaps placed an undue emphasis on the 483, and the FDA has at times leveraged this to establish regulation standards and expectations rather than through the correct process of proposal and public comment and ultimately issuance. This is clearly the case in recent 483/warning letters issued in conjunction with draft histopathology and study reports.

9.6 SUMMARY

In summary, conducting bioanalysis under GLP for multinational submission will be challenging for years to come. Work conducted in the United States will need to be completed in accordance with US FDA or EPA GLP regulations as well as OECD requirements. Work conducted outside the United States likely need only be conducted according to OECD GLP (with some minor regional accommodations), due to the MAD agreements between the 41 participating nations. Work conducted outside of China might not be accepted in China, under any circumstances, in the near future. Many quality systems are being designed to meet the United States and OECD requirements regardless of the intended registration targets, and finding ways to do this both effectively and efficiently. The proposed revision of the US FDA GLP regulations will go a long way to reducing the challenges in global harmonization, if the FDA is successful in the efforts, it is highly likely that the US EPA will follow. While international compliance remains challenging, it is slowly becoming easier and will likely continue into the future as the world continues to shrink.

REFERENCES

Environmental Protection Agency. Pesticides, Regulating Pesticides. Pesticide Registration (PR) Notices 2011-3. Jan 6, 2012. Available at www.epa.gov/PR_Notices/#2011. Accessed Apr 11, 2013.

Ertz K, Preu M. Ann. Ist. Super Sanita 2008. Vol. 44, No. 4, 390–394 provides a very comprehensive comparison of the Japanese MAFF GLP with US and OECD GLP requirements.

European Medicines Agency. Scientific Guidelines. Guideline on Bioanalytical Method Validation. Jul 21, 2011. Available at www.ema.europa.eu/docs/en_GB/document_library/Scientific_guideline/2011/08/WC500109686.pdf. Accessed Apr 11, 2013.

European Medicines Agency. Good Clinical Practices Inspectors Working Group. Feb 28, 2012. Reflection paper for laboratories that perform the analysis or evaluation of clinical trial samples. Available at www.ema.europa.eu/ema/index.jsp?curl–pages/regulation/document_listing/document_listing_000136.jsp&midamp;=WC0b01ac05800296c4. Accessed Apr 11, 2013.

European Medicines Agency. Committee for Medicinal Products for Human Use. Guideline on the Investigation of Bioequivalence. Jan 20, 2010. Available at www.emea.europa.eu/docs/en_GB/document_library/Scientific_guideline/2010/01/WC500070039.pdf. Accessed Apr 11, 2013.

Federal Authorities of the Swiss Confederation. SR 813.112.1 Ordinance on Good Laboratory Practice (OGLP). May 18, 2005. Available at http://www.admin.ch/ch/e/rs/813_112_1/index.html. Accessed Apr 11, 2013.

Nonclinical Laboratory Studies; Good Laboratory Practice Regulations, 43 Federal Register 247 (Dec 22, 1978) 59985-60020.

1987 Final Rule—Good Laboratory Practice Regulations, 52 Federal Register (Sept 4, 1987) 33768. Available at www.fda.gov/ICECI/EnforcementActions/BioresearchMonitoring/NonclinicalLaboratoriesInspectedunderGoodLaboratoryPractices/ucm072706.htm. Accessed Apr 11, 2013.

Good Laboratory Practice for Nonclinical Laboratory Studies, 75 Federal Register, 244 (Dec 21, 2010) 80011–80013. Available at www.gpo.gov/fdsys/pkg/FR-2010-12-21/pdf/2010-31888.pdf. Accessed Apr 11, 2013.

Medicines and Healthcare products Regulatory Agency-UK. Good Laboratory Practices: Background and Structure. Jul 6, 2012. Available at www.mhra.gov.uk/Howweregulate/Medicines/Inspectionandstandards/GoodLaboratoryPractice/Structure/index.htm. Accessed Apr 11, 2013.

US Food and Drug Administration, Office of Regulatory Affairs. Comparison Chart of FDA and EPA Good Laboratory Practice (GLP) Regulations and the OECD Principles of GLP. Mar 2004. Available at www.fda.gov/ICECI/EnforcementActions/BioresearchMonitoring/ucm135197.htm. Accessed Apr 11, 2013.

US Food and Drug Administration. Inspections, Compliance, Enforcement and Criminal Investigations: SNBL USA LTD. 8/9/10, Feb 17, 2011, Available at www.fda.gov/ICECI/EnforcementActions/WarningLetters/ucm222775.htm. Accessed Apr 11, 2013.

US Food and Drug Administration. Good clinical practices (GCP) is comprised of numerous federal regulations, a complete list can be found on the web site: Selected FDA GCP/Clinical Trial Guidance Documents. Mar 5, 2012. Available at www.fda.gov/ScienceResearch/SpecialTopics/RunningClinicalTrials/GuidancesInformationSheetsandNotices/ucm219433.htm. Accessed Apr 11, 2013.

Viswanathan CT, Bansal S, Booth B, et al. Workshop/Conference Report—Quantitative Bioanalytical Methods Validation and Implementation: Best Practices for Chromatographic and Ligand Binding Assays. *AAPS Journal* 2007;9(1):E30–E42. DOI: 10.1208/aapsj0901004.

10

CURRENT UNDERSTANDING OF BIOANALYSIS DATA MANAGEMENT AND TREND OF REGULATIONS ON DATA MANAGEMENT

Zhongping (John) Lin, Michael Moyer, Jianing Zeng, Joe Rajarao, and Michael Hayes

10.1 INTRODUCTION

10.1.1 Historical Overview

The essential kernel of the entire pharmaceutical discovery and development process is the data. Bioanalytical laboratories generate massive amounts of supporting data that must be organized, secured, processed, and archived. Bioanalytical data and related documentation on method validations or sample analysis for preclinical studies and clinical trials are the essential parts of new drug applications, abbreviated new drug applications, and biologics license applications submitted to the regulatory authorities such as US Food and Drug Administration (FDA) or EU European Medicines Agency (EMA). Current lifecycle data management in most bioanalytical laboratories are still a paper-based or hybrid (paper and electronic) processes, which are often inconsistent and fragmented and have been recognized as inefficient, and containing potential compliance risks for the drug development process. Paper-based and hybrid processes have been shown to slow down the rapid and efficient movement of data between the sponsors, investigators, laboratories, and contract research organizations (CROs). Figure 10.1 shows examples of the typical office (A), archive room (B), and storage room for electronic data (C). These traditional processes require about 30–40% additional administrative overhead to manage data that are not secured, searchable, not easy to share within an organization.

Current processes, shown in many cases to be inadequate, are not suitable for future business needs, and keeping them in place leads to inconsistent ways of working and failure to exploit "best practices." In this environment, there are no common processes for capturing and securing study data, no common agreed business rules for handling "raw data," and no ability to derive global performance metrics from data management to drive improvements. The traditional approaches to data capture, calculation, and verification of raw data result in a major (or serious) drain on human resources while also jeopardizing the data integrity. The administration of paper records is particularly inefficient and expensive, and data cannot be easily integrated with other technologies employed by the organization. As a result, complying with the strict principles of good laboratory practice (GLP) can prove to be very time-consuming and expensive. In the past, new technology that was implemented via "island" solutions added complexity and cost, while not taking global advantage of advanced technology solutions that could improve the efficiency and effectiveness of bioanalytical laboratory operations.

10.1.2 The Need for Regulated Bioanalysis Data Management

Almost every domain area, including the pharmaceutical industry, is moving toward applying more electronic tools and processes. "Paperless" processes that are based on standardized procedures (or business rules), and real-time acquisition and access to data have long been the "Holy Grail" of bioanalytical laboratory operations. Many

Handbook of LC-MS Bioanalysis: Best Practices, Experimental Protocols, and Regulations, First Edition. Edited by Wenkui Li, Jie Zhang, and Francis L.S. Tse.

(a)

(b)

(c)

FIGURE 10.1 The traditional paper-based or hybrid processes.

pharmaceutical companies have created business units or functional organizations that focus on driving toward a future state that leverage electronic solutions. Some bioanalytical laboratories within CROs have quickly moved toward this direction by converting their paper-based process into a truly "paperless" bioanalytical laboratory. The main drivers for implementing electronic data management system are increased efficiency and productivity, improved data quality, and better regulatory compliance. Figure 10.2 depicts the trend in the bioanalytical laboratories on data management. The integrated laboratory is provided with a scalable, integrated platform for creating, managing and sharing data in an increasingly complex global environment.

The "best of breed" approach in most pharmaceutical companies and CROs thus lead to an integrated, heterogeneous environment of separate IT systems. The business will eventually manage "e-records," instead of today's paper records, more effectively by having a common, high-quality business information system that is readily accessible.

Current practice in a bioanalytical laboratory typically involves Thermo Scientific's Watson Laboratory Information Management Systems (LIMS) for study protocol design, sample management, data processing and data statistical analysis, Pharsights WinNonlin for pharmacokinetic (PK) parameter evaluation, an Electronic Laboratory Notebook (ELN) for other sample-related data or process information and a data warehouse such as Agilent OpenLAB ECM or Waters scientific data management systems (SDMS) for the automated acquisition and storage of all types of scientific data, which facilitates all aspects of the bioanalytical workflow from scanning in received samples to scheduling runs and processing samples, to calculating results and generating reports, and to data backup and archival procedures. However, some of the study data documentation, such as the personnel records including resumes, job description and training records, are very often still paper-based. The secured web portal is designed to give access to both the internal project team and external clients (CRO sponsors) to raw and processed data (electronic raw data, result tables, study reports, etc.).

10.2 BIOANALYTICAL WORKFLOW AND DATA MANAGEMENT

10.2.1 Bioanalytical Workflow and Data Flow

Bioanalysis is the quantitative measurement of drugs, their metabolites, and biological molecules in biological samples. Automating this process to save time (increased efficiency) and reduce the number of human errors (improved quality) can be accomplished through software and hardware automation solutions.

Figures 10.3 and 10.4 depict an overall bioanalytical workflow and associated dataflow from a sample to data archiving in bioanalytical laboratories.

10.2.2 Quality Data Management Software Systems

A typical bioanalytical laboratory can produce an enormous amount of data in a very short period of time. Laboratories are constantly expanding their repertoire of instruments, quickly introducing newer and even more exquisite improvements in instrument sensitivity or reliability. As a result, the burden on ensuring and maintaining the quality and integrity of the data becomes a critical task. In today's sophisticated laboratory environment, a wide range of electronic systems are employed to capture, analyze, report, and store the data being generated. As such, adequate consideration must be given to designing the matrix that

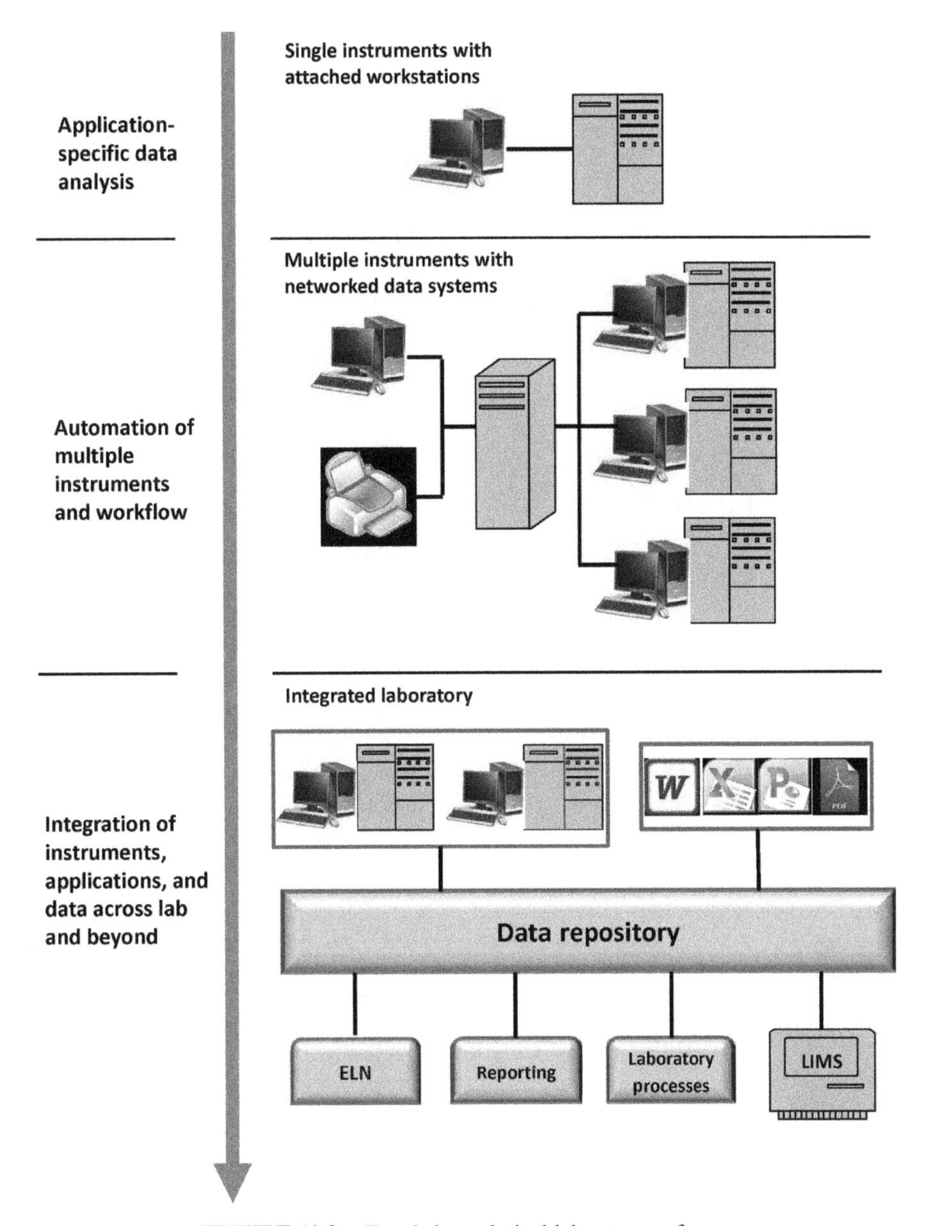

FIGURE 10.2 Trends in analytical laboratory software.

represents this "data-centric" ecosystem of the modern laboratory. Key aspects that influence this environment include reducing cost, increasing productivity, providing highest quality results, and compliance with changing regulatory landscapes.

In addition to instrument generated data, large amounts of "un-structured" data are also generated within the laboratory. These comprise hand-written notes in paper notebooks or binders, references to the preparation, and use of laboratory assets (equipment, reagents, etc.) and a deluge of word processor or spreadsheet documents. While all these activities are governed by a myriad of standard operating procedures or bioanalytical methods, monitoring the execution of these processes is a retroactive, labor intensive, and cumbersome process.

There are several systems that must interact together in order to support a bioanalytical laboratory. While each system was originally defined to perform a specific task (i.e., data capture, statistical analysis, sample management, or report generation), current trends in software capabilities suggest significant overlap across traditional boundaries. This section will review several key software components used in a bioanalytical laboratory and assess areas of redundancy between these systems.

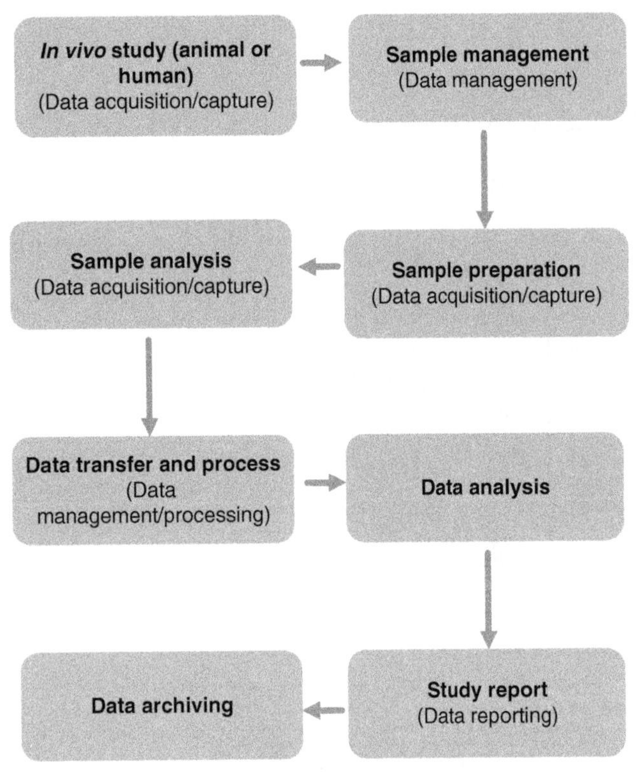

FIGURE 10.3 Bioanalytical workflow.

LIMS are an integral part of the bioanalytical laboratory. They provide a framework by which key bioanalytical processes are managed and enforced. The LIMS system is often the core system for managing the analysis of samples and the study data resulting from the analysis. The LIMS helps to ensure compliance with given procedures and provide capabilities for tracking productivity and performance. The other key competency provided by purpose-built LIMS in a bioanalytical laboratory (such as Watson LIMS) is study design and data analysis. LIMS are routinely used to capture data from instruments, compile summary results, and facilitate the generation of study reports for externalizing data.

ELNs, in their most basic state, capture details of experimental activities conducted to generate the data that is stored in the LIMS system. The information stored in current ELN systems is traditionally associated with traditional paper notebooks. Essentially, they are designed to replace paper notebooks while providing additional functionality that is not available with paper-based processes. In the bioanalytical laboratory, data can be either structured or un-structured in format. In a paper-based or hybrid environment, structured data comprise form-based worksheets or spreadsheets that normally drive a laboratory process, while unstructured data include all loose notes or observations taken during the course of an experiment. A major benefit of an ELN is the

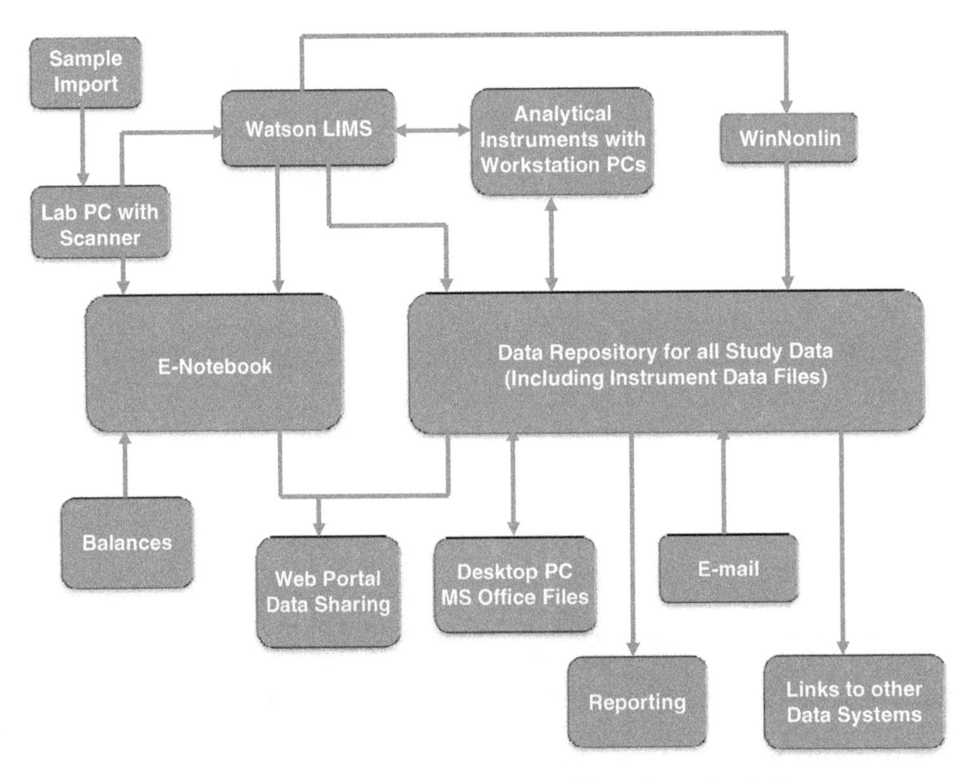

FIGURE 10.4 Bioanalytical data flows with an electronic solution.

ability to quickly find information without having to navigate bound notebooks or forms. Increasingly, ELNs have expanded their functionality into areas traditionally seen as the domain of LIMS. These include the execution of laboratory process, management of storage location and metrology needs, tracking of samples and the enforcement of business rules—all within the framework of an ELN. These systems have evolved from being a simple repository of data for protecting intellectual property to now acquiring capabilities primarily associated with LIMS.

SDMS complement ELN and LIMS by providing a very important function—the *automated* acquisition and storage of all types of scientific raw data, regardless of its source, format, and initial storage location, into a central repository. This function is distinct from that of LIMS or ELNs in that it requires no manual intervention, conferring both a level of compliance and security to raw data captured across the bioanalytical organization. These systems are exceptionally suited for handling very large data files and making them available to an ELN via direct hyperlink to its content. SDMS could also serve as a mechanism for long-term archiving of all data generated by the laboratory.

Electronic Data Management Systems (EDMS) provide support for processes relating to storage, tracking, and retrieval of electronic documents. Bioanalytical laboratories could use EDMS to replace paper records, by simply scanning them into the system, or organizing electronic records by using metadata tags to assist their filing and tracking. EDMS help control bioanalytical methods or SOPs by enforcing access restrictions, version control, check-in/check-out, and workflows for approval and sign-off. EDMS could be part of the SDMS as the documentation management system in addition to the management of scientific data.

This short overview of software systems highlights two situations—the first being an overlap between the capabilities of each application and the second being the potential effective creation of data silos across the bioanalytical enterprise. Software systems are increasingly expanding the scope of their capabilities and blur the lines between traditional categories. It is not unusual to find LIMS applications supporting ELN functionality or ELNs providing robust document management support. On the other hand, the prevalence of multiple software applications in the BA laboratory could lead to the creation of data silos, where effective organization leads to the segregation of data in "best-of-breed" software applications. This situation should be avoided, either by the appropriate deployment of more integrated software applications or by effectively leveraging each individual software solution with the end result in sight. This final deliverable could be a bioanalytical report for submission to a sponsor (as in the case of a CRO) or even a regulatory agency.

10.3 COMPUTER SYSTEMS VALIDATION

10.3.1 Overview

As the bioanalytical industry moves away from paper-based systems to electronic solutions, computer validation has an increasingly important role within the bioanalytical laboratory. Software now permeates almost all aspects of the bioanalytical workflow. When used in regulated (GLP/GCP) analysis, the computer software used for these functions must be validated. The regulations and guidance for computerized systems come from both the GLPs (21 CFR Part 58), GCPs (Computerized Systems used in Clinical Investigations, 2007), 21 CFR Part 11 as well Good Automated Manufacturing Practice (GAMP 5) guidelines. Additionally, depending on the submission, bioanalytical laboratories must also consider ex US regulations such as the OECD and Japanese GLPS.

Software developers have become increasing aware of these regulations and have incorporated many regulatory expectations into their software products such as 21 CFR Part 11 compliant E-signatures, audit trials, configurable user roles, and secure data storage. With most bioanalytical records becoming electronic, "online" archival now becomes a viable option for securely archiving data while at the same time allowing users to view (read only) data. The addition of these compliance features does not exempt a company from having to internally validate their software, but it does allow companies to more easily meet validation requirements.

Many instruments used in the bioanalytical laboratory today are controlled with computers (GC, HPLC, LC-MS) and require both instrument qualifications, as well as software validation. An integrated approach that combines both the instrument qualifications and the specific software validation into one System Validation can result in efficiencies in the validation process. Certain tests completed during instrument IQ/OQs can be referenced as part of the software validation and need not be repeated.

When developing a computer validation strategy, the bioanalytical laboratory should first categorize the software type. GAMP 5 provides a system to categorize software into several categories. The majority of bioanalytical software requiring validation would fall into two categories: category 3: off-the-shelf, nonconfigurable software and category 4: off-the-shelf, configurable software. Chromatographic instrument control, data acquisition and processing software, LIMS systems, and E-notebooks would generally fall into category 4, where the software products are purchased from an instrument/software vendor and then configured internally to meet specific business needs. Computer software developed in-house would fall into category 5. The amount of validation needed increases in direct proportion to the amount of customization required as one moves from category 3–5.

The following should be considered when developing a validation plan for a computer system:

1. *Account management*: User accounts with passwords and unique user IDs ensures controlled and documented access to the system.
2. *Generation of electronic records*: Are critical GXP electronic records created, modified, maintained, and archived by the system?
3. *Audit trail capability*: Allows study event reconstruction and tracking of modifications to study data.
4. *Data security*: Protects against intentional and accidental deletion of electronic records.
5. *Data backup and Archival*: Ensures that system data can be restored in the event of a crash and that finalized data is protected and can be retrieved in the future.

10.3.2 Outline of Validation Procedures

The major components of an "off-the-shelf" configurable computer system validation are shown in Table 10.1 and should include the following.

The first step in the validation process should be to define a validation team. The members of a validation team should include at a minimun, the system owner who has overall responsibility for the system, a system user, an IT representative, and a quality assurance (QA) representative. The specific team members and their roles should be clearly defined in the validation plan. Members of the validation team are responsible for driving the validation project and reviewing and approving the validation documents

After defining a validation team, the next recommended steps are to conduct a 21 CFR Part 11 gap analysis and a risk assessment. The gap analysis is done to evaluate the computer system against specific requirements within the 21 CFR Part 11 regulation. The risk assessment is used to assess both business and regulatory risks with the use of the computer system and define any risk mitigation strategies when necessary. Incorporating a risk assessment into the validation process allows one to focus the validation on needed critical functionalities and potentially reduce the amount of testing and thereby the validation costs. For example, Part 11 deficiencies with noncritical records or functionalities generally not used could be documented in the risk assessment and then could be excluded from validation testing.

One of the most important documents in the validation process is the user requirements specification (URS). This defines all the business and regulatory requirements that are needed for the specific system as it is intended to be used.

TABLE 10.1 The Major Components of an "Off-the-Shelf" Configurable Computer System Validation

Risk assessment	This defines the extent and focus of the validation. The more critical the GXP records generated, the more testing required. Testing may focus only on functions needed by the users. A 21 CFR Part 11 Gap analysis may also be conducted evaluating the software against 21 CFR Part 11 regulations.
User requirement specification	These define both the business and regulatory requirements of the system. Test scripts are written to test against these requirements.
Traceability matrix	This links together each user requirement with the executed test run to confirm the requirements. This is usually finalized in the Validation Summary Report.
Validation plan	Describes the overall validation strategy, roles and responsibilities and deliverables for the system. This should include a system description with data flows.
Vendor audit	A vendor audit may be conducted to assure the vendor follows quality procedures in their development of software. Confirmation of specific testing by the vendor can reduce the amount of validation tested needed.
System configuration specification	If the system is configurable (i.e., assignment of specific permissions to a various user roles in the software), this specification is used to document how the system is configured.
SOPs	Development of SOPs to define system use, administration, calibration, and maintenance.
User training	Ensuring that those validating the system have been trained, as well as defining a procedure to ensure future users of the system are appropriately trained before being given access to the system.
Component /instrument IQ/Oqs	For a system such as an LC/MS system, this would typically include a vendor installation and qualification of the LC and mass spectrometer.
User acceptance testing	(Sometimes referred to as PQ testing) Testing the entire system in its environment against the requirements defined in the URS.
Validation summary report	Summarizes the validation activities and deliverables. The report should list any deviations and incidences, as well as any needed workarounds. This may include the completed traceability matrix.
System release memo	Documentation of release of the validated to regulated production use.
Change control	Changes made to the system must follow change control procedures as defined in a change control SOP.
System development lifecycle	This includes the entire Life Cycle of the system including the initial validation, production use, periodic reviews, upgrades and eventual retirement of the system.

This would include both functionalities within the software application, as well as security processes put in place by an IT group. It must also include all phases of data management from initial acquisition through data processing to storage, archival, and restoration. User test scripts are based on testing against the URS and each URS is linked to a specific test step in a traceability matrix.

The validation plan describes the overall validation strategy, roles and responsibilities, and deliverables for the validation. The risk assessment, URSs, and user test scripts may be part of the validation plan or issued as separate documents. The validation plan should list all proposed system SOPs and documents needed to complete the validation. Any vendor IQ/OQs should also be addressed in the validation plan.

Once all testing has been completed, a validation report is prepared that summarizes all validation activities and deliverables. Any test failures during user test execution and their resolution status must be documented in the validation report. The traceability matrix is usually included in the validation report to document successful testing against each requirement specified in the URS. A system release memo may be issued separately to notify potential users that the system has been released for production use with information about any outstanding issues or workarounds.

10.3.3 Future Direction of Computer Systems Validation

In the future, as more integrated software solutions become available, validation efforts will have to move from the validation of many separate pieces to a more integrated approach where multiple system components are validated together as a single system.

10.3.4 Trend of the Regulations on Data Management

Validation of a computerized system can insure that tested procedures are in place to handle all phases of data management including acquisition, processing, reporting, and archival. However, the specific regulations that could apply to a regulated validation are many and often inadequate and inconsistent. Table 10.2 shows a summary of the regulations and guidance that could apply to GLP, GCP, and GMP analysis.

For regulated analysis within the bioanalytical laboratory, either GLP (animal samples) or GCP regulations (human samples) may apply. In addition, depending on the company location, as well as marketing intentions, ex US regulations must also be considered. The US, OECD and Japanese GLPs were written primarily for equipment use, maintenance and calibration and provide very little specific guidance for computer system validation and electronic data management.

Several additional guidance documents have been published in an effort to incorporate computer validation into GLP environments, but fail to adequately address the critical need for user requirements or are overly bureaucratic. As a consequence, many bioanalytical laboratories have implemented computer validation strategies based on a mix of the FDA clinical guidance and the GMP, GAMP 4, and GAMP 5 guidance for both GLP and GCP analysis.

TABLE 10.2 Summary of the Regulations and Guidance on Software Validation and Data Management

GLP	US GLP 21 CFR 58, 58.61 equipment design and 58.63 maintenance and calibration of equipment, 1978	Focus on equipment maintenance and calibration
	OECD GLPs—Section 4—Apparatus, Material and Reagents, 1992	
	Japanese GLPs—Article 10—Equipment, 1997	
	OECD—The Application of Principles of GLP to Computerized Systems, 1995	No reference to user requirements. Inadequate description of system life cycle.
	Drug Information Association, Computerized systems used in nonclinical safety assessment, 1988, updated 2008. (Red Apple Document)	Very bureaucratic approach. Little discussion on user requirements.
GCP	FDA—21 CFR Part 11, 1997	Addresses use of electronic records and signatures.
	FDA—Computerized Systems in Clinical Trials, 1999	Provides recommended procedures and documents for computer validation. Includes user requirements.
	FDA—Part 11, Electronic Records: Electronic Signatures—Scope and Application, 2003	Narrows scope of 21 CFR Part 11.
	FDA—Computerized Systems in Clinical Investigations, 2007	Adds Risk Assessment to the validation process.
	European Union—Pharmaceutical Inspection Cooperation Inspection Scheme—Computerized System in GXP Environments. 2007	Guidance for inspectors. Little risk management discussed.
GMP	GAMP 4, 2001	Manufacturing focused. Good process for categorizing instrument into 5 types. Includes system lifecycle.
	GAMP 5, 2008	More flexible approach to validation. Excluded category 2 instruments (firmware based) which is needed in bioanalytical laboratories.

With many bioanalytical laboratories doing both regulated GLP and clinical analysis, as well as supporting projects that may lead to products marketed outside of the United States, there is a need for a clear, standardized set of global requirements that would lead to validations that would be acceptable to any bioanalytical regulatory agency.

10.4 CHALLENGES FOR BIOANALYTICAL DATA INTEGRITY AND SECURITY

In support of clinical studies and nonclinical studies conducted in compliance with various regulations and guidance, data integrity, sharing, and security are essential to bioanalytical laboratories. In addition, the efficiency and cost of maintaining these records are equally important. This applies to data generated from different equipments and records in electronic or paper notebooks/study binders and LIMS. It also covers documentation not specific to a particular study, such as records pertaining to equipment calibration and maintenance, personnel records including resumes, job description, and training records.

To ensure data integrity, equipment software, electronic notebooks, and LIMS are validated prior to use to meet various regulations and guidance (see Section 10.3). The electronic systems also go through periodic qualification, calibration, maintenance, and change control management. The most important aspect of these exercises is to have a controlled data environment through organization, processes, and technology. In cases of data documented on paper notebooks and study binders, etc., business processes are clearly outlined in SOPs to ensure proper documentation procedures are followed.

After data collection, maintaining, and archiving raw data and other records generated by bioanalytical laboratories are critical. Moreover, storage and retrieval of information in and from archives are also quite challenging. The processes include data back up, archiving and retrieving. Business should have clear procedures to direct how, where, and how often the analytical data are backed up and archived. The official data record's location should be clear to scientists, management, and the quality control (QC) and QA units. Data must be backed up for short-term security in response to any data loss as part of a disaster recovery plan. Data backup applies for the duration of the study and must also provide appropriate data access security. Once data are archived, the archiving solution generally has its own, independent, backup, and disaster recovery procedures. Most data in the backup will be stored for 6–12 months. As a next step, data can be archived using software such as the Agilent OpenLAB ECM or Waters SDMS as a long-term data archiving plan to ensure data security. In some system designs, data backup and archiving functionality may be provided by one integrated system

General speaking, paper notebooks and supplemental notebooks are microfilmed, inventoried, and archived within 1 or 2 years of notebook issuance, then stored off-site for longer term archiving. Only an archive specialist, or one with that role, can retrieve notebooks from an archiving site. It is costly to maintain an archiving site with fire- and waterproof record storage capability. With electronic notebook and other LIMS systems, most businesses choose not to archive data out of the electronic system, relying on what is called "online" archiving via the originating application. The approval of electronic notebook entries, usually via an electronic signature or "publishing" step, or "lock" of a Watson study secures all study-related data to prevent further changes. Only those with appropriate archiving rights are given the ability to reopen (unlock) closed folders or studies. Since most studies contain records from multiple source systems, including paper, these activities should be controlled by, or closely coordinated with, the appropriate archiving specialists who have regulatory responsibilities within the organizations. The benefit of such a procedure is that all closed studies and projects are still fully accessible as "read only" that allows current problem solvers to take advantage of past work, provides improved security, and reduces the cost of access.

If instrument calibration and maintenance records, such as IQ/OQ documentation, scheduled maintenance, service reports, software/hardware upgrades and bug fixes, retirement management document, are kept in binders, they will be archived routinely according to business' polices. An electronic notebook can easily be used to maintain such instrument records, enabling efficient review, approval, and storage within a compliant system.

Any document related to computer software validation, implementation, and change control will need to be managed and archived. Similarly, standard operation procedures, training records, CVs, and job descriptions will need the same archiving and retrieving processes as well. During health authorities' inspection, these documentation are requested routinely. Recently, more and more computer software has been developed especially for this purpose. For example, the Hewlett Packard HPQC is used for managing the computer software validation process. Electronic systems also used for documentation management such as PRISM or EPIC Star, which can be used for electronic approval and storage of the documents.

It is worth mentioning that the Agilent OpenLAB ECM or Waters SDMS have been successfully used for automated capture of laboratory data generated from analytical instruments at predetermined time frames. Once implemented, all captured files are swept directly from the instrument and archived in their native format. This system allows the capture and storage of instrument raw data files and associated method files as well. The system provides role-based access to projects/views and web-based search and retrieval process.

Results files, summary files, and processed data typically are created and stored on a network drive and are archived through alternative solutions for example, LIMS. Metadata is extracted using data adapters, which makes naming conventions an important consideration for data retrieving.

The archived material requires periodic review to ensure that all the documents are archived completely and properly. Any retrieved materials needs to be documented and returned to the archive to ensure data security. The most challenging aspect of data archiving and retrieving in bioanalytical laboratories is maintaining workable computer legacy systems so retrieved data can be read to generate necessary reports over the duration of the archiving period. This is particularly important in the pharmaceutical industry due to the extended development timelines for new drugs and the need to keep records indefinitely once human trials have been initiated. This process is very costly and the potential of lost data due to corrupted or unreadable file formats is high. Alternatively, some businesses choose to print all data into portable document format (PDF) files to avoid maintaining legacy systems. In certain cases, it may be appropriate to archive the records in a nonelectronic format such as microfilm or paper. In these cases, verification of complete transfer to the new media would occur prior to destruction of the electronic records.

In final consideration, backup and archive procedures are determined based on the type of application and the business criticality of the data. For example, data may be stored "online," via the originating application, where all data are kept online and available until the application is retired, or "off-line," where electronic records are moved to a storage location that is separate from the originating application. In some cases, it may even make sense to retain paper/microfilm forms. No matter which combination of methods are used, appropriate verification must be performed to ensure that the records can continue to be read and accessed as needed throughout the retention period.

10.5 FUTURE PERSPECTIVES

The increasing pressure on bioanalytical laboratories to increase productivity, reduce cost, and ensure compliance could be viewed as a significant burden. However, this reality could drive a more efficient use of software to augment the expertise of trained laboratory professionals. While software systems cannot take the place of a trained individual in making decisions in the bioanalytical laboratory, they can enable better use of the person's time and expertise. Software has the capability to ensure high-quality work is performed consistently and to identify areas that could become issues during an audit.

The bioanalytical laboratory will have to effectively navigate the changing regulatory landscape, as well as resource constraints (both operational and financial) by actively leveraging the plethora of software solutions available. While each application has a core competency, software vendors are keen to incorporate added functionality to make their particular offering more attractive than competitor solutions. As such, some applications provide a diverse range of capabilities, possibly at the risk of adequate rigor in all aspects. New software systems have found their way into the bioanalytical laboratory to address different unresolved issues. For example, an ELN that purports to offer LIMS and EDMS functionality may not offer the same level of support as the stand-alone application. This can be considered a draw-back at the present time, but it is possible that such an application will be created and gains traction in the bioanalytical laboratory of the future. The trend toward hybrid software solutions, ones that offer more than a single category of support, is rapidly growing with every major vendor offering a range of applications—whether by acquisition of companies or by organic development of solutions that span the multiple classes described above. The more integrated the single software solution across the bioanalytical process lifecycle, the better suited it will be to reduce issues faced by current bioanalytical scientists. Today's bioanalytical laboratory, which can utilize as many as four or five software applications, is fraught with manual interventions to move data across platforms. Those core systems, together with other application systems, consist of the heterogeneous IT systems to satisfy process-related data capturing and processing in a "paperless" bioanalytical laboratory (Figure 10.5).

The ideal solution to manage this diverse ecosystem should have the following features:

1. Ability to capture and mine unstructured or free-form data—whether entered directly into the system or via a "drag-drop" function.
2. Compliance with regulations governing the capture and storage of electronic records (e.g., CFR 21 Part 11).
3. Integration with laboratory applications currently in place, as well as future systems built on industry standard protocols (LIMS, CDS, EDMS, SDMS, etc.).
4. Mechanisms for long-term storage in a format not dependent on a proprietary platform.

- Data warehouse systems, such as Agilent OpenLAB ECM or Waters SDMS, should be able to manage raw data and human readable documents of any file type, extract meta data from analytical and all word processing applications, centralize data storage, integrate with MS Office data, and provide compliance with all major regulatory guidelines, including 21CFR part 11, and provide an integrated archival and record retention solution. All bioanalytical study-related source data, such

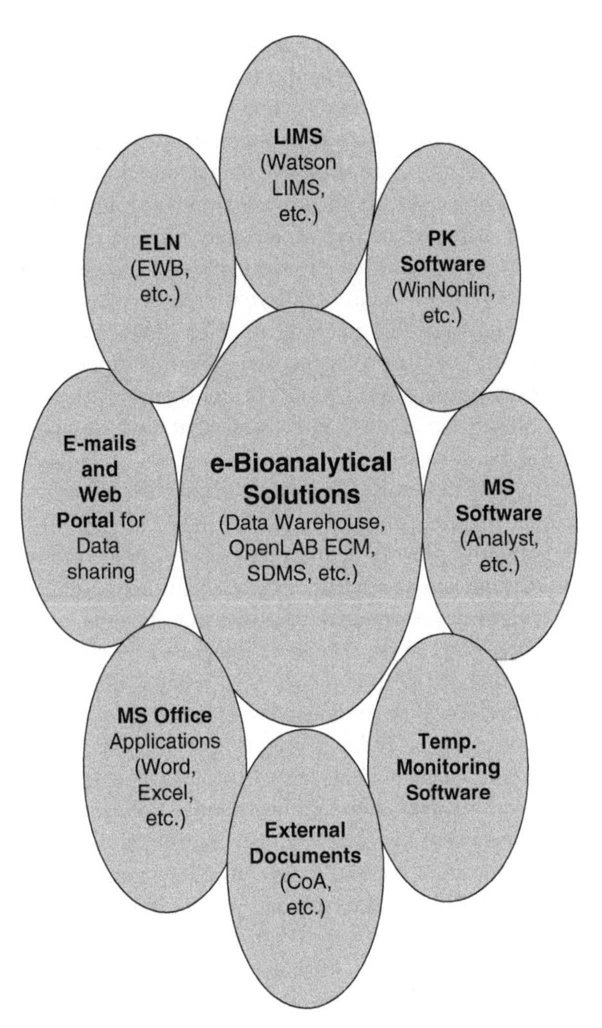

FIGURE 10.5 The heterogeneous environment of separated IT systems in a "paperless" bioanalytical laboratory.

as sample information and data generated from sample preparation and analysis, electronic data, and processed data must be able to printed into human readable file (e.g., PDF) format and documented in the eBinder of the study in the data warehouse to include the following study files:

- Responsible personnel
- Signed study protocol and all amendments
- All relevant correspondence (including relevant external reports, e-mails) and any memos to file
- Sample receipt documentation
- Method, validation report, and stability report
- Method training documentation
- Certificates of Analysis (CoA) and use records
- Analytical run preparation documentation
- Documentation of storage dates and freezer locations for samples and QCs with associated temperature documentation
- Data tables (summary of standard and QC performance, summary of analyte concentrations and PK analysis)
- Assay listing summary (table of assay dates, samples analyzed, and acceptance/rejection, etc.)
- *Each bioanalytical run:* Analytical Run Data (each contains cover sheet, acceptance criteria worksheet, sample list, results table without concentrations, regression curves, standard & QC results, Watson batch results, and chromatograms in PDF format)

- This eBinder could facilate the current requirements of the US FDA and the EU EMA for bioanalytical submission reports (e-filing) by combining information from all mentioned IT systems in a single location (eBinder) that can facilate the automatic report generation and electronic data archiving.
- In the future, technology is expected to drive toward a single global "eStorage" system for the majority of study/raw data. This will provide the ability to manage global efforts more effectively by having, in a single place, study data along with appropriate supporting records, metadata and integrated information from other laboratory systems (e.g., LIMS, ELN). Dealing with customers both internally (cross-function) and externally (sponsors) becomes easier, streamlined and more efficient due to consistencies in using a "common language" and shared business processes.
- The availability of highly integrated and bioanalytical-focused software will improve laboratory efficiency and work productivity, lower compliance risks, minimize overall operational cost, promote electronic data capture, data sharing and organizing, archiving and retrieving, enable efficient report generation, e-submission, auditing, and electronic signature, while also facilitating the audit by regulatory agencies or sponsors.

The future focus, therefore, is on building a common data model and implementing data standards for the bioanalytical laboratory as the linchpin of other data warehouses such as a clinical data warehouse that can be used across trials, across compounds, across therapeutic areas—indeed across companies and regulatory agencies as needed. It is recognized that creating a comprehensive e-bioanalytical environment is an enormous task because it involves the integration of many disparate pieces of software to manage the many facets of the bioanalytical study execution. It is very difficult for any one software developer to create such an e-bioanalytical platform in any reasonable period of time.

At present, the ability to exchange of live data across multiple systems requires enormous programming efforts in order to obtain the desired benefits. An emerging alternative to the daunting challenge of creating a single integrated data model is the creation of common software libraries that facilitate data exchange between proprietary systems. There have been numerous attempts in the past to define common data formats and these have had only limited impact on the ability to make data portable between systems. Organizations such as the Foundation for Laboratory Data Standards are working to develop standard tools, including open source class libraries, which have the potential to provide much more universal data exchange mechanisms. These libraries will identify key data fields that need to be exchanged and would not be dependent on the source system data format. Vendors would provide a pathway for their data into these libraries and would be able to continue with their proprietary data formats. Software tools would have direct access to import data from any compliant system. This approach bypasses the substantial hurdle of obtaining industry wide agreement on source data formats and the current burden of having to program custom interfaces between each combination of systems that a laboratory may be utilizing. Easier integration of data across systems would also facilitate selection of systems based on their core competencies and reduce the need for compromising functionality due to the needs of data integration.

Realizing the fully integrated laboratory environment will come about as software vendors work together with industry to integrate and standardize solutions that support the laboratory workflow. The integration of multiple core pieces of bioanalytical functionality will revolutionize the way research is conducted.

10.6 CONCLUSIONS

Regulated bioanalytical laboratories generate enormous amounts of data for regulatory submission and approval. Paper-based or hybrid lifecycle data management in the bioanalytical laboratories is common even thought it has been recognized as inefficient, and includes potential compliance risks in the drug development process. "Paperless" processes based on standardized procedures (defined business rules), real-time acquisition/storage, and access to data will be the trend for the bioanalytical laboratory operation. The computerized system validation is playing a more and more important role in this regulated environment. Detailed validation procedures are presented and future trend in the regulation of data management are discussed. In the "paperless" laboratories, the data security and integrity is critical. The best practices on data integrity, sharing, archiving, and security are recommended. Future perspective on a global quality data management software system for the bioanalytical laboratories has been discussed. As data management systems move beyond formal forms-driven processes to include a true closed loop design to manage disparate processes across the enterprise, they can provide support for collaborative processes and deliver insight into the overall state of control with the potential to close the gap between simply accomplishing regulatory compliance and delivering measurable improvements in quality and efficiency.

REFERENCES

Computerized systems used in non-clinical safety assessment. Drug Information Association, Maple Glen, PA, 1988.

Computerized Systems used in non-clinical safety assessment—current concepts in validation and compliance, Drug Information Association, Horsham, PA, 2008.

GAMP 5: A Risk-Based Approach to Compliant GxP Computerized Systems, International Society for Pharmaceutical Engineering. Available at http://www.ispe.org/gamp5. Accessed Mar 7, 2013. Retrieved Feb 28, 2012.

Guidelines for the Archiving of Electronic Raw Data in a GLP Environment, version 01. 2003. Available at http://www.bag.admin.ch/anmeldestelle/12828/12832/index.html? Accessed Mar 7, 2013.

Guidelines for the Acquisition and Processing of Electronic Raw Data in a GLP Environment, version 01. 2005. Available at http://www.bag.admin.ch/anmeldestelle/12828/12832/index.html? Accessed Mar 7, 2013.

International Society for Pharmaceutical Engineering. Good Automated Manufacturing Practice guidelines (GAMP 4). ISPE, Tampa, FL, 2001.

International Society for Pharmaceutical Engineering. Good Automated Manufacturing Practice guidelines (GAMP 5). ISPE, Tampa, FL, 2008.

Ministry of Health and Welfare. Ordinance 21, GLP standard ordinance for nonclinical laboratory studies on safety of drugs. Tokyo, 1997.

Organisation of Economic Co-operation and Development. OECD series on principles of good laboratory practice and compliance 10. The application of the principles of GLP to computerized systems. OECD, Paris, 1995.

Organisation of Economic Co-operation and Development, OECD Series on Principles of Good Laboratory Practice and Compliance 1. OECD Principles of Good Laboratory Practice, 1999.

US Food and Drug Administration. 21 Code of Federal Regualtions 58, Good Laboratory Practice for Non-clinical Laboratory Studies. Office of the Federal Register, National Archives and Records Administration, Washington, DC, 1978.

US Food and Drug Administration. 21 Code of Federal Regulations 11, Electronic Records; Electronic Signatures Final Rule. Office of the Federal Register, National Archives and Records Administration, Washington, DC, 1997.

US Food and Drug Administration. Guidance for Industry: Computerized Systems in Clinical Trials. US FDA, Silver Spring, MD, 1999.

US Food and Drug Administration. Guidance for Industry: Part 11, Electronic Records; Electronic Signatures—Scope and Application. US FDA, Silver Spring, MD, 2003.

US Food and Drug Administration. Guidance for Industry: Computerized Systems in Clinical Investigations. US FDA, Silver Spring, MD,2007.

11

REGULATORY INSPECTION TRENDS AND FINDINGS OF BIOANALYTICAL LABORATORIES

FRANK CHOW, MARTIN YAU, AND LEON LACHMAN

11.1 INTRODUCTION

Bioanalytical laboratories have always played a very important role in supporting the pharmaceutical drug development process, from discovery to preclinical (good laboratory practice (GLP)) and to the clinical (good clinical practice (GCP)) phases. While there are specific regulations written for GLP and GCP studies (21 CFR Part 320, 21 CFR Part 58, 21 CFR Part 11, 21 CFR Part 50, 21 CFR Part 56, 21 CFR Part 312), these regulations are not directly applicable to the control and operation of bioanalytical laboratories (Ocampo et al., 2007; Timmerman, 2010). To address this apparent gap, guidances have been issued by various regulatory agencies, such as the US FDA and the EMA (FDA CPGM 7348.001, 1999; FDA Guidance for Industry: Bioanalytical Method Validation, 2001; EMA Guideline on Bioanalytical Method Validation, 2011). Since the regulations only provide general information and lack many important specifics regarding bioanalytical compliance standards and testing requirements, the regulated bioanalytical laboratories have been subjected to a dynamically evolving process during the past couple of decades. Several joint industry (AAPS)-FDA Conferences have been held in Crystal City, Washington DC in 1990, 2000, 2006, and 2008 to discuss various evolving regulatory/compliance issues and to develop appropriate standards and practices. Workshop "White Papers" (also called "Consensus Paper" as it was authored by a number of experts from various regulatory agencies, CROs and pharmaceutical companies) were published (Shah et al., 1992, 2000, 2007; Miller et al., 2001; Desilva et al., 2003; Viswanathan et al., 2007; Fast et al., 2009; Savoie et al., 2010). These white papers have been used by the FDA and other regulators to revise and update their guidance documents and by CRO and pharma companies to fine-tune their control procedures to avoid regulatory citations or enforcement actions.

The general expectations of the level of compliance and controls for bioanalytical laboratories are provided in the May 2001 FDA "Guidance for Industry: Bioanalytical Method Validation." The following is an excerpt from the current FDA Guidance:

> The analytical laboratory conducting pharmacology/toxicology and other pre-clinical studies for regulatory submissions should adhere to FDA's Good Laboratory Practices (GLPs) (21 Code of Federal Regulations (CFR) Part 58) and to sound principles of quality assurance throughout the testing process. The bioanalytical method for human BA, BE, PK, and drug interaction studies must meet the criteria in 21CFR 320.29. The analytical laboratory should have a written set of standard operating procedures (SOPs) to ensure a complete system of quality control and assurance. The SOPs should cover all aspects of analysis from the time of sample is collected and reaches the laboratory until the results of the analysis are reported. The SOPs also should include record keeping, security and chain of sample custody (accountability systems that ensure integrity of test articles), sample preparation, and analytical tools such as methods, reagents, equipment, instrumentation and procedures for quality control and verification of results . . .

It should be noted that at the time of the preparation of this chapter, the FDA/OSI (Office of Scientific Investigations) is

Handbook of LC-MS Bioanalysis: Best Practices, Experimental Protocols, and Regulations, First Edition. Edited by Wenkui Li, Jie Zhang, and Francis L.S. Tse.

in the process of updating the above-referenced guidance document to include additional consensus elements as published in the 2007 and 2009 Crystal City AAPS/FDA White Paper. This updated version is designed to provide further clarification of various emerging requirements discussed during the Crystal City conferences.

In 2011, the EMA issued the European version of the guidance entitled, "Guideline on Bioanalytical Method Validation" that provides the following additional clarifications regarding their expectations on the level of compliance to GLPs for the bioanalytical laboratories as follows:

> The validation of bioanalytical methods and the analysis of samples should be performed in accordance with the principles of Good Laboratory Practice (GLP). However, as human bioanalytical studies fall outside of the scope of GLP, as defined in Directives 2004/10/EC, the sites conducting the human studies are not required to be monitored as part of a national GLP compliance programme. In addition, for clinical trials in humans, the principles of Good Clinical Practice (GCP) should be followed.

Other regulatory agencies, such as Brazil (ANVISA), China (SFDA), Japan (MHLW), and Canada (HPFB), have also recently issued or are in the process of issuing their versions of Bioanalytical Guidance.

It is evident from these guidance documents that the expectation of the FDA and other regulatory agencies is that all bioanalytical laboratories engaged in the support of human studies are required to implement proper systems, standard operating procedures (SOPs) and controls to ensure the accuracy, validity and integrity of the data reported. In the absence of specific regulations articulated by the FDA and other regulators, it is reasonable to extrapolate general principles from all applicable GLPs and good manufacturing practice (GMPs) regulations and guidances to assure appropriate level of compliance and controls.

The apparent ambiguity in applying appropriate compliance standards had resulted in many significant inspectional citations leading to severe enforcement actions. This chapter provides a detailed analysis of current regulatory guidance and expectations based on reported inspectional trends, citations, and regulatory follow-up letters. In addition, the regulatory compliance and scientific basis of these inspection citations are discussed, as applicable. Furthermore, specific recent FDA 483 observations and several current bioanalysis compliance "hot topics" that raised potential data integrity and other compliance concerns such as (a) the practices of using study samples for "preparation runs" or "equilibration runs", (b) reinjection of samples at the end of run sequences to replace nonconformance quality control (QC) or standards, (c) event investigation practices, and (d) ISR (incurred sample reanalysis) requirements, are also extensively discussed.

11.2 CURRENT REGULATORY INSPECTION TRENDS

Based on the collective review of recent FDA and other regulatory agencies inspectional citations, FDA Untitled/Warning Letters, and the collective hand-on experiences of the authors in inspecting/assisting pharmaceutical clients and CRO laboratories in the assessment and resolution of regulatory citations and enforcement actions, several general themes can be found, which are summarized below (Chow et al., 2008). Due to results of these recent FDA and other regulatory agencies' inspectional focuses, most bioanalytical laboratories have been updating their control procedures to avoid major compliance issues:

- *Data Integrity (electronic records)*: In July 2010, the FDA announced their intention to conduct inspections focusing on 21 CFR Part 11 requirements relating to human drugs in accordance with the August 2003 FDA Guidance on "Part 11, Electronic Records, Electronic Signatures, Scope and Application." The instrument and data systems used in today's bioanalytical laboratories are mostly computerized, and therefore, are subjected to the FDA 21 CFR Part 11 regulations for "Electronic Records, Electronic Signatures." The lack of adequate controls in the use of these computerized instrument and data systems has resulted in several significant FDA 483 citations, as well as enforcement actions for many bioanalytical laboratories. Examples of FDA 483 observations relative to data integrity issues are discussed in this chapter. The integrity, accuracy, and validity of the reported results could be vigorously challenged by the regulators and could result in delay and refusal of approval of the related NDA or ANDA. Recent FDA enforcement cases related to potential data integrity issues and electronic records are discussed later in detail in Section 11.5.
- Incurred sample reanalysis (*ISR*): The need to engage actual subject samples in addition to in-study QC samples to demonstrate the reproducibility of the bioanalytical method and to support the validity of the reported results was extensively discussed in the 2006 Crystal City AAPS/FDA Conference and a "White Paper" was published in 2007 (Viswanathan et al., 2007). The ISR discussion was triggered by the FDA/OSI concern that during their inspections of bioanalytical laboratories, the lack of assay reproducibility between the original and repeat results, where agreement was expected, was frequently observed. In a presentation made by Dr. Viswanathan (Associate Director of DSI[1] at the time) during the Fall 2006 GPHA meeting, FDA further

[1] OSI was formerly known as DSI (Division of Scientific Investigations) prior to the Office of Compliance reorganization into a Super Office in June 2011.

illustrated their concern related to the lack of method reproducibility by presenting results for three passing bioequivalence (BE) studies with substantially different C_{max}, T_{max}, and AUC values. A follow-up Crystal City AAPS/FDA Conference primarily focusing on the implementation of ISR requirements was held in 2008. A White Paper was published in 2009 formalizing the ISR recommendations and requirements (Fast et al., 2009). The rationale to include this new requirement was based on the fact that even if a method is demonstrated to be suitable and reproducible using spiked QC samples, it may not have the same level of performance when the actual subject samples are analyzed. Incurred samples (i.e., subject samples) are different from the spiked QC samples for several reasons. They may contain metabolites that are different than the QC samples and these metabolites can be reverted back to the parent compound and cause nonreproducible results. Other factors, such as matrix effect, sample homogeneity, and protein binding that may be unique to the incurred samples, could also play a significant role resulting in nonreproducible results. ISR is designed to provide direct evidence that the method is suitable under the actual conditions of use and is reproducible for the analysis of the subject samples. Satisfactory ISR results provide additional assurance of the accuracy and reliability of the reported results, and therefore, could facilitate and expedite the review and inspection process. During the period from 2007 to the publication of the 2009 AAPS/FDA White Paper on ISR, the implementation of ISR requirements varied significantly from laboratory to laboratory. This lack of conformance to ISR requirements due to lack of the regulatory agency's written expectations resulted in numerous FDA 483 observations. In 2011, the EMA issued a new guidance for Bioanalytical Method Validation, and ISR requirement was included. The FDA is also in the process of updating their guidance document to include the ISR requirement.

- *Adverse Trend(s) and Event Investigations*: In or about 2006–2007, as a result of the extensive FDA investigation of MDS (MDS Pharma Services) bioanalytical laboratory located in Saint Laurent, Montreal, Canada, the concern about potential data reliability issues resulted in severe enforcement action. The acceptability and validity of hundreds of approved bioequivalence/bioavailability/pharmacokinetic (BE/BA/PK) studies conducted at the MDS Saint Laurent facility were questioned by the FDA. To avoid similar compliance issues, most bioanalytical laboratories began to recognize the need to perform formal and documented investigations for any unusual trend of deviations and events observed during method validation and/or study batch runs. Formal SOPs were prepared to guide the investigation process. However, since the FDA and other regulatory agencies have not published any guidance regarding their expectations, many bioanalytical laboratories were still being cited for either not performing the required investigation or the investigations performed were inadequate. Some of the FDA/OSI and other regulators' concerns related to investigations are summarized below (Chow et al., 2008):
    - Failure to perform documented, systemic, thorough, and scientifically sound investigations for significant recurring events, anomalous results and deviations to support the exclusion of unused/unreported/rejected data.
    - Failure to identify the assignable cause/root cause of the problems and failure to perform impact assessment and/or implementation of appropriate CAPA (corrective and preventive actions).
    - As ISR becomes an FDA/OSI mandated requirement in or about 2007, bioanalytical laboratories began to face another major challenge concerning investigations of ISR failures. Bioanalytical laboratories are required, as expected by the FDA/OSI, to perform full investigations, as well as impact assessments in the event of ISR failures to support the validity of the reported results. However, due to limited experience and lack of detailed published guidance, many ISR failure investigations were found to be inadequate by the regulators.
    - The FDA/OSI has been taking the position that when an investigation determines that a certain population of the data set is inaccurate/invalid, the entire data set (including the "good data" generated by the same run sequence or related data potentially impacted by the assigned root causes) also could become questionable. Laboratories are expected to perform full investigations to justify the acceptance of the "good data." The laboratories cannot simply follow its run acceptance SOP to reject samples with apparent "bad results" and accept other samples with the apparently "good results" if the assigned root causes implicate all samples.
- *Acceptability of batch runs*: The validity/acceptability of the study batch runs need to be assessed against the performance characteristics of the bioanalytical method generated during method validation. If substantial performance gaps are observed, the validity of the BA/BE data could be questioned if current investigation does not further support the results. For example, ISR needs to be performed using a minimum number of subject samples (i.e., 5% in a study that collected $\geq$ 1000 subject samples or 10% in a study that collected $<$ 1000 subject samples) to further demonstrate the reproducibility of the methods and accuracy of the reported results (21 CFR Part 320.29). Many FDA 483s were issued due to inadequate control or poor practices in the acceptance of batch runs.

11.3 INADEQUATE INVESTIGATION—THE MDS CASE

The MDS compliance problem surfaced around 2003 following a routine inspection at the Saint Laurent facility. The FDA expressed serious concerns over the inadequacy of a wide range of laboratory control procedures and practices at their bioanalytical laboratories. One of the key FDA concerns was related to the failure of MDS to perform a systemic and thorough investigation for anomalous results, failure to identify the root cause, and failure to take proper corrective action for an apparent sample contamination incident when samples were processed using automated 96-well plate equipment (Tom-Tec). MDS followed its SOP to reject "contaminated" samples with apparent "bad results" and accepted other samples with the apparently "good results." This practice was objected to by the FDA/OSI investigators. MDS attempted to justify its approach in their responses and held meetings with the FDA to defend its position. The FDA/OSI took the position that, when an investigation determines that a certain population of the data set is considered to be inaccurate/invalid, the entire data set, including the "good data" generated by the same run sequence, is potentially impacted by the assigned root causes and can be considered questionable. Laboratories are expected to perform full investigations to justify for the acceptance of the "good data" and cannot selectively and mechanically accept "good data" and reject "bad data" using their batch run acceptance SOP criteria. The FDA rejected the MDS responses and issued two Untitled Letters to MDS in 2004 requiring MDS to review 5 years of approved BE/BA/PK studies performed at its Saint Laurent facility to assure the accuracy and reliability of the study data and the study conclusions. Several hundreds of studies in support of approved products were implicated. In a follow-up inspection in 2006, the FDA raised concern regarding the effectiveness of the MDS review and posted a letter on the FDA website in January 2007 requiring all ANDA holders to perform their own audit/remediation action to support their products that were approved based on the BE studies conducted at the MDS, Saint Laurent facility. An excerpt of the FDA letter summarizing the FDA concerns is provided here:

Dear ANDA Holder/Applicant:

We are writing to you as the sponsor of pending abbreviated new drug application(s) (ANDAs) supported by BE studies in which the bioanalytical analysis was conducted by MDS Pharma Services (MDS) at the St. Laurent (Montreal) and Blainville sites in Quebec, Canada.

FDA has conducted several comprehensive inspections of BE studies conducted by MDS since 2000. The findings of these inspections raise significant concerns about the validity of the reported results from these analytical studies conducted in support of drug applications for marketing. Our findings from these inspections include, but are not limited to, the following:

- Failure to conduct a systematic and thorough evaluation to identify and correct sources of contamination.
- Failure to investigate anomalous results.
- Lack of assay reproducibility between original and repeat results.
- Assay accuracy not assured under the conditions of sample processing.
- Biased manipulation of study data resulting in the acceptance of failed runs.
- Failure to demonstrate the accuracy of analytical methods with appropriate validation experiments and documentation.

Accordingly, with respect to these studies submitted in your applications, we recommend that within 6 months of the date of this letter you do one of the following, in order of FDA preference:

1. Repeat the BE studies.
2. Reassay the samples at a different bioanalytical facility. For this option, the integrity of the original samples must be demonstrated for the frozen storage period.
3. Commission a scientific audit by a qualified independent expert, who is knowledgeable in the area of BE studies and bioanalytical data, and selected by your company rather than by MDS, to verify the results obtained by MDS.

In addition, because one of the agency's significant findings for the inspected MDS studies was the presence of anomalous results, we are recommending for all of the above options that the blood/plasma level results obtained in the studies be compared to any published literature or other relevant information that is publicly available. If you are unable to complete one of these options within the recommended 6-month time frame, please inform us of the reason(s) and your estimated time of completion.

It is evident from the content of this letter that the FDA/OSI expressed serious concern about several laboratory controls and practices at the MDS bioanalytical laboratories. One of the key concerns was related to the failure to conduct systemic and thorough investigations for anomalous results and failure to identify the root cause. At the time, this appeared to be a new inspection focus by the OSI, and this focus appeared to run parallel with the requirement of out-of-specification (OOS) investigation for the FDA-regulated

GMP analytical laboratories. The MDS case had triggered immediate reactions from almost all other bioanalytical laboratories. New SOPs were prepared by almost every bioanalytical laboratory during this time period to assure systematic and thorough investigations would be performed. The need to perform adequate and documented investigations has now become industry standard.

11.4 DATA INTEGRITY CONCERN—THE CETERO CASE

From the review of several recent FDA/OSI Warning/Untitled letters issued to several CROs (Miller et al., 2001; Savoie et al., 2010), the concern of scientific misconduct appeared to have resurfaced again after the chaotic years during the "generic drug scandal" which erupted in late 1980s and early 1990s. During that time, several bioanalytical laboratories, such as Pharmacokinetics and Biodecision Laboratories, were discredited due to potential data integrity issues. A recent incident was related to the inspection of one of Cetero Research's bioanalytical laboratories located in Houston, Texas ("Cetero-Houston"), that resulted in the FDA questioning the integrity of the bioanalysis results generated by this laboratory. Similar to the MDS case, the FDA in an Untitled Letter to Cetero-Houston on July 26, 2011, required remediation actions to be performed for all BE/BA/PK studies conducted at Cetero-Houston from April 1, 2005 to August 31, 2009. What follows is a discussion of the Cetero-Houston Case.

In May 2010, the FDA (Dallas District Office and OSI) conducted an inspection at Cetero-Houston as a part of the FDA's Bioresearch Monitoring (BIMO) Program to confirm that data intended to support new (NDA) and generic (ANDA) applications submitted to the FDA are reliable, and to verify compliance in accordance with CFR 21 Part 320, Bioavailability and Bioequivalence Requirements. This was also a "For Cause" inspection to follow-up on a complaint received from an ex-employee at Cetero-Houston who reported to the FDA that some analysts in Cetero-Houston's bioanalytical laboratory did not follow SOPs, and there were frequent data manipulations and falsification of records that occurred in the laboratory. Prior to filing the complaint to the FDA, the complainant also informed Cetero-Houston's management regarding the alleged regulatory violations and other misconduct in the laboratory. In response to the allegations, Cetero-Houston informed the FDA that an internal investigation of the complaint was initiated, and a consultant firm was also brought in to provide an independent evaluation. One of the critical findings by Cetero-Houston during their internal investigation was the evidence for falsification of time/date of biological sample extractions. Specifically, after review of the key card electronic records for employee building entrances, Cetero-Houston noted that many analysts involved in biological sample extractions were frequently either not in the building or the entry times to the building were greater than 1 h after the documented start times of sample extractions. Due to this discrepancy, Cetero-Houston voluntarily reported to the FDA that the date/time of many sample extractions in multiple studies for multiple sponsors were falsified by their analysts due to the monetary incentive rewarded for work conducted over weekends and/or holidays. Over the time period while the internal investigation was ongoing, Cetero-Houston also forwarded periodic progress reports to the FDA. Thus, before the May 2010 FDA inspection, the FDA had received Cetero-Houston's internal investigation process reports, the final investigation report, and the third-party investigation report.

Based on the above background information, the FDA apparently waited until all internal and third-party investigations were completed before scheduling the "For Cause" inspection in May 2010 to determine (a) if the scope and conclusions of Cetero-Houston's internal investigation were sufficient to resolve concerns raised by the allegations, and (b) if the bioanalytical data generated at Cetero-Houston were reliable to support approval of applications. However, following the May 2010 inspection, the FDA decided to delay issuing an Untitled Letter to Cetero-Houston. Instead, a second "For Cause" inspection was scheduled in December 2010 to further investigate the significant findings uncovered in the May 2010 inspection and to inspect more applications pending approval at the FDA. An Untitled Letter was finally issued to Cetero-Houston on July 26, 2011, after (a) the FDA had completed all of its investigations concerning the complaint, (b) Cetero-Houston's internal investigation reports were issued, (c) the third-party investigation report was issued, and (d) the written responses to the FDA 483 observations was sent to the FDA by Cetero-Houston. In the Untitled Letter, the FDA concluded that both inspections in May and December 2010, as well as the third-party investigation in July 2009, had identified numerous significant concerns regarding the operating procedure at Cetero-Houston and that Cetero-Houston's internal investigation was deficient. What follows is a summary of the concerns raised by the FDA.

Firstly, the FDA found the scope of Cetero-Houston's internal investigation regarding falsified laboratory records with sample extractions to be insufficient, since this investigation was limited to only the date/time falsifications on the analytical procedure (AP) sheets and did not address the impact on overall study conduct, as well as other data recorded on the AP sheets. The FDA concluded that falsification of any portion of laboratory records could cast doubts upon the reliability of the remaining portions, and therefore, the true extent of the data falsification was unknown. The FDA also rejected Cetero-Houston's explanation that the date/time falsifications were due to monetary incentives for work done on weekends and/or holidays because the problems of extraction date/time were later found to occur also in weekdays. Moreover, the Untitled Letter cited examples which showed that routine audits conducted by Cetero-Houston's quality assurance (QA) department were insufficient to assure the reliability of the laboratory records.

Secondly, the FDA raised concerns that Cetero-Houston manipulated results of test samples in analytical runs to meet run acceptance criteria by conducting equilibration or "prep" runs. In the response to the Form FDA 483 observation, Cetero-Houston explained that the equilibration or "prep" runs were used to condition the LC-MS/MS systems. The FDA, however, stated that the concept of equilibrating LC-MS/MS system before performing the actual testing of official runs was not objectionable, but they objected to Cetero-Houston's routine practice of using calibration standards, QCs, blank samples, and subject samples that have not yet been officially analyzed in the equilibration or "prep" runs. In the Untitled Letter, the FDA also cited an example where an equilibration or "prep" run was run on one LC-MS/MS system, but the official run was on a different LS-MS/MS system, thus raising questions on the underlining purpose of all the equilibration or "prep" runs.

Thirdly, the FDA concluded that Cetero-Houston was not able to conduct an adequate investigation regarding the equilibration or "prep" runs because of the lack of documentation concerning the "prep" runs. The Untitled Letter pointed out that Cetero-Houston did not have an SOP for the equilibration of LC-MS/MS systems until December 2009. The lack of SOP and lack of any source records regarding the equilibration or "prep" runs limited both Cetero-Houston and the FDA to evaluate and to reconstruct how the "prep" runs were carried out, what samples were actually injected, and how they might have impacted the study results. Moreover, the possibility of sample substitution for failing calibrators, QCs, or blanks before the official runs could not be ruled out. Overall, the extent of impact of the violations related to the equilibration or "prep" runs could not be determined.

Lastly, it is interesting to note that the FDA Untitled Letter also cited findings and concerns raised in the independent third-party investigation. The below paragraph summarizes the concern raised by the independent third party that was shared by the FDA:

> Chromatography may have been acceptable, QCs may have passed and ISR data may be very good, but doubt remains regarding the integrity of the basic elements of the analytical process due to the documentation irregularities. If the foundation of the laboratory is corrupted, the data generated will be also. The investigation has uncovered practices that are clearly unacceptable. The major concern is the impact they may have had on the data generated. It is not possible to know the extent of the improprieties or their ultimate impact.

The lesson learned from the Cetero-Houston Case is that good documentation and compliance to SOPs are the keys to a good foundation in a bioanalytical laboratory. The source documentation maintained in the laboratory should allow an auditor to reconstruct the important events or steps involved in the bioanalytical process. Lack of adequate source records can also be a limitation to a thorough investigation and would prevent an investigation to identify and resolve the root cause of bioanalytical problems found in the laboratory. Moreover, if adequate source records are not maintained in the laboratory and analysts are not following SOPs, the function of the QA group to assure the reliability of the laboratory records and to detect any data falsification, data manipulation, or documentation irregularities can be limited. Although Cetero-Houston was apparently trying to handle the complaint in a proper manner by communicating with the FDA, and initiating internal and independent third-party investigations, they could not overcome the basic faults (i.e., data falsification confirmed for several analysts) over at least a period of 3–4 years, pretesting samples using equilibration or "prep" runs without SOP and failing to maintain proper documentation) that existed in the laboratory, and eventually was not able to resolve the data integrity concerns raised by the FDA.

11.5 DISCUSSION AND ANALYSIS OF SPECIFIC REGULATORY INSPECTION FINDINGS

Based on our collective experience from conducting over several hundred bioanalytical laboratory inspections worldwide and review of publicly available FDA 483 observations obtained under the Freedom of Information Act (FOIA), the most common recent FDA 483 observations have been noted in the following areas:

- Data integrity/electronic records
- Method validation
- Acceptability of batch runs
- Events/deviations investigations/resolution
- Test specimen accountability

A list of potential FDA objectionable observations, as well as the scientific and compliance basis of each of these potential observations, is presented below. This information should allow the bioanalytical laboratory management to perform appropriate risk assessment to determine the proper level of enhancement to be made to their current systems, procedures, and controls. Much of the information presented here is based on a recent 2008 article (Chow et al., 2008) published by one of the authors of this chapter with some recent regulatory citations and new focus topics added to assure the content is current.

11.5.1 Data Integrity/Electronic Records Issues

The following is a list of common observations related to the data integrity and electronic records issues:

1. Failure to include all aspects of study conduct; specifically, failure to archive or report results of "unofficial" analyses of study samples. There was no justification provided for the "unofficial" analysis/reanalysis of study samples for multiple study protocols. Furthermore, sequence files of unofficial runs or test runs were not recorded in the high-pressure liquid chromatography (HPLC) and LC-MS/MS Use Log.
2. Failure to maintain documentation and audit trails for undeclared "Test" of the reanalyses of study samples. Specifically, the firm reassayed selected samples with outlier concentration results until they obtained desirable results. The results of the "Test" runs were saved under separate "R&D" folders. Therefore, the integrity of the bioanalytical data is not assured.
3. Failure to document all aspects of study conduct:
   a. The entire run was reanalyzed and yet only the reanalysis results were reported in the final report without providing a justification for the reanalysis or not disclosing of data from the initial analyses.
4. Lack of source documents that ensure integrity of study conduct:
   a. No freezer log for removal and return of study and QC samples.
   b. Failure to maintain all sponsor's communication records (quality system)
   c. Failure to define the role of system administrator, laboratory director, and administrative authority in an SOP – laboratory director is allowed to function as system administrator who has the authority to set up and enable the audit trail, but the SOP does not clearly describe whether the laboratory director can disable the audit trail.
   d. Failure to provide adequate security for electronic records in that the computer access procedure is used to access the computer workstation and the software (e.g., "Analyst") used for analytical data acquisition; staff not involved with sample processing is allowed to access the data acquisition software.
5. Lack of SOP to describe the procedure and test samples that were used to equilibrate liquid chromatography with tandem mass spectrometry (LC-MS/MS) systems when conducting equilibration or "prep" runs.
6. Calibration standards, QCs, blank samples, and biological samples collected from study subjects that have not yet been analyzed officially were used in the equilibration or "prep" runs.

Compliance/Scientific Basis: As discussed earlier, good documentation and compliance to SOPs are the keys to a good foundation in a bioanalytical laboratory to assure the accuracy and completeness of the reported bioanalytical data can be supported. All study-related source documentation and data, including electronic records, should be fully and securely maintained in the laboratory in support of any internal QA and third-party audits, including the regulatory agencies. A complete set of records would allow an auditor to reconstruct the important events or steps involved in bioanalytical process. Lack of adequate source records can also be a limitation to a thorough investigation and would prevent an investigation to identify and resolve the root cause of bioanalytical problems found in the laboratory. Moreover, if adequate source records are not maintained in the laboratory and analysts are not following SOPs, the function of the QA group to assure reliability of the laboratory records and to detect any data falsification, data manipulation, or documentation irregularities can be limited.

11.5.2 Method Validation Issues

The requirements for validation of bioanalytical methods used in support of human BA/BE/PK studies have been an ongoing discussion topic among the bioanalytical laboratories and the FDA. This dialogue has led to the issuance of the May 2001 FDA Guidance for Industry—Bioanalytical Method Validation, which was developed based on two conferences/workshops jointly sponsored by the industry and the FDA in 1992 and 2000 (Shah et al., 1992, 2000). Several significant compliance issues related to bioanalytical method validation were extensively discussed in a recent publication by one of the authors (Chow et al., 2009). A high percentage of the recent FDA 483 observations issued to bioanalytical laboratories are related to method validation issues, as highlighted in FDA 483 observation examples provided below:

Issue 1: Many bioanalytical laboratories were cited for not including/reporting all method validation batch runs in the determination of the method suitability, accuracy, and precision. Several examples of FDA 483 observations related to this issue are provided as follows:

1. Method validation report incorrectly states the accuracy and precision of data since data from many runs performed during the prestudy validation were omitted, which otherwise would fail to demonstrate accuracy and precision of the method.
2. Failure to report all validation experiments containing valid data. Failure to include all data for statistical analysis in the method validation report. For example:
   a. 2 LOQ QC were excluded from the within-run precision and accuracy calculation.
   b. Hemolysis effect data was excluded without a valid reason.

3. Only passing long-term stability results were reported.
4. Method validation report inaccurately stated the reason for rejecting validation runs. Although the runs were rejected and aborted due to contamination peaks, they were reported as "unacceptable QC and standard curve" and "batch not reported."

Compliance/Scientific Basis: The requirement to include all pertinent accuracy and precision data in the determination of the accuracy and precision of the method is clearly stated in the May 2001 FDA Guidance for Industry—Bioanalytical Method Validation that "... Reported method validation data and the determination of accuracy and precision should include all outliers... " This requirement is also stated in FDA Compliance Program Guidance Manual, CPGM 7348.001 dated May 1, 1999, entitled, "Bioresearch Monitoring—In-vivo Bioequivalence": "... failure to include all data points, not otherwise documented as rejected for a scientifically sound reason, in determination of assay method precision, sensitivity and accuracy ... "

However, excluding failed runs from the determination of accuracy and precision of the method during validation has been a common finding at bioanalytical laboratories in the past and has resulted in many recent FDA 483 citations/observations, particularly when there is no documented rationale provided to justify these exclusions. Many bioanalytical laboratories routinely followed their procedure on chromatographic run acceptance to exclude failed method validation runs when the calibration standards and QC samples do not meet the SOP acceptance criteria. However, if these method validation runs were rejected due to inherent variability of the method, then according to the above-referenced FDA documents, this data should be included in the determination of the "actual" accuracy and precision of the method. Therefore, bioanalytical laboratories need to review all method validation runs, passed or failed, to ensure there is valid reason to support any exclusion. Unless the failed runs are due to isolated and nonmethod (inherent variability) related issues, the data should be included. The rationale for exclusion should be documented and appropriately approved to support the firm's decision, and for consideration by customers and regulatory authorities when required.

Issue 2: In the past several years, many bioanalytical laboratories were cited for collecting stability data using frozen standards. Several examples of citations relative to this issue are provided as follows:

1. The demonstration of analyte stability—recovery after freezing and thawing, postpreparative, stock solution stability, etc., did not employ an unfrozen (freshly prepared) reference (calibration standards).
2. Failure to validate postpreparation stability (i.e., autosampler stability) by comparison to freshly prepared standards.

Compliance/Scientific Basis: The May 2001 FDA Guidance for Industry for Bioanalytical Method Validation states "... All stability determinations should use a set of samples prepared from a freshly made stock solution of the analyte in the appropriate analyte-free, interference-free biological matrix... " While the FDA Guidance has been published for several years, many bioanalytical laboratories are still collecting stability data using stored frozen standards. This practice is based on the rationale that the firm only needs to demonstrate "relative stability," as their practice is to prepare calibration standards and QC samples in bulk and store them with the test specimen samples under the same conditions. This is based on the view that even if the analyte in the frozen standards degrades during storage, it would not affect the accuracy and precision of the results generated since the analyte in the test specimen samples would degrade at the same rate. However, the FDA Guidance requires the firm to demonstrate "absolute stability," so that the capability/suitability of the method can be fully understood. As a result, many FDA 483 observations issued to bioanalytical laboratories have been due to the use of frozen standards in the determination of stability.

While the traditional approach of comparing analytical results of stored standards to that of the fresh standards is recommended by the FDA Guidance, FDA also allows other valid statistical approaches using confidence limits for the evaluation of analyte stability to be used (Timm et al., 1985; Kringle et al., 2001).

Issue 3: For many bioanalytical laboratories, partial validation could be frequently performed over a period of several years to support modifications made to the method. However, there is no clear guidance provided by the FDA to define partial validation requirements, consequently resulting in regulatory citations. Examples of concerns are provided as follows:

1. The use of stability data generated from other methods/studies (i.e., HPLC/UV) to support the stability of a revised/new method (i.e., LC-MS/MS) is without proper justification.
2. Failure to generate supplemental stability data when there were significant changes in assay procedure and instrumentation.
3. Failure to revalidate the assay precision and accuracy when there was a change in LC-MS/MS instrumentation (e.g., API 5000 instead of API 4000 was used).

Compliance/Scientific Basis: Most bioanalytical methods developed and validated by commercial contract bioanalytical laboratories can remain in active use for many years in order to support generic BA/BE/PK studies. After the initial validation, methods could continue to be optimized and/or modified. As recommended by the FDA Guidance, partial validation is required to support these changes. However, the FDA Guidance does not further detail the partial validation requirements under various change conditions. The FDA Guidance only states that "... partial validation can range from as little as one intra-assay accuracy and precision determination to a nearly full validation." As a result, it has been observed that the extent of partial validation performed varies significantly among laboratories for similar changes. Many bioanalytical laboratories have considered that analyte stability would not be affected by the test method used, and therefore, stability studies would not need to be repeated. However, while the chemical stability of an analyte may not have changed but the stability duration established by using different bioanalytical methods may be different, depending on the complexity and inherent variability (e.g., different recovery) of the individual assay. Therefore, before relying on old stability data to support new methods, proper scientific justification with supporting data needs to be generated and documented.

Issue 4: Many bioanalytical laboratories had received FDA citations due to deviations from the validated method; examples are provided as follows:

1. Failure to use the same method as validated—use of manual aliquot instead of automated aliquots, changes were made in the method without validation data to support that these changes would not compromise the accuracy and precision of the data generated.
2. Failure to adequately revalidate the inter-run precision and accuracy after a procedural change of dwell times for mass transition from 150 ms to 50 ms.
3. Failure to demonstrate the accuracy of the method when using longer centrifugation time and using a plunger to push plasma through solid phase extractor.

Compliance/Scientific Basis: As discussed under Issue 3, from inception of the method, it could undergo optimization and/or modification over time. Many bioanalytical laboratories attempt to justify this, saying partial validation was not required since the change was minor. Although the justification for the immediate change may be adequate, it is important to consider the cumulative effect of all prior changes when assessing the need for revalidation. The accuracy and validity of the data generated by the modified method could be challenged by the FDA unless proper supporting validation data is generated. In order to avoid this potential compliance issue, a formal documented change control procedure should be instituted and controlled by the QA Unit to ensure the decision for revalidation will be made based on the evaluation of all changes made since the last complete validation.

Based on past experience in auditing bioanalytical laboratories, change control and change management is always a weak link in the entire quality system. Change control is a critical control element in a GMP environment, and it is evident from the current trend that this requirement is gaining more and more attention in the bioanalytical laboratory environment.

Issue 5: Many bioanalytical laboratories had received regulatory citations related to inadequate method validation. For example:

1. Failure to demonstrate the absence of matrix effect when the bioanalytical method in support of the human BA/BE study is a LC-MS/MS method.
2. Failure to evaluate matrix effects in the method validation, specifically, the matrix effect on extraction recovery and detector response.

Compliance/Scientific Basis: Matrix effect is a scientific term used to describe the effect of extracted biological matrix on the MS response of the drug analyte. Due to the complexity and inherent variability of the biological matrix, materials from the extracted matrix could affect the extent of ionization of the analyte, and consequently, may suppress and/or enhance the ionization process that could compromise the accuracy and precision of the test results generated using LC-MS/MS methods (King et al., 2000; Weng and Halls, 2002; James et al., 2004). It should be noted that matrix effect is unique to LC-MS/MS methods, and is one of the major unknown parameters that needs to be evaluated during method development and validation before the method can be considered as "suitable for its intended use." This issue has been discussed in the FDA guidance and many other publications.

Bioanalytical laboratories often evaluate the matrix effect during method development, and many have not further demonstrated the absence of matrix effect during method validation. Consequently, the development records did not adequately document efforts and data to support the absence of matrix effect. As a result, without this documentation, any event/adverse trend of highly variable recovery and/or loss of sensitivity results observed could be challenged during FDA review as potentially due to the presence of a matrix effect.

Issue 5 (cont'd):

1. The composition of the spiked QC and calibration standards are substantially different than that of the test specimen. (i.e., presence of excessive aqueous and organic solvent in the final spiked matrix samples).

Compliance/Scientific Basis: While it is not explicitly stated in the FDA Guidance, it is good science to perform method validation using analyte spiked into the same matrix as the test specimen to ensure the method validation data collected are truly representative of the actual test article conditions. It will be ideal if the spiked matrix does not contain any solvent/water. However, due to inherent solubility limitation of many analytes in biological fluid, direct spiking is impossible in many cases. Therefore, stock and diluted spiking solutions in organic/aqueous mixture are generally used. Bioanalytical laboratories need to have SOPs that define how the spike validation samples are to be prepared, and a limit of organic solvent and the aqueous content needs to be defined. Generally, the presence of organic solvent in the spiked samples should be about 2%, and in any case, should not be more than 5%.

Issue 5 (cont'd):

1. The prestudy method validation for the assay of (drug) in plasma and the calibrators and QC samples used in the studies contained a different anticoagulant-diluent from the EDTA anticoagulant used for the collecting study samples (for example, the use of FFO-fresh frozen plasma in citrate-phosphate-dextrose diluent anticoagulant).

Compliance/Scientific Basis: It was also observed for many bioanalytical laboratories located in India, control plasma with anticoagulant-diluent different than for subject specimen was common, resulting in many regulatory citations for similar concern as discussed earlier.

Issue 5 (cont'd):

1. Stability experiments were not conducted to reflect the condition of the plasma samples collected from the subjects. Specifically, the stability data was collected for (drug 1) without the presence of (drug 2) for subject samples that contained both analytes. Failure to demonstrate assay specificity of (drug 1) in the presence of (drug 2).

Compliance/Scientific Basis: For a combination drug that contains more than one drug compound, biological (e.g., plasma) samples collected from study subjects contain two or more analytes. As stability data are the basis to demonstrate integrity of subject plasma samples during sample processing (e.g., freeze/thaw stability) and during long-term storage (e.g., long-term frozen storage stability), the FDA is taking the position that test plasma samples used in stability studies should contain the same analytes as in the study plasma samples, due to possible analyte interactions. For similar reasons, assay specificity of individual analyte should also be demonstrated in the presence of other analytes contained in the biological samples.

Issue 6: Regulatory citations were made regarding inadequate validation of the size of batch run:

1. The size of the batch run during method validation is much less than that the actual study run.
2. Runs were accepted exceeding the validated run time.

Compliance/Scientific Basis: It is considered a current industry practice and FDA expectation that the method validation accuracy and precision runs should contain enough test samples that would mimic the actual size/run time of a study batch run (as defined in the test protocol for the designated BE/BA study). The ruggedness of the method can, therefore, be further demonstrated. However, this has not been the practice in the past, and this deficiency could be challenged by the FDA during study specific inspection, particularly when the ruggedness issue was observed for production runs. In order to maintain the same run time as the production runs, "dummy" extracted matrix blank samples are usually added to the method validation run sequences in order to extend the run time.

11.5.3 Batch Runs Acceptance Criteria Issues

The batch run acceptance criteria has been one of the most important topics of discussion during various FDA/Industry workshops and conferences, and many regulatory guidances have been published. However, the interpretations of the regulators' expectations and the practices varied from laboratory to laboratory and had resulted in many regulatory citations. Examples are provided as follows:

Issue 1: Batch runs were accepted when more than 50% of QC samples at each concentration level did not meet acceptance criteria.

Compliance/Scientific Basis: This has been a common FDA 483 observation for several bioanalytical laboratories. Since the requirement of having 50% of QC sample at each concentration level meeting the acceptance criteria is not explicitly defined in the May 2001 FDA Guidance, many laboratories failed to follow this requirement. Nevertheless, this observation makes good scientific sense, as adopting this additional requirement would further ensure the accuracy and precision of the data generated for the entire concentration range of the method. Many bioanalytical laboratories are revising their procedure for the Acceptance of Calibration Curves, QC samples, and batch runs to include this requirement to avoid future FDA challenges.

Issue 2:

- Analytical runs that originally failed when results of QC samples were outside the acceptance limits were reprocessed by excluding selected calibration standards until the QC standards passed.

- The order of rejection of calibration standards was not consistent. The calibration curve would not pass if proper rejection order was followed.

Compliance/Scientific Basis: As described in the FDA Guidance and various publications, it is an acceptable practice to drop some calibration standards in the construction of the regression line based on not meeting the preestablished acceptance criteria. However, the FDA Guidance does not provide the procedural details regarding the order of rejection of calibration standards. As a result, the practices of some bioanalytical laboratories became a subject of recent FDA 483 observations. Consequently, many bioanalytical laboratories have updated their SOP to provide a consistent and detailed guidance regarding the order of rejection of calibration standards.

Issue 3: Based on a review of recent FDA 483 observations issued to bioanalytical laboratories, below is a list of additional batch runs acceptance issues that could be found objectionable to the FDA during data review. Some of these objectionable conditions were due to inadequate quality systems and control, and some were due to poor laboratory practices:

1. Failure to perform or use adequate number of subject samples to assess ISR:
   a. Failure to evaluate reproducibility of the assay of (drug) in incurred samples for studies (drug). Less than 5% of samples were reanalyzed.
   b. Use acceptance criteria of +/− 30% instead of +/− 20% to assess ISR results
2. The MS detector used in the analysis of batch runs was different from the MS detector used in the prestudy validation runs that resulted in the need to modify the method in the amount of internal standard and MS operating parameters.
3. Samples were reassayed without documented reason for rejecting the original results (suspected sample interchange).
4. Failure to maintain completed records for conduct of the experiment. The identity of the analyst who aliquoted plasma samples for sample processing was not documented. The method SOPs did not specify that one analyst must perform a specific step of the analytical process for all calibrators, QCs, and study samples in a batch.
5. There is no contemporaneous record of preparation (source of plasma, anticoagulant diluent, volumes, date, preparer, etc.) of the plasma pools used to prepare the calibrator and QC samples for studies (drug) ... or the prestudy method validations of (drug).
6. The analyst who performed solid phase extraction (SPE) of calibrators for assay of (drug) was not the same analyst who performed SPEs of QCs, and study sample calibrators were extracted separately from QCs and study samples.
7. There were only 2 QC concentrations in a singlet for each extraction batch of study samples. Samples from one subject for one period and four QC concentrations in a singlet were extracted over two batches. Failure to document the QC concentrations extracted with each extraction batch of study sample.
8. QC concentration levels were not representative of the actual subject sample concentrations ... C_{max} from 2645 ng/mL to 8045 ng/mL and QCs of 300, 8500, and 20,000 ng/ml were observed.
9. Failure to use separate stocks prepared from independent weightings for preparation of calibration standards and QC samples during prestudy validation.
10. Failure to document all aspects of study conduct:
    a. Diluent used not documented; sample processing checklist cannot be clearly distinguished between sets of calibrators extracted for a specific analytical run.
    b. Failure to archive or report results of "unofficial" analysis of study samples.
    c. Entire run was a reanalysis, yet only the reanalysis results were reported, and no justification was provided for reanalysis or not disclosing of original results.
    d. The lot of blank matrix used in the preparation of calibrators was not documented at the time of preparation.
    e. Procedure required a maximum of 4–6 samples to be extracted at a time, but there was no document (a) to support that this procedure was followed or (b) to identify the samples processed in each subset in a run.
    f. Notebook entries to document validation were not contemporaneous with experiment conducted.
    g. No documentation exists to confirm the autosampler injection sequence was verified.
11. Failure to investigate the cause of anomalous concentrations in Run X. For example, subject coded as "BLQ" was flanked by samples with measurable (drug) concentrations. Subject with measurable concentration of (drug) was flanked by subject with "BLQ."
12. Runs were accepted despite significant variations in internal standards response during the run.

13. Sample residue was observed in some samples after extraction. There was no data to show that the accuracy of sample analysis was not affected by the presence of sample residue.
14. Failure to verify that samples were loaded into 96-well plates according to the preprinted plate mapping sequence; anomalous sample concentration was attributed to sample loading mistake.
15. Failure to reject run showing analyte interference >20% of LLOQ in blank samples. Run was improperly accepted by deactivating calibration standard 1.
16. When the samples with original results above the upper calibration limit were reanalyzed after dilution, the reassayed results were not in agreement with the initial results. (Method might not be reproducible for subject samples.)
17. Acceptance of batch runs data that were interrupted for a considerable period of time, and no control data shown to support accuracy of the data generated after the runs were resumed.
18. Acceptance of study data when there is an adverse trend observed. For example, the % of rejected runs, rejected QC samples, and calibration standards are excessive and not consistent with the method validation data. There is no justification to support accuracy of the accepted study data.
19. Acceptance of batch runs when anomalous results were rejected as PK outliers, and replaced by repeated assay results without proper justification to support that the outlier results were indeed PK outliers.
20. Acceptance of batch runs with significant interference observed in blanks, zero standards, and pre-dose samples for a large population of the batch run.
21. Acceptance of batch runs with numerous repeated results based on sponsor's requests without proper justifications.
22. Analytical runs were accepted when regression type was changed during analysis of study samples without proper cross-validation.
23. Acceptance of batch runs data when most repeated results were significantly different from initial results:
    - For example, the entire run was repeated due to sample out of regression range. Results of samples reassayed after dilution were significantly different from initial results. (Validity of reported data is, therefore, questionable.)
    - Acceptance of data in batch runs where many data were generated using inconsistent manual integration:
        - Inconsistencies in manual integration of chromatograms, in that identical peak of some standards and QC samples were reintegrated using different integration methods. If reintegration was performed consistently across all samples within a run, some runs would not meet the run acceptance criteria.

Compliance/Scientific Basis: It is evident from the list of objectionable issues presented above, that bioanalytical laboratories are not only expected to produce the source raw data to support the reported data, the laboratories must also be able to have the proper documentation to justify the exclusion of the unused data with scientifically sound rationale. The laboratories should also be able to demonstrate that the exclusion of these unused data would not compromise the precision and accuracy of the reported data. The bioanalytical laboratories should have proper documentation to avoid any perception of "Selective Reporting."

Due to the inherent variability of the LC-MS/MS method and the complexity of the biological matrix, the observations described above could occur during any routine analysis even if the method has been properly validated. However, if the magnitude and frequency of these events becomes significant (i.e., adverse trend), then it would become a potential FDA inspectional issue. Therefore, in order to minimize this possibility, it is highly recommended for bioanalytical laboratories to perform a thorough internal review of BA/BE/PK data generated using the "quality indicators" approach to identify and resolve potential questions regarding the accuracy and reliability of reported data. The quality indicators can be developed using the objectionable conditions presented above to ensure the study data are free of these objectionable conditions. A list of example "quality indicators" that can be used to facilitate an effective review of the LC-MS/MS method performance is provided in Table 11.2, and is further discussed in Section 11.6.

It should be noted that out of all the examples listed earlier, performing manual integration in an inconsistent manner is one of the most common FDA inspectional issues, based on our past auditing experience. Due to the need to assay an extended concentration range, the noise level at lower calibration range could be much higher than that at the higher concentration range. As a result, manual integration needs to be performed in order to improve accuracy. However, if these manual integrations are frequently performed in an inconsistent manner, it would become a potential compliance issue.

11.5.4 Events/Deviations Investigation/Resolution Issues

The MDS case, as discussed earlier, triggered immediate reactions from almost all other bioanalytical laboratories. New SOPs were prepared by almost every bioanalytical laboratory during this time period to assure that systematic and thorough investigation would be performed. The need to perform adequate and documented investigations has now become industry standard. However, since FDA and other regulatory agencies have not published any guidance regarding their expectations, many bioanalytical laboratories were still being cited for either not performing the required investigation, or the investigations performed were inadequate.

Issue 1:

- Acceptance of batch runs data when there is an adverse trend or recurring significant deviation events, and no investigation was performed to identify its root cause to support that the passing reported data were accurate and valid. For example:
  - Ten consecutive batch runs except one were rejected due to massive failures of in-study QC samples, and one batch run was accepted with marginally passed QC samples.
  - Selective acceptance of passing data generated from batch runs, when subsequent investigation concluded that the robotic sample preparation instrument used was malfunctioning at the time of use.
  - Acceptance of batch runs data when there is a significant adverse trend observed.
- Failure to perform investigation of high failure rate of analytical runs in study (drug) (33% failure rate).
- Failure to investigate when reassay of incurred samples showed significant variability, in that the original and repeat results differ significantly without proper justification.

Compliance/Scientific Basis: It is evident from the examples of objectionable laboratory practices provided above that the bioanalytical laboratories are expected to perform a thorough and well-documented investigation for recurring significant events/deviations. This had been a "hot" compliance topic in the mid-2000s and has now become an industry practice. The expectation to perform formal laboratory investigations is parallel to the cGMP requirement for pharmaceutical analytical laboratories when an OOS or aberrant test result is generated. It cannot just be replaced with repeated testing data without appropriate justification based on a formal investigation driven by a written SOP. However, performing formal or rigorous laboratory investigations for these types of issues has not been the past practice at many bioanalytical laboratories based on our past auditing experience. Due to the complexity and inherent variability of the biological matrix, many bioanalytical laboratories had allowed the replacement of "outlier" results with repeated assay of test specimens, usually following written procedures that did not require a formal documented investigation. As a result of the FDA concerns, many bioanalytical laboratories have started to review and revise their SOPs related to event investigation, resolution, and management. An effective investigation procedure should cover the control elements, as summarized in Table 11.1.

TABLE 11.1 Summary of Critical Control Elements in an Event/Deviation Investigation Procedure

Implement an SOP system, driven by QA, to document and track event investigations and to evaluate all events prior to acceptance of the BA/BE studies to include the following elements:

i. Define the events.
ii. Perform proper documented scientifically sound investigation.
iii. Identify root cause or most probable cause.
iv. Perform impact assessment for other reported data potentially implicated by these events.
v. Establish proper corrective and preventive actions (CAPA), properly implemented and tracked.
vi. Final disposition of the study data affected, and determine its impact on the accuracy and validity of the BA/BE study.
vii. Procedure to close out investigation for single event and all events at the end of the study.

Since the conclusion of the BA/BE study is not based on one data point but is based on a statistical evaluation of the totality of all data points generated, the final closure of all events should not occur until all individual events are evaluated in their totality to ensure all reported data are supportable and would not be compromised by the recurring excursions.

11.5.5 Test Specimen Accountability Issue

Issue 1:

1. Failure to maintain adequate and accurate sample records. For example, the record systems fail to document the removal or return of test samples to and from the frozen storage areas, failure to document who removed or returned the samples from the storage area, or the date and time of removal or return.
2. Lack of source document that ensure integrity of study conduct (e.g., there was no freezer log to verify identity and to record freezer removal and return of QC samples used in freeze/thaw and long-term frozen stability experiments, as well as for study plasma samples analyzed in analytical runs).

Compliance/Scientific Basis: It is a GLP requirement to document the chain-of-custody records for all test articles used. Similar control is expected for the bioanalytical laboratories to ensure the test specimen and control samples are stored and handled within the validation conditions, as defined in the stability studies during prestudy method validation. Therefore, the laboratory documentation should be designed in a way that can provide a traceable audit trail.

11.6 RECOMMENDATIONS TO SUPPORT AN EFFECTIVE FDA INSPECTION—READINESS PREPARATION

In light of the current trends seen in recent FDA 483 observations issued to bioanalytical laboratories, it is recommended that the firms should thoroughly evaluate the state of existing controls and practices to assure the accuracy and reliability of test data and conclusions therefrom. As part of inspection readiness, the firm should perform a study-specific "mock inspection" and 100% data review using the "quality indicators" approach to ensure that all systems, practices, data, and documentation are ready to support a successful FDA study-specific inspection. As discussed in Issue 3 under "Batch Run Acceptance," these quality indicators were developed based on recent FDA 483 observations to ensure all common potentially objectionable conditions can be detected and corrected prior to the FDA inspection. Examples of "quality indicators" are summarized in Table 11.2.

To support an effective mock FDA study-specific inspection, a well-planned audit strategy should be developed to ensure the scope of the audit coverage will address both study data issues and related quality system issues. The critical elements of a sample audit plan are summarized in Table 11.3.

In addition to review of hard copy raw data, review of electronic data should be performed so that potential chromatographic problems (i.e., peak tailing, peak splitting, baseline drifting, and reintegration) and unreported data not reflected in printouts can be detected and properly evaluated.

Bioanalytical laboratories should periodically review their quality systems, procedures and controls, and perform benchmarking exercises to assure they meet current standards, regulations, and expectations. Any significant gaps identified should be timely addressed and corrected. To assure a successful FDA/OSI inspection, proper preparation is the key. Below are a few additional points to be considered:

TABLE 11.2 Summary of "Quality Indicators" to be Used in Evaluating LC-MS/MS Chromatographic Data

Run acceptability: Determine whether the run was rejected or accepted with a supported valid reason; determine if the rejection rate of runs, quality controls (QCs), and/or calibration standards (STD) are excessive, identify any adverse trends, etc.

QC & STD acceptability: Determine if the calibration standard was rejected in the correct order as described in the SOP; observe to see if there is an overall trend of the % deviated from nominal when QC and calibration standards marginally meet the acceptance criteria, etc. Determine if there is any reinjection of failed QCs to produce passing QCs.

Calibration curves: Determine if all calibration curves generated in support of the study have similar slopes, intercepts and correlation coefficients.

Internal standard variability: Determine if the change of internal standard responses is indicative of potential loss in system sensitivity, instrument malfunction, sample extraction problem, recovery problem, matrix effect, etc.

System suitability: Determine if the system suitability meets the acceptance criteria. Determine if there are documented stability data to support the suitability of the system in completing the batch runs, etc. Determine if any runs were accepted with failed system suitability.

Interrupted run: Evaluate the data to ensure that there are no excessive time gaps within a batch run that exceeded the allowable limit specified by the SOP, etc. If excessive time gaps are observed, determine if the test system is requalified prior to restarting the run.

Potential interference: Evaluate the magnitude and frequency of potential interference observed in blanks, zero-standards, pre-dose samples, etc.

Poor chromatography: Determine the extent and impact of merging peaks, split peaks, spikes, traveling peaks, shift in retention time, shift in baseline, etc., on the precision and accuracy of the data generated.

Missing peak: Determine the frequency of missing peak events, as it may be indicative of potential systemic sample extraction problem, etc., that would discredit the entire batch run.

Consistency in data reporting: Detect any systemic practices that may be construed as selective reporting of data, inconsistent manual integration, rejection of calibration standards, etc.

Reported BLQ (below limit of quantitation) results: Determine the frequency and magnitude of samples with BLQ results, as this could be an indication of potential issue with analytical method and/or sample preparation error.

a. In recent years, bioanalytical sites were frequently inspected by the FDA without notifications or with a very short (1–3 days) notification. Thus, a laboratory should be ready for FDA inspection any time after its work is submitted to support an application. For a new drug application, it would be an advantage to find out from the sponsor the "User Fee" due date, as most FDA inspections are often conducted 1–3 months before this due date. This will also apply to generic drug applications, once the Generic Drugs User Fee Act is implemented.

b. To avoid unnecessary misunderstanding that would result in FDA 483 observations, QA staff, analysts, and their supervisors directly involved with the bioanalytical work should be present during the inspection to answer technical questions. In case such employees no longer work there, the laboratory should plan ahead of time and identify the staff that is most qualified to be present during the inspection.

c. If a laboratory is out of business or was moved to a new location, the location that maintains all source documents should be known, and all documentation should

TABLE 11.3 Typical Study-Specific FDA Mock Inspection Audit Plan

Study-specific documentation and records:

1. Organization, responsibility, and staffing of the Bioanalytical Laboratory operation, including QA Control in support of method validation and BE/BA/PK studies.
2. Brief overview of the control/process flow of the Bioanalytical Laboratory operation for these studies.
3. All pertinent method validation reports/protocols and their support hardcopy and electronic source data.
4. All pertinent bioanalytical methods used in support of the BE/BA/PK studies, as well as method validation and their change control records, as applicable.
5. All pertinent versions of bioanalytical reports issued in support of the BE/BA/PK studies and their supporting source data to the electronic source data level.
6. Records of any "prep runs" and/or "system equilibration runs."
7. All other related supporting records/information, such as study lead-in data, deviation reports, investigation reports, unreported/rejected/failed data, repeat analysis data, instrument/equipment calibration records, training records, pertinent internal/external communication records related to the analysis of the study, etc., in support of the studies reviewed.
8. Pertinent records for the preparation and qualification of quality control, calibration standards, and stability samples in support of method validation and BE/BA/PK studies.
9. Pertinent records to support the suitability of the reference standards, including blank control plasma / serum used.
10. Pertinent instrument calibration / qualification records / instrument use logs in support of method validation and BE/BA/PK studies.
11. All pertinent computer validation records/source data in support of the method validation and BE/BA/PK studies.
12. All pertinent training and qualification records for analysts and QA personnel engaged in the studies audited.
13. Pertinent records/logs for sample receipt, handling, storage, and control of specimens to support proper chain of custody related to the conduct of the studies, including calibration and QC samples.
14. Specimen freezer control/logs—inventory, calibration, qualification, and monitoring.
15. Recordkeeping/documentation/record retention practices.

General GLP Records, SOPs and Documentation

1. Review that there are past FDA or other regulatory agencies inspection reports and responses to assure no outstanding issues.
2. Index of SOPs related to the method validation and BE/BA/PK studies.
3. Procedure for receipt, handling, and control of specimens (freezer logs, chain-of-custody records).
4. Procedure for specimen freezer control—inventory, calibration, qualification, and monitoring.
5. Procedure to prepare and control calibration, quality control, and stability samples to support method validation and production batch runs.
6. Procedure for system suitability requirements.
7. Recordkeeping / documentation / record retention practices – study binders and form control.
8. Procedure to handle, document, and report results of repeat testing.
9. Procedure to perform testing of incurred samples.
10. Procedure to handle and document planned and unplanned deviations, and procedure to conduct proper investigation for "significant events".
11. Procedure for the transfer of electronic raw data.
12. Procedures for the review and approval of analytical raw data — laboratory check and QA check.
13. Procedure to handle PK outliers.
14. Procedures and requirements for prestudy, in-study validations, revalidation, cross-validation, and partial validation, including requirement for the validation/qualification of robotic equipments.
15. Procedures and acceptance criteria to accept calibration curves, QC samples, and batch runs.
16. Procedure to control and qualify reference standards.
17. Procedure to evaluate the stability of drug in biological fluid and stock solution.
18. Quality audit procedure for bioanalytical data and reports.
19. Instrument calibration/PM/Qualification program.
20. Computer/laboratory data system validation program.
21. Analyst training and qualification program.
22. Procedure to control the reprocessing / reintegration of samples and batch runs.
23. Instrument calibration / qualification program.
24. Reporting requirements for prestudy method validation.
25. Reporting requirements for the bioanalytical portion of the BE/BA/PK study report.

be readily retrieved for inspection. If instrumentation used in a bioanalytical method (e.g., LC-MS/MS systems) is no longer available, the laboratory should plan ahead in case additional experiments need to be conducted to resolve issues raised by the FDA during an inspection.

d. Conducting additional experiments or investigations ahead of time to resolve deficiencies identified during the mock inspection can avoid receiving Form FDA 483 citations during the FDA inspection. Any identified deficiencies that have an impact on data integrity and data accuracy must be evaluated and resolved before the FDA inspection, as reaction to these deficiencies after the FDA inspection may endanger or cause a delay in approval of an application.

REFERENCES

Chow F, Lum S, Ocampo A, Vogel P. Current challenges for FDA-Regulated bioanalytical laboratories for human (BE/BA) studies. Part II: recent FDA inspection trends for bioanalytical laboratories using LC/MS/MS methods and FDA inspection readiness preparation. Qual Assur J 2008;11:111–122.

Chow F, Ocampo A, Vogel P, Lum S, Tran N. Current challenges for FDA-regulated bioanalytical laboratories performing human BA/BE studies. Part III: selected discussion topics in bioanalytical LC/MS/MS method validation. Qual Assur J 2009;12(1): 22–30.

21 CFR Part 11, Electronic Records, Electronic Signatures. Available at http://www.accessdata.fda.gov/scripts/cdrh/cfdocs/cfCFR/CFRSearch.cfm?CFRPart=11. Accessed Jan 30, 2012.

21 CFR Part 50, Protection of Human Subjects. Available at http://www.accessdata.fda.gov/scripts/cdrh/cfdocs/cfCFR/CFRSearch.cfm?CFRPart=50. Accessed Jan 30, 2012.

21 CFR Part 54, Financial Disclosure by Clinical Investigators. Available at http://www.accessdata.fda.gov/scripts/cdrh/cfdocs/cfCFR/CFRSearch.cfm?CFRPart=54. Accessed Jan 30, 2012.

21 CFR Part 56, Institutional Review Boards. Available at http://www.accessdata.fda.gov/scripts/cdrh/cfdocs/cfCFR/CFRSearch.cfm?CFRPart=56. Accessed Jan 30, 2012.

21 CFR Part 58: Good Laboratory Practice for Nonclinical Laboratory Studies. Available at http://www.accessdata.fda.gov/scripts/cdrh/cfdocs/cfCFR/CFRSearch.cfm?CFRPart=58. Accessed Jan 30, 2012.

21 CFR Part 312, Investigational New Drug Application. Available at http://www.accessdata.fda.gov/scripts/cdrh/cfdocs/cfCFR/CFRSearch.cfm?CFRPart=312. Accessed Jan 30, 2012.

21 CFR Part 320: Bioavailability and Bioequivalence Requirements. Available at http://www.accessdata.fda.gov/scripts/cdrh/cfdocs/cfCFR/CFRSearch.cfm?CFRPart=320. Accessed Jan 30, 2012.

Desilva B, Smith W, Weiner R, et al. Recommendation for the bioanalytical method validation of ligand-binding assays to support pharmacokinetic assessments of macromolecules. Pharm Res 2003;20(11):1885–1900.

EMA Guideline on bioanalytical method validation. 2011. Available at http://www.ema.europa.eu/docs/en_GB/document_library/Scientific_guideline/2011/08/WC500109686.pdf. Accessed Jan 30, 2012.

Fast DM, Kelley M, Viswanathan CT, et al. Workshop report and follow-up—AAPS Workshop on current topics in GLP bioanalysis: assay reproducibility for incurred samples—implication of Crystal City recommendations. AAPS J 2009;11(2):238–241.

FDA Compliance Program Guidance Manual (CPGM): CPGM 7348.001 Bioresearch Monitoring—In-vivo Bioequivalence (October 1, 1999). Available at http://www.fda.gov/downloads/ICECI/EnforcementActions/BioresearchMonitoring/ucm133760.pdf. Accessed Jan 30, 2012.

FDA Guidance for Industry: Bioanalytical Method Validation. 2001. Available at http://www.fda.gov/downloads/Drugs/GuidanceComplianceRegulatoryInformation/Guidances/UCM070107.pdf. Accessed Jan 30, 2012.

FDA Guidance for Industry: Part 11, Electronic Records; Electronic Signatures-Scope and Application. 2003. Available at http://www.fda.gov/downloads/RegulatoryInformation/Guidances/ucm125125.pdf. Accessed Jan 30, 2012.

James CA, Breda M, Frigerio E. Bioanalytical method validation: a risk-based approach? J Pharm Biomed Anal 2004;35(4): 887–893.

King R, Bonfiglio R, Fernandez-Metzler C, Miller-Stein C, Olah T. Mechanistic investigation of ionization suppression in electrospray ionization. J Am Soc Mass Spec 2000;11(11): 942–950.

Kringle R, Hoffman D, Newton J, Burton R. Statistical methods for assessing stability of compounds in whole blood for clinical bioanalysis. Drug Info J 2001;35: 1261–1270.

Miller KJ, Bowsher RR, Celniker A, et al. Workshop on bioanalytical methods validation for macromolecules: Summary report. Pharm Res 2001;18(9):1373–1383.

Ocampo A, Lum S, Chow F. Current challenges for FDA-regulated bioanalytical laboratories for human (BE/BA) studies. Part 1: defining the appropriate compliance standards—application of the principles of FDA GLP and FDA GMP to bioanalytical laboratories. Qual Assu J 2007;11:3–15.

Savoie N, Garofolo F, van Amsterdam P, et al. 2009 White Paper on recent issues in regulated bioanalysis from the 3rd Calibration and Validation Group Workshop. Bioanalysis 2010;2(1): 53–68.

Shah VP, Midha KK, Findlay JWA, et al. Bioanalytical method validation: a revisit with a decade of progress. Pharm Res 2000;17(12):1551–1557.

Shah VP. The history of bioanalytical method validation and regulation: evolution of a guidance document on bioanalytical methods validation. AAPS J 2007;9(1):E43-E47.

Timm U, Wall M, Dell D. A new approach for dealing with the stability of drugs in biological fluids. J Pharm Sci 1985;74: 972–977.

Timmerman P. Regulated or GLP Bioanalysis? Proceedings of the EBF & EUFEPS Workshop, April 15–16, 2010; Brussels, Belgium.

Viswanathan CT, Cook CE, McDowall RD, et al. Analytical method validation: bioavailability, bioequivalency, and pharmacokinetics studies. Int J Pharmaceutics 1992;82(1–2):1–7.

Viswanathan CT, Bansal S, Booth B, et al. Workshop/Conference Report—Quantitative bioanalytical methods validation and implementation: best practices for chromatographic and ligand binding assays. AAPS J 2007;9(1):E30–E42.

Weng N, Halls TDJ. Systematic troubleshooting for LC/MS/MS. Pharm Tech 2002;102–120.

PART III

BEST PRACTICE IN LC-MS BIOANALYSIS

12

ASSESSMENT OF WHOLE BLOOD STABILITY AND BLOOD/PLASMA DISTRIBUTION OF DRUGS

IAIN LOVE, GRAEME T. SMITH, AND HOWARD M. HILL

12.1 ASSESSMENT OF WHOLE BLOOD STABILITY OF DRUGS

To ensure the reliable determination of drug concentrations in biological matrices and the rational interpretation of related toxicokinetic (TK) and pharmacokinetic (PK) parameters, it is critical to define the stability profile of a drug—or drug metabolites—of interest in the intended matrix.

Stability in any matrix, including working solutions, is of utmost importance in understanding the results generated from bioanalysis. Both the US Food and Drug Administration (FDA) Guidance for Industry on Bioanalytical Method Validation (FDA Guidance, 2001) and supporting White Paper (Viswanathan et al., 2007) and the European Medicines Agency (EMA) Guideline on bioanalytical method validation (EMA Guideline, 2011) have extensive sections discussing stability.

The FDA guidance document states that "Stability procedures should evaluate the stability of the analytes during sample collection and handling." This has been further emphasized by recent guidelines issued by the EMA (EMA Guideline, 2011) stating that "Sufficient attention should be paid to the stability of the analyte in the sampled matrix directly after blood sampling." In addition, the FDA guidance goes on to provide further information emphasizing the fact that the storage container may in some way impact on the stability profile of a drug: "Drug stability in a biological fluid is a function of the storage conditions, the chemical properties of the particular drug, the matrix and the container system." The stability of an analyte in a particular matrix and particular container system is relevant only to that matrix and container system, and should not be extrapolated to other scenarios.

In recent discussions within the bioanalytical community, it has been suggested that in the instances where plasma or serum is the primary matrix for analytical purposes, a stability determination in whole blood should be carried out in order to satisfy the regulatory requirements. In addition, as the relationship between a drug and the various components of blood is an important factor in the interpretation of PK profiles, it is clear that an understanding of a drugs stability in blood will be a complementary factor in that interpretation. A recent survey of its members conducted by the European Bioanalysis Forum (Freisleben et al., 2011) found that blood as a matrix for stability testing has attracted significant attention but has not been adopted as a matter of consensus. The reasons for this are numerous. It can be difficult for a bioanalytical laboratory to source blood of the required quality to perform an adequate assessment of stability if they do not have links to an animal facility or clinic. It is thought in some quarters that stability assessments carried out in plasma are an adequate indication of blood stability. Indeed this conclusion is borne out in part with published incidences of plasma and blood stability mismatches mainly limited to certain classes of compounds such as hydroxamic acids (Sugihara et al., 2000) and *N*-oxides (Kitamura et al., 1998). However, the main barrier to widespread adoption of blood stability assessments is the lack of an industry standard experimental design.

The scientific rationale for the assessment of stability of a drug in blood is to provide an animal facility or clinic with a set of working instructions to ensure that the drug is stable

Handbook of LC-MS Bioanalysis: Best Practices, Experimental Protocols, and Regulations, First Edition. Edited by Wenkui Li, Jie Zhang, and Francis L.S. Tse.

throughout the process of sampling. The stability evaluation should cover all stages of collection of the blood sample and generation of the plasma or serum sample. This includes physically removing a sample of blood from a test subject and, whenever necessary, the collection of plasma or serum fraction prior to frozen storage or transfer. Specific concerns associated with the sampling and storage of biological fluids is to be addressed in a separate chapter of this book.

The evaluation of the stability of a drug in blood is a regulatory requirement and this assessment should be carried out in the intended species matrix in which the method is validated. However, before undertaking any experimental investigations, it is advisable to take note of any existing literature surrounding the blood stability of the drug of interest in addition to any chemical stability identified during the Chemistry, Manufacturing and Controls (CMC) phase of development and the stability of any structurally related compounds, including impurities if any are known. The information gathered can give the bioanalytical laboratory vital information to focus method development experiments. It is of note, however, that a novel compound in development is unlikely to have any published stability data in the scientific literature, although in-house CMC data can be of great value.

In order to eliminate some of the technical problems associated with working with whole blood, a logical alternative can be the use of dried blood spots (DBS) as the matrix of choice. This technique is discussed elsewhere in this book. However, despite the plethora of recent interest, the use of DBS is not universally accepted in regulatory bioanalysis and some of its shortcomings have been highlighted recently by Kissinger (2011).

12.1.1 Factors Affecting the Stability of Drugs in Blood

A number of environmental and chemical factors could affect the stability of a drug in blood. For example, natural pH variance of the sampled matrix or introduced pH effects through the use of anticoagulants can impact on a drugs stability profile. Photo-oxidation can also impact on the stability of a drug. The oxygen-scavenging properties of the hemoglobin present in blood can have a stabilizing effect making blood less susceptible to photo-oxidative stability issues compared to analogous plasma samples.

To a degree these chemical and environmental factors can be easily controlled. However, biological factors can contribute significantly to instability of a drug in matrix. In the majority of cases, this instability is due to the action of plasma enzymes, although it is known that some enzymes present in red blood cells can also impact on a drug's stability. Examples of such factors include enzyme activated degradation, hydrolysis of conjugated metabolites and intermolecular conversions such as lactone/carboxylic acid conversion (Briscoe and Hage, 2009). It should also be noted that any stability assessment can be affected by the design of the evaluation experiments, for example, hydrogen/deuterium exchange of a labeled internal standard.

When a drug is found to be unstable in blood, a number of approaches can be investigated for developing a stabilization protocol. A common solution is to adopt the use of enzyme inhibitors during sample collection. For example, sodium fluoride (NaF) is a broad-spectrum esterase inhibitor, and blood tubes containing NaF are widely available. The concentration of NaF in these tubes may not be sufficient to inhibit all of the esterases present in plasma and, therefore, the use of more potent inhibitors need to be considered. A full discussion of stabilization protocols can be found elsewhere in this book. However, when it is not feasible or possible to resolve instability issues, some simple, pragmatic steps can be taken to minimize or mitigate the impact of the instability. Examples of this include immediate addition of an inactivating organic solvent such as acetonitrile or to place the blood samples on ice immediately after collection in order to arrest or slow the rate of degradation.

12.1.2 Considerations of Experimental Design for Whole Blood Stability Assessment of Drugs

Whole blood used for *in vitro* blood stability assessments should be as fresh as possible. There is a debate as to what constitutes fresh blood as a number of diagnostically significant concentration changes occur when blood reaches >4 h old as the metabolic integrity of the red cells degrade significantly. For certain parameters, such as glucose determination, concentration changes occur very shortly after sampling. It is often thought that storing blood at 4°C reduces or arrests any *ex vivo* changes but this is not the case (Richterich and Colombo, 1981). For bioanalytical applications; however, it is common for blood to be stored under refrigerated conditions for a number of days prior to use. Regardless of storage duration, it should be noted that the blood sample should be handled carefully and be homogenous prior to use. This can be achieved via gentle inversion of the blood sample or gentle mixing on an automatic rolling machine. Under no circumstances should blood samples used for stability determinations be subjected to vigorous shaking.

In order to best mimic a blood sample at the point of sampling, the fresh control blood used for any stability assessments must contain the same anticoagulant as used for the incurred samples. Whenever possible, a measure of hematocrit or other appropriate blood parameters should be employed as a qualitative check to confirm a blood source is fit for the intended purpose. In this case predefined limits of acceptance should be assigned before carrying out the assessment.

The duration of the assessment should also reflect the procedures to be employed at sampling. For example, when it is likely that a facility may require 2 h between sampling and further processing, the stability over this period of time should be evaluated. It is valid to include multiple time points covering the duration of the experiment.

In order to prepare a blood sample suitable for evaluating the stability of the drug, a pool of control blood must be fortified with the target analyte in working solutions. However, these solutions may alter the composition of the blood and, therefore, any volumes added should be kept to a minimum. Consideration may also be given to evaporating known volumes of fortifying solution before control blood is added to the sample tube. When using this approach it should be noted that potential poor solubility of dried extracts may be encountered. For example, it is generally accepted that methodologies involving an evaporation step should be avoided for peptide assays owing to solubility issues associated with the dried extract. The use of organic modifiers in fortification solutions should be kept to a minimum as blood cells will be lysed immediately upon addition of an organic solvent and the integrity of the blood sample will be compromised. It is valid to prepare the fortifying solution in plasma or protein solution, for example, albumin solution. The control blood may be fortified at ambient conditions but in order to be more reflective of the sampling process, this procedure should be performed at physiological temperature. This holds true even in the event that the stability experiment is to be carried out at ambient temperature as blood drawn from a test subject is initially at elevated temperature before processing occurs. The effect of the dosing vehicle may also need to be considered in, for example, intravenous dosing programs as a matrix effect or specificity concern may make the interpretation of the acquired stability data difficult.

12.1.3 The Choice of Analytical Matrix for Blood Stability Assessment

Bioanalytical methods that have been validated for the analysis of a drug in plasma or serum are not usually implemented to measure the drug directly in blood and, as a result, this has led to a lack of consensus about the best experimental approach to blood assessment. Broadly speaking there are two main approaches for determining the stability of a drug in blood. The first is to use the plasma derived from blood and the second is simply to use blood as the analytical matrix. Each approach has associated advantages and disadvantages.

12.1.3.1 Derived Plasma as the Analytical Matrix for Blood Stability Assessment

Considerations for Nominal Stability. Stability assessments carried out in support of drug development programs are driven by regulations and are typically carried out in plasma or serum. The analytical matrix is fortified at known low and high concentrations and processed with calibration standards and quality control (QC) samples. The results of the analysis confirm the stability of the analyte of interest if the defined nominal acceptance criteria are met. However, when using derived plasma as the analytical matrix the fortified blood samples used to conduct a stability experiment are processed in order to provide plasma for subsequent analysis. Owing to this, it is not possible to attribute a nominal value for the analyzed plasma samples. Therefore, the most commonly used approach to stability assessment in bioanalytical validation of plasma or serum assays is not appropriate for the stability assessment of a drug in whole blood. There are, however, a number of advantages associated with using derived plasma as the analytical matrix for assessment of whole blood stability. As the majority of bioanalytical methods are characterized in plasma or serum, it is reasonable to assume that a reliable plasma assay would be available or in the process of qualification/validation at the time a stability assessment in blood is to be carried out. It is less likely that a reliable blood assay would be available. Owing to this, the bioanalytical laboratory has a choice of whether to use existing assays to analyze the derived plasma in a blood stability assessment or to consider potential development of new assay conditions for the measurement of the drug in blood. In most cases, it is deemed appropriate to use existing methodologies. In addition, the use of a single analytical matrix in a bioanalytical validation intended to support regulated sample analysis makes the reporting of the study more straightforward compared to a study where results are generated from both plasma and blood.

Considerations for Experiments. The main advantages of using derived plasma as the analytical matrix for blood stability assessments have been discussed in the previous section. The derived plasma approach is dependent on the measurement of an initial value of drug in plasma (time zero) and stability is defined by comparing subsequent measurements to the time zero value. In this experiment, blood has to be initially fortified with the drug. The analyte has to be allowed to equilibrate between the various blood components before the time zero value can be measured. Therefore, the nominal concentration of drug in plasma is unknown. Although this appears to be contrary to the nominal stability approach described in the industry guidance/guideline documents, it is widely accepted as the approach maintains an adequate level of scientific validity.

It is common that multiple time points (e.g., 0, 1, 2, and 3 h) are defined in stability assessment using derived plasma. The test samples are stored under ambient conditions representative of the sampling location. The plasma fraction of the time zero blood samples is harvested after blood/plasma partitioning is completed and the plasma fraction of the stored blood samples are then harvested at known time intervals. The plasma is then stored, for example, at −20°C (this assumes that freeze-thaw and frozen stability has already been established), and analysis is performed on one analytical occasion to avoid inter batch variance. The derived plasma stability samples are commonly analyzed in replicate (e.g., $n = 6$) in a standard analytical batch with calibration standards and performance QCs. The acceptance criterion applied to each time point analyzed is the familiar $\pm 15\%$ of the concentration of drug measured at time zero. An alternative approach is to simply compare the detector response (peak area or peak area ratio where internal standard is used) measured at each time point to the detector response at time zero. The acceptance criterion is $\pm 15\%$ of the detector response of the time zero sample. The advantage of this approach is that it precludes the use of a calibration curve.

In undertaking such an experiment, care should be afforded to the preparation and handling of the plasma samples. At each time point in the experiment, centrifugation of the stability samples is required. The centrifugation processes imparts a degree of stress on the red blood cells present in the blood sample. With this stress the integrity of the red cells may become compromised and result in the liberation of biological components and target analyte from the blood cell fraction. This effect can be mitigated by applying the same centrifugation protocol to all sample time points within the experimental design.

The main disadvantage of using derived plasma for the assessment of blood stability lies principally in the definition of a time zero sample. When a blood sample is fortified with drug it will take an unknown period of time for equilibrium distribution to be reached for the drug among the various components of blood. The kinetics of such phenomena are commonly rapid but examples have been reported where the distribution may take longer (Hinderling, 1997). Moreover, the distribution can be affected by environmental factors, such as temperature variances and many biological factors (Bieri et al., 1977). Efforts to define the time-to-equilibrium are outside the scope of routine bioanalytical studies in support of drug development programs and, as a result, time zero must be arbitrarily assigned. In the event that time zero is assigned before equilibrium is reached, a false and short stability period may be defined. The phenomenon of equilibrium distribution is not a major concern when whole blood is chosen as the analytical matrix.

12.1.3.2 Blood as the Analytical Matrix for Blood Stability Assessment In some cases, it is considered more appropriate to assess drug stability in whole blood rather than derived plasma. The advantages of this are two-fold. Firstly, it affords the bioanalytical scientist a degree of flexibility in experimental design as a nominal stability assessment analogous to that described in the regulatory authorities bioanalytical guideline/guidance documents can be carried out.

The nominal experimental design using whole blood is considered to be the gold-standard approach; however, this may cause complications as a method intended for use in plasma or serum may not be appropriate for the analysis of blood samples. In this case, a simple protein precipitation methodology should be employed outside of any intended validation procedure solely for the determination of blood stability (Freisleben et al., 2011). Using this approach, stability is evaluated through the use of a qualified blood assay in conjunction with calibration standards and QCs prepared in whole blood. The blood samples used to assess stability are quantified by interpolation using a calibration curve with familiar defined acceptance criteria. However, it may be more straightforward to simply compare the response ratio at time zero with the values in samples obtained after the appropriate storage period. If these are within $\pm 15\%$ of the time zero value, then stability is confirmed.

A second advantage associated with the use of blood as the analytical matrix is that the time taken to equilibrium partitioning does not have a bearing on the results of the stability assessment. As no processing of the blood sample is carried out before extraction and analysis of the drug (e.g., no centrifugation to produce plasma), the association between blood cells or blood plasma is of no consequence.

A further approach, although in many cases impractical, is to carry out blood stability assessments using incurred blood samples. This approach removes the requirement to assume control blood fortified with analyte is analogous to the blood of the sampled test subject and is not dependent on time-to-equilibrium assumptions. Indeed it is the most robust means of measuring true blood stability but it may not be acceptable to the regulators because the initial concentration of the drug in the blood is unknown.

12.1.4 Stability of Metabolites

It should be taken into consideration that the metabolism of a drug may result in the formation of unstable metabolites that can convert back to the parent drug *ex vivo* (de Loor et al., 2008; Silvestro et al., 2011). This raises concerns that a blood stability assessment using control blood fortified with drug may not be truly reflective of an incurred blood sample. As such, it is conceivable that an incorrect sampling protocol could be provided based upon data from such a stability assessment. Owing to this, consideration should be given to the assessment of blood stability in incurred blood samples where metabolites with liabilities are known to exist. Alternatively, when a metabolite reference standard is available, it may be appropriate to carry out a parallel stability assessment to determine the influence and impact of any metabolite back-conversion.

12.1.5 Statistical Methods for Blood Stability Assessment

The acceptance criteria applied to bioanalytical stability assessments is the familiar 4-6-15 rule. That is, in a set of six replicate stability samples, at least four sample results must fall within 15% of the nominal or time zero values. Kringle et al. (2001) reported a statistical model carried out to determine the likelihood of falsely concluding stability when using the 4-6-15 rule as the means of decision making. This statistical assessment was analogous to that used in bioequivalence determinations and quantified the probability of falsely concluding stability by modeling the true percentage degradation and the precision between replicates. Kringle indicated there was a 36% risk of falsely concluding stability where a true degradation of 15% existed with an assay precision of 8%. The alternative proposal involved the use of nine replicates per concentration level analyzed at 0, 3, 6, and 24 h. The observed results assumed a number of statistical truths and were investigated by linear regression (trend analysis). Stability assessments were then carried out at each time point using a 90% confidence interval. Kringle concluded that the recommended method for assessing blood stability is to apply an equivalence test using the regression approach (e.g., a trend analysis approach comparing each stability time point with the time zero value). The main disadvantage of this approach lies in the practicality of carrying out the assessment using high numbers of replicates at each concentration level. Moreover, there is increased difficulty associated with the data interpretation and a requirement for a statistics software package.

In the event that a drug is found to be unstable in blood, an impact assessment of the extent of instability must be made. It is possible to conceive a situation where a target analyte may be stable for only a short period of time in the sampled matrix but display acceptable stability characteristics in the primary matrix. In this case, it may be necessary to process the sampled matrix within a short period of time after collection in order to ensure the integrity of the bioanalytical data. However, in many instances this may not be adequate and a modified sampling protocol or stabilizer may be required. A full discussion on sampling is given in elsewhere in this book.

12.2 BLOOD/PLASMA DISTRIBUTION OF DRUGS

When undertaking bioanalytical support of TK/PK or pharmacodynamic (PD) studies, it is important to know how the drug of interest interacts with the various components of circulating blood. Through an understanding of these interactions it is possible to interpret TK/PK parameters with reference to free or total concentrations of drug and make a rational choice of primary matrix for bioanalytical investigations.

The significance of understanding this behavior may be commonly overlooked but should be carefully considered, given the current interest in employing blood (e.g., DBS) as the primary matrix. The decision about which matrix should be chosen for the measurement of drug concentrations is important but, in most cases, is not critical. The important factors to consider are the constancy of the fraction of drug unbound in plasma and, in blood, the constancy of the hematocrit and binding to the red blood cells. Often these parameters are fairly constant but if they change then the choice of matrix becomes more important from a PK perspective (Emmons and Rowland, 2010)

The blood-to-plasma ratio of an analyte of interest is the concentration ratio of drug found in the blood and blood plasma components of whole blood. The extent of binding to blood components directly influences how a drug distributes into tissues. As the bound fraction of circulating drug is not able to cross membranes such as the blood–brain barrier or endothelial cardiac barrier, it is usually only the free, unbound fraction of drug that is available to exert a pharmacological effect (Musteata, 2011). Therefore, drugs that are highly bound are believed to have low bioavailability and distribution. It is acceptable to consider that blood is comprised of blood cells suspended in blood plasma. When a drug is administered to a subject, binding interactions between the drug and both the plasma proteins and erythrocytes occur. This lowers the concentration of free drug available to exert a pharmacological effect. However, this is often not considered and the total concentration data obtained from routine bioanalysis is adopted as a surrogate to gather PK information in lieu of determining the more appropriate free drug concentrations. An example of a thorough *in vitro* investigation of the binding relationships between the sedative profopol and various blood components was published by Mazoit and Sammi (1999) and is summarized in Figure 12.1. They found that profopol was 48% bound to plasma proteins, 16% bound to red blood cell membranes, and 35% bound to intracellular components. 1.4% of profopol was found to be unbound.

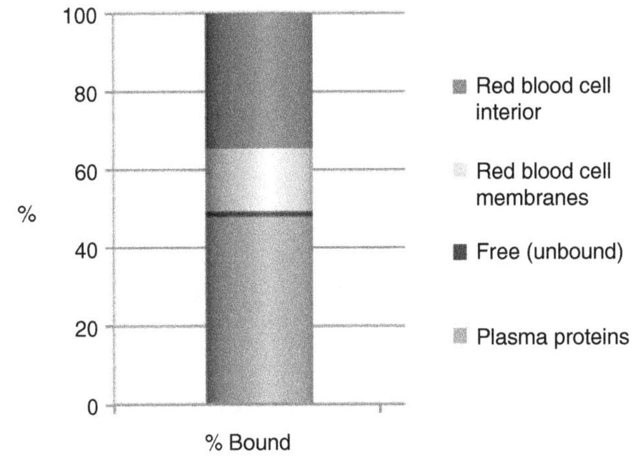

FIGURE 12.1 Distribution of propofol in blood.

12.2.1 Mechanisms of Blood/Plasma Distribution/Partitioning of Drugs

The extent of partitioning between blood and plasma is largely dependent on the lipophilicity of the drug, but the mechanism of binding to erythrocytes can be complicated by reversible binding to hemoglobin, proteins or cell membranes. The well cited review paper by Hinderling (1997) aptly summarizes a number of binding mechanisms to red blood cell components and describes the associated kinetics, see Table 12.1.

The kinetics of binding interactions can have a marked effect on the blood-to-plasma ratio defined earlier. The time taken to reach a state of equilibrium partitioning is usually rapid; for example, cyclosporine and related compounds reach equilibrium partitioning within minutes (Kawai et al., 1994). However, for some drugs this is not the case. Chlorthalidone is an example of a drug that takes hours to reach equilibrium partitioning (Fleuren and van Rossum, 1977). There are also numerous published incidences where partitioning is concentration dependent (Rose et al., 1981; Lin et al., 1991; Wong et al., 1994). This manifests as nonlinear kinetics with reference to escalating dose and occurs as the binding capacity of a given mechanism is exceeded. This is referred to as saturation. To further complicate matters, it is known that metabolites can compete and displace bound drug (Wong et al., 1996) manifesting in a dose dependent decrease in blood AUC owing to rapid plasma clearance of liberated parent drug. In addition, it is possible to observe similar effects through displacement of a weakly bound drug by endogenous components (Querol-Ferrer et al., 1991) or exogenous compounds such as coadministered drugs.

12.2.2 Measurement of the Blood-to-Plasma Partition Ratio

Determination of the blood-to-plasma ratio of a drug is traditionally carried out by incubating fortified blood at 37°C for a defined period of time. A blood aliquot is then processed to provide plasma and a further blood sample

TABLE 12.1 Drug Binding Sites in Red Blood Cells

Compound	Binding Site	Reference
Chlorpromazine		Bickel (1975)
Codeine		Mohammed et al. (1993)
Imipramine	Plasma membrane[a]	Bickel (1975)
Mefloquine		San George et al. (1984)
Pyrimethamine		Rudy and Poynor (1990)
Draflazine	Nucleoside transporter	Snoek et al. (1996)
Acetazolamide		Wallace and Riegelman (1977)
Chlorthalidone		Collste et al. (1976)
Dorzolamide	Carbonic anhydrase	Biollaz (1995)
Methazolamine		Bayne et al. (1981)
MK-927		Lin et al. (1992)
Cyclosporine A	Cyclophin	Agarwal et al. (1986)
Tacrolimus	Tacrolimus binding protein	Hooks (1994)
Aminophenzone		Hilzenbecher (1972)
Barbiturates		Hilzenbecher (1972)
Chlordiazepoxide		Hilzenbecher (1972)
Digoxin and derivatives		Hinderling (1984)
Imipramine and derivatives		Hilzenbecher (1972)
Mefloquine		San George et al. (1984)
Nitrofurantoin		Hilzenbecher (1972)
Oxyphenbutazone		Hilzenbecher (1972)
Phenothiazines	Hemoglobin	Hilzenbecher (1972)
Phenylbutazone		Hilzenbecher (1972)
Phenytoin		Hilzenbecher (1972)
Proquazone		Ross and Hinderling (1981)
Pyrimaethamine		Rudy and Poynor (1990)
Salicylic acid and congeners		Hilzenbecher (1972)
Sulfinpyrazone		Hilzenbecher (1972)
Sulfonamides		Berneis and Boguth (1976)

[a]Plasma membrane equivalent to cell membrane.

Reproduced from (Hinderling, 1997) with permission of the American Society for Pharmacology and Experimental Therapeutics.

from the fortified pool is frozen and thawed three times to facilitate lysis of the blood cells. Comparison of concentration results obtained from the lysed blood samples and plasma samples yield the blood-to-plasma ratio (Mullersman and Derendorf, 1986). Recently, Yu et al. have published a more facile methodology that requires measurement of concentrations in plasma only and uses a control sample negating the need for calibration standards and QCs. The procedure involves fortifying blood and plasma with equivalent concentrations of drug. Both matrices are incubated for a defined period of time, after which the plasma is harvested from the blood sample. The partitioning coefficient is calculated from Yu et al. (2005):

Equation 1

$$K_{\mathrm{RBC/PL}} = \frac{1}{H} x \left(\frac{I_{\mathrm{PL}}^{\mathrm{REF}}}{I_{\mathrm{PL}}} - 1 \right) + 1$$

$K_{\mathrm{RBC/PL}}$ = Partitioning coefficient
H = Hematocrit
$I^{\mathrm{REF}}{}_{\mathrm{PL}}$ = Instrument response from the plasma control sample
I_{PL} = Instrument response from the sample harvested from whole blood

A high $K_{\mathrm{RBC/PL}}$ indicates the potential for accumulation of the drug in red blood cells and the risk of subsequent hematotoxicity. This has consequences for the development of drugs intended for prolonged use. It is therefore appropriate to carry out this simple assessment in early phase drug discovery. The experimental design and interpretation of the subsequent data achieved may also be affected by the concentration(s) investigated. At an elevated concentration, it is possible for the binding capacity of blood to be exceeded and for the observed $K_{\mathrm{RBC/PL}}$ to be suppressed.

12.2.3 Factors to be Considered in Blood/Plasma Partition Assessment of Drugs

12.2.3.1 Hematocrit A full discussion of hematocrit is outside the scope of the chapter suffice to say hematocrit represents the fraction of the corpuscular elements (erythrocytes and leucocytes) in whole blood, expressed as a percentage (Richterich and Colombo, 1981). The hematocrit is traditionally measured by centrifuging a capillary of whole blood until the cells have packed to a minimum volume, the packed cell volume (PCV). The packed cell height is expressed as a percentage of the total column height. When using a centrifugation technique such as PCV, it is found that small amounts of interstitial blood plasma can affect the value obtained and as such, modern measurements of hematocrit are typically performed by automated analyzers (Stott et al., 1995). Automated analyzers use a number of mechanisms to determine hematocrit, often based on blood conductivity or red blood cell counting.

The hematocrit is an important factor in the interpretation of PK data. As the hematocrit value increases or decreases, the amount of red blood cells available to interact with any therapeutic agent is proportionally altered. This subsequently affects the partitioning coefficient in any given sample and the concentration data returned from any bioanalytical analysis. This effect is of key importance when assessing the PKs of drugs with a high affinity for red blood cells. For example, the immunosuppressive tacrolimus is a drug that exhibits a higher binding affinity for red blood cells than plasma proteins. In blood samples of elevated hematocrit, more tacrolimus is bound by the increased numbers of red blood cells (Chow et al., 1997). Accordingly, there is less free drug in the plasma fraction resulting in increased plasma clearance and lower measured plasma concentrations for the drug. Therefore, the hematocrit measurement of a parent blood sample can be important for the assessment of the plasma PK results.

12.2.3.2 Hemolysis Hemolysis is the lysis of red blood cell membranes, causing the release of hemoglobin and other cellular components into the blood plasma. Hemolysis of a plasma sample can be detected visually as a pink to red coloring of the sample if the concentration of plasma hemoglobin is $> 200\ \mu$g/ml. An example of where this visual detection limit may be raised is the case of icteric plasma drawn from jaundiced subjects. The increased levels of plasma bilirubin make the color change associated with hemolyzed plasma more difficult to detect (Richterich and Colombo 1981).

Hemolysis can arise from either of two sources: *in vivo* hemolysis or *in vitro* hemolysis. *In vivo* hemolysis may be due to pathological conditions such as hemolytic anemia or bacterial action. However, in most cases *in vitro* sources are the biggest contributing factor to hemolysis. Typical factors are prolonged use of a tourniquet, inappropriate aspiration of the blood sample (e.g., drawing blood too quickly or using too small a needle), contamination with nonphysiological fluid or excessive heating or cooling of the blood sample.

In a hemolyzed sample, the presence of lysed blood cells contaminates blood plasma. In the event that a drug is highly bound to red blood cells, analysis of a hemolyzed sample may give an artificially high drug concentration. Moreover, where LC-MS is the detection technique used in bioanalysis, it is possible that hemolysis can result in an increased plasma concentration of the biological components that affect the ionization efficiency of the target analyte. In particular, phospholipids are a class of compounds that are recognized as a cause of matrix effect in LC-MS bioanalysis (Ismaiel et al., 2010). Phospholipids are a major substituent of cell membranes and, as such, a hemolyzed plasma sample will contain a higher concentration of these components. It is now a regulatory requirement (EMA Guideline, 2011) to carry out an evaluation of matrix effects in hemolyzed plasma during the assay development/validation phase of any bioanalytical program. In the event that a drug is highly plasma bound, it is possible that the release of intracellular fluid can result in a small dilution of the blood plasma, potentially resulting in lower measured concentrations than expected. Some bioanalytical service providers take the view that hemolyzed samples should not be analyzed owing to potential unreliability of the data acquired. As a minimum it is reasonable to expect that the bioanalytical report should be appended to identify specific data points that have been generated from a hemolyzed plasma sample. A comprehensive discussion of the analysis of hemolyzed samples will be addressed in another chapter of this book.

*12.2.3.3 **Plasma Protein Binding*** In addition to the determination of blood/plasma distribution, the parameter of protein binding is of key importance in drug discovery. If a drug binds to the proteins in the blood plasma fraction, the amount of free, unbound drug available for interaction will be reduced. This binding process is saturable and reversible. As it is usually the unbound fraction of drug that is pharmacologically active, the determination of protein binding is a key parameter of interest to the interpretation of bioanalytical concentration data from a PK and PD perspective. Moreover, within the context of drug candidate design and optimization, low plasma protein binding is often desirable and investigated very early on in the drug discovery process.

High levels of protein binding can have implications with regards to drug efficacy. Small changes in the extent of protein binding can have significant effects on the concentration of free drug available to exert pharmacological effects. Changes in the level of protein binding of a drug can be caused by conditions that alter plasma protein levels or by displacement by endogenous or exogenous compounds (Lindup and Orme, 1981). One well studied example is the antiepileptic drug phenytoin. Phenytoin has a narrow therapeutic range and low bioavailability (Neuvonen et al., 1977) in part due to high binding of serum albumin. When co-administered with the anticonvulsant and mood stabilizer valproic acid, it is found that valproic acid displaces phenytoin from the albumin binding sites and raises the free phenytoin fraction significantly (Tsanaclis et al., 1984). Plasma protein binding is usually expressed as a percentage of total plasma concentration and is considered a function of the physicochemical properties of a drug. The most abundant protein in plasma, albumin, accounts for the largest proportion of available binding sites followed by glycoprotein, lipoprotein, and globulins. The *in vitro* measurement of protein binding is generally carried out using one of three experimental procedures: ultrafiltration, ultracentrifugation, or equilibrium dialysis. There are also a number of other less commonly used techniques for determining the binding and subsequent free fraction of drug such as affinity chromatography (Hage, 2002) or solid phase micro-extraction (Musteata et al., 2006). A detailed discussion of the measurement of free drug concentrations and the application of the techniques discussed earlier will be presented in another chapter of this book.

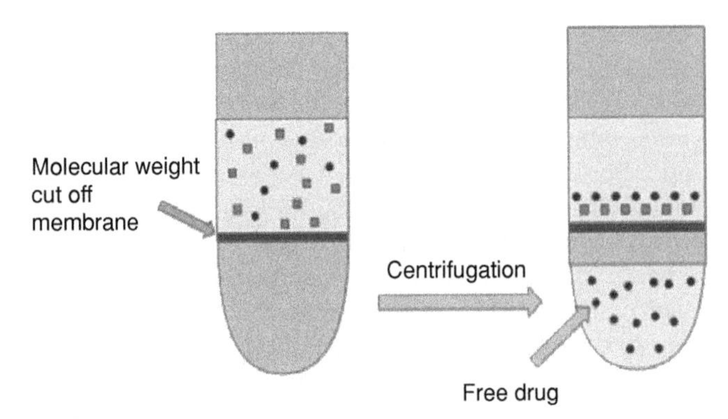

FIGURE 12.2 Illustration of ultrafiltration with molecular weight cutoff membrane.

Ultrafiltration. Ultrafiltration is used widely in the determination of plasma protein binding interactions owing to its ease of application. The technique relies on hydrostatic pressure to drive the unbound fraction of drug through a molecular weight cutoff membrane. Plasma fortified at known concentrations reflective of expected therapeutic dose is transferred to the ultrafiltration apparatus and centrifuged. Compounds of a molecular weight below the membrane cutoff perfuse easily through the membrane and into the ultrafiltrate eluate (Figure 12.2). Compounds of higher molecular weight than the membrane cutoff such as proteins and protein bound drug cannot perfuse into the eluate. An aliquot of the ultrafiltrate is then analyzed with reference to the fortified concentration. Ultrafiltration is particularly susceptible to nonspecific binding interactions with the ultrafiltration apparatus (Dow, 2006) and as such the nonspecific binding should be investigated as part of any ultrafiltration investigation.

Ultracentrifugation. Plasma protein binding may be evaluated by centrifuging a plasma sample at high g-force for a prolonged period of time. After centrifugation, the free drug concentration is measured in the supernatant with reference to the fortified concentration (drug bound to the plasma proteins is trapped within the pellet formed during centrifugation). This technique is not widely used for evaluating plasma protein binding because of the high capital cost of the equipment, the low sample throughput and the small volume of supernatant that is produced for analysis.

Equilibrium Dialysis. Equilibrium dialysis is seen as the gold standard approach for determination of plasma protein binding and is commonly employed in research laboratories. The equilibrium dialysis technique is conducted using

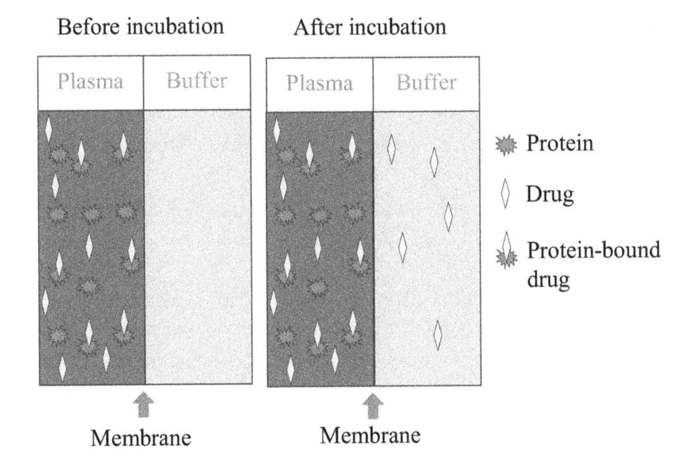

FIGURE 12.3 Illustration of equilibrium dialysis with two chambers or cells.

two chambers or cells separated by a membrane permeable to free drug but not to protein or protein bound drug (Figure 12.3). Fortified plasma is added to one chamber and buffer to the other. The dialysis apparatus is then incubated at 37°C under agitation and the unbound test article allowed to reach equilibrium distribution between the two compartments. Measurement of the test article concentration in each compartment facilitates the calculation of protein binding. Equilibrium dialysis can also be used to determine the extent of blood binding using an analogous procedure. Where blood binding is investigated by equilibrium dialysis it is common to use blood diluted with physiological saline to mitigate certain handling problems associated with whole blood.

12.3 SUMMARY

To ensure reliable determination of drug concentrations in plasma or serum, it is necessary to perform a number of stability assessments. Inclusive of this, and independent of the use of plasma or serum as the primary matrix, is the stability of the drug of interest in blood. As blood is drawn from a test subject before further processing for plasma or serum, it is of key importance to define a sampling protocol to ensure the stability of drug in blood throughout the sampling period. In order to do this, one important decision that must be made is whether to work directly with whole blood or with plasma derived from the blood sample for the stability assessment of drugs in blood. In theory, evaluating stability in blood or DBS is more straightforward than using plasma as there are no partitioning effects to consider. However, the technical aspects of developing methods to measure drugs in blood together with the persisting (at the time of writing) regulatory question marks about the DBS approach for supporting bioanlytical studies means that in most circumstances plasma is probably still the matrix of choice.

It is important to understand a drug's interactions within and between the constituent components of blood. The partitioning of a drug between blood and blood plasma as well as the nature of the binding mechanisms influences the concentration of drug available to distribute into bodily tissues. This affects the concentration of free drug available to exert a pharmacological effect or for metabolism or renal excretion. A number of simple *in vitro* techniques described in the current chapter can help probe these relationships and these investigations should be carried out as part of a carefully planned drug development program.

ACKNOWLEDGMENT

The authors acknowledge and thank Patricia Naylor for her secretarial assistance in producing this chapter.

REFERENCES

Bieri JG, Evarts RP, Thorpe S. Factors affecting the exchange of tocopherol between red blood cells and plasma. The Am J Clin Nutrition 1977;30(5):686–690.

Briscoe CJ, Hage DS. Factors affecting the stability of drugs and drug metabolites in biological matrices. Bioanalysis 2009;1(1):205–220.

Chow FS, Piekoszewski W, Jusko WJ. Effect of hematocrit and albumin concentration on hepatic clearance of tacrolimus (FK506) during rabbit liver perfusion. Drug Metab Dispos 1997;25(5):610–616.

de Loor H, Naesens M, Verbeke K, Vanrenterghem Y, Kuypers DR. Stability of mycophenolic acid and glucuronide metabolites in human plasma and the impact of deproteinization methodology. Clin Chim Acta 2008;389(1-2):87–92.

Dow N. Determination of compound binding to plasma proteins. Curr Protoc Psychopharmacol 2006;34:7.5–7.5.15.

Emmons G, Rowland M. Pharmacokinetic considerations as to when to use dried blood spot sampling. Bioanalysis 2010; 2(11):1791–1796.

European Medicines Agency. Guideline on the Validation of Bioanalytical Methods. Committee for Propietary medicinal Products for Human Use (CHPMP), London. 2011. Available at http://www.ema.europa.eu/docs/en_GB/document_library/Scientific_guideline/2011/08/WC500109686.pdf. Accessed Mar 14, 2013.

Fleuren HLJ, van Rossum JM. Pharmacokinet. Biopharm 1977, 5. Nonlinear relationship between plasma and red blood cell pharmacokinetics of chlorthalidone in man. J Pharmacokinet Pharmacodyn 1977;5(4):359–375.

Freisleben A, Brundy-Klöppel M, Mulder H, de Vries R, de Zwart M, Timmerman P. Blood stability testing: European bioanalysis forum view on current challenges for regulated bioanalysis. Bioanalysis 2011;3(12):1333–1336.

Hage DS. High performance affinity chromatography: A powerful tool for studying serum protein binding. J Chromatogr B 2002;768(1):3–30.

Hinderling PH. Red blood cells: A neglected compartment in pharmacokinetics and pharmacodynamics. Pharmacol Rev 1997;49(3):279–295.

Ismaiel OA, Zhang T, Jenkins RG, Karnes HT. Investigation of endogenous blood plasma phospholipids, cholesterol and glycerides that contribute to matrix effects in bioanalysis by liquid chromatography/mass spectrometry. J Chromatogr B Analyt Technol Biomed Life Sci 2010;878(31):3303–3316.

Kawai R, Lemaire M, Steimer JL, Bruelisauer A, Niederberger W, Rowland M. Physiologically-based pharmacokinetic study on a cyclosporine derivative, SDZ IMM 125. J Pharmacokinet Biopharm 1994;22(5):327–365.

Kissinger PT. Thinking about dried blood spots for pharmacokinetic assays and therapeutic drug monitoring. Bioanalysis 2011;3(20):2263–2266.

Kitamura S, Terada A, Inoue N, Kamio H, Ohta S, Tatsumi K. Quinone-dependent tertiary amine N-oxide reduction in rat blood. Biol Pharm Bull 1998;21(12):1344–1347.

Kringle R, Hoffman D, Newton J, Burton R. Statistical methods for assessing stability of compounds in whole blood for clinical bioanalysis. Drug Inf J 2001;35(4):1261–1270.

Lin J, Ulm EH, Los LE. Dose-dependent Stereopharmacokinetics of 5,6-dihydo-4H-4(isobutylamino)thieno(2,3-B)thiopyran-2-sulfonamide-7,7-dioxide, a Potent Carbonic Anhydrase Inhibitor, in Rats. Drug Metab Dipos 1991;19(1):233–238.

Lindup WE, Orme MC. Plasma protein binding of drugs. Br Med J (Clin Res Ed). 1981;282(6259):212–214.

Mazoit JX, Sammi K. Binding of profopol to blood components: Implications for pharmacokinetics and for pharmacodynamics. Br J Clin Pharmacol 1999;47(1):35–42.

Mullersman G, Derendorf H. Rapid analysis of ranitidine in biological fluids and determination of erythrocyte partitioning. J Chromatogr 1986; 381(2):385–391.

Musteata FM. Monitoring free drug concentrations: Challenges. Bioanalysis 2011;3(15):1753–1768

Musteata FM, Pawliszyn J, Qian MG, Wu JT, Miwa GT. Determination of drug plasma protein binding by solid phase microextraction. J Pharm Sci 2006;95(8):1712–1722.

Neuvonen PJ, Pentikäinen PJ, Elfving SM. Factors affecting the bioavailability of phenytoin. Int J Clin Pharmacol Biopharm 1977;15(2):84–89.

Querol-Ferrer V, Zini R, Tillement JP. The blood binding of cefotiam and cyclohexanol, metabolites of the prodrug cefotiam hexetil, In-Vitro. J Pharm Pharmacol 1991;43(12):863–866.

Richterich JPR, Colombo JP. *Clinical Chemistry: Theory, Practice and Interpretation.* Translated edition. Chichester: John Wiley & Sons, Ltd; 1981.

Rose JQ, Yurchak AM, Jusko WJ. Dose dependent pharmacokinetics of prednisone and prednisolone in man. J Pharmacokinet Biopharm 1981;9(4):389–417.

Silvestro L, Gheorghe M, Iordachescu A, et al. Development and validation of an HPLC-MS/MS method to quantify clopidrogel acyl glucuronide, clopidrogel acid metabolite and clopidrogel in plasma samples avoiding analyte back-conversion. Anal Bioanal Chem 2011;401(3):1023–1034.

Stott RA, Hortin GL, Wilhite TR, Miller SB, Smith CH, Landt M. Analytical artifacts in hematocrit measurements by whole-blood chemistry analyzers. Clin Chem 1995;41(2):306–311.

Sugihara K, Kitamura S, Ohta S, Tatsumi K. Reduction of hydroxamic acids to the corresponding amides catalyzed by rabbit blood. Xenobiotica 2000;30(5):457–467.

Tsanaclis LM, Allen J, Perucca E, Routledge PA, Richens A. Effect of valproate on free plasma phenytoin concentrations. Br J Clin Pharmacol 1984;18(1):17–20.

US FDA. Guidance for Drug Evaluation and Research. Guidance for Industry: Bioanalytical Method Validation. May 2001. Available at: www.fda.gov/downloads/Drugs/GuidanceComplianceRegulatoryInformation/Guidances/UCM070107.pdf.

Viswanathan CT, Bansal S, Booth B, et al. Quantitative bioanalytical methods validation and implementation: Best practices for chromatographic and ligand binding assays. Pharm Res 2007;24(10):1962–1973.

Wong BK, Bruhin PJ, Barrish A, Lin JH. Non-Linear dorzolamide pharmacokinetics in rats: Concentration dependent erythrocyte distribution and drug-metabolite displacement interaction. Drug Metab Dispos. 1996;24(6):659–663.

Wong BK, Bruhin PJ, Lin JH. Dose-dependent pharmacokinetics of L-693,612, a carbonic anhydrase inhibitor, following oral administration in rats. Pharm Res. 1994;11(3):438–441

Yu S, Li S, Yang H, Lee F, Wu J-T, Qian MG. A novel liquid chromatography/tandem mass spectrometry based depletion method for measuring red blood cell partitioning of pharmaceutical compounds in drug discovery. Rapid Commun in Mass Spectrom 2005;19(2):250–254.

13

BEST PRACTICE IN BIOLOGICAL SAMPLE COLLECTION, PROCESSING, AND STORAGE FOR LC-MS IN BIOANALYSIS OF DRUGS

MARIA PAWULA, GLEN HAWTHORNE, GRAEME T. SMITH, AND HOWARD M. HILL

13.1 INTRODUCTION

Good bioanalysis starts with good sample collection procedures. Therefore, the integrity of samples must be maintained from the time of collection to the moment of analysis, such that the determined concentration closely reflects the *in vivo* concentrations. The global and striated nature of drug discovery and development means that bioanalytical support for a particular drug or analyte will most likely occur in a number of different laboratories and test facilities around the world. To enable meaningful comparisons to be made for the data produced, detailed documents and protocols need to be developed for sharing best practice and to standardize the procedures for collecting the study samples.

How samples are collected, handled, and stored is of the utmost importance in ensuring good-quality data from the study being conducted. Although some issues with sample collection and handling may be immediately obvious, such as hemolyzed samples, others will not come to light until much later when all the samples have been analyzed. For example, discrepancies may be noted when results are compared with previous data or an incurred sample reanalysis (ISR) experiment is conducted. This may lead to further investigations, and a worst-case scenario outcome could be that the data for a study or a set of studies is rejected, necessitating repeat work at additional cost together with delays to the drug development program.

When a protocol is being drawn up for a new study that will generate biological samples for analysis, the bioanalyst should be contacted and asked for their advice on the best way to collect, handle, and store the samples. However, the difficulty is that often a comprehensive set of information is not available early on in a drug development program. Unless the samples are stored correctly, the "true" concentration of drug cannot be determined, but in order to fully assess the most appropriate collection and storage procedures, the analysis of actual incurred samples is required. This is because calibration standards and quality control samples prepared during method validation do not always reflect the complexity of an incurred sample containing parent drug and a range of metabolites, some of which may be unstable. In these situations, the bioanalyst has to use all available information (e.g., chemical structure, drug class, and the stability generated during method validation for the drug under a series of different storage conditions and possibly a limited number of metabolites, together with their personal experience) and make the best informed decision.

The problems that can potentially arise from inappropriate collection, handling, and storage of samples are numerous, but can be grouped into two broad categories, that is, physical and chemical factors, where the latter includes all the issues related to drug stability (e.g., pH, enzymatic activity, oxidation, interconversion between isomeric forms or parent drug and metabolite). However, there is a lot of overlap between these two factors as many issues are inter-related and impact on each other.

Issues due to physical factors can include the use of inappropriate containers or sample collection devices. For example, during *in vivo* sample collection, incorrect choice of needle gauge could increase cell rupture, leading to unwanted

Handbook of LC-MS Bioanalysis: Best Practices, Experimental Protocols, and Regulations, First Edition. Edited by Wenkui Li, Jie Zhang, and Francis L.S. Tse.

hemolysis. This may in turn cause matrix effects or other difficulties in sample analysis, and necessitate additional assay development work or extra validation. The choice of sample container is also significant. For example, glass is easily broken during transportation. Some drugs can adsorb to glass, while others adsorb to different plastics. If a drug is adsorbed, then the "real" circulating drug concentrations may be underestimated, leading to incorrect pharmacokinetic (PK) interpretation, and wrong decisions for dosing regimens, and so on. If a tube or sample container is too small, there may not be room for effective mixing before taking an aliquot for sample analysis. Conversely, if the tube or sample container is too large, evaporation may occur and it may be more difficult to take an aliquot for analysis, not to mention, make sample shipment more expensive. Incorrect sample storage, for example, samples are stored at too high a temperature (see Section 13.2.5) could result in the formation of fibrinogen clots in plasma, making it difficult for pipetting. Simple things can also cause problems. For example, some labels may become unreadable or detached when a sample is frozen and thawed. If a sample cannot be correctly identified, the sample should not be analyzed. This needs to be documented in the bioanalytical data report.

There are wide ranges of chemical factors that can affect sample collection. For example, plastic sample containers may leach unwanted chemical additives (used in manufacturing) into the sample and cause matrix effects. However, most often adverse chemical effects are linked to drug stability issues. If an analyte of interest is ultra-violet (UV) sensitive and the sample is exposed to sunlight, degradation may occur. If a blood sample is taken, metabolic processes, such as enzymatic reactions, may not stop immediately. Careful consideration, therefore, needs to be given about how to halt the processes as quickly as possible, allowing accurate quantitative determination to be made for both the drug and its metabolite(s). The chemical characteristics of the analyte should be taken into consideration, as the pH of the collected sample may adversely affect the stability of the analyte, possibly due to degradation and oxidization. Often stability problems lead to the underestimation of actual analyte concentrations. However, the opposite can also occur as overestimation of parent drug can happen when conjugated metabolites are hydrolyzed back to the parent drug in the sample.

When loss of a drug is observed during analysis, it should not automatically be assumed that the cause of the problem is a consequence of instability. Problems with stability can sometimes be confused with poor recovery or adsorption of the drug to the containers used in processing. If loss of drug is observed during method development, it would be the best practice to investigate the reason and then address the issue accordingly, rather than simply adding unnecessary stabilizers to the sample.

The importance of ensuring stability in sample collection and handling has been highlighted by both the Food and Drug Administration (FDA) and European Medicines Agency (EMA) in their guidance documents. The FDA document states that "Stability procedures should evaluate the stability of the analytes during sample collection and handling" (FDA Guidance, 2001). This has been further emphasized by recent guidelines issued by the EMA stating "Sufficient attention should be paid to the stability of the analyte in the sampled matrix directly after blood sampling" (EMA Guideline, 2011). Therefore, whole blood stability is also routinely assessed during method validation to ensure that stability has been evaluated for the entirety of the sample collection and handling process. The earlier any stability issues are identified, the easier it is to develop appropriate preventive measures to be incorporated into the preanalytical phase (sample collection, handling, and storage). ISR assessments have now become part of the standard bioanalytical process, but can only be performed when incurred samples are available.

Careful documentation of the procedures to be involved in the preanalytical phase is also very important. A detailed and accurate record needs to be kept as to exactly how the sample was taken, in what container it was collected and by whom, and so on, providing an audit trail, so that whatever occurred during a study can be reconstructed exactly. As well as being good laboratory practice, these records can help with troubleshooting analytical issues later on if the need arises.

This chapter attempts to provide a detailed overview of the issues and the all the different aspects of the preanalytical phase that need to be considered prior to, and during the collection, handling and subsequent storage of the study samples.

13.2 SAMPLE COLLECTION

Various different types of matrices may be collected for subsequent LC-MS/MS bioanalysis. The most common matrix is plasma; however, depending on the characteristics of the drug and its metabolic behavior, analysis of blood or serum may be more appropriate. Sometimes it is also useful to measure the concentration of the drug in urine, to help further understand the behavior of the drug under investigation, especially if a significant amount of unchanged drug is excreted by this particular route.

The drug development process is, in general, limited by constrained budgets. Therefore, scientists need to get the maximum amount of information out of the intended studies based on the limited resources; thus a new trend is emerging, with more requests to analyze tissues and/or other matrices from the organs of interest. The desire to identify and quantify the drug metabolites has moved forward, in many cases to the discovery phase. As a consequence to measure the drug in a variety of biological matrices has increased. Instead of waiting for the mass balance (ADME) studies using radiolabeled drug, LC-MS/MS analysis for the samples from the studies using the unlabeled test article may give an earlier answer as to whether the drug is reaching the target organ/tissue (distribution) and how the drug is eliminated (urine or feces), and so on.

In addition to conventional plasma, serum or whole blood, other matrices include tissues (virtually any tissue/organ in the body that the drug may distribute into), feces, other bodily fluids such as cerebrospinal fluid (CSF), saliva, sputum, bronchoalveolar lavage fluid, vaginal fluid, semen, aqueous and vitreous humor, tear fluid, and sweat. Analyzing the samples outlined above can provide answer to some specific questions raised during drug development.

In general, the key point in biological sample collection is to collect them quickly and store them at the correct temperature, stabilize an unstable drug in the matrix, and ensure that the samples are labeled correctly.

13.2.1 *In vivo* Sample Collection

There are many decisions that need to be made when designing a study in which collection and analysis of biological samples will be required. Much of the *in vivo* aspect of a study is outside the scope and input of the bioanalytical scientist, but may have an impact on sample collection. If there are stability or other issues, the bioanalytical scientist and the study director or principal investigator of the *in vivo* study should work together to ensure that the best quality sample is obtained.

When a study is set up, appropriate biological sample collection supplies should be selected, in order to provide the best quality sample for analysis. For example, when drawing blood, an appropriate gauge of needle should be selected. If too wide a needle is used, this can lead to turbulent flow and blood cells rupturing, causing hemolysis, which in turn may cause matrix effects during analysis. If too narrow a gauge is used, the slow rate of blood flow through the bore may cause clotting (WHO Guidelines, 2010).

Several clinical practices that may affect the quality of the sample taken need to be taken into account. Body posture during blood sampling generally does not affect concentrations of free drug; however, if a drug is highly protein bound, posture may have an impact. Changing from a supine to a sitting or standing position moves body water from the intravascular to the interstitial compartment. This can increase the concentration (by 5–15%) of some large molecules such as albumin, which may lead to less free drug being available. Similarly, when changing from a standing position to supine, the plasma volume increases thus causing a slight dilution effect (Guder et al., 1996). This will only be an issue if free drug concentrations, rather than total drug concentrations, are being measured. The excessive use of a tourniquet during venipuncture when collecting blood should be avoided (also taking blood samples from the same arm that has just been used for a blood pressure measurement), as constriction of veins can increase the protein and lipid content of blood (Narayanan, 1996). This may also affect the free fraction of drug if the drug is highly protein bound. Therefore, if a tourniquet is used, it should be released as soon as the needle enters the vein. If the clinical study includes intravenous infusions, blood should not be obtained from a vein close to the infusion site. Instead, blood samples should be collected from the opposite arm. Detailed guidelines and protocols for best practice in venepuncture and blood collection procedures can be found at the World Health Organization (www.who.int/publications, accessed Mar 14, 2013) or National Health Service (NHS) websites.

Over the last few years, there has been a lot of research conducted into using dried blood spots (DBS) as an alternative matrix for bioanalysis (see Chapter 31). One of the main advantages of using DBS is that the total volume required is very small, usually in the region of 30–100 μl (with accurate volumes of 10–35 μl applied per spot). Although venepuncture with subsampling of the blood sample by pipette or capillary tube can be used, the small volume required allows other alternative less invasive sampling techniques to be utilized. Finger-prick or heel-pricks have been found to be effective, and studies have shown that the patient/volunteer finds this technique less distressing than using a venous cannula. Nurses and technicians in the clinic also find taking the samples easier and quicker. Similarly, when used with animals such as mice and rats, no or less warming is required, and the sample can be taken from a tail-prick using a microcapillary tube. It has been reported that there appears to be no significant difference in the results obtained whether venous blood or capillary blood sampling is used.

For animal studies, there are various best practices that should be followed by the animal facilities and technicians taking the samples, to prevent cross contamination occurring and to ensure that the best quality data is produced. Control animals should be housed in a separate room where possible. Samples from dosed animals should be collected in the separate rooms, or collected from control animals first, then working through the remaining groups in ascending dose order. Disposable single use equipment should be used where possible to minimize the chances of cross-contamination. Similarly, gloves, aprons/overalls, and so on should be changed when moving between control and dosed animals, and if any spillages of blood, plasma, or serum occur, then general good hygiene practices should be followed (EC Guidance Document, 2006).

In certain situations having an analyst present at the animal facility or clinic is useful to ensure the correct performance of critical tasks. For example, if stabilizer, solvents, or other reagents need to be added at the exact time of sample

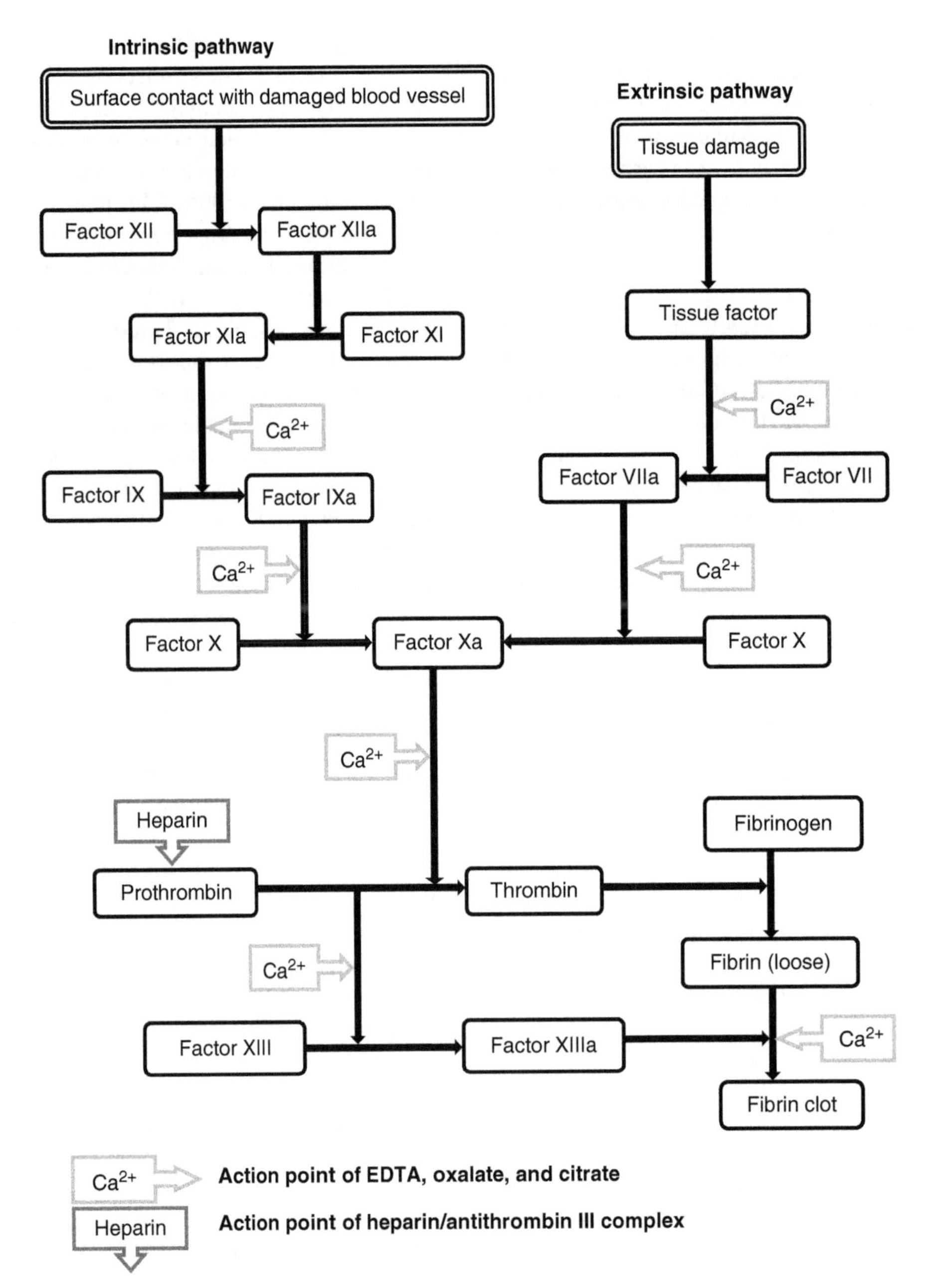

FIGURE 13.1 Anticoagulants and the coagulation cascade.

collection and the samples need to be processed immediately. This will obviate the need for training clinical staff or animal technicians, and will provide greater assurance that the integrity of the samples has been maintained during collection.

13.2.2 Choice of Anticoagulant

The *in vivo* coagulation cascade is an important homeostatic response that serves to form a fibrin and platelet clot at the site of injury of a blood vessel to prevent the loss of blood. The coagulation cascade converts fibrinogen to fibrin via the proteolytic enzyme thrombin, as shown in the simplified cascade in Figure 13.1 (Rang and Dale, 1991). Thrombin itself is produced by cleavage of prothrombin. There are many important cofactors in the clotting process, but calcium is also important. When calcium is removed by an anticoagulant chelating agent, the clotting process is halted and fibrin clots are not formed. *Ex vivo*, the benefit is that the blood and subsequent plasma produced from the whole blood remains fluid and relatively free from large clots.

Anticoagulant is normally added to the collection tube as a solution, at an appropriate concentration to inhibit clotting, but not too great as it may have unwanted effects on bioanalysis. The solution is then evaporated to dryness at a suitable temperature (i.e., so as not to decompose the salt).

Blood collection tubes are available from vendors with a wide range of anticoagulants and various counter ion options. This situation has arisen as most of the collection tubes have been developed for specific clinical chemistry tests. The choice of counter ion is more important for some clinical chemistry applications, as the quantification of different electrolytes is important and needs to be unbiased by the presence of that ion (e.g., if measuring the sodium content of blood, a sodium counter ion should not be used for the anticoagulant).

There has been debate recently in the bioanalytical community about the differences between the selection of counter ions and whether changing counter ion warrants additional validation. Anticoagulant counter ion impact was discussed and investigated by the members of the European Bioanalytical Forum (EBF). The resulting conclusion was that the EBF recommended that "in regulated bioanalysis plasma samples containing different counter ions, but the same anticoagulant, should be regarded as equal matrices, thus removing the need for partial validation (Sennbro et al., 2011)."

The selection of anticoagulant for bioanalysis can sometimes be based on what has historically been used within an animal facility and is erroneously considered to not be important. There are many benefits to selecting the right anticoagulant for an assay; for minimizing interferences, for better stabilization of a drug and/or its metabolites and/or for ease in automation (by minimizing fibrinogen clot formation, which could block pipette tips).

Commonly used anticoagulants and some of their properties are listed below. A more comprehensive summary of the anticoagulants available is presented in Appendix 2, at the end of this handbook.

13.2.2.1 Ethylenediaminetetraacetic Acid (EDTA)

EDTA is a synthetic small molecule that can be produced with a number of different counter ions, namely, sodium, dipotassium, or tripotassium, as shown in Figure 13.2. EDTA acts to prevent coagulation by forming a complex with any calcium present, and thus inhibits the formation of fibrin clots by inhibiting the coagulation cascade.

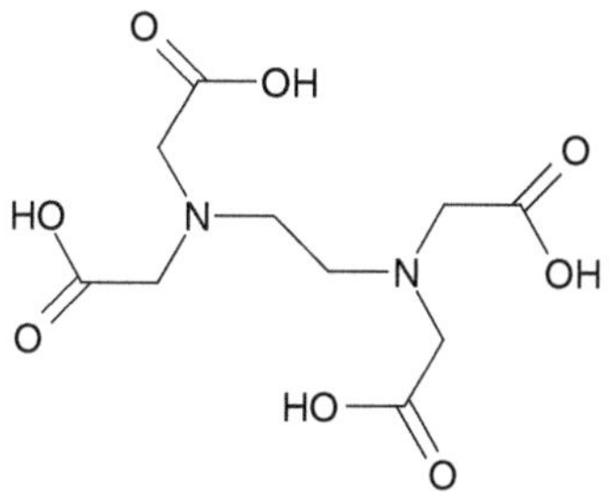

Ethylenediaminetetreacetic acid (Mw 292.24)

FIGURE 13.2 Structure of EDTA.

Both potassium and sodium salts have been used for blood sample collection, with potassium salts preferred as they are more soluble in blood. The choice of salt form also affects the pH of EDTA. In plasma, there is very little difference in pH between K_2 and K_3 EDTA with reported plasma pH of 7.1 for K_2 and pH 7.3 for K_3. However, there is a larger difference in pH of aqueous solutions. A 1% solution of K_2 EDTA has a pH of 4.8 compared with K_3 that has a pH of 7.5. It is noted that the physical form of K_3 EDTA exists as a liquid, whereas a benefit of K_2 EDTA is that it can be spray dried to coat surfaces of blood collection tube, so it does not have a dilution effect of the volume of the blood collected (Narayanan, 1996). The optimum concentration of EDTA as an anticoagulant is 1.5 mg per ml of blood.

Blood collected into EDTA tubes produces fewer plasma clots (than heparin) upon freeze thaw. This can be beneficial for automation (Sadagopan et al., 2003).

13.2.2.2 ***Heparin*** Heparin is a natural anticoagulant produced by mast cells and found in the liver and intestine, and to a lesser extent in other tissues. Commercially, heparin is isolated mainly from porcine intestine or bovine lung, and can be administered therapeutically to prevent thrombosis.

Heparin is a highly sulfated glycosaminoglycan and is widely used as an anticoagulant when collecting blood samples for bioanalytical applications. The molecule consists of variably sulfated repeating disaccharide units, and commercially available heparin typically has a molecular weight that can range from 12 to 15 kDa. The main disaccharide unit is shown in Figure 13.3. In blood or plasma, the ester and amide sulfate groups are deprotonated and attract positively charged counter ions to form a heparin salt (Linhardt and Gunay, 1999). For anticoagulation, the most commonly used salts are either lithium or sodium, but potassium and ammonium salts can also be used.

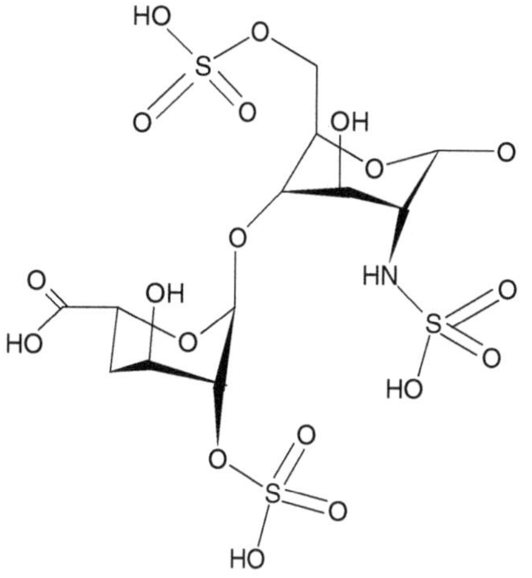

FIGURE 13.3 Structure of heparin.

Heparin is very effective as an anticoagulant, as it exerts its effect on three different stages of the coagulation cascade, inhibiting the activation of prothrombin to thrombin, inhibiting the coagulation of fibrinogen to fibrin, and also stabilizing the thrombocytes (Richterich and Columbo, 1981). The way heparin acts as anticoagulant is to change the conformation of the antithrombin III molecule, by binding to it and accelerating its action. This in turn accelerates the inhibition of the coagulation factors Xa, IX, and thrombin (Rang and Dale, 1991; Narayanan and Hamasaki, 1998), and so prevents fibrinogen from forming fibrin.

The usual concentration of heparin used is 20 units per 1 ml of blood (approximately 0.2 mg per ml of blood).

Using heparin has two main disadvantages. Firstly, it is the most expensive anticoagulant. Secondly, heparin only delays and does not prevent the clotting process. Heparinized blood will eventually start to clot in 8–12 h at room temperature (Bush, 2003). Additionally, the use of heparin should also be avoided, if trying to measure the free concentrations of triiodothyronine and thyroxine, as it can affect their binding to their carrier proteins (Burtis et al., 2001).

13.2.2.3 Oxalate Oxalate's mechanism of action is that it binds calcium and so inhibits clotting, by forming an insoluble complex with calcium ions (Narayanan and Hamasaki, 1998). Various oxalate salts can be used as anticoagulants: ammonium, lithium, potassium, and sodium. Often a combination of ammonium and potassium oxalate is used in a 3:2 ratio (known as Paul & Heller's double oxalate). Potassium oxalate has been shown to cause the shrinking of erythrocytes by drawing water out, resulting in the plasma becoming artificially diluted as cellular fluid is expelled. This in turn leads to a more diluted plasma sample, and an inaccurate determination of the drug's concentration in plasma. A concentration of 1 to 2 mg/ml of potassium oxalate is most commonly used to minimize this, as higher concentrations (e.g., 3 mg/ml and above) can lead to unwanted hemolysis and excessive shrinking of erythrocytes. Additionally, ammonium oxalate has been observed to cause erythrocytes to swell. Hence, a combined ammonium and potassium oxalate mixture causes little distortion of the erythrocytes at the recommended concentration level in the first hour after collection.

Compared to heparinized blood, using oxalate as an anticoagulant can yield a lower hemocrit value, which may lead to errors when determining drug concentrations.

Oxalate can also alter the pH of plasma considerably. This may help to stabilize some drugs, but be disadvantageous for the stability of other drugs. Oxalate has also been found to inhibit the action of acid and alkaline phosphatase enzymes which helps with stabilization of phosphatase sensitive and phosphatase labile drugs (e.g., prodrugs such as fosfluconazole and fludarabine phosphate), as hydrolysis of the phosphate group is minimized.

13.2.2.4 Citrate The use of trisodium citrate as an anticoagulant goes back to the early twentieth century, when it was developed to use as an additive in donor blood for blood transfusions.

Citrate works as an anticoagulant by chelating calcium, and forming a soluble calcium citrate complex (Narayanan and Hamasaki, 1998). This prevents the clotting cascade taking place.

Blood collection tubes have also been developed which, in addition to citrate, contain dextrose and citric acid. These are commonly called ACD (*a*cid, *c*itrate, *d*extrose) tubes. These additives help to sustain the metabolic activity of red blood cells, and improve the buffering capacity, thus prolonging the shelf life of the blood. However, ACD tubes offer no particular advantages over normal citrate tubes for bioanalytical applications.

Citrate's mode of action being similar to oxalate leads to it having similar advantages and disadvantages. Citrate can help to buffer the blood that may be a benefit, but it may also alter the pH of the blood, which could be a strength or weakness, depending on the particular drug involved. The main disadvantage of using citrate is that it has more marked effect in shrinking erythrocytes (than oxalate), leading to the dilution of plasma, and so an underestimation of the concentration of the drug in the blood.

Citrate is normally used at a concentration of 3.2% or 3.8% sodium citrate (dihydrate), and is added in a ratio of 1 part citrate solution to 9 parts blood to be effective as an anticoagulant. It is imperative that a full blood draw takes place and the blood tubes are filled to the required level (i.e., not under or over filled), so that the dilution is properly controlled. This dilution has to be born in mind when calculating plasma concentrations of drug (Narayanan, 1996).

13.2.2.5 Fluoride Sodium fluoride is a fairly weak anticoagulant, unless very high concentrations are used. It works by binding calcium to prevent clotting. Although it only has a weak anticoagulant effect, it has the advantage of being a potent enzyme inhibitor, which makes it very useful as a stabilizing agent. Sodium fluoride inhibits phosphatases and reversibly inhibits both acetylcholinesterases and butyrylcholinesterases (Heilbronn, 1965). The inhibition has been found to be even more effective if the pH is lowered. Sodium fluoride also inhibits glycolysis in erythrocytes, which makes it useful for the determination of blood glucose.

The recommended starting concentration to be used for anticoagulation is 2 mg/ml, although it is wise to optimize the concentration of fluoride for your particular drug in your specific matrix. Sodium fluoride can be added as a powder to the blood tube or as 2% solution (as it is not particularly soluble). If added as a solution, then the usual proportion is 1 part 2% sodium fluoride solution to 9 parts blood. In this case (as with sodium citrate) the dilution effect needs to be taken into account when calculating plasma concentrations of drug, and a full draw needs to be made into the sample collection tube.

Sodium fluoride is often used in combination with potassium oxalate in the same blood collection tube. This combination ensures more effective anticoagulation, as well as inhibition of a wider range of enzymes. For example, Lindegardh et al. (2007) found that fluoride-oxalate sample collection tubes were just as effective at inhibiting the hydrolysis of an oseltamivir prodrug by plasma esterases, as adding the esterase inhibitor dichlorvos, but were much safer and less toxic to use.

13.2.3 Other Sample Tube Additives

Depending on the desired result, there are many additives that can be added to blood collections tubes, usually in conjunction with one of the above anticoagulants. The general purpose for having those additives is for stabilization of the analytes of interest, whether it is to inhibit enzymes, or prevent unwanted chemical reactions such as oxidation or other types of degradation.

Commonly used additives are antioxidants (such as ascorbic acid, sodium metabisulfite, and butylated hydroxytoluene (BHT)) and enzyme inhibitors, examples of which are discussed in more detail in Section 13.6 of this chapter, and elsewhere within this publication relating to specific classes of compound. Other kinds of additives may be buffers for pH control, derivatization reagents to stabilize analytes or aid analysis, organic solvents to effectively "freeze" further chemical or enzymatic reactions, or organic solvents or buffers to help solubilize the drug in urine samples.

If additives are determined to be necessary for the intended study, it is important to establish needed logistics during the planning phase of the study: who will add the chemical(s) to the sample collection tube?; will the bioanalytical laboratory prepare the tubes in advance and supply them to the collection site?, or does the collection site have suitably trained staff who can prepare and aliquot the necessary solutions into the tubes as required? Alternatively, if the distances involved makes it feasible, could an analyst be sent to the collection site to do the preparation "on site," which could save time and eliminate the potential misunderstandings, avoid the need for training other staff, and help to generate good-quality samples.

No matter who is responsible for preparing the tubes, clear protocols and detailed procedures need to be available. It is also important to establish at which point the stabilizers are to be added; do they need to go in at the start to the blood collection tube, or can they be added to the tubes into which the separated plasma or serum is to be transferred? This can also be applied to additives used for preventing adsorption of the analyte of interest to the urine sample container, or keeping the analyte solubilized. Do they need to be added to the container prior to the urine sample collection, or can they be added postcollection?

13.2.4 Preparing Plasma from Blood

The stability assessment of drugs in blood and the associated discussion of the physiological nature of blood can be found in Chapter 12 in this publication. Despite being the principal circulating biological matrix only a minority of drug discovery and development programs are conducted with analysis of drugs in whole blood. In recent years, there has been a renewed interest in the use of DBS for quantitative bioanalysis, and more recently there has been a move toward to the use of microsampling of blood into glass capillaries for either the analysis of diluted blood or plasma. Currently, plasma or serum is still the most commonly used matrix that is acceptable from a regulatory perspective. Table 13.1 outlines a simple standard protocol for the collection of blood and the processing of it to yield a plasma sample.

TABLE 13.1 Example Protocol for the Preparation of Plasma

Step	Description
1	Select tube containing the appropriate anticoagulant.
2	Draw full amount of blood into the tube (i.e., do not partially fill, or overfill blood tubes) to ensure the correct ratio of anticoagulant to blood.
3	Gently and slowly invert tube five to eight times to ensure thorough mixing of blood and anticoagulant.
4	Centrifuge for 10–15 min at 1500–2500 g.
5	Transfer the plasma supernatant to a clean tube for storage.
6	Aliquot into smaller volumes (if duplicate samples are required).
7	Store at the specified temperature (usually −20°C or −80°C).

Unless the stability of the drug or test substance has been proven at room temperature, it is wise to store the blood tubes on crushed ice after blood draw and prior to centrifugation, and to use a chilled centrifuge for the separation of the plasma. The effect of centrifugation temperature has been investigated (Lippi et al., 2006), but no statistically significant differences were observed, with respect to the ratios of plasma/serum components obtained, thus stability of the drug or test substance would be the deciding factor whether to use chilled temperatures.

13.2.5 Minimizing Clot Formation

Minimizing clot formation is an important factor in developing a robust plasma assay. The formation of clots can lead to the clogging of pipette tips when aliquotting plasma samples, whether the pipetting is performed manually or by using a robotic system (where the presence of clots can cause a greater adverse impact). The procedures for collection and storage of plasma samples can play an important part in reducing unwanted clotting.

In general, if blood is to be collected into several different anticoagulant tubes, the preferred collection order to minimize any potential problems is citrate, heparin, EDTA, fluoride, and serum tubes.

One technique for clot reduction is to immediately freeze and store the samples at −80°C. According to Watt et al., if a sample is frozen at −20°C and then undergoes freeze–thaw cycles at this temperature, it increases the likelihood of clotting (Watt et al., 2000).

The anticoagulant is another factor that needs to be considered. The use of EDTA is commonly found to produce less clotting than heparin. Investigations into the transfer of manual methods to automated workstations showed that the failure rate for methods using heparinized plasma were much greater than using EDTA plasma (Sadagopan et al., 2003).

13.2.6 Difference between Plasma and Serum

The major difference between plasma and serum is that no anticoagulants are used in the collection of serum and all the fibrinogen and associated proteins are removed through the clotting process. Therefore, serum always contains less protein material than plasma, leading to a cleaner sample extract. Generally, it can be assumed that a plasma assay transferred to serum will perform well. However, swapping from a serum assay method to plasma may cause issues due to the inherent difference of the two matrices.

Plasma and serum are collected from blood in the same way (albeit in different tubes). Both are separated from the whole blood components such as red and white blood cells and platelets by centrifugation. The benefit of using plasma as the matrix for bioanalytical assay is that blood can be centrifuged immediately after collection with a reduced potential for hemolysis. If the drug of interest has a stability issue, using plasma is a better option than serum, as plasma samples can be processed very quickly followed by rapid frozen storage.

TABLE 13.2 Example Protocol for the Preparation of Serum

Step	Description
1	Draw a volume of blood that is 2.5 times the volume of serum required (e.g., 2 ml to produce 0.8 ml of serum).
2	Collect into tube with no anticoagulant present.
3	If using tube with a clot activator, invert tube five to six times to mix, prior to incubation.
4	Leave at room temperature (18–24°C) for approximately 30–60 min (use minimum time necessary) to allow it to clot.
5	Centrifuge for 10 min at 1500–2500 g.
6	Transfer the serum supernatant to a clean tube for storage.
7	Aliquot into smaller volumes (if duplicate samples are required).
8	Store at the specified temperature (usually −20°C or −80°C).

Table 13.2 outlines a simple standard protocol for the collection of blood and the processing of it to yield a serum sample.

13.2.7 Collection of Urine Samples

Often the first time urine is encountered as a matrix by the bioanalyst is when the drug candidate is tested for the first time in humans (first-in-human study). At a quick glance, it might seem that urine is a simpler matrix to work with than plasma, serum, or whole blood since it is less viscous, in ample supply and, has passed through a (hopefully) healthy kidney. Normal healthy urine consists of about 95% water with the remainder consisting of surplus electrolytes (ions such as Na^+, K^+, Mg^{2+}, Ca^{2+}, and Cl^-) and metabolic waste products (urea, uric acid, creatinine plus very small quantities of hormones, amino acids, etc.) (Green, 1978). However, urine lacks proteins (plasma contains about 6–8% protein material) and lipids. This can cause issues for bioanalyst as proteins and lipids perform a number of useful functions in the background of the bioanalytical process via preventing adsorption to containers (see Section 13.4.1), binding analytes and helping to solubilize them (Ji et al., 2010).

There are also number of key differences in the practical aspects between blood sample collection and handling and the way urine samples are collected and stored. This can pose additional challenges for urine sample assays.

Firstly, unlike blood sample collection where samples are collected at discrete time points following dosing of compound, urine is collected over a time interval. Typical collection times for a PK study are pre-dose (−6 to 0 h), 0–6 h, 6–12 h, 12–24 h, and 24–48 h post-dose. Based on the above sampling regime, urine will potentially be collected and stored for 24 h (for the last interval) before a subsample is taken for analysis. This immediately necessitates validation of a minimum of 24 h stability of the analyte of interest in urine. The most practical temperature for the stability evaluation would be room temperature to avoid the need for refrigerated storage of potentially numerous containers of >1 L volume.

Secondly, urine may initially be collected into an intermediate vessel before being pooled in the collection interval container. Due to the very low levels of proteins and lipids in urine, nonspecific binding or adsorption of the analyte to the walls of the collection vessels becomes more prevalent. To avoid the need for assessing adsorption to container types of different materials, the intermediate vessel should ideally be of the same material as the sample tube and collection interval container. The use of stainless steel bed pans should be avoided to prevent the need for rinsing between uses which could potentially be a source of cross-contamination.

Since urine is in ample supply relative to plasma, there is a tendency for laboratories to request large volumes for samples for their bioanalytical assays. This is not necessarily best practice for a number of reasons: (1) large sample volumes will take longer to freeze and thaw, (2) samples take up more space in refrigerators and freezers, and (3) if tubes are too full there is not enough headspace left in a tube to enable thorough vortex mixing. If this is a problem, it may be more appropriate to roller mix samples or invert to sample tube at least six times to effect mixing. Ineffective mixing has been cited in a number of cases for ISR failing for urine assays (Fu et al., 2011).

Freezing and thawing of urine samples may lead to casts (cellular membrane debris) and variable amounts of salts being precipitated, resulting in inhomogeneity of study samples. To prevent the need for performing multiple freeze–thaw cycles of urine samples, it may be appropriate to request multiple aliquots of the same collection interval (assuming the bulk sample is thoroughly mixed before sub sampling) so that a sample is analyzed after a single freeze–thaw cycle only.

As one of the major routes of elimination of the drug, urine can also contain a higher proportion of metabolites than plasma. Stability of these metabolites should also be assessed wherever practicable to ensure that reproducible results are generated for incurred samples. On the other hand, the presence of drug metabolites in urine makes it a valuable matrix for metabolite profiling. In circumstances where a urine sample will be used both qualitatively for metabolite profiling and for quantitative purposes, it may be appropriate to split the sample into two aliquots, and just add the stabilizing reagent to the aliquot that will be specifically used for the quantification of the analyte.

13.2.8 Collection of Tissue Samples

It is of benefit during drug development to form some ideas of the tissue distribution of a drug. This is normally carried out using radio-labeled drug. However, during early development labeled drug may not be available. Therefore, during preclinical development, tissues may sometimes be collected for qualitative and/or quantitative analysis of drug levels. While an assay for tissues may not always be fully quantitative, there are a number of steps that can help to minimize errors in tissue sample analysis.

Once removed, the organ or tissue should be rinsed with saline to remove any debris or blood from the surface and any major blood vessels in the tissue sample. There will of course be a residual volume of blood retained within the organ for which it may not be possible to distinguish between circulating levels and tissue associated levels of drug.

The tissue can be homogenized immediately or frozen for homogenization at a later date. Snap freezing in liquid nitrogen is a good method to ensure sample integrity, and to facilitate homogenization of tougher tissue samples. Currently, it is not possible to quantitatively analyze a drug directly in intact tissue. This can only be done in tissue homogenate, which is actually a surrogate matrix. Tissue samples should be collected into preweighed containers so the container plus tissue weight can be used to determine an accurate volume of homogenization solution to be added to the sample. While homogenization is a commonly used mechanical approach of preparing tissue samples, other methods such as enzyme digestion or acid hydrolysis can be convenient methods providing the stability of the test substance allows. An example of this is the use of the enzyme subtilisin Carlsberg to digest liver tissue for the isolation of basic drugs such as benzodiazepines (Osselton, 1977).

13.2.9 Collection of Fecal Samples

Although fecal samples are collected on a needs basis, there are a number of cases where feces collection is essential, for example, where biliary excretion is a major route of elimination, where the drug is delivered topically to the gut and fecal concentration can provide some indication of exposure. Last but not least, according to the Metabolites in Safety Testing guidance document (FDA Guidance, 2008), there is a need to identify metabolites that constitute 10% or more of the parent drug systemic exposure at steady state in humans (or 10% of total drug-related exposure as defined by the ICH M3 (R2) guidelines (ICH guidelines, 2009). The FDA guidance document states "Generally systemic exposure is assessed by measuring the concentration of the parent drug at steady state, in serum or plasma. However when measurements cannot be made in plasma of the test species for any reason, verification of adequate exposure can be made in other biological matrices such as urine, feces, or bile" (FDA Guidance, 2008).

There is a wide variety of protocols for the collection of feces from humans. Most are designed by the health services for diagnostic purposes and not for monitoring the excretion of drugs. Nevertheless, it is possible to use them as the basis for designing appropriate procedures for sampling, handling, and storing of fecal samples in drug monitoring.

The NHS website (nhs.uk/chq/Pages/how-should-I-collect-and-store-a-stool-faeces-samples.aspx?CatergoryID=69&SubCategoryID=692, accessed March 20, 2013) provides some dos and don'ts for the collection of samples:

- Do wear disposable "plastic" gloves.
- Place something in the toilet to catch the stool; the document provides some suggestions but success is largely up to the "ingenuity" of the subject.

- Do not collect urine with the sample.
- Do NOT allow the sample to touch the inside of the toilet (probably only relevant for diagnostic purposes).
- DO ensure the container lid is secured.
- DO label the container with relevant data.

While this chapter is dedicated to sample handling and storage, it may be necessary to carry out some sample processing soon after or during collection to ensure that additives, for example, "stabilizers" are homogeneously mixed in the sample. For animal species this means physical collection of fecal material followed by a "cage" wash that minimizes contamination by feces. Samples are bulked together for processing. Typically, this is then followed by addition of a buffer and homogenization using one of a variety of techniques such as Waring Blendor, Ultra Turrax homogeniser through to the use of Stomachers for large fecal samples.

13.2.10 Best Practice for Use of Centrifuges

13.2.10.1 Temperature Control Ideally, it is better to specify the need to control the temperature of centrifugation in the first place (i.e., can a protocol be simplified to use a centrifuge at ambient temperature?). This is especially important when moving into later stage clinical trials where not all laboratories or clinics involved may have a refrigerated centrifuge available.

If, due to stability issues, the samples need to be kept chilled during the collection process, the following should be observed. If a subambient temperature is required, a refrigerated centrifuge should be switched on in advance and allowed to equilibrate to the required temperature prior to placing the samples in it. It is a good practice to leave lid open following use to prevent condensation from corroding the walls of the centrifuge.

13.2.10.2 Centrifugation Force/Speed Centrifugation is used to separate the cellular material in blood such as red and white blood cells from the plasma or serum liquid. The centrifuge creates a force-field in which the denser material (cells) move faster than the material of lower density (plasma or serum). Centrifugation force is quoted in "g," the gravitational unit (acceleration due to gravity), which is equal to 9.81 m/s^2.

It is best to specify centrifugation force in g, rather than speed in revolutions per minute (rpm), since rotors of different length and different centrifuges in different laboratories will generate a different centrifugal force depending on model and rotor length.

The RCF can be calculated using the following equation (Richterich and Columbo 1981):

$$\text{RCF} = (1.118 \times 10^{-5})^* r^* (\text{rpm})^2,$$

where RCF is the centrifugal acceleration in "g," 1.118×10^{-5} is a constant derived from the angular velocity, r is the rotating radius in cm (the distance between the centrifuge axis and the center of the inserted tubes), and rpm is the speed of rotation (in revolutions per minute).

Alternatively, if the rotating radius is known, a nomogram can be used to evaluate the g force a specific speed will produce, or the speed required to give a specific g force. A nomogram can be obtained from any centrifuge manufacturer.

It is also useful to try and keep the centrifugation time to a maximum of 15 min. Spinning for longer (or faster or colder) can start to separate out lipids from blood cells or cause other problems such as hemolysis. In addition, the centrifugation conditions need to be kept consistent between batches of samples.

13.2.11 Time Factors

Sample collection and handling should be based on data obtained during bioanalytical method validation, where the stability of the drug is evaluated under different storage conditions. If there is no stability data available from bioanalytical method validation or other sources (e.g., radiolabel stability in blood combined with a radiolabeled plasma protein binding experiment), then it is best to be cautious and assume that the analyte may be unstable in blood. In which case, any blood samples collected should be placed onto wet ice immediately after collection, and all subsequent processing should be conducted on ice.

Animal technicians and clinical staff are unlikely to have a lot of awareness about sampling issues especially from a bioanalytical perspective. It is therefore the best practice to keep the sample collection conditions as simple as possible, as this will facilitate a faster sample collection and help to ensure the integrity of the samples, and consequently better quality data at the end of the study.

Consideration should also be given to the amount of time required to bleed animals, and prepare plasma, such that if a 15-min limit is applied to the plasma generation, can all samples be feasibly prepared within that time if the next bleed is scheduled 5 min later? Does the animal facility have the capacity to commit to the sampling schedule?

13.2.12 Working on Ice

Fill a deep-sided tray with wet ice and observe closely how long it takes the ice to melt, topping up with fresh ice if necessary. Tubes should be ideally be preprepared with labels printed in indelible ink, and the label should not loosen or become obliterated when it becomes wet. Hand written labels should be avoided if possible. Tubes should be kept upright and securely held in racks to prevent them from floating out of the racks during the course of sample collection.

13.3 DOCUMENTATION

As a drug moves through the development process the degree of procedural information associated with samples increases. Generic procedures in a discovery environment become more tailored methods detailed in protocols in nonclinical development. If drug development progresses further, then more rigorous procedures are documented in separate laboratory manuals for firstly single site clinical pharmacology trials moving onto global multicenter clinical trials. The need for increasingly detailed sampling collection and handling information arises as the geographic distance between where the samples are generated and where they are analyzed increases. Work instructions need to be written with utmost clarity such that all persons involved with the sample collection and handling understand their responsibilities.

Every opportunity should be taken to make documentation simple, easy to read, and readily available to all involved. Instead of sending big packages of paper documentation, all the relevant documents, instructions, templates, and so on can be put onto compact discs (CDs), memory sticks, or other electronic storage devices. Electronic documents allow opportunities for indexing and cross-referencing, thus making the documents easily searchable for the information required. Some groups have taken this concept further; for example, GlaxoSmithKline sent out DVDs with a training video of the correct procedures to take and handle DBS samples to pass on the knowledge to the phase I clinic staff involved. This enabled the very successful transfer of the technique developed in laboratories in the United Kingdom, to clinics in Australia and Korea, without anyone having to specifically to fly out to the clinic to train the staff (Evans, 2010).

13.3.1 Procedures for Nonclinical Sample Collection

A typical nonclinical protocol may contain the following text for the collection of plasma samples for a toxicokinetic study. However, the practical details collecting study samples are usually documented in specific animal handling standard operating procedures.

The protocol should ideally specify the following information, as listed in Table 13.3.

TABLE 13.3 Example of Minimum Information Required for a Nonclinical Protocol (Taken from a Rat Study)

Information item	Example
Sampling site	Lateral tail vein
Type of anesthesia if required	Isofluorane
Anticoagulant (if required)	Lithium heparin
Blood volume to be taken	0.5 ml
Special procedures/ precautions	Store on crushed ice prior to centrifugation
Volume of plasma (no. of aliquots if necessary)	0.2 ml (as 2 × 0.1 ml aliquots)
Storage conditions (of samples prior to dispatch or analysis)	−80°C

13.3.2 Procedures for Clinical Sample Collection

The same information is required for collecting clinical samples, albeit with more specific details to deal with the added complexity and regulatory requirements. As a drug passes through the development process and enters later phase studies, the complexity of sample logistics increases. For example, multicenter studies require the services of a central laboratory to provide kits for sample collection and management of sample movements from clinic to a reconciliation point to the laboratory analyzing the samples.

Table 13.4 shows an example of the information needed in a clinical protocol with regards to collection, handling, and storage of plasma samples. It is useful to provide examples of the type of labels required, or it could all be provided as a pre-prepared kit sent out to the clinics.

Preparing an example label to be reviewed by the persons handling the samples will ensure that all necessary information is captured. The receiving laboratory should be notified of the shipment prior to dispatch and delivery arranged to take place during working hours. It is the best to send out the shipment in the early part of the week, so that it can be delivered on a working day (avoiding the weekend). Delays in transit may result in unusable samples if the drug is not stable, or at a minimum extra time and expense may be required to validate the stability of the samples to mimic the shipment conditions.

If the analyte of interest is known to be unstable and there are no appropriate specific stabilizing agents that can be used (i.e., stability is reliant on temperature control), it is sensible to split the collected sample into two or more separate aliquots. This can avoid multiple freeze–thaws of the same samples for repeat analysis and/or ISR.

The use of multiple aliquots of a sample is also prudent when shipping samples between laboratories. This enables backup aliquots to be kept at the original test facility. If there is a problem with the shipment of the initial aliquot, the backup samples can then be sent to the bioanalytical laboratory.

13.3.3 Sample Identification and Randomization

The identity of a sample must be maintained through a series of events in which a number of different parties may take custody of the sample and/or the data arising from its analysis. For example, an individual may be prescreened for acceptance on a trial and given a subject accession number at a trial center. Once assigned to a trial, he or she is assigned a subject number and given a treatment based on the corresponding randomization identifier. A sample may be taken

TABLE 13.4 Information Required for a Clinical Protocol

Information required	Example
Locations of facility:	
State location of facilities where samples are being drawn and where they will be analyzed	Samples drawn at Laboratory ABC, USA, and sample analysis will be conducted at Laboratory XYZ, UK.
Analysis required:	
Type of analysis required	Determination of "test substance" in plasma for PK analysis.
Materials required:	
Details of blood tubes (material, volume and anticoagulant (ideally exact part number of tube)	BD vacutainer 3 ml plastic tubes, (Ref 367856) containing K_2EDTA as anticoagulant.
Details of plasma tubes and any additives; Material/Volume/Part number, plus any special treatment details	3 × Plasma tubes: 1.8 ml polypropylene Nunc Cryotube, Ref 375418). Plasma tubes to contain X.XX mg of ascorbic acid.
Pipette for plasma transfer	Disposable plastic pipette.
Any additional equipment	Needles, catheters, butterfly clips, etc.
Who provides materials	Laboratory XYZ, as prepared kit.
Volume to collect:	
Volume and type of blood to be drawn	3 ml of venous blood, into blood collection tube.
Volume required:	
Details of type and volume	3 × 0.4 ml of plasma.
Processing:	
Mixing details	Gently invert blood tube eight times.
Centrifugation details	Centrifuge within 30 min of collection. Centrifuge at 1500 g for 10 min at +4°C Transfer a minimum of 1.2 ml into the plasma tube, with a new disposable plastic pipette, 3 aliquots of 0.4 ml into separate tubes* (see end of this section). Freeze immediately in an upright position at −80°C (to keep sample away from the lid to minimize contamination when it is removed following thawing).
Storage:	
Details of storage conditions	Store at −80°C in a temperature monitored freezer.
Labeling:	
Instructions	The label can only be used for the intended sample. Personal information such as patient initials, gender, and date of birth should not be provided on the label (ref to GCP). Use waterproof indelible ink. Label adhesive should withstand freezing and thawing.
Information required on PK sample labels	Study code Randomization code if required Sample ID (unique number) Study day Planned protocol sampling time (e.g., 02:30) HH:mm afterdose or state pre-dose Compound name (or other drug identifier)
Packing and shipping:	
Shipping temperature	Frozen
Specific packing instructions	Samples to be packed securely in boxes, double bagged to prevent leaks, and sufficient quantity of dry ice to ensure that they remain frozen for 72 h. All applicable shipping regulations must be followed (e.g., IATA's Shipping and Dangerous Goods Regulation).
Documentation:	
Required documentation	The receiving laboratory must be notified of the shipment of any samples prior to dispatch. All mandatory fields in sample list to be completed, to provide sample inventory. Add comments about missing samples, deviations, and so on, in the comments field.

at a scheduled point and collected into a sample tube with an identity given by the central laboratory that provided the kit. This sample identity is added to the sample chain of custody and sent with the sample to the receiving laboratory that may wish to assign another identifier for the purposes of their own Laboratory Information Management System. Once the sample has been correctly identified for analysis at the laboratory, the data may be sent to a data manager, who prepares a data set for a pharmacokineticist for PK assessment. Throughout the process, the clinic, subject, and investigators for the study must remain blind to the identity of the sample treatment (known as a double blinded clinical trial). The planning and logistics for such a fragmented process must involve timely and open communication between all parties such that the aims of the trial can be met with as little manual transcription as possible.

13.4 REDUCING ADSORPTION OF ANALYTE TO CONTAINERS

The material of the container used to store a sample can affect the results obtained from analysis. Adsorption of analyte to containers is often seen when analyzing aqueous media containing less proteins or lipids, especially urine. Frequently this issue only comes to light rather late, when an existing established plasma method is applied to the analysis of urine samples, and poor recoveries or nonlinear calibration curves are observed. This is especially noticeable at low concentrations in urine when compared to low concentrations obtained in plasma.

A possible reason for adsorption of the analyte onto the container wall surface is through nonspecific binding. The container surface is likely to be negatively charged (e.g., glass or polypropylene), while the analyte may be positively charged. Also compared to plasma, there is a lack of the proteins and lipids in urine (e.g., plasma contains 6–8% proteins) that normally bind to the analyte or solubilize them. Nonspecific binding is more noticeable with lipophilic and highly protein bound drugs.

A simple experiment can be conducted to identify if adsorption is likely to occur with the selected container, so that preventive action can be taken to eliminate issues or to help resolve identified problems. An example of such an experiment is shown in Figure 13.4, and an example protocol is provided in Table 13.5.

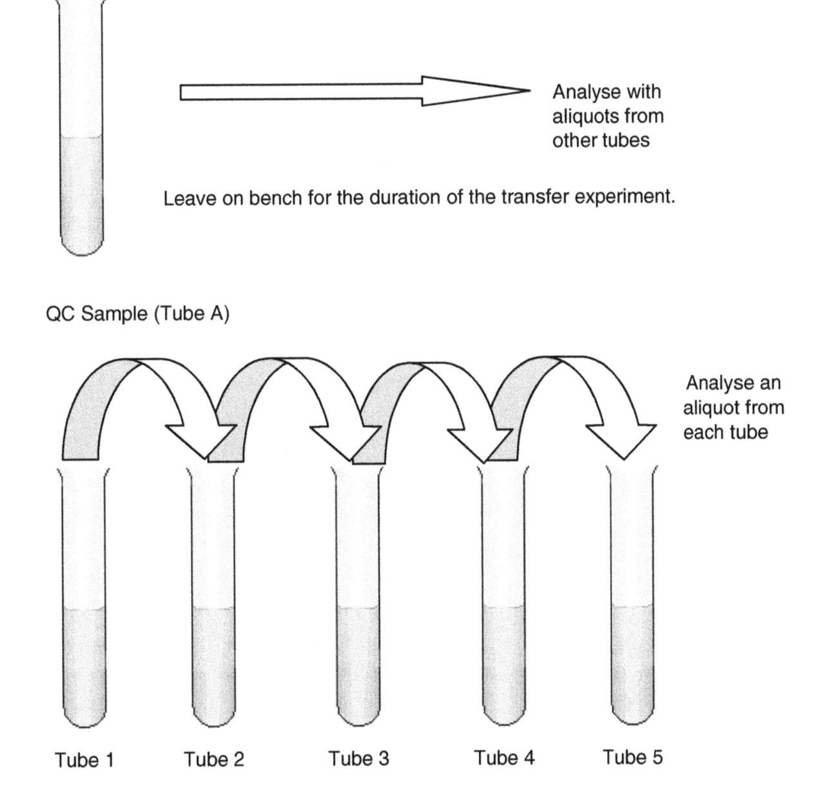

A urine sample is prepared with the test compound at the low quality control level. The sample is split in two and one portion is left on the bench (tube A), while the other is poured from one tube to the next, leaving sufficient in each tube for analysis. The concentration of test compound in each tube is then detemined.

FIGURE 13.4 Nonspecific binding adsorption test.

TABLE 13.5 Example of a Protocol to Check for Adsorption of an Analyte in a Urine Sample to Its Container

Step	Description
1	Prepare a quality control sample at a low concentration in urine.
2	Transfer an aliquot immediately into a selection of storage tubes (e.g., glass, polyethylene, and polycarbonate).
3	Leave to stand for a specific period of time (e.g., 20 min).
4	Transfer the majority of the sample into a fresh tube of the same material, leaving sufficient volume to perform the assay.
5	Repeat steps 3 and 4, three more times.
6	There should now be five tubes in total for each container material tested (as shown in Figure 13.4).
7	Take an aliquot (normal assay volume), extract and analyze using the bioanalytical assay method.

A decrease in concentrations determined across the sequence of Tube 1 to Tube 5 indicates that there is adsorption occurring (assuming that there are no stability issues).

Once analyte container surface adsorption or nonspecific binding is confirmed, there are a number of additives that can be utilized as antiadsorption reagents to help minimize/control this issue. The most commonly used ones are zwitter-ionic detergents (such as 3-[(3-cholamidopropyl)dimethylammonio]-1-propanesulfonate (CHAPS), (cholamidopropyl)dimethylammonio(hydroxyl) propanesulfonate (CHAPSO), Tween 20, Tween 80, sodium dodecyl benzenesulfonate (SDBS), bovine serum albumin (BSA) and solvents (e.g., DMSO) (Xu et al., 2005; Chen et al., 2009; Ji et al., 2010; Li et al., 2010). Phospholipids (specifically lysophosphatidylcholine (lysoPC)) have also been demonstrated to have a positive effect in reducing adsorption in human urine samples (Silvester, 2011).

Table 13.6 summarizes some suggestions as to what concentration of each agent could be used as a starting point for method development. A good way to test them is to repeat the sequential transfer experiment, adding the antiadsorptive agent to the sample. Different concentrations of promising agents should be tested to find the optimum minimum required, as many of these reagents may cause issues with the assay method, for example, ion suppression or enhancement, and if you have a choice of two effective ones, it makes sense to use the less expensive option. The use of the chosen antiadsorptive reagent can be further improved by investigating the optimum pH at which it is most effective.

TABLE 13.6 Examples of Antiadsorptive Reagents and Suggested Initial Starting Concentrations for Investigative Work

Antiadsorptive reagent	Starting concentration (%)	Comments
CHAPS	2.0	May cause interference/ matrix effects
CHAPSO	0.19	May cause interference/ matrix effects
Tween 20	0.2	May cause interference/ matrix effects
Tween 80	0.5	May cause interference/ matrix effects
SDBS	0.14	Cheap; may cause interference/ matrix effects
BSA	1	High concentration required, so can be expensive
β-Cyclodextrin	0.2	Solubility issues

13.5 SAMPLE COLLECTION FOR PEPTIDES AND PROTEINS

The development and approval of biopharmaceutical drugs is increasing significantly every year, and the bioanalyst is more frequently requested to analyze drugs that are proteins or peptides. Thus, the additional challenges these molecules present to sample collection and storage need to be addressed.

Often the distinction between a peptide and a protein is not clear. Peptides can range from two amino acids up to approximately 50 amino acids, which is often used as an arbitrary cutoff. For example, the beta amyloid peptide is considered a protein fragment, while insulin (51 amino acid monomers in size) is generally considered a large peptide. For the purposes of this chapter, we shall be looking at peptides, but many points will be applicable to proteins as well (as proteins are usually digested into peptides to facilitate analysis and quantification). The information is also applicable to peptidomimetics, which are compounds designed to mimic biologically active peptides, but due to structural differences have improved characteristics, such as increased bioavailability and resistance to enzymes.

Adsorption of peptides to the side of containers is a well-known phenomenon. Based largely on proteomic studies, that is, posthydrolysis of proteins prior to analysis, Kraut et al. (2009) have shown that the preferred tubes in which to store peptides are low adsorption plastic, followed by glass, and ordinary plastic tubes were the worst. Hydrophobic peptides, in general, show the highest adsorption (lowest recovery).

A wide variety of "protective" agents have been used to improve recovery of peptides. From a bioanalytical perspective, "loss of drug" through adsorption is likely to be worst in "standard" solutions and during the sample preparation process. Among those agents, use of BSA (1%) improves recovery of a range of peptides in glass and plastic containers for the aqueous/buffered solutions used for dosing.

However, evidence for low recovery of some peptides in biological matrices, largely plasma, serum, or blood caused by container adsorption is not as diverse, because sample processing can confound the issue. It is well known that protein–protein, protein–peptide, and peptide–peptide interactions can not only protect peptides from adsorption but also make it difficult to recover the peptide from the biological matrices.

In general, peptides are prone to instability from enzymes present in the biological matrix. The enzymes involved are peptidases, which metabolize proteins in the body. There are five main classes of proteases, that is, serine, metallo, cysteine, aspartic, and threonine proteases. Many of these are found to some extent in the blood as well as other organs/tissues, for example, insulin protease is found in erythrocytes. If peptide instability is still suspected after eliminating other causes (such as solubility, adsorption, aggregation, and poor recovery), this instability can be managed using similar techniques as for small molecule drugs in Section 13.6. Namely, effective use of temperature control (handling on ice), effective pH control (e.g., acidifying the samples with formic or hydrochloric acid), use of a suitable anticoagulant for blood collection (e.g., using EDTA to inhibit metalloproteases), and the addition of additives such a suitable protease inhibitor. Various standard protease inhibitor cocktails have been developed, but they may not be suitable for LC-MS/MS analysis. Examples of possible stabilizers for inhibiting serine proteases are diisopropyl fluorophosphate (DFP) and phenylmethylsulfonyl fluoride (PMSF). Pefabloc®, a proprietary preparation of 4-(2-aminoethyl)benzenesulfonyl fluoride hydrochloride (AEBSF), is another effective serine protease inhibitor. It is more soluble and safer (less toxic) than DFP or PMSF, and, additionally, it is more stable at low pHs.

Another issue to be aware of when working with peptides is that they can form aggregates during storage, both in solutions and in samples. Aggregation is the binding of peptide to itself (e.g., within a stock solution of the peptide) or to other peptides and proteins, in either a covalent or noncovalent manner. Covalent binding occurs when disulphide bonds or ester and amide linkages are formed, whereas noncovalent binding occurs when the peptides from charge to charge complexes. Aggregates can be small, for example, when two or three peptides associate to form dimers and trimers, respectively, which tend to be more soluble and not so affected by freezing. Bigger aggregates tend to be insoluble, and once they form they are usually impossible to resolubilize. Due to this nature, they are thought to trigger disease states in the body such as Alzheimer's disease (Parker and Reitz, 2000; Itkin et al., 2011).

Aggregates are usually formed when peptides are present in high concentrations, or under stressed conditions such as when experiencing excessive heat, pH conditions, freezing, and agitation. The latter two are of especial importance as they often encountered when biological samples are collected, handled, and stored; for example, human growth hormone (hGH) can form both soluble and insoluble aggregates on freezing, but more insoluble aggregates are formed if the sample is frozen rapidly. Fewer aggregates were also formed at pH 7.8 than at pH 7.4 when the samples were frozen. Agitation was also found to readily induce aggregation of hGH molecules (Pearlman and Bewley, 1993).

Good practice for sample collection, handling, and storage needs to be established during method development and validation in order to successfully analyze peptides in biological matrices. In particular, freezing procedure needs to be evaluated and optimized, as $\leq -20°C$, may actually be better than $\leq -80°C$ with respect to minimizing aggregation. Various other additives and optimum conditions can also be investigated. For example, the addition of surfactants, such as polysorbate, has been found to reduce the aggregation of hGH and insulin. Human serum albumin and sucrose have been found under certain conditions to improve the stability of human interleukin-1β. Insulin can be stabilized by keeping the pH neutral and the addition of extra zinc ions, which stabilizes the hexameric structure of insulin, further minimizing aggregation (Wang and Pearlman, 1993). Once suitable conditions are identified, the detailed sample collection and storage procedures can be standardized and made available to sample collection facility.

13.6 STABILIZING SAMPLES

Stability of the drug and any known metabolites is a critical factor to be considered in designing and developing good assay methods, as discussed in other chapters in this publication. Correct sample collection and handling is the first step in ensuring that valid data is generated. This can be especially challenging with drugs, for example, prodrugs, that are designed to be unstable due to their inherent mode of action of rapidly converting to an active form. There have been many recent reviews looking into the importance of preanalytical sample stability in the bioanalysis (Dell, 2004; Chen and Hsieh, 2005; Briscoe and Hage, 2009; Li et al., 2011).

The factors causing analyte instability can usefully be divided into two main groups: physical factors and structural/chemical factors (Briscoe and Hage, 2009). Physical factors can be factors such as light and temperature. The chemical structure of the drug and metabolites can be susceptible to chemical factors such as pH, oxidation, and action of enzymes all of which can lead to instability. Examples of this are if the molecule is a prodrug, lactone/acid conversion and tautomerization, acyl glucuronide migration, thiol group reactivity, *N*-oxide decomposition, and chiral interconversion.

The factors that can affect the stability of a drug can be tackled in different ways. Instability can, to some extent, be predicted by careful study of the molecular structure, and accordingly, early preventive action should be taken. Often it is not a single factor but a combination of different factors that cause instability. Therefore, each of these factors must be investigated and addressed. This section gives an overview of the issues that need to be considered during sample collection, but further information and detailed experimental protocols for dealing with specific classes of compounds are described in Part IV of this handbook.

13.6.1 Physical Factors

13.6.1.1 Photosensitivity Photochemical instability is common in molecules that contain unsaturated double bonds (e.g., alkenes where E/Z isomerization may occur) and/or heteroatom double bonds such carbonyls, nitro-aromatic groups, and aryl chloride groups (where dechlorination may occur). Some molecules such as fluoroquinilones (e.g., lomefloxacin), nisoldipine, retinoids, and Vitamin D3 may be sensitive to UV light to different extents (Moore, 2004). The extent of precautions required to ensure that stability during sample collection may vary from using ambered sample tubes, or wrapping tubes in foil, to working under special lights that emit only a specific wavelength (e.g., filtered yellow light), and storing samples in light proof boxes. If a compound is very light sensitive, it is useful to use a light meter and establish the maximum Lux level that still ensures compound stability.

13.6.1.2 Thermal Stability Increased temperatures can accelerate the speed at which analytes of interest degrade in the intended matrix or facilitate unwanted reactions. A good general strategy for handling unstable analytes is to collect and freeze blood immediately (if blood is the chosen matrix) or to process the blood sample immediately, for example, centrifuging blood for plasma at ~4°C or below, and freezing the harvested plasma sample immediately at or below −70°C. Examples of drugs well known to be temperature sensitive in matrix are simvastatin and aspirin (Buskin et al., 1982; Zhao et al., 2000; Chung et al., 2001).

Drug instability may not be immediately apparent, and therefore, it is always valuable to assess frozen stability over a longer term to monitor any slow degradation. This can be especially important for later phase clinical studies, where samples may be stored for a long period prior to analysis while waiting for more subjects to be recruited. Aspirin has been shown to degrade 20% after 11 days storage at −20°C, so needs to be stored at −80°C (Buskin et al., 1982). Similarly, a study of four different benzodiazepine drugs in whole blood has shown that a loss of 10–20% of drug occurred when stored at −20°C for period of a year, but when stored at −80°C for the same period, there was no significant loss of drug (El Mahjoub and Staub, 2000).

13.6.1.3 pH Instability Control of pH is important in cases of both chemical and enzymatic instability, as a chemical reaction may be catalyzed at a specific pH, or an enzyme may only work in a specific pH range. Research has also shown that the pH of plasma and urine can also change with time, rising from the normal physiological pH 7.4 up to pH 8.8 (Fura et al., 2003), thus precise buffering can be important for pH sensitive compounds.

Classes of compounds typically affected by pH instability are acidic compounds that form acyl glucuronide metabolites, esters, amides, lactones, and lactams. A well-known example where pH control is important is the interconversion of statins (where the lactone ring is hydrolyzed) to the β-hydroxy statin acid. At higher pHs, such as physiological pH, the interconversion is increased. Careful control of the pH, by buffering to pH 4.5 and reducing the temperature minimizes the S-interconversion (Jemal et al., 1999; Jemal and Xia, 2000; Zhang et al., 2004). A common way to control pH is to add acids, for example, citric acid, at the point of sample collection (citrate tubes are available—3.2–3.8% citrate) (Narayanan, 1996).

Acyl glucuronides are common metabolites of compounds containing carboxylic moieties (e.g., clopidogrel, salicylic acid, diclofenac, ibuprofen, valproic acid, furosemide, telmisartan, and mefenamic acid) and are also recognized for their instability at physiological pH, being very prone to hydrolysis at pH 7.4. The situation is actually more complicated as acyl glucuronides can also undergo intramolecular acyl migration, where different isomers of the drug glucuronide metabolite are produced when the acyl group migrates. Beta 1-O-acyl glucuronides are most prone to acyl migration, but the isomers produced can be much more stable. The best way to accurately measure concentrations of acylglucuronide metabolites is to "freeze" the migration, by adjusting the pH to acidic conditions with phosphoric acid (to ideally between pH 2–4), at the point of sample collection (Shipkova et al., 2003; Dell, 2004; Srinivasan et al., 2010). However, making the pH much more acidic than absolutely necessary should be avoided, since the use of strong acids could lead to unwanted hydrolysis of the analyte.

13.6.1.4 Enzymatic Instability Before attributing analyte instability to enzymatic factors it is essential to make sure that observed instability is not just a chemical hydrolysis effect, as otherwise addition of enzyme inhibitors will not be effective in stabilizing the analyte. Although there are a wide

TABLE 13.7 Examples of Enzyme Classes, Potential Inhibitors, and the Analytes That Have Been Stabilized Using the Inhibitor

Enzyme classes	Potential inhibitors	Examples of analytes
Cholinesterase	Sodium fluoride	Cefetamet pivoxil[a], Cocaine[b]
	Phenylmethylsulfonyl fluoride (PMSF)	Aspirin[c]
	Physostigmine	Pyridostigmine[d]
	Neostigmine	Physostigmine[e], Bambuterol[f]
	Paraoxon	Procaine[g]
	Dichlorvos	Oseltamivir[h]
Carboxylesterase	Sodium fluoride	Diltiazem[i], Caffeic acid[j]
	PMSF	Phenacetin[k]
	Bis-(4-nitrophenyl)-phosphate (BNPP)	Nafamostat[l], Phenacetin[k]
	Thenoyltrifluoroacetone (TTFA)	Tocopheryl succinate[m]
	Prazosin	Phenacetin[k]
Serine esterases	Diisopropyl fluorophosphate (DFP)	Nafamostat[l]
Arylesterase	5,5′-dithiobis-2-nitrobenzoic acid	Nafamostat[n], Salvinorin A[o]
Cytidine deaminase	Tetrahydrouridine (THU)	Gemcitabine[p]

[a]Wyss et al. (1988). [b]Skopp et al. (2001). [c]Liang et al. (2008). [d]Zhao et al. (2006). [e]Elsayed et al. (1989). [f]Luo et al. (2010). [g]Jewell et al. (2007). [h]Chang et al. (2009) and Lindegardh et al. (2006). [i]Kale et al. (2010). [j]Wang et al. (2007). [k]Kudo et al. (2000). [l]Cao et al. (2008). [m]Zhang and Fariss (2002). [n]Yamouri et al. (2006). [o]Tsujikawa et al. (2009). [p]Xu et al. (2004).

variety of enzyme inhibitors available, some of them are particularly toxic (e.g., paraoxon is a potent neurotoxic agent). Therefore, it is useful to explore the less toxic inhibitors first.

Sodium fluoride is commonly used in sample collection tubes to enable determination of glucose, but it can often be used to help stabilize compounds as it inhibits glucose metabolism and is a protein denaturant. Another option to consider is to collect the samples directly into organic solvent, as this will also denature all the proteins and enzymes.

The key class of enzymes that cause issues are esterases. Esterases can be further subdivided into specific classes such as cholinesterases, carboxylesterases, serine esterases, and arylesterases. These enzymes can be found in different concentrations in matrices such as blood, plasma, serum, and tissues such as liver. It is useful to note that they are most ubiquitous and potent in rodents, and generally least potent in humans (useful as implementing complicated collection procedures may be more difficult in a clinical setting). There are only four different subgroups of esterase enzymes that are present in human blood: butyrylcholinesterases, acetylcholinesterases, albumin esterases, and paraoxonases (Lindegardh et al., 2006).

It is difficult to recommend a universal enzyme inhibitor, as many enzyme inhibitors are specific to certain classes of enzyme. Therefore, a screening approach is recommended (Fung et al., 2010) to find an effective inhibitor and then customize its concentration to optimally stabilize the analyte of interest. Table 13.7 lists examples of different classes of enzyme commonly found in blood and plasma (which are known to cause stability problems), with examples of their potential enzyme inhibitors and the drugs the inhibitors have been used to stabilize. When analyzing samples it is important to remember that all blank matrix used to prepare calibration standards, QC samples, and for dilution of overrange samples must be also be treated with the same concentration of inhibitor.

13.6.2 Chemical Factors

13.6.2.1 Instability Due to Oxidation This type of instability can occur with analytes that contain phenol or alcohol groups. A common reagent used to inhibit auto-oxidation is ascorbic acid, which is effective in stabilizing mitozantrone, psilocin, and rifampin (Priston and Sewell, 1994; Chung et al., 2001). Compounds such as levodopa can be stabilized by the addition of sodium metabisulfite in combination with EDTA (Saxer et al., 2004). Another commonly used antioxidant for the analysis of lipophilic analytes such as vitamin A and vitamin E is BHT. The compound can be added at different concentrations depending on which solvent is used to dilute it (Chow and Omeye, 1983).

Another group compounds that are subjected to oxidation are thiols. Thiols can also form dimers, or form S-S bridges with endogenous molecules, such as cysteine, containing sulfur groups. A good approach to stabilize these molecules is derivatization at the sample collection stage. Methyl acrylate reagents are especially good, and commonly used for this purpose, as the derivatives they form do not produce a new chiral center. Methyl acrylate derivatization was used effectively for the analysis of omapatrilat and its metabolites, where it was added to a K_3EDTA sample collection tube at a concentration of 10 μl per 1 ml of blood (Jemal et al., 2001). Further information about derivatization reagents and auto-oxidative compounds can be found in the relevant chapters of this handbook.

13.6.2.2 ***N-oxide Reversion*** *N*-oxides are a class of labile metabolites, which are known to be potentially unstable in biological fluids. This is particularly a problem with *N*-oxides in blood before the plasma is prepared. Also the addition of some reagents to stabilize the parent molecule can in turn cause the reduction of the *N*-oxide (e.g., ascorbic acid). Care has to be taken with sample collection, so as not to cause additional problems during the extraction and analysis. If the *N*-oxide is soluble in plasma, but the parent drug is not, for example, chlorpromazine, a good potential solution is to analyze the blood and plasma separately to quantify the concentration of parent drug in blood and quantify the concentration of *N*-oxide in plasma (Dell, 2004).

13.6.2.3 ***Chiral Interconversion*** All chiral compounds can undergo interconversion (under certain specific conditions); however, it is usually only a problem if this interconversion occurs under a physiological pH and temperature. A sample can, therefore, be stabilized by adjusting the pH with the addition of a suitable buffer and storing the samples at −80°C. For example, to stabilize thalidomide, a 0.2 M, pH 2 citrate-phosphate buffer was added, and this enabled the plasma samples to be stored for more than 1 year at −80°C (Murphy-Poulton et al., 2006).

13.6.3 Stability and Dried Blood Spots

A revival of interest in the DBS technique has highlighted (in addition to requiring less blood volume, easier collection, storage, and transfer of samples) its advantage in effectively stabilizing some unstable compounds. This is because as the sample dries, it dehydrates, and the enzymes present lose their 3D structure and are inactivated. The drying time (and relative humidity of the surrounding atmosphere) can be critical to the effectiveness of this process, but the use of DBS can significantly improve the stability of some analytes, for example, oseltamivir and mycophenolate metabolites (D'Arienzo, 2010; Gataye Perez, 2010; Yapa et al., 2006).

13.7 BIOLOGICALLY HAZARDOUS SAMPLES

In general, all biological samples should be treated as though they are potentially hazardous, particularly if the samples originate from nonhuman primates or from clinical studies. Shipping, general handling procedures, sample storage, and working practices in the laboratory should be designed with this in mind. Staff performing analytical procedures with samples should wear the appropriate protective clothing such as (minimally) a laboratory coat, disposable gloves and eye protection. Any manipulations performed during sample preparation should be carried out under an appropriate extraction hood or in a biological safety cabinet in order to avoid any potentially harmful contact with the samples (e.g., aerosols can be produced during removal of sample container lids and during pipetting). Automation of the sample preparation procedure minimizes contact with toxic reagents or biologically hazardous samples.

Consideration should be given to including viral inactivation stages in the sample preparation procedure for methods that are used to analyze hazardous samples on a regular basis. Many viruses have protein or lipid coats that can be altered by physical or chemical techniques, and as a consequence, the virus becomes inactive or unable to infect. Specific inactivation procedures work most effectively against different types of virus (e.g., nonenveloped viruses (single-stranded DNA viruses and single- and double-stranded RNA viruses) or enveloped viruses (single-stranded RNA viruses and double-stranded DNA viruses)) (Sofer et al., 2003).

Some widely used procedures for inactivating viruses that may be suitable for bioanalytical applications are solvent or detergent inactivation, low or high pH treatment, and collecting samples as DBS (Sofer et al., 2003; Hersberger et al., 2004; WHO Technical report, 2004; Pelletier et al., 2006; Lakshmy, 2008).

Detergent inactivation is a common technique used in the blood plasma industry, but it is only effective against viruses with lipid coats. Most enveloped viruses cannot survive without their lipid coat, so they die when exposed to the appropriate detergent, of which the most commonly used is Triton-X 100.

Pasteurization is a simple technique that can be effective against both enveloped and nonenveloped viruses and is not dependant on the use of sophisticated equipment. The technique typically involves incubation of samples for approximately 10 h at 60°C (although longer incubation times may be required to inactivate certain types of viruses) and consequently the drug and any metabolites need to be stable under the incubation conditions.

Some viruses will denature if exposed to low or high pH, but again an important consideration when using this procedure is the stability of the drug and metabolites in acidic or alkaline conditions.

Samples collected as whole blood and stored as DBS are a lower source of potential infection, as many viruses known to be present in blood lose their ability to infect as a consequence of disruption of their envelope on drying (Lakshmy, 2008).

There are a number of different procedures that can be used to inactivate viruses in biological samples and the technique adopted should be chosen on a case-by-case basis. Consideration should be given to using a viral inactivation stage approach, particularly if biologically hazardous samples are going to be analyzed on a regular basis. Whichever technique is chosen (heat inactivation, pH control, or use of detergent) to inactivate the viruses, the impact on the sample should be fully evaluated. Additional assessments may need to be performed during the method development and validation of the bioanalytical assay to ensure that the chosen process does not impact on the determination of the drug and/or metabolite concentrations.

13.8 SAMPLE STORAGE

13.8.1 Types of Container for Sample Storage

The size of container/tube should be appropriate for the volume of sample collected. Tubes should be large enough to ensure that there is adequate headspace to allow thorough mixing. There are a number of examples in the literature that have cited insufficient mixing as a reason for assays failing ISR, particularly urine assays (Fu et al., 2011). Consideration should be given to the expansion of liquids in solid state while freezing. Conversely, a sample tube that is too big increases the likelihood of evaporation adversely affecting the volume of sample.

LC-MS instrumentation is becoming increasingly more sensitive to fulfill the needs of quantitative analysis of drug candidates and/or their metabolites in the samples collected from preclinical and/or clinical studies with very low doses. Accordingly, there is an increasing risk of detecting contamination in collected samples. For this reason, minimizing aerosols during sample handling procedure is very important. Microcentrifuge tubes often have clip on lids, which have the potential to release an aerosol when opened. Care should be taken to prevent the above scenario from occurring.

For manual sample preparation the ideal tube will be wide enough to allow a pipette to be able to access the tube and not so long to prevent a pipette tip from reaching the bottom. However, the disposable tips on handheld (lower volume) pipettes are not always narrow enough or long enough to reach the bottom of a sample tube. Automated workstations such as Hamilton Star and Tecan Freedom EVO often have sampling probes that are long enough to reach the sample contents, whichever tube size is used in a bioanalytical laboratory. Thus with automated systems, pipette heights can be optimized and set as required.

Ideally calibration standard (if not prepared fresh daily) and QC samples should be stored in the same types of tube as are used to store study samples. Where possible the volume of the stored QC aliquots should also mimic that of a typical study sample. However, freezing samples (whether QC samples or study samples) in the exact volume required for an assay increases the chances of losing sample to the walls and caps (especially if the sample is transferred to another tube for extraction).

Where sample volumes are very small, for example, from mouse PK or toxicokinetic studies, using tubes with U- or V-shaped bottom allows smaller volumes to be aspirated more easily. Low-volume tubes with the same external diameter as the regular ones but with a raised bottom are commercially available. Care should be taken when vortex mixing these tubes to ensure that a proper mixing as the movement for the top is relatively static compared to the bottom of the tube.

Thought should be given to the racking system used to store sample tubes, ensuring easy and convenient access. Ideally samples should also be stored with a small footprint, for example, 96-well cluster tubes from Matrix, Micronics, and AbGene. Such tubes can be coded with a 2D barcode to enable an entire rack of samples to be identified in seconds. Combined with an automated tube sorter such as a Bohdan, samples can be formatted to batches, subject order, and so on, as required without tedious and potentially error prone manual handling.

Accessories for sample tubes in 96-well format can considerably improve the workflow in a bioanalytical laboratory. Automated cappers and decappers can remove, hold, and reapply caps in minutes, preventing repetitive manual handling and the chance of putting a lid back on an incorrect tube. Pierceable septums are also available that allow matrix to be directly sampled without the need to remove the cap. This is particularly useful when using online extraction technique such as the Symbiosis system or Turbulent flow. In practice, the septum must allow for air to enter the tube while sample is removed to prevent a vacuum from forming in the needle that could draw a low sample volume. When handling tubes the same lid should be put back on the same tube to prevent cross-contamination from one sample to the next. For this reason, tubes with an external thread can be better since the caps have a "skirt" that prevents the thread on the cap from coming into contact with the work bench.

13.8.2 Material

Inappropriate materials used for various containers, tubes and syringes may cause unnecessary or unexpected additional issues for the bioanalyst. In general, most sample collection containers used these days are some type of plastic: polypropylene, polyethylene, polystyrene, or polycarbonate. If an issue is found (e.g., nonspecific binding) with one type of plastic, it may be resolved by switching to a different type of plastic (Stout et al., 2000). Glass tubes are rarely used these days because of their fragility during transportation and their likelihood to shatter during the freezing process. In addition, glass tubes are also more expensive to buy and to ship due to their weight. Many drugs have been found to adsorb to glass (as the surface is slightly acidic). Anecdotally, the length of time that glass is allowed to anneal can have an impact on its adsorptivity. Normally the annealing period is at least 24 h, but if the glass is cooled down too fast (in order to speed up production), it produces more active binding sites (Hill HM, personal communication). Therefore, in order to avoid problems caused by variability in glass production, silanized glassware could be used as an alternative (particularly if the drug also binds to plastics).

One of the issues found with some plastics is that a lot of different chemical additives are used in manufacturing (e.g., plasticisers such as phthalates, slip agents, and lubricants). Differences in manufacturing processes can be very important and cause issues that are very difficult to investigate. When manufacturing U-shaped tubes or 96-well plate with U-shaped wells, due to the design, the moulds can easily be removed from the finished product. With V-shaped wells or tubes the moulds are more difficult to remove, and so more lubricants may need to be used to facilitate easier removal. This process may not be monitored as closely as it should be and can lead to batch-to-batch variability from a single manufacturer, as well as variability between different manufacturers. Sometimes trace amounts of these chemical additives may leach into the sample on storage. Occasionally they may cause issues with LC-MS/MS assays by interfering background ions, or increasing matrix effects. Lists of these ions and their source chemicals can be found on the Internet (Waters' list of ESI+ Common Background Ions), which can be useful to check if a problem of this nature is suspected.

Consideration should be given to all the components of storage container, and to what conditions they will be subjected. The caps or lids used with tubes or containers are often of a slightly different material to the main vessel (e.g., if a tube has a screw cap, with a Teflon ring seal, and the sample needs to be snap frozen in liquid nitrogen, will the seal hold at −196°C?). Similarly, will the tubes be centrifuged, and if so, will they be able to withstand the *g*-force they are subjected to without deforming?

Polypropylene and polyethylene are the most commonly used materials for collection and storage of blood, plasma, and serum. They are resilient materials, which show least interaction with analytes. A wider range of tubes/containers are used to collect and store urine. In particular, polystyrene "Universal" tubes are often used in clinics for urine, but they should be used with care, as polystyrene is more prone to nonspecific binding, compared to either polypropylene or polyethylene.

13.8.3 Temperature of Samples Storage

Both the FDA and EMA guidance documents on bioanalytical method validations make frequent references to assessing stability during the sample collection processes and stability during subsequent storage. The EMA guideline infers the requirement to demonstrate stability at both −20ºC and −70/−80ºC should samples have been stored at both of these temperatures:

> The QC samples should be stored in the freezer under the same storage conditions and at least for the same duration as the study samples. For small molecules it is considered acceptable to apply a bracketing approach, i.e. in case stability has been proved for instance at −70°C and −20°C, it is not necessary to investigate the stability at temperatures in between. For large molecules (such as peptides and proteins) stability should be studied at each temperature at which study samples will be stored.

Documentation needs to track the movement of samples, from their initial receipt, any transfers between freezers, any freeze–thaw cycles they undergo, to their eventual disposal. Also the temperature during their storage needs to be monitored continually by some kind of data logger or electronic temperature monitoring system. A suitable alarm system should be in place to alert staff of temperature changes outside a prescribed range. Failure to do this for pivotal studies (e.g., bioequivalence studies) can lead to the rejection of the study.

13.8.4 Disaster Recovery Plan

Bioanalytical laboratories supporting studies for regulatory submissions should have suitable facilities for the temperature controlled storage of biological samples in their care. If anything happened to compromise the integrity of these samples, the results from their subsequent analysis would be subject to question. This may lead to a repeat study at great expense and loss of time. This could critically impact of the progress of the drug development program. Therefore, a very important aspect of sample storage for the bioanalytical laboratory is to have a disaster recovery plan in place. This plan defines the actions to be taken in the event of disastrous occurrences such as a power failure, equipment breakdown (e.g., freezer breakdown), and/or environmental factors such as flooding. The plan should specify who is responsible for each aspect and who to contact if a temperature alarm is triggered (both in and out of normal working hours). Other details in the plan include the process for evaluating the situation as to whether it is a short or long-term problem and the arrangements to address this situation. The action could include transferring the samples to a backup freezer or facility, or adding dry ice (solid carbon dioxide) to the freezer to manage the situation in a short term, until a repair made or a replacement in place. In regulatory perspective, all steps should be comprehensively documented with a full audit trail.

13.9 TRANSPORT AND SHIPPING OF SAMPLES

13.9.1 Practicalities of Transporting Samples

Samples should be collected into containers with suitable volume for easy and cost-effective transportation. On the other hand, tubes should have a sufficient air gap to allow for expansion during freezing. Appropriate subaliquot volumes should be taken of urine samples, rather than sending the whole sample, which may be over a liter for pooled urine samples.

For human feces and tissue samples, it is advisable to process the samples before incurring high transportation costs (samples should be accurately weighed prior to processing, as described in Section 13.2.8). Aliquots can then be taken and sent to the analytical laboratory while retaining the rest original sample or aliquots at the investigation site. In order to ensure the high degree of confidence in this process, it is essential that the analytical laboratory provides step-by-step sample processing protocols. The same approach can be followed for larger species studies where bulk transport may be an issue. While for smaller species studies in rat, mouse, guinea pig, and so on, the whole organ, or whole sample at a specific time point can be frozen at −20°C/−80°C and transported in dry ice.

13.9.2 Packaging

During the transportation of biological samples from an animal facility or clinic to a laboratory for analysis, or from one laboratory to another, it is important to ensure the integrity of the samples, and all respective regulations are followed (IATA Guidelines, 2005, onwards). Samples to be shipped should be packed with sufficient dry ice to last for a greater period of time than the anticipated transit period. It is a good practice to ship samples on Monday, so any issues with the shipment (e.g., flight cancelled due to bad weather, or being held up at customs) can be resolved before the weekend.

If samples are to be shipped interstate or internationally, the appropriate IATA (International Air Transport Association) Guidelines should be followed to ensure safe shipment of biological samples. The IATA guidelines specify the nature of the packaging required, both inner and outer packaging, as well any absorbent material that may be required, in order for the package to meet the requirements of the Dangerous Goods Act (or equivalent legislation).

13.9.3 Documentation

Attention should be paid to having the correct documentation in place as well as the correct packaging. This step is usually straightforward for the samples being transported within one country. For international shipping, it is important to ensure that all the relevant export and import permits been obtained and completed correctly. It is not uncommon to see a delayed shipment due to missing one or more permits, certificates, or licenses when needed. In some cases, this delay can be caused by the package not being logged with the required authorities, or essential documentation being not correctly signed. The documentation required is generally a Dangerous Goods Declaration, a Customs invoice (to grant customs clearance at the country of destination), an Import permit, an Export certificate, Sample details and an Air Waybill (AWB). The laboratory or clinic responsible for dispatching the samples should ask their courier or shipping agent for the AWB number, and inform the recipient of it, so that the shipment can be tracked.

Proper documentation is of particular importance when shipping samples collected from primates, or other wild animal species around the world. The Convention on the International Trade in Endangered Species (CITES) specifies the regulations on shipping samples (blood, plasma, or tissues) from endangered species internationally. The correct CITES permits (both import and export) need to be obtained on fee basis, in order to transport these samples legally between countries, otherwise they will be refused to entry at the receiving country. Therefore, when planning this kind of analysis, time needs to be allowed for obtaining the permits.

13.9.4 Monitoring Samples in Transit

The handling and storage of biological samples within the collection site can be carefully monitored and controlled. It is important to continue to monitor the state of the samples during shipment, up to the point that the samples are safely received and checked by the receiving analytical laboratory. Various data logger devices can be placed along with the samples in the shipment to monitor the temperature inside the package during transportation. Upon arrival, the temperature data can be downloaded, or the data logger returned to the sample dispatch site for further processing to check whether the needed temperature was maintained. If it was not, then additional stability needs to be assessed for the QC samples that reflects the conditions to which the samples were subjected to during shipping.

Various other approaches have also been used to monitor sample conditions during shipment. For example, a tube containing an aliquot of frozen water could be included with the rest of the samples. A small steel ball bearing is then placed on top of the ice inside the tube. If the samples thaw during transportation, then upon arrival, even if the dry ice was topped up and the samples refrozen, the ball bearing will be at the bottom of the tube. A similar technique is to freeze an aliquot of water in a tube in a slanted or upside-down position. This tube is then placed along with the other sample tubes in the shipping box. If the samples thaw at any point, even if later refrozen, the water/ice will be level at the bottom of the tube. However, due to significant improvement on shipment logistics these days, and the reluctance of transporters to open packages to add additional dry ice, these techniques are employed less frequently.

13.9.5 Receipt of Samples

It is important that the condition of the samples is accurately recorded upon receipt. A chain of custody form is usually supplied to record this information, to keep track of the persons handling the samples, and to provide the necessary sample storage information. This chain of custody form should be completed in a timely manner. Signatures of the sender, the receiver and any intermediary handling the samples should be entered accordingly, so that the condition of the samples/shipment can be accounted for during the entire transportation process.

Upon sample arrival, staff of receiving laboratory (through the advance notification by the sender) should be available to receive, check, and inventory the samples, and also ensure that the samples are stored correctly prior to analysis. If the samples are received out of hours (e.g., at the weekend), a minimum instruction should be available to be followed to ensure safe keeping of the samples under the correct storage conditions until the samples can formally be checked-in by the bioanalyst. A copy of the completed chain of custody form is then returned to the collection site to formally acknowledge receipt of the samples.

13.10 COMPLIANCE CHECKLIST

From a compliance perspective, the quality/integrity of conduct of a study for regulatory purposes relies on each procedure undertaken, including preanalytical phase of sample collection handling and storage, to be accurately documented. Evidence is required that all the samples were treated identically and according to the specified standard procedure. This ensures that the data produced from the analysis of the samples is consistent and can be compared across different studies (and within the same study if data is generated at more than one site) and that the correct conclusions can be drawn.

A list of questions is taken from Attachment A for the Bioequivalence Inspection Report of the FDA compliance program guidance manual (FDA, 2000).

This summarizes the factors to be considered for sample collection from clinical trials, especially for specimen handling and integrity in the analytical laboratory:

- Are receipts available for the sending/receiving of samples?
- Is a documented history of sample integrity available (e.g., the sample storage time and conditions prior to shipment)?
- Was the type of protection during shipment appropriate?
- Note conditions of the samples on arrival at the analytical laboratory, along with the identity of the person(s) receiving the samples.
- Describe the storage equipment for samples until analysis.
- Are storage procedures appropriate (e.g., protect from UV light?)
- Capacity versus number of samples in storage.
- Alarm set points correct is temperature control recording devices working correctly?
- Any evidence of sample thawing.
- Are action plans in place in case of power loss and so on?
- Are samples labeled and separated adequately in storage to prevent sample loss or mix up between studies?
- How is sample identity (ID) maintained through transfer steps?
- Is number/time of any freeze–thaw cycles, even accidental ones recorded?

With regard to sample collection, processing, handling, and storage, there are a few topics frequently picked up on during FDA audits (and audits by other monitoring authorities) of bioequivalence studies as being deficient (FDA presentation UCM182564). Therefore, it is important to have a clear link between the PK sample and its originating subject, that is, there should be no ambiguity in sample ID through the process. Also inadequate records are often seen for the blood draw time and PK sample processing time.

The handling of the samples should be within the coverage of the validated storage stability. For example, is the number of times a sample was frozen and thawed within a study less than or equal to the number of freeze–thaw cycles that were assessed during method validation? Is there sufficient stability data to cover the storage of samples prior to analysis?

The above points can be used as a checklist to ensure that the integrity of study samples is maintained during collection and storage, and that the data produced from subsequent bioanalysis is acceptable to the regulatory authorities. There should then be no reasons for the FDA or other authority to reject the study.

13.11 SUMMARY

This chapter examined the importance of collection, processing, handling, and storage of samples for bioanalytical studies. To obtain meaningful and accurate data, a good assay needs to be more than simply getting calibration standard and quality control samples to pass the required acceptance criteria. Thought needs to be given to good method development that investigates potential issues and optimizes the method to ensure that problems do not occur later. Once a method has been developed and validated, good scientific practice needs to be followed to ensure that the integrity of the samples is maintained from the time of collection to the moment of analysis.

Best practice begins with giving consideration to the matrix to be analyzed, as different matrices have different needs with regards to sample collection. Appropriate containers need to be chosen that are of a suitable material and size, for the sample matrix involved, at each transfer step of the sample collection, handling, and storage processes.

A large proportion of the samples for bioanalytical assay by LC-MS/MS will be blood, plasma, or serum. Many factors need consideration during the blood collection process, the choice of which anticoagulant to use being especially important. The blood collected needs to be processed appropriately to give plasma or serum as required. The collection of urine samples brings its own challenges as it is an aqueous medium, and so may have more issues with adsorption of the analyte to the container. The collection of more solid samples such as tissue and fecal samples has different challenges, e.g., does additional processing (homogenization) of the samples need to be performed prior to storage. Additionally, the nature of the analyte itself may impact on how samples need to be collected and stored; for example, peptides and proteins will need care to avoid adsorption and aggregation issues.

Ensuring stability of the analyte in the sample matrix is a very important aspect of the sample collection, handling, and storage process. The potential for stability issues needs to be examined prior to starting any assay development, and likely issues need to be investigated and addressed during method development. A wide variety of stabilization techniques are available to the method developer, ranging from controlling physical factors such as temperature and light to controlling chemical factors by adjusting the pH of a sample or adding of stabilization reagents such as enzyme inhibitors and antioxidants.

Once the samples have been collected and processed, care must still be taken to ensure that the storage conditions and procedures are optimal, as sometimes the samples may need to be stored for a long time prior to analysis. This includes having things like a disaster recovery plan in place, to cover problems with freezers, or unexpected issues such as flooding or power loss.

Often samples are not collected and analyzed by the same facilities or laboratories, and may be shipped around the world for analysis. Thus, the transportation and shipping of samples also needs be planned and executed carefully. This involves not only considering the basic physical aspects of ensuring the samples are maintained at the correct temperature during the journey from the collection facility to the analytical laboratory, but also ensuring that all the accompanying documentation such as sample inventories, chain of custodies, import and export permits, and any monitoring required (e.g., use of data loggers) has been addressed.

Throughout all the processes involved during sample collection, processing, storage and any transportation, the maintenance of adequate documentation of all the procedures and samples involved is essential in maintaining the integrity of the study data. This is emphasized in the guidance documents issued by the regulatory authorities. A checklist of questions is provided to help guide the analyst to record all the necessary information required to ensure that the integrity of the study samples is maintained during sample collection through to analysis. The data subsequently produced should then be acceptable to the regulatory authorities.

ACKNOWLEDGMENT

The authors acknowledge and thank Iain Love for his help in reviewing the manuscript and Patricia Naylor for her secretarial assistance.

REFERENCES

Briscoe CJ, Hage DS. Factors affecting the stability of drugs and drug metabolites in biological matrices. Bioanalysis 2009; 1(1):205–220.

Burtis CA, Ashwood RE, Bruns DE. *Tietz Fundamentals of Clinical Chemistry*. 5th ed. Philadelphia, PA: Saunders; 2001.

Bush V. Why doesn't my heparinized plasma specimen remain anticoagulated?: A discussion on latent fibrin formation in heparinized plasma. LabNotes 2003;13(2):9–14

Buskin JN, Upton RA, Williams RL. Improved liquid chromatography of aspirin, salicylate, and salicylic acid in plasma, with a modification for determining aspirin metabolites in urine. Clin Chem 1982;28:1200–1203.

Cao YG, Zhang M, Yu D, Shao JP, Chen YC, Liu XQ. A method for quantifying the unstable and highly polar drug nafamostat mesilate in human plasma with optimised solid-phase extraction and ESI-MS detection: more accurate evaluation for pharmacokinetic study. Anal Bioanal Chem 2008;391(3): 1063–1071.

Chang Q, Chow MS, Zuo Z. Studies on the influence of esterase inhibitor to the pharmacokinetic profiles of oseltamivir and oseltamivir carboxylate in rats using an improved LC/MS/MS method. Biomed Chromatogr 2009;23(8):852–857.

Chen C, Bajpai L, Mollova N, Leung K. Sensitive and cost-effective LC-MS/MS method for quantitation of CVT-6883 in human urine using sodium dodecylbezenesulfonate additive to eliminate adsorptive losses. J Chromatogr B 2009;877:943–947.

Chen J, Hsieh Y. Stabilizing drug molecules in biological samples. Ther Drug Monit 2005;27:617–624.

Chow FI, Omeye ST. Use of antioxidants in the analysis of vitamins A and E in mammalian plasma by high performance liquid chromatography. Lipids 1983;18(11):837–841.

Chung WY, Chung JK, Szeto YT, Tomlinson B, Benzie IF. Plasma ascorbic acid: measurement, stability and clinical utility revisited. Clin Biochem 2001;34(8):623–627.

D'Arienzo CJ, Ji QC, Discenza L, et al. DBS sampling can be used to stabilize prodrugs in drug discovery rodent studies without the addition of esterase inhibitors. Bioanalysis 2010;2(8):1415–1422.

Dell D. Labile metabolites. Chromatographia 2004;59:S139–S148.

EC 2006 Guidance document for GLP inspectors and GLP test facilities: Cross-contamination of control samples with test item in animal studies. Available at http://ec.europa.eu/enterprise/sectors/chemicals/files/glp/guidance_xcont_final_18_01_2006.en.pdf. Accessed Mar 14, 2013.

EMA Guideline on bioanalytical method Validation. Adopted July 2011, Effective February 2012. Available at www.ema.europa.eu/docs/en_GB/document_library/Scientific_guideline/2011/08/WC500109686.pdf. Accessed Mar 14, 2013.

El Mahjoub A, Staub C. Stability of benzodiazepines in whole blood samples stored at varying temperatures. J Pharm Biomed Anal 2000;23:1057–1063.

Elsayed NM, Ryabik JR, Ferraris S, Wheeler CR, Korte DW. Determination of physostigmine in plasma by high-performance liquid chromatography and fluorescence detection. Anal Biochem 1989;177(1):207–211.

Evans C. The application of Dried Blood Spots for quantitation of xenobiotics—a paradigm shift within Pre-clinical DMPK. Presentation given on Feb 11, 2010, at Delaware Valley, Drug Metabolism Discussion Group meeting.

FDA presentation by J.A. O'Shaughnessy. Bioequivalence and Good Laboratory Practice. Available at www.fda.gov/downloads/Drugs/NewsEvents/ucm182564.pdf. Accessed Mar 14, 2013.

FDA. Guidance for Industry: Safety Testing of Drug Metabolites. February 2008. Available at www.fda.gov/downloads/drugs/GuidanceComplianceRegulatoryInformation/Guidances/ucm079266.pdf. Accessed Mar 14, 2013.

Fu Y, Li W, Smith HT, Tse FLS. An investigation of incurred human urine sample analysis failure. Bioanalysis 2011;3(9):967–972.

Fung EN, Zheng N, Arnold ME, Zeng J. Effective screening approach to select esterase inhibitors used for stabilizing ester-containing prodrugs analyzed by LC-MS/MS. Bioanalysis 2010;2(4):733–743.

Fura A, Harper TW, Zhang H, Fung L, Shyu WC. Shift in pH of biological fluids during storage and processing: effect on Bioanalysis. J Pharm Biomed Anal 2003;32(3):513–522.

Gataye Perez A. Can dried blood spot technique be used to stabilize pro-drugs and glucuronides metabolites? EBF workshop: Connecting Strategies on Dried Blood Spots, June 17–18, 2010, Brussels, Belgium.

Green JH. *Basic Clinical Physiology*. 3rd ed. Oxford: Oxford University Press; 1978.

Guder WG, Narayanan S, Wisser H. et al. *Samples: From the Patient to the Laboratory: The Impact of Preanalytical Variables on the Quality of Laboratory Results*. Darmstadt: GIT; 1996. p 1–149.

Heilbronn E. Action of fluoride on cholinesterases. I on the mechanism of inhibition. Acta Chem Scand 1965;19:1333–1346.

Hersberger M, Nusbaumer C, Scholer A, Knopfli V, Von Eckardstein A. Influence of practicable virus inactivation procedures on tests for frequently measured analytes in plasma. Clinical Chemistry 2004;50(5):944–946.

IATA. First published in 2005. Guidelines for shipment of samples. Available at www.iata.org. Accessed Mar 14, 2013.

Itkin A, Dupres V, Dufrêne YF, et al. Calcium ions promote formation of amyloid β-peptide (1–40) oligomers causally implicated in neuronal toxicity of Alzheimer's disease. PLoS 2011;6(3):e18250. doi.10.1371/journal.pone.0018250

International conference on Harmonisation of technical requirements for registration of Pharmaceuticals for Human Use. ICH Topic M3 (R2) Guidance on Non-Clinical Safety Studies for the Conduct of Human Clinical Trials and Marketing Authorisation for Pharmaceuticals. 2009. Available at www.ichorg/LOB/media/MEDIA5544.pdf. Accessed Mar 14, 2013.

Jemal M, Ouyang Z, Chen BC, Teitz D. Quantitation of the acid and lactone forms of atorvastatin and its biotransformation products in human serum by high performance liquid chromatography with electrospray tandem mass spectrometry. Rapid comm. Mass Spec 1999;13(11):1003–1011.

Jemal M, Xia YQ. Bioanalytical method validation design for the simultaneous quantitation of analytes that may undergo interconversion during analysis. J Pharm Biomed Anal 2000;22(5):813–827.

Jemal M, Khan S, Teitz DS, McCafferty JA, Hawthorne DJ. LC-MS/MS determination of omapatrilat, a sulfhydryl-containing vasopeptidase inhibitor, and its sulfhydryl- and thioether-containing metabolites in human plasma. Analytical Chemistry 2001;73(22):5450–5456.

Jewell C, Ackermann C, Payne NA, Fate G, Voormann R, Williams FM. Specificity of procaine and ester hydrolysis by human, minipig, and rat skin and liver. Drug Metab Dispos 2007;35(11):2015–2022.

Ji JA, Jiang Z, Livson Y, Davis JA, Chu JX, Weng N. Challenges in urine bioanalytical assays: overcoming nonspecific binding. Bioanalysis 2010;2(9):1573–1586.

Kale P, Sharma R, Gupta RK, Modi S, Patidar K, Hussain S. Stability study of Diltiazem and its major metabolite in human plasma. Poster at the 2010 AAPS Annual Meeting on November 14–18, 2010, New Orleans, USA. Abstract. Available at www.AAPSJ.org/abstracts/AM_2010/T2278.pdf. Accessed Mar 14, 2013.

Kraut A, Marcellin M, Adrait A, Kuhn L, Louwagle M, Keifer-Jaquinod S, et al. Peptide storage: are you getting the best return on your investment? Defining optimal storage conditions for proteomics samples. J Proteome Res 2009;8:3778–3785

Kudo S, Umehara K, Hosokawa M, Miyamot G, Chiba K, Satouh T. Phenacetin deacetylase activity in human liver microsomes: distribution, kinetics, and chemical inhibition and stimulation. J Pharmacol Exp Ther 2000;294(1):80–87

Lakshmy R. Analysis of the use of Dried Blood Spot Measurements in Disease Screening. J Diabetes Sci Technol 2008;2(2);242–243

Liang H, Vahid M, Zhu Y, et al. Determination of Acetylsalicylic acid in Human Plasma by LC-MS/MS. Proceedings of the 57th ASMS Conference of Mass Spectrometry 1–4 June 2009 Philadelphia, PA.

Lindegardh N, Davies G, Hien TT, et al. Rapid degradation of oseltamivir phosphate in clinical samples by plasma esterases. Antimicrob Agents Chemother 2006;50(9):3197–3199.

Lindegardh N, Davies G, Hien TT, et al. Importance of collection tube during clinical studies of Oseltamivir. Antimicrob Agents Chemother 2007;51(5):1835–1836.

Linhardt RJ, Gunay NS. Production and chemical processing of low molecular weight heparins. Sem Thromb Hem 1999;3:5–16.

Lippi G, Salvangno GL, Montagnana M, Poli G, Giudi GC. Influence of centrifuge temperature on routine coagulation testing. Clin Chem 2006;52(3):537–538.

Li W, Luo S, Smith HT, Tse FLS. Quantitative determination of BAF312, a S1P-R modulator, in human urine by LC-MS/MS: Prevention and recovery of lost analyte due to container surface adsorption. J Chromatogr B 2010;878:583–589.

Luo W, Zhu L, Deng J, Lui A, Guo B, Tan W, Dai R. Simultaneous analysis of bambuterol and its active metabolite terbutaline enantiomers in rat plasma by chiral liquid chromatographyomatographs spectrometry. J Pharm Biomed Anal 2010;52(2):227–231.

Li W, Zhang J, Tse FLS. Strategies in quantitative LC-MS/MS analysis of unstable small molecules in biological matrices. Biomed Chromatogr 2011;25:258–277.

Moore DE. Photophysical and photochemical aspects of drug stability. In: Tonnesen HH, editor. *Photostability of Drugs and Drug Formulations*. Boca Raton, FL: CRC Press; 2004. Chapter 2, p 9–40.

Murphy-Poulton SF, Boyle F, Gu XQ, Mather LE. Thalidomide enantiomers: determination in biological samples by HPLC and vancomycin-CSP. J Chromatogr B 2006;831(1–2):48–56.

Narayanan S. Effect of anticoagulants used for blood collection on laboratory tests. Jpn J Clin Pathol 1996;103:73–80.

Narayanan S, Hamasaki N. Current concepts of coagulation and fibrinolysis. Adv Clin Chem 1998;33:133–168.

NHS guidance and best practice documents. Available at www.nhs.uk.

Osselton MD. The release of basic drugs by the enzymic digestion of tissues in cases of Poisoning. J Forens Sci Soc1977;17(2):189–194.

Parker MH, Reitz AB. Assembly of β-Amyloid aggregates at the molecular level. Chemtracts—Organic Chemistry 2000; 13(1):51–56.

Pearlman R, Bewley TA. Stability and characterisation of human growth hormone, Chapter 1. In: Wang YJ, Pearlman R, editors. *Stabilisation and Characterisation of Protein and Peptide Drugs. Pharmaceutical Biotechnology*. Volume 5. New York: Plenum Press; 1993.

Pelletier JPR, Transue S, Snyder EL. Pathogen inactivation techniques best practice and research. Clin Haematol 2006;19(1):205–242.

Priston MJ, Sewell GJ. Improved LC assay for the determination of mitozantrone in plasma: analytical considerations. J Pharm Biomed Anal 1994;12:1153–1162.

Rang HP, Dale MM. Haemostasis and thrombosis. Chapter 16. In: *Pharmacology*. 2nd ed. Edingburgh: Churchill Livingstone; 1991.

Richterich R, Columbo JP. *Clinical Chemistry: Theory, Practice and Interpretation*. Translated edition. Chichester: John Wiley & Sons Ltd; 1981.

Sadagopan NP, Li W, Cook JA, et al. Investigation of EDTA anticoagulant in plasma to improve the throughput of liquid chromatography/tandem mass spectrometric assays. Rapid Comm Mass Spec 2003;17(10):1065–1070.

Saxer C, Niina M, Nakashima A, Nagae Y, Masuda N. Simultaneous determination of levodopa and 3-O-methyldopa in human plasma by liquid chromatography with electrochemicaldetection. J Chromatogr B 2004;802(2):299–305.

Sennbro CJ, Knutsson M, Amsterdam P, Timmermann P. Anticoagulant counter ion impact on bioanalytical LC-MS/MS assays: results from discussion and experiments within the European Bioanalytical Forum. Bioanalysis 2011;3(21):2393–2399.

Shipkova M, Armstrong VW, Oellerich M, Wieland E. Acyl glucuronide drug metabolites: toxicological and analytical implications. Ther Drug Monit 2003;25(1):1–16.

Silvester S. Strategy for over-coming non-specific analyte adsorption issues in clinical urine samples: Consideration of bioanalysis and metabolite identification. Oral presentation at 6th Bioanalysis in Clinical Research event Feb 15–16, 2011.

Skopp G, Klingman A, Pötsch L, Mattern R. In vitro stability of cocaine in whole blood and plasma including ecgonine as a target analyte. Ther Drug Monit 2001;23(2):174–181.

Sofer G, Lister DC, Boose JA. Virus inactivation in the 1990s—and into the 21st century: Part 6, inactivation methods grouped by virus. BioPharm International 2003; 16(4), 42–52, 68.

Srinivasan K, Nouri P, Kavetskala O. Challenges in the indirect quantitation of acyl-glucuronide metabolites of a cardiovascular drug from complex biological mixtures in the absence of reference standards. Biomed Chromatogr 2010;24(7):759–767.

Stout PR, Horn CK, Lesser DR. Loss of THCCOOH from urine specimens stored in polypropylene and polyethylene containers at different temperatures. J Anal Toxicol 2000;24(7):567–571.

Tsujikawa K, Kuwayama K, Miyaguchi H, Kanamori T, Iwata YT, Inoue H. In vitro stability and metabolism of salvinorin A in rat plasma. Xenobiotica 2009;39(5):391–398.

US FDA 2000. Compliance Program Guidance Manual. Attachment A Bioequivalence Inspection Report. Available at www.fda.gov/downloads/Drugs/GuidanceComplianceRegulatoryInformation/guidances. Accessed Mar 14, 2013.

US FDA. 2001. Guidance for Drug Evaluation and Research. Guidance for Industry: Bioanalytical Method Validation. Available at www.fda.gov/downloads/Drugs/GuidanceComplianceRegulatoryInformation/Guidances/UCM070107.pdf. Accessed Mar 14, 2013.

Wang X, Bowman PD, Kerwin SM, Stavchansky S. Stability of caffeic acid phenethyl ester and its fluronated derivative in rat plasma. Biomed Chromatogr 2007;21(4):434–350.

Wang YJ, Pearlman R, editors. *Stabilisation and Characterisation of Protein and Peptide Drugs. Pharmaceutical Biotechnology Volume 5*. New York: Plenum Press; 1993.

Waters. ESI+ Common Background Ions. Available at www.waters.com/webassets/cms/support/docs/bkgrnd_ion_mstr_list.pdf. Accessed Mar 14, 2013.

Watt AP, Morrison D, Locker KL, Evans DC. Higher throughput bioanalysis by automation of a protein precipitation assay using a 96-well plate format with detection by LC-MS/MS. Anal Chem 2000;72(5):979–984.

WHO. Guidelines on viral inactivation and removal procedures intended to assure the viral safety of human blood plasma products. WHO Technical Report, Series No. 924, 2004.

WHO. Guidelines on drawing blood: best practices in phlebotomy. Geneva: WHO Press. 2010.

Wyss R, Bucheli F. Determination of of cefetamet and its orally active ester, cefetamet pivoxil, in biological fluids by high performance liquid chromatography. J Chromatogr B 1988;430:81–92.

Xu Y, Keith B, Grem JL. Measurement of the anticancer agent gemcitabine and its deaminated metabolite at low concentrations in human plasma by liquid chromatography-mass spectrometry. J Chromatogr B 2004;802(2):263–270.

Xu Y, Du L, Rose MJ, Fu I, Woolf EJ, Musson DG. Concerns in the development of an assay for determination of a highly conjugated adsorption-prone compound in human urine. J Chromatogr B 2005;818(2):241–248.

Yamouri S, Fujiyama N, Kushihara M, et al. Involvement of human blood arylesterases and liver microsomal carboxylesterase in nafamostat hydrolysis. Drug Metab Pharmacokinet 2006;21(2):147–155.

Yapa U, Ionita I, Steenwyk RC. Improve the stability of unstable compounds in blood using dry blood spotting (DBS) technique followed by LC-MS/MS analysis. Proceedings of the 58th ASMS Conference on Mass spectrometry and allied Topics, May 23–27, 2010, Salt Lake City, UT.

Zhang JG, Fariss MW. Thenoylfluoroacetone, a potent inhibitor of carboxylesterase activity. Biochem Pharmacol 2002;63:751–754.

Zhang N, Yang A, Rogers JD, Zhao JJ. Quantitation of simvastatin and its β-hydroxy acid in human plasma by using automated liquid-liquid extraction based on 96-well plate format and liquid chromatography/tandem mass spectrometry. J Pharm Biomed Anal 2004;34(1):175–187.

Zhao B, Moochhala SM, Lu J, Tan D, Lai MH. Determination of pyridostigmine bromide and its metabolites in biological samples. J Pharm Pharmacet Sci 2006;9(1):71–81.

Zhao JJ, Xie JH, Yang AY, Roadcap BA, Rodgers JD. Quantitation of simvastatin and its β-hydroxy acid in human plasma by liquid-liquid cartridge extraction and liquid chromatography/tandem mass spectrometry. J Mass Spectrometr 2000;35:1133–1143.

14

BEST PRACTICES IN BIOLOGICAL SAMPLE PREPARATION FOR LC-MS BIOANALYSIS

Guowen Liu and Anne-Françoise Aubry

14.1 WHY DO SAMPLE PREPARATION?

Liquid chromatography coupled with mass spectrometry (LC-MS) or tandem mass spectrometry (LC-MS/MS) has become the benchmark for bioanalysis for small molecules in past decades (Watt et al., 2000; Jemal and Xia, 2006; Aubry, 2011). LC-MS/MS is also now becoming an important tool for quantitative analysis of larger molecules (peptides, oligonucleotides, and proteins) in biological matrices (Ewles and Goodwin, 2011; Li et al., 2011). As powerful and sophisticated a tool as an LC-MS system can be, its performance may be compromised by the nature of the samples analyzed. There are certain physical and chemical requirements for samples that are being analyzed by LC-MS. Most biological specimens cannot be directly introduced to the LC-MS system without some pretreatment. The principles of sample pretreatment are the same regardless of the particular MS detection technology (single quadrupole, triple quadrupole, other tandem MS technology, or high-resolution MS) being used for analysis. This chapter reviews the most commonly used sample preparation techniques in the pharmaceutical and biopharmaceutical industry for LC-MS or LC-MS/MS analyses (for simplicity, we will use LC-MS in the rest of the chapter). The discussion deals primarily with small molecules but we also pointed out where some of the techniques can be applied to large molecules.

As a hybrid instrument, LC-MS needs to follow the requirements of both of its components. The LC part is designed to deal with liquid samples. The flow path from the solvent bottle, through HPLC pump, injection valve, LC column to the LC column exit is a long and delicate system. Samples injected into the LC system are required to be in liquid form, and essentially free of particles. Therefore, samples not already in liquid form, such as tissue samples (solid), blood samples (suspension), have to be processed or converted into liquid form before being loaded into the LC system. The MS is designed to generate ions into the gas phase and detect them under a high vacuum condition. Electrospray ionization (ESI) and atmospheric pressure chemical ionization (APCI) are the two most common ion sources used in an LC-MS system. To ensure good ionization efficiency, nonvolatile acid, base, or salts should not be introduced into the ESI or APCI ion source. In addition, unless a special column/mobile phase combination is used, proteins cannot be injected in the LC as they tend to precipitate at the head of the column and cause a rapid deterioration of column performance. Furthermore, matrix effect, interfering effects (ion enhancement or ion suppression) from matrix components to the analyte of interest during ionization process, is a well-described phenomenon in ESI/APCI MS (Fu et al., 1998; King et al., 2000; Dams et al., 2003; Wu et al., 2008b; Liu et al., 2009). Matrix effect is unpredictable and can lead to inaccurate measurements if not addressed properly (Wang et al., 2007; Liu et al., 2010a). Using a stable isotope labeled internal standard (SIL-IS) can compensate matrix effects in most cases. However, when matrix effect is dramatically different from sample to sample (Liu et al., 2010a) or the SIL-IS does not exactly coelute with the analyte chromatographically (Jemal et al.,

Handbook of LC-MS Bioanalysis: Best Practices, Experimental Protocols, and Regulations, First Edition. Edited by Wenkui Li, Jie Zhang, and Francis L.S. Tse.

2003; Wang et al., 2007), even a SIL-IS may not be able to compensate for all of the matrix effects. Many matrix components are hydrophilic and can be eliminated by simple extraction procedures for assays of moderately polar to hydrophobic compounds. Phospholipids have been singled out for their propensity to cause matrix effects for pharmaceuticals (Liu et al., 2009; Xia and Jemal, 2009) and much information is available regarding the removal of phospholipids from plasma (Liu et al., 2009; Pucci et al., 2009; Xia and Jemal, 2009; Jiang et al., 2012). Sensitivity enhancement is another reason for sample preparation. To detect trace amounts of analyte in a complex system, such as a biological matrix, the analyte of interest often needs to be enriched to reach the desired detection limit. In these cases, the extraction procedure is used to preconcentrate the analyte prior to LC-MS analysis. To summarize, the goals of sample preparation are to change the sample form (e.g., process a solid or heterogeneous specimen to a clear solution), simplify the sample composition (i.e., reduce matrix background), or enrich the analyte of interest (i.e., preconcentrate the sample). Sample preparation is also one of the most, if not the most, critical and time consuming part of a bioanalytical assay. A successful analytical strategy should not only achieve the goals of sample preparation but also preserve the integrity of the analyte (i.e., avoid degradation of the analyte of interest). Shown in Figure 14.1 is a schema of the most commonly used strategies by bioanalytical scientists from simply converting the specimen to a sample physically amenable to LC-MS analysis to using an analyte-specific extraction.

Many up-to-date, high quality and comprehensive reviews on the principles and advancements of sample preparation techniques for LC-MS bioanalysis have been published (Chang et al., 2007; Novakova and Vlckova, 2009; Ashri and Abdel-Rehim, 2011; Kole et al., 2011). Technique-specific or application-oriented reviews are also available: Moreno-Bondi et al. (2009) reviewed the sample preparation procedures for LC-MS analyses of antibiotics in environmental and food samples; Samanidou et al. (2011) reviewed novel strategies for sample preparation in forensic toxicology; Rudewicz (2011) reviewed the applications of turbulent flow (TFC) in drug metabolism and pharmacokinetics; Vuckovic et al. (2010) updated the status of solid-phase microextraction in bioanalysis; Chen et al. (2009) summarized the application of online SPE for liquid chromatography.

In this chapter, we are describing the commonly used sample preparation techniques in the pharmaceutical and biopharmaceutical industries, namely, protein precipitation (PPT), liquid–liquid extraction (LLE), and solid phase extraction (SPE), focusing on the need for balancing recovery and selectivity, and on the operational details of each technique. A scientifically sound bioanalytical method can still fail due to incorrect execution of the sample preparation procedures. Many details deemed trivial are, in fact, critical to the assay's successful execution: nonspecific absorption on plasticware (Ji et al., 2010; Li et al., 2010), ineffective

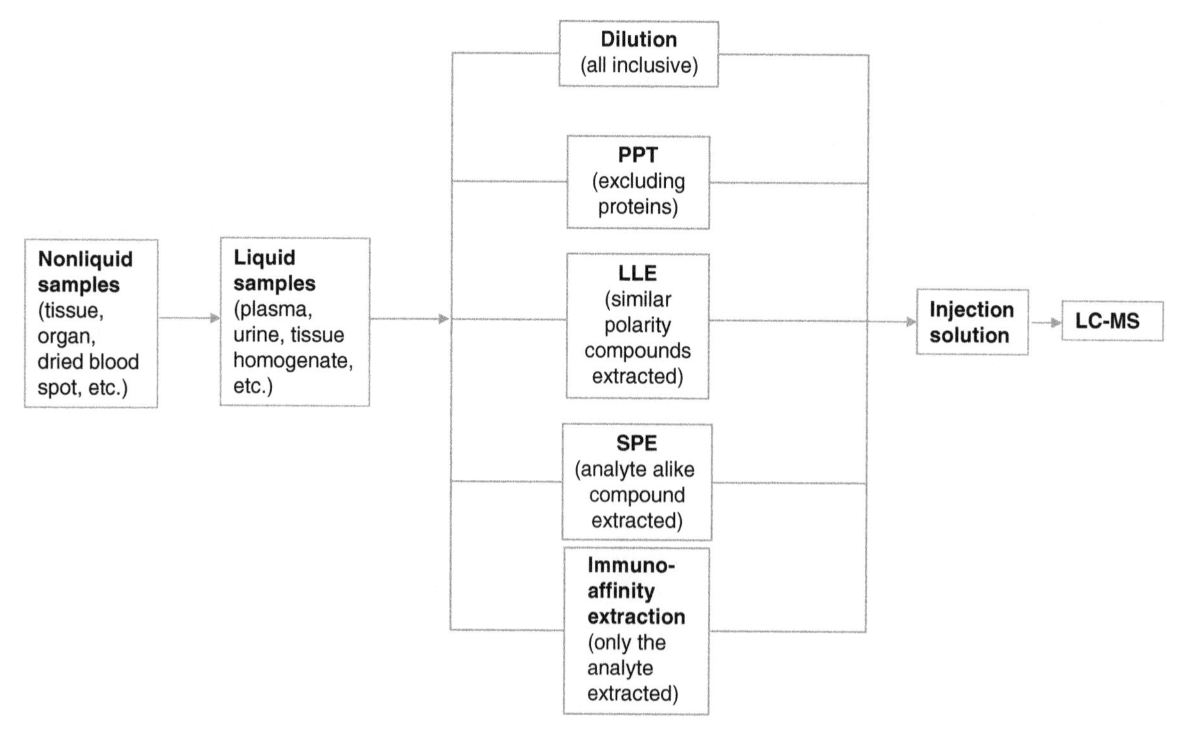

FIGURE 14.1 General overview of sample preparation of biological samples. PPT, protein precipitation; LLE, liquid–liquid extraction; SPE, solid phase extraction.

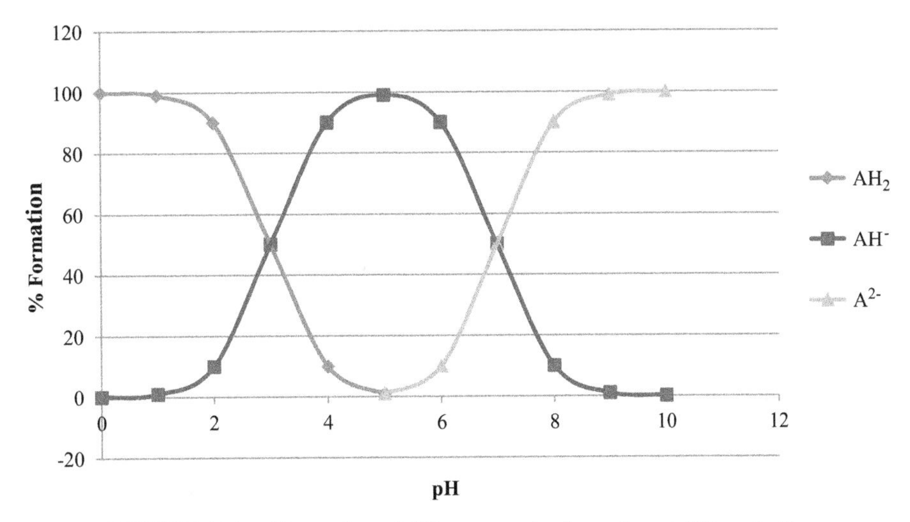

FIGURE 14.2 Illustration of different species formation at different pHs.

sample mixing prior to pipetting (Fu et al., 2011), cross-contamination during sample handling and liquid transfers, carryover from fixed-tip liquid handler, etc. In our opinion, sample preparation is all about execution of the details, especially for those new to this field. Many good bioanalytical assays fail in cross-laboratory transfers due to lack of sufficient sample preparation details in the written method. Therefore, our focus in this chapter is explaining the specificity of each sample preparation techniques and the advantages and disadvantages and the applications of each technique. Recent variations of these sample preparation techniques will also be discussed briefly. Before we discuss each specific technique, we thought it important to review the fundamental chemical principles of sample extraction.

14.2 WHAT SHOULD YOU KNOW ABOUT THE SAMPLE?

14.2.1 Analyte of Interest

Before one starts working on a sample preparation method, it is always a good practice to know the relevant chemical and physical properties of the compound(s) of interest. Knowledge of the properties of organic compounds (**log *P*, log *D*,** pKa, solubility, protein binding, stabilities, etc.) is critical to understand their compatibility with the various sample preparation techniques and the selection of experimental conditions. Therefore, it is strongly recommended to spend some time reviewing these properties and all the options in planning experiments on paper before heading to the laboratory.

14.2.1.1 pKa Many organic compounds contain acidic (H^+ donor) or basic (H^+ acceptor) functional groups. Knowing the pKa of a compound can inform the charge state of the compound at a given pH. By definition, at pH = pKa, 50% of the molecules are in the basic (protonated) form and 50% in the acidic (deprotonated) form; at pH $\ll$ pKa, almost 100% of the compounds will be unionized for acids and ionized for bases; at pH $\gg$ pKa, almost 100% of the compounds will be ionized for acids and unionized for bases. One can, therefore, map out the charge state of the molecule at various pHs according the formula **log** $[AH]/[A^-] = pKa - pH$. For example, assuming an acid (AH_2) with two pKa ($pKa_1 = 3$, $pKa_2 = 7$), the proportion of different species (AH_2, AH^-, and A^{2-}) can be mapped out at different pHs as illustrated in Figure 14.2. pH is a critical parameter in sample extraction for ionized or ionizable compounds. In LLE, the best recovery will be at a pH at which the analyte is not charged; however, in ion-exchange SPE, the analyte must be ionized to interact with the stationary phase. Compounds containing both an acidic and a basic group form zwitterions at pHs in between the two pKa. Zwitterions are particularly difficult to extract. These basic rules regarding pH are often overlooked during the initial stage of method development.

14.2.1.2 log P and log D The *partition coefficient, **P**,* is a constant to characterize the hydrophilicity–lipophilicity balance of a compound. It is the concentration ratio of the neutral form of a compound in two immiscible solvents (normally water and octanol) at equilibrium. To measure the ***P*** value of an ionizable compound, the predominant form of the compound in aqueous phase should be unionized by adjusting the pH. The "**log *P***" value of a compound is often referred to as its lipophilicity, which is the logarithm of the ratio of the concentrations of the unionized compound in the solvents (see below for a formula for **log *P***):

$$\log P = \log_{10}[C_o/C_{aq}],$$

where C_o is the concentration of the neutral compound in the water-immiscible solvent and C_{aq} is the concentration in the aqueous phase.

The *distribution coefficient*, ***D***, is another parameter to measure the lipophilicity of a compound and is defined as the ratio of the sum of the concentrations of all forms of the compound between an aqueous phase and a water-immiscible organic solvent. This parameter is both solvent and pH dependant for ionizable compounds. The "**log *D***" value is often given for a specific compound, which is derived from the following formula:

$$\log D = \log_{10} C_o/[C_{aq\,ion} + C_{aq\,neu}],$$

where C_o is the concentration of the compound in the water-immiscible organic solvent, $C_{aq\,ion}$ is the concentration of the ionized compound in water, and $C_{aq\,neu}$ is the concentration of the neutral compound in the aqueous phase.

For unionizable compounds, **log *P*** = **log *D*** at any given pH. For ionizable compounds, a relationship can be established as the following between **log *P*** and **log *D*** at a given pH with the known pKa, providing that charged molecules do not enter the organic phase:

$$\text{Acids}: \log D = \log P - \log[1 + 10^{(\text{pH}-\text{pKa})}]$$
$$\text{Bases}: \log D = \log P - \log[1 + 10^{(\text{pKa}-\text{pH})}]$$

A more detailed description of **log *D*** and **log *P*** can be found in this website: http://en.wikipedia.org/wiki/Partition_coefficient (accessed Mar 19, 2013). Unless specified, the **log *D*** and **log *P*** values of a given compound are measured using octanol as the partitioning solvent by default.

The recovery of a compound at given conditions (pH, organic solvent, volume ratio between aqueous and organic) can be predicted for LLE methods and to an extent for reversed-phase SPE methods, based on their pKa, **log *D*,** and **log *P***. The knowledge of the properties of the analyte, such as chemical stability, protein binding, solubility, and so on, will also help eliminate some obviously bad choices for sample preparation. Typically, more than one sample preparation techniques can be used for an analyte of interest and the choice is, by in large, a matter of personal preference. However, for analytes with unique physicochemical properties, some techniques or experimental parameters are obviously not good choices: for example, choosing LLE for compound with **log *D*** < 0; modifying the pH to a range where the compound is unstable; PPT for analytes that will likely coprecipitate with the proteins. In summary, understanding the analyte well is the first step of developing a sample preparation method. In addition, the expected concentration of the analyte in the matrix should be a factor to consider in choosing the extraction technique.

14.2.2 Matrix

One challenge in bioanalysis is dealing with a variety of biological matrices: plasma, serum, whole blood, urine, CSF, tissues, etc. The composition and complexity of the different matrices is significantly different. Knowing the nature of the sample is one of the key factors in deciding on a technique for sample preparation. Factors such as the normal pH of the matrix, nature and concentration of proteins and lipids, salt content, may all potentially impact the extraction. Understanding the interactions between the analyte and the matrix it resides in also helps to decide a suitable sample preparation technique. For example, when working with whole blood samples, the distribution of the compound between plasma and red blood cells (RBC) (Brockman et al., 2007), the enzymatic stability of the analyte in blood (Li et al., 2011), protein binding, etc., should be considered. A sample preparation method that is capable of releasing the analyte from the RBC is needed when the analyte distribution into RBCs is significant (Brockman et al., 2007). For tissue samples, checking that the analyte is released from the solid tissue is a must (Chng et al., 2010); for urine and CSF samples, nonspecific binding are often a concern in the absence of proteins and the procedure must be able to release the analyte from surfaces (Gu et al., 2010; Ji et al., 2010). Finally, matrix effects in the LC-MS analysis, already discussed in Section 14.1, is an important aspect of the matrix to evaluate and one that may drive the selection of the extraction method. One thing worth mentioning is that extracted matrix background for a specific technique/method is not compound dependant but sample preparation method dependant (Liu et al., 2009); this will be discussed more in Section 14.3.3.

14.3 KNOW YOUR TOOLS

Liquid handling is a predominant part of sample preparation. The steps of pipetting biological samples, adding extraction solvent, transferring intermediate solutions, and so on can be done either manually using pipettes or using automation (e.g., robotic liquid handlers). Automation is the trend and the preferred solution for liquid handling for sample preparation because of the better precision achieved, as well as ergonomic concerns with repeated motion. Many sophisticated liquid handling workstations are available, such as the Freedom EVO® from Tecan (http://www.tecan.com, accessed Mar 19, 2013), MICROLAB® STAR Liquid Handling Workstations from Hamilton (http://www.hamiltoncompany.com, accessed Mar 19, 2013), and JANUS® Automated Workstation from PerkinElmer (http://www.perkinelmer.com, accessed Mar 19, 2013). A good review by Vogeser and Kirchhoff (2011) on the progress of automation for LC-MS bioanalysis has been recently published. A chapter on the use of automation can also be found in this book.

14.3.1 Dilution

Dilution is the simplest sample preparation technique (Casetta et al., 2000; Dams et al., 2003; Rashed et al., 2005), and is also referred to as "dilute-and-shoot" (McCauley-Myers et al., 2000; Xue et al., 2007). It is an inclusive method, which yields 100% recovery of all analytes but without selectivity. All the components of a sample are loaded into the LC-MS system in this approach. The sample is simply diluted using a solution amenable for LC-MS (Wood et al., 2004; Raffaelli et al., 2006; Bishop et al., 2007; Gray et al., 2011). The applications of dilution have been mainly limited to situations in which matrix effect is not a concern and sensitivity is not an issue. The advantages of this approach include minimum sample handling, good sample integrity, and low cost. Its disadvantages include dilution of the signal and high potential for matrix effect. For these reasons, dilution is often limited to relatively simple liquid matrices containing no or few macromolecules, such as urine, tears, and CSF, and to applications that do not require high sensitivity. In reality, with the availability of more and more sensitive instrumentation and stable isotope labeled internal standards (ISs), sample preparation using dilution becomes a viable solution in bioanalysis for more applications and one that should be considered first because of its simplicity. For the most part, simple dilution is unsuitable for more complex biological samples that contain proteins or lipids.

14.3.2 Protein Precipitation

Biological samples, such as plasma and serum, which contain abundant soluble proteins, require more complex pretreatment. The process of removing the majority of proteins from the biological samples by precipitating out the proteins is known in bioanalysis as **PPT**. It is a simple, quick, and convenient sample preparation technique, favored for high-throughput, non-GLP applications. Although many approaches (Polson et al., 2003) can be used to denature proteins: such as water miscible organic solvent, acid, salts, and metal ions, PPT using an organic solvent (methanol, acetonitrile, or ethanol) seems to be dominant in bioanalytical practice (Dams et al., 2003; Flaherty et al., 2005; Xue et al., 2006b; Ma et al., 2008; Chng et al., 2010). Polson et al. (2003) reported a comprehensive study on the effectiveness of protein removal using acids (trichloroacetic acid, TCA), metal ions (zinc sulfate), salts (ammonium sulfate), and organic solvent (acetonitrile, ethanol, methanol) as the precipitants. Based on the results from this study, more than 90% of the total proteins can be removed from rat, dog, mouse, and human plasma with a precipitant-to-plasma ratio of 0.5 (10% TCA, v/v), 2 (10% zinc sulphate, w/v), 3 (saturated ammonium sulfate solution at room temperature), 1.5 (acetonitrile), 3 (ethanol), and 2.5 (methanol). A typical protocol of PPT with an organic solvent is shown later with practical details and tips provided at each step. Since 96-well plates are the most popular sample preparation format of bioanalysis in the pharmaceutical industry, all examples given in this chapter will be assuming this format is used unless otherwise stated.

14.3.2.1 A Typical Example of PPT Protocol In the example of a typical PPT method for plasma samples using acetonitrile as the organic solvent, the overall process can be completed in 1–3 h for one batch, depending on whether a dry-down and reconstitution process is included:

- Transfer 50 μl of standards (STDs), quality controls (QCs), and incurred samples into a 96-well plate.
  - This step can be performed either manually or using a robotic liquid handler. The volume transferred for each sample does not have to be accurate but has to be precise. The 96-well plate used must have a cover with perfect seal to prevent any solvent leaking and avoid well-to-well cross-contamination.
- Add 50 μl of IS working solution into each sample well.
  - The IS is added before the organic solvent. The IS may also be prepared in the organic solvent and then this step can be combined with the next step. When doing so, one must be wary of less than perfect tracking by the IS, especially if the analyte is highly protein bound. This step can easily be performed using robotic liquid handler. The volume delivered during this step needs to be precise but not necessarily accurate.
- Add 600 μl acetonitrile to each sample, seal the plate and vortex for 1 min.
  - Water-miscible organic solvent was added to the aqueous solution. The organic solvent will denature the soluble proteins and cause the proteins to precipitate out of solution. To remove a majority of the proteins (e.g., >90%) in the sample, a minimum ratio of organic to aqueous of 1.5 for acetonitrile is recommended. This step can also be performed using automation. The plate has to be sealed seamlessly to avoid cross-well contamination. Even though automation of the plate sealing and vortexing steps is possible, it is not prevailing in the bioanalytical community. This step does not require high volume precision or accuracy.
- Centrifuge the sample plate at 3000 rpm for 5 min.
  - Manual intervention is usually required: for example, the plate is manually transferred to a centrifuge for centrifugation and returned to the robotic workstation after that.
- Transfer 500 μl of the supernatant after centrifugation into a collection plate
  - Depending on the sensitivity requirement, different amounts of supernatant may be transferred out. The

supernatant contains high organic content; although this transfer step can be easily automated, caution must take to avoid liquid dripping during the transfer to avoid cross-contamination. Since the IS will compensate for any variation in the pipetting, this step does not require high precision or accuracy.

- Dry the sample plate under nitrogen flow and reconstitute each sample with 200 μl reconstitution solution.
  - There are two purposes in this step. First, the sample can be concentrated as necessary by reconstituting in a smaller volume (2.5 times in this example); second, the injection solution strength can be adjusted to match the starting LC conditions. This step may be omitted if no sample concentration is needed and high organic content solution can be injected directly into the LC system (e.g., on a HILIC column) directly. This step usually takes most of the sample preparation time because of the time required for evaporating water.
- Vortex the plate for 2 min and centrifuge the plate for 2 min and load the final sample plate to the LC-MS system for analysis.
  - This step is to make sure that all the dry residue in each well has been redissolved in the reconstitution solution and any solution on the well wall or the top is brought down to the bottom of each well. For some samples, not all the solid residue will be completely redissolved in the reconstitution solution. However, this may not be critical as long as the analyte of interest and its IS are fully dissolved in the final solution and solid particles are not in suspension.

14.3.2.2 Advantages and Limitations There are many advantages of using PPT for sample preparation: (1) it is a sample preparation method suitable for almost all types of small molecules, no matter their polarity; (2) operationally, a standard procedure can be applied to all assays; (3) it is a rapid process with good potential for automation or semiautomation (Ma et al., 2008; Tweed et al., 2010); and (4) the recovery of the analyte is literally 100% and almost all small molecules can be retained, which is, therefore, particularly advantageous for metabolite profiling (Wilson, 2011). The less attractive part of this technique lies in the fact that the processed samples after PPT still contain all small endogenous molecules (no selectivity) that may interfere with the MS ionization process, decrease the performance of the LC column, and so on. PPT has been associated with high matrix effect (Dams et al., 2003). These caveats, if not addressed properly, may significantly impact the data quality (Wang et al., 2007). PPT is very popular in drug discovery, where sample turnaround time is critical to keep up with the fast pace of discovery activities. In drug development, it was extremely popular when ESI-LC-MS was first introduced as the best tool for bioanalysis, until the issue of matrix effect became a major concern. It is regaining its popularity, especially when a stable isotope labeled IS is available, because with new instruments becoming more and more sensitive, only a small amount of biological material is injected practically eliminating concerns with matrix interference.

14.3.2.3 Other Forms of PPT To simplify the procedure and improve sample throughput further, membrane-based PPT filter plates (e.g., Unifilter® from Whatman, Strata® Impact™ from Phenomenex, Isolute® PPT+ from Biotage, Sirocco™ from Waters) are available to remove proteins in plasma or serum. In recent years, more advanced PPT plates with packed materials specifically retaining phospholipids are also available to remove both proteins and the abundant phospholipids in a plasma/serum sample (e.g., Captiva™ NDLipids from Agilent, Ostro™ from Waters, Phree™ from Phenomenex, and HybridSPE® from Sigma). By using these plates, an extracted sample free of protein and phospholipids are readily available for the subsequent LC-MS analysis. The overall PPT process can be done in a 96-well plate format without the needs for centrifugation or supernatant transfer steps, which leads to short sample preparation time and higher solvent recovery. It is a very helpful approach for low sample volume and is amenable to full automation. Each of the commercially available PPT plates comes with standardized recommended procedures. A detailed comparison of different PPT plates/tubes can be found in a recent review article by Kole et al. (2011).

14.3.3 Liquid–Liquid Extraction

LLE is another popular sample cleanup technique in bioanalysis. It is based on partition between immiscible solvents and consists in extracting compounds from the aqueous sample into a water-immiscible solvent. The distribution of the analyte between the two phases is related to its **log *P*** or **log *D*** values. When LLE approach is used, the biological sample usually is mixed with an aqueous buffer (sample extraction buffer to bring the pH of the matrix to a pH range in which an analyte of interest can be readily transferred from the aqueous phase to the organic phase). The buffer may also contain the IS. The buffered sample is then mixed with a specified volume of a water-immiscible organic solvent. Commonly used organic solvents include methyl tert-butyl ether (MTBE) (Pin et al., 2012), ethyl acetate (Kosovec et al., 2008; Cai et al., 2012), hexane (Cai et al., 2012), 1-chlorobutane (Saar et al., 2010), or a combination of two solvents and are selected based on the polarity and solubility of the analyte. By changing the organic solvent or adjusting pH and buffer ionic strength, the target analyte(s) can be favorably extracted from the aqueous phase to the organic phase, leaving most of matrix components, such as phospholipids, proteins, inorganic salts in the aqueous phase. The resulting extracted sample is much cleaner/simpler than the original biologic sample. Compared with PPT and even with SPE, LLE is a more selective sample cleanup technique. Below is a typical protocol of LLE in 96-well format. The overall process can normally be completed in less than 2 h with automation.

14.3.3.1 A Typical Example of LLE Protocol

- Transfer 50 μl of plasma samples (STDs, QCs, and incurred samples) to each well in a 96-well plate.
  - Same as PPT.
- Add 50 μl of IS working solution to each sample.
  - Operationally, this step is the same as for PPT. However, it is strongly recommended to prepare the IS working solution in a solvent that contains the minimum amount of water-miscible organic solvent, as the presence of the water-miscible solvents will greatly impact the cleanness of the LLE extracts. This step requires high pipetting precision but not necessarily high accuracy.
- After 10 min, add 100 μl sample extraction buffer (0.5 M ammonium acetate with 2% formic acid) to each sample.
  - A 10-min equilibrium time is enough to allow the IS to equilibrate with the matrix and proteins in particular to better track the analyte extraction recovery. A strong extraction buffer is to make the analyte in a neutral form so that it can be readily extracted by the organic solvent. This step can be easily automated. It (and the following steps) does not require high precision or accuracy in pipetting as the IS will compensate for any variability.
- Add 600 μl of MTBE solvent to each sample.
  - Water-immiscible organic solvent is added. This step can be easily automated, although some solvents are difficult to pipette and tend to drip. Adjustments in the setting of the liquid handler are needed based on the solvent type.
- Shake the whole plate for 15 min.
  - Automation for the shaking step is possible but difficult. However, it has been reported (Wang et al., 2006) that this step can be replaced by aspirating and dispensing multiple times, which can then be readily automated. A good seal of the plate before shaking is critical to avoid cross-well contamination. It was also reported (Aubry, 2011) that a sweet spot may exist for shaking time with regard to analyte recovery and matrix effect.
- Centrifuge the plate for 5 min at 4000 rpm.
  - This step can be automated but automation for this step is not popular.
- Transfer 500 μl of the top organic layer of each sample after centrifugation to a collection plate.
  - Since manual transfer of the organic layer may cause liquid dripping, it is recommended to perform this step using a liquid handler, which can be easily implemented.
- Dry down the plate under nitrogen flow.
  - Operationally this step is similar to PPT but usually evaporation takes less time because the organic solvent is more volatile than water.
- Reconstitute each sample with 200 μl of reconstitution buffer.
  - Operationally this step is similar to PPT. This step is almost always needed in this case, as the extraction solvent is rarely compatible with RP-LC mobile phases. However, it has been reported (Song and Naidong, 2006; Xue et al., 2006b) that the organic phase can be directly injected into the LC system when HILIC column is used.

Since LLE is a moderately selective sample preparation method, the sample preparation process can be adjusted to balance recovery and selectivity (matrix background removal) depending on the assay needs. When matrix interference is a concern, recovery may be sacrificed for selectivity. Normally, when a SIL-IS is used, the extraction process can be perfectly tracked by the IS. Therefore, near-complete recovery or even consistent recovery from sample to sample is not critical. It is, however, desirable to achieve consistent recovery, which is a good indication of the ruggedness of the procedure. Otherwise, sample preparation will always have to be focused on both the absolute recovery (the higher the better) and the consistency of the recovery from sample to sample. It is strongly recommended to use a SIL-IS or a closely related chemical analog whenever possible.

As for its applications, LLE is ideal for nonpolar to moderately polar analytes, which have favorable distribution in the water-immiscible organic solvent and, therefore, can be readily extracted from the aqueous phase. In reality, many drug candidates are nonpolar or moderately polar compounds, and a suitable LLE extraction method can often be developed. However, when a bioanalytical assay is targeting more than one analyte, for example, a drug and its metabolites, a fine tuning of the LLE conditions may be required to achieve acceptable recoveries for both analytes (Patel et al., 2008).

14.3.3.2 Advantages and Limitations One advantage for LLE is scalability, for example, a large sample volume can be used to increase the assay sensitivity if needed. Because the extracted sample is relatively clean, the signal gain from the sample volume increase will most likely surpass the signal loss from additional matrix suppression. Another practical advantage for LLE is that the extracted matrix background is predictable once the extraction conditions are determined (e.g., ethyl acetate at pH 7) for a given biological matrix

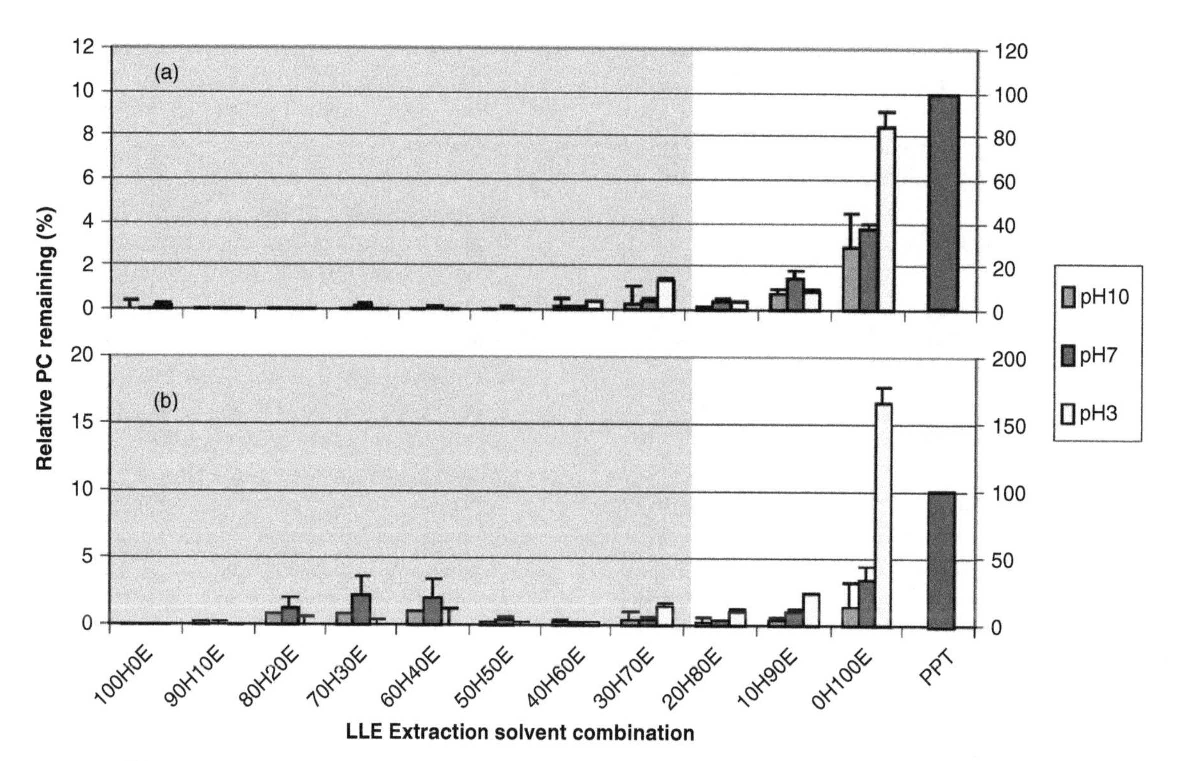

FIGURE 14.3 Phospholipids profiles for two human plasma lots after LLE at different conditions: (a) normal human plasma and (b) human plasma with high fat content. Numbers under the shadow area are multiplied by a factor of 10 for relative phospholipids response. For solvent combination, 100H0E represents hexane/ethyl acetate, 100/0, v/v; PPT represents supernatant of protein precipitation (PPT) plasma sample. Please note that PPT sample was prepared without pH adjusted (reprinted with permission from Liu et al., 2009; copyright © 2009 American Chemical Society).

(e.g., serum and plasma). Once generated, matrix background information after LLE obtained with one compound can be applied to guide method development for another. Extracted phospholipids from plasma/serum are a major concern for matrix suppression in ESI-MS (Wu et al., 2008b; Lahaie et al., 2010). Establishing a knowledge database for the extracted phospholipids under different extraction conditions has proved very useful in our laboratory. As shown in Figure 14.3 (Liu et al., 2009), the amount of extracted phospholipids differs significantly under different extraction conditions. Different combinations of hexane and ethyl acetate will also have a dramatic effect on the recovery of a specific analyte, as shown in Figure 14.4. A combination of the recovery and selectivity (matrix background) can guide the optimization of sample extraction. Taking advantage of the established data set of matrix background (phospholipids) removal, the method development process for a new compound using similar LLE can be greatly shortened (Liu et al., 2009). Other advantages for LLE include low cost, straightforward and repeatable process, ease of method transfer, and relative short method development time.

LLE is often seen in regulated bioanalytical assays for drug candidates in late stage of development. At that stage, there are usually a large number of samples to be analyzed, and automation of the sample preparation is usually desired to increase the overall sample throughput. One common misconception is that LLE is relatively difficult to automate. However, with the recent developments in robotic equipment and extraction plates (Dotsikas et al., 2006; Hussain et al., 2009; Tweed et al., 2010), most of the LLE steps can be automated as indicated in the protocol, which allows for very good throughput even though full automation is still difficult. LLE is not suitable for hydrophilic compounds, and development can be challenging when trying to extract multiple analytes with different lipophilicity.

14.3.3.3 Other Forms of LLE Other than the conventional LLE, salting-out assisted LLE (SALLE) and solid supported LLE (SLE) are the two major variations that have been used by bioanalytical scientists. SALLE is based on the "salt-induced phase separation" phenomenon to enhance the LLE efficiency. It extracts the analyte of interest using water-miscible organic solvent (such as acetonitrile) and then uses high concentration of salts (magnesium sulfate) (Zhang et al., 2009), potassium carbonate (Rustum, 1989), sodium chloride (Yoshida et al., 2004), or ammonium acetate (Wu et al., 2008a) to induce phase separation of the water miscible organic phase. By using SALLE, the extraction solvent

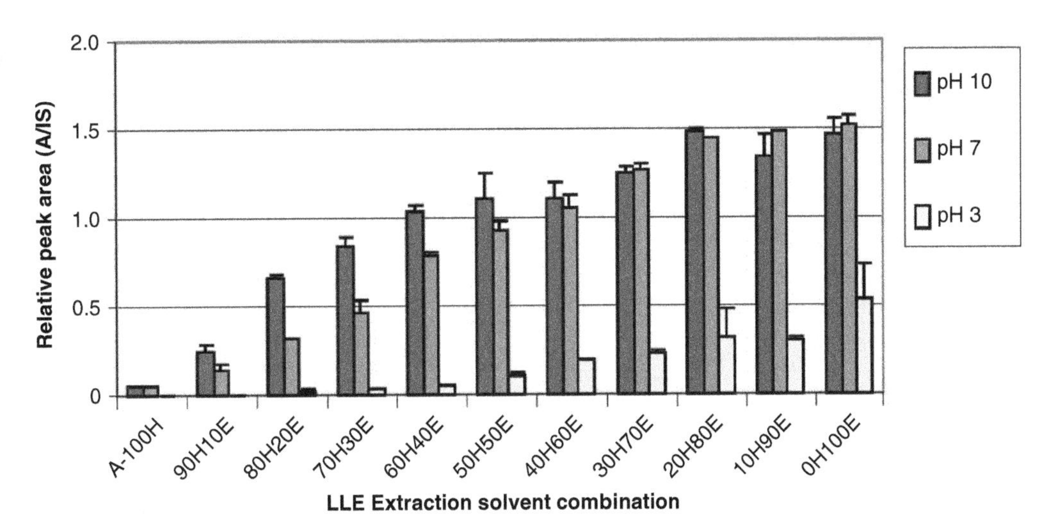

FIGURE 14.4 Recovery profiles for ketoconazole under different pH and extraction solvent combinations. The relative responses are normalized to IS response (example for solvent combination, 100H0E: hexane/ethyl acetate, 100/0, v/v) (reprinted with permission from Liu et al., 2009; copyright © 2009 American Chemical Society).

for LLE is no longer limited to water-immiscible organic solvent. Compared to conventional LLE, the advantages for SALLE include broader application (suitable for analytes from low to high lipophilicity), better analyte recovery and compatibility with RP and HILIC chromatography. Conversely, extracts typically contain more endogenous compounds and are associated with higher matrix effect. Supported liquid extraction (SLE) is a high throughput technique performed in a 96-well format that is analogous to traditional LLE. It uses an inert solid support with high surface area to improve the interface between the aqueous samples and the water-immiscible organic solvent. Briefly, biological samples are mixed with aqueous buffer and then loaded onto the solid support. Subsequently, water-immiscible solvent is added, passing through the solid support to extract the analyte of interest. The extraction mechanism involved is mainly a partition of the analyte(s) between the organic solvent and the absorbed aqueous phase on the solid support. The advantages of SLE include no emulsion formation, easy automation, and high extraction efficiency (Jiang et al., 2009, 2012).

14.3.4 Solid Phase Extraction

SPE has been used for several decades for extracting and concentrating trace amounts of analytes from biological and environmental samples (Lee and Esnaud, 1988; Peng et al., 2000; Mornar et al., 2012). The SPE process involves loading biological samples to an SPE cartridge/plate/column that is packed with solid sorbents. The analyte(s) of interest is retained by interacting with the packing material through different interaction mechanisms; the interfering matrix components are either directly passing through the sorbents during the loading step; being washed away during the washing step or retained on the stationary phase after elution, analytes of interest are retained on the stationary phase, then eluted from the sorbents using a suitable elution solvent and collected for LC-MS analysis. SPE is technically a form of low-pressure chromatography. Like for LC columns, different modes of SPE and a great variety of commercial phases are available today. When developing an SPE-based bioanalytical method, parameters needed to be determined include the type and amount of sorbent, the sample volume that can be applied without loss of recovery, the optimum loading, washing and elution conditions (time, volume, composition). SPE can be carried out either off-line (Peng et al., 2000) or online with the chromatographic system (Chen et al., 2009). Since each cartridge or each well in a multiple-well plate is essentially an individual chromatographic column, the consistency of the SPE cartridge or each well in a plate, as well as batch-to-batch variability in the SPE cartridge or plate, will have impact on the performance of an SPE method.

Based on the packing materials used, conventional SPE can be divided into three major categories. The first type is reversed-phase (RP)-SPE, which uses a nonpolar stationary phase such as the alkyl- or aryl-bonded silica; the second is normal-phase SPE using a polar stationary phase such as silica with polar functional groups (Si-CN, Si-NH2, Si-Diol, and pure silica); the third type is ion-exchange using ionic functional groups (strong or weak organic acids and bases bonded to the supporting base). Table 14.1 lists the major types of SPE, their mechanisms of retention and elution, and their applications. An SPE cartridge or plate usually comes with standard procedures for operation from the supplier. These conditions can be a very good starting point for method development. However,

TABLE 14.1 Commonly Used SPE Types and Their Mechanism of Retention and Applications

	RP-SPE	NP-SPE	Ion-exchange
Solid Phase Functional Groups	Alkyl- or aryl-bonded silica (Si-C4, Si-C8, Si-C18, Si-Ph, etc.)	Si-CN, Si-NH2, Si-Diol and pure Silica, etc.	Quaternary amine bonded silica (Strong Anion Exchange), sulfonic acid bonded silica (Strong Cation Exchange), carboxylic acid bonded silica (Weak Cation Exchange), Neutral amine (Weak Anion Exchange)
Retention Mechanism	Nonpolar–nonpolar interactions, van der waals and dispersion forces	Polar–polar interactions; Hydrogen bonding; dipole–dipole interactions; dipole-induced dipole interactions	Electrostatic attraction of the oppositely charged functional groups between the analyte and the sorbent
Loading	Samples prepared in and SPE plate conditioned with aqueous buffer with pH adjustment that analyte will not be charged	Samples prepared in and SPE plate conditioned with nonpolar solvent (e.g., hexane, dichloromethane, etc.)	An aqueous or organic, low salt solution with a pH at which the analytes and the functional groups on the stationary phase will be oppositely charged. e.g., low pH for basic compounds and strong cation exchange
Elution	Polar organic solvent such as methanol, acetonitrile, etc. with pH adjustment to the opposite of load conditions	Polar organic solvent such as methanol, acetonitrile, acetone, isopropanol	Disrupt the electrostatic interactions by pH, ionic strength and solvent modifications to neutralize groups of opposite chargeor use a more selective counter-ion to compete for ion-exchange interaction sites.
Applications	Nonpolar to moderate polar compounds: e.g., organic compound with alkyl, aromatic, alicyclic groups	Polar compounds: such as small organic compounds with hydroxyl groups, carbonyls, hetero atoms	Cation-exchange: basic compounds such as primary, secondary, tertiary and quaternary amines, etc.Anion-exchange: acidic acids such as carboxylic acids, sulphonic acids, etc.

since SPE is like any other chromatography, the optimum conditions for a specific bioanalytical method are compound specific. Fine tuning of experimental conditions for each compound is usually necessary to achieve the best results. SPE plates/cartridges are available from all the major suppliers, such as Oasis® series from Waters (www.waters.com, accessed Mar 19, 2013), HyperSep series from ThermoFisher Scientific (www.thermofisher.com, accessed Mar 19, 2013), Bond Elut series from Agilent (www.agilent.com, accessed Mar 19, 2013), EVOLUTE® series from Biotage (www.biotage.com, accessed Mar 19, 2013), Strata® series Phenomenex (www.phenomenex.com, accessed Mar 19, 2013), Discovery® and Empore™ series from Sigma-Aldrich (www.sigma-aldrich.com, accessed Mar 19, 2013). It is also worth mentioning that mixed mode (a combination of multiple retention mechanisms) SPE products are also available from almost all major vendors: Oasis® MCX/MAX from Waters, ISOLUTE® HCX/HAX from Biotage, etc. The most commonly used mixed-mode SPE is a combination of reversed-phase and ion-exchange. It utilizes dual retention mechanisms of hydrophobic and electrostatic interactions to retain basic, acidic, neutral, and zwitterionic compounds. Due to its broad applicability to a wide range of compounds, mixed-mode SPE is widely used (Shou et al., 2001; Muller et al., 2002; Jenkins et al., 2004; Ge et al., 2005; Xu et al., 2005; Yue et al., 2005; Xue et al., 2006a), however, provides less sample cleanup than other more selective SPE phases.

SPE products are available in different formats, for example, single cartridge, multiple-well plate (96-well, 384-well) or online SPE column. In the earlier days, SPE was done manually in single cartridge mode (Khan et al., 1999; van der Heeft et al., 2000; Cavaliere et al., 2003). However, with the demand for high throughput in pharmaceutical industry, 96-well format with automation is rapidly becoming the dominant platform in bioanalysis (Kaye et al., 1996; Shou et al., 2001; Shou et al., 2002; Mallet et al., 2003; Xu et al., 2007; Helle et al., 2011). No matter which format is chosen, the same principles apply. Below is a general protocol based on the 96-well format.

14.3.4.1 A Typical Sample Preparation Protocol Using Waters HLB 96-Well SPE Plate

- Pipette 50 μl of each of the blank, STDs, and QC samples to individual wells in a 96-well plate.
  - Same as PPT and LLE.

- Add 50 μl of IS working solution to each sample.
  - The IS working solution should be prepared in aqueous solution. After mixing the IS working solution with the biological samples, some time (e.g., 10 min) should be given to allow the IS to interact with the biological matrix to closely mimic the analyte's binding.
- Add 400 μl of 10 mM ammonium acetate in water to each sample.
  - This step is to buffer each sample to a given pH while maintaining a suitable solvent strength for the analyte to bind on the packing material. Protein binding concerns should be addressed during this step. Modifiers can be added to the buffer to disrupt the drug–protein binding so that the drug molecules are available for binding to the packing material during the loading step.
- Condition the 96-well SPE plate with 450 μl of methanol, followed by 450 μl of 10 mM ammonium acetate in water.
  - This step is to clean and precondition the SPE plate for sample loading. Vacuum is usually involved for efficiency and consistency.
- Transfer the sample mixtures to the preconditioned SPE plate.
  - This step is to load the samples onto the 96-well plate. Vacuum is normally needed during this step. The majority of the proteins, other macromolecules, and salts will pass through the plate to the waste. It is critical to make sure all the wells are free of previous solution before loading the samples.
- Wash SPE plate with 450 μl of water.
  - This step is to further remove residual proteins, salts, and highly polar compounds. Sometimes, the washing solution may also contain a small percentage of organic solvent to remove some less polar compounds. The desired balance of recovery and selectivity is normally decided during this step. A stronger solvent wash results in a cleaner extract but lower recovery. Vacuum is needed for this step.
- Elute SPE plate twice with 250 μl of methanol.
  - This step is to release the bound analyte of interest from the SPE packing material with strong organic solvent. Vacuum is needed for this step. It is also critical to make sure all the wells are free of previous solution before the elution solvent is added. Also the distance between the 96-well SPE plate and the collection plate has to be carefully adjusted to avoid cross-contamination. A double elution is not necessary in all cases but will results in better recovery.
- Evaporate the sample under nitrogen flow to complete dryness and reconstitute the sample with 200 μl of reconstitution solution.
  - This step is to change the solvent system for the samples to make it suitable for LC injection. Sample concentration can also be achieved during this step by reconstituting the sample in a smaller volume. This step can be omitted if the concentration and solvent are compatible with the LC-MS assay.

All liquid transfer steps in the above procedures can be routinely done using robotic liquid handler. The whole process can be easily automated using commercially available workstations. However, well-to-well and plate-to-plate differences always exist in 96-well SPE plates. It is not uncommon to observe solutions passing through one well at a different speed than another. Therefore, it is critical to make sure all the wells in the 96-well plate are free of the previous solution before applying the next solution. Unfortunately, this process normally requires visual checks by the operator, which makes walk-away automation less practical.

14.3.4.2 Advantages and Limitations The main advantages for SPE method include high-to-moderate selectivity, versatility, and suitability for full automation. Due to the availability of different types of packing material, SPE is suitable for a wide variety of compounds ranging from polar to nonpolar, acidic to basic, low to high molecular weight, and for various sample matrices. When optimized, it can achieve both high recovery and good selectivity of an analyte with very good reproducibility. Limitations of this technique include high cost, a dependence on the quality of the supplies (e.g., lot-to-lot variability of SPE plate/cartridge) and long method development time. SPE also involves many steps and requires high operational skills and good knowledge of separation science to fully actualize the advantages of SPE. A good understanding of the chemistry of the analyte and the packing material and of the retention mechanism is essential to design a good SPE method. An unoptimized SPE method is no better than a PPT method.

14.3.4.3 Other Forms of SPE The SPE format works well for other separation mechanism including immunoaffinity SPE (Delaunay-Bertoncini and Hennion, 2004), molecularly imprinted polymer (MIP) SPE (Lasakova and Jandera, 2009), and restricted access material (RAM) SPE (Souverain et al., 2004). Immunoaffinity and MIP SPE are discussed in some details in Sections 14.3.5.1 and 14.3.5.2. The RAM term was first introduced in 1991 by Desilet et al. (1991). Instead of the solid support for regular sorbents, RAM sorbents are a class of support materials with pores only accessible to small molecules. Their outer surface is coated with hydrophilic functional groups, and the pore inner surface can be coated with different functional groups, such as hydrophobic (Amini and Crescenzi, 2003) or ion exchange (Chiap et al., 2002) groups. RAM SPE takes the advantages of both regular SPE and size exclusion chromatography. Biological proteins can be easily removed from the matrix and the small organic analytes can then access the pores, and interact with the functional groups. RAM SPE is most often used in online SPE mode as it allows for direct injection of untreated plasma. A comprehensive review is available for online RAM SPE sample preparation (Souverain et al., 2004).

Operationally, other than the commonly used cartridge/plate formats of SPE, there are also many other types of SPE formats for specialized applications, such as online SPE, dispersive SPE (dSPE, also referred as QuEChERS, which stands for *Qu*ick, *E*asy, *Ch*eap, *E*ffective, *R*ugged, and *S*afe), disposable pipette extraction (DPX), microextraction SPE techniques (microextraction by packed sorbent (MEPS), solid phase microextraction (SPME), and stir bar sorptive extraction (SBSE)). Online SPE is very similar to two-dimensional chromatography, which couples an SPE cartridge to an LC column with a diverting valve in between. Two different sets of pumps are needed to deliver solvents to the SPE cartridge and the LC column. A detailed description of the hardware configuration can be found in a recent review (Chang et al., 2007). Online SPE has gained popularity for its operational advantages, in particular the minimum sample pretreatment needed before prior to LC-MS analysis. Once the system is set up and a method is developed, the production stage can be done with limited supervision. The progress, advantages, and limitations of online SPE for LC-MS bioanalysis have been extensively covered in several reviews (Xu et al., 2007; Chen et al., 2009; Novakova and Vlckova, 2009; Kole et al., 2011). In QuEChERS, the analyte of interest is first extracted from the biological matrix using organic solvent, and then sorbents are added to remove matrix components from the organic phase. It has been widely used for measuring pesticides and drug residues in food and environmental samples (Anastassiades et al., 2003; Lehotay et al., 2005; Posyniak et al., 2005). In recent years, applications were also found in the analysis of drug/metabolites in biological samples such as tissue (Fagerquist et al., 2005), milk (Whelan et al., 2010), and whole blood (Plossl et al., 2006). DPX is a variant of traditional SPE. The sorbent is loosely packed into a standard pipette tip in between two frits. Samples are loaded by aspirating through the tip from the bottom. The major advantage of DPX is its simplicity. DPX tips are commercially available with a variety of sorbents. One trend for sample preparation is miniaturization. New developments for SPE include several forms of microextraction techniques, such as SPME, MEPS, and SBSE. A common feature for these techniques is that a small amount of sorbents are used for the extraction, which allows for eluting the adsorbed analytes in a small volume, and injecting the entire volume of extract directly in the LC-MS system for analysis. The detailed description of these techniques, their advantages, and limitations are covered in several comprehensive reviews (Novakova and Vlckova, 2009; Ashri and Abdel-Rehim, 2011; Kole et al., 2011); a focused review (Vuckovic et al., 2010) for SPME is also available. No matter which operating mode is used, the principles are the same: the analyte of interest is selectively/semiselectively retained on the solid sorbent, the interfering compounds are washed away and then the analyte of interest is released and analyzed. Although there is increased interest in adopting these methods in bioanalysis, the traditional SPE in 96-well plate format is still the mainstream in bioanalysis.

14.3.5 Analyte Specific Extraction Techniques

The sample preparation techniques discussed earlier are listed in order of increasing selectivity, from the simple dilution to regular SPE methods but none of them is truly compound specific. Even though they can meet bioanalytical needs for the most part, some special cases do exist that require a more selective technique to achieve super high sensitivity or remove stubborn interferences. Matrix interferences that are most likely to interfere with the LC-MS analysis would have similar physicochemical properties to those of the analyte of interest. They are very likely to be coextracted with the analyte using the above sample extraction methods. Orthogonal or hybrid sample preparation techniques such as LLE/SPE, PPT/SPE, and PPT/LLE have sometimes been used to mitigate this issue. Compound specific extraction, such as immunoaffinity solid phase extraction (IA-SPE), SPE with MIP packing material (MIP-SPE), is a more effective approach to deal with this problem. In this section, we discuss them in detail in regard to their advantages, applications and limitations.

14.3.5.1 Immunoaffinity Extraction An immunosorbent is a selective extracting material that is prepared by immobilizing an analyte specific antibody (either polyclonal or monoclonal) on a solid support. When exposing the immunosorbent to a sample containing the analyte (antigen) of interest, a complex can form between the analyte and the immobilized antibody by selective and reversible antigen–antibody interaction. Extracting an analyte of interest using immunosorbents is called immunoaffinity extraction. It has been used for decades (Farjam et al., 1988; Haasnoot et al., 1989; Medina-Casanellas et al., 2012) to extract and enrich an analyte of interest specifically from a complex matrix. Using SPE plate/cartridge packed with immunosorbents for sample preparation is termed as IA-SPE, which can be done either online (Farjam et al., 1988) or off-line (Aranda-Rodriguez et al., 2003). Applications of IA-SPE in pharmaceutical and biomedical research have been reviewed (Delaunay-Bertoncini and Hennion, 2004).

Procedures for Immunoextraction SPE. The immunoextraction process can be divided into four steps: conditioning, percolation of the sample, washing, and elution. Detailed description of each step can be found in the review

(Delaunay-Bertoncini and Hennion, 2004) of Delaunay-Bertonicini. Briefly, the immunoextraction procedure is as follows:

Conditioning: the conditioning step is to establish an environment that favors of the interaction between the analyte and the antibody. The immunosorbents are normally stored in a PBS buffer containing a small percentage of azide to preserve the antibody activity. This buffer has to be replaced by a solution in which the specific interaction between the analyte of interest and the antibody can occur.

Loading: the biological sample after pretreatment is loaded or incubated with the immunosorbents. During this step, the capacity of the immunosorbents should be taken into account to avoid overloading the immunosorbents. The capacity of an immunosorbent is defined as the maximal amount of analyte that can be bound onto the sorbent based on the total number of accessible antibodies immobilized on the solid support.

Washing: This step serves to remove the interfering compounds that were absorbed on the sorbents through nonspecific interactions, such as hydrophobic and ionic interactions. Similarly to regular SPE procedures, the washing solution should be optimized to effectively remove the interferences but without disrupting the antigen–antibody interaction.

Elution: Any solution which can effectively disrupt the interaction between the analyte and the antibody can be used to elute the analyte and the IS, such as displacer agents, chaotropic agents, pH variations, or organic solvent; organic solvents are the most commonly used elution reagent.

The main advantage for immunoaffinity extraction is its selectivity. The immobilized antibodies on the support material are raised by immunization against a specific antigen (the analyte). Therefore, the interactions between the antibody and the antigen (analyte) are very specific. Even though antibodies can be raised against small molecules, it is much easier to raise antibodies against proteins. Immunoaffinity extraction is a powerful and efficient tool in bioanalysis. The offline (Berna et al., 2007) or online (Dufield and Radabaugh, 2012) coupling of immunoextraction with a chromatographic separation and mass spectrometric detection allows for the detection and quantification of drugs with a high sensitivity, selectivity, and reproducibility. However, compared with other sample preparation methods, such as PPT, LLE, and regular SPE, tremendous efforts have to be engaged for the preparation of immunosorbents. Producing an antibody specific to the analyte of interest and then immobilizing it on a solid support is neither a trivial nor a cheap process. Therefore, immunoextraction is often a solution of last resort for bioanalytical scientists. With the increasing interest in quantifying protein drug candidates in biological samples using LC-MS (Li et al., 2011), immunoaffinity purification of the target protein from the biological matrix (Dubois et al., 2007) or the surrogate peptides after digestion (Neubert et al., 2010) may be a necessary approach to achieve a sensitivity comparable to that of ligand binding assays, which remain the gold standard for protein drug measurements. When technology evolves to allow easy antibody generation, simple and cheap immunosorbent production for a specific analyte, or in the case when a specific analyte is widely and routinely measured, for example, biomarkers in clinical diagnostics, using standardized immunoaffinity extraction procedures (kit) for biological sample cleanup can be easily justified and also will be cost-effective.

14.3.5.2 MIP-SPE

Molecularly Imprinted Polymers. An MIP is a highly crosslinked polymer that is polymerized in the presence of a template molecule. The template molecule is removed after the polymerization, leaving complementary cavities that can then be used for molecular recognition. The cavities in the MIP specifically bind the template molecules or sometimes other molecules with very similar shape and functionality. This is very similar to the interaction between an antigen and its antibody in terms of specificity. Therefore, MIP is also referred to as a "synthetic antibody" by analogy with protein antibodies. The molecular recognition property can then be used for analyte-specific extraction in complex biological matrices. The use of MIPs as packing materials for SPE sample extraction dates back to 1994 (Sellergren, 1994). Since then, numerous applications have been reported using MIP-SPE for sample preparation in various analytical assays, including environmental sample analysis (Guan et al., 2012), food contaminants analysis (Baggiani et al., 2007; Piletska et al., 2012), and drug bioanalysis (Mullett and Lai, 1999; Yang et al., 2006; Mirmahdieh et al., 2011). Several reviews are available on the progress of using MIP-SPE for sample preparation in bioanalysis (Lasakova and Jandera, 2009; Tse Sum Bui and Haupt, 2010).

MIPs offer many advantages over protein antibodies. Proteins are difficult and expensive to purify, easily denatured (pH, heat, proteolysis), difficult to immobilize, and impractical to reuse. Antibodies to small molecules are difficult to produce as small molecules are not immunogenic by themselves. In contrast, MIPs are easy to synthesize, resistant to harsh environment (elevated temperature and pressure, extreme pHs, organic solvents), have high sample load capacity, and can be used over a long period of time. Similarly to immunoaffinity SPE, MIP-SPE can very selectively extract the target analyte from complex samples. Therefore, MIP-SPE is a practical way to concentrate the analyte of interest from a large-volume sample without retaining significant amounts of matrix interferences. This is especially helpful for an assay that needs ultrahigh sensitivity to measure trace amount of compounds in a very complex system. Limitations do exist for this technology, such as bleeding of the template molecule or heterogeneity of binding sites (nonspecific interactions). However, approaches were reported to address these concerns, making MIP-SPE a powerful analytical tool. Like any other SPE methods, MIP-SPE can be done either online or off-line.

Technically, MIPs can be prepared based on any type of template molecule. Although many reported researches on MISPE were carried using homemade MIP materials, MIP-SPE cartridges for selective compounds or a class of compounds are commercially available. Some vendors also prepare customized MIPS.

In summary, both IA-SPE and MIP-SPE use packing material specifically interacting with the analyte of interest. They offer superior selectivity, as well as the best ability to enrich the analyte of interest over other sample preparation methods. These advantages do come with a cost. In general, the initial cost for both IA-SPE and MIP-SPE is high and tremendous efforts have to be engaged to get the method up and running. Therefore, they usually tend to be a last resort in most bioanalytical laboratories. However, there is increased interest in adopting these technologies in bioanalytical applications in recent years. On one hand, with the combination of these compound-specific sample extraction techniques and availability of high sensitive instruments, many applications previously out of reach of LC-MS, such as trace analysis of biomarkers, become possible. On the other hand, although the initial cost for customized MIP or immunosorbents may be high, the overall cost may be still comparable with other techniques when a large number of samples are to be assayed using these techniques. The initial efforts and cost for customized MIPs or immunosorbents can be paid off by the efficiency of sample analysis and high quality of data.

14.3.6 Sample Preparation Techniques for Special Samples

Plasma, serum, and urine samples are the most common samples in the daily work of bioanalytical scientists. However, LC-MS bioanalysis is not limited to these ordinary matrices. LC-MS bioanalytical laboratories can be asked to measure drug/metabolites/biomarkers concentration in all kinds of samples, such as tissues, saliva, tears, bile, and most recently dried blood spots (DBS) samples. New classes of analytes such as proteins, peptides, oligonucleotides are now commonly analyzed by LC-MS. In this section, we briefly discuss sample preparation for DBS, tissue samples and protein therapeutics in plasma/serum.

14.3.6.1 Dried Blood Spots DBS, a sample collection technique that has been used for decades in clinical diagnostics, has recently emerged as a choice of microsampling technique in the pharmaceutical industry (Li and Tse, 2010). As a new sample format, sample processing of DBS sample has its own challenges and is attracting a lot of interest from bioanalytical scientists. Several sample preparation approaches have been reported (Deglon et al., 2009; Liu et al., 2010b; Abu-Rabie and Spooner, 2011). A two-step strategy was reported (Liu et al., 2010b) to best accommodate DBS samples with the approaches used for regular liquid samples. The DBS samples were first soaked with aqueous buffer to make the analyte in the blood samples more accessible to subsequent extraction. The reconstituted liquid samples can be processed in a similar way as other liquid samples using PPT, LLE, and SPE. A key factor affecting DBS sample preparation is how to efficiently and completely recover the analyte from the DBS card into the liquid phase. The term "elution efficiency" has been proposed to represent the analyte recovery from the solid DBS sample. Since this process usually is not tracked by an IS, high and consistent recovery of the analyte from the DBS sample to the liquid solution is desired. Different direct elution methods have also been reported (Deglon et al., 2009; Thomas et al., 2010; Abu-Rabie and Spooner, 2011) to elute the analyte of interest from the DBS card directly into the detection system.

14.3.6.2 Tissues and Organs Pharmaceutical drug candidates are often measured in tissue samples to better understand their distribution or for pharmacodynamic evaluation. Compared with liquid samples, sample preparation for tissue samples is more challenging. The solid tissue samples need to be converted into liquid-like samples often by homogenization (Liang et al., 2011). Organic solvent precipitation of the tissue homogenate (Gurav et al., 2012) can be done to generate a supernatant suitable for LC-MS analysis. The supernatant can also be further processed using LLE or SPE for a cleaner extract. Diluting tissue homogenate with plasma, and then processing the diluted tissue homogenate as plasma sample and quantifying it with plasma STDs was also reported (Jiang et al., 2011). A chapter devoted to tissue sample analysis using LC-MS can be found in this book.

14.3.6.3 Proteins Using LC-MS to quantitatively monitor biologics (proteins) has drawn more and more attention in recent years. Many assays have been published that use LC-MS to quantify proteins by monitoring one or several surrogate peptides after enzymatic digestion (Yang et al., 2007; Heudi et al., 2008; Ewles and Goodwin, 2011; Li et al., 2011; Mesmin et al., 2011; Wu et al., 2011). Unlike small molecules, protein drugs usually have to be digested into small peptides for sensitive quantitative analysis using LC-MS. Measuring proteins in plasma or serum using LC-MS usually involves extraction of protein drugs before digestion, such as immunoaffinity purification of target protein, or extraction of surrogate peptides after digestion.

The analysis of crude digested plasma or serum samples was also reported directly (Ouyang et al., 2012). No matter which sample preparation method is used, digestion is a critical step for protein bioanalysis using LC-MS. Many innovative approaches have been reported to quickly, effectively, and consistently perform the digestion step, such as microwave-assisted digestion (Vaezzadeh et al., 2010), organic solvent digestion (Strader et al., 2006), and pellet digestion (Ouyang et al., 2012). A digested whole plasma or serum sample is a very complex sample containing hundreds of peptides. Even though it had been demonstrated that it is feasible to quantify proteins by direct LC-MS analysis of the crude digested plasma samples, it is strongly recommended to simplify the sample composition using available techniques, such as immunoaffinity purification, MIP-SPE, or other SPE techniques.

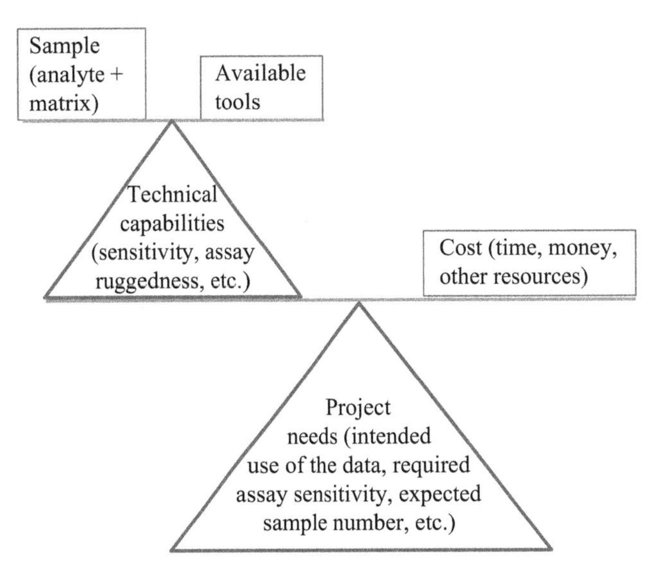

FIGURE 14.5 Illustration of bioanalytical strategy selection.

14.4 KNOW YOUR NEEDS

As a bioanalytical scientist, after examining the basic information available regarding the samples and the analyte and checking all the tools in our toolbox, the next thing is to decide what approach to use on a given project. In most cases, scientists will turn to the technique they are most familiar with or is most readily available. Sample preparation is a matter of personal preference but it is good practice to brainstorm all the options and eliminate some obviously bad choices before heading to the laboratory for experiments. Before starting a method, two questions should be asked: "what can we do?" and "what do we need to do?" A good bioanalytical strategy should derive from a comprehensive evaluation of the properties of both the analyte and the matrix, available options for sample preparation, the project needs and the cost. As illustrated in Figure 14.5, the final bioanalytical strategy should be the right balance between the technical capabilities and the cost (time, money and other resources) based on the project needs. For example, an assay that does not require high sensitivity may only need a simple "dilution" approach; a one-time-use method for a small study can settle with PPT; an assay which is expected to be used for many studies with a large number of samples should be very rugged and the initial cost of the assay may not be a concern as long as the final cost/sample is acceptable; assays for bioequivalence studies should be of the highest precision. These approaches are frequently discussed as "fit-for-purpose" or "risk-based." The best bioanalytical strategy is the most economical solution that fits the project needs.

14.5 CONCLUSION AND PERSPECTIVES

With the advancements in technology and understanding of matrix effect, sample preparation for small molecule drug candidates in ordinary matrices (e.g., plasma, urine, serum, and blood) has become straightforward in most cases. With the availability of increasingly more sensitive mass spectrometers, it is now possible to adopt a simple sample dilution strategy, in many applications, in which only a very small amount of matrix (e.g., less than 1 μl of plasma per injection) will need to be injected into the LC-MS system to achieve the required sensitivity. Also if a SIL-IS is available, matrix effects can usually be well compensated for in a bioanalytical method, which makes dilution or simple PPT even more attractive. Ideally, a sample preparation method is preferred with minimum change of the original samples to preserve their information as much as possible.

With the expanding scope of LC-MS bioanalysis, in particular in the area of biomarkers and biologics, the demand for novel and better sample preparation approaches is continuing. For routine sample analysis, greater emphasis has been placed on truly walk-away automation. The sample preparation process could be fully automated eventually. Another trend for sample preparation is the handling of small sample volumes. Microsampling technology is becoming more and more popular now that the instrument detection limits allow bioanalysis of the equivalent of 1-μl or 2-μl samples per analysis. The currently popular sample handling and processing techniques in routine bioanalysis seem to be falling behind the pace of more advanced detection techniques. In traditional sample preparation, the majority of the sample collected is not used for the final analysis but wasted. We expect that better usage of precious biological samples will be the trend and sample preparation techniques, which can effectively deal with 1-μl or 2-μl samples, will become popular in bioanalysis.

Finally, as bioanalysis-related technologies evolve, customized bioanalysis kits for a specific analyte, using technologies such as IA-SPE, MIP-SPE, with "standardized sample preparation procedures," can become a practical and cost-effective solution for the analysis of common drugs.

REFERENCES

Abu-Rabie P, Spooner N. Dried matrix spot direct analysis: evaluating the robustness of a direct elution technique for use in quantitative bioanalysis. Bioanalysis 2011;3:2769–2781.

Amini N, Crescenzi C. Feasibility of an on-line restricted access material/liquid chromatography/tandem mass spectrometry method in the rapid and sensitive determination of organophosphorus triesters in human blood plasma. J Chromatogr B Analyt Technol Biomed Life Sci 2003;795:245–256.

Anastassiades M, Lehotay SJ, Stajnbaher D, Schenck FJ. Fast and easy multiresidue method employing acetonitrile extraction/partitioning and "dispersive solid-phase extraction" for the determination of pesticide residues in produce. J AOAC Int 2003;86:412–431.

Aranda-Rodriguez R, Kubwabo C, Benoit FM. Extraction of 15 microcystins and nodularin using immunoaffinity columns. Toxicon 2003;42:587–599.

Ashri NY, Abdel-Rehim M. Sample treatment based on extraction techniques in biological matrices. Bioanalysis 2011;3:2003–2018.

Aubry AF. LC-MS/MS bioanalytical challenge: ultra-high sensitivity assays. Bioanalysis 2011;3:1819–1825.

Baggiani C, Anfossi L, Giovannoli C. Solid phase extraction of food contaminants using molecular imprinted polymers. Anal Chim Acta 2007;591:29–39.

Berna MJ, Zhen Y, Watson DE, Hale JE, Ackermann BL. Strategic use of immunoprecipitation and LC/MS/MS for trace-level protein quantification: myosin light chain 1, a biomarker of cardiac necrosis. Anal Chem 2007;79:4199–4205.

Bishop MJ, Crow BS, Kovalcik KD, George J, Bralley JA. Quantification of urinary zwitterionic organic acids using weak-anion exchange chromatography with tandem MS detection. J Chromatogr B Analyt Technol Biomed Life Sci 2007;848:303–310.

Brockman AH, Hatsis P, Paton M, Wu JT. Impact of differential recovery in bioanalysis: the example of bortezomib in whole blood. Anal Chem 2007;79:1599–1603.

Cai X, Zhong B, Su B, Xu S, Guo B. Development and validation of a rapid LC-MS/MS method for the determination of JCC76, a novel antitumor agent for breast cancer, in rat plasma and its application to a pharmacokinetics study. Biomed Chromatogr 2012;26:1118–1124.

Casetta B, Romanello M, Moro L. A rapid and simple method for quantitation of urinary hydroxylysyl glycosides, indicators of collagen turnover, using liquid chromatography/tandem mass spectrometry. Rapid Commun Mass Spectrom 2000;14:2238–2241.

Cavaliere C, Curini R, Di Corcia A, Nazzari M, Samperi R. A simple and sensitive liquid chromatography-mass spectrometry confirmatory method for analyzing sulfonamide antibacterials in milk and egg. J Agric Food Chem 2003;51:558–566.

Chang M, Ji Q, Zhang J, El-Shourbagy T. Historical review of sample preparation for chromatographic bioanalysis: pros and cons. Drug Development Research 2007;68:107–133.

Chen L, Wang H, Zeng Q, et al. On-line coupling of solid-phase extraction to liquid chromatography–a review. J Chromatogr Sci 2009;47:614–623.

Chiap P, Rbeida O, Christiaens B, et al. Use of a novel cation-exchange restricted-access material for automated sample clean-up prior to the determination of basic drugs in plasma by liquid chromatography. J Chromatogr A 2002;975:145–155.

Chng HT, New LS, Neo AH, Goh CW, Browne ER, Chan EC. A sensitive LC/MS/MS bioanalysis assay of orally administered lipoic acid in rat blood and brain tissue. J Pharm Biomed Anal 2010;51:754–757.

Dams R, Huestis MA, Lambert WE, Murphy CM. Matrix effect in bio-analysis of illicit drugs with LC-MS/MS: influence of ionization type, sample preparation, and biofluid. J Am Soc Mass Spectrom 2003;14:1290–1294.

Deglon J, Thomas A, Cataldo A, Mangin P, Staub C. On-line desorption of dried blood spot: a novel approach for the direct LC/MS analysis of micro-whole blood samples. J Pharm Biomed Anal 2009;49:1034–1039.

Delaunay-Bertoncini N, Hennion MC. Immunoaffinity solid-phase extraction for pharmaceutical and biomedical trace-analysis-coupling with HPLC and CE-perspectives. J Pharm Biomed Anal 2004;34:717–736.

Desilets CP, Rounds MA, Regnier FE. Semipermeable-surface reversed-phase media for high-performance liquid chromatography. J Chromatogr 1991;544:25–39.

Dotsikas Y, Kousoulos C, Tsatsou G, Loukas YL. Development and validation of a rapid 96-well format based liquid-liquid extraction and liquid chromatography-tandem mass spectrometry analysis method for ondansetron in human plasma. J Chromatogr B Analyt Technol Biomed Life Sci 2006;836:79–82.

Dubois M, Becher F, Herbet A, Ezan E. Immuno-mass spectrometry assay of EPI-HNE4, a recombinant protein inhibitor of human elastase. Rapid Commun Mass Spectrom 2007;21:352–358.

Dufield DR, Radabaugh MR. Online immunoaffinity LC/MS/MS. A general method to increase sensitivity and specificity: how do you do it and what do you need? Methods 2012;56:236–245.

Ewles M, Goodwin L. Bioanalytical approaches to analyzing peptides and proteins by LC–MS/MS. Bioanalysis 2011;3:1379–1397.

Fagerquist CK, Lightfield AR, Lehotay SJ. Confirmatory and quantitative analysis of beta-lactam antibiotics in bovine kidney tissue by dispersive solid-phase extraction and liquid chromatography-tandem mass spectrometry. Anal Chem 2005;77:1473–1482.

Farjam A, de Jong GJ, Frei RW, et al. Immunoaffinity pre-column for selective on-line sample pre-treatment in high-performance liquid chromatography determination of 19-nortestosterone. J Chromatogr 1988;452:419–433.

Flaherty JM, Connolly PD, Decker ER, et al. Quantitative determination of perfluorooctanoic acid in serum and plasma by liquid chromatography tandem mass spectrometry. J Chromatogr B Analyt Technol Biomed Life Sci 2005;819:329–338.

Fu I, Woolf EJ, Matuszewski BK. Effect of the sample matrix on the determination of indinavir in human urine by HPLC with turbo ion spray tandem mass spectrometric detection. J Pharm Biomed Anal 1998;18:347–357.

Fu Y, Li W, Smith HT, Tse FL. An investigation of incurred human urine sample reanalysis failure. Bioanalysis 2011;3:967–972.

Ge L, Yong JW, Tan SN, Yang XH, Ong ES. Analysis of positional isomers of hydroxylated aromatic cytokinins by micellar electrokinetic chromatography. Electrophoresis 2005;26:1768–1777.

Gray N, Musenga A, Cowan DA, Plumb R, Smith NW. A simple high pH liquid chromatography-tandem mass spectrometry method for basic compounds: application to ephedrines in doping control analysis. J Chromatogr A 2011;1218:2098–2105.

Gu H, Deng Y, Wang J, Aubry AF, Arnold ME. Development and validation of sensitive and selective LC-MS/MS methods for the determination of BMS-708163, a gamma-secretase inhibitor, in plasma and cerebrospinal fluid using deprotonated or formate adduct ions as precursor ions. J Chromatogr B Analyt Technol Biomed Life Sci 2010;878:2319–2326.

Guan W, Han C, Wang X, et al. Molecularly imprinted polymer surfaces as solid-phase extraction sorbents for the extraction of 2-nitrophenol and isomers from environmental water. J Sep Sci 2012;35:490–497.

Gurav SD, Jeniffer S, Punde R, et al. A strategy for extending the applicability of a validated plasma calibration curve to quantitative measurements in multiple tissue homogenate samples: a case study from a rat tissue distribution study of JI-101, a triple kinase inhibitor. Biomed Chromatogr 2012;26:419–424.

Haasnoot W, Schilt R, Hamers AR, et al. Determination of beta-19-nortestosterone and its metabolite alpha-19-nortestosterone in biological samples at the sub parts per billion level by high-performance liquid chromatography with on-line immunoaffinity sample pretreatment. J Chromatogr 1989;489:157–171.

Helle N, Baden M, Petersen K. Automated solid phase extraction. Methods Mol Biol 2011;747:93–129.

Heudi O, Barteau S, Zimmer D, et al. Towards absolute quantification of therapeutic monoclonal antibody in serum by LC-MS/MS using isotope-labeled antibody standard and protein cleavage isotope dilution mass spectrometry. Anal Chem 2008;80:4200–4207.

Hussain S, Patel H, Tan A. Automated liquid-liquid extraction method for high-throughput analysis of rosuvastatin in human EDTA K2 plasma by LC-MS/MS. Bioanalysis 2009;1:529–535.

Jemal M, Schuster A, Whigan DB. Liquid chromatography/tandem mass spectrometry methods for quantitation of mevalonic acid in human plasma and urine: method validation, demonstration of using a surrogate analyte, and demonstration of unacceptable matrix effect in spite of use of a stable isotope analog internal standard. Rapid Commun Mass Spectrom 2003;17:1723–1734.

Jemal M, Xia YQ. LC-MS Development strategies for quantitative bioanalysis. Curr Drug Metab 2006;7:491–502.

Jenkins KM, Young MS, Mallet CR, Elian AA. Mixed-mode solid-phase extraction procedures for the determination of MDMA and metabolites in urine using LC-MS, LC-UV, or GC-NPD. J Anal Toxicol 2004;28:50–58.

Ji AJ, Jiang Z, Livson Y, Davis JA, Chu JX, Weng N. Challenges in urine bioanalytical assays: overcoming nonspecific binding. Bioanalysis 2010;2:1573–1586.

Jiang H, Cao H, Zhang Y, Fast DM. Systematic evaluation of supported liquid extraction in reducing matrix effect and improving extraction efficiency in LC-MS/MS based bioanalysis for 10 model pharmaceutical compounds. J Chromatogr B Analyt Technol Biomed Life Sci 2012;891–892:71–80

Jiang H, Randlett C, Junga H, Jiang X, Ji QC. Using supported liquid extraction together with cellobiohydrolase chiral stationary phases-based liquid chromatography with tandem mass spectrometry for enantioselective determination of acebutolol and its active metabolite diacetolol in spiked human plasma. J Chromatogr B Analyt Technol Biomed Life Sci 2009;877:173–180.

Jiang H, Zeng J, Zheng N, et al. A convenient strategy for quantitative determination of drug concentrations in tissue homogenates using a liquid chromatography/tandem mass spectrometry assay for plasma samples. Anal Chem 2011;83:6237–6244.

Kaye B, Herron WJ, Macrae PV, et al. Rapid, solid phase extraction technique for the high-throughput assay of darifenacin in human plasma. Anal Chem 1996;68:1658–1660.

Khan JK, Bu HZ, Samarendra ZZ, Maiti N, Micetich RG. A rapid and reliable solid-phase extraction–LC/MS/MS assay for the determination of two novel human leukocyte elastase inhibitors, SYN-1390 and SYN-1396, in rat plasma. J Pharm Biomed Anal 1999;20:697–703.

King R, Bonfiglio R, Fernandez-Metzler C, Miller-Stein C, Olah T. Mechanistic investigation of ionization suppression in electrospray ionization. J Am Soc Mass Spectrom 2000;11:942–950.

Kole PL, Venkatesh G, Kotecha J, Sheshala R. Recent advances in sample preparation techniques for effective bioanalytical methods. Biomed Chromatogr 2011;25:199–217.

Kosovec JE, Egorin MJ, Gjurich S, Beumer JH. Quantitation of 5-fluorouracil (5-FU) in human plasma by liquid chromatography/electrospray ionization tandem mass spectrometry. Rapid Commun Mass Spectrom 2008;22:224–230.

Lahaie M, Mess JN, Furtado M, Garofolo F. Elimination of LC-MS/MS matrix effect due to phospholipids using specific solid-phase extraction elution conditions. Bioanalysis 2010;2(6):1011–1021.

Lasakova M, Jandera P. Molecularly imprinted polymers and their application in solid phase extraction. J Sep Sci 2009;32:799–812.

Lee CR, Esnaud H. Determination of melatonin by GC-MS: problems with solid phase extraction (SPE) columns. Biomed Environ Mass Spectrom 1988;15:677–679.

Lehotay SJ, de Kok A, Hiemstra M, Van Bodegraven P. Validation of a fast and easy method for the determination of residues from 229 pesticides in fruits and vegetables using gas and liquid chromatography and mass spectrometric detection. J AOAC Int 2005;88:595–614.

Li F, Fast D, Michael S. Absolute quantitation of protein therapeutics in biological matrices by enzymatic digestion and LC-MS. Bioanalysis 2011;3:2459–2480.

Li W, Luo S, Smith HT, Tse FL. Quantitative determination of BAF312, a S1P-R modulator, in human urine by LC-MS/MS:

prevention and recovery of lost analyte due to container surface adsorption. J Chromatogr B Analyt Technol Biomed Life Sci 2010;878:583–589.

Li W, Tse FL. Dried blood spot sampling in combination with LC-MS/MS for quantitative analysis of small molecules. Biomed Chromatogr 2010;24:49–65.

Li W, Zhang J, Tse FL. Strategies in quantitative LC-MS/MS analysis of unstable small molecules in biological matrices. Biomed Chromatogr 2011;25:258–277.

Liang X, Ubhayakar S, Liederer BM, et al. Evaluation of homogenization techniques for the preparation of mouse tissue samples to support drug discovery. Bioanalysis 2011;3:1923–1933.

Liu G, Ji QC, Arnold ME. Identifying, evaluating, and controlling bioanalytical risks resulting from nonuniform matrix ion suppression/enhancement and nonlinear liquid chromatography-mass spectrometry assay response. Anal Chem 2010a;82:9671–9677.

Liu G, Patrone L, Snapp HM, et al. Evaluating and defining sample preparation procedures for DBS LC-MS/MS assays. Bioanalysis 2010b;2:1405–1414.

Liu G, Snapp HM, Ji QC, Arnold ME. Strategy of accelerated method development for high-throughput bioanalytical assays using ultra high-performance liquid chromatography coupled with mass spectrometry. Anal Chem 2009;81:9225–9232.

Ma J, Shi J, Le H, et al. A fully automated plasma protein precipitation sample preparation method for LC-MS/MS bioanalysis. J Chromatogr B Analyt Technol Biomed Life Sci 2008;862: 219–226.

Mallet CR, Lu Z, Fisk R, Mazzeo JR, Neue UD. Performance of an ultra-low elution-volume 96-well plate: drug discovery and development applications. Rapid Commun Mass Spectrom 2003;17:163–170.

McCauley-Myers DL, Eichhold TH, Bailey RE, et al. Rapid bioanalytical determination of dextromethorphan in canine plasma by dilute-and-shoot preparation combined with one minute per sample LC-MS/MS analysis to optimize formulations for drug delivery. J Pharm Biomed Anal 2000;23:825–835.

Medina-Casanellas S, Benavente F, Barbosa J, Sanz-Nebot V. Preparation and evaluation of an immunoaffinity sorbent for the analysis of opioid peptides by on-line immunoaffinity solid-phase extraction capillary electrophoresis-mass spectrometry. Anal Chim Acta 2012;717:134–142.

Mesmin C, Fenaille F, Ezan E, Becher F. MS-based approaches for studying the pharmacokinetics of protein drugs. Bioanalysis 2011;3:477–480.

Mirmahdieh S, Mardihallaj A, Hashemian Z, Razavizadeh J, Ghaziaskar H, Khayamian T. Analysis of testosterone in human urine using molecularly imprinted solid-phase extraction and corona discharge ion mobility spectrometry. J Sep Sci 2011;34:107–112.

Moreno-Bondi MC, Marazuela MD, Herranz S, Rodriguez E. An overview of sample preparation procedures for LC-MS multiclass antibiotic determination in environmental and food samples. Anal Bioanal Chem 2009;395:921–946.

Mornar A, Sertic M, Turk N, Nigovic B, Korsic M. Simultaneous analysis of mitotane and its main metabolites in human blood and urine samples by SPE-HPLC technique. Biomed Chromatogr 2012.

Muller C, Schafer P, Stortzel M, Vogt S, Weinmann W. Ion suppression effects in liquid chromatography-electrospray-ionisation transport-region collision induced dissociation mass spectrometry with different serum extraction methods for systematic toxicological analysis with mass spectra libraries. J Chromatogr B Analyt Technol Biomed Life Sci 2002;773:47–52.

Mullett WM, Lai EP. Rapid determination of theophylline in serum by selective extraction using a heated molecularly imprinted polymer micro-column with differential pulsed elution. J Pharm Biomed Anal 1999;21:835–843.

Neubert H, Gale J, Muirhead D. Online high-flow peptide immunoaffinity enrichment and nanoflow LC-MS/MS: assay development for total salivary pepsin/pepsinogen. Clin Chem 2010;56:1413–1423.

Novakova L, Vlckova H. A review of current trends and advances in modern bio-analytical methods: chromatography and sample preparation. Anal Chim Acta 2009;656:8–35.

Ouyang Z, Furlong MT, Wu S, et al. Pellet digestion: a simple and efficient sample preparation technique for LC-MS/MS quantification of large therapeutic proteins in plasma. Bioanalysis 2012;4:17–28.

Patel BN, Sharma N, Sanyal M, Shrivastav PS. Simultaneous determination of simvastatin and simvastatin acid in human plasma by LC-MS/MS without polarity switch: application to a bioequivalence study. J Sep Sci 2008;31:301–313.

Peng SX, King SL, Bornes DM, Foltz DJ, Baker TR, Natchus MG. Automated 96-well SPE and LC-MS-MS for determination of protease inhibitors in plasma and cartilage tissues. Anal Chem 2000;72:1913–1917.

Piletska EV, Burns R, Terry LA, Piletsky SA. Application of a molecularly imprinted polymer for the extraction of kukoamine a from potato peels. J Agric Food Chem 2012;60:95–99.

Pin H, Hong-Min L, Ming Y, Qin L. A validated LC-MS/MS method for the determination of vinflunine in plasma and its application to pharmacokinetic studies. Biomed Chromatogr 2012; 26:797–801.

Plossl F, Giera M, Bracher F. Multiresidue analytical method using dispersive solid-phase extraction and gas chromatography/ion trap mass spectrometry to determine pharmaceuticals in whole blood. J Chromatogr A 2006;1135:19–26.

Polson C, Sarkar P, Incledon B, Raguvaran V, Grant R. Optimization of protein precipitation based upon effectiveness of protein removal and ionization effect in liquid chromatography-tandem mass spectrometry. J Chromatogr B Analyt Technol Biomed Life Sci 2003;785:263–275.

Posyniak A, Zmudzki J, Mitrowska K. Dispersive solid-phase extraction for the determination of sulfonamides in chicken muscle by liquid chromatography. J Chromatogr A 2005;1087:259–264.

Pucci V, Di Palma S, Alfieri A, Bonelli F, Monteagudo E. A novel strategy for reducing phospholipids-based matrix effect in LC-ESI-MS bioanalysis by means of HybridSPE. J Pharm Biomed Anal 2009;50:867–871.

Raffaelli A, Saba A, Vignali E, Marcocci C, Salvadori P. Direct determination of the ratio of tetrahydrocortisol + allo-tetrahydrocortisol to tetrahydrocortisone in urine by LC-MS-MS. J Chromatogr B Analyt Technol Biomed Life Sci 2006;830:278–285.

Rashed MS, Saadallah AA, Rahbeeni Z, et al. Determination of urinary S-sulphocysteine, xanthine and hypoxanthine by liquid chromatography-electrospray tandem mass spectrometry. Biomed Chromatogr 2005;19:223–230.

Rudewicz PJ. Turbulent flow bioanalysis in drug metabolism and pharmacokinetics. Bioanalysis 2011;3:1663–1671.

Rustum AM. Determination of cadralazine in human whole blood using reversed-phase high-performance liquid chromatography: utilizing a salting-out extraction procedure. J Chromatogr 1989;489:345–352.

Saar E, Gerostamoulos D, Drummer OH, Beyer J. Identification and quantification of 30 antipsychotics in blood using LC-MS/MS. J Mass Spectrom 2010;45:915–925.

Samanidou V, Kovatsi L, Fragou D, Rentifis K. Novel strategies for sample preparation in forensic toxicology. Bioanalysis 2011;3:2019–2046.

Sellergren B. Direct Drug Determination by Selective Sample Enrichment on an Imprinted Polymer. Anal Chem 1994;66: 1578–1582.

Shou WZ, Jiang X, Beato BD, Naidong W. A highly automated 96-well solid phase extraction and liquid chromatography/tandem mass spectrometry method for the determination of fentanyl in human plasma. Rapid Commun Mass Spectrom 2001;15: 466–476.

Shou WZ, Pelzer M, Addison T, Jiang X, Naidong W. An automatic 96-well solid phase extraction and liquid chromatography-tandem mass spectrometry method for the analysis of morphine, morphine-3-glucuronide and morphine-6-glucuronide in human plasma. J Pharm Biomed Anal 2002;27:143–152.

Song Q, Naidong W. Analysis of omeprazole and 5-OH omeprazole in human plasma using hydrophilic interaction chromatography with tandem mass spectrometry (HILIC-MS/MS)–eliminating evaporation and reconstitution steps in 96-well liquid/liquid extraction. J Chromatogr B Analyt Technol Biomed Life Sci 2006;830:135–142.

Souverain S, Rudaz S, Veuthey JL. Restricted access materials and large particle supports for on-line sample preparation: an attractive approach for biological fluids analysis. J Chromatogr B Analyt Technol Biomed Life Sci 2004;801:141–156.

Strader MB, Tabb DL, Hervey WJ, Pan C, Hurst GB. Efficient and specific trypsin digestion of microgram to nanogram quantities of proteins in organic-aqueous solvent systems. Anal Chem 2006;78:125–134.

Thomas A, Deglon J, Steimer T, Mangin P, Daali Y, Staub C. On-line desorption of dried blood spots coupled to hydrophilic interaction/reversed-phase LC/MS/MS system for the simultaneous analysis of drugs and their polar metabolites. J Sep Sci 2010;33:873–879.

Tse Sum Bui B, Haupt K. Molecularly imprinted polymers: synthetic receptors in bioanalysis. Anal Bioanal Chem 2010; 398:2481–2492.

Tweed JA, Gu Z, Xu H, et al. Automated sample preparation for regulated bioanalysis: an integrated multiple assay extraction platform using robotic liquid handling. Bioanalysis 2010;2:1023–1040.

Vaezzadeh AR, Deshusses JM, Waridel P, et al. Accelerated digestion for high-throughput proteomics analysis of whole bacterial proteomes. J Microbiol Methods 2010;80:56–62.

van der Heeft E, Dijkman E, Baumann RA, Hogendoorn EA. Comparison of various liquid chromatographic methods involving UV and atmospheric pressure chemical ionization mass spectrometric detection for the efficient trace analysis of phenylurea herbicides in various types of water samples. J Chromatogr A 2000;879:39–50.

Vogeser M, Kirchhoff F. Progress in automation of LC-MS in laboratory medicine. Clin Biochem 2011;44:4–13.

Vuckovic D, Zhang X, Cudjoe E, Pawliszyn J. Solid-phase microextraction in bioanalysis: New devices and directions. J Chromatogr A 2010;1217:4041–4060.

Wang PG, Zhang J, Gage EM, et al. A high-throughput liquid chromatography/tandem mass spectrometry method for simultaneous quantification of a hydrophobic drug candidate and its hydrophilic metabolite in human urine with a fully automated liquid/liquid extraction. Rapid Commun Mass Spectrom 2006;20:3456–3464.

Wang S, Cyronak M, Yang E. Does a stable isotopically labeled internal standard always correct analyte response? A matrix effect study on a LC/MS/MS method for the determination of carvedilol enantiomers in human plasma. J Pharm Biomed Anal 2007;43:701–707.

Watt AP, Morrison D, Locker KL, Evans DC. Higher throughput bioanalysis by automation of a protein precipitation assay using a 96-well format with detection by LC-MS/MS. Anal Chem 2000;72:979–984.

Whelan M, Kinsella B, Furey A, et al. Determination of anthelmintic drug residues in milk using ultra high performance liquid chromatography-tandem mass spectrometry with rapid polarity switching. J Chromatogr A 2010;1217:4612–4622.

Wilson ID. High-performance liquid chromatography-mass spectrometry (HPLC-MS)-based drug metabolite profiling. Methods Mol Biol 2011;708:173–190.

Wood M, Laloup M, Samyn N, et al. Simultaneous analysis of gamma-hydroxybutyric acid and its precursors in urine using liquid chromatography-tandem mass spectrometry. J Chromatogr A 2004;1056:83–90.

Wu H, Zhang J, Norem K, El-Shourbagy TA. Simultaneous determination of a hydrophobic drug candidate and its metabolite in human plasma with salting-out assisted liquid/liquid extraction using a mass spectrometry friendly salt. J Pharm Biomed Anal 2008a;48:1243–1248.

Wu ST, Ouyang Z, Olah TV, Jemal M. A strategy for liquid chromatography/tandem mass spectrometry based quantitation of pegylated protein drugs in plasma using plasma protein precipitation with water-miscible organic solvents and subsequent trypsin digestion to generate surrogate peptides for detection. Rapid Commun Mass Spectrom 2011;25: 281–290.

Wu ST, Schoener D, Jemal M. Plasma phospholipids implicated in the matrix effect observed in liquid chromatography/tandem mass spectrometry bioanalysis: evaluation of the use of colloidal silica in combination with divalent or trivalent cations for the selective removal of phospholipids from plasma. Rapid Commun Mass Spectrom 2008b;22:2873–2881.

Xia YQ, Jemal M. Phospholipids in liquid chromatography/mass spectrometry bioanalysis: comparison of three tandem mass spectrometric techniques for monitoring plasma phospholipids, the effect of mobile phase composition on phospholipids elution and the association of phospholipids with matrix effects. Rapid Commun Mass Spectrom 2009;23:2125–2138.

Xu RN, Fan L, Rieser MJ, El-Shourbagy TA. Recent advances in high-throughput quantitative bioanalysis by LC-MS/MS. J Pharm Biomed Anal 2007;44:342–355.

Xu Y, Du L, Rose MJ, Fu I, Woolf EJ, Musson DG. Concerns in the development of an assay for determination of a highly conjugated adsorption-prone compound in human urine. J Chromatogr B Analyt Technol Biomed Life Sci 2005;818: 241–248.

Xue YJ, Akinsanya JB, Liu J, Unger SE. A simplified protein precipitation/mixed-mode cation-exchange solid-phase extraction, followed by high-speed liquid chromatography/mass spectrometry, for the determination of a basic drug in human plasma. Rapid Commun Mass Spectrom 2006a;20:2660–2668.

Xue YJ, Liu J, Pursley J, Unger S. A 96-well single-pot protein precipitation, liquid chromatography/tandem mass spectrometry (LC/MS/MS) method for the determination of muraglitazar, a novel diabetes drug, in human plasma. J Chromatogr B Analyt Technol Biomed Life Sci 2006b;831:213–222.

Xue YJ, Yan JH, Arnold M, Grasela D, Unger S. Quantitative determination of BMS-378806 in human plasma and urine by high-performance liquid chromatography/tandem mass spectrometry. J Sep Sci 2007;30:1267–1275.

Yang J, Hu Y, Cai JB, Zhu XL, Su QD. A new molecularly imprinted polymer for selective extraction of cotinine from urine samples by solid-phase extraction. Anal Bioanal Chem 2006;384:761–768.

Yang Z, Hayes M, Fang X, Daley MP, Ettenberg S, Tse FL. LC-MS/MS approach for quantification of therapeutic proteins in plasma using a protein internal standard and 2D-solid-phase extraction cleanup. Anal Chem 2007;79:9294–9301.

Yoshida M, Akane A, Nishikawa M, Watabiki T, Tsuchihashi H. Extraction of thiamylal in serum using hydrophilic acetonitrile with subzero-temperature and salting-out methods. Anal Chem 2004;76:4672–4675.

Yue H, Borenstein MR, Jansen SA, Raffa RB. Liquid chromatography-mass spectrometric analysis of buprenorphine and its N-dealkylated metabolite norbuprenorphine in rat brain tissue and plasma. J Pharmacol Toxicol Methods 2005;52:314–322.

Zhang J, Wu H, Kim E, El-Shourbagy TA. Salting-out assisted liquid/liquid extraction with acetonitrile: a new high throughput sample preparation technique for good laboratory practice bioanalysis using liquid chromatography-mass spectrometry. Biomed Chromatogr 2009;23:419–425.

15

BEST PRACTICE IN LIQUID CHROMATOGRAPHY FOR LC-MS BIOANALYSIS

STEVE UNGER AND NAIDONG WENG

15.1 INTRODUCTION

This chapter discusses the best practices in liquid chromatography when performing liquid chromatography–mass spectrometry (LC-MS) bioanalysis. Chromatography plays a pivotal role for the establishment of a bioanalytical LC-MS method to support the intended studies. The life cycle of a bioanalytical method evolves, and new findings may require additional method enhancements (Figure 15.1). Typically a bioanalytical LC-MS method can be divided into four phases: method development, method validation, sample analysis, and method enhancement. In the early development stages of a drug candidate, due to the high attrition of drug candidates and typically smaller number of samples, some of the features of chromatography such as high-speed analysis, automation and multiplexing of multiple instruments, which are important considerations when analyzing large sets of clinical samples to achieve optimal throughput, may be less important. Autosampler carryover that can be quite detrimental for clinical analysis of large numbers of samples could be easily managed for small preclinical study samples by inserting extra blanks after each incurred samples. On the other hand, information regarding the stability of analytes and metabolites in biological fluids is incomplete or is still being gathered at these early stages, so additional precautions should be used when developing chromatographic conditions. Extended gradient elution and longer run time might be used to ensure the chromatographic separation of the analytes of interest from the metabolites, in-source breakdown products of the conjugated metabolites, or endogenous compounds that cause matrix effects. When using nonselective sample extraction procedures (usually generic protein precipitation methods that can be used across different preclinical species), more extensive chromatographic separation is a good choice to improve assay selectivity. Findings collected from method validation and sample analysis may lead to method reoptimization and revalidation, which can be beneficial for future studies should the drug candidates survive the early attrition.

Therefore for a new practitioner in the field of bioanalysis, a common and much simplified question is "What is good chromatography?" Responses often use examples of poor chromatography that may include peak splitting or tailing as well as insufficient retention time for the analytes of interest. While these more obvious observations may already give a good indication for whether or not the chromatographic system is under control, there are many more factors that can impact the overall quality of the results generated from LC-MS. A more comprehensive understanding of how chromatography works in the context of bioanalytical LC-MS and what are the potential pitfalls will be discussed here. In regulated bioanalysis, acceptance criteria should be predefined so that decisions are unbiased and consistently applied to all samples and runs (Briscoe et al., 2007). Analysts can reanalyze samples for assignable causes and an isolated instance of poor chromatography is one reason that can be used to justify reinjection. Provided samples have not exceeded extract stability, reinjection on a properly prepared instrument is also the most straightforward approach. When there is a consistently poor chromatography for multiple samples or runs, investigation and resolution of the root cause is required prior to further continuation of sample analysis.

Handbook of LC-MS Bioanalysis: Best Practices, Experimental Protocols, and Regulations, First Edition. Edited by Wenkui Li, Jie Zhang, and Francis L.S. Tse.

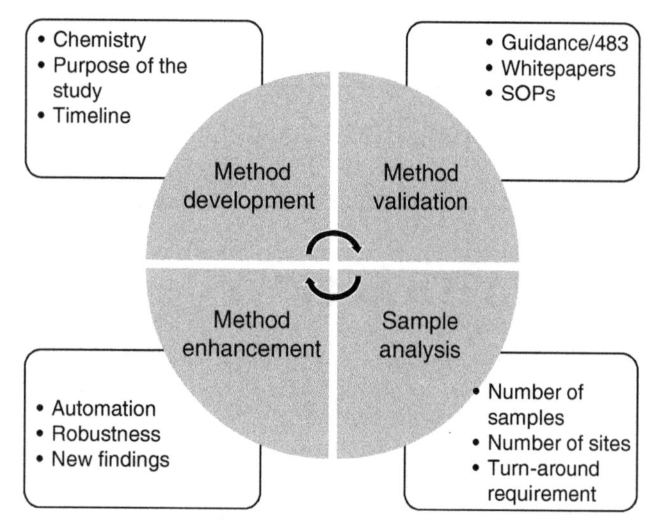

FIGURE 15.1 Life cycles of bioanalytical method.

A good separation can be described by a series of chromatographic figures of merit, including peak symmetry, resolution of critical pairs, retention capacity factor, number of plates, and other quantitative measurements, that define the requirements of the separation (Snyder et al., 2009). Chromatographic separation and detector response define an assay's ability to resolve, detect, and quantify individual components (Williams, 1991). Performance can be judged against standards and system suitability samples to determine whether the resulting separation and detector response is within acceptable requirements. Therefore, a good chromatogram is one that achieves these predefined requirements.

Historically and prior to the introduction of LC-MS, chromatography played a primary role for bioanalysis and was often the only possible tool to achieve desirable selectivity. Detectors such as UV or fluorescence are in general inadequate for discriminating different analytes, especially analytes of similar structures such as parent drug and metabolites. With the introduction of the online MS detection in the 1980s, its tremendous sensitivity and selectivity immediately opened a new avenue for bioanalytical scientists to achieve more selective, sensitive, and higher throughput analysis. Using almost the same sample preparation and chromatographic conditions, bioanalysis of warfarin enantiomers in human plasma by LC-MS is at least 25-fold more sensitive than LC-UV (Naidong and Lee, 1993; Naidong et al., 2001b). The much improved sensitivity and selectivity was clearly demonstrated. The absence of interfering peaks in LC-MS/MS assays has left many wondering about the purpose of an LC separation (van de Merbel, 2001). Testing detection specificity with normal plasma often shows few interfering components to separate. The burden of tedious method development to separate endogenous and/or metabolite interferences was greatly reduced. MS responses are often limited by electronic noise and chromatograms show few peaks. Therefore, optimization of the detector plays a more critical role in defining what defines an acceptable run. LC-MS/MS methods require an accurate assessment of peak response both prior to starting runs and during assessment of sample runs. It is also true that too frequently assignment of failed runs for poor chromatography was actually poor MS response, an elevated baseline, or the inability of a data system to properly define the beginning and end of a peak. Source cleaning or MS optimization rather than column or LC maintenance was often deemed the proper solution.

In the early days of LC-MS, chromatography was merely used to obtain acceptable peak shape. Often, generic methods consisted of a blastic gradient elution on a short C18 column. The run time was in many cases less than 1–2 min with no or minimal retention capacity. The selectivity of MS detection and its obvious advantage in speed reduced the importance of the chromatographic separation, except for a few special separation requirements that became the last stronghold for classic chromatographers. Separation of enantiomers using a chiral column, detection of endogenous biomarkers that have numerous compounds of highly related structures and chromatographic resolution of isobaric metabolites were the few areas that still were recognized where chromatographic expertise was needed. It was only after some pitfalls were learnt that the importance of chromatographic separation for general bioanalytical LC-MS resurfaced (Jemal and Xia, 2006). The bioanalytical practitioners are continuing to learn and appreciate the art of chromatography which is currently again playing an equally important roles in conjunction with sample preparation and mass spectrometer detection. Table 15.1 summarizes some of the historical perspectives. In this book chapter, theoretical and practical considerations for LC-MS will be systematically discussed.

15.2 THEORETICAL CONSIDERATIONS

The primary goal of chromatography in bioanalysis is to separate the analytes of interest from the metabolites, endogenous interferences, and any other components such as phospholipids and dosing vehicles since these compounds can negatively impact the quantitation of the analytes. While some of these compounds can be visibly viewed on the chromatogram (e.g., isobaric metabolites or in-source breakdown of phase II metabolites), most of them are invisible at the detection channel (ion suppression from phospholipids and coeluting metabolites). In an ideal world, the analyte of interest should be chromatographically well resolved from all other compounds except the internal standard (IS). Approaches for chromatographic resolution can be achieved by having better column efficiency (higher plate count or lower plate height) and/or better chromatographic resolution (higher peak capacity).

TABLE 15.1 A Brief History of Main-stream Quantitative Bioanalysis Using Liquid Chromatography

	LC-UV/FLU/EC	LC-MS/MS
Dominant time frame	1970s– ~1995	Since 1990
Throughput	24–48 samplers per 2 days	96–192 samples per day
Typical sample volume	1 ml	<0.2 ml
Typical extraction format	Tubes—manual	96-well liquid handler
Typical extraction chemistry	LLE, SPE, and combination with elegant/complicated washing/eluting strategy. Derivatization often used to enhance sensitivity. PPT seldom used or used as an integrated part of overall separation and analyte enrichment.	PPT, SPE, LLE, often with single step. Combination and derivatization seldom used.
Typical chromatography column size	250 × 4.6 mm, 5 μm and 100 × 4.6 mm, 3 μm	50 × 3 mm, 3 μm and 50 × 2 mm, sub 2 μm
Typical retention factor (k') and run time	>5 and >20 min	>2 and < 5 min
Typical mobile phase	Both volatile and nonvolatile but must not interfere the detection	Volatile (salt such as Na and K cannot be used)
Typical detection and sensitivity (LLOQ)	UV/Vis (~0.1 μg/ml) FLU (~low ng/ml) Sensitivity often limited by endogenous interference	~ pg/ml to low ng/ml Sensitivity often limited by ionization efficiency and detector noise
Primary selectivity domain	Sample extraction and chromatography	MS/MS detection
Selection of internal standard	Structure analog which tracks analyte during the extraction	Stable-labeled analyte (primarily used to tract analyte in MS)
Quantitation	Peak height	Peak area

15.2.1 Parameters Affecting Column Efficiency

In order to understand the chromatographic separation, it is important to review the basic chromatography theory on Van Deemter equation (Figure 15.2). From Figure 15.2, an optimum mobile phase velocity exists for a column at which its highest efficiency would be realized. While the A term is somewhat fixed for each column, the B term and C term play a significant role in the column efficiency. For any given column, due to the longitudinal diffusion, tremendous efficiency loss will be observed when operating under a very low velocity, something to be aware of when using nano-flow

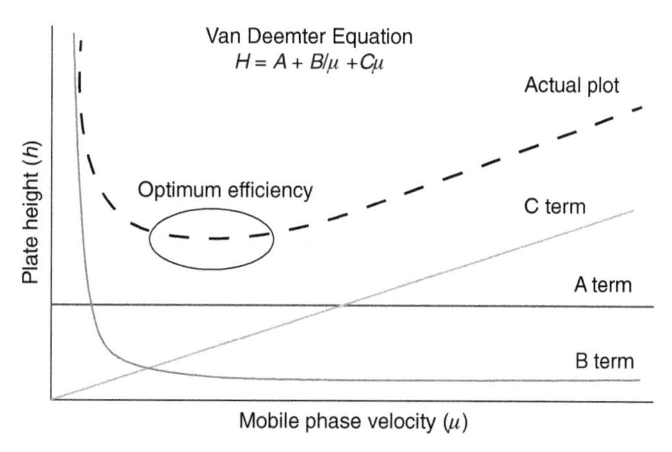

- A term represents the contribution from eddy diffusion. Eddy diffusion results from radial flow inequalities through a packed bed.
- B term (B/μ) represents the contribution from longtudinal diffusion.
- C term ($C\mu$) represents the contribution from resistnace to mass transfer in the stationary and mobile phases.
- μ is mobile phase velocity (mm/s)

FIGURE 15.2 An optimum mobile phase velocity exist for a column at which its highest efficiency would be realized.

chromatography. On the other hand, narrow bore columns frequently used in nano-flow chromatography have much less efficiency loss due to eddy diffusion. The C term is more related to the characteristics of the analytes. While low mobile phase velocity is more favorable for C term, it is counter-balanced by an unfavorable B term. At higher velocity, while the impact from longitudinal diffusion is minimized, the mass transfer in and out of the stationary phase plays a more significant role. In particular, analytes with high molecular weights such as protein and peptides can have quite unfavorable mass transfer at high velocity. For these large analytes, lower mobile phase velocity, larger pore sizes (300 Å or larger instead of typical 80–100 Å for small molecules), smaller particle sizes, and elevated column temperature (to accelerate mass transfer) are often employed. Larger pore size and smaller particles have shorter diffusion path lengths, allowing a solute to travel in and out of the particle faster. Therefore, the analyte spends less time inside the particle where peak diffusion can occur. Column with smaller particles, especially those sub-1.7 μm, will need a pumping system that can handle high column back pressure.

15.2.2 Peak Capacity and Two-Dimensional Chromatography

Another very useful measurement for the overall chromatography selectivity and efficiency is the peak capacity. Peak capacity is defined as the maximum number of peaks that can be theoretically separated on a column at given chromatographic conditions. Selectivity and column efficiency need to be optimized to maximize peak capacity in the one-dimensional separation. The highest peak capacity is obtained if all peaks are evenly spaced and column efficiency remains unchanged. Selectivity optimization can be achieved by changing mobile phase composition, changing column temperature or changing composition of stationary phase. This approach represented the classic chromatography optimization and is still a preferred way to chromatographically resolve multiple isobaric interferences or metabolites. Sufficient selectivity does not guarantee good efficiency as two well-separated peaks may have very broad peak shape. Therefore, a complementary approach is to optimize column efficiency, which can be usually realized by using smaller particle column and minimizing system extra volume. While this approach is an easier and faster way to separate analytes from each other or from matrix components, one may still need to be aware that going to smaller particles may not improve the selectivity and therefore the space between the analytes. The improved separation is solely from reduced plate height that may be compromised upon repeated injections of biological extracts onto the column. It was reported that going from 3.5 μm particle column to 1.8 μm particle only improved peak capacity by 80% with an undesirable five-fold increase on column back pressure (Gilar et al., 2004). To achieve optimal chromatographic separation, both good selectivity and high efficiency are required.

In two-dimensional chromatography, the peak capacity is the product of peak capacities from one-dimensional chromatography and tremendous overall chromatography performance enhancement by employing two-dimensional chromatography is achieved. Whereas the technology is routine in other fields such as metabonomics or proteomics research, its potential has been underutilized in bioanalysis LC-MS. Another advantage of using two-dimensional chromatography is that only a very small fraction from the one-dimensional chromatography needs to be directed (heart cut) to the second chromatography since most of the assays use stable isotope labeled analytes as the internal standard (IS). The mobile phase composition in the first dimension is also less restricted and even nonvolatile mobile phase may be used. This provides additional opportunity for method optimization. Even for bioanalytical methods that do not use two-dimensional chromatography, the concept of gaining additional separation power can still be considered in conjunction with sample preparation. For example, when a reversed-phase LC is used, reversed-phase solid phase extraction (SPE) may not be the best choice for achieving maximal selectivity. Reversed-phase SPE may be better combined with hydrophilic interaction chromatography (HILIC) or ion-exchange chromatography (IEC) since they use different separation mechanism.

15.2.3 Chromatographic Secondary Interactions

Understanding secondary interactions in reversed-phase chromatography is important to achieving a good separation. Secondary interaction refers to an additional solute-stationary phase interaction beyond the intended primary interaction. On a reversed-phase column, the secondary interaction is usually caused by the transient binding of solute and residual silanol groups on the stationary phase (Figure 15.3). This interaction can become a dominant force in a mobile phase of high organic content, leading to a U-shape curve for the retention and organic content in mobile phase (Naidong et al., 2001a). The shape of the curve depends upon the stationary phase, mobile phase composition, and the characteristic of the analytes.

15.2.4 Isocratic versus Gradient Chromatography

Both isocratic and gradient elution are routinely used for bioanalytical LC-MS methods. In isocratic elution, the mobile phase composition is kept constant. Gradient elution method has a changing composition of mobile phase, usually a linear change of organic content from low to high for a reversed-phase chromatography, while the other compositions of the mobile phases were kept unchanged. The choice of isocratic or gradient elution depends upon the compound and sample

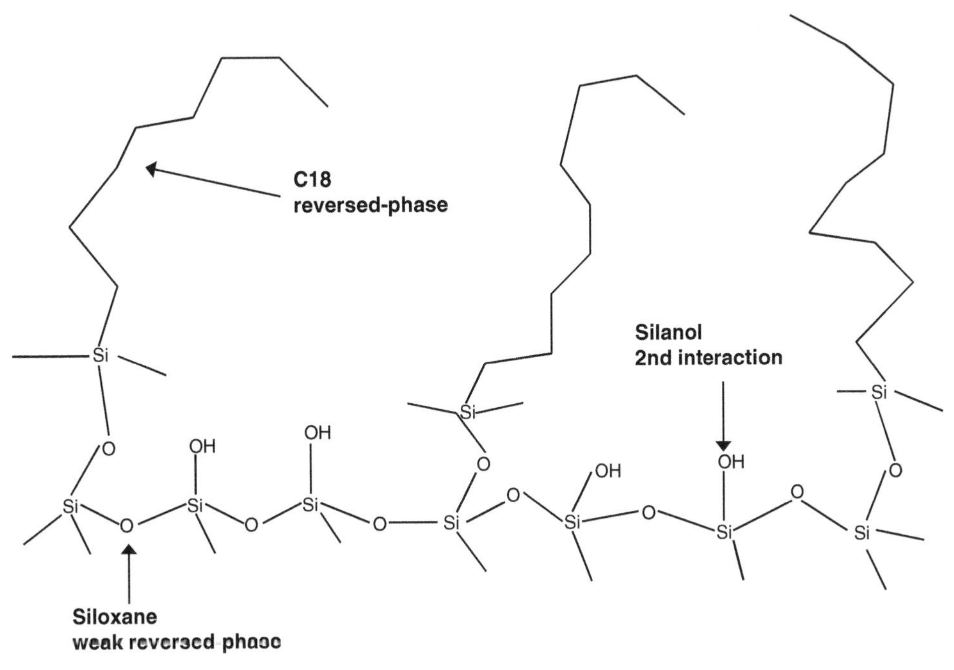

FIGURE 15.3 Bonded C18 reversed-phase surface and secondary interactions.

cleanup. Typically, sample cleanup using protein precipitation requires more extensive chromatography with gradient elution, while a more selective cleanup such as liquid-liquid or mixed-mode SPE may only need isocratic elution. While gradient elution gives a sharp peak shape for a late eluting analyte, it may have more carryover. Also, because mobile phase composition is in constant change, a good IS, preferably stable labeled analyte, should be used to compensate for any MS fluctuation.

15.2.5 Injection Solution

The selection of injection solvent is often an overlooked parameter for a LC-MS method. Injection solvent plays an important role in chromatographic efficiency, in particular with isocratic elution. The elution strength of the injection solvent should be slightly weaker than that of the mobile phase to obtain on-column stacking and focusing of the analyte band. Otherwise peak distortion and broadening can be noted (Naidong et al., 2001a). One also needs to be aware that a too weak injection solvent may cause carryover, in particular for gradient elution. In gradient elution, injection solvent at equal to or slightly stronger than the initial mobile phase elution strength is preferred. Injection volume is also critical when considering the solvent strength of the reconstitution solution. Too large an injection volume and too strong a solvent will result in a diffuse peak and poor resolution. Analysts who avoid evaporation steps but want to inject large volumes need to pay particular attention to its impact on peak shape.

15.2.6 Retention Factor

Once acceptable column efficiency and selectivity are achieved; injection solvent is optimized, and elution mode is chosen, one needs to ensure that analytes are adequately retained on column. Retention factor k' is calculated by $(t'-t_0)/t_0$, where t' is the retention time of the analyte and t_0 is retention time of an unretained compound in an isocratic elution. At least a retention capacity of 2 is required for quantitative bioanalytical LC-MS.

15.2.7 High Speed LC-MS

The sequential chromatographic run time is still the most significant limitation for high-throughput bioanalytical LC-MS. Tremendous effort has been made to shorten the run time by either reducing the mass spectrometer occupying time using multiplexing of several columns in a sequential order or by decreasing the chromatographic run time. Decreasing chromatographic run time represents the most straightforward and easiest solution since it does not require special instrument setup. However, there are several important factors to consider for high-speed analysis as both selectivity and resolution can be affected:

$$\text{Runtime } t_R = \frac{L^*A}{F}\left(1+k'\right)$$

$$\text{Resolution } R = {}^1\!/_4(\alpha-1)(L/H)^{1/2}\,\frac{k'}{1+k'},$$

where L is column length; A is column stationary phase related constant; F is flow rate; k' is capacity factor; H is plate height; L/H is plate count N.

Approaches to achieve fast LC (decrease t_R) include using a strong solvent (decrease k'), short column (decrease L), or high flow rate (increase F).

Fast LC with strong solvent will decrease retention capacity k'. Reducing the capacity factor (k') can lead to potential matrix effects (Matuszewski et al., 1998). Decreasing k' will also decrease resolution/selectivity that may result in interference among analytes (e.g., drug metabolites: glucuronides, sulphates, isomers, prodrug/drug).

Fast LC with a short column ($L\downarrow$) will reduce runtime without changing k' but will decrease the resolution (R) that may lead to potential interference or matrix effects.

Fast LC with high flow ($F\uparrow$) will reduce runtime by increasing flow rate without affecting k'. A high flow rate will increase system back-pressure. High flow rate may also reduce N (column efficiency) by increasing plate height (H). Columns that have been successfully used for high speed LC-MS/MS usually require flat Van Deemter plots at high flow rates to minimize the loss of column efficiency as well as maintain a low column back pressure. Use of a system that can handle the high system pressure is recommended.

Small particle size (sub-2 μm) columns, monolithic columns, fused core particle columns, and bare silica column have been successfully used as high speed LC-MS stationary phases.

15.2.8 Major Types of Stationary Phases

15.2.8.1 Reversed Phase Reversed-phase liquid chromatography (RPLC) is the most frequently used liquid chromatographic method in bioanalytical LC-MS. RPLC has a relatively nonpolar stationary phase and an aqueous, moderately polar mobile phase consisting of water, an organic solvent such as acetonitrile or methanol, and small amount of volatile acid, base, or buffer. Organic solvent is the stronger elution solvent than water and in the gradient elution the percentage of the organic solvent is usually linearly increased to increase the elution strength. The analytes are primarily retained by their hydrophobic interaction with the stationary phase usually consisting of organic carbon chains. The most commonly used stationary phase is silica treated with organosilane [$R(CH_3)_2SiCl$], where R represents different bonded phases (e.g., C18, phenyl, etc.) to afford specific selectivity characterization. For a good RPLC, the analytes of interest should be in one single form in the mobile phase. Use of a mobile phase with its pH close to the pKa (for acidic compounds) or pKb (for basic compounds) will lead to the different retention of charged and uncharged forms of the same compound. Since the charged form is more polar than the uncharged one, it will elute faster on the reversed-phase column. This will lead to peak splitting of the analyte. A good rule of thumb is the mobile phase pH should be at least two units away from the pKa or pKb. Secondary interaction may also play roles in enhancing the chromatographic selectivity as long as it is well controlled and is not competiting with the primary interactions to cause peak distortion, tailing, or even splitting. When this secondary interaction becomes less desirable, one may use one of more of the following strategies to overcome it:

Use of mobile phase pH to neutralize analytes: Since most secondary interactions are due to the ion interaction between the stationary phase and charged analyte, one of the most effective approaches is to reverse the mobile phase pH. For basic analytes, a basic mobile phase will keep the analytes in the uncharged form and this usually leads to an improved retention and peak shape. Even though electrospray ionization (ESI) response can be improved by detecting precharged analytes in solution, that is, acidic mobile phase for basic analytes to form protonated molecules, the increase in the amount of organic solvent in the mobile phase that is needed to elute the less polar uncharged analyte improves ionization efficiency. It should also be noted that ionization of analytes does not usually require precharged status of the analyte (Zhou et al., 2002).

Neutralizing analytes by using ion-pair reagents: Ion-pair reagents are frequently used in LC-UV but are less used for LC-MS since they can decrease the ionization efficiency of the analyte. Recently, volatile ion-pair reagents specially designed for LC-MS have also been developed though they have not been widely adopted (Gao et al., 2005). Ion-pair reagents added to the mobile phases may be particularly useful for zwitterions analytes on reversed-phase LC. Modification of mobile phase will not make zwitterions uncharged as they contain both acidic and basic function groups. One may use mobile phase pH modification to drive zwitterions analytes to carry only either positive or negative charge and then use the ion pair reagent to neutralize the remaining charge

Suppression of residual silanol activities by using trifluoroacetic acid (TFA): TFA usually improves peak shape of basic analytes if other approaches fail. TFA suppress the ionization of the residual silanol groups and make them no longer interact with the analytes. However, TFA suppress ionization of the analytes. TFA fixing includes postcolumn addition or including these acids directly in the mobile phases (Shou and Naidong, 2005). The pKa of silanols is ~2.0 so using a strongly acidic mobile phase modifier such as TFA will convert $Si–O^-$ to Si–OH and overcome the strong ionic binding of protonated amines to silanols. The addition of other amines such as ammonium acetate can result in a similar improvement by binding with and masking $Si–O^-$ sites.

15.2.8.2 HILIC HILIC has become a favored choice for separation of drugs and/or metabolites due to its superior ability to retain polar compounds (Jian et al., 2011; Naidong, 2003). In HILIC, the analyte interacts with a hydrophilic stationary phase and is eluted with a relatively hydrophobic mobile phase. Bare silica is the most commonly used stationary phase. HILIC employs water-miscible polar organic solvents such as acetonitrile, and water is the stronger eluting solvent. The gradient elution typically starts with 5–10% aqueous in mobile phase and increases to up to 50–60% for elution of the analytes. In HILIC, polar compounds are more highly retained than nonpolar compounds, and the elution order is usually the reverse of that on RPLC columns. In addition to the good retention for polar analytes, sensitivity is also improved due to the more favorable electrospray condition of a highly organic mobile phase (Naidong, 2003). Another advantage of this chromatographic mode is the amenability to high flow rate on non-UPLC (ultrahigh performance liquid chromatography) system due to low back pressure from high organic mobile phase and lack of bonded chains on the stationary phases (Shou et al., 2002). Typical organic solvents such as methyl tert-butyl ether (MTBE) and ethyl acetate have weaker elution strength than the HILIC mobile phase, and therefore, direct injection of SPE eluent onto HILIC-MS/MS is feasible, which not only increases the throughput by eliminating the tedious drying down and reconstitution steps but also avoids potential loss of analytes due to adsorption and volatility (Li et al., 2004; Naidong et al., 2004).

15.2.8.3 Normal Phase Normal phase liquid chromatography (NPLC) employs a polar stationary phase and a nonpolar, nonaqueous mobile phase. NPLC is infrequently used for bioanalytical LC-MS due to its difficulty of operation and poor ionization efficiency. One exception is chiral LC-MS/MS where normal phase condition is used frequently (Chapter 41). Even trace levels of moisture in the mobile phase can significantly change the chromatographic result (retention time, peak shape, resolution, etc.) since water is a very strong elution solvent for NPLC. Due to lack of aqueous components in typical NPLC mobile phase, the analyte ionization is very poor in a nonaqueous environment. Usually postcolumn infusion of aqueous solution is needed for enhancing the ionization and thus the sensitivity.

15.2.8.4 Ion Exchange IEC is important mode of separation for large biomolecules such as peptides, proteins, or oligonucleotides as therapeutic agents. IEC retains and separates ions or polar compounds based on their charges. The analytes (anions or cations) are retained on the stationary phase by adsorption to functional groups with opposite charge, and can be eluted by increasing the concentration of a similarly charged species that will displace the analyte ions from the stationary phase. The chromatographic elution is achieved by either increasing the mobile phase ionic strength to dislodge the analytes from the stationary phase or by changing the mobile phase pH gradually to uncharge the analytes. The uncharged analytes will no longer have ion–ion interaction with the functional groups of the opposite charge. IEC-MS/MS is a useful tool for analysis and metabolite identification of therapeutic oligonucleotides (Lin et al., 2007).

15.2.9 Choice of the Right Column Sizes

Narrow bore columns possess a decrease in eddy diffusion but possibly an increased longitudinal diffusion. Nevertheless, the overall column efficiency is improved. Given the same amount of analyte injected onto the column, the narrow bore column also has higher peak concentration. A 50% reduction in the column width would theoretically increase the peak concentration and thus the sensitivity 400%. The narrow bore column also increases ESI efficiency. There is certainly a trend in the bioanalytical quantitation of proteins and peptides by using narrow bore columns. A number of vendors also offer narrow bore columns prefabricated on chips to minimize any dead volume and to make use of narrow columns easier. However, narrow bore columns have still been less used for routine bioanalysis than packed capillary columns (~0.15 to 0.5 mm) or small microbore columns (1 mm i.d.) (Li et al., 2009; Pan et al., 2011). This is because the column loadibility is directly proportional to the column width. A doubling of column width would result in 400% increase on column loadibility, which is important for routine bioanalysis. The sensitivity gained by going to smaller column is lost due to the much decreased column loadability. System blockage is another challenge for the narrow bore column but vendors have been making steady progress on solving this problem.

15.3 PRACTICAL CONSIDERATIONS FOR METHOD DEVELOPMENT

15.3.1 Holistic View of Practical Bioanalytical LC-MS Method Development

A holistic view of method development would consider the limitations and opportunities for all three important areas which are usually divided into sample preparation, chromatography, and MS detection (Figure 15.4). With the goal of developing a robust method in mind, a good balance among all of the technical consideration would ensure a method

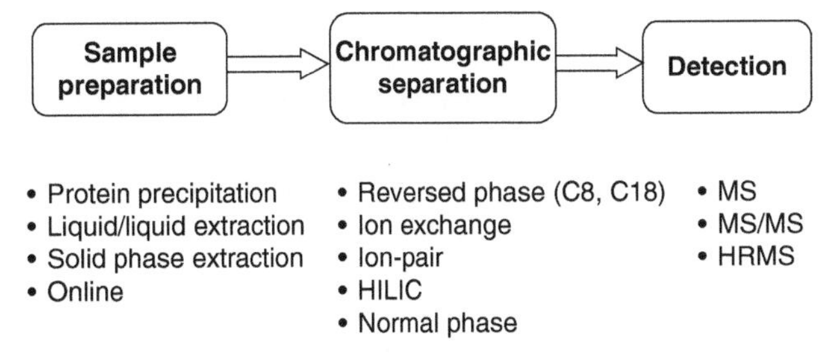

FIGURE 15.4 Holistic view on method development.

suitable for the analysis. It should be noted that the rational of choosing extraction method should be linked with chromatographic condition and MS detection. It should also be noted that usually the extraction was not optimized for matrix but rather for analytes. Chromatography serves an important link between sample preparation and MS detection. A less extensive sample cleanup usually requires a more efficient chromatographic separation by optimizing column efficiency and selectivity. A longer run time is usually required. A more elaborate sample cleanup, in particular with orthogonal separation mechanism to the LC, may lead to a shortened but acceptable chromatographic run time with lower retention capacity. In a similar way, use of a high-resolution mass spectrometer may reduce the requirement of chromatographic separation of interference peaks arising from isotopic contribution.

15.3.2 Matrix Effects

Ideally, one would prefer to separate analytes from matrix effects by extraction. In plasma, phospholipids present at high concentrations are the main cause of matrix effects when using ESI (Pulfer and Murphy, 2003). The use of phospholipid removal extraction plates, such as the Varian Captiva, Waters Ostro, and Supelco HybridSPE products, often rely upon a lanthanide metal sorbent to retain phospholipids (Wu et al., 2008). Much depends upon the extraction selectivity of analytes relative to matrix components. SPE is a low resolution form of an analytical separation; analyte is either on or off the sorbent bed. Affinity of the analyte for the extraction media and the wash solutions defines the purity of the eluent. A mixed mode SPE is an alternative means to yield a purer extract (Chambers et al., 2007). Mixed mode SPE has a high affinity to bind many drugs, allowing aggressive washing of phospholipids prior to analyte elution. The same extraction chemistry on SPE as the analytical separation does little to improve an assay. Having an extraction procedure that uses an orthogonal retention mechanism to the analytical separation gives the best overall separation. With even the best extraction, the total removal of phospholipids may not be possible due to their high abundance in plasma.

Trapping columns can be used to further reduce the presence of phospholipids and other endogenous material on analytical columns. However, once injected onto an analytical column, bioanalysts must be aware of their presence and, during method development, take appropriate steps to resolve them. MS monitoring of matrix effects, common in method development, is also an option during sample analysis. If phospholipid monitoring is used during validation or sample analysis, analysts must ensure that predefined criteria are established and used to judge runs.

Once injected onto an HPLC, the need for phospholipid separation prior to MS-detection will be apparent when the analyst determines analyte matrix factors (Buhrman et al., 1996). Each analyte, matrix and source interface is different. Atmospheric pressure chemical ionization is less prone to matrix effects as the analyte is separated from endogenous components by vaporization prior to its ionization (Huang et al., 1990). ESI requires an LC separation of analytes from matrix effects. An alternative is to compensate with a stable label IS. Inverse detection to identify unknown matrix effects or the direct monitoring of known matrix effects should be performed during method development. Inverse detection uses a postcolumn flow stream of the analytes with a blank extract injection (King et al., 2000). Regions of ion suppression or enhancement are noted in the chromatogram and serve to identify regions where analytes should not elute. A more sensitive means is the direct detection of known components such as phospholipids from plasma extracts (Little et al., 2006).

A comparison of the elution profiles of phospholipids under isocratic reverse phase conditions on a Waters XBridge C18 column (2.1 × 50 mm, 3.5 μm) using a mobile phase of 5 mm ammonium formate/5 mm formic acid in water and varying the proportion of acetonitrile illustrates the strong retention of phospholipids (Xia and Jemal, 2009). The total removal of all phospholipids required at least 80 min (240 column volumes) when using 65% acetonitrile, 20 min for 75% acetonitrile, 10 min for 85%, and 5 min for 95% acetonitrile. Comparing the phospholipid elution behavior using acetonitrile, methanol or isopropyl alcohol showed a surprising result. The major phosphatidylcholine ion at *m/z* 758 was not eluted until 220 min when methanol was used. By contrast, the total removal of phospholipids required only 20 min for acetonitrile and 3 min for isopropyl alcohol.

This finding highlights both the importance of routinely monitoring for all phospholipids during method development and the difficulty in removing them. Many investigators employ a gradient wash step to elute phospholipids. Most use just a few minutes of methanol or acetonitrile seen here to be ineffective in totally removing phospholipids. If time is critical, multiplexing (Deng et al., 2002) can be used to allow sufficient time to both flush and reequilibrate the column. Staggered injection employs two or more LC flow streams feeding into a single MS ion source (Wu, 2001). Typically, these systems are used whenever speed is needed for complex, high resolution separations. A better approach is never to allow phospholipids to accumulate on the column. Online extraction under turbulent-flow chromatography (Wu et al., 2000) has been shown to be an effective means to remove plasma proteins, but it does little to reduce phospholipids prior to LC-MS analysis. In a related but more effective 2D LC approach, trapping phospholipids on a guard column with front cutting of the analyte has reduced phospholipids in protein precipitated samples (Van Eeckhaut et al., 2009).

Phospholipids should be considered worthy of separation (Remane et al., 2010). Additional separation requirements can include IV formulations such as polyethylene glycols that can suppress ionization at various regions of elution across a chromatographic separation (Tiller and Romanyshyn, 2001). For a stable label IS assay, one may consider whether a retained phospholipid needs to be removed. One risk is that phospholipids accumulate to the point at which a bolus of saturated phospholipids elutes in many subsequent injections. Therefore, it is better to flush the column to avoid accumulation.

15.3.3 Metabolites

Regulatory agencies need to fully understand toxicity that is mediated by reactive metabolites. Guidance on Metabolites in Safety Testing requires that significant ($\geq$10%) metabolites are measured in humans and the primary toxicology species (FDA, 2008). The more recent International Conference on Harmonization: Guideline on Nonclinical Safety Studies for the Conduct of Human Clinical Trials and Marketing Authorization for Pharmaceuticals (ICH M3[R2]) describes the threshold as 10% of total drug-related exposure. Where they differ, the ICH guidance supersedes the CDER Guidance (Robison and Jacobs, 2009). For PK-PD assessments, it is common to measure parent plus active metabolites in humans and the pharmacology model. Having an accurate assessment of active or toxic metabolite exposures across species greatly aids this assignment. A reliable LC-MS method that simultaneously measures parent drug and all metabolites is a great advantage. There are bioanalytical challenges, particularly when metabolites are great in number or structural diversity. Bioanalysts will need to consider whether they can develop and validate a combination assay, and the LC separation plays a critical role. An alternative approach uses individual LC-MS assays; however, this approach can be expensive for an extensively metabolized drug.

A chromatographic best practice includes a system suitability sample prepared from metabolites that need to be measured or resolved from other interfering metabolites. Since oxidation is a common biotransformation, measurement of a selected metabolite from several isomers is a common requirement. To ensure that the separation has been achieved, isomeric or isobaric metabolites that could potentially interfere are tested. To assure metabolic specificity, preparation of the system suitability sample needs to consider the known metabolism in each species. Figures of merit such as resolution requirements for critical pairs should be predefined. Failure to maintain resolution can compromise the assay so assessment at the end is as critical as at the beginning of a run.

A second instance where metabolic specificity is important is the analysis of parent drug known to undergo isomerization or phase II conjugation. Examples of isomerization include mammalian target of rapamycin compounds undergoing lactone ring-opening and dehydration to the seco isomer, or selective estrogen receptor modulator drugs undergoing Z to E conversion (Cai et al., 2007; Kieser et al., 2010). More common is the glucuronidation or sulfation of parent drug. One might assume that any phase II conjugate would be resolved from its parent. However, if the retention characteristics are dominated by parent chemistry, there may be no selectivity for its conjugate. Interference from an isobaric metabolite with parent drug is far less likely since two metabolic processes are required that generally also results in a different product ion.

If all isomers are measured, they will be present in standard and QC samples. If not, they must be included in a system suitability sample at the beginning and at the end of the run. It is not sufficient to observe just a single peak in study samples without demonstrating that the separation had the potential to resolve other components.

Acyl glucuronides are a good example of how a chromatographic resolution requirement should be set. Acid drugs can form 1-β acyl glucuronides that have been associated with covalent binding and possible idiosyncratic toxicity (Ojingwa et al., 1994). Rearrangement is correlated with reactivity and toxicity, so the assay should measure only the 1-β isomer (Xue et al. 2006). A chromatographic requirement is, therefore, to separate the 1-β anomer from the seven other rearranged forms. The separation of each rearranged isomer from one another is not required. However, stabilization to rearrangement is required since each isomer can have different multiple reaction monitoring (MRM) response.

When metabolic specificity is in question, plasma from subjects on these medications can be purchased from suppliers that will include metabolites at physiologically relevant concentrations. An alternative test includes spiking metabolites from generating systems such as *in vitro* sources or bile duct cannulated collections.

When simultaneously measuring numerous or dissimilar metabolites, it may be necessary to use segmented scans. Such is the case when sulfate conjugates are measured using negative ion detection, while parent drug and phase I metabolites are measured using positive ion detection. The analyst needs to decide what can be effectively measured as a combination assay that meets regulatory requirements and what should be determined using tiered assays.

15.3.4 Biomarkers

Assays that determine endogenous components such as biomarkers require special attention to detection specificity. The presence of components with similar biochemistry is common to biomarker assays so the requirements of the separation are severe. Examples abound including steroids, prostaglandins, and retinols in the field of lipidomics, peptides in proteomics, or CNS biogenic amines. The challenge of biomarker assays is generally preparing accurate matrix standards and ensuring that separations are maintained across a diversity of subjects. Testing in numerous lots to accurately describe the total human population that will be seen in clinical studies is a must. Regulatory guidance requires testing of both detection specificity as well as the ability to accurately measure drug at the lower limit of quantification (LLOQ) level. Finding interference during method development is far better than finding that an assay has interference during clinical sample analysis. Analysts should be prepared for this possibility, so standard operating procedures (SOPs) must define how much revalidation is required for any change within a separation.

15.3.5 Coadministration of Drugs

When comedication therapy is needed, a combination assay may be used. Concomitant medication is common in oncology or virology indications, where synergistic pharmacology is possible or drug resistance needs to be controlled. Often, drugs are diverse in nature so the resulting method needs to deal with this complexity. Compromises may be needed to ensure the optimum recovery and separation of the group analytes. Therefore, matrix effects may need to be dealt with by LC separation and use of a stable label IS rather than a selective extraction.

Coadministration of drug candidates, described as cassette or N-in-1 pharmacokinetic studies, is used in discovery organizations (Berman et al., 1997). Typically, 10–25 drug candidates are coformulated, dosed, and their pharmacokinetic properties assessed. The study design allows for rapid screening of drug candidates for clearance, volume of distribution, and site-of-action determinations. A criticism of cassette dosing is the potential for drug interactions. A variation of this approach in clinical drug interaction studies will dose Pittsburgh, Cooperstown, Karolinska, or Glaxo cocktails (Chen et al., 2003; Palmer et al., 2001). An example P450 cocktail includes (1A2) caffeine, (2C9) tolbutamide, (2C19) omeprazole, (2D6) dextromethorphan, and (3A4) midazolam. A general difficulty with the approach is the inconsistency when deciding upon the best probe substrates for the cocktail. To reduce bioanalytical demands, use of a consistent cocktail is the best approach. A cocktail approach is becoming less used as *in vitro* assessments allow better design of drug interaction studies with individual compounds.

Interference testing ensures that detection specificity is preserved when other medications are given. A cocktail of drugs is generally screened in method development and tested in validation. A mixture of over-the-counter medications at a therapeutically relevant (low μg/ml) concentration is often used. Method development should consider the potential for interference that can be predicted based upon the molecular weight and physicochemical properties. When observed, a decision is made in method development to either revise the LC separation or use a more selective MRM.

Analysts must ensure that both exogenous and endogenous, highly related components are resolved. Variability in background is generally less a problem for nonclinical studies where animals are purpose-bred and live on a restricted diet. The human population is vastly different, so lot screening is critical to understanding the requirements of the separation. Knowing the requirements of the separation will define how well an assay can be optimized.

15.4 SAMPLE ANALYSIS

15.4.1 Batch Runs

Once the analyst has defined the requirements of the separation, it is relatively straightforward to provide guidance on what good chromatography is and what constitutes an acceptable run. The simplest requirements to start a bioanalytical run are proper retention time, signal-to-noise, and linearity. An acceptable analytical run also considers instrument drift before and during runs. System suitability tests generally require precision, as determined by the coefficient of variation, to be within 10 %CV by multiple injections of a LLOQ sample before the run. Additional requirements can include judging within run response by the consistency of IS response.

To ensure consistent inter-run performance, the analyst should avoid changing critical parameters that affect response. Tuning ion optics to maximize the transmission of ions should be performed. However, changing the collision energy, mass resolution, or extract injection volume should not. Changing collision energy (or gas thickness) can affect the response of interfering ions. The same is true when decreasing mass resolution. Either can impact detection specificity. For an electronic noise-limited assay, increasing the injection volume can improve response but will also limit batch size. European guidance requires an assessment of batch size during validation (EMA, 2011). Injecting more sample extract can result in greater matrix effects, shorter column lifetime and reduced performance on later injections (Heller, 2007).

15.4.2 Carryover

Adherence to other criteria needs to be considered. This includes an assessment of the potential for carryover to have impacted study sample quantification during an analytical run. Carryover is contamination by the preceding sample. It is most typically associated with the autosampler (Vallano et al., 2005). However, it can also be present in other devices such as fixed tip liquid handlers. Carryover is often not critical to high-throughput ADME (absorption, distribution, metabolism, and excretion) or discovery studies. However, carryover requirements for regulated studies are demanding and can limit laboratory efficiency. Separations play a large role here so be prepared to manage it.

Carryover is tested by injecting a sequence of single blank (IS only) followed by an upper limit of quantification sample, then another single blank prior to any run. When carryover is more than 20% of the LLOQ response, the analyst should reduce it prior to starting the run. If carryover increases during the run, acceptance of data must be justified. When noted during method development, the assay range may be restricted. This can be difficult when multiple doses are tested, such as in single or multiple ascending dose studies where a wide assay range is needed. During sample analysis, precautions include the proper placement of samples to minimize changes in concentration and protection of samples with washout blanks. The latter is a poor choice since it leads to long analytical runs. The best approach is to reduce carryover below the acceptable level. If not reduced or controlled, one may need to fail individual results or runs, intersperse blank samples, and state its potential impact on study results.

For these reasons, there has been considerable effort to improve autosamplers. Improvements include using more inert materials, minimizing sample contact with the syringe needles, and adding more wash solvents and cycles. Inability to control carryover will negatively impact laboratory efficiency. Selecting the best autosampler should include testing carryover and washing options as part of user requirements.

To reduce carryover, the analyst should consider the location and extent of adsorption, substitute materials to reduce it, or enhance solvent interactions to overcome it. Solvent wash substitution is preferred for speed and broader application to multiple systems. First, increase the number, volume and type of wash steps (acidic or basic pH adjustments by adding formic acid or triethylamine, solvents such as acetonitrile, isopropyl alcohol, and acetone or metal chelators such as EDTA). Select wash solvents that minimize electrostatic, silanophilic, metalophilic, and hydrophobic interactions. Consider upgrading the autosampler with a separate pump that washes the injection valve and needle.

Second, determine the location of carryover. Change to an uncontaminated column. Determine the extent of carryover in the autosampler injection valve or needle by substitution of an uncontaminated injection valve or needle. Determine if replacement of the autosampler or its materials reduces background. If possible, replace filter frit or columns with nonabsorptive materials.

During method development, the type of autosampler will have great impact on carryover. Once chosen, another model should not be substituted without being tested for equivalence. The autosampler is a critical component to a method. Be sure that autosampler washing programs do not perturb your separation and are completed in time for the next injection.

Turbulent flow uses the online trapping of analytes onto large particle columns under turbulent flow conditions (Wu et al., 2000). Plasma proteins are not retained. Trapped analytes are subsequently eluted onto an analytical separation by back-flushing the trapping column. Unfortunately, the carryover criteria for regulated bioanalysis have made the direct injection of matrix using turbulent flow impossible. The buildup of residue from absorptive components in plasma frequently results in carryover that is far in excess of 20% of the LLOQ response.

15.4.3 Peak Tailing

Peak tailing is often correlated with carryover due to an analyte's high affinity binding to both the injector and stationary phase. Additives to the mobile phase can overcome these strong associations, but must be done with care to avoid affecting MS response. Typically, a stationary phase with better protection from surface silanols is used. However, reverse phase separations that depend upon a secondary interaction for their selectivity can be affected. Compounds that are strong chelators may show both high carryover and peak tailing due to binding within the injector. A simple test is to remove the column and observe the elution profile of a pure solution under flow-injection conditions. Unless corrected, tailing may either degrade chromatographic resolution or make integration impossible.

15.4.4 Conditioning

Decreased response at lower concentrations due to reduced recovery from the LC column is often seen for absorptive compounds. A lower response factor (sensitivity) of a pure solution at lowest standard concentrations is an indicator and should be evaluated during method development. When analyzing extracts, variability in LC recovery is rarely simple due to coextractables also binding to active sites. Therefore, results using pure analyte solutions may not mimic those seen with extracts. Extraction by protein precipitation illustrates why. Unlike more selective extractions, protein precipitation results in abundant phospholipids that can act as a carrier during extraction and through the LC system. However, phospholipids also result in matrix effects. The analyst must, therefore, decide whether the speed and utility of protein precipitation is worth the risk of variability in matrix effects. For those performing discovery studies, it is a common approach. For regulated bioanalysis, particularly in clinical studies, a more selective extraction with appropriate conditions (carrier) is used.

When analyzing adsorptive compounds at lower concentrations, consistency of recovery through the LC system must be considered. When using radioactivity to quantify metabolites, the sum of individual response on-column is compared to the total off-column response. Identical experiments are difficult in LC-MS; however, LC recovery can be tested using pure standards. Adsorption losses may also be noted when a standard curve shows decreased sensitivity at lower levels. Both the absolute and relative response should be considered since high levels of IS can compensate by masking active surfaces. Adding more IS can be a solution. However, finding a stationary or mobile phase that avoids adsorption losses is a better approach.

Active sites can be made less absorptive by covering them with analyte, a procedure known as conditioning. Conditioning is achieved by repetitive injections of analyte onto a chromatographic system prior to running samples. Since passivation should be achieved as quickly as possible, injecting a high level extract is preferred. In regulated bioanalysis, this sample must not be associated with either a prior or present run. Multiple injections from a single extract are allowed. Once a steady-state signal is obtained and low-level response is confirmed by system suitability, the run may proceed. Ideally, the number of conditioning injections should be defined during method development and stated within a written SOP or test method procedure. In practice, the reuse of columns or history of LC injections affects adsorption. Having a standard practice to inject a limited number of conditioning samples with system suitability tests must be clearly defined in a SOP.

The practice of conditioning is similar to the use of a silylating agent to cover active sites within a gas chromatographic (GC) system. Unlike GC, binding is generally not covalent and will be easily overcome by the shear and solvation forces of the mobile phase. One, therefore, needs to determine how long a running LC system can remain idle before reconditioning is needed. For regulated bioanalysis, this time should be defined within an SOP. As long as injections continue to be performed at regular intervals, a portion of the injected analyte is available to cover active sites.

15.4.5 Diversion of LC Flow

Postcolumn divert valves are sometimes used in bioanalytical laboratories. The valve controls the flow of mobile phase to the mass spectrometer, reducing contamination to the ion source at times when analyte does not need to be detected. This can be to divert void material, later eluting components or both. Unless a makeup flow stream is added, the procedure causes a brief overheating of the interface and deposition of material until the flow stream is reestablished. Once the valve is redirected to the MS interface, it may take some time to flush out any deposited interference and reestablish a stable baseline. Therefore, a best practice is to ensure the flow stream is reestablished at least 30 s prior to peak elution. Analysts must also ensure that band broadening due to the extracolumn dead volume of the divert valve does not impact the separation. This is particularly important for UPLC or microbore LC. An unexpected consequence may be failing to observe nonretained analyte from an overloaded injection. This can occur when the presence of coextracted materials overload the capacity of the separation. Using a divert phase can mask these problems. If planning to use a divert valve, a best practice is to ensure testing both with and without a divert valve during method development. European guidance requires a test of batch size during validation. Using a divert valve to avoid source contamination will impact this assessment. The use of a divert valve at any part of the validation requires its consistent use during sample analysis.

15.4.6 Column Use in Sample Analysis

As one develops more complex separations or assays larger runs, column lifetime needs to be considered. Understanding when a column will fail due to pressure overload or poor chromatography is critical to understanding batch size. Ghosting is the elution of a retained interference from a previous injection cycle and has the potential to negatively affect any separation. Always use a guard column and consider frits when assaying sample extracts that could contain particulate. A good analytical column should be protected as it is both expensive and a critical part of the assay.

Temperature is critical to a separation so column heaters should be used to control it (Hao et al., 2008). Conformational forms can impact a separation, requiring the use of elevated temperature to overcome peak tailing or splitting. Examples include dipeptides such as captopril that contains proline at the C-terminus and undergo on-column cis-trans isomerization (Henderson and Horvath, 1986). Elevated temperature can also be used to improve efficiency by decreasing mobile phase viscosity. Elevated temperatures will result in faster column degradation so separation efficiency must be balanced against column lifetime.

Recording injection history has been difficult due to a lack of integrated software capable of reading different vendor's column tags. Vendors could do more here as paper recording of injection history is a tedious and time-consuming process. It is important to isolate methods on individual columns. Comingling methods on the same column can result in failures due to degradation by another procedure. To ensure an accurate understanding of column lifetime and batch size, individual assays must be run on individual columns.

When changing columns, attention must be given to ensure equivalent performance. Requirements will be part of your system suitability. If lot changes have occurred, the selectivity of the separation may have been compromised. Amines are often separated using a secondary retention on an exposed silanol that can be impacted by an improved end capping. If the separation is critical, one must return to an original lot or find and validate an equivalent column. The latter can be challenging, so best practice in method development is to assure that the method achieves the desired separation in multiple lots. If the integrity of the column is ever in question, analysis of the test mixture provided within the performance report will confirm its original specifications.

15.4.7 Instrument Communication

An unfortunate situation occurs when an autosampler and LC system communicate with the mass spectrometry data system through contact closure. When runs are interrupted, errors can result by improperly associating sample injections with data being acquired. Out-of-sequence injections can be eliminated by direct communication between systems. This appears to be a more common problem when an LC system is used from one vendor and the mass spectrometer from another. For a regulated laboratory, direct communication should be a primary requirement of any system that is being considered for purchase.

15.4.8 Data Processing

MS data systems use a number of algorithms to process quantitative data. Because data is often limited by electronic noise, digital smoothing is used. Unfortunately, the use of a moving average smooth has further complicated peak integration and data processing. Ideally, one should define an optimum number of data points for each chromatographic peak. Twelve to sixteen data points will avoid aliasing but also ensure that the maximum dwell time is used to accumulate signal and reduce noise. Too many data points will require the extensive use of digital smoothing. Post-processing can bias peak integration so regulated bioanalysts must ensure that a consistent set of digital smoothing is defined and used. A moving average smooth will affect peak response so it must be consistently applied within the run. Since day-to-day and instrument-to-instrument performance can vary, requiring exact processing conditions across multiple days is unrealistic. However, many laboratories provide an upper boundary as guidance to smoothing. Reasons for changes need to be recorded and deviations justified. Other algorithms are less inclined to bias peak integration.

Within the field of regulated bioanalysis, the ability to use automated data processing to achieve consistent peak integration is an important part of what constitutes good chromatography. Bias associated with manual integration has made it impossible to defend analyst choices. Therefore, manual integration has been rendered obsolete in regulated bioanalysis. Two factors influence whether automated processing is possible, the algorithm's ability to properly define slope changes needed to select the beginning and end of peaks and unresolved chemical components that interfere with defining a Gaussian peak. Because of this, poor chromatography is often recognized as whenever processing algorithms fail to consistently define and integrate peaks.

15.5 MAKING RUGGED METHODS

Method development includes first, recognition that all components have been separated followed by optimization of the separation. The first objective requires an assurance of peak purity and an understanding of the elution of reference compounds. The second involves iterative reanalysis under different experimental conditions until an optimum is achieved.

15.5.1 Column Selection

To assist in selecting columns and mobile phases, a multiplexed LC system has many advantages. First, it allows the orderly introduction of numerous mobile phase compositions. This allows acidic, neutral, and basic compositions to test both chromatographic and MS response characteristics under LC flow conditions. Second, columns are sequentially tested under conditions in a consistent manner. Third, the system and resulting data processing can be automated. Most vendors provide column selection systems, particularly those that market multiplexed LC pumps. Column selectivity helps most in achieving a separation but is most difficult to predict. Therefore, using an automated process gives the best global view in a minimum period of time (Wang et al., 2010).

15.5.2 Optimizing the Separation

Once a column has been selected, optimization under isocratic or gradient conditions is relatively straightforward (Snyder et al., 1988). For multicomponent mixtures under gradient conditions, experimental design software, such as ChromSword or DryLab, greatly reduces method development time. Within just a few injections, resolution maps show optimum conditions to resolve critical pairs for both gradient time and column temperature. Resolution requirements should include no interference from endogenous components, metabolites, matrix effects, and any coadministered compounds in the final optimization.

15.5.3 Changes Across Species and Matrices

Knowing how a separation can used in multiple species or matrices is critical to toxicology studies. The extracts of biological fluids and tissue homogenates are quite complex. Across mammalian species, their biochemistry is well conserved. So while abundances may vary, individual components are consistent. Therefore, analytical solutions for one matrix are generally applicable to all species. Most assays can be simultaneously developed in all toxicology species and man. There may be special requirements for analyte stabilization in some species to overcome the enhanced esterase activity in rodents or oxidase activity in dogs (Li et al., 2011).

There may also be differences in assay range or sensitivity, as a result of differences in efficacy, safety margins, pharmacokinetics, or study design. However, the separation should remain the same. Once the metabolism is fully known within each species, requirements to resolve individual isobaric or isomer metabolites may be relaxed provided specificity is not compromised.

When considering penetration or site of toxicity questions, an analyst may be asked to translate a plasma assay to another matrix such as brain or liver tissue, urine, CSF, synovial fluid, tears, or any range of biological fluids or tissue. Urine is the most common since it is readily available and serves to define renal clearance. Urine exposures address questions associated with renal toxicity or, in the case of some medications, site of action. Exposure in urine also defines a lower boundary of the bioavailability of orally dosed drugs. Urine assays have their own challenges, some of which include lack of homogeneity, different matrix effects or interferences, and nonspecific binding (Ji et al., 2010).

Differences exist for each matrix that is assayed. Therefore, it is not surprising that the same separation in one matrix may not be useful in another. Differences in exposure, recovery, and stability further complicate the ability to translate assays across matrices. Measuring remote exposures is a common expectation within discovery organizations whose primary mission is to rapidly establish PK-PD models. Site of penetration is required for target engagement, so pharmaceutical companies perform numerous studies each year that require systemic and site of action assays. These studies may employ generic analytical methodology such as protein precipitation, fast gradients on a universal column, mixed matrix calibration standards, generic stabilization, and any range of procedures that minimize method development. Quality drug candidates that have the potential for investigational new drug (IND) studies receive more attention. Within regulated bioanalysis, definitive methods must be established and fully validated to support preclinical studies. The toxicology study director and bioanalyst need to define what matrices can be assessed in good laboratory practices (GLP) studies.

15.5.4 Column Lifetime

Column lifetime can also be tested by multiple reinjections of a limited number of standards during method development. Provided column capacity is not exceeded, quickly reaching a fatigue or failure limit may also be possible by injecting a larger volume of extract. Testing large numbers of injection cycles under different conditions is one way to expose this liability. Testing batch size during method validation must include extraction variability, so reinjection alone is insufficient. However, knowing when a new column is needed by testing during method development will reduce failures in validation or sample analysis.

15.6 LESS IS MORE

15.6.1 Microseparations

Sensitivity improvements are seen when, at a constant linear flow velocity and fixed particle size, the column inner diameter (ID) is reduced. Since ESI is a concentration-dependent detector, smaller ID columns afford as much as a 20-fold improvement in concentration when moving from a 4.6 to a 1.0 mm column. However, capacity suffers and the amount that can be injected without band broadening is limited. This is not a problem when analyzing volume-limited samples. Microseparations have a large impact on the bioanalysis of small volume samples. When dead volume is properly reduced, scaling to smaller ID columns gives the analysis-increased sensitivity. Elution of an analyte in a smaller volume increases its concentration, generating a greater ESI response (Moseley et al., 1992). Therefore, there is considerable interest in the potential of microseparations to impact small volume analyses.

A best practice in microseparations is to avoid large dilutions and scale separations accordingly. When working with packed capillaries, the column is interfaced directly through the ID of the metal interface capillary so as to minimize extracolumn dead volume. Liquid and gas flow conditions as well as column alignment are important to ensure a consistent signal. Injection volume is limited for isocratic separations. Trace enrichment by injecting a larger volume that is weaker in solvent strength is an option. However, focusing of a large injection volume often requires a gradient separation that can be time-consuming at lower flow rates.

15.6.2 Microsampling

Microsampling adheres to the "less is more" principle. Improvements include a reduction in (1) interanimal variability by analysis of serial sampling in mouse pharmacokinetic studies, (2) toxicology satellite groups, (3) trauma when collecting clinical samples, and (4) risk of depleting sample when performing pediatric or complex biomarker studies where many individual assays are needed (Rainville, 2011). Dried blood spots (DBS) have analytical challenges that are readily solved using microseparations. A typical punch from a spot is only 3–5 μl and far less than a normal 100 μl aliquot. For an electronic noise-limited assay, this difference will result in a 25-fold higher LLOQ. Not diluting the sample during extraction or analysis is, therefore, critical. ESI is a concentration-dependent detector so keep the sample as concentrated as possible. Microseparations play a large role in helping make DBS possible for high potent drugs (Hooff et al., 2011). Without it, few assays would be able to measure pg/ml plasma concentrations needed for potent drugs.

Besides DBS, there are numerous other matrices where sample volume is limited. For ocular indications, tears represent a particularly useful medium for analysis. Synovial fluid for arthritic drugs, microdialysate for CNS drugs, biopsy tumor for oncology drugs, pediatric inner ear fluid for antiinfective drugs, antiviral drugs in peripheral blood mononuclear cells (PBMCs) are some of many microsampling applications. PBMCs represent a particularly attractive but analytically challenging medium for virology drugs such as reverse transcriptase inhibitors (Shi et al., 2002). Typically, only 1×10^6 cells (~0.5 μl) of cell lysate is isolated from a 1 ml blood sample. Drug levels are reported in amount per number of cells, so counting cells is as critical as the LC-MS analysis.

Microdialysis sampling allows correlation of neurotransmitter affect (directly by measuring biogenic amines such as serotonin or indirectly through receptor occupancy) and free drug concentrations in the brain. Kennedy illustrated how a single LC-MS assay using benzoyl chloride derivatization could yield levels for many common neurotransmitters and metabolites. A total of 17 analytes were separated in 8 min with a sample volume of only 5 μl (Song et al., 2012).

Much of proteomics evolved using packed capillary LC due to the need for both high-resolution separations and high-sensitivity analysis. Twenty years ago, the difficulty of performing quantitative analysis using packed capillary LC systems with slowly equilibrating syringe pumps made it impractical for high-throughput bioanalysis. Today's tools are much better and routinely peptide or protein drugs are being assayed by LC-MS in regulated studies. As protein drugs and biomarkers successfully move from discovery pharmacology to clinical studies, signature peptide assays will be translated into biomarker assays or used to provide more specific pharmacokinetic assessments of biologic drugs in support of regulated studies.

15.7 HIGH-RESOLUTION AND HIGH-SPEED ANALYSIS

15.7.1 UPLC

Separations are driven by the selectivity of the column chemistry. However, particle size is also critical to column performance. While smaller particles result in higher pressure, they also give higher efficiency and greater capacity. The Van Deemter equation defines the height equivalent of a theoretical plate (HETP) as the sum of three terms that represent eddy diffusion, molecular diffusion, and mass transfer kinetics. The C term (mass transfer) is directly proportional to the square of particle diameter. Even a small decrease in particle diameter will, therefore, have a large improvement in mass transfer and HETP. A high-resolution separation may be turned into a high-speed separation, so increasing the number of plates is a straightforward manner in which to improve any separation (Churchwell et al., 2005; Xiang et al., 2006).

Fast chiral separation of drugs using columns packed with sub-2 microm particles and ultra-high pressure has also been reported (Guillarme et al., 2010).

Twenty-five years ago, 5-μm particles were common. In the 1990s, 3.5-μm particles were manufactured. Today, sub-2-μm particles are produced with excellent uniformity and stability (Wu et al., 2006). Fused core® columns can be operated at higher temperatures, yielding faster separations by reducing mobile phase viscosity and resistance to mass transfer (Brice et al., 2009). When operated within the pressure limitation of HPLC systems, smaller particles afford many of the performance characteristics of UPLC on conventional pumping systems. When sub-2-μm columns are combined with ultrahigh pressure pumps (15,000 psi), further increases in speed and resolution are possible.

UPLC scaling is important to any established laboratory having a large number of existing HPLC assays that need to be transferred to UPLC. Because earlier MS interfaces could not accommodate high flow rates, LC-MS analysis had been performed using packed capillary column or 1-mm ID columns. Therefore, scaling assays down from a traditional HPLC assay 250 × 4.6 mm (5-μm particle) to a 100 × 2.1 mm with a 3-μm particle is mostly needed for non-MS-based assays. LC-MS users were already working with 1-mm ID microbore column and, over the years, translated from longer 150 × 1.0 mm columns to faster, higher flow 50 × 2.1 mm columns.

The commercialization of UPLC with stable sub-2-μm particles has made laboratories consider how to translate microbore HPLC assays to a 50 × 2.0 mm with a 1.7-μm particle column. Vendors provide scaling calculators on their Internet sites which takes into account particle size, column length, and injection volume. For an equivalent flow velocity, the concentration of analyte is inversely related to the square of the column ID. A twofold decrease in particle size will result in a fourfold increase in pressure, where doubling the column length simply translates into a twofold increase in pressure. Moving from a 3.0 to 1.7-μm particle, therefore, translates into a 3.1-fold increase in pressure. Commercial UPLC systems have produced pressure limits ranging from 12,000 (822 bar) to 20,000 psi (1370 bar). This is far short of what has been demonstrated in academic laboratories with 1.0-μm particle size columns (MacNair et al., 1999). Given the complexities of operating under these conditions, most laboratories are happy to translate their HPLC assays to UPLC conditions. Not only was it necessary to develop pumping systems and components that could operate under higher pressure, but significant advances in the stability and uniformity of small particle manufacturing was needed.

Analysts must be careful to avoid the introduction of dead volume when plumbing such systems. Particle size and distribution, packed bed uniformity, linear flow velocity, mass transfer, and column volume all contribute to column band broadening. To preserve column efficiency, an analyst must ensure that extra column dead volume is not introduced. Sources of dead volume include the injector, tubing (diameter and length), connector, and ion source. The same care given to interfacing MS to a packed capillary or 1 mm ID separation should be provided for UPLC; avoid dead volume. Systems must also use low volume mixers. Additional dead time can impact gradient formation across different model UPLC systems. Therefore, retention times can be different for different models. Test performance and do not assume equivalence when transferring assays across different UPLC systems.

15.7.2 Monolithic Columns

Improvements in speed are gained by reducing the resistance to mass transfer so that higher flow rates achieve faster separations. The same improvement gained by reducing particle size and increasing flow rate in UPLC is possible using monolithic HPLC columns. Monolithic columns contain porous rod structures characterized by mesopores and macropores (Jiang et al., 2011; Wu, 2001). These provide high permeability, high surface area and a large number of channels for diffusion into the phase. Unlike UPLC, monolithic columns are typically run under normal pressures and are less sensitive to dead volume and backpressure. There are, however, fewer commercial options for column chemistries.

Monolithic columns are ideally suited for large molecules. When biomolecules are separated on smaller particles, higher backpressures result due to their large molecule size. In monolithic columns, backpressures are low and dynamic binding capacities can be up to ten times greater than that for particulate packings (Ali et al., 2009; Altmaier and Cabrera, 2008). Unlike particulate columns, shear forces or eddying effects are minimized. Mesopores provide multiple paths for convective flow so mass transport through the column is unaffected by flow rate. Monolithic columns can, however, suffer from wall effects when the monolith pulls away from the cladding and flow of the mobile phase occurs around the stationary phase rod. Decladding has been reduced by advances in column construction.

15.7.3 Fused-Core Particles (Superficially Porous Silica Microspheres)

The fused-core technology utilizes a uniform porous silica layer that is grown around a spherical solid silica core. This unique combination provides chromatographic performance enhancement by increasing the rate of mass transfer by decreasing the effects of diffusion and by reducing the losses in efficiency with nearly monodispersed sub-2-μm particles. The concept of fused-core technology was introduced in the 1990s by Kirkland as >5 μm particles for the separation of macromolecules (Kirkland, 1992). Several vendors now commercialize their own superficially porous particles in a variety of stationary phases. The advantages of fused-core particles for the determination of analytes in biological fluids were evaluated by several groups (Badman et al., 2010; Cunliffe et al., 2009; Cunliffe, 2012; Hsieh et al., 2007; Song et al., 2009). One unique feature about this technology is the low flow resistance and thus the ability of achieving UPLC like efficiency without the need of using a system than can handle high column backpressure.

15.7.4 Multiplexing

LC separations are time-consuming. When using more expensive MS detection, multiplexing can offer a significant advantage. While this can be achieved using multiple flow streams into a multiplexed interface, source design was complex and suffered from reduced sensitivity (Deng et al., 2002). A more common practice is to stagger injection cycles (Wu, 2001). As one LC is injecting a sample onto the first column another HPLC is moving analyte from a second column into the mass spectrometer. When the peak from the second column has eluted, the MS flow stream switches back to capture the peak from first column. This process continues, allowing two injections to be analyzed in the same time as one.

Configuring a system is as simple as adding another pump and a switching valve. However, coordination of injection cycles and the proper recording of run times are critical in a regulated bioanalytical laboratory. Cohesive (now owned by Thermo Scientific) was first to commercially develop systems that incorporate two or more pumping systems into a common flow path and have associated software that ensures proper timing. Today, other vendors have offered staggered injection. Analysts need to demonstrate that assays can provide equivalent results when separate columns are used. Controlling the balance of pressure and flow rates across the two columns is critical. Methods should be validated in the manner samples are to be analyzed. Therefore, a multiplexed assay needs to be fully tested prior to its implementation. The analyst needs to consider what defines a batch and may wish to associate one extraction batch with an individual column flow stream. Commingling of injections has the risk of failing the entire batch.

15.8 SPECIAL CHALLENGES AND OPPORTUNITIES

15.8.1 Enantiomers

Drugs may be dosed as a racemic mixture or inversion of a chiral center may require enantiomer-specific assays be used in pharmacokinetic studies. Often, oxidative metabolism induces chirality, requiring the separation of R and S metabolites. More commonly used phases include β-cyclodextrin or covalently bound protein columns such as ovomucoid or α1-acid glycoprotein (Zhou et al., 2010; Xiao et al., 2003; Ward and Ward, 2010). Column vendors offer advice on their Internet sites, which is helpful for method development. Developing a robust bioanalytical separation can be challenging. Unless the extract is sufficiently pure, a good chiral separation can rapidly deteriorate. Chiralpak® AGP column (α1-acid glycoprotein) has broad applicability but has limits on pressure (2000 psi), organic modifier (20%), pH range (4–7), and temperature (40°C). Many chiral separations use normal phase conditions, so samples must be free of water and extracts properly sealed to avoid evaporation. Chiral separations require three point interactions in either the stationary or mobile phase. Achieving the desired enantiomer separation on the stationary phase eliminates the need to add chiral modifers to the mobile phase, which can deteriorate MS response.

15.8.2 Polar Analytes and HILIC

Reverse phase separations have been used in the majority of bioanalytical separations. Their use in the analysis of highly polar compounds has been limited in MS applications due to the general requirement to add ion-pairing reagents, which impacts MS response. The utility of performing aqueous-based separations on silica columns was largely overlooked until Alpert reported HILIC (Adamovics, 1986; Alpert, 1990). Unlike normal phase separations, higher water content is required on a HILIC separation. Vendors have offered many options including pure silica or amino and cyano columns, which can be operated under HILIC conditions. Since water is the strong solvent, a large volume of an organic extract can be loaded onto a HILIC column without band broadening. This avoids dry-down time and losses without sacrificing detection sensitivity (Nguyen and Schug, 2008). However, there are some disadvantages. Columns are more prone to overloading that can cause peak splitting or distortion. Surface adsorption is the primary retention mechanism that limits capacity more than partitioning. Pure silica columns also are more susceptible to hydrolysis under higher pH than are many reverse phase columns. Polymer-based HILIC columns have better stability.

The selectivity of a separation is not improved by simply adding another column with the same retention characteristics. A major advantage of HILIC is its complementary combination with reverse phase or weak anion exchange as an orthogonal separation (Jandera, 2011; Jian et al., 2010). Orthogonal separations are needed for complex analysis, such as proteomics or metabolomics. When a switching valve is added, heart-cutting of individual components and subsequent LC-MS analysis offers another means to further improve detection specificity and assay selectivity. Some refocusing and care to avoid overloading the secondary column is needed for a 2D LC method.

Ultrahigh pressure HILIC columns are also available. The Van Deemter curve for 1.7 μm BEH silica shows little increase in HETP profile even at higher flow rates (Grumbach et al., 2008). Due to its low backpressure, a sub-2-μm HILIC particle is useful at normal HPLC pressures affording many of the advantages of UPLC separations on an HPLC system. Separations as fast as a few seconds have been achieved. Clearly, many options are evolving that will afford very high-speed bioanalysis in the future.

15.9 CONCLUSIONS

A thorough understanding of both chromatographic theory and its practical application as well as the holistic view of sample preparation, chromatography, and mass detection will help the practitioners to develop a robust bioanalytical method. Many factors can contribute to the potential failures of any given bioanalytical method. Some of the most important factors one should consider and mitigate are matrix effects, in-source breakdown of phase II conjugated metabolites, separation of isobaric metabolites, carryover, loss due to nonspecific adsorption, and stability. Besides a good scientific understanding of LC-MS/MS, bioanalytical scientists should be aware of the current regulatory requirements as a worldwide submission needs alignment with regional requirements. In conjunction with science and regulatory compliance, throughput and cost reduction must also be considered.

In conclusion, good chromatography is a critical component of any bioanalytical method and can be described as the ability to consistently achieve separation requirements throughout an entire analytical run. A good separation is a critical component of a bioanalytical method and should aid in achieving the required specificity. Efficient separations perform their job in as little time as possible while managing to meet all assay and any regulatory requirements. Separation technology is ever advancing and serves as the foundation for a successful bioanalytical method. Requirements should be defined and tested using a system suitability sample at the beginning and end of a run. A properly cleaned, tuned, and calibrated MS system helps to ensure that detector response is sufficient for proper peak identification and integration. Together, these contribute to a successful analytical run.

REFERENCES

Adamovics J, Unger SE. Preparative liquid chromatography of pharmaceuticals using silical gel with aqueous eluents. J Liq Chrom 1986;9:141–155.

Ali I, Gaitonde VD, Aboul-Enein HY. Monolithic silica stationary phases in liquid chromatography. J Chromatogr Sci 2009;47(6):432–442.

Alpert AJ. Hydrophilic-interaction chromatography for the separation of peptides, nucleic acids and other polar compounds. J Chromatogr A 1990;499:177–196.

Altmaier S, Cabrera K. Structure and performance of silica-based monolithic HPLC columns. J Sep Sci 2008;31(14):2551–2559.

Badman ER, Beardsley RL, Liang Z, Bansal S. Accelerating high quality bioanalytical LC/MS/MS assays using fused-core columns. J Chromatogr B 2010;878(25):2307–2313.

Berman J, Halm K, Adkison K, Shaffer J. Simultaneous pharmacokinetic screening of a mixture of compounds in the dog using API LC/MS/MS analysis for increased throughput. J Med Chem 1997;40:827–829.

Brice RW, Zhang X, Colon LA. Fused-core, sub-2 μm packings, and monolithic HPLC columns: a comparative evaluation. J Sep Sci 2009;32(15-16):2723–2731.

Briscoe CJ, Stiles MR, Hage DS. System suitability in bioanalytical LC/MS/MS J. Pharm Biomed Anal 2007;44(2):484–491.

Buhrman D, Price P, Rudewicz P. Quantitation of SR 27417 in human plasma using electrospray liquid chromatography-tandem mass spectrometry: a study of ion suppression. J Am Soc Mass Spectrom 1996;7:1099–1105.

Cai P, Tsao R, Ruppen ME. In Vitro metabolic study of temsirolimus: Preparation, isolation, and identification of the metabolites. Drug Metab Dispos 2007;35(9):1554–1563.

Chambers E, Wagrowski-Diehl DM, Lu Z, Mazzeo JR. Systematic and comprehensive strategy for reducing matrix effects in LC/MS/MS analyses. J Chromatogr B 2007;852(102):22–34.

Chen Y-L, Junga H, Jiang X, Weng N. Simultaneous determination of theophylline, tolbutamide, mephenytoin, debrisoquin, and dapsone in human plasma using high-speed gradient liquid chromatography/tandem mass spectrometry on a silica-based monolithic column. J Sep Sci 2003;26(17):1509–1519.

Churchwell MI, Twaddle NC, Meeker LR, Doerge DR. Improving LC-MS sensitivity through increases in chromatographic performance: comparisons of UPLC-ES/MS/MS to HPLC-ES/MS/MS. J Chromatogr B 2005;825(2):134–43.

Cunliffe JM. Fast chromatography in the regulated bioanalytical environment: sub-2-μm versus fused-core particles. Bioanalysis 2012;4(8):861–863.

Cunliffe JM, Noren CF, Hayes RN, Clement RP, Shen JX. A high-throughput LC–MS/MS method for the quantitation of posaconazole in human plasma: implementing fused core silica liquid chromatography. J Pharm Biomed Anal 2009;50(1): 46–52.

Deng Y, Wu J-T, Lloyd TL, Chi, CL, Olah, TV, Unger SE. High-speed gradient parallel liquid chromatography/tandem mass spectrometry with fully automated sample preparation for bioanalysis: 30 seconds per sample from plasma. Rapid Commun Mass Spectrom 2002;16:1116–1123.

European Medicines Agency. Guideline on bioanalytical method validation. 2011. EMEA/CHMP/EWP/192217/2009. Availabile at http://www.ema.europa.eu/docs/en_GB/document_library/Scientific_guideline/2011/08/WC500109686.pdf. Accessed Mar 20, 2013.

European Medicines Agency. ICH Topic M 3 (R2). Non-Clinical Safety Studies for the Conduct of Human Clinical Trials and Marketing Authorization for Pharmaceuticals. Note for Guidance on Non-Clinical Safety Studies for the Conduct of Human Clinical Trials and Marketing Authorization for Pharmaceuticals (CPMP/ICH/286/95). 2009. Available at http://www.emea.europa.eu/docs/en_GB/document_library/Scientific_guideline/2009/09/WC500002720.pdf. Accessed Mar 20, 2013.

FDA. Guidance for Industry, Safety Testing of Drug Metabolites. 2008. Available at http://www.fda.gov/OHRMS/DOCKETS/98fr/FDA-2008-D-0065-GDL.pdf. Accessed Mar 20, 2013.

Gao S, Zhang Z-P, Karnes HT. Sensitivity enhancement in liquid chromatography/atmospheric pressure ionization mass spectrometry using derivatization and mobile phase additives. J Chromatogr B 2005;825: 98–110.

Gilar M, Daly AE, Kele M, Neue UD, Gebler JC. Implications of column peak capacity on the separation of complex peptide mixtures in single- and two-dimensional high-performance liquid chromatography. J Chromatogr A 2004;1061(2): 183–192.

Grumbach ES, Diehl DM, Neue UD. The application of novel 1.7 μm ethylene bridged hybrid particles for hydrophilic interaction chromatography. J Sep Sci 2008;31(9):1511–1518.

Guillarme D, Bonvin G, Badoud F, Schappler J, Rudaz S, Veuthey JL. Fast chiral separation of drugs using columns packed with sub-2 microm particles and ultra-high pressure. Chirality 2010;22(3):320–30.

Hao Z, Xiao B, Weng N. Impact of column temperature and mobile phase components on selectivity of hydrophilic interaction chromatography (HILIC). J Sep Sci 2008;31(9):1449–1464.

Heller D. Ruggedness testing of quantitative atmospheric pressure ionization mass spectrometry methods: the effect of co-injected matrix on matrix effects. Rapid Commun Mass Spectrom 2007;21:644–652.

Henderson DE, Horvath C. Low temperature high-performance liquid chromatography of cis-trans proline dipeptides. J. Chromatogr. 1986;368(2):203–213.

Hooff GP, Meesters RJW, van Kampen JJA, et al. Dried blood spot UHPLC-MS/MS analysis of oseltamivir and oseltamivircarboxylate–a validated assay for the clinic. Anal Bioanal Chem 2011;400(10):3473–3479.

Hsieh Y, Duncan CJ, Brisson JM. Fused-core silica column high-performance liquid chromatography/tandem mass spectrometric determination of rimonabant in mouse plasma. Anal Chem 2007;79(15):5668–5673.

Huang EC, Wachs T, Conboy JJ, Henion JD. Atmospheric pressure ionization mass spectrometry. Detection for the separation sciences. Anal Chem 1990;62(13):713A–725A.

Jandera P. Stationary and mobile phases in hydrophilic interaction chromatography: a review. Anal Chim Acta 2011;692(1-2): 1–25.

Jemal M, Xia Y-Q. LC-MS Development Strategies for Quantitative Bioanalysis. Current Drug Metabolism 2006;7:491–502.

Ji AJ, Jiang Z-P, Livson Y, Davis JA, Chu JX-G, Weng, N. Challenges in urine bioanalytical assays: overcoming nonspecific binding. Bioanalysis 2010;2(9):1573–1586.

Jian W, Edom RW, Xu Y, Weng N. Recent advances in application of hydrophilic interaction chromatography for quantitative bioanalysis. J Sep Sci 2010;33: 1–17.

Jian W, Xu Y, Edom RW, Weng N. Analysis of polar metabolites by hydrophilic interaction chromatography–MS/MS. Bioanalysis 2011 3(8):899–912.

Jiang Z, Smith NW, Liu Z. Preparation and application of hydrophilic monolithic columns. J Chromatogr A 2011; 1218(17):2350–2361.

Kieser KJ, Dong WK, Carlson KE, Katzenellenbogen BS, Katzenellenbogen JA. Characterization of the Pharmacophore Properties of Novel Selective Estrogen Receptor Downregulators (SERDs). J Med Chem 2010;53(8):3320–3329.

King R, Bonfiglio R, Fernandez-Metzler C, Miller-Stein C, Olah T. Mechanistic investigation of ionization suppression in electrospray ionization. J Am Soc Mass Spectrom 2000;11: 942–950.

Kirkland JJ. Superficially porous silica microspheres for the fast high-performance liquid chromatography of macromolecules. Anal Chem 1992;64(11):1239–1245.

Li AC, Junga H, Shou WZ, Bryant MS, Jiang XY, Naidong W. Direct injection of solid-phase extraction eluents onto silica columns for the analysis of polar compounds isoniazid and cetirizine in plasma using hydrophilic interaction chromatography with tandem mass spectrometry. Rapid Commun Mass Spectrom 2004;18(19):2343–2350.

Li W, Zhang J, Tse FLS. Strategies in quantitative LC-MS/MS analysis of unstable small molecules in biological matrices. Biomed Chromatogr 2011;25:258–277.

Li Z, Yao J, Zhang Z, Zhang L. Simultaneous determination of omeprazole and domperidone in dog plasma by LC-MS method. J Chromatogr Sci 2009;47(10):881–884.

Lin ZJ, Li W, Dai G. Application of LC-MS for quantitative analysis and metabolite identification of therapeutic oligonucleotides. J Pharm Biomed Anal 2007;44(2):330–341.

Little J, Wempe M, Buchanan C. Liquid chromatography-mass spectrometry/mass spectrometry method development for drug metabolism studies: Examining lipid matrix ionization effects in plasma. J Chromatogr B 2006;833:219–230.

MacNair JE, Patel KD, Jorgenson JW. Ultrahigh-pressure reversed-phase capillary liquid chromatography: Isocratic and gradient elution using columns packed with 1.0-μm particles. Anal Chem 1999;71(3):700–708.

Matuszewski BK, Constanzer ML, Chavez-Eng CM. Matrix effect in quantitative LC/MS/MS analyses of biological fluids: A method for determination of finasteride in human plasma at picogram per milliliter concentrations. Anal Chem 1998;70:882–889.

Moseley MA, Unger SE. Packed column liquid chromatography/electrospray ionization/ mass spectrometry in the pharmaceutical sciences: Characterization of protein mixtures. J Microcolumn Sep 1992;4:393–398.

Naidong W. Bioanalytical liquid chromatography tandem mass spectrometry methods on underivatized silica columns with aqueous/organic mobile phases. J Chromatogr B Analyt Technol Biomed Life Sci 2003;796(2):209–224.

Naidong W, Chen YL, Shou W, Jiang X. Importance of injection solution composition for LC-MS-MS methods. J Pharm Biomed Anal 2001a;26(5-6):753–767.

Naidong W, Lee JW. Development and validation of a high-performance liquid chromatographic method for the quantitation of warfarin enantiomers in human plasma. J Pharm & Biomed Anal 1993;11:785–792.

Naidong W, Ring PR, Midtlien C, Jiang X. Development and validation of a sensitive and robust LC-tandem MS method for the analysis of warfarin enantiomers in human plasma. J Pharm Biomed Anal 2001b;25(2):219–226.

Naidong W, Zhou W, Song Q, Zhou S. Direct injection of 96-well organic extracts onto a hydrophilic interaction chromatography/tandem mass spectrometry system using a silica stationary phase and an aqueous/organic mobile phase. Rapid Commun Mass Spectrom 2004;18(23):2963–2968.

Nguyen HP, Schug KA. The advantages of ESI-MS detection in conjunction with HILIC mode separations: fundamentals and applications. J Sep Sci 2008;31(9):1465–1480.

Ojingwa JC, Spahn-Langguth H, Benet LZ. Reversible binding of tolmetin, zomepirac, and their glucuronide conjugates to human serum albumin and plasma. J Pharmacokinet Biopharm 1994;22(1):19–40.

Palmer JL, Scott RJ, Gibson A, Dickins M, Pleasance S. An interaction between the cytochrome P450 probe substrates chlorzoxazone (CYP2E1) and midazolam (CYP3A). Br J Clin Pharmacol 2001;52:555–561.

Pan J, Fair SJ, Mao D. Quantitative analysis of skeletal symmetric chlorhexidine in rat plasma using doubly charged molecular ions in LC-MS/MS detection. Bioanalysis 2011;3(12):1357–1368.

Pulfer M, Murphy R. Electrospray mass spectrometry of phospholipids. Mass Spectrom Rev 2003;22:332–364.

Rainville P. Microfluidic LC–MS for analysis of small-volume biofluid samples: where we have been and where we need to go. Bioanalysis. 2011;3(1):1–3.

Remane D, Meyer MR, Wissenbach DK, Maurer HH. Ion suppression and enhancement effects of co-eluting analytes in multi-analyte approaches: systematic investigation using ultra-high-performance liquid chromatography/mass spectrometry with atmospheric-pressure chemical ionization or electrospray ionization. Rapid Commun Mass Spectrom 2010;24(21):3103–3108.

Robison TW, Jacobs A. Metabolites in safety testing. Bioanalysis 2009;1(7):1193–1200.

Shi G, Wu JT Li Y, Geleziunas R, Gallagher K, Emm T, Unger S. Novel direct detection method for quantitative determination of intracellular nucleoside triphosphates using weak anion exchange liquid chromatography/tandem mass spectrometry. Rapid Commun Mass Spectrom 2002;16:1092–1099.

Shou WZ, Chen YL, Eerkes A, Tang YQ, Magis L, Jiang X, Naidong W. Ultrafast liquid chromatography/tandem mass spectrometry bioanalysis of polar analytes using packed silica columns. Rapid Commun Mass Spectrom 2002;16(17):1613–1621.

Shou WZ, Naidong W. Simple means to alleviate sensitivity loss by trifluoroacetic acid (TFA) mobile phases in the hydrophilic interaction chromatography-electrospray tandem mass spectrometric

(HILIC-ESI/MS/MS) bioanalysis of basic compounds. J Chromatogr B Analyt Technol Biomed Life Sci 2005;825(2): 186–192.

Snyder LR, Glajch JL, Kirkland JJ. *Practical HPLC Method Development*. New York: John Wiley & Sons, Inc.; 1988.

Snyder LR, Kirkland JJ, Dolan JW. *Introduction to Modern Liquid Chromatography*. New York: John Wiley & Sons, Inc.; 2009.

Song P, Mabrouk OS, Hershey ND, Kennedy RT. In Vivo neurochemical monitoring using benzoyl chloride derivatization and liquid chromatography-mass spectrometry. Anal Chem 2012;84(1):412–419.

Song W, Pabbisetty D, Groeber EA, Steenwyk RC, Fast DM. Comparison of fused-core and conventional particle size columns by LC–MS/MS and UV: application to pharmacokinetic study. J Pharm Biomed Anal 2009;50(3):491–500.

Tiller PR, Romanyshyn LA. Implications of matrix effects in ultra-fast gradient or fast isocratic liquid chromatography with mass spectrometry in drug discovery Rapid Commun. Mass Spectrom 2001;16(2):92–98.

Vallano PT, Shugarts SB, Woolf EJ, Matuszewski BK. Elimination of autosampler carryover in a bioanalytical HPLC-MS/MS method: a case study. J Pharm Biomed Anal 2005;36(5):1073–1078.

van de Merbel NC. Is HPLC becoming obsolete for bioanalysis? Chromatographia 2001;55(1): S53-S57.

Van Eeckhaut A, Lanckmans KS, Smolders I, Michotte Y. Validation of bioanalytical LC-MS/MS assays: Evaluation of matrix effects. J Chromatogr B 2009;877:2198–2207.

Wang J, Aubry A-F, Cornelius G, et al. Importance of mobile phase and injection solvent selection during rapid method development and sample analysis in drug discovery bioanalysis illustrated using convenient multiplexed LC-MS/MS. Anal. Methods 2010;2(4):375–381.

Ward TJ, Ward KD. Chiral separations: A review of current topics and trends. Anal Chem 2010;84(2):626–635.

Williams RR. Fundamental limitations on the use and comparison of signal-to-noise ratios. Anal Chem 1991;63(15): 1638–1643.

Wu J-T. The development of a staggered parallel separation LC/MS/MS system with on-line extraction for high-throughput screening of drug candidates in biological fluids. Rapid Commun Mass Spectrom 2001;15(2):73–81.

Wu J-T, Zeng H, Deng Y, Unger SE. High-speed liquid chromatography/tandem mass spectrometry using a monolithic column for high-throughput bioanalysis. Rapid Commun Mass Spectrom 2001;15:1113–1119.

Wu J-T, Zeng H, Qian M, Brogdon BL, Unger SE. Direct plasma sample injection in multiple-component LC-MS-MS assays for high-throughput pharmacokinetic screening. Anal Chem 2000;72:61–67.

Wu N, Liu Y, Lee ML. Sub-2μm porous and nonporous particles for fast separation in reversed-phase high performance liquid chromatography. J Chromatogr A 2006;1131(1-2):142–150.

Wu S, Schoener D, Jemal M. Plasma phospholipids implicated in the matrix effect observed in liquid chromatography/tandem mass spectrometry bioanalysis: evaluation of the use of colloidal silica in combination with divalent or trivalent cations for the selective removal of phospholipds from plasma. Rapid Commun Mass Spectrom 2008;22:2873–2881.

Xia Y-Q, Jemal M. Phospholipids in liquid chromatography/mass spectrometry bioanalysis: comparison of three tandem mass spectrometric techniques for monitoring plasma phospholipids, the effect of mobile phase composition on phospholipids elution and the association of phospholipids with matrix effects. Rapid Commun Mass Spectrom 2009;23:2125–38.

Xiang Y, Liu Y, Lee ML. Ultrahigh pressure liquid chromatography using elevated temperature. J Chromatogr A 2006; 1104(1-2):198–202.

Xiao TL, Rozhkov RV, Larock RC, Armstrong DW. Separation of the enantiomers of substituted dihydrofurocoumarins by HPLC using macrocyclic glycopeptide chiral stationary phases. Anal Bioanal Chem 2003;377(4):639–654.

Xue Y-J, Liu J, Simmons NJ, Unger S, Anderson DF, Jenkins RG. Separation of a BMS drug candidate and acyl glucuronide from seven glucuronide positional isomers in rat plasma via HPLC with MS/MS detection. Rapid Commun Mass Spectrom 2006;20:1776–1786.

Zhou S, Prebyl BS, Cook KD. Profiling pH changes in the electrospray plume. Anal Chem 2002;74(19):4885–4888.

Zhou Z-M, Li X, Chen X-P, Fang M, Dong X. Separation performance and recognition mechanism of mono(6-deoxy-imino)-β-cyclodextrins chiral stationary phases in high-performance liquid chromatography. Talanta 2010;82(2):775–784.

16

BEST PRACTICE IN MASS SPECTROMETRY FOR LC-MS

Richard B. van Breemen and Elizabeth M. Martinez

16.1 INTRODUCTION

Although gas chromatography-mass spectrometry (GC MS) became a routine analytical technique by the late 1960s, several decades elapsed before high-performance liquid chromatography-mass spectrometry (LC-MS) reached a similar level of reliability, reproducibility, and robustness. Successful LC-MS requires selective removal of the high-performance liquid chromatography (HPLC) mobile phase and then formation of gas-phase sample ions for analysis in a high-vacuum mass spectrometer. Today, LC-MS enables routine qualitative and quantitative analyses of a variety of pharmaceuticals and biomedical compounds that is limited by solubility instead of volatility. These steps have been achieved in atmospheric pressure ionization sources that may be interfaced with a variety of mass spectrometer analyzers.

16.2 ANALYZERS MOST OFTEN USED FOR LC-MS AND LC-MS/MS BIOANALYSIS

During LC-MS, the rate at which mass spectra are recorded should match the peak width produced during LC. Although analog-to-digital signal conversion was once a limiting factor, electronics have become so fast that data acquisition rates are now analyzer dependent. Data acquisition rates from mass spectrometers today are determined by the type of analyzer, the *m/z* range being recorded, and the type of scan. Table 16.1 shows different types of analyzers and their typical scan rates when recording mass spectra and tandem mass spectra.

Ideally, at least 20 data points should be obtained across a chromatographic peak so that chromatographic resolution is not lost due to sampling. Although conventional HPLC peak widths are often 10–20 s, ultrahigh pressure liquid chromatography (UHPLC) peak widths can be 1–3 s. Therefore, time of flight (Tof) instruments, which are the only mass analyzers capable of recording $\geq$20 mass spectra per second are best for use with UHPLC, while slower scanning instruments such as quadrupole, Orbitrap, and ion trap mass spectrometers are more appropriate for use with HPLC. Examples of the effects of mass spectrometer scan rates on chromatographic resolution are shown in Figure 16.1.

During LC-MS analysis, mass spectra of eluting compounds are usually recorded continuously. A total ion chromatogram (TIC) may be used to determine when compounds eluted from the HPLC system and are detected. The TIC is produced by summing all signals detected in each mass spectrum into a single data point and plotting the signal intensity of each point versus scan number or elution time. In Figure 16.2a, the TIC is shown for the negative ion electrospray UHPLC-MS analysis of an extract of licorice (*Glycyrrhiza uralensis*) using a quadrupole mass spectrometer (see Section 16.2.1 for additional information about quadrupole mass spectrometers).

When samples contain complex mixtures as in extracts of serum, urine extract or botanicals, then the TIC may show many overlapping peaks and be less informative about each eluting species. In such cases, the computer may be used to look for ions corresponding to specific molecular ions, protonated molecules, deprotonated molecules, and so on, and then computer-reconstructed mass chromatograms may be plotted showing these signals. Note that scan-type mass spectra are not always needed or recorded during LC-MS and LC-MS/MS analyses. Instead, selected ion monitoring (SIM) or selected reaction monitoring (SRM) may be used.

Handbook of LC-MS Bioanalysis: Best Practices, Experimental Protocols, and Regulations, First Edition. Edited by Wenkui Li, Jie Zhang, and Francis L.S. Tse.

TABLE 16.1 Types of Mass Spectrometers Most Often Used for LC-MS and LC-MS/MS

Mass analyzer	Scanning rate (scans/s)	Resolving power	*m/z* Range	Tandem MS
Quadrupole	5	<4 000	4 000	None
Triple quadrupole	5	<4 000	4 000	Low resolution
Time-of-flight (Tof)	>1 000	≥15 000	>200 000	none
FTICR	<1	>200 000	<10 000	High-resolution MS^n
Orbitrap	≤10	>60 000	<6 000	High-resolution MS^n
Ion trap (IT)	≤5	<4 000	<4 000	Low-resolution MS^n
QTof	5	≥14 000	4 000	High resolution
IT-Tof	≤5	≥14 000	4 000	High resolution

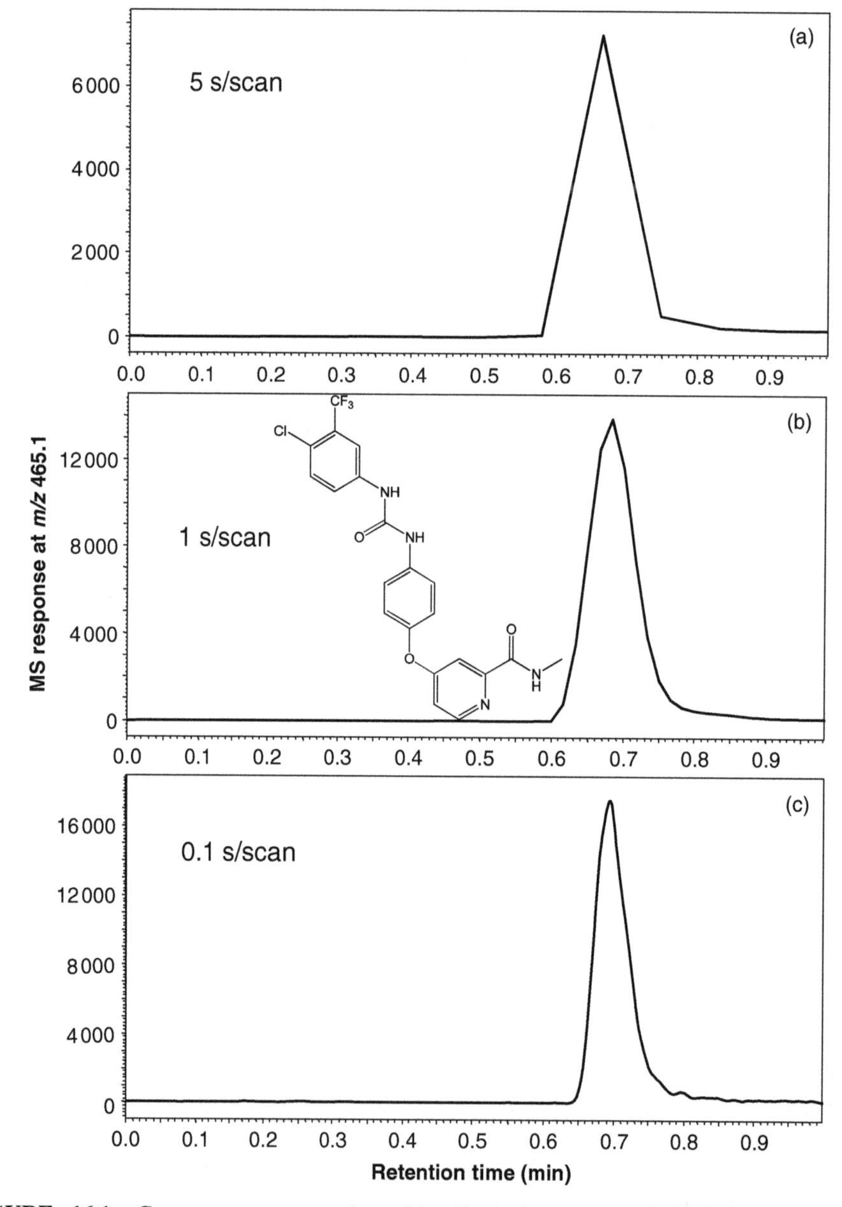

FIGURE 16.1 Computer-reconstructed positive ion electrospray UHPLC-MS mass chromatograms obtained using a quadrupole mass spectrometer showing the effect of different scan rates on the detection of protonated sorafenib at *m/z* 465.1: (a) 5 s/scan, (b) 1 s/scan, and (c) 0.1 s/scan. As the scan rate was increased, the chromatographic peak shape and signal-to-noise ratio improved. The scan range was *m/z* 50–500.

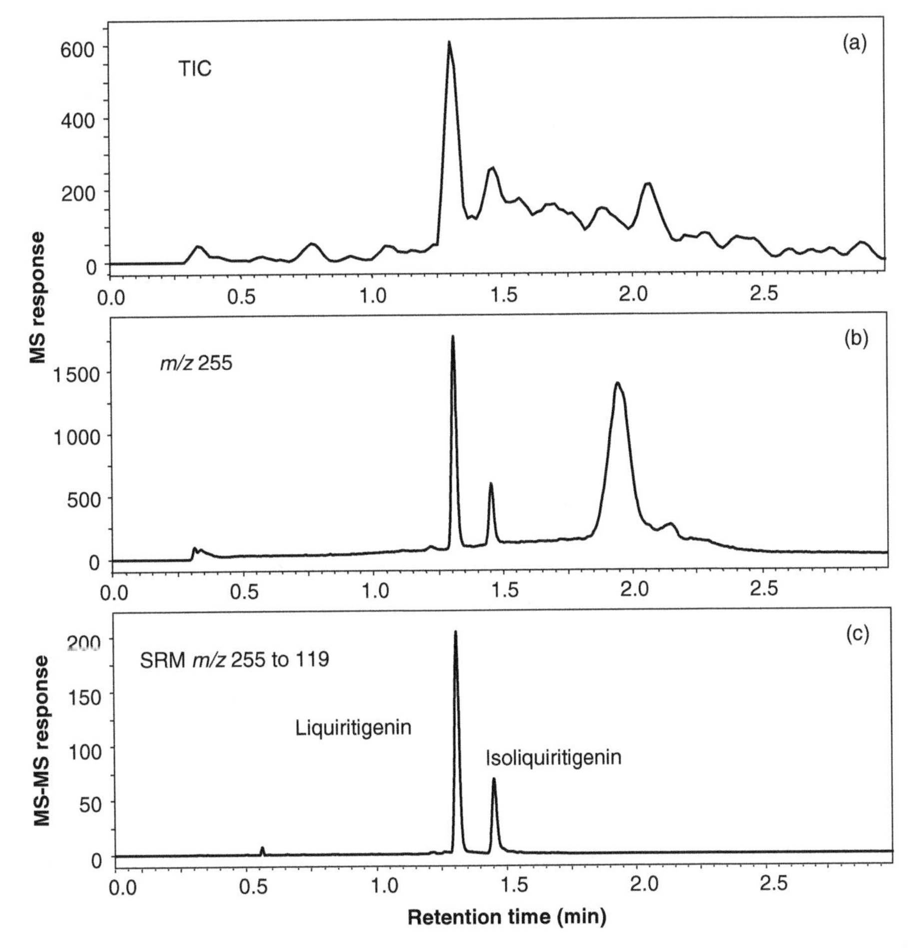

FIGURE 16.2 UHPLC-MS analysis of an extract of licorice (*Glycyrrhiza uralensis*) using negative ion electrospray and reversed-phase UHPLC: (a) TIC over the scan range *m/z* 100 to *m/z* 500; (b) Selected ion monitoring (SIM) mass chromatogram showing the intensity of the signal of *m/z* 255, which corresponds to the deprotonated molecules of liquiritigenin and isoliquiritigenin; (c) UHPLC-MS/MS analysis of the same extract using SIM of the transition of *m/z* 255 to *m/z* 119. Note the detection of liquiritigenin at a retention time of 1.30 min and its isomer, isoliquiritigenin, at 1.45 min.

If the molecular ions, protonated molecules or deprotonated molecules of specific analytes are known, then SIM may be used to detect only these ions. During SIM, signals for only a few ions are sampled on the millisecond timescale in rapid succession and recorded as data points of chromatograms. Since the sampling rate for SIM is fast (10–25 ms/ion) compared with the chromatographic timescale (peak widths 1–20 s), multiple ions may be monitored in rapid succession using SIM during a single LC-MS or UHPLC-MS analysis. In Figure 16.2b, elution of the licorice compounds liquiritigenin and isoliquiritigenin were recorded using SIM during UHPLC-MS by monitoring their deprotonated molecules of *m/z* 255. Note that the narrower peak shapes obtained using SIM, which were the result of the improved duty cycle of SIM compared with scan mode on a quadrupole mass spectrometer.

Although chromatographic separation provides a unique dimension of analyte selection and SIM provides molecular mass as another dimension of selectivity, multiple ions of identical nominal mass might still be detected during LC-MS or UHPLC-MS. For example, UHPLC-MS analysis with SIM of *m/z* 255 of the licorice extract shows peaks at 1.90 min and 2.10 min in addition to those of liquiritigenin and isoliquiritigenin at 1.30 min and 1.45 min, respectively. To distinguish among such possibilities, an additional level of selectivity may be used based on tandem MS in a technique called SRM as shown in Figure 16.2c.

During SRM LC-MS/MS, signals for one or more analytes are monitored by selecting a precursor ion such as the molecular ion during the first stage of MS, fragmenting the precursor ion using collision-induced dissociation (CID), and then recording one or more CID product ions formed during a second stage of MS. In Figure 16.2c, UHPLC-MS/MS with SRM was applied to the selective detection of liquiritigenin and its closely related structural isomer isoliquiritigenin. The precursor ion of *m/z* 255 was selected using the first quadrupole of a triple quadrupole mass spectrometer (see Section 16.1.2 for more information about triple quadrupole mass spectrometers), fragmented using CID in the next sector, and then the product ion of *m/z* 119 was selected by the final quadrupole. By recording the signal corresponding to the SRM transition of *m/z* 255 to *m/z* 119, only peaks for liquiritigen and isoliquiritigenin were detected during analysis of the licorice extract (Figure 16.2c).

SIM may be used with any type of mass analyzer, but quadrupole mass spectrometers, which can be set to transmit ions of a single *m/z* value at a time, are used most often. Since Tof mass spectrometers record multiple mass spectra each second during a chromatographic separation, SIM chromatograms may be reconstructed from these mass spectra using a computer. For LC-MS/MS analysis using SRM, triple quadrupole mass spectrometers are usually used. In fact, the duty cycle of triple quadrupole mass spectrometers is so efficient in SRM mode that they are the most popular type of mass spectrometer for quantitative analysis. Ion trap instruments may be used for SIM or SRM, but the multistep process of ion acquisition and selection is slower and less efficient than for SIM or SRM using quadrupole-type instruments. Quadrupole time-of-flight (QqTof) mass spectrometers are a type of hybrid tandem mass spectrometer used for SRM. In the QqTof, precursor ions are selected in the quadrupole, fragmented using CID and then complete product ion tandem mass spectra are recorded using the Tof analyzer. Computer-reconstructed SRM chromatograms may then be produced from the tandem mass spectra that were recorded using the QqTof.

In general, SRM is as fast as SIM (typically 10–20 ms/ion), and outstanding chromatographic fidelity may be obtained. Therefore, both SIM and SRM are ideal for use with HPLC or UHPLC. On the other hand, SIM and SRM do not contain complete mass spectra, and they are not useful for structure elucidation. Instead, SIM and SRM are usually used in combination with LC-MS or LC-MS/MS for quantitative analysis as described in Section 16.4. If additional structural information is required, then mass spectra and tandem mass spectra are usually obtained.

When a peak is observed in a TIC or a computer-reconstructed mass chromatogram such as the peak eluting at 1.30 min in Figure 16.2a, then additional information may be obtained for the eluting compound by viewing the associated mass spectrum stored in the data file for that analysis. (Note that mass spectra are only acquired during scan mode and during SIM or SRM analyses.) The mass spectrum will contain information such as the molecular ion of the analyte, molecular ions for coeluting compounds, contaminant ions from the mass spectrometer ion source, and possibly some fragment ions of the analyte if any formed in the ion source. The mass spectrum corresponding to the peak at 1.30 min is shown in Figure 16.3a, and an ion corresponding to the deprotonated molecule of liquiritigenin was observed at *m/z* 255. However, ions that are probably unrelated to liquiritigenin are also present in Figure 16.3a that might correspond to contaminants in the mobile phase or coeluting compounds from the sample extract. These background ions provide no structural information about the analyte and can confuse its identification.

One solution to the identification of the analyte would be to use high-resolution accurate mass measurement (discussed in detail under high-resolution mass analyzers below) to determine the elemental composition of the ion of *m/z* 255. Another solution would be to carry out data-dependent product ion tandem MS to obtain structurally significant fragment ions. Yet another solution would be combine high-resolution accurate mass measurement with tandem MS using an instrument such as a QqTof mass spectrometer (discussed in Section 16.2.2) to obtain the tandem mass spectrum of the precursor ion of *m/z* 255 as shown in Figure 16.3b. In this case, accurate mass measurement provided an elemental composition of $C_{15}H_{11}O_4$ that is −4.9 ppm from the theoretical mass of liquiritigenin (measured *m/z* 255.0661; theoretical *m/z* 255.0673). Examination of the tandem mass spectrum provides additional structural information about the analyte such as the characteristic fragment ions of *m/z* 119 and *m/z* 135 that were formed by cleavage of the central ring of liquiritigenin.

16.2.1 Low-Resolution Mass Spectrometers and Tandem Mass Spectrometers (Quadrupole, Triple Quadrupole, Ion Trap, and Q-trap)

Quadrupole, triple quadrupole, ion trap, and Q-trap mass spectrometers are scanning-type instruments that may be used to obtain low-resolution mass spectra. Triple quadrupole, ion trap, and Q-trap mass spectrometers are also capable of recording tandem mass spectra at low resolution. Since triple quadrupole mass spectrometers actually consist of several instruments connected in tandem, they are known as tandem-in-space tandem mass spectrometers. In contrast, ion trap mass spectrometers carry out tandem MS in multiple steps within the same analyzer and so are defined as tandem-in-time tandem mass spectrometers. Since Q-trap instruments are hybrids in which a linear ion trap has been substituted for the final quadrupole of a triple quadrupole mass spectrometer, this mass spectrometer has both tandem-in-space and tandem-in-time capabilities.

Quadrupole mass spectrometers act as mass filters and set up an oscillating electromagnetic field through which ions of a particular *m/z* value from a beam of ions follow a stable, spiral path from the ion source to an impact-type detector. Each opposing pair of rods in the quadrupole is electrically

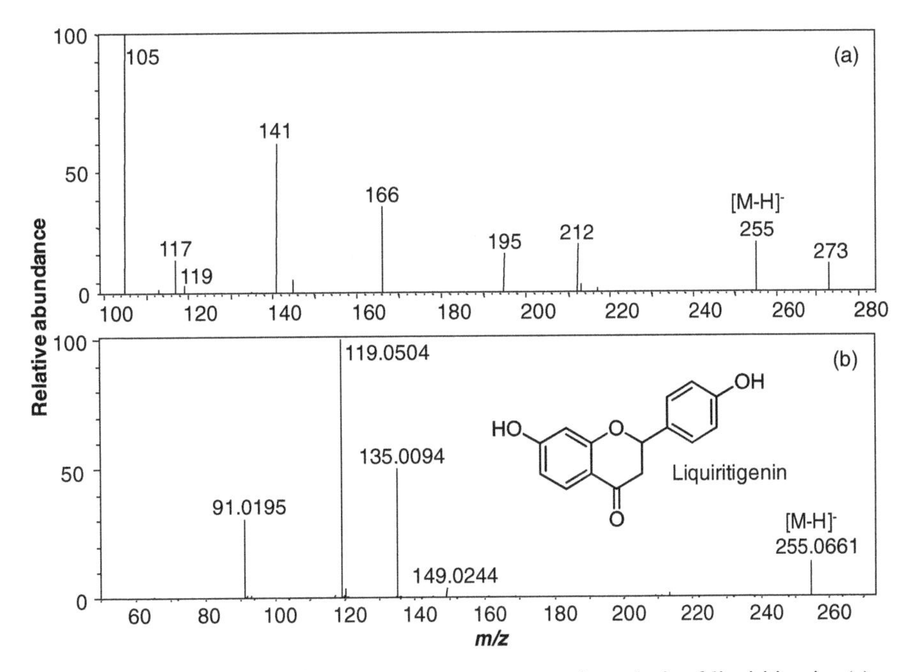

FIGURE 16.3 Negative ion electrospray mass spectrometric analysis of liquiritigenin: (a) mass spectrum recorded during the UHPLC-MS analysis shown in Figure 16.2a and (b) product ion tandem mass spectrum of the deprotonated molecule of liquiritigenin at *m/z* 255. Note that ion of *m/z* 119 is the most abundant product ion in the tandem mass spectrum and also appears as a probable in-source fragment ion of liquiritigenin in the UHPLC-MS spectrum.

connected, and each adjacent rod has an opposing charge. By controlling the voltage and rate of charge inversion on each pair of rods, the *m/z* value of ions passing through the quadrupole is controlled. The equations describing the stable trajectories of ions in a quadrupole mass spectrometer are called Mathieu equations, and the ion physics of these instruments have been described elsewhere (Dawson, 1997).

During tandem MS using a triple quadrupole instrument, two quadrupole instruments are connected via a radiofrequency-only quadrupole that serves as both a collision chamber and ion guide. The CID chamber in the middle section is filled with a collision gas such as helium, nitrogen, or argon at a pressure that is sufficient to facilitate multiple collisions between ions entering from the first quadrupole and the collision gas. These collisions result in fragmentation to produce structurally significant product ions, which continue into the final quadrupole for analysis. If the CID gas pressure is too high, the ion beam is quenched or scattered, and the signal is lost. If the CID gas pressure is too low, then insufficient fragmentation occurs.

Like other types of tandem mass spectrometers, triple quadrupole mass spectrometers may be used to obtain product ion tandem mass spectra. In this mode, the first quadrupole is set to transmit ions of a particular *m/z* value to the collision chamber, and the final quadrupole is operated in scan mode so that product ion tandem mass spectra may be recorded. Product ion tandem mass spectra provide information about fragmentation patterns of a preselected precursor ion with minimal background noise.

If the first quadrupole of a triple quadrupole mass spectrometer is scanned while the final quadrupole is set to transmit ions of a specific *m/z* value, this is called precursor scanning. Precursor ion tandem mass spectra facilitate the identification of ions within mixtures that fragment to form specific product ions. This type of MS/MS scanning is helpful for the detection of compounds in mixtures that share a common fragment ion related to a structural feature such as a sulfate group or a glutathione moiety. If the first and last quadrupoles of the triple quadrupole mass spectrometer are scanned while being offset by a specific *m/z* value, then constant neutral loss tandem mass spectra are produced. Constant neutral loss tandem mass spectra are particularly useful for identifying ions within mixtures that fragment to eliminate a common structural unit that may be small such as water or carbon dioxide or large such as glucuronic acid, palmitic acid, and so on. As mentioned in the previous section, triple quadrupole mass spectrometers are also useful for SRM studies, in which the first and last quadrupoles are not scanned but instead are set to transmit specific precursor and product ions, respectively. Triple quadrupole mass spectrometers are particularly efficient for SRM experiments since no data are lost scanning unimportant ions; although less sensitive in scan modes than in SRM mode, triple quadrupole instruments are still quite useful for constant neutral loss scanning and precursor ion scanning during LC-MS/MS.

Ion trap mass spectrometers may be cylindrical (3D) or linear (2D) in design and have similarities to quadrupole instruments except that the ions do not travel through the instrument in a continuous beam. Instead, ions are stored (trapped) within corkscrew-shaped orbits within the ion trap while the voltage and polarity of the surrounding rods are oscillated as in a quadrupole mass spectrometer. The ion trapping stage may be timed to fill the trap with a constant number of ions for optimum mass accuracy. Alternatively, maximum numbers of ions may be accumulated and stored for analysis, which helps to amplify weak ion currents and enhance sensitivity, but can lead to poor mass assignments in the mass spectrum due to space-charge effects (Qiao et al., 2011). In the cylindrical ion trap, ions are focused in a spherical volume, whereas ions are stored in a cylindrical volume of the linear ion trap, which provides for greater storage capacity before space-charge effects become significant. The ions stored in the trap are not measured, and mass spectra are not recorded, until the ions are sequentially ejected axially from the trap toward an impact detector. For additional information regarding the design and function of ion trap mass spectrometers, see March and Todd (2005).

For tandem-in-time analyses, precursor ions of the desired *m/z* value are retained in the trap while ions of higher and lower *m/z* values are ejected. CID is carried out by allowing a low pressure of helium to enter the trap. Product ions formed during CID remain within the trap until ejection and measurement by the detector. Not only can ion trap mass spectrometers be used for product ion MS/MS (MS^2) measurements, but additional stages of tandem MS (MS^n) are possible. At each stage, a precursor ion is selected by ejecting unwanted ions from the trap, CID is carried out, and product ions are stored in the trap. The product ions can be ejected and measured as a tandem mass spectrum (MS^2), or a particular product ion may be selected for another round of CID (MS^3), and so on; MS^n is particularly useful for obtaining structural information for unusually stable product ions, but is not suitable for neutral loss and precursor ion scanning.

Tandem mass spectra recorded using a triple quadrupole mass spectrometer often show more types of fragmentation and products ions of higher abundance (especially at lower *m/z* values) than those obtained using an ion trap. The reasons for these differences are twofold. First, triple quadrupole mass spectrometers use slightly higher energy CID than do ion traps (although CID in both systems is considered low energy CID (<50 eV) so that more fragmentation occurs and more types of bonds are dissociated. Second, ion traps do not trap low-mass ions as efficiently as high-mass ions, so that ions of low *m/z* value typically have lower abundance in ion trap tandem mass spectra than in corresponding triple quadrupole tandem mass spectra. Hybrid Q-trap mass spectrometers are available that can select ions in the first quadrupole like a triple quadrupole mass spectrometer and fragment them at triple quadrupole CID energy levels. In the final stage of these hybrid instruments, Q-trap mass spectrometers are able to trap and store ions in a linear ion trap for MS^n measurements. Alternatively, the linear ion trap instruments may be used as a single quadrupole, a triple quadrupole, or as an ion trap mass spectrometer.

16.2.2 High-Resolution Mass Spectrometers (Tof, Q-Tof, Tof-Tof, Ion Trap-Tof, FTICR, Orbitrap)

In MS, resolving power is defined as $M/\Delta M$ (resolution is the inverse of this term, $\Delta M/M$), where M is the *m/z* value of a singly charged ion and ΔM is the difference (measured in *m/z*) between M and the next highest ion. Alternatively, ΔM may be defined in terms of the width of the peak. High resolution is typically regarded as a resolving power of at least 10 000 such that the molecular ions of most drug-like molecules (i.e., compounds with molecular masses less than $\sim$500) can be resolved from each other. After resolving a sample ion from others in a mass spectrum, an accurate mass measurement may be carried out by comparing the *m/z* value of the unknown to that of a calibration standard. Accurate mass measurements within 10 ppm of a theoretical elemental composition are generally considered confirmative of the composition of low-mass compounds (MW $<$ 1 000). The types of mass spectrometers capable of exact mass measurements include reflectron Tof instruments, QqTof hybrid mass spectrometers, ion trap-Tof hybrid mass spectrometers, Orbitraps, and Fourier transform (FT) ion cyclotron resonance (FTICR) mass spectrometers (Table 16.1).

Tof mass spectrometers equipped with reflectron capability provide high-resolution mass spectra (Table 16.1) with mass accuracy usually within 5 ppm. Tof mass spectrometers are especially useful when chromatographic peak widths are extremely narrow as in UHPLC-MS because of their ability to obtain complete mass spectra on a microsecond timescale without conventional scanning. This feature improves the instrument duty cycle during data acquisition. Another unusual feature of Tof mass spectrometers is their theoretically unlimited mass range, which is typically limited only by the sensitivity of the detector for high-mass ions.

Tof mass spectrometers operate by accelerating a focused packet of ions to several thousand volts and then directing them out of the ion source into a field-free drift tube toward an impact-type detector. Ideally, all ions leaving the ion source have the same energy, but unfortunately, this ideal is not yet realized. Therefore, a reflectron device is included at the end of the drift tube to help reduce the energy spread of the ions and to eliminate neutrals formed by post-source decay. The reflectron directs the energy-focused ions back down the drift tube again to the detector for measurement.

Since all the ions of the packet that leaves the ion source have approximately equal momentum, and this momentum is the product of mass and velocity ($\mathbf{p} = mv$), the velocity of each ion is inversely proportional to its mass. Therefore, low-mass ions travel the fastest through the drift tube and strike the detector first, and high-mass ions arrive last. The Tof of each ion is recorded and converted to an *m/z* value according to the equation, $m = (2E/d^2)t^2$, where t is the Tof, d is the length of the flight tube, and E is the accelerating energy applied to the ions. For additional details on the design and operation of Tof mass spectrometers, see Cotter (1997).

Tandem Tof (Tof-Tof) mass spectrometers are available that provide high-energy CID for exceptionally informative product ion tandem mass spectra. However, these instruments are usually designed for use with matrix-assisted laser desorption ionization, which is a pulsed ionization source perfectly matched for the pulsed nature of Tof mass spectrometers. Although continuous-beam ionization sources such as electrospray are ideal for LC-MS, they are difficult to match with Tof mass spectrometers that require packets of ions of equal energy. However, this problem has been overcome by the use of hybrid mass spectrometers such as the QqTof MS (Chernushevich et al., 2001) and ion trap-Tof MS (Douglas et al., 2005) that combine quadrupoles or ion traps with Tof analyzers. In these hybrid instruments, LC-MS ion sources such as electrospray, atmospheric pressure chemical ionization (APCI), or atmospheric pressure photoionization (APPI) (see description of LC-MS ion sources in Section 16.3) are interfaced to a quadrupole or an ion trap mass spectrometer that is then connected to a Tof mass analyzer for high-resolution measurements (Table 16.1). For high-resolution tandem MS, CID may be carried out in the second, rf-only quadrupole of a QqTof mass spectrometer (as in a triple quadrupole instrument), and then the Tof analyzer provides high-resolution measurements of the product ions. Like an ion trap, the ion trap-Tof mass spectrometer is capable of MS^n measurements, but unlike the ion trap, the Tof mass spectrometer then provides high-resolution accurate mass measurements of the product ion. Although Tof-Tof mass spectrometers are capable of high-energy CID, QqTof and ion trap-Tof instruments can only carry out low-energy CID, which means that fragmentation will be less extensive and not all types of fragment ions can be observed.

FTICR mass spectrometers provide the highest resolving power and mass accuracy of any type of mass spectrometer that can exceed 200 000 (Table 16.1) (Marshall et al., 1998). In the FTICR mass spectrometer, ions are trapped in circular, corkscrew orbits as in the ion trap, but the orbits are defined by a strong external magnetic field produced by a superconducting magnet. The period of each orbit is determined by the *m/z* value of the ion. Instead of detecting ions as they strike an impact detector, as in quadrupole, ion trap, and Tof instruments, ions are measured in FTICR mass spectrometers while still in orbit. The ions create images of themselves as they orbit within the conductive walls of the FTICR cell, and the frequency of this image current is proportional to the *m/z* value of the ions. The amplitude of the signal is proportional to the number of ions. Fourier transformation of the complex signal, which represents the sum of multiple signals of different frequencies produced by ions of different *m/z* values, is necessary to deconvolute the signal and produce a mass spectrum. Like ion trap instruments, FTICR mass spectrometers are capable of MS^n measurements, except that each stage of MS is carried out at ultrahigh resolving power.

Since each scan requires >1 s (Table 16.1), FTICR mass spectrometers are unsuitable for use in combination with fast chromatography or UHPLC. However, FTICR mass spectrometers are useful for capillary LC-MS studies in which the chromatographic peaks are many seconds in width. Hybrid ion trap-FTICR mass spectrometers have been particularly useful in LC-MS applications involving drug metabolism and proteomics. As FTICR mass spectrometers are the most expensive instruments to purchase and maintain, many laboratories requiring ultrahigh resolution MS are turning to Orbitrap mass spectrometers as alternatives.

The Orbitrap mass spectrometer was first described by Makarov (2000). The instrument traps and measures ions without quenching them in a manner similar to the FTICR, except that a superconducting magnet is not required (Hu et al., 2005; Makarov and Scigelova, 2010). Trapped between an inner and an outer electrode using a quadro-logarithmic electrostatic potential, ions revolve around the central electrode and oscillate harmonically along its axis with a frequency characteristic of their *m/z* values. The image current signal of these oscillations is converted into a frequency using FT in a manner similar to that used in the FTICR. The Orbitrap mass analyzer has demonstrated high resolving power and high mass accuracy (Table 16.1).

The Orbitrap is capable of MS^n measurements with high mass accuracy, although the resolving power of the Orbitrap is less that of the FTICR mass spectrometer (Table 16.1). The scanning rate of the Orbitrap is $\sim$10-fold faster than the FTICR ($\geq$10 scan/s) and is compatible with online HPLC analysis. However, this scan rate is still too slow for most fast chromatography and UHPLC applications. Like the ion trap-FTICR, ion trap-Orbitrap hybrid mass spectrometers have become popular for proteomics studies and some drug metabolism applications.

16.3 IONIZATION TECHNIQUES FOR UHPLC-MS

Electrospray (Figure 16.4) and APCI (Figure 16.5) have become the most widely used ionization sources and HPLC-interfaces for drug discovery using MS. Unlike early LC-MS interfaces and ionization sources such as thermospray, particle beam, and continuous-flow fast atom bombardment, electrospray and APCI interfaces operate at atmospheric

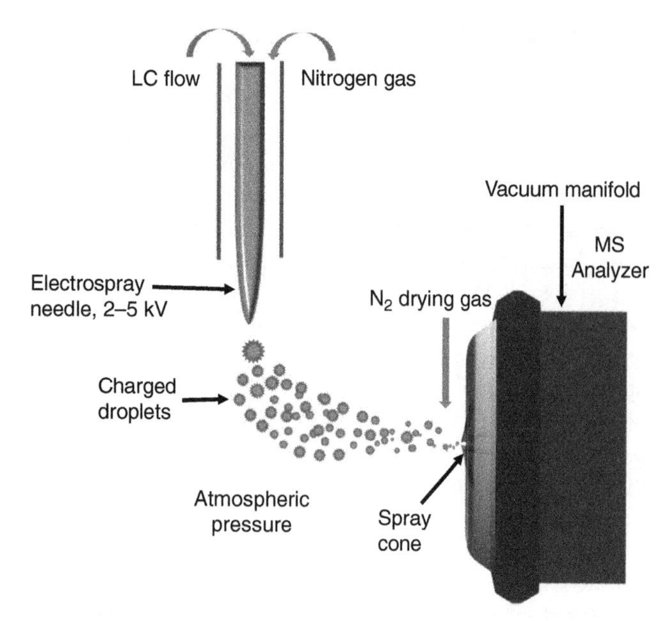

FIGURE 16.4 Schematic of an electrospray source. During LC-MS with electrospray, nitrogen is used as a nebulization and drying gas to facilitate the formation of small droplets that become charged as they emerge from a Taylor cone and electrospray needle at high potential. As the solvent evaporates, analytes becomes gas phase ions that are accelerated into the mass spectrometer through a potential gradient. In some designs, heated nitrogen is directed across the entrance to the mass spectrometer as a curtain gas to provide additional desolvation and to prevent the solvent from entering the mass spectrometer.

pressure, do not depend upon vacuum pumps to remove solvent vapor, and are compatible with a wider range of HPLC mobile phases and flow rates. Like all LC-MS systems, the solvent system for electrospray and APCI should contain only volatile solvents, buffers, or ion pair agents to reduce fouling of the mass spectrometer ion source. In general, electrospray and APCI form abundant molecular ion species. When fragment ions are formed, they are usually more abundant in APCI than electrospray mass spectra.

16.3.1 Electrospray

During electrospray, the HPLC eluate is sprayed through a capillary electrode at high potential (usually 2000–7000 V) to form a fine mist of charged droplets at atmospheric pressure (Figure 16.4). As the charged droplets migrate toward the opening of the mass spectrometer due to electrostatic attraction, they encounter a crossflow of heated nitrogen that increases solvent evaporation and prevents most of the solvent molecules from entering the mass spectrometer. Molecular ions, protonated, or deprotonated molecules, and cationized species such as $[M+Na]^+$ and $[M+K]^+$ can be formed. The relative abundance of each of these species depends upon the chemistry of the analyte, the pH, the presence of proton donating or accepting species, and the levels of trace amounts of sodium or potassium salts in the mobile phase. For additional information on electrospray ionization, see Cole (1997). An example of the C_{18} reversed phase UHPLC-negative ion electrospray mass spectrometric analysis of the natural product liquiritigenin from *G. uralensis* (licorice) is shown in Figure16.2 and 16.3. In this

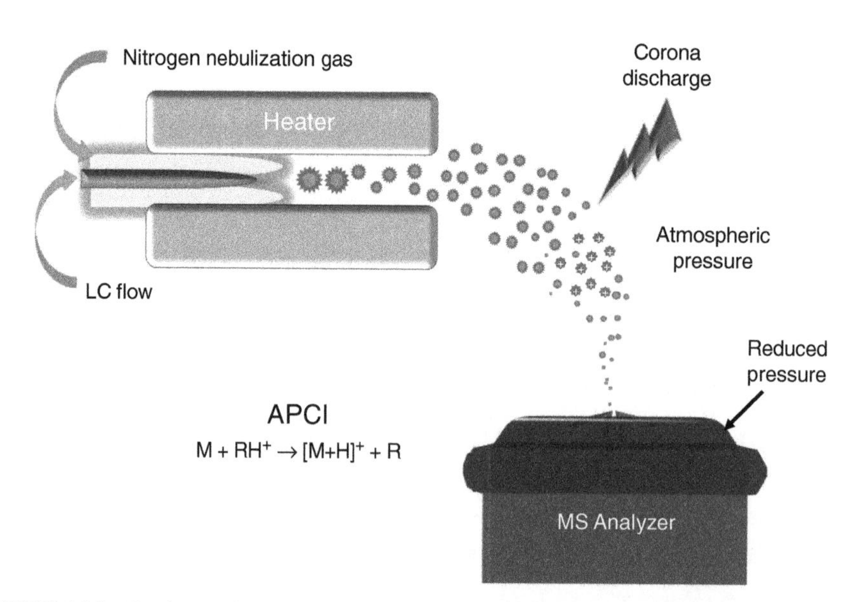

FIGURE 16.5 During APCI, eluate from the UHPLC system is sprayed through a heated capillary and desolvated using heated nitrogen. Functioning as a CI reagent gas, solvent molecules are ionized by a corona discharge and then ionize sample molecules through proton transfer or charge exchange.

example, liquiritigenin was detected as its deprotonated molecule with minimal fragmentation during electrospray UHPLC-MS (Figure 16.3a). During product ion MS-MS with CID, abundant fragment ions were observed (Figure 16.3b).

In addition to singly charged ions, electrospray is unique as an ionization technique in that multiply charged species are common and often constitute the majority of the sample ion abundance. In contrast, most other ionization techniques (including APCI, matrix-assisted laser desorption ionization and APPI) produce singly charged species. A consequence of forming multiply charged ions is that they are detected at lower *m/z* values (i.e., $z > 1$) than the corresponding singly charged species. This has the benefit of allowing mass spectrometers with modest *m/z* ranges to detect and measure ions of molecules of exceedingly high mass. For example, electrospray has been used to measure ions with molecular masses of hundreds of thousands or even millions of daltons on mass spectrometers with *m/z* ranges of only a few thousand. For additional information about theory and applications of electrospray, see Cole (1997).

16.3.2 APCI

The APCI ion source and LC interface (Figure 16.5) uses a heated nebulizer to form a fine spray of the mobile phase. Note that the droplets formed during APCI are considerably larger than those formed during electrospray. Heated nitrogen gas is used to facilitate the evaporation of solvent from the droplets. The resulting gas-phase sample molecules are ionized by collisions with solvents ions, which are formed by a corona discharge in the atmospheric pressure chamber. Molecular ions, $M^{+\cdot}$ or $M^{-\cdot}$, and/or protonated or deprotonated molecules can be formed. The relative abundance of each type of ion depends upon the sample itself, the mobile phase, and the ion source parameters. Next, ions are accelerated into the mass spectrometer analyzer for measurement through a narrow opening or skimmer that helps the vacuum pumps to maintain low pressure inside the analyzer, while the APCI source remains at atmospheric pressure. For more information on APCI and other atmospheric pressure ionization techniques, see Covey et al. (2009).

In general, APCI facilitates the ionization of nonpolar and low molecular mass species while electrospray is more useful for the ionization of polar and higher mass compounds. In this sense, APCI and electrospray are often complementary ionization techniques. However, during the analysis of large or diverse combinatorial libraries, both polar and nonpolar compounds are usually present. As a result, no one set of ionization conditions using APCI or electrospray is adequate to detect all the compounds contained in some libraries of compounds. To help address situations when neither electrospray nor APCI are effective for the ionization of a class or library of compounds, APPI is available.

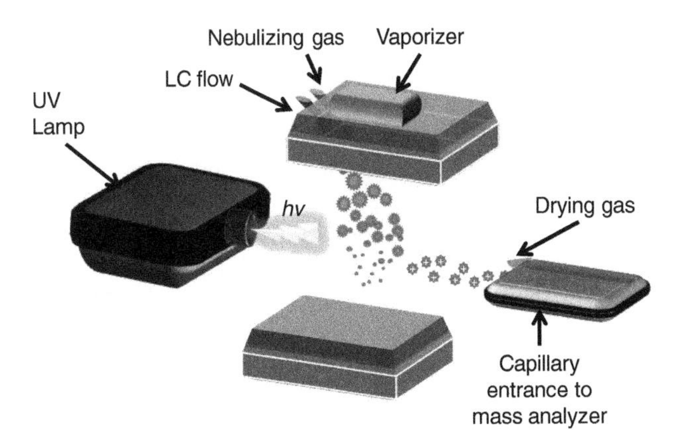

FIGURE 16.6 Schematic of an APPI source. During LC-MS with APPI, nitrogen is used as a nebulization and drying gas to facilitate the formation of small droplets as in APCI. As the solvent evaporates, analytes become gas-phase molecules and are ionized upon absorption of UV photons. If the analytes do not absorb UV light, then a solvent with a UV chromophore may be added as a dopant to the mobile phase. This dopant becomes ionized in the APPI source upon UV irradiation and then can ionize the analyte by chemical ionization.

16.3.3 APPI

To facilitate the analysis of compounds that do not ionize efficiently using electrospray or APCI, a UV ionization technique called APPI has been developed for use with LC-MS and UHPLC-MS (Raffaelli and Saba, 2003). During APPI, the HPLC eluate is sprayed at atmospheric pressure as in APCI. Instead of using a corona discharge as in APCI, ionization occurs during APPI due to irradiation of the analyte molecules by an intense UV light source (Figure 16.6). Since the analytes become ionized upon absorption of UV photon, the carrier solvent must not absorb UV light at the same wavelengths or interference would prevent sample ionization and detection. Therefore, mobile phase selection is more limited for LC-MS analysis using APPI than for either electrospray or APCI. If the analytes do not absorb UV light, then a solvent with a UV chromophore may be added, usually postcolumn, as a dopant to the mobile phase. This dopant becomes ionized in the APPI source upon UV irradiation and then can ionize the analyte through chemical ionization processes.

16.4 LC-MS IN QUANTITATIVE ANALYSIS

Therapeutic drug monitoring, pharmacokinetics studies, testing for drugs of abuse, and so on require quantitative analysis of drugs and/or drug metabolites in complex biological fluids such as serum, plasma, and urine. Since most therapeutic agents are not volatile enough for analysis using gas chromatography-MS, HPLC and UHPLC are used primarily for the quantitative analysis of therapeutic agents. Compared with UV or fluorescence as LC detectors, MS is more selective for analyte detection by providing mass as an additional level of characterization. With the availability of electrospray, APCI, or APPI for ionization (see Section 16.3), virtually all type of drugs, drug metabolites, and biological molecules may be measured using LC-MS/MS. Therefore, LC-MS (and UHPLC-MS) may be considered a universal method for quantitative analysis of therapeutic agents.

By combining the selectivity of chromatography, mass selection (MS), and product ion selection (MS/MS), LC-MS/MS is one of the most specific analytical techniques available for quantitative analysis. Note that specificity may be considered to be the measurement of an analyte without interference from other compounds. As shown in Figure 16.2, UHPLC-MS/MS using SRM provides significantly more selectivity than does UHPLC-MS. Not only does the use of tandem MS eliminate peaks from potentially interfering compounds, but it reduces background noise and thereby enhances signal-to-noise, which is essential for the quantitative analysis of trace levels of drugs and drug metabolites in serum, urine, and tissues.

Even after partial purification of analyzes using solvent/solvent extract or solid phase extraction, the complexity of these samples still usually requires chromatographic separation of the analytes from the matrix prior to quantitative analysis. When tandem MS was first being used for biomedical analysis, it was hoped that the selectivity of MS-MS would eliminate the need for chromatographic separations. Unfortunately, not all constituents of a mixture ionize with equal efficiency, and some compounds can suppress or enhance the ionization of others in the mixture in a process known as a matrix effect. Therefore, chromatography is indispensible for quantitative analysis of therapeutic agents, natural products, and biomolecules. The elimination of matrix effects is so important for quantitative analysis using LC-MS/MS that most guidelines for the development and validation of analytical methods using this technique require that matrix effects be considered and eliminated (US Department of Health and Human Services, 2001; European Medicines Agency, 2011).

Although MS and MS/MS provide exquisite selectivity, MS cannot always distinguish between isomeric compounds; in this situation, chromatography is essential. For example, Figure 16.7a shows the HPLC-MS/MS analysis of isomeric prostaglandins PGD_2 and PGE_2 extracted from murine bone marrow-derived macrophage. Both PGD_2 and PGE_2 formed abundant deprotonated molecules of *m/z* 351 during negative ion electrospray and then fragmented during CID to form abundant product ions of *m/z* 271; the transition of *m/z* 351 to *m/z* 271 was monitored during HPLC-MS/MS using SRM.

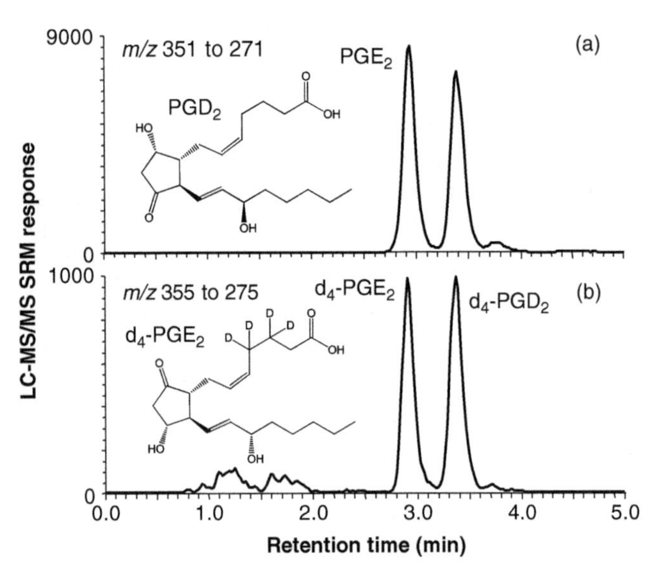

FIGURE 16.7 Negative ion electrospray HPLC-MS/MS with CID and SRM of prostaglandin E_2 (PGE_2) and PGD_2 extracted from a cell culture incubation of mouse bone marrow-derived macrophage. Deuterated analogs of PGE_2 and PGD_2 were added to the samples as surrogate standards for quantitative analysis.

The use of HPLC facilitated the baseline separation of these isomeric prostaglandins for quantitative analysis using electrospray MS-MS with SRM (Figure 16.7a).

The ions used in SRM are selected on the basis of abundance and selectivity. For example, the molecular ion, protonated molecule, or deprotonated molecule (whichever is most abundant) is typically used as the precursor ion, and the most abundant fragment ion or else another abundant ion that provides structural information is selected as the product ion for SRM. This SRM transition is called the quantifier. Often, a second SRM transition is used as a qualifier for the same analysis. The ratio of the signals for the quantifier and qualifier SRM transitions should remain constant for all analysis of a particular compound, whether they are standards or biomedical specimens. If the ratio of SRM signals should deviate significantly for a particular sample, then that analysis should be discarded and the sample reanalyzed. If the deviation remains, then the analytical method is invalid and must be modified and revalidated.

Quantitative analysis using LC-MS/MS is similar to most other quantitative approaches in that a standard curve is required (Figure 16.8). Due to matrix effects as discussed above, the standards should be prepared in the same matrix as the unknown such as blank serum or urine. As in GC-flame ionization, HPLC-UV detection, and so on, an internal standard can be added to the sample immediately before LC-MS/MS analysis to control for variations in detector response or injection volume. Alternatively, only mass spectrometers may utilize a special type of internal standard known as a

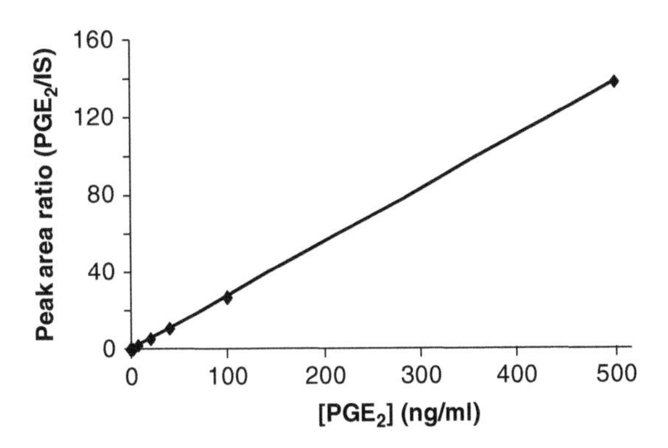

FIGURE 16.8 Standard curve for the quantitative analysis of prostaglandin-E_2 using HPLC-MS/MS with CID and SRM. A constant concentration of d_4-PGE_2 was used as a surrogate standard for PGE_2 and was added to the matrix prior to solvent/solvent extraction from blank medium. The SRM transitions for PGE_2 and d_4-PGE_2 were as shown in Figure 16.7. Note that the signals for PGE_2 were normalized to a constant signal for the surrogate standard d_4-PGE_2 that corrected for sample losses at all stages of sample preparation as well as variations in mass spectrometer response.

surrogate standard that may be added at the beginning of sample preparation.

Labeled with stable isotopes such as ^{13}C, ^{15}N, ^{18}O, and/or deuterium, surrogate standards are otherwise identical to the analyte and share the same chemical reactivity, stability, solubility, and chromatography properties. Although surrogate standards coelute with their corresponding analyte, they are distinguished in the mass spectrometer by their higher masses (Figure 16.7b). For example, the LC-MS/MS analysis of prostaglandins PGD_2 and PGE_2 and their deuterated surrogate standards d_4-PGD_2 and d_4-PGE_2 are shown in Figure 16.7. Note that the unlabeled prostaglandins and their corresponding deuterated surrogate standards coeluted during reversed phase HPLC separation. In the standard curve for the LC-MS/MS analysis of PGE_2 shown in Figure 16.8, note that the signals for PGE_2 on the y-axis were normalized to a constant signal for the surrogate standard. This normalization corrects not only for instrument response and injection variation at the time of analysis (like routine internal standards) but also corrects for sample losses or degradation at any stage of handling such as extraction or pipetting.

16.5 PERSPECTIVES

Looking ahead, LC-MS/MS analyses should continue to benefit from enhancements in chromatographic separations. During the last decade, UHPLC systems became available from multiple manufacturers and were installed in many different laboratories. As UHPLC systems were combined with MS, the speed and chromatographic resolution of UHPLC-MS/MS assays improved significantly compared with the corresponding HPLC-MS/MS assays. This trend should continue as more MS laboratories acquire UHPLC systems and as the performance of UHPLC columns improves.

In this chapter, the most common mass spectrometers used for LC-MS and MS-MS were discussed. Missing from this list were magnetic sector mass spectrometers, since LC-MS using these high-resolution instruments has become rare and is being carried out instead by QqTof, ion trap-Tof, FTICR, and Orbitrap high-performance mass spectrometers. During the next decade, expect that Orbitrap mass spectrometers will gradually replace FTICR instruments for high-resolution LC-MS^n applications. This change will be driven by increasing costs and shortages of helium that are anticipated for superconducting FTICR magnets and by improvements in Orbitrap technology.

Besides improvements in Orbitrap resolving power, speed, and sensitivity, the scan speed and sensitivity of triple quadrupole mass spectrometers should continue to be improved. Enhancement in sensitivity will benefit all modes of action of triple quadrupole mass spectrometers. Increases in scan speed will be particularly beneficial for precursor ion scanning and constant neutral loss scanning modes, for which triple quadrupole mass spectrometers are particularly well suited.

REFERENCES

Chernushevich IV, Loboda AV, Thomson BA. An introduction to quadrupole-time-of-flight mass spectrometry. J Mass Spectrom 2001;36:849–865.

Cole RB, editor. *Electrospray Ionization Mass Spectrometry, Fundamentals, Instrumentation and Applications*. John Wiley & Sons, Inc.; 1997.

Cotter RJ. *Time-of-flight Mass Spectrometry: Instrumentation and Applications in Biological Research*. American Chemical Society; 1997.

Covey RT, Thomson BA, Schneider BB. Atmospheric pressure ion sources. Mass Spectrom Rev 2009;28:870–897.

Dawson PH, editor. *Quadrupole Mass Spectrometry and Its Applications*. Springer-Verlag; 1997.

Douglas DJ, Frank AJ, Mao D. Linear ion traps in mass spectrometry. Mass Spectrom Rev 2005;24:1-29.

European Medicines Agency, Committee for Medicinal Products for Human Use (CHMP): Guideline on Bioanalytical Method Validation, London, United Kingdom, Jul 21, 2011 (Doc. Ref. EMEA/CHMP/EWP/192217/2009).

Hu Q, Noll RJ, Li H, Makarov A, Hardman M, Graham Cooks R. The Orbitrap: a new mass spectrometer. J Mass Spectrom 2005;40:430–443.

Makarov A. Electrostatic axially harmonic orbital trapping: a high-performance technique of mass analysis. Anal Chem 2000;72:1156–1162.

March RE, Todd JFJ. *Quadrupole Ion Trap Mass Spectrometry*. 2nd ed. Wiley-Interscience; 2005.

Marshall AG, Hendrickson CL, Jackson GS. Fourier transform ion cyclotron resonance mass spectrometry: a primer. Mass Spectrom Rev 1998;17:1–35.

Makarov A, Scigelova M. Coupling liquid chromatography to Orbitrap mass spectrometry. J Chromatogr A 2010;1217:3938-3945.

Qiao H, Gao C, Mao D, Konenkov N, Douglas DJ. Space-charge effects with mass-selective axial ejection from a linear quadrupole ion trap. Rapid Commun Mass Spectrom 2011;25:3509-3520.

Raffaelli A, Saba A. Atmospheric pressure photoionization mass spectrometry. Mass Spectrom Rev 2003;22:318-331.

US Department of Health and Human Services, Food and Drug Administration, Center for Drug Evaluation and Research (CDER). *Guidance for Industry: Bioanalytical Method Validation*, 2001. Available at http://www.fda.gov/cder/guidance/index.htm. Accessed Mar 18, 2013.

17

USE OF INTERNAL STANDARDS IN LC-MS BIOANALYSIS

AIMIN TAN AND KAYODE AWAIYE

17.1 INTRODUCTION

Internal standards (ISs) are commonly used in liquid chromatography–mass spectrometry (LC-MS) bioanalysis (Wieling, 2002; Stokvis et al., 2005; Bakhtiar and Majumdar, 2007; Tan et al., 2012a). The main purpose of utilizing ISs is to improve the accuracy and precision of quantitation as well as the robustness of bioanalytical methods. During the treatment and analysis of biological samples (e.g., plasma, serum, blood, urine, and tissue), variations and/or losses can occur. These include transfer loss, adsorption loss, evaporation loss, variation in injection volume and, especially, in MS response due to ion-suppression or enhancement (matrix effect). By adding an equal amount of an IS that has similar physicochemical properties as the analyte to all samples in a batch and using analyte/IS response ratios for quantitation, most of the aforementioned variations and losses can be corrected or compensated for. Therefore, the accuracy and precision of reported concentrations and the reliability of bioanalytical methods can be significantly improved through the proper use of a good IS.

Then, what is a good IS? How should its concentration be determined? When and how should it be added? Why are stable isotope labeled (SIL) ISs preferred yet one should still be cautious in their usage? Should IS responses be monitored during incurred sample analysis? What are the root causes of IS response variations? What are their potential impacts on the integrity of reported concentrations? All these questions will be addressed in this chapter with the focus on small molecules (molecular weights typically less than 1000 Da).

17.2 SELECTION AND USE OF IS

17.2.1 Selection of IS

Generally speaking, there are two types of ISs, that is, structural analog and SIL ISs. Whenever possible, SIL ISs should be used because they are most effective (Viswanathan et al., 2007).

In case a SIL IS is used in quantitative LC-MS or GC-MS analysis, it is sometimes termed as isotope dilution mass spectrometry (IDMS) (Moore and Machlan, 1972). For a SIL IS, its molecular weight should be ideally 4 or 5 Da higher than that of the analyte to reduce isotopic interference (Bakhtiar and Majumdar, 2007), though this is not an absolute necessity, that is, it's dependent on many other factors like chemical purity and concentration range. For example, norethindrone-$^{13}C_2$ was successfully used for the determination of norethindrone over the concentration ranges of 2.5–500 pg/ml and 0.05–10 ng/ml (Li et al., 2005). In addition, among SIL ISs, those labeled with ^{13}C and/or ^{15}N are usually preferable over those labeled with deuterium (^{2}H, D, or d) in terms of performance (Berg and Strand, 2011; GCC, 2011a), despite less difficulty in the synthesis of deuterated ISs and therefore less expense. Finally, the location of stable isotope atoms should be given consideration in synthesizing a deuterated IS, so that deuterium-hydrogen exchange would not occur during sample preparation (Chavez-Eng et al., 2002; Savard et al., 2010).

Despite the desirable performances of SIL ISs, they are not always available or are too expensive. Then, structural analogs can be used as ISs. It is preferable that the key

Handbook of LC-MS Bioanalysis: Best Practices, Experimental Protocols, and Regulations, First Edition. Edited by Wenkui Li, Jie Zhang, and Francis L.S. Tse.

chemical structure and functionalities (e.g., –COOH, $-SO_2$, $-NH_2$, halogens, and heteroatoms) of an analog IS are the same as those of the analyte of interest with difference only being C–H moieties (length and/or position). Modifications in key chemical structure and/or functionalities would result in significant differences in ionization efficiency and extraction recovery (Stokvis et al., 2005). In addition, a structural analog IS should not correspond to any *in vivo* biotransformed products of the analyte (e.g., hydroxylated metabolite and *N*-dealkylation metabolite). An appropriate structural analog IS can usually be found from the same therapeutic class as the analyte or by key chemical structure search. Once potential structural analogs are found, their physicochemical properties, such as log D (hydrophobicity) versus pH, can be calculated and compared with those of the analyte using computer programs (e.g., Pallas) prior to being experimentally tested (Tan et al., 2012a).

Even though an IS is expected to track the analyte of interest in all the three distinctive stages of LC-MS bioanalysis, that is, sample preparation (extraction), chromatographic separation, and mass spectrometric detection, these stages are not of the same importance as far as the selection of ISs is concerned. Furthermore, the relative importance of the above three stages could change as a different strategy is adopted during the method development. In short, all the three stages should be considered as a whole and they all affect the selection of ISs interactively. For example, when the extracts of samples contain coeluting matrix components that cause ion suppression or enhancement, then tracking the analyte during MS detection to avoid, minimize, or eliminate matrix effects becomes more important. On the other hand, different extraction methods may have different requirements for ISs. For instance, the requirement for ISs to track an analyte during a simple "dilution-and-shoot" sample preparation procedure would be less stringent than that for liquid–liquid extraction (LLE) or solid-phase extraction (SPE) methods.

Finally, when an appropriate IS cannot be found, alternative approaches should be taken. For example, in early drug discovery where less strict criteria are used (Timmerman et al., 2010) or when clean extracts are obtained and variations in sample extraction and LC-MS analysis are minimized, it is possible to develop a method without the use of any IS (Wieling, 2002). Another possibility is the use of ECHO peak technique (Zrostlíková et al., 2002; Alder et al., 2004), where the analyte itself is used as its own IS. In this case, shortly (typically 30–50 s) after the injection of an unknown sample, a standard solution is also injected, which results in two peaks for each analyte, one from the unknown sample and the other from the standard solution (an echo peak). By using their response ratio for quantitation, the matrix effect might be compensated for because the two peaks are in close proximity and they may have been affected in the same manner by the coeluted matrix components, which usually have a broader peak shape.

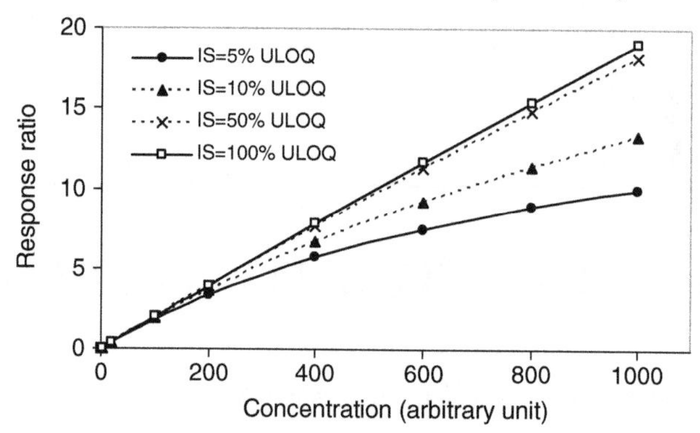

FIGURE 17.1 Simulation results demonstrate that calibration curves become progressively nonlinear with decreasing concentrations of internal standard (cross-contribution from the analyte to the internal standard is equivalent to 5% of the concentration of the analyte). IS, internal standard; ULOQ, upper limit of quantitation (reproduced from Tan et al. (2011) with permission from Elsevier).

17.2.2 Determination of IS Concentration

While no one would argue against the importance of a good IS in LC-MS bioanalysis, no consensus exists regarding the determination of its concentration (the concentration after being mixed with samples) or the amount to be added to each sample. Some prefer a low IS concentration corresponding to one-third of the upper limit of quantitation (ULOQ) (Ansermot et al., 2009), whereas others recommend half of the ULOQ (Sojo et al., 2003; Bakhtiar and Majumdar, 2007) or even higher than the ULOQ (Cuadros-Rodríguez et al., 2007). In contrast, many others simply believe that IS concentration is not important as long as the same amount is added to all samples in an assay batch. However, a recent research has demonstrated that the accuracy and linearity can be significantly impacted when an inadequate IS concentration is used (Tan et al., 2011). As shown in Figure 17.1, for a given degree of cross-signal contribution from an analyte to its IS either due to chemical impurity in reference standard or because of isotopic interference, calibration curves become increasingly nonlinear as the IS concentration decreases. Furthermore, an inappropriate IS concentration can even lead to significant systematic errors in analyzing unknown samples, for example, incurred samples, which may not be seen in quality control (QC) samples. The reason for this is that the source of analyte for the QC samples (analyte reference standard) is different from that of unknown samples (e.g., dosed medication), and they may contain different amounts of IS impurity. Therefore, determining an appropriate IS concentration is not a trivial task. On the other hand, it is, unfortunately, very difficult to set a clear-cut guideline regarding IS concentration. Based on our experience, all the following factors should be taken into consideration.

TABLE 17.1 Minimum Internal Standard Concentrations Required When an Analyte Contributes to the Internal Standard Response in LC-MS

Concentration range of analyte	Weighting factor of regression	Cross-signal contribution (%) from analyte to internal standard and the recommended minimum concentrations (% of the upper limit of quantitation) of the internal standard					
		0%	0.5%	1%	2.5%	5%	10%
1000-fold (e.g., 1–1000 ng/ml)	$1/X^2$	>0	1.6	3.2	7.9	16	32
	1/X	>0	3.5	6.9	18	35	69
	None	>0	168	339	840	1679	3358
500-fold	$1/X^2$	>0	1.6	3.1	7.7	16	31
	1/X	>0	3.3	6.6	17	33	66
	None	>0	84	168	418	836	1671
250-fold	$1/X^2$	>0	1.5	3	7.4	15	30
	1/X	>0	3	5.9	15	30	59
	None	>0	42	83	207	414	828
100-fold	$1/X^2$	>0	1.5	2.9	7.2	15	29
	1/X	>0	2.8	5.6	14	28	56
	None	>0	18	36	88	176	352

Reformatted from Tan et al. (2011) with permission from Elsevier.

The first factor to be considered is the magnitudes of cross-signal contribution between an analyte and its IS due to chemical impurity and/or isotopic interference. Although not absolutely necessary (Tan et al., 2011), the generally accepted criteria are equal to or less than 20% of the lower limit of quantitation (LLOQ) response and 5% of the IS response for IS-to-analyte and analyte-to-IS contributions, respectively, which are the same as those recommended in regulatory guidelines for method selectivity or specificity (ANVISA, 2003; EMA, 2011). Based on these criteria and the magnitudes of cross-signal contribution, the minimum IS concentration required ($C_{\text{IS-min}}$) and the maximum IS concentration allowed ($C_{\text{IS-max}}$) can be calculated using Equations (17.1) and (17.2), respectively (Tan et al. 2012a):

$$C_{\text{IS-min}} = m \times \text{ULOQ}/5, \quad (17.1)$$

$$C_{\text{IS-max}} = 20 \times \text{LLOQ}/n, \quad (17.2)$$

where m and n represent the percentages of cross-signal contributions from analyte-to-IS and IS-to-analyte, respectively. In addition, as shown in Figure 17.1, calibration curves can become progressively nonlinear when the IS concentration is lowered. Depending on the calibration range, weighting factor of calibration regression, and the degree of cross-contribution from analyte to its IS, another minimum IS concentration exists for each of the different combinations (Table 17.1), lower than which the acceptance criteria of accuracy for calibration standards (CSs), that is, bias within $\pm 20\%$ at the LLOQ and within $\pm 15\%$ for the rest, cannot be met when linear calibration is used for the given combination. Fortunately, the minimum IS concentrations listed in Table 17.1 are usually automatically satisfied when Equation (17.1) is used to determine the minimum IS concentration, except for cases that nonweighted linear regression is used. For instance, when the cross-signal contribution from analyte to IS is 2.5%, the minimum IS concentration calculated according to Equation (17.1) is 50% of the ULOQ, which is higher than the corresponding values listed under the column heading of 2.5% in Table 17.1, except those associated with nonweighted regression. Hence, if a nonweighted linear regression is ever to be used, higher IS concentration should be used and the corresponding minimum IS concentrations in Table 17.1 must be satisfied. Moreover, when there is the aforementioned systematic error in the analysis of unknown samples, high IS concentration is usually helpful to reduce the error (Tan et al., 2011).

The second factor is MS detection sensitivity toward the analyte and the IS. When the sensitivity is relatively high for the IS, then its concentration can be lowered. Otherwise, high IS concentration should be used to achieve adequate signal-to-noise ratio to reduce the impact of random detection noise in IS response.

The third factor is ion suppression or enhancement. When there is matrix effect, the more closely an IS coeluates with the analyte, the better the matrix effect would be compensated for. As the concentration of analyte in unknown samples varies while the amount of IS added is constant, a choice must be made to match which part of a calibration curve. Generally speaking, the segment from one-third to half of the ULOQ is more important because it is expected to cover the average C_{max} for most drugs and metabolites. This might be the reason why some researchers have proposed to use an IS concentration around one-third or half of the ULOQ. In case analyte signal is suppressed by a coeluting IS, low IS concentration should be used to maintain a low detection limit for the analyte. On the other hand, a high IS concentration is necessary to obtain good reproducibility when the IS signal is suppressed by the analyte (Liang et al., 2003; Sojo et al., 2003).

The fourth factor, which is often overlooked, is the linearity of IS response within the range of expected concentrations (Hewavitharana, 2011). The commonly used response ratio method of calibration is based on single-point calibration for the IS, which assumes that the IS response is linear within the range of IS concentrations expected in all samples analyzed. In cases of ion suppression/enhancement or recovery variation, the range of responses obtained for the IS can be wide. Hence, ensuring the linearity of IS response over a wide range of concentrations above and below the selected IS concentration is critical for obtaining accurate results.

In addition to the above, other factors need to be considered as well, such as solubility, loading capacity, and regression precision. For example, the IS concentration should not be so high as to cause issues of solubility or exceed the loading capacity of SPE cartridges or similar products. The ratios of analyte-to-IS responses should also be appropriate, for example, not too small at the low concentration end. Otherwise, when an inappropriate regression algorithm or calculation precision is used, unreliable or even wrong regression results may be obtained.

To conclude, by simply adding the same amount of an IS to all samples in an assay batch is not enough for ensuring good accuracy. Its concentration must be properly chosen as well because an inappropriate one could affect linearity, accuracy, and precision. On the other hand, there is no clear-cut guideline to follow. As discussed earlier, many different factors should be considered. Some have very specific criteria to be met, like cross-signal contributions, while others do not. In general, a high IS concentration is preferable in order to improve the linearity of calibration curve, to match high concentration samples (usually more important), to reduce the potential systematic error in the analysis of unknown samples, or to have adequate signal-to-noise ratios for the IS. However, some situations dictate a lower one, for example, ion suppression by a coeluting IS or high MS detection sensitivity for the IS. Therefore, the most important thing is to understand how different factors affect the selection of IS concentration and assay performance, and to know how to adjust IS concentration when desired performances are not obtained.

17.2.3 Addition of IS

Generally speaking, the earlier an IS is added, the better the correction can be for the possible variability and/or analyte losses in LC-MS analysis. This is why an IS is usually added to samples prior to extraction, for example, before the addition of a buffer and organic solvent in LLE. In this way, the IS goes through most of the sample extraction and LC-MS analysis steps together with the analyte of interest. Any variations, if not all, that occur to the analyte in these processes are expected to be compensated for by the IS. This might be the reason why the IS was initially termed "processed IS" (Wieling, 2002; Stokvis et al., 2005).

Sometimes, it may be difficult to add an IS early. For example, in the quantitation of a drug in both free and liposome-encapsulated forms in biological samples, the IS was added only after the separation of the two analyte forms by SPE to reduce the potential impact of IS addition (as well as its solvent) on the integrity of fragile liposomes (Lee et al., 2001; Viel et al., 2010).

On the other hand, an IS can be introduced even after chromatographic separation to compensate mainly for ion suppression or enhancement during MS detection (Choi et al., 1999). This approach is useful to avoid using multiple ISs in a multicomponent analytical method and to seek the benefit of matrix effect correction as a result of coelution of an analyte and the IS (even with a structural analog IS). However, one must make sure that the variability and analyte losses, if any, during sample preparation and LC separation are minimized.

In the analysis of dried blood spot (DBS) or dried matrix spot (DMS) specimens, the addition of IS can be much more complicated (Abu-Rabie et al., 2011; Meesters et al., 2011) due to the 'dry' nature of these samples. The possible options of IS addition include (a) mixing the IS with liquid samples first and then spotting their mixtures; (b) spotting the IS first followed by spotting liquid samples (both individually spotted); (c) spotting or spraying the IS after liquid samples have been spotted; and (d) adding the IS in a soaking solvent at the start of sample extraction (currently the most commonly used approach). Nevertheless, the same principle mentioned in the beginning of this section (the earlier an IS is added, the better the correction for variability and losses can be) holds true for DBS/DMS analysis. In other words, the best results in terms of recovery and accuracy are usually obtained when an IS is added to liquid samples prior to spotting (Meesters et al., 2011).

17.2.4 Chromatography and MS Detection of IS

In LC separation, coelution of an IS with the analyte of interest is usually preferred in order to reduce the impact due to matrix effect on the quantitation of the analyte in MS detection and/or to expand linearity range. This practice is particularly important when structural analog ISs are used (Kitamura et al., 2001; Shi, 2003). For MS detection, the same type of ionization and multiple reaction monitoring transitions should be used for both the analyte and its IS. In the case of SIL ISs, it is also preferable that some, if not all, of the stable isotope atoms remain in the product ions to avoid possible cross-contamination during MS/MS detection, though it is rare with modern mass spectrometers (Morin et al., 2011).

17.3 PERFORMANCE OF IS

17.3.1 SIL IS versus Structural Analog One

Many reports have demonstrated that SIL ISs outperform structural analog ones in terms of assay precision and accuracy provided that the sample extraction and LC separation are comparable (Jemal et al., 2003; Stokvis et al., 2005; Taylor et al., 2005; Lanckmans et al., 2007; O'Halloran and Ilett, 2008). Moreover, SIL ISs may enable the extension of assay dynamic ranges, use of less reliable precursor to product ion transitions (e.g., loss of water) in MS/MS detection, and prolongation of in-processing and post-preparation stabilities (Stokvis et al., 2005; Lanckmans et al., 2007).

17.3.2 Precautions when using SIL IS

Since the performance of SIL ISs is usually very good and their physicochemical properties are almost identical to those of the analyte, some people may tend to believe that SIL ISs are a magic solution to all LC-MS bioanalysis problems. In fact, due to the complexity of biological samples, some precautions should still be taken when using a SIL IS, in particular, the deuterated ones. Discussed below are some potential pitfalls or situations where the performance of SIL ISs may not be as expected. It should be pointed out that the performance of structural analog ISs can also be a concern in some of the following situations, even more serious.

The first is when there is a difference in retention time (albeit small) between the analyte and its deuterated IS. In case the elution of the analyte and IS happens to overlap with the steep rising or declining edges of matrix component(s) that can cause ion-suppression or enhancement (Figure 17.2), significant error in quantitation may occur because of differential ion-suppression or enhancement (Wang et al., 2007; Lindegardh et al., 2008; Zhang and Wujcik, 2009). This kind of differential ion-suppression or enhancement can also result in significant difference between initial concentrations and those obtained after reinjection on a different LC-MS system, though the results of all CSs and QC samples were accepted in both runs (Tan et al., 2009a). It should be advised that the difference in retention time between an analyte and its deuterated IS can be even wider when the number of deuterium substitution is increased and/or the resolving power of LC is improved, such as through the use of ultra performance liquid chromatography (Berg and Strand, 2011).

The second is possible exchange between deuterium and hydrogen atoms during sample processing, such as in water-containing solvent or human plasma (Chavez-Eng et al., 2002). The position of deuterium atoms in the molecule of

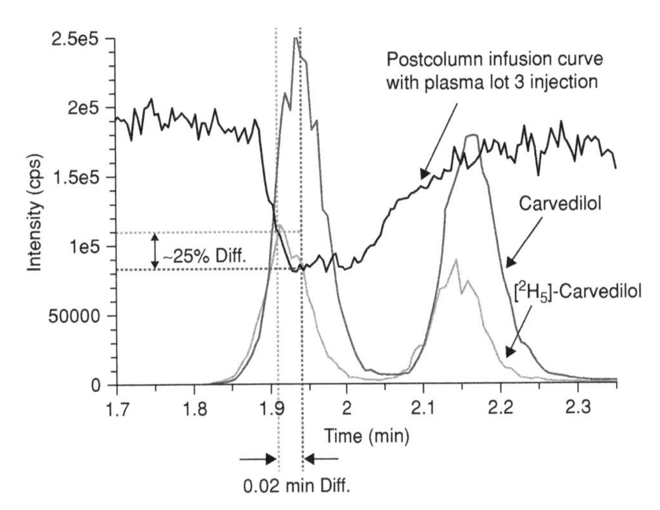

FIGURE 17.2 Profile of postcolumn infusion of carvedilol with an injection of control plasma extract (lot 3, the problematic lot) overlaid with the LC-MS/MS chromatograms of carvedilol and its deuterated internal standard (D_5-carvedilol) to demonstrate a significant difference in ion suppression (~25%) due to even a very small difference in retention time (0.02 min) between carvedilol-S (1.93 min) and its deuterated internal standard (1.91 min) (reproduced from Wang et al. (2007) with permission from Elsevier).

the IS may have an impact on this exchange. For example, it was reported that a significant loss of deuterium atoms was observed only when deuterium atoms are adjacent to a keto group of a given molecule. This is apparently due to the possible conversion between keto- and enol- forms (Savard et al., 2010).

The third is possible significant difference in extraction recovery between an analyte and its deuterated IS. For example, the extraction recoveries of haloperidol and haloperidol-d_4 were reported at 72% and 44%, respectively (Wieling, 2002). The large difference was believed due to the difference in physicochemical properties, such as p*K*a, or because of the aforementioned hydrogen-deuterium exchange during sample processing.

The fourth is potential quantitation bias due to nonuniform matrix effects between CSs and QCs/unknown samples (Liu et al., 2010). For example, when there is significant ion enhancement matrix effect for CSs, the calibration curve may be nonlinear. Hence, a quadratic regression model is used. However, the ion enhancement matrix effect in unknown samples and/or QCs may be less pronounced or even on the contrary, it is ion suppression matrix effect. Then, the analyte response vs. concentration relationship may be linear in QC and/or unknown sample matrix. Quantitation bias for QC/unknown samples can occur when they are quantified by a quadratic calibration curve. To evaluate the potential risk in this type of situation, a diagnostic factor Q (Q = ULOQ $(-A/B)$) has been proposed by Liu et al., where A and B are the parameters before X^2 and X in the quadratic model of

TABLE 17.2 Differences in Various Bioanalytical Perspectives between CSs/QCs and Incurred Samples

Item	CSs/QCs	Incurred samples
Screening criteria for matrix sources	Usually loose	Usually specific and strict, for example, age 40–50 and nonsmoker or others, dependent on the objectives of a study
No. of lots/sources	Usually more than one source (pooled)	One single source
PH	Averaged due to pooling	Variable
Extra components associated with the investigated medication	None or limited, especially after blank screening	Presence of variable levels of metabolite(s), comedication and non-active ingredients in drug formulation, depending on subject health conditions
Amount collected	Usually large, for example, 200 ml per collection	Usually small, for example, 7 ml per sampling time
Number of freeze/thaw cycles prior to being extracted	Usually 2 or more (when the prepared CSs/QCs are used for sample analysis)	Usually 1 (initial analysis) or 2 (repeat analysis or ISR)
Storage tube and preuse storage	Usually stored at −20°C and without special protection until being selected for a specific study	Could be collected under sodium light and stored at −80°C immediately after collection
Amount of anticoagulant	May be different because of different amounts collected	

CS, calibration standard; QC, quality control; ISR, incurred sample reanalysis.
Reformatted from Tan et al. (2009a) with permission from Elsevier.

$Y = AX^2 + BX + C$, respectively. This Q factor can be modified to Q = ULOQ/LLOQ ($-A/B$) for a more general application, that is, nonunit LLOQ cases. When it is larger than the critical value for a given concentration range and calibration scheme, quantitation bias can be significant for some samples in the calibration range (Tan et al., 2012b).

Finally, when a SIL IS is used for the analysis of more than one analyte, for example, a SIL drug being used as the IS for the parent drug and its metabolite, significant bias in quantitation may exist for the metabolite. This is because the ion-suppression or signal enhancement, if any, posed to the metabolite during LC-MS can be very different from those borne by the IS. The latter may be also subjected to the ion-suppression or signal enhancement by the parent drug (Jian et al., 2010; Remane et al., 2010). As a result of the effect, the response of SIL IS can vary due to the presence of the variable amount of parent drug in incurred samples. In other words, metabolite/IS response ratios are biased by parent drug concentration in unknown samples. However, it should be mentioned that the SIL IS of the parent drug is in fact a structural analog IS to the metabolite.

Since the aforementioned pitfalls or situations (except the last two) are mainly associated with deuterated ISs, it is therefore preferable to use the SIL ISs labeled with ^{13}C and/or ^{15}N (Berg and Strand, 2011; GCC, 2011a).

17.3.3 Variation in IS Response during Incurred Sample Analysis

There have been mixed expectations regarding IS response variations in LC-MS bioanalysis community. As the main purpose of utilizing ISs is to correct and compensate for the variability of the analyte(s) of interest during sample extraction and LC-MS analysis, the IS responses are expected to be variable. Considering the various differences between CSs/QCs and incurred samples (Table 17.2) as well as the difference in the number of matrix lots (e.g., 6) tested during method validation and the number of subjects (e.g., 40) enrolled in a study, the chance of encountering variable IS responses during study sample analysis would be quite high. However, excessive variation in IS response, especially those associated with study sample analysis, could cast a doubt on the integrity of the obtained results and/or the validity of the bioanalytical method. Therefore, several recommendations have been made for monitoring IS responses during study sample analysis (e.g., Jemal et al., 2003; Bakhtiar and Majumdar, 2007; Tan et al., 2009a), although no agreement has been reached yet in the bioanalytical community (Viswanathan et al., 2007; GCC, 2011b). Briefly, the recommended approaches include (1) setting the upper and lower boundaries for acceptable IS responses of unknown samples (e.g., 50–150% of the mean IS response of known samples) and (2) performing a trend analysis, that is, using the IS response variation of known samples (CSs and QCs) to define an acceptable IS response variation for unknown samples. Each of these approaches has its advantages and disadvantages. Despite disagreement on which approach should be taken and what kind of magnitude of variability can be considered acceptable, it is generally agreed in bioanalytical community that the acceptance criteria for IS response in LC-MS bioanalysis should be established with sound scientific judgment prior to sample analysis.

TABLE 17.3 Common Patterns/Trends and Causes of Internal Standard Response Variation and Associated Potential Impact on Assay Accuracy

IS response variation pattern or trend	Typical cause	Potential impact on LC-MS assay accuracy
Interbatch variation, that is, change in slope of calibration curve between batches	Difference in LC-MS instrument optimization or conditions; difference in IS amounts added; instability of IS in working solution; poor ruggedness of assay method	Usually no if the batches are otherwise acceptable
No or doubled IS responses for individual samples	Error in addition of IS working solution or other reagents	Normally yes
Continuous increase or decrease in IS signal over the course of a batch	Malfunctioning or inadequate conditioning of mass spectrometer; inadequate mixing of extracted samples	Case by case
Sudden decrease in IS response for all samples	Problem in autosampler or mass spectrometer	Usually no if adequate signal-to-noise ratio is maintained and linear calibration is used
Variable IS response for both unknown and CSs/QCs	Method issue, for example, lengthy sample processing, incomplete derivatization; improper execution of method	Case by case
Consistently high or low IS responses for the study samples, including pre-dose samples, collected from a subject or a group of subjects	Intersubject matrix difference, resulting in variable recovery and/or variable levels of ion-suppression or signal enhancement among the individual subject matrices.	Case by case. Method modification may be warranted
Variable IS responses only for post-dose samples	Suppression or enhancement of IS response by analyte and/or drug metabolites, comedications and their metabolites, drug formulation materials	Usually no when caused by the analyte because CSs and QCs are similarly affected; normally yes when caused by others, for example, comedications

IS, internal standard.

The root causes for IS response variation are numerous (Tan et al., 2009a, 2009b; Jian et al., 2010). Summarized in Table 17.3 are some common patterns/trends in IS response variation, possible root causes, and the potential impact on the accuracy of quantitation. In most cases, there is no clear "yes or no" answer to the question of whether the assay accuracy may have been impacted. Nevertheless, errors in IS addition can be easily singled out for reassays using some preset criterion, for example, 50–150% of mean IS response of known samples (CSs and QCs). For other patterns or trends such as those listed in Table 17.3, if they were not observed during method validation, evaluation or investigation is necessary.

During the investigation, upon any errors in sample processing and analysis have been ruled out, it would be helpful to ask the following series of questions: (1) Was the whole batch affected, for example, system suitability samples, extracted and nonextracted samples (neat solution sample or extracted blank reconstituted with neat solution)? (2) Were there any differences in IS response between CSs/QCs and unknown samples, between extracted and nonextracted samples, and between the samples injected at the beginning and those at the end of the batch? (3) Were the results of all subject samples analyzed in the same batch affected? (4) Did the abnormal IS responses only occur to the samples from a certain patient population? (5) Was there a difference in IS responses between pre-dose and post-dose samples? (6) Was there any comedication that was not checked in method development/validation?

To illustrate how the aforementioned enquiries are useful in identifying root cause(s), an example is presented later. In this case, consistently high IS responses (higher than those of CSs and QC samples) were observed for incurred samples from a few subjects during the LC-MS/MS analysis of repaglinide in a study (Figure 17.3a). Since the IS responses from the system suitability injections and the injections of CSs and QCs that bracketed/blanketed the entire assay sequence were all acceptable, the issue was apparently not associated with the autosampler or mass spectrometer. In addition, as this anomaly was also observed in the pre-dose sample injections, the root cause was unlikely related to drug metabolite(s). Furthermore, the IS responses were normal for the samples collected from most other subjects in the same study. Based on the earlier description, subject specific matrix effect may have contributed to this observation. Therefore, postcolumn (postelution) infusion tests were conducted using the pooled control blank (which was used for the preparation of CSs and QCs, Figure 17.3b) and the pre-dose sample from a subject whose samples showed high IS responses in LC-MS/MS analysis (Figure 17.3c). By comparing the results of the two postcolumn infusion tests, it is evident that the root cause was the difference in ion suppression between the samples from the "problematic" subject and

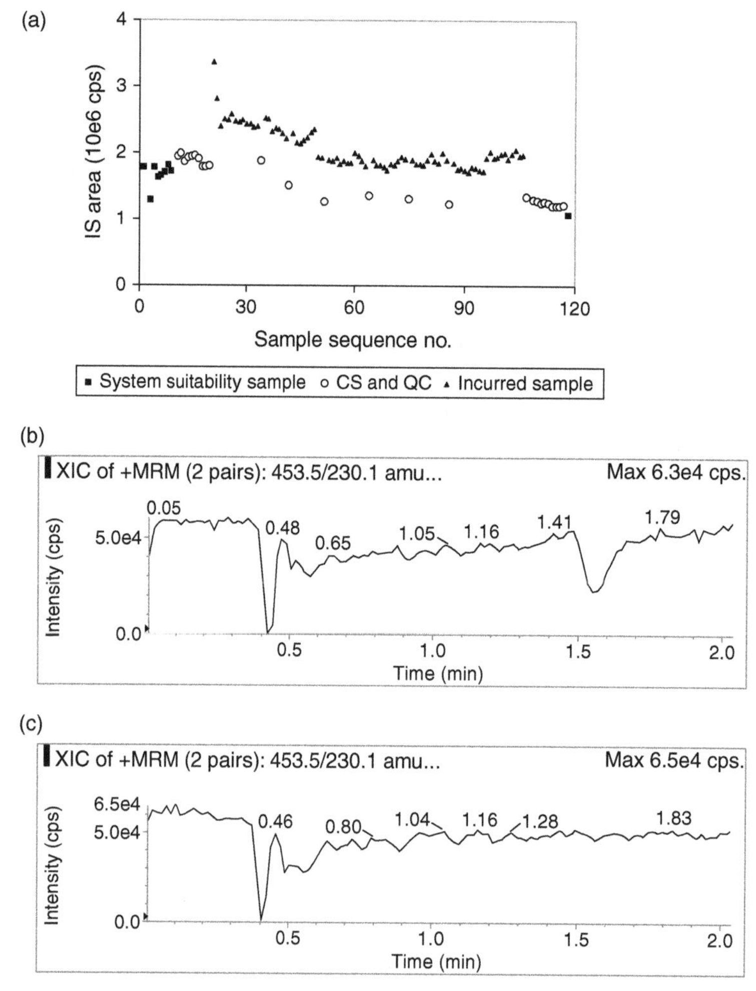

FIGURE 17.3 (a) High internal standard (IS) responses were observed for incurred samples only. Analyte: repaglinide; Extraction: automatic liquid–liquid extraction. (b) Postelution infusion results show that ion suppression existed near the retention time of the analyte (1.57 min) from the pooled control blank used for the preparation of calibration standard (CS) and quality control (QC) samples. (c) Absence of ion suppression near the retention time of the analyte in subject pre-dose sample (reproduced from Tan et al. (2009a) with permission from Elsevier).

CSs/QCs (Tan et al., 2009a). This example also highlights the importance of being open minded during troubleshooting because ion enhancement was initially thought as the cause for the observed high IS responses. At the end of investigation, it was still due to ion suppression.

Based on the results of investigation, the affected samples may be reinjected, reanalyzed, or the original results may be accepted together with scientific justification. The last approach, that is, reporting the original results with scientific justification, would not only be preferable, that is, saving time and cost, but it may be also the only option in some cases. For the example given above, when all the samples from a subject show consistently higher or lower IS responses than those of the CSs and QC samples in LC-MS/MS due to inter-lot matrix differences, the same or similar IS responses would be seen again from reassays if the same LC-MS/MS method is used. Without adequate justification, the obtained concentration values for the study samples collected from the respective subjects (due to abnormal IS responses) should not be reported. In terms of justification, it is important to demonstrate not only that the IS's ability to track the analyte had not been impacted but also that the relationship between analyte concentration and analyte/IS response ratio had not been changed, for example, from linear to quadratic or vice versa. Otherwise, the obtained concentrations may still not be reportable despite good reassay reproducibility shown in some cases. Back to the example given in the above, there are different options that can be taken to provide scientific justification. First, if a sufficient amount of matrix can be obtained from the "problematic" subject, for example, by pooling pre-dose and other low concentration samples, QC samples can be prepared in this matrix and analyzed to evaluate their accuracy. Second, standard addition test can be performed in a few samples to demonstrate the recovery of the added analyte standard. Lastly, some selected samples can be reanalyzed after appropriate dilutions with the pooled blank used for the preparation of CSs and QCs to bring their IS responses close to those of CSs and QCs. Then, the results obtained after dilutions can be compared with the initial ones to see if they are matching. Whichever option is chosen, it is critical to test at both low and high QC concentration levels.

17.4 CONCLUSION

ISs play critical roles in ensuring the accuracy of quantitation in LC-MS bioanalysis and the robustness of bioanalytical methods. In general, the physicochemical properties of an IS, particularly hydrophobicity and ionization properties, should be as close as possible to the analyte(s) of interest so that variations/losses in the bioanalytical process for the analyte(s) including sample preparation, LC separation, and MS detection can be corrected for. For this reason, SIL ISs, especially those labeled with ^{13}C and/or ^{15}N, should be used whenever possible. Nevertheless, some precautions should still be taken while using SIL ISs, especially the deuterated ones.

An IS should be added as early as feasible in LC-MS bioanalysis. Its amount to be added to study samples, that is, IS concentration, must be properly chosen based on several factors, including the magnitudes of cross-signal contribution between the analyte and its IS, assay concentration range, weighting factor to be used in regression, detection sensitivity, IS linearity range, and ion suppression/signal enhancement between the analyte and its IS. An inappropriate IS concentration can potentially affect the linearity, precision, and accuracy of an LC-MS assay.

Although IS response variations during study sample analysis are somehow expected, they still should be closely monitored. The root causes for IS response variations in an LC-MS assay run are multiple. They include difference in IS amounts added, malfunctioning of LC-MS system, improper execution of assay method, poor ruggedness of assay method, intersubject matrix difference, and ion suppression or enhancement by analyte/drug metabolites. Among these causes, errors in IS addition should be singled out for reassay without hesitation. Any pattern or trend in IS responses that were not seen during assay method validation should trigger an immediate investigation for root cause. Since the same pattern or trend in IS responses, for example, consistently higher IS responses for all samples from a subject, may be caused by totally different reasons, diligence in analyzing all possibilities can help an efficient investigation. Upon investigation, the respective sample extracts can be reinjected or the samples can be reassayed, provided that scientific justification is in place. Sometimes, the initial results can be accepted with additional supporting documents to demonstrate that the IS's ability to track the analyte and the analyte concentration–response (ratio) relationship had not been impacted despite the abnormal IS responses observed. Otherwise, the bioanalytical method should be revamped and revalidated prior to the reanalysis of the study samples.

ACKNOWLEDGMENT

The first author thanks his family (Cailin and Joyce) for their support during the preparation of this chapter.

REFERENCES

Abu-Rabie P, Denniff P, Spooner N, Brynjolffssen J, Galluzzo P, Sanders G. Method of applying internal standard to dried matrix spot samples for use in quantitative bioanalysis. Anal Chem 2011;83:8779–8786.

Alder L, Lüderitz S, Lindtner K, Stan HJ. The ECHO technique-the more effective way of data evaluation in liquid chromatography-tandem mass spectrometry analysis. J Chromatogr A 2004;1058:67–79.

Ansermot N, Rudaz S, Brawand-Amey M, Fleury-Souverain S, Veuthey JL, Eap CB. Validation and long-term evaluation of a modified on-line chiral analytical method for therapeutic drug monitoring of (R,S)-methadone in clinical samples. J Chromatogr B Analyt Technol Biomed Life Sci 2009;877:2301–2307.

ANVISA (Brazilian Sanitary Surveillance Agency). Guide for validation of analytical and bioanalytical methods. Resolution RE no. 899, 2003.

Bakhtiar R, Majumdar TK. Tracking problems and possible solutions in the quantitative determination of small molecule drugs and metabolites in biological fluids using liquid chromatography-mass spectrometry. J Pharmacol Toxicol Methods 2007;55:227–243.

Berg T, Strand DH. ^{13}C labelled internal standards—a solution to minimize ion suppression effects in liquid chromatography-tandem mass spectrometry analyses of drugs in biological samples? J Chromatogr A 2011;1218:9366–9374.

Chavez-Eng CM, Constanzer ML, Matuszewski BK. High-performance liquid chromatographic-tandem mass spectrometric evaluation and determination of stable isotope labeled analogs of rofecoxib in human plasma samples from oral bioavailability studies. J Chromatogr B Analyt Technol Biomed Life Sci 2002;767:117–129.

Choi BK, Gusev AI, Hercules DM. Postcolumn introduction of an internal standard for quantitative LC-MS analysis. Anal Chem 1999;71:4107–4110.

Cuadros-Rodríguez L, Bagur-González MG, Sánchez-Viñas M, González-Casado A, Gómez-Sáez AM. Principles of analytical calibration/quantification for the separation sciences. J Chromatogr A 2007;1158:33–46.

European Medicines Agency (EMA). Guideline on bioanalytical method validation, 2011.

GCC (Global CRO Council). The 3rd global CRO council for bioanalysis at the International Reid Bioanalytical Forum. Bioanalysis 2011a; 3:2721–2727.

GCC (Global CRO Council). Recommendations on: internal standard criteria, stability, incurred sample reanalysis and recent 483s by the Global CRO Council for Bioanalysis. Bioanalysis 2011b; 3:1323–1332.

Hewavitharana AK. Matrix matching in liquid chromatography-mass spectrometry with stable isotope labelled internal standards—is it necessary? J Chromatogr A 2011;1218:359–361.

Jemal M, Schuster A, Whigan DB. Liquid chromatography/tandem mass spectrometry methods for quantitation of mevalonic acid in human plasma and urine: method validation, demonstration of using a surrogate analyte, and demonstration of unacceptable matrix effect in spite of use of a stable isotope analog internal standard. Rapid Commun Mass Spectrom 2003;17:1723–1734.

Jian W, Edom RW, Xu Y, Gallagher J, Weng N. Potential bias and mitigations when using stable isotope labeled parent drug as internal standard for LC-MS/MS quantitation of metabolites. J Chromatogr B Analyt Technol Biomed Life Sci 2010;878:3267–3276.

Kitamura R, Matsuoka K, Matsushima E, Kawaguchi Y. Improvement in precision of the liquid chromatographic-electrospray ionization tandem mass spectrometric analysis of 3′-C-ethynylcytidine in rat plasma. J Chromatogr B Analyt Technol Biomed Life Sci 2001;754:113–119.

Lanckmans K, Sarre S, Smolders I, Michotte Y. Use of a structural analogue versus a stable isotope labeled internal standard for the quantification of angiotensin IV in rat brain dialysates using nano-liquid chromatography/tandem mass spectrometry. Rapid Commun Mass Spectrom 2007;21:1187–1195.

Lee JW, Petersen ME, Lin P, Dressler D, Bekersky I. Quantitation of free and total amphotericin B in human biologic matrices by a liquid chromatography tandem mass spectrometry method. Ther Drug Monit 2001;23:268–276.

Li W, Li YH, Li AC, Zhou S, Naidong W. Simultaneous determination of norethindrone and ethinyl estradiol in human plasma by high performance liquid chromatography with tandem mass spectrometry—experiences on developing a highly selective method using derivatization reagent for enhancing sensitivity. J Chromatogr B Analyt Technol Biomed Life Sci 2005;825: 223–232.

Liang HR, Foltz RL, Meng M, Bennett P. Ionization enhancement in atmospheric pressure chemical ionization and suppression in electrospray ionization between target drugs and stable-isotope-labeled internal standards in quantitative liquid chromatography/tandem mass spectrometry. Rapid Commun Mass Spectrom 2003;17:2815–2821.

Lindegardh N, Annerberg A, White NJ, Day NPJ. Development and validation of a liquid chromatographic-tandem mass spectrometric method for determination of piperaquine in plasma: stable isotope labeled internal standard does not always compensate for matrix effects. J Chromatogr B Analyt Technol Biomed Life Sci 2008;862:227–236.

Liu G, Ji QC, Arnold ME. Identifying, evaluating, and controlling bioanalytical risks resulting from nonuniform matrix ion suppression/enhancement and nonlinear liquid chromatography-mass spectrometry assay response. Anal Chem 2010;82:9671–9677.

Meesters R, Hooff G, van Huizen N, Gruters R, Luider T. Impact of internal standard addition on dried blood spot analysis in bioanalytical method development. Bioanalysis 2011;3:2357–2364.

Moore LJ, Machlan LA. High accuracy determination of calcium in blood serum by isotope dilution mass spectrometry. Anal Chem 1972;44:2291–2296.

Morin LP, Mess JN, Furtado M, Garofolo F. Reliable procedures to evaluate and repair crosstalk for bioanalytical MS/MS assays. Bioanalysis 2011;3:275–283.

O'Halloran S, Ilett KF. Evaluation of a deuterium-labeled internal standard for the measurement of sirolimus by high-throughput HPLC electrospray ionization tandem mass spectrometry. Clin Chem 2008;54:1386–1389.

Remane D, Wissenbach DK, Meyer MR, Maurer HH. Systematic investigation of ion suppression and enhancement effects of fourteen stable-isotope-labeled internal standards by their native analogues using atmospheric-pressure chemical ionization and electrospray ionization and the relevance for multi-analyte liquid chromatographic/mass spectrometric procedures. Rapid Commun Mass Spectrom 2010;24:859–867.

Savard C, Pelletier N, Boudreau N, Lachance S, Lévesque A, Massé R. Relative instability of deuterated internal standard under different pH conditions and according to deuterium atoms location. Proceedings of the 58th ASMS Conference on Mass Spectrometry and Allied Topics, 2010 May 23–27; Salt Lake City, UT, USA.

Shi G. Application of co-eluting structural analog internal standards for expanded linear dynamic range in liquid chromatography/electrospray mass spectrometry. Rapid Commun Mass Spectrom 2003;17:202–206.

Sojo LE, Lum G, Chee P. Internal standard signal suppression by co-eluting analyte in isotope dilution LC-ESI-MS. Analyst 2003;128:51–54.

Stokvis E, Rosing H, Beijnen JH. Stable isotopically labeled internal standards in quantitative bioanalysis using liquid chromatography/mass spectrometry: necessity or not? Rapid Commun Mass Spectrom 2005;19:401–407.

Tan A, Boudreau N, Lévesque A. Internal standards for quantitative LC-MS bioanalysis. In: Xu QA, Madden TL, editors. *LC-MS in Drug Bioanalysis*. New York: Springer; 2012a. p 1–32.

Tan A, Awaiye K, Jose B, Joshi P, Trabelsi F. Comparison of different linear calibration approaches for LC-MS bioanalysis. J Chromatogr B Analyt Technol Biomed Life Sci. 2012b; 911:192–202.

Tan A, Hussain S, Musuku A, Massé R. Internal standard response variations during incurred sample analysis by LC-MS/MS: case by case trouble-shooting. J Chromatogr B Analyt Technol Biomed Life Sci 2009a; 877:3201–3209.

Tan A, Lévesque IA, Lévesque IM, Viel F, Boudreau N, Lévesque A. Analyte and internal standard cross signal contributions and their impact on quantitation in LC-MS based bioanalysis. J Chromatogr B Analyt Technol Biomed Life Sci 2011;879: 1954–1960.

Tan A, Montminy V, Gagné S, Musuku A, Massé R. Troubleshooting of least-expected causes in bioanalytical method development and application. Proceedings of 2009 AAPS Annual Meeting and Exposition, 2009b; Nov. 8–12; Los Angeles, CA, USA.

Taylor PJ, Brown SR, Cooper DP, et al. Evaluation of 3 internal standards for the measurement of cyclosporine by HPLC-mass spectrometry. Clin Chem 2005;51:1890–1893.

Timmerman P, Kall MA, Gordon B, Laakso S, Freisleben A, Hucker R. Best practices in a tiered approach to metabolite quantification: views and recommendations of the European Bioanalysis Forum. Bioanalysis 2010;2:1185–1194.

Viel F, Santos N, Tan A, et al. Simultaneous quantitation of free and liposomal drug forms in human serum by evaporation-free extraction. Proceedings of the 58th ASMS Conference on Mass Spectrometry and Allied Topics, May 23–27, 2010; Salt Lake City, UT, USA.

Viswanathan CT, Bansal S, Booth B, et al. Workshop/conference report—Quantitative bioanalytical methods validation and implementation: best practices for chromatographic and ligand binding assays. AAPS J 2007;9(1):E30–E42.

Wang S, Cyronak M, Yang E. Does a stable isotopically labeled internal standard always correct analyte response? A matrix effect study on a LC/MS/MS method for the determination of carvedilol enantiomers in human plasma. J Pharm Biomed Anal 2007;43:701–707.

Wieling J. LC-MS-MS experiences with internal standards. Chromatographia 2002;55:S107–S113.

Zhang G, Wujcik CE. Overcoming ionization effects through chromatography: a case study for the ESI-LC-MS/MS quantitation of a hydrophobic therapeutic agent in human serum using a stable-label internal standard. J Chromatogr B Analyt Technol Biomed Life Sci 2009;877:2003–2010.

Zrostlíková J, Hajšlová J, Poustka J, Begany P. Alternative calibration approaches to compensate the effect of co-extracted matrix components in liquid chromatography-electrospray ionization tandem mass spectrometry analysis of pesticide residues in plant materials. J Chromatogr A 2002;973:13–26.

18

SYSTEM SUITABILITY IN LC-MS BIOANALYSIS

CHAD BRISCOE

18.1 OVERVIEW

Your instruments were qualified months or years ago. You have recently validated a new LC-MS/MS assay for use in an upcoming clinical study. Your time and your clinical samples are extremely valuable. How are you going to bridge the gap between instrument qualification and method implementation? The way that is done is through what is commonly called system suitability testing. No matter what type of analytical testing or any testing you are doing at all, system suitability is a critical process to know that your instrument or equipment is ready to run.

System suitability is the process of ensuring your instruments are prepared and performing properly for your analysis. Traditionally a system suitability test consists of a few injections performed prior to running a batch of samples to ensure the instrument is ready for the run. However, the system suitability test is much more than that. It is the bridge between a qualified instrument and the injection of bioanalytical test samples. The extent of that system suitability test is dependent on the amount of risk you are willing to take with the injection of your samples. Very simple, rugged assays may not justify significant system suitability testing, whereas challenging assays with limited sample volume may require less risk to be taken and thus more extensive system suitability testing. The success of the system suitability test is about the design of the test, proper qualification of the instruments and proper development of a rugged assay. Because all of these aspects are critical to successful system suitability testing, they are all discussed in this chapter.

The concept of system suitability is often made much more complicated than it needs to be. System suitability is simply a way to ensure our instruments and equipment are ready for a specific analysis that they are soon to be used for. This is what differentiates system suitability from the rest of the IQ/OQ/PQ process. Sometimes system suitability is considered the end of the PQ part of the qualification process, and sometimes it is a completely distinct activity. What is critical to understand is that it is generally the only part of the process that is considered specific to an individual assay. If you consider the application of the instrument to be tiered as shown in Figure 18.1, instrument qualification is at the base of the model. Installation qualification (IQ) is the first step. Every instrument of a similar design installed for any purpose would have the same IQ. This is followed by the operational qualification (OQ), which is generally more specific to the environment the instrument is installed in and thus is a bit more specific. It is not uncommon for the IQ and OQ tests to be performed by the manufacturer. The performance qualification (PQ) will contain tests that are more specific to the customer's intended use. Parameters are tested that are unique to the environment for use. There are many helpful references to look toward for the IQOQPQ process (Burgess et al., 1998; Bansal et al., 2004). A method must be developed and validated with a qualified instrument. Part of that process includes identifying what the system suitability process should be. System suitability is the next step to qualify an instrument for a specific assay. At the apex of the triangle, we find the most specific use of the instrument is the use of the MS for analysis of samples with embedded quality control (QC) checks.

This chapter will focus primarily on the system suitability tier of the triangle. It is in this function that system suitability is so critical to take an instrument from a complex electronic machine to an analytical instrument to support

Handbook of LC-MS Bioanalysis: Best Practices, Experimental Protocols, and Regulations, First Edition. Edited by Wenkui Li, Jie Zhang, and Francis L.S. Tse.

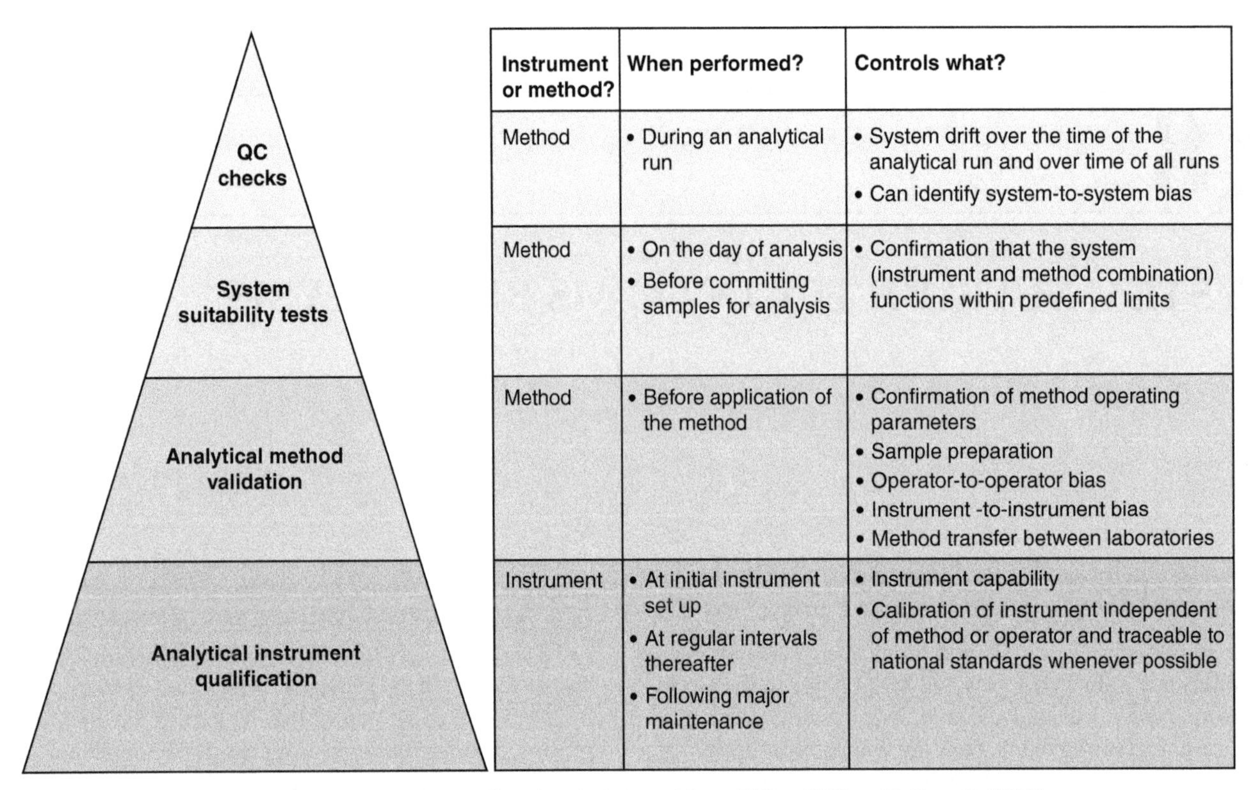

Instrument or method?	When performed?	Controls what?
Method	• During an analytical run	• System drift over the time of the analytical run and over time of all runs • Can identify system-to-system bias
Method	• On the day of analysis • Before committing samples for analysis	• Confirmation that the system (instrument and method combination) functions within predefined limits
Method	• Before application of the method	• Confirmation of method operating parameters • Sample preparation • Operator-to-operator bias • Instrument -to-instrument bias • Method transfer between laboratories
Instrument	• At initial instrument set up • At regular intervals thereafter • Following major maintenance	• Instrument capability • Calibration of instrument independent of method or operator and traceable to national standards whenever possible

FIGURE 18.1 The data quality triangle (adapted from USP <1058>; McDowall, 2010).

research. System suitability enables the researcher to accomplish two simple but critical tasks:

1. *Get it right the first time*: Research is expensive and time is often very limited. Also, in many applications of modern LC/MS/MS, research is intended to be profitable as in applications of LC/MS/MS in CROs. Therefore, when a batch of samples, a series of experiments, or a day of method development is invested by a scientist, they need to know that the instrument will perform for them as expected.
2. *Provide assurance that the instrument is not at fault*: There are always many unknowns in research. It is important for scientists to know that the instrument's performance is not a variable. In the same vein, when routine Bioanalysis is conducted, it is not uncommon for unexpected results to occur. Properly designed system suitability tests can minimize or eliminate the LC/MS/MS from question or help to identify that the LC/MS/MS was at fault.

18.2 REGULATORY REQUIREMENTS FOR SYSTEM SUITABILITY

Historically, the FDA has been the driver for defining the global expectations for Regulated Bioanalysis from GLP studies through all phases of clinical bioanalysis. These expectations were developed over the course of several meetings jointly sponsored by the American Association of Pharmaceutical Scientists (AAPS) and the FDA. These meetings are commonly referred to as "The Crystal City Meetings" because they were held in the Crystal City area of Arlington, Virginia. The FDA Guidance for Bioanalytical Method Validation from May 2001 (FDA, 2001) was a by-product of the second of these four meetings. However, the guidance only briefly mentions system suitability stating that system suitability is "based on the analyte and technique, a specific standard operating procedure (SOP) (or sample) should be identified to ensure optimum operation of the system used." The lack of specific regulations or guidances for system suitability has led to some confusion as to which regulations to apply to the application of system suitability in bioanalysis. Table 18.1 provides an overview of a number of the global regulations and guidances that include a reference to system suitability.

In February 2012 the European Medicines Agency (EMA) also released a guidance for Bioanalytical Method Validation (EMA, 2011). While that guidance does not specifically mention system suitability, it does give very good justification for the criticality of system suitability testing as it states "Re-injection of a full analytical run or of individual calibration standard samples or QC samples, simply because the calibration or QCs failed, without any identified

TABLE 18.1 Some of the Global Regulations that Specifically Discuss System Suitability and Instrument Qualification

GLP
- Swiss Public Health AGIT (SFOPH, 2007)
- OECD Application of GLP to Computerised systems (OECD, 1995)
- ANVISA (2002) *Manual for Good Bioavailability and Bioequivalence Practices.* Vol. 2 (ANVISA, 2002)

GMP
- USP ><621> (USP, 2006)
- USP <1058> (USP, 2006)
- FDA 21 CFR Part 11 (FDA, 2008)
- EMA GMP Annex 11 (EMA, 2011)
- FDA General Principles of Software Validation (FDA, 2002)

Clinical
- FDA Guidance Comp systems in Clinical Investigations (FDA, 1999)
- Japanese Guideline for Electromagnetic Records in Applications for Pharmaceuticals (MLHW, 2005)

Engineering
- ISPE GAMP 5 (ISPE, 2008)
- NIST Risk Mgmt Guide for IT systems (NIST, 2002)

analytical cause, is not acceptable." Therefore, it is critical to conduct system suitability testing in order to ensure the system performs properly as it is clearly not acceptable to reinject without evidence of poor instrument performance. It is a well-accepted concept that reinjection to get acceptable results is not allowed. This is one of the arguments that has been made for designing a system suitability test that looks at system performance at both the beginning and end of the batch (Briscoe et al., 2007). Briefly, what was described in this paper is an approach to system suitability that separates the instrument performance throughout the batch from the performance of standards and QCs. Therefore, if a batch fails and it can be demonstrated that the reason for failure is instrument drift, carryover or poor sensitivity that started during the batch, then the batch can be reinjected due to an obvious instrument failure. This may save reextraction of valuable, limited volume clinical samples by allowing reinjection under the principal that the instrument failed and thus the standards and QCs did not cause batch failure. This is discussed further later in the chapter.

The importance the FDA places on system suitability was clarified in several inspections of LC-MS/MS Bioanalysis studies as early as 2005. In one 483 finding the firm was cited for utilizing a system suitability procedure that was not documented in a predefined procedure. The process was not at question, the lack of a written procedure was the concern. More recently the FDA has cited a laboratory's implementation of system suitability as findings in multiple inspections in 2010 (FDA, 2010, 2006). In these findings, it was noted that the laboratories "prep runs" were performed without a "written procedure to describe the selection, evaluation and reporting." This looks very much like the finding from 5 years ago with regards to the need for an SOP to define the system suitability process. In this case, however, the FDA took issue with the actual process used as they also stated that it is "not recommended" to use subject samples, standards, QCs, and blanks that are to be analyzed as part of the official run. The FDA expresses a concern that by injecting these types of samples as part of prep or system suitability runs, manipulation of the final result could occur. This is a valid concern if fraud is suspected. There is another important distinction that the FDA is making that is pertinent to this chapter. They are recognizing that system suitability is a critical activity worthy of inspection and thus proper conduct. They also identify that it should be a distinct part of the bioanalytical process. In summary, several important lessons can be extracted from these FDA findings. System suitability should (1) be a distinct activity in the bioanalytical process; samples to be injected in the final batch should not be injected as system suitability (2) be written in a predefined procedure; and (3) have thoroughly documented and traceable results.

18.3 MONITORING INSTRUMENT PERFORMANCE

18.3.1 Setting Up for Success

System suitability is a holistic approach to ensure your instrument is performing properly for the analysis of your clinical samples. This starts with ensuring your instrument is ready to run samples even before samples are injected. The preinjection preparation includes proper installation, maintenance, calibration, and setup for your run. Only when all of these activities are properly performed is the instrument ready to inject system suitability samples and ultimately the samples for bioanalysis.

Proper installation of your system requires following some type of predetermined instrument qualification procedure. Various procedures have been described in the literature and regulations (Bansal et al., 2004; USP, 2008; ISPE, 2011 2011) and the practice of instrument qualification continues to evolve (Burgess and McDowall, 2012). Complete treatment of this topic is outside the scope of this chapter, but there are several critical elements of the process to highlight. First of all, the end user must be closely involved in the process to ensure that the qualification procedure is appropriate for the intended use. The manufacturer or installer of the system should conduct installation and operational qualification that has been preapproved by the end user. The end user then performs any additional testing that is unique to how they will use or configure the instrument. This is the PQ. If this is the first of this type of instrument to be installed; SOPs defining the use, maintenance, and calibration of the equipment should be written.

Once the instrument has been installed, it is necessary to maintain it properly. The maintenance required for each system will vary depending on whether it is one of the high-pressure liquid chromatography (HPLC) or mass spectrometer components. However, what is critical is that sufficient investment in time and expense is planned for as proper maintenance is required to ensure quality operation of the system for the long term. Routine maintenance should be conducted according to a fixed schedule by a qualified service person. Often the maintenance is performed by a combination of a trained factory service technician and an individual that has been trained or acquired skills for more basic maintenance tasks. The routine maintenance should also include calibration. The calibration approach should be adapted from the factory specifications in order to maintain the instrument as close to possible in its initial performance specifications. Unplanned maintenance will also be required at times when the instrument has unexpected breakdowns. This unfortunate situation should also be planned for as much as possible by approaches such as paying for a factory maintenance contract, stocking extra parts, or qualifying multiple systems for the assay being used.

Another critical factor to ensuring proper system performance is to ensure robustness in the assay during the method development process. There are various ways that this is accomplished. These include using quality reagents, parts and consumables, preparing clean samples, assessing performance with multiple columns and instruments and testing the maximum batch length. When it comes to utilizing quality materials for developing your assay and maintaining your instrument there is no doubt that you generally get what you pay for and in the case of bioanalysis you generally pay a little more to do things right the first time or pay a lot to do things over. High purity reagents with quality certificates of analysis are critical whether they are solvents or reference standards. If the supplier cannot provide manufacturing and testing details along with a detailed certificate of analysis, there is no assurance of the product quality or reproducibility of the product. The same is true of the materials used in the development of the assay whether it is SPE or HPLC columns or the HPLC and mass spectrometer replacement parts. Similarly with the reagents, suppliers should be able to provide evidence of good manufacturing practices (GMPs) and some details on the quality and reproducibility of the materials. This type of information should always be requested for all suppliers to help ensure that they have a quality system in place to generate good quality materials.

The best materials can be used in any improperly tested assay and still result in an assay that is not reproducible and robust. There are many factors that can go wrong with an assay, and those are treated in significant detail in other chapters of this book. For the sake of the instrument performance, the best samples are the cleanest samples. An assay that is both visually clean and has been optimized to minimize matrix effects will produce the best results on the instrument. For this reason, liquid-liquid extractions and solid phase extractions are preferred to protein precipitation as those approaches generally produce cleaner samples. Once an optimal sample cleanup procedure has been designed there are a few tests that can be performed to ensure that your system is suitable to analyze samples with the assay that has been developed. First of all, when possible the assay should be tested on multiple different production lots of reagents, sample preparation materials, and HPLC columns. This is necessary because even high quality supplies will have some variability. It is important to go to a reasonable extent to ensure that the variability in the materials you have chosen do not affect the assay performance. The same applies to the HPLC and mass spectrometer used. Properly maintained equipment should perform nearly equivalent but that is not always the case. This is frequently observed with the mass spectrometer when very low limits of quantitation are required. Mass spectrometer's response will vary between instruments and tends to be dependent on the compound or compound class that is being assayed, thus one system may have sufficient response for an assay but another instrument may not. A final test of robustness that is essential is to establish the maximum batch length to inject on the instruments. This is typically evaluated during method development and validated during the method validation runs. Performance of the instrument beyond this maximum batch length is not known, and therefore, reliability of the instrument with more injections than those tested cannot be assured.

If the LC-MS/MS system has just recently been set up for use or has not run any batches for a period of time, it is often a good idea to perform a series of priming injections on the LC-MS/MS system in order to assist in the equilibration of the column, the gradient, and the ionization process. The number of injections may vary from a few (2–3) up to 50 or more with some unusually challenging assays. The samples used for priming are generally extracts and should be either specifically prepared for this purpose, or be extracts from previous batches that have been pooled to eliminate the possibility for bias. The acceptance criteria for such priming activities would be documented in the study plan or SOP and is typically involving the precision of a certain number of consecutive injections. As was previously discussed, under no circumstances should the samples used for priming be from the batch that is about to be injected. This practice provides the potential for falsifying the results as samples from the batch could be replaced with others if the results do not come back as expected. Some laboratories have used this in a practice called "prepping" (US FDA, 2011), whereby the technicians would inject standards from a batch and if they did not produce the expected result; they would not inject the

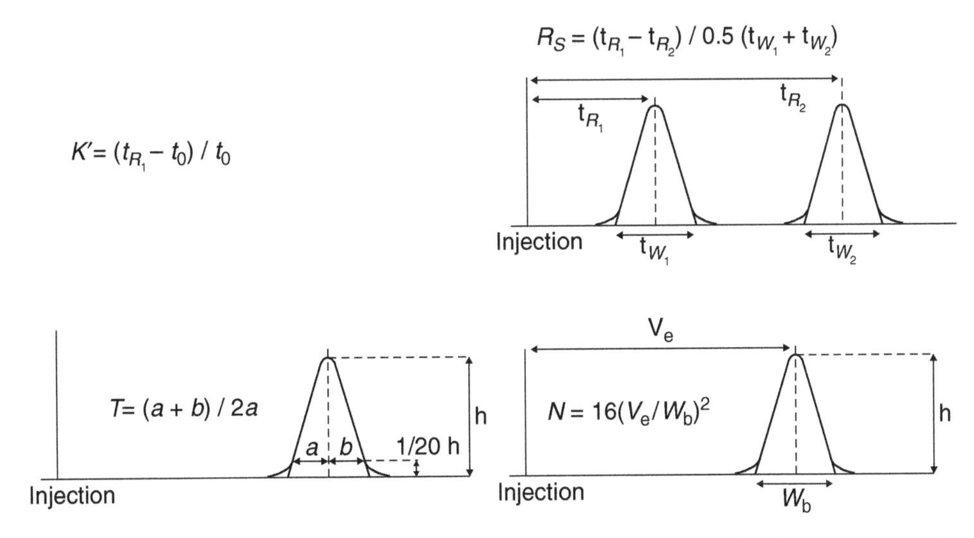

FIGURE 18.2 Typical parameters for the assessment of system suitability in chromatographic systems (Phenomenex).

batch. However, in some cases, they would replace the "bad" standards with standards from a previous batch that were known to be good so that the batch would pass acceptance criteria.

18.3.2 Chromatographic Systems

System suitability assessments are often implemented to monitor chromatographic performance. While it is not as critical in bioanalysis applications as in GMP to tightly control descriptors such as peak shape and retention time, in some applications, it can be very important. System suitability has been written about extensively for applications to HPLC, specifically for GMP use and we can apply these principals for LC-MS/MS applications in bioanalysis.

There are a number of parameters that are commonly used in assessment of system suitability for monitoring HPLC performance (Figure 18.2). These parameters are used to monitor peak shape (asymmetry, A_S, or tailing, T), retention time/capacity factor (k'), theoretical plates (N)/plate height (H), and resolution (R_S). These can be used in a variety of different ways in bioanalysis. The US Pharmacopeia Chapter 621 on Chromatography (USP, 2006) outlines the use of several of these parameters for use principally in GMP applications. In bioanalysis, these are not required tests but are often helpful depending on the assay performance and type of chromatography. The FDA's reviewer guidance for validation of chromatographic methods (FDA, 1994) suggests that k' should be >2, $N > 2000$, $A_s > 2$, and $R_s > 2$. When these parameters are implemented for use in system suitability tests, it would be expected that the performance characteristics would be very similar in sample analysis to those observed during validation.

One example of the application of resolution is an application for an assay for the caffeine metabolite paraxanthine (Havel, 2012). In this example, the chromatography (Figure 18.3) had an interference peak that was necessary to maintain separation. In this application, a system suitability sample demonstrating the separation was run at the beginning and end of the batch to ensure the interference would be separated throughout. In this case, the separation was critical to ensuring acceptable performance especially at low concentrations. It was also critical to ensure that resolution was maintained throughout the run. Therefore, the sample was run at the beginning and end of the batch. Had resolution been lost part way through the run and this system suitability sample was not run at the end, the issue would not be detected and the run results would not be reliable at the low concentrations.

18.3.3 Mass Spectrometry Issues

For implementation of system suitability testing in LC-MS bioanalysis, it is critical to also ensure the mass spectrometer performance is acceptable. There are generally two issues observed with the mass spectrometer that are monitored with system suitability tests, sensitivity, and response reproducibility. The mass spectrometer may not have the same response from batch to batch or study to study for a variety of reasons, so it is important to check the system before starting the batch. It may be that the instrument is dirty from a previous run. Another possibility is that the system has not equilibrated long enough. It may need longer to heat up further to get maximally efficient ionization. Similarly, a mass spectrometer that has not been properly equilibrated may have issues with inconsistent response or drifting response

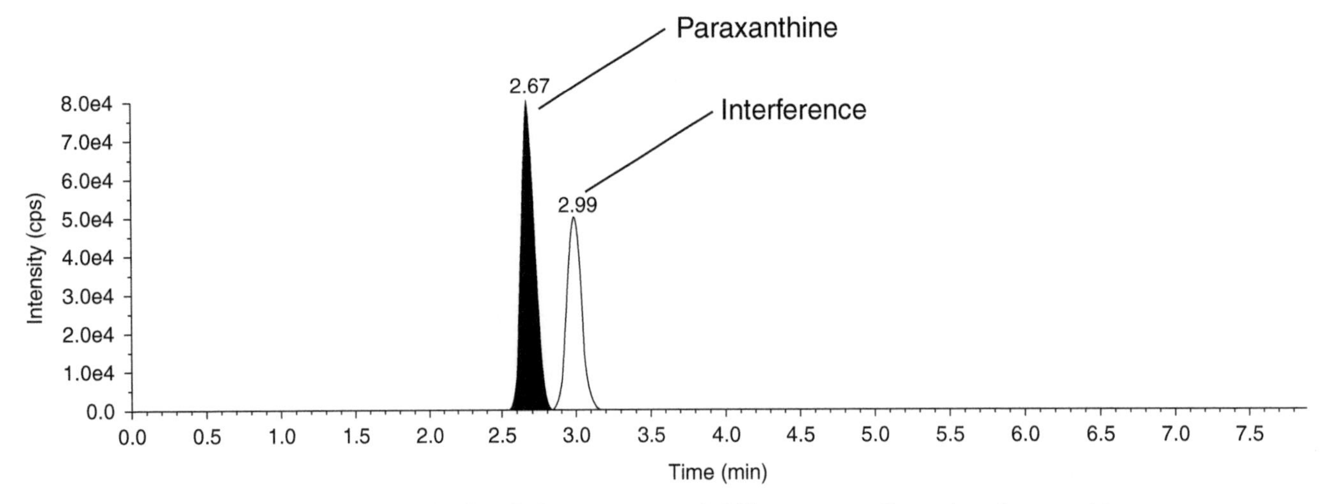

FIGURE 18.3 Example of resolution as a system suitability parameter. Separation of paraxanthine from an unknown interfering peak (Havel, 2012).

over time. Ways to monitor these performance characteristics are outlined in other sections of this chapter.

While system equilibration is one way to help the operator avoid issues with the mass spectrometer, there are other factors that can play a significant role as well. One important aspect is to remove the phospholipid and other matrix components in order to inject a clean sample extract. Many examples have been identified in the literature (Bennett et al., 2006; Chang et al., 2011; Guo and Lankmayr, 2011) and all follow a few general principals. In LC-MS/MS analysis the samples must be properly cleaned up and when developing methods, multiple approaches to monitoring matrix effect should be followed such as measurement of matrix factor and postcolumn infusion. For a detailed discussion of these approaches, refer to Chapter 21 of this book.

Another way to avoid issues with the mass spectrometer is to establish performance limits for your assays. Maximum and minimum injection volumes should be tested in method development. Injecting too much can cause issues with peak shape due to column degradation or can cause buildup of matrix in the mass spectrometer leading to signal deterioration. Injecting too little is not generally a problem unless the response is lower overall and then the limit of quantitation does not have enough response or if you are injecting below the performance limits of the HPLC injector you could get a more variable peak response. Batch length was previously discussed as a parameter that should be tested in method development and subsequently validated. Other MS parameters should be evaluated as well depending on your application and type of MS. Generally MS tuning parameters should not be changed without some degree of revalidation or consideration for their impact and an assessment using some type of system suitability test.

18.3.4 Applications in LC–MS/MS Bioanalysis

When using LC-MS/MS systems for method validation and sample analysis it is important to have tools to ensure that the system is performing properly. These include equilibration samples that have previously been discussed, utilization of system suitability specific samples and proper application of calibrators, QCs and blank samples as dictated in the various bioanalytical guidances.

System suitability is designed into all LC-MS/MS batches used in regulated bioanalysis that are designed to meet the FDA guideline for standard calibrator, QC and blank sample placement. While the use of calibrators, QCs, internal standards, and blanks are not typically considered system suitability samples, they benefit the demonstration of system suitability in a variety of ways. Calibrators are frequently utilized at both the beginning and end of a batch of samples. This enables scientists to evaluate instrument performance by looking for drift between the curves. If the response observed in one set of samples is significantly different from the other, there may be a problem with the instrument drifting. The use of two standard curves in a single run may also help to overcome issues by averaging out problems such as variability and detector drift. If a system were to drift by 15–20% over the course of a run the standards could help to average out this issue in order to achieve more acceptable results. This approach should be taken with caution as it can mask real problems. This potential issue is discussed later. QC samples are indicators of whether a drift such as this is a problem or not. With too much drift, the calibrators at the end of the batch may fail acceptance low. Blank samples are indicators of carryover. While a peak identified in a blank sample is not a definitive indicator of carryover, it is a good indicator when placed following a high concentration sample. The blank could of course be simply contaminated with drug. Blank samples may also be strategically located in validation studies to demonstrate robustness of robotic techniques that are sometimes prone to dripping and consequently contamination of neighboring wells. Internal standard is one of the best indicators of system suitability that is included as a standard part of every batch of samples. Observing the trend in internal standard response can be used to detect response drift in a batch of samples. Also, the internal standard response can indicate a problem with an autosampler injection and thus a problem with the system for an individual sample. For example, if a single sample has an internal standard response that is significantly lower than the majority of the samples, it can be an indicator that the autosampler did not inject properly for that sample. While it is not suggested in either the FDA or EMA guidance, it is common to apply some type of internal standard acceptance criteria. Typically this would be a comparison of the mean of the internal standard response of known samples (standards and QCs) with the response of individual samples to detect anomalies. Should an especially low or high internal standard response be detected, the result would be rejected. It may be that this result is due to a sample processing error or an injector error but in either case, the result is deemed unacceptable.

When designing a system suitability test with samples specifically processed for this purpose, one should consider the critical instrument related parameters to evaluate in all batches. These include response at the lower limit of quantitation (LLOQ), carryover, and signal stability (Briscoe et al., 2007). These parameters are critical as they are all directly related to the performance of the standards and QCs that batch acceptance are based on and they will relate directly to the instruments ability to properly quantify them. Additionally, the system suitability serves to bridge instrument performance from method development to validation to bioanalysis of samples to support studies. Instrument performance should be consistent throughout all phases of bioanalytical studies and ultimately through multiple studies to support clinical studies. The validated state of instrument performance is based on an instrument performing in a particular way. With challenging methods that have challenging LLOQ's, column switches or other aspects that make them difficult and maybe less rugged than is ideal, the system suitability test designed with these performance parameters can help to ensure performance is excellent.

Another factor to consider when monitoring batches is that LC-MS/MS runs are dynamic. When running long batches of samples derived from biological matrices it is not uncommon for some matrix elements to still be present even in meticulously processed samples. These matrix elements cause changes to the system over the length of the run. Often the instrument performance is monitored by using standards, QCs, and blank samples that are required to be placed throughout the run. The use of standards, QCs, blanks, and internal standard response for monitoring instrument performance has one significant flaw. These samples are all different as they have been through the extraction process independently. It is preferred to make a single system suitability sample that can be consistently applied throughout a study or series of studies so the extraction is never the question and calibration solutions do not have to be used as surrogates. It is good to know the instrument has started out well but it is just as critical to know that it finished with the same performance also. These system suitability samples are placed at the beginning and end of the batch. These samples would be extracted so they are similar to the samples on the batch but pooled and redistributed so that they are identical throughout the batch. For instance, the application of duplicate standards can help to account for issues of instrument drift but may hide problems with data accuracy as demonstrated by Briscoe et al. (2007). They presented a case study where >15% drift was detected using a system suitability test with samples placed at the beginning and end of the batch. The criteria for system suitability in this test required reinjection when drift was greater than 15% because it is demonstrating inadequate performance of the instrument. The batch was reinjected due to the 15% drift and 33 of 112 clinical sample results changed by more than 15% as well. If no system suitability test had been performed at the end of the batch, more than one-third of the samples would have been reported inaccurately. If duplicate standards were utilized, some of the drift would have been accounted for by averaging out the response from the beginning to end of the batch but in that case a much larger subset of results would likely have been inaccurately reported, only a smaller percentage of inaccuracy. In the concept of front and back system suitability, it is also critical to have an LLOQ sample at the beginning and end of the batch to ensure that the response is adequate throughout the run and that signal loss does not drop to unacceptably low levels (generally S/N of less than 5 or 10) at the end of the batch. Again, it is critical that this sample of equivalent concentration to the low standard be pooled and redistributed so that the initial and final sample are the same and extraction variability does not play a part.

Another very common issue in LC-MS/MS runs is carryover. It is common to ensure carryover is not an issue when a batch is started. However, over the course of a long run, it sometimes occurs when assaying a sticky compound that carryover will start part way through a batch. This can sometimes be identified by observing the performance of blanks in a batch. However, if a peak is observed in an extracted blank that is extracted with the rest of the batch, it may be an issue of contamination that occurs in the extraction process. Also, as the FDA has recently raised concerns about laboratories placing pretested blank samples in batches and not using samples from the batch as system suitability checks (FDA, 2006, 2010) it may be preferred to use samples specifically

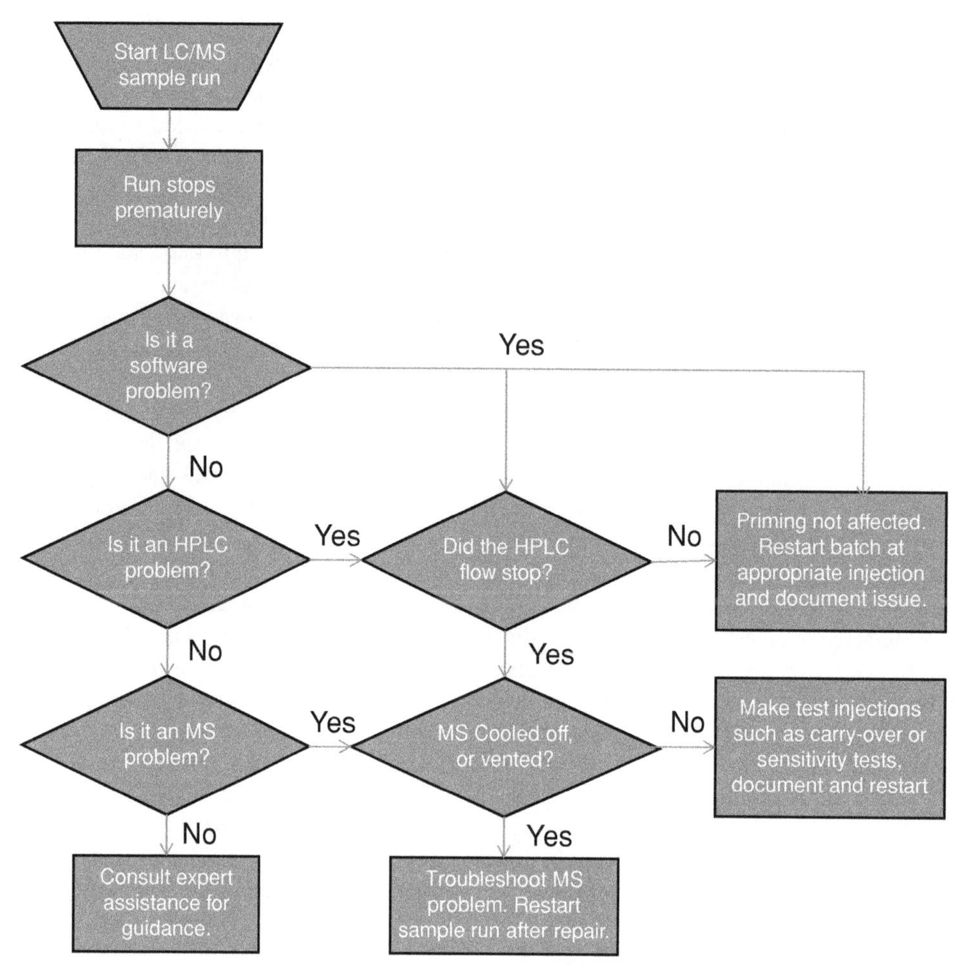

FIGURE 18.4 Approach to determining the proper course of action due to unexpected analytical run stoppage.

designed for checking carryover in order to verify this critical parameter.

18.3.5 Restarts and Reinjections

If an instrument fails part way through a batch run it is going to be necessary to conduct a requalification prior to running the samples. Requalification does not need to be an extensive procedure but must be documented in a procedure such that the approach is defined. Typically the approach taken to requalification would be to have this written into a system suitability SOP. The types of instrument failures would be categorized with a decision tree (such as in Figure 18.4) to guide the approach for the requalification. The following is one approach:

1. *No system stoppage*: This type of failure of a system would be something that affects the run integrity if no intervention is taken but does not affect the primed state of the system. For instance, occasional software errors can cause missed injections or it can be identified part way through a batch that the injection sequence is incorrect. If the point of error can be determined and the system is still injecting the error can be corrected and just documentation is needed to make the correction. If the point of error cannot be determined, system suitability may not need to be repeated but the run will typically need to start again from the beginning along with making adequate documentation of the issue.
2. *HPLC fails but not the MS*: Occasionally, issues will occur where the HPLC system will fail but not the MS. This may occur for a variety of reasons. These errors also fall into two general categories. Errors resulting in stoppage of injection (e.g., injection needle breakage) and errors resulting in stopped flow (e.g., run out of mobile phase, blown fitting, or pump failure). In the event of a stoppage of injection, typically a very simple procedure can be followed to ensure that upon starting the injection again, the response and performance is

similar. This is often accomplished by reinjecting the batch starting from the last acceptable QC sample and ensuring that response for the QC both in terms of area and area ratio are similar to the original injection. If significant change in either has occurred, additional measures should be taken. If the flow of the HPLC stops, a more extensive procedure must be taken. When flow through to the MS is interrupted, the source can become significantly hotter and the spray conditions are affected. This can also cause problems with spray electrodes due to overheating. Therefore, more than one injection is typically warranted to ensure the system is suitable to reinitiate injection. These additional injections may include sensitivity tests or a carryover check. Some autosamplers continue to inject even after flow has stopped, this can result in a buildup of compound in the injector and thus a high chance for carryover. If a fitting prior to the HPLC column is blown out, it is often due to excessive buildup on the column and the column needs to be replaced resulting in complete reinjection and repeat of the full system suitability test.

3. *Failures resulting in completely stopping the HPLC and MS system*: This would include various types of catastrophic failures with the MS such as loss of vacuum, power loss, or breakdown of HPLC causing the MS to revert to "standby" mode. In these cases, significant additional requalification will be needed possibly including recalibration of the system or even the repeat of some operational qualification if a major repair is required. In these cases, it is probably essential that the batch will be restarted from the beginning.

Whether batches are paused and restarted from the point they stopped or are stopped and restarted from the beginning, it is critical that the process be predefined to the extent possible in an SOP and that all data be saved and all actions documented.

18.4 CONCLUSIONS

Ensuring that an instrument is suitable to execute a specific analytical method to generate concentration results for support of a study is an extensive process. It starts with proper IQ/OQ/PQ, continues with the establishment of a maintenance program, and will also include specific samples that are what is traditionally called system suitability. Standards, QCs, and blank samples can also serve as indicators of system suitability. Requalification of a system after the interruption of an injection sequence also requires some type of system suitability that should be predefined in an SOP.

The system suitability process that is appropriate for your company and your application should be customized to meet your needs and the needs of your assay. If the process is well-thought out starting with establishment of the goals of the system suitability test, the expected outcomes and the process for documenting the system suitability test then the bioanalytical scientist will be successful in ensuring that their instrument is performing how it needs to for successful generation of bioanalytical results.

REFERENCES

ANVISA. *Manual for Good Bioavailability and Bioequivalence Practices*. Volume 2(21). 2002.

Bansal SK, Layloff T, Bush ED, et al. Qualification of analytical instruments for use in the pharmaceutical industry: a scientific approach. AAPSPharmSciTech 2004;5(Article 22). Available at http://www.aaps.org/PharmSciTech/. Accessed Apr 11, 2013.

Bennett PK, Meng M, Čápka V. Managing phospholipid-based matrix effects In bioanalysis. Proceedings of the 17th International Mass Spectrometry Conference, Aug 27–Sep 1, 2006; Prague, Czech Republic.

Briscoe CJ, Stiles MR, Hage DS. System suitability in bioanalytical LC/MS/MS. J Pharm Biomed Anal 2007;44:484–491.

Burgess C, McDowall RD. Stimuli to the Revision Process. An Integrated and Harmonized Approach to Analytical Instrument Qualification and Computerized System Validation—A Proposal for an Extension of Analytical Instrument Qualification 1058. Mar 9, 2012.

Burgess C, Jones D, McDowall RD. Equipment qualification for demonstrating the fitness for purpose of analytical instrumentation. Analyst 1998;123:1879–1886.

Chang M, Li Y, Angeles R, et al. Development of methods to monitor ionization modification from dosing vehicles and phospholipids in study samples. Bioanalysis 2011;3(15):1719–1739

European Medicines Agency (EMA). Guideline on Bioanalytical Method Validation, 21 July 2011.

EMA. Volume 4 Good Manufacturing Practice Medicinal Products for Human and Veterinary Use Annex 11: Computerised Systems. Jun 30, 2011.

Guo X, Lankmayr E. Phospholipid-based matrix effects in LC–MS bioanalysis. Bioanalysis 2011;3(4):349–352.

Havel J. Validation of Caffeine in Human Plasma by LC/MS/MS. Unpublished work. 2012. PRA International.

ISPE. Gamp 5: Compliant GxP Computerized Systems. Feb 2008.

ISPE. A Risk-Based Approach to GxP Process Control Systems (2nd ed.) (Gamp® Good Practice Guide). 2011.

McDowall RD, Smith P, Vosloo N. System Suitability Tests and AIQ. 2010; Available at http://www.perkinelmer.com/CMSResources/Images/44-74522WTP_WhySSTisnosubstitute forAIQ.pdf. Accessed Mar 21, 2013.

OECD. OECD Series on Principles of Good Laboratory Practice and Compliance Monitoring Number 10. GLP Consensus Document. The Application of GLP to Computerised Systems Environment Monograph. 1995. Available at http://www.

oecd.org/officialdocuments/displaydocumentpdf/?cote=ocde/gd(95)115&doclanguage=en. Accessed Mar 21, 2013.

Phenomenex HPLC Calculations. Available at http://freedownload.is/pdf/hplc-calculation. Accessed Mar 21, 2013.

Swiss Federal Office of Public Health. Working Group on Information Technology. Guidelines for the Validation of Computerised Systems v02, 2007.

Stoneburner G, Goguen A, Feringa A. NIST Risk Management Guide for Information Technology Systems. Jul 2002.

US FDA. General Principles of Software Validation; Final Guidance for Industry and Staff. Jan 11, 2002.

US FDA. Guidance for Industry: Computerized Systems Used in Clinical Trials. Apr 1999.

US FDA Form 483, Inspection Report. Dec 2010.

US FDA Form 483, Inspection Report. May 2010.

US FDA Form 483, Inspection Report. Aug 2006.

US FDA Untitled Letter addressed to Dr. Roger Hayes, Cetero Research. Reference No.: 11-HFD-45-07-02. Jul 26, 2011.

US FDA Center for Drug Evaluation and Research. Guidance for Industry, Bioanalytical Method Validation. May 2001. Available at http://www.fda.gov/downloads/Drugs/GuidanceComplianceRegulatoryInformation/Guidances/ucm070107.pdf. Accessed Mar 21, 2013.

US FDA Center for Drug Evaluation and Research. Reviewer Guidance; Validation of Chromatographic Methods. 1994. Available at http://www.fda.gov/downloads/Drugs/GuidanceComplianceRegulatoryInformation/Guidances/UCM134409.pdf. Accessed Mar 21, 2013.

US FDA Title 21 Code of Federal Regulations Part 11. Apr 1, 2008.

US FDA 21 CFR Guidance for Industry Part 11, Electronic Records; Electronic Signatures—Scope and Application. Aug 2003.

US Pharmacopeia Chapter <621> Chromatography–System Suitability. Feb 20, 2006.

US Pharmacopeia Chapter <1058> Analytical Instrument Qualification. 2008.

19

DERIVATIZATION IN LC-MS BIOANALYSIS

TOMOFUMI SANTA

19.1 INTRODUCTION

The combination of liquid chromatography with mass spectrometry (LC-MS) and tandem mass spectrometry (LC-MS/MS) are frequently utilized for the sensitive and selective determination of the trace level compounds in biological samples. In particular, LC-MS(/MS) equipped with atmospheric pressure chemical ionization (APCI) or electrospray ionization (ESI) as the ion source is most often used method.

These ionization methods generate the gas phase ions of the analytes. In the APCI mode, reagent gas ions, formed from solvent molecules by corona discharge, generate the analytes ions by gas phase reactions. In the ESI mode, gas phase ions are mainly generated by transferring the ions in solution into gas phase in the presence of a strong electrical field. Therefore, generally, APCI is useful for low to medium polarity compounds having high proton affinity atoms such as oxygen and nitrogen; on the contrary, ESI is more useful for the compounds that can form ionic species in the solution. ESI is the more frequently used method for the analysis of biomolecules, since it can ionize wider range of the compounds and is applicable to the polar compounds or large molecular weight compounds. Furthermore, ESI requires lower temperature for ionization compared with APCI, and thus, it can be used for thermally unstable compounds.

The analyte having the following properties can be sensitively analyzed by LC/ESI-MS(/MS). First, it must be in its ionic form in the solution phase or be chargeable through adduct formation in gas phase reaction (Cech and Enke, 2001). Second, it is preferable to have the volatile structure, that is, to have the appropriate hydrophobic structure, since the hydrophobic ions prefer the droplet-air interface and reside at the droplet surface generated by electrospray. Consequently, these ions enter the gas phase more readily than those in the droplet interior and show the higher signal intensities (Cech and Enke, 2000, 2001; Cech et al., 2001; Zhou and Cook, 2001). Therefore, the chargeability and hydrophobicity of the analytes are the critical factors for the ionization efficiency (Okamoto et al., 1995; Nordstrom et al., 2004; Henriksen et al., 2005; Ehrmann et al., 2008). Furthermore, the hydrophobic compounds can be well separated on the reversed-phase column from salts and interfering compounds possessing suppression effects on ESI (Nordstrom et al., 2004). And the hydrophobic compounds are eluted by the mobile phase with the higher organic solvent content. The higher organic solvent content is suitable for the generation of charged droplets by electrospray and thus gives the higher signal intensities (Cech and Enke, 2001; Zhou and Hamburger, 1995).

However, not all the compounds can be favorably analyzed by LC/ESI-MS(/MS). Thus, chemical derivatization of the analyte is often used to enhance the detection sensitivity. So far, a number of derivatization reagents for LC-MS have been reported and they are summarized in several review papers (Higashi and Shimada, 2004; Gao et al., 2005; Johnson, 2005; Iwasaki et al., 2011; Santa, 2011). The reagents for peptides and proteins were also reviewed (Leitner and Lindner, 2004; Toyo'oka, 2012). The readers should refer to these articles. This section presents an overview of the derivatization reagents that have been applied to LC/ESI-MS(/MS), focusing on the application to low molecular weight compounds.

Handbook of LC-MS Bioanalysis: Best Practices, Experimental Protocols, and Regulations, First Edition. Edited by Wenkui Li, Jie Zhang, and Francis L.S. Tse.

19.2 DERIVATIZATION REAGENTS ENHANCING IONIZATION EFFICIENCY IN LC/ESI-MS

ESI is the method to generate gas phase ions by transferring the ions in solution into gas phase in the presence of a strong electrical field. Therefore, the chargeable and volatile analytes are sensitively detectable in ESI-MS. The initial derivatization reagents are mainly focused on the improvement of the chargeability. The analytes were derivatized to structures that are ionic or can be readily ionizable in solution; that is, the derivatization reagents introduced a permanently charged moiety or a readily ionizable moiety into the analytes. The former is sometimes called "charged derivatization" and the latter is called "ionizable derivatization."

In the early stage of the derivatization in ESI-MS, charged derivatization method is often adopted. Quirke et al. (1994) reported the derivatization of alkyl halides, alcohols, phenols, thiols, and amines using a number of the reagents to achieve the enhancement of the signal intensities. A typical example is the derivatization of alcohol using 2-fluoro-1-methylpiridinium *p*-toluenesulfonate (Figure 19.1a). Cholesterol was converted to its *N*-methylpiridyl ether, and an intense [M^+] ion of the derivative was observed in positive ESI-MS.

Girard's reagent P (1-(carboxymethyl)pyridium chloride hydrazide; Gir P) and Girard's reagent T ((carboxymethyl)trimethylammonium chloride hydrazide; Gir T) are the derivatization regents for carbonyl compounds (Figure19.1b and 19.1c). They were originally used for the separation of the keto-steroids from the other steroids by the derivatization to water-soluble hydrazones. These reagents possess the quaternary amino group and are able to enhance the ionization efficiency of the analytes. Shackleton et al. (1997) adopted Gir P for the analysis of steroids, 17β-fatty acid esters of testosterone. The nonpolar steroids were derivatized to the water-soluble hydrazones and analyzed in LC/ESI-MS. Gir P and Gir T are sometimes used for the analysis of carbonyl compounds.

The phosphonium moiety was also used for the charged derivatization. Barry et al. (2003) reported *S*-pentafluorophenyl tris(2,4,6-trimethoxyphenyl)phosphonium acetate bromide (TMPP-AcPFP) for alcohols and (4-hydrazino-4-oxobutyl)[tris(2,4,6-trimethoxyphenyl)] phosphonium bromide (TMPP-PrG) for aldehydes and ketones (Figure 19.1d and e). These reagents were used for the derivatization of sugar and steroids. TMPP-PrG enhanced detection responses of carbonyl compounds, but TMPP-AcPFP did not always improved the detection responses of alcohols. The reagents having TMPP moiety for amines and carboxylic acids were also reported (Leavens et al., 2002).

A ferrocene-based electrochemically ionizable derivatization was reported (Van Berkel et al., 1998). Ferrocenoyl azide was used for the derivatization of alcohols and phenols (Figure 19.1f). The generated ferrocene carbamates were ionized during the electron spray process by one electron oxidation and formed the radical cation. Cholesterol was sensitively detected by this method.

Charged derivatization gives strong signal intensity in ESI-MS comparing with ionizable derivatization; however, the generated derivatives were sometimes too hydrophilic and troublesome for the separation on the reversed-phase column.

19.3 DERIVATIZATION REAGENTS IN LC/ESI-MS/MS

Tandem mass spectrometry (MS/MS) enables sensitive detection of the analytes, since MS/MS detection decreases the noise level and improves the signal-to-noise ratios comparing with MS detection. To be sensitively analyzed by LC/ESI-MS/MS, it is desired for the target analyte to have the suitable structure for MS/MS detection, that is, to fragment efficiently upon collision-induced dissociation (CID) and generate an intense and particular product ion. Recently, derivatization reagents suitable for LC/ESI-MS/MS have

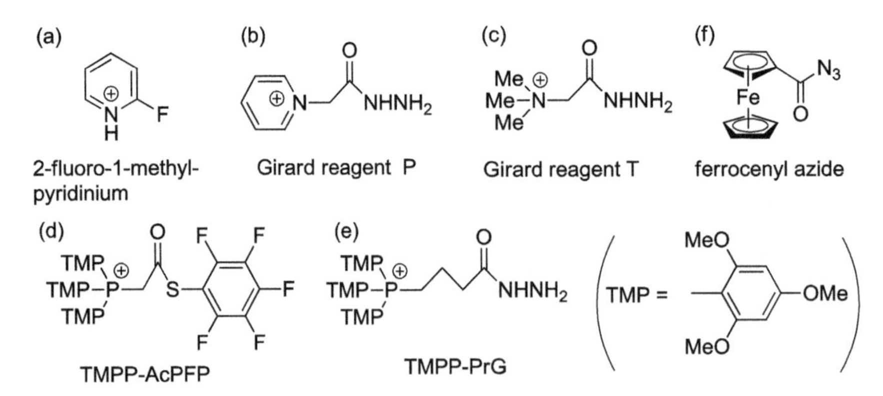

FIGURE 19.1 Chemical structures of the derivatization reagents enhancing ionization efficiency of analytes.

been reported. These reagents generate the derivatives having a structure suitable for MS/MS detection as well as enhance the ionization efficiency of the analytes. Therefore, most of the reagents carry the ionization moiety, hydrophobic moiety, and the structure suitable for MS/MS detection. Usually, esters, hydrazones, urea or thiourea, aromatic sulfonyl compounds, and alkyl quaternary ammonium compounds are easily cleaved by CID. The main reagents for each functional group are described below with a few examples of their application.

19.3.1 For Ketones and Aldehydes

Ketones and aldehydes possess neutral functional groups. The ionization efficiencies in ESI of these compounds are sometimes low. Therefore, a chargeable moiety was often introduced to enhance the ionization efficiency. In the early stage, hydroxylamine was used for derivatization of carbonyl compounds (Lampinen-Salomonsson et al., 2006). Though the detection sensitivity was increased, several product ions were produced by CID, and thus, it is not so suitable for MS/MS detection.

One of the prominent reagents for carbonyl compounds is HMP (2-hydrazino-1-methyl-pyridine) (Figure 19.2a). HMP has a quaternary ammonium group as an ionization moiety and a hydrophobic aromatic structure. Higashi et al. (2005) used HMP for the derivatization of oxo-steroids such as testosterone, and pregnenolone, progesterone. HMP reacted with oxo-steroids at 60°C within 1 h in ethanol containing 0.5% trifluoroacetic acid (TFA). The derivatives of mono-oxosteroids provided only their [M^+] ions in MS analysis. And in the MS/MS analysis of the HMP derivatives, the product ion at *m/z* 108 was observed, which was formed by the cleavage of the N-N bond of the hydrazone by CID. The HMP-oxo-steroids were separated on a reversed-phase column and detected by MS(/MS). The attained limits of detection (LOD) of the HMP derivatives of oxo-steroids were 70–1600-folds higher compared with the intact steroids. On the contrary, Gir P derivatives of oxosteroids gave many structurally informative product ions in the MS/MS analysis. The most abundant product ions was $[M\text{-}79]^+$, which was formed by the elimination of the pyridine moiety from [M^+], but the intensity was rather low. The LODs in selected reaction monitoring (SRM) mode were almost equal or inferior to those in the selected ion monitoring (SIM) mode. In addition, Gir P derivatives showed a significant peak tailing in chromatographic separation (Higashi et al., 2005). Thus, Gir P was not so suitable for the LC/ESI-MS/MS analysis of oxo-steroids. HMP is sometimes used for the analysis of oxo-steroids.

However, HMP is not effective for increasing the detection responses of di-oxo-steroids, since small molecules with a multicharge are unstable in the gas phase and provided multiple ions. To overcome these problems, 2-hydrazinopyridine (HP) was used for di-oxo-steroids such as androsterone and progesterone. The generated derivatives of androsterone and progesterone provided the intense product ions at *m/z* 322 and 348, respectively (Higashi et al., 2007a). Although the structures of these ions could not be identified, it was inferred that they were formed by the loss of one HP moiety together with a part of the A-ring. HP was suitable for the analysis of di-oxo-steroids (Shibayama et al., 2008; Hala et al., 2011).

Dansyl moiety is also suitable for MS/MS detection (Figure 19.2b). Dansyl hydrazine (5-dimethylaminonaphthalene-1-sulfonyl hydrazine; Dns-Hz) was used for the analysis of succinylacetone in dried blood spot specimens (Al-Dirbashi et al., 2006). The generated dansyl hydrazone of succinylacetone selectively gave the product ion at *m/z* 170 by CID,

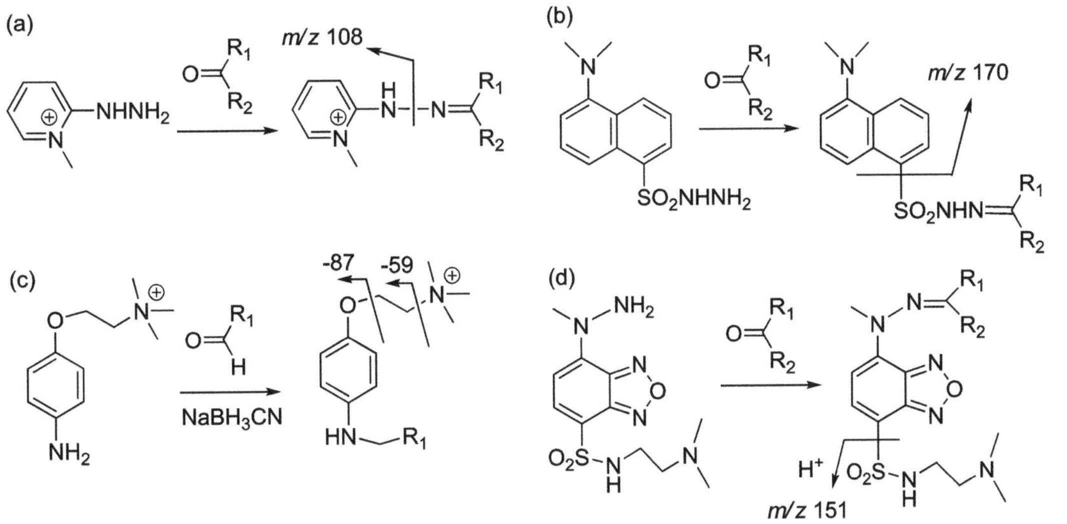

FIGURE 19.2 Derivatization reaction for ketones and aldehydes, and the product ions of the derivatives: (a) HMP, (b) Dns-Hz, (c) 4-APC, and (d) DAABD-MHz.

assigned to the cleavage of dimethyaminonaphtyl moiety originated from the reagent. The product ion spectra of the derivatives were rather simple and clear. The transition of *m/z* 462 ($[M+H]^+$) to *m/z* 170 was used for SRM.

4-(2-(Trimethylammonio)ethoxy)benzenaminium halide (4-APC) reacted with aldehydes and generated amines in the presence of $NaBH_3CN$. 4-APC possesses an aniline moiety for the reaction with the aliphatic aldehydes and a quaternary ammonium group for the improvement of sensitivity. The generated derivatives gave the product ions at *m/z* $[M\text{-}59]^+$ and $[M\text{-}87]^+$, derived from the loss of trimethylammonium moiety and trimethylaminoethyl moiety by CID (Eggink et al., 2008) (Figure 19.2c). Similarly, 4-(2-((4-bromophenethyl)dimethylammonio)ethoxy)benzenaminium dibromide (4-APEBA) was reported. It contains a bromophenyl group, and the isotopic signature of ^{79}Br and ^{81}Br isotope (100:98) provided confirmation of the presence of bromine in the derivatives (Eggink et al., 2010). DAABD-MHz (4-[2-(*N,N*-dimethylamino)ethylaminosulfonyl]-7-*N*-methylhydrazino-2,1,3-benzoxadiazole) (Santa et al., 2008) reacted with aldehydes at 50°C within 10 min in acetonitrile containing 0.5% TFA. The generated derivatives gave a predominant product ion at *m/z* 151 by CID, derived from the protonated (*N,N*-dimethylamino)ethylaminosulfonyl moiety of the reagent (Figure 19.2d). 4-Hydrazino-*N,N,N*-trimethyl-4-oxobutanaminium iodide (HTMOB) was a modified Girard's reagent. The derivatives gave rather a simple product ion spectrum, derived from the neutral loss of 59 Da (trimethylamino moiety). HTMOB was applied to the profiling of ketones, ketoacids, and ketodiacids in the children's urine samples by ESI-MS/MS without chromatographic separation (Johnson, 2007).

19.3.2 For Phenols and Alcohols

Alcohols and phenols are neutral and sometimes hydrophilic compounds. Therefore, deivatization is often used to enhance the ionization efficiency and to increase the hydrophobicity. One of the most widely used reagents is dansyl chloride (5-dimethylamino-1-naphthalenesulfonyl chloride; Dns-Cl) (Figure 19.3a). A typical example is the analysis of ethinylestradiol (EE), a potent synthetic estrogen (Anari et al., 2002). The ionization efficiencies of steroids were usually low, and thus, they cannot be sensitively analyzed by LC/ESI-MS/MS. EE possesses the hydrophobic aromatic structure and two hydroxyl groups. The estimated dissociation constants (p*K*a) of phenolic and hydroxyl group were 10.4 and 13.3, respectively. Therefore, EE is supposed to be ionized minimally in solution under acidic to neutral pH range suitable for chromatographic separation. For example, less than 0.001% of EE is estimated to be ionized in mobile phase containing 0.1% formic acid (pH 3). This may explain low ionization efficiency of EE using electrospray, since ionization in the gas phase is known to be directly affected by ionization in solution. In addition, CID of EE did not generate an intense product ion. The intensity of the product ion to the precursor ion was less than 0.1% for EE. These results suggested that intact EE was not suitable for SRM. Anari et al. (2002) used Dns-Cl for the derivatization of EE. Dns-Cl reacted with a phenolic group of EE at 60°C within several minutes in acetone and sodium bicarbonate buffer (pH 10.5). The introduction of a dansyl functional group bearing a basic nitrogen shifted the predicted p*K*a of the molecule to about 3.3, and 67% of the molecule is estimated to be ionized in the solution of pH 3. As expected, an intense protonated molecule ($[M+H]^+$) of 3-dansyl-EE at *m/z* 530 was observed in the mass spectrum. Furthermore, CID produced an intense product ion at *m/z* 171, corresponding to the protonated 5-(dimethylamino)-naphthylene moiety (Figure 19.4). SRM using the transition of protonated 3-dansyl-EE at *m/z* 530 to *m/z* 171 resulted in a sensitive detection of the derivative, and the attained limit of detection for EE was 0.2 fg/ml. Dns-Cl is widely used for the analysis of steroids having phenol structure.

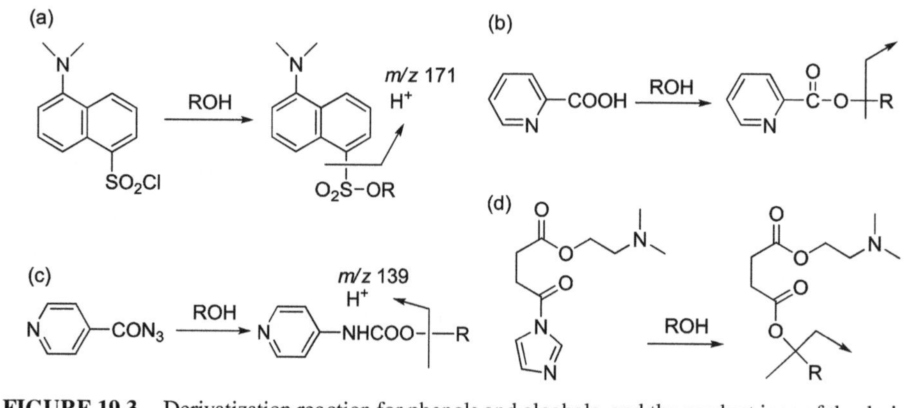

FIGURE 19.3 Derivatization reaction for phenols and alcohols, and the product ions of the derivatives: (a) Dns-Cl, (b) picolynic acid, (c) NA, and (d) MDMAES imidazole.

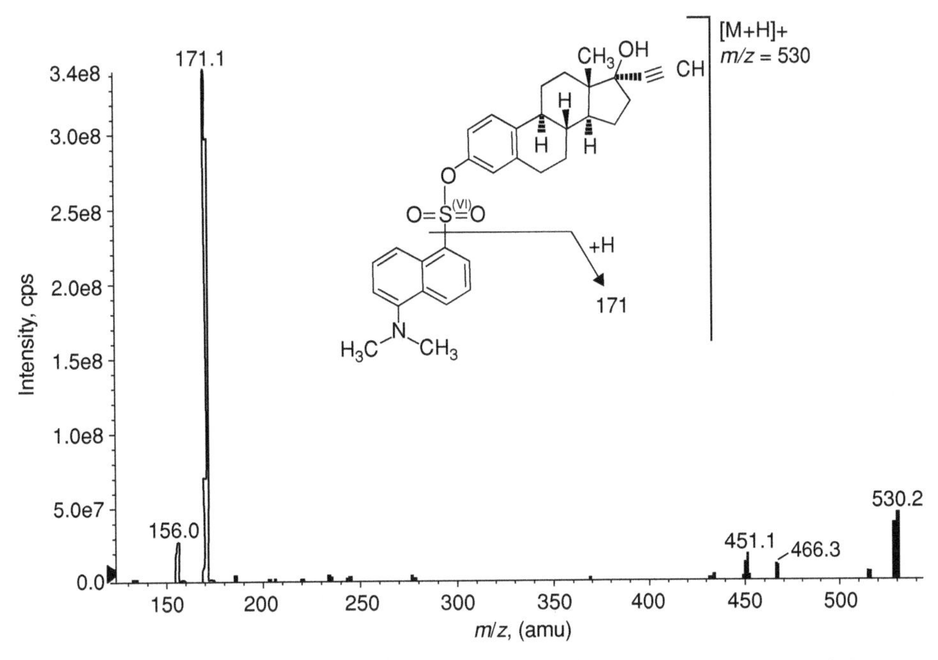

FIGURE 19.4 Product ion scan of the *m/z* 530 ($[M+H]^+$ of 3-dansyl-ethinylestradiol) (reproduced with permission from Anari et al., 2002; copyright 2002 American Chemical Society).

Dns-Cl was also useful for alcohols such as testosterone, cholesterol, retinol, cholecalciferol, and so on (Tang and Gengerich, 2010). Dns-Cl reacted with these alcohols at 65°C for 1 h in the presence of 4-(dimethylamino)-pyridine and *N*,*N*-diisopropylethylamine in dichloromethane. The dansylated alcohols gave the strong signal of the protonated molecules ($[M+H]^+$), and CID produced the product ion at *m/z* 171 and *m/z* 252. SRM using the transition of $[M+H]^+$ of the derivative to *m/z* 171 or *m/z* 251 resulted in a sensitive detection of the derivatives. The sensitivities were 17–1000-folds increased comparing with the intact alcohols. This method was applied to characterizing the P450 oxidation products in human liver extracts.

Honda et al. used picolynic acid for the derivatization of the hydroxyl group (Figure 19.3b). Picolynic acid reacted with 7α-hydroxy-4-cholestene-3-one, a biomarker for bile acid biosynthesis, in 30 min at room temperature in the presence of the condensation reagents in tetrahydrofuran. The derivative gave the product ion at *m/z* 383 by CID due to the loss of picolinic acid moiety. The transition of *m/z* 506 ($[M+H]^+$) to *m/z* 383 was used for SRM (Honda et al., 2007). Picolinic acid was sometimes used for the analysis of the steroids having hydroxyl groups.

Introducing two permanently charged group to the small molecules did not improve the detection responses, since the multicharged small molecules are unstable in the gas phase and provided multiple ions. Therefore, the reagent having a highly proton affinity group such as pyridyl group is favorable for the derivatization of di-hydroxysteroids. Higashi et al. (2007b) used isonicotinoyl azide (NA) for the derivatization of di-hydroxysteroids such as 5α-androstane-3α,17β-diol (Figure 19.3c). NA reacted with two hydroxyl groups of estradiol at 80°C for 30 min in benzene. The generated derivative gave the product ion at *m/z* 139 by CID, which was assigned to the protonated pyridyl carbamic acid. The transition of $[M+H]^+$ ion to *m/z* 139 was usable for SRM.

Johnson et al. (2001) synthesized mono-(dimethylamino-ethyl) succinyl (MDMAES) imidazole (Figure 19.3d). MDMAES imidazole reacted with hydroxyl group at 70°C for 10 min in dichloromethane containing 1% triethylamine. The generated MDMAES ester gave the product ion at *m/z* 369 by CID due to the neutral loss of MDMAES moiety (189 Da). Cholesterol and dehydrocholesterol in dried spot of plasma were analyzed by ESI-MS/MS without chromatographic separation.

Simple esterification of the analytes was sometimes useful for the hydrophilic alcohols. Propionyl and benzoyl anhydride were used for the derivatization of bases, cytokinins, ribosides, and intact nucleotides such as AMP, ADP, and AMP (Nordstrom et al., 2004). The ESI response was enhanced by the formation of hydrophobic derivatives. In addition, the retention on a reversed-phase column was greatly increased, and the derivatives were separated without the need for an ion-paring reagent, known for its unwanted suppression effects on ionization.

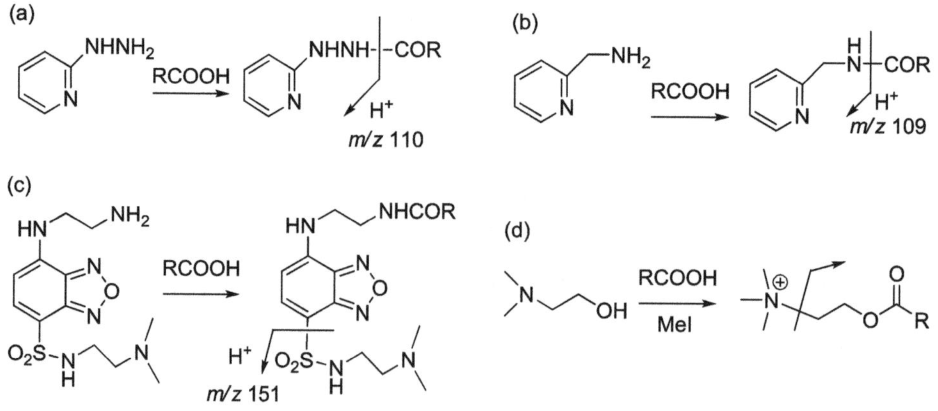

FIGURE 19.5 Derivatization reaction for carboxylic acids and the product ions of the derivatives: (a) HP, (b) PA, (c) DAABD-AE, and (d) TAME alcohol.

19.3.3 For Carboxylic Acids

Carboxylic acids are detectable in the negative ESI-MS. However, their sensitivity is rather poor. In addition, the mobile phases for the carboxylic acids separation are not always compatible with ESI-MS. Therefore, carboxylic acids were usually transformed to the hydrophobic and ionizable derivatives.

Butanolic HCl derivatization is sometimes used for carboxylic acids. The generated butyl esters are rather hydrophobic and ionizable by ESI. The generated esters often gave the product ion at *m/z* of ($[M+H\text{–}56]^+$) due to the loss of C_4H_8 (56 Da) by CID. One of the examples is the analysis of methylmalonic acid, the marker for a group of metabolic disorders caused by deficiency in methylmalonyl-CoA mutase or a defect in vitamin B_{12} metabolism. Its di-butyl ester gave the product ion at *m/z* 119 due to the loss of 2 C_4H_8. The transition of *m/z* 231 ($[M+H]^+$) to *m/z* 119 was used for SRM (Magera et al., 2000). Another example for butanolic HCl derivatization of carboxylic acids is the simultaneous analysis of amino acids and acylcarnitines in dried blood spots for newborn screening (Chace et al., 2003; Chace and Kalas, 2005; Li and Tse, 2010). The generated butyl esters of amino acids were introduced to ESI-MS/MS without chromatographic separation. Most of α-amino acids butyl esters gave the intense product ions correspond to the loss of $HCOOC_4H_8$ (102 Da) by CID. Therefore, amino acids profile in biological samples can be obtained by the neutral loss scan of 102 Da. Acylcarnitines are the marker metabolites for inherited disorders related to organic acid and fatty acid metabolism. Acylcarnitines have a quaternary ammonium group and a carboxylic group in their structure. Their butyl esters were introduced to ESI-MS/MS and gave the common product ion at *m/z* 85 by CID. Therefore, acylcarnitine profile can be obtained by precursor ion scan of *m/z* 85. These methods are widely used for the analysis of amino acids and acylcarnitines in urine, plasma, serum, or blood (dried blood spot) (Millington et al., 1990; Rashed et al., 1994).

Higashi et al. reported the simple and practical method for the analysis of carboxylic acids using 2-hydrazinopyridine (HP) and 2-picolylamine (PA) (Figure19.5a and 19.5b). Several biological important compounds having carboxylic acids, such as chenodeoxycholic acid, were derivatizated with these reagents at 60°C for 10 min in the presence of the condensation reagents. The resulting HP- and PA-derivatives were highly responsive in ESI-MS. The generated derivatives of HP gave the strong product ion at *m/z* 110, derived from the protonated 2-hydrazinopyridine moiety, and those of PA gave the strong product ion at *m/z* 109, derived from 2-picolylamine moiety, by CID. The transitions to *m/z* 109 from the $[M+H]^+$ of the PA-derivatives were used for SRM. The PA-derivatization was successfully applied to a biological sample analysis; the derivatization followed by LC/ESI-MS/MS enabled the detection of trace amount of the clinically important carboxylic acids such as bile acids, homovanillic acid in human saliva. The detection responses of PA-derivatives were increased by 9–158-folds over the intact carboxylic acids (Higashi et al., 2010). Similarly, 3-picolylamine and 3-picolylcarbinol were used for fatty acids (Li and Frank, 2011), and 3-hydroxy-1-methyl-piperidine for malonic acid (Honda et al., 2009), 3-(hydroxymethyl)-pyridine for the polar molecules in plant extracts (Kallenbach et al., 2009), and 4-dimethylaminobenzylamine for valproic acid and its metabolite (Cheng et al., 2007).

DAABD-AE (4-[2-(*N,N*-dimethylamino)ethylaminosulfonyl]-7-(2-aminoethylamino)-2,1,3-benzoxadiazole) (Santa et al., 2007) was synthesized and used for the analysis of very long chain fatty acids, the markers of peroxisomal disorders (Figure 19.5c). DAABD-AE reacted with fatty acids at 60°C for 45 min in the presence of the condensation reagents. The generated amide derivative provided almost the single product ion at *m/z* 151 by CID, derived from

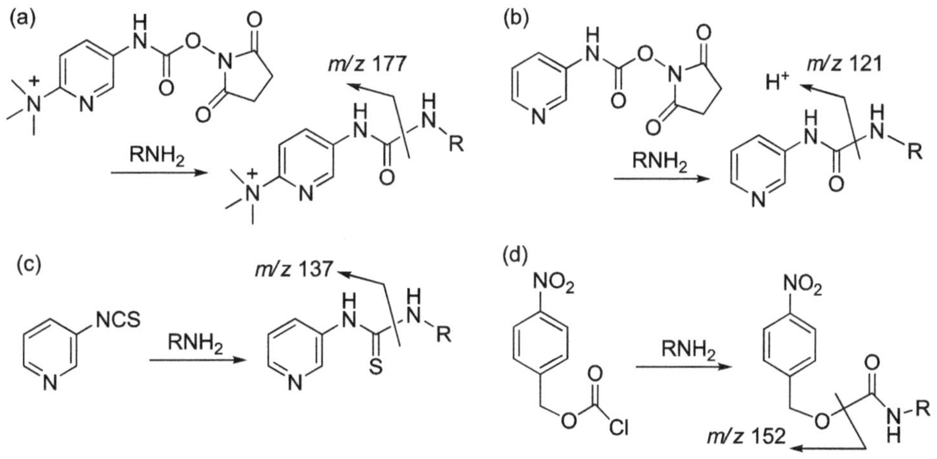

FIGURE 19.6 Derivatization reaction for amines and the product ions of the derivatives: (a) THAS, (b) APDS, (c) Py-NCS, and (d) 4-nitrobenzyl chloroformate.

the protonated (*N,N*-dimethylamino)ethylaminosulfonyl moiety of the reagent. The transition of $[M+H]^+$ ions to *m/z* 151 was used for SRM. Compared to standard gas chromatography-mass spectrometric methods routinely used for this purpose, this LC/ESI-MS/MS method is more simple, saves 75% of instrument time and requires one tenth of biological sample volume (Al-Dirbashi et al., 2008).

TMAE (trimethylaminoethyl) ester derivatives were also used for the analysis of very long chain fatty acids (Johnson et al., 2003) (Figure 19.5d). Fatty acids were treated with oxalyl chloride, dimethylaminoethanol, followed by the methylation with methyl iodide. These derivatives gave the product ion by the loss of 59 Da, derived from $(CH_3)_3N$ moiety of the derivatization reagent, and each fatty acid derivative was detected by SRM. The generated derivatives were suitable for MS/MS detection. However, the three-step derivatization reaction was tedious for routine assay. TMAE or DMAE (dimethylaminoethyl) ester derivatization of fatty acids and ESI-MS/MS analysis without chromatographic separation were reported.

19.3.4 For Amines

The compounds having amino group are easily protonated under acidic conditions and suitable for ESI-MS. However, the analysis of amines is often troublesome because of their high polarity, basicity, and high water solubility. Chemical derivatization makes amines more hydrophobic and the generated derivatives can be more easily separated from the interfering compounds on the reversed-phase column, and can be sensitively detected in ESI-MS. In addition, the increase in the molecular weight decreases in the background noise from the matrix, since the background is generally lower in the higher mass range.

THAS (*p-N,N,N*-trimethylammonioanilyl-*N*′-hydroxysuccidimidyl carbamate iodide) was the reagent designed for LC/ESI-MS/MS. It reacted with amino acids at 60°C for 10 min in borate buffer (pH 8.8) to form urea compounds. The derivatives gave the characteristic cleavage at the urea bond by CID, and produced characteristic product ions at *m/z* 177, derived from the reagent moiety. Amino acids were analyzed with the detection limits of atto-mole level (Shimbo et al., 2009b) (Figure 19.6a). APDS (3-aminopyridyl-*N*-hydroxysuccimidyl carbamate) was also designed for LC/ESI-MS/MS and used for the analysis of more than 100 compounds having amino group in biological fluid. The generated derivatives are hydrophobic compared with corresponding THAS derivatives, and thus, they are suitable for the separation on the reversed-phase column. The transition of all the protonated molecular ions to the common product ion at *m/z* 121 was monitored (Shimbo et al., 2009a) (Figure 19.6b).

As described earlier, urea and thiourea moieties are easily cleaved by CID and suitable structure for MS/MS. AQC (6-aminoquinolyl-*N*-hydroxysuccinimidyl carbamate) is the commonly used fluorescence derivatization reagent for amino acids. AQC was used for the analysis of β-*N*-methylamino-L-alanine and 2,3-diaminobutylic acid, a potential human neurotoxin and its isomer. The urea structure of the derivatives was cleaved by CID, and the product ions at *m/z* 145 and *m/z* 171 derived from the AQC moiety were observed (Spacil et al., 2010). AQC was also used for amino acids analysis. (Fiechter and Mayer, 2011). NIT (naphthylisothiocyanate) was used for the determination of eighteen kinds of primary and secondary amines in air samples. The derivatives of the primary amines gave the common base peak at *m/z* 144 and the product ion at *m/z* 127 by CID, whereas those of secondary amines gave the common base peak at *m/z* 186 and the product ion at *m/z* 128 by CID (Claeson et al., 2004). Conventional isothiocyanates, 3-pyridyl isothiocyanate (Py-NCS) (Figure 19.6c), *p*-(dimethylamino)phenyl isothiocyanate (DMAP-NCS), and *m*-nitrophenyl isothiocyanate (NP-NCS) were applied to the derivatization of amines. The generated derivatives having thiourea structures gave the intense product ions at *m/z* 137, *m/z* 179, *m/z* 137 by CID, respectively (Santa, 2010).

The reagent developed for proteomics, iTRAQ (isobaric tags for relative and absolute quantitation) (Ross et al., 2004), was also applicable for amino acid analysis by LC/ESI-MS/MS. A comparison of LC-MS/MS, GC-MS, and amino acid analyzer using iTRAQ derivatization was reported (Kaspar et al., 2009).

Dns-Cl was sometimes used for amines. The generated derivatives gave the product ion at *m/z* 171, derived from protonated dimethylaminonaphtyl moiety of the reagent. Dns-Cl was used for the analysis of musimol and ibotenic acid, bioactive compounds in mushroom. The transitions of (M^+) ions to *m/z* 171 were used for SRM (Tsujikawa et al., 2007). Chloroformates such as nitrobenzyl chloroformate (Blum et al., 2000) (Figure 19.6d), PrCl (propyl chloroformate), and FMOC (9-fluorenylmethyl chloroformate) (Uutela et al., 2009) were also used for amines. The generated derivatives were easily fragmented by CID and suitable for MS/MS detection.

19.4 CONCLUSION

The derivatization reagents in LC/ESI-MS/MS applied to the low molecular weight compounds are summarized. The derivatization reagents enhance the chargeability and hydrophobicity of the analyte to improve the ionization efficiency. Furthermore, several reagents transform the analytes to the structures suitable for MS/MS detection. The generated derivatives were fragmented easily by CID and efficiently generated the particular intense product ions. The prominent derivatization regents for LC/ESI-MS/MS are indispensable for the sensitive and selective detection of the various kinds of the compounds in the fields of biomedical analysis.

REFERENCES

Al-Dirbashi OY, Rashed MS, Ten Brink HJ, et al. Determination of succinylacetone in dried blood spots and liquid chromatography tandem mass spectrometry. J Chromatogr B 2006;831:274–280.

Al-Dirbashi OY, Santa T, Rashed MS, et al. Rapid UPLC-MS/MS method for routine analysis of plasma pristanic, phytanic, and very long chain fatty acid markers of peroxisomal disorders. J Lipid Res 2008;49:1855–1862.

Anari MR, Bakhtiar R, Zhu B, Huskey S, Franklin RB, Evans DC. Derivatization of ethinylestradiol with dansyl chloride to enhance electrospray ionization: application in trace analysis of ethinylestradiol in rhesus monkey plasma. Anal Chem 2002;74:4136–4144.

Barry SJ, Carr RM, Lane SJ, et al. Use of S-pentafluorophenyl tris(2,4,6-trimethoxyphenyl)phosphonium acetate bromide and (4-hydrazino-4-oxobutyl)[tris(2,4,6-trimethoxyphenyl)] phosphonium bromide for the derivatization of alcohols, aldehydes and ketones for detection by liquid chromatography/electrospray mass spectrometry. Rapid Commun Mass Spectrom 2003;17:484–497.

Blum W, Aichholz R, Ramstein P, Kuhnol J, Froestl W, Desrayaud S. Determination of the $GABA_B$ receptor agonist CGP 44532 (3-amino-2-hydroxypropylmethylphosphinic acid) in rat plasma after pre-column derivatization by micro-high-performance liquid chromatography combined with negative electrospray tandem mass spectrometry. J Chromatogr B 2000;748:349–359.

Cech NB, Enke CG. Relating electrospray ionization response to nonpolar character of small peptides. Anal Chem 2000;72:2717–2723.

Cech NB, Enke CG. Practical implications of some recent studies in electrospray ionization fundamentals. Mass Spectrome Rev 2001;20:362–387.

Cech NB, Krone JR, Enke CG. Predicting electrospray response from chromatographic retention time. Anal Chem 2001;73:208–213.

Chace DH, Kalas TA. A biochemical perspective on the use of tandem mass spectrometry for new born screening and clinical testing. Clin Biochem 2005;38:296–309.

Chace DH, Kalas TA, Naylor EW. Use of tandem mass spectrometry for multianalyte screening of dried blood specimens from newborns. Clin Chem 2003;49:1797–1817.

Cheng H, Liu Z, Blum W, et al. Quantification of valproic acid and its metabolite 2-propyl-4-pentenoic acid in human plasma using HPLC-MS/MS. J Chromatogr B 2007;850:206–212.

Claeson AS, Ostin A, Sunesson AL. Development of a LC-MS/MS method for the analysis of volatile primary and secondary amines as NIT (naphthylisothiocyanate) derivatives. Anal Bioanal Chem 2004;378:932–939.

Eggink M, Wijtmans M, Ekkebus R, et al. Development of a selective ESI-MS derivatization reagent: synthesis and optimization for the analysis of aldehydes in biological mixtures. Anal Chem 2008;80:9042–9051.

Eggink M, Wijtmans M, Kretschmer A, et al. Targeted LC–MS derivatization for aldehydes and carboxylic acids with a new derivatization agent 4-APEBA. Anal Bioanal Chem 2010;397:665–675.

Ehrmann BM, Henriksen T, Cech NB. Relative importance of basicity in the gas phase and in solution for determining selectivity in electrospray ionization mass spectrometry. J Am Mass Spectrom 2008;19:719–728.

Fiechter G, Mayer HK. Characterization of amino acid profiles of culture media pre-column 6-aminoquinolyl-N-hydroxysuccinimidyl carbamate derivatization and ultra performance liquid chromatography. J Chromatogr B. 2011;879: 1353–1360.

Gao S, Zhang ZP, Karnes HT. Sensitivity enhancement in liquid chromatography/atmospheric pressure ionization mass spectrometry using derivatization and mobile phase additives. J Chromatogr B 2005;825:98–110.

Hala D, Overturf MD, Peterson LH, Huggett DB. Quantification of 2-hydrazinopyridine derivatized steroid hormones in fathead minnow (Pimephales promelas) blood using LC-ESI + /MS/MS. J Chromatogr B 2011;879:591–598.

Henriksen T, Juhler RK, Svensmark B, Cech N. The relative influence of acidity and polarity on responsiveness of small organic molecules to analysis with negative ion electrospray ionization mass spectrometry (ESI-MS). J Am Mass Spectrom 2005;16:446–455.

Higashi T, Ichikawa T, Inagaki S, Min JZ, Fukushima T, Toyo'oka T. Simple and practical derivatization procedure for enhanced detection of carboxylic acids in liquid chromatography-electrospray ionization-tandem mass spectrometry. J Pharm Bioamed Anal 2010;52:809–818.

Higashi T, Nishio T, Hayashi N, Shimada K. Alternative procedure for charged derivatization to enhance detection responses of steroids in electrospray ionization-MS. Chem Pharm Bull 2007a;55:662-665.2288;

Higashi T, Nishio T, Yokoi H, Ninomiya Y, Shimada K. Studies on neurosteroids XXI. An improved liquid chromatography-tandem mass spectrometric method for determination of 5α-androstane-3α,17β-diol in rat brains. Anal Sci 2007b;23: 1015–1019.

Higashi T, Shimada K. Derivatization of neutral steroids to enhance their detection characteristics in liquid chromatography. Anal Bioanal Chem 2004;378:875–882.

Higashi T, Yamauchi A, Shimada K. 2-Hydrazino-1-methylpyridine: a highly sensitive derivatization reagent for oxosteroids in liquid chromatography-electrospray ionization mass spectrometry. J Chromatogr B 2005;825:214–222.

Honda A, Yamashita K, Ikegami T, et al. Highly sensitive quantification of serum malonate, a possible marker for de novo lipogenesis, by LC-ESI-MS/MS. J Lipid Res 2009;50:2124–2130.

Honda A, Yamashita K, Numazawa M, et al. Highly sensitive quantification of 7α-hydroxy-4-cholesten-3-one in human serum by LC-ESI-MS/MS. J Lipid Res 2007;48:458–464.

Iwasaki Y, Nakano Y, Mochizuki K, et al. A new strategy for ionization enhancement by derivatization for mass spectrometry. J Chromatogr B 2011;879:1159–1165.

Johnson DW. Contemporary clinical usage of LC/MS: analysis of biologically important carboxylic acids. Clin Biochem 2005;38:351–361.

Johnson DW. A modified Girard derivatizing reagent for universal profiling and trace analysis of aldehydes and ketones by electrospray ionization tandem mass spectrometry. Rapid Commun Mass Spectrom 2007;21:2926–2932.

Johnson DW, ten Brink HJ, Jakobs C. A rapid screening procedure for cholesterol and dehydrocholesterol by electrospray ionization tandem mass spectrometry. J Lipid Res 2001;42:1699–1705.

Johnson DW, Trinh MU, Oe T. Measurement of plasma pristanic, phytanic and very long chain fatty acids by liquid chromatography-electrospray tandem mass spectrometry for the diagnosis of peroxisomal disorders. J Chromatogr B, 2003;798:159–162.

Kallenbach M, Baldwin I, Bonaventure G. A rapid and sensitive method for the simultaneous analysis of a aliphatic and polar molecules containing free carboxyl groups in plant extract by LC-MS/MS. Plant methods 2009;5:17–27.

Kaspar H, Dettmer K, Chan Q, et al. Urinary amino acid analysis: A comparison of iTRAQ-LC-MS/MS, GC-MS, and amino acid analyzer. J Chromatogr B 2009;877:1838–1846.

Lampinen-Salomonsson M, Beckman E, Bondesson U, Hedeland M. Detection of altrenogest and its metabolites in post administration horse urine using liquid chromatography tandem mass spectrometry-increased sensitivity by chemical derivatization of glucuronic acid conjugate. J Chromatogr B 2006;833: 245–256.

Leavens WJ, Lane SJ, Carr RM, Lockie AM, Waterhous I. Derivatization for liquid chromatography/electrospray mass spectrometry: synthesis of tris(trimethoxyphenyl)phosphonium compounds and their derivatives of amine and carboxylic acids. Rapid Commun Mass Spectrom 2002;16:433–441.

Leitner A, Lindner W. Current chemical tagging strategies for proteome analysis by mass spectrometry. J Chromatogr B-Anal Technol Biomed Life Sci, 2004;813:1–26.

Li W, Tse FLS. Dried blood spots sampling in combination with LC-MS/MS for quantitative analysis of small molecules. Biomed Chromatogr 2010;24:49–65.

Li X, Frank A. Improved LC-MS method for the determination of fatty acids in red blood cells by LC-Orbitrap MS. Anal Chem 2011;83:3192–3198.

Magera MJ, Helgeson JK, Matern D, Rinaldo P. Methylmalonic acid measured in plasma and urine by stable-isotope dilution and electrospray tandem mass spectrometry. Clin Chem 2000;46:1804–1810.

Millington DS, Kodo N, Norwood DL, Roe CR. Tandem mass spectrometry: a new method for acylcarnitine profiling with potential for neonatal screening for inborn errors of metabolism. J Inherit Metab Dis 1990;13:321–324.

Nordstrom A, Tarkowski P, Tarkowska D, et al. Derivatization for LC—electrospray ionization—MS: a tool for improving reversed-phase separation and ESI response of bases, ribosides, and intact nucleotides. Anal Chem 2004:76:2869–2877.

Okamoto K, Takahashi K, Doi T. Sensitive detection and structural characterization of trimethyl(p-aminophenyl)-ammonium-derivatized oligosaccharides by electrospray ionization-mass spectrometry and tandem mass spectrometry. Rapid Commun Mass Spectrom 1995;9:641–643.

Quirke JME, Adams CL, Van Berkel GJ. Chemical derivatization for electrospray ionization mass spectrometry. 1. Alkyl halides, alcohols, phenols, thiols, and amines. Anal Chem 1994;66:1302–1315.

Rashed MS, Ozand PT, Harrison ME, Watkins PJF, Evans S. Electrospray tandem mass spectrometry in the diagnosis of organic acidemias. Rapid Commun Mass Spectrom 1994;8:129–133.

Ross PL, Huang YLN, Marchese JN, et al. Multiplexed protein quantitation in *Saccharomyces cerevisiae* using amine-reactive isobaric tagging reagents. Mol Cell Proteomics 2004;3:1154–1169.

Santa T, Al-Dirbashi OY, Ichibangase T, et al. Synthesis of benzofurazan derivatization reagents for carboxylic acids in liquid chromatography/electrospray ionization-tandem mass spectrometry (LC/ESI-MS/MS). Biomed Chromatogr 2007;21: 1207–1213.

Santa T, Al-Dirbashi OY, Ichibangase T, Rashed MS, Fukushima T, Imai K. Synthesis of 4-[2-(*N*,*N*-dimethylamino)ethylaminosulfonyl]-7-*N*-methylhydrazino-2,1,3-benzoxadiazole (DAABD-MHz) as a derivatization reagent for aldehydes in liquid chromatography/electrospray ionization-tandem mass spectrometry. Biomed Chromatogr 2008;22:115-118.

Santa T. Isothiocyanates as derivatization reagents for amines in liquid chromatography/electrospray ionization-tandem mass spectrometry (LC/ESI-MS/MS). Biomed Chromatogr 2010;24:915–918.

Santa T. Derivatization reagents in liquid chromatography/electrospray ionization tandem mass spectrometry. Biomed Chromatogr 2011;25:1–10.

Shackleton CHL, Chuang H, Kim J, de la Torre X, Segura J. Electrospray mass spectrometry of testosterone esters: Potential for use in doping control. Steroids 1997;62:523–529.

Shibayama Y, Higashi T, Shimada K, et al. Liquid chromatography-tandem mass spectrometric method for determination of salivary 17a-hydroxyprogesterone: A noninvasive tool for evaluating efficacy of hormone replacement therapy in congenital adrenal hyperplasia. J Chromatogr B 2008;867:49–56.

Shimbo K, Oonuki T, Yahashi A, Hirayama K, Miyano H. Precolumn derivatization reagents for high-speed analysis of amines and amino acids in biological fluid using liquid chromatography/electrospray ionization tandem mass spectrometry. Rapid Commun Mass Spectrom 2009a;23:1483–1492.

Shimbo K, Yahashi A, Hirayama K, Nakazawa M, Miyano H. Multifunctional and highly sensitive precolumn reagents for amino acids in liquid chromatography/tandem mass spectrometry. Anal Chem 2009b;81:5172–5179.

Spacil Z, Eriksson J, Honasson S, Rasmussen U, Ilag LL, Bergman B. Analytical protocol for identification of BMAA and DAB in biological samples. Analyst 2010;135:127–132.

Tang Z, Gegerich FP. Dansylation of unactivated alcohols for improved mass spectral sensitivity and application to analysis of cytochrome P450 oxidation products in tissue extracts. Anal Chem 2010;82:7706–7712.

Toyo'oka T. LC-MS determination of bioactive molecules based upon stable isotope-coded derivatization method. J Pharm Biomed Anal 2012;69:174–184.

Tsujikawa K, Kuwayama K, Miyaguchi H, et al. Determination of muscimol and ibotenic acid in Amanita mushrooms by high-performance liquid chromatography and liquid chromatography-tandem mass spectrometry. J Chromatogr B 2007;852:430–435.

Uutela P, Ketola RA, Piepponen P, Kostianinen R. Comparison of different amino acid derivatives and analysis of rat brain microdialysates by liquid chromatography tandem mass spectrometry. Anal Chim Acta 2009;633:223–231.

Van Berkel GJ, Quirke JME, Tigani RA, Dilley AS, Covey TR. Derivatization for electrospray ionization mass spectrometry. 3. Electrochemically ionizable derivatives. Anal Chem 1998;70:1544–1554.

Zhou S, Cook KD. A mechanistic study of elecrospray mass spectrometry: charge gradients within electrospray droplets and their influence on ion response. J Am Soc Mass Spectrom 2001;12:206–211.

Zhou S, Hamburger M. Effects of solvent composition on molecular ion response in electrospray mass spectrometry: investigation of the ionization process. Rapid Commun Mass Spectrom 1995;9:1516–1521.

20

EVALUATION AND ELIMINATION OF MATRIX EFFECTS IN LC-MS BIOANALYSIS

BERND A. BRUENNER AND CHRISTOPHER A. JAMES

20.1 INTRODUCTION

The analysis of pharmaceutical drugs in biological matrices using LC-MS–based methods can be impacted by the presence of coextracted materials that affect the ionization of the compound of interest. The consequence in terms of the instrument response observed for the analyte can be either enhancement or more typically suppression of the signal, an event which is referred to in general as the matrix effect and has been the subject of several published reviews (Bakhtiar and Majumdar, 2007; Cote et al., 2009; Trufelli et al., 2011). The degree to which signal response is modified can vary between calibration standards, quality control (QC) samples, and among different samples within a study. Of particular concern is the situation where substances affecting ionization are present in study samples only and not in the calibration standards or QCs, as a bias in the concentrations determined for the unknown samples may occur in spite of the assay fully meeting defined acceptance criteria. Many categories of compounds are known to contribute to matrix effects, including both endogenous biological components such as phospholipids in plasma as well as exogenous materials from drug formulations, drug metabolites, and anticoagulants. Although sources of these interfering compounds may seem ubiquitous, in order to affect the analytical results, the interfering compounds must both be extracted from the biological sample and then coelute with the analyte during the ionization process. Recognition of this requirement suggests means by which matrix effects can be mediated, namely modifications to the sample extraction procedures, chromatography, ionization conditions, and compensation through use of stable labeled internal standards (ISs). In this chapter, we discuss the known causes and mechanisms of matrix effects as well as approaches to evaluate and reduce the impact of matrix effects on quantitative LC-MS.

20.2 POTENTIAL IMPACT OF MATRIX EFFECTS

In the early stages of pharmacokinetic screening, the LC-MS/MS methods generally use generic extraction and chromatographic conditions, nonoptimized ISs, and are not subject to extensive validation. Consequently, the methodologies used for early drug discovery are potentially more prone to matrix effects as compared to later in the development phase. However, screening paradigms adopted during drug discovery generally strike a balance between method quality and ability to screen large numbers of compounds. Consequently, some risk of matrix effects can be tolerated in drug discovery, whereas for validated methods occurrence of matrix effects causing bias in reported results would not be acceptable. In one example illustrating the impact of matrix effects during drug discovery, ion suppression was shown to result in lower rat plasma concentrations of greater than fivefold after IV dosing (Schuhmacher et al., 2003). These differences resulted in an initial underestimation of the area under the curve (AUC) exposure and overestimation of clearance (CL) values, and could also cause overestimation of calculated bioavailability. The matrix effect was attributed to the IV dosing vehicle (PEG 400), and the methodology was subsequently modified to enable accurate quantitation of the drug. If not identified and corrected, such inaccuracies can be misleading in the selection of drug candidates.

Handbook of LC-MS Bioanalysis: Best Practices, Experimental Protocols, and Regulations, First Edition. Edited by Wenkui Li, Jie Zhang, and Francis L.S. Tse.

The observation of matrix effects is not limited to preclinical pharmacokinetic screening assays. There are numerous examples of compounds for which matrix effects impacted sensitivity, accuracy, and precision during method development and validation (Buhrman et al., 1996; Matuszewski et al., 1998; Matuszewski et al., 2003). A wide range of matrix effects, from 0% to 50%, were reported for validated LC-MS/MS methods that were published in the first half of 2008 (Van Eeckhaut et al., 2009). Nevertheless, in most of these methods, the use of an IS could compensate for the ion suppression or enhancement. Thus, the presence of a matrix effect does not necessarily preclude successful method validation, so long as appropriate acceptance criteria for accuracy, precision, and matrix effect can be met. In addition to affecting the analysis of small molecule drugs, matrix effects can also impact the analysis of larger peptide therapeutics. For example, the validation of an LC-MS/MS method for octreotide, an octapeptide used to inhibit excessive hormone secretion from tumors, required multiple approaches including the use of mixed-mode solid phase extraction (SPE), ultrahigh-pressure liquid chromatography (UHPLC), and online phospholipid removal to establish a robust and a sensitive methodology (Ismaiel et al., 2011).

20.3 COMMON CAUSES OF MATRIX EFFECTS

There are many materials present in biological samples that could potentially lead to matrix effects during LC-MS/MS analysis. To cause a matrix effect the materials must be present in the final extract of the sample (i.e., not removed during sample preparation) and also coelute in the chromatographic system with the analyte and/or the IS. However, even if coextraction and coelution of these materials occurs, the analysis may not be impacted as matrix effects are dependent on several other factors, including the type of atmospheric pressure ionization interface used, relative concentrations of the materials concerned, and compensation by calibration or stable isotope labeled (SIL) ISs. Compounds that have been identified to contribute to matrix effects may be classified as either endogenous or exogenous components depending on their origin (Antignac et al., 2005). Endogenous sources of interference are the organic and inorganic species normally present within the biological matrix itself, and may include metabolites of the analyte formed *in vivo*. Since the composition of biological tissues and fluids such as plasma, urine, or cerebrospinal fluid can vary widely, and can differ within and across species as well, a large range of potential interferences may be encountered (Dams et al., 2003). Moreover, significant intraindividual variation of a particular matrix component, such as plasma lipids, may occur as a result of changes in disease state or diet.

Plasma is by far the most common biofluid selected for sampling and analysis of pharmaceutical agents, and normally contains high concentrations of salts, proteins, lipids, and phospholipids as primary components and to a lesser extent carbohydrates, peptides, and various other organic compounds (Antignac et al., 2005). All of these constituents can contribute to matrix effects. Through the process of sample preparation, discussed in detail later, these components can be removed to varying degrees. Phospholipids have been highlighted in multiple publications as potentially one of the major causes of matrix effects, particularly as they are not effectively removed using the common practice of protein precipitation (Little et al., 2006; Ismaiel et al., 2007, 2010). Furthermore, due to the hydrophobic nature of the lipid chains, these compounds typically elute later under reversed-phase chromatographic conditions and thus are more likely to coelute with the analyte and be a source of interference. This stands in contrast to highly polar salts that are not chromatographically retained and typically emerge in the void volume prior to elution of the analyte. Consideration must also be given to exogenous materials that can lead to matrix effects. These can occur as a result of the sample preparation process or may be artifacts of dosing or subsequent sample handling. Substances known to cause matrix effects in this category include anticoagulants such as Li-heparin, plasticers from containers used for sample collection and storage (Mei et al., 2003), buffers, ion-pairing reagents (Gustavsson et al., 2001), and coextracted excipients used in intravenous drug formulations. In particular, Tween-80 and polyethylene glycol (PEG)-400 can cause matrix effects when used in intravenous and possibly oral formulations (Weaver and Riley, 2006). Although Li-heparin has been shown to cause matrix effects, the different counter ions used with EDTA anticoagulants (NaEDTA, K_2EDTA, or K_3EDTA) appear to have no impact (Bergeron et al., 2009; Sennbro et al., 2011b). In addition, comedications or over the counter drugs may also be present in clinical plasma samples and may lead to matrix effects that can affect quantitation of the analyte (Leverence et al., 2007).

20.4 MECHANISM OF MATRIX EFFECTS

Several mechanisms have been proposed to explain matrix effects, but the exact process remains uncertain (Cech and Enke, 2000; King et al., 2000). In the case of ion suppression, the mechanisms proposed are dependent on the mode of ionization. It is generally recognized that electrospray ionization (ESI) is more susceptible to matrix effects than atmospheric pressure chemical ionization (APCI) (Matuszewski et al., 2003). In ESI, ion formation is a multistep process, and compounds coeluting with the analyte can affect ionization in both the liquid and gas phases. With ESI, ions are first formed in the liquid phase as the LC effluent elutes from a capillary held at high voltage (Cole, 2000). The high electric field results in the formation of numerous small charged droplets that undergo further fission and solvent evaporation (Kebarle, 2000; Kebarle and Verkerk, 2009). At this stage, interfering compounds from the matrix may compete with the analyte for a limited number of available charges in the droplet, resulting in signal suppression (Enke, 1997). The loss of linearity in ESI at high analyte concentrations (above 10^{-5} M) (Ikonomou et al., 1990) is consistent with the competition for charge mechanism (Constantopoulos et al., 2000). The next step is transfer of the ion to the gas phase, which first requires access of the ion to the surface of the liquid drop. High concentrations of matrix components can limit access of the analyte to the surface of the drop. In addition, species with hydrophobic moieties such as lipids as well as dosing additives such as Tween 80 and PEG 400 have high surface activity and can thereby limit the number of analyte ions reaching the surface of the drop, further suppressing the ionization efficiency of the analyte (Xu et al., 2005). These matrix components can also affect the viscosity and surface tension of the drop, resulting in less efficient droplet formation and subsequent solvent evaporation leading to a decreased number of ions reaching the gas phase. As a result of the importance of surface activity on ionization, matrix effects can also be compound dependent and a function of analyte polarity. Compounds of higher polarity will tend to be concentrated in the inner aqueous phase of the drop rather than at the surface and thus have lower surface activity and are known to exhibit a greater tendency to undergo ion suppression (Bonfiglio et al., 1999). The presence of components acting as ion-pairing agents either from mobile phase additives or the matrix itself can also result in the formation of neutral complexes and cause signal suppression (Apffel et al., 1995). Once in the gas phase, the analyte ion is still susceptible to the influence of matrix components that have also been transferred to the gas phase. Neutral matrix species may then compete for protons from the charged analyte based on their relative gas phase basicities through proton transfer reactions. Those components with higher gas phase basicities will remove a proton from the analyte, neutralizing its charge resulting in a drop in signal intensity (Amad et al., 2000). Nonvolatile components originating from the matrix or mobile phase can also be detrimental through the formation of solid particles and coprecipitation of the analyte during desolvation of the charged droplets resulting in suppression of signal (King et al., 2000). In contrast to ESI, the APCI ionization process involves formation of ions from neutral species directly in the gas phase through a chemical ionization process from reagent ions generated by the corona discharge needle in the source. The presence of interfering matrix components in the liquid phase generally does not result in direct suppression as occurs in ESI, and APCI is thus less prone to matrix effects. However, as a result of the higher temperatures and rapid solvent evaporation encountered during APCI, precipitation of analyte with nonvolatile species is believed to be a more dominant suppression mechanism than in ESI (King et al., 2000). Also, once in the gas phase the various matrix species may compete with the analyte for protons from the gas phase reagent ions, thereby suppressing ionization of the analyte, similar to the gas phase ion suppression mechanism in ESI.

Although the exact mechanisms of ion suppression may not be completely defined, in general the matrix effects appear to occur due to competition for charge or neutralization of ionization processes. Conversely, ion enhancement would result when the presence of the matrix components would increase the efficiency of ion formation or the gas phase transfer processes described. Reduction of matrix effects can therefore be achieved through various strategies including decreasing the level of matrix components, improving separation of interfering materials from the analyte, and even by simple dilution of samples to reduce the overall concentrations of both analyte and coextracted materials (Schuhmacher et al., 2003; Larger et al., 2005).

20.5 METHODS TO IDENTIFY AND EVALUATE MATRIX EFFECTS

Several approaches to evaluate the extent and impact of matrix effects on quantitative LC-MS/MS assays have been developed. A widely published but qualitative evaluation of matrix effects is obtained by the postcolumn infusion approach (Bonfiglio et al., 1999; Muller et al., 2002; Mallet et al., 2004; Souverain et al., 2004). This technique identifies specific regions in the chromatographic profile where the analyte response may be susceptible to matrix effects, and is useful in evaluating modifications to the analytical method designed to minimize matrix effects. In postcolumn infusion, a constant amount of analyte is introduced to the LC column eluent using a mixing tee and syringe pump, producing a constant signal in the mass spectrometer (Figure 20.1). A solvent blank and blank matrix extract are then injected, and the two elution profiles are compared. Matrix components present in the injected samples are chromatographically separated under the same conditions as in routine sample analysis. A decrease in response indicates a region in the chromatogram where the matrix components are suppressing ionization of the analyte and an increase in response indicates regions of ion enhancement (Figure 20.2). The disadvantages of this technique are that it can be time-consuming, especially if multiple analytes are evaluated, and the concentration levels infused are generally high and not representative of the matrix effect near the LLOQ. Furthermore, the technique does not monitor matrix effects quantitatively, but only indicates in what regions of the chromatograms interfering components are eluting and are potentially causing ion suppression effects. In addition to postcolumn infusion,

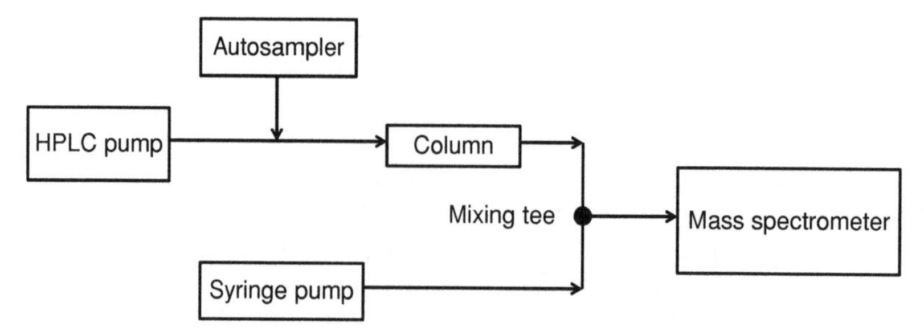

FIGURE 20.1 Postcolumn infusion configuration.

dilution of study samples with plasma free of dosing vehicle and comparison of concentrations to undiluted samples can be used to identify matrix effects (Larger et al., 2005).

More quantitative assessment of matrix effects is usually obtained by comparison of responses from various extracted samples and solutions under the standard analytical conditions for the method. The absolute effect of the matrix can be defined as the matrix factor (MF). The MF is calculated as the relative peak response of an analyte in the presence of matrix ions to the response in the absence of matrix ions, excluding variability due to recovery issues (Matuszewski et al., 2003). In practice, the MF is determined from the ratio of peak area or height of the analyte in blank extracted matrix that has been spiked postextraction with analyte, compared to the peak response of the same analyte concentration in a neat solvent solution:

$$\text{Matrix factor} = \frac{\text{Peak response in blank matrix extract spiked postextraction}}{\text{Peak response in neat solution}}$$

An MF value of 1 results when the response in neat solution is the same as in the matrix blank spiked postextraction, indicating the absence of a matrix effect. Values of the MF <1 indicate a suppression effect and values >1 suggest ion enhancement (Viswanathan et al., 2007).

The IS-normalized MF can be calculated by dividing the MF of the analyte by the MF of the IS as shown here, or by using the analyte/IS ratios in the aforementioned equation (Bansal and DeStefano, 2007).

$$\text{IS normalized MF} = \frac{\text{MF for analyte}}{\text{MF for IS}}$$

An IS-normalized MF value of one indicates that the analyte and IS are affected to the same degree by matrix suppression or enhancement. Values for IS-normalized MF ranging from 0.80–1.20 are generally considered acceptable (Kollipara et al., 2011). In the case of stable-isotope labeled ISs, the analyte and IS are typically affected by matrix effects to the same degree and IS-normalized MF values near 1 are typically observed. Indeed, the ability of a stable-isotope labeled IS to compensate for matrix effects is one of the primary benefits of their use.

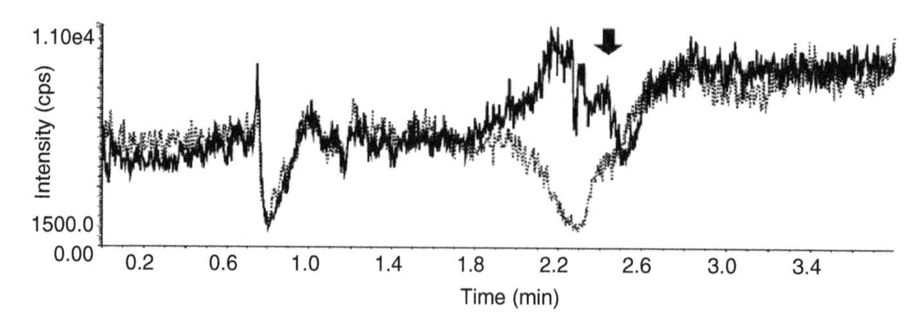

FIGURE 20.2 An example of a postcolumn infusion experiment. The solid line trace was obtained from the injection of blank control plasma extract. The dotted line trace is from injection of plasma extract obtained from animals dosed with vehicle (polysorbate 80) only. A drop in signal intensity of the infused analyte indicates chromatographic regions of ion suppression. The first region of suppression occurs near the void volume (0.8 min) in both samples due to unretained matrix components. The second region of suppression at 2.2 min is observed only for the plasma sample containing the formulation vehicle, which occurs near the retention time of the analyte (arrow) (reprinted from Larger et al. (2005); copyright (2005), with permission from Elsevier).

Alternate approaches that may serve as indicators of matrix effects in different lots of matrix have also been suggested. For example, calibration curves can be prepared in five separate matrix lots and the slopes of the regression lines compared. As a guide for method suitability, one group proposed that the precision of the slopes from five different lots have a coefficient of variation (CV) of <3–4%, based on results from methods developed in their laboratories (Matuszewski, 2006).

Monitoring the mass spectrometric response of specific matrix components such as phospholipids is another qualitative tool useful during method development (Table 20.1). Phospholipid monitoring can be performed by monitoring mass transitions specific for the various species to ensure chromatographic separation of the analyte from the phospholipid components, thereby minimizing potential matrix effects (Chambers et al., 2007). The various lipid classes present in human plasma include phospholipids, cholesterol esters, free cholesterol, and triacylglycerols. Of the phospholipids, the majority is present in the form of phosphatidylcholine, which comprises nearly 70% of the total in this class (Ismaiel et al., 2010). A more generalized approach monitors a common fragment ion of all glycerophosphocholines including phosphatidylcholine, lysophosphatidylcholine, and sphingomyelin, the trimethylammonium-ethyl phosphate ion at *m/z* 184. The technique described used in-source collision induced dissociation to form the ion at *m/z* 184 and a single selected reaction monitoring channel with a transition of *m/z* 184 → 184 (Little et al., 2006). An alternate approach uses precursor ion scanning of *m/z* 184 to identify choline containing lipids or neutral loss scanning to identify other lipid classes, which has the advantage of identifying the precursor ions, and therefore, the specific phospholipid corresponding to the individual chromatographic peaks observed (Xia and Jemal, 2009).

20.6 REGULATORY GUIDANCE ON EVALUATION AND AVOIDANCE OF MATRIX EFFECTS

The potential for matrix effects to adversely affect accuracy and precision of LC-MS/MS analyses is recognized as a critical element in the validation of bioanalytical methods used in support of regulated studies. Although the need to evaluate matrix effects has been a topic of discussion in numerous conferences and workshops on the subject of bioanalytical method validation, a harmonized approach to its evaluation has yet to be achieved. The current FDA guidance on bioanalytical method validation states that "matrix effects should be investigated to ensure that precision, selectivity, and sensitivity will not be compromised," but does not specify how these tests should be performed (FDA, 2001). In practice, many bioanalytical laboratories assess if matrix effects impact quantitation by spiking six different lots of matrix with analyte at the low and high QC concentrations and establish if the accuracy and precision criteria of the method (± 15%) are met in the various matrix lots (Kollipara et al., 2011). The recently approved European Medicines Agency (EMA) guidelines on bioanalytical method validation outline specific testing requirements for the analyte and IS (EMA, 2011). These include calculating the MF and IS-normalized MF in at least six lots of matrix from individual donors at both low and high concentrations. The IS-normalized MF should have a CV of less than 15%. In addition, if formulation excipients known to cause matrix effects are used, these should be included in the evaluation, preferably from subjects administered the excipients. The EMA guidelines also indicate matrix testing be performed for hemolyzed or hyperlipidemic plasma. The need to cross-validate methods for differing counter ions for the same anticoagulant remains unresolved from a regulatory perspective (Sennbro et al., 2011a). Regardless of the approach taken, the criteria and requirements for matrix effect testing during validation should be specified according to appropriate standard operating procedures governing the regulated laboratory.

20.7 METHODS TO AVOID OR ELIMINATE MATRIX EFFECTS

There are several approaches that can be taken to reduce and minimize matrix effects. Since coelution of the analyte with the interfering component is a requirement for the occurrence of matrix effects, modifications to chromatographic conditions can often be used to reduce their impact (Trufelli et al., 2011). Changes to gradient conditions, mobile phase strength, and pH are effective strategies to shift the retention of the analyte away from chromatographic regions affected by ion suppression (Chambers et al., 2007). The hydrophobic character of phospholipids can result in their strong retention on reversed-phase columns. When gradient conditions are not sufficient to remove these interfering components, they may build up on the column or elute in subsequent injections, resulting in inconsistent matrix effects. In one study, optimization of mobile phase conditions to reduce matrix effects for high throughput applications using protein precipitation revealed that although acetonitrile was a stronger solvent for eluting phospholipids than methanol at concentrations of ≤90% organic modifier, methanol was a more effective solvent in eluting phospholipids at concentrations >95% when using a C18 column (Ye et al., 2011). This result was attributed to the reduced solubility of phospholipids in acetonitrile. However, using gradient elution with a mixture of 1:1 methanol/acetonitrile as the organic modifier for a variety of model compounds comprised of neutral, acidic, and basic drugs was found to provide the best combination of analyte resolution, peak shape, as well as separation from phospholipids. Further improvements in resolution as well

TABLE 20.1 Monitoring Common Lipids Associated with Matrix Effects in Positive Ion Mode

Lipid	Technique	Transition (*m/z*)	Comment	Reference
All glycerophosphocholines (including lysophosphatidylcholines and sphingomyelin)	In-source CID-MRM	184 → 184 (trimethylammonium-ethyl phosphate ion)	Single MRM channel can be used to minimize sensitivity loss.	Little et al. (2006)
Lysophosphatidylcholines		104 → 104	Transition at *m/z* 184 is common for all lipids with phosphocholine head groups.	
Lysophosphatidylcholines	MRM	496 → 184 524 → 184	Multiple MRM channels can be monitored, each specific for a unique phospholipid.	Ghosh et al. (2011), Ismaiel et al. (2010), Cote et al. (2009); Pucci et al. (2009), Chambers et al. (2007)
Glycerophosphatidylcholines		704 → 184 758 → 184 786 → 184 806 → 184		
All cholesterols and cholesterol esters		369 → 369 $[M+H\text{-}H_2O]$	Transition at *m/z* 369 common for cholesterol and esters.	
All glycerophosphocholines (including lysophosphatidylcholines and sphingomyelin)	Precursor ion scan of *m/z* 184.	N/A	Separate scan function from MRM.	Ghosh et al. (2011), Xia et al. (2009), Little et al. (2006)
Glycerophosphoethanolamines	Neutral loss scan of 141	N/A		
Glycerophosphoserines	Neutral loss scan of 185	N/A		

as reductions in run time can be realized by the implementation of UHPLC, which uses columns with sub-2 μm particle diameters (Guillarme et al., 2010; Remane et al., 2010). The application of UHPLC to a set of 10 different compounds using various sample preparation techniques and chromatographic conditions was found to improve separation between analytes and coeluting species and statistically determined to reduce matrix effects (Chambers et al., 2007).

Sample preparation is another strategy that can be used to selectively remove interfering components and thereby reduce matrix effects. Of the common sample preparation techniques, protein precipitation is most often used due to simplicity of the process, applicability to a wide range of analytes, and cost-effectiveness. However, in terms of sample cleanup protein precipitation techniques are considered to be most prone to matrix effects because they result in relatively dirty extracts that still contain numerous matrix components such as salts and phospholipids. SPE and liquid–liquid extraction (LLE) techniques are both superior in this regard, but are more complex to develop and may not be suitable for certain compound classes. LLE techniques generally provide clean sample extracts; however, application is limited to compounds that readily partition into solvents that are immiscible with water, and LLE techniques are therefore not suitable for ionic or highly polar compounds. The choice of extraction solvents may be further limited due to environmental and safety concerns, and LLE also typically requires solvent dry down and reconstitution steps prior to injection. Sample preparation using SPE can also be effective in reducing matrix effects, and the wide range of chemistries available for SPE make this technique applicable to most types of drug molecules (Li et al., 2006).

A recent publication compared various sample extraction techniques for nevirapin in human plasma, including protein precipitation, LLE, and SPE, and illustrates the improvements that can be achieved by these methodologies (Ghosh et al., 2011). In this study, the greatest matrix effect was observed when protein precipitation was used, with an average MF of 0.30. Comparison of methanol and acetonitrile as extraction solvents followed by phospholipid monitoring demonstrated that acetonitrile yielded a cleaner extract that contained fewer coextracted lipids than methanol, consistent with earlier reports (Chambers et al., 2007). Using LLE with ethyl acetate or methyl tertiary butyl ether (MTBE) as the extraction solvent, the matrix effect was significantly reduced relative to protein precipitation, with an average MF of 0.80. As with protein precipitation, differences in the extent of coextracted lipids were observed depending on the extraction solvent, with the less polar solvent (MTBE) found to extract fewer phospholipids than ethyl acetate. In this example, reversed-phase SPE (Oasis® HLB) was found to be superior to LLE both in terms of reduction of matrix effect (MF of 0.99) and minimal coextraction of phospholipids. In another study, multiple SPE modes including cation exchange, reversed-phase, and mixed-modes were evaluated and compared (Chambers et al., 2007). The results demonstrated that the cleanest extracts were obtained from mixed-mode SPE, which used a combination of cation exchange and reversed-phase mechanisms. Recent new technologies in sample preparation include development of filtration plates that selectively remove phospholipids from protein precipitated plasma samples as they pass through the filter during extraction. These include the Captiva ND®Lipids from Agilent and the HybridSPE®-phospholipid protein precipitation plates from Sigma-Aldrich, and have been demonstrated to reduce matrix effects and improve assay sensitivity (Pucci et al., 2009; Kole et al., 2011).

Selection of the atmospheric pressure ionization source and ionization polarity can also impact the extent of matrix effects observed. For assays requiring high sensitivity, ESI is generally utilized; however, ESI is known to be more susceptible to matrix effects than APCI (Dams et al., 2003; Remane et al., 2010) due to the differences in ionization mechanism previously discussed. Consequently, if sensitivity is not an issue and the analyte is thermally stable, switching from ESI to APCI can be an effective strategy that can be employed to reduce matrix effects. In one example, matrix effects for chlorpheniramine in human plasma were reduced by 75% when APCI was used instead of ESI. This study also demonstrated that phospholipids themselves associated with ion suppression, phosphatidylcholine and lysophosphatidylcholine, had much lower ionization efficiencies using APCI versus ESI, which may explain why APCI tends to be less prone to matrix effect (Ismaiel et al., 2008). However, as previously mentioned, APCI is not immune to matrix effects, as noted by examples in the literature (Sangster et al., 2004). Also, when APCI is used, there is a greater risk of thermal degradation of labile metabolites back to parent compound. The ion polarity mode selected for analysis can also moderate the degree of matrix effect (Antignac et al., 2005). In general, negative ion mode is considered more selective and tends to have fewer issues related to ion suppression. However, many analytes are not amenable to analysis in the negative ion mode, and choice of ionization mode is usually dictated by assay sensitivity.

Use of ISs is one of the most common approaches to compensate for matrix effects during sample analysis. If IS is affected by the same suppression or enhancement phenomenon as the analyte, any changes in ionization conditions would only affect the absolute peak areas, but not the analyte to IS ratio and thus compensate for any matrix effects. To function most effectively in compensating for changes in ionization conditions and minimize matrix effects, ISs should have similar physicochemical properties, ionization efficiencies, and chromatographic retention times as the analyte. Structural analog ISs should also not contain the analyte as an impurity, not be a possible metabolite of the analyte, and preferably differ by C–H groups rather than heteroatom containing moieties ($-SO_2$, $-NO_2$, $-NH_2$, or halogens) to minimize differences in recovery and ionization efficiency (Bakhtiar and Majumdar, 2007). SIL compounds are generally considered to be ideal ISs due to their nearly identical chemical properties as the analyte. The similar chemical properties lead to compensation for variability during the extraction process and generally result in chromatographic coelution with the analyte. As a result, the SIL ISs are expected to undergo the same degree of matrix suppression or enhancement as the analyte. Unfortunately, SIL ISs are not readily available for all compounds due to cost and synthetic complexity, particularly during early method development, and for applications in discovery bioanalysis. In addition, examples have been cited of differential response between an analyte and its SIL analog. For example, depending on the number and location of the substitutions, replacement of hydrogen with deuterium can cause slight changes in the hydrophobic properties of the molecule, and this isotope effect can result in minor changes in chromatographic retention (Iyer et al., 2004). Under conditions where even subtle shifts in retention between analyte and IS occur, differences in response due to matrix effects have been observed. In one such example, the presence of residual triethylamine in sample extracts as a result of the SPE process resulted in a reduction of piperaquine response by 50% relative to a the corresponding D6 labeled IS (Lindegardh et al., 2008). An additional example involved the quantitation of carvedilol and a D5 IS, in which a slight difference in retention time resulted in ~20% suppression of the IS relative to the unlabeled analyte in a specific lot of plasma, resulting in a corresponding positive bias of the QC samples (Wang et al., 2007). The coelution of analyte and corresponding SIL has even been observed to result in their mutual suppression in ESI and enhancement under APCI, although this effect was minimized when appropriate IS concentrations were selected (Liang et al., 2003).

Another approach to reducing matrix effects is to decrease the amount of sample and coextracted components introduced into the LC-MS/MS system. Samples for which matrix suppression occurs will usually show an increase in apparent analyte concentration under these conditions. As long as sensitivity is not an issue, this can be readily achieved by simply reducing the injection volume or dilution of the sample extract prior to injection (Schuhmacher et al., 2003). Alternatively, dilution of the sample with matrix prior to extraction can also be performed. This technique was successfully used for discovery analysis to check for matrix effects; the earliest time points (5 min) after IV dosing were diluted 5- and 10-fold with plasma and analyzed together with the undiluted sample. If large differences in concentration between diluted and undiluted samples were observed, matrix suppression was indicated. In contrast, in the absence of a matrix effect, both the diluted and undiluted samples had similar concentrations. A decision tree approach using a 15% difference criteria was applied and used in reporting the appropriate result (Larger et al., 2005).

20.8 FUTURE PROSPECTS

Increasing regulatory expectations for bioanalytical methods used in support of regulated studies may require consideration of additional validation requirements in the future. Indeed, testing of hemolyzed plasma during validation is now routinely performed in many bioanalytical laboratories, and can be considered as a unique type of matrix effect evaluation. Hemolysis can occur due to lysis of the red blood cells during whole blood collection or processing, causing the release of heme and other intracellular components that can result in visibly red plasma and potentially affect quantitation (Hughes et al., 2009). Subsequent modification of the HPLC gradient resulted in improved separation of the interfering component from the IS and eliminated the matrix effect.

Matrix effects may have additional impact during routine analysis even if quantitation is not directly affected. Reproducibility of IS response is another emerging area in regulated bioanalysis, and suppression of IS response below predefined acceptance criteria due to matrix effects may require further investigation (Lowes et al., 2011). At a minimum, additional resources would be required for reanalysis, and in the worst-case valuable concentration data may be unobtainable if the result was rejected, demonstrating the value in minimizing such matrix effects during the method development process prior to validation.

The recent growth in the use of dried blot spot (DBS) sampling suggests that this technique may become even more commonplace in the future (Li and Tse, 2010). Although the advantages of DBS as a sample collection technique have been well documented, the potential impact of matrix effects derived from the sample collection paper has not been thoroughly investigated. The paper used for DBS cards may be pretreated with various substances to aid in the deactivation of biohazardous pathogens, and when coextracted these can lead to matrix effects. One such example is the use of FTA Elute™ cards for discovery pharmacokinetic screening, in which matrix effects were observed for a number of test compounds using a generic IS. Matrix effects were minimized to acceptable levels by a combination of 2D-HPLC-MS/MS and optimization of the reconstitution solvent and volume, which led to a generic methodology applicable to discovery PK screening (Clark and Haynes, 2011).

Since LC-MS/MS analysis involves multiple distinct processes and steps, there exist multiple opportunities to avoid or reduce the risk of matrix effects. Advances in both instrument sensitivity and design may improve robustness of LC-MS/MS methods for quantitative bioanalysis. Increases in mass spectrometer sensitivity would allow for further dilution of samples, potentially eliminating matrix effects as well as allowing the application of methods with clean extracts but low analyte recoveries. Optimization of ionization source configurations may also result in improved tolerance of coeluting matrix components. At this time, it seems unlikely that any single improvement or new technology will eliminate the potential for matrix effects during the atmospheric pressure ionization process, but rather that incremental improvements will continue to be achieved.

ACKNOWLEDGMENTS

The authors thank Philip Wong, PhD, for providing critical review and helpful comments for this manuscript.

REFERENCES

Amad MH, Cech NB, Jackson GS, Enke CG. Importance of gas-phase proton affinities in determining the electrospray ionization response for analytes and solvents. J Mass Spectrom 2000;35:784–789.

Antignac JP, de Wasch K, Monteau F, De Brabander H, Andre F, Le Bizec B. The ion suppression phenomenon in liquid chromatography-mass spectrometry and its consequences in the field of residue analysis. Anal Chim Acta 2005;529:129–136.

Apffel A, Fischer S, Goldberg G, Goodley PC, Kuhlmann FE. Enhanced sensitivity for peptide mapping with electrospray liquid chromatography mass spectrometry in the presence of signal suppression due to trifluoroacetic acid-containing mobile phases. J Chromatogr A 1995;712:177–190.

Bakhtiar R, Majumdar TK. Tracking problems and possible solutions in the quantitative determination of small molecule drugs and metabolites in biological fluids using liquid chromatography-mass spectrometry. J Pharmacol Toxicol Methods 2007;55:227–243.

Bansal S, DeStefano A. Key elements of bioanalytical method validation for small molecules. AAPS J 2007;9:E109–E114.

Bergeron M, Bergeron A, Furtado M, Garofolo F. Impact of plasma and whole-blood anticoagulant counter ion choice on drug stability and matrix effects during bioanalysis. Bioanalysis 2009;1:537–548.

Bonfiglio R, King RC, Olah TV, Merkle K. The effects of sample preparation methods on the variability of the electrospray ionization response for model drug compounds. Rapid Commun Mass Spectrom 1999;13:1175–1185.

Buhrman DL, Price PI, Rudewicz PJ. Quantitation of sr 27417 in human plasma using electrospray liquid chromatography tandem mass spectrometry—a study of ion suppression. J Am Soc Mass Spectrom 1996;7:1099–1105.

Cech NB, Enke CG. Relating electrospray ionization response to nonpolar character of small peptides. Anal Chem 2000;72:2717–2723.

Chambers E, Wagrowski-Diehl DM, Lu Z, Mazzeo JR. Systematic and comprehensive strategy for reducing matrix effects in LC/MS/MS analyses. J Chromatogr B Analyt Technol Biomed Life Sci 2007;852:22–34.

Clark GT, Haynes JJ. Utilization of DBS within drug discovery: a simple 2D-LC-MS/MS system to minimize blood- and paper-based matrix effects from FTA elute DBS. Bioanalysis 2011;3:1253–1270.

Cole RB. Some tenets pertaining to electrospray ionization mass spectrometry. J Mass Spectrom 2000;35:763–772.

Constantopoulos TL, Jackson GS, Enke CG. Challenges in achieving a fundamental model for ESI. Anal Chim Acta 2000;406: 37–52.

Cote C, Bergeron A, Mess JN, Furtado M, Garofolo F. Matrix effect elimination during LC-MS/MS bioanalytical method development. Bioanalysis 2009;1:1243–1257.

Dams R, Huestis MA, Lambert WE, Murphy CM. Matrix effect in bio-analysis of illicit drugs with LC-MS/MS: influence of ionization type, sample preparation, and biofluid. J Am Soc Mass Spectrom 2003;14:1290–1294.

EMA. 2011. Guideline on Bioanalytical Method Validation. Available at http://www.ema.europa.eu/docs/en_GB/document_library/Scientific_guideline/2011/08/WC500109686.pdf.

Enke CG. A predictive model for matrix and analyte effects in electrospray ionization of singly charged ionic analytes. Anal Chem 1997;69:4885–4893.

FDA. 2001. Center for Drug Evaluation and Research. Guidance for Industry: Bioanalytical Method Validation. Available at http://www.fda.gov/downloads/Drugs/GuidanceCompliance RegulatoryInformation/Guidances/ucm070107.pdf.

Ghosh C, Shashank G, Shinde CP, Chakraborty B. A systematic approach to overcome the matrix effect during LC-ESI-MS/MS analysis by different sample extraction techniques. J Bioequiv Bioavail 2011;3:122–127.

Guillarme D, Schappler J, Rudaz S, Veuthey JL. Coupling ultra-high-pressure liquid chromatography with mass spectrometry. Trends Anal Chem 2010;29:15–27.

Gustavsson SA, Samskog J, Markides KE, Langstrom B. Studies of signal suppression in liquid chromatography-electrospray ionization mass spectrometry using volatile ion-pairing reagents. J Chromatogr A 2001;937:41–47.

Hughes NC, Bajaj N, Fan J, Wong EY. Assessing the matrix effects of hemolyzed samples in bioanalysis. Bioanalysis 2009;1:1057–1066.

Ikonomou MG, Blades AT, Kebarle P. Investigations of the Electrospray interface for liquid chromatography/mass spectrometry. Anal Chem 1990;62:957–967.

Ismaiel OA, Halquist MS, Elmamly MY, Shalaby A, Karnes HT. Monitoring phospholipids for assessment of matrix effects in a liquid chromatography-tandem mass spectrometry method for hydrocodone and pseudoephedrine in human plasma. J Chromatogr B Analyt Technol Biomed Life Sci 2007;859:84–93.

Ismaiel OA, Halquist MS, Elmamly MY, Shalaby A, Thomas Karnes H. Monitoring phospholipids for assessment of ion enhancement and ion suppression in ESI and APCI LC/MS/MS for chlorpheniramine in human plasma and the importance of multiple source matrix effect evaluations. J Chromatogr B Analyt Technol Biomed Life Sci 2008;875:333–343.

Ismaiel OA, Zhang T, Jenkins R, Karnes HT. Determination of octreotide and assessment of matrix effects in human plasma using ultra high performance liquid chromatography-tandem mass spectrometry. J Chromatogr B Analyt Technol Biomed Life Sci 2011;879:2081–2088.

Ismaiel OA, Zhang T, Jenkins RG, Karnes HT. Investigation of endogenous blood plasma phospholipids, cholesterol and glycerides that contribute to matrix effects in bioanalysis by liquid chromatography/mass spectrometry. J Chromatogr B Analyt Technol Biomed Life Sci 2010;878: 3303–3316.

Iyer SS, Zhang ZP, Kellogg GE, Karnes HT. Evaluation of deuterium isotope effects in normal-phase LC-MS-MS separations using a molecular modeling approach. J Chromatogr Sci 2004;42:383–387.

Kebarle P. A brief overview of the present status of the mechanisms involved in electrospray mass spectrometry. J Mass Spectrom 2000;35:804–817.

Kebarle P, Verkerk UH. Electrospray: from ions in solution to ions in the gas phase, what we know now. Mass Spectrom Rev 2009;28:898–917.

King R, Bonfiglio R, Fernandez-Metzler C, Miller-Stein C, Olah T. Mechanistic investigation of ionization suppression in electrospray ionization. J Am Soc Mass Spectrom 2000;11:942–950.

Kole PL, Venkatesh G, Kotecha J, Sheshala R. Recent advances in sample preparation techniques for effective bioanalytical methods. Biomed Chromatogr 2011;25:199–217.

Kollipara S, Bende G, Agarwal N, Varshney B, Paliwal J. International guidelines for bioanalytical method validation: a comparison and discussion on current scenario. Chromatographia 2011;73:201–217.

Larger PJ, Breda M, Fraier D, Hughes H, James CA. Ion-suppression effects in liquid chromatography-tandem mass spectrometry due to a formulation agent, a case study in drug discovery bioanalysis. J Pharm Biomed Anal 2005;39:206–216.

Leverence R, Avery MJ, Kavetskaia O, Bi H, Hop C, Gusev AI. Signal suppression/enhancement in HPLC-ESI-MS/MS from concomitant medications. Biomed Chromatogr 2007;21:1143–1150.

Li KM, Rivory LP, Clarke SJ. Solid-phase extraction (SPE) techniques for sample preparation in clinical and pharmaceutical analysis: a brief overview. Curr Pharm Anal 2006;2:95–102.

Li W, Tse FLS. Dried blood spot sampling in combination with LC-MS/MS for quantitative analysis of small molecules. Biomed Chromatogr 2010;24:49–65.

Liang HR, Foltz RL, Meng M, Bennett P. Ionization enhancement in atmospheric pressure chemical ionization and suppression in electrospray ionization between target drugs and stable-isotope-labeled internal standards in quantitative liquid chromatography/tandem mass spectrometry. Rapid Commun Mass Spectrom 2003;17:2815–2821.

Lindegardh N, Annerberg A, White NJ, Day NPJ. Development and validation of a liquid chromatographic-tandem mass spectrometric method for determination of piperaquine in plasma. Stable isotope labeled internal standard does not always compensate for matrix effects. J Chromatogr B Analyt Technol Biomed Life Sci 2008;862:227–236.

Little JL, Wempe MF, Buchanan CM. Liquid chromatography-mass spectrometry/mass spectrometry method development for drug metabolism studies: examining lipid matrix ionization effects in plasma. J Chromatogr B Analyt Technol Biomed Life Sci 2006;833:219–230.

Lowes S, Jersey J, Shoup R, et al. Recommendations on: Internal standard criteria, stability, incurred sample reanalysis and recent 483s by the Global CRO Council for Bioanalysis. Bioanalysis 2011;3:1323–1332.

Mallet CR, Lu ZL, Mazzeo JR. A study of ion suppression effects in electrospray ionization from mobile phase additives and solid-phase extracts. Rapid Commun Mass Spectrom 2004;18:49–58.

Matuszewski BK. Standard line slopes as a measure of a relative matrix effect in quantitative HPLC-MS bioanalysis. J Chromatogr B Analyt Technol Biomed Life Sci 2006;830:293–300.

Matuszewski BK, Constanzer ML, Chavez-Eng CM. Matrix effect in quantitative LC/MS/MS analyses of biological fluids: a method for determination of finasteride in human plasma at picogram per milliliter concentrations. Anal Chem 1998;70: 882–889.

Matuszewski BK, Constanzer ML, Chavez-Eng CM. Strategies for the assessment of matrix effect in quantitative bioanalytical methods based on HPLC-MS/MS. Anal Chem 2003;75:3019–3030.

Mei H, Hsieh Y, Nardo C, et al. Investigation of matrix effects in bioanalytical high-performance liquid chromatography/tandem mass spectrometric assays: Application to drug discovery. Rapid Commun Mass Spectrom 2003;17:97–103.

Muller C, Schafer P, Stortzel M, Vogt S, Weinmann W. Ion suppression effects in liquid chromatography-electrospray-ionisation transport-region collision induced dissociation mass spectrometry with different serum extraction methods for systematic toxicological analysis with mass spectra libraries. J Chromatogr B 2002;773:47–52.

Pucci V, Di Palma S, Alfieri A, Bonelli F, Monteagudo E. A novel strategy for reducing phospholipids-based matrix effect in LC-ESI-MS bioanalysis by means of HybridSPE. J Pharm Biomed Anal 2009;50:867–871.

Remane D, Meyer MR, Wissenbach DK, Maurer HH. Ion suppression and enhancement effects of co-eluting analytes in multi-analyte approaches: systematic investigation using ultra-high-performance liquid chromatography/mass spectrometry with atmospheric-pressure chemical ionization or electrospray ionization. Rapid Commun Mass Spectrom 2010;24:3103–3108.

Sangster T, Spence M, Sinclair P, Payne R, Smith C. Unexpected observation of ion suppression in a liquid chromatography/atmospheric pressure chemical ionization mass spectrometric bioanalytical method. Rapid Commun Mass Spectrom 2004;18:1361–1364.

Schuhmacher J, Zimmer D, Tesche F, Pickard V. Matrix effects during analysis of plasma samples by electrospray and atmospheric pressure chemical ionization mass spectrometry: Practical approaches to their elimination. Rapid Commun Mass Spectrom 2003;17:1950–1957.

Sennbro CJ, Knutsson M, Timmerman P, Van Amsterdam P. Anticoagulant counter ion impact on bioanalytical LC-MS/MS assay performance: Additional validation required? Bioanalysis 2011a;3:2389–2391.

Sennbro CJ, Knutsson M, Van Amsterdam P, Timmerman P. Anticoagulant counter ion impact on bioanalytical LC-MS/MS assays: Results from discussions and experiments within the European Bioanalysis Forum. Bioanalysis 2011b;3:2393–2399.

Souverain S, Rudaz S, Veuthey JL. Matrix effect in LC-ESI-MS and LC-APCI-MS with off-line and on-line extraction procedures. J Chromatogr A 2004;1058:61–66.

Trufelli H, Palma P, Famiglini G, Cappiello A. An overview of matrix effects in liquid chromatography-mass spectrometry. Mass Spectrom Rev 2011;30:491–509.

Van Eeckhaut A, Lanckmans K, Sarre S, Smolders I, Michotte Y. Validation of bioanalytical LC-MS/MS assays: evaluation of matrix effects. J Chromatogr B 2009;877:2198–2207.

Viswanathan CT, Bansal S, Booth B, et al. Quantitative bioanalytical methods validation and implementation: Best practices for chromatographic and ligand binding assays. Pharm Res 2007;24:1962–1973.

Wang S, Cyronak M, Yang E. Does a stable isotopically labeled internal standard always correct analyte response? A matrix effect study on a LC/MS/MS method for the determination of carvedilol enantiomers in human plasma. J Pharm Biomed Anal 2007;43:701–707.

Weaver R, Riley RJ. Identification and reduction of ion suppression effects on pharmacokinetic parameters by polyethylene glycol 400. Rapid Commun Mass Spectrom 2006;20:2559–2564.

Xia YQ, Jemal M. Phospholipids in liquid chromatography/mass spectrometry bioanalysis: comparison of three tandem mass spectrometric techniques for monitoring plasma phospholipids, the effect of mobile phase composition on phospholipids elution and the association of phospholipids with matrix effects. Rapid Commun Mass Spectrom 2009;23:2125–2138.

Xu X, Mei H, Wang S, et al. A study of common discovery dosing formulation components and their potential for causing time-dependent matrix effects in high-performance liquid chromatography tandem mass spectrometry assays. Rapid Commun Mass Spectrom 2005;19:2643–2650.

Ye Z, Tsao H, Gao H, Brummel CL. Minimizing matrix effects while preserving throughput in LC-MS/MS bioanalysis. Bioanalysis 2011;3:1587–1601.

21

EVALUATION AND ELIMINATION OF CARRYOVER AND/OR CONTAMINATION IN LC-MS BIOANALYSIS

Howard M. Hill and Graeme T. Smith

21.1 OVERVIEW

Carryover and contamination have many definitions based on authorship and application. In the context of chromatography, carryover is used by manufacturers to define the quality of their instrumentation. It is largely a function of the engineering and design process related to the injection system, assuming that all dead spaces have been considered. Carryover is a phenomenon where the test substance from a preceding sample is introduced into the next sample injection because a small amount of the injection solvent lingers in a "dead" space of the system or as a result of adsorption of the analyte of interest to the injector system. In the latter case, the carryover is more likely to be chemistry (or compound) or solvent specific and can vary from one drug to the other. In most cases, carryover is usually seen as a consistently sized peak from one injection to the next. However, under some circumstances the amount or the "peak" may increase from one injection to the next.

Contamination can be defined as the test substance peak that appears in analytical check samples (e.g., matrix blanks) and control group, pre-dose and placebo samples. Contamination can also occur in calibration standards, quality control (QC) samples and incurred samples, but it is less easy to detect because of the presence of the test substance. Contamination may also be caused by another "peak" from an undefined source whose retention time and MS characteristics mimic the test substance.

All forms of carryover can be considered as contributors to contamination but for ease of use and definition, carryover and contamination will be seen as separate entities for this chapter. "Carry-over," "carryover," or "carry over" are all variations in "spelling" and do not represent any differences in science. Regulators seem to prefer the hyphenated form, while manufacturers prefer the combined wording.

The amount of drug carried from one injection to the other is usually small; however, it may significantly impact the integrity of the lower limit of quantification (LLOQ) of the method and measured drug concentrations that are close to the LLOQ in the study samples. There has been a trend within the Pharmaceutical Industry to develop liquid chromatography–mass spectrometry (LC-MS) bioanalytical methods that fit all intended studies for a given species matrix. This means that the dynamic range of the assay method covers the concentration range in samples from low dose early clinical studies (e.g., first-in-human (FIH) study) through to later phase studies with fixed doses. Consequently, the standard curve range can be at least three orders of magnitude, that is, 1–1000 units/ml, although the range of 1–100 units/ml is more appropriate for bioequivalence studies. The wider the calibration range, the more likely problems with carryover will occur.

Contamination, on the other hand, tends to be more random than carryover, and the problem can arise from multiple sources that can make it more difficult to diagnose and rectify.

Carryover and contamination can affect both the precision and accuracy of the method, and therefore, both must be carefully monitored and controlled.

Handbook of LC-MS Bioanalysis: Best Practices, Experimental Protocols, and Regulations, First Edition. Edited by Wenkui Li, Jie Zhang, and Francis L.S. Tse.

21.2 CURRENT UNDERSTANDING OF REGULATORY PERSPECTIVES ON CARRYOVER AND CONTAMINATION

The regulatory bodies are concerned about the impact of carryover and contamination on the interpretation of study data. In the ICH S3A Note for Guidance on Toxicokinetics: The Assessment of Systemic Exposure in Toxicology Studies (1995), it is stated that "It is often considered unnecessary to assay samples from control groups. Samples may be collected and then assayed if it is deemed that this may help in the interpretation of the toxicity findings or in the validation of the assay method." However, in the European Medicines Agency (EMA) Guideline on the Evaluation of Control Samples in Nonclinical Safety Studies: Checking for Contamination with the Test Substance (2005) comments that "A survey conducted by the European Federation of Pharmaceutical Industries Association (EFPIA) shows that contamination of controls with different levels of test substance often occurs during toxicology studies, regardless of the route of administration used, the dose levels and duration of treatment."

A survey of 30 countries in Europe and the United States was reported widely in the early 2000s with regard to the approaches for the analysis of control samples. This has been discussed recently in more detail by Olejniczak (2011) and earlier by Tse (2006). In Tse's report, only 67% of companies collected control samples and among those companies, 20% did not analyse the control samples, which was more or less in line with the ICH S3A 1995 Guidance on Toxicokinetics as summarized by Hill in a separate observation and recommendation (2004). The recommendations provided by the EMA (EMA, 2005) require that sampling of control groups and analytical procedures be integrated into pivotal studies that require toxicokinetic evaluation. At the time, some guidance as to what constituted a pivotal study was provided although they were not incorporated into the EMA Guideline. The proposals outlined in Table 21.1 are a guide only and should not be seen as regulatory opinion.

21.2.1 FDA Perspective

The FDA (US Food and Drug Administration) Guidance for industry on bioanalytical method validation (2001) did not dwell on this issue. The relevant paragraphs from the guidance have been reproduced here with comments as to their relevance to the issue(s) under discussion. Concern about contamination in its broadest definition is evident throughout the FDA Guidance documentation:

> Selectivity is the ability of an analytical method to differentiate and quantify the analyte in the presence of other components in the sample. For selectivity, analyses of blank samples of the appropriate biological matrix (plasma, urine, or other matrix) should be obtained from at least six sources. Each blank sample should be tested for interference, and selectivity should be ensured at the lower limit of quantification (LLOQ).
>
> Potential interfering substances in a biological matrix include endogenous matrix components, metabolites, decomposition products, and in the actual study, concomitant medication and other exogenous xenobiotics. If the method is intended to quantify more than one analyte, each analyte should be tested to ensure that there is no interference.

While this process highlights potential sources of interference, absence of an extraneous or unexpected peak during method validation does not guarantee freedom from contamination during study sample analysis. Indeed, variations in matrix, exogenous xenobiotics, degradation products, and metabolites may manifest themselves on a batch basis. In addition, reagents and blank plasma (processed and native) may be contaminated by the test substance.

In the aforementioned extracts, the FDA are concerned about the impact on the LLOQ, that is, "lower limit of quantification."

As per the guidance, the lowest standard on the calibration curve should be accepted as the limit of quantification if the following conditions are met:

TABLE 21.1 Requirements for Collecting and Analyzing Control Samples in Regulatory Toxicology Studies

Study type	Conditions requesting controls sampling and analysis
Single-dose toxicity	Sampling of controls may be not requested.
Repeat dose	All studies that include toxicokinetics evaluation—control samples—should be collected and preserved for potential subsequent analysis.
Genotoxicity	If controls in general toxicology studies are contaminated, consider analyzing samples from control animals.
Reproductive toxicology	Control samples from main study should be analyzed.
Carcinogenicity	Controls sampling and analysis should be performed.
Safety pharmacology	If toxicokinetics is performed, control animals should be sampled and analysis considered.
Routes of administration	Control sampling and analyses should be performed irrespective of route administration.

Note: Proposed requirements for collecting and/or analyzing samples in Regulatory Toxicology Studies. This proposal has not been incorporated into regulatory guidelines but is provided for guidance only.

- The analyte response at the LLOQ should be at least five times the response compared to blank response.
- Analyte peak (response) should be identifiable, discrete, and reproducible with a precision of 20% and accuracy of 80–120%.
- The accuracy, precision, reproducibility, response function, and selectivity of the method for endogenous substances, metabolites, and known degradation products should be established for the biological matrix. For selectivity, there should be evidence that the substance being quantified is the intended analyte.
- The specificity of the assay methodology should be established using a minimum of six independent sources of the same matrix. For hyphenated mass spectrometry-based methods, however, testing six independent matrices for interference may not be important. In the case of LC-MS and LC-MS/MS-based procedures, matrix effects should be investigated to ensure that precision, selectivity, and sensitivity will not be compromised. Method selectivity should be evaluated during method development and throughout method validation and can continue throughout application of the method to actual study samples.

The suggestion in the guidance that the MS detector will offer enough specificity not to require evaluation of multiple blank plasma samples should be considered with caution. Indeed the need to evaluate matrix effects in plasma from different sources echoes this concern. Use of the word "may" should offer sufficient warning to ensure that specificity is confirmed through experimentation. The message is, even for a highly specific methodology such as LC-MS/MS, vigilance is essential and never ASSUME!

As any interference distorting the LLOQ by 20% or more will compromise the validity of a validated method, consequently rigorous monitoring of any potential problems should be performed throughout the "in-study" analysis.

Further "guidance"/advice may be found in the Bioresearch Monitoring Compliance Program Document—"*in vivo* Bioequivalence, Compliance Program 7348.001" (1999), which requires inspectors to "ensure the analytical laboratory has scientifically sound data to support claims for the specificity of the assay employed in this study. Ascertain the laboratory's justification for noninterference, both endogenous and exogenous (e.g., metabolites and solvent contamination) in measuring the analyte (drug metabolites, etc.) studied" (page 32). In the context of sample analysis the inspectors are asked "to evaluate the source of blank biological fluids (was each subject's zero hour sample used as the blank, pooled plasma etc.?) Were interferences noted in the analytical source data for these samples? Specifications should be established to ensure blank biological fluids are as similar as possible to the biological matrix for the subject samples" (page 34). Confirmation in the report is needed in the form of "chromatograms showing Reagent Blank, Sample Blank, Internal Standard, a standard run, a quality control run, and a set of chromatograms for one subject over the entire span of the study" (page 35).

Although carryover is rarely mentioned in the 2001 Guidance for Industry on Bioanalytical Method Validation, there was a half paper overview of carryover and contamination in the White Paper of 2007 (Viswanathan et al., 2007). Furthermore, the White Paper stated "During validation the operator should assess the analyte response due to blank matrix while eliminating or minimizing other contaminations. The analyte response at the LLOQ should be at least five times the response due to the blank matrix."

21.2.2 EMA Perspective

The EMA Guideline on Bioanalytical method validation (2011) outlines the agency's position on carryover, but interestingly does not mention contamination:

- The Guideline states that "Carry-over should be addressed and minimised during method development. During validation carry-over should be assessed by injecting blank samples after a high concentration sample or calibration standard at the upper limit of quantification. Carry-over in the blank sample following the high concentration standard should not be greater than 20% of the lower limit of quantification and 5% for the internal standard. If it appears that carry-over is unavoidable, study samples should not be randomised. Specific measures should be considered, tested during the validation and applied during the analysis of the study samples, so that it does not affect accuracy and precision. This could include the injection of blank samples after samples with an expected high concentration, before the analysis of the next study sample."
- Although the EMA Guideline does not mention contamination per se, it does take a very similar line to the FDA Guidance on selectivity. The EMA states that "The analytical method must be able to differentiate the analyte(s) of interest and internal standard from endogenous components in the matrix or other components in the sample. Selectivity should be proved using at least 6 individual sources of the appropriate blank matrix, which are individually analysed and evaluated for interference. Use of fewer sources is acceptable in the case of rare matrices. Normally, absence of interfering components is accepted where the response is less than 20% of the lower limit of quantification for the analyte and 5% for the internal standard."
- During the analysis of study samples, the EMA Guideline states that an analytical run should include "blank

samples (matrix processed without analyte or internal standard) and zero samples (blank matrix processed with internal standard)." Depending on the analytical method, if the blank and zero samples are placed appropriately throughout the bioanalytical batch, then this should offer a means of monitoring carryover and contamination during routine sample analysis.

21.3 CARRYOVER—WHAT IS IT?

The earliest and complete discussions about carryover can be traced back to papers by Dolan (2001a) some 10 years ago. A more recent review of the practical implications of carryover was provided by Fluhler (2006) in a presentation at the Delaware Valley Drug Metabolism Discussion Group meeting. He defined carryover as "the analyte retained by the chromatographic system during the injection of a sample that appears in subsequent blank or unknown samples." According to Fluhler, carryover causes "systematic error (bias) that may affect the measured value of the sample." On the other hand, some forms of carryover can be cumulative (Dolan, 2001b).

Snyder and Dolan (2007) define three categories of carryover, that is, (a) classic or true carryover which is largely attributable to design issues, (b) carryover due to adsorption, and (c) carryover due to incomplete elution.

True or classical carryover is, in general, proportional to the previous sample and is constant, that is, if all samples were of the same analyte concentration, the amount of carryover would be the same for each sample. Fluhler illustrated this with some examples reproduced from Dolan's paper (2001a, 2001b) by assuming the error (Bias %) using 5% carryover from one injection to the other. As shown in Table 21.2, the error introduced to calibration standards injected in a descending order of concentration from high to low is approximately 20%. In contrast, the carryover can be reduced to approximately 2% for a calibration curve injected in an ascending order from low to high (Table 21.3). However, when samples are injected in a random concentration order, the 5% carryover can be a problem, as shown in Table 21.4. A blank injected after a 1000 ng/ml sample will have a signal that is equal to 5% of the previous sample while a further injection will have 5% carryover from the previous blank. This can significantly distort the value of a sample (depending upon its nominal concentration). Fluhler further illustrated this with examples from Zeng et al. (2006) who developed an equation to calculate the impact of carryover, that is, the estimated carryover influence (ECI) from preceding samples on the quantification of following samples:

For a fixed carryover,

$$\text{The effect of carryover ECI (\%)} = \text{RC} \times \text{CR} \times 100,$$

where

$$\text{Relative carryover (RC)} = \frac{\text{Peak area of blank}}{\text{Peak area of previous sample}} \times 100,$$

$$\text{Concentration ratio (CR)} = \frac{\text{Previous sample concentration}}{\text{Following sample concentration}} \times 100.$$

Table 21.5 (Zeng et al., 2006) shows the influence of carryover from a preceding sample on the quantification of a subsequent sample where the concentrations of the preceding samples and the following samples were increased from

TABLE 21.2 Error (%) Caused by Carryover (5%) from a Hypothetical Standard Curve Injected from High to Low

Concentration	Response	Error(%)
1000	1000.0	0
300	350.0	17
100	117.5	18
30	35.9	20
10	11.8	18
3	3.6	20
1	1.2	18
0	0.06	–

TABLE 21.3 Error (%) Caused by Carryover (5%) from a Hypothetical Standard Curve Injected from Low to High

Concentration	Response	Error(%)
0	0.0	–
1	1.0	0.0
3	3.1	1.7
10	10.2	1.5
30	30.5	1.7
100	101.5	1.5
300	305.1	1.7
1000	1015.3	1.5

TABLE 21.4 Error (%) Caused by Carryover (5%) from One Sample to the Other with Random Analyte Concentrations

Concentration	Response	Error (%)
1	1.0	0.0
1000	1000.1	0.0
0	50.0	–
3	5.5	83.3
100	100.3	0.3
30	35.0	16.7
300	301.8	0.6
10	15.1	50.9

TABLE 21.5 Influence of Carryover from Preceding Sample (N) on the Quantification of the Following Sample (N + 1) for a Fixed Concentration Ratio (CR) of 100

Conc. sample N	Conc. sample $N + 1$	Avg. conc. (ng/ml) $n = 5$	Accuracy[a] (%)	Abs. error[b] (ng/ml)	Rel. error[c] (%)	EC[d] (ng/ml)	ECI[e] (%)
1000	10	10.5	105	0.48	4.8	0.05	5
5000	50	51.0	102	1.00	2.0	2.60	5
10000	100	104	104	4.20	4.2	5.00	5
20000	200	209	105	9.10	4.6	10.0	5

[a]Expressed as (average concentration)/(nominal concentration) × 100%.
[b]Expressed as (average concentration—nominal concentration).
[c]Expressed as (absolute error/following sample concentration) × 100%.
[d]Estimated carryover (EC) expressed as (nominal concentration of preceding sample x RC), relative carryover (RC) = 0.05%.
[e]Estimated carryover influence (ECI) expressed as (RC × CR) × 100%.
Based on Zeng et al. (2006).

1000 to 20,000 ng/ml and from 10 to 200 ng/ml, respectively, in order to maintain a concentration ratio of 100. The results show that the absolute errors increase, but the relative errors remain approximately constant. This shows that the relative error is related to relative carryover and also to the concentration ratio. Zeng et al. used this data in conjunction with other experiments to demonstrate that the greater the concentration ratio, the higher the influence of carryover when the relative carryover is constant. The authors concluded that if the relative error of the method (prior to carryover effects) is less than 10%, then carryover (ECI < 5%) will not have a significant effect on the accuracy. While this provides a useful criterion for designing the validation and the in-life analysis format it is not always possible to maintain an ECI of less than 5% and consequently analysts should confirm lack of influence during method development and confirm it in validation.

Nonclassical carryover due to adsorption of the analyte of interest to the detector system appears only after "several" or more sample injections. Waters (2011) details the potential LC-MS front end components that may be contaminated and thus contribute to carryover. These components include the electrospray interface (ESI) probe, especially the probe tip, and capillary unions. In addition, the sample cone, lockspray baffle, ion source block, source enclosure, PEEK tubing connecting column to API source, components of the integral flow divert/injection valve (if fitted), throttle valve, and PEEK support block may potentially be other problematic areas. Although the aforementioned list relates primarily to Waters systems, users should understand the materials and construction of their particular instrument(s) in order to extrapolate the Waters suggestions.

Wear and tear is slightly more complex and just as "design" carryover should be evaluated at the installation qualification (IQ) stage, performance qualification (PQ) may be used to ensure that wear and tear is not contributing to carryover by incorporating a blank following a high concentration injection, and or a series of blanks following a number of sequential injections of the high concentration samples.

Comparing carryover during routine analysis with that from the IQ stage can provide some indication of wear. However, one of the bigger problems associated with wear and tear is the exposure of more active sites / surface area where the drug may be adsorbed. All the manufacturers cite this as a major problem. With increasing numbers of samples injected there is an increase in the amount of drug adsorbed. This issue is difficult to identify as the amount adsorbed must reach a critical mass before "measurable" amounts are observable in blank sample injections. This could be further complicated if samples are injected randomly.

The MS detector itself can also cause carryover as a consequence of a phenomenon known as "crosstalk." This occurs when ions are removed too slowly from the collision cell during the transition process. The problem can arise when selective reaction mode (SRM) with a short dwell time is employed for analysis of multiple analytes on "old" MS instruments. This can still be an issue as there are still many "old" machines out there (Hughes et al., 2007). Even with the newer "crosstalk free" instruments, there is evidence that crosstalk could still be present (Morin et al., 2011).

21.4 CONTAMINATION—WHAT IS IT?

Contamination can be classified into three categories:

1. Contamination of the sample with the test substance prior to sample processing/handling and storage. The term "sample processing/handling" relates to the procedure of withdrawing the sample from the subject to placing it in a sample tube.
2. Contamination with the test substance during sample preparation. This refers to preparing the samples for analysis, for example, extraction cleanup and placing on the autosampler.

3. Contamination due to the presence of unknown compound(s) that behaves like the test substance. These "contaminants" may cause ion suppression or rarely enhancement in LC-MS/MS detection of analyte of interest. In addition, contamination may be due to late eluting peaks from previous injections.

Irrespective of whether the samples are preclinical or clinical in origin, contamination might be due to atmospheric factors, errors in study sample collection, processing and transfer, unclean labware and reagents, or due to use of internal standards that are insufficiently isotopically pure. From a study conduct point of view, contamination can occur during sample collection, for example, in the clinic or animal houses or in the bioanalytical laboratory.

21.4.1 Atmospheric Contamination

Atmospheric contamination can be due to a variety of reasons. Splashes and/or aerosols can occur during evaporation of solvents used during sample preparation, although it varies depending upon the temperature of evaporation and shape of the "tube" or "well" of 96-well assay plate.

When using both manual and automated liquid handling systems, airflows can affect the transfer processes. Many factors, including the extent and direction of airflow in fume hoods and air conditioning systems, can be obviated at the laboratory design stage (Clinical Laboratory Standards Institute (CLSI) Laboratory Design). Failure to do so can result in the distribution of all manner of contaminants but especially test substance.

21.4.2 Contamination during Dosing, Sample Collection, and Storage

Contamination can occur outside the bioanalytical laboratory, for example, in the animal houses and in the clinic, as a result of misdosing or mishandling the samples that are collected.

Misdosing can occur in a number of ways. For example, control animals or volunteers scheduled to receive control dose formulation or placebo subjects can be dosed with the test substance in error. Alternatively, the control dose formulation may be contaminated with the test substance or high dose formulations are administered to animals/volunteers in low dose groups. Environmental contamination can also be a problem with rodent dietary studies where feed can accidentally be transferred between cages. Similarly, contact between control and treated animals can be an issue if there is dietary feed in their fur and the animals groom each other.

Contamination of preclinical and clinical samples can occur at all stages of collection and storage. For example, samples can be incorrectly identified (i.e., control/placebo samples switched with samples from dosed groups) or disposable pipettes are reused when collecting plasma following separation from red blood cells.

Cross-contamination can also occur between plasma samples that are not stored correctly prior to freezing (i.e., leakage may occur from sample containers stored horizontally rather than in upright position).

21.4.3 Contamination during Sample Preparation

The following sample preparation and analysis procedures are prone to contamination or carryover if the necessary precautions are not in place:

- *Sample mix-up*: Incorrect samples selected for analysis.
- *Sample transfer/aliquotting*: Aliquots from correct sample transferred to wrong tube/well prior to extraction.
- *Equipment contamination*: Test substance adsorbs to surface of analytical equipment.
- *Contamination due to reuse of disposable labware*: The ideal scenario in LC-MS bioanalysis is that the labware, for example, pipette tips and containers, is discarded after a single use. However, there are cases where the same disposable labwares may be repeatedly used.
- *Splashing and creeping*: There are now a wide range of automated dispensers, integrated workstations, and robotic liquid handlers with plate transport functionality for LC-MS bioanalysis (Lab Manager, 2011). These devices are from the Tecan, Hamilton, Gilson, Tomtec, Thermo Scientific, and so on (Caldwell, 2008). If not programmed correctly, those devices can transfer/dispense reagents, solvents, and samples inappropriately, resulting in cross-contamination (Hughes et al., 2007). A common problem is "splashing" due to the use of large distances between tips and liquid surfaces. Although air gaps can be programmed to prevent liquid samples from dripping during transfer, once again, the problem remains a common issue. This is especially true when low viscosity liquids are being transferred. Splashing can also occur during solvent evaporation using 96-well plates if the correct distance between nitrogen needles and liquid surfaces and proper nitrogen flow (pressure) is not tested for suitability. Another common problem is the creeping of solvent in the process of evaporation with volatile solvents (Chang et al., 2007) if nitrogen flow and evaporating temperature were not set up properly.

21.4.4 Reagents as a Source of Contamination

Contamination problem can occur if the test substance is accidentally added to a reagent. If this happens, the test substance will appear in all the extracts from a particular batch (i.e., blanks, calibration standards, QCs, and incurred samples) and will almost certainly invalidate the data from that batch.

Perhaps the most widely used reagent in sample preparation and chromatography is water. There are many techniques such as distillation, deionization, reverse osmosis, or other combinations for producing purified water.

The quality of water is not necessarily an indicator of its applicability to LC-MS—however, the ability to define a specific grade is essential for traceability purposes. These grades are largely based on the same parameters, that is, resistivity and conductivity, pH, total solids, silicate, particulates, organics and microbial content, which are variously defined by a range of different organizations, for example, ASTM (2009) ISO 3696, and NCCLS and Pharmacopeia standards.

Leaching of organics from storage containers, collection vessels, and pipework are another source of potential interferants from chemicals such as phthalates, detergents, lubricants, and PEG (polyethylene glycol) related materials used to calibrate mass spectrometers.

Solvents, whether they are for cleaning purposes or used as reagents in bioanalysis, should always be of the highest "purity" although there is not a definitive definition of highest purity. It is, therefore, important to check that the solvents to be used are "fit for purpose." The well-established phrase "or equivalent" should NOT be used as a reason to use what may appear to be a similar reagent. The alternative reagent should be proven to be a valid substitute.

21.4.5 Internal Standards as a Source of Contamination

Physicochemical analogs continue to be used as internal standards, but there is a growing trend to use stable isotope labeled internal standards for LC-MS bioanalysis as recommended by the EMA (EMA, 2011). It is unusual for analog internal standards to have impurities similar to, or cochromatographed with the test substance. However, these types of internal standards can cause problems when metabolites need to be quantified in the same analytical run. This is because impurities from the synthetic process for the test substance can be similar to or the same as the metabolites of the drug. Hughes et al. (2007) pointed out that there can be a significant amount of unlabeled test substance in the preparation of any stable label internal standard. It is not always a trivial task to ensure that the unlabeled test substance is removed in its entirety. Therefore, it is incumbent upon the "user" to ensure that each new batch of stable label internal standard does not have amounts of unlabeled test substance that might compromise the LLOQ of the method.

21.5 WHAT DO THE MANUFACTURERS SAY?

The design engineering features that contribute to carryover are well known and are eloquently described in the Leap technologies reference (Leap Technologies, 2005). Injector system design is pivotal to the understanding of classical carryover, for example, whether the needle takes sample extract from an assigned vial and injects the extract into a Rheodyne type valve or whether the sampling needle positions in line so that the HPLC effluent flows through it. One aspect of minimizing carryover that the manufactures strive to address is reducing the dead space within the system, but controlling adsorption and developing appropriate rinsing solvents and regimes also need to be considered.

The current trend to UHPLC and to the use of cartridge-based systems, and related Lab on a Chip systems, means that the amount of pipe work can be minimized and the connections at least standardised rather than leaving these to the mercy of the analyst plumber.

It is unlikely that manufacturers can address adsorption issues surrounding all chemical entities. Indeed, while stainless steel systems may show little adsorption with certain chemicals, it may be necessary to use polymer-based tubing and injection valves to minimize carryover with other chemical entities. While much of carryover is associated with the injection system, adsorption to the column material (e.g., packing material, frits and filters) can also be significant.

The manufacturers approach to resolving these issues are discussed as follows.

21.5.1 Shimadzu

As one of the current generations of LC systems, the Nexera with the SIL-30ACMP autosampler is claimed (amongst many others) to have "lower carryover and shorter cycle times than competitor systems." This is achieved by reducing contact area, the use of special coatings and surface treatments and a new needle seal.

21.5.2 LEAP Technologies

This manufacturer of autosamplers discusses minimizing and eliminating carryover in PAL (Prep and Load) systems in their document "A primer for Reducing Carryover in PAL Autosampler Systems" Rev 1.1. (Leap Technologies, 2005).

LEAP defines carryover as a "detector response from previously injected sample, detectable (quantifiable?) when a subsequent injection is made." While they measure carryover for acetaminophen as being 0.0004% of the upper limit of quantification, they elegantly caveat this with the statement, "This is a fantastic number but says nothing about the autosampler performance using your samples." Similar to other manufacturers, LEAP identified dead space and adsorption as the two major problems. LEAP's philosophy to washing is that two washes are better than one and that "like dissolves like" (e.g., if the solvent/solute molecules are structurally similar, then the substance will dissolve in the solvent). Addition of solvent modifiers at 0.1–1.0% may be needed.

Other components contributing to dead space include improper needles or fittings. While coating stainless steel needles with PTFE or glass can reduce adsorption, building up of test substance in the syringe is a common problem. Valves and tubing (especially incorrect fitting of ferrules) can contribute to dead volumes. While PEEK tubing produces less adsorptive carryover, the solvent flow can be easily "blocked" by excessive crimping of ferrules.

The guide provides troubleshooting tips for reducing and/or eliminating carryover in PAL systems. Many of the suggestions also have relevance to other manufacturers' equipment. The guide suggests the use of partial loop filling and sandwich loop filling for minimizing contact of the test substance with rotor and valve surfaces. In the latter case, the sample is sandwiched between solvent, air, sample, air and solvent in the needle and loop of the injector. It is important to remember that the elution strength of the injection solvent should be lower than that of the starting solvent for the chromatography gradient. This is especially important as gradient systems are commonly used in bioanalysis to separate interferants and late eluting peaks, the latter can interfere with "future" sample injections.

21.5.3 Waters

Waters provides an overview on eliminating contamination in HPLC and MS detection systems. While not specific to Waters systems, it does provide an excellent generic outline of points to consider. Although the latest brochure of Waters (Waters, 2011) did not mention the word carryover, their technology briefing notes entitled "Low sample carryover for sticky analytes with the Acquity UPLC H-Class System" (Jenkins T, Application Note P/N 720003616en) discuses a simplified washing system for terfenadine. It was shown that a single solvent could be used to clean the exterior of the needle in the injection port—which does not come in contact with the sample or the mobile phase, while the needle interior is washed by the mobile phase. The success in reducing carryover to less than 0.004% was attributed to the flow-through-needle design. As indicated in the brochure, solvent choice is extremely important—the solvent should be compatible with the drug, that is, capable of dissolving it while not capable of distorting the chromatography. For these two reasons, choice of solvent will be drug substance specific—beware use of generic wash solvents.

One of the latest Waters UPLC systems at the time of publication is the Acquity UPLC I-class system. The system is claimed to have even lower carryover than its predecessor, thus complementing MS sensitivity and helping to extend the MS linear dynamic range. Another recently developed instrument is the H-class Bio. This is an ultra-inert system that offers a solution to some of the contamination and carryover problems seen with large biological molecules.

21.5.4 Agilent

The current comments refer largely to the Agilent 1290 Infinity LC and LC-MS systems presented by Naegele et al. (2010). The presentation concentrates mainly on the use of the Agilent "Flexible Cube" that is designed to remove instrument-related carryover by needle seat back flushing. The capabilities of the cube are illustrated by a number of examples tested by Agilent. They itemise the source of carryover in the HPLC component of LC-MS system as:

- Adsorption to needle and capillary material.
- Needle design, improper sealing, or worn out parts (dead space and adsorption).
- Valve rotor stator, design, worn out (some adsorption to mainly dead space issues).
- Capillary fittings misadjusted (at all connections contaminated with sample) (attributed to dead space issues). Column-related carryover can be attributed to dead space and/or adsorption to fittings and connections of the column as well as stationary phase–sample interactions.

21.5.5 Thermo Scientific

Using the Thermo Surveyor as an illustration, Elmashni (2007) highlighted that the flush volume needs to be increased in order to remove materials that are either sticking to the wall or not displaced because of slower laminar flow next to the tubing sides. In addition, he acknowledged that the test substance may "stick" to the surface of the polymeric rotor material. It is proposed therefore "to conduct a syringe flush while rotating the valve to ensure rotor material is flushed." For the Thermo Surveyor, the calculated carryover was 0.0089% using a 400 μl flush, but this number could be improved by an order of magnitude via increasing the flush volume by more than 10-fold. Nonetheless, it is unlikely that this process will resolve the problem of sticky substances that adhere to the chromatographic system.

More recently Thermo developed the Accela Open Autosampler in 2010. According to the design of the system, the sample can be held in a holding loop between the syringe and the injection port. Using the software it is possible to wash the entire sample path using up to two solvents. In a test for chlorhexidine, the measured carryover was 0.003%.

21.6 MANAGING CARRYOVER AND CONTAMINATION IN LC-MS BIOANALYSIS

Carryover and contamination must be identified and minimized in order to avoid an adverse impact on the outcome of the intended preclinical or clinical study. Therefore, it is recommended that bioanalytical scientists receive appropriate training to be able to deal with these matters and that industry best practices are implemented in set of standard operating procedures. The philosophical approach and detailed practices may vary from one organization to another, but it is important to have an appropriate policy in place. The suggestions and observations outlined later should be read and implemented in that context.

21.6.1 Identification and Confirmation of Carryover and Contamination

Carryover should be assessed by placing matrix blanks immediately after high concentration calibration standards (e.g., ULOQ calibrator) or high QC samples. On the other hand, one or more matrix blanks should be placed at the start of the analytical batch to ensure that the system is clean before committing incurred samples for analysis. They are also placed at appropriate places throughout the batch (e.g., after high calibration standards or QCs) so that carryover can be monitored over a period of time. The reagent blanks and solvent blanks can also be used to monitor carryover. Using this strategy, it is fairly straightforward to identify if carryover is a problem in a particular batch. If a peak is present in the blank matrix injected immediately after the highest calibration standard with LC-MS response (peak area) greater than 20% of that of the LLOQ standard for the analyte or more than 5% for the internal standard, then significant carryover has been identified.

Similarly, contamination can also be assessed by monitoring extracted matrix blanks, reagent blanks, and solvent blanks for the presence of the analyte or internal standard. The presence of analyte in pre-dose, control, or placebo samples in the assay sequence must be checked. Once again, if a peak is present in any of the aforementioned samples that is greater than the LLOQ, contamination may have been identified. Although it is not always straightforward to assign the cause of contamination, using the data obtained from the blank injections may help point to the cause of the problem. For example, if peaks are present in the matrix blanks, it may indicate that contamination is occurring during sample preparation. However, if peaks are present in the reagent or solvent blanks, then it may indicate that one of the reagents is contaminated. Peaks present only in the pre-dose, placebo, or control samples but not in the blank matrix, reagent, or solvent blanks suggest that the contamination is likely to have occurred outside the bioanalytical laboratory, possibly due to misdosing or *ex vivo* contamination during sample collection.

The following examples help highlight some of these issues.

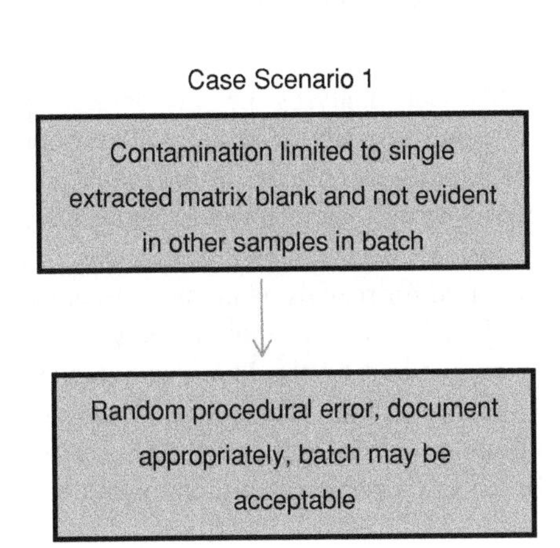

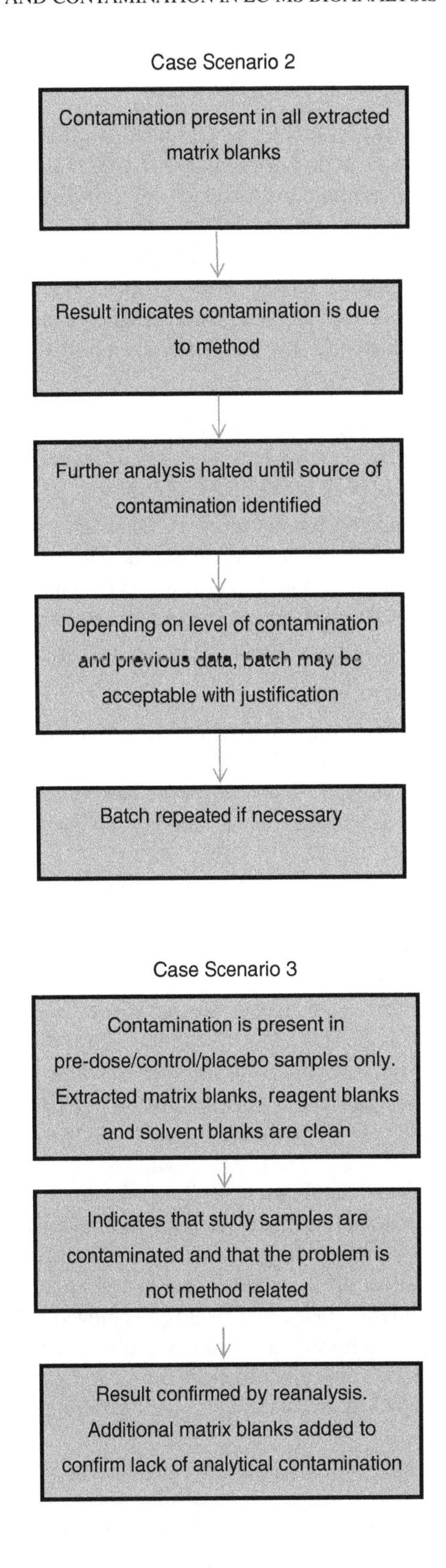

Contamination can affect all samples in a batch, not just the various blank samples that are used for diagnostic purposes. Therefore, it is important to check the entire batch and particularly for anomalies in the profiles from incurred samples.

21.6.2 Impact Assessment of Carryover and Contamination on Study Outcomes

System carryover should be evaluated by injecting blank matrix extracts after injections of ULOQ calibration standards. The maximum peak area of any carryover should be ≤ 20% for the analyte and 5.0% for the internal standard of the respective peak areas of the LLOQ calibrators in the same sample sequence. If the carryover acceptance criterion is not met, the impact of carryover on the unknown study sample must be evaluated (Zeng, 2006). For example:

- The absolute carryover (AC) can be calculated using the matrix blank with the highest carryover in the batch: carryover peak area/ULOQ calibrator.
- The concentration difference (CD) of unknown samples can be calculated as: sample E peak area/sample L peak area, where sample L and E are two samples injected sequentially and sample E precedes sample L.
- Calculate the maximum CD in the run (MCD).

The maximum impact of the carryover on unknown samples is AC × MCD × 100%.

After the impact of the carryover has been assessed, a decision has to be made on whether any sample reanalysis is required.

It may not always be possible to eliminate carryover due to the engineering of the system. While the impact of carryover on the results can be minimized through arranging the order of samples or via interspersing of blanks, there is always likely to be measurable carryover. Clouser-Roche et al. (2008) suggested a more pragmatic approach to dealing with carryover. The approach is largely driven by a combination of a risk based/fit for purpose criteria. Whether it has an impact on the study outcomes depends largely on the study objectives.

Although contamination is generally monitored and detected by including blank samples in a standard sample batch, it should not be forgotten that contamination, if occurring, may affect all the components in the batch (i.e., calibration standards, QCs, and incurred samples). Therefore, along with monitoring the blank samples for unexpected peaks, it is also important to carefully review the toxicokinetic (TK)/pharmacokinetic (PK) data generated from the study samples.

For TK studies where the concentration-time profiles are obtained through sparse sampling approaches (typically rat and mouse), there are limited sample time points available and these results may not provide adequate information for clearly identifying a sample being inconsistent within the concentration-time profile. In general, samples from TK studies should be reanalyzed only if the batch fails to meet the required acceptance criteria (i.e., if the blank samples show an obvious problem with contamination or carryover or the calibration standards/QCs are out of specification).

For clinical study samples or TK study samples collected via serial bleeding, the respective study samples may be considered for reanalysis if the observed concentration for the analyte is more than twice or below one-half of the expected concentration based upon the surrounding concentrations in the concentration versus time profile for that analyte. Sample results at or near the C_{max} may not easily be considered using this approach.

The following approach recommended can be useful in handling the unexpected analyte concentration due to contamination or other reasons:

- Reanalyze the sample in duplicate within the same batch to generate two independent concentration values for the analyte(s).
  - If the repeat values are within 20% of each other, then the median of the three (the original and two repeat values) will be reported.
  - If the values are >20% from each other, no result will be reported.
- If only one repeat reanalysis is possible due to limited sample volume, then a single repeat analysis should be performed for the suspect sample.
  - If the reassay value is within 20% of the original value, the original value is reported.
  - If the repeat analysis of the suspect sample is between 20% and 30% of the original value, the mean of the two values will be reported.
  - If the repeat analysis for the suspect sample differs from the original value by more than 30%, no result will be reported.
- Insufficient volume is available for a single repeat analysis.
  - If there is insufficient sample volume to perform even a single repeat analysis, the original value may be reported, but identified as a specific data issue in the bioanalytical report (i.e., insufficient sample volume available to confirm the value).

For trough sample analyses (i.e., pre-dose samples taken during steady-state drug administration), it may not be possible to clearly identify samples requiring reanalysis based on the surrounding trough sample values and dosing regimen (i.e., bid or tid). In general, the criterion described earlier should not be used for reanalysis in these circumstances.

For clinical studies, the decision to analyze placebo samples should be justified by the Pharmacokineticist prior to the start of analysis. For all TK studies, all samples obtained from animals in the control groups should be analyzed unless specified in the study protocol. If there is an observed drug concentration above the LLOQ for an analyte in a control/placebo group or pre-dose sample (except trough sample), then that particular sample should be reanalyzed.

If sufficient sample volume remains, the following procedures should be followed:

- A single repeat analysis should be performed with a blank (control) sample from the same biological matrix from the same species.
  - If the repeat value for the analyte in question and the blank sample result are negative (i.e., below the LLOQ), then the value will be reported at zero (below the lower limit of quantification, BLQ) for that analyte.
  - If the repeat value for the analyte in question is positive and the blank sample is negative, then the original result for the analyte will be reported.
  - If the repeat value for the analyte in question and blank sample result are both positive, then the repeat analysis is not valid and the analyst should define and eliminate the cause of the "apparent" contamination, before reanalysis, if possible

However, if there is insufficient volume remaining to perform a reanalysis then the original result will be reported, but identified as a specific data issue in the bioanalytical report (i.e., insufficient sample volume available to confirm result).

In cases where an analyte is confirmed to be present in a control, placebo or pre-dose sample (except trough sample), an investigation should be carried out to assess the impact on the overall TK evaluation or on the clinical study. This information must be communicated to the responsible scientists involved with the study (i.e., Study Director, Principal Investigator, Pharmacokineticist). An assessment of the impact on the study should be conducted:

- It is important to assess whether the contamination was due to a procedural error in the animal house or clinical site(s) (i.e., misdosing, contamination during collection, etc.) or due to error(s) in the bioanalytical laboratory. Based on the frequency, magnitude and time (post-dose) of occurrence of positive values of analyte(s) of interest and it metabolite(s), it may be possible to identify whether the contamination occurred by misdosing or *ex vivo* contamination. The results of the blank samples that are analyzed as part of the sample sequences should help confirm whether the contamination occurred in or outside the bioanalytical laboratory.
- If a definitive bioanalytical cause cannot be identified, then the laboratory should provide support to try and help identify the problem in the animal houses or clinic.
- Depending on the potential impact of the positive values, further bioanalytical investigations can include examination of the specificity of the bioanalytical method beyond typical acceptance standards, for example, confirming multiple MRM transitions for analyte identity. The observation of any positive values, the results of the investigation and the potential impact on TK/PK interpretation should be fully documented in the bioanalytical report.

If it is confirmed that peaks observed in the pre-dose/control/placebo samples have not originated in the bioanalytical laboratory, then the problem is likely to have occurred in the animal house or clinic. This may be a consequence of misdosing or because of a problem that occurred during sample collection (e.g., samples misidentified or repeat use of disposable pipettes during separation of plasma from red blood cells). Whatever the reason, an investigation needs to be initiated at the animal house or clinic facility to ascertain what actually happened so that the impact can be properly assessed.

If misdosing is confirmed, reassigning the correct dose to the right profile may require exhaustive evaluation of the paper trail and a definitive investigation involving everyone from dose preparation through to sample collection and storage. At the end of the investigation, it may require the intervention of a toxicokineticist or pharmacokineticist to determine this linkage. From a compliance point of view, the events surrounding the choice of the data should be fully documented and justified, or a complete repeat of the study may be necessary.

21.6.3 Minimizing and/or Eliminating Carryover and Contamination

This section is devoted to postvalidation approaches assuming that there is no evidence of unacceptable carryover or contamination during method development and validation and that the PQ indicates that the systems are operating as per instrument qualification results.

While many problems occur during sample analysis as random events, it is possible to minimize them based on the experience of similar issues previously identified and addressed—alternatively known as preemptive problem solving or more formally as Corrective and Preventive Action (CAPA). Some of these may have arisen and resolved during method development. However, without proper documentation into detailed method procedure, the same issues are likely to reoccur.

One of the most contentious approaches is to process control samples separately from dosed samples—geographically (on another bench and using separate equipment)—but they should be assayed on the same system.

Be aware of sample transfer processes as drips and splashes may take place. As soon as an issue occurs, it should be immediately documented and addressed properly. Whenever possible, avoid reusing materials such as tubes and cap liners. Ensure all labware is appropriately washed / cleaned. Inject samples in profile order (e.g., from lowest to highest concentration) in order to minimize carryover effects on the lowest concentration. The calibration standards should be treated likewise although some analysts run a standard curve at the start of the batch from low to high concentration and a second curve at the end of the sequence in reverse order, to assess possible carryover effects. Before using reagents, check if necessary for possible contamination, that is, a reagent blank taken through the extraction and sample concentration (where applicable) process. Use a calibration curve range consistent with the expected concentration range in the study samples—again to minimize carryover.

An understanding of the source of carryover and contamination can dictate the format and size of an analytical batch. A completely randomized approach is where standards, QCs and study samples are distributed randomly over a batch. While this has statistical validity in "minimizing" the variability due to position in the batch, it can make it difficult to evaluate variability due to carryover and is not recommended by the regulators if carryover is likely to be problematic. Listed below are the key actions that can help reduce or minimize carryover or contamination:

21.6.3.1 Cleaning Glassware and Plastic Components Glass is still an important material for use as a receptacle for preparing reagents from standards to mobile phases, but its popularity is diminishing largely due to its adsorptive properties with respect to basic drugs.

Contamination of glassware can occur as a result of the manufacturing and storage process. Therefore, new glassware should always be rinsed or washed before use, ideally with pure methanol, followed by a deionized water rinse. However, specific cleaning procedures may be needed depending upon the nature of the contaminant. It is essential that such basic articles are clean. As the old statistical saying goes "garbage in = garbage out." It is essential that the cleaning process does not compromise the "calibration" of glassware used for pivotal measurements. The essence of good washing technique is available everywhere. Among the many sources, Lab Equipment World (www.laboratoryequipmentworld.com/caring-cleaning-lab-glassware.html) (2012) provides one of the most up to date overviews. All laboratories have their standard operating procedures for washing glassware and these should be reviewed regularly. There are three major components in the washing process of laboratory glassware, that is, prewash, the washing procedure and finally drying and storage. The main point is to ensure that cross-contamination with test substance does not occur and any potential interference is removed. At the prewash stage, it important to use solvents compatible with the solubility of the test substance, this should be followed by a rinse with deionised water. If this "soft" process does not work, soaking the glassware in "chromic" acid is necessary, although this may also etch the glass surface. The washing process may be manual but more usually it is carried out with an appropriate commercial washing machine using tailored wash cycles that ends up with a deionised water rinse and drying either in the machine with gentle heat or manually transferred to an oven.

Once dried (and cooled), glassware should be stored away from any potential contamination by the test substance and should be stoppered or covered.

The ideal solution to resolving contamination is the use of disposable substitutes. While this may appear expensive, it can be a pragmatic (and cost-effective) approach to rapid problem solving of cross-contamination. Consequently, plastics are now replacing glass containers and stainless steel tubing. However, this material is not without its own problems. In reality, it is not the plastic itself but the additives used in the manufacture of plastic that contributes to the issue of contamination. Kattas et al. (2004) provided a comprehensive review of the different plastics and the additives used in manufacturing. The types of agents added have many different functions, some having multiple roles. These include antioxidants (e.g., benzophenone and benzotriazole), light stabilizers (e.g., benzophenone), antistatics, plasticisers (e.g., phthalate esters, aliphatic esters and trimellitate esters), lubricants (sometimes known as slip agents or mould release agents, for example, fatty acid amides, fatty acid esters), antiblocking agents and heat stabilizers. While they may be present in low concentrations (circa 0.1%), they can still present a problem to high sensitivity assays. Although the MS detector may be specific enough to preclude interference from chemically unrelated compounds, this not always the case especially when the MS detector is operated in single ion monitoring mode.

21.6.3.2 Removing Contamination from the LC-MS System Procedures for cleaning the analytical system can be found in the Waters document (Bergvall, 2011) "Controlling Contamination in Ultra Performance LC-MS and HPLC-MS systems (Waters 715001307 Rev D) and similar documents from other vendors. While many of the components of MS detectors are common to most systems, it is essential that the user understands the construction of the machine and is able to clean all the components appropriately. The author of the article suggests that under "ideal" circumstances each component should be removed, cleaned or replaced and tested one at a time. HOWEVER if "contamination persists it is likely that the components may be recontaminated after cleaning. In the case where contamination persists and all other possible sources have been eliminated, it may be judicious and cost-effective to clean and or replace all suspected parts simultaneously."

It is essential that the analyst understands the limitation of their remit when "dismantling" MS components for cleaning. Components should be sonicated in the appropriate solvent for between 15 and 60 min and a range of solvents are detailed in the brochure.

A simpler procedure for cleaning both the LC system and MS detection system can be found in the Agilent Technologies article "What is steam cleaning in LC/MS?—General LC/MS" (2010). In this process the vapors of specific solvents, that is, isopropanol, methanol, acetonitrile and water pass are pumped through the LC system and vaporized at the heated interface before passage into the detector. Overnight cleaning using this approach significantly reduces the background noise. For LC system cleaning, use of a solvent mixture of 25:25:50 (cyclohexane/acetonitrile/isopropanol) is recommended. It is important NOT to leave this solvent in the system for extended periods of time as cyclohexane can degrade the pump seals. When following either the Waters or Agilent cleaning procedures it is important to ensure the analytical column is NOT connected.

21.6.3.3 Cleaning the HPLC Column Historically, HPLC columns were very expensive consumable items and, as such, analysts made a lot of effort to take care of them to ensure longevity and performance. Nowadays, columns are made with packing materials based on established technology that are carefully quality controlled and are relatively cheap to purchase. As a general rule—if there is a doubt about a column's history, then discard it. If starting a new study, uses of new columns are recommended. If you have to clean a column in order to remove interferences/contaminants, then the article by Majors (2003) is recommended. The article deals mainly with Reversed Phase HPLC columns. Since the contaminants /interference may have similar physicochemical properties to the test substance (contaminant may be the test substance itself), then solvents that are suitable for the test substance can be the ideal starting points for column cleaning. It is generally recommended to use 20 column volumes of a stronger solvent run in isocratic mode. However, jumping immediately from a weak solvent to a very strong solvent can cause precipitation of buffer salts in the column. Therefore, flushing out the salts with water (containing low percentage organic) is appropriate. From the article, Majors further recommends a sequence of solvents, each with at least a 10-column volume pass. In order to ensure solvent compatibility, he recommends the following sequence: methanol, acetonitrile, 75% acetonitrile/25% isopropanol mix followed by pure isopropanol, methylene chloride and finally hexane. However before returning to the regular mobile phase, it is necessary to flush with acetonitrile.

Agilent (www.chem.agilent.com/cag/cabu/ccleaning.htm, accessed Mar 23, 2013) presents a similar approach but recommends that dimethylformamide may be better than methylene chloride and hexane for cleaning reversed phase columns. Agilent also proposes a procedure for cleaning normal phase columns. The procedure involves reversing the column and flushing with 50:50 (v/v) methanol:chloroform, followed by ethyl acetate, then reversing the column again, followed by equilibration with starting mobile phase.

Another approach is to use an isocratic mobile phase following elution of the peak of interest by increasing the proportion of the stronger solvent in mobile phase. Indeed this is now a well-accepted practice for removing late eluting peaks. If this approach fails, the more complex and protracted procedure described earlier should be employed,—or alternatively a new column should be used (Dolan, 2011). However, a continuous high organic column wash might not be as effective as cycling between high and low organic mobile phases as indicated by Williams et al (2012). Instead, a four-cycle sawtooth-wash procedure can be more practical. Nevertheless, it is necessary to evaluate and confirm the best approach on case-by-case basis.

21.6.3.4 Use of Precautions in Sample Preparation and LC-MS Injection The current trend is to carry out sample preparation using multiwell systems for protein precipitation (PPT) and solid phase extraction (SPE). For liquid–liquid extraction (LLE), although it can be carried out using this format, possible contamination of the two layers, that is, the aqueous and organic layer, needs to be monitored carefully. For this reason, supported liquid extraction (SLE) may be a better option to use in a multiwell format. In many cases, LLE is still performed at the macrolevel using glass or plastic tubes. While it is possible to use liquid handling systems, many laboratories still use manual systems especially when sample numbers are limited and do not warrant robotic method development. When using manual systems, a rapid and effective process for minimizing contamination is to freeze (dried ice and acetone) the lower aqueous layer and decant the upper organic layer. On the other hand, reducing the number of sample transfer steps will lower the chance of cross-contamination. This is an important reason why protein precipitation is the most common approach of sample preparation.

Whenever possible, suitable parameters (e.g., tip heights) should be evaluated and implemented to prevent any possible splashing in liquid transfer on automated devices. On the other hand, appropriate nitrogen flow (pressure) and temperature should be employed in the step of evaporating sample extracts using 96-well format.

For monitoring cross-contamination during sample preparation and/or carryover during LC-MS injection, use of blanks interspersed amongst the batch can be very useful. In some cases, especially when injector carryover is due to adsorption, additional blank injection after each sample injection (standards, QCs and study samples) may be the only way of minimizing carryover by washing off the adsorbed drug after each injection. With this regard, solvent blanks can be equally useful in carrying out this role. While the above of using alternate blanks/washes (to monitor and/or remove contamination/carryover) may be an expedient approach to resolving an urgent or intractable problem, it should not be seen as a substitute to ensuring that carryover and contamination, if any, are properly evaluated and addressed during the development and validation of the intended method.

21.6.3.5 Training It is important that bioanalytical scientists understand the importance of identifying contamination and carryover and their impact on bioanalytical data. Therefore, a comprehensive training program should be in place to address all key aspects related to contamination and carryover in LC-MS bioanalysis The aspects should cover as a minimum sample handling and preparation, operation of LC-MS/MS systems and an understanding of the related standard operating procedures and regulatory perspectives of bioanalysis as well. Training programs should also cover how to assess the impact of carryover and contamination on study data, how to perform a root cause investigations and how to ensure that those responsible for managing toxicity and clinical studies are made aware of potential problems.

The training program should be underpinned by a set of clear and concise standard operating procedures that outline industry best practices and up to date regulatory requirements for bioanalytical studies

21.7 SUMMARY

In summary, carryover can be defined as (1) true or classical carryover or (2) adsorptive or contamination-caused carryover. In the former case it is largely a function of the engineering design of the injection system. The latter normally requires constant monitoring.

Carryover caused by adsorption to the system can be found anywhere between the injector and the detector. Once carryover manifests itself, the amount eluted is generally constant. In order to resolve this, the adsorbed drug must be washed out of the system. This requires an understanding of where the drug is adsorbed to and what solvent or solvent mix can be used to wash it off the system. Carryover is, in general, predictable and quantifiable.

It is apparent that instrument manufacturers will continue to improve sample injection and robotic sample transfer and preparation systems as part of the product development process. With advancement of material sciences, it is likely that the problem of nonclassical or adsorptive carryover will be meaningfully addressed—if not eliminated. The problem of classical carry over may be further minimized by the implementation of Chip based or modular replaceable systems.

Contamination may be classified into three “types”: (1) contamination occurring during sample handling from sampling though to storage, (2) contamination during the in-laboratory sample preparation, and (3) contamination by nontest substance material that interferes with quantification of the test substance. These forms of contamination tend to appear randomly and can be difficult to troubleshoot. It is important to preemptively ensure that contamination does not take place via implementing proper procedures or employing proper washing solutions.

Cross-contamination and *ex vivo* contamination still remain the remit of the analyst and good laboratory practice and eternal vigilance are essential. Regardless of whatever the reasons may be, the appearance of quantifiable test substance in controlled group animal samples from preclinical studies or pre-dose human samples from single ascending dose (SAD) first-in-human (FIH) or placebo subjects in clinical studies must be investigated and the impact, if any, must be assessed.

ACKNOWLEDGMENTS

The authors acknowledge and thank Patricia Naylor for her secretarial assistance in producing this chapter.

REFERENCES

Agilent Technologies, What is Steam Cleaning in LC/MS 2010—General LC/MS. Available at www.chem.agilent.com/en-US/Technical-Support/Instrument-systems/Mass-Spectrometry/FAQ/Pages/kb002010.aspx. Accessible Apr 11, 2013.

Agilent, How to Regenerate Heavily Fouled Columns. Available at www.chem.agilent.com/cag/cabu/ccleaning.htm. Accessed Mar 23, 2013.

Bergvall S. How to maintain the performance of your MS system. Based on Brochure 715001307 Revision D Controlling Contamination in Ultraperformance LC/MS and HPLC/MS systems. Presentation Nov 9, 2011, Bastad Sweden. Available at www.waters.com/webassets/cms/support/docs/715001307d_cntrl_cntm.pdf. Accessible Apr 11, 2013.

Caldwell E. Technical Note: 07005, A wash Protocol to Determine and Eliminate Liquid Carry Over Using the Thermo Scientific Matrix PlateMate 2 x 2 with Stainless Steel Syringe. Thermo Scientific. 2008.

Chang MS, Kim EJ, El-Shourbagy TA. Evaluation of 384 well formatted sample preparation technologies for regulated bioanalysis. *Rapid Commun Mass Spectrom* 2007;21:64–72.

Clouser-Roche A, Johnson K, Fast D, Tang D. Beyond pass/fail: A procedure for evaluating the effect of carryover in bioanalytical LC/MS/MS methods. *J Pharm Biomed Anal* 2008;47:146–155.

CLSI Laboratory Design; Approved Guideline Second Edition—GP18 A2, Vol. 27, No. 7.

Dolan JW. Autosampler carry over. *LCGC* 2001a;19(2) (1):164–168.

Dolan JW. Attacking carryover problems. *LCGC* 2001b;19(10)(2):1050–1054.

Dolan JW. "Column Triage" LCGC Europe, Oct 1, 2011. Available at http://www.chromatographyonline.com/lcgc/Column%3A+LC+Troubleshooting/Column-Triage/ArticleStandard/Article/detail/747882. Accessed Mar 23, 2013.

Elmashni D. HPLC Carryover-Decreased Sample Carryover Using the Surveyor Autosampler. Application Note No. 330-2007. Available at www.thermo.com/eThermo/CMA/PDFs/Articles/articlesFile_4266pdf. Accessed Mar 23, 2013.

EMA. Note for guidance on toxicokinetics: a guidance for assessing systemic exposure in toxicology studies (CPMP/ICH/384/95) (ICH Topic S3A), June 1995.

EMA Guideline on the Evaluation of Control Samples in Non-Clinical Safety Studies: Checking for Contamination with the Test Substance. CPMP/SWP/1094/04. 2005. Available at www.ema.europa.eu. Accessed Mar 23, 2013.

European Medicines Agency. Guideline on Validation of Bioanalytical Methods EMEA/CHMP/EWP/192217/2009. 2011. Available at www.ema.europa.eu. Accessed Mar 23, 2013.

Fluhler E. In a presentation Dealing with Carryover During Validations and Beyond at the Delaware Valley Drug Metabolism Discussion Group (DVDMDG) Sheraton, Bucks County, Feb 22, 2006.

Hill HM. Contamination of Control Samples in Regulatory Toxicology Studies. PSWC-Poster, May/June 2004, Kyoto, Japan.

Hughes NC, Wong EYK, Fan J, Bajaj N. Determination of carryover and contamination for mass spectrometric-based chromatographic assays. *AAPS J* 2007;9(3): E353–E360. Article 42. Available at http: // www.aapsj.org. Accessed Mar 23, 2013.

Jenkins T. Waters Technology Brief Low Sample Carryover for Sticky Analytes with the ACQUITY UPLC H-Class system. Application Note P/N 720003616en.

Kattas L, Gastrock F, Levin I, Caccatore A. Plastic Additives Chapter 4 in Modern Plastic Handbook, Digital Engineering Library @McGraw-Hill Copyright, 2004. Available at www.digitalengineeringlibrary.com. Accessed Mar 23, 2013.

Lab Equipment World. 2012. Available at www.laboratory equipment world.com/caring-cleaning-lab-glassware.html.

Lab Manager's Independent Guide to Purchasing an Automated Liquid Handler by John Buie, May 5, 2011.

Leap Technologies, A Primer for Reducing Carryover in PAL Autosampler Systems Rev 1.1. 2005. Available at www.leaptec.com. Accessed Mar 23, 2013.

Majors RE. The Cleaning and Regeneration of Reversed Phase HPLC Columns. LC-GC Europe, p 2–6, July 2003.

Morin L-P, Mess J-N, Furtado M, Garofolo F. Reliable procedures to evaluate and repair crosstalk for bioanalytical MS/MS assays. *Bioanalysis* 2011;3(3):275–283

Naegele E, Buckenmaier S, Frank M. 2010. Achieving lowest carry-over with Agilent 1290 Infinity LC and LC/MS systems. Available at www.chem.agilent.com/Library/eseminars/Public/Achieving%20Lowest%20Carryover%20w1290_051810.pdf. Accessible Apr 11, 2013.

Olejniczak K. European guideline for evaluation of control samples from non-clinical safety evaluation studies. *Historical Perspective and Suggestions on Implementation* 2011;204, 200, 215.84/5. Available at www.ema.europa.eu/docs/en_GB/document_library/Presentation/2009/10/WC500004191.pdf. Accessible Apr 11, 2013.

Snyder LR, Dolan JW. *High Performance Gradient Elution: The practical Application of the Linear Solvent Strength Model.* John Wiley & Sons Inc.; 2007. p 203.

Standard Test Methods for Operating Performance of Continuous Electrodeionization Systems on Reverse Osmosis Permeates from 2 to 100S/cm see also for other water related standards. ASTM D6807-02. 2009.

Tse FLS. Analysis of Plasma from Control Animals in Safety Studies, Bioanalytical Considerations, Feb 2, 2006. Presented at The Toxicology Forum, 31st Winter Meeting, Washington, DC.

US FDA, Bioresearch Monitoring Program (BIMO) Compliance Programs, 7348.001 *In Vivo* Bioequivalence, Chapter 48 Bioresearch Monitoring: Human Drugs. 1999.

US FDA. Guidance for Industry: Bioanalytical Methods Validation. US Department of Health and Human Services, FDA, Center for Drug Evaluation and Research, Rockville, MD, May 2001. Available at www.fda.gov/downloads/drugs/guidance complianceregulatoryinformation/guidances/ucm070107.pdf. Accessible Apr 11, 2013.

Viswanathan CT, Bansal S, Booth B, et al. Quantitative bioanalytical methods validation: best practices for chromatographic and ligand binding assays. *AAPS J* 2007;9(1):E30–E42.

Waters. 2011. Controlling contamination in Ultra performance LC/MS and HPLC/MS Systems. 7150001307 Rev. D. Available at http://www.waters.com/webassets/cms/support/docs/715001307d_cntrl_cntm.pdf. Accessed Mar 23, 2013.

Williams JS, Donahue SH, Gao H, Brummel CL. Universal LC-MS method for minimized carryover in a discovery bioanalytical setting. *Bioanalysis* 2012;4(9):1025–1037.

Zeng W, Musson DG, Fisher AL, Quang AQ. A new approach for evaluating carryover and its influence on quantitation in high-performance liquid chromatography and tandem mass spectrometry. *Rapid Commun Mass Spectrom* 2006;20:635–640.

22

AUTOMATION IN LC-MS BIOANALYSIS

Joseph A. Tweed

22.1 INTRODUCTION

At present, one of the most widely used techniques in quantitative bioanalysis in support of pharmacokinetics, toxicokinetics, therapeutic drug monitoring, and biomarker assessment in drug discovery and development is liquid chromatography–tandem mass spectrometry (LC-MS/MS). The LC-MS/MS technique, which will be hereby referred to as LC-MS, has greatly revolutionized the bioanalytical work as evidenced by many factors, including a significantly shortened duration of drug discovery and development (Lee and Kerns, 1999; Ackermann et al., 2002). Due to the nature of complexity of various biological matrices, prior to LC-MS, biological samples need to be properly processed to minimize the impact of interfering biological matrix components that either enhance or suppress ionization of target analytes (drugs, metabolites, biomarkers) during tandem mass spectrometric detection (Kebarle and Tang 1993; Buhrman et al., 1996; Matuszewski et al., 1998). There have been abundance of comprehensive method summaries for biological sample processing used in LC-MS bioanalysis (Jemal and Xia, 2006; Chang et al., 2007a; Xu et al., 2007a). Of note, Chang provides a brief history of events related to sample preparation and detection technologies routinely used in LC-MS bioanalysis (Table 22.1).

These summaries conclude that sample preparation is the most time-consuming and labor-intensive step in LC-MS bioanalysis. With this regard, various automation systems have been developed and/or improved in the past decades to assist bioanalytical scientists in the delivery of timely quantitation results (Wells, 2003; Laycock et al., 2005; Vogeser et al., 2011).

In the practice of automation-assisted LC-MS bioanalysis, the automation systems, that is, automatic liquid handlers, can be used for many sample processing steps, including the preparation of calibration standards and quality control (QC) samples, sample aliquoting, sample transfer, solvent delivery, and other tasks as well. Nowadays, these workstations have largely replaced manual liquid transfers for parallel sample preparation (e.g., 96-well plate). This has resulted in higher sample processing throughput and has significantly reduced the monotony of repetitive tasks experienced by the bioanalytical scientist. This chapter aims to provide the reader with a comprehensive perspective regarding the use of various robotic liquid handling platforms in automation-assisted LC-MS bioanalysis. Furthermore, approaches to creating reliable and robust robotic liquid handling methods for LC-MS bioanalysis and the emerging trends and technologies in automation within LC-MS bioanalysis discipline are to be discussed.

22.2 AN OVERVIEW OF AUTOMATED SAMPLE PREPARATION IN LC-MS BIOANALYSIS

Sample preparation techniques have evolved from low-throughput methods tied to large sample volumes processed in test tubes to high-throughput approaches using smaller sample volumes in plate-based microplate formats of 24, 48, and 96 wells. The use of the 96-well plate footprint has dramatically increased laboratory throughput and allowed the simultaneous sample preparation and extraction of 96 samples in a single batch, also known as parallel processing. The new geometrical footprint became an industry standard amenable to automation through appropriately designed instrumentation, multichannel pipettes and many varieties of 96-well plate-based laboratory consumables (Majors, 2004). Since the 1980s, many bioanalytical laboratories have

Handbook of LC-MS Bioanalysis: Best Practices, Experimental Protocols, and Regulations, First Edition. Edited by Wenkui Li, Jie Zhang, and Francis L.S. Tse.

TABLE 22.1 A Brief History of Sample Preparation and Detection Technologies Used in LC-MS Bioanalysis

Era	PK requirement	Detection technology	New goals of sample preparation	Major sample preparation technology
1950–1975	Detection of metabolites; estimate exposure	Colorimetry; radioimmunoassay (RIA); gas chromatography (GC)	Bring the analyte concentration to assay range; remove interference; make analyte volatile	Dilution; LLE; PPE; TLC; column chromatography (normal phase and ion exchange); derivatization
1975–1985	Determination of exposure	RIA; enzyme-linked immunoassay (ELISA);high-pressure liquid chromatography (HPLC) with visible ultraviolet detection	Bring the analyte concentration to assay range;protein removal; remove interferences	Dilution; use of internal standard; LLE with back extraction; silica-based reverse chromatography with intention for fractionation; commercial SPE cartridge
1985–1995	GLP bioanalytical	RIA; ELISA; HPLC; GC; GC-MS; capillary zone electrophoresis	Reliability of quantitative data; validated assay with proven sample history and stability	Automation; online elution of SPE; online SPE; use of analog internal standard
1995–2000	Guidance for the industry	ELISA; HPLC; GC; GC-MS; HPLC-MS	Validated assay with proven specificity; cost reduction to compete with contract research organization	Commercial automation, high-throughput (high-density) assay based on 96-well SBS format; pre- and postextraction techniques in the SBS format
2000 to current	Biomarker and large molecule determination	HPLC-MS/MS; (HPLC) n-MS; Biacore; Mesoscare	Reduce matrix effect; improve incurred sample repeatability; reduce manual labor to compete with off-shoring	Integrated process in the SBS forma; time-sharing of MS by multiplex of HPLCs using multiple sprayers or stream selection valves online SPE

Reprinted from Chang et al. (2007a). Copyright © 2007, with permission from John Wiley & Sons.

adopted robotic automation technology in the development of high-throughput assays supporting drug discovery and development. Robotic liquid handling workstation manufacturers, such as Packard, Biomek, Tecan, Zymark, Hamilton, and a minority of others, provide channel- and multichannel-based liquid handling workstations available for routine bioanalytical assay work. A current snapshot of some commonly used robotic liquid handling workstations and their features employed for routine LC-MS bioanalysis is provided in Table 22.2.

LC-MS bioanalysis can be significantly assisted by the implementation of automation, especially in the three major areas of sample preparation and extraction: protein precipitation (PPT), solid phase extraction (SPE), and liquid–liquid extraction (LLE). The automation-assisted LC-MS bioanalysis has been widely implemented in the bioanalytical community for various purposes ranging from drug discovery to postapproval therapeutic drug monitoring. Comprehensive work focused on various automated sample preparation approaches using robotic liquid handling and online hardware or consumable technology are detailed by Wells (Wells, 2003) and more recently reviewed by Vogeser (Vogeser et al., 2011). These automated sample preparation and extraction techniques and their quality considerations when implemented within the LC-MS bioanalysis discipline are discussed in detail later within the chapter.

22.3 ROBOTIC LIQUID HANDLING PIPETTING MODES AND ASSOCIATED TECHNOLOGY

A wide variety of robotic liquid handling platforms are commercially available today to address many, if not most, of the sample preparation needs in LC-MS bioanalysis. In order to properly explain how routine manipulations of solutions, solvents, and liquids are performed through robotic liquid handling, a few key robotic liquid handling workstation platforms are highlighted.

Three modes of displacement are commonly used for liquid handling manipulations via a robotic liquid handling workstation: (1) liquid–air displacement, (2) air displacement, and (3) positive displacement. Liquid–air displacement uses the movement of system liquid (most commonly degassed water) within dedicated channel pipette tubing to aspirate and dispense desired liquids that are metered through discrete syringes for each pipetting channel. Some robotic platforms using liquid–air displacement for routine liquid manipulations are Freedom EVO (Tecan), Biomek FXp (BeckmanCoulter), and Janus (PerkinElmer). Air displacement pipetting involves the manipulation of a desired volume of liquid into and out of a disposable tip through the movement of a plunger, a process akin to a standard laboratory air displacement hand pipette. The MicroLab STAR (Hamilton) is an example of a robotic liquid handling

TABLE 22.2 Robotic Liquid Handling Workstations Commonly Employed for Routine Automated Sample Preparation in LC-MS Bioanalysis

Features		Manufacturer						
		Tecan	Perkin Elmer	Hamilton	Beckman Coulter	Tomtec	Gilson	Zinsser
General	Latest model(s)	Freedom EVO	Janus	STAR AT Plus2	Biomek FX^{P}	Quadra 4 Quadra-PLUS	Quad-Z 215	SPEEDY LISSY
	Customizable	√	√	√	√			√
	Scalable	√	√	√	√			√
	Tip configuration	1, 2, 4, 8, 96, 384	4, 8, 96, 384	8, 12, 16, 96, 384	1, 8, 96, 384	96	4	4, 8
	Variable probe spacing	√	√	√	√		√	√
Pipetting	Displacement mode	Air and liquid–air	Liquid–air	Air	Liquid–air	Air	Liquid–air	Liquid–air
	Channel pipetting	√	√	√	√		√	√
	96-well pipetting	√	√	√	√	√		
	384-well pipetting	√	√	√	√			
	Nanoliter (nl) pipetting	√	√	√	√			
	Fixed tips	√	√		√		√	√
	Conductive disposable tips	√	√	√	√			
	Nonconductive disposable tips	√	√	√	√	√	√	√
	Positive displacement; disposable tips			√				√
Technology	Capacitance	√	√	√	√			√
	Pressure monitoring	√	√	√				
	1D barcode scanning	√	√	√	√			√
	2D barcode scanning	√	√	√				
	Temperature control	√	√	√	√			√
	Vortex/agitation	√	√	√	√			√
	Gripper/arm manipulation	√	√	√	√			√
	SPE vacuum manifold	√	√	√	√	√		√
	SPE negative pressure elution	√	√	√	√	√		√
	SPE positive pressure elution					√		√

workstation using air displacement pipetting. The fundamental principles in manipulating liquids via liquid–air displacement and air displacement are essentially the same. Regardless of the mode of air displacement used, both create a change in air pressure within a channel to move a desired volume of liquid in a fixed or disposable tip. A negative pressure differential (partial vacuum) results in an aspiration sequence, and a release of the negative pressure (or applied positive pressure) results in a dispense sequence. However, a liquid–air displacement system uses a syringe–valve setup to expel a system liquid volume equal to the volume of air displaced, whereas the air displacement setup uses a piston to displace a column of air equal to the desired volume to be pipette. Because some degree of static pressure acts upon the liquid being moved in a fixed or disposable tip by a displaced column of air (gap volume), cavitations in the liquid sample can occur during routine pipetting sequences. Preventing these types of occurrences, and associated liquid dispensing challenges using air displacement pipetting, is covered in more detail later in this chapter.

Manufacturers of a respective robotic liquid handling platform can more clearly articulate the perceived strengths and limitations of the system liquid–air displacement versus air displacement technology and thus will not be covered in this review. Regardless of the mode of air displacement pipetting used, the current technology available on robotic liquid handling platforms results in liquid delivery conforming to a high degree of precision and accuracy. For example, using the Tecan Freedom Evo® platform, a typical 10 μl volume of water pipetted with disposable tips is guaranteed by the manufacturer to have a precision (%CV) value of <3.5% (Tecan, 1999). Similarly, a volume of 10 μl of Hamilton's volume verification solution (borate buffer) pipetted with 10 μl disposable tip on the Microlab STAR platform results in precision (%CV) and accuracy (%RE (relative error), trueness) values of 0.5% and 1.5%, respectively (Hamilton, 2011).

Positive displacement pipetting uses a special type of disposable tip that contains a pluger-in-tip or capillary-piston design. An example of a robotic handling workstation employing the use of positive displacement pipetting is the Hamilton ATplus2. In a similar fashion to manual positive displacement pipetting, there is no air interference in the pipetting of a desired liquid. Therefore, the piston is always in direct contact with the liquid during a pipetting sequence, which significantly limits cavitations occurring during routine transfers. The uniform pressure applied within the capillary-piston prevents aerosols, or liquid dripping from the tip, regardless of the physical properties of the sample liquid (viscosity, volatility, density).

Modern robotic liquid handling platforms offer a variety of technology and features aimed to enhance productivity and enhance operational performance. To illustrate this point, DiLorenzo provided a summary of the technology and features available in the Hamilton STAR such as compression-induced O-ring expansion (CO-RE), capacitance-/pressure-based liquid detection, and monitored air displacement (MAD) aimed to enhance automated robotic liquid handling performance (Dilorenzo et al., 2001). Nearly all manufacturers offer features such as liquid-level sensing, disposable tips, and designs incorporating both channel pipettors (available in various numbers from 1, 4, 8, 12, or 16 channels) and 96-well pipetting within one liquid handling platform. The development and introduction of the Hamilton Microlab STAR and technology more recently introduced allows for liquid manipulations in the nanoliter range as low as 50 nl with adequate precision (%CV) to $\leq$13% (PerkinElmer, 2004). Additional technology offers pressure monitoring during the liquid transfer process and can be of particular use with liquids prone to clotting such as matrix sample aliquots (i.e., plasma). For example, the Tecan Group provides the pressure monitoring pipetting (PMP) tool for the Tecan Freedom Evo® platform, which monitors and records pressure changes during liquid aspiration and dispense. Similarly, Hamilton Robotics offers total aspirate and dispense monitoring (TADM) for the Hamilton STAR liquid handling platforms (Figure 22.1). In broad terms, PMP and TADM work by comparing real-time aspiration and dispense pipetting profiles to simulated or model profiles. If an aspiration or dispense step falls outside of an acceptable range, some type of error handling is initiated by the software for remediation. Tools such as MAD, PMP, and TADM provide increased assurance that the desired volumes of liquid have been appropriately delivered during routine aspiration and dispense steps, particularly during matrix deliveries. Of considerable importance is the impact of sample clots (e.g., thrombin, fibrinogen, and viscous lipids) in plasma samples during routine aspirate and dispense cycles. The selection of anticoagulant may significantly impact the throughput, quality, and overall feasibility of assays intended for automated sample preparation (Sadagopan et al., 2003). This is mostly

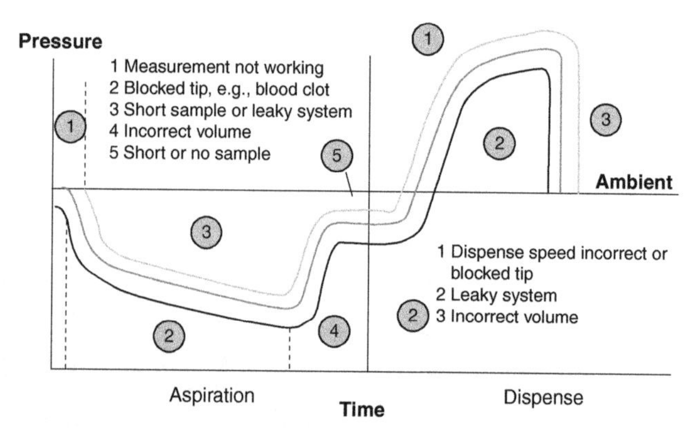

FIGURE 22.1 Total aspirate and dispense monitoring (TADM) technology used by the Hamilton Microlab STAR robotic liquid handling platform for the pressure monitoring of individual channel pipettes during the aspirate and dispense steps of a pipetting sequence (images of TADM, TADM copyright and TADM IP are the property of Hamilton Bonaduz AG, used with permission).

due to the level of clotting and turbidity that occurs over a typical freeze and thaw cycle(s) experienced by a plasma sample during routine sample processing (Zhang et al., 2000b). If ethylenediaminetetraacetic acid (EDTA) containing plasma cannot be used during automated plasma sample preparation, careful attention must be paid to mitigate the impact of "clots" during the automated aspirate and dispense steps. In addition to the more sophisticated tools detailed earlier (TADM, MAD, PMP), centrifugation, the choice of labware (cluster tube, 96-well plate), and labware dimensions (e.g., tapered, round, and V-bottom) all can play an important role in the overall success of an automated assay. More information on this topic is provided later in this chapter in the discussion regarding PPT plates.

Nearly all robotic liquid handling manufacturers offer modular platform designs or original equipment manufacturer (OEM) automation solutions amenable to a wide variety of laboratory applications. This flexibility enhances the utility of the robotic liquid handling platform for routine LC-MS bioanalytical portfolio support. Modular design provides the laboratory the capability of performing many different types of sample preparation protocols, or routine liquid manipulations, with a single-liquid handling platform. This concept was first demonstrated by Gu in the reporting of a customized software interface used to prepare bioanalytical samples via three different sample preparation techniques (PPT, SPE, or LLE) on a Tecan Genesis RSP® (Gu et al., 2006). More recent work describes the development of a multiple assay sample preparation and extraction platform using the Hamilton Microlab STAR for the same three extraction techniques (Tweed et al., 2010). In the later reference, for example, use of OEM equipment and the modular platform design allows for automated sample dilutions via the integrated plate shakers and provides the ability to perform SPE with the integrated SPE manifold. Robotic liquid handling hardware and software have also been demonstrated to improve the overall workflow in routine bioanalytical portfolio support. Many manufacturers provide the ability to read sample barcodes directly from plates or sample tubes via integrated barcode scanners. In LC-MS bioanalysis, one-dimensional (1D) or two-dimensional (2D) barcodes can be used for routine sample management logistics and downstream sample processing. Recently published work described the use of 2D barcode technology (shown in Figure 22.2) for routine bioanalytical portfolio support in a regulated setting (Zhang et al., 2010; Tweed et al., 2012). Some advantages of 2D barcode processing include improved sample chain-of-custody, limited manual manipulations for sample organization and categorization, and a reduction in the setup time required for routine batch and reassay batch sample analysis. Use of 2D barcode processing has also been described for sample storage and collection in a drug discovery setting (Laycock et al., 2005). This work also proposed a potential sample management logistics solution via radiofrequency identifier (RFID) technology. Comparatively, it would seem feasible

FIGURE 22.2 An image of a polypropylene 96-well two-dimensional (2-D) barcoded sample tracker storage tube (image adapted from Nova Biostorage+, Inc., McMurray, PA (formerly Micronic USA), used with permission).

that RFID technology could be used for bioanalytical sample logistics, as RFID sample container tubes were patented for intended use in pharmaceutical clinical trials (Veitch and Biddlecombe, 2002). However, no reports could be found within the LC-MS bioanalysis discipline describing any such applications RFID technology for sample management logistics or sample preparation.

22.4 OPTIMIZING ROBOTIC LIQUID HANDLING PERFORMANCE

Robotic liquid handling platforms often require some degree of optimization prior to routine use in a LC-MS bioanalysis laboratory. All vendors offer some type of vendor volume verification or volume QC kit to ensure the physical liquid manipulations meet vendors established guidelines and specifications. The volume verification process is typically performed following platform installation, instrument movement, scheduled maintenance calls, or on an as-needed basis depending upon the users required level of compliance and routine instrument use. For channel-based robotic liquid handling platforms, volume verification usually entails a gravimetric evaluation of specified volumes of liquid (typically water or buffer) with an established protocol to establish communication with a balance and the operating robotic liquid handler (Xie et al., 2004). A gravimetric volume verification is dependent not only upon the performance of the liquid handler but also upon the quality of the balance used and environmental factors such as temperature, humidity, residual vibrations, and pressure. In an iterative fashion, internal adjustments are made within the liquid handling software that adjusts the calibration parameters of the associated liquid being pipetted. The approach to volume verification and the precise means by which liquid aliquot volumes are optimized will differ, depending upon the internal software design that controls the physical attenuation of the displacement mode used for liquid manipulation (calibration curve, volume range adjustment, liquid class, etc). It should be noted that when routinely performed by the vendor or a trained scientist, data that meets or exceeds vendor specifications can be achieved during typical volume verification.

In a similar fashion, Artel Inc. (Estbrook, ME) has demonstrated success of a multichannel verification system (MVS) used specifically to verify the liquid transferring performance of multichannel robotic liquid handling workstations (Bradshaw et al., 2005; Artel, 2011). Broadly speaking, the MVS volume verification approach uses a plate reader to measure the absorbance values of proprietary solutions (used for specific volume ranges) under controlled protocols performed in 96-well and 384-well footprints (Knaide et al., 2006). In minutes, a scientist can determine the quality of the respective robotic liquid handling protocol and iteratively make any adjustments to the calibration parameters of liquid handler. Work at Artel demonstrated some advantages of the dual wavelength absorbance protocol when assessing the precision and accuracy of dilutions performed using robotic liquid handling platforms (Bradshaw et al., 2007). This is of particular relevance when a laboratory identifies dilution problems within their bioanalytical assay directly associated with robotic liquid handling performance (Kim et al., 2007). Careful attention must be given to the routine volume verification techniques used to ensure optimal liquid handling performance and avoid costly disruptions in workflow (Albert and Bradshaw, 2007). A generic, low-cost, and simple approach detailed by Stangegaard uses different volumes of an OrangeG stock solution measured by absorbance to assess the accuracy and precision of manual pipettes or channel-based robotic liquid handlers (Stangegaard et al., 2011). Although the technology, protocols, and volume verification tools provide a means by which the scientist can demonstrate that the robotic liquid handler physically performs with a high degree of precision and accuracy, in practical terms, additional work is required to ensure that a robotic liquid handling platform is optimized to perform routine sample preparation and extraction assays in LC-MS bioanalysis.

Gravimetric or Artel's MVS volume verification protocols use aqueous-based solutions and typically result in high quality and reproducible data sets collected during a routine preventive maintenance visit. It is quite possible that the same performance data demonstrated during the volume verification process may not translate accordingly to the robotic liquid handling protocol used for LC-MS bioanalytical sample preparation. Artel's MVS system was used to demonstrate that nonaqueous solutions such as dimenthyl sulfoxide (DMSO) (high-throughput screening), serum (bioanalysis) and nonaqueous-based reagents with detergents (molecular biology) can be measured with the MVS system directly to assess liquid handling performance (Albert et al., 2006). Protocols entailed mixing Artel's proprietary reagents with the nonaqueous-based reagent, creating customized calibration

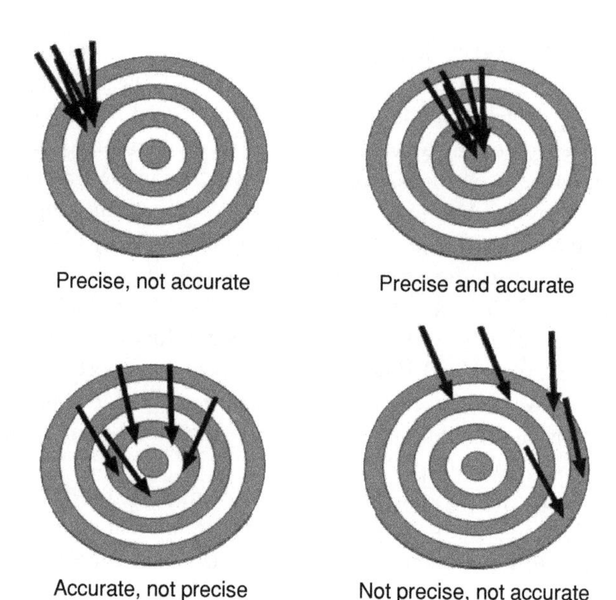

FIGURE 22.3 Schematic representation visualizing precision and accuracy when delivered by a manual pipette or a robotic liquid handler (Bradshaw and Albert, 2010) (copyright © 2010, with permission from John Wiley & Sons).

solutions that provide a more accurate representation of liquid handling performance when delivered by a robotic liquid handler. Although a number of factors have an impact on the overall quality of an automated liquid handling routine, the accuracy and precision (schematically represented in Figure 22.3) of the delivery of solvents, matrix aliquots, and solutions are of critical importance (Bradshaw and Albert, 2010). In the work performed by Xie et al. (2004), a disparity in accuracy was demonstrated during the delivery of dog plasma using nonoptimized calibration curves for the Tecan Genesis robotic liquid handler. Excellent precision is obtained in the matrix delivery (<3% CV); however, due to the different density of dog plasma as compared with water for example, the inaccuracy of the matrix delivery was found to be high (15% RE). After measuring the density with a portable density meter and correcting for the density change in the internal Tecan Genesis calibration file, the inaccuracy of the matrix delivery was found to be approximately 0.1% RE.

Some of the most common sample preparation and extraction techniques automated for use in LC-MS bioanalysis are PPT, SPE, LLE, and supported liquid extraction (SLE). A variety of matrices, solvents, and solutions are used in these techniques and, as a result, may require some degree of optimization on a liquid handling platform to ensure optimal assay performance. To illustrate this point, if the same protocol was used for volume verification process mentioned earlier for a liquid–air or air displacement pipetting and the test solution used was replaced with plasma; the likelihood of meeting vendor specifications becomes harder to achieve. These pipetting modes on the robotic liquid handlers are impacted by the viscosity and density of liquid and, as a result, impact the overall precision and accuracy of each robotic liquid handling pipetting sequence. Furthermore, for liquid–air displacement robotic liquid handlers, if adjustments are not performed to internal pipetting parameters, the overall quality of the resulting assay will be compromised producing misleading results (Dong et al., 2006). A thin layer of system liquid in the liquid path has been implicated to dilute out serially prepared calibration and QC samples during bioanalytical sample preparation (Gu and Deng, 2007). Different remedial strategies were proposed in the previous reference and by Ouyang for liquid–air displacement robotic liquid handlers using either fixed or disposable tips (Ouyang et al., 2008).

In addition to the dilution effect mentioned earlier, it is also prudent to optimize the matrices, solvents, and solutions used in any LC-MS–based bioanalytical sample preparation, especially for quantitative deliveries (calibration standard and QC samples, routine sample aliquoting, or a working internal standard solution delivery). Xie was the first to demonstrate this optimization approach in significant detail on a Tecan Genesis (liquid–air displacement mode) robotic liquid handler (Xie et al., 2004). Another particularly good example describing this process details the liquid class optimization of plasma, organic stock solutions, and tissue homogenate samples encompassing a pipetting range of 5–300 μl for a typical automated PPT method (Palandra et al., 2007). Matrices like homogenized tissues (heart, muscle, lung, liver, etc), blood products (blood, serum, plasma) and body fluids (urine, bile, saliva, etc.) along with organic solvents and other solutions (PBS, detergents) should be verified; however, all matrices may not require optimization due to similar viscosities, densities, volatility, and so on. Other groups have also taken this approach with air displacement robotic liquid handlers (Hamilton STAR) to ensure optimal assay sample preparation performance (Tweed et al., 2010; Zhang et al., 2011).

22.5 SOLID PHASE EXTRACTION

Tied with the explosion of LC-MS–based analysis, increased throughput was achieved for typical LC-MS methods using robotic liquid handling platforms most notably impacting SPE assays and their clinical trial application (Allanson et al., 1996; Kaye et al., 1996). For a brief history of SPE and its application in routine LC-MS bioanalysis, readers are encouraged to refer to the review from Venn (Venn et al., 2005). In addition to the comprehensive work for high-throughput LC-MS sample preparation cited earlier, reviews by Rossi and Kataoka focus specifically on online and offline SPE (Rossi and Zhang, 2000; Kataoka, 2003). Work by Simpson reported a fully automated SPE platform integrated with an automated vacuum system and liquid level sensing pipetting channels using a Packard Multiprobe II (Simpson et al., 1998). The expanding scope of robotic applications in LC-MS bioanalysis developed into increased SPE

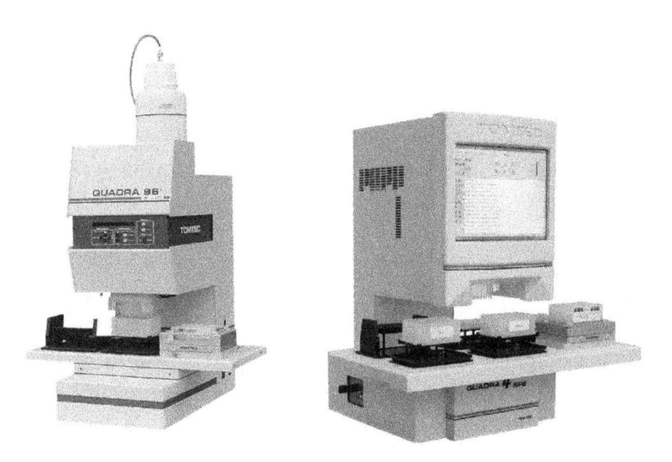

FIGURE 22.4 The Tomtec Quadra96 robotic liquid handler (left panel) and the more recent version of this robotic liquid handling platform the Tomtec Quadra4 (images adapted from Tomtec, Inc., Hamden, CT, with permission).

functionality in both hardware and software control demonstrated by the Zymark 96-well SPE robotic system (Callejas et al., 1998; Joyce et al., 1998). This robotic platform was one of the first to incorporate several disparate SPE hardware components into an integrated SPE method such as a storage carousel, robotic liquid handler, and a SPE manifold all manipulated with use of a rotating robotic arm.

An example of an early robotic liquid handling platform with 96-well parallel processing capabilities that impacted SPE LC-MS bioanalysis and is still in use today in many bioanalytical laboratories is the Tomtec Quadra96™ shown in Figure 22.4. The Tomtec Quadra96™ incorporates a six position automatic shuttle deck and a fixed position 96-well pipettor equipped with disposable polypropylene tips. This platform significantly increased throughput for SPE in a 96-well geometry (Janiszewski et al., 1997; Zhang and Henion, 1999; McMahon et al., 2000). Programming routines on the Tomtec Quadra96™ were accomplished via front panel controls or paired with proprietary computer software. The Tomtec Quadra96™ demonstrated an excellent example of how design, function, and accessibility (ease of use in routine operation) made the Tomtec Quadra96™ a laboratory workhorse within the LC-MS bioanalysis discipline.

As the technology and capabilities of the liquid handling platform evolved, so did the desire to add additional automated capabilities to routine SPE assays. Multiple robotic liquid handling workstations were serially used or paired for the various stages of routine bioanalytical SPE protocols (Shou et al., 2001; Shou et al., 2002; Song et al., 2005). In these applications, channel-based robotic liquid handling platforms were used to prepare or fortify samples for extraction by aliquoting samples from discrete sample tubes into 96-well plates followed by the addition of internal standard. The fortified matrix samples were then moved to another 96-well robotic liquid handler (typically a Tomtec Quadra 96) where the SPE protocol was performed. Another approach detailed by Deng demonstrated remarkable SPE throughput (Deng et al., 2002). In this setup, a Zymark track robot was fully integrated with a Tecan Genesis robotic liquid handler for full automation of an SPE protocol that incorporated automated pipetting, centrifugation, and refrigerated carousel operation paired with multiplexed LC-MS detection.

The ability to expand capacity was also explored in the parallel SPE processing of 384-well samples at a time. Having the same dimensional footprint as a 96-well plate but with increased well density, the logical appeal of the 384-well SPE plate would allow increased throughput with minor modifications to standard SPE protocols. Early work in this area proved 384-well SPE utility in automated LC-MS bioanalysis (Biddlecombe et al., 2001; Rule et al., 2001). A more specific application using an antibody-based immunosorbent SPE in a 384-well nonautomated format was also described (Nevanen et al., 2005) although not automated. SPE in 384-well format, however, has never attained a use level similar to that of typical 96-well plate based SPE assays used in LC-MS bioanalysis. Chang revisited the 384-well design for bioanalytical LC-MS bioanalysis in the development of a manual SPE protocol and an automated LLE protocol (Chang et al., 2007b). Both extraction techniques demonstrated high-quality results; however, the 384-well routine must be carefully optimized to address overall method quality considering the increased possibility of cross-well contamination during sample preparation and general LC-MS method ruggedness during long batch runs.

Typical SPE protocols paired with reversed phase liquid chromatography usually require evaporation (dry-down) of the eluted analyte(s). This is typically performed under a stream of heated nitrogen and final reconstitution (recon) of the dried eluent with a solvent of weaker elution strength than the mobile phase needed to maintain chromatographic resolution. These steps add significant time to an SPE protocol and drastically reduce throughput in an automated SPE protocol. Some groups devised customized SPE elution techniques and novel chromatographic approaches to avoid these tedious and time-consuming steps in order to attain higher throughput or circumvent molecule stability or nonspecific adsorption problems observed during evaporation and reconstitution steps of the SPE protocol (Naidong et al., 2002; Yang et al., 2003). Implementing these approaches can reap dividends for routine portfolio support if sensitivity can be sacrificed using a customized SPE elution that ultimately impacts recovery due to higher aqueous content in elution solvent. Furthermore, direct injection of a SPE sample eluent with a typically high organic composition requires the analyte of interest to be sufficiently polar to undergo normal phase or pseudonormal phase (silica, HILIC) chromatography.

Another automated SPE protocol implemented to avoid solvent evaporation and sample reconstitution was first demonstrated by Mallet with the introduction of the

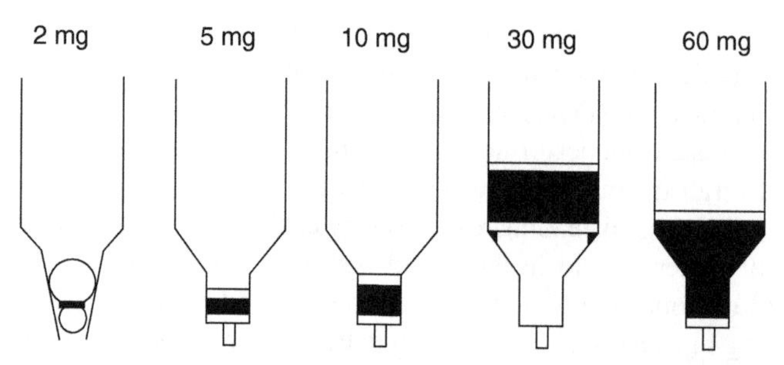

FIGURE 22.5 The novel tip design of the Waters μ-Elution plate (2 mg) using spherical frits and a cone-shaped tip design as compared with other standard cylindrical tip formats and packing volumes (reprinted from Mallet et al. (2003); copyright © 2003, with permission from John Wiley & Sons).

μElution™ plate (Mallet et al., 2003). The μElution™ plate allows up to a 15-fold increase in sample concentrating ability by allowing large matrix aliquots (375 μl) to be loaded on the μElution™ plate and be eluted with a volume as small as 25 μl. This advancement was made possible by a novel plate design incorporating spherical frits and a conical tip design as opposed to standard circular disk-shaped frits and cylindrical tip design (Figure 22.5). Applications by Yang have demonstrated its practical use in the simultaneous determination of simvastatin and simvastatin carboxylic acid in human plasma (Yang et al., 2005), and also for the determination of cortisol in human plasma (Yang et al., 2006), which utilized low-elution volumes, thereby avoiding the SPE dry-down and reconstitution steps. All of these approaches increased throughput by the avoidance of the dry-down and reconstitution steps and demonstrated their practical application in the delivery of reliable and robust quantitative LC-MS bioanalytical data.

More recently, tip-based SPE protocols have been used for selective high-throughput sample preparation in LC-MS bioanalysis. Work by Shen demonstrated the application of tip-based SPE using a Tomtec Quadra 96 robotic liquid handling platform (Shen et al., 2006, 2007). Compared to conventional SPE plate-based approaches, tip-based SPE increases throughput as it does not require the use of a vacuum manifold and has excellent flow characteristics attributed to the monolithic technology. Readers are encouraged to refer to the comprehensive work by Saunders reviewing the use of monolithic phases for both online and off-line SPE applied to LC-MS bioanalysis (Saunders et al., 2009). The SPE-tips, commercially known as OMIX™, are available in a number of different sorbent chemistries that are packed in a Tomtec 450 μl polypropylene tip and have been recently reported for use in a Hamilton Microlab STAR nonconductive polypropylene tip (Luckwell and Beal, 2011). Based upon these published reports, tip-based SPE has been shown to be a reliable and robust alternative to conventional plate-based SPE with demonstrated recoveries and selectivity comparable to plate-based methods. Furthermore, similar to the μElution™ plate mentioned earlier, the tip-based SPE allows small volume elution that may facilitate the avoidance of the dry-down and reconstitution steps providing additional gains in achieved throughput.

22.6 PROTEIN PRECIPITATION

An early report on a novel robotic platform demonstrated a high degree of customization and operational control in PPT sample preparation using HPLC-UV detection (Fouda and Schneider, 1987). The incorporation of a 96-well plate footprint resulted in significant gains in throughput for PPT sample preparation and resulted in the emergence of the 96-well filter plate (Biddlecombe and Pleasance, 1999). The filter plate PPT protocol is a popular alternative to manual PPT because it eliminates the need for sample centrifugation (Rouan et al., 2001). When compared to similar manual protocols, the filter plate PPT increases sample preparation efficiency and enhances data quality by reducing manual sample manipulations (Walter et al., 2001). Additional throughput advancements were achieved with the filter plate PPT paired with automated preparation of matrix sample dilutions (Watt et al., 2000; O'Connor et al., 2002).

PPT filter plates were also demonstrated in an application outside of its intended function to "crash" matrix sample proteins. In order to reduce the impact of matrix sample clots (e.g., thrombin and fibrinogen) during the routine pipetting of matrix samples using robotic liquid handling platforms, Berna proposed using PPT filter plates during sample collection and storage to minimize clot formation (Berna et al., 2002). This work (and others) demonstrated potential solutions to address the negative impact of matrix sample clot formation on the routine pipetting performance of robotic liquid handling platforms (Sadagopan et al., 2003; Villa et al., 2004; Peng et al., 2005). Although the proposed sample collection approaches were different, these researches also investigated potential issues associated with nonspecific binding of target analytes to material used in sample collection and sample preparation (i.e., PPT filter plates). Furthermore, internal automation programming files controlling the aspirate and dispense parameters of the syringes or pistons in a liquid handler such as a liquid-transfer performance file (Multiprobe II), liquid classes (Tecan, Hamilton) were noted to impact pipetting performance (precision and accuracy) and ultimately the quality of an automated sample preparation technique (Xie et al., 2004; Pucci et al., 2005; Palandra et al., 2007). Issues such as these draw attention to general liquid handling performance considerations and quality concerns when robotic liquid handling platforms are used for bioanalytical sample preparation. This is an area of critical importance to the LC-MS bioanalysis discipline and is discussed in detail in Section 22.4.

In a similar fashion to automated SPE assays, the serial use of channel and multichannel robotic liquid handlers were utilized for matrix sample aliquoting and 96-well parallel sample processing (Kitchen et al., 2003). Also, multichannel robotic liquid handling platforms (typically a Tomtec Quadra 96) were used in automated PPT protocols both with (Pereira and Chang, 2004) and without (Yang et al., 2004; Xu et al., 2005) the use of PPT filter plates. Throughput gains using the Tecan Genesis platform were demonstrated in the development of a single-pot PPT technique where the supernatant was directly injected from a well that contained the precipitated matrix sample proteins (Xue et al., 2006). Controlled by fine adjustments to the autosampler offset setting, only the supernatant of the crashed sample was injected leaving the pellet undisturbed.

Automated PPT was demonstrated for tissue homogenates as well. Work by Xu reported the serial use of an automated and programmable tissue homogenizing system called the Tomtec Autogizer paired with the Packard Multiprobe II robotic liquid handler (Xu et al., 2007b). This automated process started from preweighed rat brain tissues in test tubes, which were homogenized in an automated fashion using the Tomtec Autogizer. Homogenized samples were then transferred to a 96-well plate for PPT sample preparation. On the Multiprobe II, wide-bore tips were chosen to prevent tip clogging during homogenate sample transfer. This was required because of the unique physical properties of the homogenized brain tissue (increased viscosity, surface tension, etc.). Although no changes to liquid performance files were mentioned, the automated method demonstrated comparable assay precision and accuracy to the same homogenate pipetting procedure conducted manually.

Increased functionality of liquid handling workstations for automated PPT was demonstrated by advanced robotic platform integration and customized software applications. A fully automated PPT methodology with customized software tools was developed for drug discovery applications using PPT filter plates (Palandra et al., 2007). This work and the work by Gu demonstrated the use of customized software integrated with laboratory information management systems (Gu et al., 2006). In the graphical user interface approach offered by Palandra and shown in Figure 22.6, the scientist designed a PPT methodology that started from the preparation of calibration standards and QC samples and ended with filtered precipitated matrix samples or "extracts" ready for LC-MS injection. Similar to Deng's high-throughput SPE methodology described earlier, Ma integrated a variety of components (centrifuge, plate sealer, plate piercer, plate shakers, robotic arms) with a Tecan FreedomEvo200 liquid handler for fully automated and unattended PPT processing of matrix samples for LC-MS analysis shown in Figure 22.7 (Ma et al., 2008).

A hybrid technique incorporating the simplicity of PPT and SPE selectivity has been developed for routine

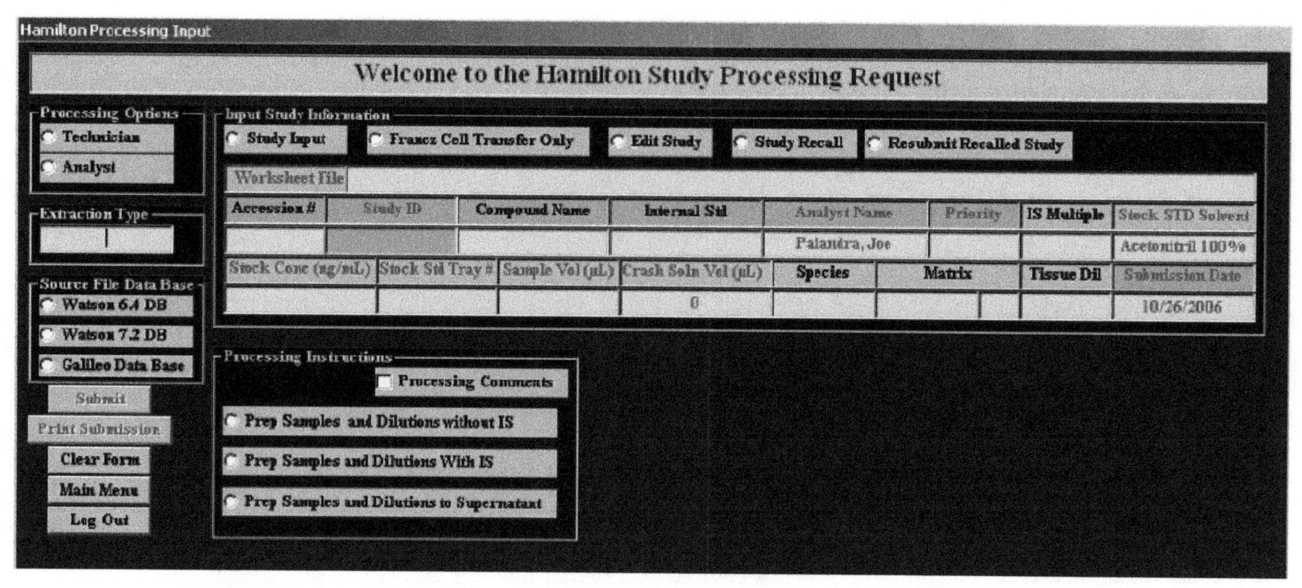

FIGURE 22.6 A customized graphical user interface (GUI) used to input project/study specific information and protein precipitation sample preparation parameters into a Hamilton STAR robotic liquid handling workstation (reprinted with permission from Palandra et al. (2007); copyright (2007) American Chemical Society).

LC-MS sample preparation, which generically depletes phospholipids on a PPT filter plate. It is widely known that phospholipids are implicated to be a major constituent in prepared or extracted samples analyzed via tandem mass spectrometric detection. This has been thoroughly reviewed in the literature (Van Eeckhaut et al., 2009; Jemal et al., 2010). Efficiency of electrospray ionization is either increased or reduced depending upon the amount of residual matrix components, salts, and phospholipids remaining in the prepared sample extract (Côté et al., 2009). This

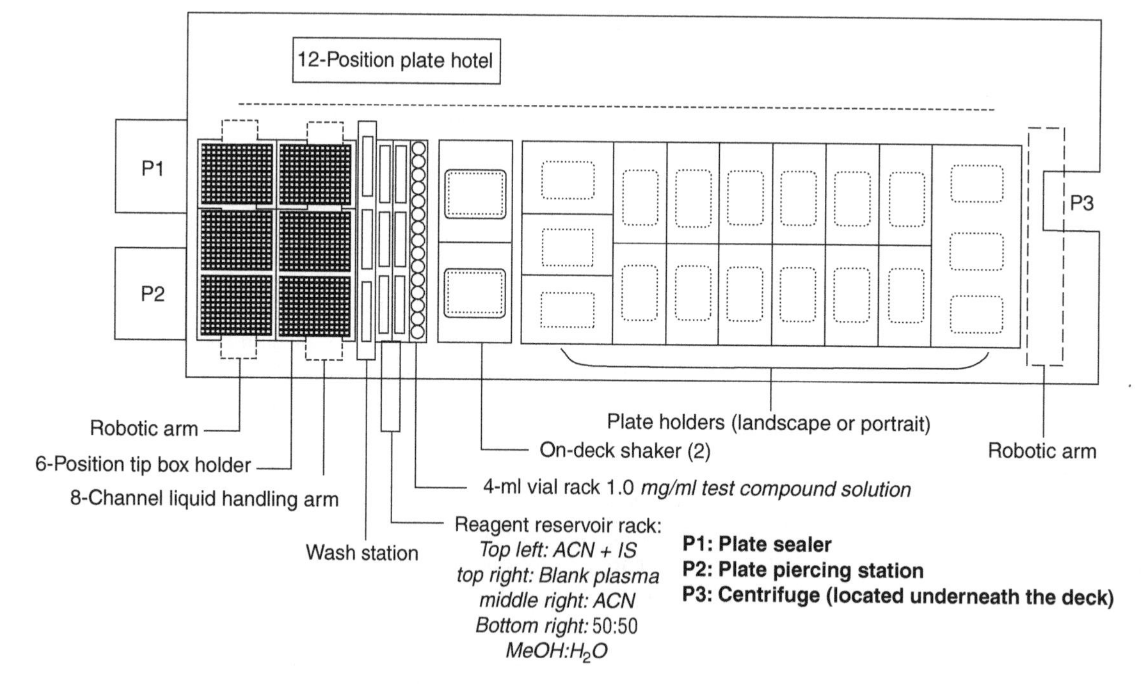

FIGURE 22.7 A schematic representation of the Tecan Freedom EVO® 200's deck layout for fully automated protein precipitation sample preparation (reprinted with permission from Ma et al. (2008); copyright (2008) Elsevier).

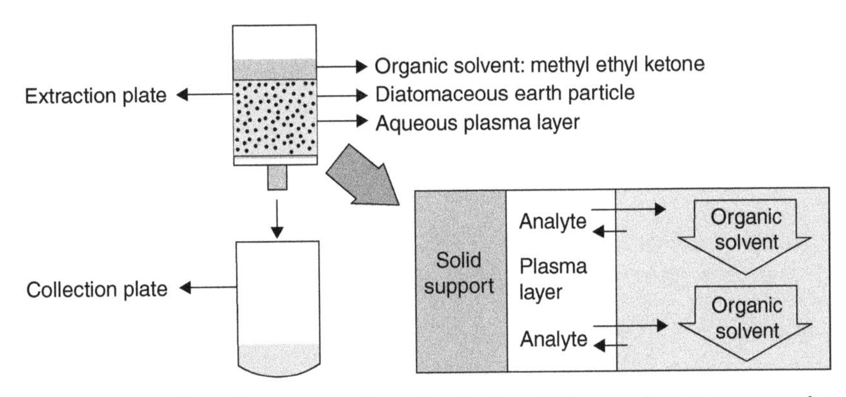

FIGURE 22.8 A schematic representation of a single well from a diatomaceous earth packed 96-well plate used for solid supported liquid–liquid extractions (reprinted with permission from Peng et al. (2000); copyright (2001) American Chemical Society).

ionization effect is routinely observed in routine LC-MS bioanalysis, especially when using less selective sample preparation techniques like PPT (Ye et al., 2011). At least two manufacturers currently offer 96-well PPT plates that selectively retain phospholipids during sample preparation, commercially available as Hybrid-SPE and Captiva ND Lipids from Sigma-Aldrich and Agilent, respectively (Agilent, 2011; Sigma-Aldrich, 2002). These phospholipids depletion plates have been recently reported in LC-MS applications (Ismaiel et al., 2010; Ardjomand-Woelkart et al., 2011) and have proven to be amenable to automation via robotic liquid handling platforms (Pucci et al., 2009; Jiang et al., 2011). Further work is expected employing this depletion technology paired with automation in order to enhance throughput and improve the overall quality of the LC-MS bioanalytical sample preparation technique.

22.7 LIQUID–LIQUID EXTRACTION

The LLE is another popular extraction technique used in LC-MS bioanalysis and is often automated via robotic liquid handling platforms. The earliest reports of automated 96-well plate LLE extractions detailed the use of a channel-based (Jemal et al., 1999) or a 96-well based (Steinborner and Henion, 1999; Zweigenbaum et al., 1999; Ramos et al., 2000) multichannel robotic liquid handler for sample manipulations with off-line sample mixing and extraction followed by organic phase separation. A variant of traditional LLE using modified diatomaceous earth packed cartridges was introduced by Varian and is commercially known as ChemElute™. This new variant of LLE offered the selectivity of LLE without requiring the need for sample mixing using vortexing, rotatation, or sonication prior to phase separation using flash-freezing or centrifugation. Shown in its application for bioanalytical support of clinical trials (Burton et al., 1997; Zhao et al., 2000), this variant of LLE was also demonstrated in a 48-well footprint that significantly increased throughput compared with cartridge-based approaches (Wang et al., 2001). Subsequently, even higher analytical speed was achieved in the 96-well format (Wang et al., 2002). Early work by Peng paired a diatomaceous earth 96-well plate (shown in Figure 22.8) with a Tomtec Quadra-96 robotic liquid handler and was reported as the first fully automated LLE extraction for LC-MS bioanalysis (Peng et al., 2000). Referred to as SLE for the purposes of this chapter, this approach offered a simple and convenient alternative to 96-well based LLE. More readily amenable to automation, SLE also offered a reduced potential for cross-contamination compared to such problems identified with 96-well LLE protocols (Basileo et al., 2003). Furthermore, the automated SLE has increased overall laboratory efficiency when applied to routine discovery and development portfolio support (Licea-Perez et al., 2007) and when compared to conventional LLE. Additionally, extraction recovery and selectivity (matrix effect profile) were demonstrated to be comparable as determined by LLE (Nguyen et al., 2010; Wu et al., 2010). The consumable costs of SLE plates are comparable to SPE plates ($180–290 per plate) and result in scalable (variable) costs to a bioanalytical laboratory, depending upon the number of experiments performed or samples extracted. Important to the overall quality of an SLE protocol, automated or not, is the controlled application of pressure (positive or negative) used to facilitate aqueous sample loading and complete solvent elution on a vacuum manifold. Negative (vacuum) pressure was used in the loading and elution steps of an automated SLE of fluoxetine and norfluoxetine in human plasma (Li et al., 2011). Li was able to demonstrate excellent throughput for the SLE portion of this protocol at just 10 min per 96 samples. Unlike negative pressure, positive pressure allows uniform pressure to be applied across all wells within a 96-well plate and facilitates better flow-through characteristics regardless of sample viscosity. Comprehensive work by Jiang and later by Pan have detailed the use of a positive pressure apparatus in automated SLE protocols (Jiang et al., 2008a,

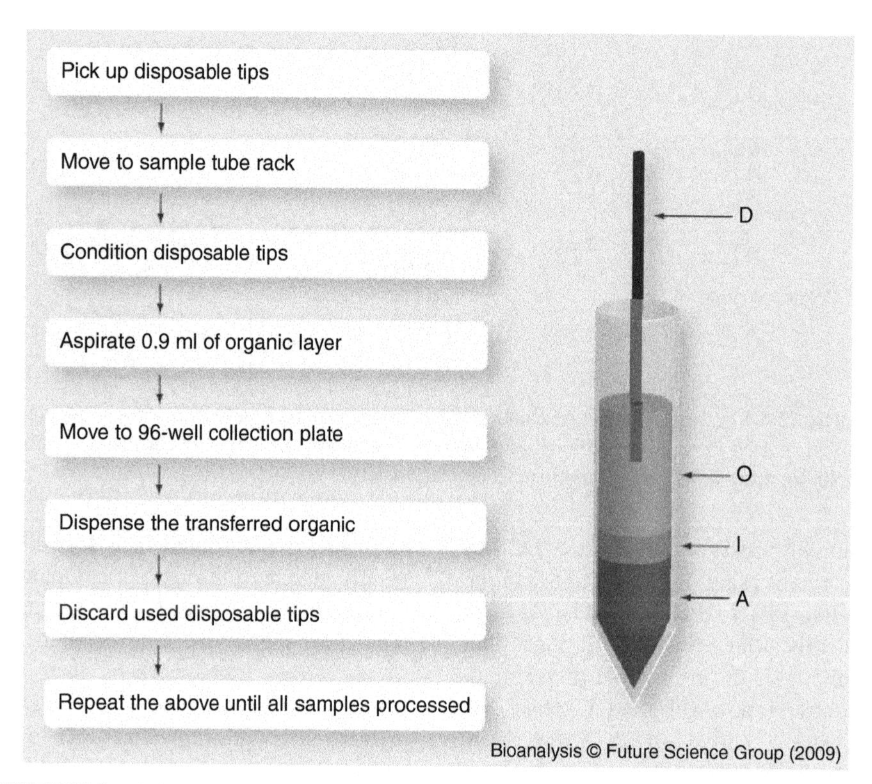

FIGURE 22.9 Schematic representation of the transfer of the organic layer (O) of the LLE extraction using a disposable tip (D) performed via a Multiprobe II robotic liquid handling platform. During pressure equilibration of the disposable tip, careful consideration was made during the organic transfer as to not disturb the intermediate layer (I) phase-separation of the aqueous (A) and organic layers (reproduced from Hussain et al. (2009) with permission of Future Science Ltd).

2008b; 2009; Pan et al., 2010, 2011). Unfortunately, use of the positive pressure apparatus, such as the Speedisk®96 (J.T. Baker, Inc.), is typically performed off-line relative to the robotic liquid handling platform, thus requiring manual intervention and decreased throughput which is arguably rate-limiting to the overall bioanalytical process. To solve this problem, Tomtec has recently made available a positive pressure tool for use on their Quadra4 robotic liquid handling platform (Figure 22.4) to perform positive pressure elution for any filter/SPE/SLE plate based assay (Tomtec, 1999).

High-throughput discovery and clinical applications for LLE have been described using the Tomtec Quadra96 and fixed height pipetting (Brignol et al., 2001; Shen et al., 2002; Dotsikas et al., 2006; Li et al., 2007; Yadav et al., 2008). Solvents used in LLE are typically highly volatile and nonconductive (methyl t-butyl ether, ethyl acetate, chloroform, etc.), thereby disallowing use of automated liquid level detection using conductive tips. Careful attention must be paid to the aspirate and dispense steps of robotic liquid handlers in the manipulation of these solvents to ensure optimal LLE performance. Adjustments to transfer stage heights for 96-well robotic liquid handlers like the Tomtec (Zhang et al., 2000a) or tip-depth positions for channel-based robotic liquid handlers (Tecan, Hamilton, Biomek) are critical to high-quality LLE performance conducted on robotic liquid handling platforms. One approach developed by Hussain used the Multiprobe EX HT II to equilibrate (pressure condition) a portion of the disposable tip headspace with the organic layer of a centrifuged phase-partitioned sample at a predetermined fixed position shown in Figure 22.9 (Hussain et al., 2009). This allowed the organic layer of the sample to be transferred to another 96-well plate for subsequent "dry-down" step and prevented any organic layer solvent from dripping during transfer due to uncontrolled expansion or contraction of the solvent vapor headspace.

Another approach implemented to improve the overall quality of an automated LLE protocol was demonstrated in the use of positive displacement tips (Hamilton MicroLab AT Plus 2) for a liquid–liquid back extraction of dextromethorphan and dextrorphan in human plasma (Bolden et al., 2002). When implemented for automated LLE, positive displacement tips improved liquid handling manipulations of highly volatile solvents and increased throughput by preventing solvent drips and eliminated the need to equilibrate the headspace of disposable tips. In an another automated LLE back extraction methodology, sequential use of two robotic liquid handlers was demonstrated in addition to in-tip LLE phase mixing by repetitive aspirate and dispense cycles of the combined fortified aqueous sample and extraction solvent (Xu et al., 2004). This team of bioanalytical scientists at Abbott Laboratories demonstrated that the Hamilton AT Plus 2 was a robust robotic liquid handling platform that delivered high-quality results from semi-automated (Ji et al., 2004; Wang et al., 2006b; Zhang et al., 2006) and fully automated (Rodila et al., 2006; Wang et al., 2006a) LLE assays.

For robotic liquid handlers using air displacement pipetting modes, the highly volatile, nonconductive LLE solvents still remained a challenge to the quality and throughput of automated LLE. The risk of dripping extraction solvent before and after phase mixing, and the fact that automated liquid level sensing technology could not be used with fixed or disposable tips, required employment of alternative methods. To achieve greater throughput and improve automated LLE assay quality, several reports demonstrated the use of the sequential use of robotic liquid handling platforms (e.g., Multiprobe II followed by Tomtec Quadra96) for specific steps (i.e., matrix sample aliquoting, LLE solvent delivery, and phase mixing) within a given automated LLE protocol (Eerkes et al., 2003; Zhang, 2004; Riffel et al., 2005; Apostolou et al., 2007).

The risk of cross-well contamination during the 96-well phase mixing is of great importance to the overall quality of the automated LLE protocol. There are advantages and disadvantages in performing extraction phase/sample mixing online (in-tip) or off-line (vortex-mixing, sonication, and rotation) relative to the robotic liquid handler. Obvious advantages of the in-tip mixing are decreased cycle times and reduced potential to cross-contaminate wells; however, if the in-tip phase mixing is not sufficiently turbulent to promote a vigorous mixing of the aqueous and organic layers reduced analyte recovery will be evident. Conversely, off-line phase mixing (following manual plate sealing) will most definitely promote adequate mixing of phases but may increase the likelihood of cross-contaminating the 96-well plate due to leakage of the sealed well during off-line phase mixing. Therefore, appropriate testing must be performed to ensure that potential sources of cross-contamination are addressed during off-line phase mixing to ensure rugged automated LLE assay performance. Optimized heat-sealing conditions (Ji et al., 2004) and special plates providing better sealing conditions (Eichhold et al., 2007) have both been described. Also explained are steps designed to prevent contamination after phase separation (access to top organic layer) as described by Wang via the direct puncture of the heat seal (Wang et al., 2006). When using the 384-well format, the burden of cross-contamination becomes an even greater challenge in automated LLE assays. However, with careful experimental design and optimized phase-mixing and phase-separation steps, the automated 384-well LLE has shown promising applicability for routine LC-MS bioanalytical support (Chang et al., 2007b).

22.8 PRACTICAL CONSIDERATIONS: STRATEGY, QUALITY, AND COMPLIANCE

Automated LC-MS bioanalytical sample preparation methods range in complexity that is highly dependent upon the objectives of the sample preparation approach and the level of automation desired. The appropriate balance between cost and complexity must be achieved by leveraging automated techniques and technologies for routine LC-MS bioanalysis. The appropriate strategy must differentiate between the "fit-for-purpose" sample preparation protocols employed during early drug discovery to the more rugged and selective sample preparation and extraction techniques often required in later regulated preclinical and clinical drug development. Accordingly, the appropriate resources must be directed to successfully implement the strategy from installation through routine production.

The development of a strategy directed toward automation implementation in a laboratory setting is not a new concept (McDowall, 1992; Vogelsanger, 1992; Mole et al., 1993). A variety of decisions must be made regarding what will be automated, how the process/technique will be implemented, and what technical direction is required regarding the associated technology (hardware, software, programming, etc.) (Kozlowski, 1996). Practical guiding concepts regarding cost and the economic impact of automation can aid in hardware and resourcing decisions, especially in today's contracting economic environment (Gurevitch, 2004). Ideally, these concepts should all be included in an encompassing validation or qualification plan supporting the implementation of the automated technique, discussed later in this chapter.

A common misconception in the development and implementation of robotic liquid handling techniques is that the instrumentation, vendor support, and appropriate functional training are all that is needed to deliver an automated solution into routine bioanalytical production. While this approach may be suitable in some cases, it certainly does not address the strategy, time, logistics, and resources supporting the hardware and software (methods) required for routine operation and maintenance, especially in a regulated setting. Alignment among the stakeholders (management, users, automation specialists, informatics group, etc.) implementing the automation platform must be attained which details the goals, purpose, and expected accountability for the automation tools/techniques proposed (McDowall, 1994). To that end, the automation investment should be monitored appropriately to justify future decisions to buy-in or change course based upon monitored use or collected metrics (Schultz et al., 2003; Benn et al., 2006).

Fortunately, vendors often provide training and services that can achieve the end goal of robotic implementation into a production environment. However, as the scope and complexity of the robotic liquid handling platforms evolve, so must the skills and competencies of the scientists performing the work. The increased scope and complexity of modern liquid handling platforms routinely used in LC-MS bioanalysis requires increased competency in a number of disciplines such as analytical chemistry, robotics, computer science, and regulatory compliance (hardware, software, and assay specific). Some of these concepts have been discussed and demonstrated in a number of automated platforms using customized robotic setups (Deng et al., 2002; Ma et al., 2008) or customized software interfaces integrated with robotic liquid handlers (Gu et al., 2006; Palandra et al., 2007; Tweed et al., 2010).

Robotic liquid handling instrumentation is typically used for the sample preparation and extraction steps of routine LC-MS bioanalytical methods supporting both nonregulated (non-GxP) drug discovery and regulated (GxP) drug development portfolios. The degree of regulatory compliance associated with the implementation and use of robotic liquid handling platforms is dependent upon several factors. In general terms, the level of compliance associated with the installation, maintenance, and routine use of robotic liquid handlers should be commensurate with the regulatory requirements of the laboratory. A regulated bioanalytical laboratory has stringent compliance criteria associated with bioanalytical method validation and sample analysis, established by FDA and EMA guidelines (US-FDA, 2001; EMEA, 2011). Accordingly, the implementation and routine use of the instrumentation in a regulated setting requires significantly more resources and effort to deliver automated assays into a routine production environment than may be needed for early exploratory assay development. User requirements for hardware maintenance, software compliance, and general assay requirements (assay acceptance, method qualification, etc.) may differ significantly in regulated versus nonregulated laboratories. Therefore, it is important to explore some key compliance considerations that must be evaluated before robotic liquid handling instrumentation is implemented in a bioanalytical laboratory.

For any regulated LC-MS bioanalysis laboratory implementing robotic liquid handling, compliance is ultimately determined by the predetermined risk level an organization is willing to bear. For the purposes of this discussion, the scope of risk encompasses both scientific and business based risk and should drive any decisions and strategy surrounding robotic liquid handling hardware, software, and subsequent automated sample preparation and extraction techniques developed for routine production use. As a result, an appropriate balance must be achieved between the business needs and compliance risk a bioanalysis laboratory can sustain when implementing robotic liquid handling instrumentation.

In a GxP laboratory, instrumentation must conform to hardware compliance standards set by the FDA (USFDA-21CFR58.63, 2001; USFDA-21CFR211.68, 2001). Chan

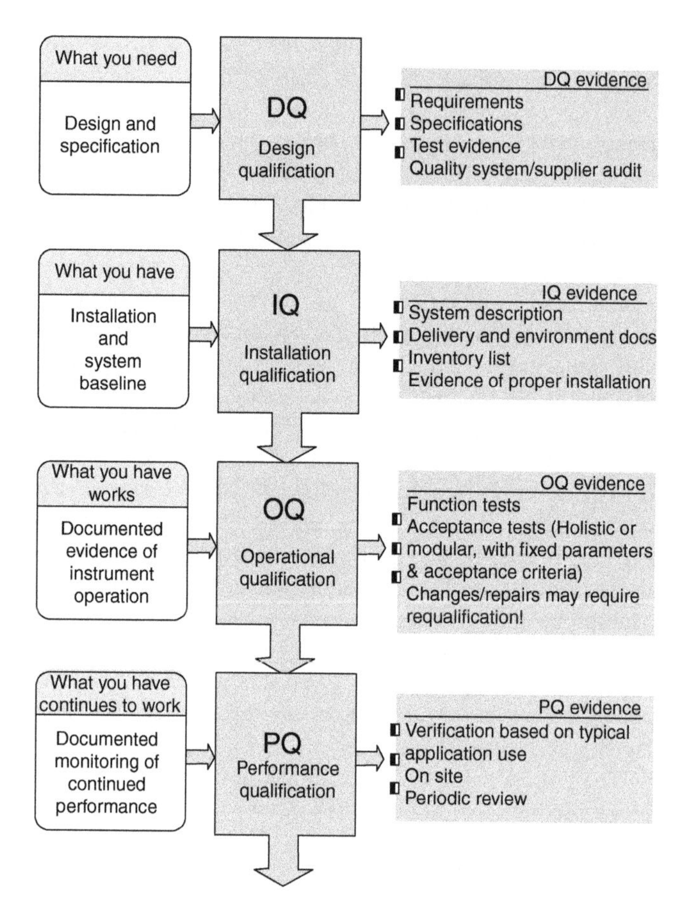

FIGURE 22.10 Instrument qualification and its key deliverables via the 4Q model (reprinted from Chan et al. (2010); copyright © 2010, with permission from John Wiley & Sons).

et al. appropriately details hardware validation and qualification concepts and recommended approaches to achieve an appropriate risk-based compliance endpoint (Chan et al., 2010). The compiled concepts for hardware validation and qualification result in a logical process map explained as the 4Qs: design qualification (DQ), installation qualification (IQ), operational qualification (OQ), and performance qualification (PQ) (Figure 22.10). These concepts can and should be applied to the validation and qualification of robotic liquid handling platform for use in an LC-MS bioanalytical laboratory and are detailed accordingly later.

A DQ, sometimes referred to as the design specification stage, establishes a viable implementation and support strategy and is fundamental to achieving success. Creating a DQ provides a foundation for making decisions upon the objectives, hardware, standard operating procedures (SOPs), internal and external resources, and other components critical for a successful rollout of any robotic liquid handling platform.

Most, if not all, manufacturers of robotic liquid handling platforms offer some type of IQ/OQ package upon purchase. The IQ establishes that the instrument has been installed as per vendor specifications and configured appropriately.

IQ documents information that includes, but is not limited to, hardware/software make and model, electrical voltages, site requirements, and firmware version. The OQ, however, establishes that the instrumentation is functionally sound and must only be initiated once the IQ is complete and meets specifications. During the OQ, a series of functional tests are performed and documented to provide evidence that the instrument meets the operational requirements established by the vendor. For example, during a typical robotic liquid handling OQ, some protocols executed would test positional movements, pipetting function (volume verification), functional platform integration (shakers, barcode readers, vacuum manifolds, etc.), and routine maintenance procedures.

To address perceived compliance gaps associated with computer software, the FDA established guidelines for electronic records and signatures commonly known as CFR Part 11compliance (USFDA-21CFR11, 2003). CFR Part 11 compliance has a significant impact on a variety of software packages employed for a variety of uses within LC-MS bioanalysis discipline. In the IQ/OQ stage of validation and qualification, the level of compliance desired to fulfill the requirements of CFR Part 11 must be implemented. It has been argued that because many robotic liquid handling platforms do not generate or collect "raw data," they may be exempt from CFR Part 11 compliance regulations. This determination is typically based on the compliance risk level assessment established by the laboratory and will thus be outside the scope of this review. In common terms, however, it is again important to stress that a tangible level of risk must be established by the laboratory or organization when deciding upon CFR Part 11 compliance and robotic liquid handling platform use. Ideally, these decisions should be decided upon in the DQ stage and carried out during OQ with subsequent resources and support infrastructure in place to facilitate a predetermined level of compliance. Regardless of the CFR Part 11 compliance decision, it is recommended that the fundamental structure and documentation of a computer validation process be followed. For routine GxP use, it is of paramount importance to ensure that the sample preparation and extraction methods used on the robotic liquid handling platform are secure and do not change without proper authorization and documentation.

The PQ is the sole responsibility of the organization and is specific to the hardware, software, and objective of the robotic liquid handling platform being implemented. For LC-MS bioanalytical applications, practical considerations of the PQ include the following: Does a specific robotic liquid handling technique satisfactorily perform routine sample preparation and extraction? How will consistent and predictable performance be evaluated? What degree of testing and level of control will the aforementioned technique be subjected?

In the PQ process, a computer validation framework can significantly enhance compliance and the scientific validity of the robotic liquid handling protocol being implemented. A typical computer validation setup includes a validation plan (test plan), validation procedures (test scripts), and a validation report (test report). This strategic approach provides structure, and more importantly, documentation to the robotic liquid handling PQ process. The test plan should formalize the scope of validation activities and establish the tests being performed. Validation test plans should detail the steps completed during protocol execution and establish acceptance criteria for the method executed via test scripts, if required. The validation report documents the results of the specific automated sample preparation method being tested that, upon approval, releases the method for routine production use. Fortunately, the FDA validation guidance for bioanalytical methods provides a framework to test automated sample preparation and extraction techniques. The SOPs of an organization and the regulatory guidelines previously mentioned provide tangible targets to which robotic liquid handling sample preparation and extraction techniques must meet for eventual use in a production setting. The performance of calibration standard and QC sample data provide a legitimate assessment of the overall quality of an automated technique. Furthermore, several published reports have detailed side-by-side comparisons of automated versus manual sample preparation techniques, which practically eliminate any possibility of systematic errors (bias) introduced by automated protocols. Fundamentally, the PQ provides the bioanalytical scientist with a level of confidence establishing that the sample preparation technique performed on a robotic liquid handling instrument is no different from manual performance. The required effort needed to achieve this endpoint is tedious and time-consuming; however, the benefits of implementing automation over time will offset this initial investment, if planned appropriately.

22.9 CONCLUDING REMARKS

It is widely accepted that robotic liquid handling instrumentation is an important component in the development of high-quality sample processing techniques in LC-MS bioanalysis. Assays that can be routinely automated via robotic liquid handling platforms include (1) SPE, (2) PPT, and (3) LLE with each having its respective variations therein. These assays can be typically performed in a high-throughput fashion using parallel sample processing with high-density 96-well plate footprint. The use of robotic liquid handling platforms in LC-MS bioanalysis has resulted in significant gains in bioanalytical laboratory efficiency and productivity with a high level of quality. Several key factors that impact the overall quality of robotic liquid handling sample preparation techniques include (1) the choice of robotic liquid handling hardware, (2) pipetting mode (air, liquid–air, positive) used for liquid displacement, and (3) settings used for liquid aspirate and dispense manipulation. Routine system qualification either by gravimetric analysis or absorbance testing has been demonstrated to be critical. With the advancement in automation technology, more sophisticated "off-the-shelf" products are expected to be available that allow for more seamless integration of laboratory software and hardware for automation in LC-MS bioanalysis.

REFERENCES

Ackermann BL, Berna MJ, Murphy AT. Recent advances in use of LC/MS/MS for quantitative high-throughput bioanalytical support of drug discovery. Curr Top Med Chem 2002;2(1):53–66.

Agilent. 2011. Available at http://www.chem.agilent.com/en-US/Products/columns-supplies/samplepreparation/spe/captivandlipids/Pages/default.aspx. Accessed Mar 24, 2013.

Albert KJ, Bradshaw JT. Importance of integrating a volume verification method for liquid handlers: applications in learning performance behavior. J Assoc Lab Autom 2007;12(3):172–180.

Albert KJ, Bradshaw JT, Knaide TR, Rogers AL. Verifying liquid-handler performance for complex or nonaqueous reagents: a new approach. J Assoc Lab Autom 2006;11(4):172–180.

Allanson JP, Biddlecombe RA, Jones AE, Pleasance S. The use of automated solid phase extraction in the'96 well'format for high throughput bioanalysis using liquid chromatography coupled to tandem mass spectrometry. Rapid Commun Mass Spectrom 1996;10(7):811–816.

Apostolou C, Dotsikas Y, Kousoulos C, Loukas YL. Quantitative determination of donepezil in human plasma by liquid chromatography/tandem mass spectrometry employing an automated liquid–liquid extraction based on 96-well format plates: Application to a bioequivalence study. J Chromatogr B 2007;848(2):239–244.

Ardjomand-Woelkart K, Kollroser M, Li L, Derendorf H, Butterweck V, Bauer R. Development and validation of a LC-MS/MS method based on a new 96-well Hybrid-SPE™-precipitation technique for quantification of CYP450 substrates/metabolites in rat plasma. Analytical and Bioanalytical Chemistry 2011;400(8):2371–2381.

Artel. 2011. MVS Multichannel verification system. Available at http://www.artel-usa.com/products/mvs_advanced.aspx. Accessed Mar 24, 2013.

Basileo G, Breda M, Fonte G, Pisano R, James CA. Quantitative determination of paclitaxel in human plasma using semi-automated liquid–liquid extraction in conjunction with liquid chromatography/tandem mass spectrometry. J Pharm Biomed Anal 2003;32(4–5):591–600.

Benn N, Turlais F, Clark V, Jones M, Clulow S. An automated metrics system to measure and improve the success of laboratory automation implementation. J Assoc Lab Autom 2006;11(1):16–22.

Berna M, Murphy AT, Wilken B, Ackermann B. Collection, storage, and filtration of in vivo study samples using 96-well filter plates to facilitate automated sample preparation and LC/MS/MS analysis. Anal Chem 2002;74(5):1197–1201.

Biddlecombe RA, Benevides C, Pleasance S. A clinical trial on a plate? The potential of 384-well format solid phase extraction for high throughput bioanalysis using liquid chromatography tandem mass spectrometry. Rapid Commun Mass Spectrom 2001;15(1):33–40.

Biddlecombe RA, Pleasance S. Automated protein precipitation by filtration in the 96-well format. J Chromatogr B 1999;734(2):257–265.

Bolden RD, Hoke Ii SH, Eichhold TH, McCauley-Myers DL, Wehmeyer KR. Semi-automated liquid–liquid back-extraction in a 96-well format to decrease sample preparation time for the determination of dextromethorphan and dextrorphan in human plasma. J Chromatogr B 2002;772(1):1–10.

Bradshaw JT, Curtis RH, Knaide TR, Spaulding BW. Determining dilution accuracy in microtiter plate assays using a quantitative dual-wavelength absorbance method. J Assoc Lab Autom 2007;12(5):260–266.

Bradshaw JT, Albert KJ. Chapter 15: Instrument qualification and performance verification for automated liquid-handling systems. In: Chan CC, Lam H, Zhang XM, editors. Practical Approaches to Method Validation and Essential Instrument Qualification. John Wiley & Sons; 2010. p. 347–375

Bradshaw JT, Knaide T, Rogers A, Curtis R. Multichannel verification system (MVS): a dual-dye ratiometric photometry system for performance verification of multichannel liquid delivery devices. J Assoc Lab Autom 2005;10(1):35–42.

Brignol N, McMahon LM, Luo S, Tse FLS. High-throughput semi-automated 96-well liquid/liquid extraction and liquid chromatography/mass spectrometric analysis of everolimus (RAD 001) and cyclosporin a (CsA) in whole blood. Rapid Commun Mass Spectrom 2001;15(12):898–907.

Buhrman DL, Price PI, Rudewicz PJ. Quantitation of SR 27417 in human plasma using electrospray liquid chromatography-tandem mass spectrometry: a study of ion suppression. J Am Soc Mass Spectrom 1996;7(11):1099–1105.

Burton R, Mummert M, Newton J, Brouard R, Wu D. Determination of SR 49059 in human plasma and urine by LC-APCI/MS/MS. J Pharm Biomed Anal 1997;15(12):1913–1922.

Callejas SL, Biddlecombe RA, Jones AE, Joyce KB, Pereira AI, Pleasance S. Determination of the glucocorticoid fluticasone propionate in plasma by automated solid-phase extraction and liquid chromatography–tandem mass spectrometry. J Chromatogr B 1998;718(2):243–250.

Chan CC, Lam H, Zhang XM, editors. Practical Approaches to Method Validation and Essential Instrument Qualification. Hoboken, NJ: John Wiley & Sons; 2010.

Chang MS, Ji Q, Zhang J, El-Shourbagy TA. Historical review of sample preparation for chromatographic bioanalysis: pros and cons. Drug Dev Res 2007a;68(3):107–133.

Chang MS, Kim EJ, El-Shourbagy TA. Evaluation of 384-well formatted sample preparation technologies for regulated bioanalysis. Rapid Commun Mass Spectrom 2007b;21(1):64–72.

Côté C, Bergeron A, Mess JN, Furtado M, Garofolo F. Matrix effect elimination during LC–MS/MS bioanalytical method development. Matrix 2009;1(7):1243–1257.

Deng Y, Wu JT, Lloyd TL, Chi CL, Olah TV, Unger SE. High-speed gradient parallel liquid chromatography/tandem mass spectrometry with fully automated sample preparation for bioanalysis: 30 seconds per sample from plasma. Rapid Commun Mass Spectrom 2002;16(11):1116–1123.

Dilorenzo ME, Timoney C, Felder RA. Technological advancements in liquid handling robotics. J Assoc Lab Autom 2001;6(2):36–40.

Dong H, Ouyang Z, Liu J, Jemal M. The use of a dual dye photometric calibration method to identify possible sample dilution from an automated multichannel liquid-handling system. J Assoc Lab Autom 2006;11(2):60–64.

Dotsikas Y, Kousoulos C, Tsatsou G, Loukas YL. Development and validation of a rapid 96-well format based liquid–liquid extraction and liquid chromatography–tandem mass spectrometry analysis method for ondansetron in human plasma. J Chromatogr B 2006;836(1–2):79–82.

Eerkes A, Shou WZ, Naidong W. Liquid/liquid extraction using 96-well plate format in conjunction with hydrophilic interaction liquid chromatography–tandem mass spectrometry method for the analysis of fluconazole in human plasma. J Pharm Biomed Anal 2003;31(5):917–928.

Eichhold TH, McCauley-Myers DL, Khambe DA, Thompson GA, Hoke SH. Simultaneous determination of dextromethorphan, dextrorphan, and guaifenesin in human plasma using semi-automated liquid/liquid extraction and gradient liquid chromatography tandem mass spectrometry. J Pharm Biomed Anal 2007;43(2):586–600.

EMEA. Guideline on Bioanalytical Method Validation. 2011.

Fouda HG, Schneider RP. Robotics for the bioanalytical laboratory a flexible system for the analysis of drugs in biological fluids. TrAC, Trends Anal Chem 1987;6(6):139–147.

Gu H, Deng Y. Dilution effect in multichannel liquid-handling system equipped with fixed tips: problems and solutions for bioanalytical sample preparation. J Assoc Lab Autom 2007;12(6):355–362.

Gu H, Unger S, Deng Y. Automated Tecan programming for bioanalytical sample preparation with EZTecan. ASSAY Drug Dev Technol. 2006;4(6):721–733.

Gurevitch D. Economic justification of laboratory automation. J Assoc Lab Autom 2004;9(1):33–43.

Hamilton. 2011. Microlab STAR Line. Available at http://www.hamiltonrobotics.com/fileadmin/user_upload/products/startour/MR-0805-03_STAR_LINE_web.pdf. Accessed Mar 24, 2013.

Hussain S, Patel H, Tan A. Automated liquid–liquid extraction method for high-throughput analysis of rosuvastatin in human EDTA K2 plasma by LC–MS/MS. Bioanalysis 2009;1(3):529–535.

Ismaiel OA, Zhang T, Jenkins RG, Karnes HT. Investigation of endogenous blood plasma phospholipids, cholesterol and glycerides that contribute to matrix effects in bioanalysis by liquid chromatography/mass spectrometry. J Chromatogr B 2010;878(31):3303–3316.

Janiszewski J, Schneider RP, Hoffmaster K, Swyden M, Wells D, Fouda H. Automated sample preparation using membrane microtiter extraction for bioanalytical mass spectrometry. Rapid Commun Mass Spectrom 1997;11(9):1033–1037.

Jemal M, Ouyang Z, Xia Y-Q. Systematic LC-MS/MS bioanalytical method development that incorporates plasma phospholipids risk avoidance, usage of incurred sample and well thought-out chromatography. Biomed Chromatogr 2010;24(1):2–19.

Jemal M, Teitz D, Ouyang Z, Khan S. Comparison of plasma sample purification by manual liquid–liquid extraction, automated 96-well liquid–liquid extraction and automated 96-well solid-phase extraction for analysis by high-performance liquid chromatography with tandem mass spectrometry. J Chromatogr B 1999;732(2):501–508.

Jemal M, Xia Y-Q. LC-MS development strategies for quantitative bioanalysis. Curr Drug Metabol 2006;7(5):491–502.

Ji QC, Todd Reimer M, El-Shourbagy TA. 96-Well liquid–liquid extraction liquid chromatography-tandem mass spectrometry method for the quantitative determination of ABT-578 in human blood samples. J Chromatogr B 2004;805(1):67–75.

Jiang H, Jiang X, Ji QC. Enantioselective determination of alprenolol in human plasma by liquid chromatography with tandem mass spectrometry using cellobiohydrolase chiral stationary phases. J Chromatogr B 2008a;872(1–2):121–127.

Jiang H, Li Y, Pelzer M, et al. Determination of molindone enantiomers in human plasma by high-performance liquid chromatography–tandem mass spectrometry using macrocyclic antibiotic chiral stationary phases. J Chromatogr A 2008b;1192(2):230–238.

Jiang H, Randlett C, Junga H, Jiang X, Ji QC. Using supported liquid extraction together with cellobiohydrolase chiral stationary phases-based liquid chromatography with tandem mass spectrometry for enantioselective determination of acebutolol and its active metabolite diacetolol in spiked human plasma. J Chromatogr B 2009;877(3):173–180.

Jiang H, Zhang Y, Ida M, LaFayette A, Fast DM. Determination of carboplatin in human plasma using hybrid SPE-precipitation along with liquid chromatography–tandem mass spectrometry. J Chromatogr B 2011;879(22):2162–2170.

Joyce KB, Jones AE, Scott RJ, Biddlecombe RA, Pleasance S. Determination of the enantiomers of salbutamol and its 4-O-sulphate metabolites in biological matrices by chiral liquid chromatography tandem mass spectrometry. Rapid Commun Mass Spectrom 1998;12(23):1899–1910.

Kataoka, H. New trends in sample preparation for clinical and pharmaceutical analysis. TrAC, Trends Anal Chem 2003;22(4):232–244.

Kaye, B, Herron WJ, Macrae PV, et al. Rapid, solid phase extraction technique for the high-throughput assay of darifenacin in human plasma. Anal Chem 1996;68(9):1658–1660.

Kebarle P, Tang L. From ions in solution to ions in the gas phase—the mechanism of electrospray mass spectrometry. Anal Chem 1993;65(22):972A–986A.

Kim J, Flick J, Reimer MT, *et al.* LC-MS/MS determination of 2-(4-((2-(2S,5R)-2-Cyano-5-ethynyl-1-pyrrolidinyl)-2-oxoethylamino)-4-methyl-1-piperidinyl)-4-pyridinecarboxylic acid (ABT-279) in dog plasma with high-throughput protein precipitation sample preparation. Biomed Chromatogr 2007;21(11):1118–1126.

Kitchen CJ, Wang AQ, Musson DG, Yang AY, Fisher AL. A semi-automated 96-well protein precipitation method for the determination of montelukast in human plasma using high performance liquid chromatography/fluorescence detection. J Pharm Biomed Anal 2003;31(4):647–654.

Knaide TR, Bradshaw JT, Rogers A, McNally C, Curtis RH, Spaulding BW. Rapid volume verification in high-density microtiter plates using dual-dye photometry. J Assoc Lab Autom 2006;11(5):319–322.

Kozlowski, MR. Problem-solving in laboratory automation. Drug Discov Today 1996;1(11):481–488.

Laycock JD, Hartmann T. Automation. Integrated Strategies for Drug Discovery Using Mass Spectrometry. John Wiley & Sons, Inc.; 2005. p 511–542.

Lee MS, Kerns EH. LC/MS applications in drug development. Mass Spectrom Rev 1999;18(3–4):187–279.

Li W, Luo S, Smith HT, Tse FLS. Simultaneous determination of midazolam and 1′-hydroxymidazolam in human plasma by liquid chromatography with tandem mass spectrometry. Biomed Chromatogr 2007;21(8):841–851.

Li Y, Emm T, Yeleswaram S. Simultaneous determination of fluoxetine and its major active metabolite norfluoxetine in human plasma by LC-MS/MS using supported liquid extraction. Biomed Chromatogr 2011;25:1245–1251.

Licea-Perez H, Wang S, Bowen CL, Yang E. A semi-automated 96-well plate method for the simultaneous determination of oral contraceptives concentrations in human plasma using ultra performance liquid chromatography coupled with tandem mass spectrometry. J Chromatogr B 2007;852(1–2): 69–76.

Luckwell J, Beal A. Automated micropipette tip-based SPE in quantitative bioanalysis. Bioanalysis 2011;3(11):1227–1239.

Ma J, Shi J, Le H, Cho R, Huang JC-J, Miao S, Wong BK. A fully automated plasma protein precipitation sample preparation method for LC–MS/MS bioanalysis. J Chromatogr B 2008;862(1–2):219–226.

Majors RE. New developments in microplates for biological assays and automated sample preparation. LC GC Mag-North Am-Solut Separat Sci 2004;22(11):1062–1072.

Mallet CR, Lu Z, Fisk R, Mazzeo JR, Neue UD. Performance of an ultra-low elution-volume 96-well plate: drug discovery and development applications. Rapid Commun Mass Spectrom 2003;17(2):163–170.

Matuszewski BK, Constanzer ML, Chavez-Eng CM. Matrix effect in quantitative LC/MS/MS analyses of biological fluids: a method for determination of finasteride in human plasma at picogram per milliliter concentrations. Anal Chem 1998;70(5):882–889.

McDowall RD. Strategic approaches to laboratory automation. Chemometr Intell Lab 1992;17(3):259–264.

McDowall RD. Laboratory automation: quo vadis? Chemometr Intell Lab 1994;26(1):37–42.

McMahon LM, Luo S, Hayes M, Tse FLS. High throughput analysis of everolimus (RAD001) and cyclosporin A (CsA) in whole blood by liquid chromatography/mass spectrometry using a semi-automated 96-well solid-phase extraction system. Rapid Commun Mass Spectrom 2000;14(21):1965–1971.

Mole D, Mason RJ, McDowall RD. The development of a strategy for the implementation of automation in a bioanalytical laboratory. J Pharm Biomed Anal 1993;11(3):183–190.

Naidong W, Shou WZ, Addison T, Maleki S, Jiang X. Liquid chromatography/tandem mass spectrometric bioanalysis using normal-phase columns with aqueous/organic mobile phases-a novel approach of eliminating evaporation and reconstitution steps in 96-well SPE. Rapid Commun Mass Spectrom 2002;16(20):1965–1975.

Nevanen TK, Simolin H, Suortti T, Koivula A, Söderlund H. Development of a high-throughput format for solid-phase extraction of enantiomers using an immunosorbent in 384-well plates. Anal Chem 2005;77(10):3038–3044.

Nguyen L, Zhong W-Z, Painter CL, Zhang C, Rahavendran SV, Shen Z. Quantitative analysis of PD 0332991 in xenograft mouse tumor tissue by a 96-well supported liquid extraction format and liquid chromatography/mass spectrometry. J Pharm Biomed Anal 2010;53(3):228–234.

O'Connor, D, Clarke DE, Morrison D, Watt AP. Determination of drug concentrations in plasma by a highly automated, generic and flexible protein precipitation and liquid chromatography/tandem mass spectrometry method applicable to the drug discovery environment. Rapid Commun Mass Spectrom 2002;16(11):1065–1071.

Ouyang Z, Federer S, Porter G, Kaufmann C, Jemal M. Strategies to maintain sample integrity using a liquid-filled automated liquid-handling system with fixed pipetting tips. J Assoc Lab Autom 2008;13(1):24–32.

Palandra J, Weller D, Hudson G, et al. Flexible automated approach for quantitative liquid handling of complex biological samples. Anal Chem 2007;79(21):8010–8015.

Pan J, Fair SJ, Mao D. Quantitative analysis of skeletal symmetric chlorhexidine in rat plasma using doubly charged molecular ions in LC–MS/MS detection. Bioanalysis 2011;3(12):1357–1368.

Pan J, Jiang X, Chen Y-L. Automatic supported liquid extraction (SLE) coupled with HILIC-MS/MS: an application to method development and validation of erlotinib in human plasma. Pharmaceutics 2010;2(2):105–118.

Peng SX, Branch TM, King SL. Fully automated 96-well liquid–liquid extraction for analysis of biological samples by liquid chromatography with tandem mass spectrometry. Anal Chem 2000;73(3):708–714.

Peng SX, Cousineau M, Juzwin SJ, Ritchie DM. A 96-well screen filter plate for high-throughput biological sample preparation and LC–MS/MS analysis. Anal Chem 2005;78(1): 343–348.

Pereira T, Chang SW. Semi-automated quantification of ivermectin in rat and human plasma using protein precipitation and filtration with liquid chromatography/tandem mass spectrometry. Rapid Commun Mass Spectrom 2004;18(12):1265–1276.

PerkinElmer. 2011. Available at: http://www.perkinelmer.com/Catalog/Category/ID/Janus.

Pucci V, Di Palma S, Alfieri A, Bonelli F, Monteagudo E. A novel strategy for reducing phospholipids-based matrix effect in LC–ESI-MS bioanalysis by means of HybridSPE. J Pharm Biomed Anal 2009;50(5):867–871.

Pucci V, Monteagudo E, Bonelli F. High sensitivity determination of valproic acid in mouse plasma using semi-automated sample preparation and liquid chromatography with tandem mass spectrometric detection. Rapid Commun Mass Spectrom 2005;19(24):3713–3718.

Ramos, L, Bakhtiar R, Tse FLS. Liquid-liquid extraction using 96-well plate format in conjunction with liquid chromatography/tandem mass spectrometry for quantitative determination of methylphenidate (Ritalin®) in human plasma. Rapid Commun Mass Spectrom 2000;14(9):740–745.

Riffel KA, Groff MA, Wenning L, Song H, Lo M-W. Fully automated liquid–liquid extraction for the determination of a novel insulin sensitizer in human plasma by heated nebulizer and turbo ionspray liquid chromatography-tandem mass spectrometry. J Chromatogr B 2005;819(2):293–300.

Rodila RC, Kim JC, Ji QC, El-Shourbagy TA. A high-throughput, fully automated liquid/liquid extraction liquid chromatography/mass spectrometry method for the quantitation of a new investigational drug ABT-869 and its metabolite A-849529 in human plasma samples. Rapid Commun Mass Spectrom 2006;20(20):3067–3075.

Rossi DT, Zhang N. Automating solid-phase extraction: current aspecs and future prospekts. J Chromatogr A 2000;885(1–2):97–113.

Rouan MC, Buffet C, Marfil F, Humbert H, Maurer G. Plasma deproteinization by precipitation and filtration in the 96-well format. J Pharm Biomed Anal 2001;25(5–6):995–1000.

Rule G, Chapple M, Henion J. A 384-well solid-phase extraction for LC/MS/MS determination of methotrexate and its 7-hydroxy metabolite in human urine and plasma. Anal Chem 2001;73(3):439–443.

Sadagopan NP, Li W, Cook JA, et al. Investigation of EDTA anticoagulant in plasma to improve the throughput of liquid chromatography/tandem mass spectrometric assays. Rapid Commun Mass Spectrom 2003;17(10):1065–1070.

Saunders KC, Ghanem A, Boon Hon W, Hilder EF, Haddad PR. Separation and sample pre-treatment in bioanalysis using monolithic phases: a review. Anal Chim Acta 2009;652(1–2):22–31.

Schultz H, Alexander J, Petersen J, et al. The automatic metric monitoring program. J Assoc Lab Autom 2003;8(1):24–27.

Shen JX, Tama CI, Hayes RN. Evaluation of automated micro solid phase extraction tips (μ-SPE) for the validation of a LC–MS/MS bioanalytical method. J Chromatogr B 2006;843(2):275–282.

Shen JX, Xu Y, Tama CI, Merka EA, Clement RP, Hayes RN. Simultaneous determination of desloratadine and pseudoephedrine in human plasma using micro solid-phase extraction tips and aqueous normal-phase liquid chromatography/tandem mass spectrometry. Rapid Commun Mass Spectrom 2007;21(18):3145–3155.

Shen Z, Wang S, Bakhtiar R. Enantiomeric separation and quantification of fluoxetine (Prozac®) in human plasma by liquid chromatography/tandem mass spectrometry using liquid-liquid extraction in 96-well plate format. Rapid Commun Mass Spectrom 2002;16(5):332–338.

Shou WZ, Jiang X, Beato BD, Naidong W. A highly automated 96-well solid phase extraction and liquid chromatography/tandem mass spectrometry method for the determination of fentanyl in human plasma. Rapid Commun Mass Spectrom 2001;15(7):466–476.

Shou WZ, Pelzer M, Addison T, Jiang X, Naidong W. An automatic 96-well solid phase extraction and liquid chromatography—tandem mass spectrometry method for the analysis of morphine, morphine-3-glucuronide and morphine-6-glucuronide in human plasma. J Pharm Biomed Anal 2002;27(1–2):143–152.

Sigma-Aldrich. 2011. Available at http://www.sigmaaldrich.com/analytical-chromatography/sample-preparation/spe/hybridspe-ppt.html. Accessed Mar 24, 2013.

Simpson H, Berthemy A, Buhrman D, et al. High throughput liquid chromatography/mass spectrometry bioanalysis using 96-well disk solid phase extraction plate for the sample preparation. Rapid Commun Mass Spectrom 1998;12(2):75–82.

Song Q, Junga H, Tang Y, et al. Automated 96-well solid phase extraction and hydrophilic interaction liquid chromatography–tandem mass spectrometric method for the analysis of cetirizine (ZYRTEC®) in human plasma—with emphasis on method ruggedness. J Chromatogr B 2005;814(1):105–114.

Stangegaard M, Hansen AJ, Frøslev TG, Morling N. A simple method for validation and verification of pipettes mounted on automated liquid handlers. J Lab Autom 2011;16(5):381–386.

Steinborner S, Henion J. Liquid–liquid extraction in the 96-well plate format with SRM LC/MS quantitative determination of methotrexate and its major metabolite in human plasma. Anal Chem 1999;71(13):2340–2345.

Tecan. 2011. Freedom EVO®—Specifications. Available at http://www.tecan.com/platform/apps/product/index.asp?MenuID=2696&ID=5273&Menu=1&Item=21.1.8.2. Accessed Mar 24, 2013.

Tomtec. 2011. Available at http://www.tomtec.com/index.htm. Accessed Mar 23, 2013.

Tweed JA, Gu Z, Xu H, et al. Automated sample preparation for regulated bioanalysis: an integrated multiple assay extraction platform using robotic liquid handling. Bioanalysis 2010;2(6):1023–1040.

Tweed JA, Walton J, Gu Z. Automated supported liquid extraction using 2D barcode processing for routine toxicokinetic portfolio support. Bioanalysis 2012;4(3):249–262.

US-FDA. 2001. Guidance for Industry: Bioanalytical Method Validation. Available at http://www.fda.gov/downloads/Drugs/GuidanceComplianceRegulatoryInformation/Guidances/ucm070107.pdf. Accessed Mar 24, 2013.

USFDA-21CFR11. Guidance for Industry Part 11, Electronic Records; Electronic Signatures—Scope and Application. Available at http://www.fda.gov/downloads/RegulatoryInformation/Guidances/ucm125125.pdf. Accessed Mar 24, 2013.

USFDA-21CFR58.63. PART 58 GOOD LABORATORY PRACTICE FOR NONCLINICAL LABORATORY STUDIES Available at: http://www.accessdata.fda.gov/scripts/

cdrh/cfdocs/cfcfr/CFRSearch.cfm?CFRPart=58&showFR=1&subpartNode=21:1.0.1.1.22.4. Accessed Mar 24, 2013.

USFDA-21CFR211.68. PART 211 – CURRENT GOOD MANUFACTURING PRACTICE FOR FINISHED PHARMACEUTICALS, Subpart D–Equipment, Sec. 211.68 Automatic, mechanical, and electronic equipment. Available at http://www.accessdata.fda.gov/scripts/cdrh/cfdocs/cfcfr/CFRSearch.cfm?fr=211.68. Accessed Mar 24, 2013.

Van Eeckhaut A, Lanckmans K, Sarre S, Smolders I, Michotte Y. Validation of bioanalytical LC–MS/MS assays: evaluation of matrix effects. J Chromatogr B 2009;877(23):2198–2207.

Veitch JD, Biddlecombe RA. Sample container with radiofrequency identifier tag, Google Patents. 2002.

Venn RF, Merson J, Cole S, Macrae P. 96-Well solid-phase extraction: a brief history of its development. J Chromatogr B 2005;817(1):77–80.

Villa JS, Cass RT, Karr DE, Adams SM, Shaw JP, Schmidt DE. Increasing the efficiency of pharmacokinetic sample procurement, preparation and analysis by liquid chromatography/tandem mass spectrometry. Rapid Commun Mass Spectrom 2004;18(10):1066–1072.

Vogelsanger M. Robots: future key elements in laboratory automation. Chemometr Intell Lab 1992;17(1):107–109.

Vogeser M, Kirchhoff F. Progress in automation of LC-MS in laboratory medicine. Clin Biochem 2011;44(1):4–13.

Walter RE, Cramer JA, Tse FLS. Comparison of manual protein precipitation (PPT) versus a new small volume PPT 96-well filter plate to decrease sample preparation time. J Pharm Biomed Anal 2001;25(2):331–337.

Wang, AQ, Fisher AL, Hsieh J, Cairns AM, Rogers JD, Musson DG. Determination of a β3-agonist in human plasma by LC/MS/MS with semi-automated 48-well diatomaceous earth plate. J Pharm Biomed Anal 2001;26(3):357–365.

Wang AQ, Zeng W, Musson DG, Rogers JD, Fisher AL. A rapid and sensitive liquid chromatography/negative ion tandem mass spectrometry method for the determination of an indolocarbazole in human plasma using internal standard (IS) 96-well diatomaceous earth plates for solid-liquid extraction. Rapid Commun Mass Spectrom 2002;16(10):975–981.

Wang, PG, Jun Z, Eric MG, et al. A high-throughput liquid chromatography/tandem mass spectrometry method for simultaneous quantification of a hydrophobic drug candidate and its hydrophilic metabolite in human urine with a fully automated liquid/liquid extraction. Rapid Commun Mass Spectrom 2006a;20(22):3456–3464.

Wang PG, Wei JS, Kim G, Chang M, El-Shourbagy T. Validation and application of a high-performance liquid chromatography–tandem mass spectrometric method for simultaneous quantification of lopinavir and ritonavir in human plasma using semi-automated 96-well liquid–liquid extraction. J Chromatogr A 2006b;1130(2):302–307.

Watt AP, Morrison D, Locker KL, Evans DC. Higher throughput bioanalysis by automation of a protein precipitation assay using a 96-well format with detection by LC–MS/MS. Anal Chem 2000;72(5):979–984.

Wells DA. High Throughput Bioanalytical Sample Preparation Methods and Automation Strategies. Amsterdam: Elsevier Science; 2003.

Wu S, Li W, Mujamdar T, Smith T, Bryant M, Tse FLS. Supported liquid extraction in combination with LC-MS/MS for high-throughput quantitative analysis of hydrocortisone in mouse serum. Biomedical Chromatography 2010;24(6):632–638.

Xie IH, Wang MH, Carpenter R, Wu HY. Automated calibration of TECAN genesis liquid handling workstation utilizing an online balance and density meter. ASSAY Drug Dev Technol 2004;2(1):71–80.

Xu N, Kim GE, Gregg H, et al. Automated 96-well liquid–liquid back extraction liquid chromatography–tandem mass spectrometry method for the determination of ABT-202 in human plasma. J Pharm Biomed Anal 2004;36(1):189–195.

Xu RN, Fan L, Rieser MJ, El-Shourbagy TA. Recent advances in high-throughput quantitative bioanalysis by LC–MS/MS. J Pharm Biomed Anal 2007a;44(2):342–355.

Xu S, Zheng S, Shen X, Yao Z, Pivnichny J, Tong X. Automated sample preparation and purification of homogenized brain tissues. J Pharm Biomed Anal 2007b;44(2):581–585.

Xu X, Zhou Q, Korfmacher WA. Development of a low volume plasma sample precipitation procedure for liquid chromatography/tandem mass spectrometry assays used for drug discovery applications. Rapid Commun Mass Spectrom 2005;19(15):2131–2136.

Xue YJ, Liu J, Pursley J, Unger S. A 96-well single-pot protein precipitation, liquid chromatography/tandem mass spectrometry (LC/MS/MS) method for the determination of muraglitazar, a novel diabetes drug, in human plasma. J Chromatogr B 2006;831(1–2):213–222.

Yadav M, Contractor P, Upadhyay V, et al. Automated liquid–liquid extraction based on 96-well plate format in conjunction with ultra-performance liquid chromatography tandem mass spectrometry (UPLC–MS/MS) for the quantitation of methoxsalen in human plasma. J Chromatogr B 2008;872(1–2):167–171.

Yang, AY, Sun L, Musson DG, Zhao JJ. Application of a novel ultra-low elution volume 96-well solid-phase extraction method to the LC/MS/MS determination of simvastatin and simvastatin acid in human plasma. J Pharm Biomed Anal 2005;38(3): 521–527.

Yang AY, Sun L, Musson DG, Zhao JJ. Determination of M + 4 stable isotope labeled cortisone and cortisol in human plasma by μElution solid-phase extraction and liquid chromatography/tandem mass spectrometry. Rapid Commun Mass Spectrom 2006;20(2):233–240.

Yang L, Clement RP, Kantesaria B, et al. Validation of a sensitive and automated 96-well solid-phase extraction liquid chromatography–tandem mass spectrometry method for the determination of desloratadine and 3-hydroxydesloratadine in human plasma. J Chromatogr B 2003;792(2):229–240.

Yang L, Wu N, Clement RP, Rudewicz PJ. Validation and application of a liquid chromatography–tandem mass spectrometric method for the determination of SCH 211803 in rat and monkey plasma using automated 96-well protein precipitation. J Chromatogr B 2004;799(2):271–280.

Ye Z, Tsao H, Gao H, Brummel CL. Minimizing matrix effects while preserving throughput in LC–MS/MS bioanalysis. Bioanalysis 2011;3(14):1587–1601.

Zhang H, Henion J. Quantitative and qualitative determination of estrogen sulfates in human urine by liquid chromatography/tandem mass spectrometry using 96-well technology. Anal Chem 1999;71(18):3955–3964.

Zhang J, Reimer MT, Nicholas EA, Qin CJ, Tawakol AE-S. Method development and validation for zotarolimus concentration determination in stented swine arteries by liquid chromatography/tandem mass spectrometry detection. Rapid Commun Mass Spectrom 2006;20(22):3427–3434.

Zhang J, Wei S, Ayres DW, Smith HT, Tse FLS. An automation-assisted generic approach for biological sample preparation and LC–MS/MS method validation. Bioanalysis 2011;3(17):1975–1986.

Zhang N, Hoffman KL, Li W, Rossi DT. Semi-automated 96-well liquid–liquid extraction for quantitation of drugs in biological fluids. J Pharm Biomed Anal 2000a;22(1):131–138.

Zhang N, Rogers K, Gajda K, Kagel JR, Rossi DT. Integrated sample collection and handling for drug discovery bioanalysis. J Pharm Biomed Anal 2000b;23(2):551–560.

Zhang N. Quantitative analysis of simvastatin and its β-hydroxy acid in human plasma using automated liquid–liquid extraction based on 96-well plate format and liquid chromatography-tandem mass spectrometry. J Pharm Biomed Anal 2004;34(1):175–187.

Zhang Y, Gu Z, Tweed JA. The automated analysis of 2-dimensional barcode toxicokinetic study samples via protein precipitation. Int Drug Discov 2010;(October/November 2010):66–73.

Zhao JJ, Xie IH, Yang AY, Roadcap BA, Rogers JD. Quantitation of simvastatin and its β-hydroxy acid in human plasma by liquid–liquid cartridge extraction and liquid chromatography/tandem mass spectrometry. J Mass Spectrom 2000;35(9):1133–1143.

Zweigenbaum J, Heinig K, Steinborner S, Wachs T, Henion J. High-throughput bioanalytical LC/MS/MS determination of benzodiazepines in human urine: 1000 samples per 12 hours. Anal Chem 1999;71(13):2294–2300.

23

LC-MS BIOANALYSIS OF DRUGS IN TISSUE SAMPLES

Hong Gao, Stacy Ho, and John Williams

23.1 INTRODUCTION

Analysis of drugs and their metabolites in tissue (commonly referred as tissue bioanalysis) plays a vital role in the pharmaceutical industry from discovery and development of new drug candidates to therapeutic monitoring of marketed drugs. The analysis result is pivotal for confirming target engagement of drug candidate for understanding its efficacy, safety, and distribution, as well as for establishing pharmacokinetics (PK)–pharmacodynamics (PD) relationship. Monitoring drugs in human tissues allow safe dosing of individualized and target-specific medicines with narrow therapeutic index (Noll et al., 2011).

Liquid chromatography–tandem mass spectrometry (LC-MS/MS) is commonly used for the analysis of drugs and metabolites in tissues and other biofluids due to its exquisite sensitivity and selectivity. However, to representatively extract the analytes of interest for LC-MS/MS analysis, samples must be first processed into a liquid format. While it is a relatively simple task for the bioanalysis of plasma, urine, and other biofluids, it is much more challenging for the bioanalysis of solid tissues. One of the major hurdles specific to tissue bioanalysis is tissue homogenization, which is the process of transforming a solid sample into a liquid form that is compatible with commonly used analyte extraction procedures. Tissue homogenization is a tedious and labor intensive process. If it is not operated under carefully controlled conditions, the homogenization process may also impact quantitation accuracy and analyte stability. Another major challenge is that it is difficult (if not impossible) to prepare standards and quality control samples that resemble the study samples since the spiked samples will not have the same drug/tissue binding as well as intra- and intercellular distribution in tissues as the incurred samples. Additionally, the residual blood, which may contain drugs and metabolites at different concentration levels from the tissue samples, will affect the quantitation accuracy, precision, and reproducibility.

Because of the unique challenges of tissue bioanalysis, developing and qualifying a tissue analysis method can be time-consuming. To balance efficiency and data quality, a "fit-for-purpose" strategy is recommended when establishing LC-MS/MS method for tissue bioanalysis.

23.2 CLASSIFICATION OF TISSUES

A tissue is comprised of different cell types that work together to carry out a specific function. Animal tissues are classified into four basic types, that is, epithelial, connective, muscular, and nervous tissues (William, 1972). Epithelial tissue typically consists of single to multiple layers of cells. They form the skin surface, linings of cavities and tubes, many glands, and surfaces of internal organs. Connective tissue is composed primarily of fibrous cells. They form tendons, inner layers of skin, and the connective framework of fibers in muscles, capsules, ligaments around joints, cartilage, and bone. Connective fibrous tissues are elastic and strong. Muscle tissue is divided into three subtypes: skeletal muscle, smooth muscle, and cardiac muscle. Skeletal muscle is made of striated long muscle fibers, which are held together by connective tissue. Smooth muscle is the major component of the inner walls of blood vessel, lymphatic vessel, urinary bladder, uterus, reproductive tract, gastrointestinal tract,

Handbook of LC-MS Bioanalysis: Best Practices, Experimental Protocols, and Regulations, First Edition. Edited by Wenkui Li, Jie Zhang, and Francis L.S. Tse.

TABLE 23.1 Tissue Classification Based Upon Physical Characteristics

Classification	Tissue and organ
Soft	Adipose, adrenal, brain, kidney, liver, lung, ocular tissue (eyes, aqueous humor, conjunctiva, choroid, cornea, iris/ciliary body, retina, sclera, vitreous humor), pancreas, parathyroid, pituitary, prostate, salivary, spinal cord, spleen, thymus, thyroid, tonsils
Tough	Artery, bladder, colon, endothelium, esophagus, fallopian tube, heart, intestine, lymph nodes, mammary gland, ovary, placenta, stent, stomach, skeletal muscle, testes, tumor, ureter, uterus
Hard	Bone, bone marrow, cartilage, hair, hard skin, nail

respiratory tract, arrector pili of skin, the ciliary muscle, and the iris of the eye. Cardiac muscle is a fatigue resistant muscle found in the heart. Nervous tissue is the main tissue comprising the brain, spinal cord, and nerves. When several different tissue types work together to perform a specific function, this functional unit is called an organ. Based upon the physical properties, most solid tissues and organs can be classified as soft, tough (resistant to force, yet flexible or stringy), or hard (resistant to force, inflexible or brittle), as shown in Table 23.1. The physical classification of a tissue often dictates the appropriate homogenization strategy required for LC-MS/MS bioanalysis.

23.3 WORKFLOW

Figure 23.1 outlines the workflow of LC-MS/MS bioanalysis of drugs in tissue. In the following sections, the current practices, their scientific rationale, and challenges are discussed for each step.

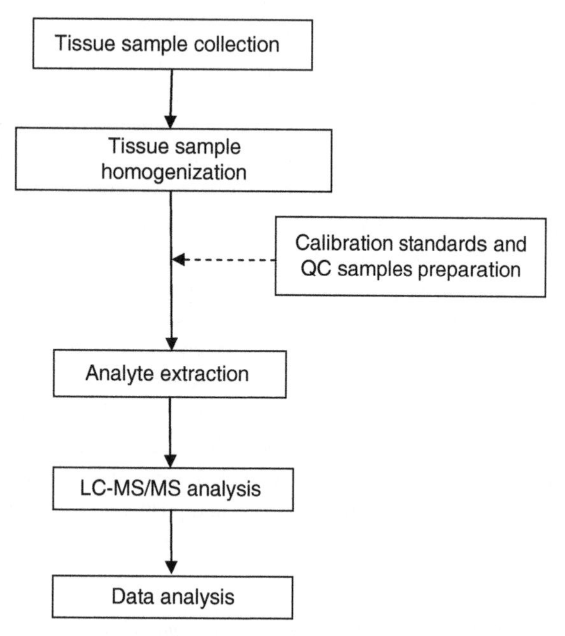

FIGURE 23.1 Tissue sample analysis workflow.

23.4 TISSUE SAMPLE COLLECTION

Tissue sample collection is critical for successful bioanalysis. Tissues must be handled properly and in a timely manner to avoid contamination and to prevent potential chemical, photo, or enzymatic degradation of the analyte of interest (Espina et al., 2009). Improper handling can result in metabolic profile changes (e.g., acylcarintines, Petucci et al., 2011) and altered analyte levels from the *in vivo* state.

Procedures for tissue sample collection and postcollection operations should be predetermined. Important questions such as what tissues are to be collected, when the tissues should be collected (time points), and whether whole or partial tissues are to be obtained for the analysis, should be addressed. If partial tissues are to be collected, sampling location (edge or center) and size of the partial tissues should be specified. The procedure should also detail the postcollection treatment if applicable. Any deviations from the predetermined tissue sample collection procedure should be documented.

There are two common approaches to collect tissue samples: postmortem tissue collection and *in vivo* biopsy. Postmortem sample collection is conducted by removing tissues from the subjects after euthanization. It is commonly used for preclinical animal studies. Whole tissues are usually collected for small tissues and organs, such as brain, heart, liver, lung, kidney, spleen, stomach, and lymph nodes from small animals (rat and mice), whereas partial tissues are usually collected for large tissues, such as skin, muscle, and hair. Partial tissue collection is also common for large animals (dog and monkey). *In vivo* biopsy is an invasive method for sampling tissues from living subjects. It is widely used for clinical diagnostics.

Tissue perfusion indicates the level of blood flow between the capillaries and the tissue vasculature. Perfusion levels differ among organs, with fat and brain are considered poorly perfused while heart, kidney, and liver are considered highly perfused. Immediately prior to tissue harvesting, blood is often replaced with a saline solution (Fenyk-Melody et al., 2004). Drug levels in saline perfused tissue will more accurately reflect the true concentration within tissue that is unadulterated with blood. The magnitude of concentration

TABLE 23.2 Currently Available Techniques for Tissue Sample Homogenization

Category	Technique	Apparatus	Mechanism
Homogenization	Grinding	Dounce Homogenizer (Fisher) and Potter-Elvehjem Homogenizer (Corning)	Disrupt tissues by mechanical force
	Cryogenic grinding	Freezer mill (SPEX CertiPrep) and CryoPrep (Covaris)	Disrupt tissues by pulverizing frozen tissues
	Rotor–stator homogenization	Waring Blender (Waring), Polytron homogenizer (Polytron), Tomtec homogenizer, and Autogizer (Tomtec)	Disperse tissues by grinding tissues with rotating blades
	Bead beating	Precellys-24 (Precellys), FastPrep-24 (MP Biomedicals), and Omni Bead Ruptor-24 (Omni)	Disrupt tissues by agitating tissues against beads at high speed
	Acoustic cell disruption	Covaris E-Series (Covaris)	Shear tissues by high frequency and energy waves
Digestion	Chemical digestion	Acid Digestion Vessels (Parr Instrument)	Break down tissue structures by acid or base
	Enzymatic digestion	Enzyme Kits (Worthington Biochemical Corporation)	Digest tissues with enzymes
Direct extraction	Accelerated solvent extraction	ASE200 and ASE350 Extractor (DIONEX and Applied Separations)	Extract analytes under controlled pressure and temperature
	Pressure cycling technology	Barocycler NEP3229 with PULSE™ Tubes (Boston Biomedica)	Extract analyte under controlled pressure and temperature

Reproduced from Xue et al. (2012) with permission of Future Science Ltd.

differences between blood-perfused and saline-perfused tissue will depend on the level of perfusion in the tissue and the distribution of the analyte. Perfusion may not be necessary for all tissues in the early drug discovery process but may be beneficial to compounds in the late stage lead optimization, especially for candidates with low tissue penetrating ability (Fenyk-Melody et al., 2004). Perfused/nonperfused drug concentrations will show the greatest difference in highly perfused tissues such as the heart (Gao, 2012; Ho, 2012).

After the tissues are removed from the animal, they should be rinsed with cold saline to remove any blood on the surface of the tissue and blotted dry with lint-free paper (Jia et al., 2010; Oliveira et al., 2011). The tissue weight is important in determining the amount of solvent needed for tissue homogenization or digestion and in calculating total drug and metabolite in the tissue. For efficient analysis, tissue weight should be recorded prior to storage. Tissue samples should be stored in appropriate containers (such as tissue homogenization containers) and labeled properly. Typical information on a label includes study ID, subject ID, tissue type, and collection time point.

For analytes that are chemically or enzymatically unstable, tissues need to be treated immediately after collection. Snap-freezing the sample either in liquid nitrogen or on dry ice and storage at −80°C is commonly employed to minimize analyte degradation by enzymatic activity. Immediate homogenization of tissue in a suitable pH adjusted buffer or with added enzyme inhibitors are also ways to minimize analyte degradation.

23.5 TISSUE SAMPLE PREPARATION

Drug and metabolite in tissue, either inside the cells (intracellular) or in the medium between cells (extracellular), must be accessible by the extraction solvent. To achieve this, solid tissue samples are processed via mechanical homogenization, ultrasonic disruption, and chemical or enzymatic digestion (Table 23.2). Among these techniques rotor–stator homogenization, bead beating and acoustic disruption are widely employed in the bioanalysis laboratories.

23.5.1 Rotor–Stator Homogenization

Rotor–stator homogenizer has a stationary cylinder (stator) and an internal blade (rotor or generator) that is rotated at high speed to finely chop tissues. The rotor–stator homogenizer usually processes a tissue sample in less than 30 seconds and is compatible with a wide variety of generator probes to satisfy application requirements. It is ideal for grinding tough tissues like heart and skeleton muscle (Yu et al., 2011). Polytron, Omni, TomTec, and other brands like Tissumizer by TekMar, Tissue Tearor by BioSpec, and Powergen by Fisher are commonly used. The handheld devices are available in many sizes that can accommodate a wide range of volumes. The efficiency of handheld homogenizer is low because only one sample can be processed at a time and thorough cleaning is required between samples. To overcome these weaknesses, some handheld models are equipped with disposable generators (e.g., Omni plastic blender). Automated devices are available to process multiple samples either with consumable probes (e.g., Omni Prep Multi-Sample Homogenizer) or with automated probe wash programs (e.g., Tomtec Autogizer, Hamden, Connecticut, USA). However, it should be noted that fibrous material caught in the rotor may require manual cleaning (Yu and Cohen, 2004). As the homogenization process generates heat, it is generally a good practice to keep samples on ice to reduce thermal degradation.

23.5.2 Bead Beating

Bead beating tissue homogenization is achieved through the high speed shaking of a tissue sample against small hard beads inside the sealed sample vial. The high speed and specific motion of the beads produce efficient grinding and can process most tissue types, including difficult samples such as tumor, muscle, heart, bone, skin, hair, cartilage, and more. Bead beating devices can also dry grind snap-frozen tissues. The duration and speed of agitation are programmable for most devices. The typical homogenization time is 30 second or less for soft tissues. Since the process is brief, the heat generated during tissue processing usually does not raise the sample temperature significantly. Beads made of ceramic or metal are commonly used. Depending on the characteristics of the tissue, a suitable bead material and an agitation program that controls duration and speed should be developed prior to the sample processing (Gao et al., 2007; Germann and Powell; 2011). Different vial sizes (i.e., 2, 7, 15, and 50 ml) can be selected to accommodate various sizes of tissue. The automated devices can easily process up to 24 samples in one run (e.g., Precellys-24, Omni Bead Ruptor-24, and Fastprep-24 beads beater) when samples are smaller than 500 mg and 2-ml vial is used. For samples larger than 500 mg, larger vials (7-ml) are available but the throughput is decreased accordingly. To prevent thermal degradation, a cooling system is used to maintain the temperature in the chamber where sample tubes are placed (e.g., Precellys-24 and Fastprep-24 beads beater). One advantage of using the bead-beating vial is that tissue samples can be directly collected in the pretared vials with selected beads, which eliminates the sample transfer step. Single use of disposable vials eliminates sample cross-contamination.

23.5.3 Acoustic Disruption

Strong acoustic energy directed at a tissue can cause pressure disturbances and intense vibration resulting agitating and lysing of tissue cells immersed in the liquid suspension. Depending on the acoustic energy and how energy is directed, three types of devices [sonication, ultrasonication, and adaptive focused acoustics (AFA)] are commonly available. Sonication ($\pm$ 1 kHz) is best suited for soft tissues like brain and ocular tissue (Lehner et al., 2011; Jiang et al., 2009). Since the energy may not always correctly be scaled to the sample, sonication may cause "hot spots" and degrade thermally labile compounds. Ultrasonication oscillates at 20 kHz and can be adapted to process several samples at the same time. Small and large ultrasonic probes are available, allowing a variety of sample volumes to be processed. To prevent excessive heating generated during the process, ultrasonic treatment is applied in multiple short bursts to the sample immersed in the coldwater bath. AFA is a high frequency ($\pm$ 1 MHz) acoustic-based process with more controllable wavelengths to avoid "hot spots" and compound degradation. With this advantage, a 96-well plate of samples can be easily processed in one batch. The process can be run isothermally and convergence of the mechanical energy is controlled by the program. A shorter wavelength, coupled with controlled wave packets, enables the energy to traverse materials, such as sample tubes, prior to achieving the peak energy density inside of the sample tube (e.g., Covaris E-series). The reduction of heat damage using isothermal processing can improve the analyte recovery (Oliveira et al., 2011). Due to the precisely controlled energy delivery, the process is standardized and results are reproducible. With AFA there is no direct contact with sample, no cleanup is needed, and no aerosol is formed during the sample handling and process, which can prevent cross-contamination. Takach et al. (2004) evaluated the AFA system with rat heart, liver, and muscle tissues and demonstrated results comparable to Polytron homogenization.

23.5.4 Other Techniques

23.5.4.1 Grinding Grinding tissue using a traditional mortar and pestle is the simplest technique to homogenize tissue samples. Multiple sizes of mortar and pestles made of a variety of different materials are available. Manually grinding tissue is labor and time demanding but is a simple and inexpensive choice for processing a small number of samples. Grinding tissue using a cryogenic grinder is a great alternative for thermally labile compounds. Freeze mill (SPEX CertiPrep) is a cryogenic grinding device using small magnetic bars to pulverize the sample in a chamber full of liquid nitrogen where low temperature (−196°C) are maintained. Cryogenic grinding is an ideal tool for hard and fibrous tissues like bone, teeth, hair, tumor, cartilage, heart, and skin. The efficiency of the grinding is improved when the sample size is small. This can be achieved by precutting the sample into smaller and more uniform pieces, or grinding for additional cycles. For instance, it is suggested to cut bone into small pieces approximately 5 mm in diameter before grinding (Ji et al., 2008). The cleanup of the components of freezer mill vials is labor intensive and the use of disposable vials is recommended.

23.5.4.2 Digestion Hard and strong fibrous tissues like bone, cartilage, and hair are difficult to homogenize when using mechanical techniques. Chemical and enzymatic digestion is often used to break down the fibrous structure. Mineral acids (HCl, HNO_3, HF, H_2SO_4), base (KOH, NaOH), and hydrogen peroxide are often used as the reagents for chemical digestion. The selection of reagents for chemical digestion depends on the tissues and the analyte stability under the digestion conditions (Henderson et al., 1995). Bone can be digested in concentrated nitric acid and nitric–perchloric acid (AOAC, 1990), and heated to different temperatures. Hair can be digested either by NaOH (Favretto et al., 2006; Marchei et al., 2005) or HCl (Girod and Staub, 2000; Heimbuck and Bower, 2002). Enzymatic digestion has recently become more commonly used for analysis of drug in tissue (Yu et al., 2004). Based on the type of tissue, a range of enzymes are available with different digestive properties and efficiencies. Collagenase, proteinase K, DNAse, elastase, glucuronidase-aryl-sulfatase, hyaluronidase, papain, and lipase are the most common enzymes used for tissue digestion. The choice of enzyme is driven by the desired tissue or component to be digested.

23.5.4.3 Direct Extraction Accelerated solvent extraction (ASE) and pressure cycling technology (PCT) are two types of technologies developed and applied to environmental and animal tissue sample preparation in the food industry. The applications of ASE and PCT for pharmaceutical analysis of large animal tissue have been reported (DIONEX application note 358). Based on the sample size, ASE and PCT can selectively disrupt membrane structures of tissue samples. For soft animal tissue, usually 50–150 mg of sample is placed inside a Pulse tube and subjected to a suitable pressure-cycling regimen. For tough or fibrous tissue such as animal muscle and skin, the sample may be pretreated with the PCT shredder directly in the tube.

23.6 CALIBRATION STANDARD AND QC SAMPLE PREPARATION

Calibration standards and QC samples can be prepared using matching blank tissue matrix or surrogate matrix. A matching blank tissue matrix is the drug-free matrix of the same tissue and species as that of the sample and it is referred to as "authentic" blank matrix by some scientists.

23.6.1 Matching Blank Tissue Matrix

To mimic the matrix of the study (incurred) samples, matching blank tissue matrix is preferred for the preparation of calibration standards and QC samples. Animals from the control group under the same husbandry environment are the ideal source. The prescreening of blank matrix to ensure they are free of endogenous analyte is recommended for tissue analysis method qualification or validation (Xue et al., 2012a). To reduce animal usage, it is suggested that only a minimum amount of blank tissue be collected based on the size of the analytical batch and numbers of standards and QC samples that are needed. Alternatively, the blank tissues can be obtained from commercial sources.

23.6.2 Surrogate Matrix

Surrogate matrix can be used for the preparation of calibration standards and QC samples in situations, where (1) authentic blank tissue is not available or insufficient; or (2) there are multiple types of tissues to be analyzed and each type of tissue has limited number of samples. For the latter, rather than preparing multiple calibration curves in each individual blank tissue, all tissue samples can be quantified using calibration standards and QCs prepared in surrogate matrix. It is worth noting that QC samples should be prepared in both surrogate matrix and matching blank tissue whenever possible, when conducting assay qualification using surrogate matrix. QC samples in the surrogate matrix provide the information on accuracy and precision of analyte in surrogate matrix, while QC samples in matching blank tissue matrix mimic the sample analysis accuracy and precision. Blank plasma from the same species is often used to serve as surrogate matrix. Other matrices such as physiological pH buffer also can be used as the surrogate matrix. Whenever the surrogate matrix is used, it is always a good practice to make sure the internal standard response cross the whole sample run (batch) is consistent; otherwise the surrogate matrix may be not appropriate for the tissue analysis under the defined method conditions (Gao, 2012).

23.6.3 Calibration Standard and QC Preparation

Calibration standards and quality control samples are normally prepared by spiking the analyte into the prescreened blank matrix. To mimic study samples as closely as possible, the percentage (v/v) of the spiking solution of analyte should be minimal (less than 5% of the total volume of the tissue homogenates). Analyte spiking solution is normally prepared in acetonitrile, methanol or 50:50 methanol:water. The analyte (drug and metabolite) can be spiked into the intact tissue before homogenization (Jimenez-Diaz et al., 2010) or into the homogenate after the tissue homogenization (Scheidweiler et al., 2008).

23.7 ANALYTE EXTRACTION

Once tissue sample is disrupted or digested, the analyte is ready to be extracted from the tissue homogenate or digestion mixture for LC-MS/MS bioanalysis. The details of biological sample extraction for LC-MS/MS analysis captured in other chapters are also applicable to tissue bioanalysis. Unlike liquid matrices such as plasma and serum, tissue homogenate mainly contains large quantities of cellular membrane particles that may affect the extraction process.

Protein precipitation by organic solvent, for example, acetonitrile, methanol or a mixture of acetonitrile and methanol, is widely applied by scientists (Wang et al., 2011). Depending on the pKa of the analyte, the pH of solvent may need to be adjusted to improve the precipitation efficiency and/or analyte stability. Liquid–liquid extraction (LLE) (Blum et al., 2008) and solid phase extraction (SPE) (Xu et al., 2011) are the two most common methods for tissue sample extraction. Salting-out–assisted LLE (Xiong et al., 2011) and supported-liquid extraction (Nguyen et al., 2010) are also effective extraction methods. When compared to protein precipitation, these methods usually yield cleaner extracts for tissue sample LC-MS/MS bioanalysis. Caution may be needed when extracting the mixture after digestion either with acid/base or enzymes. Chemical digestion employs strong acid or alkaline conditions and the reaction is stopped by neutralization. The salts that remain in the mixture can strongly affect analyte ionization during LC-MS/MS analysis if they are not removed from the extract. To remove interferences effectively, multiple extraction methods may be considered. For insistence, protein precipitation followed by either LLE (Ahmadkhaniha et al., 2009) or SPE (Korecka et al., 2010) extraction. Other extraction methods such as matrix solid-phase dispersion (Priyanthi et al., 2009) and solid-phase microextraction (Poli et al., 2009) to prepare tissue samples can make the analyte(s) extraction much easier due to their unique mechanisms. Still, further cleanup may be necessary to eliminate matrix effects as in the case of mycotoxin analysis in fish tissue (Laganà et al., 2003).

The level of drug/protein binding in tissue homogenate will affect the free fraction of analyte and should be factored into analyte concentration calculations if nondenatured sample extract is analyzed directly (Spoto et al., 2006; Wang et al., 2008). The bound fraction of analyte will be released upon denaturing the protein and analysis will represent the total analyte concentration in the sample. Therefore, at a minimum, it is recommended that the tissue homogenate be precipitated with organic solvent prior to LC-MS/MS bioanalysis.

23.8 LC-MS/MS ANALYSIS

The quantitative bioanalysis of drugs in biological samples using LC-MS/MS are specifically discussed in other chapter. In many cases, LC-MS/MS parameters used for the analyte in plasma can be used as the starting point for the further optimization of an assay for a specific tissue matrix. It is worthwhile mentioning that, when investigating multiple tissues, such as a drug tissue distribution study, the chromatographic and mass spectrometric conditions may be different for each particular tissue matrix due to tissue composition and interferences.

Compared to plasma, tissue homogenate contains a complex mixture of cellular components with varying levels of phospholipids, leading to variable matrix effects. In a LBH589 mouse PK study, it was observed that matrix effect was measured at 16.96% for liver, 16.92% for lung, and 13.56% for kidney but only 9.92% for plasma (Estella-Hermoso de Mendoza et al., 2011). Direct injection of supernatant of tissue homogenate or protein precipitation extracts usually is associated with a greater matrix effect issue. Better performance is often observed when samples are extracted with LLE and SPE techniques. Many tissues are prepared with multiple techniques to yield cleaner extracts. For instance, when determining the concentration of polybrominated diphenyl ethers in human placenta, the tissue homogenate are extracted first with acetone/hexane (7:3, v/v) and then with hexane/MTBE (9:1, v/v) to avoid matrix effects (Priyanthi et al., 2009).

One issue that may affect assay reproducibility is the pipetting error due to homogenate heterogeneity caused by foaming during homogenization. Foam is usually generated when using rotor–stator homogenizer and bead beating technique. Antifoam agents may be added to the buffer prior to homogenization (Gao, 2012). Foam will break apart over time but the waiting period may be impractical. Centrifugation of the homogenate at high speed can easily break apart the foam but may also cause severe data inaccuracy and poor reproducibility due to the heterogeneous nature of the homogenate (Gao, 2012).

23.9 FIT-FOR-PURPOSE LC-MS/MS METHOD QUALIFICATION

Because the calibration standards and quality control samples for tissue analysis do not closely mimic the incurred tissue samples, the bioanalytical method may not be considered to be fully validated. Instead of "method validation", "method qualification" based on fit-for-purpose application is an appropriate term to describe the process of method establishment.

23.9.1 Recovery

Analyte recovery measures the percentage of analyte extracted from the sample matrix. The recovery can affect the quality of LC-MS/MS bioanalytical method. Poor recovery directly affects the method sensitivity in most cases. For tissue sample analysis, the analyte recovery is usually measured by spiking analyte into blank tissue homogenate. This approach does not truly mimic the distribution of analyte in real tissue samples. Therefore, different approaches should be incorporated into the LC-MS/MS assay development to minimize the risk related to recovery. Spiking analyte onto the blank tissue or injecting analyte into the blank tissue followed by tissue homogenization and extraction is another way to determine the analyte recovery. Determination of the absolute recovery by using incurred sample containing radiolabeled compound by comparing radioactivity levels prior to homogenization and following homogenization and extraction is the most rigorous approach but is the most difficult to conduct.

For liquid biological matrices such as plasma and urine, analyte recovery can be easily evaluated by comparing a known amount of analyte spiked into the matrix before and after the extraction. A similar approach can be applied to tissue analysis by spiking the analyte into the tissue homogenate. The result will be accurate if the analyte can equally and freely distribute in the tissue homogenate. However, this is not often known a priori and therefore assumes risk (Xue et al., 2012b). Spiking a known amount of analyte into intact tissues is another approach that is used for some assays (Ji et al., 2004; Zhang et al., 2006, 2007). However, this method cannot be applied to hard tissue like hair and bone and faces the same issues of uneven analyte distribution within the tissue. Recovery evaluation using incurred samples that are from radioisotope labeled and cold material codosed animals can be more accurate method of measurement (Safarpour et al., 2009). The drawback of this approach is that the total radioactivity represents both analyte and its metabolite(s), and therefore, an investigation of metabolite present in the tissue is required. Based on the potential spatial distribution differences in tissue, the true assessment of analyte recovery is a considerable challenge. Scientists should select the approach that is most likely to be used for the method qualification.

23.9.2 Stability

The analyte stability in stock solution and working solutions should be established. The analyte stability in tissue homogenate should also be investigated during the method development and qualification. Analyte is spiked into homogenate for the evaluation of the bench-top stability, short-term storage (at −20°C and/or −80°C), and autosampler stability. Although the established analyte stability in homogenate does not mimic the incurred tissue samples, such practices provide critical information on tissue sample collection, handling, and storage. They can also prevent unnecessary time spent on method development. If the surrogate matrix is used for standards, the analyte stability in surrogate (bench-top, short term, freeze-thaw, long-term) in addition to in tissue homogenate should be evaluated.

The stability of analytes in tissue sample can be important and should be evaluated. However, it is not often done due to technical and logistical difficulties. There are two commonly practiced approaches to prepare tissue samples containing analyte: one is to spike the analyte onto the tissues and the other is to inject the analyte into the tissue. The prepared tissue samples are then homogenized to determine the analyte stability. The types of stability determination in tissue are determined by the actual practice from sample collection to sample analysis. The bench-top stability should be verified to cover the duration from tissue collection to frozen storage, and the duration for the tissues waiting to be homogenized on the bench. In most case, tissues are collected and immediately frozen and the tissue homogenization occurs after a certain period of storage. Therefore, the freeze–thaw stability for at least one cycle and long-term frozen storage stability of the analyte in tissue should be evaluated.

Analyte stability in homogenate is usually evaluated as bench top and autosampler stability during method qualification. The storage stability of analyte in tissue is a challenge because the absolute amount of the analyte and its recovery from the tissue may not be known. Use of incurred tissue samples to establish the storage stability is a practical choice but the analyte homogeneity remains a concern.

23.9.3 Linearity, Sensitivity, Accuracy, and Precision

When developing a method for tissue analysis, the linearity, sensitivity, accuracy and precision of the method should be evaluated and meet the predefined acceptance criteria prior to analyzing study samples. A one-day qualification run should be sufficient for discovery and preclinical development project (Zhang et al., 2006; Sakurada and Ohta, 2010; Zhou and Gallo, 2010). For more critical studies, three or more runs may be conducted. As discussed in previous sections, when surrogate matrix is used to prepare the standards and QC samples, the accuracy and precision of QC samples in both surrogate matrix and in matching blank tissue matrix should be evaluated. Poor QC results in surrogate matrix typically points to poor performance of standards and leads to low-quality data. On the other hand, if QC results in surrogate matrix meet the acceptance criteria but in matching blank tissue matrix is poor, it could mean poor selection of surrogate matrix (poor accuracy results) or issues in analyte extraction (poor accuracy and/or precision results).

23.9.4 Qualification and Acceptance Criteria

There is no specific bioanalytical guidance for tissue sample LC-MS/MS bioanalysis from the regulatory authorities including the US FDA and European EMA. Given the variety and complexity of solid tissue matrices, a "fit-for-purpose" qualification strategy proposed for biomarker analysis (Lee et al., 2006) with wider acceptance criteria may be used when establishing a tissue analysis method (Breda et al., 2011; DeMuth et al., 2010; Garofolo et al., 2011; Savoie et al., 2010). In general, a qualified method is deemed sufficient for tissue sample bioanalysis in most cases unless the study is

TABLE 23.3 Proposed Tiered Method Establishment for Bioanalysis of Drugs in Tissue

	Exploratory	Qualified	Validated
Reference standard	No COA needed	COA may be needed	COA needed
Internal standard	Yes or No	Yes	Yes
Accuracy and precision runs	1	1-3	3
Selectivity	No	Yes	Yes
Levels of QCs	1	2-3	3-4
Blank matrix	Surrogate	Authentic if possible or surrogate	Authentic if possible
Matrix effect	No	Yes	Yes
Recovery	No	From homogenate	From homogenate and intact tissue if feasible
Stability	No	Homogenate	Homogenate and intact tissue if feasible
ISR	No	No	Yes
Acceptance criteria (may be modified based on study purpose)	4-6-30 (4 out of 6 QCs are within 30% of nominal concentration)	4-6-20 (4 out of 6 QCs are within 20% of nominal concentration)	4-6-15 (4 out of 6 QCs are within 15% of nominal concentration)

COA, certificate of analysis.
Reproduced from Xue et al. (2012) with permission of Future Science Ltd.

critical for addressing safety issues (Xue et al., 2012a). Based on the purpose of a tissue study, the extent of method qualification may be minimal in the discovery phase and increase with the desired critical outcome. A full qualification may be performed when primary PK/PD relationships in a particular tissue are the driving force in decision making. To guide current tissue bioanalysis method qualification on the basis of "fit-for-purpose," a three-level method establishment and acceptance criteria have been proposed (Xue et al., 2012a) as shown in Table 23.3.

23.10 DATA ANALYSIS AND REPORTING

Data analysis is conducted after the LC-MS/MS bioanalysis. The measured concentrations of drug and metabolites in tissue are calculated by taking into account the proper dilution factor used in sample preparation. The measured concentration unit is weight per weight, that is, ng/g or μg/g, for solid tissue (such as brain). The total amount of drug in a certain tissue is calculated by multiplying the measured concentration in tissue by the total tissue weight.

23.11 SUMMARY

Determination of drug and metabolite levels in biological tissues by LC-MS/MS is widely used to answer critical questions. Unlike liquid samples, the key challenge to tissue bioanalysis is the structural reduction of the tissue matrix into a liquid format from which the analyte of interest can be extracted. Based upon the size, number, and type of tissue, several techniques, including mechanical homogenization and chemical/enzymatic digestion, can be used. Given the complexity of tissue composition, additional efforts for characterizing the method's analytical figures of merit are often required to provide reliable LC-MS/MS results. Use of multiple extraction techniques and/or surrogate matrices provide practical solutions to tissue bioanalysis challenges. The unknown nature of drug tissue distribution *in vivo* complicates the assessment of method recovery and analyte stability. To address the changing requirements for tissue bioanalysis at different stages of drug discovery and development, the "fit-for-purpose" strategy is a valuable tool to guide bioanalytical scientists in selecting a suitable level of method qualification to support the desired study outcome.

ACKNOWLEDGMENT

The authors wish to extend sincere thanks to Drs Naidong Weng, Wenkui Li, and Jie Zhang for their comments and discussions.

REFERENCES

Ahmadkhaniha R, Shafiee A, Rastkari N, Kobarfard F. Accurate quantification of endogenous androgenic steroids in cattle's meat by gas chromatography mass spectrometry using a surrogate analyte approach. Analytica Chimica Acta 2009;631(1):80–86.

AOAC. AOAC Official Methods of Analysis. 15th ed., Arlington, Virginia: Association of Official Analytical Chemists, 1990; p 84–85.

Blum M, Dolnikowski G, Seyoum E, et al. Vitamin D-3 in fat tissue. Endocrine 2008;33(1):90–94.

Breda M, Garofolo F, Caturla MC, et al. The 3rd global CRO council for bioanalysis at the international reid bioanalytical forum. Bioanalysis 2011;3(24):2721–2727.

DeMuth JE, Hayes MJ, Amaravadi L, et al. Summary of the eleventh annual university of wisconsin land O'Lakes bioanalytical conference. Bioanalysis 2010;2(10):1677–1681.

DIONEX. Extraction of contaminants, pollutants, and poisons from animal tissue using accelerated solvent extraction (ASE). Application Note 359. On web http://www.dionex.com/en-us/events/market/2008/lp-75513.html.

Espina V, Mueller C, Edmiston K, Sciro M, Petricoin E, Liotta L. Tissue is alive: new technologies are needed to address the problems of protein biomarker pre-analytical variability Proteomics Clin Appl 2009;3(8):874–882.

Estella-Hermoso de Mendoza A, Imbuluzqueta I, Campanero MA, et al. Development and validation of ultrahigh performance liquid chromatography—mass spectrometry method for LBH589 in mouse plasma and tissues. J Chromatogr B 2011;879:3490–3496.

Favretto D, Frison G, Vogliardi S, Ferrara SD. Potentials of ion trap collisional spectrometry for liquid chromatography/electrospray ionization tandem mass spectrometry determination of buprenorphine and nor-buprenorphine in urine, blood and hair samples. Rapid Commun Mass Spectrom 2006;20:1257–1265.

Fenyk-Melody J, Shen X, Peng Q, et al. Comparison of the effects of perfusion in determining brain penetration (brain-to-plasma ratios) of small molecules in rats. Comp Med 2004;54(4):378–381.

Gao H, Thompson S, Verollet R, Duval A. New tool refines tissue homogenization. Genet Eng Biotechnol 2007;27(19): 28.

Gao H. Tissue sample LC-MS/MS bioanalysis: method development and considerations. Presentation at 13th Annual Land O'Lakes Bioanalytical Conference. Madison, WI, July 16–20, 2012.

Garofolo F, Rocci ML Jr, Dumont I, et al. Recent issues in bioanalysis and regulatory findings from audits and inspections. Bioanalysis 2011;3(18):2081–2096.

Germann M, Powell KD. Fast quantitation of biomarkers N-acetylaspartate and N-acetylaspartylglutamate in mouse brain homogenates using HILIC and tandem mass spectrometry. Presented at The 59th ASMS Conference on Mass Spectrometry and Allied Topics. Denver, CO, June 5–9, 2011.

Girod C, Staub C. Analysis of drugs of abuse in hair by automated solid-phase extraction, GC/EI/MS and GC ion trap/CI/MS. Forensic Sci Int 2000;107(1–3):261–271.

Heimbuck CA, Bower NW. Teaching experimental design using a GC-MS analysis of cocaine on money: a cross-disciplinary laboratory. J Chem Educ 2002;79(10):1254

Henderson GL, Harkey MR, Jones RT. Analysis of hair for cocaine. International Research on Standards and Technology 1995; NIH Publication No. 95-3727:91–120.

Ho S. Overview on quantitative of drugs in tissue. Presentation at 13th Annual Land O'Lakes Bioanalytical Conference, Madison, WI, July 16–20, 2012.

Ji AJ, Saunders JP, Amorusi P, et al. A sensitive human bone assay for quantitation of tigecycline using LC/MS/MS. J Pharm Biomed Anal 2008;48(3):866–875.

Ji QC, Zhang J, Reimer MT, Watson P, El-Shourbagy T. Method development for the quantitation of ABT-578 in rabbit artery tissue by 96-well liquid-liquid extraction and liquid chromatography/tandem mass spectrometric detection. Rapid Commun Mass Spectrom 2004;18:2293–2298.

Jia Y, Xie H, Wang G, et al. Quantitative determination of helicid in rat biosamples by liquid chromatography electrospray ionization mass spectrometry. J Chromatogr B Analyt Technol Biomed Life Sci 2010;878(9–10):791–797.

Jiang S, Chappa AK, Proksch JW. A rapid and sensitive LC/MS/MS assay for the quantitation of brimonidine in ocular fluids and tissues. J Chromatogr B 2009;877:107–114.

Jimenez-Diaz I, Zafra-Gomez A, Ballesteros O et al. Determination of bisphenol a and its chlorinated derivatives in placental tissue samples by liquid chromatography-tandem mass spectrometry. J Chromatogr B Analyt Technol Biomed Life Sci 2010;878(32):3363–3369.

Korecka M, Clark CM, Lee VMY, Trojanowski JQ, Shaw LM. Simultaneous HPLCMS-MS quantification of 8-iso-PGF (2 alpha) and 8,12-iso-iPF(2 alpha) in CSF and brain tissue samples with on-line cleanup. J Chromatogr B Analyt Technol Biomed Life Sci 2010;878(24):2209–2216.

Laganà A, Bacaloni A, Castellano M, et al. Sample preparation for determination of macrocyclic lactone mycotoxins in fish tissue, based on on-line matrix solid-phase dispersion and solid-phase extraction cleanup followed by liquid chromatography/tandem mass spectrometry. J AOAC Int 2003;86(4):729–736.

Lee JW, Devanarayan V, Barrett YC, et al. Fit-for-purpose method development and validation for successful biomarker measurement. Pharm Res 2006;23(2):312–328.

Lehner A, Johnson M, Simkins T et al. Liquid chromatographic-electrospray mass spectrometric determination of 1-methyl-4-phenylpyridine (MPP(+)) in discrete regions of murine brain. Toxicol Mech Methods 2011;21(3):171–182.

Marchei E, Durgbanshi A, Rossi S, Garcia-Algar O, Zuccaro P, Pichini S. Determination of arecoline (areca nut alkaloid) and nicotine in hair by high-performance liquid chromatography quadrupole mass spectrometry. Rapid Commun Mass Spectrom 2005;19:3416–3418.

Nguyen L, Zhong WZ, Painter CL, Zhang C, Rahavendran SV, Shen ZZ. Quantitative analysis of PD 0332991 in xenograft mouse tumor tissue by a 96-well supported liquid extraction format and liquid chromatography/mass spectrometry. J Pharm Biomed Anal 2010;53(3):228–234.

Noll BD, Coller JK, Somogyi AA, et al. Measurement of cyclosporine A in rat tissues and human kidney transplant biopsies—A method suitable for small (<1 mg) samples. Ther Drug Monit 2011;33(6):688-693.

Oliveira LT, Garcia GM, Kano EK, Tedesco AC, Mosqueira VCF. HPLC-FLD methods to quantify chloroaluminum phthalocyanine in nanoparticles, plasma and tissue: application in pharmacokinetic and biodistribution studies. J Pharm Biomed Anal 2011;56(1):70–77.

Petucci C, Rojas-Betancourt S, Gardell SJ. Comparison of tissue harvest protocols for the quantitation of acylcarnitines in mouse heart and liver by mass spectrometry. Presented at The 59th

ASMS Conference on Mass Spectrometry and Allied Topics, Denver, CO, June 5–9, 2011.

Poli D, Caglieri A, Goldoni M, Coccini T, Roda E, Vitalone A. Single step determination of PCB 126 and 153 in rat tissues by using solid phase microextraction/gas chromatography—mass spectrometry: comparison with solid phase extraction and liquid/liquid extraction. J Chromatogr B 2009;877: 773–783.

Priyanthi RA, Dassanayake S, Wei H, Chen RC, Li A. Optimization of the matrix solid phase dispersion extraction procedure for the analysis of polybrominated diphenyl ethers in human placenta. Anal Chem 2009;81:9795–9801.

Safarpour H, Connolly P, Tong X, Bielawski M, Wilcox E. Overcoming extractability hurdles of a 14C labeled taxane analogue milataxel and its metabolite from xenograft mouse tumor and brain tissues. J Pharm Biomed Anal 2009;49:774–779.

Sakurada K, Ohta H. Liquid chromatography-tandem mass spectrometry method for determination of the pyridinium aldoxime 4-PAO in brain, liver, lung, and kidney. J Chromatogr B Analyt Technol Biomed Life Sci 2010;878(17–18):1414–1419.

Savoie N, Garofolo F, van Amsterdam P, et al. Recent issues in regulated bioanalysis & global harmonization of bioanalytical guidance. Bioanalysis 2010;2(12):1945–1960.

Scheidweiler KB, Barnes AJ, Huestis MA. A validated gas chromatographic electron impact ionization mass spectrometric method for methamphetamine, methylenedioxymethamphetamine (MDMA), and metabolites in mouse plasma and brain. J Chromatogr B Analyt Technol Biomed Life Sci 2008;876(2):266–276.

Spoto B, Fezza F, Parlongo G, et al. Human adipose tissue binds and metabolizes the endocannabinoids anandamide and 2-arachidonoylglycerol. Biochimie 2006;88(12):1889–1897.

Takach EJ, Zhu Q, Yu S, Qian M, Hsieh F. New Technology in tissue homogenization: using focused acoustic energy to improve extraction efficiency of drug compounds prior to LC/MS/MS analysis (poster). 52nd ASMS Conference on Mass Spectrometry, May 26, 2004. Poster: WPJ 143.

Wang C, Wang S, Chen Q, He L. A capillary gas chromatography-selected ion monitoring mass spectrometry method for the analysis of atractylenolide I in rat plasma and tissues, and application in a pharmacokinetic study. J Chromatogr B Analyt Technol Biomed Life Sci 2008;863(2):215–222.

Wang LL, Liu ZH, Liu DH, Liu CX, Juan Z, Zhang N. Docetaxel-loaded-lipid-based nano suspensions (DTX-LNS): preparation, pharmacokinetics, tissue distribution and antitumor activity. Int J Pharm 2011;413(1-2):194–201.

William T. Keeton, Biological Science. 2nd ed. W.W. Norton & Company Inc.; 1972. p 85.

Xiong H, Bi H, Wen Y, et al. Determination of SYUIQ-F5, a novel telomerase inhibitor and anti-tumor lead-compound, in tissues and plasma samples by LCMS-MS: application to a tissue distribution study in rat. Chromatographia 2011;73(11–12):1073–1080.

Xu N, Qiu C, Wang W, et al. HPLC/MS/MS for quantification of two types of neurotransmitters in rat brain and application: Myocardial ischemia and protection of Sheng-Mai-San. J Pharm Biomed Anal 2011;55(1):101–108.

Xue YJ, Gao H, Ji QC, et al. Bioanalysis of drug in tissue: current status and challenges. Bioanalysis 2012a;4(21):2637-2653.

Xue YJ, Melo B, Vallejo M, et al. An integrated bioanalytical method development and validation approach: case studies. Biomed Chrom 2012b; Doi:10.1002/bmc.2682.

Yu C, Cohen LH. Tissue sample preparation — not the same old grind. LCGC Europe 2004;17(2):96–101.

Yu C, Penn LD, Hollembaek J, Li W, Cohen LH. Enzymatic tissue digestion as an alternative sample preparation approach for quantitative analysis using liquid chromatography-tandem mass spectrometry. Analytical Chem 2004;76(6):1761–1767.

Yu D, Rummel N, Shaikh B. Development of a method to determine albendazole and its metabolites in the muscle tissue of yellow perch using high-performance liquid chromatography with fluorescence detection. J AOAC Int 2011;94(2):446–452.

Zhang J, Reimer MT, Alexander NE, et al. Method development and validation for zotarolimus concentration determination in stented swine arteries by liquid chromatography/tandem mass spectrometry detection. Rapid Commun Mass Spectrom 2006;20:3427–3434.

Zhang J, Reimer MT, Ji QC, et al. Accurate determination of an immunosuppressant in stented swine tissues with LC–MS/MS. Anal Bioanal Chem 2007;387:2745–2756.

Zhou Q, Gallo JM. Quantification of sunitinib in mouse plasma, brain tumor and normal brain using liquid chromatography-electrospray ionization-tandem mass spectrometry and pharmacokinetic application. J Pharm Biomed Anal 2010;51(4):958–964.

24

LC-MS BIOANALYSIS OF DRUGS IN URINE

ALLENA J. JI

24.1 INTRODUCTION

Information of urinary drug excretion is important for understanding the route of its metabolism and elimination. Most of preclinical urine sample quantization methods are based on the measurement of radiation activities on the radio-labeled compounds, while the majority of clinical urine assays are based on liquid chromatography–mass spectrometry (LC-MS) measurements. As per the current pharmaceutical industrial practice, if urine excretion of a drug candidate and/or its active metabolite(s) exceeds 10% of the administered dose, their concentrations in urine will need to be determined. In most cases, bioanalytical work on urine samples are often conducted after the completion of the analysis for the plasma (or whole blood/serum) samples. It has been often done that the method for urine samples employs the same or similar sample preparation approaches as that for the plasma assay. However, urine presents unique challenges for the method development. Urine samples have variable pH, variable salt concentrations (ion strength), variable buffer capacity, endogenous components with similar structures and polarity as analytes of interest, and high possibility of loss of analytes due to their nonspecific bindings to the container surfaces. The methods of urine sample collection are, in general, more complex than those for other biological matrices. Therefore, some necessary evaluations should be conducted before a suitable procedure is in place for the clinical urine sample collection. In this chapter, approaches of identifying and resolving analyte nonspecific binding issues in developing bioanalytical method for urine samples and the associated urine assay troubleshootings are described.

24.2 BEST PRACTICE IN DEVELOPING A ROBUST URINE LC-MS QUANTITATION METHOD

Urine is an aqueous solution of greater than 95% water, with the remaining constituents, in the order of decreasing concentrations, urea 9.3 g/l, chloride 1.87 g/l, sodium 1.17 g/l, potassium 0.750 g/l, and creatinine 0.670 g/l as well as other inorganic and organic compounds. The pH of urine can vary between 4.6 and 8, with neutral being the norm (Wikipedia 2012). Urine salt concentrations can also vary from subject to subject. Urine does not normally contain protein and lipids that are present in whole blood, plasma, or serum. In quantitative analysis of urine samples, lack of protein and lipids in urine samples can be associated with the issue of nonspecific binding or container surface adsorption of drug molecules, especially those lipophilic and highly protein bound drugs. A practical approach to identify nonspecific binding in the method development and to establish a robust urine bioanalytical method is discussed later.

24.2.1 Identification and Resolution of Analyte Loss due to Urine Nonspecific Binding

1. *Understanding general urine sample collection procedures at clinical sites*: It is important to understand the general clinical procedure for urine sample collection as the container surface area and number of transfers during the collection can affect the severity of adsorption if it is present. A general urine sample collection procedure in clinical setting is summarized in

Handbook of LC-MS Bioanalysis: Best Practices, Experimental Protocols, and Regulations, First Edition. Edited by Wenkui Li, Jie Zhang, and Francis L.S. Tse.

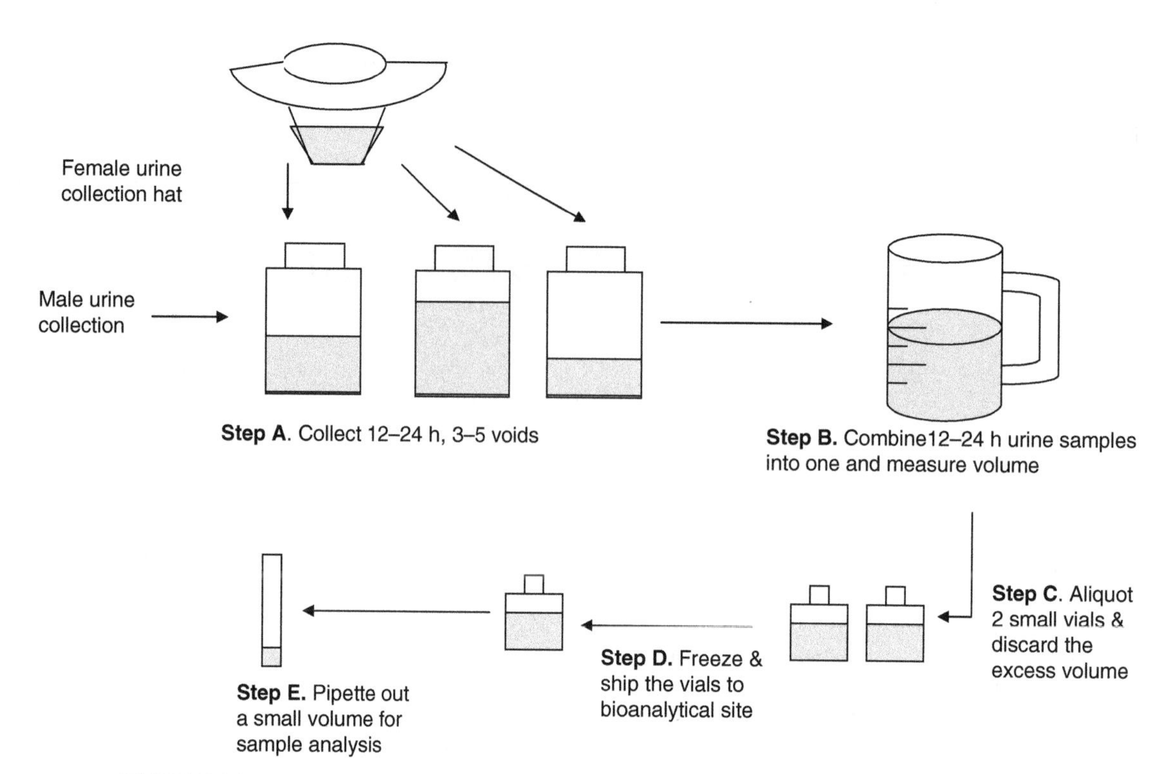

FIGURE 24.1 A general urine collection procedure for dose recovery studies (reproduced from Ji et al. (2010) with permission of Future Science Ltd.).

Figure 24.1. The collection hats are often used by females, elderly males, and disabled subjects, while the collection bottles are often used by male healthy subjects. All collected urine bottles are placed in a refrigerator for 1–3 days, depending on a study design. When multiple urine voids occur within a single interval, the individual void is collected in a separate bottle and then all voids within a defined interval are to be combined and mixed thoroughly. A portion of the combined urine sample is to be transferred to a smaller container (20-ml or 5-ml polypropylene vial with screw cap). The urine samples are then frozen and shipped to the designated bioanalytical site(s) for analysis. Since the entire urine sample collection involves multiple steps of urine surface area exposure to the container inner-wall, the final analyte concentration can be greatly affected by its adsorption to the surface area, generally referred to as nonspecific binding. The nonspecific binding results in inaccuracy of quality control (QC) samples and nonlinear standard curve response, due to the decreasing concentrations after sequential tube transfer of the analyte aqueous solution. Therefore, during the urine method development it is important for bioanalytical scientists to identify analyte nonspecific binding by mimicking clinical urine sample collection situation for a maximum container surface exposure.

2. *Identification of nonspecific binding:* To determine nonspecific binding for an analyte of interest, a scheme chart of a sequential approach for a urine method development to overcome the nonspecific binding is depicted in Figure 24.2. There are several ways to identify nonspecific binding (Ji et al., 2010), the simplest approach is sequential transfer test as shown in Figure 24.3. The test can be performed by preparing a drug solution at a concentration of about three times the lower limit of quantification (3 × LLOQ) in aqueous solution or urine. Four or five test tubes that are dry, clean, and comparable in terms of size and composition (e.g., polypropylene or polystyrene) to the intended collection containers are used for the sequential transfer test. A portion of the solution is transferred into the first tube and then from the first tube to the second. The same procedure is repeated for the remaining tubes. During each transfer, the solution in the donor tube should be left at room temperature for 5–10 min to allow analyte adsorption, if any, to take place. Also, keeping an appropriate volume from each test tube before the transfer is needed for the determination of analyte response using an established LC–MS/MS

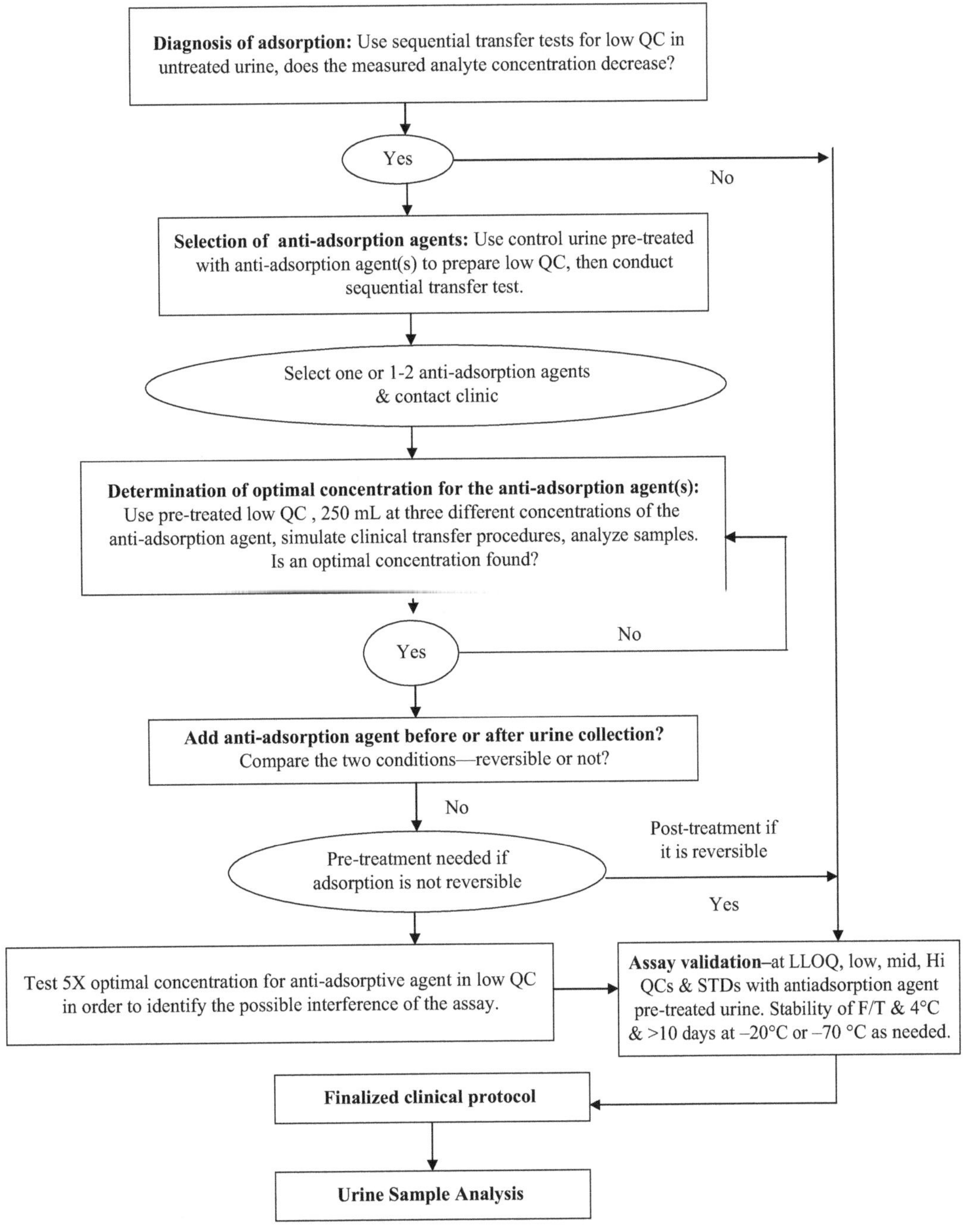

FIGURE 24.2 Method development scheme for overcoming nonspecific binding for urine sample analysis (reproduced from Ji et al. (2010) with permission of Future Science Ltd.).

method. If the analyte concentrations (or response) from Tube 1 through Tube 5 quickly decrease, strong adsorption of the analyte to the surface of the test tubes can be concluded.

3. *Commonly used antiadsorption agents and their selection*: Once it was determined that nonspecific adsorption will adversely affect the quantitation of drugs and metabolites in urine samples, appropriate approaches should be taken to minimize and mitigate this issue. In certain cases, adsorption can be simply prevented by controlling the pH of urine. Depending on the individual's diet and other physiological conditions, the pH value of human urine varies from pH 4.5 to 8. The pH of the urine can directly affect the degree of adsorption for some drugs to the surface of the sampling container as a result of a change of ionization form of the analyte(s) at certain pH. For acidic drugs, when the urine pH is below their pK_a values, the drug molecules

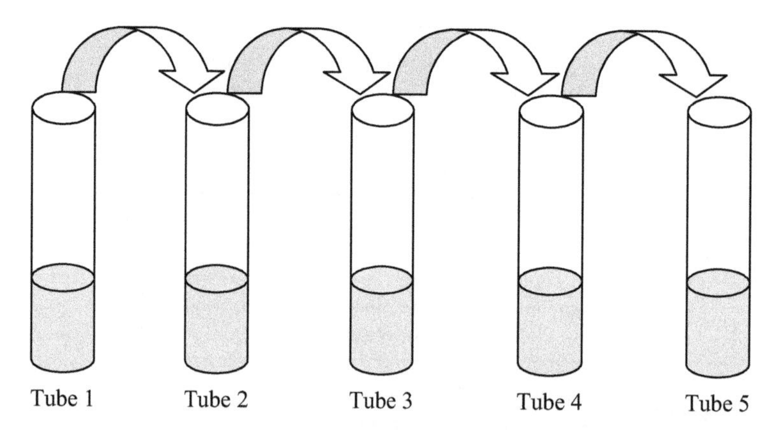

FIGURE 24.3 Sequential transfer method for identifying nonspecific binding in urine assay. The method involves three steps: Step A: prepare analyte at the low QC level in animal or human control urine or aqueous solution (organic component should be <3%). Step B: pour urine solution from one tube to the next and leave enough urine in each tube for the urine assay (do not use a pipette). Step C: pipette 100 μl of the urine or appropriate volume from each tube for determination of the analyte concentration using a bioanalytical method (reproduced from Ji et al. (2010) with permission of Future Science Ltd.).

appear to be neutral in the matrix. Therefore, the drugs are less subjected to the adsorption to the negatively charged container surface (e.g., polypropylene, polyethylene, polystyrene, and glass) due to the low binding affinity (Amshey and Donn, 2000). This approach is feasible at the clinical site since small amount of acid such as phosphoric acid can be easily added to the collection container without affecting the total urine volume. Similarly, the same procedure should also be used during the preparation of QC samples and calibration standards. One, of course, will need to assess, in the incurred samples, the potential impact to the analyte stability or breakdown of phase II conjugated metabolites to the parent compounds.

In some cases, addition of organic solvent(s) to the urine samples can also solve the adsorption issue (Palmgren et al., 2006). Organic solvents such as methanol, acetonitrile, DMSO, isopropanol can prevent adsorption when a large volume of the solvents (>10% of volume of the urine, or compound dependent) are added to the urine samples. The mechanism of this antiadsorption action is the factor that those small and polar organic molecules quickly interact with the binding sites on the internal surface of the container, resulting in a blockage of any electrostatic interaction of the analyte(s) with the container surface. Another possible explanation is that the polarity of the urine is decreased upon the addition of those organic solvents, resulting in an enhanced solubility of the analyte(s) in the urine. For single urine sampling, the organic solvent can be added to each sample tube at a one-to-one ratio (>10%, or compound dependent) during the collection. This approach can be easily adopted at clinical sites. However, for serial urine sampling to determine the dose recovery of the drug(s), this approach might not be practical due to the volume change.

When a drug molecule has the issue of nonspecific binding in urine samples that cannot be easily resolved by pH change or addition of an organic solvent, an antiadsorption agent needs to be identified and added to the containers before urine sample collection and to the control urine as well for preparation of the calibration standards and QC samples. Several types of antiadsorption agents are commercially available. They include BSA (bovine serum albumin) (Hoffman, et al., 2006), human plasma (Dubbelman et al., 2012), CHAPS(3-[3-cholamidopropyl)-dimethylammonio]-1-propane sulfonate) (Li et al., 2010; Silvester and Zang 2012), SDBS(sodium dodecyl benzene sulfonate) (Chen et al., 2009; Silvester and Zang 2012), Tween 20 (Xu et al., 2005), Tween 80 (Li et al., 2010; Silvester and Zang 2012), DMSO (dimethyl sulfoxide) (Schwartz et al., 1997), β-cyclodextrin (Li et al., 2010), and quaternary ammonium salts (Amshey and Donn, 2000). These agents have been reported for blocking adsorption in urine assays. The mechanism of the antiadsorption agent, advantages and disadvantages, suggested concentrations, and cost per subject for commonly used antiadsorption agents are summarized in Table 24.1. The advantage is that these antiadsorption agents can be used with minimal impact to the urine volume. Each antiadsorption solution can be added to the control urine to prepare a urine solution at low QC level in a small volume (such as 10 ml). Then a sequential transfer test is conducted for each urine solution containing different types of antiadsorption agent. The samples from each step of transfer are collected, extracted, and analyzed using LC-MS/MS. The results of peak area or ratio of analyte to internal standard and the chromatograms will be reviewed and compared. If more than one antiadsorption agents show negligible decrease of tendency for peak response, the one with less chromatographic interference and lower cost should then be selected as the candidate for the further test described later.

4. *Determining the optimal concentration of the antiadsorption agent:* When the antiadsorption agent is selected, the optimal concentration of this agent needs to be determined for both the urine assay and clinical urine collection procedure. In general, a primary urine container such as a 800-ml collection hat or a 500-ml squared polyethylene (or polypropylene) bottle should be used to prepare 3 sets of low urine QCs, each set with three different antiadsorption agent concentrations. One antiadsorption agent should be at the one recommended concentration in literature (Table 24.1), the other two concentrations can be chosen based on user's needs. For example, one is approximately two times lower and the other one is approximately two times higher than the recommended concentration. The QCs need to be prepared with a total volume of 250 ml each. This 250 ml volume is an approximate mean volume of common human urine voids. To simulate the primary urine collection condition, each QC should be prepared directly in a primary urine collection container. For example, the 250 ml low QC can be prepared by mixing 2.5 ml of the analyte stock solution with ~247.5 ml of antiadsorption agent treated control urine at one of the three concentrations. Then these three low QCs (250 ml each) should go through a few transfers that are similar to the sample collection procedure shown in Figure 24.1. In each step of the transfer, a few aliquots ($n = 3$) are taken for the determination of analyte concentration. The urine samples from the primary collection container (collection hat, or square bottle), and final small test tube (or 20 ml vial) are analyzed in triplicate. All of these samples should be analyzed on the same day when the transfer steps occur. If the analyte concentration measured in the urine from the final tube does not have a trend of decreasing (Figure 24.1, Step A to Step E) for one of the three low QCs, then the optimal concentration of antiadsorption agent is found. If all three low QCs with different concentrations of the antiadsorption agent show similar LC-MS/MS responses (peak area, peak ratios, or concentrations), then the lowest concentration of the antiadsorption agent should be chosen as the optimal antiadsorption agent concentration.

Data analysis can guide the modification of the urine sample collection procedure. For example, in Figure 24.1, if analyte concentration results from Steps A to D are equivalent or similar but sample from Step E is about 20% lower, it indicates insufficient amount of antiadsorption agent in the final sample transfer step (Step E). There may not be a need to add more antiadsorption agent to the primary urine sample collection bottle or hat (Step A). Instead, by adding suitable volume (e.g., 100 μl) of the antiadsorption agent in the last 5-ml tube (Step E), the problem can be solved. In general, the higher the ratio of surface area to urine volume, the more adsorption is anticipated. For example, a 3-l jar has about one eighth of the surface area to volume ratio of 20-ml collection vials. To save cost, low concentration of antiadsorption agent may be used in the jar due to low ratio of surface.

5. *Time of adding antiadsorption agent in primary collection container*: After a suitable antiadsorption agent is selected and its optimal concentration is determined, when (before or after primary urine collection) to add the antiadsorption agent solution should then be investigated. Xu et al. (2005) reported that Tween 20 (used for preventing nonspecific adsorption for a specific drug) can be added to the urine sample at the time after initial collection. This was determined by weighing the urine in container, followed by determining the volume of antiadsorption agent needed to be added. In another example, Li et al. (2010) compared the recoveries of an analyte of interest in urine samples treated with antiadsorption agents added pre- and post-urine collection. The results show that adding 1% (w/v) CHAPS before urine collection leads to a 100% recovery of the analyte, whereas adding the same antiadsorption agent after primary urine collection can only recover ~ 55% of the analyte. This highlights the importance of determining whether adsorption is reversible or not. If the adsorption process is reversible, then the antiadsorption agent can be added to a urine sample after its collection. Based on volume of collected urine, a suitable amount of antiadsorption agent can be added to each urine sample to reach its optimal antiadsorption agent concentration. If the adsorption process is not reversible, the antiadsorption agent must be added to a urine sample at the time of collection. In this case, if a certain volume of antiadsorption agent solution is added to the urine collection container before urine is collected, then the final concentration of the antiadsorption agent will vary due to the variability of the urine volume collected. Therefore, a concentration range of the antiadsorption agent should be tested during method development. As noted in another report (Ji et al., 2010), antiadsorption agent at a concentration of five times of the optimal concentration in the low QCs was evaluated against a standard curve prepared with urine treated with the same agent at the optimal concentration.
6. *Compiling urine sample collection procedure for clinical studies*: The urine collection manual is complied after the completion of the method development. The manual would contain detailed information about the materials, chemicals, suppliers, procedures for preparation of all solutions, and final urine collection tubes.

TABLE 24.1 Summary of Commonly Used Antiadsorption Agents in Urine Sample Collection

Agent name	Mechanism of action	Advantages	Disadvantages	Recommended conc. in urine[a]	Estimated cost (estimated amount) per Subject[b]
Bovine serum albumin (BSA) or plasma	It interacts with negatively charged surface of polymer (polypropylene and polystyrene), and also strongly binds to the analyte, leading to the blockage of adsorption of the analytes to the container surface.	Works for almost all analytes in urine.	May require a high concentration (>1%, w/v); May interfere with chromatographic separation. Cost is relatively high. Considered to be biohazardous for subjects.	1% (w/v)	~$154.40 (~ 60 g)
CHAPS(3-[3-cholamidopropyl)-dimethylammonio]-1-propane sulfonate)	It is a zwitterionic detergent. Its hydrophilic group binds to analyte's polar group so that it prevents the adsorption of an analyte to the urine container. Its nonpolar group interacts with the nonpolar portion of the analyte, further inhibiting the adsorption of the analyte to the container wall.	Requires a low concentration in urine. Its solid form is easy to handle.	May not work for some drugs. The cost is very high.	0.18% (w/v or 3 mM)	~$131.22 (~11 g)
SDBS (sodium dodecyl benzene sulfonate)	It is a zwitterionic detergent and its mechanism is similar to CHAPS.	Requires a very low concentration in urine. Its solid form is easy to handle. Cost is the lowest among the various agents.	Only works for certain drugs. It can cause ion suppression in mass spectrometric detection.	0.14% (w/v or 4 mM)	~$0.07 (3 g)
Tween 80 and Tween 20	It is zwitterionic detergent. Its hydrophilic groups binds to the analyte's polar group so that it prevents the adsorption of the analyte to the container surface.	Works for many drugs. It requires a low concentration in urine. In some cases, it can be added to urine after urine collection to retrieve the adsorbed analyte. The amount added to urine can be determined by weighing the urine specimen. The cost is low.	May not work for certain drugs. It has high viscosity as a liquid and may not be easy to handle.	0.5% (v/v) for Tween 80; 0.2% (v/v) for Tween 20.	~$1.80 (30 ml) for Tween 80; ~$0.64 (12 ml) for Tween 20.
β-cyclodextrin	A group of cyclic oligomers of glucose that can form water-soluble inclusion complex with small molecule drugs and a portion of large-molecule compounds. Therefore, it can block the adsorption of the analyte to the container wall.	Blocks adsorption for certain drugs. The cost is reasonably low.	Only works for certain drugs. The solid form has a large flake-type form. It has low solubility in methanol and water.	0.2% (w/v)	~$18.24 (12 g)

[a]The recommended concentrations were based on publications. It can be utilized as the starting concentration for selection/optimization.
[b]The price was based on the optimal concentration for one subject in a clinical study, 12 bottles of urine samples (500 ml each) are counted from 0 to 72 h (0–4 h, 4–8 h, 8–12 h, 12–24 h, 24–48 h, 48–72 h) post-dose. The listed prices are based on Sigma-Aldrich online catalog of January 2012.

The manual would provide information of procedure for shipping final urine sample specimens to the bioanalytical sites. Table 24.2 provides a typical example of urine sample collection.

24.2.2 Sample Preparation Approaches for Urine Bioanalysis

The common methods of urine sample preparation for LC-MS/MS analysis include direct dilution, liquid–liquid extraction (LLE), solid-supported liquid extraction (SLE), and solid phase extraction (SPE).

1. *Direct dilution*: Direct dilution method involves mixing the urine sample with an appropriate solution. The organic component in the final urine solution should be similar to the initial LC mobile phase composition. This method is the first choice and is commonly used for the analytes with relatively high urine concentrations. It is important to control the pH and the ionic strength of the diluted urine samples. A proper chromatographic condition should be implemented to separate analyte from all interfering components and to avoid matrix effect in mass spectrometric detection.
2. *LLE and SLE*: LLE is the most popular approach to urine extraction. The principle of LLE is to separate compounds based on their relative solubility in two different immiscible liquids, usually an aqueous buffer and an organic solvent. It is an extraction of a substance from one liquid phase into another liquid phase. In LLE, pH adjustment of the urine sample is highly necessary to ensure that drug and the internal standard appears as uncharged for a better extraction recovery using immiscible organic solvent. In general, the pH of urine samples should be adjusted to 2 units below the pK_a when analyzing acidic drugs or to 2 units above the pK_b when dealing with basic molecules. The commonly used organic LLE solvents include methyl-*t*-butyl ether (MTBE), ethyl acetate, methylene chloride, and hexane. Among these solvents, MTBE is the most commonly used not only because it is suitable for extraction of many nonpolar compounds from urine but also it can be quickly dried down. The drying time for 10 ml MTBE is less than 10 min at 37°C or 15 min at room temperature. In addition, MTBE is lighter than urine so the bottom aqueous phase can be frozen using a dry ice methanol or dry ice acetone bath, while top organic layer remains unfrozen. Using the dry ice methanol bath (or dry ice acetone bath) allows the top organic layer to be easily poured into a set of clean tubes.

In addition to the above manual method using test tube for extraction, LLE can be also conducted in 96-well plate using automated liquid handlers. Followed by addition of the aqueous urine samples, internal standard, buffer, and organic solvent, the plate should be are capped with a plate mat or heat sealer foil. After proper mixing and centrifuging, the top organic layer from the wells is transferred to another clean 96-well plate using an automated liquid handler (Tomtec, Janus, Hamiton, Multiprobe, or Tecan etc.). The organic solution is then evaporated and the sample residues are reconstituted with an appropriate reconstitution solution. The extraction recovery using a 96-well plate is usually about 10–20% lower than that using the tubes. This is possibly due to the limited space in each well of the plate for efficient mixing and much smaller volume ratio of the organic solvent to urine sample for partitioning. It is important to leave 30–40% of the space in each well for efficient mixing of the organic with the sample matrix. Table 24.3 lists an example of a typical procedure for LLE using tube-format and 96-well plate-format for a compound.

SLE is, in general, similar to LLE (Biotage, 2009) in terms of extraction mechanism. However, SLE employs solid sorbent, specially treated diatomaceous earth materials, as a solid support for LLE. In SLE, the sorbent in cartridges or 96-well plate adsorbs aqueous analyte solution (Note: do not add excess volume of aqueous phase to overflow the sorbent), organic solvent (such as MTBE) is then added to extract and elute the analytes from the sorbent. The extract containing analyte is collected, dried, and reconstituted for LC-MS/MS analysis. Compared to conventional LLE, the major advantage of SLE is its speed of the extraction and, in general, less matrix effect (Mulvana, 2010), and is easily automated. The major disadvantage of SLE is more expensive than regular LLE.

3. *SPE*: SPE is commonly used for the extraction of drugs at low concentrations with a decreased matrix components in the sample extract prior to LC-MS/MS analysis. In general, pH adjustment is necessary for SPE to ensure the highest extraction recovery of the analyte(s) of interest and the lowest unwanted interfering components. Three major advantages of SPE over other methods are (a) it has better specificity for extracting the drug; (b) it enriches drug concentration since unwanted component can be washed out and analyte(s) can be eluted, evaporated, and reconstituted in a small volume; and (c) it is easily to be automated using various liquid handlers.

24.2.3 Sample Analysis and Calculation of Percent Dose Recovered in Urine

Urine sample analysis can be initiated after method validation is completed. During urine sample analysis, sample

TABLE 24.2 A Representative Urine Sample Collection Procedure

Supplies
1. 500-ml squared bottle, high-density polyethylene, wide mouth, NalGene, VWR Catalog # 16121-060.
2. 800-ml specimen collection hat. Medical Supply Depot, M.C. Johnson, Item # KND4014/CS
3. 30% bovine serum albumin (BSA), Sigma-Aldrich, Catalog # A8577-1L.
4. 15-ml centrifuge tubes, Polypropylene, VWR, Catalog# 21008-105.
5. 5-ml culture tube, disposable, polypropylene, VWR, Catalog# 60818-486.

Preparation of anti-adsorption agent solutions
1. Add 15 ml of 30% BSA in water to a 15-ml centrifuge tube, cap tube and store the solution in a refrigerator until use. Expiration: 1 year from preparation at 4°C.

Urine collection procedure
1. Before subjects use a collection bottle or collection hat, pour the above antiadsorption agent solution into each 500-ml squared polypropylene bottle (for male subjects) or each 800-ml polypropylene collection hat.
2. For male subject urine samples, upon completion of urine collection, cap the bottles, followed by proper mixing via inverting the bottle upside down for four to five times. Place the urine specimen bottle(s) in a refrigerator. Urine samples from each interval needs to be combined into a 2000-ml graduate cylinder cup for volume measurement. For female subject urines samples, properly mixing the samples via rotating the hat for one to two times. Pour urine from each hat into a 500-ml squared polypropylene collection bottle, followed by capping and mixing via inverting the bottle upside down for four to five times, place the bottle(s) in a refrigerator until all urine samples collected from the assigned interval can be combined into a 2000-ml graduate cylinder for volume measurement.
3. Pool all urine specimens from the same collection interval into a 2000-ml graduate cylinder. The volume per interval per subject is recorded in unit ml. Then 3.0 ml of the pooled urine sample from each interval is pipetted into a 5-ml cryotube for primary urine specimen and in another 5-ml cryotube for backup urine specimen. The rest of excess urine is discarded. The graduate cylinder is rinsed with deionized water for three times, then it can be used for next measurement.
4. Two 5-ml urine sample tubes (primary and backup specimens) are capped and stored in a −20°C freezer until shipment with dry ice to bioanalytical sites in two different shipment dates.

mixing is a key step. If urine sample volume is more than 70% of the tube space, the tubes should be inverted for several times, and then followed by vortex-mixing before analysis. Urine concentrations cannot be used for directly calculation of dose recovery. Unlike plasma concentrations, the urine volume collected for different individuals can be significantly different within the same collection period. Therefore, dose recovery should be calculated using the amount of drug in urine over the respective time intervals by multiplying the analyte urine concentrations with sample volume. The cumulative amount of drug/metabolite(s) in a dose collection time course is added from the amount of drug/metabolite(s) from all intervals. Finally, the dose recovery is calculated as follows:

Amount of drug from each interval (ng or μg) = urine volume (ml) × urine concentrations (ng/ml or μg/ml).

For example, urine sampling intervals are 0–4 h, 4–8 h, 8–12 h, 12–24 h, 24–48 h, and 48–72 h.

Total cumulative amount of drug in urine = sum of all the amounts at all intervals (ng or μg, convert to mg).

Conversion of units: 1 ng = 0.001 μg; 1 μg = 0.001 mg.

$$\text{\% Dose recovery} = \frac{\text{Cumulative amount of drug in urine (mg)}}{\text{Amount of an oral or IV dose (mg)}} \times 100.$$

For the urine samples containing preservatives or an anti-adsorption agent, no correction in volume is necessary. This is because the volume of these agents has been included in consideration of the total sample volume.

24.2.4 Summary of TroubleShooting Urine Sample Analysis

Common troubleshootings in urine assays include the following:

1. Nonspecific binding can result in one or more of the following phenomena (Ji et al., 2010): (1) nonlinear response of standard curve; (2) poor reproducibility of the assay (especially at the lower concentration level) and the large bias compared to the nominal values; (3) poor extraction recovery of the analyte after one or more freeze–thaw cycle. Adding antiadsorption agent, or organic solvent to urine samples prior to the collection or adjusting pH of the urine samples can prevent nonspecific binding or container surface adsorption.
2. Variable pH or ionic strength of urine can lead to inconsistent extraction recoveries and nonreproducible analyte LC-MS/MS response between runs. The issue can be addressed by adding an acid (such as acetic acid,

TABLE 24.3 A Typical LLE Procedure for Urine Sample Extraction Using Conical Tube Format and 96-Well Plate Format

A: Using conical tubes

1. Place urine sample tubes in a tap water batch (~ 22°C) for approximately 30 min. If urine tube is almost full, invert tubes for four to seven times, followed by vortex-mixing for approximately 2 min.
2. Pipette 100 μl urine sample into a 10-ml conical glass tube with screw cap.
3. Add 50 μl of internal standard working solution. For urine blank sample without IS, add 50 μl of IS solvent.
4. Add 50 μl 0.1 M sodium acetate (pH 5.00) to each urine sample tube.
5. Add 5 ml of methyl-*t*-butyl ether (MTBE) to all tubes.
6. Cap the tube, followed by vortex-mixing using a multivortexer for approximately 10 min.
7. Centrifuge the tubes at approximately 3000 rpm for 5 min at room temperature.
8. Place conical tube into a dry ice acetone bath for 1 min, the aqueous layer is frozen.
9. Uncap the tubes and pour the top MTBE layer into another set of clean 10-ml conical tube. Discard the urine tubes containing aqueous phase.
10. Evaporate the MTBE extract tubes at 37°C under nitrogen flow in a TurboVap for ~10 min.
11. Reconstitute dried sample residues with 500 μl of reconstitution solution. Vortex the conical tubes for 3 min.
12. Transfer reconstituted sample extract into 1-ml HPLC ambient vials and submit the samples for LC-MS/MS analysis.

B: Using 96-well plate

1. Place urine sample tubes in a water bath (~ 22°C) for approximately 30 min. If urine tube is almost full, invert tubes for four to five times and pause-vortex each sample tube for approximately 2 min.
2. Pipette 100 μl of urine sample into a 2-ml 96-well plate (or 1.4 ml microtubes in 96-well plate format).[a]
3. Add 50 μl of internal standard working solution. For blank sample without IS, add 50 μl solvent.[a]
4. Using Tomtec, add 50 μl 100 mM sodium acetate (pH 5.0) into each well of the urine sample.
5. Using Tomtec, add 400 μl (× 2 times) of MTBE to each well of the plate.
6. Cap the plate with a plate mat tightly (or use caps to seal 96-well format microtubes).
7. Vortex the plate for approximately 10 min at a low speed.
8. Centrifuge the plate at 3000 rpm for 5 min at room temperature.
9. Remove the plate mat (or uncap the microtubes).
10. Using Tomtec, transfer 300 μl (× 2 times) of MTBE layer to another clean plate.
11. Evaporate the sample extracts in the plate at 37°C under nitrogen flow using Turbo Vap96 for ~ 5 min.
12. Using Tomtec, add 200 μl of reconstitution solution to each well of sample residue.
13. Cap the plate and vortex plate at a low speed (setting 3) for ~ 3 min.
14. Submit the plate for LC-MS/MS analysis.

[a]The procedure can be performed with a pipette or with an automated liquid handler such as Hamilton, Janus, MultiProbe, or Tican.

hydrochloric acid, and citric acid) during sample collection to control the pH of urine samples, or adding high concentration buffer (0.2 M to 1.0 M) to adjust urine ionic strength before extraction.

3. Urine sample volume for the collection should not be greater than two thirds of the tube volume to ensure enough space for proper mixing.
4. Inconsistent extraction recovery due to nonadsorption issues can be also solved by adding a small volume of plasma or bovine serum albumin solution as a reagent to the urine samples, then extract the urine samples. The plasma sample preparation approaches such as protein precipitation method (PPT), LLE, SLE, or SPE can be used for urine sample after adding plasma or BSA. The reason is that plasma is not only a good buffer to control urine pH, but it is also a good reagent to adsorb salts in urine so that the ionic strength is controlled.

24.3 SUMMARY

In summary, quantitation of drugs in urine presents technical challenges. Nonspecific binding of drugs to the urine sample container is an important factor to be considered and evaluated at a very early stage of method development. Appropriate antiadsorption agents should be investigated for preventing or retrieving the analyte(s) of interest from the loss due to nonspecific binding. In addition, pH and ions strength adjustment should be considered during urine assay development.

REFERENCES

Amshey J, Donn R. US patent 606020-methods for reducing adsorption in an assay. PatentStorm 2000.

Biotage AB. Comparison of Liquid-Liquid Extraction (LLE) and Supported Liquid Extraction (SLE): Equivalent Limits of Quantitation with Smaller Sample Volumes. The Application Notebook, 2009, September 1. Available at http://chromatographyonline.findanalytichem.com/lcgc/Application+Notes/Comparison-of-Liquid-Liquid-Extraction-LLE-and-Sup/ArticleStandard/Article/detail/623987. Accessed Mar 28, 2013.

Chen C, Bajpai L, Mollova N, Leung K. Sensitive and cost-effective LC/MS/MS method for quantitation of CVT-6883 in human urine using sodium dodecylbenzenesulfonate additive to eliminate adsorptive losses. J Chromatogr B: Analyt Technol Biomed Life Sci 2009;877(10):943–947.

Dubbelman AC, Tibben M, Rosing H, et al. Development and validation of LC-MS/MS assays for the quantification of bendamustine and its metabolites in human plasma and urine. J Chromatogr B 2012;893–894:92–100.

Hoffman B, Bhadresa S, Zhao L, Weng N: Adsorptive losses in a bioanalytical LC/MS/MS urine assay. Poster of the 54th ASMS Conference on Mass Spectrometry, May 28–June 2, 2006, Seattle, WA, USA.

Ji AJ, Jiang Z, Livson Y, Davis JA, Chu JX, Weng N. Challenges in urine bioanalytical assays: overcoming nonspecific binding. Bioanalysis 2010;2(9):1573–1586.

Li W, Luo S, Smith H, Tse F. Quantitative determination of BAF312, ASLP-R modulator, in human urine by LC-MS/MS: prevention and recovery of lost analyte due to container surface adsorption. J Chromatogr B: Analyt Technol Biomed Life Sci 2010;878(5–6):583–589.

Mulvana DE. Critical topics in ensuring data quality in bioanalytical LC–MS method development. Bioanalysis 2010;2(6):1050–1072.

Palmgren JJ, Monkkonen J, Korjamo T, Hassinen A, Auriola S. Drug adsorption to plastic containers and retention of drugs in cultured cells under in vitro conditions. Eur J Pharm Biopharm 2006;64(3):369–378.

Schwartz M, Kline W, Matuszewski B: Determination of a cyclic hexapeptide (l-743 872), a novel pneumocandin antifungal agent in human plasma and urine by high-performance liquid chromatography with fluorescence detection. Anal Chim Acta 1997;352(1–3):299–307.

Silvester S, Zang F. Overcoming non-specific adsorption issues for AZD9164 in human urine samples: consideration of bioanalytical and metabolite identification procedures. J Chromatogr B 2012;893-894:134–143.

Wikipedia, the free encyclopedia. Available at http://en.wikipedia.org/wiki/Human_urine, 2012. Accessed mar 28, 2013.

Xu Y, Du L, Rose MJ, Fu I, Woolf EJ, Musson DG. Concerns in the development of an assay for determination of a highly conjugated adsorption-prone compound in human urine. J Chromatogr B: Analyt Technol Biomed Life Sci 2005;818(2):241–248.

25

LC-MS BIOANALYSIS OF UNBOUND DRUGS IN PLASMA AND SERUM

THEO DE BOER AND JAAP WIELING

25.1 INTRODUCTION

In today's bioanalysis, usually the total concentration of a drug in plasma and serum is evaluated. In those cases where the drug of interest is bound to plasma proteins, applying a sample purification step will liberate the drug from the proteins and the total concentration can be quantitated. If a drug is extensively distributed over the tissues and organs, tissue sample analyses using fit-for-purpose methodologies may be applicable. On the other hand, if a drug is mainly distributed in the erythrocytes, whole blood analysis would be the method of choice. However, it is well known that in almost all cases the free (intracellular) concentration of a drug is responsible for its biological activity and pharmacological/toxicological processes, including membrane permeation, (metabolic) clearance, and receptor binding. Measuring the free or unbound fraction of a drug is an important tool for understanding the relationships between plasma/serum levels of the drug and patient responses (efficacy/safety) (Li et al., 2011; Rakhila et al., 2011; Streit et al., 2011). In addition, plasma protein binding (PB) and preferential partitioning of drugs into the erythrocytes are important factors to be considered in setting the initial dose of the drug in phase I clinical trials.

Due to the diversity of the binding properties of the investigational drug(s) to a variety of proteins (albumin, α1-glycoproteins and lipoproteins), the PB studies are usually performed in plasma rather than laboratory solutions. For quantitative analysis of the free or unbound fraction of a drug, several methodologies, such as equilibrium dialysis, ultracentrifugation, and ultrafiltration, have been implemented (Musteata, 2011) using either the "hot" (radioactive) or "cold" (stable isotope labeled) compounds. Each of the methods has some advantages over the others. This chapter covers two major sections: in Section 25.4, different techniques for separating the bound and free (unbound) fraction of the drugs is to be discussed and the individual case studies are to be captured in Section 25.5.

25.2 PROTEIN BINDING

When a drug is absorbed in the bloodstream (systemic circulation), it will be distributed over the blood, the tissues and the organs, where the ratio of the distribution is dictated by the physicochemical properties (pK_a, log P) of the drug and by both the pH values of the body fluids (pH 1–8) and the perfusion properties of the membranes that separate the plasma from the tissue. In the blood stream, the drug can be present in its free form, it can either enter the erythrocytes and/or bind to erythrocyte membranes, or it can interact with proteins (e.g., albumin and α-1 acid glycoprotein (AGP)) present in the plasma. In general, acidic drugs mainly bind to human serum albumin, whereas basic drugs bind to lipoproteins and α-1 AGP. To modulate fluctuations in the free concentration, endogenous compounds such as vitamins and steroid hormones (e.g., testosterone and estradiol) tend to bind to specific globulins, for example, the sex hormone binding globulin (SHBG), as well. The binding to the proteins is always reversible. When the concentration of the drug in the plasma increases, usually the free fraction of the drug will increase. This is indeed the case for α-1 AGP binding compounds, but uncommon for albumin binding compounds (concentration of albumin is approximately

Handbook of LC-MS Bioanalysis: Best Practices, Experimental Protocols, and Regulations, First Edition. Edited by Wenkui Li, Jie Zhang, and Francis L.S. Tse.

25–100 times higher than that of α-1 AGP). Hence, for albumin binding compounds, the free fraction is rather constant, which is especially critical for steroid hormones. A special form of PB is the stereoselective binding of a racemic mixture of a chiral drug. In several cases, significant differences between the interaction of the enantiomers (eutomer and distomer) with the proteins have been observed (see also Section 25.5.4).

25.3 REGULATORY REQUIREMENTS REGARDING UNBOUND DRUG CONCENTRATIONS IN SPECIAL POPULATIONS

Unbound drug concentrations should correlate better to safety and efficacy than the total plasma concentrations when the free fraction of the drug in plasma is subjected to change as for instance in special clinical populations, for example, patients suffering from comorbid hepatic and renal insufficiency, cancer, immune diseases (Shackleton, 2011). Both FDA and EMA have provided guidelines for assessment of potential differences in plasma PB between healthy subjects and special populations (EMA, 2004; US FDA, 1998). The guidelines recommend that the unbound fraction is determined *ex vivo* at least at trough and maximum plasma concentration levels. However, when plasma PB of a drug candidate is known to be independent of its plasma concentration, the presence of metabolites and other time-varying factors (meals, circadian rhythms, and concomitant medications), the results from *in vitro* studies are considered acceptable.

25.4 TECHNIQUES FOR THE BIOANALYSIS OF UNBOUND DRUGS IN PLASMA AND SERUM

As both temperature and pH are the two important parameters that influence the binding of small molecules to proteins, the control of temperature (37°C) and pH (7.4) is of utmost importance. During the entire sample handling process that encompasses the time from blood collection to actual analysis, changes in physiological pH can occur. This may unequivocally influence the PB properties of small molecules. Other parameters that affect the plasma PB are the concentrations of the drug and proteins and the solubility, stability, and displacement of the drug. Drug displacement from the drug–protein complex can occur upon direct competition of two drugs at the same binding site. Therefore, for highly protein-bound (>95%) drugs, a small displacement can result in a significant increase in the free drug concentration in plasma (Lee et al., 2003). On the other hand, while freezing and thawing of plasma samples seems not to affect the free concentration, special attention should be given to the possible impact of long-term storage of plasma on the PB capacities of small molecules as lipolysis can occur during the storage period. Lipolysis can lead to increase in the free fatty acid levels in plasma, resulting in fatty-acid induced protein conformational changes that may influence the nonspecific binding (NSB) of small molecules to proteins (Howard et al., 2010).

The most popular approaches for assessment of free drug concentrations in complex biological samples have been to date: equilibrium dialysis and ultrafiltration, the latter seems nowadays to be the more popular one due to its simplicity and high-throughput potential. Alternative approaches include analytical ultracentrifugation, solid phase microextraction (SPME), and optical biosensors (surface plasmon resonance), and they are gaining more and more interest. Although several techniques have been described that can measure free drug concentrations without separating the drug from the binding protein (e.g., calorimetry, chromatography, electrophoresis, and spectrometry), they are not suitable for the analysis of complex biological samples (Musteata, 2011).

Specific attention should be paid to the NSB of analyte(s) of interest to the membrane filter or apparatus. Without necessary evaluation and the associated prevention, the unbound drug concentration can be underestimated. NSB can be significantly reduced via using filters/membranes made of Teflon (Wells, 2003). Nevertheless, if there is still considerable NSB upon employing these materials, analytical ultracentrifugation should be considered.

Alternatively Schuhmacher et al. (2000) proposed a methodology that measures the partitioning of a drug between plasma and blood cells as well as the partitioning between buffer and blood cells. The free fraction of the drug can be calculated by dividing the two partition coefficients.

25.4.1 (Rapid) Equilibrium Dialysis

Equilibrium dialysis separates free from protein-bound analytes in small volumes of complex biological matrices (serum, plasma, whole blood) according to diffusion through the pores of a selective and semipermeable membrane in a concentration driven process (Figure 25.1). While ultrafiltration is probably the most popular method, especially in clinical settings, equilibrium dialysis is still considered as

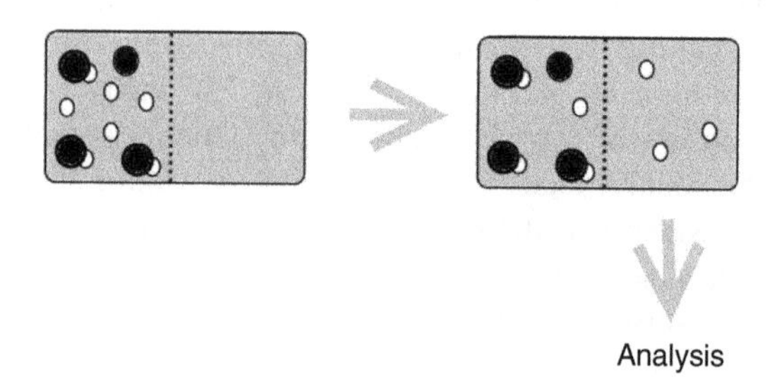

FIGURE 25.1 Schematic representation of the equilibrium dialysis methodology.

the reference method ("gold standard") for monitoring free drug concentrations (Musteata, 2011). The method allows studying the true nature of the interaction under equilibrium conditions and is therefore believed to be less susceptible to experimental artifacts. The process is as follows: in one chamber of the equilibrium dialysis device an accurate volume of sample is added. In the other chamber an equivalent volume of a buffer solution is pipetted. In between the two chambers a membrane is located. The molecular weight cut-off (MWCO) of the semipermeable membrane (5 KDa, 10 KDa, 30 KDa) is chosen such that it will retain the proteins and allows the free analyte(s) to pass the membrane until equilibrium is achieved. As full equilibrium is reached after incubation at 37°C under stirring or mixing conditions for 3 h up to 2 days (depending on drugs, membrane materials, and devices), the method is, in spite of its accuracy, rather time-consuming. Consequently, equilibrium dialysis is not suitable for unstable compounds.

Several equilibrium dialysis devices are available: Dispo-Equilibrium dialyzer™ (for single use), Micro-Equilibrium Dialyzer™ (reusable), Equilibrium Dialyzer-96™ (96-wells) as well as a multi-Equilibrium Dialyzer™ that is suitable for 20 parallel assays (Harvard Bioscience, Holliston, MA, USA).

Recently, high-throughput parallel sample processing devices have been introduced by Thermo Scientific/Pierce (Rockford, IL), HTdialysis LLC (Gales Ferry, CT) and BD Biosciences (San Jose, CA). Using these rapid equilibrium dialysis (RED) devices, potential complications like volume shifts and protein leakage can be minimized by reducing the equilibration time considerably. For example, Waters et al. (2008) and Howard et al. (2010) reported equilibrium times within 6 h at 37°C, where classic equilibrium dialysis often required equilibrium times up to 24 h or longer. Furthermore, RED offers the possibility of analyzing both the plasma fraction and the buffer fraction, which allows studying the true nature of the interaction under equilibrium conditions. It is believed that RED is nowadays a better choice of method than others for assessment of unbound fractions of drugs in plasma and serum.

The percentage of PB (%PB) can be calculated using Equation (25.1):

$$\%\text{PB} = 100 \times (1 - f_\text{u}) = 100\left(\frac{C_\text{ET} - C_\text{EF}}{C_\text{ET}}\right). \quad (25.1)$$

In Equation (25.1), f_u is the free fraction, C_ET is the total concentration of the drug, that is, the sum of the free drug concentration at equilibrium (C_EF) and the protein-bound concentration at equilibrium (C_EB). The NSB ($\%\text{NSB}_\text{ED}$) can be calculated using Equation (25.2):

$$\%\text{NSB}_\text{ED} = 100 \times \left(\frac{C_\text{I} - C_\text{D} - C_\text{R}}{C_\text{I}}\right). \quad (25.2)$$

In Equation (25.2), C_I is the measured initial concentration, C_D is the concentration of the donor well, and C_R is the concentration of the receiver well.

25.4.2 Ultrafiltration

Ultrafiltration separates free from protein-bound analytes in small volumes of complex biological matrices (serum, plasma, whole blood) according to molecular weight and size using centrifugal force. Unlike equilibrium dialysis, a diffusion and concentration driven process, ultrafiltration separates molecules based on the MWCO value of the low-adsorptive hydrophilic membranes (Figure 25.2). Hence, ultrafiltration is a much faster technique than equilibrium dialysis and much more suitable for high-throughput analysis.

Ultrafiltration devices, such as the Microcon®, Centricon®, Centriplus®, and Centrifree® devices from Millipore (Bedford, MA, USA), are available in several MWCO values (3 KDa, 10 KDa, 30 KDa, 50 KDa, and 100 KDa). These devices are also available in several volumes (from 10 μl up to 15 ml). Typically, 0.15–1 ml plasma or serum is added to the device after which the device is placed in to a centrifuge. A higher yield of ultrafiltrate is claimed when a fixed angle is used instead of a swinging bucket (Wells, 2003). In the course of ultrafiltration, all nonfilterable components present in the plasma compartment become increasingly concentrated. This may result in a so-called sieve effect (Ekins, 1992; Kwong, 1985) because the membrane may not act as a perfect molecular sieve but instead discriminates between water molecules and drug molecules if the drug molecule is relatively large, for example, >500 Da. As a result of the "sieve effect," the obtained ultrafiltrate from the later stage of ultrafiltration may have been diluted compared to those obtained from the early stage. Therefore, centrifugation of study (plasma/serum) samples at 1500 × g using a fixed angle rotor for 20 min is generally recommended, which should result in approximately 30% (v/v) ultrafiltrate. (Wells, 2003).

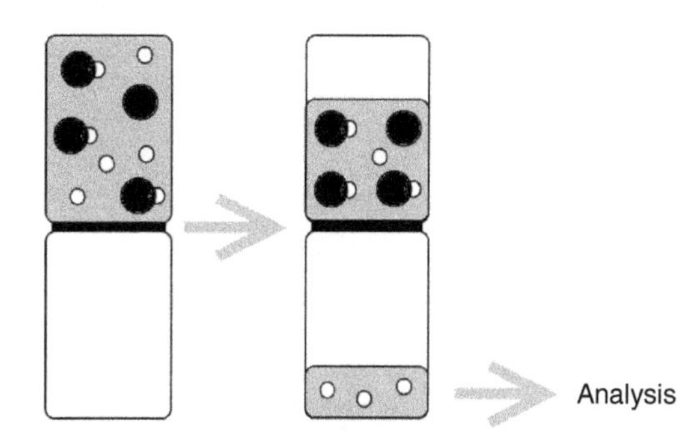

FIGURE 25.2 Schematic representation of the ultrafiltration methodology.

NSB of the analytes to the ultrafiltration device (membranes and devices) is a commonly encountered issue that leads to an underestimation of the free-drug concentration if proper prevention is not in place. Lee et al. (2003) reported the NSB can be minimized by pretreating the filter membranes with Tween 80 for neutral/acidic compounds and benzalkonium chloride for basic compounds. In another case, Taylor and Harker (2006) added control plasma retentate (the part of a solution that does not cross the membrane) to the obtained ultrafiltrate to prevent NSB of corticosteroids.

The NSB of the ultrafiltrate unit can be calculated using equation (25.3):

$$NSB_{UF} = \frac{C_{BD} - C_{BF}}{C_{BD}}. \quad (25.3)$$

In Equation (25.3), C_{BD} is the total drug concentration in buffer solution (usually phosphate buffered saline pH 7.4) before centrifugation and C_{BF} is the drug concentration in the filtrate after centrifugation. When $C_{BF} = C_{BD}$, NSB is 0 and there is no need of NSB correction for the %PB calculation. When $C_{BF} < C_{BD}$, it can be assumed that a fraction of drug disappeared. The NSB correction of %PB can be calculated using Equation (25.4):

$$\%PB = 100 \times (1 - f_u) = 100\left(1 - \frac{C_{SF}}{C_{SD}(1 - NSB_{UF})}\right). \quad (25.4)$$

In Equation (25.4), f_u is the free fraction, C_{SF} is the drug concentration in the serum–plasma filtrate, and C_{SD} is the nominal serum–plasma donor concentration.

25.4.3 Analytical Ultracentrifugation

The ultracentrifuge is a centrifuge that is optimized for spinning a rotor at very high speeds, capable of separating particles ("large molecules" from "small molecules") from a solution when subjected to gravitational forces according to their size, shape, density, viscosity of the medium, and rotor speed (Figure 25.3). There are two kinds of ultracentrifuges, the preparative and the analytical ultracentrifuge. Both classes of instruments are commonly used in molecular biology, biochemistry, polymer science, and drug discovery and development.

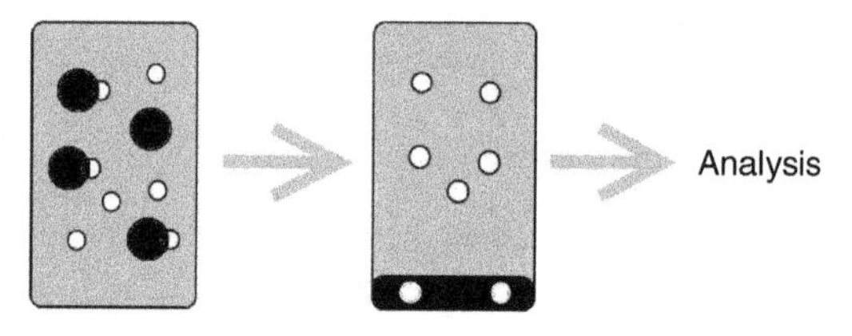

FIGURE 25.3 Schematic representation of the ultracentrifugation methodology.

Although equilibrium dialysis and ultrafiltration may have been most commonly used for the assessment of PB, analytical ultracentrifugation can be seen as a complementary technique with this regard. For analytical ultracentrifugation, fixed angle rotors or swinging buckets are used. The use of a swinging bucket results in a longer precipitation time than that with fixed angle rotors, due to a lower relative centrifugal force. Typically, plasma is added to special designed tubes, for example, cellulose propionate, polycarbonate, polyethylene, polypropylene, or polyallomer (Beckman Coulter, Brea CA, USA). These tubes can typically tolerate centrifugation force up to 2,000,000 × g, depending on the centrifugation time, size, and material of the tubes. Nitrocellulose tubes are inexpensive and intended for single use, while polyallomer tubes are reusable but more expensive. One of the advantages of the ultracentrifigation technique is the fewer issues of NSB compared to the ultrafiltration experiments. However, physical phenomena related to ultracentrifugation such as sedimentation, back diffusion, viscosity, and binding to plasma lipoproteins in the supernatant fluid should be carefully addressed. The NSB and the PB can be calculated using Equations (25.3) and (25.4).

25.4.4 Solid Phase Microextraction

SPME was introduced more than 20 years ago as a promising extraction technique in analytical chemistry (Arthur and Pawliszyn, 1990). In short, SPME uses a small fused-silica fiber coated with a suitable polymeric phase mounted in a syringe-like protective holder. During extraction the fiber is exposed to the sample by suppressing the plunger. Sorption of the analytes on the fiber takes place in either the sample by direct-immersion or the headspace of the sample tube. After equilibrium the fiber is withdrawn in the needle and introduced into the analytical instrument where the analytes are either thermally desorbed or redissolved in a proper solvent (Figure 25.4). The applicability of SPME for determining drug binding to proteins in human plasma has been previously described (Musteata et al., 2006; Musteata et al., 2007; Zhan et al., 2011). Howard et al. (2010) acknowledged that because the free drug is never partitioned into a solution that is void of protein, very hydrophobic compounds with limited water solubility may be analyzed by this methodology. Commercial SPME fibers coated with nonspecific polymers are available from Sigma-Aldrich (formerly: Supelco), while more tailor-made approaches using fibers coated with molecularly imprinted polymers (Prasad et al., 2009) still remain in the academic environment.

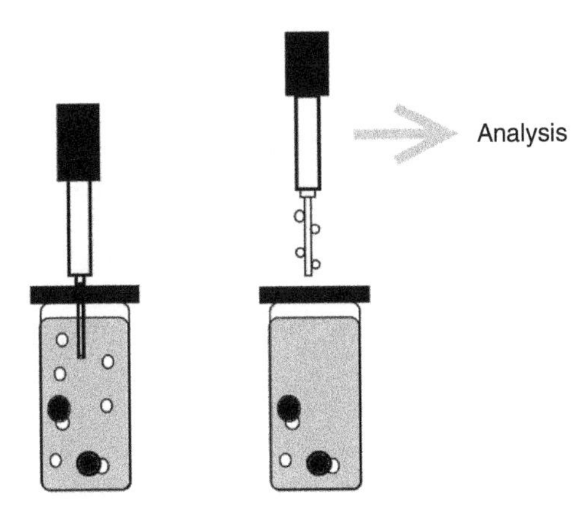

FIGURE 25.4 Schematic representation of the SPME methodology.

The percentage PB can be calculated using Equation (25.1), where in this case the free plasma concentration at equilibrium (C_{EF}) can be calculated using Equation (25.5):

$$C_{EF} = \frac{m_{plasma}}{fc} = \frac{V(C_{E0} - C_{ET})}{fc}. \qquad (25.5)$$

In Equation (25.5), fc is the fiber constant and m_{plasma} is the amount of drug that is extracted by the fiber and equals $V\,(C_{EO} - C_{ET})$, where V is the plasma volume, C_{EO} is the concentration at $t = 0$ h and C_{ET} is the total concentration in plasma at equilibrium. The fiber constant fc represents the product of the partition coefficient of the drug (between fiber and solution) and the volume of the fiber (for liquid coatings) or the active surface of the fiber (for solid coatings) (Musteata et al., 2006). The fiber constant may be easily determined by extraction from standard solutions of the drug in PBS or "plasma water," when the total drug concentration is considered to be equal to the free concentration C_{free} as shown in Equation (25.6):

$$fc = \frac{m_{standard}}{C_{freestandard}} = \frac{m_{standard}}{\left(C_{0standard} - \frac{m_{standard}}{V_{standard}}\right)}. \qquad (25.6)$$

In Equation (25.6), $C_{0\,standard}$ represents the initial concentration in a volume ($V_{standard}$) of standard solution before extraction, and $m_{standard}$ is the amount of drug extracted.

25.4.5 Considerations in Developing and Validating a Robust LC-MS/MS Method for the Analysis of Unbound Drugs in Biological Matrices

In today's quantitative bioanalysis, RED and ultrafiltration seem to be the most favorable approaches for assessment of unbound fractions of drugs in plasma and serum as shown in recently published papers (Deng et al., 2011; Larsen et al., 2011; Li et al., 2011; Rakhila et al., 2011; Streit et al., 2011).

Regardless which sample processing method is to be used, several key evaluations have to be conducted to ensure the robustness of an intended LC-MS methodology for measuring the unbound fraction:

- Selection of devices (brand, type).
- Stability of analyte during incubation (dialysis) or ultrafiltration/ultracentrifugation.
- Assessment of NSB.
- Centrifugation temperature, time, and speed (ultrafiltration).
- Any impact of bench top stay to the measured free value.
- Any impact of *F/T* on the measured free values.
- Any impact of long-term storage on the measured free values.
- Any impact of the presence of organic solvent (spiking solution) in the STD/QCs.

More specific, when ultrafiltration is used, considerable attention should be paid on the impact of the ultrafiltrate volume on the free drug concentration ("edge effect") (Zhang and Musson, 2006). In addition to this "edge effect" also the "sieve effect" (Section 25.4.2) should be studied.

A final remark regarding the validation of the final method procedure should be made. It should be kept in mind that because no real plasma sample (incurred sample) with a known *in vivo* concentration of a free analyte is available, the value of the actual method validation is some-how limited. During their validation of a method for the determination of unbound vadimazan in human plasma using ultrafiltration, Li et al. (2011) acknowledged that the precision and accuracy determination during the method validation was simply the analytical result of spiked plasma ultrafiltrate samples that pass through the filter membrane to mimic the plasma ultrafiltration process.

25.5 EXAMPLES

25.5.1 General Applications

Barré et al. (1985) compared equilibrium dialysis (Teflon microcells, Dianorm®; Diachema, Ruschlikon, Zurich, Switzerland), ultrafiltration (Emit® system, Syva Co., Palo Alto, CA) and ultracentrifugation (Polyallomer tubes, Beckman) for determining the plasma-protein-binding characteristics of valproic acid. They observed a difference between the results obtained by ultracentrifugation and those obtained by equilibrium dialysis and ultrafiltration (that correlated very well with each other). These differences may be due to some physical phenomena related to ultracentrifugation (Section 25.4.3). Lee et al. (2003) compared an ultrafiltrate unit

(Millipore) and a 96-well equilibrium dialyzer (Harvard Bioscience) in a study that was set up to reduce or prevent NSB of ten drugs to the testing units. Severe NSB was observed for etoposide, hydrocortisone, propranolol, and vinblastine when using the ultrafiltrate unit. The NSB for the above compounds was reduced from 87–95% to 13–64% via pretreatment with Tween 80 (neutral or acidic compounds) or benzalkonium chloride (basic compounds). Interestingly, when NSB was below 50%, the PB data obtained via ultrafiltration were comparable to those obtained via equilibrium dialysis. Fung et al. (2003) used the Microcon-96 ultrafiltrate assembly (Millipore) with a molecular cutoff of 30 KDa to determine the PB for 32 compounds in one single experiment conducted at 3000 $\times$ g for 45 min. They compared the calculated percentages of unbound drugs using the 96-well approach with the individual values measured or reported in literature. Excellent correlations were observed ($R^2 = 0.94$ (Millipore) and $R^2 = 0.92$ (Harvard Bioscience), respectively).

Some very recent studies (Deng et al., 2011; Rakhila et al., 2011) evaluated the use of RED for the assessment of unbound fractions of vismodegib and teriflunomide, respectively. Typically, 400 μl plasma (or less) was dialyzed against an equal amount of buffer solution in a RED plate. Completion of the dialysis was achieved within 6 h at 37°C.

25.5.2 Special Applications I: Steroids

In men, approximately 44–65% of the circulating testosterone is specifically bound to SHBG, sometimes referred to as sex steroid binding globulin (SSBG), while 33–54% is nonspecifically bound to albumin. In women, 66–78% of the circulating testosterone is bound to SHBG and 20–32% to albumin. Only 2–3% of total testosterone is available in its pharmacologically active free form (Emadi-Konjin et al., 2003). As testosterone is only weakly (nonspecifically) bound to albumin and to cortisol-binding globulin (~3%), essentially all albumin (and CBG (corticosteroid binding globulin)) bound testosterone is available for interaction with classical nuclear steroid receptors, whereas the testosterone bound to SHBG is tightly bound and considered to be inactive. For that reason free testosterone (FT) and the sum of free and albumin bound testosterone, usually referred to as "bioavailable testosterone (BioT)" or non-SHBG bound testosterone are much better markers for hyperandrogenism and hypogonadism than the total testosterone concentration (Cumming and Wall, 1985; De Ronde et al., 2006). The most popular technique to determine bioavailable testosterone is by precipitating SHBG (together with the specifically bound sex steroids) with a saturated ammonium sulfate solution. In short, serum is diluted 1:1 with a saturated solution of ammonium sulfate after which the testosterone concentration is determined directly in the supernatant or indirectly by reconstituting the precipitate with a buffer solution (Tremblay et al., 1974; Morley et al., 2002). The gold standard for assessment of FT is isotope dilution (e.g., using [^{3}H]-labeled testosterone) equilibrium dialysis. However, this technique is much more laborious and costly compared to ultrafiltration (Chen et al., 2010; Hackbarth et al., 2011) and SPME (Zhan et al., 2011). To overcome the issues related to the use of nonautomated, time-consuming techniques in a (high-throughput) laboratory and to avoid the use of extremely sensitive and hence expensive chromatographic techniques (FT levels in women and hypogonadal men can be as low as 1 pg/ml), several authors have published mathematic algorithms for the calculation of FT and BioT that have been compared in a recent paper by De Ronde et al. (2006). These equations are based on the total testosterone concentration and the binding affinities of testosterone with SHBG and albumin. As the affinity constants used in the equations are arbitrarily chosen, assuming a constant value for all individual samples, and as none of the suggested algorithms are validated against a reference method in using a significant number of samples over- or underestimation of the calculated FT and BioT concentrations may occur. Moreover, although usually the calculated FT correlates well with FT measured by the reference equilibrium dialysis method, the calculated FT concentration is highly dependent on the accuracy of total testosterone, SHBG, and albumin quantification (DeVan et al., 2008).

The most commonly used algorithm for the calculation of BioT and FT was originally described by Sodergard et al. (1982) and derived by Vermeulen et al. (1999). This algorithm assumes that FT and BioT concentrations can be calculated when the concentrations of total testosterone, SHBG, and albumin are known. Critical parameters for a proper estimation of FT and BioT are the association constants for binding of testosterone to SHBG (K_{SHBG}) and to albumin (K_{ALB}), as summarized by De Ronde et al. (2006). The formulas are presented in Table 25.1 for the calculation of FT and BioT. Alternative equations for bioT (Morris et al., 2004) and FT (Ly and Handelsman, 2005) were published as well. However, there is currently no consensus on which constants are considered ideal (Hackbarth et al., 2011). Dhinsa et al. (2011) calculated free and bioavailable testosterone and estradiol concentrations using the equations in Table 25.1. They made an adjustment to the formula by multiplying the SHBG concentration by two (i.e., $b = N + K_{SHBG} \times \{2[\text{SHBG}] - [\text{TT}]\}$), because 1 mol of SHBG homodimer binds to 2 mol of testosterone. With this adjustment for the number of SHBG binding sites, FT measured by equilibrium dialysis were similar to the concentrations calculated with the suggested algorithms ($r^2 > 0.98$). Free estradiol was calculated using the same formula but with a weaker association constant for estradiol for SHBG (0.6×10^9 l/mol) and for albumin (0.61×10^4 l/mol).

For estimation of the unbound prednisolone concentration, Ruiter et al. (2012) compared their ultrafiltrate results

TABLE 25.1 Equations for the Calculation of FT and BioT

FT (mol/l)	$N = 1 + K_{ALB} \times [\text{Albumin}]$
	$a = N \times K_{SHBG}$
	$b = N + K_{SHBG} \times ([\text{SHBG}] - [\text{TT}])$
	$c = -[\text{TT}]$
	$[\text{FT}] = \left(\frac{-b + \sqrt{b^2 - 4ac}}{2a}\right)$
bioT (mol/l)	$[\text{BioT}] = N \times [\text{FT}]$

SHBG, sex hormone binding globulin; TT, total testosterone; FT, free testosterone; BioT = bioavailable testosterone.
Assumptions: All concentrations ([Albumin]; [SHBG], and [TT]) should be given in mol/l.
Molar weight albumin is 69,000 g/mol; molar weight SHBG is 90,000 g/mol.
Affinity constants:
$K_{SHBG} = 1 \times 10^9$ l/mol; $K_{ALB} = 3.6 \times 10^4$ l/mol (Vermeulen et al., 1999)
$K_{SHBG} = 5.97 \times 10^8$ l/mol; $K_{ALB} = 4.06 \times 10^4$ l/mol (Sodergard et al., 1982)
$K_{SHBG} = 1.4 \times 10^9$ l/mol; $K_{ALB} = 1.3 \times 10^4$ l/mol (Emadi-Konjin et al., 2003)

with equations that were previously reported (Miller et al., 1990; Shibasaki et al., 2008). Both equations yielded lower concentrations, which can be due to the fact that the specific binding to CBG was not included into the equations.

25.5.3 Special Applications II: Drugs Affecting the Central Nervous System

Drugs for the treatment of conditions related to the central nervous system (CNS) need to pass the blood–brain barrier as well as the blood-cerebrospinal barrier that separates the systemic circulation and the brain. The process of transporting drugs into brain can be described by complex equilibrium models, in which the rate of transportation depends on the rate of influx, the dissociation of drug bound to plasma proteins as well as the rates of elimination or metabolism of the drug in the brain compartment (Howard et al., 2010). Lipophilic, unbound and nonionized drugs may undergo passive transcellular diffusion from blood to the CNS (He et al., 2009). As cerebrospinal fluid (CSF), produced in the brain, contains only trace levels of proteins (15–45 mg/dl), drug concentrations measured in the CSF may be similar to the actual unbound concentration of the drug in plasma.

25.5.4 Special Applications III: Chiral Drugs

The enantiomers of a racemic drug can enantioselectively bind to plasma proteins, resulting in independent pharmacokinetic properties for each enantiomer. Jensen et al. (2011) evaluated seven ultrafiltration devices for the separation of free warfarin (>99% is bound) from plasma proteins. A chiral LC–MS/MS method was developed for the analysis of both *S*-warfarin (eutomer) and *R*-warfarin (distomer) in ultrafiltrate. The Centrifree device from Millipore (Section 25.4.2) was chosen for preparation of the ultrafiltrate. The device was reported to be reproducible (<1.6%) and free of NSB free concentrations in ultrafiltrate samples from 72 patients were between 2.2 and 17.5 ng/ml for *R*-warfarin and between 0.8 and 6.5 ng/ml for *S*-warfarin. The mean values obtained for the PB were 99.1% and 99.4%, for *R*-warfarin and *S*-warfarin, respectively.

REFERENCES

Arthur CL, Pawliszyn Solid phase microextraction with thermal desorption using fused silica optical fibers. J Anal Chem 1990;62(19):2145–2148.

Barré J, Chamouard JM, Houin G, Tillement JP. Equilibrium dialysis, ultrafiltration, and untracentrifugation compared for determining the plasma-protein-binding characteristics of valproic acid. Clin Chem 1985;31(1):60–64.

Chen Y, Yazdanpanah M, Wang XY, Hoffman BR, Diamandis EP, Wong P-Y. Direct measurement of serum free testosterone by ultrafiltration followed by liquid chromatography tandem mass spectrometry. Clin Biochem 2010;43:490–496.

Cumming DC, Wall SR. Non-sex hormone-binding globulin-bound testosterone as a marker for hyperandrogenism. J Clin Endocrinol Metab 1985;61:873–876.

De Ronde W, Van der Schouw YT, Pols HAP, et al. Calculation of bioavailable and free testosterone in men: a comparison of 5 published algorithms. Clin Chem 2006;52(9):1777–1784.

Deng Y, Wong H, Graham RA, et al. Determination of unbound vismodegib (GDC-0449) concentration in human plasma using rapid equilibrium dialysis followed by solid phase extraction and high-performance liquid chromatography coupled to mass spectrometry. J Chromatogr B Analyt Technol Biomed Life Sci 2011;879:2119–2126.

DeVan ML, Bankson DD, Abadie JM. To what extent are free testosterone (FT) values reproducible between the two Washingtons, and can calculated FT be used in lieu of expensive direct measurements? Am J Clin Pathol 2008;129:459–463.

Dhinsa S, Furlanetto R, Vora M, Ghanim H, Chaudhuri A, Dandona P. Low estradiol concentrations in men with subnormal testosterone concentrations and type 2 diabetes. Diabetes Care 2011;34(8):1854–1859.

Ekins R. The free hormone hypothesis and measurement of free hormones [Editorial]. Clin Chem 1992;38:1289–1293.

Emadi-Konjin P, Bain J, Bromberg IL. Evaluation of an algorithm for calculation of serum "bioavailable"testosterone (BAT). Clin Biochem 2003;36:591–596.

European Medicines Agency (EMA). Committee for medicinal products for human use (CHMP). Note for guidance on the evaluation of the pharmacokinetics of medicinal products in patients with impaired renal function. (Adopted by CHMP: June 22–23, 2004). Available at http://www.ema.europa.eu/ema/. Accessed Mar 28, 2013.

Fung EN, Chen Y-H, Lau YY. Semi-automatic high-throughput determination of plasma protein binding using a 96-well plate filtrate assembly and fast liquid chromatography-tandem mass spectrometry. J Chromatogr B Analyt Technol Biomed Life Sci 2003;795:187–194.

Hackbarth JS, Hoyne JB, Grebe SK, Singh RJ. Accuracy of calculated free testosterone differs between equations and depends on gender and SHBG concentration. Steroids 2011;76:48–55.

He H, Lyons KA, Shen X, et al. Utility of unbound plasma drug levels and P-glycoprotein transport data in prediction of central nervous system exposure. Xenobiotics 2009;39(09):687–693.

Howard ML, Hill JJ, Galluppi GR, McLean MA. Plasma protein binding in drug discovery and development. Comb Chem High Throughput Screen 2010;13(2):170–187.

Jensen BP, Chin PKL, Begg EJ. Quantification of total and free concentrations of R- and S-warfarin in human plasma by ultrafiltration and LC-MS/MS. Anal Bioanal Chem 2011;401:2187–2193.

Kwong TC. Free drug measurements: methodology and clinical significance. Clin Chim Acta 1985;151(3):192–216.

Larsen HS, Chin PK, Begg EJ, Jensen BP. Quantification of total and unbound concentrations of lorazepam, oxazepam and temazepam in human plasma by ultrafiltration and LC-MS/MS. Bioanalysis 2011;3(8):843–852.

Lee KJ, Mower R, Hollenbeck T, et al. Modulation of nonspecific binding in ultrafiltration protein binding studies. Pharm Res 2003;20(7):1015–1021.

Li W, Lin H, Smith HT, Tse FLS. Developing a robust ultrafiltration-LC-MS/MS method for quantitative analysis of unbound vadimezan (ASA404) in human plasma. J Chromatogr B Analyt Technol Biomed Life Sci 2011;879:1927–1933.

Ly LP, Handelsman DJ. Empirical estimation of free testosterone from testosterone and sex hormone-binding globulin immunoassays. Eur J Endocrinol 2005;152:471–478.

Miller PF, Bowmer CJ, Wheeldon J, Brocklebank JT. Pharmacokinetics of prednisolone in children with nephrosis. Arch Dis Child 1990;65:196–200.

Morley JE, Patrick P, Perry HM3rd. Evaluations of assays available to measure free testosterone. Metabolism 2002;51(5):554–559.

Morris PD, Malkin CJ, Channer KS, Jones TH. Mathematical comparison of techniques to predict biologically available testosterone in a cohort of 1072 men. Eur J Endocrinol 2004;151:241–249.

Musteata FM. Monitoring free drug concentrations: challenges. Bioanalysis 2011;3(15):1753–1768.

Musteata FM, Pawliszyn J, Qian MG, Wu JT, Miwa GT. Determination of drug plasma protein binding by solid-phase microextraction. J Pharm Sci 2006;95:1712–1722.

Musteata ML, Musteata FM, Pawliszyn J. Biocompatible solid phase microextraction coatings based on polyacylonitrile and SPE phases. Anal Chem 2007;79:6903–6911.

Prasad BB, Tiwari K, Singh M, Sharma PS, Patel AK, Srivastava S. Zwitterionic molecularly imprinted polymer-based solid-phase micro-extraction coupled with molecularly imprinted polymer sensor for ultra-trace sensing of L-histidine. J Sep Sci 2009;32(7):1096–1105.

Rakhila H, Rozek T, Hopkins A, et al. Quantitation of total and free teriflunomide (A77 1726) in human plasma by LC-MS/MS. J Pharm Biomed Anal 2011;55:325–331.

Ruiter AFC, Teeninga N, Nauta J, Endert E, Ackermans MT. Determination of unbound prednisolone, prednisone and cortisol in human serum and saliva by on-line solid-phase extraction liquid chromatography tandem mass spectrometry and potential implications for drug monitoring of prednisolone and prednisone in saliva. Biomed Chromatogr 2012;26(7):789–796.

Schuhmacher J, Buhner K, Witt-Laido A. Determination of the free fraction and relative free fraction of drugs strongly bound to plasma proteins. J Pharm Sci 2000;89:1008–1021.

Shackleton G. Special populations: protein binding aspects. In: Vogel HG, Maas J, Gebauer A, editors. Drug Discovery and Evaluation: Methods in Clinical Pharmacology. 1st ed. Berlin Heidelberg: Springer-Verlag; 2011. p 67–71.

Shibasaki H, Nakayama H, Furuta T, et al. Simultaneous determination of prednisolone, prednisone, cortisol, and cortisone in plasma by GC-MS: estimating unbound prednisolone concentration in patients with nephrotic syndrome during oral prednisolone therapy. J Chromatogr B Analyt Technol Biomed Life Sci 2008;870:164–169.

Sodergard R, Backstrom T, Shanbhag V, Carstensen H. Calculation of free and bound fractions of testosterone and estradiol-17ß to human plasma proteins at body temperature. J Steroid Biochem 1982;16:801–810.

Streit F, Binder L, Hafke A, et al. Use of total and unbound Imatinib and metabolite LC-MS/MS assay to understand individual responses in CML and GIST patients. Ther Drug Monit 2011;33(5):632–643.

Taylor S, Harker A. Modification of the ultrafiltration technique to overcome solubility and non-specific binding challenges associated with the measurement of plasma protein binding of corticosteroids. J Pharm Biomed Anal 2006;41(1):299–203.

Tremblay RR, Dube JY. Plasma concentrations of free and non-TeBG bound testosterone in women on oral contraceptives. Contraception 1974;10:599–605.

US FDA—Guidance for Industry. Pharmacokinetics in patients with impaired renal function – study design, data analysis, and impact on dosing and labeling. US Department of Health and Human Services, Center for Drug Evaluation and Research and Center for Veterinary Medicine (1998). Available at www.fda.gov/drugs/guidancecomplianceregulatoryinformation/guidances/default.htm. Accessed Mar 28, 2013.

Vermeulen A, Verdonck L, Kaufman JM. A critical evaluation of simple methods for the estimation of free testosterone in serum. J Clin Endocrinol Metab 1999;84:3666–3672.

Waters NJ, Jones R, Williams G, Sohal B. Validation of a rapid equilibrium dialysis approach for the measurement of plasma protein binding. J Pharm Sci 2008;97:4586–4595.

Wells DA. High throughput bioanalytical sample preparation. Methods and Automation Strategies, 1st ed. Amsterdam, The Netherlands: Elsevier; 2003.

Zhan Y, Musteata FM, Basset FA, Pawliszyn J. Determination of free and deconjugated testosterone and epitestosterone in urine using SPME and LC-MS/MS. Bioanalysis 2011;3(1): 23–30.

Zhang J, Musson DG. Investigation of high-throughput ultrafiltration for the determination of an unbound compound in human plasma using liquid chromatography and tandem mass spectrometry with electrospray ionization. J Chromatogr B Analyt Technol Biomed Life Sci 2006;843(1):47–56.

26

LC-MS BIOANALYSIS OF DRUGS IN BILE

HONG GAO AND JOHN WILLIAMS

26.1 BILE AND BILIARY EXCRETION

Bile analysis plays an important role in the optimization of drug candidates due to bile's integral role in the absorption, distribution, metabolism, and excretion (ADME) pathways associated with drug disposition. The qualitative and quantitative assessment of exogenous components in bile is critical to understanding the interplay between pharmacokinetic and pharmacological effects initiated after drugs are administered. Due to the complex composition of bile, its analysis by LC-MS must be carefully conducted to ensure accurate, precise, and reproducible results.

Bile is a biofluid composed primarily of solubilized lipids that are secreted by liver hepatocytes and stored in the gallbladder for most vertebrates (Figure 26.1). Bile is released from the gall bladder into the duodenum during digestion to aid the absorption of lipids and lipid soluble nutrients. The components of bile, as a percentage of total solutes, are bile acids (61%), fatty acids, phospholipids (15%), cholesterol (9%), protein (7%), bilirubin (3%), and other components (5%) (Kristiansen et al., 2004). From the gastrointestinal functioning point of view, the bile acids are the most important components in the bile. Bile acids are bioactive molecules derived from cholesterol by multiple pathways of cytochrome-mediated metabolism in the liver. Their synthesis is initiated by α-hydroxylation at the 7 position of cholesterol followed by side chain shortening and conjugation (Norlin and Wikvall, 2007). In addition to aiding the absorption of dietary lipids, bile acids are important signaling molecules in the regulation of cholesterol homeostasis, liver function, and enterohepatic circulation (Chiang, 2004; Eloranta and Kullak-Ublick, 2005; Thomas et al., 2008). In humans, the primary bile acids formed in hepatocytes are cholic acid and chenodeoxycholic acid. These acids are extensively conjugated with taurine or glycine prior to release in the small intestine and ultimately pumped back into the portal circulation to be recycled in the liver in a process known as enterohepatic circulation. Bile conjugates that are not recycled are deconjugated and further dehydroxylated by gut flora in the colon to produce secondary bile acids, notably deoxycholic and lithocholic acids. Conjugation increases the solubility of bile acids in the small intestine by lowering the pK_a. The amphipathic nature of conjugated bile salts promotes the formation of micelles in the intestinal lumen, which emulsifies lipophilic compounds. Bile acids are also essential cofactors in the degradation of lipids by pancreatic lipase; thus facilitating lipophilic absorption by intestinal cells (Hofmann, 1999). The multitude of bile acids, fatty acids, phospholipids, and other endogenous components that compose bile are the main sources of interference in the LC-MS analysis of bile. The matrix effects caused by these sources must be minimized and accounted for to properly analyze drugs and metabolites in bile.

Biliary excretion is a critically important function of bile. Bile flow serves as the medium through which endogenous and xenobiotic compounds can be excreted into the feces for direct elimination or reabsorption and elimination by the kidneys through urine. Endogenous molecules that undergo biliary excretion include bile salts, cholesterol, steroid hormones, bilirubin, and others. Bilirubin is a toxic breakdown product of hemoglobin that is continually generated from the recycling of red blood cells. Accumulation of bilirubin leads to the telltale signs of jaundice, yellowing of the skin and sclera of the eyes. Elevated levels of bilirubin in the body may indicate certain diseases that affect the liver such as hepatitis C or liver cancer. Xenobiotics are often

Handbook of LC-MS Bioanalysis: Best Practices, Experimental Protocols, and Regulations, First Edition. Edited by Wenkui Li, Jie Zhang, and Francis L.S. Tse.

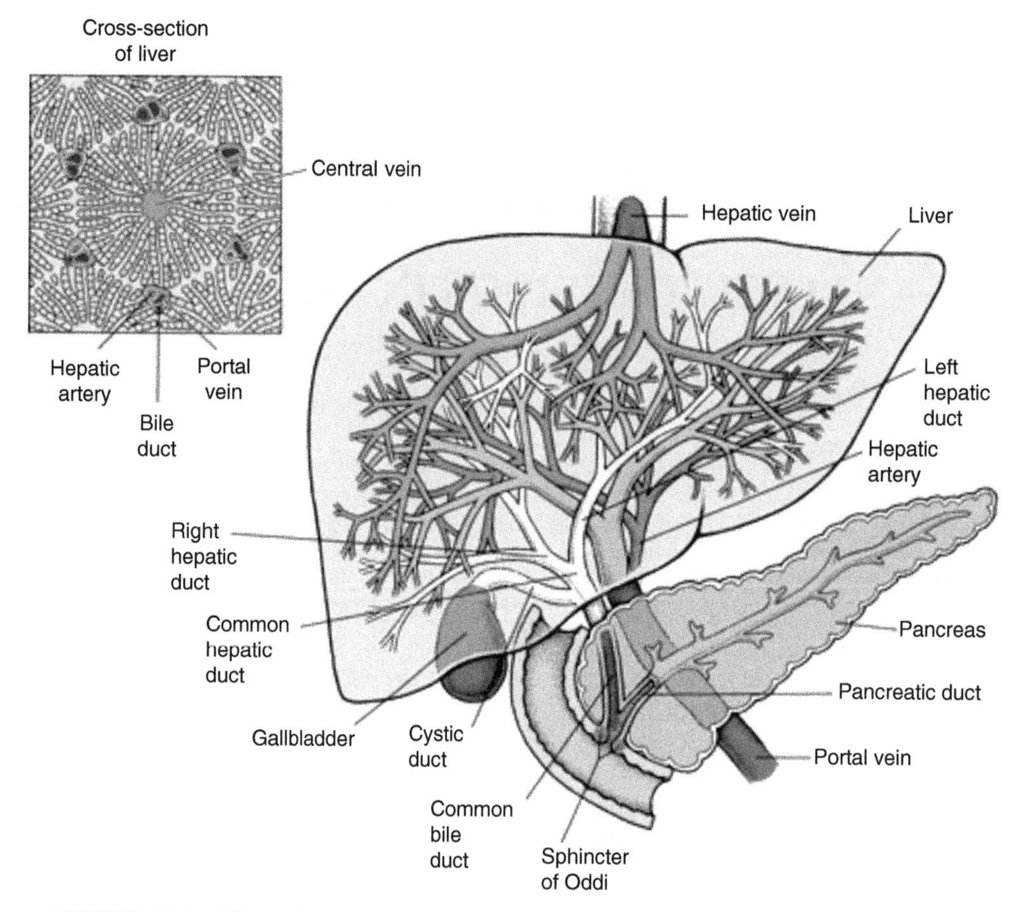

FIGURE 26.1 View of the liver, biliary ducts, and gallbladder where bile is stored and released. Source: http://www.merckmanuals.com/home/liver_and_gallbladder_disorders/biology_of_the_liver_and_gallbladder/overview_of_the_liver_and_gallbladder.html, accessed Mar 28, 2013 (used with permission of Merck & Co., Inc.).

biotransformed in the liver into phase I or II metabolites to increase solubility of lipophilic species and increase their clearance in bile. The tendency for a compound to be eliminated by bile is affected by many physicochemical factors, such as the polarity, charge, and size of the molecule and the elimination can proceed by either passive diffusion or active transporter-mediated uptake (Roberts et al., 2002). In humans, drugs with a molecular weight of greater than 500 g/mole and with both polar and lipophilic groups are more likely to be excreted in bile while smaller molecules are generally excreted at insignificant levels. Conjugation, particularly with glucuronic acid, facilitates biliary excretion by increasing aqueous solubility (Wang et al., 2006).

Once in the bile flow, compounds can be eliminated directly through the feces. Often, however, compounds excreted in bile will be reabsorbed into the portal blood flow from the gastrointestinal tract. Molecules can be either intact or as a phase I or II metabolites that have reverted to their unmetabolized forms due to bacterial or enzymatic action in the gut. This absorption and reabsorption process is defined as enterohepatic circulation (Kaye, 1976; Rollins and Klaassen, 1979; Klaassen and Watkins, 1984) shown in Figure 26.2. When enterohepatic circulation occurs, it may significantly extend the half-life of a drug and prolong exposure of the drug to the body, which can lead to toxicity due to over exposure. As shown in the pharmacokinetic profile of estrone in Figure 26.3, several enterohepatic recycling phases give rise to secondary peaks that more than doubles the half-life of

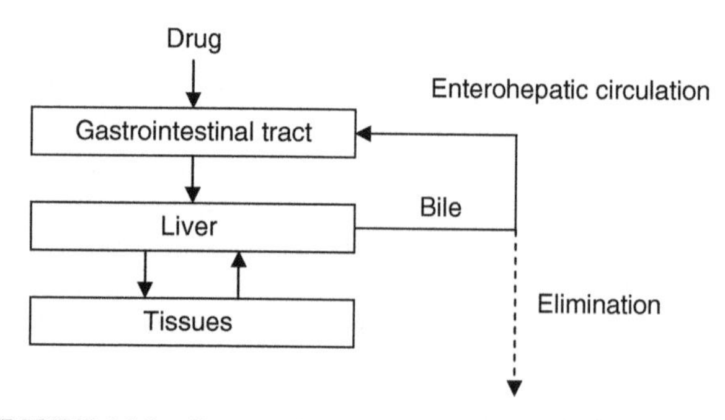

FIGURE 26.2 Drug reabsorption from gastrointestinal tract in the enterohepatic cycle.

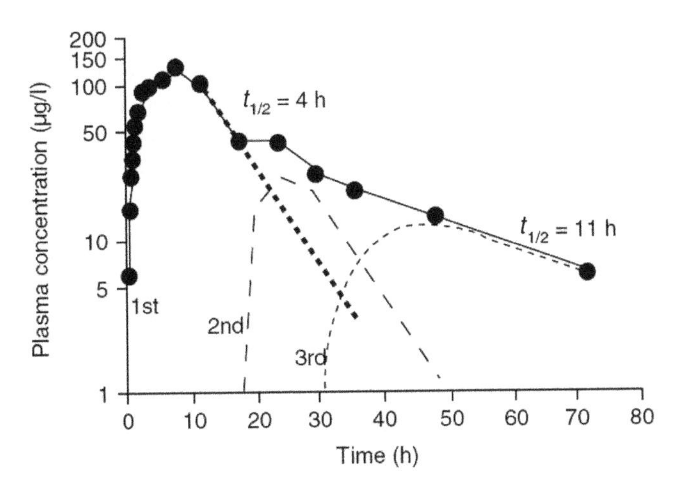

FIGURE 26.3 Plasma pharmacokinetic profile of estrone after oral dosing of estradiol 1.5 mg in a desogestrel combination to 18 healthy postmenopausal women. Also shown are the second and third enterohepatic recycling phases leading to secondary peaks and an increase in half-life of estrone (Vree and Timmer (1998), used with permission of John Wiley & Sons).

the drug (Vree and Timmer, 1998). The effects of enterohepatic circulation on a drug's disposition have been described and can be accurately predicted (Colburn and Lucek, 1988; Peris-Ribera et al., 1992; Wang and Reuning, 1992).

The impairment of biliary excretion is known as cholestasis and can arise from physical obstruction of the bile pathway or in response to inhibition of transporters associated with bile flow. Obstructive cholestasis can occur due to gallstone formation (Phemister et al., 1939), tumor growth (Yang et al., 2011), or inflammation and scarring of the bile ducts (Aller et al., 2008). Metabolic cholestasis typically arises from inhibition or down regulation of transporter proteins located in the hepatocytes and cholangiocytes. Over 14 transporters have been associated with bile formation including the ATP-dependent export pumps MDR3, MRP2, and BSEP, responsible for translocation of bile salts, phospholipids, and organic anions through the bile canaliculi (Trauner et al., 1998). Upon competitive inhibition of these transporters by xenobiotics such as cyclosporin A and rifampicin, accumulation of bile salts will lead to cholestasis and eventual liver disease (Stieger et al., 2000).

Since biliary excretion plays an important role in drug metabolism and disposition, bioanalysis of drugs and their metabolite(s) in bile using LC-MS has become more widely used to answer critical questions in drug discovery and development. For instance, prior to preclinical and early clinical mass balance (ADME) studies, which provide toxicity and therapeutically effective dose range information, complementary tissue distribution studies in rats and bile-duct-cannulated animals are usually conducted to provide more pharmacokinetic information. The detection of metabolites present in bile can facilitate the elucidation of metabolic pathways of a test compound. Quantitative and qualitative changes in metabolic pathways for a compound can be assessed *in vivo* by measuring the amount of the dosed drug excreted in the bile and by determining possible shifts in the profile of metabolites. Transgenic mice expressing biotransformation enzymes or knockout mice lacking biotransformation enzymes or membrane-associated transporters are now commonly used to address the *in vivo* role in the metabolism and excretion of xenobiotics. These studies provide tissue distribution and biliary elimination profiles of the drug candidate. Evaluation of the metabolites in bile also provides important information at this stage (Sakaguchi et al., 2006). Data obtained from the animal metabolism and excretory pathways of a drug are useful to design a clinical mass balance study (Zhang and Comezoglu, 2008). Observations have revealed that in addition to plasma matrix pharmacokinetic characterization, bile matrix interaction helps to refine the pharmacokinetic disposition of drug and the associated metabolite(s) (Srinivas and Mullangi, 2011). It is important to characterize hepatobiliary elimination of drug(s) in bile because bile can be a potential site of drug–drug interaction that may result in significant alterations in systemic or hepatic exposure. Utilizing bile samples in conjunction with plasma samples to probe mechanistic action of the drug disposition, including the potential for drug–drug interaction in preclinical species, have been reported (Liu and Tsai, 2002; Tsai, 2005; Huang et al., 2009) and are drawing more attentions. In forensic investigations, bile samples are often collected in postmortem cases for the determination of possible poisoning or drug abuse since drug-induced deaths are commonly reported cases (Agarwal and Lemos, 1996; Kintz et al., 1997; Tracqui et al., 1998; Gaulier et al., 2000). For example, insulin, a macromolecule of interest in forensic toxicology, is normally present in bile at lower concentrations than those found in plasma (Bailey et al., 1976) and is often measured in bile to confirmation if an over dose has been administered.

26.2 BILE SAMPLE COLLECTION

Proper sample collection is the first step to accurately determine reliable drug levels in bile. The bile sample collection and pretreatment prior to bioanalysis can directly impact the data integrity. To ensure that measured drug levels accurately reflect the concentration at the time of sampling, it is important to prevent any degradation of the parent drug and back conversion of conjugated metabolites. The ratio of the parent drug and its metabolite must be maintained in the bile matrix during bile sampling. Therefore, the method of bile sample collection, handling, and storage must be well designed and defined in the protocol prior to the investigation.

For studies in nonclinical species, mouse and rat bile samples are typically collected via bile duct cannulation of animals over a defined time interval, which varies from 10 min to 6 h, up to 24 h or more, after drug administration (Zhou and Chowbay, 2002; Ma et al., 2006; Lin et al., 2008). Bile from cannulated animals can be collected either manually or automatically via the cannulate in programmed intervals rather than in a single continuous drain. For some drugs, first pass metabolism in the liver and excretion into the bile occurs rapidly after dosing, necessitating a prompt initial collection of bile sample. Many researchers have chosen to sample in brief time intervals immediately after dose administration. As time lapse from initial dose increases, so does the time interval of subsequent sample collection (Zhou and Chowbay, 2002; Ma et al., 2006). To facilitate bile sample collection, the animals are anesthetized by intraperitoneal injection and an abdominal incision is made for common bile duct cannulation. The animal body temperature is usually maintained under a heating lamp during the experimental procedures to prevent hypothermic alterations of the bile flow (Li et al., 2011a). In many cases, the time interval chosen for collection of bile samples is equally important as the technique used when the data is used to derive the maximum benefits in the calculation of the bile related pharmacokinetic parameters in conjunction with plasma pharmacokinetics (Srinivas and Mullangi, 2011).

For animal pharmacokinetic and pharmacology studies, microdialysis is another option for timed intervals that allows bile perfusate to be collected into refrigerated fraction collectors. Microdialysis has proven to be a beneficial technique for collecting the unbound fraction of a drug from biological fluids. The advantages of microdialysis included minimal sample collection volume, no apparent interference with physiological functions, sampling for extended periods of time and minimal sample loss. A major limitation that can lead to poor recovery of the test compound is the choice of an unsuitable probe. Therefore, choosing the proper microdialysis probe will ensure the maximum recovery. There are several vital parameters that must be considered when selecting a probe including membrane composition, probe shaft length, and the molecular weight cutoff (Srinivas and Mullangi, 2011). The molecular weight cutoff of commercially available microdialysis probes covers a wide range of approximately 20–100 kD. While water-soluble compounds generally diffuse freely across the microdialysis membrane, diffusion of highly lipophilic analytes is more limited and variable. The microdialysis for evaluation of estradiol and fusidic acid was successful but not for corticosteroids (Stahl et al., 2002). The recovery of water-soluble compounds usually decreases rapidly if the molecular weight of the analyte exceeds 25% of the membrane's molecular weight cutoff. The shunt microdialysis probe is widely used for microdialysis (Huff et al., 1999) enabling the monitoring of the concentration of low molecular weight species in the bile of conscious, freely moving rats (Scott and Lunte, 1993; Gunaratna et al., 1994; Hadwiger et al., 1994). The shunt, which carries the bile flow, is of appropriate dimension for implanting into the bile duct of an adult rat. Suspended inside the shunt is a linear microdialysis probe that continuously samples low molecular weight compounds from the bile. In anesthetized rats, a shunt probe can be inserted at the upstream end toward the liver to easily collect the bile (Huff et al., 1999).

In clinical studies, bile can be collected by direct aspiration of the gallbladder, by duodenal intubation or by T-tube drainage. The Meltzer–Lyon method (Brown, 1920) is a commonly used method for bile sample collection due to ease of operation. With this method, bile samples are collected via the drainage of the biliary tract through a duodenal tube after relaxation of the sphincter of Oddi by the local application to the duodenum of a concentrated solution of magnesium sulfate (Brown, 1920). Bile samples are also collected from patients by puncture of the gallbladder or bile duct at surgery or immediately after percutaneous trans-hepatic cholangiographic drainage. Many studies of drug excretion in human bile have been performed by means of T-tubes from the bile ducts of patients undergoing surgery for obstructive jaundice (Nishijima et al., 1997). This method of bile collection is not ideal because bile flow and composition is often severely altered during the study period if not all bile is collected and enterohepatic circulation is partially interrupted. Loc-I-Gut is a single-pass perfusion technique with multichannel tubing (Bergman et al., 2006; Persson et al., 2006) and is an option for accessing human bile. Using Loc-I-Gut, bile samples are withdrawn via the Loc-I-Gut device from the proximal jejunum. In contrast with clinical investigation, bile sample is directly collected from the postmortem body in the case for comprehensive forensic toxicology analysis (Vanbinst et al., 2002).

Research demonstrates that some conjugated drug metabolite(s), especially the acyl glucuronide metabolite, can easily undergo hydrolysis and intramolecular acyl migration; therefore, the storage condition following bile sample collection needs to be well defined and followed closely to avoid any potential degradation (Khan et al., 1998). Several major factors such as pH, sample handling temperature, and storage conditions should be considered. To facilitate a suitable pH environment, prefilled tubes with buffer are normally used to collect bile samples. Some conjugated molecules are sensitive to unsuitable pH. Acyl glucuronidated molecules are easily hydrolyzed at basic pH. Many buffers at different pH can be used based on the chemical structure of the analyte and its conjugate. Ammonium acetate, sodium acetate, phosphoric buffer, and other acidic solutions are a few examples widely used for stabilizing conjugates in bile fluids (Fura et al., 2003). Additionally, it is of critical importance to cool down the bile samples immediately on an ice bath following collection (Scherer et al., 2009). Once a time point collection is completed, bile samples should be immediately transferred into a freezer maintained at −20°C or −80°C. Any failure of these three steps may cause inaccurate measurements of the analyte.

26.3 BILE SAMPLE PREPARATION

Compared to other biofluid samples (plasma, serum, CSF, and urine), bile samples contain a large amount of acids, salts, and endogenous molecules. Those small molecular entities often interfere with the bioanalysis of drugs and their metabolites leading to severe matrix effects, particularly when mass spectrometry method is used. As discussed in other chapters, those interfering components must be removed or minimized during the sample preparation prior to LC-MS analysis. Simple dilution-and-shoot (D-S) of supernatant after centrifugation, protein precipitation (P-P) followed by centrifugation, liquid–liquid extraction (LLE) followed by evaporation and reconstitution, solid phase extraction (SPE), column-switching techniques, and injection of protein-free fraction following microdialysis are the numerous extraction techniques commonly used for bile sample preparation for bioanalysis.

For drug discovery studies, where robustness is often traded for speed, sample D-S (Van Asperen et al., 1998; Chen et al., 2008) and direct injection of the supernatant following centrifugation (Wang et al., 2006) of the bile sample may be a suitable choice. In the analysis of the bile marker compound indocyanine green dosed in dogs, the bile samples were simply diluted with water and injected onto LC-MS for analysis after centrifugation (Chen et al., 2008). The method was validated for sample analysis for the pharmacokinetics study. P-P with iced acetonitrile (Zhou and Chowbay, 2002; Dhananjeyan et al., 2006), methanol (Li et al., 2011b), or 50:50 acetonitrile/methanol is also popularly applied in laboratories (Bi et al., 2006; Moon et al., 2006; Li et al., 2011b). For samples collected from microdialysis, protein-free fraction can be directly injected following the microdialysis procedures (Tsai et al., 1999; Tsai, 2005). It is suggested that D-S and P-P method should be used cautiously as bile sample composition varies from time point to time point when bile samples are collected for pharmacokinetic purposes. The variation may cause inaccuracy of the concentration data when severe interference exists. Diluting bile samples with drug-free blank plasma at a high ratio (e.g., >2), then performing P-P may overcome some variation in bile composition and yield more consistent results (Lee et al., 1996; Van Asperen et al., 1998). This process has shown to minimize matrix differences in recovery, selectivity, and specificity by converting the bile matrix into a plasma matrix allowing it to be analyzed using validated methods for plasma analysis.

LLE and SPE methods are more advanced sample preparation techniques that can remove most interfering components from the bile matrix. The organic solvent eluent may be directly injected onto LC-MS system (Cheng et al., 2007) for analysis when the solvent is compatible with the mobile phases and the sensitivity is not sacrificed. Otherwise, the organic fraction is usually evaporated and the sample is reconstituted with mobile phase before injection onto LC-MS system (Tietz et al., 1984; Singh et al., 2003; Zhang et al., 2005; Ma et al., 2006; Bansal et al., 2008). Single organic solvents such as ethyl ether, methyl tert-butyl ether (MTBE), hexane, dichloromethane (DCM), or mixtures of several solvents are selected for LLE extraction based on the physicochemical properties of the analyte. For a better recovery and less matrix interference, Ma et al. (2006) extracted naringenin with ethyl acetate after diluting the bile tenfold with saline and Zhang et al. (2009) extracted felbinac using ethyl ether:dichloromethane (60:40, v/v) after acidifying the bile matrix. Both methods have been validated. LLE can now be automated using a 96-well format liquid handler to more efficiently handle bile sample extraction. Also, SPE using different cartridge chemistries (e.g., HPB, MCX, and C18) are widely used for bile sample preparation. Bile samples are usually loaded onto SPE cartridges without further dilution yet some drug stabilizers or pH adjustment may be necessary (Hasler et al., 1993; Alnouti et al., 2007). Many types of SPE cartridges are available on the market. The selection of a suitable extraction cartridge depends on the physicochemical properties of the test compound. Usually, single extraction works for the majority of drug-like compounds. However, multiple or combinations of different extraction techniques should be considered when single extraction is inadequate. In some cases, bile samples may be precipitated and followed by either LLE or SPE, or multiple LLE with different solvents, or multiple SPE extraction with different sorbents.

Column switching can be used as an online cleanup technique that can reduce off-line sample extraction steps and accelerate bioanalysis (Baek et al., 1999). This technique utilizes many components including a six-port automatic switching valve, a cleanup column, concentrating column, and an analytical column. Although the column switching is a complex system setup, it allows direct inject of bile sample without significant interference from endogenous components. Sakaguchi (Sakaguchi et al., 2006) successfully validated a column-switching method for quantitative analysis of 8-hydroxyquinoline and its glucuronide conjugate using two-dimensional extraction LC. The two dimensions of chromatography were reversed-phase chromatography and cation-exchange chromatography. The endogenous components were removed by the cleanup column while the analytes were retained on the analytical column for analysis.

26.4 BILE SAMPLE LC-MS BIOANALYSIS

The endogenous components such as bile acids, salts, cholesterol, bilirubin and phospholipids in bile (Bickel et al., 1970) present big challenges for bioanalytical scientists. Those components often interfere with the detection and quantitation of small molecule drug entities and their metabolites via coelution and ion suppression or enhancement. It is the best to remove those interfering compounds from the samples prior to LC-MS analysis. Sample extraction using suitable technique discussed earlier plays an important role in sample analysis because it improves the sample cleanliness significantly. However, many interfering components can still remain in the extract and their impacts to the LC-MS analysis are not non-negligible. Therefore, in addition to the sample extraction, high-resolution chromatography, which separates analytes from the interferences chromatographically, is a further important step to improve the data quality. To achieve this, a high-resolution analytical column with high plate number and small particle size should be used and mobile phase, flow rate, and column oven temperature should be optimized. In addition to selecting proper sample extraction and optimizing LC conditions, optimization of mass spectrometry parameters such as gas flow, temperature, voltages, suitable ionization mode (ESI or APCI) and polarity (positive or negative) is often used to minimize the ion suppression or enhancement from the matrix. For instance, Felbinac (or biphenylylacetic acid, molecular weight: 212.2 g/mol) is an active metabolite of fenbufen, which is used to treat muscle inflammation and arthritis (Kohler et al., 1980). With such a low molecule weight, felbinac is strongly interfered with by the components in bile matrix (Zhang et al., 2009). To support a preclinical pharmacokinetic study dosed with felbinac, felbinac trometamol and probenecid, Zhang et al. (2009) developed and validated a method capable of separating those three drugs and analyzed the bile samples successfully. They also determined that negative ionization mode was more suitable for febinac, felbinac trometamol, and probenecid compared to the positive mode. Morphine (molecular weight: 285.3 g/mol) is also a low molecular weight compound, which makes quantitation of morphine and its metabolite, morphine-3-glucuronide (M3G) in bile a challenge. In a study for investigating the influence of phenobarbital on morphine metabolism and disposition, Alnouti et al. (2007) used a long Luna C18 column (150 $\times$ 2.1 mm, 5 μm) with acetonitrile and 7.5 mM ammonium format at pH 9.3 as mobile phases to achieve the separation of morphine from M3G and other interferences in bile samples.

As discussed in Section 26.2, many drugs are metabolized to form glucuronic acid conjugates that are excreted into bile and then eventually eliminated from the body. Compared to plasma, bile samples may contain more glucuronide metabolites. This can make LC-MS bioanalysis more challenging because many glucuronides are labile and tend to deconjugate during sample collection, preparation, or ionization in the mass spectrometer. If deconjugation occurs, the concentration of glucuronide metabolites will be under estimated while the parent drug will be overestimated, particularly if the metabolite and parent drug are coeluted. Generally, the conjugated metabolite is more polar and therefore elutes earlier than its parent drug under reversed-phase chromatographic conditions. Still, the polarity difference between parent drug and metabolite may not be sufficient to enable a good separation. It is always a good practice to confirm if the parent drug is glucuronidated and present in the bile. The presence of a glucuronide metabolite can be determined by monitoring the neutral loss of m/z 176, the (M + H + 176) peak in a full scan mass spectrum, or by adding m/z 176 to the parent transition and monitoring for both the parent and product transition in multiple reaction monitoring mode. Once the glucuronide peak is observed in the bile sample, the bioanalytical method should be developed accordingly to avoid potential problems. Measures such as sample stabilization, chromatographic separation, and mass spectrometer ion source condition optimization can be helpful. As discussed in Section 26.2, bile samples should be acidified to pH 3 to prevent the degradation of glucuronide conjugates. To separate the parent drug from its glucuronide metabolite, chromatographic conditions such as analytical column selection, mobile phase composition, elution gradient, flow rate, and temperature should be investigated. Also, the ion source gas flow, temperature, and declustering potential should be optimized to keep the glucuronide conjugate intact. Depending on the functional group and position of the parent drug that is conjugated with glucuronic acid, there may be several different types of glucuronides that behave distinctly differently from one to another during the LC-MS bioanalysis. For example, compared to a phenolic glucuronide, an acyl glucuronide is much less stable, particularly at physiological pH, and may require more attention when quantitative bioanalysis is considered.

Although there are no regulatory guidelines for bile matrix sample analysis, many researchers have developed and validated their LC-MS bioanalytical methods for analyzing drugs and metabolites in bile. In these cases, the same criteria for plasma methods are used for validating the bile matrix method. Full validation of bile matrix methods may not be necessary if the concentration of the test compound is negligible (Srinivas and Mullangi, 2011). For analysis of drugs and the associated metabolites in bile, a fit-for-purpose strategy is commonly employed in many laboratories (Sporkert et al., 2007). Before extensive method development, it is important to determine if the drug or its metabolites are present in the bile by spot check. Validated plasma assays may be employed to determine if the bile contains high levels of analyte and metabolite(s) (Singh et al., 2003). When the concentration of drug and its metabolite(s) in bile are predicted to be high it becomes more critical to drug discovery and development decision making and full method validations are recommended. Many validated methods have evaluated calibration ranges, recovery, matrix effects, selectivity, stability, accuracy, and precision (Ma et al., 2006; Chen et al., 2008;

Zhang et al., 2009). In cases of some well-behaved drugs and metabolites, partial validation may be sufficient (Alnouti et al., 2007; Lin et al., 2008). Partial validation should include the evaluation of selectivity/specificity because selectivity/specificity in bile matrix may differ from the plasma or serum matrix. For instance, bile may present a host of endogenous materials that may cause interference, affect recovery, and cause anomalies in the chromatography. Since bile samples may manifest higher levels of phase II metabolites in addition to the parent drug, the applicability of partial method validation may be difficult to determine. This is often the case in situations where the plasma/serum matrix method has not been validated for phase II metabolite since it does not carry the same level of phase II metabolite(s) (Srinivas and Mullangi, 2011). In such cases, the quantitative method of measuring phase II metabolites in bile should be validated.

For bile sample analysis, calibration standards and quality control samples should be prepared in the same matrix if possible. Commercially available drug-free blank bile is often chosen for this purpose. But one needs to be mindful of the possible inconsistency between purchased blank and the incurred bile samples, especially when D-S and P-P extraction methods are used. It is recommended to use high ratio of dilution to minimize the difference between the matrix sources in this scenario. It may become less problematic when the samples are thoroughly cleaned via LLE and SPE extraction, but the matrix effect should still be evaluated.

Including internal standard for bile sample preparation and analysis is a common practice to overcome systematic experimental errors and compensate for the recovery losses of analytes. There are two main types of internal standards that are used: the structurally close analogs and isotopic-labeled compounds. Isotopic-labeled internal standards are superior due to their ability to match extractability, matrix effect, and coelution with the analyte. However, isotopic-labeled compounds are not readily available for purchase and are difficult to synthesize. Structurally similar analogs can be beneficial by matching chromatographic and ionization characteristics on mass spectrometry but can be affected by different interferences in bile extracts than the analyte. The selection of a suitable internal standard really depends on the stage of project (discovery vs. development) and availability of the isotope-labeled compound.

26.5 SUMMARY

Bioanalysis of drugs and their associated metabolite(s) in bile provides important insights into drug disposition. Bile analysis can also help elucidate ADME issues encountered during drug optimization and development, especially when used in conjunction with other matrix data. Due to the instability of analytes and their conjugates in bile, appropriate bile sample collection should be considered as an important part of the sample analysis. Due to the complexity of the bile composition, the method development and incurred sample analysis present a great analytical challenge. Matrix effect and inconsistency of sample matrix are the major issues associated with bile sample analysis by LC-MS. To achieve optimal bioanalytical results, bile samples should be extracted via suitable preparation techniques. The bioanalytical method for quantifying glucuronide-conjugated metabolite needs to be well developed to prevent the degradation and underestimation during quantification. Depending on the goal of the investigation, fit-for-purpose strategy can be used to meet the needs of the study.

REFERENCES

Agarwal A, Lemos N. Significance of bile analysis in drug-induced deaths. J Anal Toxicol 1996;20(1):61–63.

Aller MA, Arias JL, Garcia-Dominguez J, Arias JI, Duran M, Arias J. Experimental obstructive cholestasis: the wound-like inflammatory liver response. Fibrogenesis Tissue Repair 2008;1(1):6.

Alnouti YM, Shelby MK, Chen C, Klaassen CD. Influence of phenobarbital on morphine metabolism and disposition: LC-MS/MS determination of morphine (M) and morphine-3-glucuronide (M3G) in Wistar-Kyoto rat serum, bile, and urine. Curr Drug metab 2007;8(1):79–89.

Baek M, Rho YS, Kim DH. Column-switching high-performance liquid chromatographic assay for determination of asiaticoside in rat plasma and bile with ultraviolet absorbance detection. J Chromatogr 1999;732(2):357–363.

Bailey CJ, Flatt PR, Atkins TW, Matty AJ. Immunoreactive insulin in bile and pancreatic juice of rat. Endocrinol Exp 1976;10(2):101–111.

Bansal T, Awasthi A, Jaggi M, Khar RK, Talegaonkar S. Development and validation of reversed phase liquid chromatographic method utilizing ultraviolet detection for quantification of irinotecan (CPT-11) and its active metabolite, SN-38, in rat plasma and bile samples: application to pharmacokinetic studies. Talanta 2008;76(5):1015–1021.

Bergman, E, Forsell, P, Tevell, A, Persson, E M, Hedeland, M, Bondesson, U, Knutson, L, and Lennernas, H. Biliary secretion of rosuvastatin and bile acids in humans during the absorption phase. Eur J Pharm Sci 2006;29(3–4):205–214.

Bi, Y A, Kazolias, D, Duignan DB. Use of cryopreserved human hepatocytes in sandwich culture to measure hepatobiliary transport. Drug Metab Dispos 2006;34(9):1658–1665.

Bickel MH, et al. Metabolism and biliary excretion of the lipophilic drug molecules, imipramine and desmethylimimpramine in rat, experiments in vivo and with isolated perfused livers. Biochem Pharmacol 1970;19:2425–2435.

Brown G. The Meltzer-Lyon method in the diagnosis of the biliary tract. J Am Med Assoc 1920;75(21):1414–1416.

Chen CY, Fancher RM, Ruan Q, Marathe P, Rodrigues AD, Yang Z. A liquid chromatography tandem mass spectrometry method for the quantification of indocyanine green in dog plasma and bile. J Pharmaceut Biomed 2008;47(2):351–359.

Cheng CL, Kang GJ, Chou CH. Development and validation of a high-performance liquid chromatographic method using fluorescence detection for the determination of vardenafil in small volumes of rat plasma and bile. J Chromatogr A 2007;1154(1–2):222–229.

Chiang JY. Regulation of bile acid synthesis: pathways, nuclear receptors, and mechanisms. J Hepatol 2004;40(3):539–551.

Colburn WA, Lucek RW. Noncompartmental area under the curve determinations for drugs that cycle in the bile. Biopharm Drug Dispos 1988;9(5):465–475.

Dhananjeyan MR, Erhardt PW, Corbitt C. Simultaneous determination of vinclozolin and detection of its degradation products in mouse plasma, serum and urine, and from rabbit bile, by high-performance liquid chromatography. J Chromatogr A 2006;1115(1–2):8–18.

Eloranta JJ, Kullak-Ublick GA. Coordinate transcriptional regulation of bile acid homeostasis and drug metabolism. Arch Biochem Biophys 2005;433(2):397–412.

Fura A, Harper TW, Zhang H, Fung L, Shyu WC. Shift in pH of biological fluids during storage and processing: effect on bioanalysis. J Pharmaceut Biomed 2003;32(3):513–522.

Gaulier JM, Marquet P, Lacassie E, Dupuy JL, Lachatre G. Fatal intoxication following self-administration of a massive dose of buprenorphine. J Forensic Sci 2000;45(1):226–228.

Gunaratna PC, Wilson GS, Slavik M. Pharmacokinetic studies of alpha-difluoromethylornithine in rabbits using an enzyme-linked immunosorbent assay. J Pharmaceut Biomed 1994;12(10):1249–1257.

Hadwiger ME, Telting-Diaz M, Lunte CE. Liquid chromatographic determination of tacrine and its metabolites in rat bile microdialysates. J Chromatogr B Biomed Appl 1994;655(2):235–241.

Hasler F, Krapf R, Brenneisen R, Bourquin D, Krahenbuhl S. Determination of 18 beta-glycyrrhetinic acid in biological fluids from humans and rats by solid-phase extraction and high-performance liquid chromatography. J Chromatogr 1993;620(1):73–82.

Hofmann AF. Bile acids: the good, the bad, and the ugly. News Physiol Sci 1999;14:24–29.

Huang SP, Lin LC, Wu YT, Tsai TH. Pharmacokinetics of kadsurenone and its interaction with cyclosporin A in rats using a combined HPLC and microdialysis system. J Chromatogr B Analyt Technol Biomed Life Sci 2009;877(3):247–252.

Huff J, Heppert K, Davies M. The microdialysis shunt probe: profile of analytes in rats with erratic bile flow of rapid changes in analyte concentration in the bile. Current Sep 1999;18(3):85–90.

Kaye CM. The biliary excretion of acebutolol in man. J Pharm Pharmacol 1976;28(5):449–450.

Khan S, Teitz D S, Jemal M. Kinetic analysis by HPLC-electrospray mass spectrometry of the pH-dependent acyl migration and solvolysis as the decomposition pathways of ifetroban 1-O-acyl glucuronide. Anal Chem 1998;70(8):1622–1628.

Kintz P, Jamey C, Tracqui A, Mangin P. Colchicine poisoning: report of a fatal case and presentation of an HPLC procedure for body fluid and tissue analyses. J Anal Toxicol 1997;21(1):70–72.

Klaassen CD, Watkins JB, 3rd. Mechanisms of bile formation, hepatic uptake, and biliary excretion. Pharmacol Rev 1984;36(1):1–67.

Kohler C, Tolman E, Wooding W, Ellenbogen L. A review of the effects of fenbufen and a metabolite, biphenylacetic acid, on platelet biochemistry and function. Arzneimittel-Forschung 1980;30(4A):702–707.

Kristiansen TZ, Bunkenborg J, Gronborg M, et al. A proteomic analysis of human bile. Mol Cell Proteomics 2004;3(7):715–728.

Lee ED, Lee SD, Kim WB, Yang J, Kim SH, Lee MG. Determination of a new carbapenem antibiotic, DA-1131, in rat plasma, urine, and bile by column-switching high-performance liquid chromatography. Res Commun Mol Path 1996;94(2):171–180.

Li CY, Qi LW, Li P. Correlative analysis of metabolite profiling of Danggui Buxue Tang in rat biological fluids by rapid resolution LC-TOF/MS. J Pharmaceut Biomed 2011a;55(1):146–160.

Li X, Delzer J, Voorman R, De Morais SM, Lao Y. Disposition and drug-drug interaction potential of veliparib (ABT-888), a novel and potent inhibitor of poly(ADP-ribose) polymerase. Drug Metab Dispos 2011b;39(7):1161–1169.

Lin LC, Chen YF, Lee WC, Wu YT, Tsai TH. Pharmacokinetics of gastrodin and its metabolite p-hydroxybenzyl alcohol in rat blood, brain and bile by microdialysis coupled to LC-MS/MS. J Pharmaceut Biomed 2008;48(3):909–917.

Liu SC, Tsai TH. Determination of diclofenac in rat bile and its interaction with cyclosporin A using on-line microdialysis coupled to liquid chromatography. J Chromatogr B Analyt Technol Biomed Life Sci 2002;769(2):351–356.

Ma Y, Li P, Chen D, Fang T, Li H, Su W. LC/MS/MS quantitation assay for pharmacokinetics of naringenin and double peaks phenomenon in rats plasma. Int J Pharm 2006;307(2):292–299.

Moon YJ, Sagawa K, Frederick K, Zhang S, Morris ME. Pharmacokinetics and bioavailability of the isoflavone biochanin A in rats. AAPS J 2006;8(3):E433–42.

Nishijima T, Nishina M, Fujiwara K. Measurement of lactate levels in serum and bile using proton nuclear magnetic resonance in patients with hepatobiliary diseases: its utility in detection of malignancies. Jpn J Clin Oncol 1997;27(1):13–17.

Norlin M, Wikvall K. Enzymes in the conversion of cholesterol into bile acids. Curr Mol Med 2007;7(2):199–218.

Peris-Ribera JE, Torres-Molina F, Garcia-Carbonell MC, Aristorena JC, Granero L. General treatment of the enterohepatic recirculation of drugs and its influence on the area under the plasma level curves, bioavailability, and clearance. Pharmaceut Res 1992;9(10):1306–1313.

Persson EM, Nilsson RG, Hansson GI, et al. A clinical single-pass perfusion investigation of the dynamic *in vivo* secretory response to a dietary meal in human proximal small intestine. Pharmaceut Res 2006;23(4):742–751.

Phemister DB, Aronsohn HG, Pepinsky R. Variation in the cholesterol, bile pigment and calcium salts contents of gallstones formed in gallbladder and in bile ducts with the degree of associated obstruction. Ann Surg 1939;109(2):161–186.

Roberts MS, Magnusson BM, Burczynski FJ, Weiss M. Enterohepatic circulation: physiological, pharmacokinetic and clinical implications. Clin Pharmacokinet 2002;41(10):751–790.

Rollins DE, Klaassen CD. Biliary excretion of drugs in man. Clin Pharmacokinet 1979;4(5):368–379.

Sakaguchi T, Yamamoto E, Kushida I, Kajima T, Asakawa N. Effective on-line purification for cationic compounds in rat bile using a column-switching LC technique. J Pharmaceut Biomed 2006;40(2):345–352.

Scherer M, Gnewuch C, Schmitz G, Liebisch G. Rapid quantification of bile acids and their conjugates in serum by liquid chromatography-tandem mass spectrometry. J Chromatogr B Analyt Technol Biomed Life Sci 2009;877(30):3920–3925.

Scott DO, Lunte CE. *In vivo* microdialysis sampling in the bile, blood, and liver of rats to study the disposition of phenol. Pharmaceut Res 1993;10(3):335–342.

Singh SK, Mehrotra N, Sabarinath S, Gupta RC. HPLC-UV method development and validation for 16-dehydropregnenolone, a novel oral hypolipidaemic agent, in rat biological matrices for application to pharmacokinetic studies. J Pharmaceut Biomed 2003;33(4):755–764.

Sporkert F, Augsburger M, Giroud C, Brossard C, Eap CB, Mangin P. Determination and distribution of clotiapine (Entumine) in human plasma, post-mortem blood and tissue samples from clotiapine-treated patients and from autopsy cases. Forensic Sci Int 2007;170(2–3):193–199.

Srinivas NR, Mullangi R. An overview of various validated HPLC and LC-MS/MS methods for quantitation of drugs in bile: challenges and considerations. Biomed Chromatogr 2011;25(1–2):65–81.

Stahl M, Bouw R, Jackson A, Pay V. Human microdialysis. Curr Pharm Biotechno 2002;3(2):165–178.

Stieger B, Fattinger K, Madon J, Kullak-Ublick GA, Meier PJ. Drug- and estrogen-induced cholestasis through inhibition of the hepatocellular bile salt export pump (Bsep) of rat liver. Gastroenterology 2000;118(2):422–430.

Thomas C, Pellicciari R, Pruzanski M, Auwerx J, Schoonjans K. Targeting bile-acid signalling for metabolic diseases. Nat Rev 2008;7(8):678–693.

Tietz PS, Thistle JL, Miller LJ, Larusso NF. Development and validation of a method for measuring the glycine and taurine conjugates of bile acids in bile by high-performance liquid chromatography. J Chromatogr 1984;336(2):249–257.

Tracqui A, Kintz P, Ludes B. Buprenorphine-related deaths among drug addicts in France: a report on 20 fatalities. J Anal Toxicol 1998;22(6):430–434.

Trauner M, Meier PJ, Boyer JL. Molecular pathogenesis of cholestasis. New Engl J Med 1998;339(17):1217–1227.

Tsai TH. Concurrent measurement of unbound genistein in the blood, brain and bile of anesthetized rats using microdialysis and its pharmacokinetic application. J Chromatogr A 2005;1073(1–2):317–322.

Tsai TH, Tsai TR, Chen YF, Chou CJ, Chen CF. Determination of unbound 20(S)-camptothecin in rat bile by on-line microdialysis coupled to microbore liquid chromatography with fluorescence detection. J Chromatogr 1999;732(1):221–225.

Van Asperen J, Van Tellingen O, Beijnen JH. Determination of doxorubicin and metabolites in murine specimens by high-performance liquid chromatography. J Chromatogr 1998;712(1–2):129–143.

Vanbinst R, Koenig J, Di Fazio V, Hassoun A. Bile analysis of drugs in postmortem cases. Forensic Sci Int 2002;128(1–2):35–40.

Vree TB, Timmer CJ. Enterohepatic cycling and pharmacokinetics of oestradiol in postmenopausal women. J Pharm Pharmacol 1998;50(8):857–864.

Wang J, Nation RL, Evans AM, Cox S, Li J. Determination of antiviral nucleoside analogues AM365 and AM188 in perfusate and bile of the isolated perfused rat liver using HPLC. Biomed Chromatogr 2006;20(3):244–250.

Wang YM, Reuning RH. An experimental design strategy for quantitating complex pharmacokinetic models: enterohepatic circulation with time-varying gallbladder emptying as an example. Pharmaceut Res 1992;9(2):169–177.

Yang H, Li TW, Peng J, et al. A mouse model of cholestasis-associated cholangiocarcinoma and transcription factors involved in progression. Gastroenterology 2011;141(1):378–388, 388 e1–4.

Zhang C, Wang L, Yang W, et al. Validated LC-MS/MS assay for the determination of felbinac: application to a preclinical pharmacokinetics study of felbinac trometamol injection in rat. J Pharmaceut Biomed 2009;50(1):41–45.

Zhang D, Comezoglu SN. ADME Studies in Animals and Humans: Experimental Design, Metabolite Profilinf and Identification, and Data Presentation, Drug Metabolism and Drug Design and Development. John Wiley & Sons, Inc.; 2008.

Zhang W, Zhang C, Liu R, et al. Quantitative determination of Astragaloside IV, a natural product with cardioprotective activity, in plasma, urine and other biological samples by HPLC coupled with tandem mass spectrometry. J Chromatogr B Analyt Technol Biomed Life Sci 2005;822(1–2):170–177.

Zhou Q, Chowbay B. Determination of doxorubicin and its metabolites in rat serum and bile by LC: application to preclinical pharmacokinetic studies. J Pharmaceut Biomed 2002;30(4):1063–1074.

27

LC-MS BIOANALYSIS OF INTRACELLULAR DRUGS

Fagen Zhang and Michael J. Bartels

27.1 INTRODUCTION

In recent years, evidence has shown that the efficacy and toxicity of many drugs (especially some of intracellular drugs such as anticancer and antiretroviral drugs) are related to the concentrations of parent drug and/or active metabolites within in cell (Becher et al., 2002a). Therefore, during drug development, it is crucial to understand the pharmacokinetics (PK) of those drugs at intracellular level. Also, for some drugs with narrow therapeutic ranges, but high inter- and intrapatient variability in PK, monitoring intracellular drug and/or active metabolite concentrations in routine clinical is necessary to achieve safe and efficacious therapeutic treatment. For those purposes, appropriate quantitation methods for determining the intracellular concentrations of drugs and/or drug metabolites are critical. Traditionally, methods used for quantifying drugs or drug metabolites mainly include reversed phase high-performance liquid chromatography (RP-HPLC) in combination with UV absorbance (El-Gindy et al., 2000; Lal et al., 2003; Kaji et al., 2005; Park et al., 2008), fluorescence (Nirogi et al., 2006; Raghavamenon et al., 2009; Konda et al., 2010; Chen et al., 2011), and electrochemical detection (Chen et al., 1996; Reynolds et al., 1992). For drugs with no appropriate ultraviolet or fluorescent chromophores, derivatization followed by HPLC-UV/FL analysis, HPLC-ELSD (Forget and Spagnoli, 2006), or radioimmunoassay (RIA) (Kaul et al., 1996; Kominami et al., 1996, 1997, 1999; Zhou et al., 1996) techniques have been often used. However, in most cases, these traditional methods are not sensitive or reproducible enough for the analysis of trace-level intracellular concentration of drugs and/or their metabolites. Since the introduction of electrospray ionization (ESI), atmospheric pressure chemical ionization (APCI), and atmospheric pressure photoionization (APPI) techniques for atmospheric pressure ionization (API), liquid chromatographic separation coupled with API mass spectrometric (LC-API/MS) detection has been widely used for the bioanalysis of intracellular drugs and/or metabolites in support of toxicokinetics (TK) and PK in drug development. Compared to the above traditional methods, liquid chromatography–mass spectrometry (LC-MS) is excellent in terms of assay specificity and analysis speed and sensitivity (Covey et al., 1986; Xu et al., 2007). The number of publications on quantitative bioanalysis of intracellular drugs via LC-MS has increased exponentially over the last few years (Claire, 2000; Chi et al., 2001; Pruvost et al., 2001; Becher et al., 2002a, 2002b, 2003; Hennere et al., 2003; Huang et al., 2004; Rouzes et al., 2004; Colombo et al., 2005; Ehrhardt et al., 2007; Pruvost et al., 2008; Jansen et al., 2009a, 2009b, 2011; Bushman et al., 2011; Coulier et al., 2011).

The general procedures involved in LC-MS bioanalysis of intracellular drugs are three steps: (a) cell isolation, counting, and lysis, (b) cleanup of lysed cell, and (c) LC-MS analysis (Figure 27.1). In this chapter, sample treatment techniques (including cell isolation, cell counting, and cell lysis), LC-MS instrumental setups including the chromatographic separation and detection mode and application examples will be summarized.

27.2 SAMPLE PREPARATION

In most cases, the intracellular level of drugs *in vivo* or *in vitro* cannot be quantified directly without proper sample pretreatment. The sample pretreatment includes isolation, counting,

Handbook of LC-MS Bioanalysis: Best Practices, Experimental Protocols, and Regulations, First Edition. Edited by Wenkui Li, Jie Zhang, and Francis L.S. Tse.

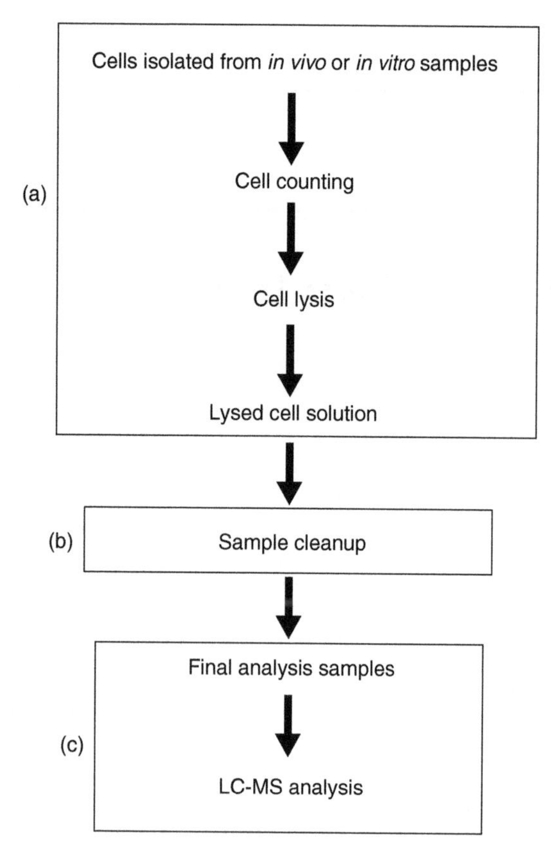

FIGURE 27.1 The general procedures involved in LC-MS bioanalysis of intracellular drugs are three steps: (a) cell isolation, counting, and lysis; (b) cleanup of lysed cell; and (c) LC-MS analysis.

and lysis of the cells, and cleanup of the harvested subcellular fraction of cells to remove endogenous proteins, carbohydrates, salts, lipids, and other endogenous compounds. Although sample pretreatment for LC-MS quantitation does not need to be as elaborate as for other LC assays (such as LC-UV quantitation), it is critical to remove matrix components that might interfere with the MS detection of the analytes of interest via ion suppression. Therefore, protein precipitation (PPT), solid phase extraction (SPE), and liquid–liquid extraction (LLE) are commonly employed.

27.2.1 Cell Separation and Counting

During preclinical or clinical development of drugs, blood, plasma, or serum are the common matrices for the determination of TK or PK of the test compounds, while peripheral blood mononuclear cells (PBMC) is the most commonly used matrix for intracellular drug bioanalysis. PBMC is a combination of both lymphocytes and monocytes in the blood. PBMC makes up about one-third of the white blood cells (WBC), also known as leukocytes. PBMC can be usually prepared from relatively large volumes (2–25 ml) of collected whole blood by density gradient techniques such as Ficoll-pague (Boltz et al., 1987; Slusher et al., 1992; Kawashima et al., 2000; Gahan et al., 2001; Nilsson et al., 2008) or directly drawn into commercial available tubes containing a density gradient (Jansen et al., 2011). Special precaution should be used to prevent erythrocyte contamination (Jansen et al., 2011) during PBMC preparation. PBMC with pink or red color, an indication of erythrocyte contamination, showed a larger and more variable matrix effect during LC-MS analysis (Shi et al., 2002) than the PBBC without color. To prevent contamination, an extra erythrocyte lysis procedure (such as lysis with an ammonium chloride solution) should be applied (Durand-Gasselin et al., 2007). Due to continuing cell integrity during the isolation process, any ongoing analyte metabolism *ex vivo* can be prevented by quick isolation process on ice (Pruvost et al., 2001).

To express the final intracellular concentration of drug per cell, the number of isolated cells should be counted and quantified. This can be done by traditional quick methods such as a hemocytometer (Jemal et al., 2003), microscope (Becher et al., 2003), or flow cytometry (Ahmed et al., 2001; Mascola et al., 2002). To test cell viability, the trypan blue method where a vital stain (a diazo dye) was used to selectively color dead cells blue and live cells with intact cell membranes were not colored (Baran et al., 2011) can be used. In some occasions, total protein or DNA in the isolated cells can also be used to express the final intracellular drug/or metabolite concentrations normalized to cell quantity (Jansen et al., 2009a, 2011) (Table 27.1).

Although most published intracellular drug bioanalysis methods are focused on PBMC, intracellular drug bioanalysis have also been reported for other *in vitro* cell types such as CTM-4 cells (Cahours et al., 2001), MCF7 and human kidney HEK 293T cells (Cahours et al., 2001; Kim et al., 2005, 2009; Luo et al., 2007; Seifar et al., 2008; Chen et al., 2009; Jauhiainen et al., 2009; Jansen et al., 2009a; Furugen et al., 2011; Huang et al., 2011; Lan et al., 2011; Mlejnek et al., 2011). In those cases, cells either treated with drugs or not were then washed with PBS buffer (or other related buffer dependent on the cell types) and trypsinized, and cell number is usually counted by hemocytometer (Thomsen et al., 2004; Kim et al., 2009) (Table 27.2).

27.2.2 Cell Lysis

After cells are isolated, a critical step is cell lysis to recover the drug or drug metabolites from the cells. Methods used in cell lysis include use of organic solvent, organic acid, basic conditions, and lysis instrumentation (such as sonicator) (Tables 27.1 and 27.2). For example, to quantify the cyclosporine level in PMBC, PMBC was lysed with organic solvent (MeOH) containing internal standard (Ansermot et al., 2007). Intracellular delavirdine (a nonn-ucleoside reverse transcriptase inhibitor) in PMBC was quantified

TABLE 27.1 Quantitation of Intracellular Drugs in PMBC by LC-MS

Analyte's name	Class of drugs	Cell preparation methods	Cell lysis procedure	Sample preparation	Chromatographic type	Column type	Mobile phase	MS type (polarity)	LOQ	Validation	References
Neviparine; delavidine; efavirenz	Non-nucleoside reverse transcriptase inhibitors	Whole blood was centrifuged by conventional Ficoll gradient	Cells were lysed with tris-HCl/MeOH (30/70; v/v)	Lysed cell solutions were centrifuged and supernatant was evaporated and final residue reconstituted in tris solution (0.05 M, pH 7.4) for analysis	RPLC	Nova Pak C_{18}	(a) 10% ACN + 90% 25 mM ammonium acetate with 0.1% AA; (b) 90% ACN + 10% 25 mM ammonium acetate with 0.1% AA	QqQ (+)	0.5 ng/ml	Yes	Pelerin et al. (2005)
Stavudine	Nucleotide reverse transcriptor inhibitor	Whole blood was centrifuged by conventional Ficoll gradient	Cells were lysed with tris-HCl/MeOH (30/70; v/v)	Lysed cell solution was centrifuged and supernatant was evaporated to certain level (120–150 μl) for analysis	RPLC	SMT-C_{18}, OD 5 μm, 100 Å (150 mm × 2.1 mm)	(a) DMH (10 mM) + ammonium formate (3 mM, pH 11.5); (b) DMH (20 mM) + ammonium formate (6 mM)/ACN (1:1, v/v)	**QqQ (−)**	138 fmol/7 ml Blood (9.8 fmol /10^6 cells)	Yes	Pruvost et al. (2001)
Cyclosporine A	Immuno-suppressive drug	Whole blood was centrifuged by conventional Ficoll gradient	Cells lysed with MeOH containing internal standards (IS)	Lysed cell solution was centrifuged and supernatant was evaporated and final residue was reconstituted in MeOH for analysis	RPLC with online cleanup (column switch)	Clean Column: Xterra MS C_8, 5 μm, 2.1 mm × 10 mm; Analytica Column: Xterra MS C_{18}, 5 μm, 2.1 × 50 mm	(a) MeOH; (b) water	**QqQ (+)**	5 ng/ml (0.5 fg/cell)	Yes	Ansermot et al. (2007)
Zidovudine triphosphate	Nucleotide reverse transcriptor inhibitor	Whole blood was centrifuged by conventional Ficoll gradient	Cells were lysed with tris-HCl/MeOH (30/70; v/v)	Lysed cell solution was centrifuged and supernatant was first dephosphorylation by acid phosphatase, then the resulting solution was cleaned up by waters oasis HLB SPE and final solution was evaporated and reconstituted in MeOH for analysis	RPLC	Waters XTerraTM RP_{18} 3.5 μm, 2.1 mm × 150 mm	(a) Water 0.1% AA; (b) water with 10% ACN;	**QqQ (+)**	5 fmol/10^6 cells	Yes	King et al. (2006a)

(continued)

TABLE 27.1 ***(Continued)***

Analyte's name	Class of drugs	Cell preparation methods	Cell lysis procedure	Sample preparation	Chromatographic type	Column type	Mobile phase	MS type (polarity)	LOQ	Validation	References
Didanosine and stavudine	Nucleotide reverse transcriptor inhibitor	Commercial PBMC	Cells were lysed by instrument (Fischer dismembrator)	Lysed cell solution was centrifuged and supernatant was cleaned up SPE with Bond Elut cartridge	RPLC	Keystone, BDS C_{18} column (5 μm, 4.6 mm × 150 mm)	(a) Methanol–water (16:84 v/v) containing 0.05% TFA, 1 mM ammonium formate; (b) methanol–water (80:20, v/v) containing 0.05% TFA, 1 mM ammonium formate	**QqQ** (+)	0.4 ng/ml	Yes	Huang et al. (2004)
Amprenavir (APV), atazanavir (ATZ), efavirenz (EFV), indinavir (IDV), lopinavir (LPV), nelfinavir (NFV), nevirapine (NVP), ritonavir (RTV), saquinavir (SQV), and tipranavir (TPV). APV and ATZ	Antiretroviral drugs	Whole blood was centrifuged by conventional Ficoll gradient	Cell were lysed by instrument (ultrasound bath)	Lysed cell solution was centrifuged and supernatant was directly analyzed	RPLC	Waters Symmetry shield RP_{18}, 3.5 μm (2.1 mm × 50 mm)	(a) 10 mM ammonium acetate/10mM formic acid; (b) ACN/10 mM formic acid	**QqQ** (+)	0.0125–0.2 ng per ml of cell extract	Yes	Elens et al. (2009)
Lamivudine (3TC), zidovudine (AZT), stavudine (d4T), abacavir (ABC), etc.	Antiretroviral drugs	Whole blood was centrifuged by conventional Ficoll gradient	Cells were with MeOH/water (1:1, v/v) overnight at 5°C	Lysed cell solution was cleaned up by LLE extract with MeOH/water (1:1, v/v) and extract was diluted by water and analyzed	RPLC (HPLC and UPLC)	Different columns: Polymeric column 5 μm (2.1 mm × 150 mm); X-Terra Column MS 5 μm (2.1 mm × 150 mm); Supercogel ODP-50, 5 μm (2.1 mm × 150 mm)	(a) Water with 0.1 % formic acid; (b) ACN with 0.1% formic acid	**QqQ** (+)	Not reported	Partially validated	Becher et al. (2002a)

Lopinavir and ritonavir	HIV protease inhibitor	Whole blood was cen-trifuged by conven-tional Ficoll gradient	Cells were first alkalinized with 2 mM K_3PO_4, then *t*-butylmethy-lether was added and whole solution followed by utltrasonica-tion.	LLE extract was evaporated to dryness and reconstituted in mobile phase and analyzed	RPLC	Phenomenex® JupiterTM Proteo column (C_{12}, 4 μm, 100 mm × 2.1 mm)	**(a)** Water with 0.1% acetic acid including 20 mM ammonium acetate; (b) acetonitrile	QqQ (+)	0.1 ng/cell pellet (~3 × 106 cells)	Partially validated	(Ehrhardt et al. (2007)
Deoxyribonu cleotide triphosphate	Antiretroviral drugs	Whole blood was cen-trifuged by conven-tional Ficoll gradient	Cells were lysed with 0.05 M tris-HCl/MeOH (30/70, v/v)	Lysed cell solution was centrifuged and supernatant was transferred to tube and evaporated to certain level (120–150 μl)	RPLC	Supelcogel, ODP-50 (5 μm 2.1 mm × 150 mm)	(a) Buffer A was composed of 50% 6 mM DMH in 20 mM ammonium formate (pH 5); (b) 50% A + 50% acetonitrile	**QqQ (-)**	0.03–0.04 pmol/10^6 cells to	Partially validated	Hennere et al. (2003)
Nucleosides and nucleotides	Antiviral-retroviral drugs	Whole blood was cen-trifuged by conven-tional Ficoll gradient	Cells were lysed with perchloric acid	Lysed cell solution was treated KOH, then centrifuged and supernatant was used for analysis	RPLC	Hypercarb column (5 μm, 2.1 mm × 100 mm)	(a) 1 mM ammonium acetate in ACN/H_2O (15:85, v/v) with pH 5; (b) 25 mM ammonium bicarbonate in ACN/H_2O(15:85, v/v)	**QqQ** (+)	4.29–52.7 nM	Validated	Jansen et al. (2009b)
Thymidine triphosphate	Antiviral-retroviral drugs	Whole blood was cen-trifuged by conven-tional Ficoll gradient	Cells were lysed with 70% MeOH at −20°C	Lysed cell solution was centrifuged and supernatant was transferred to tube for direct analysis	RPLC	C_8 Waters Sentry (5 μm, 3.9 mm × 20 mm)	(a) 2 mM ammonium buffer; (b) acetonitrile	**QqQ** (+)	1.4 ng/ml	Validated	Chi et al. (2001)
Indinavir, amprenavir, saquinavir, ritonavir, nelfinavir, lopinavir, atazanavir, efavirenz	Anti-HIV drugs	Whole blood was cen-trifuged by conven-tional Ficoll gradient	Cells were lysed by sonicating in 50% MeOH	Lysed cell solution was centrifuged and supernatant was transferred to tube for direct analysis	RPLC	C_{18} Symme-tryShieldTM (3.5 μm 2.1 mm × 30 mm)	(a) 2 mM ammonium buffer containing 0.1% formic acid; (b) acetonitrile containing 0.1% formic acid	**QqQ** (+)	0.2–0.4 ng/ml	Validated	Colombo et al. (2005)

(continued)

TABLE 27.1 (*Continued*)

Analyte's name	Class of drugs	Cell preparation methods	Cell lysis procedure	Sample preparation	Chromatographic type	Column type	Mobile phase	MS type (polarity)	LOQ	Validation	References
Nucleotides	Reverse transcriptase inhibitor	Commercial PBMC	Cells were lysed with a mixing solvent containing 10% 10 mM ammonium phosphate, 20% deionized water, and 70% MeOH at −80°C	Lysed cell solution was cleaned up with C_{18} SPE column	IP-HPLC	Xterra2 RP18, (3.5 μm, 1.0 mm × 100 mm) and Xterra2 MS, (3.5 μm, 1.0 mm × 100 mm)	10 mM ammonium phosphate, pH 6.4, with 2 mM TBAH (tetrabutylammonium hydroxide) and 15% acetonitrile	**QqQ (-)**	0.08 picomol/10(6) cells	Partially validated	Claire (2000)
Didanosine and stavudine	Antiretroviral drugs	Whole blood was centrifuged by conventional Ficoll gradient	Cells were lysed with tris-HCl/MeOH (30/70; v/v))	Lysed cell solution was centrifuged and supernatant was evaporated and reconstituted in MeOH for analysis	IP-HPLC	Supelcogel, ODP-50 (5 μm, 2.1 mm × 150 mm)	Ion-pairing reagent (dimethylhexylamine) mixed with an ammonium formate buffer and acetonitrile	**QqQ** (−)	6.1 and 5.3 fmol/106 cells	Validated	Becher et al. (2003)
Atazanavir	HIV protease inhibitor	Commercial PBMC	Cells were lysed by sonicator	Lysed cell solution was centrifuged and supernatant was evaporated and reconstituted in water with 0.1% acetic acid and followed by SPE work station cleanup	RPLC	YMC Basic (5 μm, 50 mm × 2 mm)	Isocratic mobile phase containing water/acetonitrile with 0.025% formic acid	**QqQ** (+)	5 fmol/106 cells	Validated	Jemal et al. (2003)
Nucleosides and nucleotides	Antiretroviral drug	Whole blood was centrifuged by conventional Ficoll gradient	Cells were lysed with 100 l PBS at −80°C	Lysed cell solution was extracted by LLE with MeOH/Water	RPLC and IP-HPLC	different C_{18} columns	(a) Water with 0.1% formic acid; (b) MeOH with 0.1% formic acid (RPLC); and (a) 5 mM HA in water (pH 6.3); (b) 5 mM HA in 90% acetonitrile/10% water (pH 8.5) (IP-HPLC)	**QqQ** (+) **(RPLC); QqQ** (−) **(IP-HPLC)**	1–2 nM (RPLC); 1–5 nM (IP-HPLC)	Validated	Coulier et al. (2011)

Nucleoside analog	Antiviral drugs	Commercial PBMC	Cells were lysed with MeOH (30/70; v/v))	Lysed cell solution was centrifuged and supernatant was clean up by SPE cartridges	RPLC	Synergi Polar RP (2.5 μm, 2.0 mm × 100 mm)	(a) 2% acetonitrile and 0.1% formic acid in ultrapure water; (b) 6% 2-propanol and 0.1% acetic acid in ultrapure water	**QqQ** (+)	2.5 fmol to 0.1 p mol	Partially validated	Bushman et al. (2011)
Nucleoside triphosphate (D-D4FC-TP)	Anti-HIV drug	Commercial PBMC	Cells were lysed with MeOH (30/70; v/v))	Lysed cell solution was cleaned up by protein precipitation procedure	Anion-exchange HPLC (WAX-LC)	Keystone BioBasic (5 μm, 1 mm × 20 mm)	Acetonitrile/Water (30/70, V/V) containing 2 mM ammonium acetate and different pH	**QqQ** (+)	5 fmol/10(6) cells	Partially validated	Shi et al. (2002)
Didanosine and stavudine	Anti-HIV drugs	Commercial PBMC	Cells were lysed by Fisher 550 dismembrator	Lysed cell solution was mixed with acetonitrile, centrifuged, and supernatant was cleaned up by SPE cartridges	RPLC	Keystone, BDS C18 column (5 μm, 4.6 mm × 150 mm)	Methanol/water (16:84, v/v) containing 0.05% trifluoroacetic acid, and 1 mM ammonium formate	**QqQ** (+)	0.4 ng/ml	Partially validated	Huang et al. (2004)
Amprenavir, lopinavir, ritonavir, saquinavir, and efavirenz	Antiretroviral drugs	Whole blood was cen-trifuged by conven-tional Ficoll gradient	Cells were lysed in a solution of 1-acid glycoprotein (1 mg/ml) in sodium azide (0.1%, w/v)	Lysed cell solution was extracted by LLE procedure with 200 ul of 0.05 M Na_2CO_3, 1.2 ml of a mixture of *n*-pentane and ethyl acetate (50:50, v/v)	RPLC	X-TERRATM MS C18 column (5 μm, 4.6 mm × 100 mm)	Acetonitrile/water (50:50, v/v) containing 0.04% formic acid	**Q**(+)	1 ng/3 × 106 to 2 ng/3 × 106 cells	Partially validated	Rouzes et al. (2004)

TABLE 27.2 Quantitation of Intracellular Drugs in Other Cells by LC-MS

Analyte's name	Class of drugs	Cell type	Cell lysis procedure	Sample pretreatment	Chromatographic type	Column type	Mobile phase	MS type (polarity)	LOQ	Validation	References
Metabolites from tricarboxylic acid cycle	Endogenous metabolites	Wild type *Escherichia coli* K12 W3110 strain	Cells were lysed with Water/MeOH (30/60; v/v)	Lysed cell was extracted with 0.3 M KOH (dissolved in 25% ethanol) centrifuged and supernatant was transferred to tube after filtration for analysis	RPLC	Synergi Hydro-RP (C_{18}) (4 μm, 150 mm × 2.1 mm)	(a) 10 mM tributylamine adjusted with 15 mM acetic acid; (b)methanol	Q (−)	4.2 to 1260.2 nM	Partially validated	Luo et al. (2007)
Imatinib and its main metabolites	Therapeutic drug for chronic myelogenous leukemia	Bcr-Abl positive cells	Cells were lysed by one of the following ways: (a) 4% (w/v) trichloroacetic acid; (b) 4% (w/v) formic acid; (c) 100% methanol; (d) 50% (v/v) methanol in water; (e) 1% (w/v) trichloroacetic acid + 50% (v/v) methanol in water; (f) 1% (w/v) formic acid + 50% (v/v) methanol in water	Lysed cell solution was centrifuged and supernatant was used for analysis	RPLC	Polaris C_{18} (5 μm, 250 × 2.0 mm)	Methanol/water (65:35, v/v) containing 7 mM ammonium acetate	Q (−)	1 nmol/l for imatinib and 2 nmol/l for CGP 74588	Partially validated	Mlejnek et al. (2011)
Uric acid	Endogenous metabolites	Human umbilical vein endothelial cells	Cells were lysed by sonicator in 0.3 M KOH	Lysed cell solution was centrifuged and supernatant was filtered and transferred to HPLC vial for analysis	RPLC	Phenomenex Luna C_{18}(2) (5 μm, 150 mm × 4.6 mm)	(a) 5 mM ammonium acetate/0.1% acetic acid; (b) MeOH	QqQ (−)	Not reported	Yes	Kim et al. (2009)
Estrogens and estrogen metabolites	Hormones	MCF-7 breast cancer cells	Cells were dissolved in buffer and lysed by sonicator	Lysed cell solution was extracted by ethyl acetate (LLE) and followed by derivatization with dansyl chloride	RPLC	C_{18} thermo scientific (1.9 μm, 30 mm × 2.1 mm)	(a) 0.1% formic acid in water; (b) 0.1% formic acid in acetonitrile	QqQ (+)	88 pM to 9770 pM	Partially validated	Huang et al. (2011)
Sphingosine and sphingosine 1-phosphate	Cell inducer and inhibitors	HEK 293 cells	Cell lysed by MeOH/water	Lysed cell solution was centrifuged and supernatant was transferred to clean glass vial for analysis	RPLC	Luna-RP column (5 μm, 150 mm L × 2 mm)	Methanol/water (95:5 v/v) containing 0.1% formic acid	QqQ (+)	1 ng/ml and 0.1 ng/ml	Partially validated	Lan et al. (2011)

Prostanoids	Bioactive lipid mediator	Epithelial Calu-3 cells and human lung adeno-carcinoma epithelial A549 cells	Cells were dissolved in buffer and lysed by sonicator	Lysed cell solution was cleaned up by SPE (Bond Elut® C18 disk cartridge)	RPLC (HPLC)	A Shiseido CAPCELL PAK C_{18} MGII column	Acetonitrile/water/acetic acid (40:60:0.1, v/v/v).	QqQ (−)	0.0125 to 0.2 ng per ml of cell extract	Not reported	Furugen et al. (2011)
Didanosine triphosphate	Anti-HIV drug	CEM-T4 cells	Cells were lysed by 70% MeOH buffered to pH 7.4 on ice	Lysed cell solution supernatant was dephosphorylation followed by SPE cleanup procedure	RPLC	Purospher RP-18e column (3 μm, 30 mm × 2. mm)	Mobile phase CH_3OH/H_2O (25/75) containing 1% formic acid	QqQ (−)	0.1 ng/ml	Partially validated	Cahours et al. (2001)
Nucleoside triphosphate (NTP) and deoxynucleo-side triphosphate (dNTP)	Anticancer drug	Human leukemia cell lines K562, NB4, ML-1, MV4-11, and THP-1	Cells were deproteinized with 60% MeOH then lysed by sonicator on ice bath	Lysed cell solution was centrifuged and supernatant was used for analysis	RPLC	Supelcogel ODF-50 (5 μm, 150 × 2.1 mm)	(a) 5 mM DMHA in ultrapure water buffered to pH 7 by 90% formic acid; (b) 5 mM DMHA in ACN (50:50, v/v)	LCQ ion Trap (−)	5 nM	Partially validated	Chen et al. (2009)
ATP analogs	Anticancer drug	Breast cancer cell line (MCF-7)	Cells were lysed with water/acetonitrile (2/3, v/v)	Lysed cell solution was centrifuged and supernatant was transferred to tube or evaporated to certain level	RPLC	Phenomenex Gemini C_{18} (5 μm, 50 mm × 2.00 mm)	(a) 20 mM DMHA formate with pH adjusted to 6.8 with formic acid; (b) 80% methanol containing 2 mM DMHA formate	QqQ (−)	0.02–0.03 μM	Validated	Jauhiainen et al. (2009)
3′-azido-3′-deoxythymidine (AZT) and metabolites	Anti-HIV drugs	Human T-lymphoblastoid leukemia (CEM) cells	Cells lysed with ice-cold 60% methanol-40% 15 mM ammonium acetate buffer (pH 6.65) at −20°C overnight	Lysed cell solution was dried in vacuo then submitted to online cleanup procedure via column-switch	Capillary ion-pairing LC	Capillary Zorbax XDB-C18 column (5 μm, 150 mm × 0.5 mm)	(a) 15 mM ammonium acetate, pH 6.65; (b) methanol	QqQ (−)	Not reported	Partially validated	Kim et al. (2005)
Penicillin G	Antibiotics	Aerobic glucose-limited chemostat cultures of *Penicillium chrysogenum*	Cells were lysed with 75% EtOH at 95°C	Lysed cell solution was centrifuged to certain level and then diluted in water and centrifuged again and supernatant was used for analysis	Ion-pair reversed-phase liquid chromatography	XTerraMSC_{18} column (3.5 μm, 150 mm × 2.1 mm)	(a) 2 mM DBAA with 5% (v/v) acetonitrile; (b) 2 mM DBAA with 84% (v/v) acetonitrile	QqQ (−)	92 nM	Partially validated	Seifar et al. (2008)

by LC-MS through the tris-HCl/MeOH basic lysis process (Pelerin et al., 2005). The HIV protease inhibitor, atazanavir in PMBC was quantified by LC-MS through the cell lysis *via* a sonicator (Jemal et al., 2003). More cell lysis methods and applications used in intracellular LC-MS quantitation are summarized in Tables 27.1 and 27.2).

27.2.3 Extraction and Isolation Methods

After cell lysis, samples usually cannot be assayed directly without a proper pretreatment to remove endogenous compounds such as lipids and salts or to increase the drug concentration in the final sample extracts prior to LC-MS analysis. PPT, SPE, and LLE are the main procedures used for sample preparation (Tables 27.1 and 27.2).

27.2.3.1 Protein Precipitation PPT is a quick and simple method of sample pretreatment for intracellular drug bioanalysis. Due to the high selectivity and specificity of the multiple reaction monitoring (MRM) MS detection, simple PPT using organic solvent is often the first choice of sample preparation to quickly remove proteins and other precipitable components in the lysed cell solutions (Stewart et al., 1998; Shi et al., 2002). However, many intracellular drugs (such as nucleotide antiviral compounds) are very polar and difficult to be extracted from lysed cell solutions with organic solvents due to the hydrophilicity of those drugs.

27.2.3.2 Solid Phase Extraction SPE is a very useful technique for sample preparation. Compared to the PPT, SPE method can reduce matrix background to a greater degree, resulting in an increased MS/MS detection sensitivity (Tables 27.1 and 27.2). In SPE process, lysed cell solution is routinely centrifuged and the resulting supernatant transferred to a SPE column/plate. SPE can be performed off-line manually, semiautomated, or online. Use of the 96-well format for high-throughput quantitation can be used for the sample pretreatment to shorten sample process times. The extraction efficiency of SPE depends on analyte types, the type of sorbents, the sample volume and pH, the content of organic modifier, and elution volume. The SPE cartridges that have been used in the publications include C_{18}, C_8, and other cartridges (Tables 27.1 and 27.2).

27.2.3.3 Liquid–Liquid Extraction LLE is also one of sample pretreatment methods used for intracellular LC-MS analysis. The organic solvent (extraction solvent) can be added to the cell pellet before lysis (Ehrhardt et al., 2007) or after cell lysis (Rouzes et al., 2004; Huang et al., 2011). The selection of extraction solvent is dependent on the types of analytes. After extraction, the organic solvents are evaporated and sample residues reconstituted in mobile phase and injected onto the LC-MS system.

27.3 LC-MS INTRACELLULAR DRUG BIOANALYSIS

Various interface and MS detection techniques for intracellular drug LC-MS bioanalysis are summarized this section along with optimization of experimental LC-MS conditions and method validation.

27.3.1 LC-MS Interfaces

Currently, the many LC-MS interfaces that have been used in intracellular bioanalysis for drugs and/or drug metabolites share one common characteristic: they all employ API technique, a soft-ionization process for MS detection. ESI, APCI, and APPI are all common API techniques and among them. ESI has been heavily used in intracellular bioanalysis of drugs or drug metabolites (Slusher et al., 1992; Shi et al., 2002; Staines et al., 2005; Stevens et al., 2008; Serdar et al., 2011; van Haandel et al., 2011; Turnpenny et al., 2011) (Tables 27.1 and 27.2).

27.3.2 MS Detection Techniques

Several combinations of tandem MS scan techniques have been used in intracellular bioanalysis. Among them, single quadruple (Q), the triple quadrupole (QqQ), hybrid quadrupole-time of flight (Q-TOF), and linear ion trap (LIT) mass spectrometers are the most utilized in intracellular bioanalysis (Jansen et al., 2011) (Tables 27.1 and 27.2).

Single-stage MS can be used in combination with other detection techniques (such as UV) to facilitate the identification of intracellular drugs or drug metabolites with comparison against the reference standards or reference data. However, there are still some bioanalytical drawbacks derived from the characteristic full-scan mass spectrum of some drugs (or drug metabolites), which often gives only a molecular adduct or weak fragment ion(s). In particular, the matrix components often enhance or suppress the MS detector response, leading to great variances between the relative abundance of different ions in the MS spectra (Pico et al., 2004). Therefore, during LC-MS analysis, the matrix of interest and the corresponding sample preparation procedure should be considered together with the chromatographic separation and MS detection sensitivity. For this purpose, LC-MS/MS is often used for intracellular drug quantification, while liquid chromatography multistage mass spectrometry (LC-MS^n) used for identification of the unknowns in the matrix of interest.

Tandem mass spectrometer is one of the most widely used instruments for LC-MS bioanalysis. Compared to single-quadrupole instrument, the tandem MS provides various selective screening strategies (such as full-scan, neutral-loss, precursor-ion, and product-ion scan). In particular, MRM provides beatable selectivity, specificity, and sensitivity that are required for quantitative intracellular drug quantification (Jemal et al., 2003; Huang et al., 2004; Kim et al., 2005, 2009; King et al., 2006a, 2006b; Jauregui et al., 2007; Jauhiainen et al., 2009; Jansen et al., 2009a, 2009b) (Tables 27.1 and 27.2).

27.3.3 Liquid Chromatography

For LC-MS quantitative analysis of intracellular drugs or drug metabolites, no matter which MS detection (scan, SIM, MRM) is to be used, good chromatographic separation of analytes of interest from matrix components is always critical.

27.3.3.1 Reversed-Phase Liquid Chromatography

Reversed-phase liquid chromatography (RPLC) is commonly used for quantitative analysis of intracellular analytes (Becher et al., 2003; Hennere et al., 2003; Huang et al., 2004; Rouzes et al., 2004; Ehrhardt et al., 2007; Elens et al., 2009; Jansen et al., 2009b; Jauhiainen et al., 2009; Coulier et al., 2011; Furugen et al., 2011; Huang et al., 2011) (Tables 27.1 and 27.2). In RPLC, most stationary phases are based on silica that has been chemically modified with octadecyl (C_{18} or ODS) or octyl (C_8 or ODS). Using a short and small particle size, HPLC column can reduce the chromatography runtime and facilitate the high-throughput analysis. Simple gradients starting with a low percentage of methanol (MeOH) or acetonitrile (ACN) in ammonium acetate or formate buffer (Lynch et al., 2001; Kim et al., 2004, 2005, 2009; Bousquet et al., 2008). Although those chromatographic conditions are relatively straightforward for the analysis of most intracellular drugs, the lack of retention for nucleotides limits the applicability of conventional RPLC (Jansen et al., 2011).

UPLC (ultrahigh performance liquid chromatography) provides significantly better selectivity and chromatographic resolution than the traditional HPLC method and enables much shorter cycle time for the LC-MS bioanalysis of compounds of interest (Ciric et al., 2010; Ji et al., 2010; Michopoulos et al., 2011; Pedersen et al., 2011; Rao et al., 2011; Cheng et al., 2012; Sauve et al., 2012). In general, it can be expected that UPLC-MS-MS will be much more widely used.

27.3.3.2 Ion-Pairing Chromatography Due to the high polarity and charged nature of some intracellular drugs, such as nucleotide analogs, regular reversed-phase chromatography techniques are deemed unsuitable for the chromatographic separation of these compounds with matrix components unless other means can be employed to help the stationary phase of the column to retain the compounds during the separation. In this case, ion-pairing (IP) chromatography is considered ideal chromatographic method in term of compatibility with MS detection for analysis of charged compounds such as intracellular nucleotides (Seifar et al., 2008). The separation is based on the formation of ion pairs between the negatively charged analyte(s) and positively charged IP reagents. The stationary phase is a conventional C_{18} or C_8 (Cai et al., 2004; Qian et al., 2004; Cordell et al., 2008; Seifar et al., 2009; Jansen et al., 2011) (Tables 27.1 and 27.2). The most common IP reagents used in mobile phases are a variety of cationic reagents, such as alkylamines. The concentration of IP agent and the pH in IP mobile phases is critical and needs to be optimized to ensure good retention and peak shape with a minimum ion suppression in MS detection.

27.3.3.3 Ion Exchange Chromatography Most intracellular drugs and/or their metabolites can behave very well either on IP or RPLC during LC-MS analysis. However, the two chromatography techniques have very limited applications to the very polar compounds, such as intracellular nucleoside triphosphates (Shi et al., 2002). An alternative chromatographic technique is ion exchange chromatography (IEC). IEC involves the use of more concentrated or more selective competing ions. Traditionally, high salt concentration was used in the mobile phase. However, the high concentration of salt in mobile phase can suppress ionization and quickly the salts deposit at the ion source inlet of the MS instrument, making it impossible for direct ion exchange LC-MS bioanalysis. However, if a pH gradient (pH 6–10.5) is applied to a WAX column, the charges of the basic functional groups (pK_a ~8) of the column were changed. As a result, the capacity of the column decreased at a higher pH, which leads to the elution of anionic nucleotide phosphates (Shi et al., 2002). In this way, the high salt concentration (ammonium acetate) in mobile phase can be reduced greatly and IEC LC-MS or LC-MS-MS can directly used for those intracellular drug analysis (Table 27.1).

27.3.3.4 Porous Graphitic Carbon Chromatography

Porous graphitic carbon (PGC) chromatography consists of flat carbon sheets of hexagonally arranged carbon atoms that retain both polar and ionic compounds on its surface. Polar or polarizable molecules can cause a charge-induced dipole at the graphite surface, resulting in retention of negatively charged analytes such as nucleotides. Charged analytes can be retained well on a PGC column without using IP agents. Accordingly, the molecules can be eluted from the column without the need for concentrated salts in mobile phase. Because of this unique feature, PGC is very suitable for LC-MS analysis of certain drugs such as a nucleotides (Jansen et al., 2011). Also, it has been found that a remarkable selectivity exists in structurally similar compounds, due to the difference caused by contact area with the planar carbon sheets. Remarkable selectivity can be obtained for the analysis of structurally similar compounds.

27.3.3.5 Capillary Electrophoresis Chromatography Capillary electrophoresis chromatography (CEP) has been widely used to separate ionic analytes with high resolution, the CEP hyphenated with MS has been widely used for the intracellular bioanalysis of some antiviral nucleosides (Agrofoglio et al., 2007) and other nucleotide drugs by using MS compatible running buffer (mobile phase) (Jansen et al., 2011). In most cases, a volatile buffer such as ammonium bicarbonate or acetate are used for those bioanalyses in order to obtain good sensitivity (Tables 27.1 and 27.2).

27.3.3.6 Hydrophilic Interaction Liquid Chromatography As a newer form of normal-phase chromatography, hydrophilic interaction liquid chromatography (HILIC) is a useful technique for the retention of polar analytes. In HILIC, polar columns are used in combination with mobile phases rich in organic solvents. Polar analytes are eluted from the column by increasing the water content of the mobile phases. Due to the highly volatility of the applied mobile phases, hyphenation with MS is favorable and used for intracellular bioanalysis (Jansen et al., 2011). Although some column vendors are marketing column specific for HILIC, most columns used with normal phase HPLC such as pure silica columns or cyano columns can operate in HILIC conditions (Xu et al., 2007; Jian et al., 2010). The typical mobile phase for HILIC includes acetonitrile with a small amount of water content. It is commonly believed that in HILIC, the mobile phase forms a water rich layer on the surface of the polar stationary phase creating a LLE system. With HILIC, polar compounds have longer retention times than nonpolar compounds. This method does not require IP reagents or high water content for separation of very polar compounds and provides better peak shapes for basic compounds. The high concentration of organic solvent in the mobile phase pushes up the sensitivity of LC-MS analysis (Hsieh, 2008; Jian et al., 2011). Due to its unique property, HILIC-MS has been used for quantifying intracellular nucleotides (Pucci et al., 2009; Preinerstorfer et al., 2010).

27.3.4 Method Validations

According to the current health authority guidance (FDA, 2001; EMEA, 2011) and industrial practice, an LC-MS method for the analysis of intracellular drugs should be validated prior to use in support preclinical or clinical studies. The validation is essential to ensure the robustness of the intended assay method. Most of published intracellular LC-MS bioanalysis were based this important guideline and the LC-MS quantitation methods were validated or partially validated (Tables 27.1 and 27.2). A few other sources of validation criteria can be found in white papers by groups such as the nonprofit Calibration and Validation Group (Agency, 2011; Garofolo et al., 2011).

27.4 APPLICATION

To date, many papers have been published on LC-MS bioanalyis of intracellular drugs or drug metabolites, primarily in peripheral blood mononucleous cells (PMBC) and other cells such as cultured cells. In this section, application examples were summarized in Tables 27.1 and 27.2 according to cell types, and method information, such as LC conditions, MS conditions, validation status, limit of quantitation (LOQ), and limit of detection (LOD), was also included.

27.4.1 LC-MS Bioanalysis of Intracellular Drugs in PMBC

A number of LC-MS bioanalysis methods for the determination of drugs in PMBC have been published. Most of these published methods are focused on antitumor, antiviral, and immunosuppressive drugs (Table 27.1). Nucleotide analogs are a primary example of this class of pharmaceuticals. The metabolism and mechanism of action of nucleotide analogs were summarized by Cohen et al. (2010). A review on intracellular bioanalysis of therapeutic nucleotide analogs via mass spectrometry has also been published recently (Jansen et al., 2011). Some of LC-MS intracellular bioanalysis examples were summarized in Table 27.1.

27.4.2 LC-MS Bioanalysis of Intracellular Drugs in Cultured Cells

LC-MS bioanalysis methods have also been successfully applied in cultured cells to quantify intracellular drug or drug metabolite concentrations. Some of the application examples were summarized in Table 27.2.

27.5 CONCLUSIONS

Quantitative analysis of intracellular level of drugs and/or drug metabolites is critical for the assessment of their efficacy and toxicity during preclinical and clinical drug development. LC-MS quantitation method has been widely used for quantitative analysis of intracellular concentrations of drugs and/or drug metabolites. Different chromatographic techniques, MS ionization techniques, sample treatment techniques including cell lysis and further treatment techniques were explored and used in intracellular LC-MS bioanalysis according to the Health Authority guidance and industrial practice.

REFERENCES

Agency EM. Guideline on bioanalytical method validation. 2011. Available at http://www.ema.europa.eu/docs/en_GB/document_library/Scientific_guideline/2011/08/WC500109686.pdf. Accessed Apr 13, 2013.

Agrofoglio LA, Bezy V, Chaimbault P, Delepee R, Rhourri B, Morin P. Mass spectrometry based methods for analysis of nucleosides as antiviral drugs and potential tumor biomarkers. Nucleos Nucleot Nucl 2007;26:1523–1527.

Ahmed M, Venkataraman R, Logar AJ, et al. Quantitation of immunosuppression by tacrolimus using flow cytometric analysis of interleukin-2 and interferon-gamma inhibition in CD8(−) and CD8(+) peripheral blood T cells. Ther Drug Monit 2001;23:354–362.

Ansermot N, Fathi M, Veuthey JL, Desmeules J, Hochstrasser D, Rudaz S. Quantification of cyclosporine A in peripheral blood mononuclear cells by liquid chromatography-electrospray mass spectrometry using a column-switching approach. J Chromatogr B Analyt Technol Biomed Life Sci 2007;857:92–99.

Baran Y, Bielawski J, Gunduz U, Ogretmen B. Targeting glucosylceramide synthase sensitizes imatinib-resistant chronic myeloid leukemia cells via endogenous ceramide accumulation. J Cancer Res Clin Oncol 2011;137:1535–1544.

Becher F, Pruvost A, Gale J, et al. A strategy for liquid chromatography/tandem mass spectrometric assays of intracellular drugs: application to the validation of the triphosphorylated anabolite of antiretrovirals in peripheral blood mononuclear cells. J Mass Spectrom 2003;38:879–890.

Becher F, Pruvost A, Goujard C, et al. Improved method for the simultaneous determination of d4T, 3TC and ddl intracellular phosphorylated anabolites in human peripheral-blood mononuclear cells using high-performance liquid chromatography/tandem mass spectrometry. Rapid Commun Mass Spectrom 2002a;16:555–565.

Becher F, Schlemmer D, Pruvost A, et al. Development of a direct assay for measuring intracellular AZT triphosphate in humans peripheral blood mononuclear cells. Anal Chem 2002b;74:4220–4227.

Boltz G, Penner E, Holzinger C, et al. Surface phenotypes of human peripheral blood mononuclear cells from patients with gastrointestinal carcinoma. J Cancer Res Clin Oncol 1987;113: 291–297.

Bousquet L, Pruvost A, Didier N, Farinotti R, Mabondzo A. Emtricitabine: Inhibitor and substrate of multidrug resistance associated protein. Eur J Pharm Sci 2008;35:247–256.

Bushman LR, Kiser JJ, Rower JE, et al. Determination of nucleoside analog mono-, di-, and tri-phosphates in cellular matrix by solid phase extraction and ultra-sensitive LC-MS/MS detection. J Pharm Biomed Anal 2011;56:390–401.

Cahours X, Tran TT, Mesplet N, Kieda C, Morin P, Agrofoglio LA. Analysis of intracellular didanosine triphosphate at sub-ppb level using LC-MS/MS. J Pharm Biomed Anal 2001;26:819–827.

Cai Z, Qian T, Yang MS. Ion-pairing liquid chromatography coupled with mass spectrometry for the simultaneous determination of nucleosides and nucleotides. Se Pu 2004;22:358–360.

Chen LS, Fujitaki JM, Dixon R. A sensitive assay for the aminoimidazole-containing drug GP531 in plasma using liquid chromatography with amperometric electrochemical detection: a new class of electroactive compounds. J Pharm Biomed Anal 1996;14:1535–1538.

Chen P, Liu Z, Liu S, et al. A LC-MS/MS method for the analysis of intracellular nucleoside triphosphate levels. Pharm Res 2009;26:1504–1515.

Chen Q, Zeng Y, Kuang J, et al. Quantification of aesculin in rabbit plasma and ocular tissues by high performance liquid chromatography using fluorescent detection: application to a pharmacokinetic study. J Pharm Biomed Anal 2011;55:161–167.

Cheng XL, Wei F, Xiao XY, et al. Identification of five gelatins by ultra performance liquid chromatography/time-of-flight mass spectrometry (UPLC/Q-TOF-MS) using principal component analysis. J Pharm Biomed Anal 2012;62:191–195.

Chi J, Jayewardene A, Stone J, Gambertoglio JG, Aweeka FT. A direct determination of thymidine triphosphate concentrations without dephosphorylation in peripheral blood mononuclear cells by LC/MS/MS. J Pharm Biomed Anal 2001;26:829–836.

Ciric B, Jandric D, Kilibarda V, Jovic-Stosic J, Dragojevic-Simic V, Vucinic S. Simultaneous determination of amoxicillin and clavulanic acid in the human plasma by high performance liquid chromatography-mass spectrometry (UPLC/MS). Vojnosanit Pregl 2010;67:887–892.

Claire RL, 3rd. Positive ion electrospray ionization tandem mass spectrometry coupled to ion-pairing high-performance liquid chromatography with a phosphate buffer for the quantitative analysis of intracellular nucleotides. Rapid Commun Mass Spectrom 2000;14:1625–1634.

Cohen S, Jordheim LP, Megherbi M, Dumontet C, Guitton J. Liquid chromatographic methods for the determination of endogenous nucleotides and nucleotide analogs used in cancer therapy: a review. J Chromatogr B Analyt Technol Biomed Life Sci 2010;878:1912–1928.

Colombo S, Beguin A, Telenti A, et al. Intracellular measurements of anti-HIV drugs indinavir, amprenavir, saquinavir, ritonavir, nelfinavir, lopinavir, atazanavir, efavirenz and nevirapine in peripheral blood mononuclear cells by liquid chromatography coupled to tandem mass spectrometry. J Chromatogr B Analyt Technol Biomed Life Sci 2005;819:259–276.

Cordell RL, Hill SJ, Ortori CA, Barrett DA. Quantitative profiling of nucleotides and related phosphate-containing metabolites in cultured mammalian cells by liquid chromatography tandem electrospray mass spectrometry. J Chromatogr B Analyt Technol Biomed Life Sci 2008;871:115–124.

Coulier L, Gerritsen H, van Kampen JJ, et al. Comprehensive analysis of the intracellular metabolism of antiretroviral nucleosides and nucleotides using liquid chromatography-tandem mass spectrometry and method improvement by using ultra performance liquid chromatography. J Chromatogr B Analyt Technol Biomed Life Sci 2011;879:2772–2782.

Covey TR, Lee ED, Henion JD. High-speed liquid chromatography/tandem mass spectrometry for the determination of drugs in biological samples. Anal Chem 1986;58:2453–2460.

Durand-Gasselin L, Da Silva D, Benech H, Pruvost A, Grassi J. Evidence and possible consequences of the phosphorylation of nucleoside reverse transcriptase inhibitors in human red blood cells. Antimicrob Agents Chemother 2007;51:2105–2111.

Ehrhardt M, Mock M, Haefeli WE, Mikus G, Burhenne J. Monitoring of lopinavir and ritonavir in peripheral blood mononuclear cells, plasma, and ultrafiltrate using a selective and highly sensitive LC/MS/MS assay. J Chromatogr B Analyt Technol Biomed Life Sci 2007;850:249–258.

El-Gindy A, El Walily AF, Bedair MF. First-derivative spectrophotometric and LC determination of cefuroxime and cefadroxil in urine. J Pharm Biomed Anal 2000;23:341–352.

Elens L, Veriter S, Yombi JC, et al. Validation and clinical application of a high performance liquid chromatography tandem mass spectrometry (LC-MS/MS) method for the quantitative determination of 10 anti-retrovirals in human peripheral blood mononuclear cells. J Chromatogr B Analyt Technol Biomed Life Sci 2009;877:1805–1814.

EMEA. 2011. Guideline on Bioanalytical Method Validation. Available at http://www.ema.europa.eu/docs/en_GB/document_library/Scientific_guideline/2011/08/WC500109686.pdf. Accessed Apr 13, 2013.

FDA. 2001. Guidance for Industry (Bioanalytical Method Validation. Available at wwwfdagov/downloads/Drugs//Guidances/ucm070107pdfSimilar.

Forget R, Spagnoli S. Excipient quantitation and drug distribution during formulation optimization. J Pharm Biomed Anal 2006;41:1051–1055.

Furugen A, Yamaguchi H, Tanaka N, et al. Quantification of intracellular and extracellular prostanoids stimula btedy A23187 by liquid chromatography/electrospray ionization tandem mass spectrometry. J Chromatogr B Analyt Technol Biomed Life Sci 2011;879:3378–3385.

Gahan ME, Miller F, Lewin SR, et al. Quantification of mitochondrial DNA in peripheral blood mononuclear cells and subcutaneous fat using real-time polymerase chain reaction. J Clin Virol 2001;22:241–247.

Garofolo F, Rocci ML, Jr, Dumont I, et al. 2011 White paper on recent issues in bioanalysis and regulatory findings from audits and inspections. *Bioanalysis* 2011;3:2081–2096.

Hennere G, Becher F, Pruvost A, Goujard C, Grassi J, Benech H. Liquid chromatography-tandem mass spectrometry assays for intracellular deoxyribonucleotide triphosphate competitors of nucleoside antiretrovirals. J Chromatogr B Analyt Technol Biomed Life Sci 2003;789:273–281.

Hsieh Y. Potential of HILIC-MS in quantitative bioanalysis of drugs and drug metabolites. J Sep Sci 2008;31:1481–1491.

Huang HJ, Chiang PH, Chen SH. Quantitative analysis of estrogens and estrogen metabolites in endogenous MCF-7 breast cancer cells by liquid chromatography-tandem mass spectrometry. J Chromatogr B Analyt Technol Biomed Life Sci 2011;879:1748–1756.

Huang Y, Zurlinden E, Lin E, et al. Liquid chromatographic-tandem mass spectrometric assay for the simultaneous determination of didanosine and stavudine in human plasma, bronchoalveolar lavage fluid, alveolar cells, peripheral blood mononuclear cells, seminal plasma, cerebrospinal fluid and tonsil tissue. J Chromatogr B Analyt Technol Biomed Life Sci 2004;799: 51–61.

Jansen RS, Rosing H, Schellens JH, Beijnen JH. Protein versus DNA as a marker for peripheral blood mononuclear cell counting. Anal Bioanal Chem 2009a;395:863–867.

Jansen RS, Rosing H, Schellens JH, Beijnen JH. Simultaneous quantification of 2′,2′-difluorodeoxycytidine and 2′,2′-difluorodeoxyuridine nucleosides and nucleotides in white blood cells using porous graphitic carbon chromatography coupled with tandem mass spectrometry. Rapid Commun Mass Spectrom 2009b;23:3040–3050.

Jansen RS, Rosing H, Schellens JHM, Beijnen JH. Mass spectrometry in the quantitative analysis of therapeutic intracellular nucleotide analogs. Mass Spectrometry Reviews 2011;30:321–343.

Jauhiainen M, Monkkonen H, Raikkonen J, Monkkonen J, Auriola S. Analysis of endogenous ATP analogs and mevalonate pathway metabolites in cancer cell cultures using liquid chromatography-electrospray ionization mass spectrometry. J Chromatogr B Analyt Technol Biomed Life Sci 2009;877:2967–2975.

Jauregui O, Sierra AY, Carrasco P, Gratacos E, Hegardt FG, Casals N. A new LC-ESI-MS/MS method to measure long-chain acylcarnitine levels in cultured cells. Anal Chim Acta 2007;599: 1–6.

Jemal M, Rao S, Gatz M, Whigan D. Liquid chromatography-tandem mass spectrometric quantitative determination of the HIV protease inhibitor atazanavir (BMS-232632) in human peripheral blood mononuclear cells (PBMC): practical approaches to PBMC preparation and PBMC assay design for high-throughput analysis. J Chromatogr B Analyt Technol Biomed Life Sci 2003;795:273–289.

Ji C, Walton J, Su Y, Tella M. Simultaneous determination of plasma epinephrine and norepinephrine using an integrated strategy of a fully automated protein precipitation technique, reductive ethylation labeling and UPLC-MS/MS. Anal Chim Acta. 2010;670:84–91.

Jian W, Edom RW, Xu Y, Weng N. Recent advances in application of hydrophilic interaction chromatography for quantitative bioanalysis. J Sep Sci 2010;33:681–697.

Jian W, Xu Y, Edom RW, Weng N. Analysis of polar metabolites by hydrophilic interaction chromatography–MS/MS. Bioanalysis 2011;3:899–912.

Kaji H, Maiguma T, Inukai Y, et al. A simple determination of mizoribine in human plasma by liquid chromatography with UV detection. J AOAC Int 2005;88:1114–1117.

Kaul S, Stouffer B, Mummaneni V, et al. Specific radioimmunoassays for the measurement of stavudine in human plasma and urine. J Pharm Biomed Anal 1996;15:165–174.

Kawashima H, Mori T, Kashiwagi Y, Takekuma K, Hoshika A, Wakefield A. Detection and sequencing of measles virus from peripheral mononuclear cells from patients with inflammatory bowel disease and autism. Dig Dis Sci 2000;45:723–729.

Kim J, Chou TF, Griesgraber GW, Wagner CR. Direct measurement of nucleoside monophosphate delivery from a phosphoramidate pronucleotide by s isotope labeling and LC-ESI(−)-MS/MS. Mol Pharm 2004;1:102–111.

Kim J, Park S, Tretyakova NY, Wagner CR. A method for quantitating the intracellular metabolism of AZT amino acid phosphoramidate pronucleotides by capillary high-performance liquid chromatography-electrospray ionization mass spectrometry. Mol Pharm 2005;2:233–241.

Kim KM, Henderson GN, Ouyang X, et al. A sensitive and specific liquid chromatography-tandem mass spectrometry method for the determination of intracellular and extracellular uric acid. J Chromatogr B Analyt Technol Biomed Life Sci 2009;877:2032–2038.

King T, Bushman L, Anderson PL, Delahunty T, Ray M, Fletcher CV. Quantitation of zidovudine triphosphate concentrations from human peripheral blood mononuclear cells by anion exchange solid phase extraction and liquid chromatography-tandem mass spectroscopy; an indirect quantitation methodology. J Chromatogr B Analyt Technol Biomed Life Sci 2006a;831:248–257.

King T, Bushman L, Kiser J, et al. Liquid chromatography-tandem mass spectrometric determination of tenofovir-diphosphate in human peripheral blood mononuclear cells. J Chromatogr B Analyt Technol Biomed Life Sci 2006b;843:147–156.

Kominami G, Nakamura M, Chomei N, Takada S. Radioimmunoassay for a novel benzodiazepine inverse agonist, S-8510, in human plasma and urine. J Pharm Biomed Anal 1999;20: 145–153.

Kominami G, Nakamura M, Mizobuchi M, et al. Radioimmunoassay and gas chromatography/mass spectrometry for a novel antiglaucoma medication of a prostaglandin derivative, S-1033, in plasma. J Pharm Biomed Anal 1996;15:175–182.

Kominami G, Ueda A, Sakai K, Misaki A. Radioimmunoassay for a novel lignan-related hypocholesterolemic agent, S-8921, in human plasma after high-performance liquid chromatography purification and in human urine after immunoaffinity extraction. J Chromatogr B Biomed Sci Appl 1997;704:243–250.

Konda A, Soma M, Ito T, et al. Stereoselective analysis of ritodrine diastereomers in human serum using HPLC. J Chromatogr Sci 2010;48:503–506.

Lal J, Mehrotra N, Gupta RC. Analysis and pharmacokinetics of bulaquine and its major metabolite primaquine in rabbits using an LC-UV method—a pilot study. J Pharm Biomed Anal 2003;32:141–150.

Lan T, Bi H, Liu W, Xie X, Xu S, Huang H. Simultaneous determination of sphingosine and sphingosine 1-phosphate in biological samples by liquid chromatography-tandem mass spectrometry. J Chromatogr B Analyt Technol Biomed Life Sci 2011;879:520–526.

Luo B, Groenke K, Takors R, Wandrey C, Oldiges M. Simultaneous determination of multiple intracellular metabolites in glycolysis, pentose phosphate pathway and tricarboxylic acid cycle by liquid chromatography-mass spectrometry. J Chromatogr A 2007;1147:153–164.

Lynch T, Eisenberg G, Kernan M. LC/MS determination of the intracellular concentration of two novel aryl phosphoramidate prodrugs of PMPA and their metabolites in dog PBMC. Nucleosides Nucleotides Nucleic Acids 2001;20:1415–1419.

Mascola JR, Louder MK, Winter C, et al. Human immunodeficiency virus type 1 neutralization measured by flow cytometric quantitation of single-round infection of primary human T cells. J Virol 2002;76:4810–4821.

Michopoulos F, Theodoridis G, Smith CJ, Wilson ID. Metabolite profiles from dried blood spots for metabonomic studies using UPLC combined with orthogonal acceleration ToF-MS: effects of different papers and sample storage stability. Bioanalysis 2011;3:2757–2767.

Mlejnek P, Novak O, Dolezel P. A non-radioactive assay for precise determination of intracellular levels of imatinib and its main metabolite in Bcr-Abl positive cells. Talanta 2011;83:1466–1471.

Nilsson C, Aboud S, Karlen K, Hejdeman B, Urassa W, Biberfeld G. Optimal blood mononuclear cell isolation procedures for gamma interferon enzyme-linked immunospot testing of healthy Swedish and Tanzanian subjects. Clin Vaccine Immunol 2008;15:585–589.

Nirogi RV, Kandikere VN, Mudigonda K. Quantitation of zopiclone and desmethylzopiclone in human plasma by high-performance liquid chromatography using fluorescence detection. Biomed Chromatogr 2006;20:794–799.

Park CW, Rhee YS, Go BW, et al. High performance liquid chromatographic analysis of rabeprazole in human plasma and its pharmacokinetic application. Arch Pharm Res 2008;31:1195–1199.

Pedersen TL, Keyes WR, Shahab-Ferdows S, Allen LH, Newman JW. Methylmalonic acid quantification in low serum volumes by UPLC-MS/MS. J Chromatogr B Analyt Technol Biomed Life Sci 2011;879:1502–1506.

Pelerin H, Compain S, Duval X, Gimenez F, Benech H, Mabondzo A. Development of an assay method for the detection and quantification of protease and non-nucleoside reverse transcriptase inhibitors in plasma and in peripherical blood mononuclear cells by liquid chromatography coupled with ultraviolet or tandem mass spectrometry detection. J Chromatogr B Analyt Technol Biomed Life Sci 2005;819:47–57.

Pico Y, Blasco C, Font G. Environmental and food applications of LC-tandem mass spectrometry in pesticide-residue analysis: an overview. Mass Spectrom Rev 2004;23:45–85.

Preinerstorfer B, Schiesel S, Lammerhofer M, Lindner W. Metabolic profiling of intracellular metabolites in fermentation broths from beta-lactam antibiotics production by liquid chromatography-tandem mass spectrometry methods. J Chromatogr A 2010;1217:312–328.

Pruvost A, Becher F, Bardouille P, et al. Direct determination of phosphorylated intracellular anabolites of stavudine (d4T) by liquid chromatography/tandem mass spectrometry. Rapid Commun Mass Spectrom. 2001;15:1401–1408.

Pruvost A, Theodoro F, Agrofoglio L, Negredo E, Benech H. Specificity enhancement with LC-positive ESI-MS/MS for the measurement of nucleotides: application to the quantitative determination of carbovir triphosphate, lamivudine triphosphate and tenofovir diphosphate in human peripheral blood mononuclear cells. J Mass Spectrom 2008;43:224–233.

Pucci V, Giuliano C, Zhang R, et al. HILIC LC-MS for the determination of 2′-C-methyl-cytidine-triphosphate in rat liver. J Sep Sci 2009;32:1275–1283.

Qian T, Cai Z, Yang MS. Determination of adenosine nucleotides in cultured cells by ion-pairing liquid chromatography-electrospray ionization mass spectrometry. Anal Biochem 2004;325:77–84.

Raghavamenon AC, Dupard-Julien CL, Kandlakunta B, Uppu RM. Determination of alloxan by fluorometric high-performance liquid chromatography. Toxicol Mech Methods 2009;19:498–502.

Rao DD, Sait SS, Mukkanti K. Development and validation of an UPLC method for rapid determination of ibuprofen and diphenhydramine citrate in the presence of impurities in combined dosage form. J Chromatogr Sci 2011;49:281–286.

Reynolds DL, Eichmeier LS, Giesing DH. Determination of MDL 201,012 at femtomole/millilitre levels in human plasma by liquid chromatography with electrochemical detection. Biomed Chromatogr 1992;6:295–299.

Rouzes A, Berthoin K, Xuereb F, et al. Simultaneous determination of the antiretroviral agents: amprenavir, lopinavir, ritonavir, saquinavir and efavirenz in human peripheral blood mononuclear cells by high-performance liquid chromatography-mass spectrometry. J Chromatogr B Analyt Technol Biomed Life Sci 2004;813:209–216.

Sauve EN, Langodegard M, Ekeberg D, Oiestad AM. Determination of benzodiazepines in ante-mortem and post-mortem whole blood by solid-supported liquid-liquid extraction and UPLC-MS/MS. J Chromatogr B Analyt Technol Biomed Life Sci 2012;883–884:177–188.

Seifar RM, Ras C, van Dam JC, van Gulik WM, Heijnen JJ, van Winden WA. Simultaneous quantification of free nucleotides in complex biological samples using ion pair reversed phase liquid chromatography isotope dilution tandem mass spectrometry. Anal Biochem 2009;388:213–219.

Seifar RM, Zhao Z, van Dam J, van Winden W, van Gulik W, Heijnen JJ. Quantitative analysis of metabolites in complex biological samples using ion-pair reversed-phase liquid chromatography-isotope dilution tandem mass spectrometry. J Chromatogr A 2008;1187:103–110.

Serdar MA, Sertoglu E, Uyanik M, Tapan S, Akin O, Cihan M. Determination of 5-fluorouracil and dihydrofluorouracil levels by using a liquid chromatography-tandem mass spectrometry method for evaluation of dihydropyrimidine dehydrogenase enzyme activity. Cancer Chemother Pharmacol 2011;68:525–529.

Shi G, Wu JT, Li Y, et al. Novel direct detection method for quantitative determination of intracellular nucleoside triphosphates using weak anion exchange liquid chromatography/tandem mass spectrometry. Rapid Commun Mass Spectrom 2002;16:1092–1099.

Slusher JT, Kuwahara SK, Hamzeh FM, Lewis LD, Kornhauser DM, Lietman PS. Intracellular zidovudine (ZDV) and ZDV phosphates as measured by a validated combined high-pressure liquid chromatography-radioimmunoassay procedure. Antimicrob Agents Chemother 1992;36:2473–2477.

Staines AG, Burchell B, Banhegyi G, Mandl J, Csala M. Application of high-performance liquid chromatography-electrospray ionization-mass spectrometry to measure microsomal membrane transport of glucuronides. Anal Biochem 2005;342:45–52.

Stevens AP, Dettmer K, Wallner S, Bosserhoff AK, Oefner PJ. Quantitative analysis of 5′-deoxy-5′-methylthioadenosine in melanoma cells by liquid chromatography-stable isotope ratio tandem mass spectrometry. J Chromatogr B Analyt Technol Biomed Life Sci 2008;876:123–128.

Stewart BH, Chung FY, Tait B, Blankley CJ, Chan OH. Hydrophobicity of HIV protease inhibitors by immobilized artificial membrane chromatography: application and significance to drug transport. Pharm Res 1998;15:1401–1406.

Thomsen AE, Christensen MS, Bagger MA, Steffansen B. Acyclovir prodrug for the intestinal di/tri-peptide transporter PEPT1: comparison of in vivo bioavailability in rats and transport in Caco-2 cells. Eur J Pharm Sci 2004;23:319–325.

Turnpenny P, Rawal J, Schardt T, et al. Quantitation of locked nucleic acid antisense oligonucleotides in mouse tissue using a liquid-liquid extraction LC-MS/MS analytical approach. Bioanalysis 2011;3:1911–1921.

van Haandel L, Becker ML, Williams T, Leeder JS, Stobaugh JF. Measurement of methotrexate polyglutamates in human erythrocytes by ion-pair UPLC-MS/MS. Bioanalysis 2011;3:2783–2796.

Xu RN, Fan L, Rieser MJ, El-Shourbagy TA. Recent advances in high-throughput quantitative bioanalysis by LC-MS/MS. J Pharm Biomed Anal 2007;44:342–355.

Zhou XJ, Chakboub H, Ferrua B, Moravek J, Guedj R, Sommadossi JP. Radioimmunoassay for quantitation of 2′,3′-didehydro-3′-deoxythymidine (D4T) in human plasma. Antimicrob Agents Chemother 1996;40:1472–1475.

28

LC-MS BIOANALYSIS OF ENDOGENOUS COMPOUNDS AS BIOMARKERS

WENYING JIAN, RICHARD EDOM, AND NAIDONG WENG

28.1 INTRODUCTION

For many years, endogenous compounds such as catabolic/anabolic products, bioactive small molecules, peptides, and proteins have been measured in clinical laboratories as biomarkers using a variety of techniques. Biomarker, as defined by the US National Institute of Health (NIH) sponsored Biomarker Definitions Working Group, refers to a characteristic that is objectively measured and evaluated as an indicator of normal biological processes, pathogenic processes, or pharmacologic responses to a therapeutic intervention (NIH, 2001). Recently, biomarkers have been recognized and accepted as useful tools for drug discovery and development process, which is being shifted from a "trial-and-error" practice to a more mechanistic-based and target-driven paradigm (Katz, 2004; Goodsaid and Frueh, 2007; Wagner et al., 2007). As demonstrated in Figure 28.1 (Lee and Hall, 2009), appropriate application of biomarkers to preclinical and clinical drug discovery and development can facilitate activities such as target and candidate selection, risk assessment, study design, dose escalation, patient stratification, and safety surveillance, thus improving the productivity of drug development and reducing the time and cost of bringing new therapies to patients.

The biomarkers analyzed in clinical chemistry laboratories as diagnostic tools, such as the panels of cardiac, liver, or kidney function markers, are considered as routine biomarkers (Lee et al., 2005). Routine biomarkers are often measured using well-established methods such as FDA approved commercial kits. In contrast, endogenous molecules being explored as research tools in the drug discovery and development process are often referred to as "novel biomarkers" that may require specialized reagents and techniques for analysis, which are not readily available in a clinical chemistry setting. LC-MS technology, due to its intrinsic sensitivity and selectivity, and its capability to provide absolute quantitation, is being increasingly utilized in quantitative analysis of novel biomarkers. This chapter discusses the challenges involved in biomarker assay development using LC-MS and the approaches to overcome the issues with an emphasis on fit-for-purpose method development and validation.

Assay development for a biomarker is significantly more complex than in the case for exogenous drugs, as depicted in Table 28.1. The complexity associated with biomarker quantitation can be summarized into the following points:

1. The fundamental challenges of biomarker measurement originate from their endogenous nature. Preexisting analyte in the biological matrix used to prepare calibration standards and quality control (QC) samples may complicate the bioanalytical assay, leading to compromised assay accuracy and precision. Therefore, one of the most important aspects of a successful biomarker assay is the ability to overcome interference of the endogenous levels and their variations.
2. Unlike exogenous drugs that can be purified and well characterized for various development programs, biomarker standards are often not readily available. Batch-to-batch variability in the standards may introduce error into the assay results. This is especially true for peptide or protein biomarkers because these

Handbook of LC-MS Bioanalysis: Best Practices, Experimental Protocols, and Regulations, First Edition. Edited by Wenkui Li, Jie Zhang, and Francis L.S. Tse.

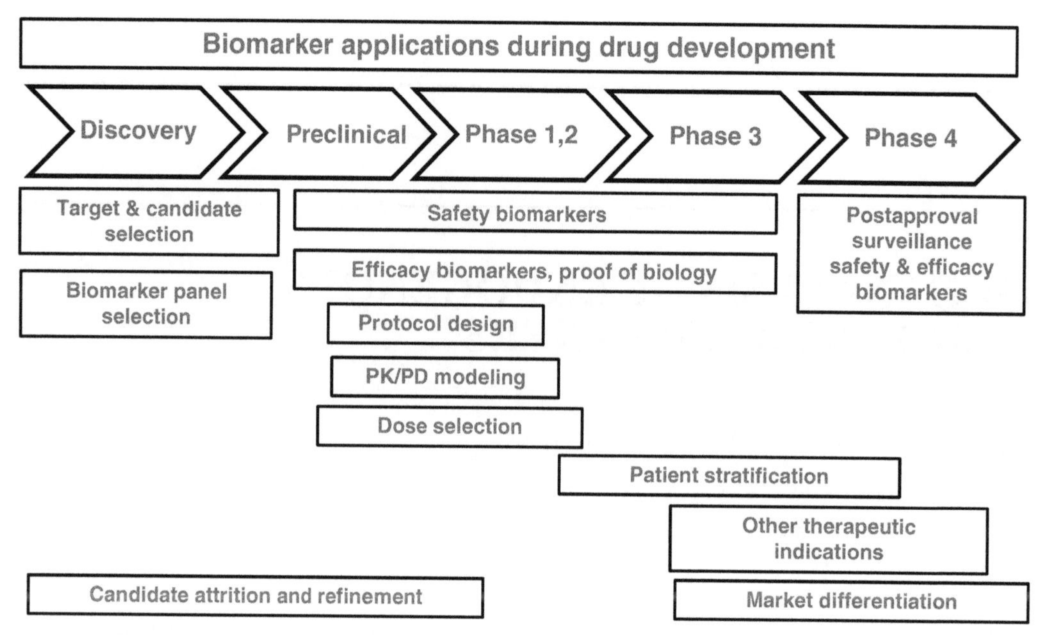

FIGURE 28.1 Application of biomarkers in drug discovery and development (Lee and Hall, 2009) (reproduced with permission).

molecules are generally heterogeneous in nature and their characterization is challenging.

3. Specificity of an LC-MS-based assay for an exogenous drug can be confirmed by the absence of an interference peak at the expected retention time of the target analyte in blank matrix samples. However, specificity for endogenous compounds is very difficult to assess due to the lack of a true blank matrix. In addition, naturally occurring analogs, such as those generated from the same biosynthetic pathway of the target biomarker, could cause interference due to the similarity in the structures.
4. While the concentrations of exogenous drugs in preclinical or clinical study samples often fall within an acceptable detection range of the LC-MS technique, biomarkers may exist in extremely low abundance. More seriously, they may be significantly downregulated by a therapeutic agent, thus requiring extensive assay development in order to achieve the desired assay sensitivity.

TABLE 28.1 Comparison of Biomarker Assays to Toxicokinetic/Pharmacokinetic (TK/PK) Drug Assays

	Drug bioanalysis	Biomarker bioanalysis
Analyte	Exogenous, not present in the matrix blank	Endogenous, usually present in the matrix blank
Matrix	Analyte-free	Contains the analyte, causing accuracy and specificity issues
Specificity	Usually easy to confirm	Difficult to confirm due to the endogenous nature of analyte; often more subject to interference from endogenous analogs
Reference standard	Certified, well characterized	Certified standard usually not available, often purchased from vendors, not well qualified with batch-to-batch variability
Sensitivity	Usually adequate	May require very high sensitivity due to the low abundance of the analyte
Range	Easily defined based on dose and TK/PK behavior of the analyte	Not easily defined due to intersubject and intrasubject variability in baseline and response to drug treatment, disease, or biological modulation
Stability	Easily established	Often poor and not easily established, may be generated *ex vivo* from precursors
Sample collection procedure	Usually straightforward	Extra caution is needed due to biological instability, diurnal change, food intake, and so on

5. The calibration curve range should cover the expected levels that result from both the control and the drug treatment. However, it may be difficult to do so for biomarkers due to the often inconsistent literature reports of baseline endogenous concentrations, inter- and intrasubject variability, and the potential disparity between healthy and afflicted populations. The assay development should be conducted in the context of the biological system being modulated. If a biomarker is expected to be dramatically upregulated, the task of developing a reliable assay that is free from impact of the baseline analyte may be relatively easier to accomplish, as opposed to detection of a relatively moderate change that could be masked by biological variations.
6. Concentrations of endogenous molecules, especially bioactive ones, are often tightly mediated by the body through regulatory mechanisms such as metabolism or protease degradation. Therefore, the stability of biomarkers is often intrinsically poor, posing a serious challenge for assay development. One also needs to keep in mind the endogenous nature of biomarkers and be aware that they could be generated from their precursors that may be present in the biological matrix. If *ex vivo* production is observed, it is necessary to separate the target analyte from its biological precursor as quickly as possible and/or to take certain precautions to prevent the conversion, such as addition of an enzyme inhibitor.
7. It is crucial to maintain sample integrity from collection through analysis, so-called vein-to-vessel stability. Issues such as instability or adsorptive loss may impact analyte concentrations from the point of collection, which requires extra caution in assay development and study protocol design. In addition, variability caused by biological factors such as diurnal change, food effect, and emotional change should also be considered in protocol design to ensure confident detection of the changes caused by drug treatment on top of these variations.

Overall, due to their endogenous nature, measurement of biomarkers presents a series of challenges to bioanalytical laboratories and often requires extensive method development for optimal assay performance.

28.2 APPROACHES FOR BIOMARKER QUANTITATION

For biomarker quantification, it is essential to select an appropriate approach to prepare calibration standard and QC samples in order to mimic the incurred samples (the real study samples) and to achieve accurate measurement of the analyte. Ideally, these samples are prepared by spiking known amounts of analyte into the intended biological matrix. However, for endogenous compounds, the authentic biological matrix usually contains an unknown amount of the analyte, making it unsuitable for preparation of standards and QCs. Different approaches have been developed to overcome this challenge based upon the nature of the target analyte and the intended biological matrix (van de Merbel, 2008; Houghton et al., 2009; Ciccimaro and Blair, 2010).

28.2.1 Authentic Analyte in Authentic Matrix

A conventional approach to overcome the challenge due to the presence of endogenous analyte is standard addition. Analyte with increasing concentrations is added to individual aliquots of a biological sample to construct serial calibration standards. The obtained intercept of the calibration curve is equal to the endogenous level of the analyte.

In some rare situations, analyte levels in authentic matrix can be essentially negligible due to species, gender, age, diurnal changes, disease conditions, and so on. For example, the concentrations of sex hormones are considerably different between male and female, as well as in populations of different ages. In another example of quantifying metalloproteinase-9 in mouse plasma, rat plasma (but not mouse plasma) was found analyte-free and was successfully used to construct the calibration curve (Ocana and Neubert, 2010). For the measurement of environmental or occupational exposure biomarkers, the biological matrix collected from nonexposed subjects can serve as blank matrix to prepare calibration standard samples (Pan et al., 2004; Li et al., 2006). Occasionally, the target analyte may be readily subject to chemical or enzymatic degradation, and therefore, it can be present at very low or zero concentrations in the authentic matrix if a stabilization procedure is not employed. Examples include ascorbic acid, catecholamines, esters, and lactones (Boomsma et al., 1993; Karlsen et al., 2005; Li et al., 2005). In these cases, conditions such as elevated temperature and exposure to oxygen in air can accelerate the degradation of analyte for generating analyte-free matrix. The resulting matrix needs to be treated with stabilizers, for example, antioxidant or enzyme inhibitor, before it can be spiked with the analyte for reliable preparation of calibration standard and QC samples.

There are also situations where analytes are actually generated *ex vivo* instead of *in vivo*, and therefore, the baseline in unstimulated biological matrix is zero. For example, in a clinical study supported by our laboratory, the plasma samples for analysis of leukotriene B_4 (LTB_4) were generated by stimulating the human whole blood with calcium ionophore to mobilize the LTB_4 biosynthesis, followed by plasma collection. Because the baseline concentration of LTB_4 in unstimulated human plasma is negligible, calibration standard and QC samples can be directly prepared in authentic matrix.

Although the "authentic analyte in authentic matrix" approach is considered to be the best for reliable assay performance, it is not very often applicable to biomarker analysis due to the difficulty in finding true blank matrix or in completely depleting the endogenous analytes from the matrix. Furthermore, if the matrix is prepared by depleting an unstable analyte, caution needs to be taken to ensure the stability of the spiked analyte by using stabilizers or certain procedures such as temperature or pH control. It is also necessary to confirm that there is no back-conversion of the reacted endogenous analyte. For example, for an endogenous lactone compound that was depleted via hydrolysis, the hydroxy-carboxylic acid could react again to reform the lactone ring, elevating the concentrations of the analyte in the standards/QCs. In addition, it is worth mentioning that the "authentic matrix" referred to in this approach may not be completely equivalent to a true matrix. For example, stimulation versus nonstimulation may cause some modification of the matrix components. The same is also true for rat versus mouse plasma, male versus female plasma, stabilization versus not stabilized, and so on. The potential nonequivalence may need to be considered in assay development, especially during troubleshooting.

28.2.2 Authentic Analyte in Surrogate Matrix

So far, the most often-used approach for endogenous analyte quantitation is to prepare standards/QCs in analyte-free surrogate matrix, which can be practical and cost-efficient. Different types of surrogate matrix are available:

1. *Buffer*: Phosphate-buffered saline (PBS) is often used due to its similar pH (7.4) and ionic strength (150 mM) to plasma. Alternatively, protein buffer such human serum albumin (HSA) or bovine serum albumin at concentration of 4–6% in PBS to mimic the protein content of human plasma can be used. Protein(s) in buffer help to increase the solubility of analytes and to prevent analyte losses due to nonspecific adsorption.
2. *Stripped matrix*: Many endogenous analytes in a biological matrix can be depleted by activated charcoal. However, the disadvantages of this approach include, but is not limited to, incomplete removal of the analyte and the inability to remove compounds that are bound to lipoproteins (such as cholesterol) (van de Merbel, 2008).
3. *Immunoaffinity extracted matrix*: Analyte can be selectively removed by using a specific antibody that is bound to sorbent on a column or on magnetic beads. The matrix generated via this approach is virtually intact and very close to the authentic one in nature. However, the procedure is expensive and time-consuming.
4. *Commercially available, artificial, biological matrix*: Examples include SeraSub synthetic serum and UriSub synthetic urine provided by CST Technologies (Great Neck, NY), and the serum substitute from STEMCELL Technologies (Vancouver, Canada) that contains HSA, insulin, transferrin, and buffer.

In this approach, analyte concentrations in incurred samples are measured against calibration curves prepared in surrogate matrix based on the assumption that the method does not exhibit significant bias. However, the potential differences between surrogate and authentic matrices might lead to a systematic bias between the standard and incurred samples during sample analysis in the following areas:

1. *Matrix effects*: The components in surrogate matrix can differ significantly from those in the authentic matrix, resulting in a varying extent of ion suppression/enhancement in LC-MS assays.
2. *Recovery*: Protein binding can impact analyte recovery during sample preparation. The analyte may have specific protein binding in authentic matrix that is absent in surrogate matrix. For example, vitamin D is tightly bound to vitamin D binding protein (DBP) in serum (Vogeser, 2010). If denaturing conditions are not strong enough to completely dissociate vitamin D from its protein binding, the extraction recovery of the analyte in incurred samples will be lower than that in surrogate matrix that does not have DBP.
3. *Stability*: Surrogate matrix may lack certain enzyme systems, such as proteases or esterases, which mediate degradation of certain types of analytes in authentic matrix.
4. *Solubility*: In the absence of carrier proteins and lipids, the solubility of some analytes in aqueous synthetic matrix may be lower than that in authentic matrix.
5. *Specificity*: Certain components in authentic matrix may interfere with the detection of the analyte of interest. However, this interference may not be present in the standard samples prepared in surrogate matrix. An example may be the quantification of endogenous hydroxy-sterols. Many isobaric, positional isomers exist in authentic biological fluids such as plasma or serum. However, they may be absent in surrogate matrices.

Use of a suitable internal standard (IS), especially a stable isotope labeled one which behaves similarly to the analyte of interest in LC-MS, can help overcome the issue of matrix effects mentioned earlier. However, none of the other issues such as protein binding, stability, solubility, and specificity can be completely solved by using a stable isotope labeled IS, as they have already impacted sample integrity before the addition of IS. Therefore, it is critical to fully evaluate the method by using QC samples prepared at least at one concentration level using authentic matrix to demonstrate the absence of those potential issues with the assay (DeSilva et al., 2003).

The baseline level of the endogenous analyte, the expected assay range, and the availability of the authentic matrix need to be considered prior to preparing the QC samples. In a generic approach for small molecule biomarker quantitation, Houghton et al. (2009) recommend the use of pooled control matrix for the preparation of the mid QC, the initially measured concentration of which is designated as the nominal concentration (Houghton et al., 2009). The high QC is prepared by spiking the control matrix with known amounts of analyte with the resulting concentration near the upper limit of the calibration curve. Preparation of the low QC can be achieved by dilution of the control matrix with surrogate matrix with the final concentration of the analyte within threefold of the lower limit of quantitation (LLOQ). The LLOQ QC is prepared by spiking the surrogate matrix. In another case for quantitation of vitamin D in human serum in our laboratory, a different scheme of QC sample preparation was employed. The LLOQ QC was prepared in a surrogate matrix (protein buffer). The horse serum, which contains marginal levels of vitamin D (Harmeyer and Schlumbohm, 2004), was screened for its baseline concentration prior to being spiked with a known amount of analyte to reach the expected low QC concentration. Pooled human serum was spiked on top of the measured endogenous level to reach the designated mid and high QC concentrations. In another case for quantitation of LTB_4 in human sputum, an alternative approach had to be taken due to the limited supply of control sputum, which was prepared by a special protocol at the clinical site. The volume the control sputum was only sufficient for the preparation of QC samples at a single level. Since the baseline level of LTB_4 was below the expected low QC concentration, the low QC was prepared by spiking the control sputum, while the mid and high QCs were prepared in PBS containing HSA. The capability to accurately measure the baseline levels and to achieve the designated concentration of the low QC, verified the performance of the assay at the low end of the calibration curve, which is a confirmation of reliable assay performance. In any case, it is preferable to secure a large enough control matrix pool to prepare all the QC samples needed for assay validation and sample analysis in order to minimize the variations caused by variable baseline analyte levels.

28.2.3 Surrogate Analyte in Authentic Matrix

A novel and scientifically elegant approach to overcome the issues of analyte baseline levels in control matrix is to employ stable isotope labeled compounds as surrogate analytes (Jemal et al., 2003; Li and Cohen, 2003). The stable isotope labeled analytes, such as those labeled by ^{13}C, ^{15}N, or deuterium, theoretically exhibit identical physical-chemical properties and LC-MS behavior to the natural compounds, but they do not exist in the authentic matrix. Therefore, they can be used to prepare calibration standard and QC samples in authentic matrix without any endogenous interference. The concentrations of the authentic analyte can be calculated by applying its mass responses to the calibration curve constructed using the surrogate analyte. Usually, a differently labeled stable isotope analyte is used as IS to ensure the ruggedness of the assay, which is common in current LC-MS-based bioanalytical practice. If a different version of stable label is not readily available, a nonendogenous structural analog may be employed as IS. As demonstrated in Figure 28.2, for analysis of endogenous fatty acid amides, namely, arachidonyl ethanolamide (AEA), oleoyl ethanolamide (OEA), and palmitoyl ethanolamide (PEA), D_4-AEA, D_4-OEA, and $^{13}C_2$-PEA were used as surrogate analytes, and D_8-AEA, D_2-OEA, and D_4-PEA were used as IS (Jian et al., 2010). The authentic analytes AEA, OEA and PEA are endogenous and were constantly present in all the samples, while the surrogate analytes were spiked at different concentrations to construct the calibration curve. The ISs were spiked at one concentration to compensate for variations in sample extraction and analysis. In another recently published article measuring 4-β-hydroxy-cholesterol (4-βHC) in human plasma as a biomarker of CYP3A activity, D_7-4-βHC and D_4-4-βHC were used as surrogate analyte and IS, respectively (Goodenough et al., 2011).

For QC preparation, different approaches have been adopted. For quantitation of AEA, OEA, and PEA in our laboratory, all levels of QCs were prepared using surrogate analytes spiked in authentic matrix, and assay validation was conducted based on concentrations of the surrogate analytes (Jian et al., 2010). In addition, the endogenous analytes in the QC samples were also determined for their concentrations in the same assay validation runs (such as accuracy/precision and stability evaluation). The reproducibility and stability of the authentic analytes demonstrated in these experiments further confirmed the assay performance. The same calibration and QC sample approach was used for quantitation of 4-βHC in human plasma as biomarker of CYP3A activity, where freeze–thaw and long-term storage stability were evaluated using a human plasma pool containing only the endogenous analyte (Goodenough et al., 2011). The stability evaluation was based on comparison of calculated concentrations after freeze–thaw or long-term storage to those determined before. Alternatively, some research laboratories prefer to keep the use of surrogate analyte in QC samples at a minimum by preparing any QC above baseline levels using authentic analyte and only those below baseline using a surrogate. In this case, it is important to secure a large control matrix pool for reliable preparation of the authentic analyte QCs. Any of the above methods would be considered acceptable as long

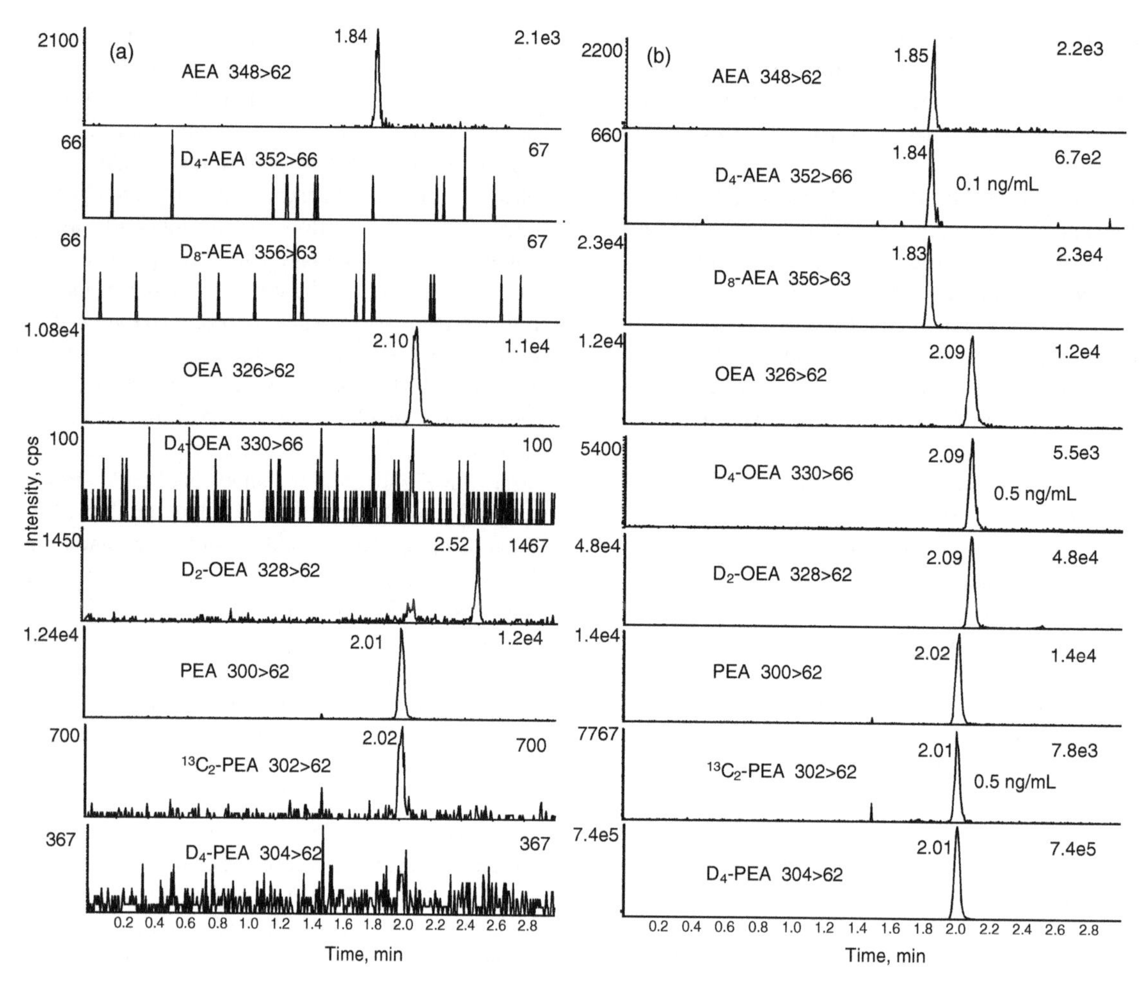

FIGURE 28.2 LC-MS/MS chromatograms of (a) an extracted blank human plasma sample without internal standards; (b) extracted human plasma spiked with D_4-AEA (0.1 ng/ml), D_4-OEA (0.5 ng/ml), and $^{13}C_2$-PEA (0.5 ng/ml) at LLOQ with internal standards. Authentic analytes: AEA, OEA, PEA; surrogate analytes: D_4-AEA, D_4-OEA, $^{13}C_2$-PEA; internal standards: D_8-AEA, D_2-OEA, D_4-PEA (reproduced with permission from (Jian et al., 2010)).

as the validation experiments prove the scientific soundness of the assay for reliably and reproducibly measuring analyte concentrations in the intended, true matrix.

The surrogate analyte approach is completely free from interference caused by the endogenous analyte so that calibration standard samples can be prepared in the intended matrix. In addition, it is considered the only reliable way to measure a decrease in the biomarker level, as it is the only approach that a QC level below the normally observed baseline can be prepared in true matrix.

Due to the above advantages, the surrogate analyte approach has gained popularity and is being increasingly applied for measurement of endogenous analytes. Nonetheless, for successful application, a number of precautions should be taken:

1. Mass spectrometric response can be different between the authentic and surrogate analyte, which could lead to biased quantitation. In most circumstances, the difference is very minor and a correction is not necessary, especially given the fact that a majority of biomarker applications fall into the category of "relative comparison" (i.e., determining fold changes pre- and post-treatment). In the case of absolute quantitation, a correction factor should be applied, which is determined by comparing the responses of equal-molar concentrations of surrogate analyte and authentic analyte.
2. Stable isotope labeled compounds need to be carefully characterized for isotopic purity. Nonlabeled impurity in the stable isotope material may interfere with the authentic analyte. Partial incorporation of the label can also lead to biased results. For example, if a D_4-compound contains only 70% full incorporation, with the remainder being D_3-, or D_2-, and so on, this would produce systematic bias in the surrogate analyte

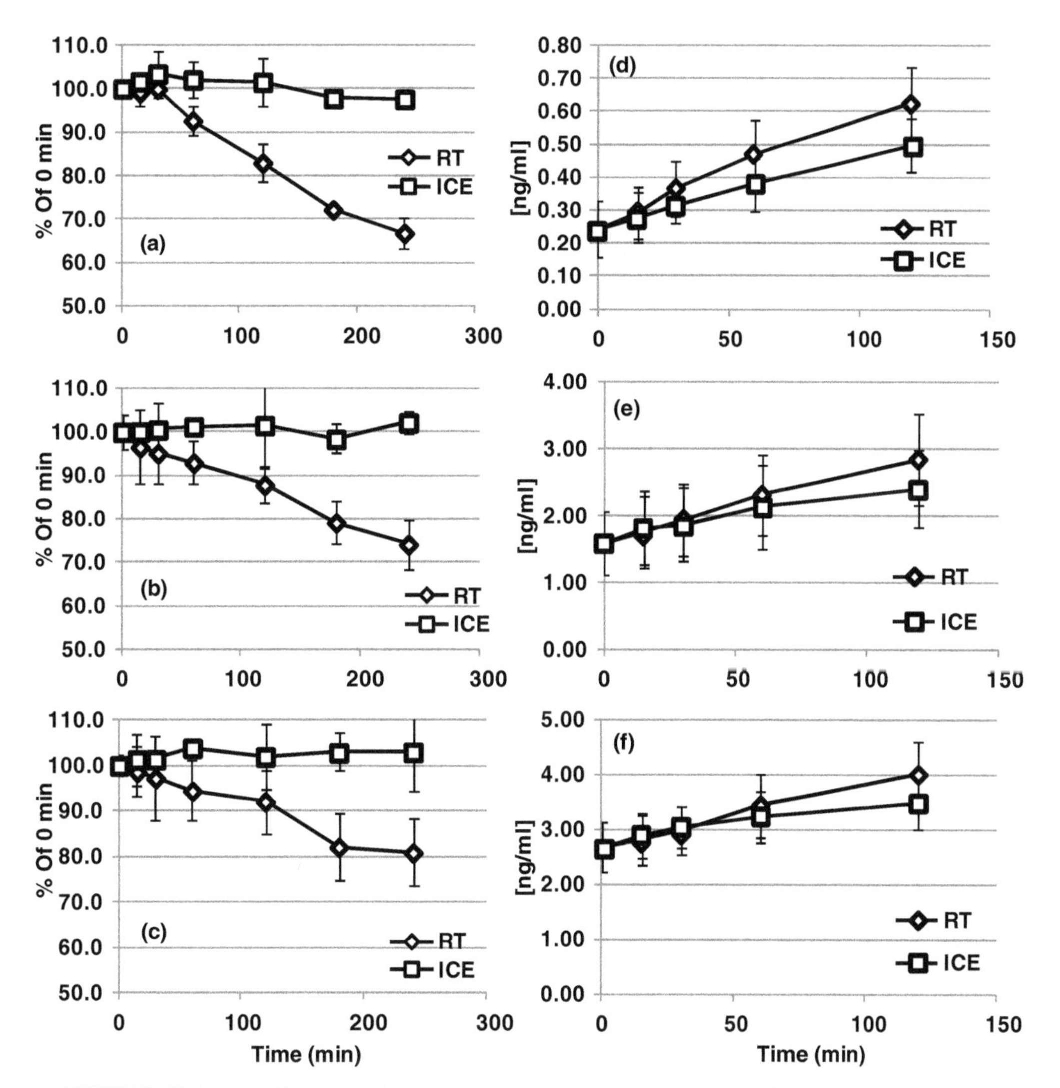

FIGURE 28.3 Stability evaluation of (a) D_4-AEA, (b) D_2-OEA, (c) D_4-PEA spiked in purchased human whole blood; mean concentrations of endogenous (d) AEA, (e) OEA, (f) PEA in fresh human whole blood ($n = 6$). RT, room temperature; ICE, on ice (modified with permission from (Jian et al., 2010)).

approach if the isotopic purity is not incorporated in calculations.

3. It is critical to ensure that the multiple reaction monitoring (MRM) channels of the authentic analytes, surrogate analytes, and ISs do not produce isotopic contribution to each other. The assay scheme needs to be strategically designed to choose suitable surrogate analyte and IS to avoid interference. For compounds containing naturally abundant heavy isotopes such as Cl or S, additional labels will be needed.

A potential pitfall when using this approach is that surrogate analytes may generate misleading stability data due to their exogenous nature. Fatty acid amides are typical examples. The surrogate analytes demonstrated time- and temperature-dependent loss in whole blood (Figure 28.3a–c). In comparison, the endogenous analyte showed rapid elevation over the time (Figure 28.3d–f). While both surrogate and authentic analytes are subject to hydrolase activity, authentic analytes are also being constantly released from their precursors, *N*-acyl-phosphatidylethanolamines (NAPE), leading to an overall increase in the measured concentrations (Natarajan et al., 1984; Cravatt et al., 1996; Hillard and Jarrahian, 2003). Similarly, back-conversion of certain metabolites to the parent compound could also generate authentic analyte, while having no impact on the exogenous surrogate analyte. Therefore, for a throughout assessment of stability, it is recommended to study the authentic analytes in true biological matrix, preferably in incurred samples, especially if there is potential for generation from a precursor or back-conversion from a metabolite.

28.2.4 Enzyme Activity Assays

All approaches discussed earlier are based on measurement of molecules involved in biological pathways, such as quantitation of enzyme substrates or products as indicators of enzyme activation/inhibition. Alternatively, for biological machineries that are present in plasma or blood cells, their activity can be directly measured *ex vivo* by incubating exogenously spiked substrates into the collected biological matrix. Enzyme activity can be expressed as the decrease of substrate concentration or the increase of product concentration over the incubation period, often normalized to the protein content of the sample. To differentiate from endogenous compounds, radiolabeled or stable isotope labeled substrates are used so that endogenous interference is not expected to be an issue. For an activity assay to work for an *in vivo* system, the enzyme involved in the target biological process has to be readily present in collectable biological matrix, which limits its use. However, in cases where it is applicable, enzyme activity assessment provides a direct measurement of target engagement for drugs targeting enzymatic systems. The data can generate valuable information for pharmacokinetics (PK)–pharmacodynamics (PD) correlation.

An enzyme activity experiment needs to be carefully designed with respect to the following considerations:

1. *Incubation conditions*: Substrate concentration should be at or below the Km value to avoid saturation of the enzyme and to keep the reaction velocity within the linear range. The pH and ionic strength (salt concentration) of the reaction buffer should be optimized for the target enzyme. Enzyme reaction rate increases with temperature, although enzyme protein denaturation at high temperature can decrease product formation. It is desirable to use a temperature that gives a suitable reaction rate so that turnover is large enough to be reliably measured, but not too high to deplete the substrate, which compromises the ability of the assay to differentiate among different samples. The same is true for incubation time in that a large enough of change in substrate or product concentration without substrate depletion should be generated within the incubation period. In most cases, an enzyme assay has already been established and the substrate concentration, buffers, and other parameters reported previously can be used.
2. *Biological matrix for bioanalytical assay:* For an enzyme activity assay, the natural biological matrix contains enzyme activity, and therefore, it cannot be directly used for preparation of standard and QC samples. Alternatively, a substitute matrix such as protein buffer or heat-deactivated matrix can be used. It is important to ensure the matrix is completely free from enzyme activity so that the measurement of substrate or product concentration is accurate and reproducible.
3. *Stability*: Since the incubation samples are usually generated fresh at the test site immediately before analysis, it is not necessary to evaluate the bench-top stability, freeze–thaw stability, or long-term freezer storage stability of the substrate or product in the incubation samples. The QC samples can be prepared daily and incubated along with the incurred samples. However, it is important to test the collected biological samples (e.g., blood cells) for enzyme activity to evaluate if activity has been maintained during shipping and storage. In this sense, the stability evaluation needs to be done for the biological matrix, rather than for the substrate or product. For example, for an enzyme activity assay in white blood cells developed in our laboratory, the stability evaluation was conducted by measuring the enzyme activity in white blood cells stored in −80°C for certain periods of time and comparing to that determined on day 0 in same lot of cells from same donor. The purpose of this experiment was to mimic the storage of the incurred white blood cell samples and to ensure that there was no activity loss before the test.

28.3 BIOMARKER ASSAY VALIDATION AND STUDY CONDUCT

28.3.1 Fit-for-Purpose

Currently, the consensus of the bioanalytical community is that a "fit-for-purpose" approach should be used for biomarker quantitation (Lee et al., 2005, 2006; Chau et al., 2008; Cummings et al., 2008, 2010). A position paper has provided extensive discussion on this topic and defined the key component of fit-for-purpose: "assay validation should be tailored to meet the intended purpose of the biomarker study, with a level of rigor commensurate with the intended use of the data" (Lee et al., 2006). The position of the biomarker assay in the spectrum from research tool to clinical end-point indicator dictates the stringency of performance verification. In addition, the nature of the bioanalytical technology, ranging from qualitative (categorical), quasi-quantitative (possesses certain attributes), relative quantitative, to definitive quantitative, also influences the level of performance verification required (Cummings et al., 2010). Biomarker measurement using LC-MS technology falls into the category of definitive quantitation and is often performed with a relatively high level of assay verification. The USFDA Guidance for Bioanalytical Methods Validation for generation of toxicokinetic (TK)/PK data for exogenous analytes is often adopted by researchers in biomarker quantitation with some extent of modification to address the challenges associated with endogenous analytes (USFDA, 2001; Viswanathan et al., 2007; Fast et al., 2009).

Recently, more clarification regarding biomarker assay validation was provided (Booth, 2011a, 2011b). For biomarkers used in early development stages, such as phase 0 or phase I clinical trials, the data are generally used internally to make decisions about candidate selection or continuing the development of the drug. The extent of method validation is not considered of critical regulatory importance and the sponsor should incorporate whatever level of validation deemed appropriate. On the other hand, "For biomarker assays intended to support the determination of safety and efficacy (approval) of a new drug or biological product, and/or to support labeled dosing instructions or patent treatment, the bioanalytical assay should be fully validated" (Booth, 2011a). In these cases, it is recommended that assay validation parameters and acceptance criteria should be same as the assay intended to measure drug concentrations. If broader acceptance criteria are needed, the criteria must be specified and rationale/justification should be provided (Booth, 2011a). It is anticipated that the next version of Bioanalytical Method Validation guidance to be issued by USFDA will include endogenous compound assays.

In Europe, when biomarker measurements are performed on samples collected in clinical trials, the laboratories conducting these analyses are subject to clinical trial regulations (Fontaine and Rosengren, 2001; HMSO, 2004, 2006). All aspects of sample analysis are governed by standard operating procedures (SOPs) and are subject to full quality assurance audits.

28.3.2 Prevalidation Considerations

Depending on the specific biomarker being monitored and the intended biological matrix, an appropriate strategy needs to be selected before any other bioanalytical parameters can be evaluated. As discussed in the previous sections, in some cases, an authentic analyte in authentic matrix approach can be applied to the analysis of endogenous analytes. Alternatively, it is often preferable to use the surrogate analyte approach if two versions of stable isotope labeled analyte can be readily obtained. However, if one or no stable isotope labeled analytes are available, surrogate matrix will have to be used to avoid endogenous interference. In addition, for control matrix such as cerebral spinal fluid (CSF) that is difficult to obtain, it is also more practical to use surrogate matrix. If the target enzyme is readily present in collectable biological matrices, an enzyme activity assay can be considered for a direct measurement.

Regardless of the approach to be taken, it is very critical to use a suitable IS. Ideally, a stable isotope labeled analyte can be used to ensure sufficient compensation of assay variation, especially matrix effects, which often significantly impact assay accuracy and precision. If only one version of stable isotope labeled analyte is available, it is often preferable to take the surrogate matrix approach and use it as IS, rather than taking a surrogate analyte approach using it as a surrogate analyte in authentic matrix while choosing a structural analog as IS. In general, the matrix effects in different matrices can be effectively compensated for by a stable isotope labeled IS. However, as discussed in Section 28.2, it is very important to always include at least one level of QC in true matrix because some other aspects of bioanalytical assays such as recovery and stability could be different in the surrogate and authentic matrix.

Unlike drug candidates that often have a reliable supply of reference compounds from internal resources, reference standard and IS of biomarkers are often obtained from outside vendors. In this case, the quality of the materials and reliability of the vendor should be carefully evaluated. Enough quantity of consistent quality/purity to cover entire validation and study support should be obtained upfront. For consistent preparation of QC samples, a large enough pool of control matrix also needs to be secured to cover the validation and study.

Very often, a biomarker of interest has been extensively studied. Therefore, reviewing literature can provide useful background information for setting a starting point for assay development. However, one needs to be cautious in attempting to fully adopt a literature method as it might have been used for research purposes only, and it may have not gone through rigorous method validation.

To choose an appropriate calibration curve range, one needs to consider the baseline level of the analyte, the inter- and intrasubject variability, as well as the analyte concentration changes due to the effect of drug treatment. Therefore, it is recommended to screen multiple matrix lots early in method development to assess the baseline level and intersubject variations. Data from previous studies and the scientific literature can serve as references. The LLOQ should be low enough to cover the baseline, as well as the potential changes in analyte concentration if there is drug-induced downregulation. Similarly, the upper limit of quantitation (ULOQ) should be high enough to cover the projected peak concentration as a result of maximum upregulation by the drug, although dilutions can always be done.

Endogenous analytes are often highly susceptible to natural regulatory mechanisms in the body. This can present a challenge for *ex vivo* stability assessment of an intended bioanalytical assay. Stability should be evaluated using authentic analyte in the authentic matrix during method development. It is recommended to use freshly collected matrix for stability evaluation if potential instability is suspected for the analyte, as the matrix components that mediate the instability, such as enzymes, may not be as active in aged matrix. If any instability is observed, appropriate measures need to be taken, such as the use of stabilizers, or extra caution in sample shipping/handling, or a change in sample collection procedures. Due to the endogenous nature of biomarkers, it is also recommended to evaluate potential *ex vivo* biosynthesis in biological matrix by conducting the stability evaluation in the presence of their precursors, if the precursors are expected to be present in the incurred samples.

28.3.3 Assay Validation

Unlike drugs, for which validation in full compliance with regulatory guidelines needs to be conducted to support preclinical good laboratory practice (GLP) studies and clinical studies, biomarker assay validation is often tailored to include only selected components that are judged by the researchers as appropriate, for the purpose of expediting the process and preserving resources. Depending on the intended purpose of the biomarker, the scope of assay validation can range from basic experiments, such as accuracy, precision, and stability evaluation, to a full scope of components recommended in the regulatory guidance for drug assay validation (USFDA, 2001; Viswanathan et al., 2007; Fast et al., 2009). As discussed in Section 28.3.1, in most cases, a "fit-for-purpose" biomarker assay is acceptable as long as it demonstrates scientific soundness and traceability to meet the study objective in a defined context of its use. The term "qualification" is often used to refer to an assay that is not fully validated and contains "fit-for-purpose" elements. In this chapter, "validation" is used generically to refer to any type of activities for assay performance verification, therefore covering both concepts of qualification and validation. Many aspects of method validation experiments are similar for endogenous biomarkers and exogenous drugs. The features specific for endogenous compounds and containing "fit-for-purpose" elements are discussed below.

28.3.3.1 Validation Plan and Acceptance Criteria Before starting a validation, the final analytical procedure and the anticipated validation experiments should be documented. Appropriate documentation should be maintained to support the intended use of the data throughout the validation and study support. The acceptance criteria also need to be determined prior to initiating validation. So far, no guidance or consensus has been given for acceptance criteria of biomarker assays (Lee, 2009; Lee and Hall, 2009). For LC-MS assays of small molecular biomarkers, the 4-6-X rule for drug TK/PK assays is commonly adopted, which requires that out of six replicates of QCs at low, mid, and high concentration levels, the results for at least four must be within ±X% (±15% for drug, ±20% at LLOQ) of the nominal values for the analytical run to be accepted. In general, an "X" value of 20% can be used for biomarker assays, given the fact that analytical variability is often higher for biomarker assays than for drugs. Other than choosing acceptance criteria based on expected analytical assay performance, factors such as the intended use of the data and the expected modulation caused by drug treatment should be also considered. For example, for a biomarker expected to change by only 30% upon drug treatment, more stringent criteria need to be used in order to increase the predictive power of the data, while for a 10-fold treatment effect, more lenient criteria is acceptable because of the large magnitude of change in biomarker levels.

28.3.3.2 Accuracy, Precision, Sensitivity, and Linearity Accuracy and precision is measured by analysis of QC samples over separate analytical batches, each containing QC levels of at least three concentrations: low, mid, and high. Assessment of precision and accuracy at the LLOQ and above the ULOQ may not be important, as these concentrations are usually outside the physiologically relevant range for a well-designed assay. If a surrogate analyte or a surrogate matrix approach is used, it is important to evaluate the quality of the data for the authentic analyte in authentic matrix. As demonstrated in the example of fatty acid amide quantitation, in addition to the determination of intraassay and interassay accuracy and precision of the surrogate analytes (Table 28.2a), the concentrations of the endogenous analytes in the QC samples were also calculated (Table 28.2b). The endogenous analytes were found to be quantified in a highly reproducible manner on a daily basis, demonstrating excellent reproducibility.

Similar to drug assays, calibration curve fitting is determined by applying the simplest model that best describes the concentration–response relationship. Weighting ($1/x$ or $1/x^2$) is usually used. Fundamentally, assays based on LC-MS technology should produce a linear concentration–response relationship. Therefore, linear regression analysis is preferred. When LC-MS assays produce significantly nonlinear calibration results, an investigation should be conducted. Fitting the data to an exponential regression may mask an underlying problem with assay, such as adsorptive losses, poor solubility, or instability of the analyte.

28.3.3.3 Stability Stability of analyte in biological matrix should be evaluated for short-term (bench-top), freeze–thaw cycles, and long-term storage (frozen) to mimic the conditions that are expected to be experienced by the incurred samples. Stock solution stability should be evaluated to cover the daily use (bench-top) and long-term storage. In addition, stability of processed samples in the autosampler should be evaluated. Finally, as discussed previously, if an enzyme activity approach is taken, it is recommended to evaluate the "stability" of the biological matrix for conservation of the enzyme activity during handling and storage.

In cases of *ex vivo* production of an endogenous biomarker, use of a surrogate analyte for stability testing might result in a different profile. Likewise, surrogate matrix may lack the components (e.g., enzymes) that mediate the

TABLE 28.2 (a) Accuracy and Precision of Analysis of AEA, OEA, and PEA in Human Plasma (ANOVA Analysis of Three Analytical Runs). QC Samples Are Prepared Using Surrogate Analytes ($n = 6$). (b) Mean ($n = 6$) Calculated Concentrations (ng/ml) of Endogenous Analytes AEA, OEA, and PEA in QC Samples in Three Analytical Runs.

(a)

		Nominal conc. (ng/ml)	Mean observed conc. (ng/ml)	%Bias	Between run precision (%CV)	Within run precision (%CV)	Total variation (%CV)
D_4-AEA	LQC	0.300	0.321	7.0	5.3	7.1	8.8
	MQC	5.00	4.93	−1.4	2.0	5.5	5.8
	HQC	8.00	8.31	3.9	1.4	3.2	3.5
D_4-OEA	LQC	3.00	2.94	−2.0	1.3	3.0	3.3
	MQC	50.0	49.4	−1.2	0.0	4.9	4.6
	HQC	80.0	80.6	0.8	0.0	3.6	3.3
$^{13}C_2$-PEA	LQC	1.50	1.49	−0.7	3.5	5.6	6.6
	MQC	25.0	25.2	0.8	2.1	3.4	4.0
	HQC	40.0	41.5	3.8	0.0	2.9	2.8

(b)

		Run 1	Run 2	Run 3	Average	%CV
AEA	LQC	0.275	0.276	0.290	0.280	3.0
	MQC	0.264	0.274	0.304	0.281	7.4
	HQC	0.269	0.264	0.304	0.279	7.8
OEA	LQC	1.735	1.790	1.849	1.791	3.2
	MQC	1.719	1.766	1.832	1.772	3.2
	HQC	1.720	1.723	1.862	1.768	4.6
PEA	LQC	1.832	1.687	1.920	1.813	6.5
	MQC	1.826	1.670	1.779	1.758	4.6
	HQC	1.815	1.617	1.846	1.759	7.1

potential instability, and therefore, its use could generate misleading stability results. Therefore, it is critical to test stability using authentic analyte in authentic matrix, either unspiked or spiked on top of the baseline level, preferably at low and high QC concentrations. The stability samples are subjected to the conventional tests and measured concentrations are compared to nominal concentrations. Since the nominal concentrations for endogenous analytes includes the measured (endogenous) component, an initial measurement needs to be conducted to obtain the day 0 concentrations so that stability results can be compared against the initial values.

There is no regulatory guideline for acceptance criteria on stability evaluation. It is common to see a ±15% rule used (compared to the nominal values) as per current practice for drug assays. Considering the complexity of biomarker assays, a more relaxed criteria of ±20% could be applied for stability assessments. In reality, endogenous compounds may be more susceptible to natural degradation processes than exogenous compounds, even when preventive measures have been taken. In these cases, the instability information, including the test conditions, the time frame, and extent of instability loss, should be used to estimate the potential impact on the obtained results from incurred samples.

28.3.3.4 Specificity Endogenous analytes are often subjected to interference from isobaric compounds or structural analogs in the biological matrix. Specificity of the assay for the target analyte in the presence of those interfering components should be evaluated in method development, and the results documented during validation. For a regular LC-MS assay for exogenous compounds, specificity can be demonstrated with no difficulty by the absence of an interfering peak at the expected retention time of the analyte in an extracted blank matrix sample. The same evaluation can be done for surrogate analytes in authentic matrix or authentic analyte in surrogate matrix. However, for the authentic analytes in authentic matrix, it is challenging to determine specificity, since the analytes are always present in the samples due to their endogenous nature. Therefore, based on the existing knowledge about the analyte, a more flexible experimental design should be applied.

Several MS techniques can be explored to evaluate specificity:

1. Comparison of chromatographic peak shape between injections of the neat analyte and extracted, authentic samples. Any difference in peak shapes may suggest coeluting isobaric interferences.

2. Several MRM channels can be simultaneously monitored for each analyte to compare the relative intensities of different product ions between the neat and extracted samples.
3. Product ion or MS^3 spectra generated from the peak at the expected retention time in extracted matrix samples and neat standard solutions can be compared to unveil the presence of any coeluting components.
4. High-resolution MS provides the capability to differentiate coeluting isobaric compounds that are of different elemental compositions.

Specificity can also be evaluated by using alternative chromatographic conditions. Shallower gradients, longer analytical runs, modification of mobile phase additives, or change of stationary phase may facilitate the separation of coeluting components and reveal the interference that was buried under the analyte peak. Separation power can be maximized by using two-dimensional LC based on orthogonal mechanisms, such as reversed phase and hydrophilic interaction chromatography (HILIC), for better elucidation of coeluting components from one dimension alone. In some circumstances, it may also be necessary to conduct chiral chromatography for separation of naturally occurring enantiomers. For example, three isobaric isomers of LTB_4 may be generated from nonenzymatic hydrolysis of the precursor Leukotriene A_4 (LTA_4) (Borgeat and Samuelsson, 1979b, 1979a; Lee et al., 1984; Fretland and Anglin, 1991; Sala et al., 1996). They share identical MRM transitions with authentic LTB_4 and may cause overestimation of its concentrations if not chromatographically separated. The specificity evaluation in our laboratory involved injecting the mixture of LTB_4 and the three isomers and demonstrating their baseline separation.

Chemical derivatization of functional groups, for example, phenols, alcohols or amines of biomarker molecules of interest, can alter their chromatographic behavior and enhance the sensitivity in MS detection, thus further improving assay specificity (Quirke et al., 1994). In addition, data obtained from complimentary assay platforms, such as Enzyme Linked Immuno Sandwich Assay (ELISA), LC-UV, and LC-fluorescence, can be compared with those of LC-MS to elucidate if there is a potential specificity issue.

28.3.3.5 ***Matrix Effects*** In general, stable isotope labeled IS can effectively compensate for ionization variation caused by matrix components in LC-MS-based assays. Therefore, matrix effects evaluation may not be necessary for biomarker assay validation if a stable isotope labeled IS is in use. Nevertheless, several approaches have been popular for evaluating of matrix effects in LC-MS assays for endogenous analytes:

1. If a stable isotope labeled analyte is available, it can be used in the place of authentic analyte for the assessment of matrix effects via a conventional approach to compare instrument response in post-spiked matrix samples to that in neat solutions.
2. For evaluation of the impact of matrix on authentic analytes, a dilution linearity experiment can be conducted. If the same results can be obtained for an authentic analyte in the intended matrix before and after dilution using a matrix-free buffer (or surrogate matrix), the assay is considered to be free from matrix effect liability (Houghton et al., 2009).
3. Significant variability of standard curve slopes (run to run) in different matrix lots (or surrogate matrix lots) may serve as a good indicator of relative matrix effects. It is suggested the %CV of slopes from five individual matrix lots not exceed 3–4% for a method to be considered free from matrix effects liability (Matuszewski, 2006).
4. If the recovery of the analyte has been determined, matrix effects for an authentic analyte can be calculated by comparing instrument response from an extracted sample of known concentration to that of a corresponding neat solution that was spiked with an equal amount of the analyte expected in the matrix sample (amount × recovery).

28.3.3.6 ***Recovery*** Evaluation of extraction efficiency of endogenous analytes from biological matrix might not be necessary if a stable isotope labeled IS is used. Nonetheless, it can be evaluated via different approaches:

1. If a stable isotope labeled analyte is available, it can be used in place of the authentic analyte to measure recovery. The chromatographic peak area in extracted samples is compared to that from blank matrix spiked postextraction, or the relative response of the stable isotope labeled analyte to the endogenous analyte is compared, as the endogenous analyte is supposed to be present in same amount in extracted and postspiked samples.
2. The following equation can be used to calculate recovery of authentic analyte, assuming the baseline concentration has been determined in the lots of biological matrix used for the recovery experiment:

$$\frac{(\text{Baseline concentration} + \text{spiked concentration}) \times \text{recovery}}{\text{Baseline concentration} \times \text{recovery} + \text{spiked concentration}} = \frac{\text{Peak area of prespiked sample}}{\text{Peak area of postspike sample}}$$

This approach requires that the spiked concentration is in sufficient amount relative to the baseline concentration for accurate and reproducible calculation. Therefore, recovery evaluation at lower concentrations is usually not conducted if the baseline concentration is relatively high. It will be up to the judgment of the researchers for each individual case on how to select the appropriate concentrations for this experiment.

28.3.4 Study Support

28.3.4.1 Study Design Concentrations of endogenous analytes are usually subject to regulation in the body such as diurnal changes and food effects. These biological variations may complicate the interpretation of data in a preclinical or clinical study for investigation of drug effect on biomarker concentrations. This is especially true for new or novel biomarkers, for which historical information on normal concentrations, biological variations, and the expected extent of change due to disease conditions or drug treatment are largely unknown. Taking the above factors into consideration, a study protocol must be strategically designed to avoid generating confounding or uninterpretable data:

1. Intersubject variations often preclude comparison between placebo versus treated subjects for detection of changes in biomarkers caused by drug treatment, unless the change is dramatic.
2. Intrasubject variations may preclude comparisons between pretreatment (0 h) versus post-treatment. Otherwise, a crossover design is necessary to compare the biomarker concentration profiles from the same subject dosed with drug and placebo. Alternatively, the biomarker levels on the day before drug treatment (Day-1) can be used as baseline to compare with those obtained on the day of treatment (Day 1).
3. If the time frame and magnitude of changes in biomarker levels caused by drug treatment are unknown, the probability to detect changes reliably could be increased by serial blood sampling rather than selected time points. After trends are elucidated, it may be possible to reduce the blood sampling to fewer points taken at key times post-dose.

A phase 0 clinical study can be conducted for better understanding of biological variations of biomarkers. The phase 0 study simulates typical single ascending dose or multiple ascending dose studies in population, diet, activity restrictions, sample collection time points, and sample processing procedure, with exception that it does not involve administration of any novel or marketed pharmaceutical agent. Biological samples are collected for measurement of baseline biomarker levels over different time points. The intersubject and intrasubject variations of the targeted biomarker elucidated in a phase 0 study can be taken into consideration for future clinical study design and data interpretation.

28.3.4.2 Preanalytical Sample Integrity Prior to conduct of a study, it is critical to evaluate the integrity of the analyte during sample collection and to establish a reliable processing procedure for minimal *ex vivo* changes. The evaluation experiment should mimic the conditions to be used in the intended study as much as possible in terms of freshness of the matrix, the processing time, and the processing conditions. For example, in order to establish the plasma collection procedure for fatty acid amide measurement for a clinical study, stability experiments were conducted in fresh blood. The blood samples were either immediately processed for plasma (within 5 min of collection) or incubated at room temperature or on wet ice for different periods of time. It was found that the endogenous fatty acid amide levels were rapidly elevated after incubation due to their release from the precursors located in blood cells (Figure 28.3d–f). To prevent *ex vivo* change of fatty acid amide levels, it was critical to separate the plasma from blood cells as soon as possible so that the enzymes/transporters involved in the biosynthesis and anabolism are removed. Accordingly, a sample collection that required separation of plasma from blood within 3 min of collection was established and applied to support the clinical studies (Jian et al., 2010).

For analytes that are subject to losses due to instability, such as protease degradation, precautions such as temperature control, pH adjustment, or stabilizer addition should be taken. In some cases, it is necessary to precharge the vacutainers with a stabilizer in order to prevent any rapid analyte loss upon sample collection. This entails confirmation of the preservation of the stabilizer and maintenance of vacuum during the storage of the vacutainers.

For biological fluids of relatively low-protein content such as urine and CSF, collection tubes, pipettes, and storage containers must be evaluated to minimize nonspecific adsorption of the biomarker. Certain procedures such as addition of surfactant or organic solvent, or use of a specially treated low-binding container, may be necessary to prevent adsorption.

It is important to standardize the procedure for all sample collection and handling, and to keep them consistent throughout the entire study, which is especially true for multisite studies. Specific instructions such as sample volume, type of venipuncture, type of collection tube, *g*-force, temperature and duration of centrifugation, and so on, should be detailed in the study protocol, and appropriate training should be conducted for the laboratory personnel, if necessary.

28.3.4.3 Sample Analysis Verification of method performance of ongoing sample analysis involves analysis of QC samples at low, mid, and high concentrations, including at least one level of QC samples in authentic matrix. Ideally, the

QC samples should be the same batch used for method validation to ensure assay performance. Again, the often adopted acceptance criteria is 4-6-X, with X determined *a priori* based on biological variability, the expected change caused by drug treatment, and the intended use of the data.

It is desirable during sample analysis that a complete set of samples to be compared (e.g., all the samples from same dose group, or all the samples from same subject) are analyzed within the same batch to reduce analytical bias. If this is not possible, extra care is then required to ensure that inter-batch variations do not mask true differences between the sample sets to be compared. This would be especially true with methods where acceptance criteria have been loosened beyond the conventional $\pm 15\%$ (bias) window, or when a small degree of change in the biomarker is measured.

Incurred sample reproducibility (ISR) evaluation has recently been introduced as a mandatory requirement for regulated bioanalysis to ensure the assay has adequately addressed potential issues associated with incurred samples. These issues include, but are not limited to, metabolite conversion, protein-binding differences in patient samples, recovery problems, sample in-homogeneity, analyte instability, and matrix effects (Fast et al., 2009). For endogenous analytes, repeat analyses of QC samples prepared using authentic analyte in authentic matrix during validation mimic ISR evaluation. Formal measurement of ISR using incurred samples from a preclinical or clinical study also offers a good opportunity to evaluate assay performance over the full range of analyte concentrations in biological matrices. It can also help to evaluate the impact of other factors, such as the presence of drug, metabolites and concomitant medications in the incurred samples, on the measurement of biomarkers of interest. However, despite the fact that ISR can offer the insight into the performance of an intended method, conducting ISR is not mandatory for biomarker analysis at the present time.

28.4 SUMMARY AND PROSPECTIVE

Endogenous analytes are increasingly utilized as biomarkers in preclinical and clinical studies in support of drug discovery and development. Due to the endogenous nature of biomarkers, bioanalysis of those compounds usually encounters a series of challenges. Among the various challenges, dealing with endogenous levels, maintaining analyte integrity from collection to analysis, achieving specificity, and obtaining sufficient sensitivity are the most difficult ones. Those challenges entail special consideration and meticulous experiment design in method development, validation, and study conduct.

LC-MS provides definitive quantitation with high specificity and sensitivity, and therefore, it has become one of the most popular platforms for biomarker quantitation, especially for small molecule biomarkers. Four major approaches for biomarker analysis, and their pros and cons, have been discussed in detail in this chapter: (1) authentic analyte in authentic matrix, (2) authentic analyte in surrogate matrix, (3) surrogate analyte in authentic matrix, and (4) enzyme activity assay. Among the four approaches, the second and third are the ones often applied. In particular, the surrogate analyte in authentic matrix approach provides a simple way to avoid endogenous interference at all concentration levels within the calibration range, also allowing reliable measurement of a decrease in biomarker levels below its endogenous concentration. Nevertheless, it is recommended to evaluate assay performance for the authentic analyte in true matrix whenever possible.

Currently, there is no regulatory guidance specific for biomarker bioanalysis. The "fit-for-purpose" approach, which tailors the burden of assay verification according to the drug development stage, the intended use of the data, and the nature of the assay platform, has gained consent within bioanalysis community. For small molecular biomarkers analyzed using LC-MS, researchers often adopt the existing regulatory guidance for drug TK/PK assay validation. The term "fit-for-purpose" reflects flexible inclusion/exclusion of validation experiments, experiment design, and acceptance criteria. In general, assessment of accuracy, precision, and stability is considered the essential part of assay validation, while others, for example, matrix effect and recovery, are considered optional, especially when a stable isotope labeled IS is used. It is important to set the acceptance criteria prior to start of validation or sample analysis, as well as to maintain proper documentation. Scientific soundness of the assay and reconstructability of the study conduct are crucial factors to consider in the whole process of method development, validation, and study conduct.

Recently, more clarification regarding biomarker assay validation has been provided by FDA which specified that data generated for regulatory actions should be supported by a validated assay (Booth, 2011a, 2011b). It is expected that bioanalytical guidance concerning biomarker assay validation will be released in the near future.

Another emerging trend in biomarker quantitation is the LC-MS bioanalysis of peptide or protein biomarkers (Carr and Anderson, 2005; Addona et al., 2009; Ciccimaro and Blair, 2010). Large numbers of biomarkers fall into the categories of peptides and proteins, and they have been traditionally analyzed using ligand-binding assays (LBA) such as ELISA. LC-MS affords the ability to measure the molecules with high specificity at the primary structure level. As LC-MS assay development is relatively fast with no need to raise antibodies, the approach has been increasingly explored as an alternative to conventional LBA in support of drug discovery and development. The discussion of LC-MS analysis of peptides and proteins is out of the scope of this chapter.

Nevertheless, it is worth mentioning that the basic approaches for measuring endogenous compounds can also be applied to quantitation of endogenous peptides and proteins.

REFERENCES

Addona TA, Abbatiello SE, Schilling B, et al. Multi-site assessment of the precision and reproducibility of multiple reaction monitoring-based measurements of proteins in plasma. *Nat Biotechnol* 2009;27:633–641.

Boomsma F, Alberts G, Van Eijk L, Man In't Veld Aj, Schalekamp Ma. Optimal collection and storage conditions for catecholamine measurements in human plasma and urine. Clin Chem 1993;39:2503–2508.

Booth B. Biomarkers Where we are going with method validation. Applied Pharmaceutical Analysis Meeting, Sept 11–14, 2011a, Boston, MA.

Booth B. When do you need a validated assay? Bioanalysis 2011b;3:2729–2730.

Borgeat P, Samuelsson B. Arachidonic acid metabolism in polymorphonuclear leukocytes: unstable intermediate in formation of dihydroxy acids. Proc Natl Acad Sci USA 1979a;76:3213–3217.

Borgeat P, Samuelsson B. Metabolism of arachidonic acid in polymorphonuclear leukocytes: structural analysis of novel hydroxylated compounds. J Biol Chem 1979b;254:7865–7869.

Carr SA, Anderson L. Protein quantitation through targeted mass spectrometry: the way out of biomarker purgatory? Clin Chem 2005;54:1749–1752.

Chau CH, Rixe O, Mcleod H, Figg WD. Validation of analytical methods for biomarkers used in drug development. Clin Cancer Res 2008;14:5967–5976.

Ciccimaro E, Blair IA. Stable-isotope dilution LC-MS for quantitative biomarker analysis. Bioanalysis 2010;2:311–341.

Cravatt BF, Giang DK, Mayfield SP, Boger DL, Lerner RA, Gilula NB. Molecular characterization of an enzyme that degrades neuromodulatory fatty-acid amides. Nature 1996;384:83–87.

Cummings J, Raynaud F, Jones L, Sugar R, Dive C. Fit-for-purpose biomarker method validation for application in clinical trials of anticancer drugs. Br J Cancer 2010;103:1313–1317.

Cummings J, Ward TH, Greystoke A, Ranson M, Dive C. Biomarker method validation in anticancer drug development. Br J Pharmacol 2008;153:646–656.

Desilva B, Smith W, Weiner R, et al. Recommendations for the bioanalytical method validation of ligand-binding assays to support pharmacokinetic assessments of macromolecules. Pharm Res 2003;20:1885–1900.

Fast DM, Kelley M, Viswanathan CT, et al. Workshop report and follow-up—AAPS Workshop on current topics in GLP bioanalysis: Assay reproducibility for incurred samples—implications of Crystal City recommendations. AAPS J 2009;11:238–241.

Fontaine N, Rosengren B. Directive 2001/20/EC of the European Parliament and of the Council: on the approximation of the laws, regulations and administrative provisions of the Member States relating to the implementation of good clinical practice in the conduct of clinical trials on medicinal products for human use. Off J Eur Commun 2001;L121:34–44.

Fretland DJW, Widomski DL, Anglin CP. 6-trans-Leukotriene B4 is a neutrophil chemotoxin in the guinea pig dermis. J Leukoc Biol 1991;49:283–288.

Goodenough AK, Onorato JM, Ouyang Z, et al. Quantification of 4-beta-hydroxycholesterol in human plasma using automated sample preparation and LC-ESI-MS/MS analysis. Chem Res Toxicol 2011;24:1575–1585.

Goodsaid F, Frueh F. Biomarker qualification pilot process at the US food and drug administration. AAPS J 2007;9:E105–108.

Harmeyer J, Schlumbohm C. Effects of pharmacological doses of vitamin D3 on mineral balance and profiles of plasma vitamin D3 metabolites in horses. J Steroid Biochem Mol Biol 2004;89–90:595–600.

Hillard CJ, Jarrahian A. Cellular accumulation of anandamide: consensus and controversy. Br J Pharmacol 2003;140:802–808.

HMSO. US Statutory Instrument 2004 No. 1031. The Medicines for Human Use (Clinical Trials) Regulations. 2004.

HMSO. US Statutory Instrument 2006 No. 1928. The Medicines for Human Use (Clinical Trials) Regulations. 2006.

Houghton R, Horro Pita C, Ward I, Macarthur R. Generic approach to validation of small-molecule LC-MS/MS biomarker assays. Bioanalysis 2009;1:1365–1374.

Jemal M, Schuster A, Whigan DB. Liquid chromatography/tandem mass spectrometry methods for quantitation of mevalonic acid in human plasma and urine: method validation, demonstration of using a surrogate analyte, and demonstration of unacceptable matrix effect in spite of use of a stable isotope analog internal standard. Rapid Commun Mass Spectrom 2003;17:1723–1734.

Jian W, Edom R, Weng N, Zannikos P, Zhang Z, Wang H. Validation and application of an LC-MS/MS method for quantitation of three fatty acid ethanolamides as biomarkers for fatty acid hydrolase inhibition in human plasma. J Chromatogr B Analyt Technol Biomed Life Sci 2010;878:1687–1699.

Karlsen A, Blomhoff R, Gundersen TE. High-throughput analysis of vitamin C in human plasma with the use of HPLC with monolithic column and UV-detection. J Chromatogr B Analyt Technol Biomed Life Sci 2005;824:132–138.

Katz R. Biomarkers and surrogate markers: an FDA perspective. NeuroRx 2004;1:189–195.

Lee JW. Method validation and application of protein biomarkers: basic similarities and differences from biotherapeutics. Bioanalysis 2009;1:1461–1474.

Lee JW, Devanarayan V, Barrett YC, et al. Fit-for-purpose method development and validation for successful biomarker measurement. Pharm Res 2006;23:312–328.

Lee JW, Hall M. Method validation of protein biomarkers in support of drug development or clinical diagnosis/prognosis. J Chromatogr B Analyt Technol Biomed Life Sci 2009;877:1259–1271.

Lee JW, Weiner RS, Sailstad JM, et al. Method validation and measurement of biomarkers in nonclinical and clinical samples in drug development: a conference report. Pharm Res 2005;22:499–511.

Lee TH, Menica-Huerta JM, Shih C, Corey EJ, Lewis RA, Austen KF. Characterization and biologic properties of 5,12-dihydroxy derivatives of eicosapentaenoic acid, including leukotriene B5 and the double lipoxygenase product. J Biol Chem 1984;259:2383–2389.

Li B, Sedlacek M, Manoharan I, et al. Butyrylcholinesterase, paraoxonase, and albumin esterase, but not carboxylesterase, are present in human plasma. Biochem Pharmacol 2005;70:1673–1684.

Li W, Cohen LH. Quantitation of endogenous analytes in biofluid without a true blank matrix. Anal Chem 2003;75: 5854–5859.

Li Y, Li Ac, Shi H, et al. Determination of S-phenylmercapturic acid in human urine using an automated sample extraction and fast liquid chromatography-tandem mass spectrometric method. Biomed Chromatogr 2006;20:597–604.

Matuszewski BK. Standard line slopes as a measure of a relative matrix effect in quantitative HPLC-MS bioanalysis. J Chromatogr B Analyt Technol Biomed Life Sci 2006;830:293–300.

Natarajan V, Schmid PC, Reddy PV, Schmid HH. Catabolism of N-acylethanolamine phospholipids by dog brain preparations. J Neurochem 1984;42:1613–1619.

NIH. Biomarkers and surrogate endpoints: preferred definitions and conceptual framework. Clin Pharmacol Ther 2001;69: 89–95.

Ocana MF, Neubert H. An immunoaffinity liquid chromatography-tandem mass spectrometry assay for the quantitation of matrix metalloproteinase 9 in mouse serum. Anal Biochem 2010;399:202–210.

Pan J, Song Q, Shi H, et al. Development, validation and transfer of a hydrophilic interaction liquid chromatography/tandem mass spectrometric method for the analysis of the tobacco-specific nitrosamine metabolite NNAL in human plasma at low picogram per milliliter concentrations. Rapid Commun Mass Spectrom 2004;18:2549–2557.

Quirke Jmecl, Adams CL, Van Berkel GV. Chemical derivatization for electrospray ionization mass spectrometry. 1. Alkyl halides, alcohols, phenols, thiols, and amines. Anal Chem 1994;66:1302–1315.

Sala A, Bolla M, Zarini S, Muller-Peddinghaus R, Folco G. Release of leukotriene A4 versus leukotriene B4 from human polymorphonuclear leukocytes. J Biol Chem 1996;271:17944–17948.

USFDA. Guidance for Industry: Bioanalytical validation. Guidance for Industry: Bioanalytical validation 2001;

Van De Merbel NC. Quantitative determination of endogenous compounds in biological samples using chromatographic techniques. Trends in Analytical Chemistry 2008;27:924–933.

Viswanathan CT, Bansal S, Booth B, et al. Quantitative bioanalytical methods validation and implementation: best practices for chromatographic and ligand binding assays. Pharm Res 2007;24:1962–1973.

Vogeser M. Quantification of circulating 25-hydroxyvitamin D by liquid chromatography-tandem mass spectrometry. J Steroid Biochem Mol Biol 2010;121:565–573.

Wagner JA, Williams SA, Webster CJ. Biomarkers and surrogate end points for fit-for-purpose development and regulatory evaluation of new drugs. Clin Pharmacol Ther 2007;81:104–107.

29

LC-MS BIOANALYSIS OF DRUGS IN HEMOLYZED AND LIPEMIC SAMPLES

MIN MENG, SPENCER CARTER, AND PATRICK BENNETT

29.1 INTRODUCTION

Over the last two decades, liquid chromatography–tandem mass spectrometry (LC-MS/MS) has become extremely popular in drug development due to its inherent advantages such as specificity, sensitivity, and high throughput. While LC-MS/MS became the mainstream analytical technique for bioanalysis, the phenomenon of "matrix effects" was observed and reported in numerous publications (Matuszewski et al., 1998; Cote et al., 2009). The term "matrix effect" can be broadly applied to any change in the behavior of the compound(s) of interest and/or internal standard (ISs) due to the presence of coeluting compound(s). More specifically, the term refers to the impact on the ionization process, either positively (ionization enhancement) or negatively (ionization suppression) (Liang et al., 2003). The coeluting compounds may include salts, endogenous components, dosing vehicles, anticoagulants, constituents from hemolysis or lipemia, or from the laboratory analytical process (e.g., reagents and HPLC column bleed). Significant matrix effects would affect the precision, sensitivity, and accuracy for the concentration of analyte(s) of the interest.

A common contributor of matrix effects are endogenous phospholipids. This has been studied extensively since it was first reported in 2003 (Bennett and Van Horne, 2003; Ismaiel et al., 2008; Xia and Jemal, 2009). In contrast, matrix effects caused by hemolysis or lipemia have seldom been reported or discussed. This chapter details the current issues, the best practice, and recommendation for assessing and handling any analytical impact that may occur from hemolysis and lipemia using LC-MS/MS technique.

29.2 HEMOLYSIS

Hemolysis is caused by the rupture of red blood cell (RBC) membranes, resulting in the release of hemoglobin and other internal cellular components into plasma or serum sample. Depending on the extent of hemolysis, the hemolyzed plasma or serum samples physically appear anywhere from a pink to red color (Figure 29.1). Hemolysis is more pronounced in small rodents because of the difficulty in drawing blood. Hemolysis can occur *in vivo* or *in vitro* with the latter being the more common cause. *In vitro* hemolysis could be caused by improper specimen collection, processing, and transport, such as an improper choice in the venipuncture site; cleansing the venipuncture site with alcohol and not allowing the site to dry; an improper venipuncture indicated by a slow blood flow; the use of a small-bore needle resulting in a large vacuum force applied to the blood, causing shear stress on the RBCs; the use of a large bore needle resulting in a much faster and more forceful flow of blood through the needle; vigorous mixing or shaking of a specimen; prolonged contact of plasma with cells; exposure to excessive heat or cold; length, speed, and number of times the specimen is transported and the number of turns of the container. *In vivo* hemolysis is caused by pathological conditions including autoimmue hemolytic anemia or a transfusion reaction. *In vitro* hemolysis may be caused by improper specimen collection, processing, or transportation (Carraro et al., 2000).

In hemolyzed plasma or serum samples, the concentration of hemoglobin, bilirubin, and potassium are increased. The impact of hemolysis on general clinical data using optical technology was reported decades ago in which hemolysis

Handbook of LC-MS Bioanalysis: Best Practices, Experimental Protocols, and Regulations, First Edition. Edited by Wenkui Li, Jie Zhang, and Francis L.S. Tse.

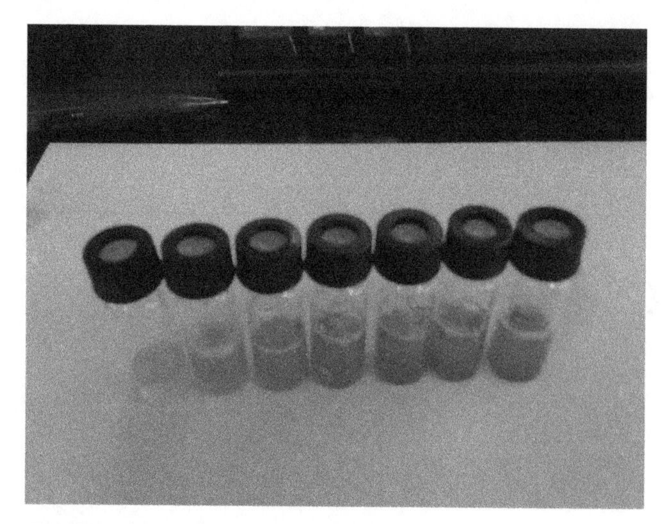

FIGURE 29.1 Full array of hemolyzed plasma samples (from left to right: 0%, 0.5%, 1%, 2%, 3%, 4%, and 5%).

resulted in certain analytes to be increased (Frank et al., 1978) or caused interference in the test method (Sonntag, 1986). The impact of hemolysis on quantitation for optical methods is easily understandable due to visible color and density changes. Typically, the data from hemolyzed sample using optical methods is rejected (Carraro et al., 2000; Laessig et al., 1976).

29.2.1 Current Status Regarding Hemolysis for LC-MS/MS Analysis

The impact of hemolysis on quantitative bioanalysis using LC-MS/MS is more complicated because LC-MS/MS technique uses a combination of the compound mass, its ability to be ionized and the appropriate chromatography resolution as a means of detection. While color and density change is not an issue, the extra amount of RBC components present in hemolyzed plasma samples may cause matrix effects depending on the chemical properties of the analytes, the extraction procedure, and/or chromatographic conditions. Bioanalytical methods for new drug development are often performed within a high regulatory compliance environment. Typically, bioanalytical methods are specifically developed, optimized, and validated for a given species and matrix, and sometimes to a certain strain or gender. In this context, it could be argued that hemolyzed plasma samples are an altered and different matrix compared to the validated matrix. However, it is also not practical to validate plasma with each percentage of the increased level of hemolysis. Unlike general clinical data in which the easiest approach is to reject data, it is very costly to reject data from hemolyzed sample for PK study during new drug development. It may result in incomplete concentration-time profiles and incorrect reporting of the C_{max}. Based on a recent publication by Tang and Thomas (2012), hemolysis may have negative impact on following assays: (1) compounds not stable due to the enzyme released from RBC; (2) compounds with known blood: plasma partitioning issue in which the drug has high affinity to RBCs; and (3) compounds with matrix effect by the endogenous interference from hemolyzed samples.

Within a commercial bioanalytical contract research organization (CRO), it is very common to receive hemolyzed plasma samples for preclinical studies. Hemolysis in human samples is much less frequent and of less magnitude. However, the percentage of hemolyzed samples is higher in phase II or III than phase I studies because the subject populations are distinctly different, that is, healthy volunteer versus patient. Because laboratories are consistently required to analyze hemolyzed samples based on a validated plasma method and the option to reject hemolyzed sample is not recommended, the only solution is to evaluate hemolysis effects and establish an acceptable threshold of hemolysis during method development (MD) and method validation (MV).

Currently, there is no consensus for conducting hemolysis test during the MD/MV stages. There are no established procedures or regulatory guidance for evaluating the effect of hemolysis on analytical methods in the United States. Globally, the only agency requires hemolysis is Agência Nacional de Vigilância Sanitária (National Health Surveillance Agency Brazil) (ANVISA, 2002). However, the experiment requirement is rather simple. It is incorporated into specificity test in which one of the six lots is hemolyzed sample. Recently, the FDA has issued deficiency letters while conducting routine audits for failure to conduct hemolysis evaluations during validations. Specific procedures for evaluating the effect of hemolysis are expected to be produced from the global bioanalytical consortium.

29.2.2 Analysis and Interpretation of Case Studies

The topic of hemolysis appears in several white papers proposing concepts to determine the impact of hemolysis on quantitation (Savoie et al., 2009, 2010). Over the last 2 years, there were three reports detailing the failure of hemolysis test and subsequent investigation and solutions (Hughes et al., 2009; Bérubé et al., 2011; Carter et al., 2011). A summary of the case studies from these three publications are included in Table 29.1 and discussed later. In general, the extent of hemolysis, the root cause, and the ultimate solution vary from analyte to analyte. The analysis and interpretation of these case studies would help us further understand the mechanism and the impact of hemolysis effect, and ultimately facilitate future MD.

Report 1 (Hughes et al., 2009)

For the atorvastatin assay, there was a visible interfering peak eluting just prior to the analyte peak in hemolyzed sample.

TABLE 29.1 Summary of the Case Studies of Failure of Hemolysis in Regulated Bioanalysis

Reference	Analyte name[a]	Original method	Hemolysis assessment	Method modification
Hughes et al. (2009)	Atorvastatin	LLE	Coeluting peak from hemolyzed sample	Change mobile phases
Hughes et al. (2009)	Phenylephrine	SPE	Lost sensitivity in hemolyzed sample	Change to stable isotope labeled IS
Hughes et al. (2009)	Carvedilol	SPE	Lost sensitivity for the metabolites in hemolyzed sample	Change to PPE/SPE extraction
Hughes et al. (2009)	Olanzapine	SPE	Lost sensitivity in hemolyzed sample	Change to LLE extraction
Bérubé et al. (2011)	Fluvoxamine	SPE	Not stable at −20°C for hemolyzed sample	Change storage temperature from −20°C to −80°C
Bérubé et al. (2011)	Morphine	SPE	Extract from hemolyzed sample not stable in the reconstitute solvent	Change recon. solvent from basic condition to neutral condition
Carter et al. (2011)	Mesalamine	SPE	Lost sensitivity in hemolyzed sample	Change collection procedure.
Carter et al. (2011)	Albuterol	SPE	Hemolysis passed at 1% and 2% but failed at 5%	Keep the method as it. Using relative hemolysis scale image to grade samples. Require to assess hemolysis impact for samples >2%
Carter et al. (2011)	Asenapine	SPE	Hemolysis passed at 0.5%, but failed at 5%	Inconclusive. Need more investigation

[a]Only parent drug listed.

A simple solution was utilized in which the LC conditions were modified and the interfering peak was separated from the analyte peak.

The failure of phenylephrine assay is slightly more complicated because only 1% hemolysis met the acceptance criteria. A simple solution to resolve this issue was to dilute hemolyzed samples with plasma. The disadvantage of this approach was that it resulted in many sample with nonreportable value due to overdilution of severely hemolyzed sample. The ultimate solution was to replace the IS with a stable isotope labeled IS.

For the case study of carvedilol, it was found that the extent of hemolysis effect was concentration dependent and only related to the metabolite. The bias also correlated with the color of the extract suggesting that hemoglobin or bilirubin played a certain role. This issue was resolved by changing the extraction from solid phase extraction (SPE) to protein precipitation extraction (PPE) + SPE.

A similar approach was taken when dealing with the failure of the hemolysis evaluation test for olanzapine in which the extraction procedure was changed from *hydrophilic-lipophilic-balanced* SPE (HLB) to PPE + SPE (MCX), then to liquid–liquid extraction (LLE) to eventually resolve this issue.

Report 2 (Bérubé et al., 2011)

In contrast to the aforementioned scenarios in which the failure of the hemolysis test was related to matrix effects or extraction or chromatography, the following two examples are due to instability associated with hemolyzed samples.

For the case of fluxoamine, failure of hemolysis test was only noticed while analyzing frozen hemolyzed QCs. Based on report from Briscoe and Hage (2009), storage temperature and anticoagulant are two main factors affecting matrix stability. The instability was eventually resolved by storing samples at lower temperature (−80°C). It was also found that hemolyzed samples were more stable in K_2EDTA plasma than in sodium heparin plasma.

For the case of morphine, it was found that the failure of the hemolysis test samples was associated with the pH of the reconstitution solvent. According to Yeh and Lach, 1961, the degradation of morphine increases at high pH conditions and with the presence of oxygen. Because hemoglobin is an O_2 carrier, it can be assumed in this case that hemoglobin induced oxidation is playing a role.

Report 3 (Carter et al., 2011)

In our laboratory, we have encountered three failures of hemolysis test samples, that is, mesalamine, albuterol, and asenapine.

As the active metabolite of sulfasalazine, mesalamine is a strong oxidizing agent. The addition of the reducing reagent, $Na_2S_3O_5$, can stabilize mesalamine in human plasma. Although whole blood stability experiments showed a trend of mesalamine degrading at longer durations and higher temperatures in whole blood, these tests met the acceptance criteria at 1–8°C. The method was initially developed in $Na_2S_2O_5$-treated human plasma, which met the acceptance criteria for all validation tests (accuracy and precision, selectivity, freeze/thaw, bench top stability, etc.) with the exception of hemolysis test (Table 29.2). The failure was

TABLE 29.2 Mesalmine Hemolysis Evaluation (Instrument Response)

Run number	Low QC level 0.0% hemolysis 6.00 ng/ml	Low QC level 0.5% hemolysis 6.00 ng/ml	Low QC level 5.0% hemolysis 6.00 ng/ml
6	0.004721	0.00597	0
	0.003208	0.005473	0
	0.005633	0.006836	0
	0.00467	0.006051	0
	0.004371	0.006184	0
	0.003673	0.00414	0
Mean	0.004	0.006	0.000
SD	0.001	0.001	0.000
%CV	19.5	15.8	N/A
%Difference	N/A	31.88	−100.00
n	6	6	6

TABLE 29.3(a) Albuterol Hemolysis Evaluation (Instrument Response)

Run number	Low level 0.0% hemolysis 3.00 pg/ml	Low level 0.5% hemolysis 3.00 pg/ml	Low level 5.0% hemolysis 3.00 pg/ml
2	0.006474	0.006025	0.010534
	0.006752	0.006292	0.009239
	0.006438	0.005905	0.008679
	0.006204	0.005930	0.008704
	0.006358	0.006885	0.008637
	0.006925	0.006284	0.009592
Mean	0.00653	0.00622	0.00923
SD	0.000266	0.000367	0.000743
%CV	4.1	5.9	8.1
%Difference	N/A	−4.67	41.47
n	6	6	6

so excessive that, at 5.0% hemolysis, mesalamine and its deuterated IS were virtually undetectable. Thus, the validation was halted and further investigation was conducted. It was found that human plasma samples must be treated with acetonitrile (MeCN) immediately after collection. This treatment completely alters the matrix and stabilizes mesalamine. This method was revalidated in MeCN extract from human plasma and passed all validation tests including long-term matrix stability at −70°C and −20°C for the extent of time tested (98 days).

The failure of hemolysis test for albuterol was less significant. During initial MD, albuterol showed acceptable storage stability in human plasma and neat solutions under various pH, temperature, and light exposure conditions. Because of these results, the validation for albuterol in human plasma proceeded. However, during MV, whole blood collection stability tests demonstrated that albuterol was only stable for up to 0.75 h in wet ice (data not shown). Similarly, hemolysis test samples passed at a level of 0.5% hemolysis but failed at 5.0% (Table 29.3(a)). This experiment was repeated at 1% and 2% and met the acceptance criteria (Table 29.3(b)). The method was validated with a special notation that any samples with a hemolysis level >2% would require a documented impact assessment.

Asenapine is completely stable in whole blood for up to 2 h at room temperature. The hemolysis test samples barely passed at 0.5% but failed at 5.0% using the initial extraction and chromatographic conditions (MCX extraction, C_{18} HPLC column and a neutral mobile phase). The following troubleshooting experiments were conducted but no conclusive results were obtained: (1) same MCX extraction + different chromatography (IBD Ultra column and acidic mobile phase)-hemolysis test failed; (2) PPE extraction + different chromatography (IBD Ultra column and acidic mobile phase)-hemolysis test failed; (3) ion suppression experiment by infusion—no suppression observed (Figure 29.2); (4) freshly prepared hemolyzed sample versus frozen hemolyzed sample-hemolysis test passed for both groups; and (5) original SRM transitioned versus new SRM transitioned-hemolysis test samples passed for both transitions. The specific reason(s) for failure of the asenapine method in hemolyzed samples could not be identified and the method was not validated.

29.2.3 Pros and Cons of the Known Hemolysis Experiment Procedures

Because there is no current formal guidance for how to conduct hemolysis tests, experimental procedures vary from laboratory to laboratory. Table 29.4 summarizes four different

TABLE 29.3(b) Albuterol Repeated Hemolysis Evaluation for Lower Percentage of Hemolysis (Instrument Response)

Run Number	Low level 0.0% hemolysis 3.00 pg/ml	Low level 1.0% hemolysis 3.00 pg/ml	Low level 2.0% hemolysis 3.00 pg/ml
3	0.005861	◊	◊
	0.005956	0.005814	0.005706
	0.006075	0.006036	0.005973
	0.006221	0.006058	0.005975
	0.006689	0.00614	0.006383
	0.007397	0.006282	0.006759
Mean	0.00637	0.00607	0.00616
SD	0.000582	0.000171	0.000413
%CV	9.1	2.8	6.7
%Difference	N/A	−4.72	−3.26
n	6	6	6

◊Sample deactivated due to analytical reasons.

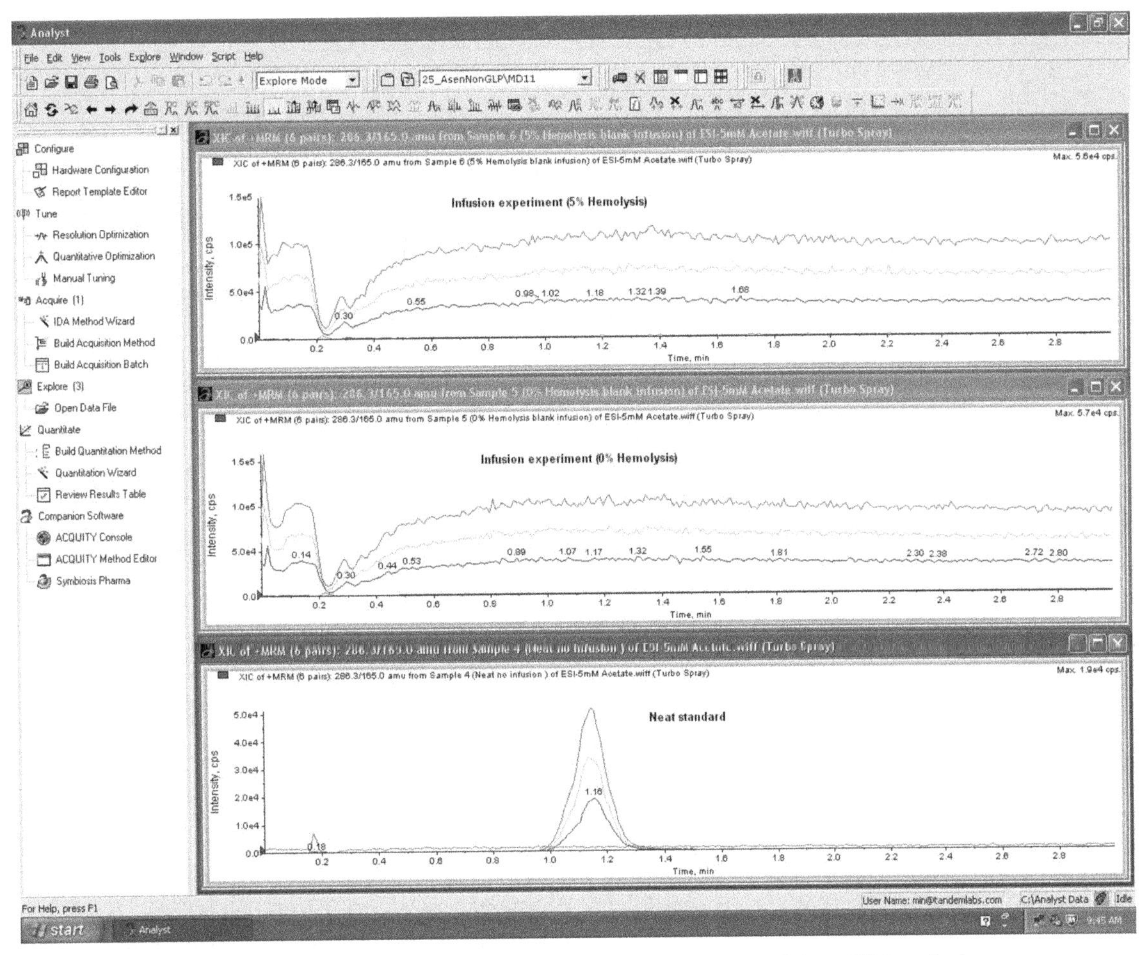

FIGURE 29.2 Asenapine Ion suppression evaluation between 0% vs. 5% hemolysis.

procedures from three CROs and one pharmaceutical company on how to conduct hemolysis experiment. Procedures 1 and 2 both compare the control versus the test group using instrument response. Also, only the low QC level was tested. The only difference is the highest level of hemolysis test group. As shown in Figure 29.1, it is impossible to visually distinguish the level of hemolysis when it is >2% hemolysis. Based on our experience, it is very rare that plasma samples for analysis appear to be hemolyzed >2%. Therefore, the evaluation at 2% hemolysis is sufficient for this evaluation purposes. Procedures 3 and 4 both quantitate hemolyzed samples against calibration curves. The extent of the hemolysis was calculated using nominal concentrations. Additionally, hemolyses at both low and high QC levels were evaluated. For Procedure 3, this test was performed using frozen hemolyzed QC. Compared to Procedures 1 and 2, Procedures 3 and 4 are very thorough and comprehensive. The drawback is the complexity to execute in the laboratory (preparation of standard calibrators) and the potential biases that may result from preparing hemolyzed and control samples separately.

29.2.4 Recommended Hemolysis Experiment Procedures

Based on the earlier discussions regarding known hemolysis experiment procedures, we recommend using Procedure 1 as the standard for hemolysis evaluation during MD and validation but using Protocols 3 and 4 for troubleshooting or investigations. The following describes the step-by-step procedure based on Procedure 1.

Experimental design:

1. Prepare hemolyzed whole blood. Aliquot appropriate amount of fresh whole blood into clean polypropylene tube. Place in −20°C or −70°C freezer for at least 30 min.
2. Prepare hemolyzed plasma or serum. Thaw above hemolyzed whole blood samples. Spike an appropriate amount of hemolyzed whole blood into blank plasma (v/v). Generate three test groups at 0% (100% normal plasma/serum), 0.5% (i.e., spike 50.0 μl of hemolyzed

TABLE 29.4 Summary of the Known Hemolysis Procedures

Protocol	Reference	Percent of hemolysis	Level and replicates	Procedure	Acceptance criteria
Protocol 1	In house procedure	0.5% and 2%	Low QC ($n = 6$)	Test at 0% (100% normal), 0.5% and 2%. Analyze six (6) replicates for each group. Use instrument response for calculation.	The %CV for each group must be ≤15%. The mean response for each group must be ≤15%. If the initial experiment fails, repeat at lower percent of hemolysis.
Protocol 2	A big pharma company	0.5% and 5%	Low QC ($n = 6$)	Test at 0% (100% normal), 0.5% and 5%. Analyze six (6) replicates for each group. Use instrument response for calculation.	The %CV for each group must be ≤ 15%. The mean response for each group must be ≤ 15%. If the initial experiment fails, repeat at lower percent of hemolysis.
Protocol 3	Bérubé et al. (2011)	7.5%	Low QC ($n = 3$) High QC ($n = 3$)	Spike the analyte at low and high QC level at 7.5% hemolyzed sample and storage at designated temperature. Analyze three (3) replicates for each level. Quantify against with standard calibration curves.	At least two out of three of the three (3) replicates at each level should have a percent deviation within ± 15% from the nominal.
Protocol 4	Hughes et al. (2009)	2%	Low QC ($n = 6$) High QC ($n = 6$)	Spike the analyte at low and high QC level at 2% hemolyzed sample and storage at designated temperature. Analyze six (6) replicates. Quantify against with standard calibration curves.	The difference of the hemolyzed sample to nonhemolyzed samples must be ≤15%.

blood into 9.95 ml of blank plasma) and 2% hemolyzed plasma/serum (i.e., spike 200 μl of hemolyzed blood into 9.8 ml of blank plasma).

3. Prepare test samples at the low QC concentration using the above three types of plasma. Use the same pipette, spiking solution, and spiking volume for each group. This procedure minimizes potential preparation error.
4. Aliquot $n = 6$ replicates for each group. There are a total of 18 samples. Because there is no calibration curves to bracket the test samples place test samples in the sequence of 0%, 0.5%, 2%, 0%, 0.5%, 2%, 0%, 0.5%, 2%, and so on.
5. Use the intended method for extraction and analysis.
6. Use instrument response (ratio of analyte peak area vs. IS peak area) for calculation.
7. If the initial experiment fails, repeat at lower percent of hemolysis such as 0.2% or 1%.

Acceptance criteria: The %CV of each group must be ≤15%. The mean response for each test group must be ≤ 15% from the mean response of the control group.

29.2.5 Best Practice and Strategy to Evaluate Hemolysis Effect

From the above case studies, it can be concluded that the root causes for the failure of hemolysis test can vary from case to case. The approaches for troubleshooting are also different from laboratory to laboratory. Typically, the troubleshooting was conducted based on the preference and experience of the scientist. Based on our experience and analysis of the known case studies, the following best practice and strategy are highly recommended for the evaluation of hemolysis effect:

1. If hemolysis test fails, the first action is to evaluate whole blood stability. Whole blood stability should be conducted during MD. Whole-blood stability may provide insight into the outcome of hemolysis testing and can facilitate troubleshooting in a failed hemolysis test.
2. If analyte is unstable in whole blood at 1–8°C within a reasonable collection time (i.e., 1 h) and hemolysis test fails, it suggests that instability is presumably related to the induced oxidation from hemoglobin. Consequently, the method would need major modification as the matrix will be changed as a result of altering the collection procedure to enable stabilization of the analyte in matrix. The following is recommended:
   a. Add antioxidant such as ascorbic acid into the biological matrix.
   b. Add enzyme inhibitors into the biological matrix.

c. Add organic solvent to precipitate proteins in the whole blood, centrifuge, and collect the supernatant at collection site.

3. If the analyte is stable in whole blood, but failed for hemolysis testing, it suggests that the failure is either related to matrix effects or a binding issue. The following investigation and method modifications are recommended:
   a. Evaluate recovery.
   b. Change chromatography or ionization mode if applicable.
   c. Evaluate extraction procedure. The best extraction procedure is to remove hemoglobin from extract. In general, hemoglobin cannot be extracted by LLE, and therefore, this extraction type should be considered. In contrast, certain SPE extractions conditions actually coextract the analyte of interest and hemoglobin. Thus, SPE should be avoided if hemolysis is an issue.
   d. Conduct postcolumn infusion experiments. If there is ion suppression or enhancement from the hemolyzed sample, it is recommended to change LC condition and/or ionization source.
   e. Conduct matrix effects experiment in which the analyte and IS are postextraction spiked into the extracts from 0%, 0.5%, and 2% hemolysis. If peak areas are consistent, it indicates that the failure is due to binding issue rather than ion suppression.
   f. To resolve binding issue, evaluate the addition of urea, high concentration or saturated salts, strong acids or bases.
   g. If structural analog is used as IS and it is affected by hemolysis, switch to stable isotope labeled IS.
4. If the instability in whole blood and hemolyzed sample is mild or insignificant and can be managed by shorter duration and reduced temperature, it is acceptable to leave the method as it is, but establish the threshold for the affected samples; that is, whole blood need to be prepared on ice bath within 1 h only and only <1% hemolyzed sample can be reported, and so on.

29.3 LIPEMIA

Similar to hemolysis, lipemia can interfere with laboratory tests that use optical methods (Kroll, 2004). The interference mechanism from lipemia is fundamentally different than that of hemolysis. In hyperlipemic sample, the amount of chylomicrons and VLDL (very low density lipoproteins) are higher than in the normal sample. Chylomicrons and VLDL are suspended particles, producing cloudiness or turbidity. Common causes of lipemia are diabetes mellitus, ethanol use, chronic renal failure, hypothyroidism, medications such as protease inhibitors (for HIV infection), estrogen, and steroids et al. (Creer and Ladenson, 1983).

29.3.1 Current Status Regarding Lipemia for LC-MS/MS Analysis

The evaluation of the quantitative impact of hemolysis is relatively easy, because the test samples can be prepared by adding hemoglobin or whole blood and the hemolyzed samples possess distinguished color. In contrast, the evaluation of the impact of lipemia on quantitation is more complicated, because there is no readily available standardized materials to produce lipemic samples, and the lipemic samples is not readily notable. It has been reported to add IntraLipid, a synthetical emulsion for intravenous administration, to serum to simulate lipemic samples. However, it was also found that samples with added IntraLipid do not perfectly mimic lipemic sample using immunoturbidimetry (Bornhorst et al., 2004).

In 2004, Chin et al. presented a research paper regarding the impact of lipemia on the quantitation of olanzapine and its metabolite desmethyl olanzapine (Chin et al., 2004). In this chapter, four categories were created for the visual inspection of lipemia sample:

1. *Category 1*: Transparent when inspected from the side and top.
2. *Category 2*: Transparent to slightly turbid when inspected from the side and not transparent when inspected from the top.
3. *Category 3*: Barely transparent to slightly turbid when inspected from the side and not transparent when inspected from the top.
4. *Category 4*: Not transparent when inspected from the side or top.

Although useful, this categorization has intrinsic bias, as it is difficult to distinguish turbidity due to fibrin particles or lipid contents. The conclusion from this case study was that lipemia can significantly influence ionization and extraction recovery.

Currently, there is no consensus for conducting lipemia tests during the MD/MV stages. There are no established procedures or regulatory guidance for evaluating the effect of lipemia on analytical methods in the United States. Globally, the only agency requires lipemia test is Agência Nacional de Vigilância Sanitária (National Health Surveillance Agency Brazil) (ANVISA, 2002). However, the experiment requirement is rather simple in which only one lipemic sample is tested as part of the specificity test.

29.3.2 Recommended Lipemia Experiment Procedures

Because the case study regarding lipemia experiment using LC-MS/MS is rarely seen in journals, two separate procedures are recommended in this chapter depending on the discretion of the readers. For both procedures, the lipemic matrix can be purchased through commercial vendors.

Procedure 1 (Williard, 2012; procedure from PharmaNet/i3 with permission)

Experimental design

1. Prepare three replicates of low (≤3 × LLOQ (lower limit of quantitation)) and high (≥75% of ULOQ) QC samples and blanks without IS in lipemic matrix. The lipemic matrix should contain the same type of anticoagulant used for MV.
2. Process and analyze the lipemic samples with the validation calibration curves and QC samples.

Acceptance criteria: Lipemic low and high QC sample results are expected to meet precision and accuracy acceptance criteria stated for intraassay validation QCs (≤15% precision and mean accuracy vs. nominal concentration).

Procedure 2

Experimental design

1. Aliquot equal amount of normal plasma (control) and lipemic plasma (Lipemia).
2. Prepare test samples at the low QC concentration using the above two types of plasma. Use the same pipette, spiking solution and spiking volume for each group. This procedure minimizes potential preparation error.
3. Aliquot $n = 6$ replicates for each group. There are a total of 12 samples. Because there is no calibration curves to bracket the test samples place test samples in the sequence of control, lipemia, control, lipemia, control, lipemia, and so on.
4. Use the intended method for extraction and analysis.
5. Use instrument response (ratio of analyte peak area vs. IS peak area) for calculation.

Acceptance criteria: The %CV of each group must be ≤15%. The mean response for lipemia group must be ≤15% from the mean response of the control group.

29.3.3 Best Practice and Strategy to Evaluate Lipemia

Based on limited information, the above two procedures provide new initiatives and guidance for future exploration of lipemia effect. There is no unambiguous if the initial experiment meets the acceptance criteria. However, if the test fails, follow-up action is required. If the analyte of the interest is targeted for a hyperlipemic population, that is, a new drug for obesity, the failure of lipemia test is not acceptable. Unlike hemolyzed samples that possess a unique color, hyperlipemic samples cannot be visually distinguished. Therefore, the intended method should be redeveloped if the test fails and the samples are targeted for a hyperlipemic population.

29.4 CONCLUSION

The failure of hemolysis testing for regulated LC-MS/MS bioanalysis assay is uncommon. For example, in our laboratory, out of hundreds of new methods developed annually, less than five assays would fail hemolysis test. However, a hemolysis failure is an indication that with the assay being used, inaccurate results may be obtained when unknown samples are hemolyzed. Thus, we believe that the assessment of hemolysis effects during MD and validation is necessary and important. The current validation guidance requires selectivity testing, that is, evaluating the accuracy and precision at LLOQ or low level from individual donors (FDA/CDER, 2001). However, this test is insufficient to account for hemolyzed samples. Although hemolysis test is required by Agência Nacional de Vigilância Sanitária (National Health Surveillance Agency Brazil) (ANVISA, 2002) and as part of the specificity test, it is considered insufficient. Although the hemolysis test is now accepted within the industry, the procedures are quite different. In order to ensure consistency, the regulatory guidance needs to be published detailing experiment design and acceptance criteria.

Although the only agency requires lipemia test is Agência Nacional de Vigilância Sanitária (National Health Surveillance Agency Brazil) (ANVISA, 2002), the experiment requirement is rather simple. At this stage, the incorporation of the evaluation of lipemia effect into regulated LC-MS/MS bioanalysis MV appears too early, but can be performed when requested in some laboratories (PharmaNet/i3). More publications of case studies and problem resolution are needed before any concrete recommendation can be issued.

ACKNOWLEDGMENT

The authors thank Clark Williard for providing Lipemic Experiment Procedure implemented at PharmaNet/i3, Princeton, NJ.

REFERENCES

ANVISA, Manual for good bioavailability and bioequivalence practice. Brazilian Sanitary Surveillance Agency, Brazil; 2002.

Bennett PK, Van Horne KC. Identification of the major endogenous and persistent compounds in plasma, serum, and tissue that cause matrix effects with electrospray LC/MS techniques. Presented at the 2003 AAPS Annual Meeting and Exposition, Salt Lake City, Utah, 2003.

Bérubé ER, Taillon MP, Furtado M, Garofolo F. Impact of sample hemolysis on drug stability in regulated bioanalysis. Bioanalysis 2011;3:2097–2105.

Bornhorst JA, Roberts RF, Roberts WL. Assay-specific differences in lipemic interference in native and Intralipid-supplemented samples. Clin Chem 2004;50:2197–2201.

Briscoe C, Hage D. Factors affecting the stability of drugs and drug metabolites in biological matrices. Bioanalysis 2009;1:205–220.

Carraro P, Servidio G, Plebani M. Hemolyzed specimens: a reason for rejection or a clinical challenge? Clin Chem 2000;46:306–307.

Carter S, Yuan WW, Zhao Y, Bessette B, Meng M. Impact of Hemolysis on the Quantitation of LC/MS/MS Assays: Case Studies of Mesalamine, Albuterol, and Asenapine in Human Plasma. Presented at ASMS 58th Conference in Denver, Colorado, 2011.

Creer MH, Ladenson J. Analytical errors due to lipemia. Lab Med 1983;14:351–355.

Chin C, Zhang ZP, Karnes HT. A study of matrix effects on an LC/MS/MS assay for olanzapine and desmethyl olanzapine. J Pharm Biomed Anal 2004;35:1149–1167.

Cote C, Bergeron A, Mess JN, et al. Matrix effect elimination during LC-MS/MS bioanalytical method development. Bioanalysis 2009;1:1243–1257.

FDA/CDER. Guidance for the Industry. Bioanalytical Method Validation, May 2001 (US Department of Health and Human Services, FDA (CDER) and (CVM), Rockville, MD).

Frank JJ, Bermes EW, Bickel MJ, Watkins BF. Effect of in vitro hemolysis on clinical values for serum. Clin Chem 1978;24:1966–1970.

Hughes NC, Bajaj N, Fan J, Wong EYK. Assessing the matrix effects of hemolyzed samples in bioanalysis. Bioanalysis 2009;1:1057–1066.

Ismaiel OA, Halquist MS, Elmanly MY, et al. Monitoring phospholipids for assessment of ion enhancement and ion suppression in ESI and APCI LC/MS/MS for chlorpheniramine in human plasma and the importance of multiple source matrix effect evaluation. J Chromatogr B 2008;875:333–343.

Kroll MH. Evaluating interference caused by lipemia. Clin Chem 2004;50:1968–1969.

Laessig RH, Hassemer DJ, Paskey TA., Schwartz TH, The effects of 0.1 % and 1.0 % erythrocytes and hemolysis on serum chemistry values. Am J Clin Pathol 1976;66:639–644.

Liang HR, Foltz RL, Meng M, Bennett P. Ionization enhancement in atmospheric pressure chemical ionization and suppression in electrospray ionization between target drugs and stable-isotope-labeled internal standards in quantitative liquid chromatography/tandem mass spectrometry. Rapid Commun Mass Spectrom 2003;17(24):2815–2821.

Matuszewski BK, Constnzer ML, Chavez-Eng CM. Matrix effect in quantitative LC/MS/MS analyses of biological Fluids: A method for determination of finasteride in human plasma at pictogram per milliliter concentrations. Anal Chem 1998;70:882–889.

Sonntag O. Haemolysis as an interference factor in clinical chemistry. J Clin Chem Clin Biochem 1986;24:127–139.

Savoie N, Booth BP, Bradley T, et al. The second calibration and validation group workshop on recent issues in good laboratory practice bioanalysis. Bioanalysis 2009;1:19–30.

Savoie N, Garofolo F, Amsterdam, PV, et al. 2009 White paper on recent issues in regulated bioanalysis from the 3rd calibration and validation group workshop. Bioanalysis 2010;2:53–68.

Tang D, Thomas E. Strategies for dealing with hemolyzed samples in regulated LC-MS/MS bioanalysis. Bioanalysis 2012;4(22):2715–2724.

Williard C. Lipemic Experiment Procedure. Courtesy of PharmaNet/i3, Princeton, NJ; 2012

Xia YQ, Jemal M. Phospholipids in liquid chromatography/mass spectrometry bioanalysis: comparison of three tandem mass spectrometric techniques for monitoring plasma phospholipids, the effect of mobile phase composition on phospholipids elution and the association of phospholipids with matrix effects. Rapid Commun Mass Spectrom 2009;23:2125–2138.

Yeh SY, Lach J. Stability of morphine in aqueous solution III. Kinetics of morphine degradation in aqueous solution. J Pharm Sci 1961;50:35–42.

30

BEST PRACTICES IN LC-MS METHOD DEVELOPMENT AND VALIDATION FOR DRIED BLOOD SPOTS

Jie Zhang, Tapan K. Majumdar, Jimmy Flarakos, and Francis L.S. Tse

30.1 INTRODUCTION

A major challenge for the pharmaceutical industry is finding new ways of increasing efficiency while reducing the cost of drug development. One area of opportunity in the research and development of new chemical entities is the sampling method for exposure assessments. In recent years, there has been a significant increase in the number of published reports related to the dried blood spot (DBS) sampling technique in drug development (Li and Tse, 2010; Majumdar and Howard, 2011). Some of the key features of DBS technology are the use of low blood volume potentially leading to decreased animal use, the ease of sample collection that does not require accurate blood volume measurements, and the low cost of sample storage and shipment obviating the need for refrigeration or freezing. In addition, a dry matrix is also considered less pathogenic than its corresponding liquid. As a result, DBS provides much desired ethical and possibly financial advantages over conventional blood sample handling.

Significant progress has been made by the bioanalytical community to better understand the DBS technology (Ji et al., 2012; Viswanathan, 2012). Aspects related to the unique DBS procedures or parameters such as DBS assay sensitivity, drying process, storage and transportation at ambient temperature, dilution integrity, and incurred sample reanalysis (ISR) have been explored and well managed. However, some technical issues including DBS homogeneity, blood hematocrit (HCT), and extraction recovery remain to be resolved before the technique can be universally applied. The US FDA currently does not accept stand-alone DBS data as a replacement for liquid matrices in any registration studies.

Although the approaches used in method development and validation of dry matrices are similar to those used for liquid biomatrices, the overall experimental protocol will need to be adapted to manage a unique set of scientific challenges posed by DBS. DBS methods must be developed and validated to meet the same acceptance criteria such as precision, selectivity, sensitivity, reproducibility, and stability, which are mandated by the health regulatory authorities for liquid matrices. In addition, many DBS specific parameters including HCT and homogeneity must be evaluated. There are extensive ongoing discussions within the bioanalytical community regarding how DBS method validation should be performed. In 2011, European Bioanalysis Forum (EBF) published its recommendations on the validation of bioanalytical methods for DBS in a white paper (Timmerman et al., 2011). This chapter provides an overview on the best practices in method development and validation for DBS as applied today.

It is worth stating that although whole blood is the most commonly used matrix when referring to DBS technology, other matrices such as plasma, serum, and urine have also been evaluated in the dry form. Therefore, the general term DMS (dried matrix spot) is also used when referring to the technology. In addition, automation based, direct elution, and direct surface desorption techniques have been developed in order to manage the tedious manual punching and off-line extraction steps that are potential impediments to the analytical benefits provided by DBS techniques. DMS and direct analysis of DBS are out of the scope in this chapter.

Handbook of LC-MS Bioanalysis: Best Practices, Experimental Protocols, and Regulations, First Edition. Edited by Wenkui Li, Jie Zhang, and Francis L.S. Tse.

30.2 METHOD DEVELOPMENT

Method development is critical in establishing a solid foundation for a validated bioassay, and it plays a key role in promoting DBS technology. Method development for DBS bioanalysis begins with the collection of basic physiochemical information about the compound (analyte) such as molecular weight, polarity, ionic character, pK_a values and solubility. If a bioanalytical method for an analyte is already developed for a liquid matrix, one can usually modify and apply it to DBS.

A standard bioanalytical approach for working with DBS typically consists of three parts: (1) preparation of spiked blood samples (i.e., calibration standards and quality controls (QCs)) in the liquid form using fresh blood with controlled HCT at a normal level; (2) spotting blood sample onto a DBS card or paper substrate. In this step, a small volume of blood sample between 15 and 30 μl is spotted onto a predefined type of DBS card followed by drying for about 2 h or more under controlled ambient temperature and humidity in the laboratory; (3) after drying, a disk, typically 3–8 mm in diameter, is punched out of the card, and sample extraction from the punched disks is performed with an organic solvent in the presence of water. The volume of extraction solvent is relatively small, ca. 100 μl. Internal standard (IS) is normally added to DBS samples during extraction. The extract is then analyzed by LC-MS/MS.

30.2.1 Preparation of Calibration Standard and QC Samples in Whole Blood

The analyte(s) is spiked into the blood matrix with an appropriate amount of stock or working solution. To reduce the likelihood of variation between the incurred samples, spiked calibration standards and QC samples due to the presence of nonendogenous organic and/or aqueous solvents, it is recommended to prepare a working intermediate in plasma, that is, a mixture of analyte stock solution and plasma. This is typically accomplished by spiking a stock solution in plasma at a 50-fold higher concentration than the target ULOQ (upper limit of quantification) of the assay. Calibration standards and QC samples are then derived by serial dilutions of the intermediate with blood to achieve the target concentrations for the DBS samples. The amount of external components, for example, solvents added to the blood should be as little as possible (typically $< 5\%$ of the final volume) to prevent solvent effects creating differences between spiked versus incurred samples. In addition, inappropriate dilution to blood matrix with external components may lead to changes in the nature of the spot formation, distribution of the compound on the filter paper, hemolysis of blood cells prior to applying to the filter paper, or the drying time.

Only fresh blood matrix is used for the preparation of calibration standards and QC samples. Fresh blood means that it is harvested on the day of use. However, as this is not feasible for many bioanalytical laboratories, it is also commonly acceptable to use blood within 2 weeks of collection and stored in a refrigerator without freezing. It is suggested to inspect the stored blood for appropriate quality prior to use, as clotting will begin over time and can have an impact on spot size and appearance of blood on DBS substrates. Use of hemolyzed blood for preparation of calibration standards and QC samples should be avoided as its potential impact on the quantification of DBS is difficult to predict.

To ensure robustness and reproducibility in bioanalytical assays, attention should be given to the integrity of spiked blood samples. Blood should be handled gently during sample dilution. Vigorous shaking of blood may induce hemolysis and should always be avoided. An appropriate equilibration period may be critical during serial dilution in blood and prior to application of the spiked sample onto the card matrix.

DBS results are subject to influence by the HCT value of the blood matrix. HCT, packed cell volume or erythrocyte volume fraction is the percentage of blood volume that is occupied by red and white blood cells. HCT level is normally about 40–45% for men and women and changes with age, sex, and general health condition. A range of 28–67% HCT generally covers the majority of juvenile and adult human blood samples likely to be encountered by an analytical laboratory, except for samples from subjects with certain medical conditions such as polycythemia and anemia (Shander et al., 2011). Differences in blood HCT level are reflected in different blood viscosity values, leading to differences in flux and diffusion properties of blood through different substrates used for DBS sample collection. The distribution of blood with a high HCT through the paper substrate is less than that of blood with a low HCT, resulting in a smaller blood spot. Therefore, DBS samples of the same punch size but prepared using blood samples with different HCT levels can generate a noticeable bias in analytical results. In addition, different HCT levels may impact the aging of the DBS spots and also the recovery of analyte extraction from DBS cards. It is, therefore, important to define the HCT level during method development and validation. The HCT level should be within the range anticipated in the target population of the planned study. When purchasing blood from vendors for the preparation of calibration standard and QC samples, the HCT levels should be noted. Adjustments can be made in the bioanalytical laboratory by adding washed and packed blood cells or plasma.

30.2.2 Selection of DBS Card

Commercially available DBS cards can be grouped according to the card material and chemical treatment as follows:

1. *Untreated cards*: Whatman DMPK-C (GE Healthcare Bio-sciences, NJ, USA) and Ahlstrom 226 (ID

Biological systems, currently part of Perkin Elmer, SC, USA).

2. *Chemically treated cards*: Whatman FTA DMPK A and FTA DMPK B. The chemical(s) in the cards are the protein denaturing agents that allow inactivation of endogenous enzymes present in biological matrices. The treatments also prevent the growth of bacteria and other microorganisms. The difference between DMPK A and DMPK B is in the blood spot area (DMPK A card is ~20% smaller than DMPK B card).
3. *Noncellulose cards*: Agilent Bond Elute noncellulose matrix spotting cards (Agilent Technologies, Inc., CA, USA). According to the manufacturer, the Bond Elute DBS card can reduce the variability of blood spot size caused by varying HCT levels.

The impact of the HCT and punch location effect were different depending on the type of DBS card. In a test using five compounds with different physicochemical characteristics, O'Mara et al. (2011) found that when comparing HCT effect on bias of individual compounds across card types, the Whatman FTA DMPK B card type produced a notably smaller number of acceptable biases for all compounds tested, particularly when compared with the DMPK A card type. In a plot of percentage bias against HCT level for individual compounds on a specific card type, the slopes were found to be dependent on not only the compound but also the card type. This compound dependency was more obvious on untreated card Ahlstrom 226 and Whatman DMPK C. At a HCT level of 0.45, for all compounds on three of the card types (Ahlstrom 226, DMPK A and DMPK C) tested, the concentration of compound measured in the perimeter punches was greater than that measured in the center punch. In contrast, all compounds tested on DMPK B card were almost homogeneously distributed across the spot. For compounds spotted on untreated cards (Ahlstrom 226 and Whatman DMPK C), the range of bias related to heterogeneous distribution of compound across the spot was greater than that compared with chemically treated card types (DMPK A and B).

Additional criteria to be considered for DBS card selection include the matrix effect, ion suppression, and recovery of the method. Using acetaminophen and its major metabolites as model compounds, Li et al. (2012b) determined the matrix effect and recovery results from five types of DBS cards (Ahlstrom 226, FTA Elute micro, FTA DMPK A, FTA DMPK B and Agilent Bond Elute cards). After applying blank blood on each card type followed by a drying period, a 3-mm disk (in three replicates at three concentration levels) was taken from each card type for extraction. Neat solutions of the analytes at known concentrations (same as the low, mid, and high QC samples) were spiked, in three replicates, into the blank DBS sample extracts. Subsequent LC-MS/MS analysis showed Ahlstrom 226 card to have the best consistency in extraction recovery.

30.2.3 Spotting Volume, Spotting Techniques, and Punch Size

The accurate determination of analyte concentration in DBS depends on the homogeneity of blood spreading across DBS card sampling area, spotting techniques of laboratory personnel, and consistency of the punch taken from the spot. Liang et al. (2011b) reported on the impact of blood spotting volume on the bioanalysis of dextromethorphan and dextrorphan using DMPK B card. Concentration differences up to ~19% were observed for the two analytes when the spotting volume was changed from 10 to 50 μl. On the other hand, using the central portion of the blood spots on FTA Elute cards, Clark et al. (2010) demonstrated an even distribution of ^{14}C-labeled compound UK-414495 in spotting volumes between 15 and 45 μl. These results suggest that the impact of spotting volume on the DBS assay may be compound dependent.

A small sample volume, typically 15–20 μl, is a good starting volume in method development. The reproducibility of the assay is then assessed at varying volumes, typically $\pm$50% of the target volume. Overloading of blood on DBS cards could lead to heterogeneous spot formation. To provide guidance on DBS sample collection to the clinical laboratories, it would be helpful to determine a spot size boundary, especially if spotting on card is done without accurate pipetting.

30.2.4 Spot Homogeneity

Spot homogeneity refers to the distribution profile of analyte across the DBS. Since it can have a significant impact on assay precision, accuracy, and reproducibility, spot homogeneity should be evaluated during method development. The factors that govern spot homogeneity include the type of filter paper (card) and the physiological parameters of the blood sample such as HCT and drying conditions. In addition, chromatographic effect during spotting can change the way an analyte moves in the filter paper, with paper acting as the stationary phase and the liquid matrix as the mobile phase. Depending on the structure of a given analyte, a chromatographic effect can drive the analyte to the edge of the spot, or concentrate it in the center of the spot. Using an autoradiogram technique, Ren et al. (2010) found uneven distribution patterns for compounds across the DBS, particularly for relatively large spot volumes (100 μl). This effect can cause bias if the same portion is not reproducibly removed between spots or when two samples from different locations on the same spot were removed for analysis. Therefore, one should aim to take larger punch samples from the same location of a spot whenever possible.

The experimental design for determining spot homogeneity is to compare the concentrations of the same analyte in the punches from different locations of the card containing the same QC samples at several concentration levels (typically low, mid, and high QC levels). Alternatively, in cases where a card size is not sufficiently large for multiple punches, concentration determinations of the same analyte can be made by comparing punches of the same QC samples from different cards. A good reproducibility (<15 %CV) derived from analysis of the QC samples along with a set of calibration standards would demonstrate the comparability of each punch and card.

30.2.5 DBS Sample Drying, Storage, and Transportation

After spotting on a DBS card, the spotted card is placed on a drying rack at ambient temperature and normal humidity. The drying time depends on the type of card, blood volume applied, and humidity. Based on a thorough evaluation on DBS storage conditions with the Whatman 903, FTA and FTA Elute cards using weight and appearance of DBS as references, Denniff and Spooner (2010) suggested that blood spots should be left to dry for at least 90 min under ambient laboratory conditions prior to further handling or analysis. It was noticed that the three types of cards examined behaved differently when exposed to conditions of high relative humidity and temperature. The exposure of FTA and Whatman 903 substrates to high relative humidity and temperature did not adversely affect the blood spot, whereas the spots on the FTA Elute expanded and diffused through the substrate over 24 h under the same conditions, suggesting that the integrity of the DBS samples has been compromised.

Storage of biological samples on DBS is more convenient than those in liquid matrices. Generally, DBS card storage at room temperature would suffice, although a lower temperature may be necessary for extending stability coverage for some analytes. Typical degradation reactions (reduction, oxidation or hydrolysis) are expected to be 10 times slower when the temperature is decreased from 22°C to 0°C (Chen and Hsieh, 2005). Nonetheless, many analytes unstable in liquid formats upon storage have shown enhanced stability in DBS (Li and Tse, 2010). Water in liquid matrices plays a critical role in enzymatic and chemical hydrolysis reactions that cleave drug molecules (Alfazil and Anderson, 2008). Water may also induce bacterial growth on DBS substrates that can alter extraction efficiency during analysis. DBS cards should be packed in zip-closure bags with desiccant packages and humidity indicator cards to ensure protection from moisture.

With generally established stability at ambient and higher temperatures, DBS samples can be shipped via conventional carriers without dry ice or refrigeration. The savings in shipping costs could be significant particularly during late-phase clinical trials that can generate a large number of samples.

30.2.6 Internal Standard

A structurally related compound, a structurally similar compound or a stable isotope labeled compound at a specific concentration is added to all samples in an analytical run as an IS to correct for any variability during sample preparation and analysis using mass spectrometric detection. When analyzing liquid samples such as plasma or serum, IS is simply added by spiking into the sample prior to extraction. In contrast, the addition of an IS in the analysis of DBS samples is somewhat complicated.

There are generally four options for introducing an IS in DBS method development as shown in Figure 30.1:

a. Introduction of IS via an extraction solvent. This simple technique is widely used and works reproducibly by compensating for matrix effects and differences in extraction efficiency, losses due to sample handling and variations in instrument sensitivity. A potential flaw of this method is that the IS may not be fully incorporated into the matrix components and sample card, thus unable to compensate for fluctuations in extraction from the DBS filter punch. In addition, this approach may not reveal changes in extraction recovery due to storage.

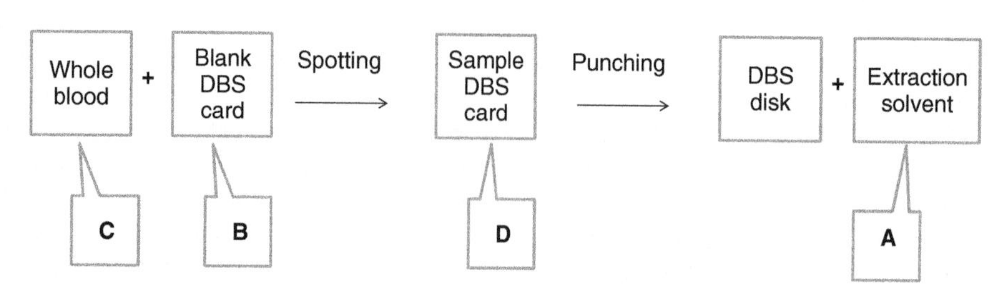

FIGURE 30.1 Four points where an internal standard can be added to DBS samples in LC-MS/MS analysis: (a) Introduction of IS via an extraction solvent; (b) DBS card pretreated with IS; (c) IS added to blood before spotting onto paper; (d) IS applied to DBS matrix prior to extraction.

b. DBS card pretreated with IS. By incorporating blank DBS cards with IS prior to applying the blood samples, the IS is integrated and extracted with the analyte. However, pretreating the DBS cards prior to sample collection is tedious and costly. The potential impact on IS migration or distribution on the DBS card during matrix spotting also needs to be clearly characterized (Meesters et al., 2011).
c. IS added to blood before spotting onto paper. The advantage of this approach is that the IS can be fully associated with blood components together with the analyte. However, this procedure may be too complicated to implement at most clinical sites. It will also require extra safety measures when sampling from patients with infectious diseases.
d. IS applied to DBS matrix prior to extraction. This is a novel technique pioneered by Abu-Rabie et al. (2011) using the TouchSpray technology to apply IS to DBS samples prior to analysis. The IS must be given sufficient time prior to extraction to bind with matrix components and paper substrate and must not adversely affect the distribution of the analyte. There was no significant difference in accuracy and precision obtained from this procedure compared with procedures A and C. This procedure can be easily configured to a fully automated method for DBS sample analysis.

Different procedures used to introduce the IS can result in markedly different recoveries. Meesters et al. (2011) recently reported that the relative recovery of a model compound nevirapine ranged between 11.4% and 108%, highlighting the need for careful evaluation of IS procedures during method development and validation in order to ensure assay integrity. On the other hand, the procedure for IS application should not be so complicated as to negate the advantages of DBS sampling.

30.2.7 Extraction Solvent, Procedure, and Recovery

The objectives of sample preparation for any bioanalytical assay are to maximally remove the matrix background and interferences, efficiently recover the analytes of interest and ISs, while maintaining an adequate sensitivity of the assay method. Changing the sample format from liquid to solid phase is accompanied with a unique set of challenges in the sample extraction process.

A common extraction procedure for DBS samples is to punch one or more DBS disks from the DBS card into tubes or 96-well plates followed by adding a finite amount of extraction solvent containing the ISs. The analyte of interest is then extracted with gentle shaking or vortex mixing. Sonication may be necessary to enhance extraction efficiency. After centrifugation, the resulting extracts are transferred manually or by an automated liquid handler to fresh tubes or 96-well plates for LC-MS/MS analysis. The extraction solutions should be able to interrupt the binding of analyte to the matrix proteins and the paper material. Several organic solvents or their mixtures with water may be considered initially, for example, methanol, methanol:water, or acetonitrile:water, at various ratios. The water in the extraction solution may be helpful for effectively eluting the analytes off the DBS card. Pure acetonitrile may not dissolve the dried blood crust completely, resulting in poor extraction of the analyte and low elution efficiency (Liu et al., 2010). Depending on the structure of the analyte, a pH modifier or buffer may be added to the extraction solvent to improve its efficiency. Several iterations may be needed to reach the optimal organic solvent:water ratio in the extraction solution that will provide maximum extraction efficiency and minimum matrix effect.

An alternative sample preparation approach is to add the organic and aqueous portions of the extraction solution in two steps. First, an aqueous solution is added to DBS disks to dissolve the dried blood, and then acetonitrile or methanol to precipitate the proteins that were eluted from the DBS cards. This approach is similar to the protein precipitation method commonly used for liquid sample analysis. If a water-immiscible organic solvent such as methyl tert-butyl ether or ethyl acetate is added to the aqueous solution, the procedure would be comparable to the liquid–liquid extraction (LLE) approach for liquid samples. Of the above, LLE appears to be the most effective extraction method in removing the matrix background introduced by DBS cards (Liu et al., 2010).

Determination of extraction recovery of analytes from blood samples spotted on a DBS card is a challenging task. Understanding the nature of analyte interactions with the substrate is prerequisite to establishing the procedure for optimal extraction. Other factors that may impact analyte recovery include HCT level, age of paper substrate, temperature, and humidity. In most DBS assays, the IS is added during sample extraction, thus the elution of the analyte from the DBS card (solid phase) to the extraction solution (liquid phase) is not monitored by the IS. It is, therefore, important to elucidate the efficiency of the extraction process during method development. If available, a radiolabeled drug substance may be used to monitor analyte extraction from the DBS card.

The current FDA guidance for liquid sample assays states that "the extent of recovery of an analyte and of the internal standard should be consistent, precise and reproducible. Recovery experiments should be performed by comparing the analytical results for extracted samples at three concentrations (low, medium and high) with unextracted standards that represent 100% recovery" (FDA, 2001). EBF recommended evaluating and documenting recovery more thoroughly for DBS assays than liquid sample assays. In particular, EBF believes that it is important to evaluate the potential impact of card storage on extraction recovery from the punched disks as part of the test for long-term stability (EBF, 2010).

30.2.8 Matrix Effects

The evaluation of matrix effect for DBS assays is similar to that used in liquid assays. Typically, a punch from a DBS card spotted with blank matrix is made followed by extraction. The analyte is then spiked into the extract at a specified concentration. The extract is analyzed along with a neat solution of the analyte to determine the matrix effect.

Some blood components bound to DBS paper may not be dissolved in aqueous or organic solvents, thus the extracts of DBS may be cleaner (i.e. less matrix effect) than that derived from liquid samples. However, DBS sample extracts may contain certain endogenous materials such as leachable chemicals that may cause high background noises or interferences during LC-MS analysis. Furthermore, different types of DBS cards may introduce different interferences into the sample extract. It is well known that treated card materials and/or the chemicals used to pretreat of DBS cards may contribute to ion suppression of the analyte causing matrix effects (Clark and Haynes, 2011)

30.2.9 Assay Sensitivity

Compared with liquid sample analysis, the sensitivity of DBS assay is likely to be lower due to the relatively small sample punch size. One can try to enhance the signal intensity of analyte in DBS samples by extracting from multiple punches or larger punched disks, and by optimizing the extraction solvent. Recent advancements in DBS technology include the use of microbore column and microflow in LC–MS/MS that have shown promising success in increasing DBS assay sensitivity (Rahavendran et al., 2012; Rainville, 2011).

30.2.10 On-Card Stabilization for Unstable Compounds

Different approaches to ensure on-card stabilization should be evaluated early in method development for analytes suspected to be unstable. Stabilizing unstable compounds is a major challenge in bioanalytical method development for liquid based samples, and a variety of stabilization techniques have been used in an attempt to protect the compound in blood, plasma, and serum after collection (Li et al., 2011). Adding a chemical reagent to blood samples immediately after collection can stop the degradation or structural changes of a drug, an example being the use of organic acids to stabilize esters and lactones. Samples containing light sensitive compounds should be handled and stored in dark chambers, and those with heat-sensitive compounds stored in deep freezers. The same principles for handling liquid matrices also apply to DBS samples. However, DBS has an advantage over liquid-based matrices; the water in the latter plays a key role in enzymatic and chemical hydrolysis reactions that can alter the molecular structure of a drug (Alfazil and Anderson, 2008). One study showed that DBS sampling stabilized two unstable prodrugs in rat whole blood for at least 21 days at room temperature without the addition of esterase inhibitors (D'Arienzo et al., 2010). Interestingly, Whatman DMPK A and B cards, despite having been treated with reagents that lyse cells upon contact, showed no advantage in terms of added stability for the investigated compounds compared to the untreated Whatman 903 Protein Saver cards. This suggests that the on-card stabilizing effect could be compound dependent. Sometimes, it may be necessary to modify the commercially available DBS cards to meet the stabilization needs of otherwise unstable compounds. For example, Liu et al. (2011b) pretreated DBS cards with citric acid to lower the pH of the spotted blood sample containing an unstable drug candidate KAI-9803 which consisted of two peptides linked by a disulfide bond. The result was an improved compound stability in DBS to at least 48 days at room temperature.

In another study, a photosensitive compound omeprazole exhibited increased stability when spotted and stored on DBS papers (Bowen et al., 2010). This compound degraded 40-90% in water, plasma or whole blood, whereas photodegradation was negligible in DBS.

30.3 METHOD VALIDATION

Method validation is conducted to confirm that an analytical procedure established during method development is suitable for its intended use. A well-developed method should be easily validated. Failure to meet the preset criteria during method validation requires a thorough investigation to understand the root cause of the problem. The knowledge gained from working with liquid matrix samples also applies to DBS.

At the current stage of DBS technology development, method validation is usually performed based on a fit-for-purpose concept since no official guidance is available. Similar to the comprehensive evaluation for a liquid matrix, DBS assay validation includes precision, accuracy, selectivity, specificity, extraction recovery, matrix effects, dilution, incurred sample reanalysis, and stabilities including whole-blood collection stability. The characterization of these parameters is made according to the existing guidelines on bioanalytical method validation (EMA, 2011; FDA, 2001). Additional validation for DBS specific parameters such as spot drying stability, HCT impact, inter- and intracard variability must be performed.

30.3.1 Selectivity, Sensitivity, and Linearity

The selectivity of a DBS method is usually assessed by analyzing DBS samples prepared from fresh blank blood collected from at least six individual human subjects or at least two individual animals. The lower limit of quantification (LLOQ) must exhibit an accuracy of within ±20% bias and precision of CV ≤20%. At least six nonzero calibration standards are analyzed in duplicates (typically one in the beginning and the other at the end of the assay sequence) in three separate validation runs. The analyte/IS peak area ratio against nominal analyte concentration is employed for calibration regression with an appropriate weighting factor.

30.3.2 Inter- and Intraday Accuracy and Precision

The inter- and intraday accuracy and precision of the DBS method are demonstrated from the analysis of 6 replicates of QCs at a minimum of four concentrations levels including one at the lowest concentration of the calibration standards on each of the three validation runs. The accuracy is expressed as the difference of the measured analyte concentrations from the nominal values (bias %) and the precision as the coefficient of variation (CV %). A bias of within ±15% and CV of ≤15% is required at all concentration levels except the LLOQ, for which a ±20% bias and ≤20% CV are considered acceptable.

30.3.3 HCT and Its Effects on the Assay

As stated in the EBF recommendation on DBS (Timmerman et al. 2011), "hematocrit is currently identified as the single most important parameter influencing the spread of blood on DBS cards, which could impact the validity of the results generated by DBS methods, affecting the spot formation, spot size, drying time, homogeneity and, ultimately, the robustness and reproducibility of the assays."

The EBF recommends that "for bioanalytical method validation, the impact of variations of HCT on the spot size and homogeneity should be understood and their impact on assay performance documented during validation. For that, clinically relevant variations of HCT (e.g., from 30–35% to 55–60%) should be evaluated during validation. Patients with physiological conditions or under medical treatment affecting the HCT beyond normal values (e.g., renal impairment and oncology patients) may require additional validation as they occur, such as including calibration standards and quality control samples prepared using matrix beyond normal HCT values." Viswanathan (2012) suggested that subsequent to acceptable homogeneity determinations, efforts need to be made to avoid or minimize the assay bias with different ranges of HCT values between calibration standards and quality control samples. An optimal range in this regard may be established during validation such that the range in the study samples can be accommodated. Thus, the investigation and the evaluation of the above parameters on the overall impact of DBS will be considered as the central part of the validation.

To determine the influence of HCT on the assay performance, fresh blood with adjusted HCT values of 30%, 40%, 50%, and 60% may be obtained either from a commercial source or prepared in the bioanalytical laboratory. These fresh blood samples are each used to prepare DBS QCs at concentrations of low and high concentration levels ($n = 6$ QCs at each level). The QC samples are analyzed in the validation runs along with the calibration standards and QC samples prepared using fresh blood with a HCT value of ~ 35%. A difference beyond ±15% of the nominal analyte concentrations in the QC samples would suggest a significant HCT effect. It is worth noting that the HCT in some disease states may be out of the range (30–60%) tested.

A significant HCT effect can be managed by the following:

1. *Correcting concentration results using each subject's HCT value*: To do this, correction factors must be established for each HCT value relative to the HCT of the blood used to prepare the standard curve during method validation. The HCT value for each study subject needs to be determined and this could add a significant burden to the clinical program.
2. *Analyzing the entire spot*: This approach yields more consistent DBS concentrations regardless of HCT levels by eliminating the variations due to spreading and nonhomogeneity. The challenge to this approach is that it requires the accurate spotting of a defined blood volume. Li et al. (2012a) described a novel procedure called perforated dried blood spot (PDBS), in which an accurate amount of blood (5–10 μl) was added, using either a Micro Safe pipette or a Drummond incremental pipette on the PDBS disks prepared from regular filter paper (6.35 mm in diameter and a 0.83 mm in thickness). PDBS samples are simply pushed by single-use pipette tips into 96-well plates for analysis. This technique provides a promising solution to the adverse impact of HCT on DBS analysis by using 100% of the sample.

30.3.4 Impact of Blood Volume and Spot Size on the Accuracy of Determination

During DBS method validation, the accuracy of determination for an analyte from DBS samples with variable sampling spot sizes is examined by spotting increasing volumes of DBS QC samples (typically 10, 20, and 40 μl) at three concentrations (low, median, and high) onto the cards. After drying, three replicates of 3-mm disks are taken from the center of each DBS QC sample and analyzed along with calibration standards. A bias within ±15% of the nominal values would suggest no apparent difference for the DBS samples made with different blood volumes. This would indicate an even distribution of blood on the card so that precise sampling on card (over the range of blood volume evaluated) may not be necessary.

30.3.5 Impact of Homogeneity on the Accuracy of Determination

The possible effect on blood diffusion by interactions of blood and/or the analyte with the DBS card materials is assessed by punching the DBS disks from the center and edge area of DBS QC samples at low, mid, and high concentrations, followed by analysis along with calibration standards. The measured analyte concentrations from both the center and edge disks of the above QCs were compared against each other and also with the nominal values. Bias within ± 15% of the nominal concentration and within ± 15% of each other would suggest no apparent chromatographic effect (i.e., no impact on spreadability).

30.3.6 Temperature Impact

DBS QC samples at low, mid, and high concentrations are analyzed each in three replicates following storage at room temperature and 2–8°C. The measured analyte concentrations are compared with the nominal values. The calculated bias (%) from the stability QCs should be within ± 15% of the nominal values. To mimic the possible situation where the DBS samples are collected and/or transported at a high temperature, a set of DBS QCs may be stored at a temperature up to 70°C for several hours (e.g., 4 h) followed by analysis with calibration standards and regular QCs. Bias within ± 15% of the nominal values would suggest the analyte is stable in the DBS sample at elevated temperatures.

30.3.7 Dilution Integrity

Dilution integrity needs to be evaluated during method validation in order to ascertain that a method is suitable for the bioanalysis of DBS samples containing analyte concentrations higher than the ULOQ. While the principles of dilution in the traditional liquid sample analysis apply to DBS, the dilution procedures for DBS samples are more complicated due to its solid format. Three common approaches to DBS sample dilution are as follows:

1. *Dilution with blank DBS extract*: In this procedure, the extract of a DBS sample from the punched disk is diluted with one or multiple extracts of blank DBS samples (the number of blank extracts used = the dilution factor − 1). An IS can be added either to the extraction solvent prior to the dilution or to the diluted extract. This method requires a great deal of blank matrix and processing additional DBS cards in each analytical run. It works well when a relatively small dilution factor is needed for a few samples, but can be costly or impractical for samples requiring large dilution factors.
2. *IS-tracked dilution*: In this approach introduced by Liu et al. (2011a), a dilution factor-adjusted IS working solution is added to the sample requiring dilution prior to sample extraction. Subsequently, the processed sample is approximately diluted to the assay linear response range for LC-MS/MS analysis. As shown in Figure 30.2, the dilution factor-adjusted IS working solution is an IS working solution at a concentration that is 10 times (i.e. the intended dilution factor) higher than that used for regular samples. The advantage of this approach is that the dilution is tracked by the IS and is no longer a volume-critical step once the concentrated IS working solution is added to a sample. The main disadvantage of this method is that standards, QCs, and study samples not requiring dilution are treated differently from diluted study samples. Therefore, it is important to closely monitor the performance of dilution QC samples in the analytical run.
3. *Subpunch dilution*: In this procedure pioneered by Alturas Analytics (Christianson et al., 2011), there are three critical steps in punching the DBS sample card and DBS blank card. (i) A fixed diameter subpunch is collected from a DBS sample to be diluted. (ii) An identical size of subpunch in a blank DBS card is taken and discarded. (iii) A regular-sized punch is taken from the blank DBS card. The punch from the blank card is then extracted together with the subpunch from the sample. A dilution factor is derived using mathematical calculations based on the sizes of the regular punch and subpunch.

30.3.8 Intercard Variability

DBS cards are manufactured under well-defined procedures, and the physical characteristics and chemical additives are expected to be identical among different lots of cards. There is consensus in the bioanalytical community that one single DBS is considered one sample. Any additional spot originating from the same liquid sample and spotted either on the same card or on a different card from the same type/manufacturer may be treated as an identical replicate sample, provided the handling and storage conditions are identical. EBF recommended in their white paper that when using cards from the same type/manufacturer, intercard variability does not need to be investigated as a discrete method validation parameter. However, it is a good practice to spread calibration standards and QC samples over multiple cards during validation, in order to identify or preclude intercard

(a) Conventional Dilution

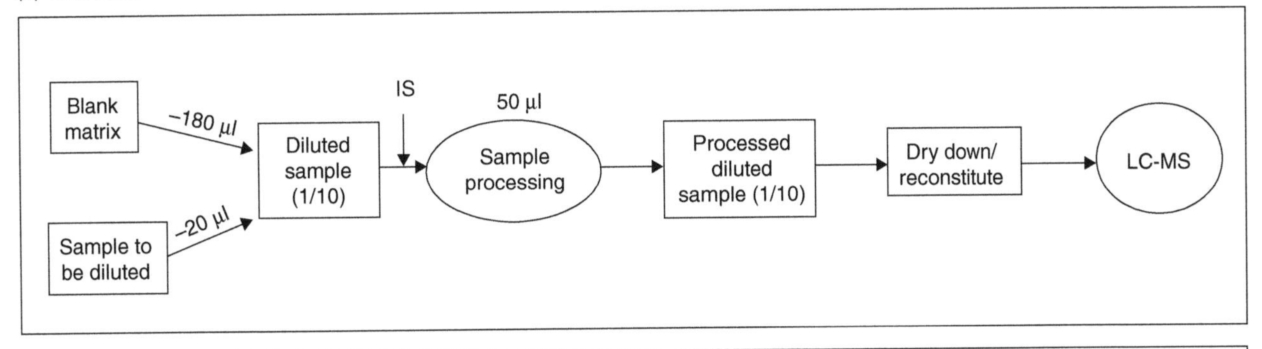

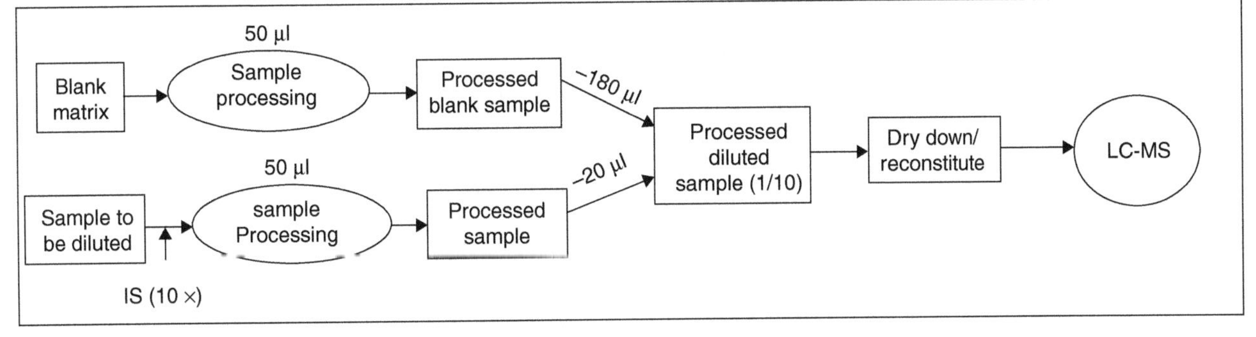

(b) IS-Tracked dilution

FIGURE 30.2 Two sample dilution processes for DBS sample analysis using LC/MS/MS: (a) Conventional sample dilution; (b) IS-tracked dilution. Shown here is an example for a dilution factor of 10 (reproduced from Liu et al., 2011a, with permission from Wiley).

variability issues when a validation run fails to meet predefined acceptance criteria (Timmerman et al., 2011).

Changing a DBS card type/manufacturer requires a partial validation of the bioanalytical method. The validation parameters recommended by EBF are linearity and sample dilution, accuracy and precision, extraction recovery, matrix effects, drying conditions (i.e., drying time and temperature), and oncard storage stability.

30.3.9 Stability Determination

The principles of stability assessment for liquid samples are applicable to DBS, which means that "evaluation of stability should be carried out to ensure that every step taken during sample preparation and sample analysis, as well as the storage conditions used do not affect the concentration of the analyte" (EMA, 2011). DBS stability evaluation is generally focused on the following three areas in the validation:

1. Whole blood stability during collection and handling of blood samples prior to spotting on DBS cards
2. *Stability on DBS cards during and after spotting*: Stability at room temperature, under frozen conditions over a certain period of time and during shipping. Evaluation of freeze–thaw stability is not necessary, as it is not relevant for DBS samples.
3. Stability during sample preparation for LC-MS analysis.

The stability of the analyte during blood collection and further handling prior to spotting on DBS cards must be established to ensure that the analytical method will yield concentration results that reflect the concentrations of the analyte in the subject at the moment of sampling. The blood used for the validation experiments should be fresh (within two weeks of collection), and the experiments should be performed at body temperature (37°C) to reflect the real-life situation. The analytical procedures for determining whole blood stability in liquid format can be used for DBS. Similarly, the stability of the analyte in stock and working solutions must be established in the method validation.

The analyte stability on DBS cards after spotting is assessed according to the intended storage conditions and following various storage durations by comparing the measured analyte concentrations in QC samples (at least three replicates each of low and high QC) with the nominal values. The bias should be within ±15% of the nominal values. Analyte in a matrix can be caused by hydrolysis of the intact conjugated molecules or other unstable conjugated molecules (e.g., acylglucuronide, *O*-sulfate/glucuronide, and *N*-oxide/glucuronide), or interaction of the analyte with endogenous components. Therefore, DBS prepared from incurred samples should be used for stability evaluation whenever possible. Accordingly, the analyte concentrations measured from the study samples (minimum 20 samples in single determinations) initially and after various storage periods are compared. The difference between the repeated measurement and the mean of two measurements (first and repeat) should be within $\pm 20\%$ for at least two-thirds of the selected samples.

As mentioned earlier, stability evaluations need to mimic the conditions of sample storage. A unique characteristic of DBS samples is that they are likely to be exposed to uncontrolled conditions such as extreme temperatures and humidity during sampling, sample shipment and storage, for example, conducting a clinical study in hot climates and at sites where freezers are not readily available. The stability profile of the analyte under these conditions needs to be evaluated during validation. Li et al. (2012b) described a validation protocol for evaluating the possible impact of humidity and high temperature on acetaminophen and its major metabolites in DBS samples. A set of DBS low and high QC samples, after drying at ambient temperature in an open laboratory (humidity $\sim$40%), was placed in sealed containers with inside humidity maintained at $\sim$80% or $\sim$0%. Another set of the QCs was placed in a Shimadzu HPLC column oven with temperature set to $\sim$60°C. At 5 and 24 h, three replicates of 3-mm disks were taken from the center of these QCs and analyzed along with the calibration standards and regular QCs that were exposed to regular humidity ($\sim$40%) at ambient temperature ($\sim$22°C). The measured analyte concentrations were compared with the nominal values.

For compounds with previously identified stability issues in a liquid matrix, appropriate caution must be taken when evaluating the analyte stability in DBS cards. For new compounds without prior stability data, studying the compound structure may help to predict potential stability issues (Li et al., 2011).

If lower-than-expected analyte concentrations are observed in DBS samples, it is important to discern whether it is due to poor extraction recovery or actual compound degradation. An interesting approach to evaluating drug stability during drying on DBS cards has been reported recently (Liu et al., 2011b).

30.3.10 Carryover

Carryover is evaluated by injecting two extracted blank DBS samples sequentially following the injection of a sample containing the analyte at ULOQ. In the first blank matrix injection, the response at the retention time region of the analyte or IS should be less than 20% of the mean response of the LLOQ samples for the analyte and less than 5% of the mean response for the IS from the same assay sequence. The response in the second blank injection serves to provide clues that may be needed for troubleshooting.

30.4 CONCLUSIONS

DBS is being explored as an important sampling tool in bioanalytics due to its many potential benefits. In order to gain widespread acceptance for use in pharmaceutical development, however, DBS has yet to improve its reliability in delivering accurate and reproducible results over time. Continued efforts are needed to overcome technological hurdles especially in minimizing hematocrit effects, ensuring spot homogeneity, enhancing extraction recoveries of analytes from DBS substrates and maximizing the stability of unstable compounds.

Despite its perceived advantages as a sample collection method, DBS sample processing in the bioanalytical laboratory today tends to be more labor intensive compared to conventional, liquid matrices. No doubt there is room for process improvements that, together with advances in automation, are expected to make DBS bioanalysis the recognized method of choice in the foreseeable future.

REFERENCES

Abu-Rabie P, Denniff P, Spooner N, Brynjolffssen J, Galluzzo P, Sanders G. Method of applying internal standard to dried matrix spot samples for use in quantitative bioanalysis. Anal Chem 2011;83(22):8779–8786.

Agilent: Hematocrit and its Impact on Quantitative Bioanalysis using Dried Blood Spot Technology. 2013. Available at http://www.chem.agilent.com/Library/applications/5991-0099EN.pdf. Accessed Apr 1, 2013.

Alfazil AA, Anderson RA. Stability of benzodiazepines and cocaine in blood spots stored on filter paper. J Anal Toxicol 2008;32(7):511–515.

Bowen CL, Hemberger MD, Kehler JR, Evans CA. Utility of dried blood spot sampling and storage for increased stability of photosensitive compounds. Bioanalysis 2010;2(11):1823–1828.

Chen J, Hsieh Y. Stabilizing drug molecules in biological samples. Ther Drug Monit 2005;27(5):617–624.

Christianson C, Johnson C, Sheaff C, Laine D, Zimmer J, Needham S. (2011) Overcoming the Obstacles of Performing Dilutions and Internal Standard Addition to DBS Analysis Using HPLC/MS/MS. Proceedings of the 50th ASMS Conference on Mass Spectrometry and Allied Topics, Orlando, FL, USA, June 1–6, 2002.

Clark GT, Haynes JJ, Bayliss MA, Burrows L. Utilization of DBS within drug discovery: development of a serial microsampling pharmacokinetic study in mice. Bioanalysis 2010;2(8):1477–1488.

Clark GT, Haynes JJ. Utilization of DBS within drug discovery: a simple 2D-LC-MS/MS system to minimize blood- and paper-based matrix effects from FTA elute^{TM} DBS. Bioanalysis 2011;3(11):1253–1270.

D'Arienzo CJ, Ji QC, and Discenza L, et al. DBS sampling can be used to stabilize prodrugs in drug discovery rodent studies without the addition of esterase inhibitors. Bioanalysis 2010;2(8):1415–1422.

Denniff P, Spooner N. Effect of storage conditions on the weight and appearance of dried blood spot samples on various cellulose-based substrates. Bioanalysis 2010;2(11):1817–1822.

EMA. 2011. Guideline on Bioanalytical Method Validation. European Medicines Agency, London. Available at http://www.ema.europa.eu/docs/en_GB/document_library/Scientific_guideline/2011/08/WC500109686.pdf. Accessed Apr 1, 2013.

FDA. 2001. Guidance for Industry: Bioanalytical Method Validation. Food and Drug Administration, Rockville, MD.

Ji QC, Liu G, D'Arienzo CJ, Olah TV, Arnold ME. What is next for dried bloodspots? Bioanalysis 2012;4(16):2059–2065.

Li F, Ploch S, Fast D, Michael S. Perforated dried blood spot accurate microsampling: the concept and its applications in toxicokinetic sample collection. J Mass Spectrom 2012a;47(5):655–667.

Li W, Doherty JP, Kulmatycki K, Smith HT, Tse FL. Simultaneous LC-MS/MS quantitation of acetaminophen and its glucuronide and sulfate metabolites in human dried blood spot samples collected by subjects in a pilot clinical study. Bioanalysis 2012b;4(12):1429–1443.

Li W, Tse FL. Dried blood spot sampling in combination with LC-MS/MS for quantitative analysis of small molecules. Biomed Chromatogr 2010;24(1):49–65.

Li W, Zhang J, Tse FL. Strategies in quantitative LC-MS/MS analysis of unstable small molecules in biological matrices. Biomed Chromatogr 2011;25(1-2):258–277.

Liang X, Li Y, Barfield M, Ji QC. Study of dried blood spots technique for the determination of dextromethorphan and its metabolite dextrorphan in human whole blood by LC-MS/MS. J Chromatogr B Analyt Technol Biomed Life Sci 2009;877(8-9):799–806.

Liu G, Ji QC, Jemal M, Tymiak AA, Arnold ME. Approach to evaluating dried blood spot sample stability during drying process and discovery of a treated card to maintain analyte stability by rapid on-card pH modification. Anal Chem 2011b;83(23):9033–9038.

Liu G, Patrone L, Snapp HM, et al. Evaluating and defining sample preparation procedures for DBS LC-MS/MS assays. Bioanalysis 2010;2(8):1405–1414.

Liu G, Snapp HM, Ji QC. Internal standard tracked dilution to overcome challenges in dried blood spots and robotic sample preparation for liquid chromatography/tandem mass spectrometry assays. Rapid Commun Mass Spectrom 2011a;25(9):1250–1256. doi: 10.1002/rcm.4990.

Majumdar TK, Howard DR. The use of dried blood spots for concentration assessment in pharmacokinetic evaluations. In: Bonate PL, Howard DR, editors. *Pharmacokinetics in Drug Development: Regulatory and Development Paradigms*. Springer; 2011. p 91–115.

Meesters R, Hooff G, van Huizen N, Gruters R, Luider T. Impact of internal standard addition on dried blood spot analysis in bioanalytical method development. Bioanalysis 2011;3(20):2357–2364.

O'Mara M, Hudson-Curtis B, Olson K, Yueh Y, Dunn J, Spooner N. The effect of hematocrit and punch location on assay bias during quantitative bioanalysis of dried blood spot samples. Bioanalysis. 2011;3(20):2335–2347.

Rahavendran SV, Vekich S, Skor H, et al. Discovery pharmacokinetic studies in mice using serial microsampling, dried blood spots and microbore LC-MS/MS. Bioanalysis 2012;4(9):1077–1095.

Rainville P. Microfluidic LC-MS for analysis of small-volume biofluid samples: where we have been and where we need to go. Bioanalysis 2011;3(1):1--3.

Ren X, Paehler T, Zimmer M, Guo Z, Zane P, Emmons GT. Impact of various factors on radioactivity distribution in different DBS papers. Bioanalysis 2010;2(8):1469–1475.

Shander A, Javidroozi M, Ashton ME. Drug-induced anemia and other red cell disorders: a guide in the age of polypharmacy. Curr Clin Pharmacol 2011;6(4):295–303.

Timmerman P, White S, Globig S, Lüdtke S, Brunet L, Smeraglia J. EBF recommendation on the validation of bioanalytical methods for dried blood spots. Bioanalysis 2011;3(14):1567–1575.

Viswanathan C. Perspectives on microsampling: DBS. Bioanalysis 2012;4(12):1417–1419.

31

LC-MS METHOD DEVELOPMENT STRATEGIES FOR ENHANCING MASS SPECTROMETRIC DETECTION

Yuan-Qing Xia and Jeffrey D. Miller

31.1 INTRODUCTION

Robustness of liquid chromatography–tandem mass spectrometry (LC-MS/MS) bioanalytical methods is critical in providing support for accurate assessment of toxicokinetics and pharmacokinetics of drug candidates in various stages of discovery and development (Jemal and Xia, 2006; Jemal et al., 2010). Many aspects are involved in developing and validating a robust LC-MS/MS bioanalytical method and most of those aspects have been captured in respective chapters of this book. In the present chapter, we would like to discuss signal enhancement in mass spectrometric detection in LC-MS/MS bioanalysis. In particular, we would like to discuss differential mobility spectrometry (DMS) and multiple reaction monitoring cubed (MRM^3) for reducing high chemical background noise, eliminating matrix interferences and separating isomeric and metabolites in quantitative LC-MS/MS bioanalysis. Furthermore, we would like to illustrate strategies for enhancing the mass spectrometric signal intensity and selectivity by utilizing either atmospheric pressure photoionization (APPI) or mobile phase additives and employing anionic and cationic adducts as analytical precursor ions in the LC-MS/MS bioanalysis process.

31.2 DIFFERENTIAL MOBILITY SPECTROMETRY

DMS, also referred to as high field-asymmetric waveform ion mobility spectrometry (FAIMS), is a variant ion mobility spectrometry (IMS). Conventional IMS separates ions according to their mobility through gas phase under the constant electrostatic field of low field strength (Purves and Guevremont, 1999; Guevremont, 2004; Kolakowski and Mester, 2007; Schneider et al., 2010). For DMS, ions are pulsed into the flight tube and their flight times are recorded. The drift time, the ion mobility, is a function of the reduced mass, charge state, and shape of an ion through its interactions with the background gas. DMS differs from IMS in the geometry of the instrumentation and separates ions via employing the field dependence of the coefficient of ion mobility. In brief, the separation of ions in DMS is based on analyte-specific differences in ion mobility under the influence of a high electric field (RF voltage), known as the separation voltage (SV), and a low electric field of a waveform applied to the two electrodes. Due to the difference between high and low field ion mobility coefficients, ions will migrate toward the walls and leave the flight path unless their trajectory is corrected by a counterbalancing DC voltage, known as the compensation voltage (COV). The COV thus controls which ions are transmitted through DMS, and gas is used to propel the ions through the electrodes. Thus, the drift of an ion toward either electrode is based on the difference in the mobility of that ion in high and low fields and is analyte specific parameter. The term DMS is used to distinguish this technique from IMS. Thus, the resolution of ions by DMS is based on the difference between the COV values of the ions. DMS, when used in conjunction with LC-MS/MS, acts as a postcolumn and premass spectrometer ion filter in which only selected ions generated from electrospray ionization (ESI) or atmospheric pressure chemical ionization (APCI) sources are transmitted. Ideally, when the selected COV is optimum for the analyte, the ions of the analyte pass through the DMS cell into the orifice of a mass spectrometer, while

Handbook of LC-MS Bioanalysis: Best Practices, Experimental Protocols, and Regulations, First Edition. Edited by Wenkui Li, Jie Zhang, and Francis L.S. Tse.

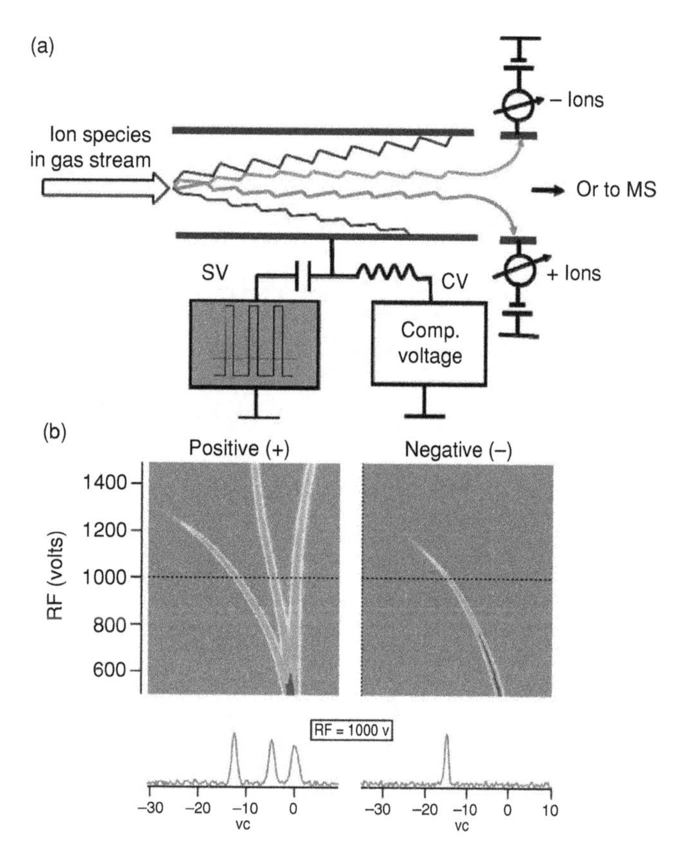

FIGURE 31.1 Schematic of DMS ion filter and sensor operation. (a) Ions generated in an atmospheric pressure ion source are carried in a transport gas through a DMS analysis region of the applied fields (SV and CV), and then detected at Faraday plate detectors. (b) Dispersion plots, recorded for dimethyl methylphosphonate (DMMP). In this experiment, SV (vertical) was scanned between 500 and 1500 V and CV (horizontal) between −40 and +10 V. Positive ions (left) and negative ions (right) were recorded simultaneously. Positive ions are separated into three ion species: reactant ions peak (RIP), DMMP monomer, and DMMP dimer. Right frame shows negative background ion behavior (reproduced from Schneider et al. (2010) with permission of Elsevier).

the ions of the background chemical ions or interferences are filtered away before reaching the orifice. Consequently, DMS provides separation between the analytes and the background interferences even in the absence of chromatographic separation. Only the ions with specific mobility dependence will pass through the electrodes of the device. This process typically occurs in the order of tens of milliseconds. The above combined feature allows bioanalytical scientists to significantly reduce interferences or background matrix ions due to the presence of isobaric compounds, coeluting species, or endogenous components in LC-MS/MS bioanalysis.

As shown in Figure 31.1, the DMS mobility cell is composed of two flat plates that are parallel to one another and define the mobility region. The ions are drawn by the transport gas flow toward the mass spectrometer. SelexION™ technology (AB Sciex) couples the DMS device to the QTRAP® 5500 or QTRAP® 6500 mass spectrometer, in which the two DMS electrodes are sealed to the inlet orifice with a juncture chamber in between. A gas port in the chamber allows full control of the residence time and resolution within the DMS device. The ion mobility coefficient is encoded in the compensation voltage and is used to correct the tilt in ion trajectory of each SV amplitude. The compensation voltage is scanned serially to pass ions according to their differential mobility, or set to a fixed value to pass only the ion species with a particular differential mobility. SelexION™ technology has short residence times, which enable rapid voltage changes for MRM operation. Due to the short MRM cycle times (20 ms pause time), DMS can couple with ultrahigh pressure liquid chromatography for fast LC analysis. The DMS device shows minimal diffusion losses and improved resolution at high voltages. In addition, it can simply be used in transparent mode (MRM mode) allowing all ions to be transmitted by turning off SV and COV voltages without the need to physically remove the DMS device. Figure 31.2 shows the comparison of LC-MS/MS chromatograms with and without DMS for the analysis of testosterone in plasma samples prepared via protein precipitation using acetonitrile. With DMS off, both MRM transitions of *m/z* 289 to *m/z* 109 and *m/z* 289 to *m/z* 97 show significant endogenous matrix interference peaks coeluting with testosterone. With DMS on, the endogenous matrix interference peaks were eliminated and the chromatograms of the same two MRM transitions show an excellent analyte peak with very clean baseline. As a result, the peak integration for testosterone was much easier for quantitation. Figure 31.3 illustrates an excellent example using DMS to enhance the full scan MS selectivity for safranin by eliminating chemical background of polyethylene glycol. With DMS off, when acquiring full scan Q1MS via infusing safranin containing PEG 400, the target mass of safranin (*m/z* 315) was not detectable due to the high background PEG 400 ions. However, with DMS on, an enhanced MS spectrum of safranin (*m/z* 315) was clearly detected while the PEG 400 ions were filtered out before entering the mass spectrometer. By using DMS, safranin was selected passing through the DMS cell, but not the PEG 400 ions, due to the different characteristic ion mobility of the analytes.

In LC-MS/MS bioanalysis, one of the potential pitfalls is the MS in-source conversion of metabolites to parent drug due to in-source fragmentation or conversion. In-source conversion of metabolites, especially phase II metabolites (glucuronide and sulfate), can generate the same drug molecular ions, which can hamper the accurate quantitative bioanalysis of the drug if there is no LC separation between the drug and its phase II conjugates (Jemal and Xia, 1999). Xia and Jemal (2009) applied FAIMS to separate ifetroban from its acyl glucuronide metabolite. The optimized COV value was determined as −13.7 V for ifetroban and −10.7 V for its acyl glucuronide, respectively. When injecting the acyl

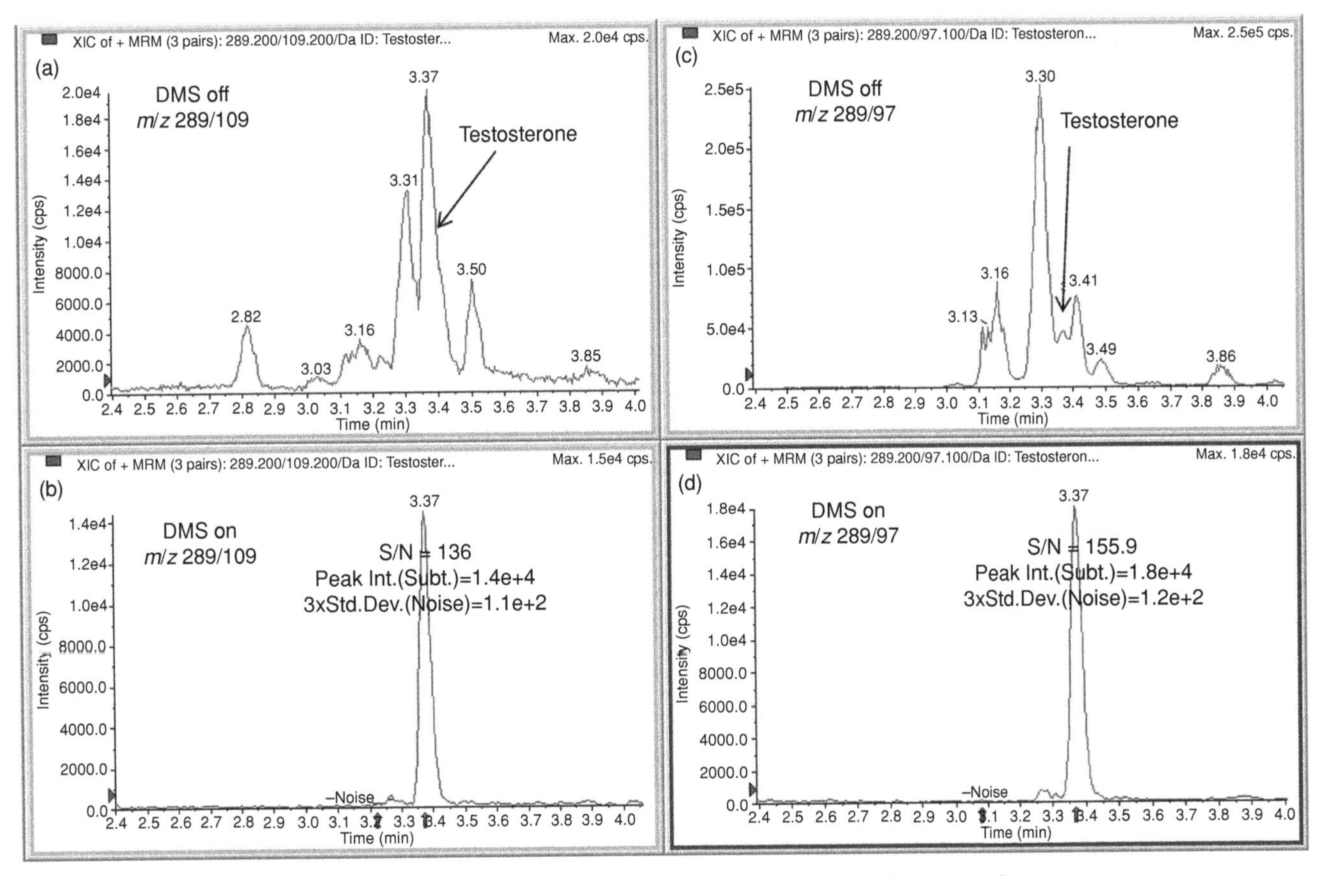

FIGURE 31.2 LC-MS/MS chromatograms of testosterone in human plasma sample extract prepared via protein precipitation using acetonitrile: (a) Deferential mobility spectrometry (DMS) off with MRM of *m/z* 289→*m/z* 109, (b) DMS on with MRM of *m/z* 289→*m/z* 109, (c) DMS off with MRM of *m/z* 289→*m/z* 97, and (d) DMS on with MRM of *m/z* 289→*m/z* 97.

glucuronide metabolite with the optimized ifetroban COV value of −13.7 V, a small ifetroban (Figure 31.4a) peak and acyl glucuronide peak were observed (Figure 31.4b). This demonstrated that the acyl glucuronide metabolite ions were filtered away at COV of −13.7 V. Thus, the in-source conversion of the acyl glucuronide metabolite to its parent drug occurred almost entirely after the orifice of the mass spectrometer, with the little conversion observed in the source chamber.

31.3 MULTIPLE REACTION MONITORING CUBED

A typical MRM process is to select a precursor ion in Q1MS, and fragment the precursor ion in the collision cell in the presence of collision gas with a suitable collision energy applied. Then a product or daughter ion is selected in Q3MS for detection. Sometimes, matrix and/or endogenous interference may share the same MRM transition as the analyte of interest. If the interference peaks coelute with the analyte, the selectivity of LC-MS/MS is hampered. This can lead to overestimation of the analyte of interest in the biological samples from the intended study. Several approaches have to be taken to address the issues. These approaches include (1) revamping sample preparation method to eliminate the interference in the sample extract prior to LC-MS/MS injection, (2) chromatographic separation of the analyte of interest from the interference, and/or (3) employing additional fragmentation in MS/MS detection with a linear ion trap (Figure 31.5). Apparently, the last one is the easiest if assay sensitivity is reachable. This additional fragmentation is referred to a MRM^3 scan, which utilizes the selection of a precursor ion in Q1MS, fragmentation of the precursor ion in the collision cell, and selection of a fragment ion in Q3 (linear ion trap, LIT) to further fragmentation to generate secondary fragment ions. An MRM^3 ion chromatogram is reconstructed from selected specific fragment ions produced from a primary product ion trapped in the Q3 linear ion trap and subsequently activated by resonant excitation. Similar to DMS, linear-ion-trap instrument with MRM^3 can be very useful for LC-MS quantitation of analytes in complex biological matrixes with greatly reduced background noise and/or endogenous interferences (Fortine et al., 2009). In

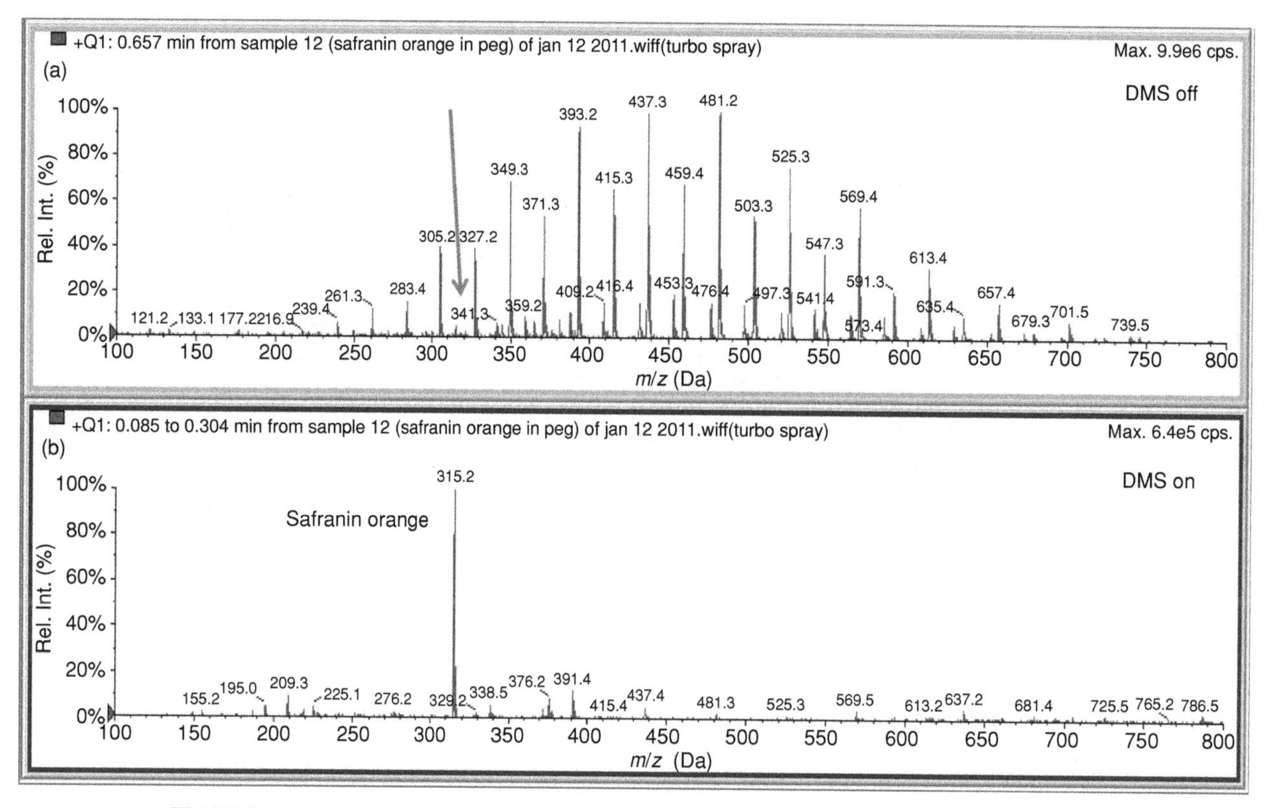

FIGURE 31.3 Positive full scan mass spectra of safranin containing polyethylene 400 (PEG 400) acquired using QTRAP® 5500 with scan range of m/z 100–800 with (a) DMS off and (b) DMS on.

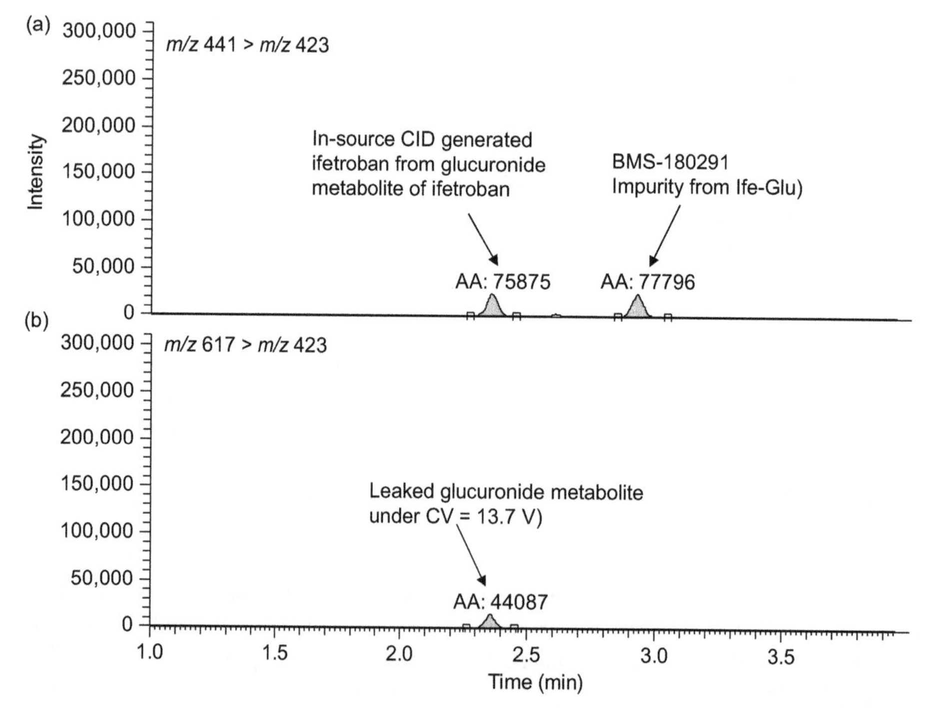

FIGURE 31.4 LC-FAIMS-SRM chromatograms obtained from the injection of a sample containing only ifetroban acylglucuronide (Ife-Glu) with the CV set at −13.7 V (optimum CV value for ifetroban): (a) ifetroban SRM channel (m/z 441 → m/z 423); (b) Ife-Glu SRM channel (m/z 617 → m/z 423) (reproduced from Xia and Jemal (2009) with permission of John Wiley & Sons, Ltd).

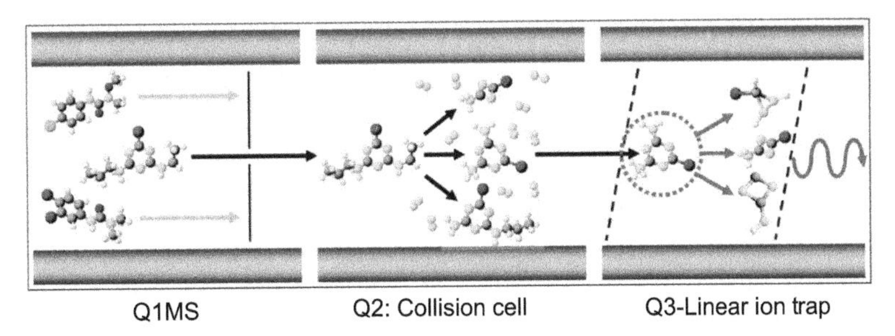

FIGURE 31.5 Schematic of multiple reaction monitoring cubed (MRM³) scan: selection of a precursor ion in Q1MS, fragmentation of precursor ion in the collision cell, and selection of a fragment ion in Q3 (linear ion trap) for further fragmentation to generate secondary fragment ions.

practice, suitable speed of MRM³ scan is necessary to allow for the second generation product ions to be extracted and integrated for quantitative bioanalysis. LIT instruments, such as QTRAP® 5500, 6500, and 4500, are commercially available to meet these needs. Xu et al. (2009) compared MRM versus MRM³ for the quantitation of exenatide in plasma. With MRM transition of *m/z* 838 to *m/z* 396, several significant peaks of plasma matrix interferences were coeluted with exenatide (Figure 31.6a), demanding an elevation of the lower limit of quantitation (LLOQ) from 5 ng/ml to 25 ng/ml for the robustness of the assay. With MRM³ (LC-MS/MS/MS) of *m/z* 838→*m/z* 396→*m/z* 202, the coeluting interference peaks were eliminated and an excellent signal-to-noise ratio was seen for the analyte at the LLOQ of 5 ng/ml (Figure 31.6b). On the other hand, due to the interference of the coeluting components, a nonlinear calibration regression has to be used for exenatide with MRM method (Figure 31.6c). With MRM³ method, a linear calibration standard curve was achieved for the analyte with no difficulty (Figure 31.6d).

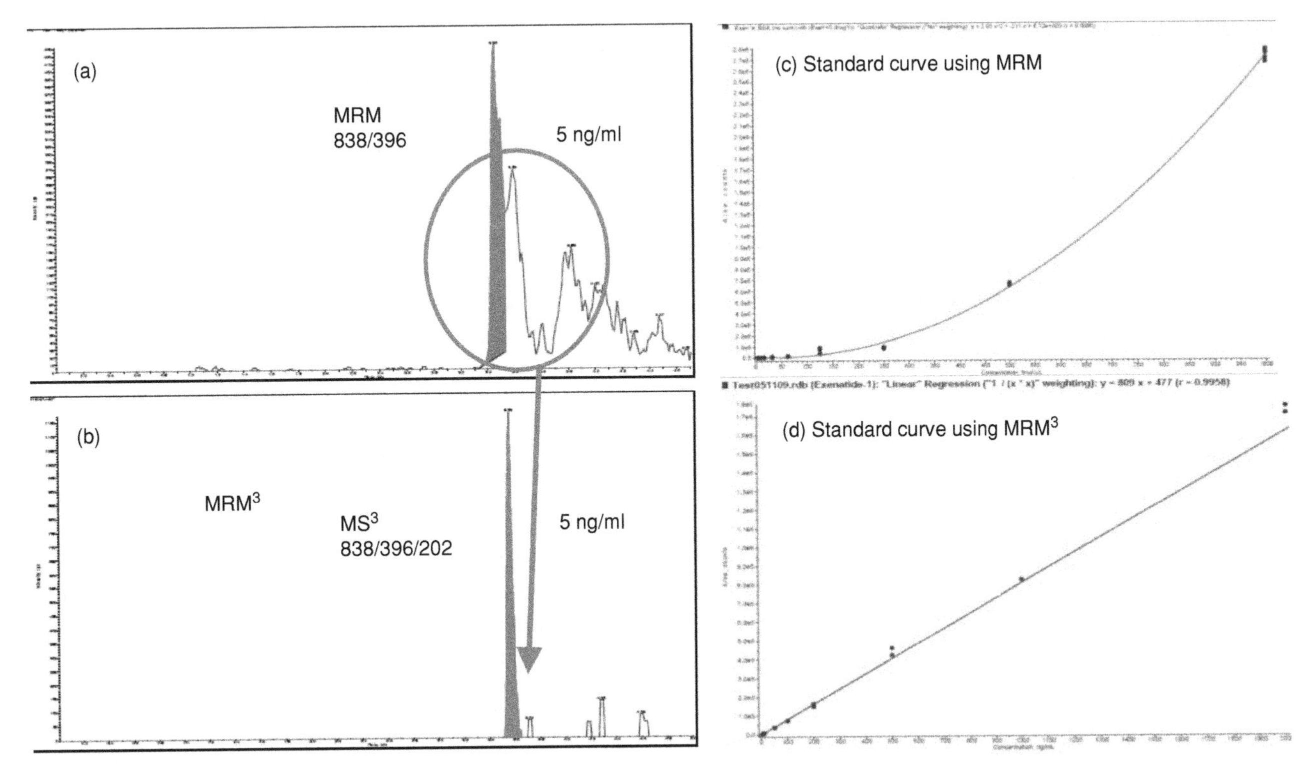

FIGURE 31.6 LC-MRM chromatograms of exenatide in plasma sample extract: (a) MRM of *m/z* 838 to *m/z* 396, (b) MRM³ of *m/z* 838→*m/z* 396→*m/z* 202, (c) with a quadratic calibration curve for the standards using MRM of *m/z* 838 to *m/z* 396, and (d) with a linear calibration curve for the standards using MRM³ of *m/z* 838→*m/z* 396→*m/z* 202.

31.4 ATMOSPHERIC PRESSURE PHOTOIONIZATION

In LC-MS/MS bioanalysis, the most widely used ionization techniques are ESI and APCI, which have been used in analysis of polar and easily ionizable compounds but may not applicable for nonpolar or less ionizable molecules. As a complement to ESI and APCI techniques, APPI can be used to expand the range and classes of compounds that can be analyzed by LC-MS toward less polar compounds (Kostiainen and Kauppila, 2009). The principal benefit of APPI, as compared to other ionization sources, is the efficient ionization of classes of nonpolar compounds, such as polychlorinated biphenyls (PCBs) and compounds with naphthalene moieties, which are not easily ionizable under conventional ESI or APCI. In APPI, a solvent is vaporized with a heated nebulizer, but the ionization process is initiated by using a vacuum ultraviolet lamp instead of a corona discharge needle as seen in APCI. APPI is less susceptible to ion suppression and salt buffer effects than APCI and ESI. The ionization process of APPI can be direct ionization or solvent-mediated reaction. The former takes place without addition of a solvent, while a dopant (such as toluene) is needed for the latter. The dopant serves as an extra solvent for initiating and enhancing the ionization. In direct APPI, the initial reaction is the formation of a molecular ion ($M + \bullet$) by photoionization of the analyte, which must possess ionization energy below the energy of the photons. In the presence of a protic solvent (methanol, water, 2-propanol, cyclohexane), the molecular ion of the analyte abstracts a hydrogen atom from the solvent to form a protonated molecule (Hanold et al., 2004; Kauppila et al., 2002). In the dopant-assisted mode, a large excess of a dopant, such as toluene or acetone, is added. The dopant molecule first undergoes photoionization and then acts as a charge carrier for subsequent ionization of trace levels of the analyte. Hanold et al. (2004) conducted comparison of ESI-CE-MS and APPI-CE-MS in analysis of a drug mixture consisting of terbutaline, salbutamol, and labetalol. They concluded that APPI gives significantly better detectability (e.g., signal-to-noise ratio) than ESI primarily due to reduced noise by the APPI source.

31.5 MS SIGNAL ENHANCEMENT VIA MOBILE PHASE ADDITIVES AND ANIONIC AND CATIONIC ADDUCTS PRECURSOR IONS

The effect of mobile phase composition on the ionization efficiency of an analyte of interest in ESI and APCI may not be easily predicted. Therefore, achieving optimal LC-MS/MS conditions may take a great effort due to the complexity of ionization processes and many factors that affect the LC-MS/MS bioanalytical assay. In LC-MS, characteristics of solvents and solvent additives, including volatility, surface tension, viscosity, conductivity, ionic strength, pH and gas phase ion–molecule reactions, may all contribute to ionization and signal responses of analyte of interest. MS detection sensitivity is also influenced by chemical and physical properties of the analyte, including p*K*a, hydrophobicity, surface activity, and proton affinity. The effect of eluent composition on the ionization efficiency of ESI, APCI, and APPI in LC-MS has been reviewed by Kostiainen and Kauppila (2009). In general, nonvolatile buffers such as phosphate and borate tend to cause increased background, signal suppression, and contamination of the ion source, resulting in decreased sensitivity and reproducibility. The most widely used aqueous mobile phases contain acetic acid, formic acid, trifluoroacetic acid (TFA), ammonium bicarbonate, ammonium hydroxide, ammonium acetate, and/or ammonium formate. In practice, the concentration of these additives should not exceed 20 mM in order to avoid suppression of ionization and reduction of sensitivity. Typical MS "friendly" organic mobile phases include acetonitrile and methanol. In most cases, mobile phase additives and buffered mobile phase are needed for the resolution, peak shape, and/or retention of analyte of interest in liquid chromatography. On the other hand, the chemical properties and concentration of the mobile phase additives have a significant effect on analyte response in ESI, APCI, and APPI.

Mallet et al. (2004) studied the influence of several additives and their concentrations on the ESI responses of acidic and basic drugs. The results showed a clear decrease in the response when the concentration of the additive such as formic acid, acetic acid, TFA, ammonium formate, and ammonium bicarbonate was increased from 0.05% to 1%. They also reported that ammonium formate and ammonium bicarbonate have a stronger suppression effect than that of acidic (formic and acetic acid) and basic buffers (ammonium hydroxide) on the ESI response of selected model drugs. Kamel et al. (1999) reported that the addition of 1% acetic acid resulted in a good LC separation and enhanced MS detection sensitivity in positive ion mode for a series of nucleosides, while the addition of 50 mM ammonium hydroxide resulted in a decreased sensitivity in negative ion mode. Duderstadta and Fischer (2008) reported that the signal intensity for polyalkene additive compounds in APCI has been found to be highly dependent upon the type of the organic solvent used in the LC separations. When employing a water/methanol gradient in place of a water/acetonitrile or a water/acetone gradient, the analyte signal intensities were increased between 2.3-fold and 52-fold.

TFA has frequently been used in the LC analyses of basic compounds. TFA not only controls the pH of the mobile phases but also acts as an ion-pair agent to improve peak shapes of basic compounds on silica-based columns. TFA is volatile and, therefore, can be used in LC-MS of proteins and peptides as well as small molecules (García, 2005; Shou and Naidong, 2005). The major drawback of using TFA in

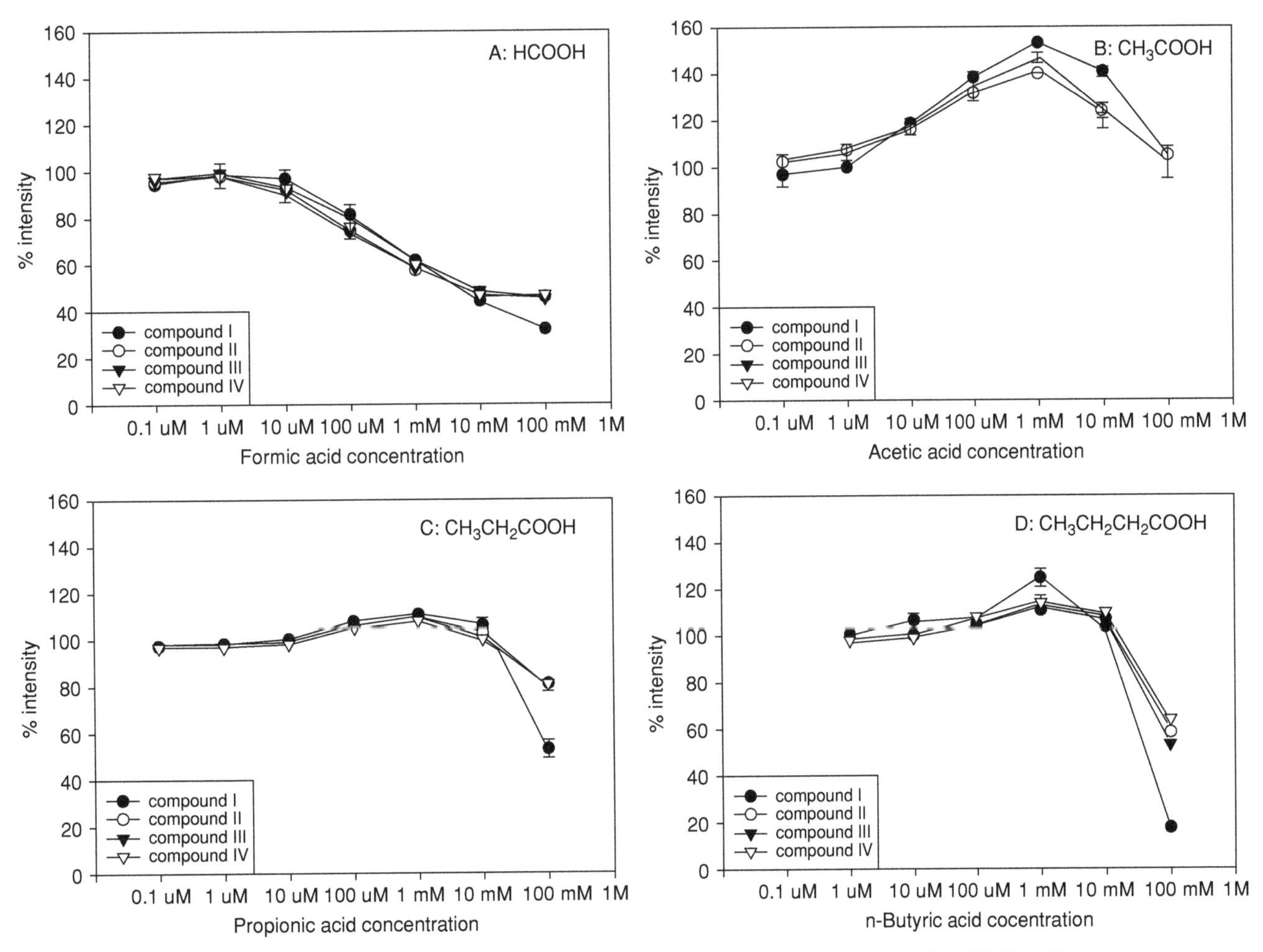

FIGURE 31.7 Effects of carboxylic acids on the negative-ion ESI responses of four SARMs. The horizontal axis represents the final concentration of modifier in the flow before entering the ESI source. The vertical axis represents the mean (SD, $n = 3$) ratio of peak area of each compound in the presence of modifier to the peak area of each compound in the absence of modifier multiplied by 100% (reproduced from Wu et al. (2004) with permission of John Wiley & Sons, Ltd).

LC-MS, however, is that TFA suppresses the ESI signals of analytes and reduces assay sensitivity. This is primarily due to the ability of TFA to form gas-phase ion pairs with positively charged analyte ions (Shou and Naidong, 2005). The suppression of ionization by TFA resulting in unstable spray has been demonstrated in several studies. The spray instability and signal reduction were due to the high conductivity and surface tension of the aqueous eluent including TFA (Chowdhury and Chair, 1991) or strong ion pairing between the TFA anion and the protonated molecule. The ion-pairing process is described as masking the protonated molecules and thereby decreasing the efficiency of the ESI droplet to emit protonated molecules to the gas phase (Kuhlmann et al., 1995). Ion pairing may also lead to reduced charge separation at the tip of the ESI sprayer and thereby to decrease ionization efficiency (Storm et al., 1999).

The most common method to overcome this issue involves the postcolumn addition of a mixture of propionic acid and isopropanol (Kuhlmann et al., 1995). However, the postcolumn addition setup requires additional pumps and is not desirable for continuous analysis of large amounts of samples. Shou and Naidong (2005) reported a simple yet very effective means of minimizing the negative effect of TFA in LC-MS/MS bioanalysis by direct addition of 0.5% acetic acid or 1% propionic acid to mobile phases containing either 0.025% or 0.05% TFA. A factor of two- to fivefold signal enhancement was achieved for eight basic model compounds. Furthermore, chromatography integrity was maintained even with the addition of acetic acid and propionic acid to existing TFA mobile phases.

Wu et al. (2004) studied the effects of various mobile phase modifiers on the negative ESI response of four

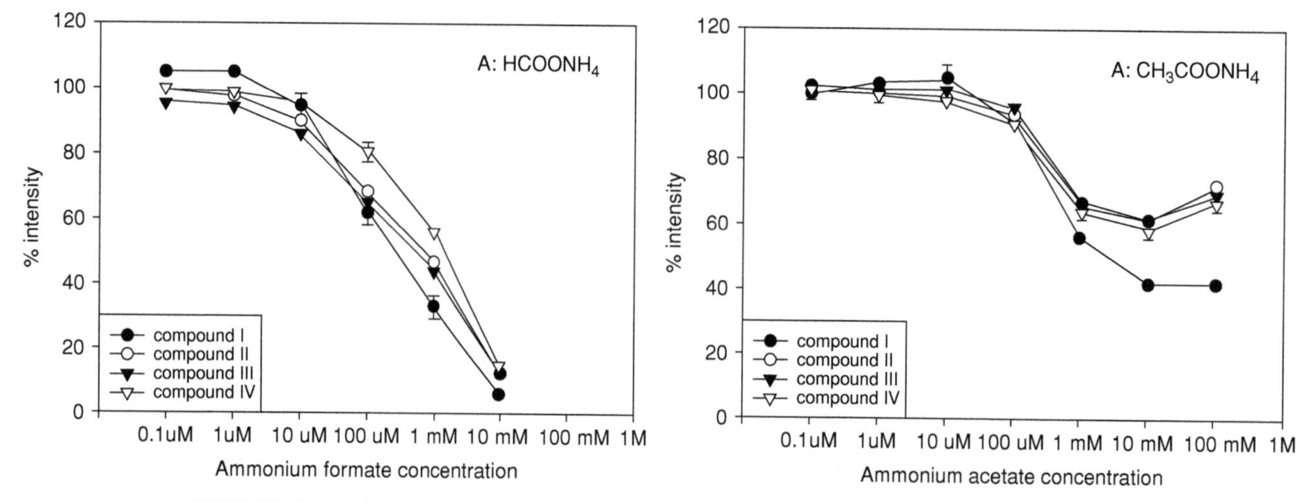

FIGURE 31.8 Effect of neutral salts on the negative-ion ESI responses of four SARMs. The horizontal axis represents the final concentration of modifier in the flow before entering the ESI source. The vertical axis represents the mean (SD, $n = 3$) ratio of peak area of each compound in the presence of modifier to the peak area of each compound in the absence of modifier multiplied by 100% (reproduced from Wu et al. (2004) with permission of John Wiley & Sons, Ltd).

selective androgen receptor modulators that do not have carboxylic acid or any other strongly acidic group. They found that acetic, propionic, and butyric acid at low concentrations from 0.1 μM to 10 mM improved the responses of the analytes to varying degrees (Figure 31.7). In contrast, formic acid decreased the MS response of the test compounds (Figure 31.7), as did ammonium formate, ammonium acetate, ammonium hydroxide, and triethylamine (Figure 31.8).

Molecules with strong acidic/basic functional groups, permanent charge moieties and/or suitable surface activity usually exhibit intense ESI signals. In contrast, molecules without presence of these structural attributes tend to respond poorly in ESI detection and are therefore difficult to be analyzed by LC-ESI-MS/MS, especially in complex biological matrices. One method to improve the signal response of those analytes is the use of solvent adduct ions as precursors for MS/MS detection. For example, anionic adduct ions can be employed as a precursor ion in negative ESI mode to increase the LC-MS/MS sensitivity. Cai and Cole (2002) conducted a systematic investigation on the attachment of small anions, such as halides, to neutral molecules to generate stable anionic adducts, $[M+X]^-$, as a means of increasing the negative ion ESI response. Kumar et al. (2004) extended the investigation using different model compounds, including dicarboxylic acids and benzoic acids. They showed that the collision-induced dissociation (CID) mass spectra of $[M+X]^-$ ions reflected the gas phase basicity of both the halide ion and $[M\text{-}H]^-$ ion of the analyte. The analytes showed a greater tendency to form adduct ions with Cl^- under ESI conditions compared with the other halide ions, such as F^-, Br^-, and I^-. Sheen and Her (2004) later reported an excellent application of fluoride adduct formation (Figure 31.9) for the sensitive quantitation of neutral drugs in human plasma. They found that the fluoride, chloride, and bromide adduct of the neutral drugs exhibited intense signals in negative ion ESI. Under CID, the major product ions of bromide and chloride adducts were the nonspecific bromide and chloride anions, respectively. In contrast, fluoride adducts produced strong $[M\text{-}H]^-$ ions, as well as $[M\text{-}H]^-$ product ions with good intensity and reproducibility (Figure 31.10).

Cationic adducts, such as ammonium or sodium, have been used as precursor ions in positive ESI mode to increase ESI responses of neutral molecules (Said et al., 2012). Amine additives have been used to suppress the formation of other adduct ions and increase the ESI sensitivity of the cholesterol-lowering drug, such as simvastatin, in human plasma. Zhao et al. (2002) compared the effect of mobile phases made from ammonium acetate and alkyl (methyl, ethyl, dimethyl, or trimethyl)-substituted ammonium acetate on the positive ESI response of simvastatin ammonium adduct. When alkyl ammonium buffer was used, it was observed that simvastatin alkyl ammonium-adduct ion was the only major molecular ion, while the formation of other adduct ions ($[M+H]^+$, $[M+Na]^+$, and $[M+K]^+$) was successfully suppressed. Among the various alkyl-substituted ammonium acetate buffers, methylammonium acetate buffer was the best one for improving the sensitivity (up to several-fold) for the simvastatin LC-MS/MS quantitation compared with that obtained using ammonium acetate buffer.

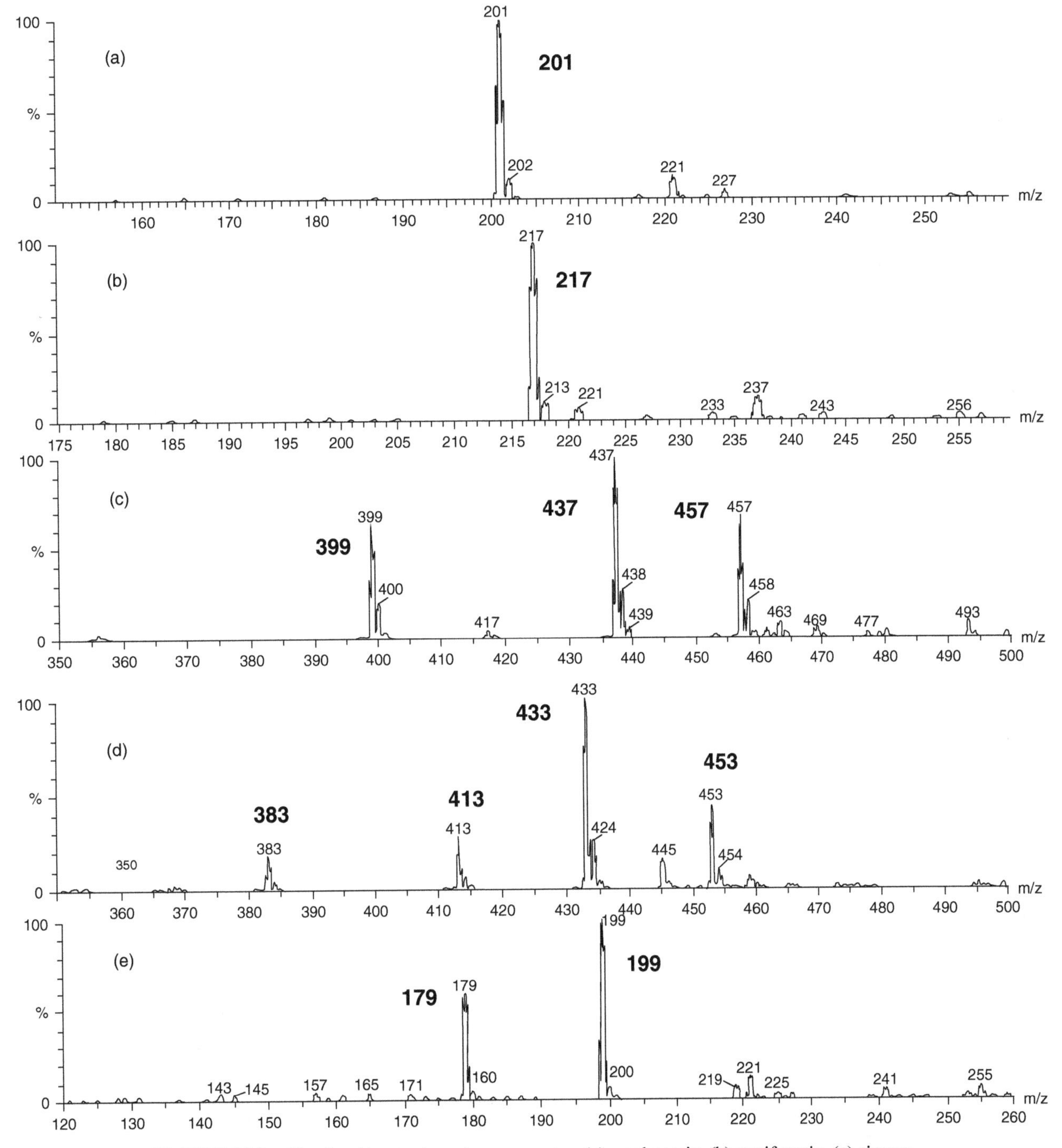

FIGURE 31.9 The fluoride attachment mass spectra: (a) mephenesin, (b) guaifenesin, (c) simvastatin, (d) podophyllotoxin, and (e) inositol (reproduced from Sheen and Her (2004) with permission of John Wiley & Sons, Ltd).

31.6 CONCLUSIONS

There are many challenging issues in LC-MS/MS bioanalysis and those challenges are different from a molecule to the other. The technologies and concepts covered in this chapter can be implemented to resolve some challenging selectivity and sensitivity issues in quantitative LC-MS/MS bioanalysis. DMS and MRM^3 can be considered as a method of choice for reducing high chemical background noise, eliminating matrix interferences and separating isomeric and metabolites in quantitative LC-MS/MS bioanalysis. The ionization processes of any compound under ESI, APCI, or APPI are

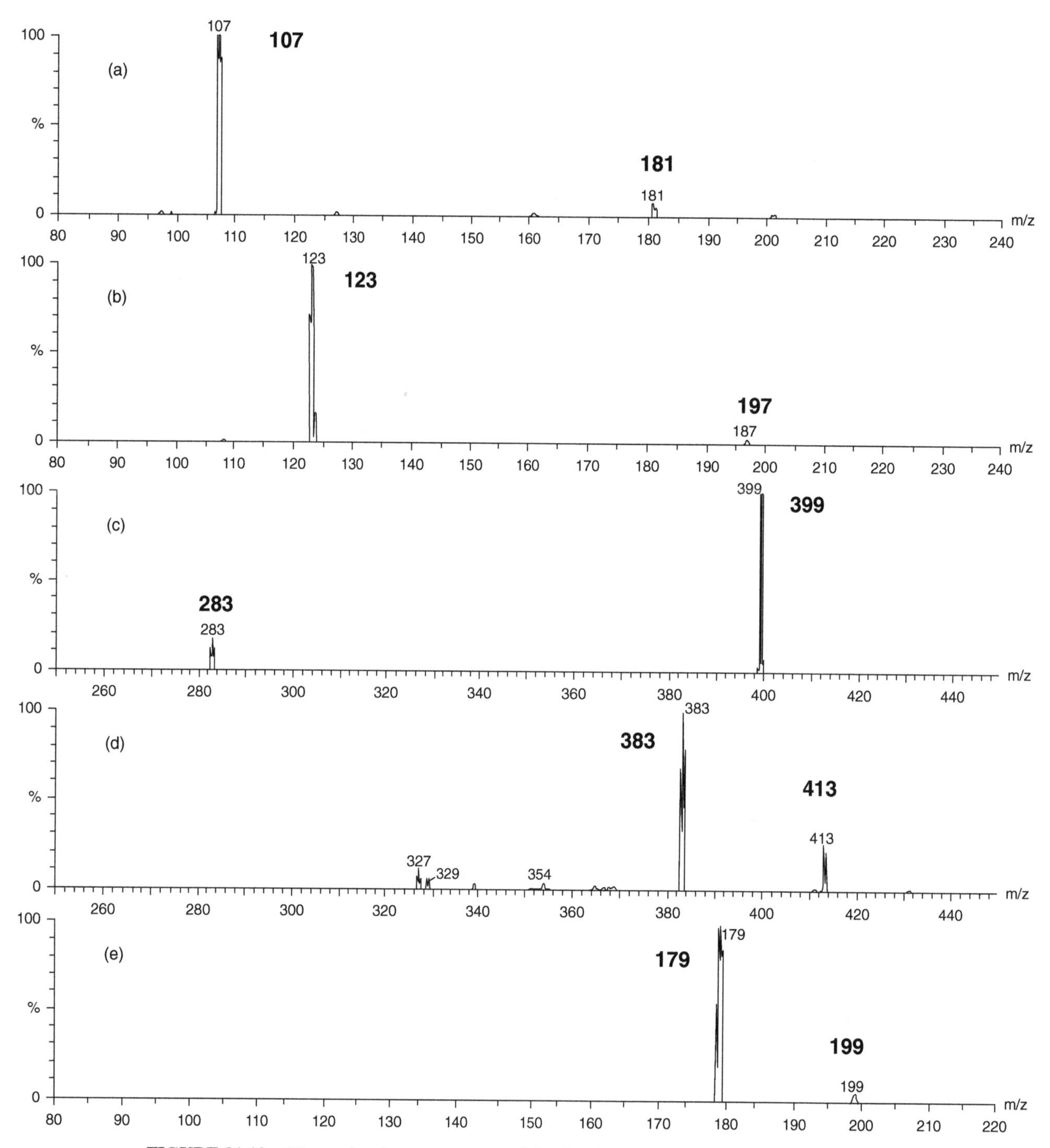

FIGURE 31.10 The product ion mass spectra of the fluoride adducts: (a) mephenesin, (b) guaifenesin, (c) simvastatin, (d) podophyllotoxin, and (e) inositol (reproduced from Sheen and Her (2004) with permission of John Wiley & Sons, Ltd).

highly complex. The composition of a mobile phase can has a significant influence on the ionization efficiency of analyte of interest in any of the above ionization modes. Therefore, it is important to optimize the mobile phase composition, including additives and their concentrations, in LC-MS/MS bioanalytical method development. For some neutral analytes, cationic or anionic adduct precursor ions can be considered as precursor ions for desired assay selectivity and sensitivity. Nevertheless, it should be emphasized that well thought-out strategies are essential in dealing with the issues of assay selectively and sensitivity in LC-MS/MS bioanalysis.

REFERENCES

Cai Y, Cole RB. Stabilization of anionic adducts in negative ion electrospray mass spectrometry. Anal Chem 2002;74:985–991.

Chowdhury SK, Chair BT. Method for the electrospray ionization of highly conductive aqueous solutions. Anal Chem 1991;63:1660–1664.

Duderstadta RE, Fischer SM. Effect of organic mobile phase composition on signal responses for selected polyalkene additive compounds by liquid chromatography–mass spectrometry. J Chromatogr A 2008;1193:70–78.

Fortin T, Salvador A, Charrier JP, et al. Multiple reaction monitoring cubed for protein quantification at the low nanogram/milliliter level in nondepleted human serum. Anal Chem 2009;81:9343–9352.

García MC. The effect of the mobile phase additives on sensitivity in the analysis of peptides and proteins by high-performance liquid chromatography–electrospray mass spectrometry. J Chromatogr B 2005;825:111–123.

Guevremont R. High-field asymmetric waveform ion mobility spectrometry: a new tool for mass spectrometer. J Chromatogr A 2004;1058(1-2):3–19.

Hanold KA, Fischer SM, Cormia PH, Miller CE, Syage JA. Atmospheric pressure photoionization. 1. general properties for LC/MS. Anal Chem 2004;76:2842–2851.

Jemal M, Ouyang Z, Xia Y-Q. Systematic LC-MS/MS bioanalytical method development. Biomed Chromatogr 2010;24:2–19.

Jemal M, Xia Y-Q. The need for adequate chromatographic separation in the quantitative determination of drugs in biological samples by high performance liquid chromatography with tandem mass spectrometry. Rapid Commun Mass Spectrom 1999;13:97–106.

Jemal M, Xia Y-Q. LC-MS Development strategies for quantitative bioanalysis. Curr Drug Metab 2006;7:491–502.

Kamel AM, Brown PR, Munson B. Effects of mobile-phase additives, solution pH, ionization constant, and analyte concentration on the sensitivities and electrospray ionization mass spectra of nucleoside antiviral agents. Anal Chem 1999;71:5481–5492.

Kauppila TJ, Kuuranne T, Meurer EC, Eberlin MN, Kotiaho T, Kostiainen R. Atmospheric pressure photoionization mass spectrometry. Ionization mechanism and the effect of solvent on the ionization of naphthalenes. Anal Chem 2002;74:5470–5479.

Kolakowski BM, Mester Z. Review of applications of high-field asymmetric waveform ion mobility spectrometry (FAIMS) and differential mobility spectrometry (DMS). Analyst 2007;132(9):842–64.

Kostiainen R, Kauppila TJ. Effect of eluent on the ionization process in liquid chromatography–mass spectrometry. J Chromatogr A 2009;1216:685–699.

Kuhlmann FE, Apffel A, Fisher SM, Goldberg G, Goodley PC. Signal enhancement for gradient reverse-phase high-performance liquid chromatography-electrospray ionization mass spectrometry analysis with trifluoroacetic and other strong acid modifiers by postcolumn addition of propionic acid and isopropanol. J Am Mass Spectrom 1995;6:1221–1225.

Kumar MR, Prabhakar S, Kumar MK, Reddy TJ, Vairamani M. Negative ion electrospray ionization mass spectral study of dicarboxylic acids in the presence of halide ions. Rapid Commun Mass Spectrom 2004;18:1109–1115.

Mallet CR, Lu Z, Mazzeo JR. A study of ion suppression effects in electrospray ionization from mobile phase additives and solid-phase extracts. Rapid Commun Mass Spectrom 2004;18: 49–58.

Purves RW, Guevremont R. Mass-spectrometric characterization of a high field asymmetric waveform ion mobility spectrometer. Rev Sci Instrum 1999;69:4094–4105.

Said R, Pohankab A, Abdel-Rehimc M, Becka O. Determination of four immunosuppressive drugs in whole blood using MEPS and LC–MS/MS allowing automated sample work-up and analysis. J Chromatogr A 2012;897:42–49.

Schneider BB, Covey TR, Coy SL, Krylov EV, Nazarov EG. Planar differential mobility spectrometer as a pre-filter for atmospheric pressure ionization mass spectrometry. Int J Mass Spectrom 2010;298:45–54.

Sheen JF, Her GR. Analysis of neutral drugs in human plasma by fluoride attachment in liquid chromatography/negative ion electrospray tandem mass spectrometry. Rapid Commun Mass Spectrom 2004;18:1911–1918.

Shou WZ, Naidong W. Simple means to alleviate sensitivity loss by trifluoroacetic acid (TFA) mobile phases in the hydrophilic interaction chromatography-electrospray tandem mass spectrometric (HILIC-ESI/MS/MS) bioanalysis of basic compounds. J Chromatogr B Analyt Technol Biomed Life Sci 2005;825(2): 186–92.

Storm T, Reemtsma T, Jekel M. Use of volatile amines as ion-pairing agents for the high-performance liquid chromatographic–tandem mass spectrometric determination of aromatic sulfonates in industrial wastewater. J Chromatogr A 1999;854:175–185.

Wu Z, Gao W, Phelps MA, Wu D, Miller DD, Dalton JT. Favorable effects of weak acids on negative-ion electrospray ionization mass spectrometry. Anal Chem 2004;76:839–847.

Xia Y-Q, Jemal M. High-field asymmetric waveform ion mobility spectrometry for determining the location of in-source collision-induced dissociation in electrospray ionization mass spectrometry. Anal Chem 2009;81:7839–7843.

Xu Y, Gutierrez JP, Lu T-S, et al. 2009. Quantification of the Therapeutic Peptide Exenatide in Human Plasma—MRM3 Quantitation for Highest Selectivity in Complex Mixtures on the AB SCIEX QTRAP® 5500 System. Available at http://www.absciex.com/Documents/Downloads/Literature/mass-spectrometry-cms_074674.pdf. Accessed Apr 1, 2013.

Zhao JJ, Yang AY, Rogers JD. Effects of liquid chromatography mobile phase buffer contents on the ionization and fragmentation of analytes in liquid chromatographic/ionspray tandem mass spectrometric determination. J Mass Spectrom 2002;37:421–433.

32

LC-MS BIOANALYSIS-RELATED STATISTICS

DAVID HOFFMAN

32.1 INTRODUCTION

The proper use of statistical techniques in bioanalytical method development, validation, and monitoring is a key tool for ensuring adequate method performance. While regulatory guidance documents and common industry practices often allow for the use of ad hoc, nonstatistical methods for evaluating method performance, the application of rigorous statistical methodology can control the risk of failing to identify methods with inadequate performance. This chapter presents a brief overview of basic statistical concepts and the application of statistical techniques to address selected issues in bioanalytical method development, validation, and monitoring.

32.2 BASIC STATISTICS

This section presents basic statistical concepts in regression analysis and the analysis of variance. Both regression and analysis of variance techniques form the basis for many statistical approaches to evaluating bioanalytical method performance.

32.2.1 Regression

Regression analysis is a widely used statistical technique to model the relationship between two (or more) variables. The objective is to estimate how a response variable (y) varies as a function of one (or more) predictor variables (x). A simple linear regression model is a model with a single predictor variable x that has a relationship with a single response variable y, given by the following function (Draper and Smith, 1998):

$$y = \beta_0 + \beta_1 x + \varepsilon,$$

where the intercept β_0 and slope β_1 are unknown constants and ε is a random error term.

The unknown constants β_0 and β_1 are typically estimated from sample data, say, $(y_1, x_1), (y_2, x_2), \ldots, (y_n, x_n)$ via the method of least squares. The least squares method estimates the unknown constants by minimizing the sum of the squared deviations between the sample data and the fitted regression line. For a simple linear regression model, the least squares estimates for β_0 and β_1 are given by

$$\hat{\beta}_0 = \bar{y} - \hat{\beta}_1 \bar{x}$$

and

$$\hat{\beta}_1 = \frac{\sum_{i=1}^{n} (x_i - \bar{x})(y_i - \bar{y})}{\sum_{i=1}^{n} (x_i - \bar{x})^2},$$

where $\bar{y} = \frac{1}{n}\sum_{i=1}^{n} y_i$ and $\bar{x} = \frac{1}{n}\sum_{i=1}^{n} x_i$.

The fitted linear regression model is then $\hat{y} = \hat{\beta}_0 + \hat{\beta}_1 x$.

The simple linear regression model above (and corresponding estimates for β_0 and β_1) assumes a linear relationship between y and x, and further assumes that the random error terms are uncorrelated and normally distributed with mean zero and constant variance (say, σ^2). The validity of these assumptions and the adequacy of the fitted model should be examined in any regression application. Violations of these assumptions may lead to poor regression predictions and invalid statistical tests. Regression model adequacy is

Handbook of LC-MS Bioanalysis: Best Practices, Experimental Protocols, and Regulations, First Edition. Edited by Wenkui Li, Jie Zhang, and Francis L.S. Tse.

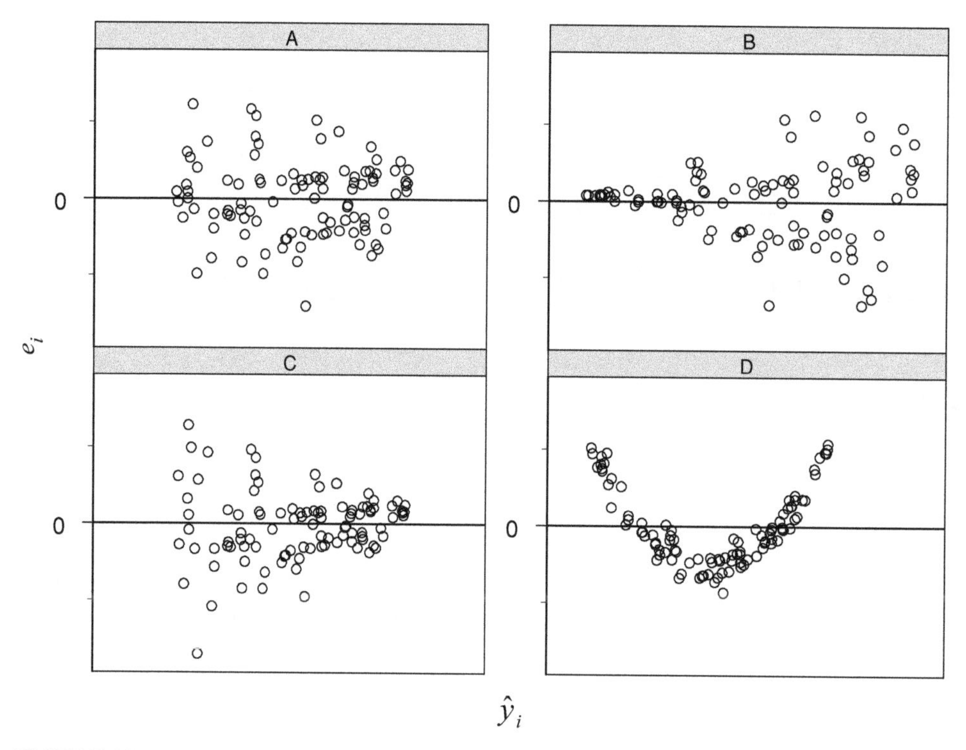

FIGURE 32.1 Various patterns for residual plots. Panel A is satisfactory. Panels B and C exhibit nonconstant variance. Panel D exhibits nonlinearity.

typically assessed via evaluation of the residuals from the fitted model, defined by

$$e_i = y_i - \hat{y}_i,$$

where y_i is an observed response and $\hat{y}_i$ is the corresponding predicted value from the fitted regression line. A particularly useful diagnostic tool is a plot of the residuals e_i versus the corresponding predicted values $\hat{y}_i$. This plot can be useful to detect nonconstant variance, nonlinearity, or potential outliers. Figure 32.1 gives various examples. Panel A illustrates residuals with no pattern and which can be contained within a horizontal band. This is a satisfactory residual plot and indicates no obvious model deficiency. Panels B and C illustrate a funnel pattern, which indicate the variance of the error terms is not constant. The outward opening funnel in panel B implies that the variance is increasing as a function of y, while the inward opening funnel in panel C implies that the variance is decreasing as a function of y. Panel D illustrates curvature in the residuals. This indicates nonlinearity and may be caused by the omission of other important predictor variables (e.g., a squared term x^2).

Other plots of residual values may also be examined. These may include normal probability plots (to assess normality), plots of residuals versus the corresponding predictor values x_i (to assess nonconstant variance and/or nonlinearity), plots of residuals versus time order (to assess correlation of error terms), and plots of residuals versus omitted predictor variables (to assess whether the omitted variable may improve the model).

As briefly described earlier, one basic assumption of the simple linear regression model is that the error terms have constant variance. In practice, it is not uncommon to encounter violations of this assumption. One method for correcting nonconstant error variance is to apply a variance-stabilizing transformation to the response variable y. Commonly used transformations include, among others, the square root, natural logarithm, reciprocal, and reciprocal square root transformations. Another approach for fitting regression models with nonconstant error variance is the method of weighted least squares. The weighted least squares method estimates the unknown regression constants by minimizing the sum of the squared deviations between the sample data and the fitted regression line, where the squared deviations are multiplied by a weight w_i chosen to be inversely proportional to the variance of y_i. For a simple linear regression model, the weighted least squares estimates for β_0 and β_1 are given by

$$\hat{\beta}_0 = \frac{\sum_{i=1}^{n} w_i y_i - \hat{\beta}_1 \sum_{i=1}^{n} w_i x_i}{\sum_{i=1}^{n} w_i}$$

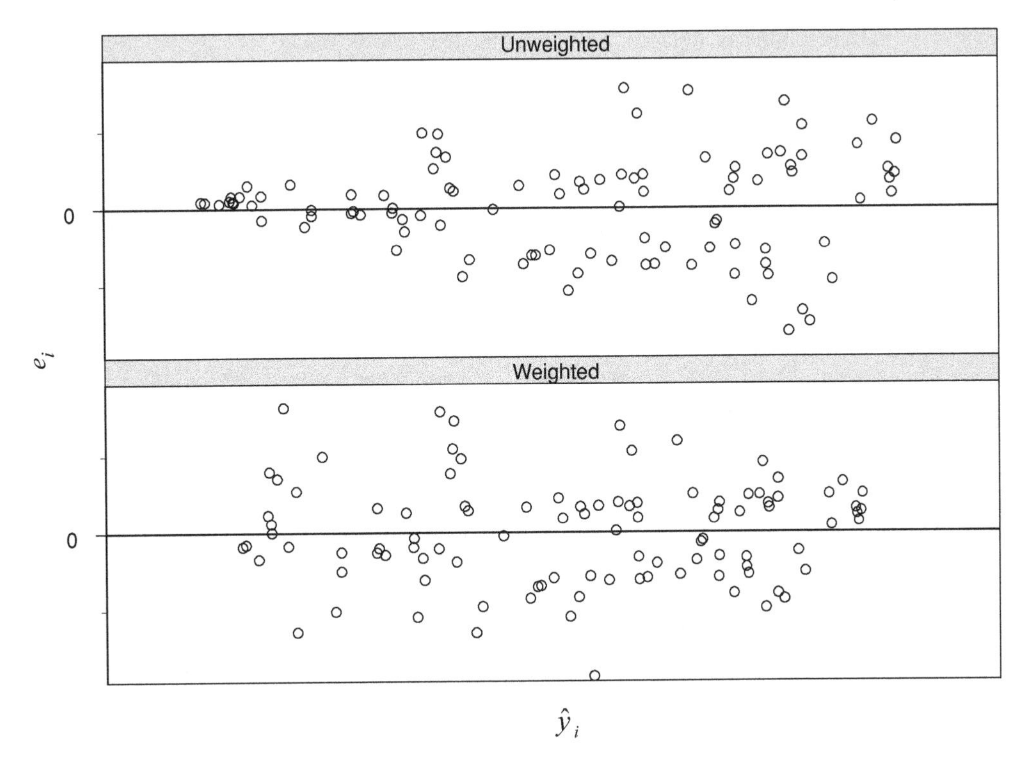

FIGURE 32.2 Illustration of unweighted versus weighted least squares in presence of nonconstant variance. Top panel gives residual plot for regression model fit by unweighted least squares. Bottom panel gives residual plot for regression model fit by weighted least squares with weight $1/x^2$.

and

$$\hat{\beta}_1 = \frac{\sum_{i=1}^{n} w_i \sum_{i=1}^{n} w_i x_i y_i - \sum_{i=1}^{n} w_i y_i \sum_{i=1}^{n} w_i x_i}{\sum_{i=1}^{n} w_i \sum_{i=1}^{n} w_i x_i^2 - \left(\sum_{i=1}^{n} w_i x_i\right)^2},$$

where w_i is the weighting factor for observed response y_i, and $\bar{y}$ and $\bar{x}$ are as before.

Figure 32.2 gives an illustration of weighted least squares in the presence of nonconstant error variance. The top panel gives the residual plot from a regression model fit by unweighted least squares. Note the outward opening funnel pattern of the residuals, indicating increasing variance as a function of y. The bottom panel gives the residual plot from a regression model fit by weighted least squares, with weight $1/x^2$. These residuals show no pattern and can be contained within a horizontal band, indicating no obvious model deficiency.

Finally, note that all of the above considerations for a simple linear regression model are also easily extended to a multiple linear regression model with k predictor variables $(x_1, \ldots, x_k)$:

$$y = \beta_0 + \beta_1 x_1 + \cdots + \beta_k x_k + \varepsilon,$$

where the parameters β_j ($j = 1, \ldots, k$) are unknown regression coefficients and ε is a random error term. Explicit formulae are not given here, but numerous software programs and packages are widely available to facilitate computation of such models.

32.2.2 Analysis of Variance

Analysis of variance is a statistical procedure that partitions the variability of a response into component parts. Analysis of variance techniques have wide utility, allowing for comparisons of the average response across various treatments or experimental conditions and estimation of the magnitude of relevant components of variance.

In the LC-MS bioanalysis context, the primary interest is typically in estimation of components of variance. Consider prestudy method validation, in which measurements are made over multiple independent assay runs with replicate determinations within each run. A statistical model to describe the measured values is given by

$$Y_{ij} = \mu + b_i + \varepsilon_{ij},$$

where Y_{ij} is the jth ($j = 1, 2, \ldots, J$) replicate observation from the ith ($i = 1, 2, \ldots, I$) assay run, μ is the true (unknown) analytical mean for the method, b_i is the random error for the ith assay run, and ε_{ij} is the random error for the jth replicate

TABLE 32.1 Analysis of Variance Table for Balanced One-Way Random Effects Model

Source	Degrees of freedom	Sums of squares	Mean square	EMS
Between-run	$df_B = I - 1$	$SS_B = J\sum_{i=1}^{I}(\bar{Y}_i - \bar{Y})^2$	$MS_B = SS_B/df_B$	$J\sigma_B^2 + \sigma_E^2$
Within-run	$df_E = I(J-1)$	$SS_E = \sum_{i=1}^{I}\sum_{j=1}^{J}(Y_{ij} - \bar{Y}_i)^2$	$MS_E = SS_E/df_E$	σ_E^2
Total	$df_T = IJ - 1$	$SS_T = \sum_{i=1}^{I}\sum_{j=1}^{J}(Y_{ij} - \bar{Y})^2$		

observation from the ith assay run. The random errors b_i and ε_{ij} are assumed to be normally and independently distributed with means zero and variances σ_B^2 and σ_E^2, respectively. These variances, σ_B^2 and σ_E^2, correspond to the between-run (inter-batch) and within-run (intra-batch) variability of the method. The total analytical variability of the method is then given by $\sigma_{TOT}^2 = \sigma_B^2 + \sigma_E^2$.

The above is commonly referred to as a one-way random effects model (Burdick and Graybill, 1992). For convenience, we assume that the data are balanced (i.e., there are J replicates in each of the I runs). Denote the overall mean of the measurements by $\bar{Y} = \sum_{i=1}^{I}\sum_{j=1}^{J} Y_{ij}/IJ$, and the mean for the ith assay run by $\bar{Y}_i = \sum_{j=1}^{J} Y_{ij}/J$. Table 32.1 gives the analysis of variance (ANOVA) table for the balanced one-way random effects model (where EMS denotes the expected mean square).

The mean squares MS_B and MS_E from Table 32.1 can be used to obtain estimates of the method within-run, between-run, and total variances. Table 32.2 gives the variance estimates obtained from the ANOVA mean squares.

32.3 CALIBRATION

The calibration function is the relationship between the instrument response (y) and the analyte concentration (x) in a sample. This relationship can be characterized by a function f such that

$$f(x) = y,$$

The inverse of this function, f^{-1}, is then used to back-calculate the analyte concentration from the instrument response (i.e., $f^{-1}(y) = x$). The function f may be linear or nonlinear. However, in LC-MS bioanalysis the calibration function f is typically assumed to be one of the two linear functions given in Table 32.3.

TABLE 32.2 Estimates of Within-Run, Between-Run, and Total Variance

Variance component	Estimate
Within-run	$\hat{\sigma}_E^2 = MS_E$
Between-run	$\hat{\sigma}_B^2 = (MS_B - MS_E)/J$
Total	$\hat{\sigma}_{TOT}^2 = \hat{\sigma}_B^2 + \hat{\sigma}_E^2$

The calibration functions given in Table 32.3 are generally estimated via the regression techniques described earlier. Commonly, the variance of the instrument response y increases as a function of the analyte concentration x. As such, the calibration function is typically fitted via weighted least squares.

One simple approach for determining the appropriate weights w is to model the variance of the response y at each level of the analyte concentration x. A simple linear regression line is then fit between the natural logarithm of the response variance $\hat{\sigma}^2$ and the natural logarithm of the analyte concentration:

$$\log(\hat{\sigma}^2) = \beta_0 + \beta_1 \log(x).$$

The weighting factor is then taken to be $1/x^{\hat{\beta}_1}$, where $\hat{\beta}_1$ is the estimated slope from the simple linear regression of log ($\hat{\sigma}^2$) on log (x). For convenience, the slope is often rounded to the nearest integer when determining the weighting factor.

Consider the example calibrations standards in Table 32.4. Table 32.4 gives the instrument response, across three independent runs, at each of six analyte concentrations prepared in triplicate (denoted Rep 1, Rep 2, and Rep 3). For each concentration, the within-run variance can be calculated via one-way ANOVA as shown previously in Table 32.1. Figure 32.3 shows the log-transformed variance versus the log-transformed concentration, along with the fitted simple linear regression line. The fitted simple linear regression line has an estimated slope of $\hat{\beta}_1 = 1.78$. Thus, after rounding the estimated slope to the nearest integer, the weighting factor for the calibration curve can be taken as $w = 1/x^2$.

Selection of the appropriate calibration function is generally based on evaluation of the back-calculated concentrations obtained from each function. The back-calculated concentrations are obtained via the inverse calibration functions shown in Table 32.3, using the estimated regression coefficients (i.e., $\hat{\beta}_0$, $\hat{\beta}_1$, and $\hat{\beta}_2$) from each fitted calibration function. Consider again the calibration standards given in Table 32.4. Using the weighting factor $w = 1/x^2$, both the simple linear and quadratic calibration functions shown in

TABLE 32.3 Typical Calibration Functions and Inverse Function

Type	Function	Inverse function
Simple linear	$y = \beta_0 + \beta_1 x$	$x = (y - \beta_0)/\beta_1$
Quadratic	$y = \beta_0 + \beta_1 x + \beta_2 x^2$	$x = (-\beta_1 + \sqrt{\beta_1^2 - 4\beta_2(\beta_0 - y)})/2\beta_2$

TABLE 32.4 Calibration Standards

Conc (ng/ml)	Run 1			Run 2			Run 3		
	Rep 1	Rep 2	Rep 3	Rep 1	Rep 2	Rep 3	Rep 1	Rep 2	Rep 3
1	0.117	0.107	0.103	0.103	0.0964	0.0922	0.114	0.105	0.0975
5	0.553	0.548	0.485	0.495	0.463	0.457	0.528	0.517	0.473
20	1.91	1.94	1.88	1.92	1.84	1.77	2.06	2.00	1.93
50	4.97	4.78	4.79	5.22	5.01	5.44	5.14	4.86	4.58
100	9.67	9.40	9.35	9.25	9.11	8.51	10.4	9.71	8.97
200	19.6	18.9	17.7	17.6	17.0	16.7	19.7	18.8	18.3

Table 32.3 were fit to the data, for each of the three runs separately.

For each calibration function, the back-calculated concentrations are then compared to the nominal concentration and expressed as a percent bias. Assessment of the average percent bias in each calibration function may be used to select an appropriate calibration function. Selection of the calibration function may be based on comparison of the average percent bias at each nominal concentration to a prespecified threshold (say, ±5%). Figure 32.4 gives the average percent bias for each calibration function fitted to the calibration standards given in Table 32.4. Note that the average bias in the simple linear calibration function (−4.5% bias) is much larger than that for the quadratic calibration function (+0.1% bias) at the 200 ng/ml nominal concentration, though still within ±5% across all nominal concentrations. Other criteria may also be examined to justify the selection of the calibration model and may include, among others: inspection of residual plots, statistical tests of significance for regression coefficients, and variability of back-calculated concentrations.

32.4 BIAS AND PRECISION

An essential part of bioanalytical method validation is the evaluation of the method bias and precision. Bias is a measure of systematic error in the analytical method and is expressed as the difference between the average concentration obtained by the method and the true concentration. Precision is a measure of random error in the analytical method and is expressed

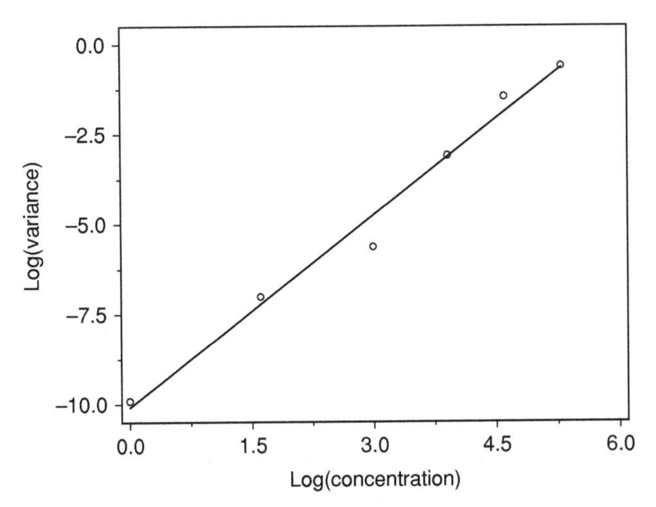

FIGURE 32.3 Log-transformed variance versus log-transformed concentration with fitted simple linear regression line for calibration standards given in Table 33.4.

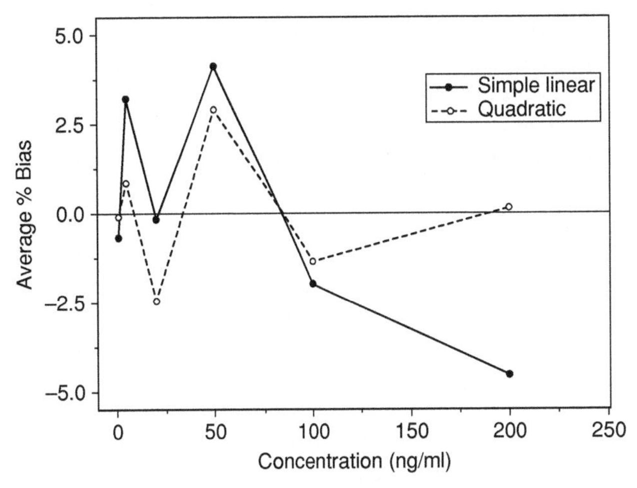

FIGURE 32.4 Average percent bias from nominal concentration for simple linear and quadratic calibration functions.

as the variability of individual measurements obtained from multiple sampling of the same homogenous sample under prescribed conditions.

Method bias and precision are typically estimated during prestudy method validation, and measurements are made over multiple independent assay runs with replicate determinations within each run. A statistical model to describe the measured concentrations is given by the one-way random effects model referenced in Section 32.2.2.

Common prestudy acceptance criteria for bioanalytical methods require the observed mean to be within $\pm 15\%$ of the nominal value and the observed precision to be $\leq 15\%$ coefficient of variation (%CV), though these limits are both 20% at the lower limit of quantification (European Medicines Agency, 2012; US Food and Drug Administration, 2001). That is, $\bar{Y}$ must be within $\pm 15\%$ of the known nominal value, and $\hat{\sigma}_{\mathrm{TOT}}$ must be $\leq 15\%$ relative to the mean value.

While acceptance criteria based solely on the point estimates $\bar{Y}$ and $\hat{\sigma}_{\mathrm{TOT}}$ are widely used, these criteria yield uncontrolled risks of rejecting suitable bioanalytical methods and accepting unsuitable bioanalytical methods (Kringle and Khan-Malek, 1994). Alternatively, the use of total error is a statistically and scientifically sound approach that incorporates both systematic and random errors. A total error approach reflects how large a measurement error can be and is easily understood by analysts. Moreover, it is a single comprehensive measure of method performance, rather than an assessment of method bias and variability individually.

Ideal acceptance criteria would ensure that a high proportion (say $\beta\%$) of future observations lie within acceptable limits (say $\pm 15\%$ of nominal), with a high degree of confidence (say $\gamma\%$). A two-sided β-content tolerance interval is a statistical interval (L, U) such that at least a proportion β of a population will lie within the interval (L, U) with $\gamma\%$ confidence. Two-sided β-content tolerance intervals provide lower (L) and upper (U) limits such that we can claim a specified proportion β of measured assay values will lie within the interval (L, U), with specified confidence coefficient γ.

For any analytical method, performance characteristics that constitute method suitability for its intended use can be defined by appropriate choice of the proportion β and acceptable limits (A, B). That is, a method is suitable for its intended use if at least a proportion β of measured assay values lie within the specified acceptance limits (A, B). Two-sided β-content tolerance intervals provide a statistical framework for controlling the risk of incorrectly accepting methods that do not fulfill these suitability requirements.

The total error approach is as follows (Hoffman and Kringle, 2007):

1. Construct a two-sided β-content tolerance interval (L, U) with desired confidence level γ (say, 90%).
2. Compare the interval (L, U) to the acceptance limits (A, B).
3. If (L, U) falls completely within (A, B), the method is accepted; otherwise, the method is not accepted.

The construction of two-sided β-content tolerance intervals for the balanced one-way random effects model is straightforward and requires only the calculation of the quantities previously described in Tables 32.1 and 32.2, as well as quantiles of the standard normal and chi-square distributions. Let $Z_{(1+\beta)/2}$ be the upper $(1+\beta)/2$ quantile of the standard normal distribution and $\chi^2_{1-\gamma,\mathrm{df}}$ be the lower γ quantile of the chi-square distribution with df degrees of freedom.

A two-sided β-content tolerance interval with confidence coefficient γ is given by Hoffman and Kringle (2005):

$$\bar{Y} \pm Z_{(1+\beta)/2}\sqrt{1+N_e^{-1}}\sqrt{\hat{\sigma}^2_{\mathrm{TOT}} + \{H_1^2(1/J)^2\mathrm{MS}_{\mathrm{B}}^2 + H_2^2((J-1)/J)^2\mathrm{MS}_{\mathrm{E}}^2\}^{1/2}}, \tag{32.1}$$

where

$$N_e = \frac{I(\mathrm{MS_B} + (J-1)\mathrm{MS_E})}{\mathrm{MS_B}},$$

$$H_1 = \frac{\mathrm{df_B}}{\chi^2_{1-\gamma,\mathrm{df_B}}} - 1, \text{ and}$$

$$H_2 = \frac{\mathrm{df_E}}{\chi^2_{1-\gamma,\mathrm{df_E}}} - 1.$$

Implementation of the total error approach requires appropriate choices of content level (β), confidence level (γ), and acceptance limits (A, B). For bioanalytical assays, 66.7% content, 90% confidence, and $\pm 15\%$ acceptance limits are logical choices. That is, the total error approach consists of constructing a two-sided $\beta = 66.7\%$ content, $\gamma = 90\%$ confidence tolerance interval. If the resulting tolerance limits are completely within $\pm 15\%$ of nominal, the assay is accepted; otherwise, it is not. The selection of $\beta = 66.7\%$ content and $\pm 15\%$ acceptance limits is intended to be consistent with typical acceptance criteria for in-study monitoring: that at least four of every six QC samples be within 15% of their respective nominal concentration.

The total error approach is illustrated by application to data from a prestudy validation experiment (Hoffman and Kringle, 2007). The data are calculated concentrations (ng/ml) of an analyte in human plasma and are shown in Table 32.5. The nominal concentration is 1 ng/ml. The sampling design consisted of six independent runs with three replicates per run. The method will be judged suitable if the entire two-sided $\beta = 66.7\%$ content, $\gamma = 90\%$ confidence tolerance interval is within (0.85, 1.15) ng/ml (i.e., $\pm$ 15% of the nominal concentration).

TABLE 32.5 Calculated Concentrations (ng/ml)

Replicate	Run 1	2	3	4	5	6
1	0.969	0.952	0.989	1.000	0.959	1.020
2	0.976	0.993	0.883	0.969	0.989	1.090
3	0.938	0.956	0.981	0.954	0.998	1.020

To calculate the interval, we construct the analysis of variance table as previously shown in Table 32.1. We have $I = 6$ runs, $J = 3$ replicates per run, and overall mean concentration of $\bar{Y} = 0.9798$ ng/ml. Table 32.6 gives the analysis of variance.

From the mean squares in Table 32.6, we have that $\hat{\sigma}^2_{\text{TOT}} = 0.00186$ and $N_e = 10.308$. The appropriate standard normal and chi-square quantiles can be easily obtained from tabulated values or from a statistical software package, and are as follows: $Z_{0.8335} = 0.96809$, $\chi^2_{0.10,5} = 1.61031$, and $\chi^2_{0.10,12} = 6.30380$. From the degrees of freedom in Table 32.6 and the chi-square quantiles above, we have $H_1 = 2.1050$ and $H_2 = 0.9036$. A two-sided β-content tolerance interval can then be calculated using Equation (32.1):

$$0.9798 \pm 0.96809\sqrt{1 + 10.308^{-1}}$$
$$\sqrt{\begin{array}{l}0.00186 + \{2.1050^2(1/3)^2 0.003251^2 \\ \quad +0.9036^2(2/3)^2 0.001167^2\}^{1/2}\end{array}}$$

The resulting two-sided β-content tolerance interval is given by (0.914, 1.046) ng/ml. Equivalently, the interval is (−8.6%, 4.6%) from the nominal concentration. Thus, the assay performance is judged suitable at this nominal concentration.

Note that the observed estimates of the bias and total %CV are −2.02% and 4.40%, respectively. Thus, the assay also passes the common acceptance criteria that $\bar{Y}$ must be within ±15% of the known nominal value, and $\hat{\sigma}_{\text{TOT}}$ must be ≤15% relative to the mean value.

32.5 STABILITY

Another key aspect of bioanalytical method validation is evaluation of analyte stability. Various stability assessments are performed during method validation and typically includes, among others, freeze-thaw stability, processed sample stability, stock solution stability, short-term (temperature or bench-top) stability, and long-term stability (US Food and Drug Administration, 2001; Nowatzke and Woolf, 2007; European Medicines Agency, 2012). The assessment of long-term analyte stability presents particular difficulties and is the focus of this section.

TABLE 32.6 Analysis of Variance Table for Data in Table 32.5

Source	Degrees of freedom	Sums of Squares	Mean Square
Between-run	$df_B = 5$	$SS_B = 0.016254$	$MS_B = 0.003251$
Within-run	$df_E = 12$	$SS_E = 0.014009$	$MS_E = 0.001167$
Total	$df_T = 17$	$SS_T = 0.030263$	

The objective of a long-term stability study is to assess potential analyte degradation encompassing the duration of time from sample collection to sample analysis. Long-term analyte stability assessment is performed by preparing stability samples at two or more nominal concentrations (US Food and Drug Administration, 2001). Typically, these stability samples are prepared by spiking control (blank) biological matrix with the analyte of interest. These stability pools are then transferred into individual storage tubes representative of those intended for the long-term storage of study samples, and are stored (frozen) under the conditions that will be used for the study samples. Long-term stability is then assessed by analysis of the stability samples over an appropriate time frame (i.e., sufficient to encompass or exceed the storage time anticipated for study samples).

Consider two possible experimental designs for long-term stability assessment: a "standard" design and a "concurrent control" design. The standard design is defined as follows. Stability samples are prepared as described earlier. Immediately following sample preparation or shortly thereafter, replicate samples are analyzed against a freshly prepared calibration curve. This can assess the accuracy of the spiked sample preparations (i.e., confirm the nominal concentration). The remaining samples are then stored as described earlier. At prespecified time points, replicate frozen stability samples are thawed and analyzed against freshly prepared calibration curves.

The concurrent control design is identical to that of the standard design, with one modification: at each prespecified time point, replicate "control" samples are analyzed concurrently with the thawed stability samples against the same freshly prepared calibration curve. The concurrent control samples can be prepared in one of two manners:

I. At each prespecified time point, replicate control samples are freshly prepared at the same nominal concentration and analyzed against the same freshly prepared calibration curve.

II. At the time of initial stability sample preparation, the samples are divided into two subsets. The first subset (i.e., stability samples) is stored at the temperature intended for study samples, as described previously. The second subset (i.e., control samples) is stored under temperatures less than −130 °C (e.g., in liquid nitrogen or other suitable freezer). At each prespecified time point, replicates of both the stability samples and control samples are analyzed against the same freshly prepared calibration curve.

With either the standard or concurrent control design, calculated analyte concentrations are subject to both within-run (intrabatch) and between-run (inter-batch) random variability intrinsic to the analytical method. The use of concurrent controls is intended to eliminate or minimize sources of between-run variability (e.g., calibration error) by including control samples in the same analytical run as the stability samples (Timm et al., 1985).

Note that the stability of such concurrent control samples should be (at minimum) informally verified via graphical inspection and/or descriptive statistics. The use of concurrent controls that exhibit degradation similar to that of stability samples over the storage time is improper and will result in an inflated risk of falsely concluding stability. Freshly prepared control samples (as described in (i) above), by definition, will not exhibit degradation; however, this introduces random variability arising from the preparation of different fresh control samples at each time point.

Various linear regression techniques may be considered for assessing long-term analyte stability (Hoffman et al., 2009). One approach for assessing long-term analyte stability is to regress the calculated stability sample analyte concentrations on storage time via the simple linear regression model considered earlier in Section 32.2.1. However, the simple linear regression model ignores the between-run (inter-batch) random errors induced by assay calibration at each time point. The simple linear model assumes that all calculated analyte concentrations are statistically independent of each other. This assumption will not be satisfied with long-term stability data, as calculated concentrations obtained against a common calibration curve will be correlated (i.e., by the between-run random error at each time point).

A nested errors linear regression model can appropriately account for both between-run and within-run random errors inherent in long-term stability data. The nested errors regression approach consists of regressing the calculated stability sample analyte concentrations on storage time via the following model:

$$y_{ij} = \beta_0 + \beta_1 x_i + \gamma_i + \varepsilon_{ij},$$

where y_{ij} is the calculated analyte concentration for the jth stability sample replicate at the ith time point, x_i is the ith time point, γ_i is the random error associated with the ith time point, and ε_{ij} is the random error for y_{ij}. The random errors γ_i and ε_{ij} are assumed to be independently and normally distributed with means zero and variances σ_B^2 and σ_E^2, respectively. These variances, σ_B^2 and σ_E^2, correspond to the between-run and within-run variability of the analytical method, respectively.

At any fixed time point x, a 90% two-sided confidence interval for the mean analyte concentration is constructed from the fitted regression model: $\hat{y} = \hat{\beta}_0 + \hat{\beta}_1 x$. The analyte is considered stable at a given time point if the 90% two-sided confidence interval lies entirely within the prespecified acceptance limits. The power of the nested error regression approach to correctly conclude stability decreases as the proportion of total variability due to between-run variability (denote by $\lambda = \sigma_B^2/(\sigma_B^2 + \sigma_E^2)$) increases.

Note that neither a simple linear nor nested errors regression approach allows for inclusion of data from concurrent control samples, which may minimize or eliminate the impact of between-run errors in the stability assessment.

One simple approach to incorporate data from concurrent control samples would be to "normalize" the stability sample analyte concentrations by the mean control sample analyte concentration at each time point. However, if the between-run and within-run random errors are assumed to follow a normal distribution, then the "normalized" random errors (i.e., ratio of errors) will be non-normally distributed. Furthermore, this simple approach presupposes a high degree of correlation between the stability and control samples at each time point. While this should be typically expected (and is the ideal outcome), it is possible that the stability and control samples may exhibit poor correlation (e.g., due to poor precision when spiking fresh control samples at each time point or possible matrix effects arising from storage at temperatures below 130 °C). In such cases, simple normalization will be detrimental, causing inflated variability and resulting in poorer precision of stability estimates.

A more flexible approach to incorporate data from concurrent control samples is to jointly model the stability sample and control sample data in a bivariate mixed model. The bivariate mixed model regression approach consists of jointly regressing the calculated stability sample and control sample analyte concentrations on storage time via the following model:

$$y_{ij} = \beta_0 + \beta_1 x_i + \gamma_i + \varepsilon_{ij},$$

$$z_{ik} = \beta_0 + \delta_i + \xi_{ik},$$

where y_{ij} is the calculated analyte concentration for the jth stability sample replicate at the ith time point, z_{ik} is the calculated analyte concentration for the kth control sample replicate at the ith time point, x_i is the ith time point, γ_i is the random error associated with the ith time point for the stability samples, δ_i is the random error associated with the ith time point for the control samples, ε_{ij} is the random error for y_{ij}, and ξ_{ik} is the random error for z_{ij}.

The within-run random errors ε_{ij} and ξ_{ik} are assumed to be independently and normally distributed with means zero and

TABLE 32.7 Calculated Concentrations (ng/ml)

Month	Rep1	Rep2	Rep3	Rep4	Rep5	Rep6						
	Fresh samples											
0	192	204	196	204	208	202						
	Stability samples (−20°C)						Control samples (<−130°C)					
Month	Rep1	Rep2	Rep3	Rep4	Rep5	Rep6	Rep1	Rep2	Rep3	Rep4	Rep5	Rep6
1	220	223	214	219	209	217	221	219	222	219	210	215
3	188	185	192	187	185	194	190	200	194	196	194	191
6	167	147	141	180	–	–	172	176	177	174	175	172
9	188	200	183	183	189	196	198	193	191	194	195	196
12	179	180	173	197	183	182	189	182	179	176	176	–
18	183	179	188	192	188	193	198	197	195	195	194	201
24	210	200	199	201	203	207	199	201	198	193	199	–

variances σ_{E1}^2 and σ_{E2}^2, respectively. These variances, σ_{E1}^2 and σ_{E2}^2, correspond to the within-run variability of the stability and control samples, respectively.

The between-run random errors γ_i and δ_i are assumed to follow a bivariate normal distribution with means zero and covariance matrix Σ given by

$$\Sigma = \begin{pmatrix} \sigma_{B1}^2 & \rho\sigma_{B1}\sigma_{B2} \\ \rho\sigma_{B1}\sigma_{B2} & \sigma_{B2}^2 \end{pmatrix}.$$

The variances σ_{B1}^2 and σ_{B2}^2 correspond to the between-run variability of the stability and control samples, respectively. The correlation parameter ρ corresponds to the correlation of the between-run random errors for the stability and controls samples analyzed at a given time point (i.e., against a common calibration curve). As the correlation parameter ρ increases, the power of the bivariate mixed model regression approach to correctly conclude stability increases accordingly.

Note that the model could also be simplified by reasonably assuming the within-run and between-run variances to be identical for both the stability and control samples (i.e., $\sigma_{E1}^2 = \sigma_{E2}^2 = \sigma_E^2$ and $\sigma_{B1}^2 = \sigma_{B2}^2 = \sigma_B^2$).

As with the simple linear and nested error approaches, a 90% two-sided confidence interval for the mean analyte concentration at any fixed time point x is constructed from the fitted regression model: $\hat{y} = \hat{\beta}_0 + \hat{\beta}_1 x$. The analyte is considered stable at a given time point if the 90% two-sided confidence interval lies entirely within the prespecified acceptance limits.

The nested errors and bivariate mixed model regression approaches are illustrated by application to data from an actual long-term stability experiment utilizing a concurrent control experimental design (Hoffman et al., 2009).

A plasma pool was spiked at 200 ng/ml of analyte and six samples were analyzed immediately following pool preparation. The remaining plasma pool was divided into two subsets (stability samples and control samples). Stability samples were stored at −20°C and control samples at less than −130°C. Six stability sample replicates and six control sample replicates were then thawed and analyzed against a freshly prepared calibration curve after 1, 3, 6, 9, 12, 18, and 24 months of storage. The raw concentration data are given in Table 32.7 (note that four observations were missing due to analytical issues and are indicated by "-" in the table).

Both the nested errors and bivariate mixed model regression approaches were applied to the data. The nested errors model was fit using only calculated concentrations from stability samples, while the bivariate mixed model was fit using calculated concentrations from both stability and control samples. Figure 32.5 shows the fitted nested errors regression model with two-sided 90% confidence interval for the mean analyte concentration. Figure 32.6 shows the fitted bivariate mixed model with two-sided 90% confidence interval for the mean analyte concentration. Note that ±15% acceptance limits correspond to (170, 230) ng/ml.

Figure 32.5 indicates substantial between-run variability in the calculated concentrations of the stability samples. This variability is reflected in the width of the two-sided 90% confidence interval about the fitted regression line. The estimated proportion of variability due to between-run variability based on the fitted nested errors regression model is $\hat{\lambda} = 0.85$, with an estimated total %CV of 9.8%. At the 24-month time point, the two-sided 90% confidence interval for the mean analyte concentration is (162, 216) ng/ml, which falls slightly outside the acceptance limits of (170, 230) ng/ml. Thus, with the nested errors regression approach, we cannot conclude the analyte is stable at 24 months.

Figure 32.6 shows good correlation between the stability and control samples. The estimated correlation of the

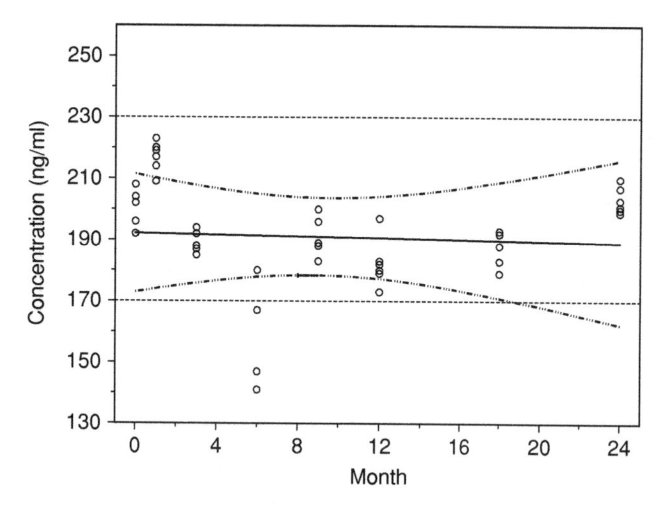

FIGURE 32.5 Fitted nested errors regression model with two-sided 90% confidence interval. Calculated concentrations for stability samples given by open circles. Acceptance limits shown at (170, 230) ng/ml.

stability sample and control sample between-run random errors based on the fitted bivariate mixed model is $\hat{\rho} = 0.93$. This strong correlation dramatically reduces the impact of the between-run random variability on the precision of the stability estimates and is reflected in the narrow confidence bounds about the fitted regression line. Note that at the 24-month time point, the two-sided 90% confidence interval for the mean analyte concentration is (178, 208) ng/ml. This interval lies entirely within the acceptance limits (170, 230) ng/ml and we can conclude the analyte is stable at 24 months.

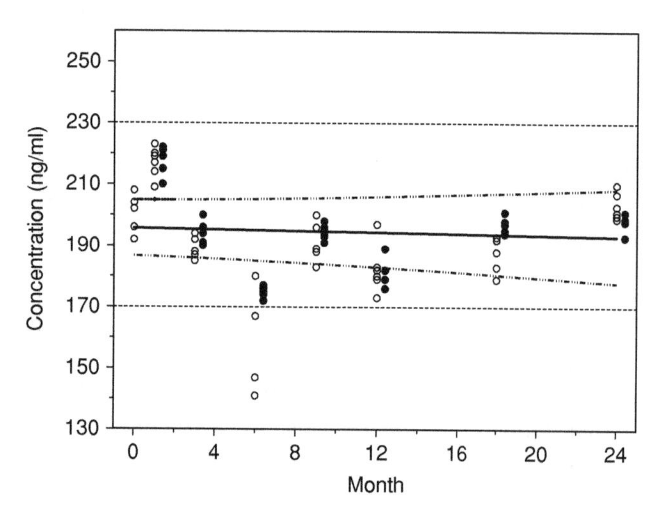

FIGURE 32.6 Fitted bivariate mixed model with two-sided 90% confidence interval. Calculated concentrations for stability samples given by open circles and for control samples by closed circles. Acceptance limits shown at (170, 230) ng/ml.

32.6 INCURRED SAMPLE REPRODUCIBILITY

The objective of incurred sample reproducibility (ISR) testing is to demonstrate that a bioanalytical method will produce consistent results from study samples when reanalyzed on a separate occasion (Viswanathan et al., 2007; Fast et al., 2009, European Medicines Agency, 2012). The term reproducibility is often used to refer to the precision of a bioanalytical method between two laboratories. In the context of ISR testing, reproducibility refers to the agreement of results obtained from the analysis of incurred samples on two (or more) separate occasions within the same laboratory.

Common acceptance criteria for ISR testing are based on the so-called 4-6-X rule, with ±20% acceptance limits (i.e., 4-6-20 rule) for small molecules or ±30% acceptance limits (i.e., 4-6-30 rule) for large molecules (Fast et al., 2009; European Medicines Agency, 2012). That is, at least 66.7% of the reanalyzed incurred samples must agree within ±20% of the original result (or ±30% for large molecules).

The acceptance limits of ±20% were likely chosen with reference to the acceptance criteria for in-study monitoring contained in the FDA bioanalytical guidance that at least 66.7% of QC samples must be within ±15% of their respective nominal concentration. The expansion from ±15% acceptance limits to ±20% acceptance limits is an apparent attempt to account for the variability in the original result.

However, the deficiencies of ad hoc approaches such as the 4-6-20 rule have been well documented (Kringle, 1994; Kringle et al., 2001). Unlike the 4-6-20 rule (or similar approaches), a statistical approach based on tolerance intervals provides strict control over the risk of incorrectly accepting truly nonreproducible bioanalytical methods.

The tolerance interval approach described later is based on the assumption that the underlying data are independent and normally distributed, though minor departures from this assumption are generally of little practical consequence. Noting that the incurred samples selected should span a wide range of concentrations and that bioanalytical precision is generally proportional to the true concentration, it is suggested that the calculated repeat and original concentrations be log-transformed prior to application of the tolerance interval and containment proportion approaches described later. In practice, the differences (repeat – original) in log-transformed concentrations will likely approximate a normal distribution, though gross departures from this assumption can be assessed via graphical techniques or statistical hypothesis tests. Further, the assumption of independence may be reasonably satisfied by appropriate choice for the number of analytical runs (i.e., incurred samples should be analyzed over as many analytical runs as practicable within a laboratory). Gross departures from the assumption of independence may inflate the risk of ISR test failure.

In the context of ISR, a two-sided β-content, γ-confidence tolerance interval may be used to determine an interval

TABLE 32.8 Repeat and Original Concentrations (ng/ml) with Percentage Difference

Sample	Original	Repeat	% Difference	Sample	Original	Repeat	% Difference
1	4030	4070	1.0	19	3240	3250	0.3
2	14100	12600	−10.6	20	11000	12200	10.9
3	2120	1530	−27.8	21	879	799	−9.1
4	21.6	19.9	−7.9	22	6.68	6.24	−6.6
5	859	761	−11.4	23	384	313	−18.5
6	192	215	12.0	24	57.0	64.8	13.7
7	2710	2790	3.0	25	2930	3150	7.5
8	13000	10000	−23.1	26	11400	11600	1.8
9	886	787	−11.2	27	2310	2270	−1.7
10	9.46	9.01	−4.8	28	6.11	6.22	1.8
11	401	381	−5.0	29	881	900	2.2
12	123	122	−0.8	30	193	169	−12.4
13	3520	3370	−4.3	31	3870	4340	12.1
14	3430	3410	−0.6	32	16600	16200	−2.4
15	2580	2630	1.9	33	1250	1180	−5.6
16	13.7	13.2	−3.6	34	10.3	9.62	−6.6
17	1410	1320	−6.4	35	581	673	15.8
18	437	433	−0.9	36	196	188	−4.1

(*L*, *U*) such that a proportion β of the (repeat − original) measurement differences lie within the interval, with a specified confidence coefficient γ. This interval (*L*, *U*) can then be compared to appropriately chosen acceptance limits (*A*, *B*). Such an approach provides a statistical framework for controlling the risk of incorrectly accepting bioanalytical methods for which less than a proportion β of the (repeat − original) measurement differences lie within acceptance limits (*A*, *B*).

A proposed tolerance interval approach is as follows (Hoffman, 2009):

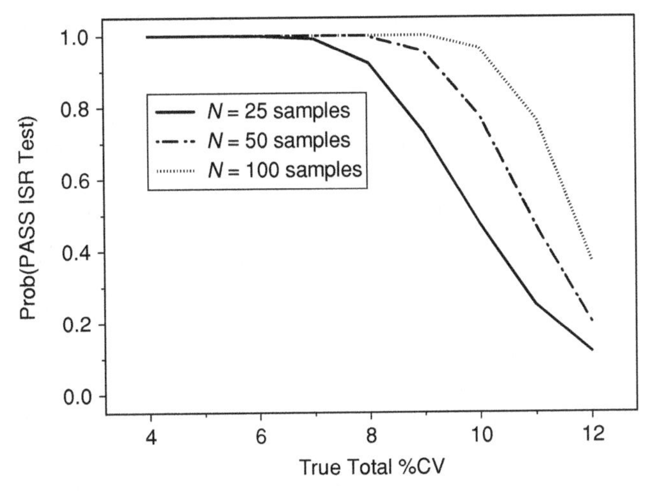

FIGURE 32.7 Probability of passing ISR tolerance interval test versus assay true total %CV, for $N = 25$, 50, or 100 incurred samples. True relative bias is 0%.

1. Construct a two-sided β-content tolerance interval (*L*, *U*) with desired confidence level γ (say, 90%).
2. Compare the interval (*L*, *U*) to the acceptance limits (*A*, *B*).
3. If (*L*, *U*) falls completely within (*A*, *B*), the ISR test is passed; otherwise, the ISR test is failed.

Let Y_i^O be the original concentration for the *i*th ($i = 1, 2, \ldots, N$) incurred sample and Y_i^R be the repeat concentration for the *i*th incurred sample. Then let

$$\Delta_i = \log(Y_i^R) - \log(Y_i^O),$$

$$\bar{\Delta} = \frac{1}{N}\sum_{i=1}^{N} \Delta_i,$$

$$\hat{\sigma}_\Delta^2 = \frac{1}{N-1}\sum_{i=1}^{N} (\Delta_i - \bar{\Delta})^2,$$

where Δ_i is the (repeat − original) difference in log-transformed concentration for the *i*th ($i = 1,2, \ldots, N$) incurred sample, $\bar{\Delta}$ is the mean of the differences in log-transformation concentration, and $\hat{\sigma}_\Delta^2$ is the variance of the differences in log-transformed concentration.

A two-sided β-content tolerance interval with confidence coefficient γ is then given by

$$\bar{\Delta} \pm Z_{(1+\beta)/2}\sqrt{1 + N^{-1}}\sqrt{(N-1)\hat{\sigma}_\Delta^2 / \chi^2_{N-1,1-\gamma}},$$

where $Z_{(1+\beta)/2}$ is the upper $(1 + \beta)/2$ quantile of the standard normal distribution and $\chi^2_{N-1,1-\gamma}$ is the lower γ quantile of the chi-square distribution with $N - 1$ degrees of freedom.

As noted earlier, implementation of a tolerance interval approach requires appropriate choices of content level (β), confidence level (γ), and acceptance limits (A, B). For assessment of ISR, 66.7% content and 90% confidence are logical choices and are consistent with typical acceptance criteria for in-study monitoring of QC samples (i.e., 4-6-15 rule). Noting that the coefficient of variation for a difference (repeat – original) of measurements is larger than that for an individual measurement by a factor of $\sqrt{2}$, limits of $\pm(15\sqrt{2})\% = 21.2\%$ are suggested. For log-transformed data, this corresponds to acceptance limits of $\pm\log(1.212)$. Thus, the proposed tolerance interval approach consists of constructing a two-sided $\beta = 66.7\%$ content, $\gamma = 90\%$ confidence tolerance interval on the differences of log-transformed measurements. If the resulting tolerance limits are completely within the $\pm\log(1.212)$ acceptance limits, the ISR test is passed; otherwise, the ISR test is failed.

Unlike the 4-6-20 rule, the tolerance interval approach proposed above strictly controls the risk of incorrectly accepting a truly nonreproducible method. Regardless of the sample size chosen, this risk is no greater than 5% (i.e., $\frac{(100-\gamma)\%}{2}$) for the tolerance interval approach.

While the risk of incorrectly accepting a truly nonreproducible method is strictly controlled, the risk of incorrectly rejecting a truly reproducible method with the tolerance interval approach must be controlled by appropriate choice of sample size. Figure 32.7 gives the probability of passing the ISR tolerance interval as a function of the true total %CV of the method, for several choices of sample size ($N = 25$, 50, or 100 incurred samples). The probabilities given in Figure 32.7 assume that the true difference (bias) between the repeat and original concentrations is 0%. For methods with small to moderate total %CV (say, $\leq 8\%$), the probability of passing the ISR tolerance interval test is high ($\geq 90\%$) with a sample size of N = 25 incurred samples. A sample size of N = 100 incurred samples will yield a high probability ($\geq 90\%$) of passing the ISR tolerance interval test for methods with total %CV as large as 10%. For methods with total %CV larger than 10%, larger sample sizes are required to maintain a high probability of passing the ISR test.

The tolerance interval approach to ISR testing is illustrated by application to data from an actual ISR experiment (Hoffman, 2009). Table 32.8 gives the repeat and original concentrations (ng/ml), as well as the simple percentage difference calculated by (repeat – original)/original × 100%.

Note that 34 of the 36 repeat concentrations (94.4%) are within ±20% of the original concentration. Thus, the ISR test passes the 4-6-20 rule acceptance criteria.

To apply the tolerance interval approach, the repeat and original concentrations are log-transformed. After log-transformation, the following statistics can be calculated: $\bar{\Delta} = -0.03350$ and $\hat{\sigma}^2_{\Delta} = 0.01034$. The appropriate standard normal and chi-square quantiles are as follows: $Z_{0.8335} = 0.96809$ and $\chi^2_{35,0.10} = 24.7966$. A two-sided $\beta = 66.7\%$ content, $\gamma = 90\%$ confidence tolerance interval is then given by

$$-0.03350 \pm 0.96809\sqrt{1 + 36^{-1}}\sqrt{(35)0.01034\ 24.7966}.$$

The resulting two-sided tolerance interval is given by $(-0.1521, 0.0851)$. The interval is entirely contained within the acceptance limits $\pm\log(1.212) = \pm 0.1923$. Thus, the ISR test passes the tolerance interval acceptance criteria.

REFERENCES

Burdick R, Graybill F. Confidence Intervals on Variance Components. New York: Marcel-Dekker; 1992.

Draper N, Smith H. Applied Regression Analysis, 3rd ed. New York: John Wiley & Sons, Inc.; 1998.

European Medicines Agency. Guideline on Bioanalytical Method Validation. February, 2012.

Fast D, Kelley M, Viswanathan CT, et al. Workshop report and follow-up—AAPS workshop on current topics in GLP bioanalysis: assay reproducibility for incurred samples— implications of crystal city recommendations. AAPS J 2009;11(2):238–241.

Hoffman D, Kringle R. Two-sided tolerance intervals for balanced and unbalanced random effects models. J Biopharm Stat 2005;15:283–293.

Hoffman D, Kringle R. A total error approach for the validation of quantitative analytical methods. Pharm Res 2007;24(6):1157–1164.

Hoffman D. Statistical considerations for assessment of bioanalytical incurred sample reproducibility. AAPS J 2009;11(3):570–580.

Hoffman D, Kringle R, Singer J, McDougall S. Statistical methods for assessing long-term analyte stability in biological matrices. J Chrom B 2009;877:2262–2269.

Kringle R. An assessment of the 4-6-20 rule for acceptance of analytical runs in bioavailability, bioequivalence, and pharmacokinetic studies. Pharm Res 1994;11:556–560.

Kringle R, Hoffman D, Newton J, Burton R. Statistical methods for assessing the stability of compounds in whole blood for clinical bioanalysis. Drug Inf J 2001;35:1261–1270.

Kringle R, Khan-Malek R. A Statistical Assessment of the Recommendations from a Conference on Analytical Methods Validation in Bioavailability, Bioequivalence, and Pharmacokinetic Studies. Proceedings of the Biopharmaceutical Section of the American Statistical Association, 1994; Alexandria, VA, USA.

Nowatzke W, Woolf E. Best practices during bioanalytical method validation for the characterization of assay reagents and evaluation of analyte stability in assay standards, quality controls, and study samples. AAPS J 2007;9(2):117–122.

Timm U, Wall M, Dell D. A new approach for dealing with the stability of drugs in biological fluids. J Pharm Sci 1985;74(9):972–977.

US Food and Drug Administration. Guidance for Industry: Bioanalytical Method Validation. May, 2001.

Viswanathan CT, Bansal S, Booth B, et al. Workshop/conference report—quantitative bioanalytical methods validation and implementation: best practices for chromatographic and ligand binding assays. AAPS J 2007;9(1): 30–42.

33

SIMULTANEOUS LC-MS QUANTITATION AND METABOLITE IDENTIFICATION IN DRUG METABOLISM AND PHARMACOKINETICS

PATRICK J. RUDEWICZ

33.1 INTRODUCTION

Pharmaceutical scientists routinely screen compounds *in vitro* for metabolic stability using incubations with liver microsomes that contain cytochrome P450s and other drug metabolizing enzymes. Drug candidates are typically classified in bins of high, moderate or low stability that indicates their susceptibility to metabolism. One strategy that is often employed is to select compounds with low or poor metabolic stability and identify soft spots of metabolism through separate *in vitro* metabolite identification (ID) experiments using LC-MS/MS. The goal is to rationally alter chemical structures and ultimately generate a lead candidate that possesses adequate *in vitro* metabolic stability while still maintaining other desirable properties like enzyme potency and selectivity. Typically, compounds that have the best *in vitro* properties, including microsomal stability, are further screened in rodent pharmacokinetic (PK) studies in which parameters such as half-life, clearance, volume of distribution, and bioavailability are determined.

For both *in vitro* metabolic stability and *in vivo* PK studies, the mass spectrometer that is used most often for quantitative determination is the triple quadrupole in the selected reaction monitoring (SRM) mode with either electrospray or atmospheric pressure chemical ionization (Bruins et al., 1987; Covey et al., 1986). In the SRM mode, a precursor ion is selected in quadrupole 1, collisionally dissociated in quadrupole 2 and a particular product ion is selected with quadrupole 3 for detection. The SRM mode is very sensitive and selective and the triple quadrupole mass spectrometer is still the workhorse for most DMPK laboratories. Metabolite ID is usually done separately using longer LC gradients and high-resolution mass spectrometers. For *in vitro* work, a separate incubation is often required at a higher substrate concentration. This decoupling of experiments for quantitation and metabolite ID creates a delay in the drug discovery process.

Recent developments in mass spectrometry are drastically changing the way bioanalysis is occurring for drug candidate optimization. Quantitation and qualitative ID (quant-qual) is beginning to merge into one experiment. The expectations of very short data turnaround times of drug discovery scientists are in part driving this change. Bioanalytical scientists are expected to increase both the speed and the efficiency of their experiments providing feedback on not only the rate of metabolism but also the sites of metabolic lability within a few days. The goals of this chapter are to describe the developments in mass spectrometry that have led to this paradigm shift in bioanalysis and to give illustrative examples of how this technology may be used for drug discovery support.

33.2 UNIT RESOLUTION MASS SPECTROMETERS IN DRUG METABOLISM AND PHARMACOKINETICS

Starting in the mid-1980s, the first online LC-MS/MS experiments for ID were done using triple quadrupole mass spectrometers (Rudewicz and Straub, 1986, 1987). With triple quadrupoles, neutral loss (NL) scans are utilized to search

Handbook of LC-MS Bioanalysis: Best Practices, Experimental Protocols, and Regulations, First Edition. Edited by Wenkui Li, Jie Zhang, and Francis L.S. Tse.

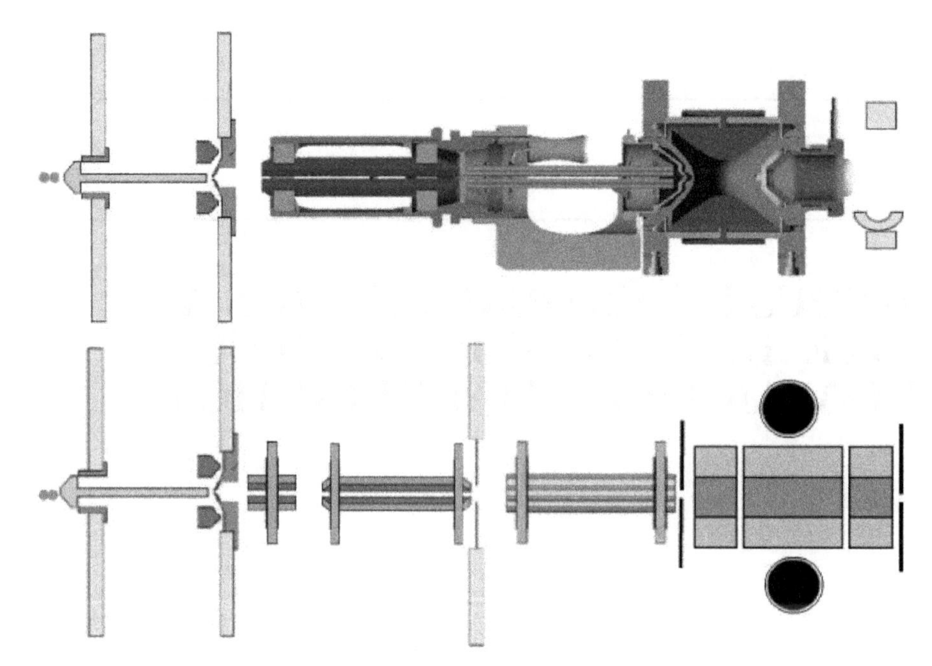

FIGURE 33.1 Three-dimensional (upper panel) and two-dimensional (lower panel) quadrupole ion trap mass spectrometers (figure is provided courtesy of Thermo Fisher Scientific).

biological matrices for drug metabolites that contain substructures such as glucuronic acid (NL 176) or aryl sulfate esters (NL 80). Precursor ion scans are also employed to identify metabolites that contain common substructural features. Although triple quadrupoles are still used today for ID, the technology has certain limitations. A triple quadrupole is a scanning mass spectrometer: as the RF/DC ratio changes, ions of a particular *m/z* reach the detector in a serial manner. Consequently, the duty cycle, or the fraction of the scan time that ions of a particular *m/z* actually reach the detector, is low. Depending upon the scan range, the duty cycle of a triple quadrupole can be less than a few percent, limiting their sensitivity and usefulness in certain ID experiments.

During the past 20 years, there have been several innovations in mass spectrometry that have advanced the field of ID significantly. One of these was the introduction of the three-dimensional (3D) quadrupole ion trap mass analyzer (March, 1997). A 3D ion trap consists of two end cap electrodes and a ring electrode and they serve as an ion storing as well as scanning mass spectrometer (Figure 33.1, upper panel). The ion storage capability results in more sensitive full scan and product ion modes relative to triple quadrupoles. 3D ion traps, however, suffer from certain disadvantages including space charging within the trap. This can compromise quantitative performance when compared to triple quadrupole mass spectrometers. In addition, in the product ion mode, only ions with a mass higher than one-third of the precursor ion are stable, resulting in the loss of structurally informative product ions in the low mass range of the spectrum. The two-dimensional (2D) or linear quadrupole ion trap consists of 4 round or hyperbolic rods with trapping lenses at each end (Figure 33.1 lower panel). 2D ion traps generally have a larger internal volume than a 3D trap meaning more ions can be stored and they suffer less from limitations due to space charging effects. Both 2D and 3D ion traps provide tandem mass spectrometry in time (MS^n) resulting in additional structural detail.

Linear ion traps are also capable of axial ion ejection along the axis of the linear quadrupole rods. In 2002, Hager and coworkers took advantage of this feature and designed the hybrid linear ion trap based on a QqQ_{LIT} rail design where the third quadrupole region is a 2D linear ion trap (Figure 33.2) (Hager, 2002). The hybrid quadrupole linear ion trap has been used successfully for both quantitative analysis and ID. This instrument is capable of all the scanning functions of a triple quadrupole mass spectrometer including NL and precursor ion scanning. It also has fast acquisition rates that are similar to a triple quadrupole mass spectrometer. However, since the third quadrupole region is actually a linear ion trap with very high trapping efficiencies, it has excellent sensitivity in the full scan product ion mode.

33.3 HIGH-RESOLUTION MASS SPECTROMETRY IN DRUG METABOLISM AND PHARMACOKINETICS

High-resolution mass spectrometry was first used for ID using double focusing magnetic and electric sector instruments. In these experiments, metabolites needed to be isolated using chromatographic techniques and then purified

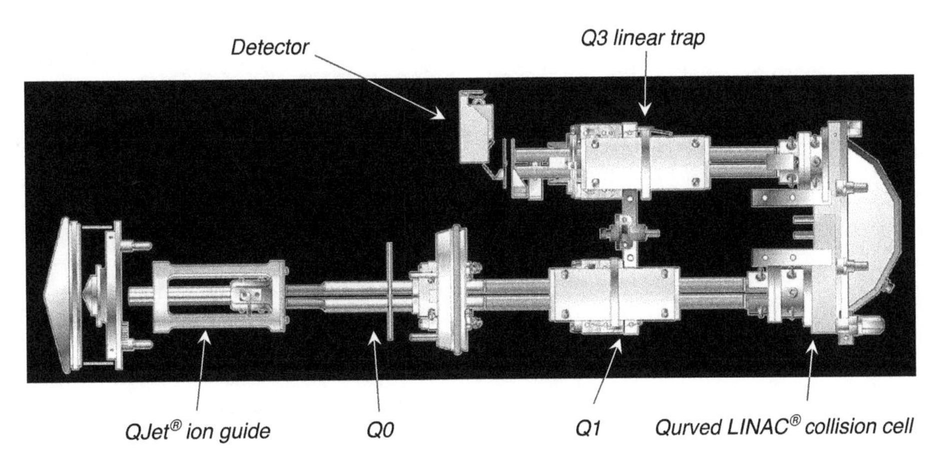

FIGURE 33.2 Quadrupole linear ion trap mass spectrometer (figure is provided courtesy of AB Sciex Pte. Ltd).

before being introduced into the mass spectrometer using gas chromatography or a direct insertion probe. Drug conjugates would often need to be hydrolyzed either chemically or enzymatically followed by high-resolution MS analysis of the primary metabolite. Derivatization was often required. With the introduction of fast atom bombardment, thermally labile drug conjugates could be analyzed without the need for hydrolysis or derivatization. Sector instruments were not easily implemented into most DMPK laboratories due to their relatively complex operational and maintenance requirements. Furthermore, because the ion source potential of a sector instrument is in the kV range, these instruments were not easily interfaced with liquid chromatography systems making online analysis problematic.

With the introduction of the hybrid quadrupole time-of-flight (QqTOF) mass spectrometer, high-resolution LC-MS/MS experiments became more accessible to DMPK scientists (Chernushevich et al., 2001). Commercially available QqTOF instruments have the quadrupole rail (Qq) orthogonal to the time-of-flight (TOF) analyzer (Figure 33.3)

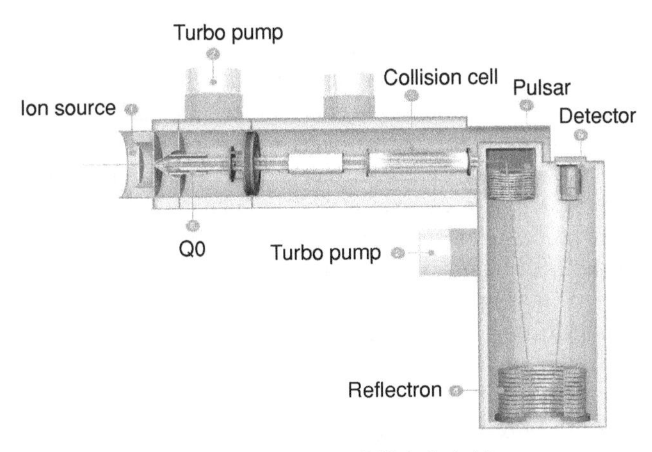

FIGURE 33.3 Quadrupole time-of-flight hybrid mass spectrometer (figure is provided courtesy of AB Sciex Pte. Ltd).

and are generally capable of resolving powers of 20,000 to 30,000 full width at half maximum (FWHM) in the reflectron mode. Initial applications of the QqTOF in DMPK laboratories focused on qualitative ID experiments. Although duty cycle limitations existed, these instruments possessed very fast data acquisition rates and were capable of mass accuracy lower than 5 ppm with the introduction of an external calibrant.

Early quantitative applications of QqTOF instruments were limited. One group reported a quantitative application using a QqTOF in the product ion mode (Yang et al., 2001). The authors performed a one-day validation for loratadine and descarboethoxyloratadine in dog plasma. The dynamic range was 1–1000 ng/ml. The sensitivity was adequate with 5.56 pg detected for each analyte on the high-pressure liquid chromatography column at the lower limit of quantitation (LLOQ). Nevertheless, at that time, QqTOF instruments suffered from limited dynamic range due to slow detector refresh rates of about 1 ns. Furthermore, LLOQs were limited due to an inefficient duty cycle.

Another high-resolution hybrid mass spectrometer that has made large inroads into DMPK laboratories is the LTQ-Orbitrap (Figure 33.4) (Hu et al., 2005; Peterman et al., 2006). This instrument contains a linear ion trap that possesses all of the capabilities of a conventional 2D trap. The ions that are stored within the trap as part of a full scan or product ion scan may either be detected using a channel electron multiplier at unit resolution or passed to a C-Trap where they are focused and then transferred to an Orbitrap mass analyzer. The Orbitrap mass analyzer consists of an inner spindle electrode and an outer barrel electrode. Using a combination of DC voltages, ions rotate with characteristic frequencies within the Orbitrap. The radial frequency is directly proportional to mass to charge ratio. Ions are detected at high resolution using a Fourier transformation.

The latest version of the Orbitrap that is currently being used for simultaneous quant-qual applications is the hybrid

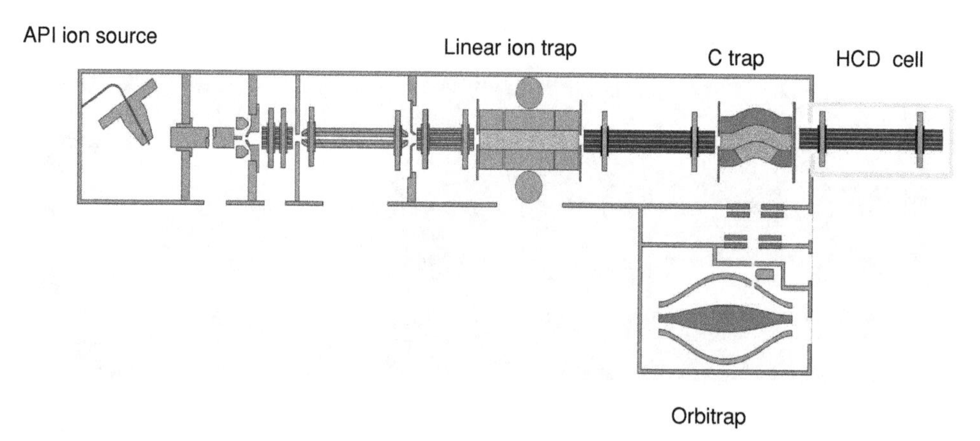

FIGURE 33.4 LTQ Orbitrap mass spectrometer (figure is provided courtesy of Thermo Fisher Scientific).

quadrupole Orbitrap (Figure 33.5). This instrument allows for quadrupole selection of precursor ions for MS/MS functionality in the higher energy collisionally activated dissociation (HCD) collision cell. The Quadrupole-Orbitrap (Q-Exactive) has a maximum resolution of 140,000; however, the resolution is dependent upon the data acquisition rate. At an acquisition rate of 4 Hz, the maximum resolution is about 70,000, whereas at 12 Hz, the resolution is approximately 17,500. The mass accuracy is excellent: 1–3 ppm with an external calibration and it typically holds for several days without using an internal lock mass.

33.4 MASS DEFECT FILTER IN LC-MS

The concept of mass defect filter is based upon the fact that, apart from carbon that by convention has an exact mass of 12, all other elements have non-integer exact masses. Most common biotransformations result in a mass defect (non-integer difference from nominal) within 50 mDa from the parent drug mass (Zhang et al., 2003, 2007). As an example, oxygenation results in an exact mass increase of 15.9949 amu and the mass defect is −5 mDa. Similarly glucuronidation results in an exact mass increase of 176.0321 or a mass defect of + 32 mDa. The mass defect of sulfation is −43 mDa, demethylation is −23 mDa, and dehydration is −16 mDA. A notable common biotransformation pathway that falls outside the 50-mDa window is glutathione conjugation that results in a mass defect of + 68 mDa. However, since this can be predicted, one may set a filter around this mass defect. Mass defect filter software simplifies high resolution full scan ion chromatograms by displaying ions within a particular mass defect range and effectively separating drug related metabolites from matrix components.

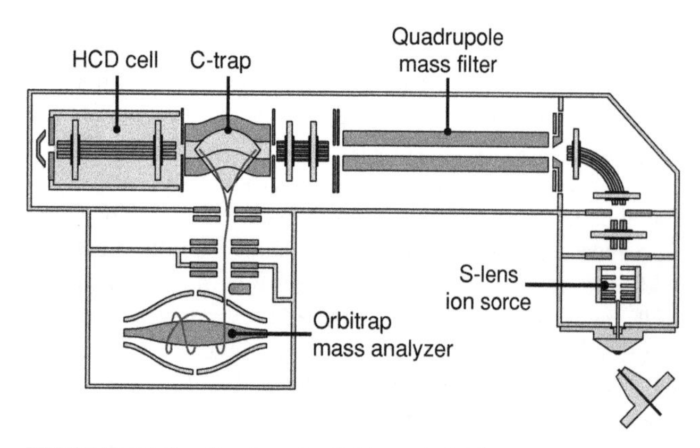

FIGURE 33.5 Quadrupole Orbitrap hybrid mass spectrometer (figure is provided courtesy of Thermo Fisher Scientific).

33.5 SIMULTANEOUS LC-MS QUANTITATION/ METABOLITE IDENTIFICATION

As ion trap and high-resolution mass spectrometry became assimilated into DMPK laboratories, several groups began to experiment with integrating quan-qual workflows for both *in vitro* and *in vivo* study support. In 2000, the use of an LC TOF instrument for the quantitation of drug candidates in rat plasma was described (Zhang et al., 2000). The compounds were dosed as a cassette at 5–10 mg/kg. Plasma samples were processed using liquid–liquid extraction. LLOQs were in the 5–25 ng/ml range with an upper limit of quantitation of 1000 ng/ml. Preliminary metabolite information was obtained using accurate mass full scan spectra with mass errors less than 10 ppm. Also in 2000, Cai et al. reported using LC-MS/MS with a 3D ion trap for the quantitation of 19 α-1a receptor antagonists and determination of their major metabolites formed *in vitro* (Cai et al., 2000). The test compounds were incubated in dog liver microsomes at a concentration of 10 μM. Samples were pooled post-incubation for cassette analysis. The parent molecule was detected in the full scan mode using nominal mass. Structural characterization of metabolites was done using MS^n tandem mass spectrometry. Cai et al. also reported the same approach for the quantitation of drug candidates with determination of their major metabolites from pooled mouse plasma samples (Cai et al., 2001). The *in vivo* metabolism data was compared with *in vitro* mouse S9 metabolism to better understand drug clearance mechanisms. Kantharaj et al. described using a 3D ion trap for the quantitation of drug candidates in the hit-to-lead stage of drug discovery (Kantharaj et al., 2003). Model compounds were incubated at a concentration of 5 μM with human liver microsomes. Quantitation of the parent compound and structural characterization of metabolites was achieved using data-dependent MS^n scan functions within a single analytical run.

Following this early work, DMPK scientists began to utilize hybrid mass spectrometers for simultaneously quantifying parent drug and screening for metabolites. Hopfgartner et al. first described the advantages of a quadrupole-linear ion trap (QqQ_{LIT}) for rapid characterization of drug metabolites in biological samples (Hopfgartner et al., 2003). They described the ability to use all of the scan functions of a triple quadrupole mass spectrometer and the added capability of obtaining high sensitivity product ion spectra with the linear ion trap. Information-dependent acquisition (IDA) experiments were used to obtain MS^2 and MS^3 spectra in the ion trap for rapid structural characterization of metabolites. In 2005, Li and coworkers published a study describing a quadrupole-linear ion trap for the simultaneous quantitation of the parent drug and ID using plasma samples from a monkey PK study (Li et al., 2005). The authors used a predefined biotransformation table with multiple reaction monitoring (MRM)-triggered IDA product ion scanning to rapidly identify metabolites in one analytical run. Excellent agreement was demonstrated between PK profiles of parent drug generated by the MRM and the MRM-IDA methods. Li et al. also detailed using LC-MS on a quadrupole-linear ion trap for the simultaneous quantitation and metabolite characterization for nefazodone using *in vitro* human liver microsome incubations at a substrate concentration of 10 μM (Li et al., 2007). Twenty-two metabolites were identified in a single LC-MS/MS run.

More recently several publications have appeared that describe the use of Orbitrap mass analyzers for quant-qual data acquisition. Zhang et al. used a hybrid LTQ-Orbitrap at a mass resolution of 15,000 (FWHM) for the quantitative analysis of drugs in plasma samples (Zhang et al., 2009). Only ions within a predefined mass range were trapped. This waveform scan function was found to improve sensitivity. In addition, automatic gain control was performed by using a prescan to determine the ion injection time for the analytical scan to eliminate space charge effects within the Orbitrap mass analyzer. The quantitative data obtained using the LTQ-Orbitrap in the full scan mode were shown to be similar to data generated on an API 4000 in the MRM mode with LLOQs between 1 and 2 ng/ml. A glucuronide metabolite for one drug candidate was also identified in rat plasma using accurate mass measurement of the protonated molecular ion for confirmation.

Bateman and coworkers described using high-resolution LC-MS with a single-stage Orbitrap mass analyzer (Exactive-MS) for collecting quant-qual data for *in vitro* stability and *in vivo* PK studies (Bateman et al., 2009). A fast generic UPLC gradient was employed using a 2.1 × 100 mm C18 column with 1.9-micron particles. With a 10 Hz acquisition rate, 30–40 data points were acquired across 3–4 s wide chromatographic peaks while maintaining mass accuracies of less than 1 ppm at a resolution of 7400. Resolution could be increased to 67,100 by lowering the acquisition rate to 1 Hz. At this acquisition rate the mass accuracy was still less than 3 ppm; however, fewer scans were acquired across each chromatographic peak. Higher energy collision decomposition spectra were obtained in a multiple collision cell without precursor selection, thus requiring chromatographic resolution and peak deconvolution to link collision spectra with precursor ions. Henry et al. compared a single-stage Orbitrap mass analyzer and a triple quadrupole mass spectrometer for the quantitation of drugs in plasma (Henry et al., 2012). The Orbitrap was operated at a resolution of 50,000 (FWHM) at *m/z* 200 with a 5 ppm mass extraction window. The high-resolution quantitative results including sensitivity, linearity, precision, and accuracy were reportedly similar to those obtained in the SRM mode on a triple-stage quadrupole. Tuning was simplified with the single-stage Orbitrap since compound specific SRM tuning was not required.

33.6 LC-MS WORKFLOW: QqQ_{LIT}

With the recent advancements in mass spectrometry technology, we are witnessing a paradigm shift in drug discovery workflows. Quantitation of the parent drug molecule may now be done simultaneously with ID and structural characterization within the same analytical run. Furthermore, because of the excellent sensitivity and fast data acquisition rates of these mass spectrometers, they may be interfaced to UPLC systems allowing for short (<3 min) run times. Two types of mass spectrometers are generally used for simultaneous quantitative-qualitative analysis: quadrupole linear ion trap and high-resolution instruments (QqTOF or Orbitrap). Each has their unique advantages and limitations.

The most attractive feature of the quadrupole linear ion trap instrument is that it functions as a triple quadrupole with the added features of a linear ion trap including enhanced sensitivity in the product ion mode. This increased sensitivity allows for metabolite characterization at low levels in

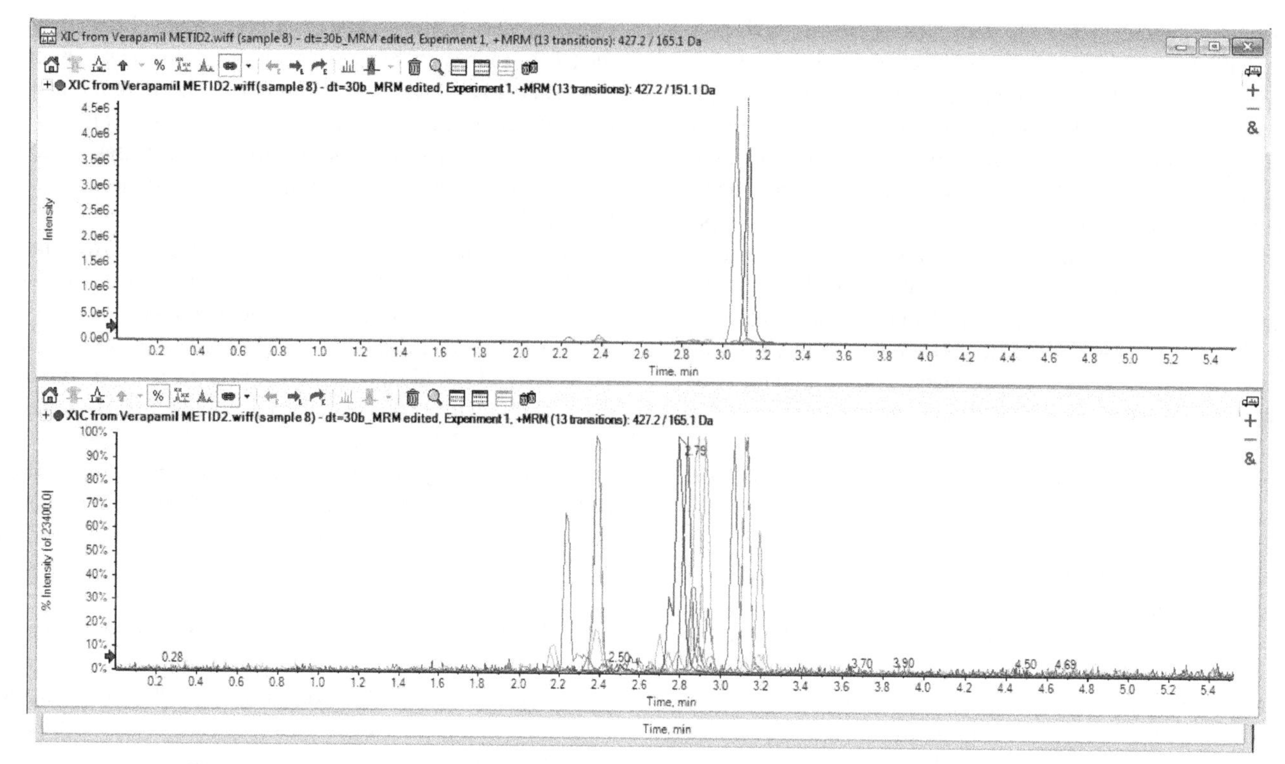

FIGURE 33.6 LC-MS/MS MRM chromatograms of verapamil and major desmethyl metabolite (upper panel) and lower abundance verapamil metabolites (lower panel).

plasma and other biological matrices such as bile, urine, and feces. In the typical quadrupole linear trap workflow, what is known as a predictive multiple reaction monitoring (pMRM) table is constructed using biotransformation sets included in the instrument software package that are derived from a list of phase I and phase II metabolic pathways. During the analysis, the instrument is set to scan in the MRM mode for the parent drug and internal standard as well as for metabolites predicted in the biotransformation table. Using an IDA algorithm, if a signal is observed for a predicted metabolite above a predefined number of ion counts, an enhanced product ion scan is obtained using the linear ion trap. The quadrupole linear trap can scan fast with 2 ms dwell times and usually typical biotransformation pathways (oxidation, reduction, hydrolysis, glucuronidation, etc.) may be covered in one analytical run. In practice, typically longer dwell times of 30–50 ms are used for the parent compound and shorter dwell times of 5–10 ms are used for pMRMs for unknown metabolites. Robust and sensitive quantitation typical of triple quadrupoles is achievable and results are similar in the MRM and pMRM modes (Rudewicz et al., 2009).

These points are illustrated for the analysis of a sample from a microsomal incubation of verapamil. For this experiment, verapamil was incubated at a concentration of 1 μM in the presence of 0.5 mg/ml rat liver microsome protein, 1 mM NADPH, 1 mM UDPGA, 3 mM $MgCl_2$, and 25 μg alamethicin/mg microsomal protein. Incubation time points were 0, 5, 15, and 30, and 60 min. Using a pMRM biotransformation table, the ion chromatogram from an analysis of a human liver microsome incubation for verapamil is shown in Figure 33.6. The upper panel contains the ion chromatogram for verapimil and a major desmethyl metabolite with a slightly shorter retention time. The lower panel shows less abundant metabolites including oxygenation, demethylation, and glucuronidation. The two chromatographic peaks at retention times of 2.2 and 2.4 min correspond to demethylated glucuronide conjugates. Their enhanced product ion spectra that contain the characteristic NL of 176 amu are shown in Figure 33.7. Quantitation for verapamil was achieved as well as simultaneous detection and structural characterization of 13 metabolites.

33.7 LC-MS WORKFLOW: HIGH-RESOLUTION MASS SPECTROMETRY

The use of high-resolution mass spectrometry for simultaneous quantitation and ID has the obvious advantages inherent in accurate mass measurements. For quantitative purposes one can separate isobaric ions with the same nominal mass but different elemental compositions. As an example, one may use high resolution for mass separation of a drug and

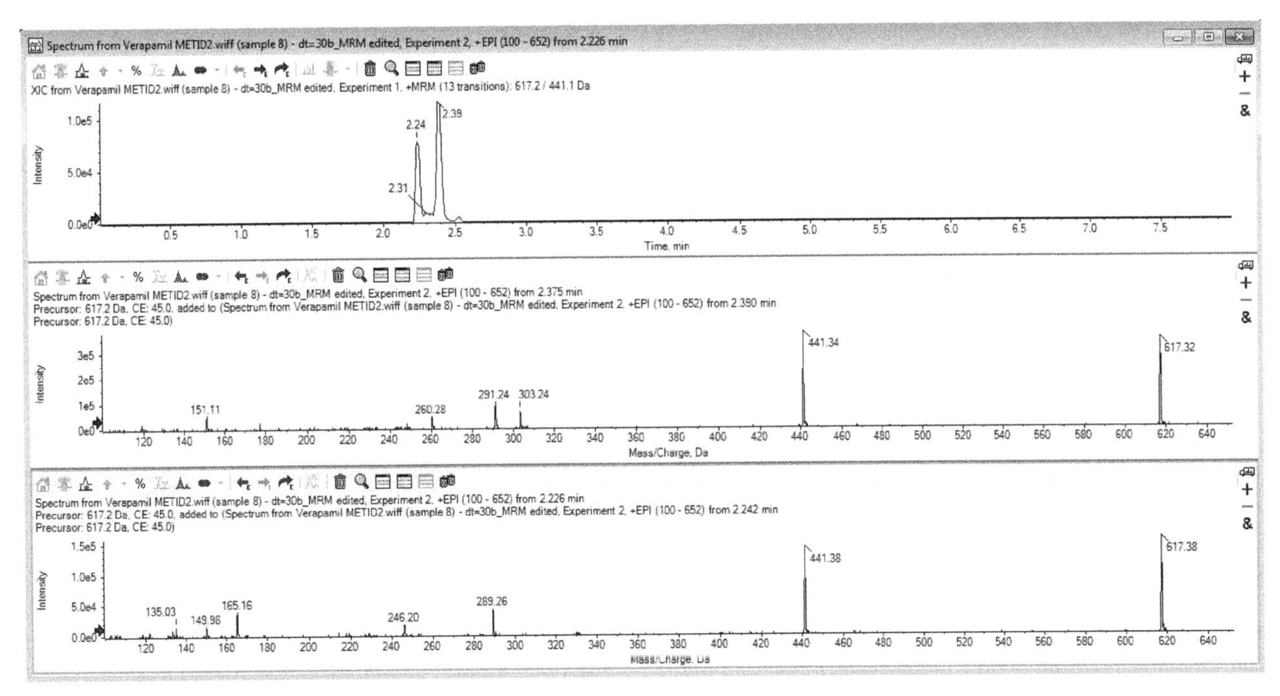

FIGURE 33.7 LC-MS/MS MRM chromatograms of major desmethyl glucuronides of verapamil and their enhanced product ion spectra.

metabolite that have the same nominal mass but different exact masses. The separation of an analyte from coeluting isobaric matrix components can also be achieved. For ID, accurate mass measurements provide elemental composition of product ions and consequently more certainty in structural assignments. Another advantage of quantitation of the parent drug in the full scan high-resolution mode is that full scan spectra of metabolites are also acquired simultaneously. Metabolites can be detected using the elemental composition from full scan exact mass measurements. Structural characterization of the metabolites may be achieved by obtaining accurate mass product ion spectra within the same run or in a subsequent analysis. Additionally, with full scan TOF quantitation, compound specific SRM tuning is eliminated.

The QqTOF that we use presently in our laboratory for quant-qual, the AB SCIEX TripleTOF™ 5600 has several improvements in technology over previous generations of QqTOFs. Ions are pulsed into the TOF at a higher energy (15 kV) and at a higher frequency (30 kHz). In addition the four channel time to digital converter has an increased frequency of 40 GHz so that the refresh rate is faster and a data point can be acquired every 25 ps (25 ps bin size). Advancements have also been made in the reflectron optics design. Taken together, these improvements result in very fast data acquisition times of 100 Hz and a higher resolving power of approximately 30,000 (FWHM). For quantitation, calibration curves generally have linearity of 3–4 orders of magnitude. Excellent mass accuracy is maintained by means of a 2-min calibrant infusion between runs using a second probe in the ion source. In addition, the data acquisition software has internal algorithms for adjusting the calibration from scan to scan.

An example of using high resolution on the latest generation of quadrupole-TOF instrumentation for quant-qual drug discovery support is illustrated in the following example. In this experiment, six drugs were incubated in a 96-well format at 1 μM initial substrate concentration using rat liver microsomes, 1 mg/ml protein. For chromatographic separation, a Synergi Polar-RP 2.5 μm, 2 × 50 mm, column was used with a generic gradient of water/acetonitrile/0.1% formic acid at a flow rate of 0.8 ml/min. The total run time was 3.7 min. The mass spectrometer was an AB SCIEX TripleTOF™ 5600 system. One generic IDA method was used for all compounds. The survey scan was a TOF full scan from 100 to 1000 Da with a 100 ms accumulation time. For the IDA MS/MS scans, two dependent scans were acquired with dwell times of 100 ms each and a collision energy of 35 V. The IDA product ion spectra were triggered in real time ("on-the-fly") using a mass defect filter.

The percent remaining of the parent drug as a function of incubation time is plotted in Figure 33.8 for the six drugs. Quantitation for each drug was done in the full scan TOF mode. An example of a full scan high-resolution spectrum for one of the drugs, haloperidol is shown in Figure 33.9. These data were acquired using a resolution of 31,042 (FWHM). Excellent mass accuracy of 0.6 ppm was obtained for the protonated molecular ion of *m/z* 376.1476. An ion chromatogram for one of the compounds, midazolam, is shown

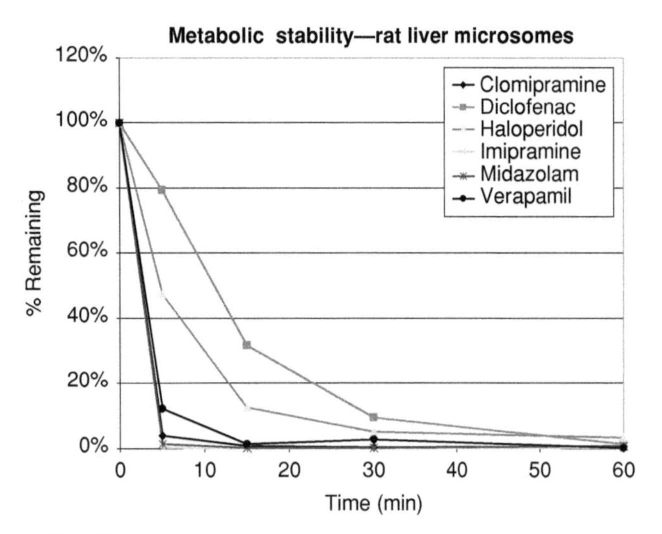

FIGURE 33.8 Metabolic stability as a function of rat liver microsome incubation times for six drugs.

in Figure 33.10. The signal to noise is more than adequate at the 30-min timepoint where there is only 0.1% of the parent remaining. Even with a full survey scan and two IDA-triggered MS/MS scans, 12–15 scans were acquired across each chromatographic peak.

Real-time, on-the-fly, IDA-triggered product ion spectra acquired by TOF obviate the need for sample reinjection for ID and structural elucidation. Figure 33.11 is a product ion spectrum of a mono-hydroxylated metabolite of midazolam that was triggered during the course of the chromatographic run by the mass defect filter. Resolution of greater than 30,000 and mass accuracy of less than 2 ppm was achieved. The MDF-based data acquisition provided coverage of the major oxidative phase I metabolites for all six compounds. As an example, Figure 33.12 contains the ion chromatograms for the phase I metabolites of imipramine and include, mono and dihydroxy imipramine, desmethyl imipramine, monohydroxy desmethyl imipramine, and imipramine *N*-oxide.

33.8 CONCLUSIONS

Recent innovations in mass spectrometry instrumentation and software are improving DMPK workflows and accelerating the pace of drug discovery. Of the two approaches described in this chapter, high-resolution mass spectrometry will have the biggest impact and is leading to a paradigm shift in bioanalysis (Korfmacher, 2011; Ramanathan and Korfmacher, 2012). Scientists are currently implementing both

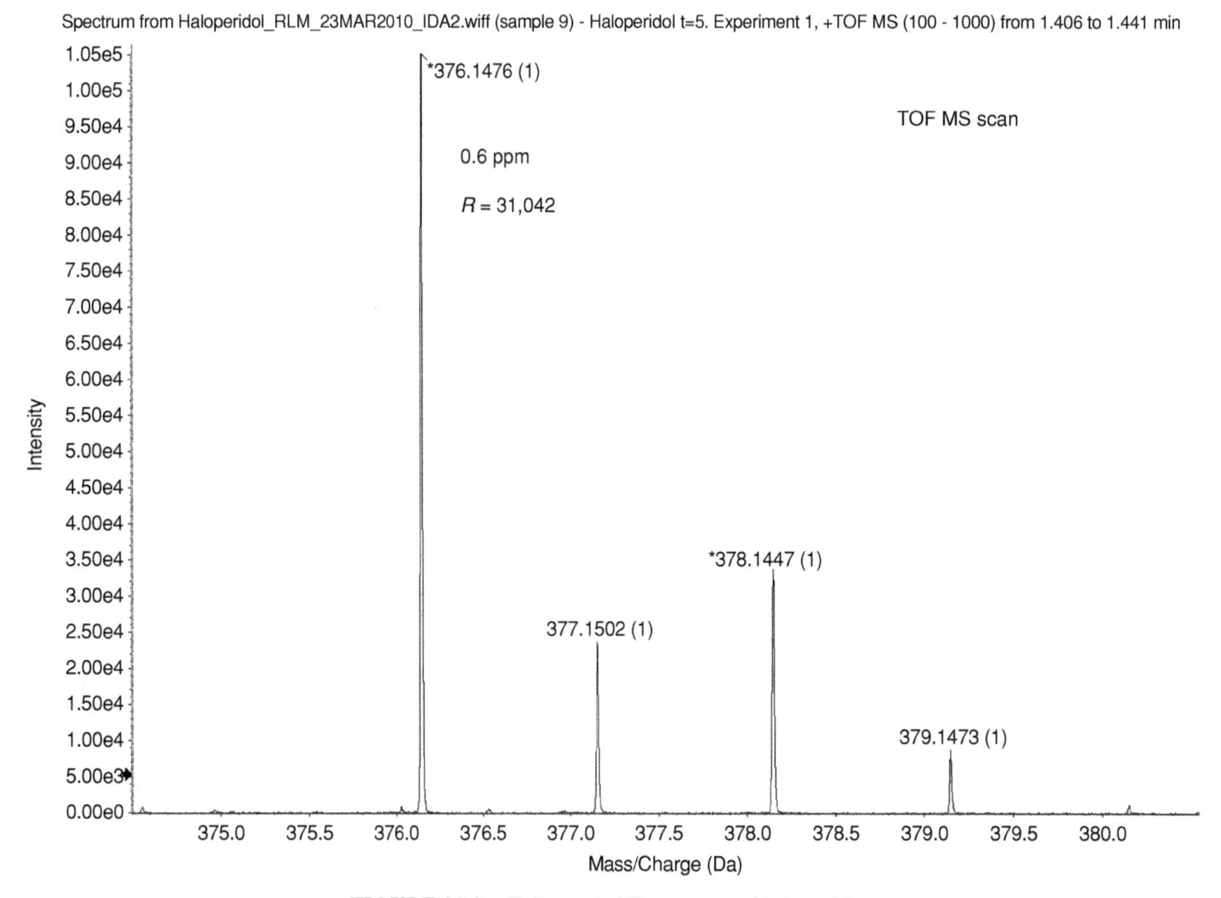

FIGURE 33.9 Full scan TOF spectrum of haloperidol.

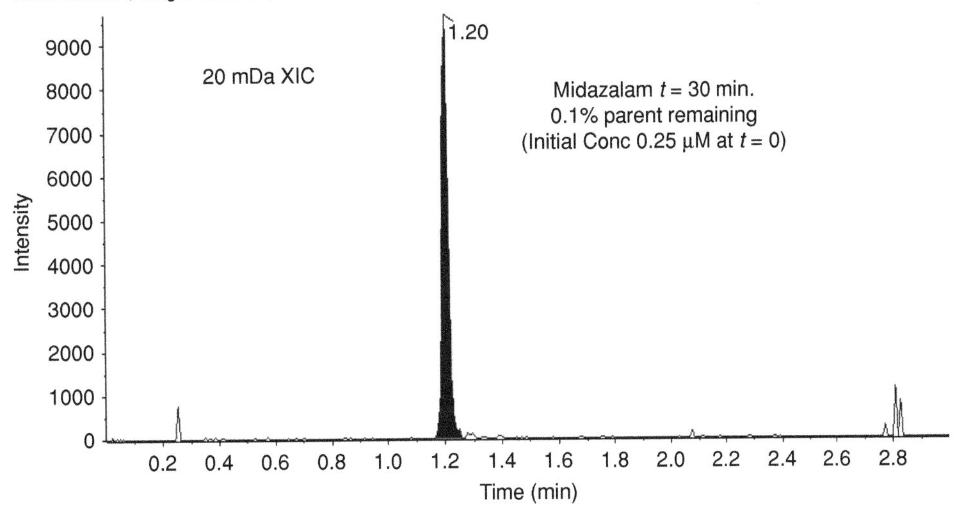

FIGURE 33.10 Ion chromatogram for midazolam in the TOF MS mode.

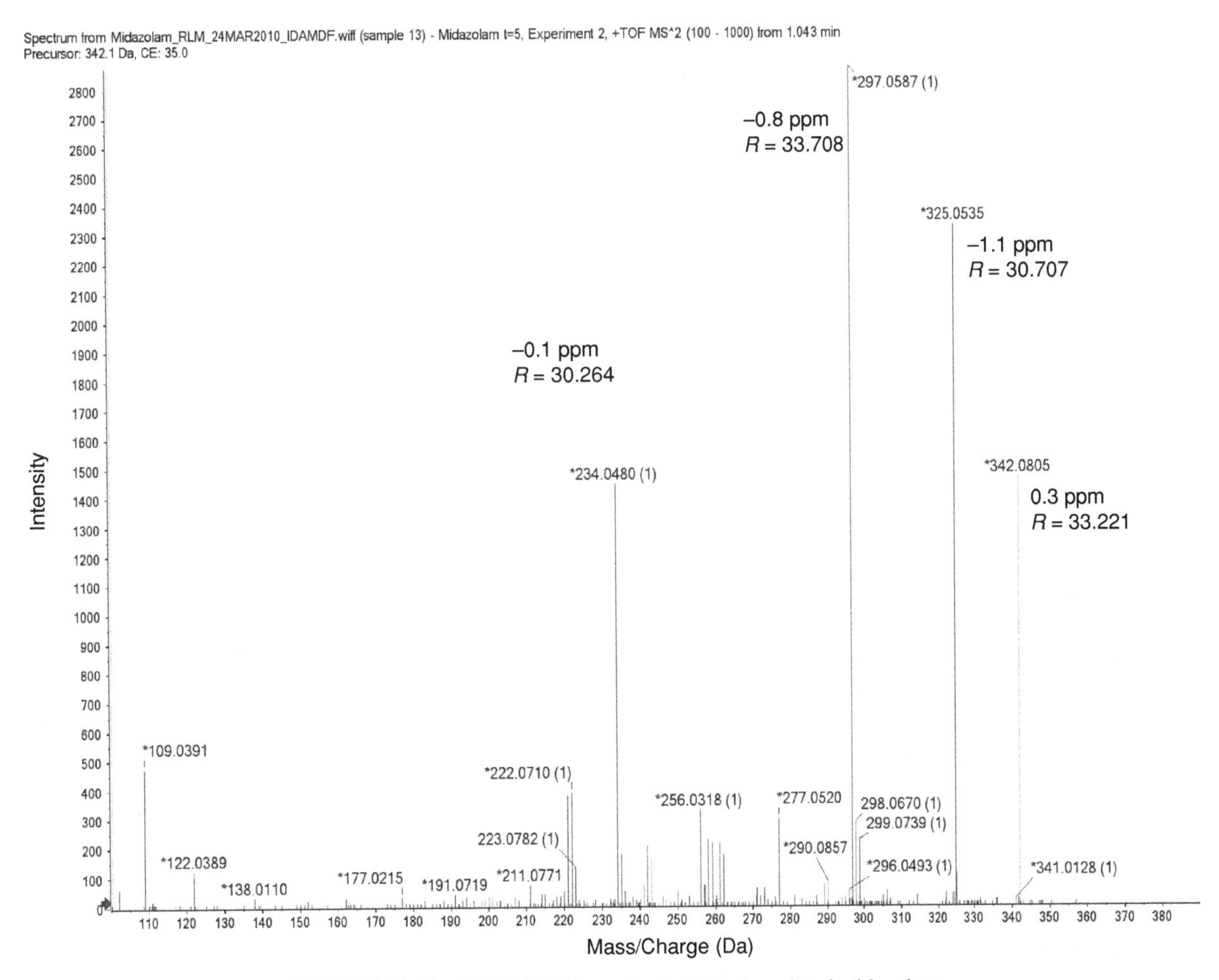

FIGURE 33.11 TOF MS/MS spectrum for hydroxylated midazolam.

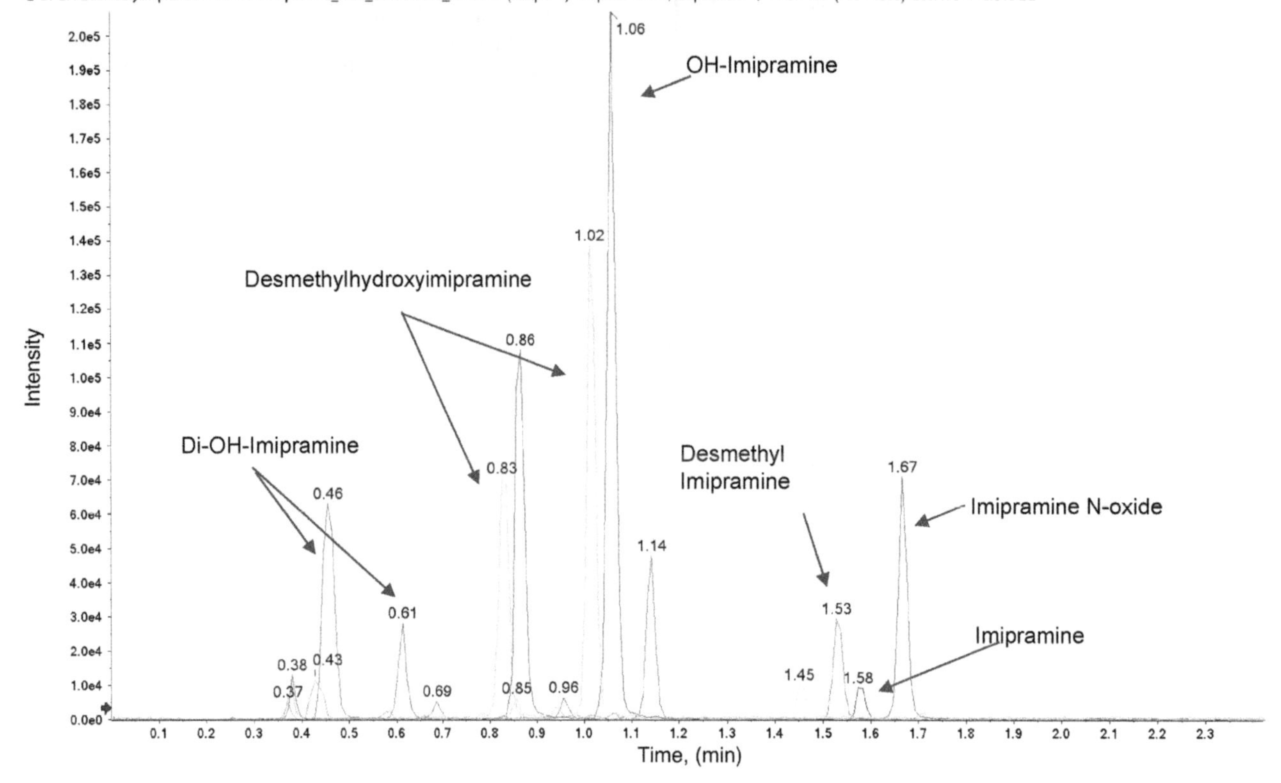

FIGURE 33.12 Ion chromatograms for metabolites of imipramine in the TOF MS mode.

Q-TOF and Orbitrap mass spectrometers for *in vitro* stability experiments as well as *in vivo* PK studies. Certain challenges exist. For high throughput *in vitro* ADME laboratories in which compounds are run in large batch sizes, speed, and efficiency of analysis must be maintained, stretching the limits of data acquisition times. Simultaneous quantitation-ID analysis for every compound is not necessarily warranted since usually only those compounds with high intrinsic clearance are of most interest to the medicinal chemist. For each quantitation – qualitative analysis, as illustrated for imipramine, a large number of product ion spectra are generated that will require interpretation that can be time and labor intensive. Mass spectrometer software can assist with interpretation of accurate mass product ion spectra. A logical extension being explored by several investigators combines predictive site-of-metabolism software, Mass Metasite with accurate mass product ion spectra for automated structural identification in the batch mode (Trunzer et al., 2009). The uncertainty of metabolite levels exists since authentic standards are generally not available. Assumptions need to be made, including equal ionization efficiencies and levels of ion suppression, to compare relative levels of metabolites in various matrices. Nevertheless, early information on the *in vitro* and *in vivo* major biotransformation pathways and the establishment of structure-metabolism relationships are invaluable for the medicinal chemist; consequently, simultaneous quantitation and ID will undoubtedly become more prevalent in DMPK laboratories, increasing the speed of the drug discovery process.

ACKNOWLEDGMENTS

The author acknowledge the contributions of Loren Olson and Hesham Ghobarah from AB SCIEX for data generation as well as David Monk from AB SCIEX and Yan Chen from Thermo Fisher Scientific for helpful discussions.

REFERENCES

Bateman KP, Kellmann M, Muenster H, Papp R, Taylor L. Quantitative-qualitative data acquisition using a Benchtop Orbitrap Mass Spectrometer. J Am Soc Mass Spectrom 2009;20:1441–1450.

Bruins AP, Covey TR, Henion JD. Ion spray interface for combined liquid chromatography/atmospheric pressure ionization mass spectrometry. Anal Chem 1987;59:2642–2646.

Cai Z, Han C, Harrelson S, Fung E, Sinhababu AK. High-throughput analysis in drug-discovery: application of liquid chromatography/ion-trap mass spectrometry for simultaneous cassette analysis of α-1a antagonists and their metabolites in mouse plasma. Rapid Commun Mass Spectrom 2001;15:546–550.

Cai Z, Sinhababu A, Harrelson, S. Simultaneous quantitative cassette analysis of drugs and detection of their metabolites by high performance liquid chromatography/ion trap mass spectrometry. Rapid Commun Mass Spectrom 2000;14:1637–1643.

Chernushevich IG, Loboda AV, Thomson BA, An introduction to quadrupole-time-of-flight. J Mass Spectrom 2001;36:849–865.

Covey TR, Lee ED, Henion JD. High-speed liquid chromatography/tandem mass spectrometry for the determination of drugs in biological samples. Anal Chem 1986;58:2453–2460.

Hager JW. A new linear ion trap mass spectrometer. Rapid Commun Mass Spectrom 2002;16: 512–526.

Henry H, Sobhi HR, Scheibner O, Bromirski M, Nimkar SB, Rochat B. Comparison between high-resolution single-stage Orbitrap and a triple quadrupole mass spectrometer for quantitative analyses of drugs. Rapid Commun Mass Spectrom 2012;2: 499–509.

Hopfgartner G, Husser C, Zell, M. Rapid Screening and characterization of drug metabolites using a new quadrupole-linear ion trap mass spectrometer. J Mass Spectrom 2003;38:138–150.

Hu Q, Noll R, Li H, Makarov A, Hardman M, Cooks RG. The Orbitrap: a new mass spectrometer. J Mass Spectrom 2005;40:430–443.

Kantharaj E, Tuytelaars A, Proost PEA, Ongel Z, van Assouw HP, Gilissen RAHJ. Simultaneous measurement of drug metabolic stability and identification of metabolites using ion-trap mass spectrometry. Rapid Commun Mass Spectrom 2003;17:2661–2668.

Korfmacher W. High-resolution mass spectrometry will dramatically change our drug-discovery bioanalysis procedures. Boanalysis 2011;3(11):1169–1171.

Li AC, Alton D, Bryant MS, Shou WZ. Simultaneously quantifying parent drugs and screening for metabolites in plasma pharmacokinetic samples using selected reaction monitoring information-dependent acquisition on a QTrap instrument. Rapid Commun Mass Spectrom 2005;19:1943–1950.

Li AC, Gohdes MA, Shou WZ. 'N-in-one' strategy for metabolite identification using a liquid chromatography/hybrid triple quadrupole linear ion trap instrument using multiple dependent product ion scans triggered with full mass scan. Rapid Commun Mass Spectrom 2007;21:1421–1430.

March RE. An introduction to quadrupole ion trap mass spectrometry. J Mass Spectrom 1997;32:351–369.

Peterman SM, Duczak N, Kalgutkar AS, Lame ME, Soglia JR. Application of a linear ion trap/orbitrap mass spectrometer in metabolite characterization studies: examination of the human liver microsomal metabolism of the non-tricyclic anti-depressant nefazodone using data-dependent accurate mass measurements. J Am Mass Spectrom 2006;17:363–375.

Ramanathan R, Korfmacher W. The emergence of high-resolution MS as the premier analytical tool in the pharmaceutical bioanalysis area. Bioanalysis 2012;4(5):467–469.

Rudewicz PJ, Straub KM. Rapid structure elucidation of catecholamine conjugates with tandem mass spectrometry. Anal Chem 1986;58:2928–2934.

Rudewicz PJ, Straub KM. Rapid metabolic profiling using thermospray LC/MS/MS. In: Benford D, Gibson GG, Bridges JW, editors. *Drug Metabolism from Molecules to Man*. New York: Taylor and Francis; 1987.

Rudewicz PJ, Yue Q, Shin Y. High-throughput strategies for metabolite identification in drug discovery. In: Wang P, editor. *High-Throughput Analysis in the Pharmaceutical Industry*. Boca Raton, FL: Taylor and Francis Group; 2009. p 141–154.

Trunzer M, Faller B, Zimmerlin A. Metabolic soft spot identification and compound optimization in early discovery phases using MetaSite and LC-MS/MS validation. J Med Chem 2009;52:329–335.

Yang L, Wu N, Rudewicz PJ. Applications of new liquid chromatography-tandem mass spectrometry technologies for drug development support. J of Chromatogr A 2001;926:43–55.

Zhang D, Cheng PT, Zhang H. Mass defect filtering on high resolution LC/MS data as a methodology for detecting metabolites with unpredictable structures: identification of oxazole-ring opened metabolites of muraglitazar. Drug Met Letters 2007;1:287–292.

Zhang H, Zhang D, Ray K. A software filter to remove interference ions from drug metabolites in accurate mass liquid chromatography/mass spectrometric analyses. J Mass Spectrom 2003;38:1110–1112.

Zhang N, Fountain ST, Honggang B, Rossi DT. Quantification and rapid metabolite identification in drug discovery using API time-of-flight LC/MS. Anal Chem 2000;72:800–806.

Zhang NR, Yu S, Tiller P, Yeh S, Mahan E, Emary WB. Quantitation of small molecules using high-resolution accurate mass spectrometers—a different approach for analysis of biological samples. Rapid Commun Mass Spectrom 2009;23:1085–1094.

PART IV

REPRESENTATIVE GUIDELINES AND/OR EXPERIMENTAL PROTOCOLS OF LC-MS BIOANALYSIS

34

LC-MS BIOANALYSIS OF ESTER PRODRUGS AND OTHER ESTERASE LABILE MOLECULES

WENKUI LI, YUNLIN FU, JIMMY FLARAKOS, AND DUXI ZHANG

34.1 INTRODUCTION

A prodrug is a "precursor" to an intended drug. Upon administration, the prodrug is converted to the intended drug, the active pharmacological agent, through normal *in vivo* bioactivation processes (metabolism). A prodrug can be used to improve oral bioavailability of the intended drug that is poorly absorbed in gastrointestinal tract. A prodrug can also be used to improve the interaction selectivity of the intended drug with target cells (Stella et al., 1985; Rautio et al., 2008; Peterson and McKenna, 2009) in treatment like chemotherapy, which is often accompanied with some unwanted adverse effects (Rooseboom et al., 2004). From a bioconversion perspective, a prodrug can be bioactivated intracellularly (e.g., in therapeutic target tissues/cells or metabolic tissues) or extracellularly (e.g., in the milieu of GI fluids, systemic circulation, and metabolic tissues) or both. The most common prodrugs are the ester-based ones that are formed by derivatizing a phenol, hydroxyl, or carboxyl group of the intended drug molecule (Stella et al., 1985; Rautio et al., 2008; Peterson and McKenna, 2009).

The cleavage process of an ester prodrug (Figure 34.1) is catalyzed by various esterases *in vivo* to form its active molecule (Rooseboom et al., 2004; Liederer and Borchardt, 2006). However, the same hydrolysis can occur *ex vivo* post-sample collection (including sample collection, processing, and storage and thawing prior to extraction). From a bioanalysis point of view, any *ex vivo* cleavage of prodrug molecules in biological matrix is unwanted (Li et al., 2011). This *ex vivo* cleavage, if not controlled, can lead to overestimation of the active drug molecule and underestimation of the prodrug. As a result, the safety and/or efficacy assessment of the prodrug and its active form in the intended studies can be inaccurate.

Over the past decades, bioanalytical scientists have made significant progress in preventing and/or blocking unwanted *ex vivo* enzyme-catalyzed cleavage of ester prodrugs in LC-MS bioanalysis. This chapter highlights some important procedures for identifying and addressing the instability issue of an ester prodrug. The same approaches can be applied for the analysis of other enzyme labile molecules that contain amides, lactams, peptides, and lactones, and so on (Li et al., 2011). LC-MS of sulfhydryl (thiol) containing molecules is not covered in this chapter.

34.2 COMMON ESTERASES THAT CATALYZE HYDROLYSIS OF ESTER PRODRUGS AND OTHER MOLECULES

Esterases belong to a heterogeneous enzyme family consisting of three subgroups (esterases A, B, and C) (Aldrige, 1952; Bergmann et al., 1957). The most important esterases include carboxylesterase, acetylcholinesterase, butyrylcholinesterase, paraoxonase, cholinesterase, and arylesterase. These enzymes are widely distributed throughout the body, including systemic circulation (blood/plasma) and tissues (e.g., liver, brain, kidney, and lung) (Liederer and Borchardt, 2006).

Research has shown that different species have different amounts and types of esterases with different substrate specificities and different rates of hydrolysis. Within a given

Handbook of LC-MS Bioanalysis: Best Practices, Experimental Protocols, and Regulations, First Edition. Edited by Wenkui Li, Jie Zhang, and Francis L.S. Tse.

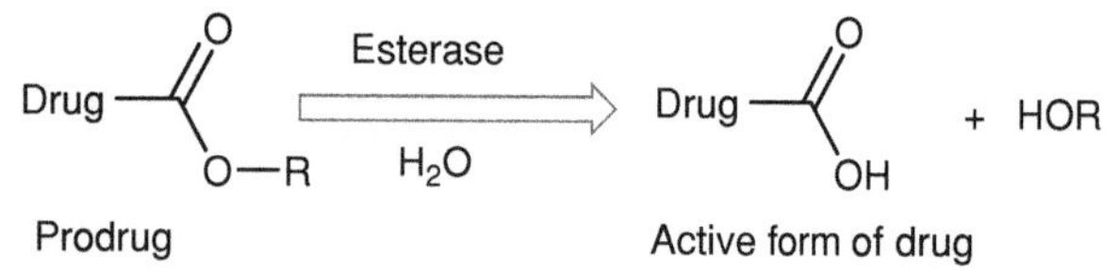

Drug–C(=O)–O–R (Prodrug) —Esterase, H_2O→ Drug–C(=O)–OH (Active form of drug) + HOR

FIGURE 34.1 A representative scheme of esterase-catalyzed hydrolysis of an ester prodrug.

species, the enzyme activities can vary from one subject to the other. There may be more than a single esterase involved in the hydrolysis of a given ester-containing compound. A given ester-containing compound may be subjected to hydrolysis catalyzed by different esterase(s) in different species. In general, esterase activity in rodent (e.g., rat and mouse) blood is much higher than that in nonrodent (e.g., dog and human) blood (Minagawa et al., 1995; Liederer and Borchardt, 2006; Koitka et al., 2010). For example, hydrolysis of cisatracurium by rat plasma esterases is more rapid than that in human plasma (Welch et al., 1995). While human plasma showed no significant issues associated with the decomposition of Ro 64-0796 and Ro 64-0802, rodent plasma must be pretreated with dichlorvos, an esterase inhibitor, to prevent hydrolysis (Wiltshire et al., 2000). The high activity in rodent species but low activity in nonrodent species of esterases often render difficulty for drug discovery and development in generating accurate preclinical model for the evaluation of prodrug candidates. This difference can also be translated to that a simple stabilization approach may be sufficient in dealing with sample analysis of human subjects, but a more sophisticated approach may be necessary for handling study samples from rodent species.

34.3 METHODOLOGY AND APPROACHES

From an LC-MS bioanalysis perspective, the major challenge for ester-containing prodrugs is the management of their instability in the intended biological matrices during biological sample collection, processing, and storage and sample preparation prior to LC-MS quantification.

34.3.1 Common Precautions

1. *Temperature control*: Temperature control plays a crucial role in assuring a reliable analysis of unstable analytes in biological samples (Chen and Hsieh, 2005; Tokumura et al., 2005; Briscoe and Hage, 2009). In general, the enzyme-catalyzed degradation of an analyte in biological matrix is significantly slower when the sample is processed on wet ice (~0°C to 4°C) compared to at room temperature (~22°C) (Chen and Hsieh, 2005). One general approach is to freeze the blood sample right after collection if blood is the matrix of choice, or to process the blood sample at a reduced temperature (i.e., centrifugation at ~4°C) for plasma, followed by immediate storage in a freezer. The duration between sample collection and storage in a freezer should be as short as possible to minimize possible degradation of unstable molecules (Li et al., 2011).
2. *Selection of anticoagulant*: Anticoagulant and/or anticoagulant counter ions can have an impact on compound stability (Evans et al., 2001; Bergeron et al., 2009). Heparin and ethylenediaminetetraacetic acid (EDTA) are common anticoagulants with different mechanisms of action. Heparin, as a polysaccharide, accelerates the inactivation of thrombin (an enzyme that promotes clotting), whereas EDTA chelates calcium ions and interrupts the clotting cascade at multiple points. EDTA can prevent the activity of calcium-dependent phospholipases and ester hydrolases, while heparin may not offer the same effect. In general, EDTA is preferred to heparin as anticoagulant in plasma samples for ester prodrugs (Li et al., 2011).

34.3.2 Stability Assessment of Prodrugs and/or Other Esterase Labile Compounds

A quick stability assessment of an ester prodrug or other esterase sensitive compound in an intended biological matrix is the first important step to determine whether one or more stabilization measures, in addition to the above (low temperature and proper anticoagulant), need to be evaluated and optimized in the early stage of bioanalytical method development.

34.3.2.1 Stability Assessment in Plasma Plasma is the most common matrix for bioanalysis. In most cases, analyte stability in plasma serves as a good indicator of its stability in whole blood (Freisleben et al., 2011). Compared to blood, plasma is more homogeneous, less viscous, and easier to handle. Stability results using plasma is less subjected to the possible impact due to blood/plasma partition of the analyte of interest if liquid–liquid extraction other than protein precipitation is the method of choice for sample preparation.

In plasma stability assessment, one (middle), two (low and high), or three (low, middle, and high) concentration levels of ester prodrug quality control (QC) samples are prepared in the untreated plasma of the intended species. Although using three concentration levels may help generate a rich data set on analyte stability, assessment using the middle QC concentration level can be very efficient and the results should serve well for the follow-up step in LC-MS bioanalytical development process. A relatively large volume of sample may be needed. The prepared QC samples are placed on a laboratory bench at room temperature (to mimic the most common situation of study sample analysis) and on wet ice (for comparison). At various time points (e.g., 0, 0.5, 1, 2, 6, and 24 h), two or more aliquots of each QC sample are taken and placed in a deeper freezer (i.e., $<-60°C$, to prevent or slow down any further degradation) or immediately treated with a fixed volume of organic (e.g., acetonirile or methanol, to stop enzyme activity) followed by vortex-mixing and storage in a deep freezer. Upon all the needed QC aliquots at the intended time points are taken and/or treated, stability assessment can be conducted using one of the following approaches:

- *Comparison of analyte/internal standard LC-MS response ratios over time*: Upon removal of stability samples (untreated or treated with organic) from the freezer and thawing, a fixed volume of internal standard (preferably stable-isotope labeled) working solution in organic solvent is added to each sample. Additional organic solvent may be needed to ensure the same amount of organic or nonorganic in the sample mixture prior to further processing. After proper vortex-mixing and centrifugation, the supernatants are transferred and evaporated. The resulting residues are reconstituted prior to injection onto LC-MS system. The LC-MS response ratios of the analyte to the internal standard for the QC samples at various time points are plotted against that of QC samples at time zero. The time zero QC samples may be freshly prepared prior to the above experiment to ensure accurate comparison. Other surrogate matrices (e.g., urine or organic solvent) that does not contain any esterase have been used (Zeng et al., 2007). However, the possible difference in matrix effect or signal suppression between the matrices needs to be taken into account for the evaluation of the stability assessment outcomes when a surrogate matrix is used as control, otherwise surrogate matrix should not be used.
- *Quantification of analyte using freshly spiked calibration standards*: The stability QC samples prepared above can be analyzed against freshly spiked calibration standards. In this process, an equal volume of each appropriate (lower limit of quantitation (LLOQ) to upper limit of quantification (ULOQ)) standard working solution (normally in 50% aqueous organic) is mixed with blank plasma in the assigned well of the assay plate. As such, there is ~25% of organic in the wells when the analyte is mixed with plasma and enzyme activity, if any, is expected to be inhibited. Upon addition of internal standard working solution and/or other reagents for further processing prior to LC-MS analysis, the measured analyte concentrations in the stability QC samples at various time points are compared to the nominal values or the values measured at time zero. To ensure the consistency of matrix effect and extraction recovery between the QC samples and the freshly spiked calibration standards, the final volume of matrix, organic and other nonmatrix components in all samples should be the same and sample mixture should be vortex-mixed well prior to any supernatant transfer for further processing.
- *Monitoring formation of hydrolysis product over time*: This approach is similar to the above with exception of that, instead of employing the freshly spiked calibration standards of ester prodrug, the active form of the drug is used to prepare the calibration standards for the assessment (Fung et al., 2010). In this case, the active drug concentration serves as a surrogate indicator of instability of the prodrug. The stability or instability of the prodrug can be unveiled by plotting the measured active drug concentration over time. The lower the active drug concentration, the better the stability of the prodrug can be. From an operational point of view, it is much easier to monitor the formation of the active drug in the intended matrix than the prodrug itself as the active drug molecule is, in general, much more stable than the prodrug.

34.3.2.2 Stability Assessment in Blood Although stability in plasma is considered as a good surrogate indicator of analyte stability in blood, there are exceptions (Freisleben et al., 2011) for certain chemical classes for which stability behavior of the drug in blood versus plasma is expected to be different based on the chemical structure. Therefore, performing blood stability assessment is necessary not only to confirm the stability coverage but also to provide adequate sampling instructions for preclinical or clinical facilities.

All three approaches listed for plasma stability assessment can be implemented for assessing the stability of analyte in blood. The analyte of interest is spiked into fresh blood of the intended species at the intended middle QC concentration level. It is worth noting that fresh blood should be used as aged blood with reduced enzyme activities may result in false outcomes. Upon a gentle mixing to ensure a proper equilibrium (~15 min) of the analyte of interest in the blood at room temperature or 37°C. The samples are incubated at 4°C (on wet ice), room temperature (~22°C) and/or 37°C as shown in Figure 34.2 for an investigational ester prodrug compound A. At various time points, at least two aliquots of blood sample are taken. The blood samples can be centrifuged for plasma or directly mixed with internal standard, followed by a conventional protein precipitation or liquid–liquid extraction process prior to LC-MS analysis.

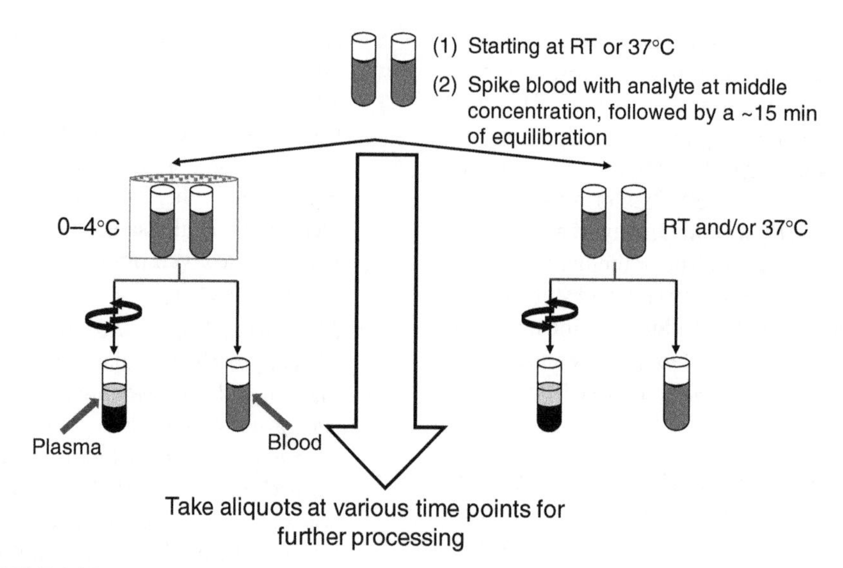

FIGURE 34.2 A representative scheme of stability assessment of prodrugs and/or other enzyme labile compounds using whole blood.

Considering the apparent difference in terms of sample handling between plasma and blood samples, comparing LC-MS response ratios of the analyte to the internal standard across the various time points might be more convenient than other means in obtaining first-hand stability information. As shown in Figures 34.3a–d, compound A appears to be stable in human blood and plasma but not stable in rat blood and plasma. Within 2 h of incubation, there are 20–40% losses of the compound in rat blood/plasma. On the other hand, as shown in Figures 34.3c and d, the enzyme-catalyzed instability of the compound in rat blood/plasma at a lower temperature is generally better when the sample is placed on wet ice (~0°C to 4°C) than at room temperature (~22°C) (Chen and Hsieh, 2005).

It is worth noting that if an analyte is stable in blood at room temperature for more than 2 h, which is the typical duration needed for harvesting plasma from blood, pretreatment of blood collection tubes with inhibitors and/or pH modifier might not be necessary. These agents can be added to the harvested plasma and then properly mixed prior to further processing. In general, treatment of plasma sample is much easier than that of blood sample.

34.3.3 Evaluation and Optimization of Stabilization Measures

As soon as instability is identified from any of the above processes, stabilization measures in addition to selection of right anticoagulant and processing temperature have to be evaluated, optimized, and implemented for the best bioanlaytical outcomes. The key stabilization measures include (1) pH control, (2) use of esterase inhibitors, and (3) combination of pH control and use of enzyme inhibitors.

34.3.3.1 pH Control It is well known that pH plays an essential role in acid-/base-catalyzed enzymatic reactions and most enzymatic reactions have a narrow working pH window. A simple treatment of study samples with pH modifiers can be very effective in stabilizing drug molecules (Li et al., 2011). In fact, a low pH can be much more effective than a lower sample processing temperature in reducing esterase activities. From a bioanalytical operation perspective, processing study samples at a low temperature is less desirable as it requires wet ice or chilled conditions, which creates challenges in using robotic liquid handler (Fung et al., 2010). The most commonly used reagents for pH control are formic acid, acetic acid, phosphoric acid, citric acid or other buffers, including Tris buffer, at a variety of concentrations.

In practice, the strength and amount of the above pH modifiers need to be adjusted to reduce the viscosity of study samples. Therefore, it is desirable to use mild condition (e.g., pH 4–5). At a lower pH (e.g., pH 3 or lower), the treated plasma samples frequently become very viscous, making pipetting difficult and unreliable (Fung et al., 2010). A general recommendation is to treat the harvested plasma samples with 0.5% formic acid (unpublished results), 0.5% phosphoric acid (unpublished results), 0.1 M HCl (Kamei et al., 2011), or treat blood samples with 0.5% citric acid (w/v) (Kamei et al., 2011) or 1 M citric acid at a final citric acid concentration of 20 mM (Fung et al., 2010). Otherwise, blood collection tubes with acid additives are commercially available for this purpose.

34.3.3.2 Use of Enzyme Inhibitors A variety of enzyme inhibitors (Table 34.1) can be used for stabilizing ester prodrugs of interest and other esterase sensitive compounds.

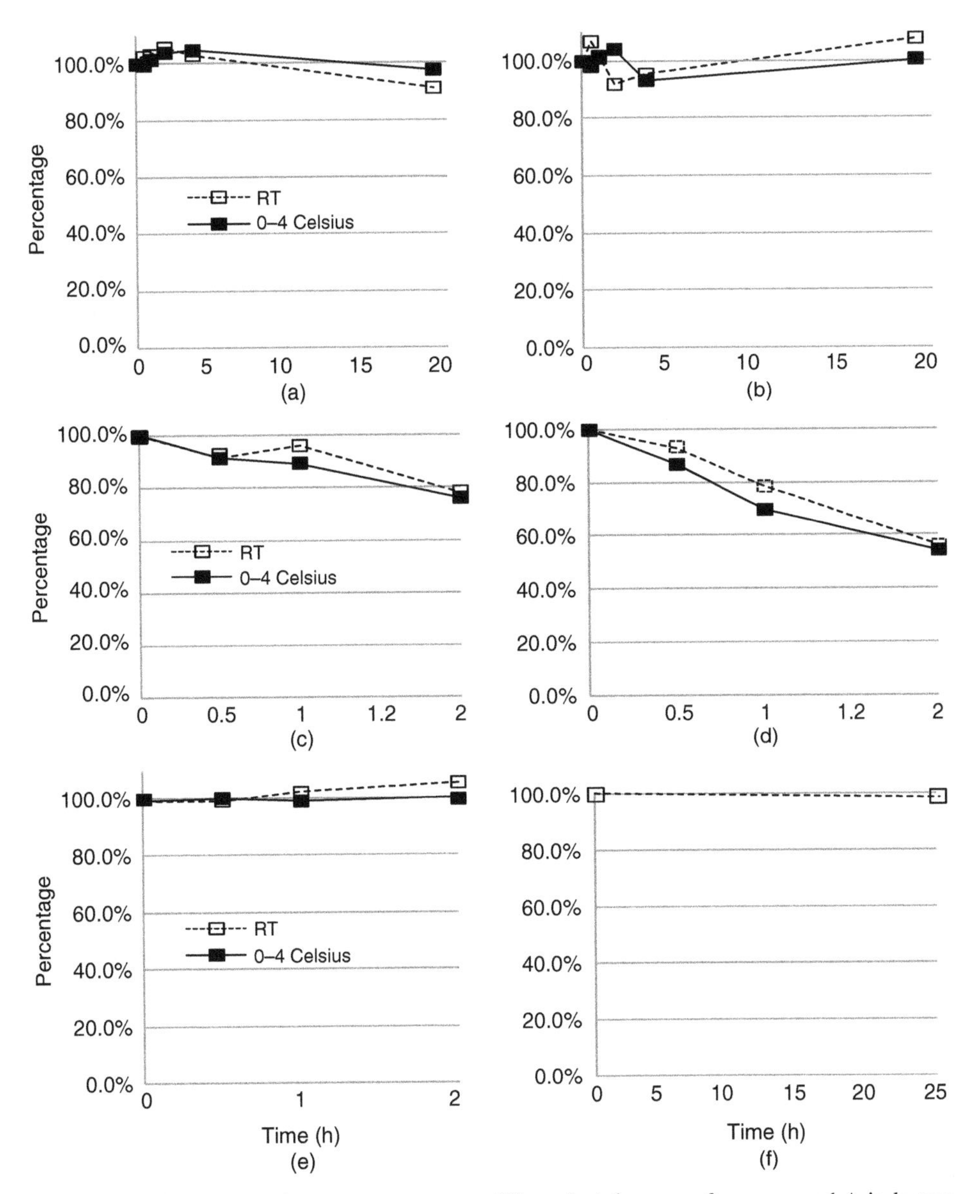

FIGURE 34.3 Measured percentage recovery (%) against time zero for compound A in human plasma (a), human blood (b), rat plasma (c) and rat blood (d) without addition of stabilizer, and in rat blood treated with ~140 mM NaF (e) and the harvested rat plasma treated with ~1% phosphoric acid (v/v) (f).

Considering the fact that not all enzyme labile compounds exhibit the same type of esterase sensitivity in the matrices of different species (Liederer and Borchardt, 2005, 2006; Minagawa et al., 1995), it is not possible to have a "universal inhibitor" for all compounds in all species and matrices. Furthermore, it is not always feasible to identify which esterases are responsible for hydrolyzing a compound and which inhibitor(s) will work the best in stabilizing the molecules in the intended matrix of a given species. For example, in rat plasma, the degradation of salvinorin A, the main active psychoactive ingredient from a medicinal plant *Salvia divinorum*, can be markedly inhibited by the addition of sodium fluoride (esterase inhibitor), phenylmethylsulfonyl fluoride (serine esterase inhibitor), and bis-p-nitrophenyl phosphate (carboxylesterase inhibitor), but little or no suppression of the degradation was observed with other functionally similar inhibitors such as 5,5′-dithiobis-2-nitrobenzoic acid (arylesterase inhibitor), ethopropazine (butylcholinesterase inhibitor), or BW284c51 (acetylcholinesterase inhibitor) (Tsujikawa et al., 2009). In another case for the identification of enzymes involved in hydrolysis of nafamostat, the hydrolytic activity in the erythrocytes and plasma

TABLE 34.1 Commonly Used Esterase Inhibitors in Bioanalysis

Names	Target enzymes	References
Prazosin	Carboxylesterase (CarbE) inhibitor	Kudo et al. (2000)
Thenoyltrifluoroacetone (TTFA)	Carboxylesterase (also serving as a chelating agent)	Fung et al. (2010) and Zhang and Fariss (2002)
Acetylcholine (Ach)	Competitive CarbE inhibitor	Yoshigae et al. (1999)
Bis(4-nitrophenyl)-phosphate (BNPP)	Carboxylesterase inhibitor, phosphodisterase inhibitor	Fung et al. (2010), Ishizuka et al. (2010), Yoshigae et al. (1999), Kudo et al. (2000), Yamaori et al. (2006), and Tsujikawa et al. (2009)
5, 5′-dithiobis-(2-nitrobenzoic acid) (DTNB, Ellman's reagent)	Carboxylesterase inhibitor	Yamaori et al. (2006), Tsujikawa et al. (2009), and Minagawa et al. (1995)
Ethopropazine (profenamine, parsidol, parsidan, parkin)	Butylcarboxylesterase inhibitor	Yamaori et al. (2006) and Tsujikawa et al. (2009)
Phenylmethylsulfonyl fluoride (PMSF)	Carboxylesterase inhibitor, Serine protease inhibitor, but does not inhibitor all serine protease, rapidly degraded in water	Fung et al. (2010), Ishizuka et al. (2010), Kim et al. (2011), Kudo et al. (2000), Yamaori et al. (2006), and Tsujikawa et al. (2009)
Diisopropyl fluorophosphate (DIFP, DFP)	Irreversible serine hydrolase inhibitor	Fung et al. (2010), Ishizuka et al. (2010), Yamaori et al. (2006), Minagawa et al. (1995), and Zeng et al. (2007)
2,2-Dichlorovinyl dimethyl phosphate (Dichlorvos, DDVP)	Carboxylesterase inhibitor, acetylcholinesterase inhibitor	Fung et al. (2010), Wiltshire et al. (2000), and Chang et al. (2009)
p-Nitrophenyl acetate (PNPA)		Yoshigae et al. (1999)
Paraoxon	Acetylcholinesterase inhibitor	Fung et al. (2010), Liederer and Borchardt (2005), and Yang et al. (2002)
Eserine (physostigmine, erserine)	Cholinesterase inhibitor, Acetylcholinesterase inhibitor	Fung et al. (2010), Ishizuka et al. (2010), Yoshigae et al. (1999), Minagawa et al. (1995), and Zhang et al. (2011)
BW284c51	Acetylcholinesterase inhibitor	Olivera-Bravo et al. (2005), Tsujikawa et al. (2009), and Yamaori et al. (2006)
Sodium fluoride (NaF), Potassium fluoride (KF)	Phosphodiesterase inhibitor, Carboxylesterase inhibitor	Fung et al. (2010), Kim et al. (2011), Kromdijk et al. (2012), Salem et al. (2011), Skopp et al. (2001), and Wang et al. (2007)
Sodium arsenate	Esterase inhibitor	Salem et al. (2011)

Reproduced from Li et al. (2011) with permission of John Wiley & Sons Ltd.

was inhibited by 5, 5′-dithiobis (2-nitrobenzoic acid), an arylesterase inhibitor, in a concentration-dependent manner. By contrast, little or no suppression of the activity was seen with phenylmethylsulfonyl fluoride, diisopropyl fluorophosphate, bis(p-nitrophenyl)phosphate, BW284c51, or ethopropazine (Yamaori et al., 2006).

As discussed, identification and use of the most suitable esterase inhibitor can be very critical. In this process, one or more esterase inhibitor candidates at two or more concentrations (e.g., 0.1, 1, and 10 mM) are screened in the intended matrix against the ester prodrug. Considering the amount of work and numbers of inhibitor candidates to be screened, one single concentration level (i.e., middle QC level) for the analyte should be sufficient as a starting point (Fung et al., 2010).

The screening process is similar to the plasma/blood stability assessment described earlier. To the plasma or blood treated with inhibitor candidates, the ester prodrug is spiked and mixed. The samples are placed at room temperature (to mimic study sample analysis situation) and on wet ice (for comparison). At various time points up to 24 h post-preparation, two or more aliquots of the samples are taken and processed for LC-MS analysis. The same approach for plasma stability assessment (refer to Section 34.3.2.1) can be employed for assessing the effectiveness of inhibitor candidates (Zeng et al., 2007; Fung et al., 2010). One needs to keep in mind that some organophosphates, for example, paraoxon, are potent neurotoxic reagents and require special handling at toxicology assessment or clinical study sites. Whenever possible, less toxic inhibitors should be used.

34.3.3.3 Combination of Addition of Enzyme Inhibitors and Use of pH Modifiers In most cases, a simple adjustment of matrix pH (Kamei et al., 2011) or addition of an enzyme inhibitor (Chang et al., 2009; Kim et al., 2011; Lindegardh et al., 2006; Yang et al., 2002; Zeng et al., 2007) can be sufficient and effective in stopping esterase-catalyzed hydrolysis of an ester prodrug and/or other esterase labile compound. However, there have been many cases where combined stabilization measures have to be utilized. This combined measure includes pH control and use of inhibitor (Tang and Sojinu, 2012; Wang et al., 2007) or use of two inhibitors (Salem et al., 2011), in addition to common precautions, for example, low temperature and proper anticoagulant.

Blood collection tubes containing certain enzyme inhibitor (e.g., NaF) or pH modifier (e.g., citric acid) are commercially available. This makes the stabilization process much easier as only one more necessary step is the treatment of the harvested plasma samples with acid or inhibitor. In practice, regardless of whether the blood tubes containing inhibitor or acid is to be used in the first place, whole-blood stability of the prodrug should be evaluated under optimized plasma pH and processing temperature to establish the procedures and conditions that are to be employed in the study sample collection and processing in preclinical or clinical studies (Fung et al., 2010).

Figures 34.3e and f depict the stability of compound A at middle QC concentration level in fresh rat blood treated with sodium fluoride (~140 mM) and in the harvested rat plasma treated with ~1% phosphoric acid. Clearly, the compound is stable for at least 2 h in rat blood and for at least 25 h in the harvested plasma. Further evaluation of rat plasma QC samples treated with NaF (~140 mM) and phosphoric acid (~1%) shows that compound A is stable after three freeze–thaw cycles and for at least 36 days following storage at $\leq -60°C$ (figure not shown). The established storage stability covers the duration from the first sample collection to the last sample analysis for a toxicokinetic study.

34.3.4 Confirmation of Effectiveness of Stabilization Measures in Method Validation and Sample Analysis

There is no exception that the health authority requirements (FDA, EMA, etc.) and the current industry practice should be followed in validating an LC-MS method for quantitative analysis of both the prodrug and its active form in the intended matrix. It is also common that pooled QCs (prodrug and drug) are employed for assay validation and associated study sample analysis. However, to confirm whether the prodrug is stable in the QC samples, QC samples containing the prodrug only need to be included in assay validation, where the prodrug alone QC samples go through the needed freeze–thaw cycles, room temperature stay, and long-term storage before analysis using freshly prepared calibration standards of both the prodrug and the active drug using the same matrix. By monitoring changes of both the parent and active drug molecule in those QC samples, a higher confidence should be gained. Furthermore, using freshly prepared calibration standards can for sure help gain this confidence.

One should always keep in mind that QC samples are not incurred samples. The QC samples, prepared by individually spiking or serial dilution, might contain ≤5% of nonmatrix component (Li et al., 2011). This small amount of non-matrix component, usually organic solvents, may provide some unexpected stabilization effect to the prodrug. On the other hand, although a large portion (>95%) of the QC sample matrix is the same as the incurred samples, QC samples do not contain the various drug metabolites, coadministered drug(s) and their metabolites, and/or dosing vehicles as found in incurred samples (Li et al., 2011). In particular, enzyme activity may be different between the incurred samples and the blank matrix that is used for preparation of calibration and QC samples. The slow denaturation of esterases in the aged matrix (whole blood, plasma, serum, etc.) often leads to a decreased hydrolysis activity. It is not unusual that a validated LC-MS method with an enzyme inhibitor concentration optimized using the standards/QCs prepared from aged matrix does not work well for the incurred samples. Heightened awareness among bioanalytical scientists of the differences between the QC samples and incurred samples would help reduce the risk of subsequent bioanalytical failure (Li et al., 2011; Meng et al., 2011). The use of representative incurred samples is necessary in order to confirm that the enzyme inhibitor(s) and other stabilization measures work well for both the calibration standards/QC samples and the incurred samples.

As soon as optimized stabilization measures are implemented in the first-in-animal or first-in-human studies, incurred sample reanalysis (ISR) and/or incurred sample stability (ISS) assessment should be conducted. The difference in the measured prodrug concentrations between the repeat analysis and initial analysis or the mean of two measurements (first and repeat) should be within ±20% for two-thirds of the incurred samples selected. Any marked decrease in the prodrug concentrations would indicate prodrug instability in the incurred samples. Under such circumstance, the stabilization measures will need to be modified prior to revalidation of the method and potential repeat of the preclinical study, which can be costly. Table 34.2 summarized the ISR results for compound A in 21 randomly selected incurred samples from a 2-week toxicokinetic study in rats. The results met the acceptance criteria, demonstrating collection of rat blood into NaF/EDTA tubes and addition of phosphoric acid (~1.0%, v/v) to the harvested plasma is effective in stabilizing the prodrug.

TABLE 34.2 Summary of Incurred Sample Reanalysis for Ester-prodrug Compound A in Rat Plasma

Sample index	Sample ID	Original concentration (ng/ml)	Repeat concentration (ng/ml)	%Bias
1	2502 10 plasma-1 Day 10 1h	94.6	92.7	−2.0
2	2502 10 plasma-1 Day 1 1h	159	150	−5.3
3	2508 10 plasma-1 Day 10 3h	8.53	7.42	−13.0
4	2001 10 plasma-1 Day 10 .5h	121	116	−3.9
5	2006 10 plasma-1 Day 1 .5h	142	132	−6.5
6	2008 10 plasma-1 Day 1 3h	3.27	2.97	−9.1
7	2009 10 plasma-1 Day 1 7h	4.38	4.47	2.1
8	3001 100 plasma-1 Day 1 .5h	1060	980	−7.8
9	3004 100 plasma-1 Day 1 7h	190	168	−11.7
10	3007 100 plasma-1 Day 10 1h	1840	1810	−1.8
11	3007 100 plasma-1 Day 1 1h	1880	1800	−3.9
12	3501 100 plasma-1 Day 10 .5h	539	531	−1.4
13	3503 100 plasma-1 Day 10 3h	185	182	−1.4
14	3503 100 plasma-1 Day 1 3h	719	777	8.1
15	4006 300 plasma-1 Day 10 .5h	803	947	18.0
16	4006 1000 plasma-1 Day 1 .5h	7280	8820	21.2[a]
17	4008 300 plasma-1 Day 10 3h	1840	1660	−9.7
18	4008 1000 plasma-1 Day 1 3h	2280	2370	4.1
19	4010 300 plasma-1 Day 10 24h	3.37	BLQ < 20.0	NA
20	4504 300 plasma-1 Day 10 7h	1870	1850	−1.3
21	4504 1000 plasma-1 Day 1 7h	2890	2920	1.1

NA = Not, applicable as the repeat result was below the adjusted LLOQ of 20.0 ng/ml after a 20-fold dilution.
[a]Denotes an ISR result that was not within the ±20% bias window (compared to the original value) due to an unconfirmed reason. However, more than 2/3 of the overall ISR sample results were within the above window.

34.3.5 Use of DBS for Stabilization of Ester Prodrugs and Other Enzyme Labile Compounds

The collection of whole blood samples on cellulose cards, known as dried blood spots (DBS), has a number of advantages over conventional blood collection. As a less invasive sampling method, DBS offers simpler sample collection and storage and easier sample transfer, with reduced infection risk of various pathogens, and requires a smaller blood volume (Li and Tse, 2010). The technique provides a totally different dimension in stabilizing unstable compounds via storing the blood samples in dried form. For a 10–40 μl of blood spot, the regular drying time is about 2–3 h in an open atmosphere at room temperature. After the spot becomes dried, the enzymes that are responsible for the instability of enzyme sensitive molecules lose their three dimensional structures, a key element for enzymatic activity in blood. Therefore, the stability of many unstable analytes can be improved (Li and Tse, 2010). Collection of DBS samples for ester prodrug and other enzyme labile compounds may be particularly meaningful. This is because most of esterase inhibitors employed for stabilization are toxic and must be handled with care. DBS may offer an alternative stabilization method for sample collection of whole blood from the intended species.

A recent research by Heinig et al. for the stability assessment of oseltamivir (Tamiflu®) demonstrated the utility of collecting DBS samples for the stabilization of ester prodrug (Heinig et al., 2011). Oseltamivir phosphate (OP) was unstable in rat EDTA blood with 35% conversion into oseltamivir carboxylate (OC) within 15 min, whereas enhanced stability was shown in DBS. During the 2 h drying process of the DBS samples at ambient temperature, OP did not degrade in rat blood on the chemically treated DMPK-A and DMPK-B cards (conversion below 3%), whereas 49% of OP was converted to OC on untreated Ahlstrom 226 cards. It was also noticed that drying the blood samples on untreated DBS cards under a stream of cold air within 30 min did not significantly reduce the conversion. Although no significant further OP degradation occurred on untreated cards after 3 days of storage after the complete dryness of the blood spots on untreated cards, the initial loss of approximately 50% presents a serious challenge for collecting 'real time' study samples on untreated DBS cards. Apparently, the observed instability of OP in untreated 226 cards is due to enzyme activity during the drying period of the blood spot. Unfortunately, it is difficult to mix a small volume (10–40 μl) of fresh blood sample with enzyme inhibitors and/or other stabilizers, such as pH modifier, prior to blood spotting to prevent the enzymatic, and/or chemical degradation.

Based on the aforementioned description, chemically treated cards appear to offer some stabilizing effect for the ester prodrug. However, this effect is apparently compound dependent as the treated Whatman FTA Elute cards were reported not offering any better stability coverage for the investigated compounds than the untreated Whatman 903 Protein Saver cards (D'Arienzo et al., 2010).

34.3.6 Representative Examples

Many research articles have been published on addressing the stability issues of LC-MS bioanalysis of ester-based prodrugs and other enzyme labile molecules. Some of the representative cases are captured in Table 34.3. These case studies highlight both general and compound specific approaches.

34.4 SUMMARY

Stability is a fundamental parameter that must be assessed and controlled in quantitative LC-MS bioanalysis of ester-based prodrugs and other enzyme labile molecules. In general, the instability of an ester-based prodrug is readily predictable because of the presence of enzyme sensitive ester bonds in the molecule. However, the instability of many other enzyme labile analytes may not be readily predictable. Furthermore, the instability of enzyme labile molecules within the same chemo-type can vary significantly in the same intended matrix and the instability of a given enzyme labile molecule can differ from one matrix to the other. Here are some general recommendations:

- Conduct necessary search for stability information of analyte of interest or its analogs. Assume analyte is unstable when its stability in the intended biological matrix is unknown.
- Conduct proper stability assessment of the analyte of interest in fresh plasma and/or blood of the intended species.
- If instability is identified, a simple adjustment of plasma pH using 0.5% or 1% acid (formic acid, phosphoric acid or citric acid) should be tested.
- Screen and optimize the use of enzyme inhibitor(s).
- If neither pH control nor inhibitor(s) alone is deemed effective, a combination of pH modifier(s) and inhibitor(s) together with other means (e.g., temperature control) should be evaluated.
- Minimize duration of study sample exposure to ambient temperature during collection, processing, and extraction for LC-MS analysis.
- Avoid using aged plasma or blood for stability assessment and/or the preparation of standards and QCs. Prepare QC samples with prodrug alone and monitor the possible change in the prodrug concentration and/or formation of the active drug molecule after freeze–thaw, at room temperature and after long-term storage.
- Avoid extreme conditions (strong base, strong acid, and high temperature) during sample extraction and/or chromatography.
- Ensure baseline separation of the analyte of interest from its metabolites and other potential interfering components.
- Include incurred sample reanalysis and/or incurred sample short-term stability assessment in method development and validation. This approach can help unveil any "hidden" instability issues that, if left unaddressed, could eventually lead to the invalidation of an otherwise "validated" method.

34.5 A REPRESENTATIVE PROTOCOL-LC-MS BIOANALYSIS OF BMS-068645 AND ITS ACID METABOLITE IN HUMAN PLASMA

Background

BMS-068645 is a selective adenosine 2A agonist that contains a methyl ester group that undergoes esterase hydrolysis to its acid metabolite, BMS-068645-acid (Zeng et al. 2007). A method is needed for simultaneous determination of both the parent (BMS-068645) and its active acid (BMS-068645-acid), respectively in dynamic range of 0.02–10 ng/ml and 0.05–10 ng/ml in human plasma.

BMS-068645 → Hydrolysis → BMS-068645-acid

TABLE 34.3 Representative Applications of Stabilization of Ester-based Prodrugs and Other Esterase Sensitive Molecules in Biological Matrices

Compound	Structure	Type of instability	Approach of stabilization	Brief stabilization procedure	Specimen	Reference
Betamethasone acetate and betamethasone phosphate	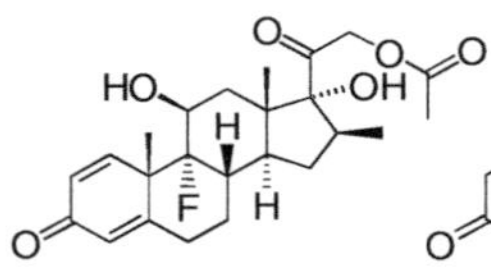 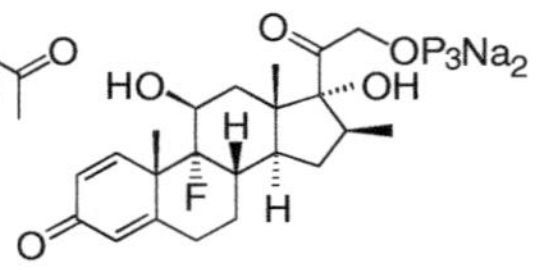	Hydrolysis of ester	Addition of inhibitors	Blood samples were collected into prechilled plastic heparinized tubes containing 10 μl 2 M sodium arsenate solution per ml blood. The harvested plasma was siphoned into prechilled plastic tubes containing 10 μl of 50% (w/v) potassium fluoride solution per ml plasma.	Human plasma	Salem et al. (2011)
Caffeic acid phenethyl ester and its fluorinated derivative	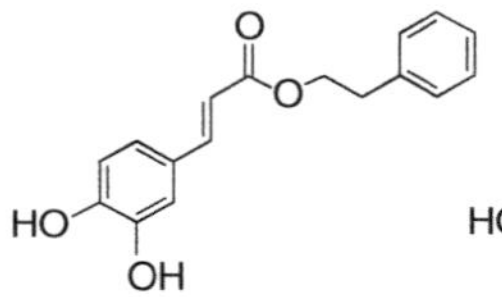 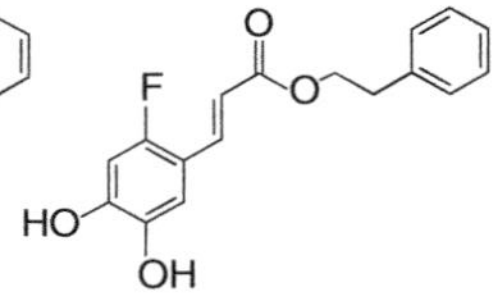	Esterase-induced degradation	Addition of NaF and pH adjustment	Addition of 0.4% sodium fluoride and pH was adjusted to 6.	Rat plasma	Wang et al. (2007)
			Collecting study blood samples into stabilizer treated tubes	100 μl aliquot of thenoyltrifluoroacetone solution (5 mg/ml in MeOH/ACN, 1:1, v/v) and 100 μl aliquot of citric acid solution (50 mg/ml in MeOH) were placed in a polypropylene tube and evaporated to dryness under a gentle nitrogen stream at 40°C. The stabilizer-treated tubes were stored at −20°C before use for collection of 400 μl dog blood. The harvested plasma sample (100 μl) was treated with 300 μl methanol and stored at −80°C prior to further processing.	Dog plasma	Tang and Sojinu (2012)

KR-62980	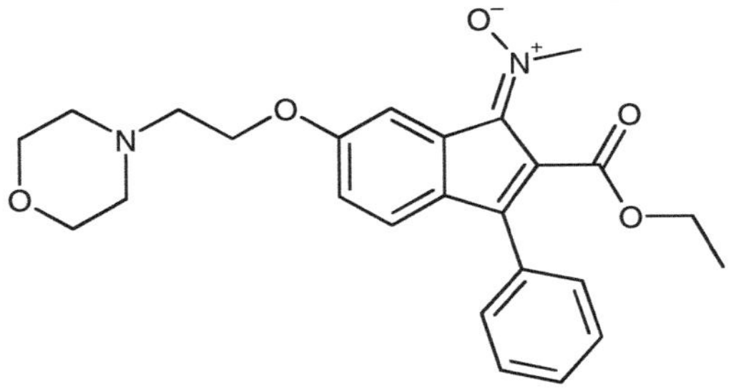	Esterase-induced instability	Store samples at very low temperature (−80°C) or add esterase inhibitors	Addition of NaF (10 mg/ml) or PMSF (0.18 mg/ml) significantly decreased the degradation rate.	Rat plasma	Kim et al. (2011)
TP300	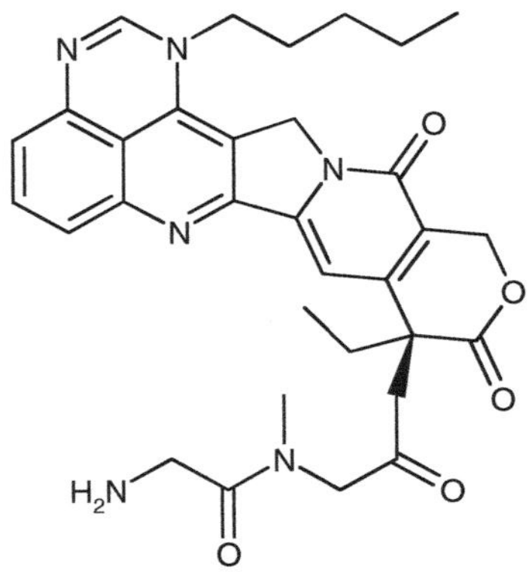	Enzyme induced degradation	Acidification of human plasma to prevent degradation of the prodrug and conversion of the active drug from lactone to carboxylate form	Immediately after collection, the blood was transferred into tubes containing heparin and citric acid (~0.5%, w/v) and stored in a refrigerator or on wet ice until centrifugation. The harvested plasma was then mixed with 1 M HCl at a ratio of 10:1 (v:v) and stored at −70°C until analysis.	Human plasma	Kamei et al. (2011)
Opioid peptide	H-Tyr-D-Ala-Gly-Phe-D-Leu-OH (DADLE)	Esterase-induced instability	Addition of esterase inhibitor	The biological matrix was preincubated with paraoxon (final concentration of 1 mM) for 15 min at 37°C before use.	Human blood, rat blood, rat tissue	Yang et al. (2002)
Nafamostat	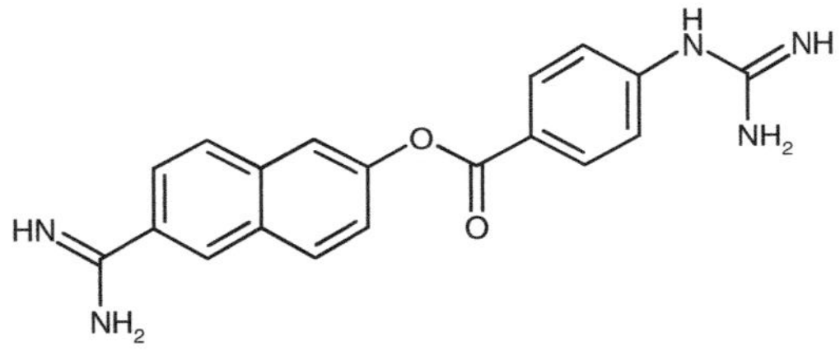	Enzymatic or nonenzymatic degradation	Quick plasma preparation	Immediate separation of plasma at 4°C and store the plasma samples at a low temperature.	Human plasma	Cao et al. (2008)
Acetylcholine	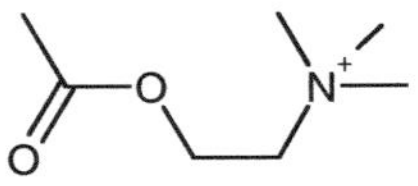	Acetyl cholinesterase induced degradation	Addition of inhibitor	Treat sample with 25 mM of eserine	Rat CSF	Zhang et al. (2011)

(*continued*)

TABLE 34.3 ***(Continued)***

Compound	Structure	Type of instability	Approach of stabilization	Brief stabilization procedure	Specimen	Reference
Oseltamivir	HN, O, O, NH$_2$, O, O	Esterase catalyzed degradation	Addition of inhibitor to sample tube prior to sample collection	Prior to blood sample (200 μl) collection, to the tubes was added 2.5 μl of dichlorvos in normal saline solution (8 mg/ml) to a final concentration of 200 μg/ml	Rat blood	Chang et al. (2009)
				Collect human blood into 4 ml NaF/Na_2EDTA (6.0 mg) tubes for further processing	Human plasma	Kromdijk et al. (2012)
				Plasma (5 ml) was transferred into a tube containing dichlorvos at 200 μg/ml (*Note*: significant inter-individual variation was observed for the samples collected without addition of the inhibitor)	Human blood/plasma	Lindegardh et al. (2006)

Reproduced from Li et al. (2011) with permission of John Wiley & Sons Ltd.

Inhibitor Screening

The instability of BMS-068645 in human blood with or without inhibitors, urine, and methanol (control) was evaluated:

- Esterase inhibitors, diisopropyl fluorophosphate (DFP) and paraoxon in water, phenylmethylsulfonyl fluoride (PMSF) and eserine in DMSO, were added to fresh human blood (heparin treated) to a final concentration of 10 mM of each inhibitor.
- BMS-068645 was added to (1) blood only, (2) blood with 10 mM inhibitors, (3) human urine, and (4) methanol to a final concentration of 25.0 ng/ml. The blood and methanol samples were maintained for 4 h at room temperature and the urine samples were maintained at room temperature for 96 h.
- Two milliliters of acetonitrile containing internal standard were added to completely quench any esterase activity ending the incubation.
- The samples were vortex mixed, centrifuged, and the supernatant was removed and dried under nitrogen at 37°C. The samples were reconstituted with 0.1% formic acid in water and analyzed by LC-MS.

Clinical Study Sample Collection Procedure

Based on the aforementioned preliminary assessment, DFP was chosen as the stabilizer for the compound in human plasma. A detailed clinical sample collection procedure was generated as below. This was followed by (1) stability assessment of the ester prodrug and the active drug in both plasma and whole blood treated as below and (2) stability assessment of the inhibitor in the blood collection tubes:

- The plastic vacutainer tubes (6 ml, K_2EDTA) were treated by the injection of 21 μl of DFP with a syringe with a sharp, thin needle (needle gauge around 22–26).
- After the addition of the inhibitor, the tubes were stored at −30°C and only warmed to room temperature less than half an hour prior to use.
- Blood samples (approximately 6 ml per sample) were collected in the tubes resulting in a final DFP concentration in blood of 20 mM.
- Immediately after collection, each blood sample was gently inverted several times to ensure complete mixing with the anticoagulant (K_2EDTA) and DFP and then placed in chipped ice.
- The blood samples were centrifuged for 15 min at 1000 $\times$ g at 4°C to obtain plasma.
- The separated plasma samples were stored at −30°C until analyzed.

Preparation of Cs/QCs in the Human Plasma

- A 40 μl portion of the 1000 ng/ml stock solution was diluted to 4.0 ml with control human K_2EDTA plasma containing 20 mM DFP to yield a combined stock solution of 10.0 ng/ml for each analyte.
- The plasma pool was diluted with plasma containing 20 mM DFP to obtain the appropriate final concentrations for both the standards and QC samples.
- Calibration standards were freshly prepared daily.
- Additional QC samples with the ester prodrug alone were prepared for monitoring the extent of the possible conversion to the active drug.

Sample Processing Procedure for Human Plasma

- Isotope-labeled internal standards, D5-BMS-068645 and D5-BMS-068645-acid, respectively at 1.0 and 5.0 ng/ml in methanol/water (50/50, v/v) were employed.
- After the addition of 100 μl of the IS working solution and 0.5 ml of 0.1N HCl solution, 4 ml of methyl-*tert*-butyl ether were added to 0.5 ml of each calibration standard, QC sample, and clinical sample.
- The samples were shaken for 20 min, and then centrifuged to separate the liquid phases.
- The organic layer from each sample was transferred to a clean tube and evaporated to dryness.
- The dried extracts were redissolved in 100 μl of the reconstitution solution containing 0.1% formic acid in acetonitrile/water (40/60, v/v) and transferred to injection vials.
- A 10 μl aliquot of the reconstituted samples was injected into the LC-MS system.

REFERENCES

Aldrige WN. Serum esterases. Biochem J 1952;53:110–117.

Bergeron M, Bergeron A, Furtado M, Garofolo F. Impact of plasma and whole-blood anticoagulant counter ion choice on drug stability and matrix effects during bioanalysis. Bioanalysis 2009;1(3):537–548.

Bergmann F, Segal R, Rimon S. A new type of esterase in hog-kidney extract. Biochem J 1957;67(3):481–486.

Briscoe CJ, Hage DS. Factors affecting the stability of drugs and drug metabolites in biological matrices. Bioanalysis 2009;1(1):205–220.

Cao YG, Zhang M, Yu D, Shao JP, Chen YC, Liu XQ. A method for quantifying the unstable and highly polar drug nafamostat mesilate in human plasma with optimized solid-phase extraction and ESI-MS detection: more accurate evaluation for pharmacokinetic study. Anal Bioanal Chem 2008;391(3):1063–1071.

Chang Q, Chow MS, Zuo Z. Studies on the influence of esterase inhibitor to the pharmacokinetic profiles of oseltamivir and oseltamivir carboxylate in rats using an improved LC/MS/MS method. Biomed Chromatogr 2009;23(8):852–857.

Chen J, Hsieh Y. Stabilizing drug molecules in biological samples. Ther Drug Monit 2005;27(5):617–624.

D'Arienzo CJ, Ji QC, Discenza L, et al. DBS sampling can be used to stabilize prodrugs in drug discovery rodent studies without the addition of esterase inhibitors. Bioanalysis 2010;2(8):1415–1422.

Evans MJ, Livesey JH, Ellis MJ, Yandle TG. Effect of anticoagulants and storage temperatures on stability of plasma and serum hormones. Clin Biochem 2001;34(2):107–112.

Freisleben A, Brudny-Klöppel M, Mulder H, de Vries R, de Zwart M, Timmerman P. Blood stability testing: European bioanalysis forum view on current challenges for regulated bioanalysis. Bioanalysis 2011;3(12):1333–1336.

Fung EN, Zheng N, Arnold ME, Zeng J. Effective screening approach to select esterase inhibitors used for stabilizing ester-containing prodrugs analyzed by LC-MS/MS. Bioanalysis 2010;2(4):733–743.

Heinig K, Wirz T, Bucheli F, Gajate-Perez A. Determination of oseltamivir (Tamiflu®) and oseltamivir carboxylate in dried blood spots using offline or online extraction. Bioanalysis 2011;3(4):421–437.

Ishizuka T, Fujimori I, Kato M, et al. Human carboxymethylenebutenolidase as a bioactivating hydrolase of olmesartan medoxomil in liver and intestine. J Biol Chem 2010;285(16):11892–11902.

Kamei T, Uchimura T, Nishimiya K, Kawanishi T. Method development and validation of the simultaneous determination of a novel topoisomerase 1 inhibitor, the prodrug, and the active metabolite in human plasma using column-switching LC-MS/MS, and its application in a clinical trial. J Chromatogr B Analyt Technol Biomed Life Sci 2011;879(30):3415–3422.

Kim MS, Song JS, Roh H, et al. Determination of a peroxisome proliferator-activated receptor γ agonist, 1-(trans-methylimino-N-oxy)-6-(2-morpholinoethoxy-3-phenyl-1H-indene-2-carboxylic acid ethyl ester (KR-62980) in rat plasma by liquid chromatography-tandem mass spectrometry. J Pharm Biomed Anal 2011;54(1):121–126.

Koitka M, Höchel J, Gieschen H, Borchert HH. Improving the *ex vivo* stability of drug ester compounds in rat and dog serum: inhibition of the specific esterases and implications on their identity. J Pharm Biomed Anal 2010;51(3):664–678.

Kromdijk W, Rosing H, van den Broek MP, Beijnen JH, Huitema AD. Quantitative determination of oseltamivir and oseltamivir carboxylate in human fluoride EDTA plasma including the *ex vivo* stability using high-performance liquid chromatography coupled with electrospray ionization tandem mass spectrometry. J Chromatogr B Analyt Technol Biomed Life Sci 2012;891-892:57–63.

Kudo S, Umehara K, Hosokawa M, Miyamoto G, Chiba K, Satouh T. Phenacetin deacetylase activity in human liver microsomes: distribution, kinetics, and chemical inhibition and stimulation. J Pharmacol Exp Ther 2000;294(1):80–87.

Li W, Tse FL. Dried blood spot sampling in combination with LC-MS/MS for quantitative analysis of small molecules. Biomed Chromatogr 2010;24(1):49–65.

Li W, Zhang J, Tse FL. Strategies in quantitative LC-MS/MS analysis of unstable small molecules in biological matrices. Biomed Chromatogr 2011;25(1-2):258–277.

Liederer BM, Borchardt RT. Stability of oxymethyl-modified coumarinic acid cyclic prodrugs of diasteromeric opioid peptides in biological media from various animal species including human. J Pharm Sci 2005;94(10):2198–2206.

Liederer BM, Borchardt RT. Enzymes involved in the bioconversion of ester-based prodrugs. J Pharm Sci 2006;95(6):1177–1195.

Lindegardh N, Davies GR, Tran TH, et al. Rapid degradation of oseltamivir phosphate in clinical samples by plasma esterases. Antimicrob Agents Chemother 2006;50(9):3197–3199.

Meng M, Reuschel S, Bennett P. Identifying trends and developing solutions for incurred sample reanalysis failure investigations in a bioanalytical CRO. Bioanalysis 2011;3(4):449–465.

Minagawa T, Kohno Y, Suwa T, Tsuji A. Species differences in hydrolysis of isocarbacyclin methyl ester (TEI-9090) by blood esterases. Biochem Pharmacol 1995;49(10):1361–1365.

Olivera-Bravo S, Ivorra I, Morales A. The acetylcholinesterase inhibitor BW284c51 is a potent blocker of Torpedo nicotinic AchRs incorporated into the Xenopus oocyte membrane. Br J Pharmacol 2005;144(1):88–97.

Peterson LW, McKenna CE. Prodrug approaches to improving the oral absorption of antiviral nucleotide analogues. Expert Opin Drug Deliv 2009;6(4):405–420.

Rautio J, Kumpulainen H, Heimbach T, et al. Prodrugs: design and clinical applications. Nat Rev Drug Discov 2008;7(3):255–270.

Rooseboom M, Commandeur JN, Vermeulen NP. Enzyme-catalyzed activation of anticancer prodrugs. Pharmacol Rev 2004;56(1):53–102.

Salem II, Alkhatib M, Najib N. LC–MS/MS determination of betamethasone and its phosphate and acetate esters in human plasma after sample stabilization. J Pharm Biomed Anal 2011;56(5):983–991.

Skopp G, Klingmann A, Pötsch L, Mattern R. *In vitro* stability of cocaine in whole blood and plasma including ecgonine as a target analyte. Ther Drug Monit 2001;23(2):174–181.

Stella VJ, Charman WN, Naringrekar VH. Prodrugs: Do they have advantages in clinical practice? Drugs 1985;29(5):455–473.

Tang C, Sojinu OS. Simultaneous determination of caffeic acid phenethyl ester and its metabolite caffeic acid in dog plasma using liquid chromatography tandem mass spectrometry. Talanta 2012;94:232–239.

Tokumura T, Muraoka A, Masutomi T, Machida Y. Stability of spironolactone in rat plasma: strict temperature control of blood and plasma samples is required in rat pharmacokinetic studies. Biol Pharm Bull 2005;28(6):1126–1128.

Tsujikawa K, Kuwayama K, Miyaguchi H, Kanamori T, Iwata YT, Inoue H. *In vitro* stability and metabolism of salvinorin A in rat plasma. Xenobiotica 2009;39(5):391–398.

Wang X, Bowman PD, Kerwin SM, Stavchansky S. Stability of caffeic acid phenethyl ester and its fluorinated derivative in rat plasma. Biomed Chromatogr 2007;21(4):343–350.

Welch RM, Brown A, Ravitch J, Dahl R. The *in vitro* degradation of cisatracurium, the R, cis-R′-isomer of atracurium, in human and rat plasma. Clin Pharmacol Ther 1995;58(2):132–142.

Wiltshire H, Wiltshire B, Citron A, et al. Development of a high-performance liquid chromatographic-mass spectrometric assay for the specific and sensitive quantification of Ro 64-0802, an anti-influenza drug, and its pro-drug, oseltamivir, in human and animal plasma and urine. J Chromatogr B Biomed Sci Appl 2000;745(2):373–388.

Yamaori S, Fujiyama N, Kushihara M, et al. Involvement of human blood arylesterase and liver microsomal carboxylesterase in nafamostat hydrolysis. Drug Metab Pharmacokinet 2006;21(2):147–155.

Yang JZ, Chen W, Borchardt RT. *In vitro* stability and *in vivo* pharmacokinetic studies of a model opioid peptide, H-Tyr-D-Ala-Gly-Phe-D-Leu-OH (DADLE), and its cyclic prodrugs. J Pharmacol Exp Ther 2002;303(2):840–848.

Yoshigae Y, Imai T, Taketani M, Otagiri M. Characterization of esterases involved in the steroselective hydrolysis of ester-type prodrugs of propranolol in rat liver and plasma. Chirality 1999;11:10–13.

Zeng J, Onthank D, Crane P, et al. Simultaneous determination of a selective adenosine 2A agonist, BMS-068645, and its acid metabolite in human plasma by liquid chromatography-tandem mass spectrometry—evaluation of the esterase inhibitor, diisopropyl fluorophosphate, in the stabilization of a labile ester-containing drug. J Chromatogr B Analyt Technol Biomed Life Sci 2007;852(1–2):77–84.

Zhang JG, Fariss MW. Thenoyltrifluoroacetone, a potent inhibitor of carboxylesterase activity. Biochem Pharmacol 2002;63:751–754.

Zhang Y, Tingley FD 3rd, Tseng E, et al. Development and validation of a sample stabilization strategy and a UPLC-MS/MS method for the simultaneous quantitation of acetylcholine (ACh), histamine (HA), and its metabolites in rat cerebrospinal fluid (CSF). J Chromatogr B Analyt Technol Biomed Life Sci 2011;879(22):2023–2033.

35

LC-MS BIOANALYSIS OF ACYL GLUCURONIDES

JIN ZHOU, FENG (FRANK) LI, AND JEFFREY X. DUGGAN

35.1 INTRODUCTION

Acyl glucuronidation is a metabolic pathway for carboxylic acid containing drugs, including many nonsteroidal antiinflammatory drugs (e.g., diclofenac and ketoprofen), fibrate (e.g., clofibrate), anticonvulsants (e.g., valproic acid), diuretics (e.g., frusemide), and others (Regan et al., 2010). In this pathway, uridine diphosphoglucuronyl transferases catalyze the transfer of a glucuronic acid moiety from cosubstrate, uridine 5′-diphosphoglucuronic acid to the –COOH group of the drug molecule, leading to the formation of an ester glucuronide. Unlike the more common ether glucuronides, acyl glucuronides (AGs) are considered reactive metabolites. They are susceptible to nucleophilic substitution and are able to covalently bind to proteins and other macromolecules (Faed, 1984). About 25% of drugs that have been withdrawn from the market due to severe adverse drug events are carboxylic acid-containing drugs (zomepirac, suprofen, alclofenac, indoprofen, etc.), and the formation of AGs were associated with the toxicities (Bailey and Dickinson, 2003). In the FDA's guidance on Metabolites in Safety Testing (FDA, 2008), AGs were listed as a class of metabolites that may require additional safety assessments during the drug development process. Because of these issues, there has been a growing interest in the bioanalysis of AGs in drug discovery and development. In the early drug discovery stage, the measurement of AGs in liver preparations provides a better understanding of compounds' liabilities, assisting with lead compound selection. In the drug development stage, quantification of AGs allows for establishing toxicity–exposure correlations. AGs tend to be more unstable than ether glucuronides. They may undergo hydrolysis to yield free glucuronic acid and parent drugs, intramolecular acyl migration, and intermolecular transacylation. To preserve the integrity of AGs, the process of their bioanalysis requires precautions. This chapter highlights the chemical reactivity and the associated experimental procedures for LC-MS bioanalysis of AGs.

35.2 CHEMICAL REACTIVITY AND BIOANALYTICAL IMPLICATIONS

The chemical reactivity of AGs essentially derives from the electrophilicity of their ester carbonyl group (Faed, 1984). As illustrated in Figure 35.1, AGs may undergo hydrolysis, intramolecular acyl migration, and intermolecular transacylation. The biological consequences of the AG transformation were extensively discussed in several reviews (Bailey and Dickinson, 2003; Regan et al., 2010; Shipkova et al., 2003; Zhou et al., 2001).

AG hydrolysis is the simplest form of nucleophilic substitution. In this process, the carbonyl group is attacked by the hydroxyl group in water. Upon hydrolysis, AGs release a molecule of glucuronic acid to become the parent aglycones. Hydrolysis *ex vivo* results in underestimation of AGs and overestimation of their parent aglycones in their quantitative analysis. Thus, to accurately quantify either an AG or its parent aglycone, it is essential to prevent AG hydrolysis. AG hydrolysis is pH dependent. In general, higher pH accelerates hydrolysis reaction. The presence of esterase or β-glucuronidase in biological matrices can also lead to enzymatic hydrolysis. In addition, plasma proteins may have differential effects on the rate of AG hydrolysis. For example, it was shown that the hydrolysis of oxaprozin glucuronide was accelerated in the presence of plasma proteins and albumin

Handbook of LC-MS Bioanalysis: Best Practices, Experimental Protocols, and Regulations, First Edition. Edited by Wenkui Li, Jie Zhang, and Francis L.S. Tse.

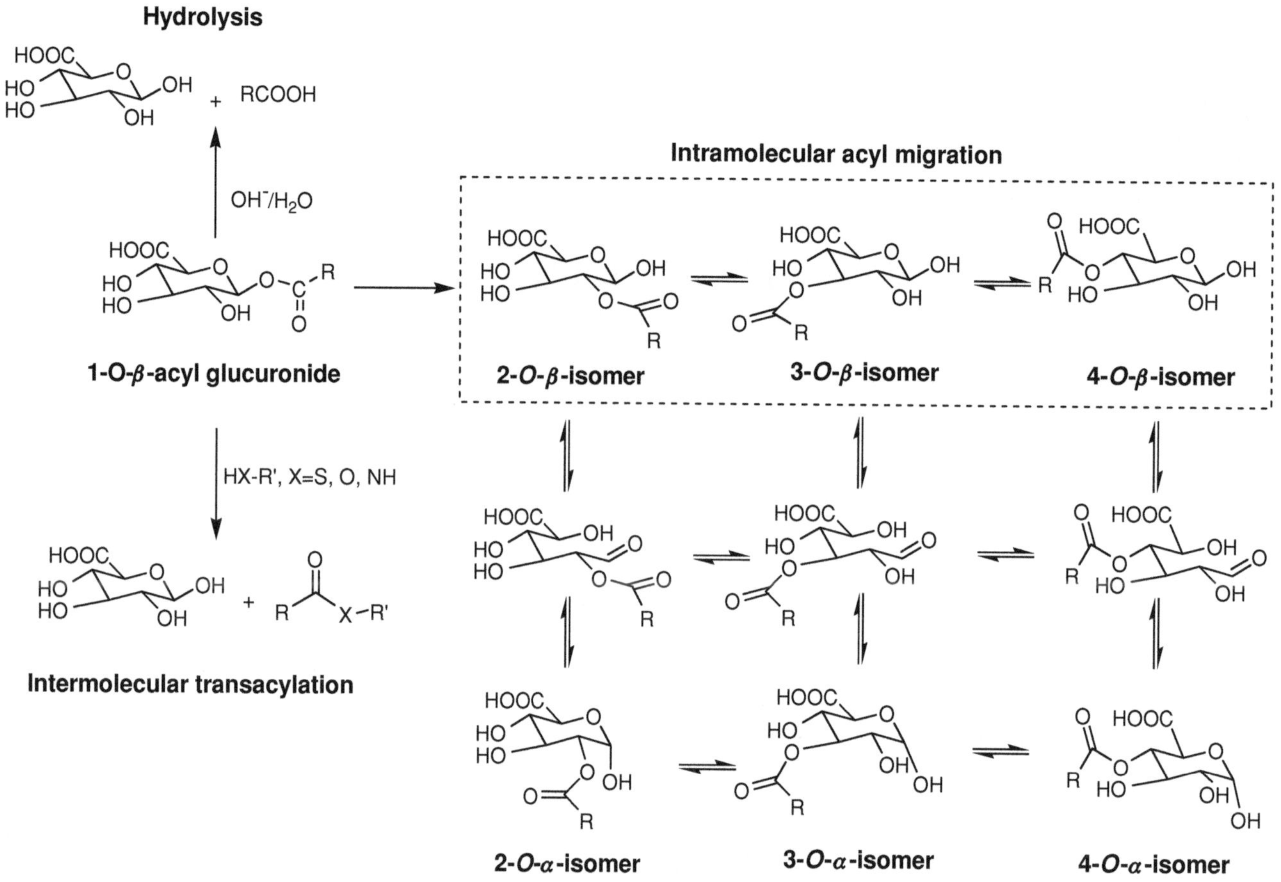

FIGURE 35.1 Summary of AG chemical reactivity.

(Ruelius et al., 1986), whereas human serum albumin had no effect on the degradation rate of S-carprofen glucuronide (van Breemen et al., 1986), and human serum albumin stabilized tolmetin glucuronide (Munafo et al., 1990). It is possible that protein sequestration or surface binding is involved in stabilizing or destabilizing AGs by plasma proteins. If the AG is internally sequestered, it may be protected from degradation. On the other hand, if it is surface bound, and the labile ester bond is exposed, the AG may degrade more rapidly.

Acyl migration describes a process in which the acyl group migrates from the hydroxyl group on the C1 position of the glucuronic acid moiety to the hydroxyl group on C2, and subsequently, C3 and C4 position (Figure 35.1). These reactions also involve nucleophilic substitution, in which the carbonyl group is attacked by the adjacent hydroxyl group in the glucuronic acid moiety. The formed positional isomers are called isoglucuronides (Dickinson, 2011). They are β-glucuronidase resistant but are subject to hydrolysis under an alkaline pH. Similar to AG hydrolysis, acyl migration is also pH dependent. Under basic conditions, isoglucuronides are more easily formed. Isoglucuronides are less liable to direct nucleophilic substitution; however, they can convert to aldehydes through transient ring opening. Subsequent cyclization of the aldehyde intermediates leads to the formation of both α and β anomers (Figure 35.1). The aldehyde intermediates may also react with the amine groups on proteins or DNA nucleosides, resulting in covalent modification of these macromolecules (one proposed mechanism for covalent modification of macromolecules by carboxylic acid containing drugs).

Additionally, AGs may react with $-NH_2$, –OH, or –SH containing molecules through intermolecular transacylation. This reaction is the other proposed mechanism for covalent modification of macromolecules by carboxylic acid containing drugs. Through intermolecular transacylation, AGs may also react with solvents, such as MeOH and EtOH (Cote et al., 2011; Silvestro et al., 2011). As these solvents are popular for sample preparation and high-pressure liquid chromatography (HPLC) analysis, solvolysis may be an issue in AG bioanalysis. For example, Silvestro et al. (2011) reported that reactions with some solvents such as MeOH and EtOH resulted in back-conversion of clopidogrel AG to its parent aglycone. Solvolysis can be prevented by using an inert solvent, such as acetonitrile (ACN). If an alcohol, such as methanol, is

TABLE 35.1 Half-Lives of AGs of Various Drugs in pH 7.4 Aqueous Buffer, at 37°C

Drugs	Half-life (h)
Tolmetin	0.26
Isoxepac	0.29
Probenecid	0.40
Zenarestat	0.42
Zomepirac	0.45
Diclofenac	0.51
Diflunisal	0.67
(*R*)-Naproxen	0.92
(*R*)-Fenoprofen	0.98
Salicylic acid	1.3
DMXAA	1.3
Indomethacin	1.4
(*R*)-Carprofen	1.73
(*S*)-Naproxen	1.8
(*S*)-Fenoprofen	1.93
(*R*)-Benoxaprofen	2.0
(*S*)-Carprofen	3.09
Ibuprofen	3.3
(*S*)-Benoxaprofen	4.1
Bilirubin	4.4
(*R*)-Flunoxaprofen	4.5
Furosemide	5.3
Flufenamic acid	7
Clofibric acid	7.3
(*S*)-Flunoxaprofen	8.0
Mefenamic acid	16.5
(*R*)-Beclobric acid	22.4
(*S*)-Beclobric acid	25.7
Telmisartan	26
Gemfibrozil	44
Valproic acid	79

Reproduced with permission from (Stachulski et al., 2006).

needed in the process, potential solvolysis should be considered. Lowering pH can also protect against solvolysis.

The chemical reactivity of an AG highly depends on the structure of its parent aglycone. One way to assess the chemical reactivity of an AG is to determine its stability in aqueous buffers at pH 7.4 and 37°C. Those AGs yielding the shortest half-lives are considered the least stable and most reactive. Table 35.1 lists the half-lives of AGs of various drugs. It is obvious that the stabilities of AGs vary significantly, from extremely unstable compounds, such as tolmetin AG (half-life = 15 min), to relatively stable compounds, such as valproic acid AG (half-life = 79 h) (Stachulski et al., 2006).

35.3 SAMPLE COLLECTION AND STORAGE

The instability of AGs makes careful sample handling crucial. As described earlier, hydrolysis and acyl migration are likely to occur at neutral to basic pH. To prevent or slow down these processes, *ex vivo* samples (blood, plasma, sample extracts) are usually handled at low temperatures (e.g., on ice) and maintained at acidic pH. Samples should also be stored at ≤−20°C. The common practice is to immediately cool samples on ice after collection and adjust the pH to 2–4 with acids (Shipkova et al., 2003). Because mineral acids have undesirable effects on biological matrices such as blood hemolysis and plasma protein precipitation, organic acids such as citric acid are often used. It is also worth mentioning that even if the samples are treated with acids before stored at below −20°C, degradation of AGs may still occur with long-term storage. As an example, it was reported that mycophenolic acid AG (AcMPAG) remained stable in acidified plasma for up to 5 months at −20°C and −80°C; however, significant reduction in AcMPAG concentration was observed thereafter (de et al., 2008). Therefore, samples containing AGs should be analyzed as soon as possible.

Moreover, additional precautions may be taken when analyzing AGs in different biological matrices. EDTA (Stachulski et al., 2006) and citrate are often used as an anticoagulants in plasma samples, since they are good inhibitors of plasma esterases. It is also prudent to separate plasma within 10 min after blood collection using a cooled centrifuge (0–4°C) (Shipkova et al., 2003). In some cases, blood may be collected in standard citrate tubes, and then additional citric acid may be added to the resulting plasma in order to lower the pH and to provide further protection of the AGs during the frozen storage. This procedure has the distinct advantage in that there would not be any severe hemolysis, as it would occur with citric acid being added directly to collected blood. Bile is slightly alkaline making it an ideal medium for AG degradation. To preserve AG integrity in bile, samples should be collected on ice and in tubes containing pH modifiers such as acetic acid (1–2 M, pH 4–5) or ammonium acetate (1 M, pH 4–5) (Mullangi et al., 2005). Urine usually has slightly acid pH, at which AG is relatively stable; however, urine pH may vary, and it may change due to the presence of drugs (Vree et al., 1994b) or during handling or storage. It is crucial to maintain acidic urine pH during sample collection, storage, and processing. For tissue samples, enzyme-catalyzed hydrolysis may be a concern. An esterase inhibitor (such as phenylmethyl sulfonyl fluoride) and a β-glucuronidase inhibitor (1,4-saccharolactone) are usually added to tissue samples prior to tissue homogenization (Mullangi et al., 2005).

Although most AGs are stable under acidic conditions, there are exceptions. For example, more than 400% deviation was observed for the concentrations of clopidogrel acid AG after the sample extracts were reconstituted in a pH 3 solution and stored at 4°C for 96 h prior to LC-MS/MS analysis, whereas under the same conditions, only 3% deviation was seen when sample extracts were reconstituted in a pH 7 solution (Bergeron et al., 2009). Flumequine AG was also found to be stable at pH 5–8, instead of at a more acidic pH (Vree et al., 1992). Thus, the pH conditions to stabilize AG during bioanalysis cannot be standardized. To establish the sample collection, storage, and processing conditions, extensive stability evaluation must be performed during method development.

35.4 SAMPLE PREPARATION

In general, sample preparation procedures for AGs should be kept as simple as possible to minimize the potential risk of AG migration or hydrolysis (Xue et al., 2006). Because it is a simple and fast procedure, protein precipitation with organic solvents, acids, or combinations of organic solvents and acids is the most commonly used approach to extract AGs from biological matrices. (Vree et al., 1993, 1994a, 1994b; Liu and Smith, 1996; Shipkova et al., 2003; Khoschsorur and Erwa, 2004; Brandhorst et al., 2006; Xue et al., 2008; Klepacki et al., 2012; Silvestro et al., 2011). However, samples prepared with protein precipitation may have higher matrix effects. Protein precipitation with organic solvents would also increase samples' organic solvent contents, which may worsen the peak shapes of AGs in certain chromatographic systems and require further dilution of samples with aqueous solvents. The efficiency of different protein precipitation methods also varies and should be evaluated during the method development. For instance, it was reported that protein precipitation with 2 M perchloric acid is not a good method for extraction of mycophenolic acid acyl glucuronide (AcMPAG) from plasma (Shipkova et al., 2003). Only 63% of AcMPAG was recovered with this method, suggesting degradation and/or incomplete release of AcMPAG. In comparison, a recovery of 102% was achieved when AcMPAG was extracted with 15% metaphosphoric acid (Shipkova et al., 2003).

Besides protein precipitation, solid-phase extraction (SPE) (Castillo and Smith, 1993; Schwartz et al., 2006), and liquid-liquid extraction (LLE) (Hermening et al., 2000) have also been used in AG sample preparation. LLE is relatively simpler than SPE; however, because AGs are polar and hydrophilic, extraction of AGs to nonpolar organic solvents such as cyclohexane, methyl-*tert*-butyl ether, or *n*-butyl chloride can be difficult (low recovery). Due to the largely different liphophilicity between AGs and their parent aglycones, it has been recommended to use a combination of polar and less polar organic solvents such as ethyl acetate/diethyl ether (1:1, v:v) to simultaneously extract both AGs and parent aglycones via LLE (Trontelj, 2012). In addition, to make sure that the glucuronic acid moieties (pK_a of 3.1–3.2) in AGs are protonated, acidic conditions are required during LLE.

SPE is also a widely used extraction technique. Depending on the chemical structure and hydrolipophilicity of the parent aglycones, the retention of AGs on SPE cartridge can be facilitated via a number of mechanisms from hydrophobic to ion pairing. It is important to choose the right combination of solid phases and eluents for the best performance. If a reverse phase mechanism is used to retain AGs, the organic solvent percentage of eluents in the washing step should be carefully selected. Too high a percentage of organic solvent may wash the hydrophilic AGs to waste, whereas too low a percentage of organic solvent may fail to wash away potentially interfering phospholipids in the matrix, resulting in higher matrix effects during LC-MS analysis. In addition, depending on the stability of AGs, acid may need to be added to the SPE eluents in order to prevent AG degradation (Castillo and Smith, 1993; Annesley and Clayton, 2005). For extremely unstable AGs, analysts can also consider online preparation with a column switching technique (Li et al., 2011; Mano et al., 2002).

35.5 LC-MS/MS QUANTIFICATION

35.5.1 Direct Analysis of AG

Quantitative analysis of AGs using LC-MS/MS can be achieved via either a direct or an indirect approach. Direct quantification allows simultaneous measurement of both AGs and their parent aglycones; however, this approach is only applicable when AG reference standards are available and their stability during sample analysis is manageable. This section focuses on discussion of direct analysis of AGs. The indirect approach is presented in Section 35.5.3.

In direct quantification, one often needs to separate the AGs from their parent aglycones and rearrangement isomers. Because the lipophilicities of AGs and their parent aglycones are largely different, their separation on HPLC can be easily achieved. Chromatographic separation of AGs from their rearrangement isomers, on the other hand, is challenging and requires well-defined chromatographic conditions. Usually, ionic strength and the pH of the mobile phase are the most important factors in separating carboxylic acid containing compounds (Andersen and Hansen, 1992; Khan et al., 1998; Xue et al., 2006). Manipulating solvent strength and mobile phase flow rate can also help further improve the resolution (Khan et al., 1998). By optimizing these four parameters, Khan et al. (1998) successfully separated ifetroban 1-*O*-β-AG from its six rearrangement isomers and its α-anomer (synthetic impurity). Figure 35.2 shows the effect of pH and ionic strength on the resolutions between ifetroban1-*O*-β-AG and its isomers. Lowering the pH increased the retention of ifetroban1-*O*-β-AG, but decreased the resolution between the AG and its isomers. Ifetroban 1-*O*-β-AG and its rearrangement isomers were best separated at pH 5-5.5; whereas, increasing ammonium acetate concentration resulted in an increase in both retention and resolution. The final mobile phase for Ifetroban 1-*O*-β-AG separation was 10 mM ammonium acetate, pH 5, containing 30% ACN.

Under the above mobile phase conditions, ifetroban 1-*O*-β-AG and its isomers eluted in the following order: 4-*O*-acyl isomers $>$1-*O*-α-anomer$>$ 1-*O*-β-anomer$>$ 3-*O*-acyl

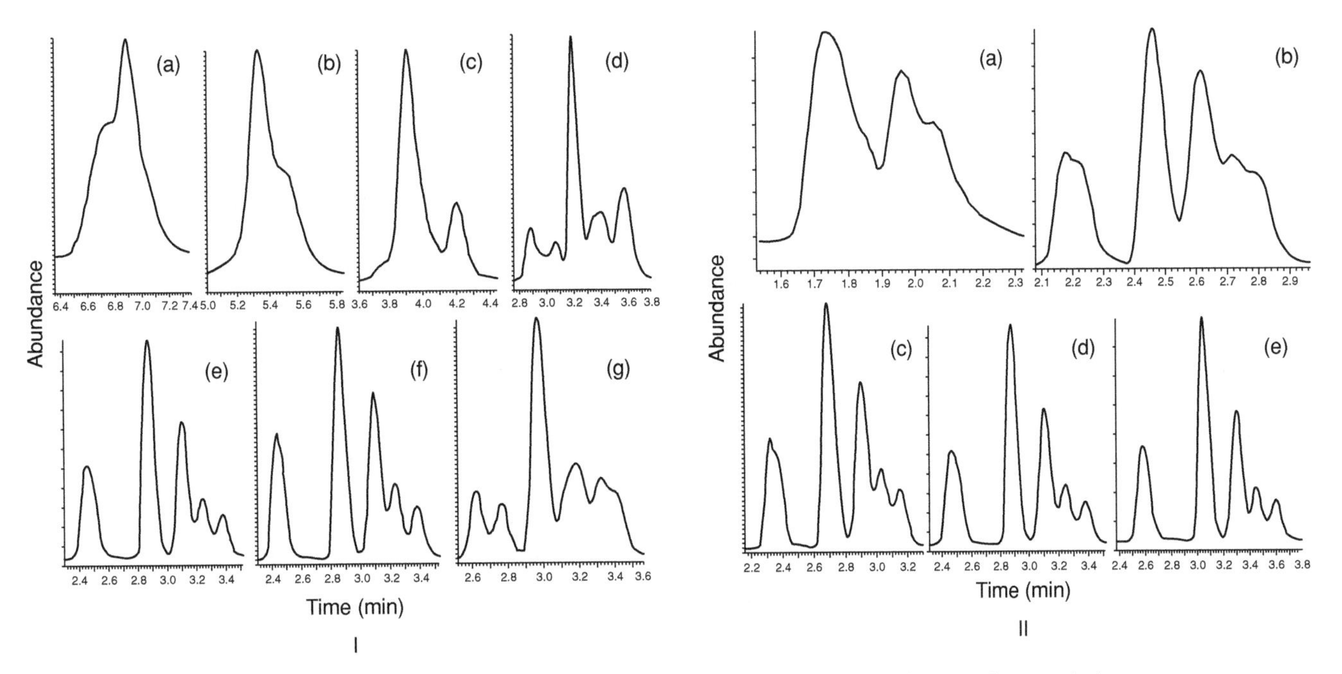

FIGURE 35.2 LC-MS chromatograms depicting the effects of pH and ionic strength on resolutions between ifetroban 1-*O*-*β*-AG and its rearrangement isomers. I. Resolution vs. pH with 7.5 mM ammonium acetate and 35% ACN and a flow rate of 0.25 ml/min: (a) pH 3.5; (b) pH 4.0; (c) pH 4.5; (d) pH 5.0; (e) pH 5.5; (f) pH 6.0; (g) pH 6.5. II. Resolution vs. ammonium acetate concentration at pH 5.5 with 35% ACN and a flow rate of 0.25 ml/min: (a) 1.0 mM; (b) 2.5 mM; (c) 5.0 mM; (d) 7.5 mM; (e) 10.0 mM (reproduced with permission from Khan et al., 1998).

isomers > 2-*O*-acyl isomers (Figure 35.3). It has been noted that the order of elution of AG isomers is the same regardless of the structures of the aglycones (Corcoran et al., 2001; Farrant et al., 1995; Mortensen et al., 2001; Sidelmann et al., 1996a; Lenz et al., 1996; Sidelmann et al., 1996b).

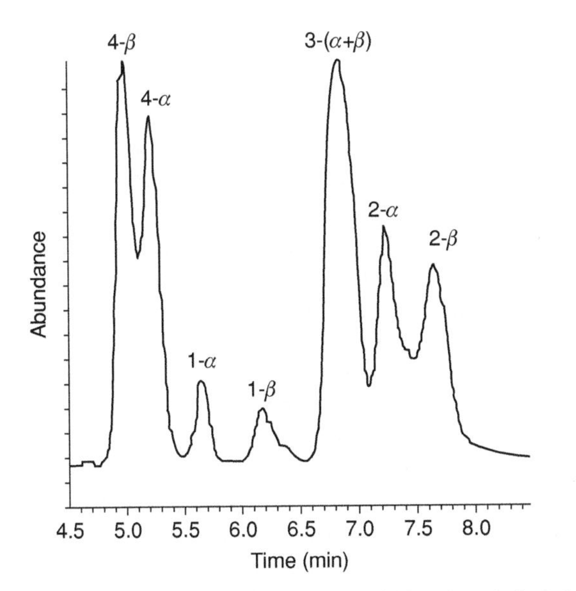

FIGURE 35.3 LC-MS chromatogram of ifetroban 1-*O*-*β*-AG and its rearrangement isomers Mobile phase conditions: pH 5.0, 10 mM ammonium acetate, and 30% ACN (reproduced with permission from Khan et al., 1998).

In addition, because the α and β anomers of 2, 3, or 4-isoglucuronides are in rapid equilibrium and can interconvert on the HPLC timescale, their complete separation is usually impossible, and distorted peak shapes are often observed (Stachulski et al., 2006). Because of this, very often, the peaks of these isomers are integrated as one peak.

Chromatographic separation of AG isomers often requires long HPLC run times. To decrease analysis time and increase the throughput in AG sample analysis, Xue et al. (2008) examined the feasibility of resolving the AGs from their rearrangement isomers by using a specific selected reaction monitoring (SRM) transition. Their findings on a model compound, muraglitazar AG, suggested that 1-*O*-*α*/*β*-AG was more prone to in-source fragmentation to the parent aglycone than the migration isomers. Under certain ion-source conditions, negligible fragmentation of the migration isomers to the parent aglycone was observed. As 1-*O*-*α*-AG is merely a synthetic impurity and does not exist *in vivo*, the SRM transition with the loss of the glucuronide moiety (−176 Da) is considered specific for 1-*O*-*β*-AG in biological matrices. By optimizing the declustering potential and the collision energy, Xue et al. (2008) developed a specific and high-throughput (2.5 min) method to quantify muraglitazar 1-*O*-*β*-AG in rat, mouse, monkey, and human plasma. The results they obtained with the high-throughput method were equivalent to the results obtained with a long LC method (Xue et al., 2006). They suggested that this approach might be applicable for other AGs.

35.5.2 In-source Fragmentation

The ester bonds of AGs are thermally unstable. During the ionization process and AGs entering the Q1 chamber, the ester bonds may break down even when the soft ionization techniques such as APCI or ESI are used. In-source fragmentation leads to formation of ions that are identical to the parent aglycones and causes overestimation of the aglycone concentrations if the aglycones are not chromatographically separated from the AGs. In general, in-source fragmentation tends to be more severe in APCI than ESI (Wainhaus, 2005). The settings of ion source temperature, cone voltage and declustering potential are critical to control the extent of in-source fragmentation (Liu and Pereira, 2002; Yan et al., 2003; Xue et al., 2008; An et al., 2010). Although analyte-specific tuning of ionization conditions can help minimize in-source fragmentation, chromatographic separation of AGs from their parents is the ultimate solution.

In addition, Mess et al. (2011) proposed to monitor the $[M\text{-}H]^-$ or $[M+NH_4]^+$ ions for the parent aglycones to minimize the impact of in-source fragmentation of AGs. As shown in Figure 35.4, AGs exhibited intense in-source fragmentation to the parent drug $[M+H]^+$ ions under positive ESI. However, only minimal in-source fragmentation to the $[M+NH4]^+$ adducts and absolutely no in-source fragmentation under negative ESI were observed. The explanation given for the minimal in-source fragmentation to the $[M+NH4]^+$ adducts is that the in-source fragmentation of AGs mostly occurs after the ammonium adducts are formed. As to the reason why AGs do not show in-source fragmentation under negative ESI conditions, the authors proposed that under negative ESI conditions, the negative charge was located at the carboxylic acid group in the glucuronic acid moiety, and therefore, the fragmentation of AGs to parent aclycones, if any, was a charge-remote event, which is much more difficult to take place than a charge-driven in-source fragmentation event occurred under positive ESI conditions when the site of protonation was located at the carbonyl oxygen of the ester.

35.5.3 Quantification of AG in the Absence of a Reference Standard (Indirect Quantification)

In many situations, especially during drug discovery stage, the authentic standards of AGs are not available. Quantification of AGs can be achieved by an indirect difference method. In this method, a sample is divided into two aliquots: one aliquot is treated with acid to preserve the integrity of the AG and the other aliquot is treated with alkaline (Loewen et al., 1989; Grubb et al., 1996; Hermening et al., 2000; Srinivasan et al., 2010) or β-glucuronidase (Vree et al., 1994bl; Stass and Kubitza, 1999; Zhao et al., 2001; Zhou et al., 2001) to convert the AG to its parent aglycone. The concentrations of the parent aglycone are then determined for both aliquots and their difference corresponds to the concentration of the AG in the sample. In samples containing both the AG and its rearrangement isomers, β-glucuronidase can only catalyze the hydrolysis of the AG, not the rearrangement isomers. Alkaline treatment, on the other hand, releases the aglycone from all isomers. Thus, if the samples are divided into three aliquots, one for acid, one for β-glucuronidase and one for alkaline treatment, the concentration of an AG and the total concentration of its rearrangement isomers can both be determined. As an example, Figure 35.5 shows the LC-MS/MS chromatograms of zomepirac, its AG, and rearrangement isomers after different treatments as well as the calculations for the concentration of the AG and total rearrangement isomers.

To quantitate AGs with the difference method, it is important to achieve complete AG deconjugation. In general, β-glucuronidases from different sources, such as bovine liver, Helix pomatia, Ampullari, and *Escherichia coli*, have different hydrolytic activity and pH optima (Vree et al., 1992; Vree et al., 1994b; Kamata et al., 2003). To hydrolyze a particular AG, β-glucuronidases from several sources are usually examined in order to find one with the highest activity toward the AG of interest. The pH conditions for the enzymatic reaction should also be evaluated and optimized. In addition, because the ionization of the aglycone in different aliquots may be different, in order to accurately determine the concentration of the aglycone in each aliquot, separate standard curves should be prepared.

The indirect quantification of AGs with the difference method is not as accurate as the direct quantification, especially when the concentrations of AGs are much lower than the parent aglycones. Because the concentration is determined based on the difference of two measurements, the additive effect of errors can sometimes result in a negative value. An alternative way to quantify AGs without reference standards is to utilize nanospray mass spectrometry. The method is based on the principle that at very low flow rates where pure electrospray takes place, there is no differential droplet desorption effect, such that the differential responses seen for molecules of different polarities and charges disappear and a polar AG exhibits the same MS response as its nonpolar parent aglycone. In a method developed by Valaskovic et al. (2006), the nanospray MS was coupled with the conventional LC-MS/MS. A calibration factor was calculated based on the relative peak area between the parent and the metabolite from nanospray MS and the relative peak area ratio from the conventional LC-MS/MS. Samples were then analyzed with LC-MS/MS and the metabolite was quantified against the standard curve of the parent compound after normalization with the calibration factor. Figure 35.6 summarizes the overall scheme of this approach. Valaskovic et al. (2006) applied this approach to quantify the AG of a proprietary compound and obtained comparable results to those yielded with an authentic standard. The drawback of this method is that it requires special instrument setup. In addition, more compounds need to be tested with this method to prove its applicability.

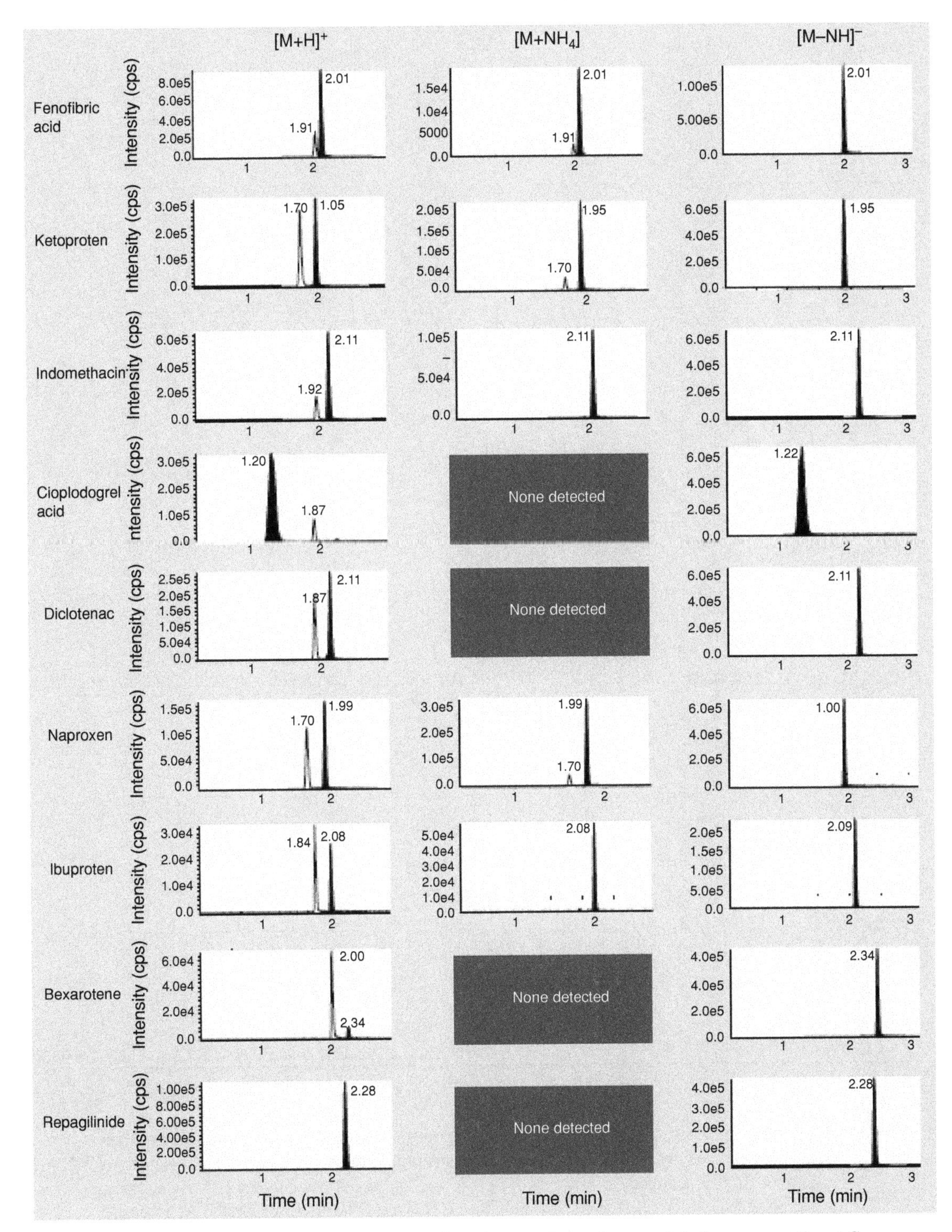

FIGURE 35.4 LC-MS chromatograms of a mixture of nine drugs and their corresponding AGs monitored under different detection modes. The drug peaks are solid and the peaks from in-source fragmentation of the AGs are open (reproduced with permission from Mess et al., 2011).

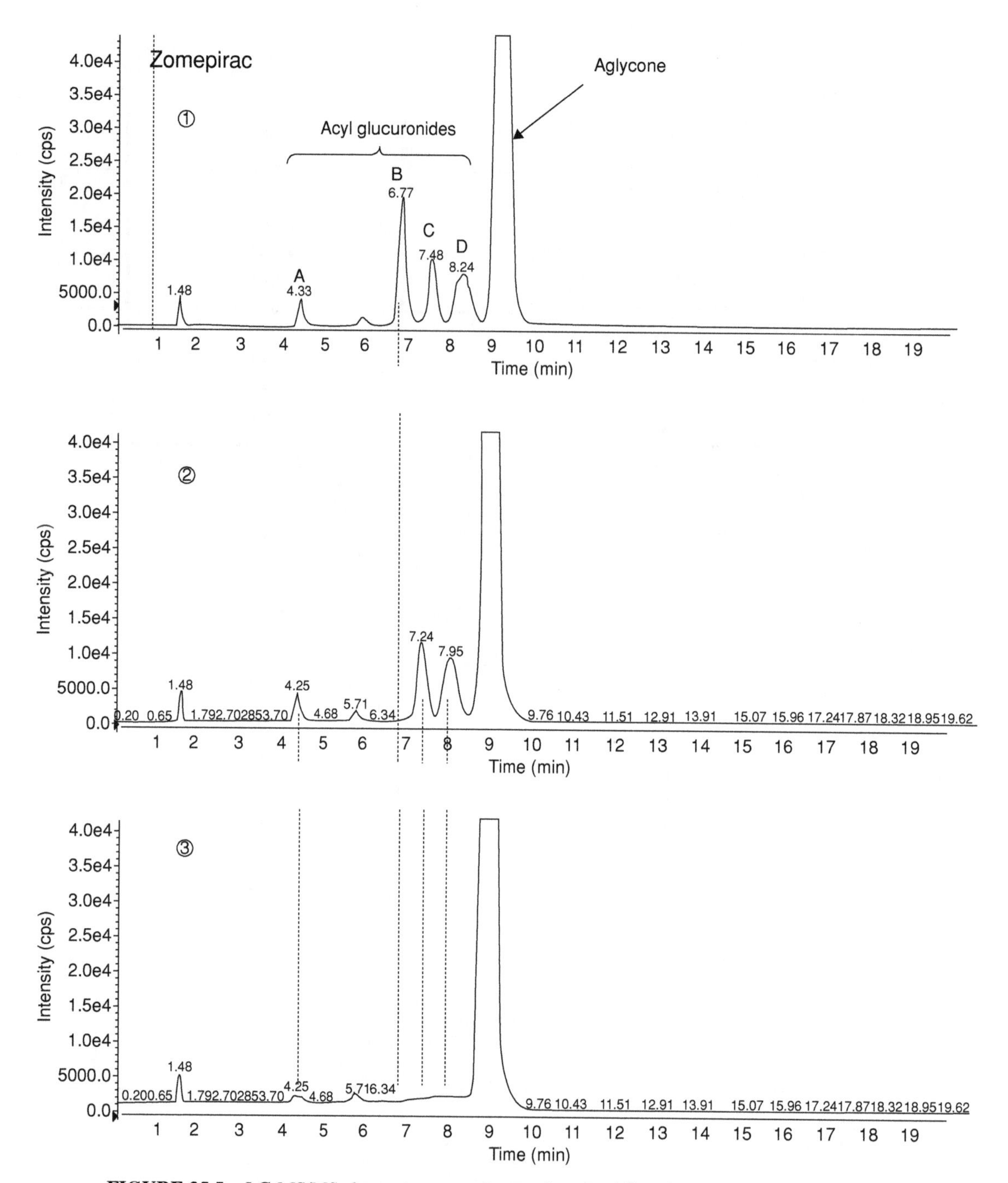

FIGURE 35.5 LC-MS/MS chromatograms of zomepirac, its AG and rearrangement isomers after different treatments: (1) obtained after zomepirac incubation with microsomes; (2) samples were hydrolyzed with β-glucuronidase estimate ([1-*O*-AG]=([zomepirac]$_2$-zomepirac]$_1$); (3) samples were hydrolyzed with alkaline ([rearrangement isomers]=[zomepirac]$_3$-zomepirac]$_2$) (reproduced with permission from Bolze et al., 2002).

35.6 INCURRED SAMPLE REANALYSIS OF AGs

Incurred sample reanalysis (ISR) plays an important role in confirming the quality of sample analysis. Readers can refer to Chapter 5 for a detailed discussion on the general procedures and acceptance criteria for ISR. For samples containing AGs, ISR should be conducted as early as possible to identify any problems associated with AG instability. ISR is not only important to ensure the quality of AG quantification but also useful in identifying the potential interference of AGs on the quantification of their parent aglycones. As noted previously, AG hydrolysis artificially increases the concentration of the

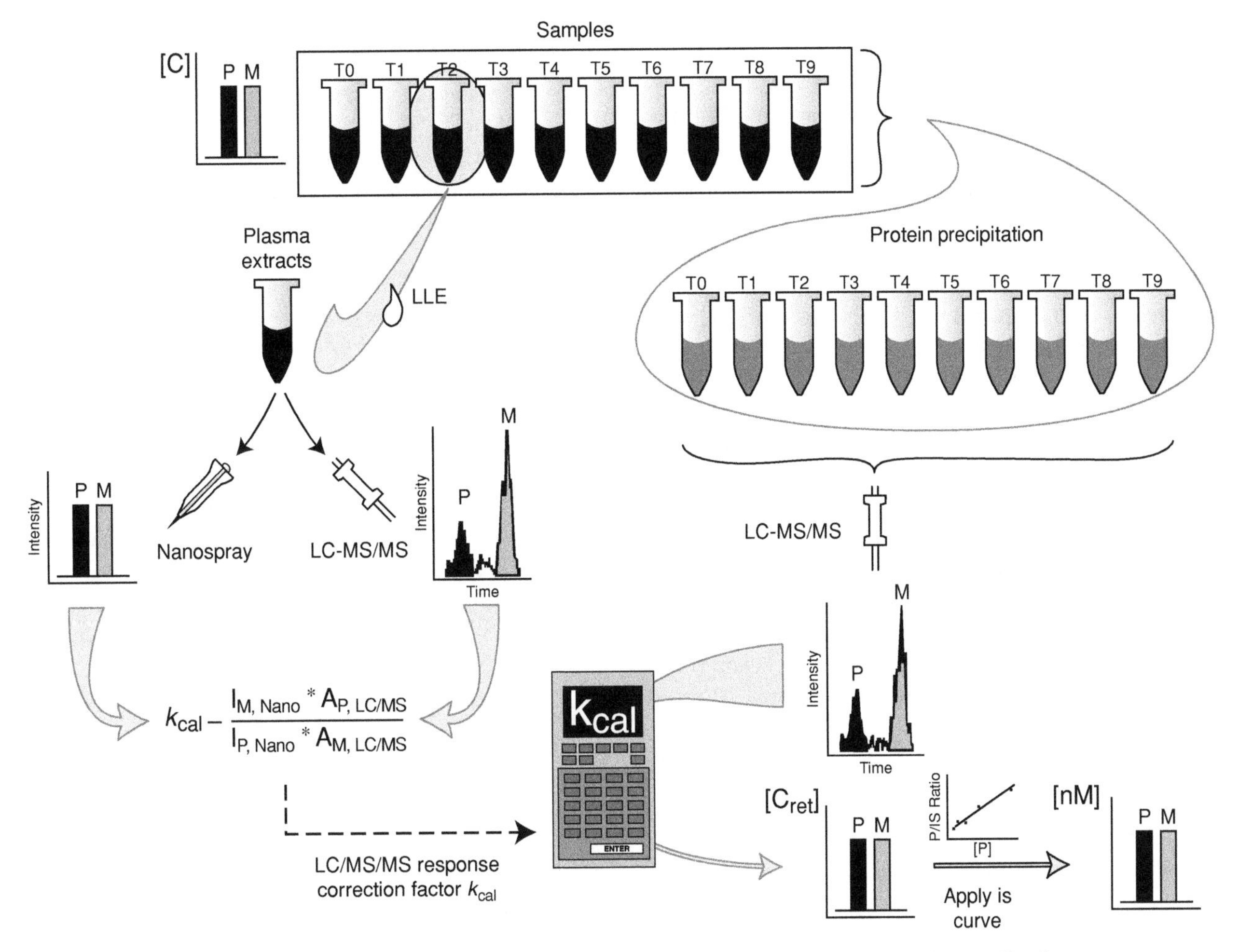

FIGURE 35.6 Summary of coupling nanospray MS with conventional LC-MS/MS in quantification of metabolites without reference standards (reproduced with permission from Valaskovic et al., 2006).

aglycone in *ex vivo* samples, affecting the accurate quantification of the aglycone. This process is especially problematic when the concentration of the aglycone is much smaller than the AG. ISR can help indentify the interference of AG hydrolysis on the aglycone quantification. In cases when the agycone is the only analyte of interest, ISR results with consistent and positive changes in the aglycone's concentrations suggest that AG hydrolysis should be examined, and actions to further prevent AG hydrolysis may be needed.

35.7 SUMMARY

LC-MS (/MS) has become a commonly used approach in AG bioanalysis. Depending on the availability and stability of AG reference standards, quantification can be achieved using direct or indirect methods. The direct methods require authentic AG standards, representing a more convenient and accurate approach. Indirect methods, which involve enzymatic or alkaline cleavage of AG's ester bond and determination of the aglycone concentration, are less accurate; however, they can be used as an alternative approach to quantify AGs when the authentic standards are not available. The major challenge in AG bioanalysis is to deal with the compound instability. Well-defined sample collection, storage, and handling procedures are needed to protect AGs from hydrolysis and acyl migration. AGs may interfere with the bioanalysis of their parent aglycones via in-source fragmentation in the LC-MS inlet. The best way to eliminate the interference is to chromatographically separate AGs from their parent aglycones so that each enters the LC-MS inlet separately. Monitoring the parent aglycones with $[M\text{-}H]^-$ or $[M+NH_4]^+$ ion could also minimize the in-source fragmentation of some AGs. In addition to in-source fragmentation, AG hydrolysis, which librates parent aglycones, is also problematic for aglycone quantification. Measures should be taken to prevent AG hydrolysis even though the parent aglycone is the only analyte of interest. Using incurred samples in addition to regular quality control samples can help identify the interference of AGs.

35.8 REPRESENTATIVE PROTOCOLS

35.8.1 Direct Quantification of a BMS Drug Candidate and Its AG in Rat Plasma (Xue et al., 2006)

Equipment and reagents

- Reference standards of the aglycone and the AG.
- 0.5 M citric acid.
- Rat plasma.
- Column: Phenomenex Luna C18, 3μ, 3 × 150 mm (Torrance, CA, USA).
- Shimadzu LC-10ADvp HPLC system (Columbia, MD, USA).
- Sciex API 4000 Triple quadrupole mass spectrometer (Foster city, CA, USA).
- ACN.
- Acetic acid.

Methods

1. Add 1 ml of 0.5 M citric acid to 5 ml plasma (final pH ~3.7) and store treated samples at −20°C or colder.
2. Mix a 50 μl aliquot of sample with 150 μl ACN containing 0.1% acetic acid and internal standard.
3. Vortex samples for 1 min and then centrifuge for 10 min.
4. Transfer the supernatant to a 96-well plate and inject into a LC-MS/MS system.[a]
5. LC-MS/MS conditions

   LC conditions[b]*:* isocratic elution with 70:30 (v/v) ACN/water containing 0.075% formic acid, pH ~2.9. A chromatogram depicting the separation of the aglycone, its AG from the AG rearrangement isomers is shown in Figure 35.7.

   MS conditions: ESI-MS/MS positive, *m/z* 531→306 for the aglycone and *m/z* 707→186 for AG. Product ion spectra of $[M+H]^+$ for the parent compound and the AG are shown in Figure 35.8.
6. Method validation
   - Prepare the parent aglycone and its AG stock solutions in dimethyl sulfoxide (DMSO) at 1 mg/ml.[c]
   - Prepare calibration standards in citric acid treated rat plasma at the following concentrations: 5, 10, 25, 50, 100, 500, 1000, 2500, and 5000 ng/ml for both AG and its parent aglycone.[d]
   - Prepare five levels of QC (5, 15, 2000, 4000, and 50,000 ng/ml) in rat plasma with dilution method.[c]
   - Validate the method by examining the standard curve linearity, QC accuracy and precision, specificity, limit of quantitation, and stability.

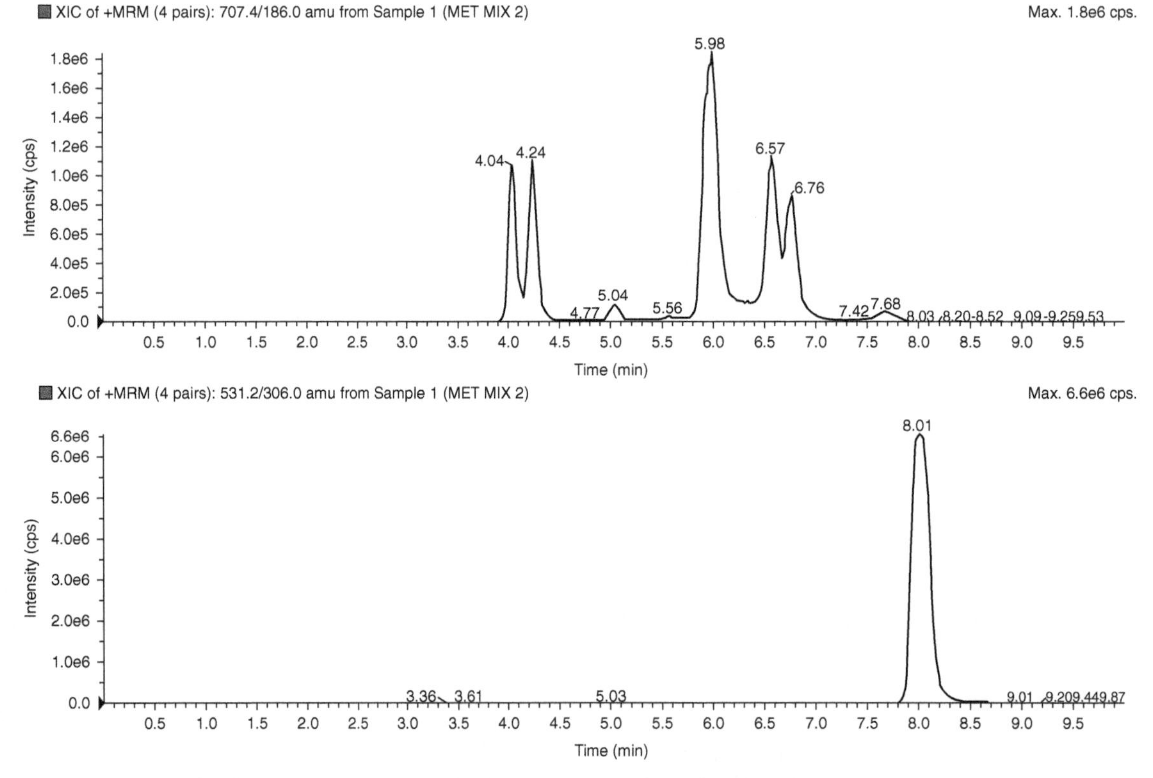

FIGURE 35.7 A chromatogram depicting the separation of a BMS drug candidate and its AG from AG rearrangement isomers. The parent drug and the AG were eluted at 8.01 min and 5.56 min, respectively (reproduced with permission from Xue et al., 2006).

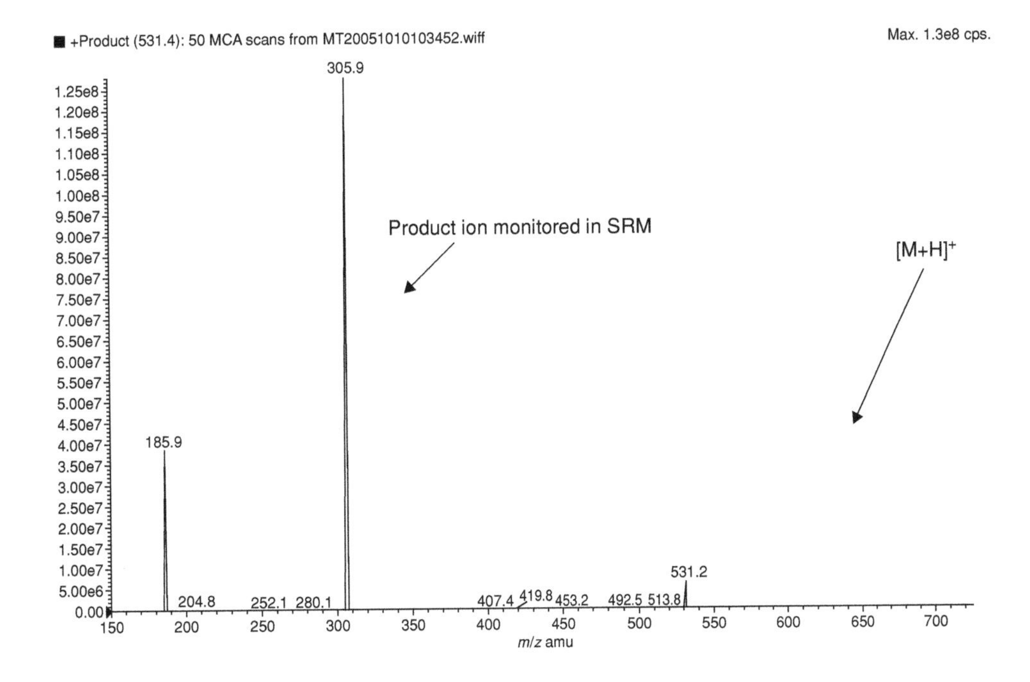

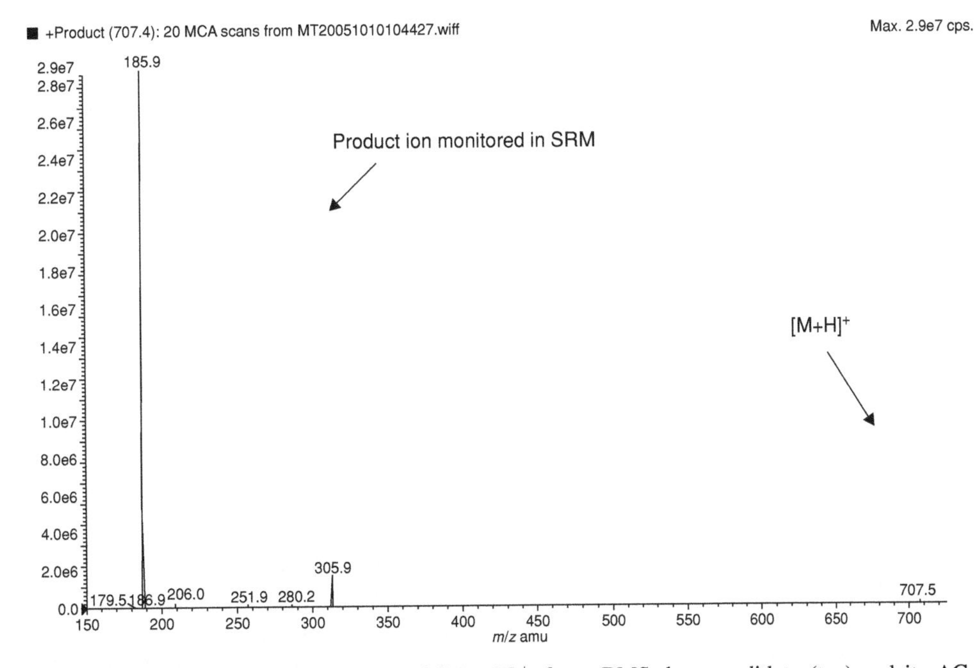

FIGURE 35.8 Product ion spectra of $[M+H]^+$ for a BMS drug candidate (top) and its AG (bottom) (reproduced with permission from Xue et al., 2006).

Notes

[a]Keep autosampler at 4°C during sample analysis to minimize possible conversion of AG.

[b]To achieve the desired separation, mobile phase pH, buffer concentration, and the organic content of the mobile phase are optimized.

[c]The DMSO stock solutions and the QC samples are stored at −20°C or colder.

[d]The calibration standards are prepared freshly.

35.8.2 Quantification of Zomepirac AG and Its Rearrangement Isomers in Human Liver Microsomal Incubation in the Absence of Their Reference Standards (Bolze et al., 2002)

Equipment and reagents

- 4% Trifluoroacetic acid in ACN.
- Zomepirac (Sigma).
- *Analytical column*: Hypersil BDS (125 X 4-mm i.d.; Thermoquest; Thermo Finnigan MAT; San Jose, CA).

- *Mass spectrometer*: API 365 (Applied Biosystems, Toronto, ON, Canada).
- *Mobile phase*: A, ACN/10 mM acetate ammonium buffer (70:30, v/v) containing 0.5% acid acetic; B, ACN/10 mM acetate ammonium buffer (4:96, v/v).
- Bovine-β-glucuronidase (Sigma).
- Phenolphthalein-1-*O*-glucuronide (Sigma).
- 1 N KOH.

Methods

1. Stop the incubation reaction by adding 1 ml of 4% trifluoroacetic acid in ACN (pH ~3–4).
2. Centrifuge the incubation samples at 1500 rpm for 10 min and collect the supernatants.
3. Store the supernatants at −80°C until analysis.
4. Develop a LC-MS/MS method for zomepirac quantification. Gradient elution at flow at of 1 ml/min is used to separate the AG isomers from the aglycone. The representative chromatogram is shown in Figure 35.5. The analytical column, mass spectrometer equipment, and mobile phase conditions are listed above.
5. Determine the free aglycone concentration in the first aliquot with LC-MS/MS ($[\text{aglycone}]_1$).[a]
6. Verify bovine-β-glucuronidase activity with a positive control, phenolphthalein-1-*O*-glucuronide.
7. Incubate the second aliquot with 1000 units of bovine-β-glucuronidase at 37°C for 2 h to cleave β_1-conjugates and liberate the corresponding aglycone part.[b]
8. Determine the free aglycone concentration in the second aliquot with LC-MS/MS ($[\text{aglycone}]_2$).[a] The concentration of 1-*O*-β-AG can be calculated with the following equation: $[1\text{-}O\text{-}\beta\text{-AG}] = [\text{aglycone}]_2 - [\text{aglycone}]_1$.
9. Incubate the third aliquot with 1 N KOH at 80°C for 3 h.
10. Determine the free aglycone concentration in the third aliquot with LC-MS/MS ($[\text{aglycone}]_3$).[a] The concentration of AG isomers can be calculated with the following equation: $[\text{AG isomers}] = [\text{aglycone}]_3 - [\text{aglycone}]_2$.

Notes

[a]Samples for the determination of free aglycone and the 1-*O*-β-AG concentration are diluted before injection into the analytical system. Calibration curves from 50 to 10,000 ng/ml, are prepared by spiking the adequate amount of zomepirac into blank matrix. Samples for determination of AG isomers are extracted by a SPE method using Oasis HLB cartridges (Waters, Saint Quentin en Yvelines, France). A separate calibration curve from 5 to 1000 ng/ml, is prepared.

[b]A preliminary experiment needs to be performed to ensure complete hydrolysis of 1-*O*-β-AG and the chemical stability of the AGs isomers during the 2-h incubation period with bovine-β-glucuronidase. The evaluation can be performed by direct graphic assessment of peak areas of corresponding compounds.

REFERENCES

An G, Ruszaj DM, Morris ME. Interference of a sulfate conjugate in quantitative liquid chromatography/tandem mass spectrometry through in-source dissociation. Rapid Commun Mass Spectrom 2010;24(12):1817–1819.

Andersen JV, Hansen SH. Simultaneous quantitative determination of naproxen, its metabolite 6-O-desmethylnaproxen and their five conjugates in plasma and urine samples by high-performance liquid chromatography on dynamically modified silica. J Chromatogr 1992;577(2):325–333.

Annesley TM, Clayton LT. Quantification of mycophenolic acid and glucuronide metabolite in human serum by HPLC-tandem mass spectrometry. Clin Chem 2005;51(5):872–877.

Bailey MJ, Dickinson RG. Acyl glucuronide reactivity in perspective: biological consequences. Chem Biol Interact 2003;145(2):117–137.

Bergeron M, Furtado M, Garofolo F, Mess JN. Evaluation of acyl glucuronide metabolites during drug quantification in bioanalysis by LC-MS/MS: from sample collection to autosampler stability. Proceedings of AAPS Annual Meeting and Exposition, Los Angeles, CA, 2009.

Bolze S, Bromet N, Gay-Feutry C, Massiere F, Boulieu R, Hulot T. Development of an in vitro screening model for the biosynthesis of acyl glucuronide metabolites and the assessment of their reactivity toward human serum albumin. Drug Metab Dispos 2002;30(4):404–413.

Brandhorst G, Streit F, Goetze S, Oellerich M, Armstrong VW. Quantification by liquid chromatography tandem mass spectrometry of mycophenolic acid and its phenol and acyl glucuronide metabolites. Clin Chem 2006;52(10):1962–1964.

Castillo M, Smith PC. Direct determination of ibuprofen and ibuprofen acyl glucuronide in plasma by high-performance liquid chromatography using solid-phase extraction. J Chromatogr 1993;614(1):109–116.

Corcoran O, Mortensen RW, Hansen SH, Troke J, Nicholson JK. HPLC/1H NMR spectroscopic studies of the reactive alpha-1-O-acyl isomer formed during acyl migration of S-naproxen beta-1-O-acyl glucuronide. Chem Res Toxicol. 2001;14(10):1363–1370.

Cote C, Lahaie M, Latour S, et al. Impact of methylation of acyl glucuronide metabolites on incurred sample reanalysis evaluation: ramiprilat case study. Bioanalysis 2011;3(9):951–965.

de LH, Naesens M, Verbeke K, Vanrenterghem Y, Kuypers DR. Stability of mycophenolic acid and glucuronide metabolites in human plasma and the impact of deproteinization methodology. Clin Chim Acta 2008;389(1-2):87–92.

Dickinson RG. Iso-glucuronides. Curr Drug Metab 2011;12(3): 222–228.

Faed EM. Properties of acyl glucuronides: implications for studies of the pharmacokinetics and metabolism of acidic drugs. Drug Metab Rev 1984;15(5-6):1213–1249.

Farrant RD, Spraul M, Wilson ID, Nicholls AW, Nicholson JK, Lindon JC. Assignment of the 750 MHz 1H NMR resonances from a mixture of transacylated ester glucuronic acid conjugates with the aid of oversampling and digital filtering during acquisition. J Pharm Biomed Anal 1995;13(8):971–977.

Grubb NG, Rudy DW, Hall SD. Stereoselective high-performance liquid chromatographic analysis of ketoprofen and its acyl glucuronides in chronic renal insufficiency. J Chromatogr B Biomed Appl 1996;678(2):237–244.

Hermening A, Grafe AK, Baktir G, Mutschler E, Spahn-Langguth H. Gemfibrozil and its oxidative metabolites: quantification of aglycones, acyl glucuronides, and covalent adducts in samples from preclinical and clinical kinetic studies. J Chromatogr B Biomed Sci Appl 2000;741(2):129–144.

Kamata T, Nishikawa M, Katagi M, Tsuchihashi H. Optimized glucuronide hydrolysis for the detection of psilocin in human urine samples. J Chromatogr B Analyt Technol Biomed Life Sci 2003;796(2):421–427.

Khan S, Teitz DS, Jemal M. Kinetic analysis by HPLC-electrospray mass spectrometry of the pH-dependent acyl migration and solvolysis as the decomposition pathways of ifetroban 1-O-acyl glucuronide. Anal Chem 1998;70(8):1622–1628.

Khoschsorur G, Erwa W. Liquid chromatographic method for simultaneous determination of mycophenolic acid and its phenol- and acylglucuronide metabolites in plasma. J Chromatogr B Analyt Technol Biomed Life Sci 2004;799(2):355–360.

Klepacki J, Klawitter J, Bendrick-Peart J, et al. A high-throughput U-HPLC-MS/MS assay for the quantification of mycophenolic acid and its major metabolites mycophenolic acid glucuronide and mycophenolic acid acyl-glucuronide in human plasma and urine. J Chromatogr B Analyt Technol Biomed Life Sci 2012;883-884:113–119.

Lenz EM, Greatbanks D, Wilson ID, et al. Direct characterization of drug glucuronide isomers in human urine by HPLC-NMR spectroscopy: application to the positional isomers of 6,11-dihydro-11-oxodibenz[b,e]oxepin-2-acetic acid glucuronide. Anal Chem 1996;68(17):2832–2837.

Li W, Zhang J, Tse FL. Strategies in quantitative LC-MS/MS analysis of unstable small molecules in biological matrices. Biomed Chromatogr 2011;25(1-2):258–277.

Liu DQ, Pereira T. Interference of a carbamoyl glucuronide metabolite in quantitative liquid chromatography/tandem mass spectrometry. Rapid Commun Mass Spectrom 2002;16(2):142–146.

Liu JH, Smith PC. Direct analysis of salicylic acid, salicyl acyl glucuronide, salicyluric acid and gentisic acid in human plasma and urine by high-performance liquid chromatography. J Chromatogr B Biomed Appl 1996;675(1):61–70.

Loewen GR, Macdonald JI, Verbeeck RK. High-performance liquid chromatographic method for the simultaneous quantitation of diflunisal and its glucuronide and sulfate conjugates in human urine. J Pharm Sci 1989;78(3):250–255.

Mano N, Nikaido A, Narui T, Yamasaki D, Goto J. Rapid and simple quantitative assay method for diastereomeric flurbiprofen glucuronides in the incubation mixture. J Chromatogr B Analyt Technol Biomed Life Sci 2002;776(1):125–131.

Mess JN, Berube ER, Furtado M, Garofolo F. A practical approach to reduce interference due to in-source collision-induced dissociation of acylglucuronides in LC-MS/MS. Bioanalysis 2011;3(15):1741–1751.

Mortensen RW, Corcoran O, Cornett C, et al. S-naproxen-beta-1-O-acyl glucuronide degradation kinetic studies by stopped-flow high-performance liquid chromatography-1H NMR and high-performance liquid chromatography-UV. Drug Metab Dispos 2001;29(4 Pt 1):375–380.

Mullangi R, Bhamidipati RK, Srinivas NR. Bioanalytical aspects in characterization and quantification of glucuronide conjugates in various biological matrices. Current Pharmaceutical Analysis 2005;1(3):251–264.

Munafo A, McDonagh A F, Smith P C, Benet L Z. Irreversible binding of tolmetin glucuronic acid esters to albumin in vitro. Pharm Res 1990;7(1):21–27.

Regan SL, Maggs JL, Hammond TG, Lambert C, Williams DP, Park BK. Acyl glucuronides: the good, the bad and the ugly. Biopharm Drug Dispos 2010;31(7):367–395.

Ruelius HW, Kirkman SK, Young EM, Janssen FW. Reactions of oxaprozin-1-O-acyl glucuronide in solutions of human plasma and albumin. Adv Exp Med Biol 1986;197:431–441.

Schwartz MS, Desai RB, Bi S, Miller AR, Matuszewski BK. Determination of a prostaglandin D2 antagonist and its acyl glucuronide metabolite in human plasma by high performance liquid chromatography with tandem mass spectrometric detection–a lack of MS/MS selectivity between a glucuronide conjugate and a phase I metabolite. J Chromatogr B Analyt Technol Biomed Life Sci 2006;837(1-2):116–124.

Shipkova M, Armstrong VW, Oellerich M, Wieland E. Acyl glucuronide drug metabolites: toxicological and analytical implications. Ther Drug Monit 2003;25(1):1–16.

Sidelmann UG, Hansen SH, Gavaghan C, et al. Measurement of Internal Acyl Migration Reaction Kinetics Using Directly Coupled HPLC-NMR: Application for the Positional Isomers of Synthetic (2-Fluorobenzoyl)-d-glucopyranuronic Acid. Anal Chem 1996a;68(15):2564–2572.

Sidelmann UG, Lenz EM, Spraul M, et al. 750 MHz HPLC-NMR spectroscopic studies on the separation and characterization of the positional isomers of the glucuronides of 6,11-dihydro-11-oxodibenz[b,e]oxepin-2-acetic acid. Anal Chem 1996b;68(1):106–110.

Silvestro L, Gheorghe M, Iordachescu A, et al. Development and validation of an HPLC-MS/MS method to quantify clopidogrel acyl glucuronide, clopidogrel acid metabolite, and clopidogrel in plasma samples avoiding analyte back-conversion. Anal Bioanal Chem 2011;401(3):1023–1034.

Srinivasan K, Nouri P, Kavetskaia O. Challenges in the indirect quantitation of acyl-glucuronide metabolites of a cardiovascular drug from complex biological mixtures in the absence of reference standards. Biomed Chromatogr 2010;24(7):759–767.

Stachulski AV, Harding J R, Lindon JC, Maggs JL, Park BK, Wilson ID. Acyl glucuronides: biological activity, chemical reactivity, and chemical synthesis. J Med Chem 2006;49(24):6931–6945.

Stass H, Kubitza D. Pharmacokinetics and elimination of moxifloxacin after oral and intravenous administration in man. J Antimicrob Chemother 1999;43(Suppl B):83–90.

Trontelj J. Quantification of Glucuronide Metabolites in Biological Matrices by LC-MS/MS. In: Prasain JK, editor. Tandem Mass Spectrometry—Applications and Principles. InTech; 2012. p 550–576.

US FDA. Guidance for industry: Satetey testing of drug metabolites. 2008.

Valaskovic GA, Utley L, Lee MS, Wu JT. Ultra-low flow nanospray for the normalization of conventional liquid chromatography/mass spectrometry through equimolar response: standard-free quantitative estimation of metabolite levels in drug discovery. Rapid Commun Mass Spectrom 2006;20(7):1087–1096.

van Breemen RB, Fenselau CC, Dulik DM. Activated phase II metabolites: comparison of alkylation by 1-O-acyl glucuronides and acyl sulfates. Adv Exp Med Biol 1986;197:423–429.

Vree TB, Biggelaar-Martea M, Verwey-van Wissen CP. Determination of indomethacin, its metabolites and their glucuronides in human plasma and urine by means of direct gradient high-performance liquid chromatographic analysis. Preliminary pharmacokinetics and effect of probenecid. J Chromatogr 1993;616(2):271–282.

Vree TB, Biggelaar-Martea M, Verwey-van Wissen CP. Determination of furosemide with its acyl glucuronide in human plasma and urine by means of direct gradient high-performance liquid chromatographic analysis with fluorescence detection. Preliminary pharmacokinetics and effect of probenecid. J Chromatogr B Biomed Appl 1994a;655(1):53–62.

Vree TB, van Ewijk-Beneken Kolmer EW, Nouws JF. Direct-gradient high-performance liquid chromatographic analysis and preliminary pharmacokinetics of flumequine and flumequine acyl glucuronide in humans: effect of probenecid. J Chromatogr 1992;579(1):131–141.

Vree TB, van Ewijk-Beneken Kolmer EW, Verwey-van Wissen CP, Hekster YA. Direct gradient reversed-phase high-performance liquid chromatographic determination of salicylic acid, with the corresponding glycine and glucuronide conjugates in human plasma and urine. J Chromatogr 1994b;652(2):161–170.

Wainhaus S. Acyl glucuronides: assays and issues. In: Korfmacher WA, editor, Using Mass Spectrometry for Drug Metabolism Studies. CRC Press; 2005. p 175–202.

Xue YJ, Akinsanya JB, Raghavan N, Zhang D. Optimization to eliminate the interference of migration isomers for measuring 1-O-beta-acyl glucuronide without extensive chromatographic separation. Rapid Commun Mass Spectrom 2008;22(2):109–120.

Xue YJ, Simmons NJ, Liu J, Unger SE, Anderson DF, Jenkins R G. Separation of a BMS drug candidate and acyl glucuronide from seven glucuronide positional isomers in rat plasma via high-performance liquid chromatography with tandem mass spectrometric detection. Rapid Commun Mass Spectrom 2006;20(11):1776–1786.

Yan Z, Caldwell GW, Jones WJ, Masucci JA. Cone voltage induced in-source dissociation of glucuronides in electrospray and implications in biological analyses. Rapid Commun Mass Spectrom 2003;17(13):1433–1442.

Zhao Y, Yang C Y, Haznedar J, Antonian L. Simultaneous determination of SU5416 and its phase I and phase II metabolites in rat and dog plasma by LC/MS/MS. J Pharm Biomed Anal 2001;25(5-6):821–832.

Zhou SF, Paxton JW, Tingle MD, Kestell P, Jameson MB, Thompson PI, Baguley BC. Identification and reactivity of the major metabolite (beta-1-glucuronide) of the anti-tumour agent 5,6-dimethylxanthenone-4-acetic acid (DMXAA) in humans. Xenobiotica 2001;31(5):277–293.

36

REGULATED BIOASSAY OF *N*-OXIDE METABOLITES USING LC-MS: DEALING WITH POTENTIAL INSTABILITY ISSUES

Tapan K. Majumdar

36.1 INTRODUCTION

Identification and characterization of metabolites are very important in the drug development process. The impact of biologically active metabolites on safety and efficacy is typically assessed by the drug metabolism and pharmacokinetics department in pharmaceutical companies. Formations of metabolites are enzymatic reactions and dependent on the chemical structure of the drug molecule. Compounds having tertiary amine functional groups can produce *N*-oxide metabolites *in vivo* and *in vitro* (Bickel, 1969). Drug molecules containing functional groups with a lone pair of electrons on the nitrogen atom (to share with an oxygen atom) such as tertiary amine, pyridine, and piperidine groups typically produce *N*-oxide metabolites as oxidation products of liver microsomal enzymes. Jenner (1971) described four types of nitrogen containing functional groups that may form *N*-oxide metabolites. The four types are (1) aliphatic amines (e.g., dimethylamphetamine, imipramine, and orphenadrine), (2) alicyclic amines (e.g., phendimetrazine, nicotine, and guanethine), (3) aromatic amines (e.g., nicotinamide and tripellenamine), and (4) aliphatic amines adjacent to aromatic ring (e.g., dimethylaniline). The *N*-oxides were described as a new class of compounds by Dunstan and Goulding (1899). In these molecules, the lone pair electrons of the nitrogen atom are shared with the oxygen atom forming a semi-polar *N*-oxide bond that is shown by a short arrow (N $\rightarrow$ O). The chemical nature of the *N*-oxides is less basic than the corresponding amines.

$$\mathrm{R{-}\overset{\displaystyle R_1}{\underset{\displaystyle R_2}{N}}{:}} + \mathrm{\ddot{O}{:}} \longrightarrow \mathrm{R{-}\overset{\displaystyle R_1}{\underset{\displaystyle R_2}{N}}\rightarrow O}$$

Tertiary amine — *N*-Oxide

36.2 FORMATION OF *N*-OXIDE METABOLITES

According to *in vitro* data, formation of *N*-oxide metabolites takes place in the hepatic microsomes through oxidative reactions catalyzed by NADPH-dependent microsomal electron transfer chains. The mechanism was described for compounds such as nicotine (Papadopoulos, 1964; Rangiah et al., 2011), procainamide (Li et al., 2012), clozapine (Bun et al., 1999; Zhang et al., 2008), *N*,*N*-dimethylaniline (Ziegler and Pettit, 1964), 1-propoxyphene (McMahon and Sullivan, 1964), imipramine (Bickel and Baggiolini, 1966), chlorpromazine (Beckett and Hewick, 1967), azo compounds (Koh and Gorrod, 1989), *N*-benzyl-*N*-ethylaniline and *N*-benzyl-*N*-ethyl-*p*-toluidine (Ulgen et al., 1997). The formation of *N*-oxide metabolites is one of the two major metabolic pathways of metabolism of tertiary nitrogen containing drugs by hepatic microsomal enzymes. The second pathway involves dealkylation of the tertiary nitrogen to a secondary amine. These two pathways are catalyzed by NADPH-dependent microsomal electron transfer chains; the first (*N*-oxide formation) involving NADPH-cytochrome c

Handbook of LC-MS Bioanalysis: Best Practices, Experimental Protocols, and Regulations, First Edition. Edited by Wenkui Li, Jie Zhang, and Francis L.S. Tse.

reductase and cytochrome P-450, the latter (*N*-dealkylation) a different flavo-protein without cytochrome P-450 (Bickel, 1971). The individual rates are highly dependent on the animal species and experimental conditions. The hepatic microsomal electron transport chain that oxidizes drugs plays a role in *N*-oxide formation using the cofactor NADPH and oxygen. It has been experimentally proven that compounds having tertiary nitrogen functional groups are readily metabolized to their *N*-oxides *in vivo* by flavin-containing monooxygenese (Ziegler, 1988). Formation of *N*-oxide metabolites in different species has been described for chlorpromazine (Harinath and Odell, 1968), chlorcyclizine (Kuntzman et al., 1967), dimethylaniline (Ziegler and Pettit, 1964; Machinist et al., 1968), and trimethylamine (Baker et al., 1963).

36.3 DISTRIBUTION AND EXCRETION OF *N*-OXIDES

The *N*-oxide metabolites are primarily excreted in the urine. Following administration of nitrogen containing drugs such as imipramine (Bickel and Weder, 1968), chlorcyclizine (Kuntzman et al., 1967), and chlorpromazine (Forrest et al., 1968) in animals or human subjects, *N*-oxides were only detected in urine and not in the body tissues. When rodents were treated with imipramine, the *N*-oxide metabolite was present in plasma, bile, intestine, feces, and urine, but not in body tissues (Bickel and Weder, 1968). However, in experiments by the same group where imipramine-*N*-oxide was directly administered the compound was detected in liver, kidney, and other tissues except brain.

36.4 METABOLISM OF *N*-OXIDES

One of the metabolic routes of *N*-oxides is that they can undergo subsequent metabolism by *N*-dealkylation, resulting in the formation of a secondary amine and an aldehyde. It is important to keep in mind that tertiary amines can directly produce secondary amines and aldehydes as metabolites through *N*-dealkylation without having *N*-oxide as an intermediary metabolite (Bickel, 1971). The second metabolic, less favored, route of *N*-oxide is formation of the parent drug (tertiary amine) through *N*-oxide reduction, which is the reversal of the *N*-oxidation.

36.5 BIOLOGICAL ACTIVITIES OF *N*-OXIDES

The systemic exposure to *N*-oxide metabolites is an area of concern due to their potentially high levels of biological activity relative to other metabolites. In several cases, *N*-oxides are equally or more active than their corresponding parent compounds. Hence, some *N*-oxides are important as pharmacological or toxicological agents, for example, *N*-oxides of alkaloids that are used in therapy as chemotherapeutic agents, antibiotics, and psychotropic drugs. Certain oncogenic *N*-oxides (e.g., purine *N*-oxides) act as antimetabolites and have been postulated as inducers of spontaneous cancer. Due to the potent pharmacological and toxicological activities of the *N*-oxide metabolites, they are routinely monitored in patients undergoing treatment along with their tertiary amine group containing parent drugs. Some of these drugs and their *N*-oxide metabolites are discussed in Section 36.8.

36.6 EXPERIMENTAL PROTOCOL

Prior to the introduction of soft ionization mass spectrometric techniques, accurate determination of *N*-oxide metabolites was a challenge. This was due to insufficient sensitivity of the instruments such as high-pressure liquid chromatography (HPLC) with UV or PDA (photodiode array) detectors or due to instability of the thermo labile *N*-oxides in the heated gas chromatography (GC)-inlet and oven. Hard ionization technique such as electron impact in GC/MS instruments also caused the conversion of *N*-oxide metabolites to parent drug. Introduction of soft ionization methods such as electrospray ionization (ESI) and atmospheric pressure chemical ionization (APCI) and availability of sensitive mass spectrometers helped low level and accurate quantification of the *N*-oxide metabolites these days. The *N*-oxides are unstable molecules and can loose the oxygen atom to revert back to the parent drug molecule under experimental conditions used during bioanalytical analyses. These conditions include high temperature, strong acidic or basic conditions, presence of antioxidants in the biomatrices, and in-source conversation during mass spectral analyses. These unfavorable conditions are abundant in bioanalytical processes such as sample collection, processing, chromatographic separation, analysis in the mass spectrometric ion source and during the collision process in tandem mass spectrometry.

In this article, only LC-MS methods are described and reviewed because of their high sensitivity and selectivity. The quantitation of *N*-oxide metabolites is described in Section 36.6. The protocol consists of several steps such as sample collection, preparation, extraction, chromatographic separation, and MS/MS analysis. These steps are briefly discussed with the possible danger of *N*-oxide conversions in each step are included here. Although soft ionization techniques have been around for over two decades, published regulated methods of *N*-oxides are still limited. Some examples from the literature are provided at the end of this discussion.

36.6.1 Sample Collection

Evaluate the stability of the *N*-oxide metabolites under the conditions (e.g., anticoagulant, stabilizer, and temperature) used for the sample collection. This is done by spiking a solution of known concentrations in the matrix followed by gentle mixing. *N*-oxides are generally more stable in plasma than in whole blood. Therefore, it is advisable to perform the evaluations in both blood and plasma. Changing the pH of blood to stabilize the parent drug may render the *N*-oxide metabolites unstable. During sample collection, conversion of *N*-oxide to parent drug can happen easily in whole blood in the presence of stabilizers. Ascorbic acid is one of the antioxidants used to prevent oxidation of the parent drug during collection of blood sample. However, ascorbic acid can cause the reduction of *N*-oxide metabolites (Dell, 2004).

36.6.2 Sample Preparation

Study sample (e.g., 0.1 ml of plasma or urine), double blanks zero samples, calibrators, in the same matrix are added to designated wells of a 96-well plate and brought to room temperature. The internal standard solution (e.g., 0.05 ml) is added to each well except to the wells containing the double blanks where the same volume (e.g., 0.05 ml) of solvent (used for internal standard solution) is added. Stable isotope labeled internal standard (SIL-IS) is strongly recommended over structural analog for a rugged method involving *N*-oxides. The SIL-IS undergoes the same changes like the analyte during extraction, ionization, and MS/MS reactions, thereby providing constant ratio of analyte/IS at a specific concentration of the analyte. The 96-well plate is mixed briefly and equilibrated at room temperature for 10 min.

36.6.3 Extraction

This step can be performed using protein precipitation (PPT), solid phase extraction (SPE), supported liquid extraction (SLE), or liquid–liquid extraction (LLE). The last three extraction methods (SPE, SLE, and LLE) will produce clean extracts with minimal matrix interferences. Extraction of biological samples under alkaline conditions may lead to the loss of the oxygen atom from the *N*-oxide metabolite. An example is the conversion of chlorpromazine *N*-oxide to the parent drug in strongly alkaline solutions (Dell, 2004). The use of sodium carbonate rather than sodium hydroxide avoided such conversion of chlorpromazine *N*-oxide during the extraction process. The use of heat during the extract drying step can also convert *N*-oxide metabolites to the parent drug.

For PPT, 0.2 ml of acetonitrile is added to each well containing 0.1 mL of plasma and the plate is vortexed for 5 min. Subsequently, the plate is sealed and centrifuged at approximately 3000 X *g* to separate the supernatant or filtrate through 96-well filtration plate to separate the supernatant. Appropriate volume of supernatant can be directly injected to the LC-MS/MS system. However, for a clean extract the supernatant is typically dried, reconstituted in selected solvent and then an appropriate volume is injected to the LC-MS/MS system. In some cases, it is a common practice to use sonication or strong acid treatment for protein bound analytes during the PPT procedure. Both sonication and acid treatment may impact stability of the *N*-oxides.

For SPE, the sample may be pretreated with acid to disrupt protein binding. After that the content of the 96-well plate is loaded on a 96-well SPE cartridge (e.g., Oasis HLB cartridge from Waters Corporation; www.waters.com), previously conditioned with 0.5–1 ml methanol followed by the same volume of water. Each well of the extraction cartridge is washed with 0.5–1 ml of 5% methanol in water (v/v). Depending on the pK_a of the analytes, a subsequent wash can be added using 2% formic acid and 2% ammonium hydroxide without affecting the stability of the *N*-oxides. The compounds are typically eluted using an appropriate volume (e.g., 0.1–0.2 ml) of selective solvent mixture (e.g., methanol:water containing 0.05% formic acid). The elution volume can be adjusted based on the sensitivity needs of the method. Mixed mode SPE stationary phases may be used for some compounds and acid and base treatment is typically needed for good extraction recovery. Care must be taken to preserve the stability of the *N*-oxides during the pH changes in the SPE procedure.

SLE on a 96-well plate (e.g., Isolute SLE+ plate from Biotage (www.biotage.com, accessed Apr 8, 2013)) is a high throughput alternative to LLE discussed later. This technique uses an inert support such as diatomaceous earth to mimic the LLE process. The sample (e.g., 0.1 ml) is mixed 1:1 (v/v) with aqueous buffer prior to loading on the SLE plate. For acidic, neutral, or basic analytes the buffer pH of acidic, neutral, and basic is maintained to render the analytes uncharged during the extraction process. This sample pretreatment step should be carefully evaluated to maintain stability of the *N*-oxides, as they are vulnerable (unstable) under strong acidic and basic conditions. The diluted sample is loaded on to the 96-well SLE plate and allowed to adsorb to the inert support for 5 min. An appropriate extraction solvent (1 ml) such as MTBE, ethyl acetate, or *n*-hexane is added and allowed to flow for 5 min without any external pressure with the remaining solvent being eluted under mild vacuum (−0.5 bar). The extract is dried and reconstituted in a suitable solvent and an appropriate volume is subsequently injected into the LC-MS/MS system.

For LLE, 0.5–1.0 ml of and appropriate extraction solvent (e.g., MTBE, ethyl acetate, and *n*-hexane) is added to each well of the 96-well plate. For acidic, neutral, or basic analytes, the pH of the aqueous phase may be adjusted to keep the analytes neutral for efficient extraction without appropriate evaluation of the *N*-oxide stability under the pH changes. The plate is sealed, vortex mixed for 5 min and centrifuged at 3000 × *g*. Subsequently, a certain volume of the organic layer from the supernatant is transferred to a clean 96-plate block. The extract is dried, reconstituted in selected solvent and then an appropriate volume is injected into the LC-MS/MS system.

36.6.4 Chromatographic Separation

Chromatographic separation of *N*-oxide metabolites is typically performed using reverse phase chromatography using C_8 or C_{18} stationary phases. The columns containing polar stationary phases, for example, HILIC, Synergy polar RP, and so on, has also been reported in the literature and discussed in Section 36.8. Typical column dimension is 50 × 2.1 mm for this type of analysis. The mobile phase is either a gradient or isocratic flow of acetonitrile (or methanol) and water containing 0.01 M ammonium acetate (or formate). The pH of the buffer may need some adjustment to obtain good chromatographic separation. Since the stability of the *N*-oxide is pH dependent, a careful evaluation of the separation method is needed for accurate quantitation of the *N*-oxides with minimal conversion.

36.6.5 MS/MS Analysis

Ionization of *N*-oxide metabolites is typically done using electrospray (ESI) or APCI. The protonated molecular ions $[M + H]^+$ formed during ionization are subjected to selected reaction monitoring (SRM) process (MS/MS) for enhanced selectivity of the method. *N*-oxides are thermally labile and in-source back conversion to the parent drug is commonly observed at the high source temperatures. When the conversion occurs in the ion source, the fraction of parent drug that is converted from the N-oxide metabolite molecule can be seen as a separate peak in the ion-chromatogram of the parent drug at the same retention time as that of the *N*-oxide molecule. Among the soft ionization techniques, only ESI can minimize the conversion of *N*-oxide. APCI converts a fraction of *N*-oxide molecules to the parent drug according to our experience (Majumdar et al., 2001). In-source conversion of *N*-oxides is severe in some of the mass spectrometer platforms. Additional fragmentation may occur by the loss of the oxygen atom while the ions are subjected to collision-induced dissociation (CID) during MS/MS process, leading to additional product ions. It is strongly advisable to use stable isotope labeled *N*-oxide as the internal standard to minimize the deleterious effects of mass spectrometric conversions during quantitative analysis of *N*-oxide metabolites.

36.7 REGULATORY CONSIDERATIONS

Regulated bioassay of *N*-oxide metabolites is performed using FDA GLP guidelines (Guidance for Industry, May 2001). The method validation is performed to evaluate six fundamental parameters such as (1) selectivity, (2) accuracy, (3) precision, (4) sensitivity, (5) reproducibility, and (6) stability.

The selectivity is established by analyzing blank biological matrix (plasma, blood, urine, or other matrix) from at least six lots or different sources. The signal intensity from matrix background or interference should not exceed 20% of the lowest level of quantification (LLOQ).

The accuracy is determined by analyzing at least five replicates of quality control (QC) samples at three different concentration levels on each day during three days of validation. The QC samples are prepared by initially spiking a known level of *N*-oxide metabolite in a blank biological matrix. The first QC level is within 3 × of LLOQ, the second level is in midcalibration range, and the third level is very close to the upper level of quantitation (ULOQ). The acceptable mean value at each level is within 15% of the theoretical concentration.

The precision is determined from the coefficient of variation (%) generated at each level of QC samples during the sensitivity and accuracy determinations discussed earlier. The acceptable level of precision is 20% at the LLOQ level and 15% at concentrations above LLOQ. The reproducibility is determined from the intra- and interday precision of the QC samples during the 3 days of validation.

The sensitivity is determined by analyzing five replicates of LLOQ samples (signal intensity is at least 5 × the response from the blank matrices). The acceptable accuracy and precision are both 20%.

The recoveries are calculated at all three levels of QC samples by comparing the concentrations to the unextracted concentrations. Recoveries should be consistent in a regulated method.

Stability is a critical factor for *N*-oxide metabolites because these metabolites are chemically unstable and the duration of stability depends on the chemistry of the molecule. Stabilities of QC samples are determined under different conditions such as room temperature (or bench-top) stability, freeze–thaw stability, short-term storage stability, long-term storage stability, post-preparative stability of sample extract, stock solution stability, and incurred sample stability (ISS).

Incurred sample reproducibility (ISR) must be established by reanalyzing 5% of the study samples. Both ISR and ISS for unstable compounds like *N*-oxide metabolites are critical for a good bioassay because these experiments are indicative of stability and reproducibility of the *N*-oxide metabolites during storage and assay conditions.

36.8 EXAMPLES

36.8.1 Nicotine and Metabolites Including *N*-Oxides

Quantitative determination of nicotine and its major metabolites (e.g., cotinine, trans-3′-hydroxycotinine, nicotine-*N*′-oxide, cotinine-*N*-oxide, and nornicotine) and their pharmacokinetics provide a useful tool for establishing uptake of nicotine and tobacco-related toxic compounds, for understanding the pharmacological effects of nicotine and nicotine

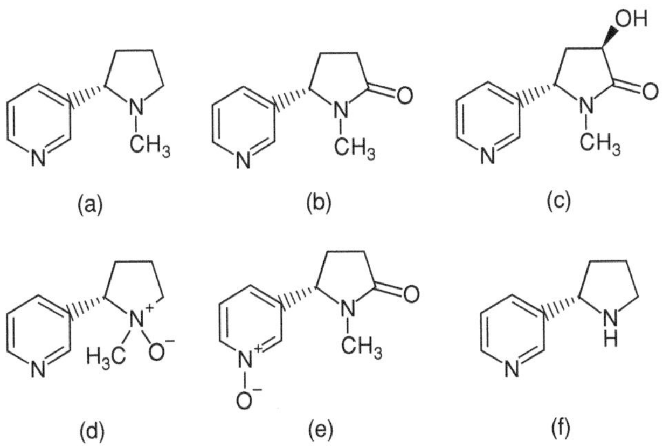

FIGURE 36.1 (a) Nicotine, (b) cotinine, (c) trans-3′-hydroxycotinine, (d) nicotine-*N*′-oxide, (e) cotinine-N-oxide, and (f) nornicotine.

addiction, and for optimizing nicotine dependency treatment. The structures of nicotine and its major metabolites are shown in Figure 36.1.

Several validated LC-MS/MS methods have been published over last few years for the regulated bioassay of nicotine and metabolites (including the *N*-oxides) in biological fluids (Meger et al., 2002; Xu et al., 2005; Pellegrini et al., 2007; Xie et al., 2008; Marclay and Saugy, 2010).

Marclay and Saugy (2010) developed an LC-MS/MS method for the detection and quantification of nicotine and its principal metabolites cotinine, trans-3-hydroxycotinine, nicotine-*N*′-oxide and cotinine-*N*-oxide in human urine. The method was developed to identify the abuse of tobacco and nicotine containing products among professional athletes. In this method, sample preparation was performed by LLE. Analysis of the extract was performed using hydrophilic interaction chromatography-tandem mass spectrometry (HILIC-MS/MS). The tandem mass spectrometer was operated in positive ESI mode with selective reaction monitoring (SRM) data acquisition. The method was validated over linear calibration ranges of 10–10,000 ng/ml for nicotine, cotinine, trans-3-hydroxycotinine, and 10–5000 ng/ml for nicotine-N′-oxide and cotinine-*N*-oxide, with correlation coefficient (r^2) greater than 0.95. The extraction efficiency (%) was concentration dependent and ranged from 70.4% to 100.4%. The LLOQ for all analytes was 10 ng/ml. The method bias and precision were between 9.4% and 9.9%, respectively. In order to measure the prevalence of nicotine exposure during the 2009 Ice Hockey World Championships, 72 urine samples from the players were collected and analyzed. Nicotine and/or metabolites were detected in every urine sample, while concentration measurements indicated an exposure within the last 3 days for eight specimens out of ten. Concentrations of nicotine, cotinine, trans-3-hydroxycotinine, nicotine-N′-oxide and cotinine-*N*-oxide were found to range between 11 and 19,750, 13 and 10,475, 10 and 8217, 11 and 3396, and 13 and 1640 ng/ml, respectively. When proposing conservative limits for nicotine consumption prior and/or during the games (50 ng/ml for nicotine, cotinine and trans-3-hydroxycotinine and 25 ng/ml for nicotine-*N*′-oxide and cotinine-*N*-oxide), about half of the hockey players were qualified as consumers.

The method published by Fan et al., 2008 was for the direct determination of nicotine, cotinine, trans-3′-hydroxycotinine, their corresponding glucuronide conjugates as well as nornicotine, norcotinine, cotinine-*N*-oxide and nicotine-*N*′-oxide in the urine of smokers for the detection of nicotine dose in large-scale human biomonitoring studies. The assay was simple and involved centrifugation and filtration of diluted urine samples (dilute and inject method). The chromatographic separation was performed on a C_{18} reversed-phase column using a gradient of 10 mM ammonium acetate, pH 6.8, and methanol as mobile phase at a flow rate of 1 ml/min. Nicotine-methyl-d3, cotinine-methyl-d3, and trans-3′-hydroxycotinine-methyl-d3 were used as internal standards. Precisions (%CV) for all the analytes at three levels of QC samples were between 2.1% and 17.0%. The recoveries for nicotine and nine nicotine metabolites ranged from 78.4% to 115.6%.

An LC-MS/MS bioassay was reported for the simultaneous quantification of biomarkers of three of the addictive materials: tobacco, caffeine, and areca nut in human breast milk (Pellegrini et al., 2007). Nicotine and its principal metabolites cotinine, trans-3-hydroxycotinine and cotinine-*N*-oxide, caffeine and arecoline were quantified in this method. The assay employed LLE using chloroform:isopropanol (95:5, v/v) under neutral condition for nicotine, cotinine, trans-3-hydroxycotinine, cotinine-*N*-oxide, and caffeine. Basic conditions were used in the same solvent for extraction of arecoline. Chromatographic separation was performed on a C_8 reversed-phase column using a gradient of 50 mM ammonium formate, pH 5.0, and acetonitrile as a mobile phase at a flow rate of 0.5 ml/min. Detection was performed by positive electrospray ionization tandem mass spectrometry using multiple reaction monitoring (MRM) (LC-ESI/MS/MS). Lower limits of quantification were 5 μg/l for nicotine, cotinine, trans-3-hydroxycotinine, cotinine-*N*-oxide and caffeine, and 50 μg/l for arecoline using 1 ml human milk per assay. The method recoveries ranged from 71.8% to 77.4% for different analytes. This method was applied for the quantitative determination of analytes in human milk to assess substance exposure in breast-fed infants in relation to clinical outcomes.

Xu et al. (2005) published a method for the quantitation of nicotine and its five major metabolites, including cotinine, trans-3′-hydroxycotinine, nicotine-N′-oxide, cotinine-*N*-oxide, and nornicotine, in human urine by liquid chromatography coupled with a TSQ Quantum triple quadrupole (Thermo Fisher Scientific, San Jose, CA) tandem mass spectrometry. Anabasine, a minor tobacco alkaloid, was also quantified in this method. SPE was employed to extract urine samples spiked with deuterium-labeled internal standards. The quantification limits of the method were 0.1–0.2 μg/l for all the analytes except for nicotine that was 1 μg/l. The method recoveries for cotinine-*N*-oxide, trans-3′-hydroxycotinine, nicotine, and anabasine in urine were close to 100%, whereas the mean recoveries of nicotine-N′-oxide, cotinine, and nornicotine were 51.4%, 78.6%, and 78.8%, respectively. The linear calibration range was 0.2–400 μg/l for nicotine-N′-oxide, cotinine-*N*-oxide, and anabasine; 0.2–4000 μg/l for cotinine, nornicotine, and trans-3′-hydroxycotinine; and 1.0–4000 μg/l for nicotine. The overall interday method bias and recovery were 2.5–18% and 92–109%, respectively.

A bioassay based on LC-MS/MS for the direct determination of nicotine and its metabolites cotinine, trans-3′-hydroxycotinine, their corresponding glucuronide conjugates as well as cotinine-*N*-oxide, norcotinine, and nicotine-*N*′-oxide in the urine of smokers has been published by Meger et al. (2002). In this method, urine samples were filtered prior to liquid chromatographic separation on a reverse phase column, the compounds were ionized using APCI method and mass specific detection was performed using selective MS/MS transitions. Deuterium-labeled nicotine, cotinine, and trans-3′-hydroxycotinine were used as internal standards. Precision (%CV) for the major nicotine analytes at levels observable in urine of smokers was better than 10%. Method accuracy for nicotine, cotinine, trans-3′-hydroxycotinine, and cotinine-*N*-glucuronide ranged from 87% to 113%. Quantitative results for the three glucuronide conjugates in urine samples of 15 smokers were compared to an indirect method in which the aglycons were detected with GC and nitrogen-selective detection (GC-NPD) before and after enzymatic splitting of the conjugates. Good agreement was found for cotinine-*N*-glucuronide (CV = 9%) and trans-3′-hydroxycotinine-O-glucuronide (CV = 20%), whereas the agreement between both methods was moderate for nicotine-*N*-glucuronide (CV = 33%). The LC-MS/MS method allowed the simultaneous determination of nicotine and eight of its major metabolites in urine of smokers with good precision and accuracy.

36.8.2 Clozapine and Metabolites

Clozapine is a dibenzene derivative well known as an atypical antipsychotic drug used for the treatment of schizophrenia and bipolar disorder. Orally administered clozapine is extensively metabolized by hepatic enzymes. Two of the major circulating and pharmacologically active metabolites are norclozapine (*N*-desmethyl clozapine) and clozapine-*N*-oxide. The structures are shown in Figure 36.2. These metabolites are routinely monitored in psychotic patients treated with clozapine.

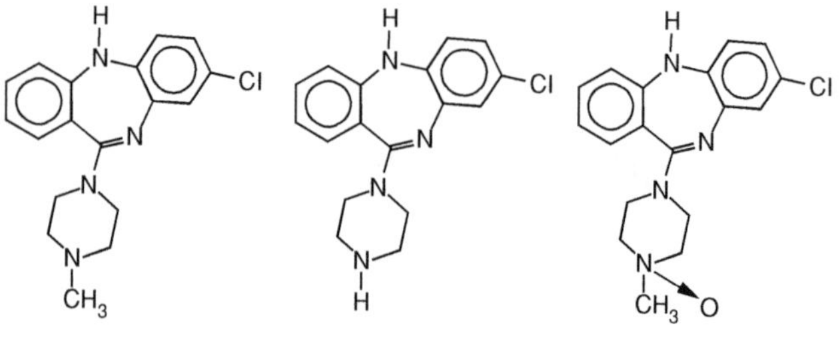

Clozapine Norclozapine Clozapine-N-oxide

FIGURE 36.2 Structures of clozapine and its metabolites norclozapine and clozapine-*N*-oxide.

A number of validated LC-MS methods have been published over last 10 years on the bioanalysis of clozapine, norclozapine, and clozapine-*N*-oxide in biological fluids (Aravagiri and Marder, 2001; Niederlaender et al., 2006; Wohlfarth et al., 2011).

Wohlfarth et al. (2011) published a validated LC-MS/MS method on the regulated bioanalysis of clozapine and its two main metabolites norclozapine and clozapine-*N*-oxide in human serum and urine. The method involved a single-step LLE of the analytes using ethyl acetate under alkaline conditions. Chromatographic separation was performed using a Synergi Polar RP column from Phenomenex (Torrance, CA, USA) using gradient elution with 1 mM ammonium formate and methanol. Data acquisition was performed on a Sciex QTRAP 2000 tandem mass spectrometer in SRM mode using positive ESI interface. Two MRM transitions were monitored for each analyte for confirmatory identification. The validation included the determination of the lower limits of quantification (1.0 ng/ml for all analytes in serum and 2.0 ng/ml for all analytes in urine), assessment of matrix effects and the determination of extraction efficiencies (52–85% in serum, 59–88% in urine) and accuracy data. The method accuracy was higher than ±90%. Dilution integrity was established for both clozapine and norclozapine in both matrices and for clozapine-*N*-oxide in serum only. For quantification of clozapine-*N*-oxide in urine a calibration with diluted calibrators had to be used. Calibration curves were linear from the 1.0 (LLOQ) to 2000 ng/ml (ULOQ) with correlation coefficients higher than 0.98. The method was used to measure the analytes in several serum and urine samples and a cerebrospinal fluid sample of an intoxicated teen age subject.

A high throughput online method has been reported by Niederlaender et al. (2006) for the determination of clozapine and metabolites (norclozapine and clozapine-*N*-oxide) in serum for therapeutic drug monitoring applications. Method development, optimization, and validation were described and a comparison with existing methods for the bioassay of clozapine and metabolites in serum and plasma was done. The method involved Prospekt-2 automated SPE system

(Spark Holland, Emmen, The Netherlands) coupled with an Agilent 1100 LC system (Agilent Technologies) in front of an API2000 mass spectrometer (Applied Biosystems, Foster City, CA). A reversed phase C_{18} SPE cartridge was used for online extraction and the chromatographic separation was performed using a reversed phase C_{18} column. Atmospheric pressure ionization (both APCI and ESI) with mass scanning in selected ion monitoring (SIM) was used for the mass spectrometric detection. Optimization of chromatographic and SPE conditions for increased throughput resulted in SPE-LC-MS cycle times of approximately 2.2 min. Limits of detection were varied from 0.15 to 0.3 ng/ml, depending on the ionization source used. A quadratic calibration curve was used for clozapine and it's *N*-oxide and a linear curve was used for the desmethyl metabolite. The correlation coefficient was higher than 0.99 for both curves. The method bias was less than 10%. Precision (both intra- and interassay) ranged from 5% RSD at the high end of the therapeutic range (700–1000 ng/ml) to 20% RSD (Organization for Economic Cooperation and Development defined limit) at the lower limit of quantitation of 50 ng/ml.

An LC-MS/MS assay method for the simultaneous determination of clozapine and its *N*-desmethyl (norclozapine) and *N*-oxide metabolites in human plasma was reported by Aravagiri and Marder (2001). A single-step LLE was used for extraction of the analytes from human plasma. Reversed phase chromatographic conditions using a C_{18} column in isocratic mode. Detection was performed using positive ESI method followed by MRM in an API2000 triple quadrupole instrument (Applied Biosystems). The monitored MRM ion transitions were m/z 327 → 270 for clozapine, m/z 313 → 192 for norclozapine, m/z 343 → 256 for clozapine-*N*-oxide and m/z 421 → 201 for the internal standard. The linear calibration range for all the analytes was 1 ng/ml to 1000 ng/ml ($r^2 > 0.998$) using 0.5 ml of human plasma. Three pooled plasma samples collected from patients administered with clozapine were used as long-term QC samples to check the validity of spiked standard curve samples made at various times. The intra- and interassay variations (%CV) for the spiked standard curve and QC samples were less than 14%. The LC-MS/MS assay was specific, sensitive, accurate and rapid. This method has been used for the plasma level monitoring of clozapine and its *N*-desmethyl and *N*-oxide metabolites in patients treated with clozapine. The plasma levels of clozapine, norclozapine and clozapine-*N*-oxide varied widely based on inter- and intrasubject data which revealed that the norclozapine and clozapine *N*-oxide metabolites were present at 58 ± 14% and 17 ± 6% of clozapine concentrations, respectively.

36.8.3 Propiverine and Its *N*-Oxide Metabolite

Propiverine hydrochloride is an anticholinergic and antimuscarinc drug used for the treatment of overactive bladder symptoms such as urinary urgency, frequency, and urge incontinence. Propiverine *N*-oxide is the major circulating metabolite that is pharmacologically active and needed to be monitored in patient under treatment.

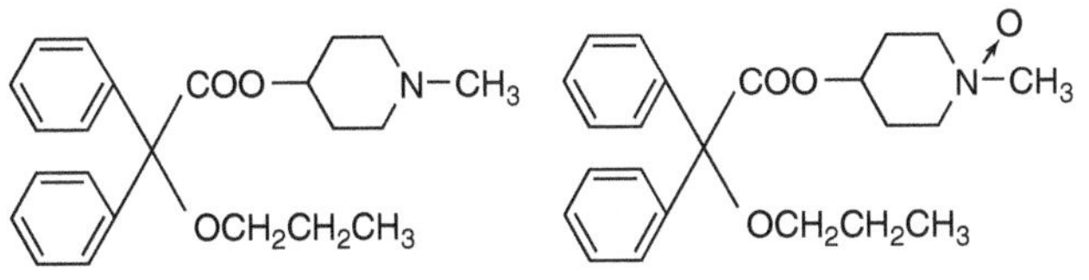

Propiverine   Propiverine *N*-oxide

The published LC-MS/MS methods were designed and validated for the quantitative determination of propiverine and its *N*-oxide metabolite in human plasma (Komoto et al., 2004; Yoon et al., 2005).

Yoon et al. (2005) reported a simple, rapid, and sensitive LC-MS/MS method for the determination of propiverine and propiverine *N*-oxide metabolite in human plasma. The extraction was done by PPT using acetonitrile. Oxybutynin was used as internal standard. The extract was directly injected into the LC-MS/MS system. The extract was subjected to reversed-phase chromatographic separation using a C_8 column. The mobile phase was 0.1% formic acid in water/acetonitrile (25:75, v/v). The mass spectrometric detection was done using ESI-MS/MS. The MRM transitions were m/z 368 → 116, m/z 384 → 183, and m/z 358 → 142 for propiverine, propiverine *N*-oxide, and oxybutynin (IStd), respectively. An alternative LC-MS method, using MTBE LLE and SIM, was validated over the calibration range of 1–250 ng/ml for propiverine and 2–500 ng/ml for the *N*-oxide metabolite, and successfully applied in a pharmacokinetic study. The lower limit of quantitation was 1 ng/ml for propiverine and 2 ng/ml for *N*-oxide in both methods. Precision (%RSD (relative standard deviation)) and accuracy values of the methods (LC-MS and LC-MS/MS) were <10.2% and >93.9%, respectively, for both analytes.

Komoto et al. (2004) published an LC-MS/MS method for the determination of propiverine hydrochloride and its metabolite, propiverine *N*-oxide in human plasma. Stable isotope labeled internal standards: propiverine hydrochloride-d_{10} and propiverine *N*-oxide-d_{10} were used in the method. LLE with dichloromethane under neutral pH of 7 was used. The chromatographic separation was performed on a C_{18} reversed phase column using methanol:1% acetic acid (50:50, v/v) as a mobile phase. Detection was performed using positive ESI in SRM mode. The method was validated over a concentration range of 2–500 ng/ml for propiverine hydrochloride and 4–1000 ng/ml for the *N*-oxide metabolite using 0.2 ml of human plasma. The intraassay precision (%RSD) and accuracy values of were <8.7% (15.2% for propiverine at LLOQ) and >88.5%, respectively, for both analytes. The interassay precision and accuracy values were <12.4% and >97.5%, respectively, for both analytes. Stability data were validated for three freeze–thaw cycles, at room temperature (4 h) and below −20°C (6 months). The method was successfully applied for the determination of propiverine hydrochloride and the *N*-oxide metabolite in clinical studies.

36.8.4 Other Basic Drugs and Their Metabolites

Roflumilast is a selective inhibitor of phosphodiesterase 4 and registered for the treatment of severe chronic obstructive pulmonary disease. Knebel et al. (2012) reported a high-throughput quantitative method for the determination of roflumilast and its *N*-oxide metabolite in human blood, plasma, and serum. The method used semiautomated LLE using ethyl acetate/*n*-heptane (1:1; v/v). Chromatographic separation was performed using a C_{18} reversed-phase column. Chromatographic peak detection was performed using tandem mass spectrometry (MS/MS) with ESI. The calibration range for assay was 0.1 ng/ml (LLOQ) and 50 ng/ml (ULOQ) for both analytes. The mean inter- and intraassay accuracies for the drug and the *N*-oxide metabolite after 3 days of validation were both higher than 96% and the mean precision for both analytes was less than 6%.

Tamoxifen (Tam) used for the treatment of estrogen receptor-positive breast cancer is a prodrug that is transformed to its major active metabolites, endoxifen, and 4-hydroxy-tamoxifen (4-OH-Tam) by various biotransformation enzymes. A novel LC-MS/MS method has been developed and published by Jaremko et al. (2010) for the quantitative determination of Tam, *N*-desmethyl-tamoxifen (ND-Tam), and tamoxifen-*N*-oxide (Tam-*N*-oxide), and the E, Z, and Z′ isomers of Endoxifen and 4-OH-Tam. The linear calibration range for the quantitation of the metabolites in plasma was 0.6 to 2000 nM. Intra- and interassay reproducibility values were 0.2–8.4% and 0.6–6.3%, respectively. The method recoveries varied from 86% to 103%. The long-term stability of endoxifen, 4-OH-Tam, and their isomers was established in fresh frozen plasma for at least 6 months. This method provided the first sensitive, specific, accurate, and reproducible quantitation of Tam and its metabolite isomers for monitoring Tam-treated breast cancer patients.

Zopiclone is a short-acting hypnosedactive agent used in the treatment of insomnia. *N*-desmethyl zopiclone and zopiclone-*N*-oxide are two of the major metabolites of zopiclone that are routinely monitored in subjects using the drug. A simple, selective, and sensitive LC-MS/MS method has been developed and validated for simultaneous quantification of zopiclone and its metabolites in human plasma (Mistri et al., 2008). The method involved SPE of the analytes followed by chromatographic separation on Waters Symmetry shield RP8 column (Phenomemex, Torrance, CA) 150 mm × 4.6 mm i.d., 3.5-μm particle size. The detection was performed by tandem mass spectrometry with a turbo ion spray interface. Metaxalone was used as an internal standard. The method had a chromatographic run time of 4.5 min and linear calibration curves over the concentration range of 0.5–150 ng/ml for both zopiclone and *N*-desmethyl zopiclone and 1–150 ng/ml for zopiclone-*N*-oxide. The intraassay and interassay accuracy and precision (%CV) evaluated at LLOQ and QC levels were within 89.5–109.1% and 3.0–14.7%, respectively, for all the compounds. The method recoveries calculated for the analytes and internal standard were ≥90% in spiked human plasma samples. The validated method was successfully used in a comparative bioavailability study after oral administration of 7.5 mg zopiclone (test and reference) to 16 healthy volunteers under fasted conditions.

Xu et al. (2005) validated a selective and sensitive LC-MS/MS method for the simultaneous determination of a novel KDR kinase inhibitor (1) and its active *N*-oxide metabolite (2) in human plasma to support pharmacokinetic studies. A Packard MultiPROBE II system (Parkin Elmer, San Diego, CA) and a TomTec Quadra 96 (Hamden, CT) liquid handling workstation were used to perform sample preparation and solid-phase extraction (SPE) in 96-well format. Following the extraction on a mixed-mode SPE using Oasis MCX 96-well plate, the analytes were separated reverse C_{18} column using a acetonitrile/ammonium acetate buffer (5 mM, pH 5.0) (60/40, v/v) as the mobile phase at a flow rate of 0.25 ml/min. A tandem mass spectrometric detection was applied using MRM under the positive ion mode with a turbo ion-spray interface. The linear ranges of the calibration curves were 0.05–400 ng/ml for 1 and 0.1–400 ng/ml for 2 on a PE Sciex API 4000 LC–MS/MS instrument. The LLOQ of the assay were 0.05 and 0.1 ng/ml for 1 and 2, respectively, using 0.4 ml of plasma sample. The intraassay precision (using five standard curves prepared by spiking compounds to five lots of plasma) was less than 4.9% for 1 and less than 9.6% for 2 at each concentration level. Assay accuracy was 95.1–104.6% of nominal values for 1 standards and 93.5–105.6% for two standards. The spiked QC samples were stable at room temperature for at least 4 h, at −70 °C for 10 days, and after three freeze–thaw cycles. The extraction recoveries were 80%, 83%, and 84% for 1 and 2 and I.S., respectively. No significant matrix effects were observed during the validation experiments. The method was successfully applied to plasma samples from clinical studies after oral administration of compound 1.

36.9 CONCLUSIONS

The *N*-oxide metabolites are formed during the metabolism of compounds containing tertiary nitrogen atom in the parent molecule. *N*-oxide metabolites are shown to exhibit potent biological activities and in some cases they possess equal or even greater biological activity than the corresponding parent compound. Therefore, it is important to monitor the concentrations of the *N*-oxide metabolites with the parent drug especially when the systemic levels are significantly high. These metabolites are excreted mostly through urine. *N*-oxide metabolites are less stable and easily convert back to the parent compound via loss of the oxygen atom under experimental conditions used in bioanalytical laboratories. This type of conversion adds to the complexities encountered in developing a good bioanalytical method for quantitation of *N*-oxide metabolites. This is the primary reason *N*-oxide metabolites eluded detection until the soft ionization technique such as ESI and APCI were available with sensitive mass spectrometric detectors. Determination of *N*-oxides is performed using the same regulatory guidelines as any other small molecule drugs. Stability of *N*-oxides are impacted at every step of the bioassay process, such as during sample collection, sample processing, extraction, chromatographic separation, in the heated ion source of mass spectrometers and during the tandem mass spectrometric collision process. For a rugged LC-MS/MS quantitation of *N*-oxides using health authority guidelines it is strongly advisable to adhere to certain experimental conditions such as: avoiding antioxidant for sample collection, using neutral pH or mild acidic and basic conditions, avoid high heat during the drying of extract, employing stable isotope labeled internal standards, and the use of ESI as the ionization method.

REFERENCES

Aravagiri M, Marder SR. Simultaneous determination of clozapine and its N-desmethyl and N-oxide metabolites in plasma by liquid chromatography/electrospray tandem mass spectrometry and its application to plasma level monitoring in schizophrenic patients. J Pharm Biomed Anal 2001;26(2):301–311.

Baker JR, Strumpler A, Chaykin S. A comparative study of trimethylamine-N-oxide biosynthesis. Biochim Biophys Acta 1963;71:58–64.

Beckett AH, Hewick DS. The N-oxidation of chlorpromazine in vitro—the major metabolic route using rat liver microsomes. J Pharm Pharmacol 1967;19:134–136.

Bickel MH. The pharmacology and biochemistry of N-oxides. Pharmacol Rev 1969;21(4):325–355.

Bickel MH. N-oxide formation and related reactions in drug metabolism. Xenobiotica 1971;1(4/5):313–319.

Bickel MH, Baggiolini M. The metabolism of imipramine and its metabolites by rat liver microsomes. Biochem Pharmacol 1966;15:1155–1169.

Bickel MH, Weder HJ. The total fate of a drug: kinetics of distribution, excretion and formation of 14 metabolites in rats treated with imipramine. Arch Int Pharmacodyn Ther 1968;173:433–468.

Bun H, Disdier B, Aubert C, Catalin J. Interspecies variability and drug interactions of clozapine metabolism by microsomes. Fundamental Clin Pharmacol 1999;13(5):577–581.

Dell D. Labile metabolites. Chromatographia Suppl 2004;59:S139–S148.

Dunstan WR, Goulding E. The action of alkyl halides on hydroxylamine. Formation of substituted hydroxylamines and oxamines. J Chem Soc (London) 1899;75:792–807.

Fan Z, Xie F, Xia Q, Wang S, Ding L, Liu H. Simultaneous determination of nicotine and its nine metabolites in human urine by LC-MS-MS. Chromatographia 2008;68(7/8):623–627.

Forrest IS, Bolt AG, Serra MT. Distribution of chlorpromazine metabolites in selected organs of psychiatric patients chronically dosed up to the time of death. Biochem Pharmacol 1968;17:2061–2070.

Harinath BC, Odell GV. Chlorpromazine-N-oxide formation by subcellular liver fractions. Biochem Pharmacol 1968;17:167–171.

Jenner P. The role of nitrogen oxidation in the excretion of drugs and foreign compounds. Xenobiotica 1971;1(4–5):399–418.

Jaremko M, Kasai Y, Barginear MF, Raptis G, Desnick RJ, Yu C. Tamoxifen metabolite isomer separation and quantification by liquid chromatography-tandem mass spectrometry. Anal Chem 2010;82(4):10186–10193.

Katagi M, Tatsuno M, Miki A, Nishikawa M, Nakajima K, Tsuchihashi H. Simultaneous determination of selegiline-N-oxide, a new indicator for selegiline administration, and other metabolites in urine by high-performance liquid chromatography–electrospray ionization mass spectrometry. J Chromatogr B 2001;759:125–133.

Knebel NG, Herzog R, Reutter F, Zech K. Sensitive quantification of romflumilast N-oxide in human plasma by LC-MS/MS employing parallel chromatography and electrospray ionization. J Chromatogr B 2012;893–894:82–91.

Koh MH, Gorrod JW. In vitro metabolic N-oxidation of azo compounds. I. Evidence for formation of azo N-oxides (azoxy compounds). Drug Metabol Drug Interac 1989;7(4):253–272.

Komoto I, Yoshida K-I, Matsushima E, Yamashita K, Aikawa T, Akashi S. Validation of a simple liquid chromatography–tandem mass spectrometric method for the determination of propiverine hydrochloride and its N-oxide metabolite in human plasma. J Chromatogr B 2004;799:141–147.

Kuntzman R, Phillips A, Tsai I, Klutch A, Burns JJ. N-oxide formation: A new route for inactivation of the antihistaminic chlorcyclizine. J Pharmacol Exp Ther 1967;155:337–344.

Li F, Patterson AD, Krausz KW, Dick B, Frey FJ, Idle JR. Metabolomics reveals the metabolic map of procainamide in humans and mice. Biochem Pharmacol 2012;83(10):1435–1444.

Machinist JM, Dehner EW, Ziegler DM. Microsomal oxidases. III. Comparison of species and organ distribution of dialkylarylamine-N-oxide dealkylase and dialkylamine-N-oxidase. Arch Biochem 1968;125:858–864.

Majumdar T, Bakhtiar T, Wu S, Winn, D, Tse F. Troubleshooting LC-MS/MS Methods for the bioanalysis of drugs: Some typical problems and solutions. Advances Mass Spectrom 2001;15:681–682.

Marclay F, Saugy M. Determination of nicotine and nicotine metabolites in urine by hydrophilic interaction chromatography-tandem mass spectrometry: Potential use of smokeless tobacco products by ice hockey players. J Chromatogr A 2010;1217(48):7528–7538.

McMahon RE, Sullivan HR. The oxidative demethylation of 1-propoxyphene and its N-oxide by rat liver microsomes. Life Sci 1964;3:1167–1174.

Meger M, Meger-Kossien I, Schuler-Metz A, Janket D, Scherer G. Simultaneous determination of nicotine and eight nicotine metabolites in urine of smokers using liquid chromatography-tandem mass spectrometry. J Chromatogr B 2002;778(1–2):251–261.

Mistri HN, Jangid AG, Pudage A, Shrivastav P. HPLC–ESI-MS/MS validated method for simultaneous quantification of zopiclone and its metabolites, N-desmethyl zopiclone and zopiclone-N-oxide in human plasma. J Chromatogr B 2008;864:137–148.

Niederlaender HAG, Koster EH, Hilhorst MJ, et al. High throughput therapeutic drug monitoring of clozapine and metabolites in serum by on-line coupling of solid phase extraction with liquid chromatography-mass spectrometry. J Chromatogr B 2006;834(1–2):98–107.

Papadopoulos NM. Nicotine-1-oxide: A metabolite of nicotine in animal tissues. Arch Biochem 1964;106:182–185.

Pellegrini M, Marchei E, Rossi S, et al. Liquid chromatography/electrospray ionization tandem mass spectrometry assay for determination of nicotine and metabolites, caffeine and arecoline in breast milk. Rapid Commun Mass Spectrom 2007;21(16):2693–2703.

Rangiah K, Hwang W-T, Mesaros C, Vachani A, Blair IA. Nicotine exposure and metabolizer phenotypes from analysis of urinary nicotine and its 15 metabolites by LC-MS. Bioanalysis 2011;3(7):745–761.

Ulgen M, Ozer U, Kucukguzel I, Gorrod JW. Microsomal metabolism of N-benzyl-N-ethylaniline and N-benzyl-N-ethyl-p-toluidine. Drug Metabol Drug Interac 1997;14(2): 83–98.

US Food and Drug Administration. Guidance for Industry on Bioanalytical Method Validation. C. F. R. 66 (100), 28526, 2001.

Wohlfarth A, Toepfner N, Hermanns-Clausen M, Auwearter V. Sensitive quantification of clozapine and its main metabolites norclozapine and clozapine-N-oxide in serum and urine using LC-MS/MS after simple liquid-liquid extraction work-up. Anal Bioanal Chem 2011;400(3):737–746.

Xu Y, Du L, Soli ED, Braun MP, Dean DC, Musson DG. Simultaneous determination of a novel KDR kinase inhibitor and its N-oxide metabolite in human plasma using 96-well solid-phase extraction and liquid chromatography/tandem mass spectrometry. J Chromatogr B 2005;817(2):287–296.

Yoon K-H, Lee S-Y, Jang M, et al. A rapid determination of propiverine and its N-oxide metabolite in human plasma by high performance liquid chromatography-electrospray ionization tandem mass spectrometry. Talanta 2005;66(4):831–836.

Zhang WV, D'Esposito F, Fabrizio ERJ, Ramzan I, Murray M. Interindividual variation in relative CYP1A2/3A4 phenotype influences susceptibility of clozapine oxidation to cytochrome P450-specific inhibition in human hepatic microsomes. Drug Metabol Dispos 2008;36(12):2547–2555.

Ziegler DM, Pettit F. Formation of an intermediate N-oxide in the oxidative demethylation of N,N-dimethylaniline catalysed by liver microsomes. Biochem Biophys Res Commun 1964;15:188–193.

37

HYDROLYSIS OF PHASE II CONJUGATES FOR LC-MS BIOANALYSIS OF TOTAL PARENT DRUGS

LAIXIN WANG, WEIWEI YUAN, SCOTT REUSCHEL, AND MIN MENG

37.1 INTRODUCTION

Drugs are metabolized by two different types of reactions: phase I and II. Phase I metabolic reactions introduce or unmask a functional group within a molecule. These reactions include hydroxylation, epoxidation, deamination, oxidation, reduction, and hydrolysis. Phase II metabolic reactions create conjugation between parent molecules (or phase I metabolites) and hydrophilic molecules, such as sulfuric acid, glucuronic acid, or other highly polar groups (e.g., glycosides and phosphate). Phase II reactions are catalyzed by conjugative enzymes, such as uridine 5′-diphospho-glucuronosyltransferase (UDP-glucuronyltransferase or UGT), sulfotransferase, glutathione *S*-transferase (GST), *N*-acetyl transferase (NAT), and methyl transferase (*N*-methyl, thiomethyl-, and thiopurinemethyl-). Glutathione conjugates are further metabolized to cysteine and *N*-acetyl cysteine adducts (i.e., mercapturic acid synthesis). These reactions result in the synthesis of hydrophilic compounds that increase their elimination efficiency via normal renal and intestinal pathways. Products of conjugation reactions have increased molecular weight and are usually inactive unlike phase I reactions that often produce active metabolites. However, as there are exceptions to most rules, morphine-glucuronides are examples of pharmacologically active phase II metabolites (Ing Lorenzini, 2012).

Worldwide regulatory agencies (US Food and Drug Administration– (FDA); European Medicines Agency– (EMA); Brazil's National Health Surveillance Agency (*ANVISA)*; etc.) require a thorough understanding of drug safety, which routinely includes monitoring exposure levels for the parent drug and all relevant metabolites. Metabolite quantitation is always required when the metabolite is toxic or pharmacologically active or when metabolite concentrations reach or exceed the parent drug concentration in biological matrices. The challenge for drug metabolism and bioanalytical scientists is to provide exposure data that is sufficiently precise and accurate to satisfy these regulatory requirements. Liquid chromatography coupled with tandem mass spectrometry (LC-MS/MS) has been widely accepted as the primary tool in the quantitative analysis of parent drugs and metabolites owing to its superior sensitivity, specificity, and efficiency (Kostiainen et al., 2003). Because the analysis must be carried out in complex biological matrices, extensive sample preparation and liquid chromatography separation techniques are often required to achieve adequate specificity and sensitivity for analysis.

37.2 METHODS AND APPROACHES

37.2.1 Principles and Methodology

Since the introduction of LC-MS/MS, there has been a common expectation that multicomponent bioanalytical assays can be easily and quickly developed for quantitative analysis. While this may generally be true for nonregulated "discovery" studies in the early phases of drug development to which expanded acceptance criteria may be applied to the analytical results, this is not always the case for regulated studies executed under good laboratory practices (GLP) that are intended for submission to worldwide regulatory agencies

Handbook of LC-MS Bioanalysis: Best Practices, Experimental Protocols, and Regulations, First Edition. Edited by Wenkui Li, Jie Zhang, and Francis L.S. Tse.

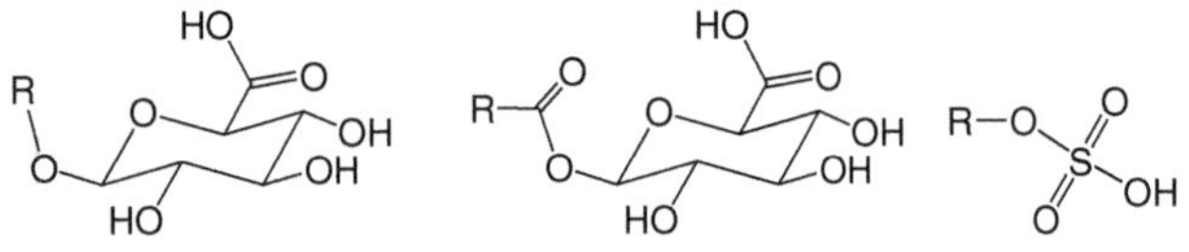

Either glucuronide (acetal) Ester glucuronide (acylal) Sulfate ester

FIGURE 37.1 Chemical structures of common phase II metabolites.

and to which much stricter acceptance criteria are applied. For GLP studies with a very stringent acceptance criteria, the likelihood of failure will statistically increase as the number of metabolites being quantified is increased. This means that multianalyte methods must be very rugged and reliable for acceptable long-term performance. Considering that parent drugs and their metabolites can have very different polarities affecting their extraction efficiency, chromatographic behavior and MS response, developing a common analytical procedure to quantify all analytes equally well often requires compromise during optimization. One of the more complex issues in measuring drugs and their metabolites together is the possible interconversion from one form to another. For example, glucuronide metabolites are unstable under certain circumstances, and can easily convert back and forth to their conjugated and nonconjugated forms. *N*-Glucuronide conjugates of primary, secondary, or *N*-hydroxylated amines are hydrolyzed to their parent compound and glucuronic acid under mild acidic conditions (Kadlubar et al., 1977), while quaternary ammonium glucuronides are hydrolyzed under basic conditions. Both *O*- and *N*-glucuronides are hydrolyzed by the enzyme β-glucuronidase, but acyl glucuronides are relatively unstable and undergo acyl group migration to form isomers that are stable to hydrolysis; however, these isomers are hydrolyzed by sodium hydroxide (Shipkova et al., 1999; Wen et al., 2006). Additionally, reference standards for these metabolites are often not available in adequate volumes or of sufficient quality (i.e., purity) for GLP quantitation. Due to these analytical considerations, it is sometimes not feasible to accurately determine the concentration of the parent drug and/or individual metabolites in certain biological matrices. An alternative quantitative approach is to convert all conjugated metabolites to the parent drug and/or to a stable intermediate metabolite in order to determine the total drug concentration in the biological matrix.

The hydrolysis of phase II metabolites such as glucuronic acid or sulfuric acid conjugates can be achieved by either acidic or enzymatic hydrolysis. In the case of acidic hydrolysis, samples are incubated with acids at elevated temperatures, leading to cleavage of acetalic and acylalic glucuronides and sulfuric esters (Figure 37.1). A typical methodology for urine sample deconjugation is to incubate the sample for 1 h at 100°C with addition of acid at a final concentration of 0.1 M. The deconjugation is strongly influenced by the choice of acid (hydrochloric or sulfuric), acid concentration, temperature, and duration of the reaction (Venturelli et al., 1995). The reactions are usually conducted in aqueous solutions; however, other solvents can sometimes assist the hydrolysis. Acid-catalyzed methanolysis is an efficient procedure for the simultaneous cleavage of steroid glucuro- and sulfo-conjugates. The procedure was first reported by Tang and Crone, who used a mixture of 1.0 M acetylchloride in methanol, thus producing anhydrous hydrochloric acid by a strong exothermic reaction (Tang and Crone, 1989). The method was further improved by Delhennin et al. who replaced the acetylchloride with trimethylchorosilane that can be mixed with methanol without special care and yet release a sufficient amount of hydrochloric acid under the experimental conditions. The reaction usually takes about 1 h at 55°C to complete (Dehennin et al., 1996). Alternatively, urine samples can be dried down and then incubated in approximately 0.04% sulfuric acid in ethyl acetate at 55°C for 1 h to deconjugate the steroid sulfates (Hauser et al., 2008). These aggressive acid hydrolysis cleavage procedures are simple and cost-effective; however, they can often be associated with the formation of unwanted artifacts or with the degradation of target analytes (Beyer et al., 2005; Kamata et al., 2003).

Enzymatic cleavage, which is usually conducted in slightly acidic to neutral conditions, is considerably gentler than the acid hydrolysis techniques. Therefore, although enzymatic hydrolysis is relatively expensive compared to acid hydrolysis, it is typically the preferred method for deconjugation. The enzymatic digestion can be performed using β-glucuronidase (GRD) and/or arylsulfatase (ARS) enzymes from various species (Gomes et al., 2009). Enzymes from mammalian sources (extracted from beef liver) and bacterial sources (extracted from *Escherichia coli*) contain β-glucuronidase activity, allowing cleavage of the glucuronide moiety. Enzymes from mollusk sources (usually extracted from *Helix pomatia*) contain both β-glucuronidase activity and sulfatase activity, the latter being responsible for sulfate hydrolysis. Although all enzyme sources possess some degree of β-glucuronidase activity, the amount of activity and the specificity and efficiency of the deconjugation process may vary significantly. The glucuronidase enzyme activity and sulfatase enzyme activity of a particular preparation are expressed as Fishman units and Roy units, respectively. One Fishman unit releases 1 μg of phenolphthalein from phenolphthalein-β-glucuronide in 1 h at 38°C. One Roy unit releases 1 μg of 2-hydroxy-5-nitrophenyl sulfate in 1 h at 38°C.

37.2.2 Method Optimization

Optimization of the hydrolysis conditions is of vital importance to ensure complete deconjugation of the metabolites. The conditions that influence enzymatic hydrolysis are

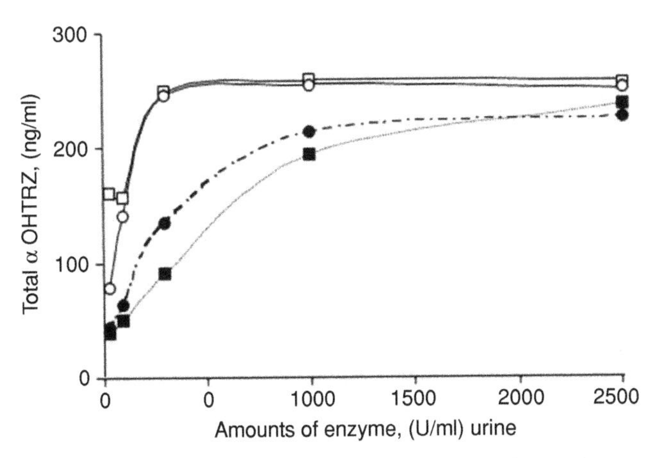

FIGURE 37.2 Dose-dependent increase of total α-hydroxytriazolam concentration with β-glucuronidases from *E. coli* (□), bovine liver (○), *P. vulgata* (●), and *H. pomatia* (■). Urine samples were incubated at 45°C for 2 h at "Sigma recommended pH" for each enzyme: pH 6.8 for *E. coli*, pH 5.0 for bovine liver, and *P. vulgate*, pH 3.8 for *H. pomatia* (data provided by Kenji Tsujikawa (Tsujikawa et al., 2004)).

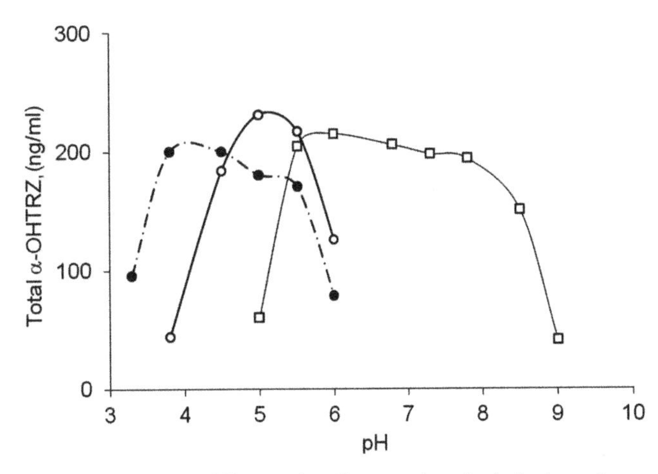

FIGURE 37.3 Effect of pH on the hydrolysis of α-hydroxytriazolam-glucuronide with β-glucuronidases from *E. coli* (□), bovine liver (○), and *P. vulgata* (●). Each sample was incubated for 2 h under the respective "optimized" conditions for each enzyme: *E. coli*. 100 U/ml in urine at 37°C; bovine liver: 100 U/ml in urine at 45°C; *P. vulgate*: 300 U/ml in urine at 60°C (data provided by Kenji Tsujikawa (Tsujikawa et al., 2004)).

amount of enzyme, temperature, duration of incubation, and pH of the reaction buffer. The enzyme activity must be sufficient to enable complete hydrolysis of the conjugates present in samples. However, the amount of enzyme required can be influenced by the type and concentration of the metabolite and other hydrolysis conditions. Additionally, the selectivity and the reactivity of the enzyme preparation can vary depending on the source and/or batch of enzyme. Consequently, it is necessary to test the enzyme activity of the preparation before carrying out preliminary studies to determine the optimum amount of enzyme for complete hydrolysis (Kamata et al., 2003; Tsujikawa et al., 2004).

An analytical method needs to be established prior to optimizing the hydrolysis conditions. Pooled samples or samples with relatively high analyte concentrations are recommended for determination of the optimal conditions. The enzyme source and concentration are among the first two variables to be evaluated. Some of the frequently used β-glucuronidases from different sources including bovine liver, *H. pomatia*, *Nepenthes ampullaria*, and *E. coli* can be evaluated at concentration ranges from 10 to 5000 units/ml. These reactions can be conducted under the respective vendor recommended conditions for 1–2 h (Figure 37.2). The measured concentrations of the conjugate (or the total concentration of the hydrolysis product) are plotted against the final enzyme concentrations in each samples. The example in Figure 37.2 indicates that 500 U/ml of glucuronidase from *E. coli* or bovine liver is sufficient to completely digest α-hydroxytriazolam-glucuronide to α-hydroxytriazolam under the experimental conditions (Tsujikawa et al., 2004).

The pH of common hydrolysis conditions is generally slightly acidic to neutral, as glucuronides are more resistant to hydrolysis under alkaline conditions. The pH dependency of the hydrolysis varies based on the enzyme preparation utilized for a particular substrate. For the enzymatic hydrolysis of α-hydroxytriazolam-glucuronide in human urine, the optimal working pH levels were 3.8–4.5 for *Patella vulgate,* 5.0–5.5 for bovine liver, and 5.5–7.8 for *E. coli* sourced β-glucuronidase enzymes (Figure 37.3). Because the pH of urine samples typically varies from approximately 4–8.5 (Bilobrov et al., 1990), it is important to adjust the pH of the samples with an appropriate buffer to obtain optimal enzymatic digestion.

The choice of temperature and duration of incubation is also compound dependent. Cleavage of dihydroepiandrosterone sulfates is improved at higher temperatures, while androsterone and etiocholanolone sulfate deconjugation favors a low temperature (Shackleton, 1986). For incubation times, durations from 30 min to 20 h have been utilized for different hydrolysis reactions (Gomes et al., 2009; Tsujikawa et al., 2004), and must be tested to select the optimal time based on the productivity and the cost-effectiveness of the overall procedure.

The sample extraction methodology and LC-MS/MS conditions are relatively easy to develop once the polar phase II metabolites are deconjugated. Simple protein precipitation extraction or sample dilution ("dilute and shoot") may be sufficient if adequate assay sensitivity can be achieved and appropriate LC column cleanup procedures (e.g., column back-flush or forward flush) are followed. However, it is more likely that the hydrolyzed samples will need to be concentrated and further purified using more extensive liquid–liquid extraction (LLE) or solid-phase extraction (SPE) cleanup procedures to achieve the required sensitivity and specificity.

SPE is easily automated in a 96-well plate format and is therefore a good option for high-throughput sample analysis. However, in addition to the ease of use and ability to automate, SPE plates are very costly to use, which may limit their viability as a sample preparation technique. Traditional manual LLE is much more cost-effective and there is no limit to how much sample volume may be used (as opposed to SPE which does have a theoretical upper limit based upon the size of the SPE vessel and amount of sorbent material), but LLE is also relatively labor intensive and therefore relatively low throughput. More recently, a technique known as supported-liquid extraction (SLE) that replicates the performance of a traditional LLE extraction has also gained popularity. SLE functions very similar to traditional LLE and is available in a 96-well format that is also easy to automate (Nguyen et al., 2010). However, the maximum sample size for SLE is somewhat limited (less than 400 μl), and the use of a stable isotope-labeled internal standard is also recommended due to the somewhat inconsistent extraction recoveries observed for some analytes, further limiting its applicability.

Final quantitative analysis can be accomplished on a reverse-phase high-pressure liquid chromatography (HPLC) column coupled with a triple-quadruple tandem mass spectrometer operated in an atmospheric-pressure chemical ionization (APCI) or electrospray ionization (ESI) mode. It has been demonstrated that matrix effects (e.g., ionization suppression and ionization enhancement) are a greater concern for ESI than for APCI and that the degree of the effect varies by instrument platform and manufacturer (Mei et al., 2003). Typical matrix effects can manifest themselves by drifting or fluctuating analyte or internal standard response during analysis, which can cause inaccuracy and imprecision of quantitative results. This may indicate that sample matrix components are being inadvertently retained on the LC column, and the uncontrolled "buildup and elution" of these matrix components on and off the LC column is preventing consistent ionization of the analytes and/or internal standards as they enter the MS system. A simple, yet effective corrective action to reduce these matrix effects is to incorporate either a forward-flush or a back-flush with a high proportion of organic solvent (acetone, acetonitrile, methanol) into the LC program between sample injections to remove these unwanted matrix components from the LC column. In order to keep the impact of matrix effects to a minimum; however, the APCI technique is preferred over the ESI technique if APCI provides adequate sensitivity for all analytes and metabolites of interest.

37.3 EXAMPLE PROTOCOL

Quantitative Determination of the Total Concentration of Dextromethorphan and Dextrorphan in Human Urine Using LC-MS/MS.

Dextromethorphan is an antitussive (cough suppressant) drug and is an active ingredients in many over-the-counter cold and cough medicines. Dextromethorphan also has other medicinal uses ranging from pain relief to psychological applications. Dextromethorphan is converted into the active metabolite dextrorphan (Figure 37.4) in the liver by the cytochrome P450 enzyme CYP2D6 (Lutz et al., 2008;

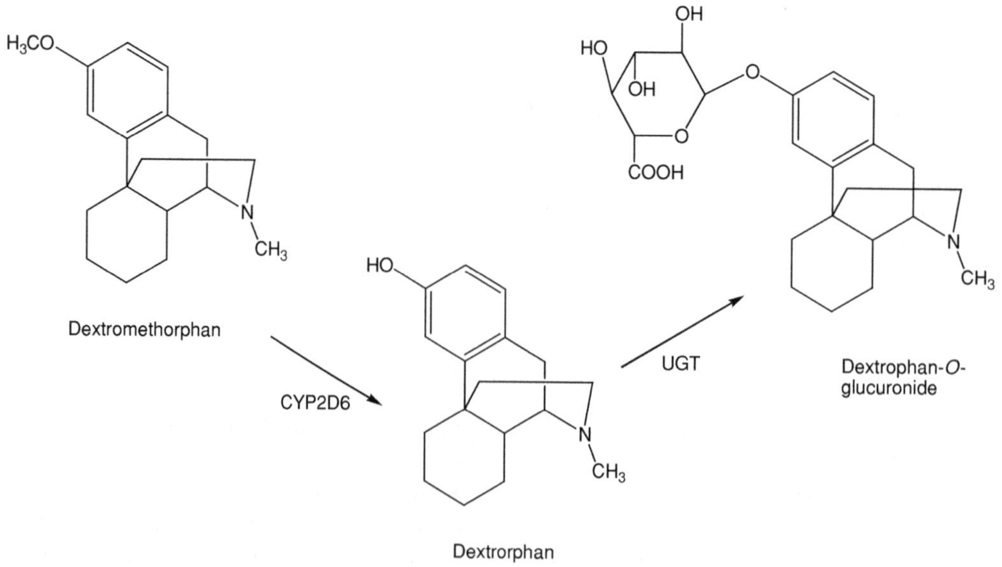

FIGURE 37.4 Main pathways for dextromethorphan metabolism in humans. CYP2D6: Cytochrome P450 enzyme 2D6; UGT: uridine diphosphoglucuronic acid and glucuronyl transferase (Lutz et al., 2008; Takashima et al., 2005).

Takashima et al., 2005). The therapeutic activity of dextromethorphan is believed to be caused by both the drug and this metabolite. Dextrorphan is then primarily metabolized to inactive glucuronide conjutates in plasma and urine by glucuronyl transferase in the presence of uridine diphosphoglucuronic acid (Capon et al., 1996; Chládek et al., 1999). Therefore, it is important to measure the total concentrations of dextromethorphan and dextrorphan in study samples for toxicology and pharmacology evaluations of dextromethorphan.

Test material and matrix

- Dextromethorphan, purchased from Sigma.
- Dextrorphan, purchased from Sigma.
- Dextromethorphan-d3, purchased from Sigma.
- Dextrorphan-d3, purchased from Sigma.
- Urine, collected in house.

Equipment and Reagents

- *Liquid handling system*: Microlab Nimbus 96, Hamilton Robotics, Inc.
- *HPLC system*: LC10AD HPLC Pumps and SCL-10AVP System Controller, Shimazu Scientific Instruments.
- *HPLC column*: Xbridge Phenyl, 5 μm, 2.1 × 50 mm, Waters Corporation.
- *Autosampler*: CTC PAL Workstation, LEAP Technologies.
- *Standard 6-port switching valve*: VICI Cheminet 10U-0263H.
- *Mass spectrometer*: Sciex API365 Triple Quadrupole MS/MS, Applied Biosystems, Inc.
- Captiva 96-well filter plate, Varian.
- β-Glucuronidase (from *E. coli*), Type IX-A, lyphilized powder, 1,000,000 to 5,000,000 units/g protein, Sigma.
- *N,N*-Dimethylformamide (DMF), HPLC grade, Burdick & Jackson.
- Ammonium hydroxide (NH_4OH) (~28–30% purity) ACS grade, Sigma-Aldrich
- Methyl *t*-butyl ether (MTBE), High purity grade, EMD.
- Phosphoric acid, HPLC grade, EMD.
- Methylene Chloride, HPLC grade, EMD.
- Acetonitrile (MeCN), HPLC grade, EMD.
- Formic acid, ACS grade, EMD.
- Methanol (MeOH), HPLC grade, EMD.
- Acetone, ACS grade, EMD.
- Water, deionized, Type 1, (typically 18.2 MΩ cm) or equivalent.
- *Needle wash 1*: 10/90 MeOH/water.
- *Needle wash 2*: 50/50 Water/DMF.
- *Mobile phase A*: 0.1% FA in water.
- *Mobile phase B*: 0.1% FA in MeOH.
- *Mobile phase C*: 50/50 MeOH/MeCN (to be used for column back-flush at 1.00 ml/min flow rate).

Method

Solution and sample preparation

1. Weigh an appropriate amount of dextromethorphan, dextrorphan, dextromethorphan-d3 and dextrorphan-d3 into the required amount of MeOH to obtain individual stock solutions at 0.500 mg/ml each.
2. Prepare a combination solution in MeOH containing 50,000 ng/ml each of dextromethorphan and dextrorphan from corresponding stock solutions in step 1.
3. Prepare standard calibrators and quality control (QC) samples in human urine from the combination solution or other appropriate intermediate solutions/samples. The standard curve consists of calibrators at eight concentrations: 1.00, 2.00, 5.00, 10.0, 50.0, 100, 250, and 500 ng/ml. The QCs are prepared at five levels: 1.00 (LLOQ), 3.00 (low), 200 (medium), 400 (high), and 1000 (dilution) ng/ml.
4. Prepare an internal standard (IS) working solution in 50/50 water/MeOH containing 100 ng/ml each of dextromethorphan-d3 and dextrorphan-d3 from corresponding stock solutions in step 1.

Sample Extraction

1. Manually aliquot 200 μl of each sample into 13 × 100 mm polypropylene tubes. For Dilution QC and samples requiring dilution, adjust the sample aliquot volume according to the dilution factor.
2. Add 50.0 μl of internal standard working solution [100 ng/ml each of dextromethorphan-d3 and dextrorphan-d3 in 50/50 water/MeOH] to all samples except matrix blanks.
3. To the matrix blanks, add 50.0 μl of 50/50 water/MeOH.
4. Add 50.0 μl of β-glucoronidase (~6250 units/ml) in 200 mM phosphate buffer (pH= ~6.8) to all samples.
5. Add 50.0 μl of 200 mM phosphate buffer (pH = ~6.8) to all samples.
6. Centrifuge all samples at approximately 3000 rpm for approximately 1 min.
7. Vortex-mix all samples at the low setting on the multitube vortexer for approximately 2 min.
8. Incubate all samples in a ~37°C water bath for approximately1 h.

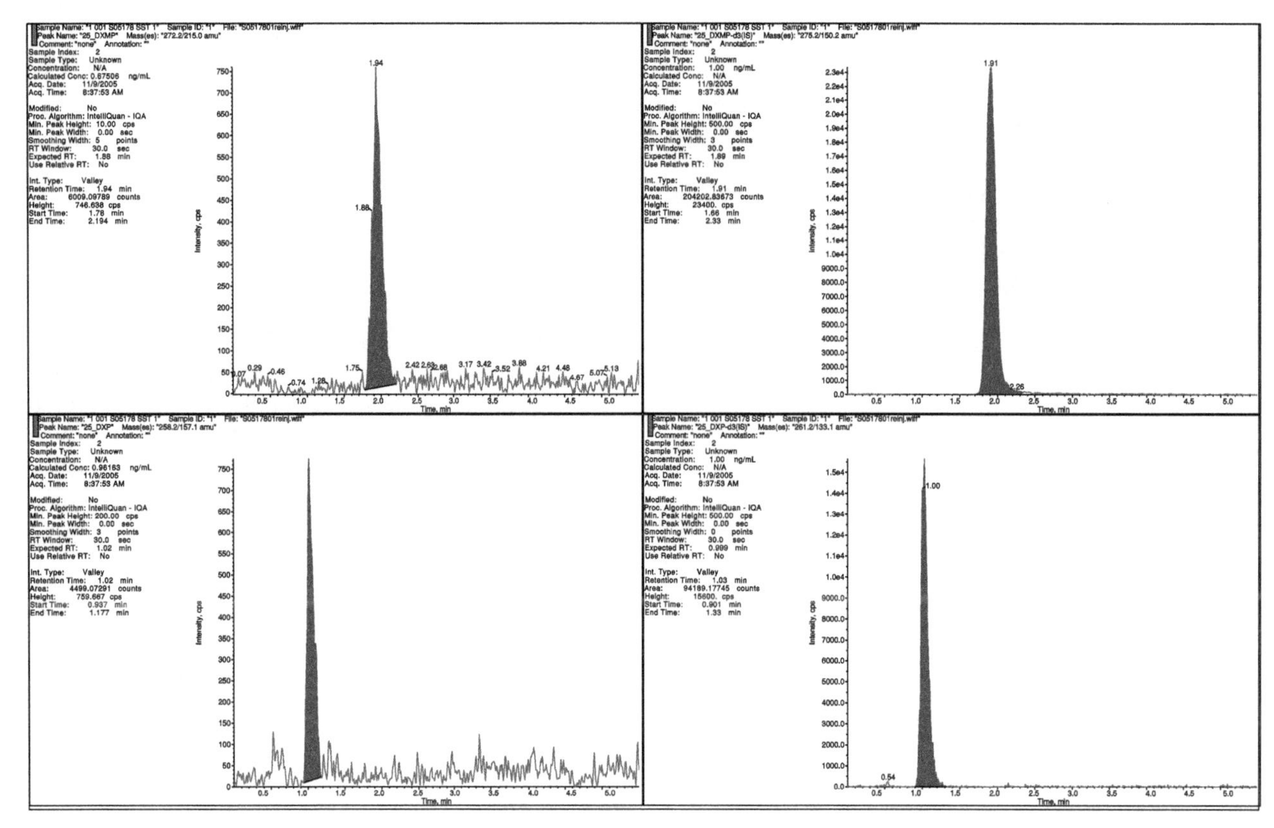

FIGURE 37.5 Chromatogram for dextromethorphan and dextrorphan at lower limit of quantitation (LLOQ; 1.00 ng/ml) in a typical validation run. Top left: dextromethorphan; top right: dextromethorphan-d3; bottom left: dextrorphan; bottom right: dextrorphan-d3.

9. Add 700 μl of 5/95 ammonium hydroxide/water (prepared fresh) to all samples. Vortex briefly.
10. Add 4.00 ml 75/25 MTBE/methylene chloride to all samples and cap the tubes.
11. Place tubes horizontally in shaker and shake on high for approximately 10 min.
12. Centrifuge at approximately 3000 rpm for approximately 5 min.
13. Freeze the aqueous layer in dry ice/acetone (or methanol) bath and decant the organic layer into clean 13 × 100 polypropylene tubes.
14. Evaporate the samples to dryness at approximately 40°C under nitrogen in the Turbovap for approximately 40 min.
15. Reconstitute the samples with 200 μl of 0.1% FA in [70/30 water/MeOH].
16. Vortex-mix the samples at the low setting on the multitube vortexer for approximately 1 min.
17. Manually transfer samples into a Captiva 96-well filter plate (Part No. A5960045) placed on top of a clean tapered 96-well collection plate. Collect the filtered sample extracts using a positive pressure manifold.
18. Centrifuge the sample plate at approximately 3000 rpm for approximately 5 min.

LC-MS/MS analysis

1. The LC system was equipped with a standard six-port switching valve to facilitate the back-flush of the LC column between injections. The retention times for dextromethorphan and dextrorphan are approximately 1.9 and 1.0 min, respectively, under the following LC conditions:
   a. *Flow rate (mobile phase A/B)*: 0.400 ml/min.
   b. *Column temperature*: 35°C.
   c. *Gradient*:

Time (min)	0.0'	3.1′	3.5′	4.0′	4.5′	5.5′
% B	38	Event 1[a]	55	38	Event 2[b]	Stop

[a]Event 1 = mobile phase C to LC column in reverse direction (mobile phase A/B to waste).
[b]Event 2 = mobile phase A/B to LC column (mobile phase C to waste).

2. The mass spectrometer was operated in positive ionization mode with an ESI source (Figure 37.5).

The mass to charge (*m/z*) transitions monitored in the multiple reaction monitoring (MRM) mode for dextromethorphan, dextrorphan, dextromethorphan-d3, and dextrorphan-d3 were 272 → 215, 258 → 157, 275 → 150, and 261 → 133, respectively. The typical MS conditions were dwell time, 100 ms; source temperature, 450°C; IS voltage, 4000 V; declustering potential (DP), 50; curtain gas, 20 psi; nebulizer gas, 8 psi; collision energy for dextromethorphan, 41 eV; and collision energy for dextrorphan, 48 eV.

REFERENCES

Beyer J, Bierl A, Peters FT, Maurer HH. Screening procedure for detection of diuretics and uricosurics and/or their metabolites in human urine using gas chromatography-mass spectrometry after extractive methylation. Ther Drug Monit 2005;27(4):509–520.

Bilobrov VM, Chugaj AV, Bessarabov VI. Urine pH variation dynamics in healthy individuals and stone formers. Urol Int 1990;45(6):326–331.

Capon DA, Bochner F, Kerry N, Mikus G, Danz C, Somogyi AA. The influence of CYP2D6 polymorphism and quinidine on the disposition and antitussive effect of dextromethorphan in humans. Clin Pharmacol Ther 1996;60(3):295–307.

Chládek J, Zimová G, Martínková J, Tůma I. Intra-individual variability and influence of urine collection period on dextromethorphan metabolic ratios in healthy subjects. Fundam Clin Pharmacol 1999;13(4):508–15.

Dehennin L, Lafarge P, Dailly P, Bailloux D, Lafarge JP. Combined profile of androgen glucuro- and sulfoconjugates in post-competition urine of sportsmen: a simple screening procedure using gas chromatography-mass spectrometry. J Chromatogr B Biomed Appl 1996;687(1):85–91.

Gomes RL, Meredith W, Snape CE, Sephton MA. Analysis of conjugated steroid androgens: deconjugation, derivatisation and associated issues. J Pharm Biomed Anal 2009;49(5):1133–1140.

Hauser B, Deschner T, Boesch C. Development of a liquid chromatography-tandem mass spectrometry method for the determination of 23 endogenous steroids in small quantities of primate urine. J Chromatogr B Analyt Technol Biomed Life Sci 2008;862(1-2):100–112.

Ing Lorenzini K, Daali Y, Dayer P, Desmeules J. Pharmacokinetic-pharmacodynamic modelling of opioids in healthy human volunteers. a minireview. Basic Clin Pharmacol Toxicol 2012;110(3):219–226.

Kadlubar FF, Miller JA, Miller EC. Hepatic microsomal N-glucuronidation and nucleic acid binding of N-hydroxy arylamines in relation to urinary bladder carcinogenesis. Cancer Res 1977;37(3):805–814.

Kamata T, Nishikawa M, Katagi M, Tsuchihashi H. Optimized glucuronide hydrolysis for the detection of psilocin in human urine samples. J Chromatogr B Analyt Technol Biomed Life Sci 2003;796(2):421–427.

Kostiainen R, Kotiaho T, Kuuranne T, Auriola S. Liquid chromatography/atmospheric pressure ionization-mass spectrometry in drug metabolism studies. J Mass Spectrom 2003;38(4):357–372.

Lutz U, Bittner N, Lutz RW, Lutz WK. Metabolite profiling in human urine by LC-MS/MS: method optimization and application for glucuronides from dextromethorphan metabolism. J Chromatogr B Analyt Technol Biomed Life Sci 2008;871(2):349–356.

Mei H, Hsieh Y, Nardo C, et al. Investigation of matrix effects in bioanalytical high-performance liquid chromatography/tandem mass spectrometric assays: application to drug discovery. Rapid Commun Mass Spectrom 2003;17(1):97–103.

Nguyen L, Zhong WZ, Painter CL, Zhang C, Rahavendran SV, Shen Z. Quantitative analysis of PD 0332991 in xenograft mouse tumor tissue by a 96-well supported liquid extraction format and liquid chromatography/mass spectrometry. J Pharm Biomed Anal 2010;53(3):228–234.

Shackleton CH. Profiling steroid hormones and urinary steroids. J Chromatogr 1986;379:91–156.

Shipkova M, Armstrong VW, Wieland E, et al. Identification of glucoside and carboxyl-linked glucuronide conjugates of mycophenolic acid in plasma of transplant recipients treated with mycophenolate mofetil. Br J Pharmacol 1999;126(5):1075–1082.

Takashima T, Murase S, Iwasaki K, Shimada K. Evaluation of dextromethorphan metabolism using hepatocytes from CYP2D6 poor and extensive metabolizers. Drug Metab Pharmacokinet 2005;20(3):177–182.

Tang PW, Crone DL. A new method for hydrolyzing sulfate and glucuronyl conjugates of steroids. Anal Biochem. 1989;182(2):289–94.

Tsujikawa K, Kuwayama K, Kanamori T, et al. Optimized conditions for the enzymatic hydrolysis of α-hydroxytriazolam-glucuronide in human urine. J of Health Science. 2004;50(3):286–289.

Venturelli E, Cavalleri A, Secreto G. Methods for urinary testosterone analysis. J Chromatogr B Biomed Appl. 1995;671(1-2):363–380.

Wen Z, Stern ST, Martin DE, Lee KH, Smith PC. Structural characterization of anti-HIV drug candidate PA-457 [3-O-(3′,3′-dimethylsuccinyl)-betulinic acid] and its acyl glucuronides in rat bile and evaluation of in vitro stability in human and animal liver microsomes and plasma. Drug Metab Dispos. 2006;34(9):1436–42.

38

LC-MS BIOANALYSIS OF REACTIVE COMPOUNDS

HERMES LICEA-PEREZ, CHRISTOPHER A. EVANS, AND YI (ERIC) YANG

38.1 INTRODUCTION

Generally metabolites are less toxic and more polar than the parent xenobiotic and thus can be easily excreted from the body. However, one of the potential consequences of xenobiotic metabolism is the formation of reactive metabolites. These reactive metabolites are typically electrophiles in nature, and have the ability of interacting with nucleophilic sites on macromolecules (DNA and proteins) to form covalently bound products (Figure 38.1; Miller and Miller, 1947; Singer and Grunberger, 1983; Hemminki et al., 1994; Dipple, 1995). The generation of reactive metabolites has been well characterized, and throughout the years, a great number of functional groups have become considered "suspicious" as potential areas for bioactivation. New chemical entities (NCE) containing these groups deserve special consideration and characterization for their bioactivation potential. A comprehensive and thorough review on structural bioactivation pathways of organic functional groups is described by Kalgutkar et al. (2005).

DNA alkylation is believed to lead to genotoxicity and hence represent a carcinogenic risk to humans (e.g., DNA damage and mutation; Lutz, 1986; Hemminki, 1993; Hemminki et al., 1994; Otteneder and Lutz, 1999). Some reactive agents have the ability to directly alkylate DNA and proteins, while others can exert their electrophilic reactivity only as a result of biotransformation or metabolism (Miller and Miller, 1947). Therefore, evaluation of their potential formation and monitoring of the exposure of putative reactive compounds with genotoxicity potential is often conducted early in drug development (Doss and Baillie, 2006). This strategy attends to minimize potential risk of idiosyncratic responses and assess any potential risk to humans (Walgren et al., 2005).

The presence of reactive compounds *in vivo* after drug intake can also be attributed to impurities in the dose, as a result of carryover from the chemical synthesis (synthetic intermediates, reaction byproducts, or degradants) and/or as result of biotransformation of parent compound to smaller reactive fragments (e.g., formation of aldehydes after enzymatic cleavage of an amide bond).

Control of pharmaceutical impurities continues to receive attention from regulatory agencies around the world. Even when proper care is taken during manufacturing of active pharmaceutical ingredients (API), a chance always remains that a synthetic residue may negatively affect safety of the API either as a result of direct toxicity of the impurity or through subsequent biotransformation. Complete removal of impurities remains a difficult challenge; regulatory authorities mandate their levels are kept to that representing minimal risk to human health (ICH Q3A(R2), 2006; ICH Q3B(R2), 2006; ICH Q3C(R5), 2006; EMEA, 2006).

In addition, the potential genotoxicity of metabolites is normally assessed in standard battery of genotoxicity assays conducted on parent drug or chemical impurities to support first-in-human (FIH) studies, which may include the Ames, mammalian cell assays, or *in vivo* micronucleus assay (Clive and Spector, 1975; Tennant et al., 1987; Cartwright and Matthews, 1994). However, additional genotoxicity studies may be required to test in isolation on potentially human specific or disproportionate metabolites exposure. Other *in vitro* screening assays such as glutathione (GSH) and cyanide trapping procedures may help to identify and characterize potential reactive metabolites and are typically completed early in the drug development process (Evans et al., 2004; Ma and Zhu, 2009).

Handbook of LC-MS Bioanalysis: Best Practices, Experimental Protocols, and Regulations, First Edition. Edited by Wenkui Li, Jie Zhang, and Francis L.S. Tse.

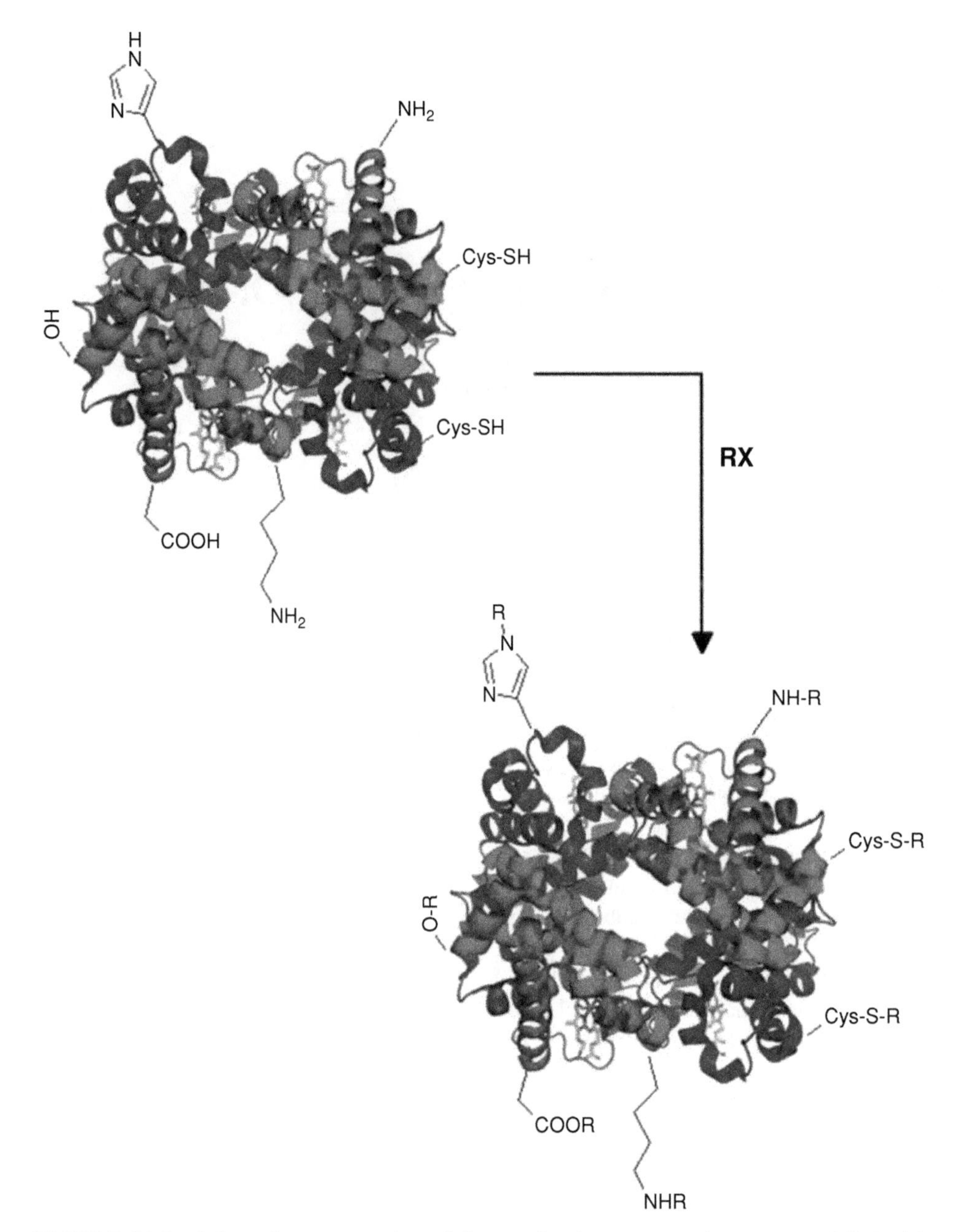

FIGURE 38.1 Schematic representation of the reaction between reactive compound RX with representative nucleophilic groups on hemoglobin.

38.2 DETERMINATION OF REACTIVE COMPOUND EXPOSURE *IN VIVO*

As the term "reactive" implies a certain degree of instability, these reactive compounds are typically short lived in *in vivo* settings and as such are difficult to analyze. In general, there are two approaches to determine potential exposure of reactive compounds in biological matrices: (1) measurement of adducts with macromolecules as an indirect marker of exposure and (2) the use of excess amount of trapping reagents (nucleophiles) to capture the unreacted fraction of these compounds. Both approaches are discussed in detail in the following text.

Electrophiles, $R^{\delta+}X^{\delta-}$ are compounds where the electron density is asymmetrical; thus the probability of finding electrons in the proximity of X is higher than R, as X is more electronegative than R. In this instance, atom R has resulting electron deficiency and is subjected to nucleophilic attack. The differential distribution of electrons in electrophilic compounds makes them unstable and the $R^{\delta+}X^{\delta-}$ bond may be broken by nucleophilic attack from another compound with excess of electrons (a nucleophile). Nucleophilic atoms such

TABLE 38.1 Nucleophilic Sites in Hemoglobin (Adapted from Ehrenberg and Osterman-Golkar, 1980)

Nucleophilic amino acid	Reactive form	Nonreactive form	pK_a
Cysteine	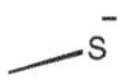 $-S^-$	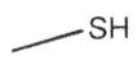 $-SH$	~8
Lysine	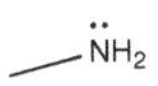 $-\ddot{N}H_2$	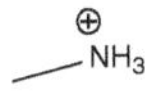 $-NH_3^{\oplus}$	~10
N-Terminal Valine	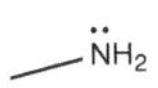 $-\ddot{N}H_2$	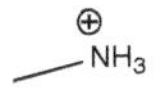 $-NH_3^{\oplus}$	7–8
Methionine	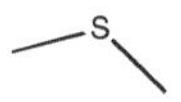 $-S-$	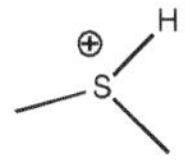 $-S^{\oplus}(H)-$	Low
Histidine	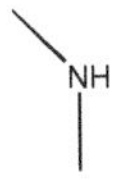 $-NH-$	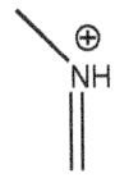 $=NH^{\oplus}-$	6–7
Serine and threonine	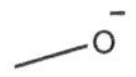 $-O^-$	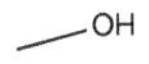 $-OH$	9–10
Aspartic and glutamic acid	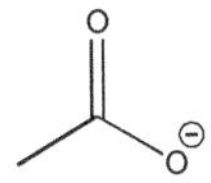 $-C(=O)O^{\ominus}$	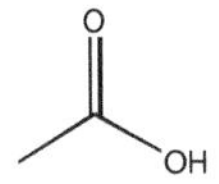 $-C(=O)OH$	~4

oxygen and nitrogen in DNA bases are particularly subjective to adduct formation. The most common DNA adducts observed are those at N7-gluanine; however, alkylation of O^6-guanine and O^2 and O^4-thymine are believed to be responsible for the premutagenic DNA lesions as they may alter the hydrogen bonding during DNA synthesis (van Zeeland et al., 1995).

38.2.1 Determination of Adduct Formation from Genotoxic Compounds with Macromolecules

As mentioned earlier, there is enough evidence to suggest that formation of DNA adducts can be responsible for induction of mutations following DNA replication or misrepair (Hemminki, 1993; Hemminki et al., 1994; Lutz, 1986; Otteneder and Lutz, 1999). As DNA is considered the target molecule for genotoxic effects, approaches have been developed to measure DNA adducts to estimate the exposure of genotoxic compounds in cells. Highly sensitive methods have been reported in the literature for measurements of DNA adduct formation including ^{32}P-postlabeling, immunoassays, fluorescence, and gas chromatography mass spectrometric methods (GS-MS) (Phillips et al., 2000). However, the practice of measuring DNA adduct formation to estimate exposure of genotoxic compounds is often limited by factors such as chemical instability of DNA adducts, the variable rate of DNA repair, and cell apoptosis (Farmer et al., 1987; La and Swenberg, 1996). Therefore, alternative methods has been developed to measure hemoglobin adduct formation as a surrogate for *in vivo* dosimetry of genotoxic compounds (Ehrenberg and Osterman-Golkar, 1980; Neumann, 1984; Osterman-Golkar et al., 1976; Tornqvist et al., 1986). A great number of hemoglobin adducts have been shown to be stable during the life span of erythrocytes (126 days in human) and are not subjected to repair. In addition, hemoglobin can be easily obtained in large quantities from both animals and humans and there are several methods available with high selectivity and sensitivity. Several nucleophilic groups in hemoglobin have the potential for adduct formation including cysteine-S^-, methionine-S, O^- in threonine and serine, histidine ring nitrogens, NH_2 in N-terminal valine, and carboxylic moiety of aspartic and glutamic acids (Ehrenberg and Osterman-Golkar, 1980; Table 38.1). The rate of the reaction between these nucleophilic groups and alkylating agents typically decreases in following order: cysteine S->amino-N, imidazol-N> carboxyl O^- (Ehrenberg and Osterman-Golkar, 1980).

Typically, *in vivo* exposure of genotoxic compounds is established as the area under the concentration curve (AUC) and has been used as the basis for risk assessment (Ehrenberg and Osterman-Golkar, 1980; Osterman-Golkar et al., 1976). There are a few methods for determination of hemoglobin adduct levels including total acid hydrolysis, Ra-Ni method, enzymatic digestion, mild basic hydrolysis (adducts from aromatic amines), as well as the more common modified Edman degradation method. Total acid hydrolysis involves boiling samples in a solution of 6 M hydrochloric acid at high temperatures (100–120°C) under vacuum. Using this procedure, proteins and adducted proteins are hydrolyzed to single amino acids and adducted amino acids, respectively. Separation of adducted amino acids from regular amino acids can be performed using a Dowex (50 W × 4) or Aminex ion exchange resins. This procedure is labor intensive with a limited sample throughput. The Ra-Ni method has been successfully used for analysis of cysteine adducts (Rappaport et al., 1993; Ting et al., 1990) as it cleaves the carbon–sulfur

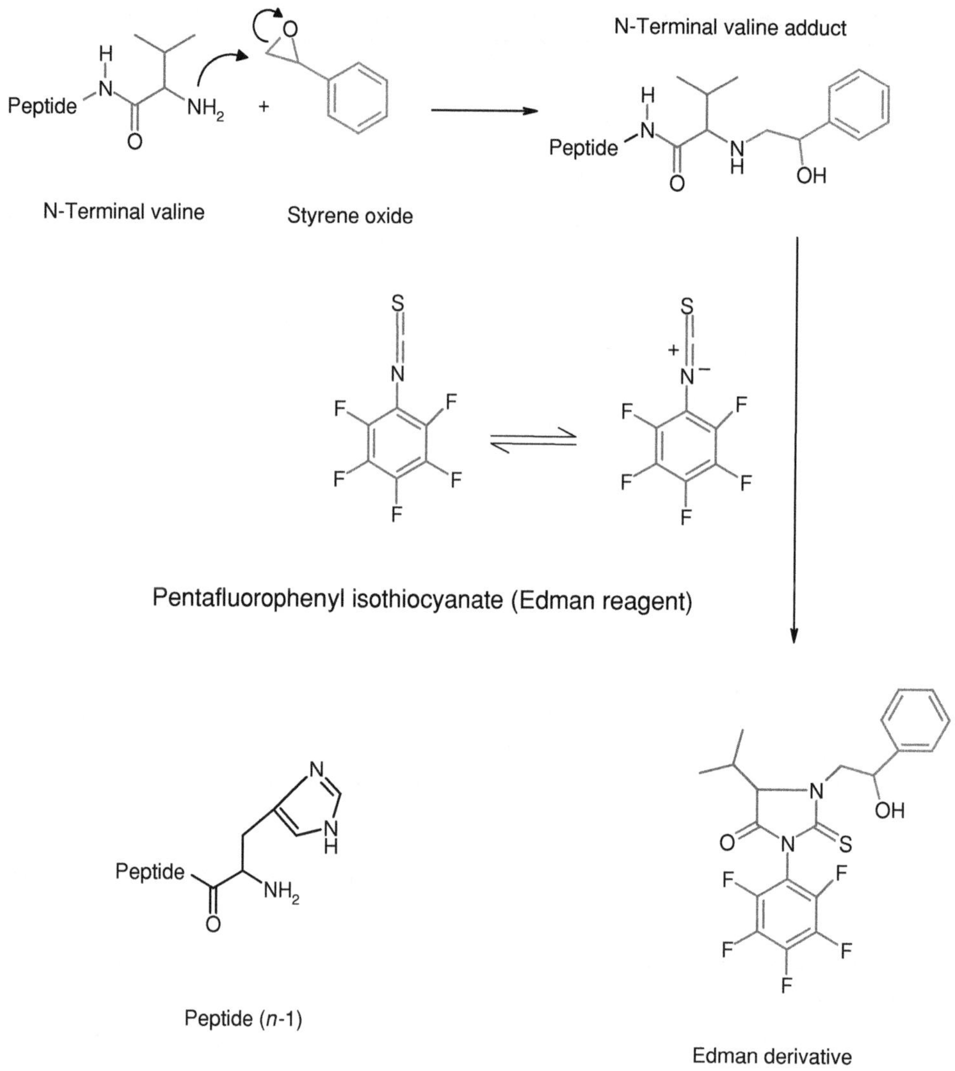

FIGURE 38.2 Formation of hemoglobin adducts of styrene oxide and cleavage using the modified Edman method.

bonds. After the carbon–sulfur cleavage, the adducted part (RH) is extracted using liquid–liquid extraction (LLE) or solid phase extraction (SPE). Ra-Ni is a nickel-aluminum alloy (1:1) saturated with hydrogen. The alloy is commercially available; however, better recovery is obtained when prepared fresh (Ohmori et al., 1981).

The formation of hydroxylamines from carcinogenic aromatic amines and their subsequent reactions with DNA are considered to be critical events in the initiation of bladder tumors (Bryant et al., 1987). The major adduct of aromatic amines with hemoglobin is a sulfinic acid amide, resulting from the reaction of arylnitroso derivatives with SH-groups. The method of analysis typically employs mild basic hydrolysis of hemoglobin to release the parent amine (Bryant et al., 1987; Neumann, 1984). The cleavage of parent amine occurs in presence of 0.1 M sodium hydroxide at ambient temperature for approximately 3 h followed by extraction with LLE or SPE for sample cleanup.

The most widely used method for quantification of hemoglobin adducts is the so-called modified Edman degradation (see Figure 38.2). The method was first introduced by Jensen et al. (1984) and improved by Tornqvist et al. (1986), and due to its sensitivity, selectivity, and improved sample throughput it has been the gold standard for determination of hemoglobin adducts with either GC-MS/MS or LC-MS/MS detection. This method employs a novel reagent "pentafluorophenyl isothiocyanate (PFPITC)" to cleave N-terminal valine adducts in hemoglobin. The reagent reacts with all primary and secondary amines in the proteins; however, only the adducted N-terminal adducts are cleaved allowing the extraction of adducted derivative using LLE or SPE. The reaction is carried out in the presence of formamide at ambient temperature, overnight incubation in slightly basic pH. The mechanism of formation of Edman-adduct derivatives is presented in Figure 38.2. Traditionally, the pioneers of these methods used GC-MS at a detection methodology; however, the use of LC-MS/MS is becoming increasingly more popular (Chevolleau et al., 2007; Fennell et al., 2000, Fennell et al., 2003; Fennell et al., 2005; Vesper et al., 2006). The modified Edman method has been successfully used for quantification of a number of aliphatic hemoglobin adducts; however, it cannot be directly applied to adduct from aldehyde products unless they are reduced prior to Edman derivatization. Aldehydes react reversibly with primary amines, for example, with N-terminal valine in hemoglobin, forming Schiff bases. Schiff bases can be transferred to stable secondary amines through reduction with sodium borohydride (NaBH4) (Kautiainen, 1992).

The modified Edman method does not work well for analysis of valine adducts to the benzene ring moiety. In this case, measurement of cysteine adducts and in particular adducts to cysteine-34 in human serum albumin (HSA) is a better choice. Cysteine-34 in HSA has an unusually low pK_a (~6.7 vs. 8.0–8.5; the average value for a free cysteine side chain in most proteins); therefore, it exists primarily in the highly nucleophilic form (Stewart et al., 2005). These adducts are typically analyzed after enzymatic digestion with pronase. Since the adduct levels are considerably lower than that for HSA, enrichment of cysteinyl adducts using thiol-affinity resins (Funk et al., 2010) or online combination of immunoaffinity chromatography (Hoos et al., 2007) is often performed prior to pronase digestion to remove unadducted HSA.

38.2.2 Determination of Free Reactive Compounds in Biological Matrices

Reactive compounds/metabolites are generally short lived and are typically eliminated by conjugation with nucleophilic groups in proteins or by hydrolysis with water molecules (e.g., epoxides and alkyl chlorides). Therefore, it is a difficult challenge to measure them in the reactive form. Other factors contributing their analytical difficulty of measurement include low molecular weight, poor ionization, and sometimes, low boiling points (volatile reactive compounds). To overcome these potential detection issues, researchers often employ trapping reagents to convert these chemicals to more stable derivatives while also improving chromatography and ionization.

Another aspect to consider during the development of methods to detect reactive metabolites is prevention of analyte degradation from the time of biological sample (blood, plasma, urine, etc.) collection to sample analysis. As these compounds can be quite unstable, sample analysis should ideally be performed immediately after blood collection or as soon as possible.

This approach may be used in preclinical studies where the in-life experiments and the bioanalysis are typically performed in the same site; however, it is impractical for clinical studies. This can be further complicated for multisite clinical studies, as bioanalytical scientists may be required to travel to the clinical sites and some special equipment required in the method may not be available at the clinical site. Also, hosting bioanalytical scientists in the proximity of the clinical sample collection site can be inconvenient and expensive. To overcome these obstacles, training of clinical personal can be provided to perform certain steps of methods such as addition of internal standard or the use of trapping reagents to capture and stabilize the reactive compound. In these instances, it is important to evaluate the stability of the investigated analyte in water and in biological matrices. For compounds that are mainly eliminated via conjugation with proteins and the rate of hydrolysis by water is low (e.g., aldehydes), it is acceptable to provide training to the clinical personnel to remove the proteins by protein precipitation with commonly used organic solvents such as acetonitrile, methanol, acetone, and so on. Short- and long-term (if possible) stability of the reactive analyte in the supernatant should be established at the freezer temperature specified in the method. If adequate stability data of the internal standard in the precipitation solution is established, the internal standard can be prepared at the bioanalytical site and then shipped to the clinical site. Practical considerations must be given to supplies, training, equipment, and shipping. However, for compounds that are also subjected to water hydrolysis (e.g., alkyl chlorides and epoxides), it is recommended to add trapping reagents immediately upon collection for stabilization purpose. The addition of internal standard can be performed later at the clinical site or bioanalytical site as described by Yang et al. (2005).

The selection of the most appropriate trapping reagent for the study is dependent on factors such as high reactivity toward the reactive compound, potential improvement of chromatographic properties and ionization, and removal of excess reagent. Once a suitable trapping reagent has been identified, it is important to optimize the reaction steps of the stabilization such as temperature, time, pH optimization, and the concentration of the trapping reagent required to achieve the highest reaction yield possible and minimize analyte losses. To minimize the reaction time, high concentrations of the trapping reagent are often employed to force the analyte to react with the trapping reagent forming a stable derivative.

An example where these challenges were addressed is described by Yang et al. (2005) for the analysis of 4-fluorobenzyl chloride (4FBC) in human plasma. Since 4FBC is very unstable in plasma; therefore, a method was optimized where 4-dimethylaminopyridine (DMAP) was used as a trapping reagent to prevent losses of 4FBC trough reaction with nucleophilic groups in plasma proteins, hydrolysis, or evaporation. Additionally, the DMAP derivative contains a quaternary amine, which was found to improve ionization resulting in an assay with excellent sensitivity (LLQ of 0.5 ng/ml). Several experiments were conducted during method development to optimize reaction parameters. Incubation of plasma containing 4FBC with an excess of DMAP in alkaline solution (100 mg/ml in methanol with 10 mM sodium hydroxide) at 37°C was found to be optimum. Interestingly, the authors discovered that when the concentration of DMAP exceeded 500 mg/ml, gel formation was observed in plasma treated with DMAP. Also, lower recovery of the DMAP derivative was observed at higher temperatures, where this was attributed to higher rate of hydrolysis of 4FBC. The method was employed to measure exposure of 4FBC present as an impurity in the drug substance. The addition of DMAP solution was performed at the clinical site and since the DMAP derivative was found to be stable at −80°C it allowed sample shipment to bioanalytical laboratory.

Another interesting approach is described by Tornero-Velez et al. (2001) for analysis of styrene oxide in human blood. The method takes advantage of favorable features of the "modified Edman method" earlier described for analysis of adducts bond to N-terminal "valine" in hemoglobin. The amino acid valine at 0.22 M in the presence of 0.22 M triethylamine is used a trapping reagent to form hydroxyphenethylvaline followed by derivatization with PFPITC to form the same Edman-derivative as described by Christakopoulos et al., (1993) for analysis of styrene oxide hemoglobin adducts. The method was used to confirm the validity of a direct method for analysis of styrene oxide in blood without derivatization. This so-called direct method employs LLE with pentane followed by analysis of styrene oxide by GC-MS in the positive chemical ionization mode. Both methods have been employed in the same samples to estimate the exposure of styrene oxide in industrial workers. Although the two methods showed comparable results, trapping valine method showed superior sensitivity and precision. Similar approaches were developed with great success for analysis of acrylamide in drinking water, coffee, and aqueous extracts of snuff (Licea Perez et al., 2003).

Many anticancer agents contain reactive alkyl moieties capable of covalent adduct formation with biological macromolecules (DNA and proteins). These compounds tend to be highly reactive, and therefore, the quantification of these compounds can present an analytical challenge. Nitrogen mustards, in particular cyclophosphamide and its isomer ifosfamide, are commonly used in the therapy of various types of cancer. Both compounds are noncytotoxic prodrugs; however, they are metabolically activated by P450 enzymes to yield a cascade of unstable metabolites. Cyclophosphamide is initially converted to 4-hydroxycyclophosphamide followed by the formation of the tautomer metabolite aldocyclophosphamide that subsequently lead to the formation of two alkylating agents, acrolein and phosphoramide mustard. Ifosfamide is metabolized in a similar way. Analysis of phosphoramide mustard is particularly challenging because it is located mainly inside cells; therefore, 4-hydroxycyclophosphamide or 4-hydroxyifosfamide concentrations are often measured in plasma. Due to its instability, assessing 4-hydroxy compounds in plasma requires the use of trapping reagents. Due to the method simplicity, the most frequently reagents used are methyl hydroxylamine (Baumann et al., 1999; Sadagopan et al., 2001) and semicarbazide (de Jonge et al., 2004). The reaction of the 4-hydroxy metabolite with methylhydroxylamine yields the stable methyloxime and is completed in 5 min at 50°C (Sadagopan et al., 2001), while the reaction with semicarbazide, yielding 4-hydroxycyclo-phosphamidesemicarbazide (de Jonge et al., 2004) requires 2 h incubation at 35°C.

38.3 MEASUREMENT OF SULFHYDRYL-CONTAINING COMPOUNDS

While many functional groups are known to potentially be reactive and have risks of unwanted health outcomes to humans, some reactive compounds (e.g., sulfhydryl-containing compounds) are essential for life and widely used in biochemical processes in metabolic reactions, control of gene expression and receptor signaling, and as antioxidants (Mitchell, 1996). Sulfhydryl containing compounds or thiols, in their reduced form, are great scavengers of free radicals and offer increase protection against oxidative DNA damage. Some examples of sulfhydryl containing compounds are cysteine, homocysteine, and GSH. Disulfide bonds (oxidized form) formed between pairs of cysteines are important structural features of many proteins. Irregular levels of sulfhydryl containing compounds *in vivo* have been associated with several human diseases. Therefore, it is of great importance to monitor the levels of these compounds in biological matrices. Immediate chemical derivatization using trapping reagents has been the method of choice for analysis of thiols in biological matrices. The selection of the reagent is based on a few factors such as selectivity and reactivity of the reagent toward thiols, stability of the derivative, chromatographic separation and detection, solubility of the reagent and derivative in biological matrices. A number of derivatization reagents have been made and successfully employed for analysis of reactive sulfhydryl containing compounds followed by ultraviolet (UV), fluorescence (FL), and LC-MS detection. Due to relatively simple derivatization (e.g., a few minutes in neutral pH at ambient temperature) and favorable absorption property (around 355 nm), 2-chloro-1-methylquinolinium tetrafluoroborate (CMQT) is one of the most popular trapping reagents for determination of thiols in human plasma and urine using UV detection (Toyo'oka, 2009). Due to increased specificity and sensitivity, labeling reagents for which derivatives have FL properties are most popular for detection of thiols in biological samples. Among the most popular FL reagents are halogeno-benzofurazans; ammonium 7-fluoro-2,1,3-benzoxadiazole-4-sulfonate (SBD-F), 4-(aminosulfonyl)-7-fluoro-2,1,3-benzo-xadiazole (ABD-F), 4-(N,N-dimethylaminosulfonyl)-7-fluoro-2,1,3-benzoxadiazole (DBD) (Toyo'oka, 2009; Uchiyama et al., 2001). The reactivity toward thiols decrease in the following order: DBD-F>ABD-F>SBD-F; however, the solubility in aqueous solutions goes in the opposite direction. A solution of DBD-F or ABD-F in organic solvents such as acetonitrile or acetone is often used to precipitate proteins and to derivatize small molecular weight thiols such as cysteine and GSH. More water-soluble reagent such as SBD-F is often used for labeling cysteine and cysteine (after reduction) residues in peptides.

Among nonfluorescence reagents, methyl acrylate is one of the most commonly used for derivatization of thiols in biological matrices. An interesting and practical approach is described by Jemal et al. (2001) for analysis of omapatrilat and its four metabolites. For stabilization purposes, methyl acrylate was added to vacutainer collection EDTA tubes (10 μl to 1 ml blood) prior to sample collection. These fortified vacutainer tubes were kept in the freezer until the day of blood collection without breaking the vacuum. After gently mixing the analytes were allowed to react on ice for 10 min followed by centrifugation for 15 min for plasma isolation. Treated plasma samples were stored at −70°C before extraction.

The most well-known reagent for labeling cysteine residues in peptides and proteins is iodoacetamide. Iodoacetamide is commonly used to bind covalently with the thiol groups of cysteine to prevent formation of disulfide bonds (Smythe, 1936). Also it is used in ubiquitin studies as an inhibitor of deubiquitinase enzymes (DUBs) because it alkylates the cysteine residues at the DUB active site.

38.4 CONCLUSION

Evaluation and monitoring of exposure of reactive compounds with genotoxicity potential is crucial during drug discovery and development to demonstrate safety to humans. In this chapter, general approaches (the use of trapping reagents and measurement of adducts with macromolecules) to assess the exposure of reactive compounds in biological matrices were discussed. While many functional groups have been shown to be reactive, due to their prevelanve specific approaches used to monitor bioactivation of thiol containing compounds in biological matrices was highlighted. Representative protocols for the characterization of protein-adducts through a modified Edman method, as well as trapping of two reactive species in biological matrices are provided in the following text.

38.5 REPRESENTATIVE PROTOCOLS

Protocol 1: The modified Edman method for determination of hemoglobin adducts. Summarized from Fennell et al. (2003) and Chevolleau et al. (2007).

Reagents

- 1 M of sodium hydroxide (NaOH)
- PFPITC
- Formamide
- Ethyl acetate
- Pentane
- Sodium chloride
- 2-Propanol
- Hydrochloric acid

Extraction Procedure

Globin Isolation

Globin is extracted from whole blood according to Mowrer et al. (1986). A summary of this procedure is described as follows:

1. Separate erythrocytes and plasma by centrifugation of whole blood at approximately 1000 × *g* for approximately 10 min.
2. Wash erythrocytes three times with ice-cold sodium chloride (0.9%) and lyze the cells by adding one volume of cold water.
3. Add six volumes of 2-propanol containing 50 ml 6 M hydrochloric acid per litter to the hemolyzate and vortex-mix it vigorously. This treatment is to detach the heme group from hemoglobin and to precipitate cell membranes.
4. Centrifuge the samples at 4–10°C at approximately 30,000 × *g* for approximately 45 min.
5. Transfer the supernatant to a new vial and then add 4 volumes of ethyl acetate to precipitate globin.
6. Wash the precipitate (globin) twice with ethyl acetate and once with pentane.
7. Allow globin samples to dry at ambient temperature and store it at −20°C until the day of analysis

Derivatization of Globin Samples Using the Modified Edman Method

1. Dissolve globin samples in formamide to make a 50 mg/1.5 ml solution.
2. Add 40 μl 1 M sodium hydroxide per 50 mg globin.
3. Add an appropriate amount of internal standard. Note: if the internal standard is an alkylated peptide or alkylate globin, add it at this point. If the internal

standard is an Edman derivative, add it after point number 5.

4. Add 10 μl of PFPITC per 50 mg globin.
5. Allow the reaction to proceed at ambient temperature under gently mixing overnight followed by incubation at 45°C for 1.5 h.
6. Extract the Edman derivative using LLE with diethyl ether (3 × 3 ml).
7. Combine the ether extract and evaporate under a steady stream of nitrogen at 50°C.

Edman Derivatives Are Further Cleaned Up Using SPE

1. Dissolve the extract in 2 ml water or a mixture of methanol water depending on the polarity of the derivative.
2. Condition the SPE cartridge (Oasis HLB, 60 mg from Waters co.) with 2 ml methanol and then 2 ml water.
3. Load the samples into the SPE cartridge.
4. Wash the wells with 5 ml water. A mixture of methanol and water can be used depending on the polarity of the derivative.
5. Elute the derivative with 5 ml methanol.
6. Evaporate the organic solution and reconstitute in the appropriate volume of methanol water solution (1:1; v/v).
7. Analyze the samples using LC-MS/MS.

Protocol 2: Trapping 4-fluorobenzyl chloride in human plasma with chemical derivatization followed by quantitative bioanalysis using high-performance liquid chromatography/tandem mass spectrometry. Summarized from Yang et al. (2005).

Reagents

- 1 M of sodium hydroxide (NaOH).
- Methanol (MeOH) containing 10 mM sodium hydroxide (NaOH).
- 100 mg/ml of 4-dimethylamino pyridine (DMAP) in MeOH (containing 10 mM NaOH). This reagent is prepared as follow: Weigh 50 g of DMAP into a 100-ml volumetric flask and add MeOH to the mark to make 500 mg/ml solution. Dilute five times with MeOH containing 10 mM NaOH to make 100 mg/ml DMAP.
- 10 mM ammonium formate (pH 3).
- 10 mM ammonium acetate (pH 7).
- 15% ammonium hydroxide (NH_4OH) aqueous solution.
- 15% NH_4OH in MeOH.
- 50/50 aqueous (15% NH_4OH)/MeOH (15% NH_4OH).
- MeOH containing 5% formic acid.

Extraction Procedure

The DMAP-containing plasma is prepared by premixing EDTA control human plasma with the derivatization reagent (100 mg/ml of DMAP in MeOH containing 10 mM sodium hydroxide) at a ratio of 90/10 (plasma/derivatization reagent, v/v). The treated plasma is issued to prepare calibration standard and quality control samples. Clinical samples are specially handled at sample collection site by adding 225 μl of plasma to a tube containing 25 μl of the derivatization reagent (100 mg/ml of DMAP in methanol containing 10 mM sodium hydroxide). Samples are vortex mixed for approximately 2 min and four replicates of 50 μl DMAP-containing plasma are aliquoted to appropriate tubes. Samples are then incubated at nominally 37°C for 1 h and then stored at nominally −80°C.

The stock internal standard solution was derivatized by mixing 990 μl of 100 mg/ml DMAP (in MeOH with 10 mM NaOH) with 10 μl of the internal standard stock solution. This solution was incubated at 37°C for 1 h. This solution is used to prepare working internal standard solution (50 ng/ml) in methanol:

1. Aliquot 50 μl of calibration standards or blanks into appropriate tubes.
2. Cap tubes and vortex mix for approximately 2 min.
3. Incubate them at nominally 37°C for 1 h.
4. Remove frozen quality control and clinical samples from −80°C freezer and allow them to thaw at ambient temperature.
5. Add 90 μl 15% ammonium hydroxide aqueous solution into the tubes.
6. Add 25 μl MeOH to double blank.
7. Add 25 μl derivatized internal standard working solution (50 ng/ml) to all other tubes.
8. Vortex mix briefly and centrifuge the tubes for approximately 1 min (approximately 3220 × *g*).
9. Condition the Varian Bond Elute CBA 96-well SPE plate with 500 μl of methanol.
10. Equilibrate the SPE plate with 500 μl of 10 mM ammonium acetate (pH 7).
11. Load 120 μl of samples and slowly apply the vacuum not greater than 5 inch Hg.
12. Wash the wells three times with 1 ml of 15% NH_4OH aqueous solution.
13. Wash the wells three times with 1 ml of 50/50 aqueous (15% NH_4OH) /methanol (15% NH_4OH).
14. Place and secure the SPE plate on a NUNC MicroWell 0.4 ml plate and centrifuge them at approximately 3220 × *g* for approximately 5 min to remove any residual washing solvent (after last 1 ml wash step).

15. Elute the samples with 150 μl of methanol containing 5% formic acid twice.
16. Cap and vortex mix briefly.
17. Inject onto HPLC-MS/MS system for analysis.

Protocol 3: LC/MS/MS Determination of omapatrilat, a sulfhydryl-containing vasopeptidase inhibitor and its sulfhydryl and thioether-containing metabolites in human plasma. Summarized from Jemal et al. (2001).

Reagents

- Methyl acrylate
- Ethyl acetate
- Hydrochloric acid

Extraction Procedure

Derivatization with Methyl Acrylate during Blood Collection

Blood samples from subjects administered omapatrilat were collected into vacutainer tubes containing EDTA and methyl acrylate, described as follows:

1. Add 10 μl methyl acrylate (per one ml blood) into each of the EDTA vacutainer tubes by piercing through the tube septum using a syringe with a noncoring stainless needle
2. Store the fortified vacutainer tubes at −20°C until the day of blood collection. The vacuum in the tubes is expected to hold for at least one year at −20°C.
3. On the day of blood collection, remove the treated tubes out of the freezer and place them on crushed ice.
4. Right before drawing the blood sample, rotate each vacutainer tube gently on its side to wet the tube walls with methyl acrylate.
5. Draw blood sample directly into the sealed vacutainer tube.
6. Gently invert the tube several times to thoroughly mix blood with the EDTA and methyl acrylate.
7. Place the blood samples on crushed ice for approximately 10 min followed by centrifugation for approximately 15 min at 1000 × *g* at 4°C to isolate plasma.
8. Transfer plasma samples to a polypropylene storage tube in a fume hood.
9. Store plasma samples at −70°C until the day of analysis.

Concentrations of omapatrilat and its metabolites in plasma were determined by LLE with ethyl acetate in the presence of 0.1 M hydrochloric acid solution followed by LC-MS/MS analysis.

REFERENCES

Baumann F, Lorenz C, Jaehde U, Preiss R. Determination of cyclophosphamide and its metabolites in human plasma by high-performance liquid chromatography-mass spectrometry. J Chromatogr B 1999;729:297–305.

Bryant MS, Skipper PL, Tannenbaum SR, Maclure M. Hemoglobin adducts of 4-Aminobiphenyl in smokers and nonsmokers. Cancer Res 1987;47:602–608.

Cartwright AC, Matthews BR. International Pharmaceutical Product Registration: Aspects of quality, safety and efficacy. New York: Ellis Horwood Limitted; 1994. p 1–873.

Chevolleau S, Jacques C, Canlet C, Tulliez J, Debrauwer L. Analysis of hemoglobin adducts of acrylamide and glycidamide by liquid chromatography-electrospray ionization tandem mass spectrometry, as exposure biomarkers in French population. J of Chromatography A 2007;1167:125–134.

Christakopoulos A, Bergmark E, Zorcec V, Norppa H, Maki-Paakkanen J, Osterman-Golkar S. Monitoring occupational exposure to styrene from hemoglobin adducts and metabolites in blood. Scand J Work Environ Health 1993;19:255–263.

Clive D, Spector JS. Laboratory procedure for assessing specific locus mutations at the Tk locus in cultured L5178Y mouse lymphoma cells. Mutat Res 1975;31:17–29.

de Jonge ME, van Dam SM, Hillebrand MJX, et al. Simu ltaneous quantification of cyclophosphamide, 4-hydroxy cyclophosphamide, N,N′,N″-triethylenethiophosphoramide (thiotepa) and N,N′,N″-triethylenephosphoramide (tepa) in human plasma by high-performance liquid chromatography coupled with electrospray ionization tandem mass spectrometry (LC-MS/MS). J Mass Spectrom 2004;39:262–271.

Dipple A. DNA adducts of chemical carcinogens. Carcinogenesis 1995;16:437–441.

Doss GA, Baillie TA. Addressing metabolic activation as an integral component of drug design. Drug Metab Rev 2006;38:641–649.

EMEA. Guideline on the limits of Genotoxic Impurities. Committee For Medicinal Products for Human Use, The European Medicines Evaluation Agency, London, 2006. CPMP/SWP/5199/02, EMEA/CHMP/QWP/251344/2006.

Ehrenberg L, Osterman-Golkar S. Alkylation of macromolecules for detecting mutagenic agents. Teratog Mutag Carcinog 1980;1:105–127.

Evans DC, Watt AP, Nicoll-Griffith DA, Baillie TA. Drug-Protein adducts: an industry perspective on minimizing the potential for drug bioactivation in drug discovery and development. Chem Res Toxicol 2004;17:3–16.

Farmer PB, Neumann H-G, Henschler D. Estimation of exposure of man to substances reacting covalently with macromolecules. Arch Toxicol 1987;60:251–260.

Fennell TR, MacNeela JP, Morris RW, Watson M, Thompson CL, Bell DA. Hemoglobin adducts from acrylonitrile and ethylene oxide in cigarette smokers: effects of glutathione s-transferase T1-null and M1-null genotypes. Cancer Epidemiol Biomar Prev 2000;9:705–712.

Fennell TR, Snyder RW, Krol WL, Sumner SCJ. Comparison of the hemoglobin adducts formed by administration of

N-Methylolacrylamide and acrylamide to rats. Toxicol Sci 2003;71:164–175.

Fennell TR, Sumner SCJ, Snyder RW, et al. Metabolism and hemoglobin adduct formation of acrylamide in humans. Toxicol Sci 2005;85:447–459.

Funk WE, Li H, Iavarone AT, Williams ER, Riby J, Rappaport SM. Enrichment of cysteinyl adducts of human serum albumin. Anal Biochem 2010;400:61–68.

Hemminki K, Dipple A, Shuker DEG, Kadlubar FF, Segerbäck D, Bartsch H. DNA adducts: identification and biological significance. IARC Scientific Publications 1994, No. 125, IARC, Lyon.

Hemminki K. DNA adducts, mutations and cancer. Carcinogenesis 1993;14:2007–2012.

Hoos JS, Damsten MC, Vlieger JSB, et al. Automated detection of covalent adducts to human serum albumin by immunoaffinity chromatography, on-line solution phase digestion and liquid chromatography–mass spectrometry. J Chromatogr B 2007;859:147–156.

ICH Q3A(R2). Impurities in New Drug Substances. In International Conference on Harmoni-sation Harmonised Tripartite Guideline. Current Step 4 version dated October 25, 2006. Available at http://www.ich.org/fileadmin/Public_Web_Site/ICH_Products/Guidelines/Quality/Q3A_R2/Step4/Q3A_R2__Guideline.pdf. Accessed Apr 13, 2013.

ICH Q3B(R2). Impurities in New Drug Products. In International Conference on Harmonisation Harmonised Tripartite Guideline. Current Step 4 version dated October 25, 2006. Available at http://www.ich.org/fileadmin/Public_Web_Site/ICH_Products/Guidelines/Quality/Q3B_R2/Step4/Q3B_R2__Guideline.pdf. Accessed Apr 9, 2013.

ICH Q3C(R5). Guideline for Residual Solvents. In International Conference on Harmonisation Harmonised Tripartite Guideline. Current Step 4 version dated October 25, 2006 Available at http://www.ich.org/fileadmin/Public_Web_Site/ICH_Products/Guidelines/Quality/Q3C/Step4/Q3C_R5_Step4.pdf. Accessed Apr 9, 2013.

Jemal M, Khan S, Teitz DS, McCafferty JA, Hawthorne DJ. LC/MS/MS Determination of omapatrilat, a sulfhydryl-containing vasopeptidase inhibitor and its sulfhydryl- and thioether-containing metabolites in human plasma. Anal Chem 2001;73:5450–5456.

Jensen S, Tornqvist M, Ehrenberg L. Hemoglobin as dose monitor of alkylating agents: determination of alkylating products of N-terminal valine. Individual susceptibility to genotoxic agents in human population. In: de Serres F, Pero R, editors. Environmental Science Research. Vol. 30. New York: Plenum Publishing Corp.; 1984. p 315–320.

Kalgutkar AS, Gardner I, Obach RS, et al. A comprehensive listing of bioactivation pathways of organic functional groups. Curr Drug Metab 2005;6:161–225.

Kautiainen A. Determination of hemoglobin adducts from aldehydes formed during lipid peroxidation in vitro. Chem-Biol Interactions 1992;83:55–63.

La DK, Swenberg, JA. DNA adducts: biological markers of exposure and potential applications to risk assessment. Mutat Res 1996;365(1–3):129–146.

Licea Perez H, Osterman-Golkar S. Sensitive gas chromatographic-tandem mass spectrometric method for detection of alkylating agents in water: Application to acrylamide in drinking water, coffee and snuff. The Analyst 2003;128:1033–1036.

Lutz WK. Quantitative evaluation of DNA binding data for risk estimation and for classification of direct and indirect carcinogens. J Cancer Res Clin Oncol 1986;112:85–91.

Ma S, Zhu M. Recent advances in applications of liquid chromatography–tandem mass spectrometry to the analysis of reactive drug metabolites. Chem Biol Interact 2009;179:25–37.

Miller EC and Miller JA. The presence and significance of bound aminoazo dyes in the livers of rats fed p-dimethylaminoazobenzene. Cancer Res 1947;7:469–480.

Mitchell S. Biological interactions of sulfur compounds. UK Taylor & Francis Ltd, 1 Gunpowder Square, London EC4 3DE; 1996. p 1–226.

Mowrer J, Tornqvist M, Jensen S, Ehrenberg L. Modified Edman Degradation applied to hemoglobin for monitoring occupational exposure to alkylating agents. Toxicol Env Chem 1986;11:215–231.

Neumann HG. Analysis of hemoglobin as a dose monitor for alkylating and arylating agents. Arch Toxicol 1984;56:1–6.

Ohmori S, Takahashi K, Ikeda M. A fundamental study of quantitative desulfuration of sulfur containing amino acids by Raney Nickel and its character. Naturforsch 1981;36b:370–374.

Osterman-Golkar S, Ehrenberg L, Segerbäck D, Hällström I. Evaluation of genetic risks of alkylating agents. II. Haemoglobin as a dose monitor. Mutat Res 1976;34(1):1–10.

Otteneder M, Lutz WK. Correlation of DNA adduct levels with tumor incidence: carcinogenic potency of DNA adducts. Mutat Res 1999;424:237–247.

Phillips DH, Farmer PB, Beland FA, et al. Methods of DNA adduct determination and their application to testing compounds for genotoxicity. Environ Mol Mutagen 2000;35(3):222–233.

Rappaport SM, Ting D, Jin Z, Yeowell-O'Connel K, Waidyanatha S, McDonaldt T. Application of Raney nickel to measure adducts of styrene oxide with hemoglobin and albumin chem. Res Toxicol 1993;6:238–244.

Sadagopan N, Cohen L, Roberts B, Collard W, Omer C. Liquid chromatography–tandem mass spectrometric quantitation of cyclophosphamide and its hydroxy metabolite in plasma and tissue for determination of tissue distribution. J Chromatogr B 2001;759:277–284.

Singer B, Grunberger D. Molecular Biology of Mutagens and Carcinogens. New York: Plenum Press; 1983.

Smythe CV. The reactions of Iodoacetate and of Iodoacetamide with various Sulfhydryl groups, with Urease, and with Yeast preparations. J Biol Chem 1936;114 (3):601–612.

Stewart AJ, Blindauer CA, Berezenko S, Sleep D, Tooth D, Sadler PJ. Role of Tyr84 in controlling the reactivity of Cys34 of human albumin. FEBS J 2005;272:353–362.

Tennant RW, Margolin BH, Shelby MD, et al. Prediction of chemical carcinogenicity in rodents from in vitro genetic toxicity assays. Science 1987;236:933–941.

Ting D, Smith MT, Doane-Setzer P, Rappaport SM. Analysis of styrene oxide-globin adducts based upon reaction with Raney nickel. Carcinogenesis 1990;11:755–760.

Tornero-Velez R, Waidyanatha S, Licea Pérez H, Osterman-Golkar S, Echeverria D, Rappaport SM. Determination of styrene and styrene-7,8-oxide in human blood by gas chromatography–mass spectrometry. J Chromatogr B Biomed Sci Appl 2001;757(1):59–68.

Tornqvist M, Mowrer J, Jensen S, Ehrenberg L. Monitoring of environmental cancer initiators through hemoglobin adducts by a modified Edman degradation method. Anal Biochem 1986;154(1):255–266.

Toyo'oka T. Recent advances in separation and detection methods for thiol compounds in biological samples. J Chromatogr B 2009;877(28):3318–3330.

Uchiyama S, Santa T, Okiyama N, Fukushima T, Imai K. Fluorogenic and fluorescent labeling reagents with a benzofurazan skeleton. Biomed Chromatogr 2001;15(5):295–318.

van Zeeland AA, Jansen JG, de Groot A, et al. Mechanisms and biomarkers of genotoxicity: molecular dosimetry of chemical mutagens. Toxicol Lett 1995;77:49–54.

Vesper HW, Ospina M, Tunde T, et al. Automated method for measuring globin adducts of acrylamide and glycidamide at optimized Edman reaction conditions Rapid Communication. Mass Spectrom 2006;20:959–964.

Walgren JL, Mitchell MD, Thompson DC. Role of Metabolism in Drug-Induced Idiosyncratic Hepatotoxicity. Crit Rev Toxicol 2005;35:325–361.

Yang E, Wang S, Bowen C, Kratz J, Cyronak MJ, Dunbar JR. Trapping 4-fluorobenzyl chloride in human plasma with chemical derivatization followed by quantitative bioanalysis using high-performance liquid chromatography/tandem mass spectrometry. Rapid Commun Mass Spectrom 2005;19 (6):759–766.

39

LC-MS BIOANALYSIS OF PHOTOSENSITIVE AND OXIDATIVELY LABILE COMPOUNDS

COREY M. OHNMACHT

39.1 INTRODUCTION

Maintaining the stability of analyte(s) in a matrix, whether of biological origin (i.e., blood, serum, plasma, and urine) or prepared in pure solution (water, methanol, etc.), is imperative for obtaining reliable results during bioanalysis. There are many factors that can adversely affect analyte stability and may lead to an over- or underestimation of the results. Failure to recognize and address these factors during method development and validation activities can be a very costly problem. It is important to remember that the results obtained from bioanalytical studies may be used to make critical decisions in establishing therapeutic dosing levels, efficacy, and other parameters used by healthcare providers for the treatment of patients. It is for these reasons that the primary responsibility of the bioanalytical laboratory is to develop and validate accurate, precise, and robust bioanalytical methods.

There are many worldwide agencies that govern the food and drug industries such as the Food and Drug Administration (FDA), the European Medicines Agency (EMA), the Agência Nacional de Vigilância Sanitária (ANVISA), the Ministry of Health, Labour and Welfare (MHLW), and the State Food and Drug Administration (SFDA) to ensure consumer safety. Many of these agencies have specific guidelines to follow when validating methods for bioanalysis. For example, the FDA, EMA, and ANVISA indicate that stabilities should be evaluated for all steps of sample preparation, analysis, and storage to ensure an accurate value for the measured concentrations (Food and Drug Administration, 2001; Agência Nacional de Vigilância Sanitária, 2003; European Medicines Agency, 2011). While it is clearly stated in the industry guidance that analyte stabilities should be evaluated, the exact manner is not clear on how to conduct the testing and what should be done if stability testing fails. Fortunately, many peer-reviewed research articles, reviews, and white papers are available that provide some assistance in dealing with unstable compounds (Bansal and DeStefano, 2007; Chandran and Singh, 2007; Nowatzke and Woolf, 2007; Savoie et al., 2010; Smith, 2010; Garofolo et al., 2011; Lowes et al., 2011).

A recent review by Li et al. (2011) provides strategies for the analysis of unstable small molecules in biological matrices by LC-MS/MS. Various reasons including enzymatic degradation, hydrolysis of conjugated metabolites, photochemical degradation, and auto-oxidation, to name a few, for analyte instability was discussed along with recommendations for the handling of these compounds. For example, protecting the analyte from light if its light sensitivity is unknown and avoiding extreme conditions (e.g., strong base, strong acid, and high temperatures) during sample extraction and/or chromatography (Li et al., 2011).

Specific challenges arise during bioanalysis when preventive measures are taken and yet the analyte displays characteristics of instability (e.g., short-term stability and autosampler stability failures). Two processes that could account for the lack of stability in the example above are photochemical and oxidative reactions. The products of the two processes can sometimes be the same, making it difficult to ascertain which one or whether both processes are the cause of the instability. In addition, a photochemical reaction can catalyze an oxidation reaction, compounding the problem.

The goal of this chapter is to provide useful tools in the identification and handling of photosensitive and oxidatively labile analytes. This goal will be accomplished by presenting the basic theory behind the mechanisms of analyte instability

Handbook of LC-MS Bioanalysis: Best Practices, Experimental Protocols, and Regulations, First Edition. Edited by Wenkui Li, Jie Zhang, and Francis L.S. Tse.

due to the effects of light and oxygen. A basic understanding of the theory behind these two mechanisms will enhance the ability to predict or identify problematic compounds early in assay development or troubleshooting. Preventive care and handling techniques will also be provided along with a generalized approach for developing methods for photosensitive and oxidatively labile compounds.

39.2 PHOTOSENSITIVE COMPOUNDS

39.2.1 Background

A photosensitive compound is one whose stability can be affected in the presence of light. The term "light" has many meanings depending on the respective study discipline. For example, light may be defined as the wavelengths in the electromagnetic spectrum that allow visual perception. In general, the electromagnetic spectrum may be separated into regions as follows beginning with the more energetic wavelengths: violet, ~380–450 nm; blue, ~450–495 nm; green, ~520–570 nm; yellow, ~570–590 nm; orange, ~590–620 nm; and red, ~620–740 nm. For the practical purposes of this chapter, light is defined as any part of the electromagnetic spectrum that has the potential to induce a photochemical reaction for a given compound.

When a photosensitive compound is irradiated with light, the chemical reactions can be complex and are essentially irreversible. In some instances, light can directly react with the analyte. On the other hand, light may react with sensitizers, and in turn, the excited sensitizer may react with the analyte. The severities of these reactions are often dependent on the analyte, the light source, and the components of the sample matrix. For example, when vitamin D was exposed to UV light in pure ethanol versus serum or plasma, it degraded in ethanol within a few minutes, while stability was maintained in biological matrices (Hollis, 2008). The author proposed that the high degree of protein binding of vitamin D in these matrices provided a protective effect from UV light.

Light can adversely affect the results obtained during bioanalysis of photosensitive compounds. Generally, photodegradation of an analyte will result in a negative bias in the measured value. Additional complications occur when conjugate metabolites can photodegradate back to the parent compound, resulting in an overestimate of the measured parent analyte. Under some circumstances, the amount of degradation may not be noticeable. However, the degradation product may create potential interferences in the assay that may also result in unreliable measurements. Employing methods like LC-MS/MS can greatly reduce the chance of this interference since resolution of the analyte from the by-product can be accomplished by chromatography as well as by mass/charge.

39.2.2 Photochemical Processes

The first law of photochemistry states that light must be absorbed by a molecule in order for a photochemical reaction to occur. During bioanalytical activities, the analyte may be exposed to both natural and artificially produced light. The energy (E) of the photon that is produced by the light is indirectly proportional to the wavelength (λ) of the incident radiation as given by the Planck–Einstein equation

$$E = hc/\lambda \tag{39.1}$$

where h is the Planck constant and c is the speed of light. Based on this equation, the ultraviolet region (i.e., 10–400 nm) of the spectrum is relatively more energetic than the visible region (i.e., 400–700 nm). When the analyte is exposed to enough energy to exceed the activation energy barrier necessary to promote an electron to a higher electron orbital, a series of reactions will follow. Common examples of reactions that can occur due to photochemical events include: hydrolysis, N-dealkylation, oxidation, dehalogenation, and isomerization (Hamann and McAllister, 1983; Le Bot et al., 1988; Wood et al., 1990; Lau et al., 1996; Hollis, 2008). Table 39.1 provides a list of photosensitive compounds in specific matrices along with examples of preventive care of handling of these compounds.

39.2.2.1 Direct Photochemical Reactions A direct photochemical reaction begins with the absorption of light by the

TABLE 39.1 Photosensitive Compounds in Matrix with Examples of Preventive Care and Handling

Compound	Matrix	Reference	Examples of preventive care and handling
Nifedipine	Human plasma	Hamann and McAllister (1983)	"Gold" fluorescent light, subdued light
Bendroflumethiazide	Human urine	Ruiz et al. (2005)	Aluminum-foil-wrapped containers
Doxorubicin	Human plasma, urine, and cell culture	Le Bot et al. (1988)	Dark
Daunorubicin and epirubicin	Nonbiological solutions	Wood et al. (1990)	Protected from light
Sunitinib	Human plasma	de Bruijn et al. (2010)	Sodium light and amber vials
Vitamin D	Ethanol	Hollis (2008)	Protein binding
5-*S*-Cysteinyldopa	Human plasma and urine	Hartleb et al. (1999)	Aluminum foil wrapped racks, dark

analyte, followed by reactions with sample matrix components or degrades to other species. The direct photochemical reaction of an analyte (A) is initiated by the absorption of a photon with sufficient energy ($h\nu$) to elevate an electron to a higher electron orbital resulting in an excited state of the analyte (A^*) as shown in Equation 39.2:

$$A \overset{h\nu}{\longleftrightarrow} [A^*] \longrightarrow P_A \quad (39.2)$$

There are at least two potential fates of the excited molecule: (1) the incident radiation is removed and the analyte returns to its ground state with the loss or transfer of energy through a physical process, or (2) the excited molecule may undergo numerous chemical reactions including fragmentation, hydrogen abstraction, intramolecular rearrangement, dimerization, or electron transfers to yield various products (P_A).

A feature of a direct photochemical reaction is that the incident light is energetic enough to photochemically excite the analyte. For example, the antianginal drug, nifedipine has an absorption wavelength maximum (λ_{max}) at approximately 235 nm. When plasma containing nifedipine is exposed to standard lighting (i.e., unprotected from light), nifedipine will decompose to a nitro- and nitrosopyridine derivative (Abou-Auda et al., 2000). Upon further evaluation, it was shown that UV light is responsible for the production of the nitropyridine metabolite while visible light (Vis) is capable of producing the nitrosopyridine metabolite (Abou-Auda et al., 2000; Offer et al., 2007).

39.2.2.2 Indirect Photochemical Reactions Besides direct photochemical reactions, indirect photochemical reactions are also possible when insufficient energy is available to directly excite the analyte. In this scenario, a matrix component known as a photosensitizer (PS), which is usually a reactive form of oxygen or a pigment, is capable of participating in a direct photochemical reaction, resulting in an intermediate excited state (PS^*) as shown in Equation 39.3:

$$PS \overset{h\nu}{\longleftrightarrow} [PS^*] \longrightarrow P_{PS} \quad (39.3)$$

$$PS^* + A \longrightarrow A^* + PS \quad (39.4)$$

$$A \overset{h\nu}{\longleftrightarrow} [A^*] \longrightarrow P_A \quad (39.2)$$

Next, the energy of the excited photosensitizer is transferred to the analyte (Equation 39.4). The analyte is now capable of undergoing similar reactions to those of direct photochemical reactions (Equation 39.2). The photosensitizer that transfers its energy to the analyte is returned to its ground state, is decomposed (P_{PS}), or is involved with subsequent chemical reactions.

Indirect photochemical reactions differ from direct reactions because they do not require the analyte absorption spectrum to overlap with the incident light. To illustrate this concept, Offer et al. investigated the photostability of levomefolic acid in the presence of rose bengal (Li et al., 2007). Levomefolic acid is quickly degraded when rose bengal, exposed to light >540 nm, is at an excited state. Besides returning to the ground state, the excited rose bengal is capable of catalyzing the metabolism of triplet state oxygen to the more reactive singlet state oxygen that is also able to degrade levomefolic acid.

39.2.3 Preventive Care and Handling of Photosensitive Compounds

There are many approaches that can be used to reduce the impact of light on the stability of photosensitive compounds. Protection of photosensitive compounds against light is most commonly achieved by controlling the light source. Physical protection can also be used that places a light impermeable barrier between the light source and the sample. Opaque storage containers and light-insulated protected rooms can also provide some protection against light.

39.2.3.1 Lighting Options In a sealed dark room, light is presumably removed, thus eliminating the root cause of photochemical instability. However, this option is not practical in most laboratories since visual perception is often required to perform many bioanalytical tasks. Subdued lighting (i.e., dimmed lighting) can be utilized but this situation is not ideal either since photons that carry sufficient energy for photochemical reactions may still be present.

Many established bioanalytical laboratories are equipped with low UV output lighting that emits wavelengths typically greater than 450 nm. For example, low-pressure sodium lamps (e.g., yellow/gold/amber light) are commonly used because they produce virtually monochromatic light at approximately 589 nm (e.g., GE covRguard® Gold; F32T8CVG). Red lamps that have been used in "darkrooms" for photographic film development can also be used to provide light protection as the light produced is approximately 700 nm.

Protection from light may also be accomplished using specialized light filters. These filters are designed to cover less-expensive conventional lighting and prevent the transmittance of certain wavelengths. Yellow shields are available that can be placed over common fluorescent lamps for minimizing the emission of light below approximately 540 nm. Shields are also available to provide for a more natural lighting environment (e.g., GE covRguard 46216; blocks UV lamp emissions from 380 to 180 nm). It is advisable to check that the light in the refrigerator goes off when the door is closed. Ideally, a filtered light source should be installed.

Natural light produced by the sun can be leaked into the laboratory by windows, doors, or unprotected hallways. The amount of sunlight leaked into the laboratory fluctuates depending on the time of day, the weather, and the season, and in turn, may cause inaccurate measurements for photosensitive compounds. Ideally, the laboratory should be void of windows and should be equipped with shielded lighting; however, this is either impractical or very costly. Alternatively a "darkroom" that is void of windows and equipped with protective lighting could be used in cases that require light protection. In situations where windows are unavoidable, window films similar to the ones used in the automotive industry can be purchased and installed (e.g., 3M PR 40; Vis transmission 39%, UV light rejected >99.9%).

39.2.3.2 Sample Storage When light comes into contact with a physical barrier, reflection, refraction, and transmittance of the light may occur. Materials that are transparent and allow the transmittance of light provide minimal protection for photosensitive compounds. The judicious selection of the storage container color and material type can provide a relatively inexpensive means of light protection.

Color-infused glassware or plastics as storage containers are commonly used to prevent light from reaching the sample. Color-infused glass (e.g., amber) is routinely used for storage of photosensitive compounds because it can block a considerable amount of UV light from reaching the contents of the vial. Colored polypropylene is also available and may provide similar light protection to amber glass with the added advantages of ease of use, less prone to breakage, and generally less susceptible to analyte absorption to the container. In the event that colored glassware is not an option, a secondary light containment system is advisable. A simple technique to protect the samples, when in transit, is to keep them stored in a cardboard box. Alternatively, wrapping the solution containers with aluminum foil is an inexpensive means of protection.

It is not unusual that photosensitive compounds are otherwise labile. For example, the temperature may affect stability. In this case, the samples may require handling below room temperature (e.g., on wet ice). Careful consideration should be given to these situations since thawing times on ice are generally increased and so is exposure to light. To minimize the influence of light in this situation, samples may be thawed on ice in a dark place before processing. On the other hand, keeping the samples on wet ice can provide some light protection if the level of the samples in the containers are kept below the surface of the ice.

39.2.4 Bioanalytical Consequences of Photosensitive Compounds

In order to obtain reliable measurements for photosensitive compounds, it is crucial that exposure to light is kept at a minimum to prevent photochemical processes. Providing protection from light in a well-controlled laboratory setting as discussed in Section 39.2.3 is relatively simple. However, it is not always possible that every stage of the bioanalytical process that will handle photosensitive samples will have the same conveniences. The impact of light on analyte stability should therefore be evaluated for all phases of the bioanalysis.

The first point of exposure of the analyte of interest to light will be during sample collection. The concentration of *S*-nitrosothiols (RSNOs), for example, is difficult to be accurately measured due to the numerous stability issues associated with RSNOs when exposed to heat, light, and trace metals (Wu et al., 2008). Wu et al. investigated the light stability of RSNOs in whole blood by drawing porcine blood through a butterfly needle/tubing either with or without protection from light (i.e., aluminum foil covering of the plastic tubing that connects the needle with the syringe) (Wu et al., 2008). When the two sets of samples were analyzed by an amperometric biosensor and data compared, it was reported that samples exposed to light during the blood draw were approximately 23.6% of those that were protected from light. Results from previous studies determined that the photodecomposition of RSNOs is primarily from the absorption of light in the 550–600 nm region (Frost and Meyerhoff, 2004).

Light exposure during the preparation of stocks, working solutions, and matrix quality controls is a potential risk for photosensitive compounds. Photostability studies of bendroflumethiazide (BMFT) in pharmaceutical formulations and urine samples have been examined (Ruiz et al., 2005). Stability studies of BMFT in buffered solutions at pH 3 and pH 7 with or without light protection were compared by HPLC and UV detection at 274 nm. During analysis of these solutions, a peak that was not BMFT, presumably a hydrolysis metabolite, was observed to increase during extended storage times. It was found that stability was better maintained for pure solutions at lower pH (i.e., pH 3) while protected from light. Photostability studies of BMFT were also conducted in urine titrated to pH 3 with or without light protection. The results of those experiments agreed with the testing performed in buffered solutions at pH 3 protected from light.

During bioanalysis, light exposure time is the greatest during the sample processing since it usually includes an extraction procedure (e.g., protein precipitation, liquid–liquid extraction, SPE, and online extraction). Nifedipine is light sensitive, and decomposes following first-order kinetics when exposed to light with an apparent half-life of approximately 2.7 h (Bach, 1983; Hamann and McAllister, 1983). To reduce the time the samples are exposed to laboratory lighting, Yriti et al. developed an online sample extraction procedure with UV detection at 338 nm for the measurement of nifedipine in human plasma (Yriti et al., 2000). The online extraction approach reduced the light exposure time of nifedipine samples to laboratory lighting to approximately 30 s.

The effect that light has on photosensitive compounds during the separation and detection phase by LC-MS/MS is

TABLE 39.2 Examples of Preventive Care and Handling of Oxidatively Labile Compounds in Various Matrices

Compound	Matrix	Examples of preventive care and handling	Reference
Rifampin	Human plasma	Ascorbic acid	Lau et al. (1996)
Hydroxycinnamates and catechins	Human urine	HCl and ascorbic acid	Nielsen and Sandstrom (2003)
Dopamine	Human plasma	Sodium metabisulfite	van de Merbel et al. (2011b)
Cysteine and cystine	Human urine	Sulfosalicylic acid	Birwé and Hesse (1991)
F4-neuroprostanes	Human brain tissue	Butylated hydroxytoluene	Musiek et al. (2004)
3′O-demethyletoposide	Human plasma	Ascorbic acid	Stremetzne et al. (1997)
Catecholamines	Human urine	HCl, EDTA, and ascorbic acid	Zhu and Kok (1997)
Olanzapine	Human plasma	Ascorbic acid	Catlow et al. (1995)
Ascorbic acid	Human plasma	Metaphosphoric acid	Karlsen et al. (2005)

generally negligible since the fluidic tubing of the system blocks light. However, thoughtful development of the LC-MS/MS component is necessary since photochemical reaction products formed during the sample collection, storage, and processing could interfere with the analysis. Since MS/MS detection is capable of providing an additional degree of separation by mass/charge, most unwanted products will not be observed unless desired. In situations where separation by mass/charge is not sufficient to provide reliable results, additional resolution will be required by chromatographic separation. For example, SU5416 is a potent antiangiogenic that is thermodynamically stable as its Z-isomer in solid form (Zhao and Sukbuntherng, 2005). When in solution and exposed to light, photoisomerization occurs forming the unstable E-isomer (i.e., SU5886). Detection of the Z- and E-isomers by MS results in identical MRM transitions (Zhao and Sukbuntherng, 2005; de Bruijn et al., 2010). For this reason, adequate chromatographic resolution of the two isomers are required to achieve accurate quantification.

39.3 OXIDATION OF COMPOUNDS

39.3.1 Background

Oxygen is one of the most abundant elements in the universe and is a vital component to life on the earth. As important as it is to support life, various species of oxygen can cause significant challenges for bioanalysis assay development due to its role in oxidation reactions. Oxidation of an analyte of interest in matrix not only reduces the amount of the analyte to be measured but can also result in the formation of unexpected metabolites and/or degradants. The latter two might interfere with the analysis.

Oxygen in its most stable ground state known as triplet state oxygen ($^3O_2$) is fairly nonreactive with respect to oxidations (Frankel, 1980). It is not until the $^3O_2$ is subjected to certain enzymes, heat, light, metals and so on, that the more reactive oxygen species (ROS) are formed (Stadtman, 1990; Kristensen et al., 1998). These ROS can be classified as oxygen-centered radicals and oxygen-centered nonradicals. Oxygen-centered radicals include superoxide anion (O_2–), hydroxyl radical (OH), alkoxyl radical (RO), and peroxyl radical (ROO). The oxygen-centered nonradicals include hydrogen peroxide (H_2O_2) and singlet oxygen ($^1O_2$).

The mechanism of oxidation can follow various types, including photo-oxidation, auto-oxidation, and nonradical oxidation. As an example, dopamine is an important neurotransmitter in the central nervous system with several hormonal functions (van de Merbel et al., 2011b). Natural and synthetic compounds capable of impeding oxidation reactions are known as antioxidants. A property of most antioxidants is that in the process of providing protection against oxidation, they in turn are oxidized (i.e., reducing agents) (Okezie, 1996; Tsuji et al., 2005). The measurement of dopamine, for example, in biological fluids is challenging partly due to the ease of oxidation via the catechol moiety, particularly at alkaline pH (van de Merbel et al., 2011b). To protect the analyte against the effects of oxidation during bioanalysis, the antioxidant sodium metabisulfite was added to human plasma and the samples handled on melting ice (van de Merbel et al., 2011b). Table 39.2 provides a list of oxidatively labile compounds along with antioxidants or other preservatives used to prevent their oxidation.

39.3.2 Oxidation Process

39.3.2.1 Auto-oxidation Auto-oxidation is the formation of crosslinked structures as a result of a free radical chain reaction in the presence of oxygen (Jensen, 2001; Hou et al., 2005). The free radical chain reaction can be divided into three steps: (1) initiation, (2) propagation, and (3) termination. During the initiation phase, a hydrogen atom is abstracted from the analyte producing a radical (R•) as shown in Equation 39.5:

$$\text{RH} \xrightarrow{\text{E,catalyst}} \text{R}\bullet + \text{H}\bullet \quad \text{(Initiation)} \qquad (39.5)$$

Abstraction of the hydrogen can be accomplished by light, heat, enzymes, and metals. The ease and rate of hydrogen removal is dependent on the number, position, and geometry

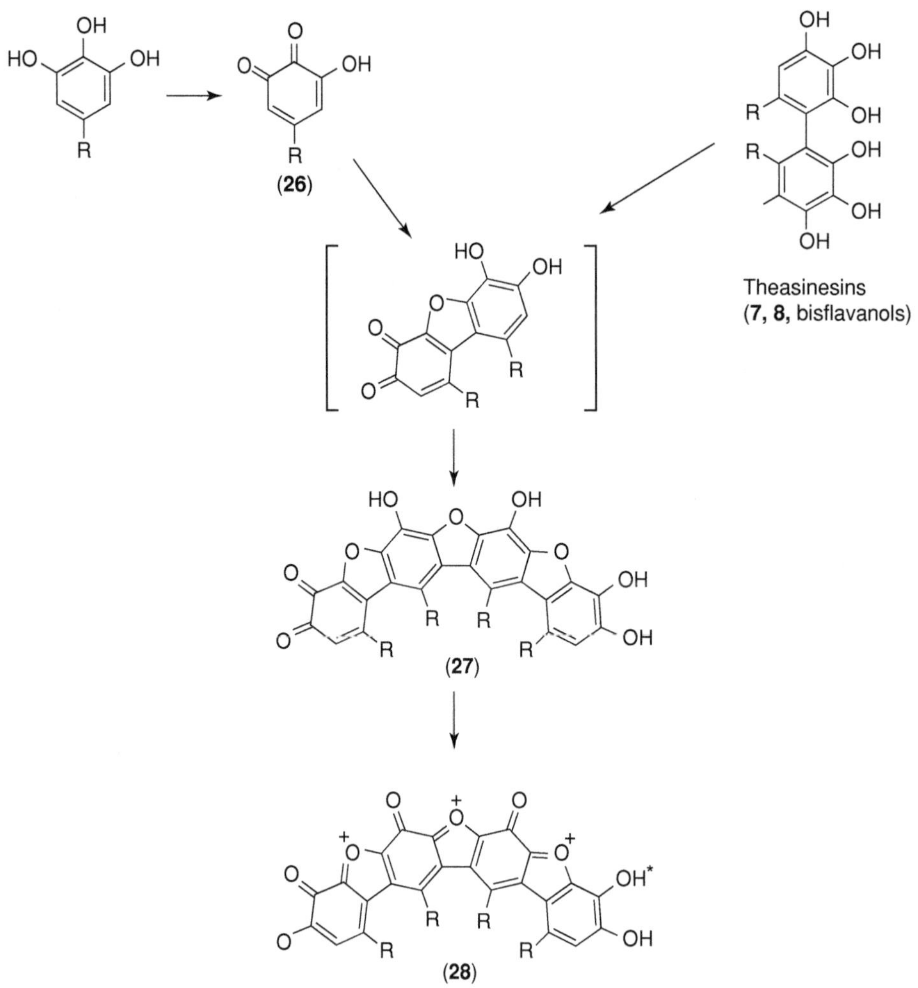

FIGURE 39.1 An example of an auto-oxidation mechanism for the self-condensation of a hydroxyquinone to form a tetramer (reproduced from Haslam, 2003) (*Continued*).

of the bonds. Compounds that have nonconjugated double bonds are typically more reactive than compounds containing conjugated double bonds. This is because the hydrogen of the methylene carbon between double bonds is susceptible to abstraction (Frankel, 1980). Next a chain reaction ensues of free alkyl radicals and peroxyl radicals to propagate the formation of more radicals as shown in Equations 39.6, 39.7, 39.8, and 39.9:

$$R\bullet + O_2 \longrightarrow ROO\bullet \quad \text{(Propagation)} \quad (39.6)$$

$$ROO\bullet + RH \longrightarrow ROOH + R\bullet \quad \text{(Propagation)} \quad (39.7)$$

$$ROOH \longrightarrow RO\bullet + \bullet OH \quad \text{(Propagation)} \quad (39.8)$$

$$RO\bullet + RH \longrightarrow ROH + R\bullet \quad \text{(Propagation)} \quad (39.9)$$

The termination phase of auto-oxidation is represented by the formation of nonradical products as shown in Equations 39.10, 39.11, 39.12, 39.13, and 39.14.

$$ROO\bullet + ROO\bullet \longrightarrow ROOR + O_2 \quad \text{(Termination)} \quad (39.10)$$

$$R\bullet + R\bullet \longrightarrow RR \quad \text{(Termination)} \quad (39.11)$$

$$RO\bullet + R\bullet \longrightarrow ROR \quad \text{(Termination)} \quad (39.12)$$

$$RO\bullet + RO\bullet \longrightarrow ROOR \quad \text{(Termination)} \quad (39.13)$$

$$ROO\bullet + R\bullet \longrightarrow ROOR \quad \text{(Termination)} \quad (39.14)$$

Since the radical electron is delocalized, various crosslinked products are possible depending on the instantaneous location of the radical electron at the time of reaction. Compounds containing unsaturated bonds, polyenes, or phenols are subject to the effects of auto-oxidation (Briscoe and Hage, 2009; Li et al., 2011). Figure 39.1 depicts an example of an auto-oxidation mechanism for the self-condensation of a hydroxyquinone (**26**) to form a tetramer (**27**). Further intramolecular oxidation may occur to yield a highly conjugation cationic structure (**28**) (Haslam, 2003).

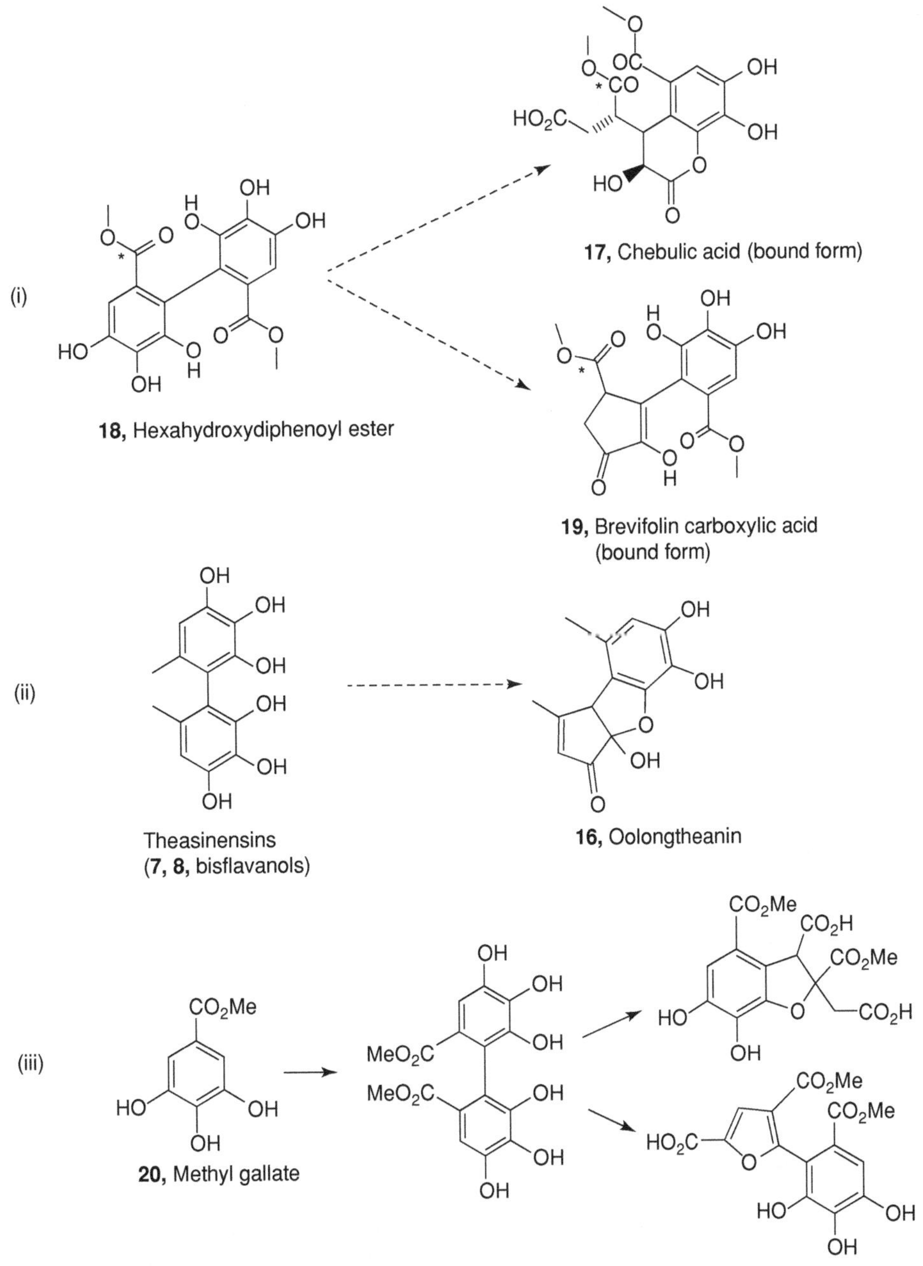

(i)

18, Hexahydroxydiphenoyl ester

17, Chebulic acid (bound form)

19, Brevifolin carboxylic acid (bound form)

(ii)

Theasinensins (7, 8, bisflavanols)

16, Oolongtheanin

(iii)

20, Methyl gallate

FIGURE 39.1 *(Continued).*

39.3.2.2 Photo-oxidation and Reactive Oxygen Species

Photo-oxidation refers to oxidation reactions that occur in the presence of light. In some cases, oxygen can act as the photosensitizer while in other cases either a sample matrix component or the analyte itself can act as the photosensitizer to catalyze the oxidation. At least three different photo-oxidation mechanisms are possible and involve an ROS.

The first photo-oxidation mechanism involves the excitation of $^3O_2$ to its ROS by light as described in Equation 39.15. The ROS then reacts with the analyte in its ground state to form the oxidation product (OP_A) as shown in Equation 39.16:

$$O_2 \xrightarrow{h\nu} \text{ROS} \tag{39.15}$$

$$\text{A} + \text{ROS} \longrightarrow \text{OP}_\text{A} \tag{39.16}$$

In this reaction, oxygen acts as a photosensitizer to catalyze the oxidation of the analyte.

The second mechanism involves excitation of the analyte by light after which the excited analyte can return to its ground state, degrade, and participate in the formation of an ROS as described in Equations 39.2 and 39.17. Next, the ROS reacts with either the ground state or excited state analyte to form the final oxidation product(s) in Equation 39.18:

$$A \overset{hv}{\longleftrightarrow} [A^*] \longrightarrow P_A \quad (39.2)$$

$$A^* + O_2 \longrightarrow A + ROS \quad (39.17)$$

$$A^*/A + ROS \longrightarrow OP_{A/A^*} \quad (39.18)$$

In this reaction, the analyte is the photosensitizer and is also the catalyst of its own oxidation.

The third mechanism involves the excitation of a component of the sample matrix (S) to its excited state by light after which its energy is transferred to oxygen to form the ROS as shown in Equations 39.19 and 39.20. Next, the ROS reacts with either the ground state or the excited state analyte to form the final oxidation product(s) described in Equation 39.18:

$$S \overset{hv}{\longleftrightarrow} [S^*] \longrightarrow P_S \quad (39.19)$$

$$S^* + O_2 \longrightarrow S + ROS \quad (39.20)$$

$$A^*/A + ROS \longrightarrow OP_{A/A^*} \quad (39.21)$$

39.3.3 Preventive Care and Handling of Oxidatively Labile Compounds

An understanding of the type of oxidation process occurring can allow for a more effective approach when exploring methods of prevention. Approaches to eliminate and/or minimize oxidation reactions of the analyte in matrix can be both chemical and physical in nature.

Chemical approaches predominantly involve the addition of reagents known as antioxidants that inhibit the free radical formation of the analyte and reacts with other reactive radicals including ROS. A list of common antioxidants and other types of oxidation-inhibiting compounds are provided in Table 39.3. Antioxidants such as ascorbic acid are typically good reducing agents, whereby in the process of preventing the analyte from being oxidized, the antioxidant is itself oxidized (Tsuji et al., 2005). A property of most antioxidants is that they are good proton donors and their radical intermediates are relatively stable due to resonance delocalization. This property leaves the analyte less prone to attack by oxygen. Strong acids such as hydrochloric acid and phosphoric acid are excellent proton donors and are commonly used to hinder oxidation (Zhu and Kok, 1997; Nielsen and Sandstrom, 2003; Waidyanatha et al., 2004).

Oxidation can be facilitated in the presence of metals such as copper, cobalt, and iron. Metals have the ability to react with molecular oxygen to form the more reactive $^1O_2$ or a peroxyl radical that is capable of chain initiation (Stadtman, 1990). In addition, heme compounds can also be responsible for auto-oxidation (Jensen, 2001). This point is of particular interest since hemolytic plasma integrity testing is becoming increasingly common in regulated bioanalysis. Hemolytic plasma samples may, in such cases, not have the same stability profile as the control matrix used for the preparation of quality controls. Removal of free reactive metals from the samples can be accomplished with compounds capable of chelation with metals such as EDTA, a common anticoagulant used for blood collection (Boomsma et al., 1993).

The selection of preservative or the amount of the selected preservative for inhibition of oxidation reactions is case by case. For example, an antioxidant that is more easily oxidized than the analyte is ideal because it can act as a sacrificial preservative. The amount of preservative (i.e., concentration, v/v%, w/v%, pH) can be established by adding various amounts of the preservative (including native conditions as a control) followed by subjecting the samples to stability-indicating tests. A plot of the measured analyte concentration or analyte LC-MS/MS response versus the amount of preservative added can be a useful tool for visually determining the amount of preservative to be added. A plot prepared in this manner will resemble a saturation curve, which will not further increase after a certain amount of preservative is added. In general, an excess amount of the preservative should be used.

Due to the nature of certain studies, addition of additives during sample collection may not be possible. In these cases, alternative approaches should be evaluated. For example, keeping the samples cold by holding on wet ice can slow the rate of oxidation. In addition, light protection procedures as discussed in Section 39.2.3 can also be employed to reduce the risk of photo-oxidation.

39.3.4 Bioanalytical Consequences of Oxidatively Labile Compounds

In order to obtain reliable measurements for oxidatively labile compounds, it is crucial that conditions in the bioanalysis process are controlled to prevent the oxidation. Protection may be accomplished with antioxidants, acids/bases, and metal chelators that can quench the ROS or prevent abstraction of labile hydrogen during the initiation phase of oxidations. In addition, control of temperature can also provide some protection by slowing the oxidation reaction kinetics.

The first stage of bioanalysis to prevent oxidation is during the sample collection. At this stage, preservatives may be added. Catecholamines, for example, often employ the addition of "cocktails" containing antioxidants, acids, and salts to the collected sample to prevent oxidation (Boomsma et al., 1993; Zhu and Kok, 1997). A method for the quantification of free and total dopamine in human plasma by LC-MS/MS was recently published by van de Merbel et al. to address sample collection stability (van de Merbel et al., 2011b).

TABLE 39.3 List of Antioxidants and Preservatives Used to Inhibit Oxidation

Antioxidant or preservative	Abbreviation	Reference
α-Tocotrienols	α-T	Tang (2003)
Butylated hydroxyanisole	BHA	Tsuji et al. (2005)
Tertiary butylhydroquinone	TBHQ	Tang (2003)
Propyl gallate	PG	Tsuji et al. (2005)
Butylated hydroxytoluene	BHT	Frankel (1980)
Ascorbic acid	AA	Nielsen and Sandstrom (2003)
Hydrochloric acid	HCl	Waidyanatha et al. (2004)
Ethylenediaminetetraacetic acid	EDTA	Karlsen et al. (2005)
Heparin	Hep	Karlsen et al. (2005)
2,5-Di-tert-butyl-hydroquinone	DTBHQ	Tang (2003)
Perchloric acid	PA	Karimi et al. (2006)
Diethylenetriaminepentaacetic acid	DTPA	Suh et al. (2009)
Glutathione	GSH	Boomsma et al. (1993
Metaphosphoric acid	MPA	Karlsen et al. (2005)
Mercaptoethanol	2-ME	Yang et al. (2006)

Sample collection and handling was performed by collecting approximately 3 ml of whole blood into K_3-EDTA tubes and immediately placed on melting ice. Plasma was harvested within 20 min after collection, then mixed with a 5% (v/v) solution containing 10% (w/v) sodium metabisulfite. The samples were then stored at −70°C until analysis. Using this collection approach, the free dopamine concentrations found after whole blood storage deviated by no more than 5% from the corresponding $t = 0$ concentrations. At room temperature, an increased bias ranging from 12% to 25% was realized after 30 min up to 1 h.

Phenolic compounds, such as catechins and hydroxycinnamates, are dietary antioxidants that have been shown to have antiatherogenic and anticarcinogenic properties (Nielsen and Sandstrom, 2003). The analysis of these compounds in urine is often difficult due to the oxidation during the urine sample collection period, which can be up to 24 h. To minimize the effects of oxidation, the urine was collected in 2.5 l containers to which 50 ml of 1 M HCl and 10 ml of 10% aqueous ascorbic acid was added. After collection of the 24 h urine, the pH of the total sample was adjusted to pH 4 with 1 M HCl before storage at −80°C. Following this protocol, hydroxycinnamates and catechins were stable for more than 7 months in the freezer with all analytes quantifying within 85% their nominal concentrations (Nielsen and Sandstrom, 2003).

A proper sample storage temperature should also be considered to ensure that stability of the analyte of interest is maintained in the intended matrix. For example, during prolonged storage of 4β-hydroxycholesterol in a −20°C walk-in freezer, a considerable increase of both the analyte and interfering oxysterols were observed (Figure 39.2) (van de Merbel et al., 2011a). It was believed that the observed increase is most likely due to the auto-oxidation of excess cholesterol. However, the same effect was not observed for the samples stored at −70°C. The lower freezer temperature may have provided the additional protection. Furthermore, the −20°C walk-in freezer had constant illumination, while the −70°C one is void of light and may have prevented the photo-induced oxidation of the cholesterols. Based on these findings, the samples were stored at −70°C in the dark to maintain sample stability for the duration of the study (van de Merbel et al., 2011a).

The prevention of oxidation during the preparation of stocks, working solutions, and matrix quality controls should also be evaluated for all phases of the bioanalysis. Rifampin, used for the treatment of tuberculosis, can be oxidized when exposed to air (Lau et al., 1996; Prueksaritanont et al., 2006). Therefore, its stock and working solutions were prepared in methanol containing 1 mg/ml ascorbic acid. Solutions prepared in this manner were shown to be stable for more than 6 months when held at −20°C and protected from light (Lau et al., 1996; Prueksaritanont et al., 2006).

The quality of the analytical samples can have a negative impact on analyte stability especially for oxidatively labile compounds. Ascorbic acid (i.e., vitamin C) is typically used as an antioxidant for analytical purposes but is also commonly ingested as a dietary supplement (Karlsen et al., 2005). Analysis of ascorbic acid is challenging because of its ability to act as a radical scavenger. It was reported that when small amounts of hemolyzed blood were added to serum, considerable acceleration of oxidation was observed (Mystkowski and Lasocka, 1939). This is not surprising since one of the products of hemoglobin oxidation is the superoxide anion (O_2−), which can take part in further oxidations. For this reason, proper training and inspection of study samples should be conducted so that additional attention can be given to visibly atypical samples.

Sample preparation prior to analysis is another potential area of concern for oxidation sensitive compounds. Extractions that are commonly employed for sample preparation often require adjustment of sample pH (e.g.,

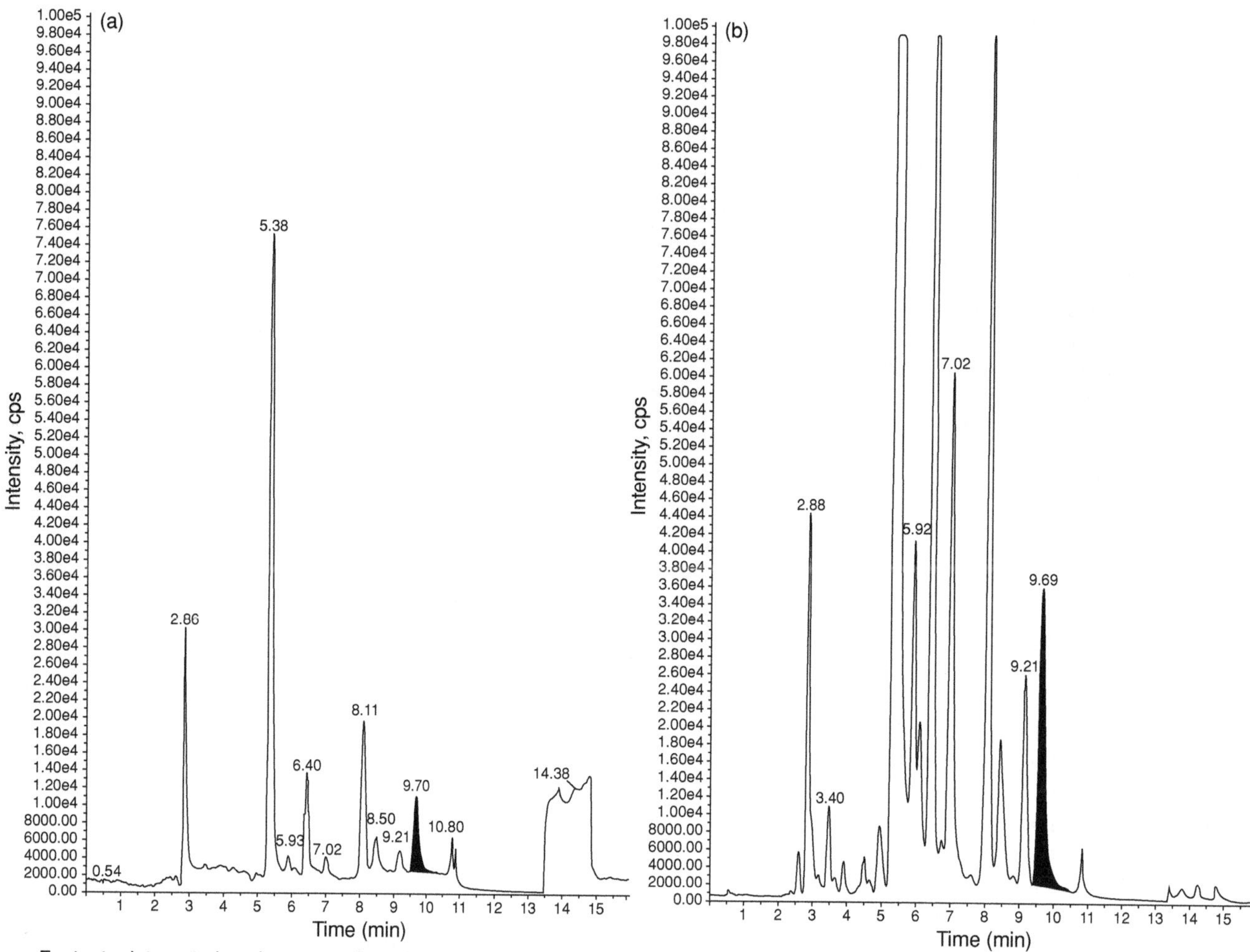

Footnote: Integrated peak at approximately 9.7 min is 4β-hydroxycholesterol. The cumulative peak at approximately 2.86 min represents 24(S), 25, and 27-hydroxycholesterol. Other peaks of interest include 7β-hydroxycholesterol (~5.4 min), 4α and 5,6β-hydroxycholesterol (~8.1 min), and 5,6β-hydroxycholesterol (~8.5 min).

FIGURE 39.2 Increased oxysterols during prolonged storage of 4*β*-hydroxycholesterol in a −20°C walk-in freezer (reproduced from van de Merbel et al., 2011a).

SPE), ionic strength, and amount of organic solvent added (e.g., liquid–liquid extraction) in order to obtain adequate extraction recoveries. It is important to ensure that the sample adjustment, and for that matter, the extraction technique, does not exacerbate analyte oxidation. Catlow et al. developed a method for the measurement of olanzapine in human plasma by SPE and HPLC-ECD (Catlow et al., 1995). From an initial evaluation, the recovery of olanzapine at levels below 10 ng/ml in matrix was variable. Further investigation determined that analyte oxidation was occurring during the sample extraction phase and prior to the analysis via HPLC-ECD. The authors found that the assay result variability was decreased while the extraction efficiency was increased upon the addition of ascorbic acid to the extraction solutions.

The sample extracts while stored in an autosampler need to be stable enough to survive the total run time of the assay sequence. Samples of ascorbic acid in heparin containing human plasma were processed via a protein precipitation method using a 10% solution of metaphosphoric acid (Karlsen et al., 2005). The supernatant was diluted with the aqueous mobile phase (i.e., 2.5 mM NaH_2PO_4, 2.5 mM dodecyltrimethylammonium chloride, and 1.25 mM Na_2EDTA in water) prior to injection for analysis. The extracts stored in an autosampler at 4°C for up to 6 h were stable within 95% of the initial concentration. However, when the autosampler temperature controller was turned off (i.e., ambient temperature), rapid degradation of the extracts was observed after 6 h with measured levels of ascorbic acid between 85% and 90% of initial concentrations. The difference in the temperature setting of the autosampler had a large impact on the rate of oxidation for the processed ascorbic acid samples. These findings emphasize the importance of temperature control when developing methods of analysis. In addition, developing methods with fast sample analysis times and small batch sizes may also allow for acceptable results.

39.4 SUMMARY

A well-planned and executed bioanalytical method development, validation, and sample analysis can reduce the likelihood of encountering sample instability issues. Many factors causing analyte instability (e.g., light, oxygen, time, temperature, pH) can usually be eliminated and/or controlled following the guidance provided in this handbook.

Since most photochemical and oxidative reactions are nonreversible, evaluation of stability or instability of the analyte of interest in the intended matrix is relatively straightforward. In this evaluation, quality control samples are subjected to various stability-indicating conditions (e.g., freeze and thawing, benchtop stability, autosampler stability). The results are compared against freshly prepared calibration standards in the same assay sequence. Results that fall outside of an acceptable range are a direct indicator of analyte instability. A general protocol for developing methods for potentially photosensitive and oxidatively labile compounds is captured in Section 39.5.

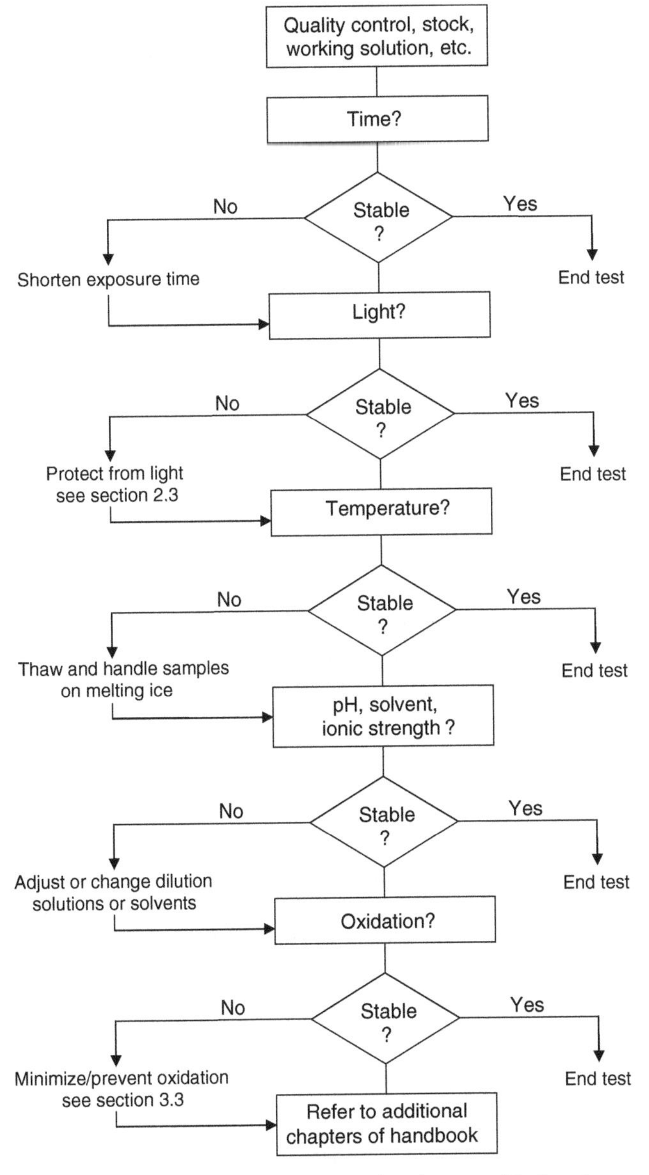

FIGURE 39.3 A stepwise approach when troubleshooting a potential instable sample.

In conjunction with the general protocol in Section 39.5, Figure 39.3 depicts a stepwise approach when troubleshooting a potential instable sample. In most cases, a specific test should be performed to identify the factor(s) that contributes to the instability. In the case of temperature, sample stability can be assessed by comparing the results of quality control samples held at ambient temperature and on melting ice.

The effect of lighting on the stability of the analyte of interest in the intended matrix can be assessed by placing samples under different lighting environments (e.g., unprotected from light, yellow light, red light, dark), followed by necessary comparison. A sample incubated in an environment completely void of light (e.g., cardboard box) can act as a negative control. If there is no statistical difference between a sample held in the dark versus one exposed to light, then the influence of light may be eliminated with a high degree of confidence. On the other hand, if light was discovered to account for the instability, then the preventive care and handling as discussed in Section 39.2.3 should be revisited.

Analyte instability associated with oxidation reactions is more difficult to identify and control than photochemical reactions. In practice, if the oxidation product is known and/or available, then monitoring the presence or absence of the product via LC-MS/MS during stability testing is feasible. By any means, if an oxidation reaction is confirmed to be responsible for the instability of the analyte of interest, a preventive procedure and handling as suggested in Section 39.3.3 should be implemented.

39.5 PROTOCOL FOR DEVELOPING LC-MS/MS METHOD FOR BIOANALYSIS OF PHOTOSENSITIVE AND OXIDATIVELY LABILE COMPOUNDS

General

Perform all experiments under defined lighting and temperature conditions.

Chromatography and Mass Spectrometry

Develop an LC-MS/MS method that is accurate, precise, selective, and rugged for the analyte.

- Begin optimization with neat analyte solutions that are protected from light.

- Ensure that repeated injections of the test solution provide consistent response and chromatographic performance (i.e., retention time, peak shape, column backpressure) before progressing to samples of biological origin.
- Ensure the LC-MS/MS method is selective for the analyte while providing adequate chromatographic and/or mass spectral resolution from potential interferences of the sample matrix.
- If reference materials are available, tune the mass spectrometer to monitor any known photochemical or oxidative products of the analyte to provide for a visual indicator of stability.
- Ensure that the components of the mobile phase (e.g., methanol, acetonitrile, buffers) do not promote oxidation reactions.
- Optimize the mobile phase pH.
- Optimize the column temperature.
- Keep autosampler light off if applicable.

Sample Preparation

Develop a sample extraction procedure with defined parameters to ensure reproducible execution.

- Optimize the light condition to minimize photochemical reactions (e.g., UV-shielded lights, amber glass vials)
- Optimize the sample processing procedure to minimize benchtop exposure time.
- Optimize the temperature for the handling and processing of the samples.
- Optimize the pH of the samples during processing.
- Investigate the impact of any evaporation steps during sample processing.

Stability Assessment

Demonstrate that the analyte is photochemical and oxidatively stable throughout the entire bioanalytical process. Analyze the samples using the optimized sample processing and LC-MS/MS method.

- Assess sample collection and handling procedure.
  a. Use as fresh of matrix as feasible to minimize loss of biological activity of the matrix (e.g., collected and used within 8 h).
  b. Begin assessment unprotected from light in an unaltered matrix (i.e., no preservatives) to stress the samples more than to be expected during the collection of actual study samples.
  c. Optimize the time, temperature, lighting conditions, and storage material.
  d. Optimize the pH, antioxidant, and antioxidant concentration.
- Assess benchtop stability of prepared quality control samples in the matrix.
  a. Begin assessment unprotected from light in unaltered matrix (i.e., no preservatives) to stress the samples more than to be expected during analysis.
  b. Perform assessment in pure solution as well as in biological matrix. Use antioxidant as necessary.
  c. Conduct a nominal assessment against a fresh standard curve as well as a comparative assessment against a sample from the same preparation pool but not subjected to stability cycle.
  d. Optimize the time, temperature, lighting condition, and storage material.
  e. Optimize the pH, antioxidant, and antioxidant concentration.
- Assess the stability during long-term storage conditions.
  a. When using a refrigerator or freezer equipped with a light, ensure that the light is UV-shielded or that the light is turned off when the door is closed.
  b. Optimize the time, temperature, lighting conditions, and storage material.
  c. Optimize the pH, antioxidant, and antioxidant concentration.
- Assess autosampler stability and autosampler reinjection reproducibility.
  a. Optimize time, temperature, lighting conditions, and storage material.
  b. Use colored glass injection vials or a 96-well injection plate that is covered with a light protective film.
  c. Turn off autosampler light if applicable if photooxidation is suspected.
  d. Optimize injection solvent to minimize oxidation reactions.

REFERENCES

Abou-Auda HS, Najjar TA, Al-Khamis KI, Al-Hadiya BM, Ghilzai NM, Al-Fawzan NF. Liquid chromatographic assay of nifedipine in human plasma and its application to pharmacokinetic studies. J Pharm Biomed Anal 2000;22(2):241–249.

Agência Nacional de Vigilância Sanitária, 2003. Guide for validation of analytical and bioanalytical methods.

Bach PR. Determination of nifedipine in serum or plasma by reversed-phase liquid chromatography. Clin Chem 1983;29(7):1344–1348.

Bansal S, DeStefano A. Key elements of bioanalytical method validation for small molecules. AAPS J 2007;9(1):E109–E114.

Birwé H, Hesse A. High-performance liquid chromatographic determination of urinary cysteine and cystine. Clin Chim Acta 1991;199(1):33–42.

Boomsma F, Alberts G, van Eijk L, Man in 't Veld AJ, Schalekamp MA. Optimal collection and storage conditions for catecholamine measurements in human plasma and urine. Clin Chem 1993;39(12):2503–2508.

Briscoe CJ, Hage DS. Factors affecting the stability of drugs and drug metabolites in biological matrices. Bioanalysis 2009;1(1):205–220.

Catlow JT, Barton RD, Clemens M, Gillespie TA, Goodwin M, Swanson SP. Analysis of olanzapine in human plasma utilizing reversed-phase high-performance liquid chromatography with electrochemical detection. J Chromatogr B 1995;668(1):85–90.

Chandran S, Singh RS. Comparison of various international guidelines for analytical method validation. Pharmazie 2007;62(1):4–14.

de Bruijn P, Sleijfer S, Lam MH, Mathijssen RH, Wiemer EA, Loos WJ. Bioanalytical method for the quantification of sunitinib and its n-desethyl metabolite SU12662 in human plasma by ultra performance liquid chromatography/tandem triple-quadrupole mass spectrometry. J Pharm Biomed Anal 2010;51(4):934–941.

European Medicines Agency. Guideline on bioanalytical method validation. July, 2011. Available at www.ema.europa.eu/docs/en_GB/document_library/Scientific_guideline/2011/08/WC500109686.pdf. Accessed Apr 4, 2013.

Frankel EN. Lipid oxidation. Prog Lipid Res 1980;19(1-2):1–22.

Frost MC, Meyerhoff ME. Controlled photoinitiated release of nitric oxide from polymer films containing S-nitroso-N-acetyl-DL-penicillamine derivatized fumed silica filler. J Am Chem Soc 2004;126(5):1348–1349.

Garofolo F, Rocci ML, Jr, Dumont I, et al. White paper on recent issues in bioanalysis and regulatory findings from audits and inspections. Bioanalysis 2011;3(18):2081–2096.

Hamann SR, McAllister RG, Jr. Measurement of nifedipine in plasma by gas–liquid chromatography and electron-capture detection. Clin Chem 1983;29(1):158–160.

Hartleb J, Damm Y, Arndt R, Christophers E, Stockfleth E. Determination of 5-*S*-cysteinyldopa in plasma and urine using a fully automated solid-phase extraction–high-performance liquid chromatographic method for an improvement of specificity and sensitivity of this prognostic marker of malignant melanoma. J Chromatogr B 1999;727(1-2):31–42.

Haslam E. Thoughts on thearubigins. Phytochemistry 2003;64(1):61–73.

Hollis BW. Measuring 25-hydroxyvitamin D in a clinical environment: challenges and needs. Am J Clin Nutr 2008;88(2):507S–510S.

Hou Z, Sang S, You H, et al. Mechanism of action of (−)-epigallocatechin-3-gallate: auto-oxidation-dependent inactivation of epidermal growth factor receptor and direct effects on growth inhibition in human esophageal cancer KYSE 150 cells. Cancer Res 2005;65(17):8049–8056.

Jensen FB. Comparative analysis of autoxidation of haemoglobin. J Exp Biol 2001;204(Pt 11):2029–2033.

Karimi M, Carl JL, Loftin S, Perlmutter JS. Modified high-performance liquid chromatography with electrochemical detection method for plasma measurement of levodopa, 3-*O*-methyldopa, dopamine, carbidopa and 3,4-dihydroxyphenyl acetic acid. J Chromatogr B 2006;836(1–2):120–123.

Karlsen A, Blomhoff R, Gundersen TE. High-throughput analysis of vitamin C in human plasma with the use of HPLC with monolithic column and UV-detection. J Chromatogr B 2005;824(1-2):132–138.

Kristensen S, Nord K, Orsteen AL, Tonnesen HH. Photoreactivity of biologically active compounds, XIV: influence of oxygen on light induced reactions of primaquine. Pharmazie 1998;53(2):98–103.

Lau YY, Hanson GD, Carel BJ. Determination of rifampin in human plasma by high-performance liquid chromatography with ultraviolet detection. J Chromatogr B 1996;676(1):147–152.

Le Bot MA, Riche C, Guedes Y, et al. Study of doxorubicin photodegradation in plasma, urine and cell culture medium by HPLC. Biomed Chromatogr 1988;2(6):242–244.

Li H, Luo W, Zeng Q, Lin Z, Luo H, Zhang Y. Method for the determination of blood methotrexate by high performance liquid chromatography with online post-column electrochemical oxidation and fluorescence detection. J Chromatogr B 2007;845:164–168.

Li W, Zhang J, Tse FL. Strategies in quantitative LC-MS/MS analysis of unstable small molecules in biological matrices. Biomed Chromatogr 2011;25(1–2):258–277.

Lowes S, Jersey J, Shoup R, et al. Recommendations on: internal standard criteria, stability, incurred sample reanalysis and recent 483s by the Global CRO Council for Bioanalysis. Bioanalysis 2011;3(12):1323–1332.

Musiek ES, Cha JK, Yin H, et al. Quantification of F-ring isoprostane-like compounds (F4-neuroprostanes) derived from docosahexaenoic acid in vivo in humans by a stable isotope dilution mass spectrometric assay. J Chromatogr B 2004;799(1):95–102.

Mystkowski EM, Lasocka D. Factors preventing oxidation of ascorbic acid in blood serum. Biochem J 1939;33(9):1460–1464.

Nielsen SE, Sandstrom B. Simultaneous determination of hydroxycinnamates and catechins in human urine samples by column switching liquid chromatography coupled to atmospheric pressure chemical ionization mass spectrometry. J Chromatogr B 2003;787(2):369–379.

Nowatzke W, Woolf E. Best practices during bioanalytical method validation for the characterization of assay reagents and the evaluation of analyte stability in assay standards, quality controls, and study samples. AAPS J 2007;9(2):E117–E122.

Offer T, Ames BN, Bailey SW, Sabens EA, Nozawa M, Ayling JE. 5-Methyltetrahydrofolate inhibits photosensitization reactions and strand breaks in DNA. FASEB J 2007;21(9):2101–2107.

Okezie IA. Characterization of drugs as antioxidant prophylactics. Free Radic Biol Med 1996;20(5):675–705.

Prueksaritanont T, Li C, Tang C, Kuo Y, Strong-Basalyga K, Carr B. Rifampin induces the in vitro oxidative metabolism, but not the in vivo clearance of diclofenac in rhesus monkeys. Drug Metab Dispos 2006;34(11):1806–1810.

Ruiz AM, Gil AM, Esteve RJ, Carda BS. Photodegradation and photostability studies of bendroflumethiazide in pharmaceutical formulations and urine samples by micellar liquid chromatography. LC GC North America 2005;23(2):182–199.

Savoie N, Garofolo F, van Amsterdam P, et al. White paper on recent issues in regulated bioanalysis & global harmonization of bioanalytical guidance. Bioanalysis 2010;2(12):1945–1960.

Smith G. Bioanalytical method validation: notable points in the 2009 draft EMA Guideline and differences with the 2001 FDA Guidance. Bioanalysis 2010;2(5):929–935.

Stadtman ER. Metal ion-catalyzed oxidation of proteins: biochemical mechanism and biological consequences. Free Radic Biol Med 1990;9(4):315–325.

Stremetzne S, Jaehde U, Schunack W. Determination of the cytotoxic catechol metabolite of etoposide (3′O-demethyletoposide) in human plasma by high-performance liquid chromatography. J Chromatogr B 1997;703(1-2):209–215.

Suh JH, Kim R, Yavuz B, et al. Clinical assay of four thiol amino acid redox couples by LC-MS/MS: utility in thalassemia. J Chromatogr B 2009;877(28):3418–3427.

Tang W. The metabolism of diclofenac–enzymology and toxicology perspectives. Curr Drug Metab 2003;4(4):319–329.

Tsuji S, Nakanoi M, Terada H, Tamura Y, Tonogai Y. Determination and confirmation of five phenolic antioxidants in foods by LC/MS and GC/MS. Shokuhin Eiseigaku Zasshi 2005;46(3):63–71.

US FDA. Guidance for Drug Evaluation and Research. Guidance for Industry: Bioanalytical Method Validation. May, 2001. Available at www.fda.gov/downloads/Drugs/GuidanceCompliance RegulatoryInformation/Guidances/UCM070107.pdf. Accessed Apr 4, 2013.

van de Merbel NC, Bronsema KJ, van Hout MW, Nilsson R, Sillen H. A validated liquid chromatography-tandem mass spectrometry method for the quantitative determination of 4beta-hydroxycholesterol in human plasma. J Pharm Biomed Anal 2011a;55(5):1089–1095.

van de Merbel NC, Hendriks G, Imbos R, Tuunainen J, Rouru J, Nikkanen H. Quantitative determination of free and total dopamine in human plasma by LC-MS/MS: the importance of sample preparation. Bioanalysis 2011b;3(17):1949–1961.

Waidyanatha S, Rothman N, Li G, Smith MT, Yin S, Rappaport SM. Rapid determination of six urinary benzene metabolites in occupationally exposed and unexposed subjects. Anal Biochem 2004;327(2):184–199.

Wood MJ, Irwin WJ, Scott DK. Photodegradation of doxorubicin, daunorubicin and epirubicin measured by high-performance liquid chromatography. J Clin Pharm Ther 1990;15(4):291–300.

Wu Y, Zhang F, Wang Y, et al. Photoinstability of S-nitrosothiols during sampling of whole blood: a likely source of error and variability in S-nitrosothiol measurements. Clin Chem 2008;54(5):916–918.

Yang B, Zhu JB, Deng CH, Duan GL. Development of a sensitive and rapid liquid chromatography/tandem mass spectrometry method for the determination of apomorphine in canine plasma. Rapid Commun Mass Spectrom 2006;20(12):1883–1888.

Yriti M, Parra P, Iglesias E, Barbanoj JM. Quantitation of nifedipine in human plasma by on-line solid-phase extraction and high-performance liquid chromatography. J Chromatogr A 2000;870(1-2):115–119.

Zhao Y, Sukbuntherng J. Simultaneous determination of Z-3-[(2,4-dimethylpyrrol-5-yl)methylidenyl]-2-indolinone (SU5416) and its interconvertible geometric isomer (SU5886) in rat plasma by LC/MS/MS. J Pharm Biomed Anal 2005;38(3):479–486.

Zhu R, Kok WT. Determination of catecholamines and related compounds by capillary electrophoresis with postcolumn terbium complexation and sensitized luminescence detection. Anal Chem 1997;69(19):4010–4016.

40

LC-MS BIOANALYSIS OF INTERCONVERTIBLE COMPOUNDS

Nico van de Merbel

40.1 INTRODUCTION

In order to obtain reliable quantitative results in LC-MS bioanalysis, as indeed in any type of analysis, it is essential that the analytes of interest be kept stable during all phases of the analytical process: from sampling and storage to extraction and instrumental analysis. Therefore, an important aspect of the development of new bioanalytical methods is the careful optimization of experimental conditions to guarantee that analyte concentrations do not change from the time of sampling until obtaining the result (Chen and Hsieh, 2005; Briscoe and Hage, 2009; Li et al., 2011). For the same reason, stability assessment is key in the validation of bioanalytical methods and many guidelines, white papers, and other articles have appeared on this subject (e.g., FDA, 2001; Nowatzke and Woolf, 2007; Viswanathan et al., 2007; EMA, 2011). For most applications, the determination of stability is quite straightforward. The analytes are added to a blank matrix sample, which is subjected to a particular storage condition and subsequently analyzed against a fresh calibration curve. Any unacceptable instability is directly apparent from the analytical result and seen as a decrease in the analyte concentration.

While many drugs, metabolites, and endogenous compounds are sufficiently stable during the entire analytical process to be handled without special precautions, a number of classes of compounds need special attention. For those compounds, there will be a need for stabilizing measures to prevent analyte degradation, such as pH adjustment, the use of additives, and/or handling at low temperatures. Several chapters in this handbook describe important classes of (potentially) unstable compounds and discuss the bioanalytical consequences of their instability: prodrugs and other enzyme-labile molecules, reactive compounds, and photosensitive and auto-oxidative compounds.

A special situation arises, when the analyte of interest itself is stable during the bioanalytical process, but one or more of its metabolites are not. Often, these metabolites are not quantified themselves, but in case they are converted back to the analyte, they can have a profound effect on the quantitative reliability of the bioanalytical results. When present in relatively high concentrations in the biological sample and when readily back-converted to the analyte during storage and analysis, they may give rise to a significant overestimation of analyte concentrations (Dell, 2004). Also in this case, stabilizing precautions are essential and stability assessment in the presence of relevant levels of unstable metabolites during method validation is just as important as the assessment of analyte stability itself. Well-known examples of potentially unstable metabolites are described in other chapters of this handbook.

Instability can be caused by enzymatic, chemical, oxidative, or photoinduced reactions and most of these reactions are essentially irreversible under typical laboratory conditions. There are, however, a limited number of compounds, which are converted to other molecules in a reversible way. This complicates the analytical approach even further, since in such a case the conversion of a compound to a product has to be prevented at the same time as the back-conversion of the product to the original molecule. The quantitative bioanalysis of this type of compounds, usually referred to as interconvertible compounds, is the topic of this chapter. Attention

Handbook of LC-MS Bioanalysis: Best Practices, Experimental Protocols, and Regulations, First Edition. Edited by Wenkui Li, Jie Zhang, and Francis L.S. Tse.

will be paid to two major classes: (1) lactones, which can be reversibly converted to their corresponding hydroxy acids, and (2) the stereoisomeric forms of a molecule such as enantiomers and diastereomers, which can be converted to one another. The chemical background of the different interconversion reactions will be discussed, as well as different approaches to stabilize samples for bioanalysis and a suitable way to validate and apply methods for interconvertible compounds. Method details that are not directly related to the interconversion processes are beyond the scope of this chapter and the reader is referred to the original papers for this information.

40.2 INTERCONVERSION OF LACTONES AND HYDROXY ACIDS

40.2.1 Background

Chemically, lactones are cyclic esters, which can be seen as the condensation products of an alcohol and a carboxylic acid within the same molecule. A lactone is characterized by a closed ring consisting of several carbon atoms and a single oxygen atom, with a ketone group at one of the carbons adjacent to the oxygen atom in the ring structure. Lactones are susceptible to hydrolysis to form the corresponding straight-chain hydroxy acids, which is a reversible reaction just as the hydrolysis of "normal" noncyclic esters (Figure 40.1). The equilibrium constant of this reaction is usually pH-dependent. At neutral and alkaline pH, the lactone form is relatively unstable and the equilibrium favors hydrolysis to the open hydroxy acid, while under acidic conditions the equilibrium is shifted toward lactonization of the hydroxy acid.

40.2.2 Statin Drugs

40.2.2.1 General A well-known class of drugs, for which the interconversion between lactone and hydroxy acid plays an important role, is the statin family. Statin drugs all inhibit HMG-CoA reductase—a crucial enzyme in the biosynthesis of cholesterol—and they are used widely to lower cholesterol levels and prevent cardiovascular diseases. After the isolation of the fungal product mevastatin in the 1970s, a compound which was never marketed, several semisynthetic and fully synthetic statins were developed and currently seven different statin drugs are available on the market (Figure 40.2). Most statins are dosed as the pharmacologically active hydroxy acid, with only lovastatin and simvastatin being inactive lactones, which can be regarded as prodrugs that need *in vivo* conversion to their hydroxy acid forms for exerting their pharmacological effect.

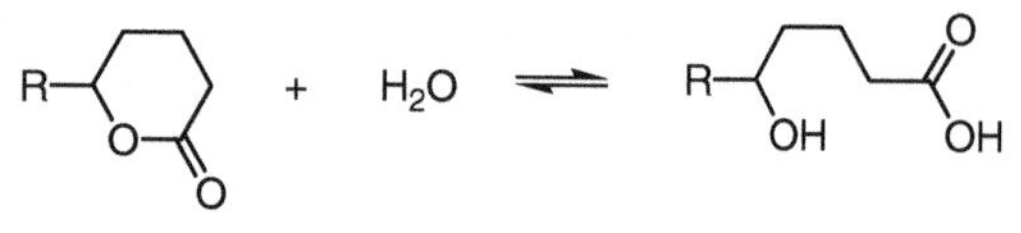

R + H_2O ⇌ R OH OH

FIGURE 40.1 General interconversion between lactones and hydroxy acids, exemplified for a six-membered lactone ring.

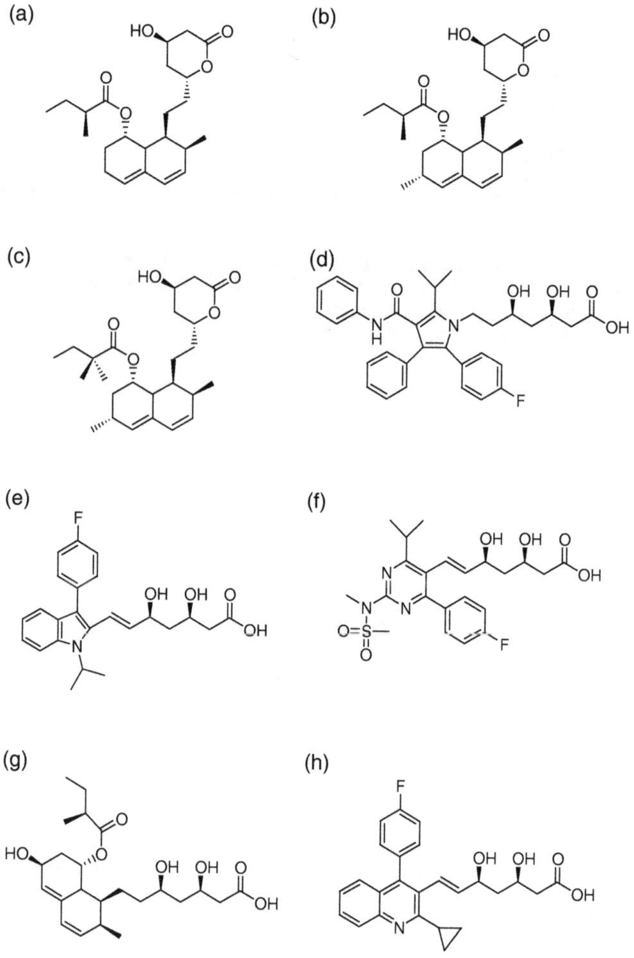

(a) (b) (c) (d) (e) (f) (g) (h)

FIGURE 40.2 Molecular structures of mevastatin (a), lovastatin (b), simvastatin (c), atorvastatin (d), fluvastatin (e), rosuvastatin (f), pravastatin (g), and pitavastatin (h).

40.2.2.2 Interconversion In general, the lactone and hydroxy acid forms of all statins coexist in equilibrium *in vivo*. Chemical conversion of one form to the other occurs in a pH-dependent way. A theoretical investigation of the mechanism of this interconversion under both acidic and alkaline conditions, using fluvastatin as a model compound (Grabarkiewicz et al., 2006), supported the general empiric finding that hydrolysis of the lactone form occurs more readily than lactonization of the hydroxy acid, particularly at high pH. Under basic conditions, the activation barrier for the hydrolysis reaction of the lactone was found to be relatively low and the overall energy gain high. This renders the

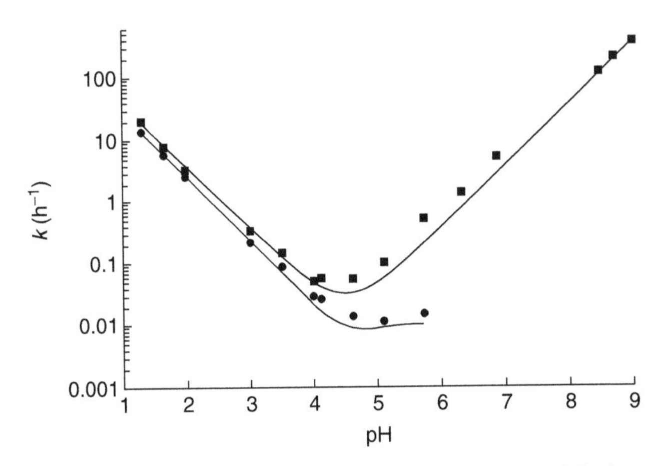

FIGURE 40.3 pH dependence of the rate constant k of the lactonization (•) and hydrolysis (■) of atorvastatin at 80°C (reproduced with permission from Kearney et al. (1993)).

lactone form unstable under basic conditions and favors the reaction toward the hydroxy acid. The reverse reaction from acid to lactone has a very high energy barrier and a negative energy gain, which makes this reaction unfavorable under basic conditions. At acidic pH, the energy barriers of both reactions are of comparable height, making the occurrence of both forms equally probable. In practice, minimum chemical interconversion is generally observed at a pH between 4 and 5. Increasing the pH above 6 greatly enhances hydroxy acid formation, while lowering the pH facilitates lactonization, as illustrated in Figure 40.3 for atorvastatin.

Next to this spontaneous chemical conversion, lactones are also susceptible to enzymatic breakdown to their hydroxy acids, by the action of esterases, of which different forms are present in plasma as well as in the liver and other tissues (Vree et al., 2001). In addition, lactonization has been described to occur in an indirect way, via acyl glucuronide conjugates of the hydroxy acids, which undergo spontaneous cyclization (Prueksaritanont et al., 2002).

40.2.2.3 Bioanalytical Consequences When developing a quantitative bioanalytical method for the lactone and hydroxy acid forms of a statin drug, these conversion and interconversion mechanisms have to be taken into account, as the reactions may typically also occur *ex vivo* during sampling, storage, and analysis. The bioanalytical literature for statins has been reviewed (Nirogi et al., 2007; Nováková et al., 2008) with some attention for the interconversion problems these molecules bring about.

Most of the information about interconversion has been obtained for simvastatin and its hydroxy acid. Plasma is typically buffered at a pH of 4 to 5 as soon as possible after thawing, to avoid interconversion during sample preparation; extracts are adjusted to the same pH to minimize interconversion during storage in the autosampler and even the mobile phase pH is often brought to pH 4.5 to avoid interconversion during the chromatographic run. In this way, sufficient stability is generally obtained during the analytical process with conversion from lactone to acid of $<1\%$ and negligible conversion from acid to lactone (Yang et al., 2005; Barett et al., 2006; Apostolou et al., 2008). If samples are stored for longer periods of time, large deviations may be observed if the sample pH is not controlled, as demonstrated by Jemal et al. (2000). For simvastatin in unbuffered plasma, a 20–40% decrease in concentration due to the conversion was found after storage for 24 h at room temperature. This conversion was reduced to insignificant levels after pH adjustment to 4.9. The use of lower storage temperatures during analysis may be helpful to further minimize interconversion effects and some authors recommend storing samples on ice/water for this purpose (Zhang et al., 2004). For the same reason, samples are typically kept at ≤ -70°C rather than at ≤ -20°C for long-term storage.

The interconversion kinetics is compound-dependent and optimum stabilizing conditions should be defined for each new analyte. The lactone form of atorvastatin, for example, showed a very fast conversion: it was nearly completely hydrolyzed in unbuffered serum after storage for 24 h at room temperature and this was reduced to 12% by lowering the serum pH to 6.0. At 4°C, hydrolysis was slower and found to be 40% after 24 h in unbuffered serum and 3% at pH 6.0. To obtain reliable results, the working temperature was therefore lowered to 4°C and the serum pH adjusted to 6.0 during the bioanalytical process (Jemal et al., 1999). For rosuvastatin, the lactone form showed a conversion to the hydroxy acid of 16% in human plasma at neutral pH and of 6.5% in plasma buffered 1:1 with a 0.1 M sodium acetate buffer (pH 4.0).The hydroxy acid was stable for 24 h at room temperature (Hull et al., 2002). Still, this was considered acceptable, because the lactone levels in post-dose samples are typically much lower than those of the hydroxy acid; which means that a few percent of lactone hydrolysis will not influence the reliability of the hydroxy acid results. Similarly, pravastatin lactone was hydrolyzed for about 30% after 24 h at 0°C in unbuffered serum, while $<5\%$ conversion occurred for serum buffered at pH 4.5 (Mulvana et al., 2000).

As the conversion of lactone to hydroxy acid occurs more readily than the reverse reaction, instability is especially problematic for samples containing high concentrations of the lactone form relative to the acid, because even when only a few percent of the lactone is hydrolyzed, hydroxy acid levels can be significantly overestimated. This situation typically occurs after dosing the lactone form of the drug, that is, for lovastatin and simvastatin. For the other statins, plasma concentrations of the lactones are usually much lower than those of the corresponding acids and a small percentage of hydrolyzed lactones will usually not affect the results for the hydroxy acid forms.

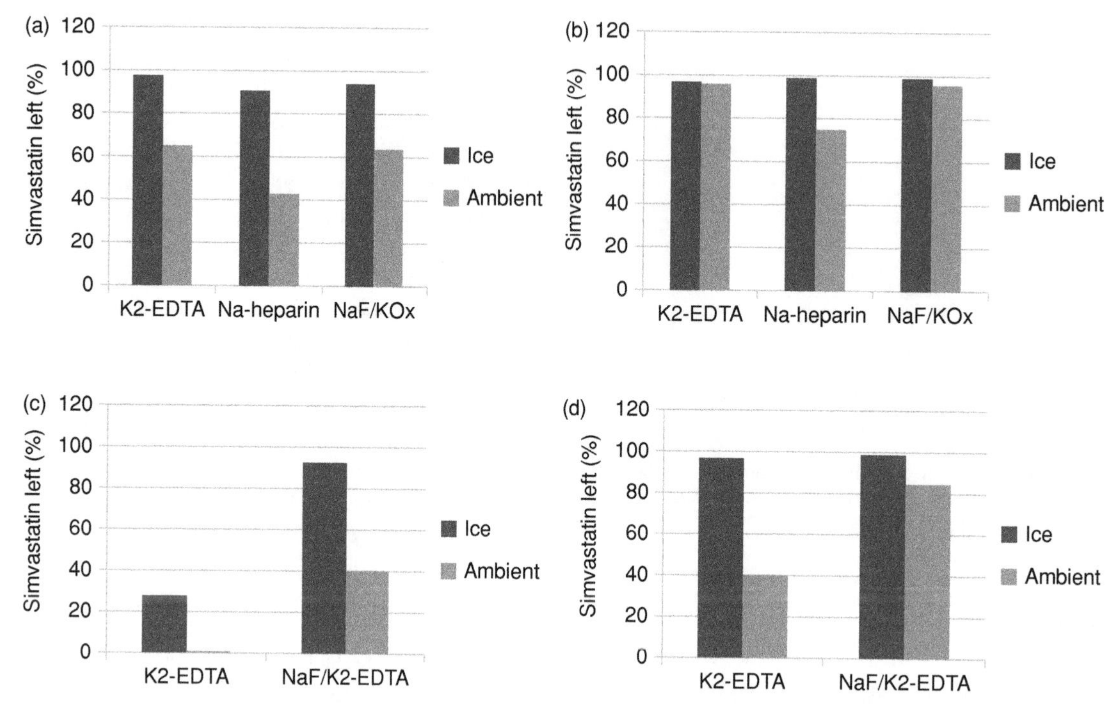

FIGURE 40.4 Percentages of simvastatin remaining after 20 h of incubation on wet ice or at ambient temperature in different types of plasma spiked with 50 ng/ml of simvastatin: human plasma pH 7.4 (a), human plasma pH 4.5 (b), rat plasma pH 7.4 (c), and rat plasma pH 4.5 (d) (PRA, unpublished results).

The choice of the anticoagulant can also have a large effect on the extent of interconversion. As shown in Figure 40.4, the percentage of simvastatin, spiked at 50 ng/ml in different types of plasma, remaining after 20 h of incubation not only depends on pH and temperature as previously discussed, but also on the anticoagulant used and, particularly, on the species. For human plasma, K_2-EDTA and a combination of sodium fluoride (NaF) and potassium oxalate (KOx) as anticoagulants give comparable results. More than 95% of simvastatin remains after storage on ice and about 65% after storage at room temperature. Apparently, the use of the esterase inhibitor, NaF does not reduce the conversion of simvastatin. This is in line with the observation that the enzymatic conversion of statins in human plasma is primarily mediated by the paraoxonase PON3 (Suchocka et al., 2006). As paraoxonases are calcium-dependent enzymes, it is advantageous to use EDTA that will inhibit the enzyme activity by chelating calcium. As a consequence, the extent of simvastatin conversion is smaller in EDTA plasma (65% unchanged at room temperature and pH 7.4 and 96% at room temperature and pH 4.5) than in heparinized plasma (43% and 75%, respectively). In contrast, the use of NaF as anticoagulant in rat plasma has an enormous positive effect on the reduction of simvastatin conversion. At pH 7.4, just 23% of the original simvastatin concentration is left after storage on ice in K_2-EDTA plasma while upon addition of NaF, 93% of the original analyte concentration remains. After storage at ambient temperature, the remaining proportion of simvastatin improves from 1% to 40% upon addition of NaF. Clearly, simvastatin conversion in rat plasma primarily occurs enzymatically by the action of plasma esterases, while in human plasma it is mainly due to a spontaneous chemical reaction. Because of the high esterase activity of rat plasma, the presence of NaF in the collection tubes is essential, next to acidification of the sample. For human plasma, the use of low storage temperatures appears to be sufficient, although acidification also improves stability, particularly at higher temperatures.

A final bioanalytical consequence is related to the possible conversion of hydroxy acid to lactone in the mass spectrometer. For example, lovastatin acid, after elution off the LC column, was partially converted to lovastatin in the ion source, either by lactonization or another form of loss of water, giving rise to a response in the mass transition for lovastatin (Figure 40.5). As this is a general risk for the simultaneous determination of lactones and hydroxy acids, care should always be taken to chromatographically separate the two forms to avoid interference with each other's quantification. Even in the absence of in-source conversion of hydroxy acid to lactone, chromatographic separation may still be necessary because the 18-mass-unit difference between lactone and acid differs by only one mass unit from the 17-mass-unit

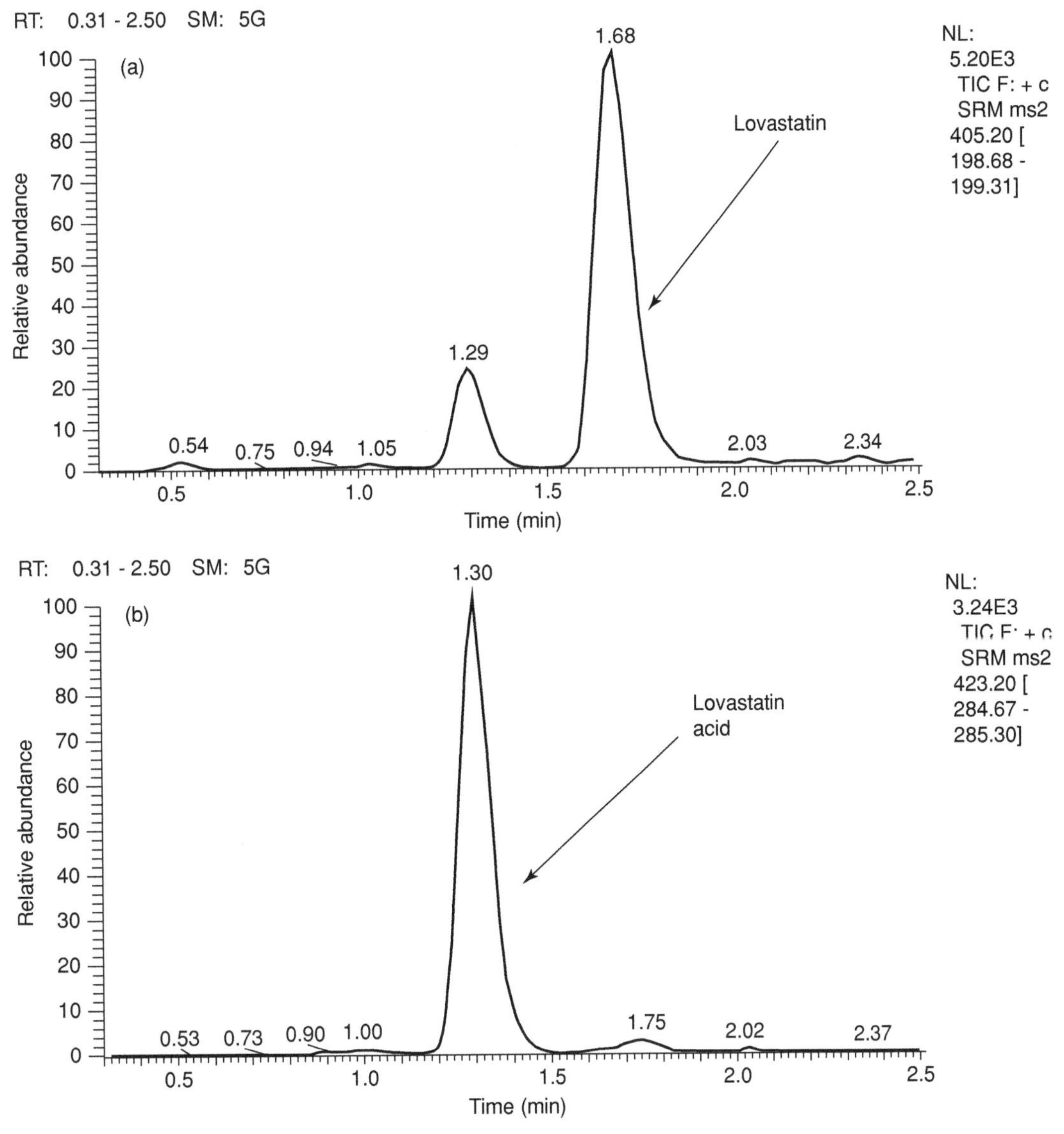

FIGURE 40.5 Selected reaction monitoring LC-MS/MS chromatograms for lovastatin (a) and lovastatin acid (b), showing a response originating from lovastatin acid in the lovastatin mass transition (reproduced with permission from Jemal et al. (2000)).

difference between the $[M + H]^+$ and the $[M + NH_4]^+$ ions of the analytes. A method using the $[M + H]^+$ ion of a hydroxy acid as the precursor ion may, therefore, suffer from interference by the A + 1 isotope response from the $[M + NH_4]^+$ ion of the lactone (Jemal and Ouyang, 2000).

40.2.3 Camptothecin Drugs

40.2.3.1 General Irinotecan and topotecan (Figure 40.6) are semisynthetic analogues of the natural alkaloid camptothecin and the only two members of their class that have received market approval as chemotherapy agents. Their administration ultimately leads to cancer cell death as a result of inhibition of topoisomerase I, an enzyme playing a role in DNA replication. Topotecan is mainly used for the treatment of ovarian and lung cancer, while the main use of irinotecan is for the treatment of colon cancer. Both drugs contain a six-membered lactone ring, which needs to remain intact for exerting its pharmacological activity, as the hydroxy acid resulting from the hydrolysis of the lactone ring is devoid of such activity. Irinotecan is cleaved *in vivo* to the extremely active metabolite SN-38, in which the lactone ring is preserved (Figure 40.6).

40.2.3.2 Interconversion In a detailed kinetic and mechanistic study with camptothecin and a number of its analogues, the interconversion equilibrium of the lactone and hydroxy acid forms was found to be pH-dependent

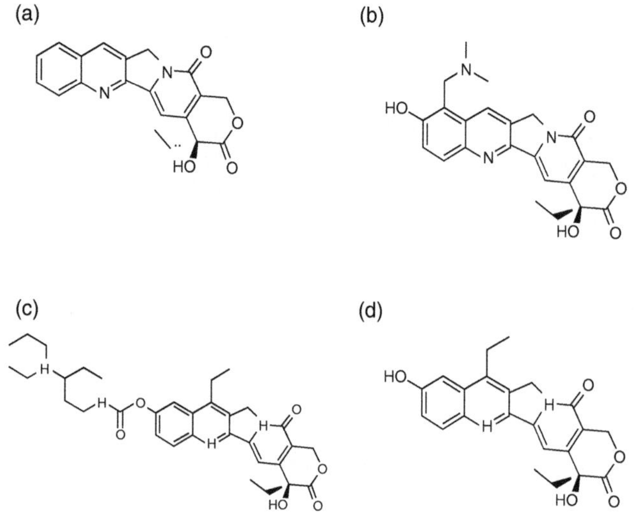

FIGURE 40.6 Molecular structures of camptothecin (a), topotecan (b), irinotecan (c), and SN-38 (d).

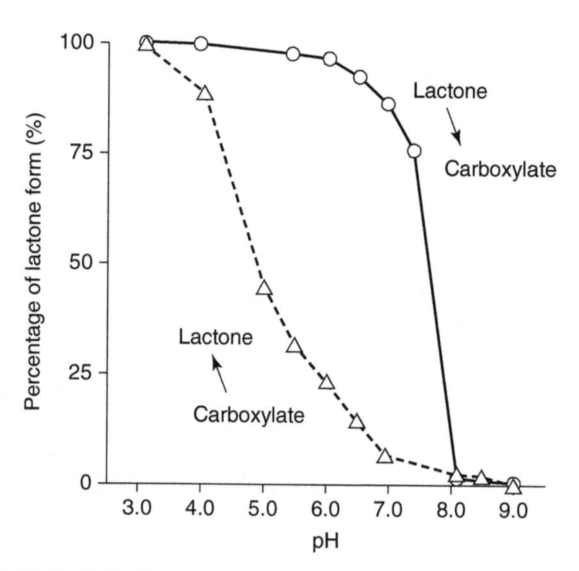

FIGURE 40.7 Percentage of the lactone form as part of the total SN-38 concentration as function of the pH in 10 mM phosphate buffer after 30 min of incubation at room temperature of lactone (○) or hydroxy acid (Δ) (reproduced with permission from Sano et al. (2003)).

(Fassberg and Stella, 1992). The conversion of lactone to acid occurred under neutral and basic conditions, with the compounds being mainly in their hydroxy acid forms at pH 8 and above. The reverse reaction was favored under neutral and acidic conditions and the compounds were found to be largely in their lactone forms at a pH below 5. The optimal pH for stabilization of both forms was between 6 and 7 and no major differences were observed for the different camptothecin analogs. Nevertheless, even at optimal pH, some conversion from lactone to acid and *vice versa* still occurs, as illustrated in Figure 40.7 for SN-38. The percentage of the compounds present in their lactone forms at equilibrium in phosphate buffered saline (PBS) at pH 7.4 is typically 15–20% (Loos et al., 2000). The presence of human serum albumin (HSA) significantly affects the interconversion equilibrium of camptothecin. Due to the high-affinity binding of the hydroxy acid form to HSA, the equilibrium shifts toward the hydroxy acid. As a result, the percentage of the lactone form of camptothecin decreased from 13% to less than 1% upon the addition of HSA at a physiological concentration to PBS (Burke and Mi, 1993). In contrast, no change in the proportion of the lactone form of topotecan was observed upon addition of HSA. Furthermore, the percentage of the lactone forms of irinotecan and SN-38 even increased to about 25% and 35%, respectively, upon addition of HSA. This is attributed to the presence of the bulky substituents in these molecules, which hinder the binding of the hydroxy acid forms to HSA and thus stabilize the lactone forms (Burke et al., 1995).

40.2.3.3 Bioanalytical Consequences The interconversion of the lactone and hydroxy acid forms of camptothecin drugs has been discussed to some extent in a number of review papers (Loos et al., 2000; Zufia et al., 2001; Mullangi et al., 2010; Ramesh et al., 2010). Most investigations found that the conversion of the lactone to hydroxy acid form of camptothecin analogs was relatively rapid, even at optimal pH. Therefore, a general recommendation is to work with these compounds at low temperatures and restrict the storage time of samples to a minimum, next to adjusting the pH to its optimal value. For example, about 15% of the lactone form of irinotecan was converted to the hydroxy acid at pH 6.4 and 10°C after just 1 h, while more than 40% of the compound was converted at pH 7.4 and 10°C (Chollet et al., 1998). At 4°C, only ~5% of irinotecan lactone was reported to be hydrolyzed at pH 6.4 in 2–3 h (Rivory and Robert, 1994). For this reason, much effort has to be put into the proper stabilization of the lactone and hydroxy acid of camptothecin drugs during analysis and this may be very challenging from a practical point of view. A way around this is to measure the total drug (lactone plus hydroxy acid) concentrations. These significantly correlate with the lactone concentrations and consequently have essentially the same clinical significance as the concentrations of the separate forms. Therefore, some bioanalytical methods employ an acidic treatment of the sample prior to extraction. In this way, any hydroxy acid in the sample is converted to the corresponding lactone for the quantification of total drug without the need for special precautions (e.g., Ragot et al., 1999). Alternatively, lactone can be converted to acid by incubation of the study samples at alkaline pH, followed by quantitation of the hydroxy acid concentration as a measure of total drug (Schoemaker et al., 2003).

Whenever it is necessary to determine the concentrations of lactone and hydroxy acid separately, a number of measures need to be taken to stabilize the ratio of lactone to acid in the study samples (Rivory and Robert, 1994; Chollet et al., 1998; Owens et al., 2003; Yang et al., 2005). First of all, plasma should be prepared as soon as possible after blood collection. Typically, blood samples should be chilled on wet ice or briefly immersed into an acetone/dry ice bath, followed by immediate centrifugation at low temperature (10°C or below). By increasing the centrifugal force to 8000 g, the duration of the centrifugation step can be limited to just 2 min. An aliquot of plasma is subsequently subjected to protein precipitation by the addition of ice-cold methanol and, after a second centrifugation step, the supernatant is frozen at ≤ -70°C. In this manner, the concentrations of lactone and hydroxy acid can be claimed to be preserved, although no detailed stability results have been reported.

After thawing, the sample extract is either transferred directly to an autosampler for injection or mixed with a solution to adjust the pH to between 6 and 7 to increase the storage stability. The autosampler temperature is kept as low as possible to further minimize interconversion. Despite all precautions, due to the limited stability of the lactones and hydroxy acids, it is often necessary to process only a small number of samples at a time to avoid prolonged storage of sample extracts prior to injection onto the chromatographic system. For example, in a report for the quantification of camptothecin and SN-38, Boyd et al. (2001) recommended to store extracted samples for no longer than 4 h in the autosampler at 4°C in order to preserve the ratio of lactone to hydroxy acid. Because of the chromatographic run time of 23 min, this enabled the analysis of batches of no more than 10 samples at a time. In another case reported by Kaneda et al. (1997), each sample extract was injected directly after preparation because of the extreme instability observed for the lactone and acid forms of SN-38 in rat plasma.

40.2.4 Other Lactone Drugs

Although a number of other lactone-containing drugs exist, not much study has been conducted and/or reported on the interconversion between their lactone and hydroxy acid forms. One example is for eplerenone, a new antihypertension drug. Similar to the lactone drugs discussed earlier, eplerenone is hydrolyzed to the pharmacologically inactive hydroxy acid under basic conditions while the hydroxy acid is converted back to the lactone at acidic pH. However, no interconversion was observed at a pH between 6.5 and 8. When maintaining a neutral pH (7.4), both forms were sufficiently stable during all stages of the bioanalytical process. In whole blood, both eplerenone lactone and its hydroxy acid were stable for 1 h on wet ice and at room temperature. In human plasma they were stable after three freeze/thaw cycles and in plasma extract for 24 h at room temperature (Zhang et al., 2003a). Similar results were obtained for human urine, in which both forms were stable for 37 days at ≤ -70°C, after three freeze/thaw cycles and for 24 h at room temperature (Zhang et al., 2003b).

40.3 STEREOISOMERIC INTERCONVERSION

40.3.1 Background

Stereoisomers are molecules that are identical in atomic composition and bonding, but different in the three-dimensional orientation of their atoms. There are different types of stereoisomers, including (1) geometric (or E- and Z-) isomers, which have a different orientation of the substituents on both sides of a double bond or a bond with a greatly restricted rotation, (2) enantiomers, which are pairs of stereoisomers with one or more asymmetric (chiral) centers and are mirror images of each other, and (3) diastereomers, which have more than one chiral center and are not mirror images of one another (with diastereomers that differ in the configuration of only one stereogenic center being called epimers).

In general, stereoisomers differ in their physicochemical properties, except for enantiomers that are identical to each other in all physicochemical aspects but the ability to rotate plane-polarized light. From a pharmacology and/or toxicology perspective, different stereoisomers of a drug candidate may have quite different properties when administered to a living organism. For this reason, drug candidates with stereoisomeric forms need to be properly evaluated to see whether a difference exists in the pharmacological and/or toxicological effects between the isomers. If so, development of the single, active stereoisomer is recommended.

A complicating factor is that a pure stereoisomeric form of a molecule may be *in vivo* converted to a mixture of the stereoisomers. This process, called racemization for enantiomers and epimerization for epimers, may lead to decreased efficacy or even toxic side effects of a drug (Ali et al., 2007). Therefore, the study of racemization or epimerization is very important and the availability of bioanalytical techniques for accurate quantification of these compounds in various biological matrices is essential.

40.3.2 Interconversion

The interconversion of stereoisomers can be the result of a variety of reactions. Probably most common is the spontaneous, base-catalyzed inversion of the chiral centers. This reaction involves deprotonation of a chiral carbon atom of the form R_3,R_2,R_1-CH to the carbanion R_3,R_2,R_1-C^- as an intermediate, which is subsequently randomly reprotonated to both stereoisomeric forms. The reaction depends on pH and temperature, but also very much on the functional groups present on the chiral carbon atom of the drug molecule. Some substituents on the chiral carbon may have a stabilizing effect

TABLE 40.1 List of Functional Groups Affecting the Configurational Stability of Chirally Substituted Carbon Atoms of the Type R_3,R_2,R_1-CH

Groups decreasing configurational stability	Neutral groups	Groups increasing configurational stability
–CO–O–R (strong)	$-CH_3$	$-COO^-$
–CO–aryl (strong)	$-CH_2CH_3$	$-SO_3$ (?)
$-CO-NR_1,R_2$		
–OH		
Halogens (?)		
Pseudohalogens (?)		
$-NR_1,R_2$		
–N=R		
–aryl		
$-CH_2$–aryl		
$-CH_2OH$ (?)		

Source: Adapted with permission from Testa et al. (1993).

while others have a destabilizing impact on the carbanion intermediate (Testa et al., 1993). Based on available experimental evidence, a list of such groups is summarized in Table 40.1. It shows that configurational instability appears to require the presence of either three carbanion-stabilizing groups or two carbanion-stabilizing groups (one of which must be strong) and one neutral group.

One example of a drug displaying extreme chiral instability is the anxiolytic agent oxazepam. The enantiomers of oxazepam rapidly racemize, even at ambient temperature and neutral pH, with an estimated half-life of racemization of just 10 min (Aso et al., 1988). Another example is thalidomide, of which the enantiomers are also readily interconverted under physiological conditions (Eriksson et al., 1995). As shown in Figure 40.8, the half-life of thalidomide racemization was ~2.3 h in human blood at 37°C. Because thalidomide is hydrolyzed *in vivo* to a number of products, a decrease in the total concentration over time (half-life ~4 h) of the compound is also apparent next to the racemization. Similar to all chemical reactions, chiral interconversion slows down at a lower temperature. For example, both enantiomers of the investigational drug MK-0767 showed a conversion of ~10% to the other form after 2 h at room temperature in human plasma. This was reduced to <3% when samples were processed on wet ice (Yan et al., 2004). Acidification of a sample may help reduce the speed of a base-catalyzed racemization. This was shown for the chiral antidiabetic drugs, pioglitazone and rosiglitazone, for which the racemization was relatively rapid at a pH 9.3 (half-life 2 h) and pH 7.4 (half-life 4 h), respectively. The racemization was considerably reduced at pH 2.5 (half-life >300 h) (Jamali et al., 2008).

Another, but less common mechanism of racemization involves the acid-catalyzed loss of a hydroxy group on a chiral carbon atom (R_3,R_2,R_1-COH) to form a carbocation

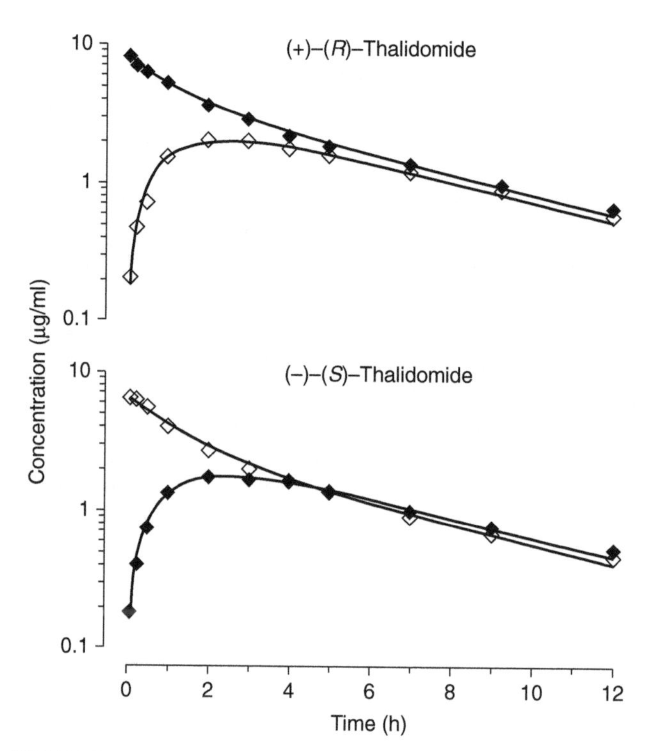

FIGURE 40.8 *In vitro* concentration curves of (*R*)-thalidomide (♦) and (*S*)-thalidomide (◊) during incubation of the indicated enantiomer in human blood at 37°C (reproduced with permission from Eriksson et al. (1995)).

intermediate R_3,R_2,R_1-C^+. Chlorthalidone is an example of a drug that racemizes extremely rapidly in this way, with a half-life of just a few minutes (Severin, 1992). Similar to the above base- or acid-catalyzed racemization, other isomerization reactions have also been described to be pH-dependent. In the case of the conversion of the E- to the Z- (geometric) isomers of a drug candidate containing an O-methylated oxime, the rate of conversion increased with decreasing pH. The isomerization at low pH was attributed to the protonation of the nitrogen atom of the oxime moiety, leading to the formation of the Z-isomer in the subsequent deprotonation phase. At pH values above 6, the conversion was negligible (Xia et al., 1999).

In addition, the exposure to light has been reported to induce the isomerization of the E- and Z-forms of drugs. The cytostatic compound SU5416 exists solely in its Z-form in the solid state, but is spontaneously converted to the E-isomer in alkaline solution when exposed to light. The E-isomer itself is unstable in solution and readily converts back to the Z-isomer, also when protected from light. In rat plasma, both isomers were stable for up to 50 min at 37°C *in vitro* in the absence of light, whereas in human plasma, a decrease of more than 90% of the E-enantiomer was observed after a 60-min incubation under the same conditions (Zhao et al., 2004). The exact cause for this significant species-related

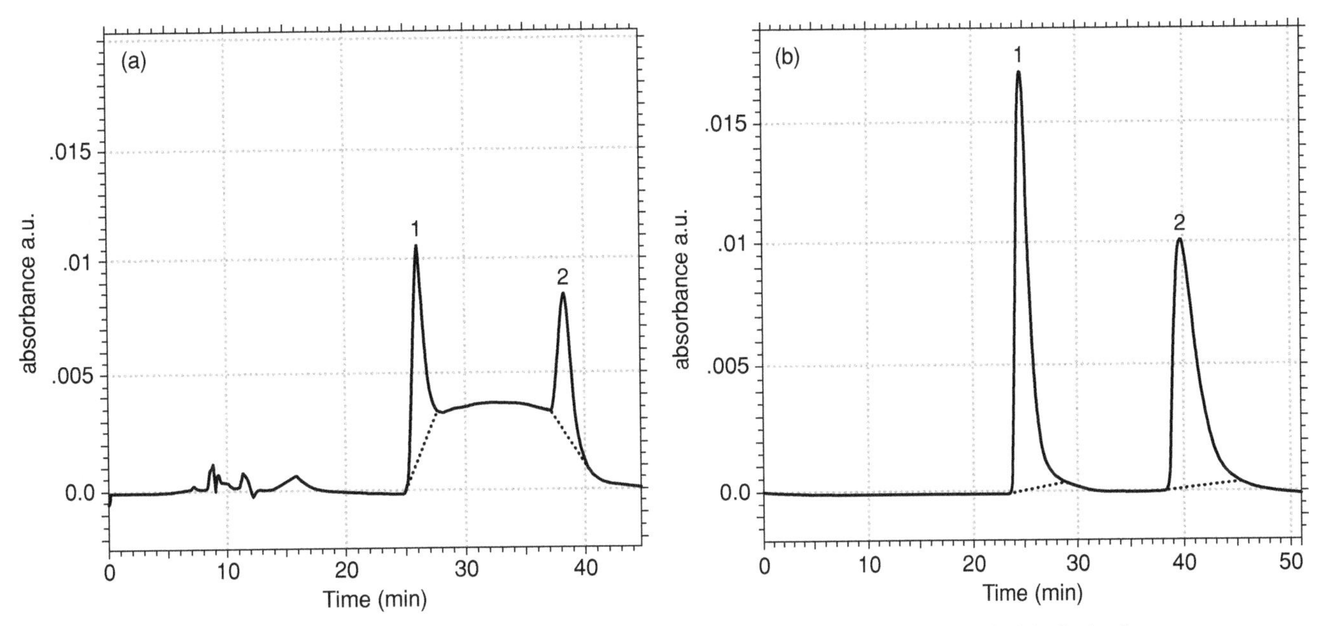

FIGURE 40.9 Chromatograms of (*R*)- and (*S*)-oxazepam (peaks 1 and 2, respectively) obtained by chiral HPLC at a column temperature of 20°C (a) and 12°C (b) (reproduced with permission from Pham-Huy et al. (2002)).

difference remains unclear, but an enzymatic reaction might have played a role.

The addition of the thiol-blocking agent N-ethylmaleimide (NEM) in combination with the antioxidant ascorbic acid was found to prevent the conversion in plasma of the dermatologic drug, all-*trans* retinoic acid to its geometric isomer 13-*cis* retinoic acid and vice versa (Wang et al., 2003). In this study, decreasing the storage temperature from 25°C to 0°C helped reduce the interconversion rate approximately threefold, but the interconversion could not be completely stopped without the addition of NEM. This suggests the involvement of a thiol-containing compound and possibly an enzyme in isomerization.

40.3.3 Bioanalytical Consequences

It is essential to ensure no further *in vitro* interconversion of stereoisomers during the bioanalytical process in order to precisely understand the *in vivo* interconversion. As most interconversion reactions of stereoisomers are dependent on pH, a proper adjustment of sample pH can usually help stabilize even highly unstable analytes. For example, the concentrations of (*S*)- and (*R*)-oxazepam could be accurately determined in rabbit plasma if the corresponding blood sample was immediately centrifuged at 0–2°C, the resulting plasma directly frozen and mixed with an equal volume of 1 N hydrochloric acid upon thawing before sample processing for analysis (Pham-Huy et al., 2002). Because of the relatively rapid interconversion at pH values above ~2 and the need for a mobile phase of pH 4.5 for an effective chiral separation, some interconversion was observed during the 50-min chromatographic run (column temperature 20°C). As a result of this on-column interconversion, the overall retention time of part of the analyte molecules was in between that of the pure enantiomers. Figure 40.9a shows a typical chromatogram illustrating the effect of on-column interconversion, which could be overcome by lowering the column temperature to 12°C (Figure 40.9b).

A number of other methods for the quantification of stereoisomers with base-catalyzed interconversion have been described. For example, the enantiomers of thalidomide were stabilized by adjustment of the serum pH to below 4 by mixing the serum sample with an equal volume of 0.2 M citrate buffer of pH 2. In this way, stability of the enantiomers at ≤ −20°C was demonstrated for at least 1 year and for at least 5 h at ambient temperature (Murphy-Poulton et al., 2006). The enantiomers of ibuprofen were quantified in human plasma after acidification of the sample with a twofold excess of 1 N hydrochloric acid, directly after thawing (Szeitz et al., 2010). For isomers with a relatively slow interconversion rate, pH adjustment may not be necessary and storage at a reduced temperature may be sufficient for stabilization. For example, the enantiomers of MK-0767 were quantified in plasma by storage of the samples at ≤ −70°C and keeping the samples on wet ice after thawing. Stability of the analyte after five freeze/thaw cycles and for at least 24 h in the autosampler at 4°C was demonstrated (Yan et al., 2004).

For analytes prone to acid-catalyzed conversion, the analysis of study samples at physiological pH may be sufficient to keep the stereoisomeric forms stable. For example, the E- and Z-isomers of the O-methylated oxime candidate drug mentioned earlier were stable in untreated human plasma for at least 41 days at ≤ -20°C and for at least 24 h at 4°C and room temperature (Xia et al., 1999). In case of photoinduced interconversion, protection of the analyte from light is obviously required. For example, the analysis of E- and Z-SU5416 in rat plasma was performed under light-protected conditions (yellow laboratory lighting) with no further pH adjustment or cooling necessary for stabilization (Zhao et al., 2004). To prevent interconversion of all-*trans* retinoic acid and 13-*cis* retinoic acid, more rigorous measures were needed, including handling study samples under red light, on wet ice, and in the presence of 0.05 mM NEM (Wang et al., 2003).

40.4 CONSIDERATIONS FOR DEVELOPING BIOANALYTICAL METHODS FOR INTERCONVERTIBLE COMPOUNDS

As discussed earlier, the speed and extent of interconversion reactions can vary considerably between compounds and, therefore, the necessary stabilizing precautions will also vary from one application to the other. For lactones/hydroxy acids and stereoisomers, pH and temperature are the main parameters that influence the interconversion kinetics. It is advisable to thoroughly study the effect of these parameters, when setting up an analytical method. The protocol listed later describes a general approach for the development of methods for interconvertible compounds. From the results of these tests, information about the appropriate handling of biological samples will be obtained and the need for pH adjustment and/or sample processing on wet ice and the required turnaround time of study sample collection and analysis can be defined.

40.5 VALIDATION AND QUALITY CONTROL OF METHODS FOR INTERCONVERTIBLE COMPOUNDS

Even after optimizing the sampling, storage, and analysis conditions for interconvertible drugs, some interconversion may be unavoidable. In such a case, it is essential to monitor the extent of the unavoidable interconversion, if any. For multianalyte assays, stability assessment during method validation is typically performed using equal concentrations of the analytes. However, in case of two analytes that can interconvert, this approach is inadequate as the relative concentrations of the analytes in study samples may differ considerably and their proportion may also change as a function of the time after dosing.

This was convincingly demonstrated by Jemal and Xia (2000) via spiking samples with the lactone and hydroxy acid forms of pravastatin in different ratios and subjecting these samples to three different storage conditions. Under condition 1, all samples were prepared in an ice bath and analyzed immediately after preparation by adjusting the pH to 4.2 to minimize interconversion. Under condition 2, the samples were analyzed after storage for 4 h at physiological pH and ambient temperature to promote the hydrolysis of lactone to hydroxy acid. Under condition 3, the pH of all samples was adjusted to 1.8 and stored for 2 h at ambient temperature to induce lactonization of the hydroxy acid. They found acceptable accuracies for all stored samples that had been spiked with equal concentrations of both forms, suggesting adequate stability under all conditions. However, the samples containing a 3- or 10-fold excess of the lactone relative to the hydroxy acid, showed an increase in the hydroxy acid concentration by 30–160% after storage at condition 2. For the samples with a 10-fold excess of the hydroxy acid, an increase of >70% in the lactone concentration was found following storage under condition 3. The evaluation clearly demonstrates that interconversion of the analytes at unknown concentrations in study samples may not be picked up by spiking samples at equal concentrations. Therefore, it is strongly recommended to prepare quality control (QC) samples in such a way that the relative concentrations of the analytes mimic the anticipated concentrations in study samples. This is especially important for assessing analyte stability during assay method validation. For routine bioanalysis, inclusion of QCs with appropriate analyte concentrations relative to study samples is important to continuously demonstrate the robustness of the bioanalytical method.

40.6 SUMMARY

In this chapter, several aspects of LC-MS/MS bioanalysis of interconvertible compounds are reviewed. Two classes of interconvertible analytes are discussed: lactones/hydroxy acids and stereoisomers. The chemical background and bioanalytical consequences of the interconversion reactions of these analyte classes are evaluated and different approaches to stabilize samples for bioanalysis are discussed. A practical protocol is suggested for the development of LC-MS/MS methods for interconvertible compounds and attention is also devoted to appropriate ways to validate and apply such bioanalytical methods.

40.7 REPRESENTATIVE PROTOCOL FOR LC-MS/MS METHOD DEVELOPMENT FOR BIOANALYSIS OF INTERCONVERTIBLE COMPOUNDS

General

Perform all experiments with samples containing just one but not both analytes, as in that way any conversion from one analyte to the other will be directly visible.

Chromatography and Mass Spectrometry

Set up an LC-MS/MS method that enables the determination of both analytes without any on-column or in-source interconversion.

1. Optimize the detection conditions in such a way that interconversion in the mass spectrometer is minimized, by fine-tuning source temperature, ion spray voltage, etc.
2. Inject the analytes separately; the pair of the analytes should be eluted as single peaks with baseline separation.
3. Optimize the mobile phase pH.[a]
4. Optimize the column temperature.[b]
5. Optimize the chromatographic run time.[c]

Sample Preparation

Set up a sample preparation procedure with good and consistent extraction recovery and sufficient selectivity with minimal interconversion during the process. Analyze sample extracts using the optimized LC-MS/MS method.

1. Optimize the pH of the different extraction steps.[a]
2. Optimize the storage temperature of samples and reagents (room temperature and stored on ice).[b]
3. Optimize the temperature of a possible evaporation of extraction solvent.[b]
4. Optimize the sample preparation throughput.[c]
5. Using the optimized procedure, define the number of samples that can be extracted in a single batch without any unacceptable interconversion.

Stability Assessment

Demonstrate the absence of interconversion under the various relevant sample storage conditions of the analytical process. Analyze samples using the optimized sample extraction procedure and LC-MS/MS method.

1. Assess analyte stability in extracts in the autosampler. Optimize the pH for sample extract reconstitution and autosampler temperature.[a,b]
2. Assess analyte stability in the intended biological sample placed on benchtop (benchtop stability) or after multiple freeze/thaw cycles (freeze/thaw stability). Optimize the sample pH and storage temperature (at room temperature and on wet ice).[a,b]
3. Assess analyte stability in whole blood, in case of methods for plasma and serum.
4. Assess analyte stability in frozen samples. Optimize the sample pH and storage temperature (−20°C and −70°C).[a,b]

Notes

[a]Lactones and hydroxy acids typically show minimal interconversion at slightly acidic pH (between 4 and 7, depending on the molecular structure) and drug enantiomers at either acidic or alkaline pH (depending on the molecular structure).

[b]Interconversion decreases at a lower temperature.

[c]Interconversion is reduced in case of short runs/extractions/storage periods.

REFERENCES

Ali I, Gupta VK, Aboul-Enein HY, Singh P, Sharma B. Role of racemization in optically active development. Chirality 2007;19:453–463.

Apostolou C, Kousoulos C, Dotsikas Y, et al. An improved and fully validated LC-MS/MS method for the simultaneous quantification of simvastatin and simvastatin acid in human plasma. J Pharm Biomed Anal 2008; 46:771–779.

Aso Y, Yoshioka S, Shibazaki T, Uchiyama M. The kinetics of the racemization of oxazepam in aqueous solution. Chem Pharm Bull 1988;36:1834–1840.

Barett B, Huclová J, Bořek-Dohalský V, Nemec B, Jelínek I. Validated HPLC-MS/MS method for simultaneous determination of simvastatin and simvastatin hydroxy acid in human plasma. J Pharm Biomed Anal 2006;41:517–526.

Boyd G, Smyth JF, Jodrell DI, Cummings J. High-performance liquid chromatographic technique for the simultaneous determination of lactone and hydroxy acid forms of camptothecin and SN-38 in tissue culture media and cancer cells. Anal Biochem 2001;297:15–24.

Briscoe CJ, Hage DS. Factors affecting the stability of drugs and drug metabolites in biological matrices. Bioanalysis 2009;1:205–220.

Burke TG, Mi Z. Preferential binding of the carboxylate form of camptothecin by human serum albumin. Anal Biochem 1993; 212:285–287.

Burke TG, Munshi CB, Mi Z, Jiang Y. The important role of albumin in determining the relative human blood stabilities of the camptothecin anticancer drugs. J Pharm Sci 1995;84:518–519.

Chen J, Hsieh Y. Stabilizing drug molecules in biological samples. Ther Drug Monit 2005;27:617–624.

Chollet DF, Goumaz L, Renard A, et al. Simultaneous determination of the lactone and carboxylate forms of the camptothecin derivative CPT-11 and its metabolite SN-38 in plasma by high-performance liquid chromatography. J Chromatogr B 1998;718:163–175.

Dell D. Labile metabolites. Chromatographia 2004;59:S139–S148.

Eriksson T, Björkman S, Roth B, Fyge Å, Höglund P. Stereospecific determination, chiral inversion in vitro and pharmacokinetics in humans of the enantiomers of thalidomide. Chirality 1995;7: 44–52.

European Medicines Agency (EMA). Guideline on bioanalytical method validation. July, 2011. Available at http://www.ema.europa.eu/docs/en_GB/document_library/Scientific_guideline/2011/08/WC500109686.pdf. Accessed Apr 11, 2013.

Fassberg J, Stella VJ. A kinetic and mechanistic study of the hydrolysis of camptothecin and some analogues. J Pharm Sci 1992;81:676–684.

Grabarkiewicz T, Grobelny P, Hoffmann M, Mielcarek J. DFT study on hydroxy acid–lactone interconversion of statins: the case of fluvastatin. Org Biomol Chem 2006;4:4299–4306.

Hull CK, Penman AD, Smith CK, Martin PD. Quantification of rosuvastatin in human plasma by automated solid-phase extraction using tandem mass spectrometric detection. J Chromatogr B 2002;772:219–228.

Jamali B, Bjørnsdottir I, Nordfang O, Hansen SH. Investigation of racemisation of the enantiomers of glitazone drug compounds at different pH using chiral HPLC and chiral CE. J Pharm Biomed Anal 2008;46:82–87.

Jemal M, Ouyang Z. The need for chromatographic and mass resolution in liquid chromatography/tandem mass spectrometric methods used for quantitation of lactones and corresponding hydroxy acids in biological samples. Rapid Commun Mass Spectrom 2000;14:1757–1765.

Jemal M, Ouyang Z, Chen BC, Teitz D. Quantitation of the acid and lactone forms of atorvastatin and its biotransformation products in human serum by high-performance liquid chromatography with electrospray tandem mass spectrometry. Rapid Commun Mass Spectrom 1999;13:1003–1015.

Jemal M, Ouyang Z, Powell ML. Direct-injection LC-MS-MS method for high-throughput simultaneous quantitation of simvastatin and simvastatin acid in human plasma. J Pharm Biomed Anal 2000;23:323–340.

Jemal M, Xia YQ. Bioanalytical method validation design for the simultaneous quantitation of analytes that may undergo interconversion during analysis. J Pharm Biomed Anal 2000;22: 813–827.

Kaneda N, Hosokawa Y, Yokokura T. Simultaneous determination of the lactone and carboxylate forms of 7-ethyl-10-hydroxycamptothecin (SN-38), the active metabolite of irinotecan (CPT-11) in rat plasma by high performance liquid chromatography. Biol Pharm Bull 1997;20:815–819.

Kearney AS, Crawford LF, Mehta SC, Radebaugh GW. The interconversion kinetics, equilibrium, and solubilities of the lactone and hydroxy acid forms of the HMG-CoA reductase inhibitor CI-981. Pharm Res 1993;10:1461–1465.

Li W, Zhang J, Tse FL. Strategies in quantitative LC-MS/MS analysis of unstable small molecules in biological matrices. Biomed Chromatogr 2011;25:258–277.

Loos WJ, de Bruijn P, Verweij J, Sparreboom A. Determination of camptothecin analogs in biological matrices by high-performance liquid chromatography. Anticancer Drugs 2000;11:315–324.

Mullangi R, Ahlawat P, Srinivas NR. Irinotecan and its active metabolite, SN-38: review of bioanalytical methods and recent update from clinical pharmacology perspectives. Biomed Chromatogr 2010;24:104–123.

Mulvana D, Jemal M, Pulver SC. Quantitative determination of pravastatin and its biotransformation products in human serum by turbo ionspray LC/MS/MS. J Pharm Biomed Anal 2000;23:851–866.

Murphy-Poulton SF, Boyle F, Gu XQ, Mather LE. Thalidomide enantiomers: determination in biological samples by HPLC and vancomycin-CSP. J Chromatogr B 2006;831:48–56.

Nirogi R, Mudigonda K, Kandikere V. Chromatography-mass spectrometry methods for the quantitation of statins in biological samples. J Pharm Biomed Anal 2007;44:379–387.

Nováková L, Šatinský D, Solich P. HPLC methods for the determination of simvastatin and atorvastatin. Trends Anal Chem 2008;27:352–367.

Nowatzke W, Woolf E. Best practices during bioanalytical method validation for the characterization of assay reagents and the evaluation of analyte stability in assay standards, quality controls, and study samples. AAPS J 2007;9:E117–E122.

Owens TS, Dodds H, Fricke K, Hanna SK, Crews KR. High-performance liquid chromatographic assay with fluorescence detection for the simultaneous measurement of the carboxylate and lactone forms of irinotecan and three metabolites in human plasma. J Chromatogr B 2003;788:65–74.

Pham-Huy C, Villain-Pautet G, Hua H, et al. Separation of oxazepam, lorazepam, and temazepam enantiomers by HPLC on a derivatized cyclodextrin-bonded phase: application to the determination of oxazepam in plasma. J Biochem Biophys Meth 2002;54:287–299.

Prueksaritanont T, Subramanian R, Fang X, et al. Glucuronidation of statins in animals and humans: a novel mechanism of statin lactonization. Drug Metab Disp 2002;30:505–512.

Ragot S, Marquet P, Lachâtre F, et al. Sensitive determination of irinotecan (CPT-11) and its active metabolite SN-38 in human serum using liquid chromatography–electrospray mass spectrometry. J Chromatogr B 1999;736:175–184.

Ramesh M, Ahlawat P, Srinivas NR. Irinotecan and its active metabolite, SN-38: review of bioanalytical methods and recent update from clinical pharmacology perspectives. Biomed Chromatogr 2010;24:104–123.

Rivory LP, Robert J. Reversed-phase high-performance liquid chromatographic method for the simultaneous quantitation of the carboxylate and lactone forms of the camptothecin derivative irinotecan, CPT-11, and its metabolite SN-38 in plasma. J Chromatogr B 1994;661:133–141.

Sano K, Yoshikawa M, Hayasaka S, et al. Simple non-ion-paired high-performance liquid chromatographic method for simultaneous quantitation of carboxylate and lactone forms of 14 new camptothecin derivatives. J Chromatogr B 2003;795:25–34.

Schoemaker NE, Rosing H, Jansen S, Schellens JH, Beijnen JH. High-performance liquid chromatographic analysis of the anticancer drug irinotecan (CPT-11) and its active metabolite SN-38 in human plasma. Ther Drug Monit 2003;25:120–124.

Severin G. Spontaneous racemization of chlorthalidone: kinetics and activation parameters. Chirality 1992;4:222–226.

Suchocka Z, Swatowska J, Pachecka J, Suchocki P. RP-HPLC determination of paraoxonase 3 activity in human blood serum. J Pharm Biomed Anal 2006;42:113–119.

Szeitz A, Edginton AN, Peng HT, Cheung B, Riggs KW. A validated enantioselective assay for the determination of ibuprofen enantiomers in human plasma using ultra performance liquid chromatography with tandem mass spectrometry (UPLC-MS/MS). Am J Anal Chem 2010;2:47–58.

Testa B, Carrupt PA, Gal J. The so-called "interconversion" of stereoisomeric drugs: an attempt at clarification. Chirality 1993;5:105–111.

US Food and Drug Administration. Guidance for Industry—Bioanalytical Method Validation. May, 2001. Available at www.fda.gov/downloads/Drugs/GuidanceComplianceRegulatoryInformation/Guidances/UCM070107.pdf. Accessed Apr 11, 2013.

Viswanathan CT, Bansal S, Booth B, et al. Quantitative bioanalytical methods validation and implementation: best practices for chromatographic and ligand binding assays. Pharm Res 2007;24:1962–1973.

Vree TB, Dammers E, Ulc I, Horkovics-Kovats S, Ryska M, Merkx IJ. Variable plasma/liver and tissue esterase hydrolysis of simvastatin in healthy volunteers after a single oral dose. Clin Drug Invest 2001;21(9):643–652.

Wang CJ, Pao LH, Hsiong CH, Wu CY, Whang-Peng JJ, Hu OYP. Novel inhibition of *cis*/*trans* retinoic acid interconversion in biological fluids—an accurate method for the determination of *trans* and 13-*cis* retinoic acid in biological fluids. J Chromatogr B 2003;796:283–291.

Xia YQ, Whigan DB, Jemal M. A simple liquid-liquid extraction with hexane for low-picogram determination of drugs and their metabolites in plasma by high-performance liquid chromatography with positive ion electrospray tandem mass spectrometry. Rapid Commun Mass Spectrom 1999;13:1611–1621.

Yan KX, Song H, Lo MW. Determination of MK-0767 enantiomers in human plasma by normal-phase LC-MS/MS. J Chromatogr B 2004;813:95–102.

Yang AY, Sun L, Musson DG, Zhao JJ. Application of a novel ultra-low elution volume 96-well solid-phase extraction method to the LC/MS/MS determination of simvastatin and simvastatin acid in human plasma. J Pharm Biomed Anal 2005;38:521–527.

Zhang JY, Fast DM, Breau AP. Development and validation of a liquid chromatography–tandem mass spectrometric assay for eplerenone and its hydrolyzed metabolite in human plasma. J Chromatogr B 2003a;787:333–344.

Zhang JY, Fast DM, Breau AP. A validated SPE-LC-MS/MS assay for eplerenone and its hydrolyzed metabolite in human urine. J Pharm Biomed Anal 2003b;31:103–115.

Zhang N, Yang A, Rogers JD, Zhao JJ. Quantitative analysis of simvastatin and its β-hydroxy acid in human plasma using automated liquid–liquid extraction based on 96-well plate format and liquid chromatography–tandem mass spectrometry. J Pharm Biomed Anal 2004;34:175–187.

Zhao Y, Sukbuntherng J, Antonian L. Simultaneous determination of Z-SU5416 and its interconvertible geometric E-isomer in rat plasma by LC/MS/MS. J Pharm Biomed Anal 2004;35:513–522.

Zufia L, Aldaz A, Giráldez J. Separation methods for camptothecin and related compounds. J Chromatogr B 2001;764:141–159.

41

LC-MS BIOANALYSIS OF CHIRAL COMPOUNDS

Naidong Weng

41.1 INTRODUCTION

Stereoisomers are molecules that are identical in atomic constitution and bonding, but differ in the three-dimensional arrangement of the atoms. Stereoisomers include diastereomers and enantiomers. Diastereomers are stereoisomers that are not mirror images of each other. These include meso compounds, *cis–trans* (*E-Z*) isomers, and nonenantiomeric optical isomers. They are physically and chemically distinct and they are also pharmacologically different (unless they are interconverted *in vivo*). From a bioanalytical point of view, they are readily separated without using chiral methods. Enantiomers are two stereoisomers that are related to each other by a reflection: they are mirror images of each other that are nonsuperimposable (Figure 41.1). Enantiomers have identical physical (except optical rotation) and chemical (except in a chiral environment) properties in an achiral environment. Many drugs, drug candidates, and their metabolites have one or more chiral centers and could have one or more pairs of enantiomers. Many of the biological macromolecules such as receptors, transporters, and enzymes prefer binding to one of the two enantiomers, leading to potential difference in their absorption, distribution, metabolism, and excretion. These macromolecules are abundant in chiral centers of defined configuration, for example, D-sugars and L-amino acids. It is well established that enantiomers may have different biological and pharmacologic activities. One enantiomer may have the desired effect (eutomer) while the other enantiomer may have unwanted side or even toxic effects (distomer) as exemplified by the toxicity associated with thalidomide, benoxaprofen, and terodoline (Srinivas et al., 2001). Enantiomers should be treated as different chemical compounds. Issues, considerations, and regulatory requirements for enantiomeric drug development were discussed in the aforementioned review article (Srinivas et al., 2001). Strategies for chiral drug development include development of a racemate, a single enantiomer, fixed ratio (nonracemic) of enantiomers, and racemic switches. Pros and cons of each strategy were also outlined. For example, in the strategy of racemic switches, the use of levofloxacin, the active enantiomer of ofloxacin, is beneficial because of the improved therapeutic and safety index. Regulatory control began in the United States with the publication in 1992 of guidelines on the development of chiral drugs (Food and Drug Administration, 1992), followed by the European Union (Committee for Proprietary Medical Products, 1993).

Metabolic and pharmacokinetic data on racemic mixtures obtained from nonstereoselective assays can be highly misleading when attempting to relate plasma concentrations to pharmacological effect or therapeutic benefit. Early evaluation of the stereoselective pharmacological effects and pharmacodynamics of enantiomers is essential in drug development. Species-related differences in stereoselectivity need to be evaluated carefully before the data are extrapolated to humans. Stereoselective plasma protein binding and metabolism can differ considerably among species (Lin and Lu, 1997; Srinivas, 2006). In a more recent review article titled "Stereoselectivity in Drug Metabolism," a comprehensive appraisal of stereochemical aspects of drug metabolism (i.e., enantioselective metabolism and first-pass effect, enzyme-selective inhibition or induction and drug interaction, species differences, and polymorphic metabolism) is reviewed (Lu, 2007). If the metabolism of the different enantiomers is handled by different enzymes, which

Handbook of LC-MS Bioanalysis: Best Practices, Experimental Protocols, and Regulations, First Edition. Edited by Wenkui Li, Jie Zhang, and Francis L.S. Tse.

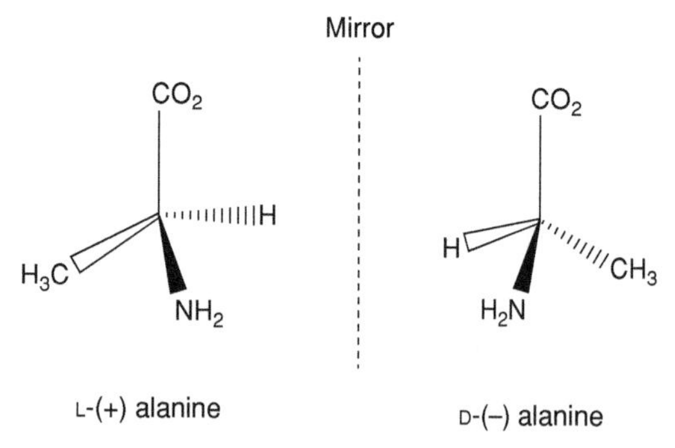

FIGURE 41.1 Mirror image of a chiral compound.

are either polymorphic or can be induced or inhibited, and if their pharmacodynamic effects have differences either in strength or in quality, an enantiospecific analysis is urgently needed (Rentsch, 2002). We propose a decision tree on the potential need of using chiral assays to measure parent and metabolites (Figure 41.2). It should be especially noted that stereoselective metabolism can occur for the nonchiral drug, leading to a need for using chiral assays for the measurement of a metabolite.

Bioanalytical support plays a key role in drug development. The importance of stereospecific bioanalytical monitoring in drug development was emphasized in an excellent review article dating back to 1996 (Caldwell, 1996), while as early as in the early 1980s LC-MS (liquid chromatography–mass spectrometry) determination of optically active drugs was reported by Henion and coauthors (Henion and Maylin, 1980; Crowther et al., 1984). LC is the method of choice for chiral separation. LC methods devised for chiral separation can be classified into three categories: (1) direct method using chiral stationary phase; (2) addition of a chiral reagent to the mobile phase, which then forms adducts with the enantiomers, and separation on an achiral stationary phase; (3) separation of the diastereomers formed by precolumn derivatization with a chiral derivatization reagent (Mišl'anová and Hutta, 2003). Of the above approaches, (1) and (3) are most commonly used for chiral quantitative bioanalysis. Methods of adding chiral reagents to the mobile phase are typically poor on chiral chromatographic separation and are usually not amenable to LC-MS due to the nonvolatile chiral reagents. Nevertheless, this is the preferred approach for the chiral capillary zone electrophoresis (CZE), thanks to the tremendous separation efficiency. Chiral capillary zone electrophoresis in general has limited usage in quantitative bioanalysis due to the low injection volume (low nl) and therefore poor sensitivity.

In the last 10 years, a number of excellent review articles focusing on quantitation of enantiomers have appeared in literatures (Table 41.1). These literatures, though not entirely focused on the chiral bioanalytical LC-MS/MS analysis, provide wealthy information about fundamental chiral separation using various approaches.

In contrast to the wealthy information of reviews focusing on general chiral separation, only a few LC-MS/MS methods for chiral analysis of pharmaceutical compounds and their metabolites in biological fluids are reviewed (Chen et al., 2005; Erny and Cifuentes, 2006; Liu et al., 2009),

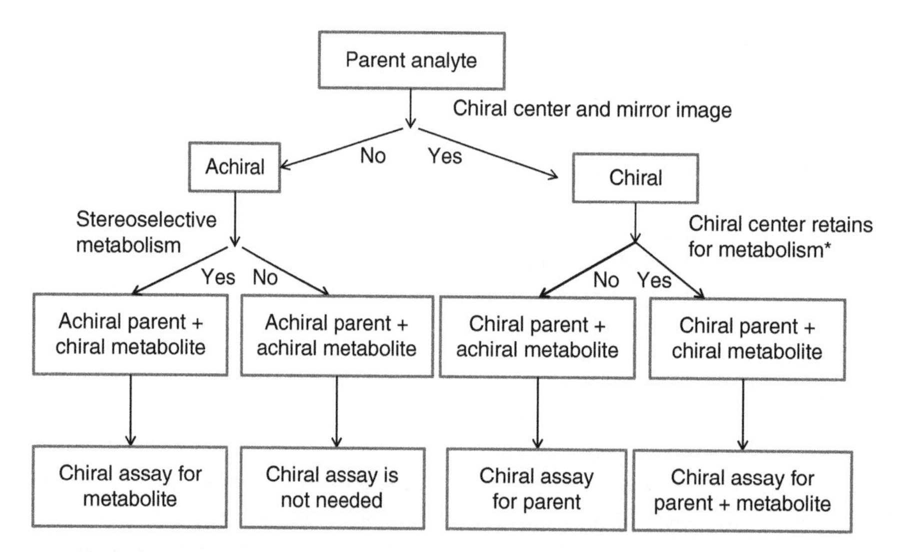

FIGURE 41.2 Decision tree on the potential need of using chiral assay to measure parent and metabolites.

TABLE 41.1 Important Review Articles Focusing on Quantitation of Enantiomers

Authors and year	Title
Ilisz et al. (2008)	Application of chiral derivatizing agents in the high-performance chromatographic separation of amino acid enantiomers: a review
Toyo'oka (2002)	Resolution of chiral drugs by liquid chromatography based upon diastereomer formation with chiral derivatization reagents
Sun et al. (2001)	Chiral derivatization reagents for drug enantioseparation by high-performance liquid chromatography based upon precolumn derivatization and formation of diastereomers: enantioselectivity and related structure
Bhushan (2011)	Enantiomeric purity of chiral derivatization reagents for enantioresolution
Millot (2003)	Separation of drug enantiomers by liquid chromatography and capillary electrophoresis, using immobilized proteins as chiral selectors
Haginaka (2008)	Recent progress in protein-based chiral stationary phases for enantioseparations in liquid chromatography
Ikai et al. (2008)	Immobilized-type chiral packing materials for HPLC based on polysaccharide derivatives
Beesley and Lee (2009)	Method development strategy and applications update for CHIROBIOTIC chiral stationary phases
Lämmerhofer (2010)	Chiral recognition by enantioselective liquid chromatography: mechanisms and modern chiral stationary phases
Schmid and Gübitz (2011)	Enantioseparation by chromatographic and electromigration techniques using ligand-exchange as chiral separation principle
Wang et al. (2008b)	Recent development of enantioseparation techniques for adrenergic drugs using liquid chromatography and capillary electrophoresis: a review
Fried and Wainer (1997)	Column-switching techniques in the biomedical analysis of stereoisomeric drugs: why, how and when

even though LC-MS/MS analysis of enantiomers in biological fluids using chiral stationary phases has been reported as early as in the early 1980s (Henion and Maylin, 1980). A comprehensive and critical review of chiral bioanalytical literatures as well as a proposal for the method validation strategy is needed to timely address the increasing requirements for chiral bioanalysis.

Based on our experiences, there are some significant differences between chiral and achiral bioanalytical LC-MS/MS method development. Due to the superior specificity of the tandem mass spectrometry, a bioanalytical scientist seldom has any difficulty in finding a reversed-phase C18 or a hydrophilic-interaction chromatography (HILIC) column that will give a good retention, column efficiency, satisfactory peak shape, and chromatographic resolution from matrix suppression or interferences. Stable-labeled isotope of the analyte is usually available as the internal standard to compensate for any variability during extraction, chromatography, ionization, and detection. One can often use gradient elution to further improve the peak shape, eliminate down-field interference, and reduce the run time. The column reproducibility, even from different column manufacturers, is now quite good. In-depth chromatographic knowledge is helpful but is not considered essential anymore. The method development scientist would focus his or her energy on streamlining and optimizing the process from sample extraction to data processing, from the use of automation to further enhance productivity to explore new technology such as dried-blood spot for sample volume reduction or high-resolution mass spectrometry for improving selectivity. Chiral bioanalytical methods, on the other hand, would require extensive chromatographic know-how and experiences. For every chiral method, the primary and the most time-consuming task is to find a way to chromatographically separate the enantiomers. Of course, one may argue that precolumn derivatization with a chiral reagent may eliminate this difficult task but, as will be discussed below, this approach is also not without its own challenges, including the inadequate purity of the chiral derivatization reagents, potential racemization of the enantiomers, and conversion of the metabolites back to the parent analyte under the harsh derivatization condition. Direct chiral analysis should still be the first choice. The most frequently used chiral stationary phases are cyclodextrin and Chirobiotic columns in the macrocyclic CPS, polysaccharide-based chiral column, and protein-based chiral column such as chiral AGP. Applications of these chiral columns will be discussed further.

41.2 APPLICATIONS

41.2.1 Chiral Derivatization

Even though chiral derivatization is less used for the bioanalytical chiral LC-MS method development, it still offers some advantages over direct chiral separation and sometimes it is the last resort available if the direct separation of enantiomers is unsuccessful. Figure 41.3 highlights some of the most frequently used chiral derivatization reactions. There are several advantages of using this approach.

- May improve sensitivity by introducing a more ionizable function group, by having better on-column

Function group	Product	Reactions
Amino	Amides	RR′NH + XCOR″ → RR′NCOR″
	Carbamates	RR′NH + ClCOOR″ → RR′NCOOR″
	Ureas	RR′NH +O=C=NR″ → RR′NCONHR″
Hydroxy	Esters	ROH + XOCOR′ → ROOCR′
	Carbonates	ROH + ClCOOR′ → ROCOOR′
	Carbamates	ROH + O=C=NR′ → ROCONHR′
Carboxy	Esters	RCOOH + R′OH → RCOOR′
	Amides	RCOOH + R′NH2 → RCONHR′

FIGURE 41.3 Highlights of some of the most frequently used chiral derivatization reactions.

retention, and by using UHPLC column to gain column efficiency.

- May improve selectivity by using selective reaction and by employing more extensive sample cleanup in conjunction with the derivatization process.
- May reduce cost by using an achiral column since typical achiral columns are less expensive, possessing longer column life and usually having shorter run time.

However, there are also several disadvantages:

- May introduce quantitation bias if the chiral reagent is not pure (Figure 41.4).
- May have potential racemization.
- Cost of the chiral reagent can be high.
- Excessive reagent may cause degradation of the analytical column.
- May have length sample extraction and derivatization.
- May have concern for breakdown of other metabolites, for example, phase II conjugates, which can cause quantitation bias.
- May need to perform extraction prior to derivatization (to remove labile metabolites and matrix components, which may interfere with reaction) and extraction after derivatization (to remove excessive reagent and/or to extract analytes to a more stable condition).
- In order to ensure selectivity, need to optimize MS condition so that the fragmentation used for the detection is not the loss of derivatization reagent.

The possibility of converting metabolites to the parent or other metabolites during the derivation step should be carefully evaluated. Potential racemization of the reagent or enantiomers should also be thoroughly investigated and excluded by using high-purity reagents and optically pure enantiomers for method development. In this way, any racemization can be easily detected and the reaction condition may be modified to avoid racemization. Any elevation of the other enantiomer's level in the sample after derivatization, over the sample before derivatization may indicate potential racemization. One common mistake is to use a racemate to optimize the reaction condition, resulting in the failure to detect racemization.

Stereoselective analysis of tiopronin enantiomers in rat plasma using LC-MS after chiral derivatization was reported by Wang et al. (2009). Tiopronin has a sulfhydryl group and a precolumn chiral derivatization with 2,3,4,6-tetra-*O*-acetyl-β-glucopyranosyl isothiocyanate (GITC) in acetonitrile was employed. The resulting diastereomers were separated on a C18 column using a mobile phase of methanol/water containing 5.3 mM formic acid with a gradient elution. The two diastereomers are well resolved ($R = 2.2$). To exclude the possibility of racemization during sample preparation, plasma samples spiked with a single enantiomer were analyzed separately and no formation of other enantiomers was observed. Because of the stability concern for the thiol compound, blood was drawn, processed to plasma immediately, and transferred to tubes containing HCl solution to prevent tiopronin from forming disulphide in biological samples. The extraction and derivatization of the samples should be immediately completed after sample collection. One added advantage of precolumn derivatization is the further stabilization of the thiol group after derivatization. This method could be further improved by using stable-labeled tiopronin instead of *N*-isobutyryl-D-cysteine (IBDC) as the internal standard

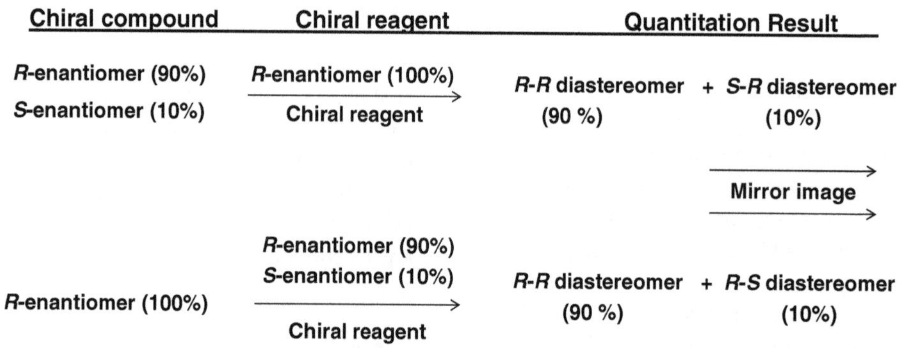

Chiral compound | Chiral reagent | Quantitation Result

R-enantiomer (90%)
S-enantiomer (10%)
R-enantiomer (100%) → Chiral reagent
R-R diastereomer (90 %) + *S-R* diastereomer (10%)

→ Mirror image →

R-enantiomer (100%)
R-enantiomer (90%)
S-enantiomer (10%) → Chiral reagent
R-R diastereomer (90 %) + *R-S* diastereomer (10%)

FIGURE 41.4 Influence of purity of chiral reagent on the quantitation result.

to improve the method ruggedness as IBDC has a different chemical structure from tiopronin and may not track well during extraction and derivatization. IBDC also elutes quite later than the derivatized tiopronin and may not compensate well for the matrix effect.

GITC was also used in the chiral LC-MS/MS analysis of carvedilol enantiomers in human plasma (Yang et al., 2004). The derivatization reaction is on the chiral alcohol group. Carvedilol racemate was extracted from human plasma by protein precipitation using acetonitrile containing D_5-carvedilol as the internal standard. Extracts were then derivatized with GITC and analyzed by LC-MS/MS. A slight difference in retention time between the analyte and the stable-labeled internal standard, caused by the deuterium isotope effect, has resulted in a different degree of ion suppression. The difference was significant enough to change the analyte to internal standard peak area ratio and affect the accuracy of the method (Wang et al., 2007).

The use of columns packed with sub-2 μm particles in liquid chromatography with very high pressure conditions (UHPLC) was investigated for the fast enantioseparation of drugs (Guillarme et al., 2010). The analysis of enantiomers was carried out after a derivatization procedure using two different reagents, 2,3,4-tri-*O*-acetyl-α-D-arabinopyranosyl isothiocyanate (AITC) and *N*-α-(2,4-dinitro-5-fluorophenyl)-L-alaninamide (Marfey's reagent). AITC works well in the presence of hydroxyl and amino groups, while Marfey's reagent can be used with compounds that contain amino groups with low steric hindrance. For Marfey's derivatization procedure, the pH and temperature should be set to a maximal value of 9 and 1 h at 55°C, respectively, to avoid compound racemization. For AITC derivatization procedure, the samples also need to be heated for 1 h at 55°C. The reaction time can be reduced by increasing the temperature, but the risk of racemization becomes significant. Separation of several amphetamine derivatives was achieved in 2–5 min and similar results were obtained with β-blockers whereas the individual separation of several pairs of enantiomers was performed in 1 min or less on a Waters Acquity BEH C18 column (50 × 2.1 mm I.D., 1.7 μm). While the separation efficiency is very impressive, it should be pointed out that the derivatization (high temperature, high pH, and length derivatization) may not be conducive for measuring drug candidates in biological fluids as metabolites, particularly unstable phase II metabolites such as acyl glucuronides, will convert back to the parent and cause overestimation.

41.2.2 Direct Separation

The biggest advantage of using direct chiral column is the removal of the derivatization step. Potential quantitation bias due to impure or racemized chiral derivatization reagent is eliminated. Even if there is some racemization of chiral

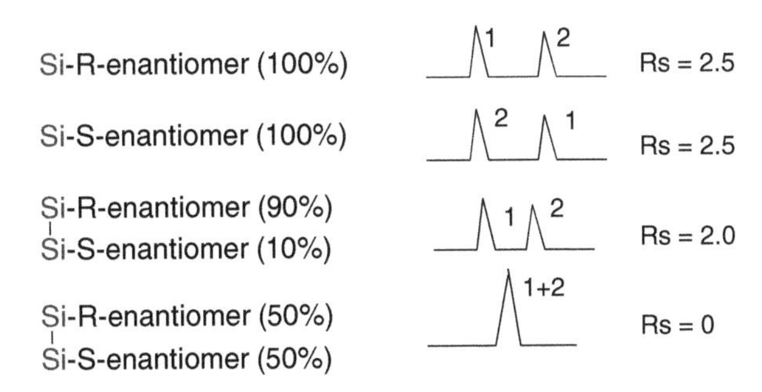

FIGURE 41.5 Influence of the chiral moiety purity of stationary phase on the separation of enantiomers.

selector on stationary phase, it usually does not significantly affect the enantiomeric separation (Figure 41.5).

Nevertheless, it is still a challenge to develop and validate robust quantitative bioanalytical chiral LC-MS methods because (1) the success of chiral separation largely depends on experiences as well as trial and error; an experienced chromatographer is needed; chiral method development is difficult and time-consuming; (2) the chiral columns are very expensive and it is not feasible to stock many different types of chiral columns; (3) the chiral columns in general are less stable and less reproducible than the typical achiral columns; (4) in order to separate enantiomers, chiral chromatographic run time is usually long; (5) many chiral separation conditions are at their best under normal phase, which is not conducible for the ionization of ESI or APCI (atmospheric pressure chemical ionization); HPLC setup for normal phase is also more difficult; streamlining of sample preparation with normal-phase LC is also a challenge; (6) chiral columns usually lack chromatographic resolution power for nonchiral separation and there are few parameters that can be played to improve the separation, leading to potential interferences or ion suppression from metabolites; (7) gradient elution on chiral column may be a challenge; (8) chiral columns are also usually lower on column efficiency and sub-2 μm chiral stationary phase is unavailable.

Based on the classification proposed by Armstrong (Armstrong and Zhang, 2001), the common chiral stationary phases are

- Macrocyclic chiral columns, which include the two most commonly used chiral columns for bioanalysis (cyclodextrins and glycopeptides) and chiral crown ethers columns, which are used for chiral analytes with primary amine on the chiral center, for example, amino acids;
- Polymeric chiral columns, such as derivatized carbohydrates which are most frequently used under normal phase conditions;
- Protein chiral stationary phases, which are most frequently used under highly aqueous conditions, and synthetic polymers;

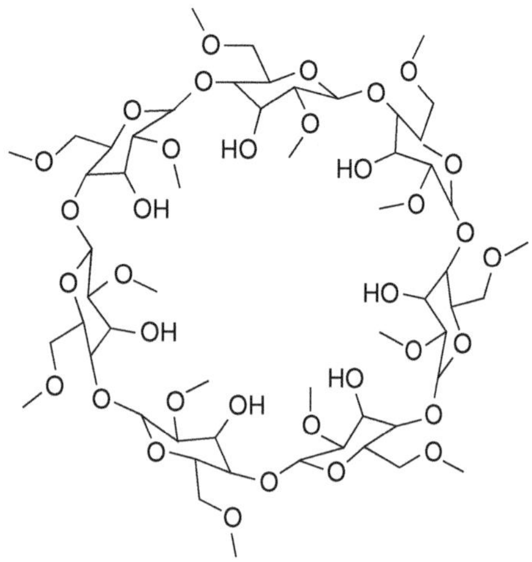

FIGURE 41.6 Chemical structure of β-cyclodextrin.

- π–π association chiral columns, which can be π-Electron acceptor, π-Electron donor or a combination of π-Electron acceptor and donor. These types of columns are less used for bioanalysis;
- Ligand exchange chiral columns;
- Miscellaneous and hybrid chiral columns.

41.2.3 Cyclodextrins Stationary Phase-Based Chiral LC-MS/MS Bioanalysis

Cyclodextrin-bonded chiral stationary phases, introduced by Armstrong (Armstrong and Jin, 1989), are based on α-, β-, or γ-cyclodextrins, which are macrocycles with 6, 7, and 8 glucose units connected via α-1,4-linkages that adopt the shape of a truncated cone with cavity diameters of 0.57, 0.78, and 0.95 nm, respectively. Figure 41.6 shows the chemical structure of β-cyclodextrin. The internal surface of the cyclodextrin cavity is hydrophobic, while the rim surfaces are hydrophilic due to the presence of hydroxyl groups. Derivatized cyclodextrins modulate the conformational flexibility of the cyclodextrins and provide additional binding sites that may alter the chiral-recognition profiles. Nevertheless, the native cyclodextrins (in particularly the β-cyclodextrin), which are bonded to silica gel at the narrower ring hydroxyls via ether linkage, are still the most frequently used one for chiral bioanalysis.

Cyclodextrin-based chiral stationary phases can be used under normal phase, polar organic phase, reversed-phase, and supercritical fluid chromatography (SFC) conditions. The most useful condition for bioanalysis is the polar organic phase mode. Under polar organic phase mode, hydrophilic interactions may be reinforced and chiral analytes interact with hydrophilic groups bound to the polar chiral surface of the cyclodextrin, resulting in chromatographic separation of the enantiomers. Polar enantiomers with more than one polar functional group, one of which is at or close to the chiral center, were found to be good candidates for chiral separation under polar organic mode on the cyclodextrin chiral stationary phase. This mode is particularly useful for the bioanalysis of drugs and metabolites that are both polar and have more than one polar group (Wang et al., 2008a).

An LC-MS/MS method was developed and validated for the quantitation of warfarin enantiomers in human plasma using polar organic mobile phase on a β-cyclodextrin column (Weng et al., 2001). Warfarin is a common anticoagulant agent by inhibiting vitamin K-dependent coagulation factors. *S*-warfarin is two to five times more potent than *R*-warfarin. *S*-warfarin is a substrate of 2C9, while *R*-warfarin is a substrate of 1A2. Up to 23% of Caucasians and 2% African-Americans may have one or more altered 2C9 genes, possibly contributing to abnormally decreased enzyme activity. *S*-Warfarin is more quickly metabolized than *R*-warfarin. The polar organic mobile phase consists of acetonitrile–acetic acid–triethylamine (1000:3:2.5) and column is a β-cyclodextrin column of 250 × 4.6 mm ID from ASTEC. The sample preparation is a simple liquid–liquid extraction with ethyl ether. *S*-*p*-Chlorowarfarin was used as the internal standard. Baseline chromatographic separation between the *S*- and *R*-warfarin is achieved within 10 min. Under this polar organic mobile phase condition, hydroxyl metabolites of warfarin are well resolved from the warfarin enantiomers (Weng and Lee, 1993). The β-CD column also demonstrated excellent column stability and lot-to-lot reproducibility.

41.2.4 Chirobiotic Stationary Phase-Based Chiral LC-MS/MS Bioanalysis

Chirobiotic chiral stationary phases offer six different types of molecular interactions: ionic, H-bond, π–π, dipole stacking, hydrophobic, and steric. The most commonly used Chirobiotic chiral stationary phases are teicoplanin and vancomycin, namely Chirobiotic T and Chirobiotic V, respectively, from ASTEC. They were introduced by Armstrong in the mid-1990s (Armstrong et al., 1994). The structure of vancomycin is shown in Figure 41.7. They also possess multiple inclusion sites that influence selectivity based on the molecular shape of the analyte. The optimization of enantiomer resolution is achieved by changing the mobile phase to leverage the types and relative strengths of the various interactions. These interactions are unique to Chirobiotic chiral stationary phases and are responsible in a large part for their desirable retention characteristics toward polar and ionizable analytes in aqueous and nonaqueous solvents.

Enantiomers of various analytes with diversified chemical structures have been successfully separated on the macrocyclic chiral stationary phases. As early as in 1999, a rapid and sensitive reversed-phase chiral LC-MS/MS assay for the determination of D- and L-*threo*-methylphenidate (Ritalin) in human plasma was developed using a vancomycin-bonded

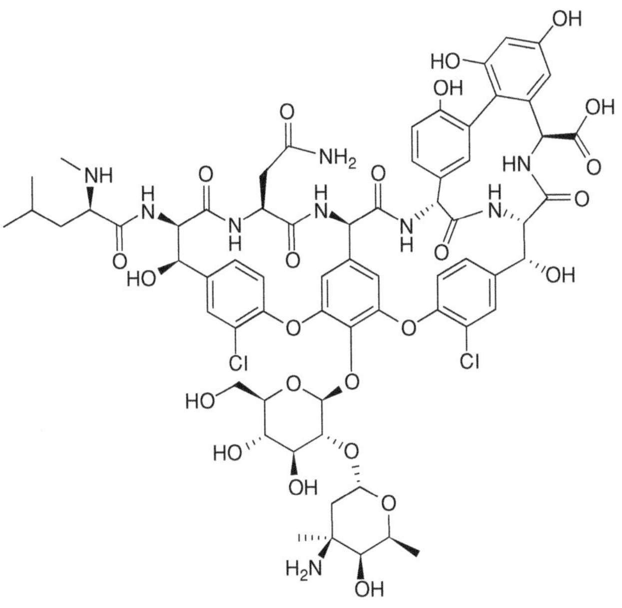

FIGURE 41.7 Chemical structure of vancomycin.

column. Ritalin is used for the treatment of attention deficit hyperactivity disorder (ADHD). Methylphenidate has two chiral centers and is marketed as a racemic mixture. It is known that D-*threo*-methylphenidate is pharmacologically more active than the L-*threo*-methylphenidate. This chiral stationary phase exhibited excellent performance with no separation deterioration observed after ~2500 injections (Ramos et al., 1999). As observed also for the cyclodextrin column (Weng et al., 2001), this kind of column stability is probably attributed to the polar organic mobile phase consisting of methanol and 0.05% ammonium trifluoroacetate. The plasma concentration-time profile for a child with ADHD after an oral administration of racemic methylphenidate indicates that the plasma levels of D-*threo*-methylphenidate were substantially higher than those of the L-*threo*-methylphenidate. One interesting finding of this study is that the stability of methylphenidate exhibits enantioselective differences. The D-enantiomer showed considerable degradation after storage at room temperature for 24 h, whereas the l-enantiomer appeared to be stable for >24 h. Chiral LC-MS/MS using Chirobiotic stationary phases were further explored for the analysis of eight model compounds including ritalinic acid, pindolol, fluoxetine, oxazepam, propranolol, terbutaline, metoprolol, and nicardipine (Bakhtiar and Tse, 2000).

Stereoselective analysis of labetalol in human plasma by LC-MS/MS using Chirobiotic V column and a mobile phase of methanol, acetic acid, and diethylamine (100:0.3:0.1%) was reported (Carvalho et al., 2009). Labetalol is a first-choice antihypertensive agent for pregnant women and is clinically available as a mixture of two racemates. The (*R*,*R*)-isomer is responsible mainly for all β1-antagonistic activity. The (*S*,*R*) isomer is highly selective for the α1 adrenoceptor, whereas the (*S*,*S*)-isomer is a less potent adrenoceptor antagonist and the (*R*,*S*)-isomer is inactive. The chromatographic resolutions of the enantiomers (*S*,*R* and *R*,*S* as well as *S*,*S* and *R*,*R*) were excellent while the resolution between *S*,*R* and *S*,*S* isomers was incomplete.

Simultaneous determination of warfarin enantiomers and 7-OH-warfarin enantiomers in human plasma on a Chirobiotic V column were reported (Zuo et al., 2010). Plasma samples were extracted with mixed-mode cation-exchange (MCX) cartridge. It should be noted that even though 7-OH warfarin enantiomers are chromatographically resolved from other hydroxyl metabolites such as 6-OH, 8-OH, 10-OH, and 4′-OH warfarin enantiomers, the major drawback of this method is the lack of separation between the warfarin and 7-OH-warfarin enantiomers.

A stereoselective method was described for the simultaneous determination of the S- and R-enantiomers of venlafaxin and its three metabolites in human plasma and whole blood samples (Kingbäck et al., 2010). Venlafaxine is used to treat psychiatric disorder and belongs to the class of dual serotonin and noradrenaline reuptake inhibitors. It is administered as a racemate. The *R*-enantiomer is a potent inhibitor of both serotonin and noradrenaline reuptake, while the *S*-enantiomer is more selective in inhibiting serotonin. Venlafaxin is metabolized mainly by the cytochrome P-450 system in the liver to its main metabolites by either or both *O*- and *N*-demethylation. Stereoselective metabolism was observed where CYP2D6 shows an appreciable stereoselectivity toward the *R*-enantiomer. The chiral separation was achieved by using a 250 × 2.1 mm Chirobiotic V column with a mobile phase of tetrahydrofuran and ammonium acetate (10 mM) pH 6.0 (10:90, v/v). A postcolumn infusion of 0.05% formic acid in acetonitrile was employed to increase the sensitivity of the method. A reversed-phase C8 solid-phase extraction (SPE) was used to extract the analytes of interest.

LC-MS/MS analysis of salbutamol (albuterol) and its 4-*O*-metabolite using Chirobiotic T column was reported (Joyce et al., 1998). Chirobiotic T, also a macrocyclic glycopeptide, exhibits complementary stereoselectivity to Chirobiotic V. Chiral analysis of albuterol enantiomers in dog plasma using online sample extraction and polar organic mode chiral LC-MS/MS on a Chirobiotic T column was reported (Wu et al., 2004). The polar mobile phase consists of methanol, 0.02% formic acid, and 0.1% ammonium formate. The two enantiomers were chromatographically separated with retention times of 5.1 and 5.6 min, respectively.

Chiral LC-MS/MS based on Chirobiotic T columns was also successfully used to analyzing pindolol enantiomers in human plasma. Pindolol is a nonselective β-adrenergic antagonist (β-blocker) for the treatment of cardiovascular diseases (Wang and Shen, 2006). S-Pindolol is considered more potent than *R*-Pindolol. To increase throughput, staggered sample injection was employed using a CTC Trio Valve system on a CTC HTS PAL autosampler.

Chirobiotic T column was also used in the simultaneous chiral LC-MS/MS analysis of prodrug bambuterol and terbutaline enantiomers in rat plasma (Luo et al., 2010) and molindone enantiomers in human plasma (Jiang et al., 2008b).

41.2.5 Protein Stationary Phase-Based Chiral LC-MS/MS Bioanalysis

This class of chiral stationary phase includes human and bovine serum albumin (HAS and BSA), α_1-acid glycoprotein (AGP), ovomucoid (OVM), cellobiohydrolase (CBH), avidin, and pepsin. Among these, AGP seems to be the most frequently applied chiral protein stationary phase. CBH is an effective complement to the AGP column for the separation of basic drugs of many compound types, while HSA can be used to separate acidic compounds directly. Though the separation mechanism is yet to be understood, its broad scope of applicability and aqueous–organic mobile phases, which are compatible directly with mass spectrometry, are its greatest advantages, and oftentimes it is used as one of the first choices for chiral separation. Enantioselectivity can be controlled or improved by changes in the mobile phase composition: pH, buffer, and organic modifier types and concentrations. However, protein columns are very expensive and they have limited chemical stabilities with restrictions in organic modifier content and temperature as well as typically inferior chromatographic efficiencies. For an AGP column, column temperature is a very important parameter to be optimized. Dramatic enantioselectivity change can be observed even by changing a few degrees of column temperature. Successful chiral separations of various analytes were achieved on protein chiral stationary phases, particularly on an AGP chiral column.

In 1999, Zhong and Chen published an enantioselective LC-MS/MS method for the determination of antiarrhythmic agent propafenone and its 5-OH metabolite in human plasma. They used a Chiral AGP column (150 × 4 mm, 5 μm from Chrom Tech) with a mobile phase of 10 mM aqueous ammonium acetate (pH 5.96) and 1-propanol (100:9) for the parent and 10 mM aqueous ammonium acetate (pH 4.0) and 2-propanol (100:9) for the metabolite (Zhong and Chen, 1999). The column was maintained at 20°C.

Racemate bupropion is commonly used for the treatment of depression and is extensively metabolized. When hydroxylated on the *N-t*-butyl carbon, bupropion rapidly undergoes ring closure, forming hydroxybupropion with two chiral centers. Only the *R,R*- and *S,S*-hydroxybupropion are formed due to the steric hindrance for the *R,S*- and *S,R*-hydroxybupropion. Unlike bupropion, which is rapidly racemized in plasma, hydroxybupropion has a much slower rate of racemization and therefore the plasma concentrations of the hydroxybupropion enantiomers accurately reflect the *in vivo* stereospecificity. Prevention of racemization during the sample preparation was achieved by acidifying the solution and keeping the samples stored at −20°C until analysis. A Chiral AGP column (100 × 2 mm, 5 μm) was used with a gradient elution containing 20 mM formate buffer and methanol at a flow rate of 0.22 ml/min (Coles and Kharasch, 2007). This method was validated and applied to analyzing samples from a subject. The predominant enantiomers in both plasma and urine are *R*-bupropion and *R,R*-hydroxybupropion.

A Chiral LC-MS/MS method using a Chiral AGP column (150 × 4 mm, 5 μm) was used to analyze the *R*- and *S*-enantiomers of the metabolites, 2- and 3-dechloroethylifosfamides, of the ifosfamide, which contains an asymmetrically substituted phosphorous atom and is a chemotherapeutical agent used to treat solid tumors (Aleksa et al., 2009). The mobile phase is a pH and ionic strength gradient elution using 10 mM aqueous ammonium acetate (pH 7) and 30 mM aqueous ammonium acetate (pH 4). While the separation for the pair of enantiomers for the 3-dechloroethylifosfamide is within 10 min (*R*-enantiomer elutes at 6 min and *S*-enantiomer at 8.3 min), there is only about 50–60% separation for the pair of enantiomers for the 2-dechloroethylifosfamide eluting between 5.5 min and 6.5 min.

Chiral LC-MS/MS using AGP column was also used for the quantitation of eszopiclone, a drug for the treatment of both transient and chronic insomnia, in human plasma (Meng et al., 2010). A computer software application of ACD Lab® was used to facilitate the development and optimization of the chiral separation. A Chiral AGP column (50 × 2 mm, 5 μm) was maintained at 30°C and a mobile phase of 10 mM aqueous ammonium acetate and methanol (85:15) was used at a flow rate of 0.5 ml/min. The baseline chiral separation was achieved within 3 min.

Bioanalytical LC-MS/MS methods using AGP column was also applied to analyze enantiomers of azelnidipine in human plasma (Kawabata et al., 2007), hyoscyamine in human serum and rabbit serum, fluoxetine and its metabolite norfluoxetine in ovine plasma (Chow et al., 2011), methadone and its metabolite 2-ethylidine-1,5-dimethyl-3,3-diphenylpyrrolidine in human serum (Etter et al., 2005), and reboxetine in rat plasma and brain (Turnpenny and Fraier, 2009).

More recently, CBH-based stationary phase is more frequently used. CBH has been widely used as a chiral selector for the separation of basic drugs, especially for β-blockers such as alprenolol (Jiang et al., 2008a). The key chiral recognition mechanism of CBH has been attributed to the electrostatic and hydrophobic interactions between CBH and the chiral analytes. A high-throughput, sensitive, and enantioselective LC-MS/MS-based bioanalytical method was developed and validated for the simultaneous determination of individual enantiomers of metoprolol and its metabolite

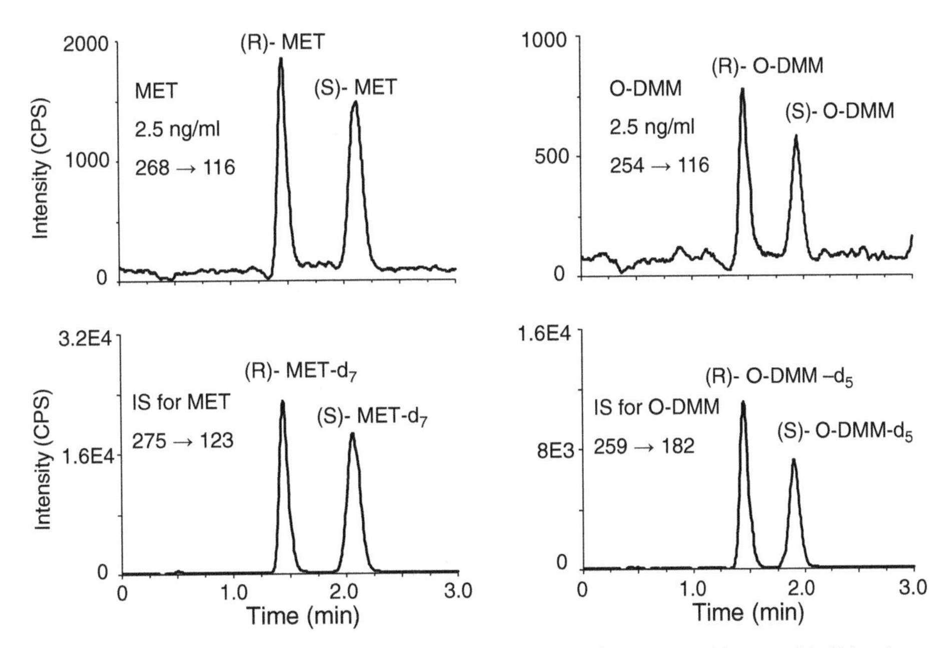

FIGURE 41.8 Representative LC-MS/MS chromatograms of an extracted human dried blood spots (DBS) whole blood sample with a mixture of MET and O-DMM enantiomers at 2.50 ng/ml (LLOQ) with internal standards of MET-d7 and O-DMM-d5.MET: Metoprolol; O-DMM: O-desmethyl metoprolol (adapted from Liang et al. (2010), with permission from Future Science Ltd).

O-desmethylmetoprolol (*O*-DMM) in human whole blood using the DBS sampling technique (Liang et al., 2010). Representative LC-MS/MS chromatograms of an extracted human dried blood spots (DBS) whole blood sample with a mixture of MET and *O*-DMM enantiomers are shown in Figure 41.8.

41.2.6 Polysaccharide Stationary Phase-Based Chiral LC-MS/MS Bioanalysis

Coated or immobilized polysaccharide chiral stationary phases are mostly operated in normal phase (alkane-alcohol) but can also be used in polar organic mode and reversed-phase mode. The normal phase mode tends to provide the best chiral separation for the polysaccharide CSPs, among which Cellulose tris(3,5-dimethylphenylcarbamate) such as Chiralcel OD and Amylose tris(3,5-dimethylphenylcarbamate) such as Chiralcel AD and Chiralpak IA are the most frequently used CSPs in this class. The chemical structures of cellulose and amylase are shown in Figure 41.9. Different groups can be attached to the cellulose and amylase backbones to form various stationary phases.

While they are enormously popular for the LC in conjunction with non-MS detections such as UV and fluorescence, their use in bioanalytical LC-MS/MS is more limited, probably due to the use of normal phase condition, which is not conducive for ionization. Postcolumn addition of polar solvents/salt is needed to achieve ionization of the analytes (Zavitsanos and Alebic-Kolbah, 1998). A chiral LC-MS/MS method on a Chiral OJ-H column operated under normal phase with postcolumn addition of ammonium hydroxide was used to measure unconjugated and total S-equol in human plasma and urine (Plomley et al., 2011). Tramadol, *O*-desmethyltramadol, and *N*-desmethyltramadol in human plasma were assayed on a Chiralpak AD column with the postcolumn addition of ethanol and 10 mM ammonium acetate (95:5) (De Moraes et al., 2012). Efforts on overcoming this drawback have been made. Cai et al. (2007) reported the use of atmospheric pressure photo-ionization (APPI) for improving sensitivity for normal-phase LC-MS chiral analysis of pharmaceuticals. Atmospheric pressure photo-ionization sensitivity was approximately 2–530 times higher than APCI. Unlike APCI and ESI, APPI has no potential explosion hazard, as APPI involves neither APCI corona needle discharge nor ESI high voltage discharge.

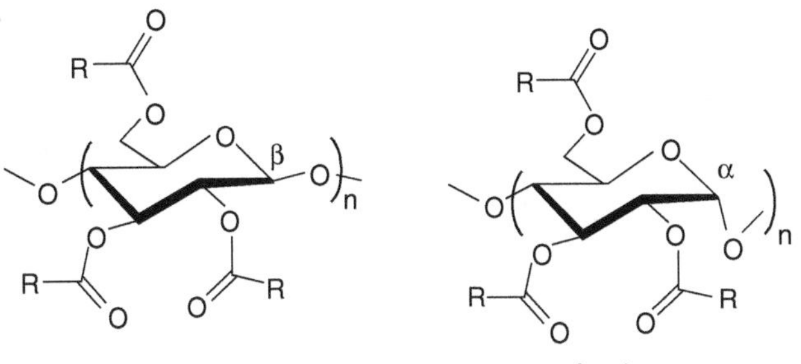

Cellulose Amylase

FIGURE 41.9 Chemical structures of cellulose and amylase.

A nonflammable solvent ethoxynonafluorobutane (ENFB) was used to replace flammable *n*-hexane or *n*-heptane in the mobile phase for a chiral LC-MS/MS method for the determination of intrinsic metabolic clearance in human liver microsomes (Zhang et al., 2008).

Chiralpak IA column was successfully used to separate enantiomers of hesperetin (Lévèques et al., 2012). Hesperetin is a flavonoid affecting vascular function and is abundant in citrus fruits. Hesperetin enantiomers have different transport and bioactivity. Biological matrices were incubated with β-glucuronidase/sulfatase, and hesperetin was isolated by SPE using a 96-well plate mixed-mode reversed-phase and anion-exchange SPE. Hesperetin enantiomers were analyzed using a Chiralpak IA-3 column using a mobile phase of water, formic acid, and acetonitrile. Baseline chiral separation was achieved within 7 min.

41.2.7 Two-Dimensional Enantioselective LC-MS/MS Bioanalysis

Chiral columns usually lack the resolution power to separate structurally similar compounds such as metabolites from the parent drug. For example, the Chiral AGP column showed an excellent separation of the two pairs of enantiomers of DU-124884, 5-HT_1-like receptor agonist, and its *N*-demethylated metabolite KC-9048, but one enantiomer of DU-124884 overlapped with one enantiomer of KC-9048 (Weng et al., 1996). Figure 41.8 also demonstrates the lack of separation between the enantiomers of metoprolol and the enantiomers of its metabolite O-desmethylmetoprolol. Lack of nonchiral separation power may lead to interference from the metabolites, if the metabolite can be converted to the parent in the source. Therefore, a combination of achiral and chiral stationary phases may be used to add the additional separation power. Two-dimensional chromatography also allows the possibility of reducing the total chromatographic run time by allowing the simultaneous of achiral and chiral chromatographic separation from two consecutive injections. The achiral column is always placed in front of the chiral column using either heart-cut column switch or direct front/or back elution. The achiral column also serves as the guard column to protect the expensive but more fragile chiral column. This also allows to trap the late eluting interference on the achiral column, subsequently being removed using back-flushing or gradient elution. Chiral columns are seldom used under gradient elution due to the slow re-equilibrium and limited compatibility with a wide range of mobile phases of different elution power. In order to use this achiral/chiral column switch technique, one would need a thorough understanding of both chiral and achiral chromatography.

A two-dimensional LC-MS/MS was used to profile warfarin and its hydroxyl metabolites in human plasma (Jones et al., 2011). Hydroxyl metabolites were better resolved from warfarin and from each other on a phenyl-based reversed-phase chromatography using an Acquity UPLC BEH Phenyl column (2.1 × 150 mm, 1.7 μm) operated at 60°C. The second column in tandem is an Astec Chirobiotic V column (2.1 × 150 mm, 5 μm) operated at room temperature. The mobile phase was methanol—water with 0.01% formic acid (55:45) at a flow rate of 0.3 ml/min. While the achiral resolution for the metabolites is excellent, the chiral separations for 8-OH warfarin and warfarin are compromised due to higher than ideal organic content in the mobile phase for the chiral resolution. The chiral separation would be much better at 20% methanol in the mobile phase but the chromatographic retention on the phenyl column would be too long.

Online SPE with LC-MS/MS method was used for the determination of enantiomers of warfarin, phenprocoumon, and acenocoumarol in human plasma (Vecchione et al., 2007). The first-dimensional column (online SPE) is a Perfusion column Poros R2 20 (2 × 30 mm) and the chiral separation was achieved on a Chira-Grom-2 column (1 × 250 mm, 8 μm) maintained at 40°C. In the online SPE, the injected sample was flushed through the perfusion Poros column with a high flow rate (1.9 ml/min) of 0.5% aqueous formic acid. The mobile phase for the chiral column is acetonitrile–methanol–formic acid (33:33:0.4) at 0.1 ml/min. With the valve switching, the analytes trapped by the Poros column are eluted to the chiral column for the enantioselective separation. Excellent chiral separation was achieved for all three analytes within 10 min.

Sertraline is a selective serotonin reuptake inhibitor (SSRI) and is used to treat major depression. Achiral–chiral column switching LC-MS/MS was developed for the determination of sertraline enantiomers in rat plasma (Rao et al., 2010). A Hisep RAM column (50 × 4.0 mm, 5 μm) and a mobile phase, containing 0.02 M aqueous ammonium acetate (pH 8)–acetonitrile (86:14) at a flow rate of 1 ml/min was used to trap the analytes and remove the proteins. The RAM column has the hydrophilic outer and hydrophobic inner phases that exclude large molecules such as proteins in the void volume while retaining selectively the small hydrophobic analytes. At 2.5 min, the analyte was eluted from RAM to a CYCLOBOND I 2000 DM (250 × 4.6 mm, 5 μm) (derivatized β-cyclodextrin) with a mobile phase of 0.1% aqueous trifluoroacetic acid–acetonitrile (86:14) at a flow rate of 0.8 ml/min. Chromatographic baseline resolution of the two pairs of enantiomers was achieved with a retention time of about 14.0 min for 1*S*,4*S*, 14.8 min for 1*R*,4*R*, 17.2 min for 1*S*,4*R*, and 20.1 min for 1*R*,4*S*.

A similar column-switching setup was also used by another group to analyze pantoprazole and lansoprazole enantiomers, respectively (Gomes et al., 2010; Barreiro et al., 2011). A bovine serum albumin octyl column (RAM-BAS C8) was used in the first dimension for the extraction, while a noncommercial polysaccharide-based chiral column was used in the second dimension for the chiral separation.

41.2.8 Chiral Supercritical Fluid Chromatography–MS/MS for Bioanalysis

More recently, SFC on packed chiral column in conjunction with mass spectrometry has also become a valuable tool for quantitative chiral bioanalysis. It employs carbon dioxide above or near its critical temperature of 31°C and pressure of 73 bars, combined with an organic modifier such as methanol or ethanol (Taylor, 2008). SFC–MS has been successfully used for chiral bioanalysis of propranolol (Chen et al., 2006), warfarin (Coe et al., 2006), and ketoprofen (Hoke et al., 2000). The addition of a significant amount of a viscosity-lowering agent such as carbon dioxide to the mobile phase provides a number of advantages for chiral separation including the fast analysis time. Higher flow rates can be attained without a concomitant loss in chromatographic efficiency, resulting in typically a several-fold reduction in analysis time.

41.3 CURRENT CONSIDERATIONS AND PROTOCOLS IN DEVELOPING A ROBUST CHIRAL LC-MS/MS BIOANALYTICAL METHOD

41.3.1 Define Project Purpose

The first step of developing a robust chiral LC-MS/MS bioanalytical method is to fully understand what is needed from the project perspective. In many cases, a qualified assay is sufficient to confirm there is lack of *in vivo* interconversion of the single metabolite enantiomer or to confirm the formed metabolite is a racemate or a single enantiomer. This type of exploratory work may lead to a decision whether or not a fully validated chiral LC-MS/MS bioanalytical method is needed. In this type of exploratory work, it may be sufficient to separate the enantiomers under the nonoptimal condition, that is, long chromatographic elution etc. As long as the enantiomers are chromatographically resolved and some level of confidence on the lack of interconversion of enantiomers in biological fluids is established, the method can be then qualified with the "fit-for-purpose" approach to address specific questions by analyzing selected and usually limited numbers of samples. On the other hand, if the chiral LC-MS/MS bioanalytical method is going to be used in pivotal studies to address safety concerns or the single enantiomer is administered with a known polymorphism *in vivo* interconversion of enantiomer, a robust and fully validated chiral LC-MS/MS bioanalytical method must be established to meet the challenges of regulatory scrutinies. Since the assays will be repeatedly used for analyzing thousands of samples from multiple studies and possibly at multiple analytical sites, these methods should be vigorously tested and should be simple to use. Additionally, throughput may be also critical to support fast turnaround studies and therefore a long chromatographic run may be less desirable.

41.3.2 Establish Chiral Chromatographic Separation

A unique challenge for developing any chiral LC-MS/MS bioanalytical method is the chromatographic separation of the enantiomers. Chiral stationary phases are expensive and therefore it is impossible to stock many different types of commercially available chiral columns. There are many literature methods published and one should always first try to find whether there are available literature methods for the analyte of interest or for analytes with similar structures. However, it should be made aware that it is not always straightforward to transfer a literature method, even for the same analyte. It would be even harder to predict whether modifying a method for a structurally similar analyte will be successful. There are so many parameters that can dramatically influence the chiral separation. Besides, not all chiral methods published in literature are suitable for bioanalysis, for which one has to consider the compatibility of the method with the sample preparation, ionization, and MS detection. Reviewing all chiral bioanalytical LC-MS/MS methods published in the last 10 years reveal that almost all of the quantitative bioanalytical chiral assays were performed on the following chiral stationary phases: Chirobiotic V or T, β-Cyclodextrin, Chiral AGP, Chiralcel OD, Chiralcel AD, or Chiralpak-IA. The most successful operation mode is polar organic mobile phase for Chirobiotic V or T and β-Cyclodextrin, reversed-phase mobile phase for Chiral AGP, and normal-phase mobile phase for the last stationary phases belonging to polysaccharide chiral stationary phases. The use of normal phase is a drawback for LC-MS/MS but there are mitigations that can be used to overcome this issue as already discussed previously. Chiral SFC–MS/MS is an area that is underdeveloped but has shown a great potential for quantitative bioanalytical LC-MS/MS, particularly for less polar chiral analytes. Chiral derivatization can always be used as the last resort but should be used with caution against racemization. A typical protocol for chiral separation is as follows.

41.3.2.1 Determine the Required Sensitivity for the Assay

This may dictate which type of chiral stationary phase (if separation can be achieved on multiple types of chiral stationary phases) one may use. Our experiences show the sensitivity in the following order (decreasing): polar-organic mobile phases (i.e., mixture of acetonitrile and methanol with some volatile organic acid or base as modifiers) such as those used for Chirobiotic V or T and β-Cyclodextrin; reversed-phase mobile phases (mixture of acetonitrile, water, and small amounts of volatile acid, base, or buffer), typically used for protein-based chiral columns or polysaccharide-based chiral columns; normal-phase mobile phases (mixture of hexane, ethanol, or isopropanol with or without volatile acid or base) typically used for polysaccharide-based chiral columns. If the separation of the enantiomers can only be achieved on one type of chiral column but the sensitivity is inadequate, other approaches, such as postcolumn addition of modifiers or derivatization to enhance sensitivity, can be explored. However, one needs to be aware that these additional steps may also introduce potential irreproducibility, which may lead to poor method ruggedness.

41.3.2.2 Separation of the Enantiomers of the Analyte of Interest This may be the single most difficult challenge in developing a robust chiral bioanalytical LC-MS/MS method. Ideally, the enantiomers should be baseline resolved chromatographically with narrow, sharp, and symmetrical peaks. This is very important, especially when the smaller enantiomer elutes after the larger one. If the peak is severely tailed, the smaller enantiomer may be buried in the larger enantiomer peak if the separation is inadequate. The chiral resolution achieved using a racemate represents the best scenario and could be misleading. If the enantiomer with disproportionately low concentration (minor peak) elutes on the tail of the isomer of much higher concentration (major peak), integration of the minor peak can be quite challenging. In this case, one may use the chiral column of the chiral selector with the opposite optical rotation or a different chiral column to achieve either, a complete resolution of the two enantiomers even at disproportionate concentrations, or elution of the minor peak in front of the major peak.

On the other hand, chiral separation can be sometimes too good, that is, the elution of the two enantiomers can be several minutes apart. This can also lead to not only a long run time but also to a less robust method, especially when stable-labeled enantiomeric internal standards are unavailable. The late eluting enantiomer may also suffer poorer sensitivity because of the usually broader peak shape.

The other challenge for chiral separation, as already mentioned before, is the separation of other related substances such as metabolites. In comparison with typical achiral columns, chiral stationary phases typically have less resolution powers on the separation of related substances. Coelution of several compounds together may lead to ionization suppression and quantitation bias if this is not compensated by using a stable-labeled internal standard. Effort should be made to resolve them if possible. Sometimes, a tandem addition of an achiral column that can be operated under a similar chromatographic condition may boost the achiral resolution. As previously mentioned, the achiral column, when placed in front of the chiral column, can also serve as a protector for the expensive chiral column and extend its usage life.

To support routine sample analysis, the chiral column should also demonstrate sufficient stability and lot-to-lot reproducibility.

41.3.3 Selection of Internal Standard

An ideal internal standard will track the analyte of interest through the entire sample extraction, chromatography, and mass spectrometry. A stable-labeled internal standard is an ideal candidate. The C_{13} label is better than the H_2 label as the latter may be chromatographically resolved from the analyte. The rule of thumb is that for every 100 Dalton, one additional label should be placed. For example, for an analyte with 300 Dalton, ideally three heavy labels (H_2 or C_{13} or a combination) should be used to make the stable-labeled internal standard. Of course, if the product ion also carries the heavy labels, one can use even fewer labels. Since the enantiomers are resolved chromatographically, stable-labeled internal standard of the single enantiomer will have a different retention from the other enantiomer and may not track it very well. Therefore, one needs stable-labeled internal standards for both enantiomers and oftentimes this can be done by using the stable-labeled internal standard for the racemate.

41.3.4 Sample Extraction

Extracting analytes from biological samples is not unique to chiral bioanalytical LC-MS methods. The extraction itself is nonchiral and both enantiomers should be equally extracted by using typical extraction protocols such as protein precipitation, liquid/liquid extraction with water-immiscible organic solvents, or SPE. One unique feature for the extraction of enantiomers is that for very highly protein-bound enantiomers, distribution of the protein binding prior to extraction is needed to ensure reproducible extraction for both enantiomers since the protein binding can be enantiomer specific, meaning that both enantiomers are not equally bound to the protein. Typical extraction usually does not cause interconversion of the enantiomers but this may need to be tested out during the method development. Of course, one still needs to minimize any potential significant loss of analyte during the extraction due to adsorption, degradation, or evaporation. Matching of the injection solvent with the chromatographic condition is another important consideration.

41.4 CURRENT METHOD VALIDATION CONSIDERATIONS AND PROTOCOLS FOR CHIRAL BIOANALYTICAL LC-MS/MS

41.4.1 Reference Standards

Detailed guidelines for bioanalytical analysis of enantiomers are not available but a few issues on chiral bioanalytical LC-MS/MS validation should be addressed with the emphasis on establishment of stereospecificity of the method.

During the early stages of drug development, pure enantiomers may not be as easily available as the racemic compound. The purity of the racemate is usually better defined than the enantiomers. For an LC-MS/MS method, if enantiomers are sufficiently resolved to allow the quantitation of small amounts of one enantiomer in the presence of its isomer at a disproportionate concentration, it is of advantage to validate part of the method parameters. They include accuracy, precision, sensitivity, linearity, dilution, carry-over, recovery, and matrix effect using the calibration standards and quality control samples prepared from the racemates. On the other hand, in order to investigate the potential racemization of the enantiomers during sample storage and preparation, additional QCs should also be prepared from the single enantiomers for any stability experiments (storage, freeze/thaw, bench-top, reinjection, and autosampler stabilities).

If chiral derivatization is employed, the use of a single pure enantiomer to verify the stereospecificity becomes even more important as many derivatization reactions occur at an elevated temperature or at an extreme pH. Enantiomers of high enantiomeric purity are desirable for these experiments but their absolute potencies may be less important. If a single enantiomer is unavailable, one should try to isolate a small quantity by using chiral chromatography, preferably under normal phase condition for the ease of removing eluents (Weng et al., 1994). Those purified enantiomers can then be fortified with the biological matrix for stability experiments.

41.4.2 Stability

For enantiomers, stability should be established for both the lack of racemization of the enantiomers and the lack of degradation. If the exact potencies other than the enantiomeric purity of single enantiomers are not defined, the theoretical concentrations of single enantiomer QCs are not calculable. This is not an issue since the purpose of the experiments is to determine whether there is any racemization and degradation, which can be both stereoselective and nonstereoselective, compared with the initial control. The observed concentrations were calculated using the calibration standards on the condition that the normal racemate QCs and calibration standards pass the acceptance criteria. Instead of monitoring the decrease of the fortified enantiomer concentration, it is much easier to follow the racemization by monitoring the formation of the opposite enantiomer. The absence of significant increase of the opposite enantiomer indicates that there is no conversion. In the absence of knowledge on the enantiomer ratios in the incurred samples, this experiment needs to be performed separately using each enantiomer:

$$\text{Racemization}\,(\%) = 100 \times [(C_{\text{opposite enantiomer}}/C_{\text{fortified enantiomer}})_{\text{final}} - (C_{\text{opposite enantiomer}}/C_{\text{fortified enantiomer}})_{\text{initial}}],$$

$$\text{Degradation}\,(\%) = 100 - 100 \times [(C_{\text{fortified enantiomer}})_{\text{initial}} - (C_{\text{fortified enantiomer}})_{\text{final}}]/(C_{\text{fortified enantiomer}})_{\text{initial}}.$$

TABLE 41.2 Examples of Racemization and Degradation

	A: QC (ng/ml)		B: QC (ng/ml)		C: QC (ng/ml)	
	(+)	(−)	(+)	(−)	(+)	(−)
Initial	122	2.2	122	2.2	122	2.2
After 43 days	120	1.3	98	2.0	107	15.9
Racemization (%)		0		0		13.1
Degradation (%)	1.8		20		12.3	

Three different cases are discussed (Table 41.2). Case A indicates that both racemization and degradation are minimal after 43 days of storage in a −20°C freezer. For Case B, while the racemization is minimal, the level of degradation is too high to be acceptable. For Case C, there is some level of degradation but the racemization is quite significant. If the (−) enantiomer is about 1/10 of the (+) enantiomer in the incurred samples, 13.1% racemization from (+) to (−) will cause >100% overestimation.

41.4.3 Accuracy, Precision, Linearity, and Sensitivity

Accuracy, precision, linearity, and sensitivity for the chiral LC-MS method should be established in a similar fashion as the achiral assay. Quality control samples should be prepared in mimicking the incurred samples. If a single enantiomer was dosed and inversion of enantiomer occurs *in vivo*, one may need to prepare quality control samples in the lopsided approach (different ratio of two enantiomers) to mimic the concentrations of the two enantiomers in incurred samples.

41.4.4 Selectivity, Specificity, Recovery, and Matrix Effect

Method selectivity, specificity, recovery, and matrix effect should be established for both enantiomers. While a different recovery is not expected for enantiomers, the selectivity, specificity, and matrix effect can very much depend upon the individual enantiomer since enantiomers are chromatographically resolved.

41.4.5 Incurred Sample Reanalysis

Incurred sample reanalysis should be performed in the same fashion as for the achiral assays. One should select sufficient quantities of incurred samples that can be representative of incurred samplesthat may have different ratios of the two enantiomers.

41.5 CONCLUSION

Quantitative bioanalytical chiral LC-MS/MS methods are essential for drug development. Direct chiral separation using chiral stationary phases is the preferred approach but chiral derivatization can also be a valuable tool. Two-dimensional achiral–chiral combination can be used to gain additional separation power. Supercritical fluid chromatography in conjunction with MS should be further explored since it provides fast chiral separation. For the method development and validation for robust chiral bioanalytical methods, potential racemization during storage, extraction, and chromatography must be tested and excluded.

REFERENCES

Aleksa K, Nava-Ocampo A, Koren G. Detection and quantification of (R) and (S)-dechloroethylifosfamide metabolites in plasma from children by enantioselective LC/MS/MS. Chirality 2009;21:674–680.

Armstrong DW, Jin HL. Liquid chromatographic separation of anomeric forms of saccharides with cyclodextrin bonded phases. Chirality 1989;1(1):27–37.

Armstrong DW, Rundlett K, Reid III GL. Use of a macrocyclic antibiotic, rifamycin B, and indirect detection for the resolution of racemic amino alcohols by CE. Anal Chem 1994;66:1690–1695.

Armstrong DW, Zhang B. Chiral stationary phases for HPLC. Anal Chem 2001;73:557A–561A.

Bakhtiar R, Tse FL. High-throughput chiral liquid chromatography/tandem mass spectrometry. Rapid Commun Mass Spectrom 2000;14:1128–1135.

Barreiro JC, Vanzolini KL, Cass QB. Direct injection of native aqueous matrices by achiral–chiral chromatography ion trap mass spectrometry for simultaneous quantification of pantoprazole and lansoprazole enantiomers fractions. J Chromatogr A 2011;1218:2865–2870.

Beesley TE, Lee J-T. Method development strategy and applications update for CHIROBIOTIC chiral stationary phases. J Liq Chromatogr Related Technol 2009;32:1733–1767.

Bhushan R. Enantiomeric purity of chiral derivatization reagents for enantioresolution. Bioanalysis 2011;3(18):2057–2060.

Cai S-S, Hanold KA, Syage JA. Comparison of atmospheric pressure photoionization and atmospheric pressure chemical ionization for normal-phase LC/MS chiral analysis of pharmaceuticals. Anal Chem 2007;79:2491–2498.

Caldwell J. Importance of stereospecific bioanalytical monitoring in drug development. J Chromatogr A 1996;719:3–13.

Carvalho T, Cavalli R, Marques M, Da Cunha S, Baraldi C, Lanchote V. Stereoselective analysis of labetalol in human plasma by LC-MS/MS: application to pharmacokinetics. Chirality 2009;21:738–744.

Chen J, Korfmacher WA, Hsieh Y. Chiral liquid chromatography-tandem mass spectrometric methods for stereoisomeric pharmaceutical determinations. J Chromatogr B 2005;820:1–8.

Chen J, Hsieh Y, Cook J, Morrison R, Korfmacher WA. Supercritical fluid chromatography-tandem mass spectrometry for the enantioselective determination of propranolol and pindolol in mouse blood by serial sampling. Anal Chem 2006;78:1212–1217.

Chow TW, Szeitz A, Rurak DW, Riggs KW. A validated enantioselective assay for the simultaneous quantitation of (R)-, (S)-fluoxetine and (R)-, (S)-norfluoxetine in ovine plasma using liquid chromatography with tandem mass spectrometry (LC/MS/MS). J Chromatogr B 2011;879:349–358.

Coe RA, Rathe JO, Lee JW. Supercritical fluid chromatography-tandem mass spectrometry for fast bioanalysis of R/S-warfarin in human plasma. J Pharm Biomed Anal 2006;42:573–580.

Coles R, Kharasch ED. Stereoselective analysis of bupropion and hydroxybupropion in human plasma and urine by LC/MS/MS. J Chromatogr B 2007;857:67–75.

Committee for Proprietary Medical Products. Working parties on quality, safety and efficacy of medical products. Note for guidance: investigation of chiral active substances 1993; III/3501/91.

Crowther JB, Covey TR, Dewey EA, Henion JD. Liquid chromatographic/mass spectrometric determination of optically active drugs. Anal Chem 1984;56:2921–2926.

De Moraes MV, Lauretti GR, Napolitano MN, Santos NR, Godoy ALPC, Lanchote VL. Enantioselective analysis of unbound tramadol, O-desmethyltramadol and N-desmethyltramadol in plasma by ultrafiltration and LC-MS/MS: application to clinical pharmacokinetics. J Chromatogr B 2012;880:140–147.

Erny GL, Cifuentes A. Liquid separation techniques coupled with mass spectrometry for chiral analysis of pharmaceuticals compounds and their metabolites in biological fluids. J Pharm Biomed Anal 2006;40:509–515.

Etter ML, George S, Graybiel K, Eichhorst J, Lehotay DC. Determination of free and protein-bound methadone and its major metabolite EDDP: enantiomeric separation and quantitation by LC/MS/MS. Clin Biochem 2005;38:1095–1102.

Fried K, Wainer IW. Column-switching techniques in the biomedical analysis of stereoisomeric drugs: why, how and when. J Chromatogr B 1997;689:91–104.

Gomes RF, Cassiano NM, Pedrazzoli Jr J, Cass QB. Two-dimensional chromatography method applied to the enantiomeric determination of lansoprazole in human plasma by direct sample injection. Chirality 2010;22:35–41.

Guillarme D, Bonvin G, Badoud F, Schappler J, Rudaz S, Veuthey J. Fast chiral separation of drugs using columns packed with sub-2 microm particles and ultra-high pressure. Chirality 2010;22:320–330.

Haginaka J. Recent progress in protein-based chiral stationary phases for enantioseparations in liquid chromatography. J Chromatogr B 2008;875:12–19.

Henion JD, Maylin GA. Drug analysis by direct liquid introduction micro liquid chromatography mass spectrometry. Biomed Mass Spectrom 1980;7:115–121.

Hoke II SH, Pinkston JD, Bailey RE, Tanguay SL, Eichhold TH. Comparison of packed-column supercritical fluid chromatography–tandem mass spectrometry with liquid chromatography–tandem mass spectrometry for bioanalytical determination of (R)- and (S)-ketoprofen in human plasma following automated 96-well solid-phase extraction. Anal Chem 2000;72:4235–4241.

Ikai T, Yamamoto C, Kamigaito M, Okamoto Y. Immobilized-type chiral packing materials for HPLC based on polysaccharide derivatives. J Chromatogr B 2008;875:2–11.

Ilisz I, Berkecz R, Péter A. Application of chiral derivatizing agents in the high-performance chromatographic separation of amino acid enantiomers: a review. J Pharm Biomed Anal 2008;47: 1–15.

Jiang H, Jiang X, Ji QC. Enantioselective determination of alprenolol in human plasma by liquid chromatography with tandem mass spectrometry using cellobiohydrolase chiral stationary phases. J Chromatogr B 2008a;872:121–127.

Jiang H, Li Y, Pelzer M et al. Determination of molindone enantiomers in human plasma by high-performance liquid chromatography-tandem mass spectrometry using macrocyclic antibiotic chiral stationary phases. J Chromatogr A 2008b;1192:230–238.

John H, Eyer F, Zilker T, Thiermann H. High-performance liquid-chromatographic tandem-mass spectrometric methods for atropinesterase-mediated enantioselective and chiral determination of R- and S-hyoscyamine in plasma. Anal Chim Acta 2010;680:32–40.

Jones DR, Boysen G, Miller GR. Novel multi-mode ultra performance liquid chromatography-tandem mass spectrometry assay for profiling enantiomeric hydroxywarfarins and warfarin in human plasma. J Chromatogr B 2011;879:1056–1062.

Joyce KB, Jones AE, Scott RJ, Biddlecombe RA, Pleasance S. Determination of the enantiomers of salbutamol and its 4-O-sulphate metabolites in biological matrices by chiral liquid chromatography tandem mass spectrometry. Rapid Commun Mass Spectrom 1998;12:1899–1910.

Kawabata K, Samata N, Urasaki Y et al. Enantioselective determination of azelnidipine in human plasma using liquid chromatography-tandem mass spectrometry. J Chromatogr B 2007;852:389–397.

Kingbäck M, Josefsson M, Karlsson L et al. Stereoselective determination of venlafaxine and its three demethylated metabolites in human plasma and whole blood by liquid chromatography with electrospray tandem mass spectrometric detection and solid phase extraction. J Pharm Biomed Anal 2010;53:583–590.

Lämmerhofer M. Chiral recognition by enantioselective liquid chromatography: mechanisms and modern chiral stationary phases. J Chromatogr A 2010;1217:814–856.

Lévèques A, Actis-Goretta L, Rein MJ, Williamson G, Dionisi F, Giuffrida F. UPLC-MS/MS quantification of total hesperetin and hesperetin enantiomers in biological matrices. J Pharm Biomed Anal 2012;57:1–6.

Liang X, Jiang Y, Chen X. Human DBS sampling with LC-MS/MS for enantioselective determination of metoprolol and its metabolite O-desmethyl metoprolol. Bioanalysis 2010;2:1437–1448.

Lin JH, Lu AY. Role of pharmacokinetics and metabolism in drug discovery and development. Pharmacol Rev 1997;49:403–449.

Liu K, Zhong D, Chen X. Enantioselective quantification of chiral drugs in human plasma with LC-MS/MS. Bioanalysis 2009;1:561–576.

Lu H. Stereoselectivity in drug metabolism. Expert Opin Drug Metab Toxicol 2007;3:149–158.

Luo W, Zhu L, Deng J et al. Simultaneous analysis of bambuterol and its active metabolite terbutaline enantiomers in rat plasma by chiral liquid chromatography-tandem mass spectrometry. J Pharm Biomed Anal 2010;52:227–231.

Meng M, Rohde L, Čápka V, Carter SJ, Bennett PK. Fast chiral chromatographic method development and validation for the quantitation of eszopiclone in human plasma using LC/MS/MS. J Pharm Biomed Anal 2010;53:973–982.

Millot MC. Separation of drug enantiomers by liquid chromatography and capillary electrophoresis, using immobilized proteins as chiral selectors. J Chromtogr B 2003;797:131–159.

Mišl'anová C, Hutta M. Role of biological matrices during the analysis of chiral drugs by liquid chromatography. J Chromatogr B 2003;797:91–109.

Plomley JB, Jackson RL, Schwen RJ, Greiwe JS. Development of chiral liquid chromatography-tandem mass spectrometry isotope dilution methods for the determination of unconjugated and total S-equol in human plasma and urine. J Pharm Biomed Anal 2011;55:125–134.

Policy Statement for the Development of New Stereoisomeric Drugs, Food and Drug Administration (1992) 57 Fed Reg 22 249.

Ramos L, Bakhtiar R, Majumdar T, Hayes M, Tse F. Liquid chromatography/atmospheric pressure chemical ionization tandem mass spectrometry enantiomeric separation of dl-threo-methylphenidate, (Ritalin) using a macrocyclic antibiotic as the chiral selector. Rapid Commun Mass Spectrom 1999;13:2054–2062.

Rao RN, Kumar KN, Shinde DD. Determination of rat plasma levels of sertraline enantiomers using direct injection with achiral-chiral column switching by LC-ESI/MS/MS. J Pharm Biomed Anal 2010;52:398–405.

Rentsch KM. The importance of stereoselective determination of drugs in the clinical laboratory. J Biochem Biophys Methods 2002;54:1–9.

Schmid MG, Gübitz G. Enantioseparation by chromatographic and electromigration techniques using ligand-exchange as chiral separation principle. Anal Bioanal Chem 2011;400:2305–2316.

Srinivas NR. Drug disposition of chiral and achiral drug substrates metabolized by cytochrome P450 2D6 isozyme: case studies, analytical perspectives and developmental implications. Biomed Chromatogr 2006;20:466–491.

Srinivas NR, Barbhaiya RG, Midha KK. Enantiomeric drug development: issues, considerations, and regulatory requirements. J Pharm Sci 2001;90:1205–1215.

Sun XX, Sun LZ, Aboul-Enein HY. Chiral derivatization reagents for drug enantioseparation by high-performance liquid chromatography based upon pre-column derivatization and formation of diastereomers: enantioselectivity and related structure. Biomed Chromatogr 2001;15:116–132.

Taylor LT. Supercritical fluid chromatography. Anal Chem 2008;80:4285–4294.

Toyo'oka T. Resolution of chiral drugs by liquid chromatography based upon diastereomer formation with chiral derivatization reagents. J Biochem Biophys Methods 2002;54:25–56.

Turnpenny P, Fraier D. Sensitive quantitation of reboxetine enantiomers in rat plasma and brain, using an optimised reverse phase chiral LC-MS/MS method. J Pharm Biomed Anal 2009;49:133–139.

Vecchione G, Casetta B, Tomaiuolo M, Grandone E, Margaglione M. A rapid method for the quantification of the enantiomers of Warfarin, Phenprocoumon and Acenocoumarol by two-dimensional-enantioselective liquid chromatography/electrospray tandem mass spectrometry. J Chromatogr B 2007;850:507–514.

Wang H, Shen Z. Enantiomeric separation and quantification of pindolol in human plasma by chiral liquid chromatography/tandem mass spectrometry using staggered injection with a CTC Trio Valve system. Rapid Commun Mass Spectrom 2006;20:291–297.

Wang S, Cyronak M, Yang E. Does a stable isotopically labeled internal standard always correct analyte response? A matrix effect study on a LC/MS/MS method for the determination of carvedilol enantiomers in human plasma. J Pharm Biomed Anal 2007;43:701–707.

Wang C, Jiang C, Armstrong DW. Considerations on HILIC and polar organic solvent-based separations: use of cyclodextrin and macrocyclic glycopetide stationary phases. J Sep Sci 2008a;31:1980–1990.

Wang Z, Ouyang J, Baeyens WRG. Recent development of enantioseparation techniques for adrenergic drugs using liquid chromatography and capillary electrophoresis: a review. J Chromatogr B 2008b;862:1–14.

Wang H, Ma C, Zhou J, Liu XQ. Stereoselective analysis of tiopronin enantiomers in rat plasma using high-performance liquid chromatography-electrospray ionization mass spectrometry after chiral derivatization. Chirality 2009;21:531–538.

Weng N, Lee JW. Development and validation of a high-performance liquid chromatographic method for the quantitation of warfarin enantiomers in human plasma. J Pharm Biomed Anal 1993;11:785–792.

Weng N, Lee JW, Hulse JD. Development and validation of a chiral HPLC method for the quantitation of methocarbamol enantiomers in human plasma. J Liq Chromatogr Related Technol 1994;17:3747–3758.

Weng N, Pullen RH, Arrendale RF, Brennan JJ, Hulse JD, Lee JW. Stereospecific determinations of (+/−)-DU-124884 and its metabolites (+/−)-KC-9048 in human plasma by liquid chromatography. J Pharm Biomed Anal 1996;14:325–337.

Weng N, Ring PR, Midtlien C, Jiang X. Development and validation of a sensitive and robust LC-tandem MS method for the analysis of warfarin enantiomers in human plasma. J Pharm Biomed Anal 2001;25:219–226.

Wu ST, Xing J, Apedo A, Wang-Iverson DB, Olah TV, Tymiak AA, Zhao N. High-throughput chiral analysis of albuterol enantiomers in dog plasma using on-line sample extraction/polar organic mode chiral liquid chromatography with tandem mass spectrometric detection. Rapid Commun Mass Spectrom 2004;18:2531–2536.

Yang E, Wang S, Kratz J, Cyronak M. Stereoselective analysis of carvedilol in human plasma using HPLC/MS/MS after chiral derivatization. J Pharm Biomed Anal 2004;36:609–615.

Zavitsanos AP, Alebic-Kolbah T. Enantioselective determination of terazosin in human plasma by normal phase high-performance liquid chromatography-electrospray mass spectrometry. J Chromatogr A 1998;794:45–56.

Zhang Y, Caporuscio C, Dai J et al. Development and implementation of a stereoselective normal-phase liquid chromatography-tandem mass spectrometry method for the determination of intrinsic metabolic clearance in human liver microsomes. J Chromatogr B 2008;875:154–160.

Zhong D, Chen X. Enantioselective determination of propafenone and its metabolites in human plasma by liquid chromatography-mass spectrometry. J Chromatogr B 1999;721:67–75.

Zuo Z, Wo SK, Lo CMY, Zhou L, Cheng G, You JHS. Simultaneous measurement of S-warfarin, R-warfarin, S-7-hydroxywarfarin and R-7-hydroxywarfarin in human plasma by liquid chromatography-tandem mass spectrometry. J Pharm Biomed Anal 2010;52:305–310.

42

LC-MS BIOANALYSIS OF PEPTIDES AND POLYPEPTIDES

HONGYAN LI AND CHRISTOPHER A. JAMES

42.1 INTRODUCTION

Peptides are biopolymers of amino acids. By convention, they are distinguished from proteins on the basis of size, typically containing less than 100 amino acid residues (Lien and Lowman, 2003). They often have a high degree of secondary structure but lack tertiary structure.

The development of analytical methods for the quantitative analysis of peptides in biological matrices has become increasingly important due to the growing interest in novel peptide therapeutics and peptide biomarkers in biopharmaceutical research. Bioactive peptides from natural sources can help to identify novel pharmacophores and can be used as templates for synthetic peptide therapeutics (Sato et al., 2006; Miranda et al., 2008a, 2008b). Endogenous peptides are of interest as biomarkers and can be used as diagnostic tools for disease progression and to aid in the development of therapeutic treatments (De Kock et al., 2001; Dalluge, 2002; Shushan, 2010; Van Den Broek et al., 2010). Peptide analysis is also becoming important to measure enzymatically generated signature peptides, such as tryptic peptides as quantitative surrogates for peptide/protein bioanalysis (Halquist and Karnes, 2010).

Liquid chromatography–mass spectrometry detection (LC-MS, LC-MS/MS) has evolved as the method of choice for peptide bioanalysis over the past decade (John et al., 2004; Van Den Broek et al., 2008; Cutillas, 2010). It combines chromatographic separation in the liquid phase and the mass spectrometric separation in the gas phase, which offers unparalleled assay specificity for targeted peptide quantitative analysis of target peptides. LC-MS/MS has been widely used for small molecule bioanalysis. The well-established methodologies and the advantages in specificity, sensitivity, wide dynamic range, accuracy, and precision for small molecule bioanalysis have generally been found for LC-MS/MS bioanalysis of peptides, but with some adaption of approaches to meet the unique challenges due to the analytical characteristics of peptides, and the complexity of the biomatrices associated with peptide analytes. One of the significant advantages of an LC-MS/MS approach over the ligand binding assays is that it does not require specific antibody reagents to achieve adequate selectivity to distinguish structurally similar peptides. Therefore, LC-MS/MS assay development is much faster, and the assay results are definitive and comparable across different research settings. However, methods using peptide-specific antibodies for immunoaffinity extraction prior to LC-MS/MS analysis may have additional advantages when low pg/ml quantitation limits are required for some endogenous bioactive peptides or peptide biomarkers (Berna et al., 2006, 2008).

With the fast growing number of research papers and reviews on LC-MS and LC-MS/MS bioanalysis for peptides, this chapter provides some highlights describing the methods/approaches for peptide bioanalysis, the key challenges associated with peptide analysis, and providing example assay protocols to help scientists developing robust peptide bioanalytical methods for their intended research.

Handbook of LC-MS Bioanalysis: Best Practices, Experimental Protocols, and Regulations, First Edition. Edited by Wenkui Li, Jie Zhang, and Francis L.S. Tse.

42.2 METHODS AND APPROACHES

42.2.1 Peptide Characteristics and Peptide Handling

42.2.1.1 Adsorption Peptides, especially large peptides, often tend to adsorb to solid surfaces including adsorption to storage tubes, pipette tips, injection syringes, etc., potentially resulting in erroneous measurement of peptide concentrations. Peptide adsorption should be carefully evaluated at the beginning of the method development (Song et al., 2002; John et al., 2004; Wilson et al., 2010). Factors contributing to adsorption include peptide amino acid composition, surface materials, solvents, biomatrix, pH, etc. Approaches for preventing adsorption include using organic-aqueous solvents, adding blocking reagents or structurally related peptide analogs to block nonspecific binding sites, and using appropriate container materials (e.g., silanized glass or low-adsorption polypropylene vials) for solution transfer and storage. Generally, aqueous solutions with low peptide concentrations should be avoided. Adsorption from plasma/serum samples or solutions prepared in plasma/serum are often far less significant due to the plasma proteins suppressing the nonspecific binding. For low protein content biomatrices such as CSF, urine, and saliva, biological sample collection/handling procedures should carefully address the adsorption liability of the target peptide analytes.

42.2.1.2 Peptide Stability Lyophilized peptides are generally very stable; however, when they are placed in solutions or in biological matrices, peptides may be subject to chemical and enzymatic degradation (Mesmin et al., 2010). Therefore, it is important to ensure appropriate peptide stability when performing bioanalytical measurements. There are various protease inhibitors and inhibitor cocktails that can help to minimize or eliminate proteolytic degradation (Wolf et al., 2001). Sometimes simply acidifying the samples, for instance using trifluoroacetic acid (TFA), can stabilize peptides. The stability characteristics are peptide dependent. It is crucial to perform stability studies to verify if stability treatments can provide sufficient peptide stability for reliable quantitative analysis. It is also possible that occasionally the peptide degradation kinetics are so fast, that the peptide cannot be stabilized effectively; for example, no reliable bioanalytical method has yet been established for Apelin-13 in human plasma (Mesmin et al., 2010).

42.2.1.3 Peptide Size Another critical characteristic for peptide bioanalysis by LC-MS/MS is the peptide size (or peptide molecular weight), which can range from several hundred to 10 kDa. In contrast to most small molecules, a distinct feature of peptides under electrospray ionization (ESI) conditions is the formation of multiply charged molecular ions. Multiple charging usefully results in the formation of peptide ions that are within the mass range of most commercial quadrupole mass spectrometer instruments. However, significant multiple charging of large peptides (e.g., peptide molecular weight >5 kDa) also results in a loss of sensitivity due to distribution of ions among multiple-charge states. Different physicochemical properties associated with peptide size will affect not only the choice of MS detection, but also the choice of LC separation and extraction procedures for biological sample preparation. For peptides below 5 kDa, the approaches for bioanalytical method development are generally similar to the approaches followed for small molecule bioanalysis and the detection can be based on the intact peptide (Chang et al., 2005; Van Den Broek et al., 2006). However, the selection of the precursor molecular ion from different charge states, and the selection of the product ion for MS/MS detection should be carefully evaluated for sensitivity and specificity. For peptides above 5 KDa, these choices become increasingly complicated and it can be difficult to achieve appropriate sensitivity due to significant charge distribution and the collision-induced dissociation (CID) fragmentation characteristics of large peptide ions (Ji et al., 2003). Analyzing large peptides by well-controlled enzymatic digestion (e.g., trypsin digestion, similar to the bottom-up approach in proteomics research) has proven to be a feasible approach, in which a quantitative surrogate peptide is identified and quantitated; the surrogate peptide concentration serves to represent the concentration of the intact peptide (Van Den Broek et al., 2007; Berna et al., 2008). It is very important to carefully evaluate and select the unique surrogate peptides. Special attention should be paid to the formation of the same tryptic peptides from potential metabolites, which may compromise the assay selectivity for the intact peptide. It is also important to optimize the digestion procedure so that the surrogate peptides are formed reproducibly and quantitatively.

42.2.1.4 Therapeutic Peptide Conjugates Peptides generally have a very short circulatory half-life preventing them from being viable therapeutics in their native form. One of the strategies to overcome this limitation is to conjugate the therapeutic peptide to polyethylene glycol polymer (PEG) or a carrier protein (e.g., FC and mAb) via stable covalent linkages. Analysis of therapeutic peptide conjugates has predominantly been performed by ligand binding assays (e.g., enzyme-linked immunosorbant assay [ELISA]). However, LC-MS/MS combined with immunoaffinity capture is now emerging as an approach with unique advantages and great potential to be the method of choice for the bioanalysis (Xu et al., 2010). As demonstrated in protocol 3 for a 40 kDa PEGylated therapeutic peptide bioanalysis (Xu et al., 2010), an anti-PEG antibody was utilized to target the PEG moiety for immunoaffinity capture, followed by enzymatic digestion and MS/MS detection of the surrogate peptide. This methodology provided a highly specific measurement of the PEG-peptide conjugate in human plasma. As an alternative, we recently published a direct quantitative bioanalytical method for a 20 kDa PEGylated calcitonin gene-related peptide (CGRP) peptide using LC-MS/MS, in which the quantitative surrogate peptides were generated by in-source fragmentation instead of by solution phase digestion from the intact analyte molecules (Li et al., 2011). The readily applicable and specific MS/MS detection of the therapeutic peptide warhead by LC-MS/MS enables fast and cost-effective method development compared to that by ligand binding assays.

42.2.2 Methods for MS Detection

42.2.2.1 MS/MS Detection Using Triple Quadrupole Mass Spectrometer LC-MS/MS on triple quadrupole mass spectrometers are almost exclusively used for quantitative analysis of peptides in biological samples (John et al., 2004; Li et al., 2009; Cutillas, 2010; Wilson et al., 2010; Ewles and Goodwin, 2011), particularly with MS/MS detection performed in the selected reaction monitoring (SRM) mode. In this mode, a specific *m/z* that corresponds to the peptide of interest is filtered through the first quadrupole (Q1). The peptide is then fragmented in the second quadrupole (Q2) by CID. Subsequently, the third quadrupole (Q3) is set to allow only one specific *m/z* corresponding to a unique peptide fragment ions to reach the detector. A signal is registered only when a predefined fragment ion arises from the predefined precursor. This SRM detection based on unique ion transitions for the peptides (precursor ion *m/z* → product ion *m/z*) is thus highly specific. The SRM detection also has a high duty cycle, high ion transmission efficiency, and large dynamic range because both quadrupole analyzers (Q1 and Q3) are operated in a mass filtering mode. As a result, LC-MS/MS using triple quadrupole mass spectrometers is generally a superior choice for highly sensitive and selective methods for the measurement of peptides in biological matrices.

CID of peptides in a triple quadrupole mass spectrometer produces several types of fragment ions with specific nomenclature (Roepstorff and Fohlman, 1984; Johnson et al., 1987); the N-terminus containing b-ions and the C-terminus containing y-ions are typically found to be the most abundant product ions, and they are widely used for peptide quantitative analysis. In general, immonium ions, small b-ions, and y-ions with *m/z* less than 200 should be avoided due to potential high chemical noise from the coeluting matrix components even though these small ions may yield very good sensitivity. In addition, unlike small molecules, it is not unusual for a product ion to have an *m/z* higher than that of the precursor ion due to lower number of charges, and in some circumstances, this may provide some advantages in terms of selectivity (Berna et al., 2008).

As stated in the previous section, peptides form multiply charged molecular ions when ionized with ESI, and the charge state distribution depends on peptide amino acid composition, mobile phase composition, and pH. Peptide molecular ions with higher charge state generally have higher fragmentation efficiency but they may not be the most abundant among the molecular ion species. Therefore, evaluation and optimization of each precursor ions and their product ions in terms of sensitivity and specificity is important for method development especially for large peptides.

42.2.2.2 MS/MS Detection Using Ion Trap Mass Spectrometer While triple quadrupole mass spectrometers are considered the instrument of choice for peptide quantitative bioanalysis, it has been found that the ion trap instrument, especially the linear ion trap mass spectrometer, can be a good alternative for the LC-MS/MS bioanalysis of some large peptides due to the difference in the peptide fragmentation process by CID (Shipkova et al., 2008). In triple quadrupole mass spectrometer, precursor ions are accelerated by a dc voltage through a pressurized collision cell (Q2). Multiple-step fragmentation can take place in the collision cell where initial peptide fragments can further fragment in the collision cascade resulting in many product ions with relatively low abundance. This becomes more significant for large peptides since they require more energy for the onset of the CID. Therefore, the lack of abundant product ions for some large peptides can lead to a reduction in the sensitivity of LC-MS/MS methods.

In an ion trap mass spectrometer, precursor ions are excited by *m/z* specific resonant excitation frequency and fragmented by the collision with helium gas in the ion trap (Schwartz et al., 2002). Since the product ions have different *m/z* to that of the precursor ions, the resultant peptide fragment ions are off resonance and do not further fragment, leading to fewer product ions with higher relative intensity and thus can offer better peptide MS/MS detection sensitivity. Recent advances in faster MS/MS full scan and larger ion trapping capacity in linear ion trap instruments, further improve their duty cycle, limits of quantitation and dynamic range (Schwartz et al., 2002; Hager and Le Blanc, 2003; Londry and Hager, 2003). However, in most cases, triple quadrupole mass spectrometers still give superior bioanalytical performance for most peptides. LC-MS/MS using ion trap mass spectrometers can serve as a good alternative approach for peptide bioanalysis and may hold advantages for some large peptides (Shipkova et al., 2008).

42.2.2.3 MS Detection Using HR/AM Mass Spectrometer Single ion monitoring (SIM) has been used for peptide bioanalysis performed on single quadrupole mass spectrometers or ion trap instruments (Yamaguchi et al., 2000; Wolf et al., 2001). The lack of selectivity compared to MS/MS detection certainly requires adequate sample preparation and extensive chromatographic separation to minimize matrix interferences and other chemical noises, which makes this approach much less attractive with instruments offering unit mass resolution. However, recent introduction of the mass spectrometry instruments capable of both high resolution/accurate mass (HR/AM) measurement and quantitative analysis, such as the AB SCIEX TripleTOF™ 5600 System and the Thermo Scientific Q Exactive mass spectrometer (Michalski et al., 2011), may offer alternative approaches for peptide analysis. With these instruments, peptides of interest are first selected by the quadrupole mass filter (Q) and followed by HR/AM mass detection. The ultra high resolution, for example, up to 140,000 offered by the Orbitrap mass spectrometer, provides high selectivity for target peptides from interfering components from the biological matrix. SIM detection for peptide molecular ions does not require CID fragmentation and hence allows for simple and fast method development. This provides advantages especially for large peptides that have poor CID fragmentation characteristics. The use of multiplexed SIMs in the Q Exactive instrument is also very effective using multiple peptide molecular ions of different charge states simultaneously for MS detection, providing highly selective and sensitive quantitative analysis of large peptides, for instance, 250 pg/ml lower limits of quantitation (LLOQ) for insulin in plasma.

42.2.3 Methods for LC Separation

ESI is mostly used to interface LC separation and MS detection for peptide bioanalysis. It is well known that coeluting matrix components from LC separation can cause ion suppression and lead to poor analytical performance in terms of sensitivity, accuracy, and precision (King et al., 2000; Shen et al., 2005).

Improving chromatographic separation is one of the most effective and convenient ways of circumventing ion suppression. Reversed-phase chromatography, using nonpolar hydrophobic stationary phases and polar aqueous mobile phases, is predominantly used for peptide bioanalysis using LC-MS and LC-MS/MS. Peptide chromatography is mainly based on adsorption–desorption mechanisms (Geng and Regnier, 1984). Peptides interact with hydrophobic stationary phases by adsorption, and they are desorbed at a critical organic solvent concentration in the mobile phases. The critical organic solvent concentration is generally less than 50% (v/v) for many peptides. Therefore, most reversed-phase high-performance liquid chromatography (HPLC) methods for peptides use shallow increasing gradients of organic solvents such as methanol and acetonitrile to achieve optimum peptide separation and sharp peaks. For good separation and ESI, 0.1% formic acid in mobile phases is usually used and is suitable for many peptides. Further improvement of the chromatography can be achieved by using ultra-high-performance liquid chromatography (UHPLC) with columns packed with sub-2 μm particles (e.g., ACQUITY BEH C18, 1.7 μm) in combination with a mobile-phase delivery system that can operate at high backpressures of up to 17,000 PSI. The major advantages of UHPLC over conventional HPLC utilizing columns packed with 3–5 μm particles include significantly improved column efficiency, sensitivity, resolution, peak capacity, and speed (Churchwell et al., 2005). Moreover, UHPLC are normally operated at higher column temperatures, which not only help reduce mobile phase viscosity but also help LC separation of peptides, especially large peptides.

Occasionally, two-dimensional (2D)-HPLC is used in order to achieve good separation of the target peptides from the background matrix components. Two-dimensional HPLC utilizes two orthogonal chromatographic separations (Motoyama et al., 2006; Motoyama et al., 2007; Xu et al., 2010); for instance, SCX/RP-HPLC for peptide separation has been widely used in the proteomics area. However, it requires special instrumentation and time to develop 2D HPLC methods. It might have some advantages for particular peptides, but traditional 1D HPLC or UHPLC separations, combined with the high resolving power of modern mass spectrometry, can usually meet the specific analytical requirements for LC-MS/MS peptide bioanalysis.

42.2.4 Biological Sample Preparation

Preparation of biological samples is often a key part of successful methods for peptide analysis, and is often critical for method selectivity, sensitivity, and to avoid ion suppression. Sample preparation typically removes interfering matrix components, such as serum proteins, lipids, salts, and particulates, and also serves to preconcentrate peptide analytes from biological samples and make the samples more suitable for LC-MS/MS analysis. Commonly used techniques of sample preparation for peptide bioanalysis are solid phase extraction, immunoaffinity extraction, and protein precipitation.

42.2.4.1 Solid Phase Extraction Solid phase extraction (SPE) has frequently been utilized for peptide extraction from biological fluids (Thevis et al., 2006; Li et al., 2009; Wilson et al., 2010) and tissue homogenates, as well as for sample cleanup after enzymatic digestion. The extraction is based on the interaction of the peptide with the functional groups embedded on the surface of the SPE sorbents. A wide range of SPE sorbent chemistries are available, providing versatile and selective peptide extraction utilizing specific chemical interactions such as hydrophobicity, hydrophilicity, ion-exchange, and mixed-mode properties. Ion-exchange SPE generally is more selective and provides cleaner extracts while mixed-mode SPE offers good recovery for a broad range of peptides. SPE can also eliminate large serum proteins through size-exclusion mechanism using sorbents with pore size less than 80 Å while retaining peptides and small proteins. In addition, the ability to use 96-well and micro-elution format makes it attractive for high-throughput sample preparation with small final extraction volumes. The correct choice of SPE sorbent and procedure is mainly peptide dependent, and is important to carefully evaluate peptide load/wash/elute profiles on the selected SPE platforms to optimize analyte recovery and the selectivity of the extraction.

42.2.4.2 Immunoaffinity Extraction Immunoaffinity extraction is an emerging technique for large molecule bioanalysis by LC-MS/MS (Wolf et al., 2001, 2004; Berna et al., 2006, 2008). It utilizes the same mechanism of capture antibodies as ligand binding assays (ELISA) using highly specific antibody–antigen interactions to isolate and enrich analytes of interest from the biological matrix.

Immunoaffinity extraction can be performed by online capture using immunoaffinity columns or by off-line capture using pipette tips or beads containing immobilized capture antibodies. Immunoaffinity extraction using magnetic beads is becoming increasingly popular because it allows flexible extraction/elution or downstream enzymatic digestion in 96-well plate format for high-throughput bioanalysis. It is also compatible with many robotic systems for automation. Due to the requirement to have a specific capture antibody, the applicability of this technique for peptide bioanalysis will be dependent on the purpose of the analytical task and the reagent availability. For example, immunoaffinity extraction may be essential to achieve low pg/ml LLOQs for an endogenous bioactive peptides in the biological matrix (Li et al., 2007; Berna et al., 2008), while immunoaffinity extraction might not be needed for therapeutic peptides for which 1 ng/ml LLOQ is generally sufficient, and can be achieved by other extraction methods without the need of costly peptide-specific antibodies.

42.2.4.3 Protein Precipitation Protein precipitation (PPT) of plasma/serum using organic solvents or acids represents the simplest sample extraction technique, which has been frequently used for small molecule bioanalysis. However, it is not widely applicable for peptides (especially for large peptides) due to significant interferences from the many matrix components that are not removed by this technique. However, PPT can be useful in combination with another sample extraction procedure such as SPE (Xue et al., 2006), or for peptides not presenting major analytical challenges.

42.2.4.4 Tryptic Digestion Analyzing large peptides and peptide conjugates by well-controlled enzymatic digestion has proven a feasible approach (Van Den Broek et al., 2007; Xu et al., 2010; Wu et al., 2011). Peptides generated by trypsin digestion are generally preferred for LC-MS/MS analysis since tryptic peptides often have the appropriate size/mass (7–20 amino acid residues), good ionization efficiency, and they produce predictable MS/MS fragment ions. However, trypsin digestion of intact peptide results in the formation of multiple tryptic peptides, and has the risk of compromising assay selectivity since potential peptide metabolites may form the same tryptic peptides. Careful selection of the surrogate peptide is very important for it to serve as an accurate representation of the concentration of the intact peptide. Other important considerations are internal standard (IS) choice, digestion parameters, and digest cleanup prior to LC-MS/MS analysis.

42.3 EXAMPLE EXPERIMENTAL PROTOCOLS

Protocol 1: LC-MS/MS bioanalysis of a human bradykinin B1 receptor antagonist peptide in human plasma (Wilson et al., 2010).

Peptide Analyte D-Orn-Lys-Arg-Pro-Hyp-Gly-Cpg-Ser-D-Tic-Cpg is a human B1 receptor antagonist peptide for the treatment of chronic pain (MW 1180 Da).

Biomatrix Human plasma

Equipment and Reagents

- *LC-MS/MS*: Shimadzu LC-20 HPLC coupled to an Applied Biosystems API 5000 mass spectrometer, ESI interface, positive mode
- *Analytical RP-HPLC column*: Varian Polaris C18 column, 75 mm × 2.1 mm, 5 μm
- *Mobile phase A*: 5:95:0.1 methanol/water/formic acid (v/v/v)
- *Mobile phase B*: 95:5:0.1 methanol/water/formic acid (v/v/v)
- *IS*: Stable isotope labeled peptide (SIL-peptide), D-Orn-*Lys-Arg-Pro-Hyp-Gly-Cpg-Ser-D-Tic-Cpg, Lys is labeled with $[^{13}C_6,^{15}N_2]$, $\Delta m = 8$
- *SPE*: Waters Oasis MAX SPE plate in 96-well format (10 mg, 30 μm)

Method

- *Peptide stock solution and plasma standard preparation*: Primary standard stock solutions (1 mg/ml) of peptide and SIL-peptide IS were prepared in 25:75 MeOH/water in polypropylene vials. Secondary standard spiking stock solutions were prepared in human plasma at concentrations of 1, 10, and 100 μg/ml in polypropylene tubes. Peptide standards for the calibration curve were prepared in polypropylene tubes using the secondary standard spiking stock solutions by serial dilution to yield concentrations of 1, 2, 2.5, 5, 10, 20, 25, and 50 ng/ml. Working IS solution (1000 ng/ml) was prepared in 25:75 MeOH/water.

- *SPE*: Waters Oasis MAX SPE plate was preconditioned with 1 ml of methanol followed by 1 ml of water. After plasma samples were loaded onto the plate under vacuum, the plate was washed sequentially with 1 ml of 2:98 ammonium hydroxide/water and then 1 ml of 50:50 acetonitrile, methanol. After washing, the samples were eluted into a 96-well collection plate using 0.5 ml of 2:90:8 acetic acid/methanol/water. The samples were dried down and reconstituted with 200 μl of 5:95:0.1 methanol/water/formic acid.
- *Mass spectrometry detection*: Full scan MS of the peptide identified two molecular ions (M^{2+} and M^{3+}) at *m/z* at 590.9 and 394.4. The M^{2+} ion was chosen for SRM detection based on experiments in which ion transitions for both M^{2+} and M^{3+} as precursor ions were monitored. The M^{2+} precursor resulted in better analytical performance in terms of sensitivity, accuracy, and precision. D-Tic immonium ion at *m/z* 132.2 was used as the product ion in the SRM ion transition; it was found to be the most abundant product ion as shown in the MS/MS spectrum of M^{2+} (Figure 42.1). Unlike immonium ions from native amino acids, for instance proline immonium ion at *m/z* 70, which are typically avoided as production ions for SRM detection, D-Tic immonium ion demonstrated good selectivity due to the fact it is from a non-natural amino acid. After optimization, the SRM at *m/z* 591 $\rightarrow$ 132 (B1 peptide) and *m/z* 595 $\rightarrow$ 132 (IS) were used for quantification of B1 peptide in human serum.
- *LC-MS/MS analysis*: Chromatographic separation of SPE-extracted samples were performed using gradient elution at a mobile phase flow rate of 0.3 ml/min. The initial eluent composition was 10% B. The eluent was kept at 10% B for 0.8 min and increased to 95% B in 1.2 min, and held at 95% B for 2.5 min. It was then reduced to 10% B in 0.3 min and allowed to equilibrate at 10% B for 0.7 min. The total run time was 5.5 min. The injection volume was 25 μl. Chromatograms of a control human plasma extract and 1 ng/ml LLOQ are shown in Figure 42.2.

Key Analytical Method Features

The method was developed and validated for the determination of a therapeutic B1 peptide in human plasma over the concentration range of 1–50 ng/ml. SIL-peptide was used as an IS. Since a small D-Tic immonium ion (*m/z* 132.2) was used as the product ion for SRM detection, to avoid significant chemical noise, a mixed-mode ion-exchange SPE extraction was implemented to provide a cleaner SPE extract. Intraday precision and accuracy determined from quality control (QC) samples were 7.3% CV or less and within 6.0% bias, respectively. Interday precision and accuracy were 5.2% CV or less and within 2.4% bias, respectively. Precautions were taken to prevent peptide adsorption, in particular all dilutions of standard solutions were prepared in human plasma and stock solutions were prepared using polypropylene containers and organic-aqueous solvent. The validated assay was used in support of human clinical trials.

Protocol 2: UPLC-MS/MS bioanalysis of endogenous hepcidin peptides in mouse serum.

Peptide Analyte

Intact mouse hepcidin (hepcidin-25) is an endogenous 25-amino acid peptide hormone with sequence: DTNFPICIFCCKCCNNSQCGICCKT (MW 2754.5 Da). Eight Cys residues are crosslinked with four internal disulfide bonds. Hepcin-24, a peptide lacking an Asp from the N-terminus of hepcidin-25, is also found in mouse serum with sequence: TNFPICIFCCKCCNNSQCGICCKT (MW 2639.4 Da). Hepcidin has been shown to be an important mediator of iron homeostasis. Analytical methods for measuring serum hepcidin peptides (both hepcidin-25 and hepcidin-24) are intended to support experiments to understand their biological role in anemia of inflammation (AI) in the mouse animal model.

Biomatrix

Rabbit control serum was used as a surrogate matrix for preparing calibration standards. Prescreened mouse control serum with negligible levels of endogenous hepcidin peptides was used for preparing QC samples to evaluate method performance with rabbit control serum.

Equipment and Reagents

- *UPLC-MS/MS*: Waters ACQUITY UPLC coupled to an Applied Biosystems API 4000 mass spectrometer, ESI interface, positive mode
- *Analytical RP-UPLC column*: ACQUITY BEH C18 column, 50 mm $\times$ 1.0 mm, 1.7μm
- *Mobile phase A:* 5:95 ACN/water (v/v) with 0.1% formic acid
- *Mobile phase B*: 95:5 ACN/water (v/v) with 0.1% formic acid
- *IS*: Human hepcidin, a structurally similar peptide analog is used as the IS. It has the sequence: DTHFPICIFCCGCCHRSKCGMCCKT.
- *SPE*: Waters Oasis HLB μElution SPE plate in 96-well format (30 μm)

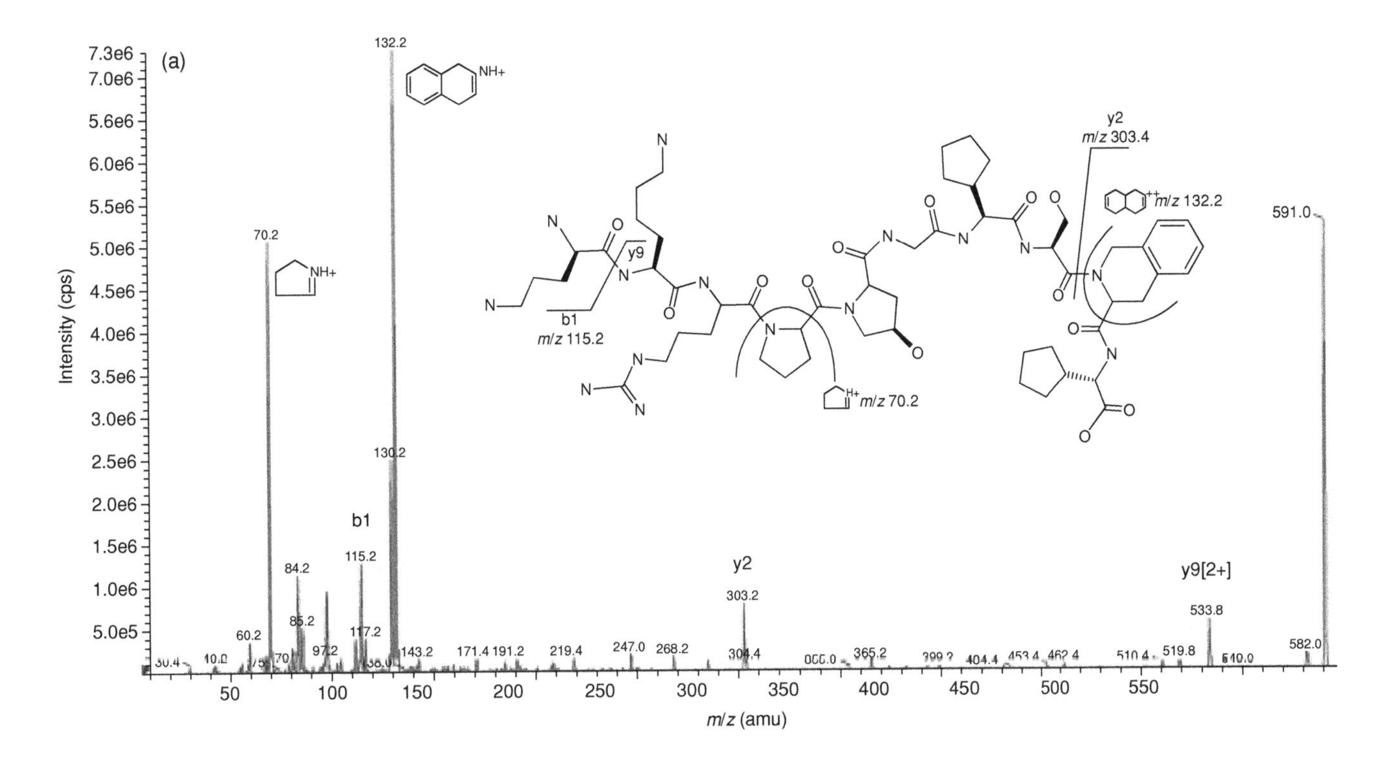

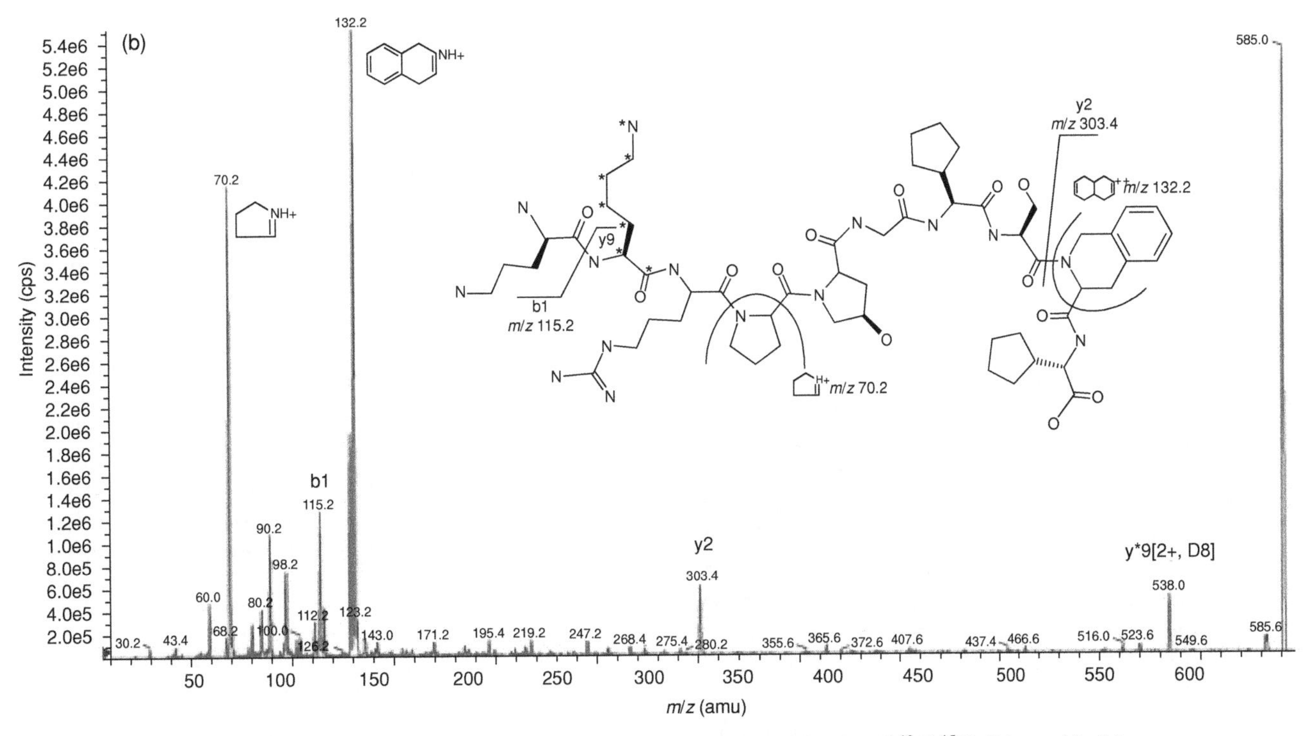

FIGURE 42.1 Product ion mass spectra (Q3) of the B1 peptide (a) and $^{13}C_6{}^{15}N_2$-B1 peptide (b) for molecular ions of 591 and 595, respectively, and the proposed fragment structures (reprinted with permission from Wilson et al. (2010); copyright 2010 Elsevier B.V.).

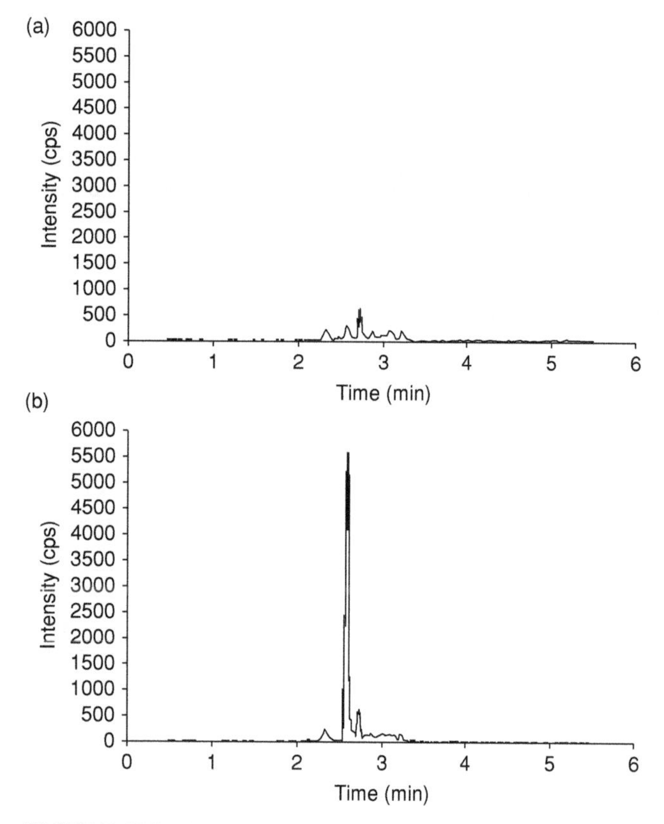

FIGURE 42.2 Chromatogram of control plasma (a) and a standard at the LLOQ of 1 ng/ml (b), prepared using solid phase extraction and analyzed using HPLC-MS/MS (reprinted with permission from Wilson et al. (2010); copyright 2010 Elsevier B.V.).

Method

- *Peptide stock solution and plasma standard preparation*: Primary stock solutions (1 mg/ml) made from mouse hepcidin-25 and hepcidin-24 reference standards were used to prepare 100 μg/ml secondary stock solutions. Both stock solutions were prepared in 50/50 (v/v) methanol/water and stored in a refrigerator at 2 to 8 °C prior to use. Standard samples were prepared in rabbit serum, standards at concentrations of 1, 2.5, 5, 10, 25, 50, 100, 250, and 500 ng/ml were prepared by serial dilution of a freshly prepared 5000 ng/ml solution in rabbit serum. IS solution (100 ng/ml of human hepcidin) was prepared in 30:70 MeOH/water (v/v).
- *SPE*: 50 μl of each serum sample was aliquoted into the appropriate well of a 96-well plate, followed by the addition of 50 μl of IS solution (100 ng/ml human hepcidin) and 100 μl of 0.1 M TFA and vortex mixed. These pretreated samples were then loaded onto a preconditioned Oasis HLB μElution 96-well SPE plate under vacuum, and washed sequentially with water (100 μl), 30/75 (v/v) methanol/water with 2% ammonium hydroxide (100 μl), and water (100 μl). The plate was then eluted with 25 μl of 90/10 (v/v) methanol/water with 2% acetic acid into a 96-well collection plate, followed by a 25 μl of water dilution to each well and the samples were vortex mixed.
- *Mass spectrometry detection*: Full scan MS of the mouse hepcidin-25 identified three molecular ions (M^{2+}, M^{3+} and M^{4+}) at *m/z* at 1378.4, 919.0, and 621.7. The M^{3+} ion was chosen for SRM detection after evaluating SRM ion transitions using M^{2+}, M^{3+}, and M^{4+} as precursor ions. MS/MS of M^{3+} ion (*m/z* 919) resulted in an abundant and selective product ion ${y_{21}}^{2+}$ at *m/z* 1139. Similar MS and MS/MS were observed for Hepcidin-24 and the M^{3+} ion at *m/z* 880 and its ${y_{21}}^{2+}$ product ion at *m/z* 1139 was chosen for SRM detection as shown in Figure 42.3.
- *LC-MS/MS analysis*: Chromatographic separation of SPE-extracted samples were performed using gradient elution at a mobile phase flow rate of 0.25 ml/min. The LC gradient profile was as follows (min/% of mobile phase B): 0.0/15, 0.1/15, 1.0/50, 1.1/95, 1.6/95, 1.65/15, and 2.0/15. Total runtime was 2 min. The column temperature was 50 °C. The injection volume was 10 μl. Chromatograms of mouse hepcidin-24, hepcidin-25 at 1 ng/ml LLOQ, and IS human hepcidin-25 are shown in Figure 42.4.

Key Analytical Method Features

The UHPLC-MS/MS method can definitively measure mouse hepcidin-25 and its hepcidin-24 isoform simultaneously, hence providing a better understanding of their biological roles in the AI mouse model. The UPLC-MS/MS method is highly sensitive and has LLOQ of 1 ng/ml for both mouse hepcidin-25 and hepcidin-24 from 50 μl serum. The linear range of the method is 1–500 ng/ml. The 2-min run time enables high-throughput analysis with accuracy and precision less than or equal to $\pm$ 15%. Hepcidin occurs endogenously but varies in amino acid sequence between species; consequently, serum from different species can be used as analyte-free matrix to prepare QC and calibration standards for other species.

Protocol 3: Bioanalysis of a PEGylated therapeutic peptide using immunoaffinity LC-MS/MS and trypsin digestion (Xu et al., 2010).

Peptide Analyte

MK-2662 is a 38 AA peptide with a 40 kDa branched PEG covalently linked to the C-terminus: H[Aib] DGTFTSDYSKYLDSRRAQDVQWLMNTKRNRNNIAC-PEG. It is a GLP-1 receptor agonist developed for the treatment of type 2 diabetes. It contains an aminobutyric

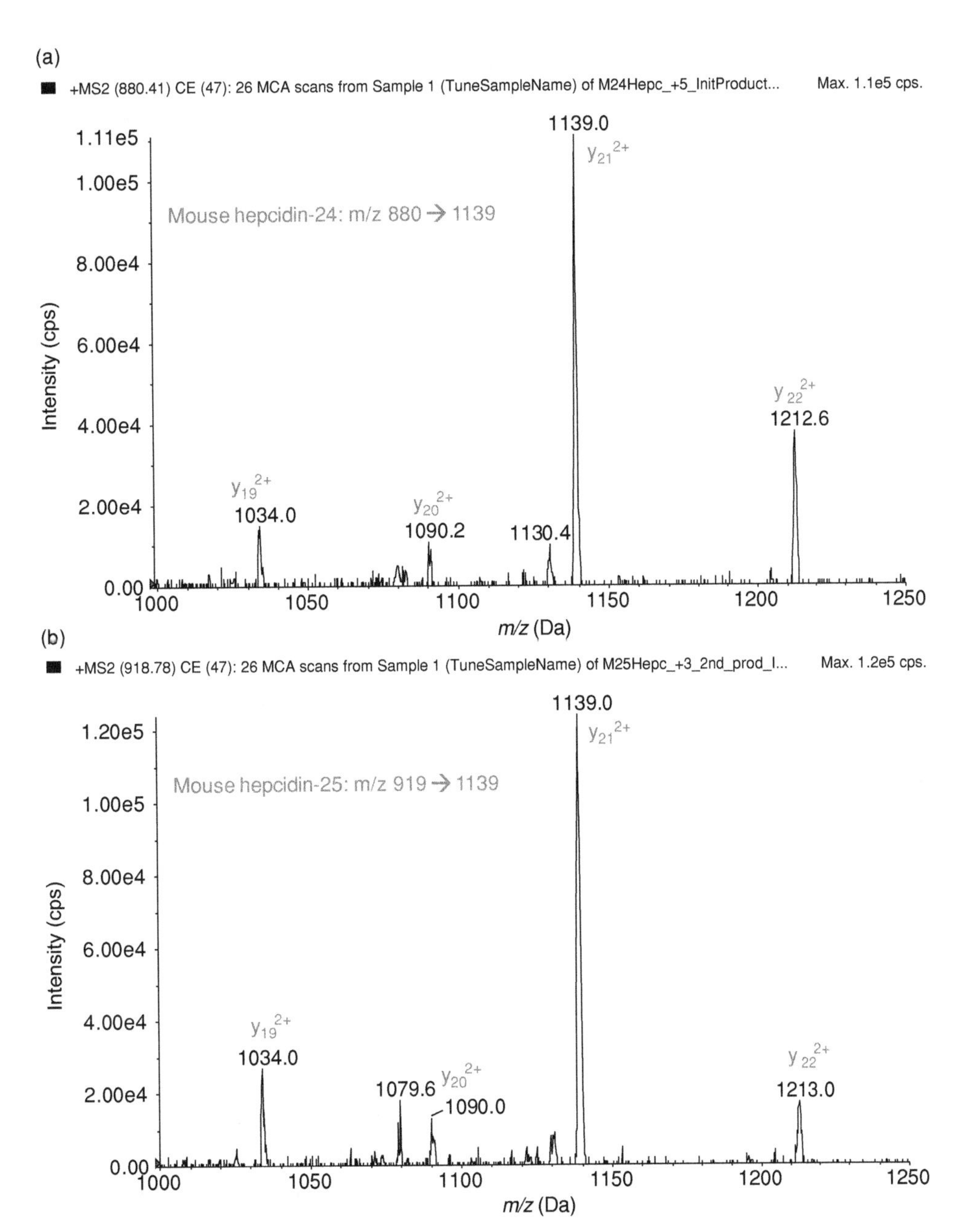

FIGURE 42.3 Product ion mass spectra of mouse hepcidin peptides: (a) hepcidin-24, M^{2+} (*m/z* 880) and (b) hepcidin-25, M^{2+} (*m/z* 919).

acid (Aib) residue, which is a non-natural amino acid providing significant proteolytic resistance.

Biomatrix

Human plasma. Clinical samples were collected in blood collection tubes (BD-P700) containing proprietary DPP-IV inhibitors. Control human plasma for preparing standards and QCs was stabilized with Linco DPP-IV inhibitor (20 μl/ml plasma).

Equipment and Reagents

LC-MS/MS

- A Cohesive Flux 2300 (2D HPLC) coupled to an Applied Biosystems API 5000 mass spectrometer, ESI interface, positive mode
- *Analytical cation-exchange column (1D)*: Thermo BioBasic SCX, 50 mm × 2.1 mm, 5 μm
- *Analytical RP-HPLC column (2D)*: Thermo Hypersil Gold PFP, 50 mm × 2.1 mm, 5 μm

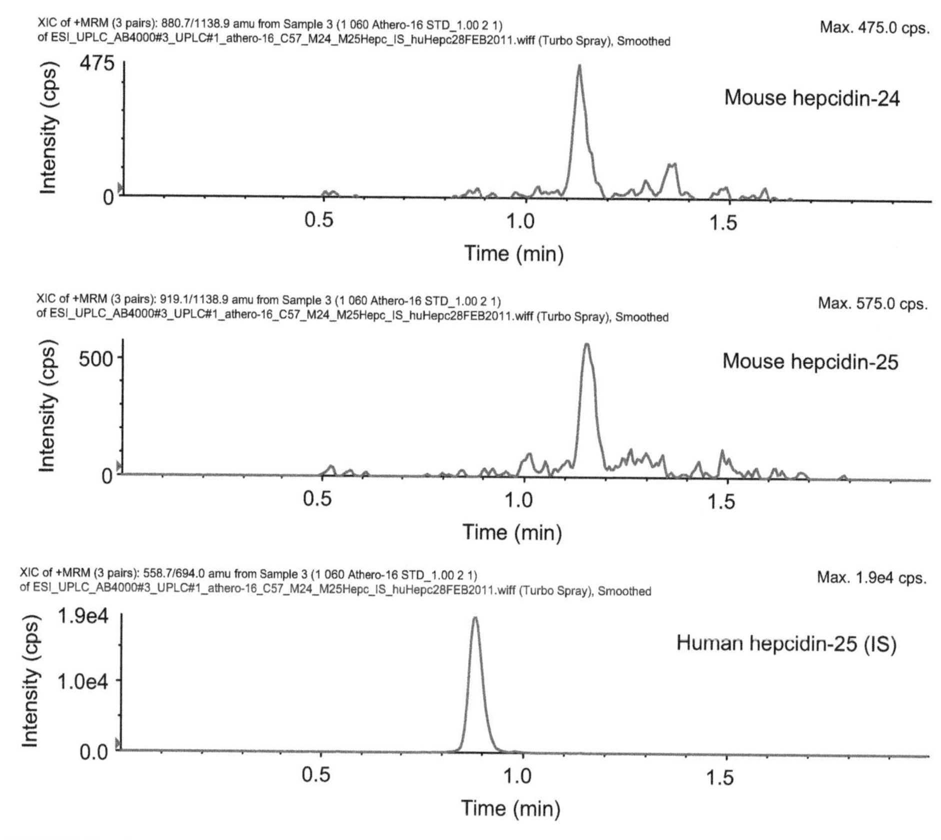

FIGURE 42.4 Chromatogram of mouse hepcidin-24 [*m/z* 880 → 1139] and mouse hepcidin-25 [*m/z* 919 → 1139] at the LLOQ of 1 ng/ml in plasma analyzed using UPLC-MS/MS.

- *1D HPLC mobile phase A*: 50 mM ammonium formate (pH 3) in 10% acetonitrile
- *1D HPLC mobile phase B*: 200 mM ammonium formate (pH 3) in 10% acetonitrile
- *2D HPLC mobile phase A*: 10 mM ammonium formate in 0.1% formic acid
- *2D HPLC mobile phase B*: 10 mM ammonium formate in 90% acetonitrile with 0.1% formic acid
- *IS*: H[Aib]DGT*FTSDYSKYLDSRRAQD*FVQWLM NTKRNRNNIAC-PEG was used as IS. *F is labeled phenylalanine with $[^{13}C_9,^{15}N_1]$, $\Delta m = 10$
- *Immunoaffinity extraction*: Biotinylated anti-PEG rabbit monoclonal antibody conjugated to the streptavidin-coated magnetic beads
- *Enzymatic digestion*: Trypsin

Method

- *Peptide stock solution and plasma standard preparation*: MK-2662 stock solutions (40 μM) were prepared in acetonitrile/water (30/70, v/v). Working standards, containing MK-2662 at the different concentration levels, were prepared by serial dilutions of MK-2662 stock solution with 1% bovine serum albumin (BSA), aliquoted to 1.5 ml polypropylene microcentrifuge tubes and stored at −70 °C. A 500-nM IS working solution was prepared in 1% BSA. Plasma calibration standards were prepared freshly by adding 40 μl of working standard and 40 μl of 500 nM IS into 200 μl of control plasma to provide final concentrations of MK-2662 ranging from 2 to 200 nM for immunoaffinity extraction.
- *Immunoaffinity extraction using immunoaffinity beads (containing 20 ug Ab/ml) and trypsin digestion*: 200 μl of each plasma sample was aliquoted into the appropriate well of a 96-well plate, followed by the addition of 40 μl of 1% BSA, 40 μl of 500 nM IS, and 50 μl immunoaffinity beads. Immunoaffinity beads (50 μl) were added to the freshly prepared plasma standards. The plate was then incubated at 30 °C for 2 h. After washing with PBST (phosphate-buffered saline solution with Tween 20) (pH 7.4) and ammonium bicarbonate (pH 8), 140 μl of 67 μg/ml trypsin in 50 mM ammonium bicarbonate (pH 8) was added into each well and incubated at 37 °C for about 3 h. The supernatant was transferred into a 96-well plate and quenched with formic acid. After mixing and centrifugation, 20 μl of the sample was injected on LC/LC-MS/MS for analysis.

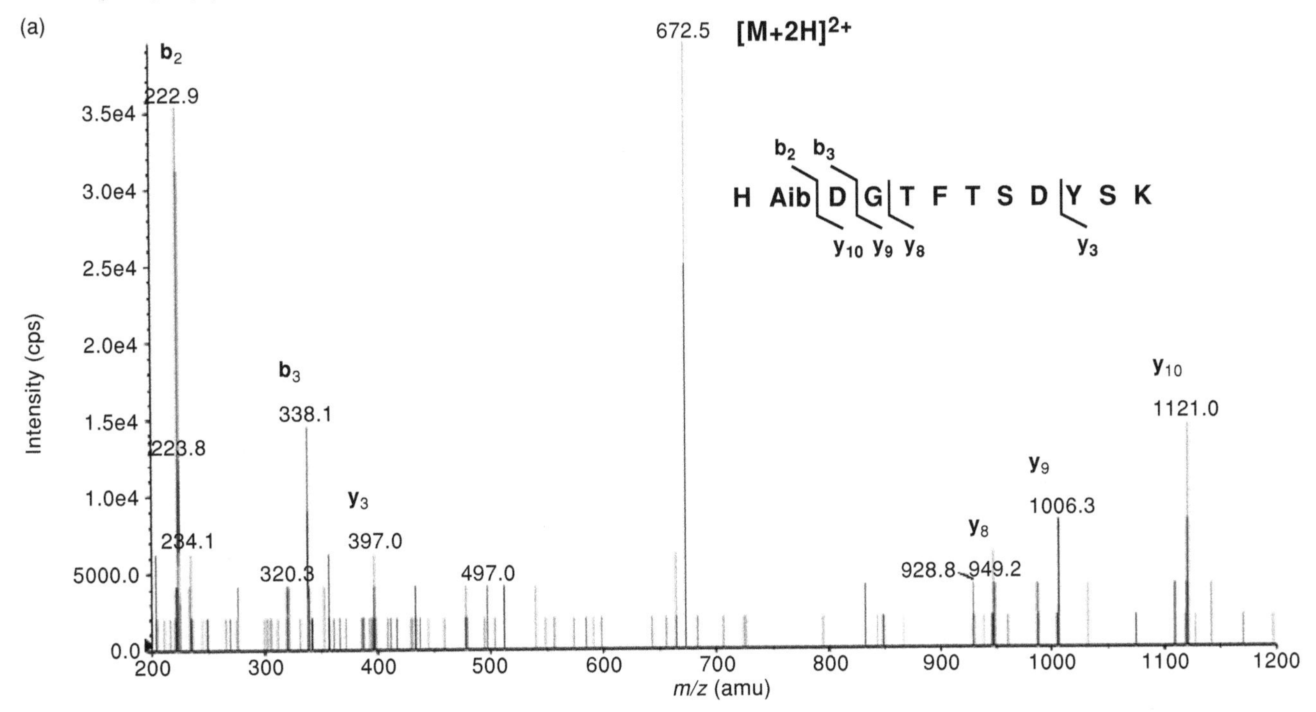

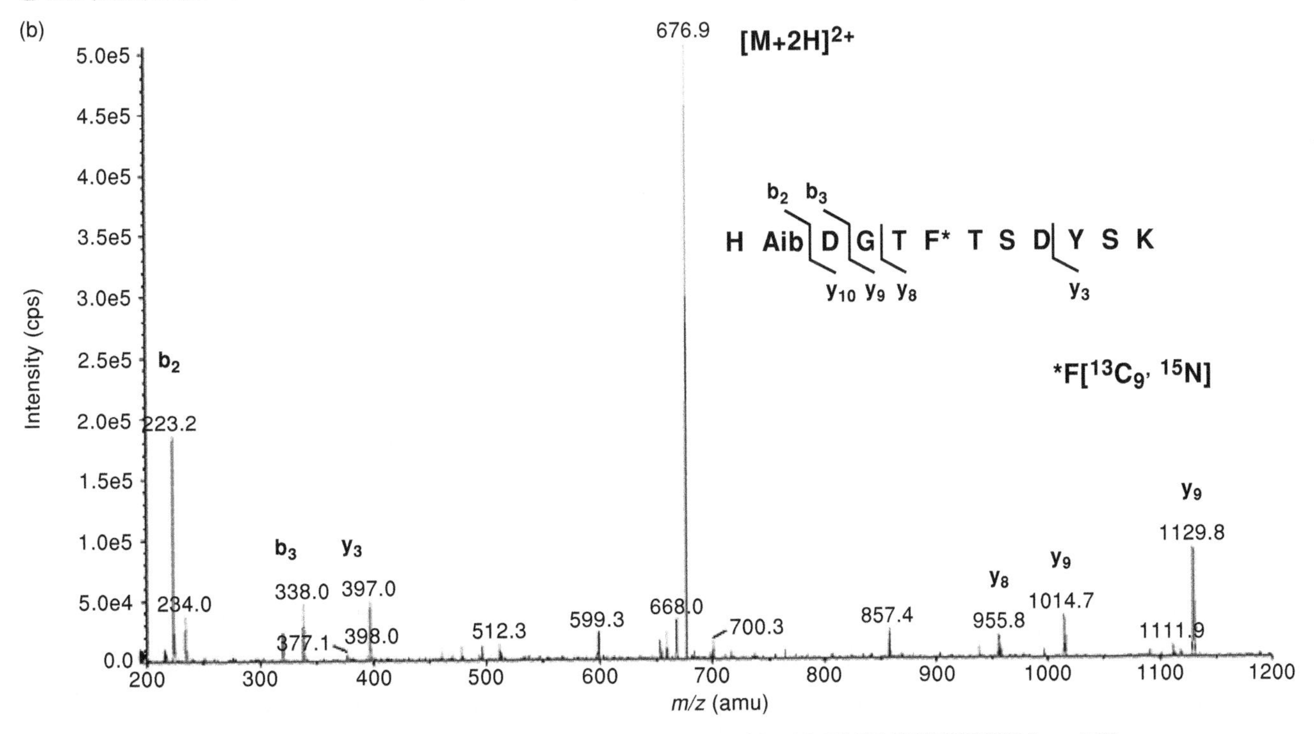

FIGURE 42.5 Product ion mass spectra of tryptic peptides (a) HAibDGTFTSDYSK from MK-2662, M^{2+}, and (b) HAibDGTF[$^{13}C_9$,^{15}N]TSDYSK from [$^{13}C_{18}$,$^{15}N_2$] MK-2662 (IS), M^{2+} (reprinted with permission from Xu et al. (2010); copyright 2010 American Chemical Society).

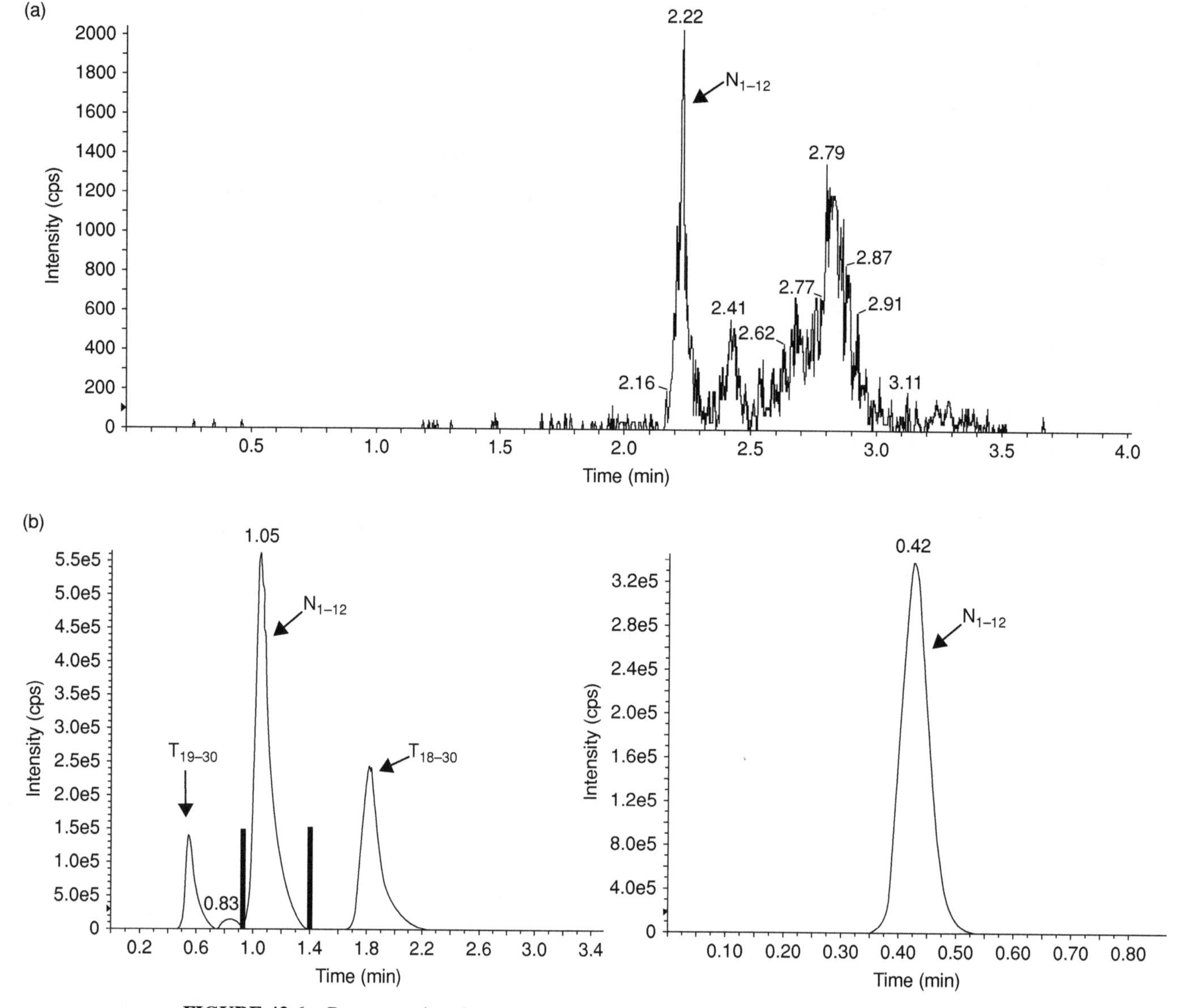

FIGURE 42.6 Representative chromatograms of the N_{1-12}-mer [*m/z* 672 → 223] from MK-2662 plasma sample analyzed using (a) 1D LC–MS/MS or (b) 2D LC–MS/MS. The left panel of (b) is the chromatogram from the SCX column (the first dimension), where T_{18-30} and T_{19-30} were the peptides obtained from trypsin digestion of MK-2662. The right panel of (b) is from the RP column (the second dimension). The total run time for 2D LC-MS/MS (reprinted with permission from Xu et al. (2010); copyright 2010 American Chemical Society).

- *Mass spectrometry detection*: Digestion of MK-2662 and IS with trypsin resulted in 12-AA N-terminal peptides: H[Aib]DGTFTSDYSK and H[Aib]DGT*FTSDYSK. They were selected as the surrogate quantitative peptides for MK-2662 and the IS. Precursor ions for the surrogate peptides were monitored as the doubly charged molecular ions, $[M]^{2+}$ at *m/z* 672.1 and 676.9, respectively. The MS/MS product ion spectra showed singly charged b- and y-ions (Figure 42.5). The b_2 ion at *m/z* 223 was selected due to its intensity and specificity, where a non-natural amino acid [Aib] in b_2 could allow MS differentiation of MK-2662 from the endogenous oxyntomodulin fragments. After optimization, the SRM at *m/z* 672 → 223 (MK-2662) and *m/z* 677 → 223 (IS) were used for quantification of MK-2662.
- *LC-MS/MS analysis*: 2D HPLC using two orthogonal separating chromatographic mechanisms to separate the target surrogate peptides from the background matrix components. Cation-exchange chromatography was performed on a BioBasic SCX (50 mm × 2.1 mm, 5 μm) column and was used as the first dimension with

a salt gradient from 50 to 200 mM ammonium formate (pH 3) at a flow rate of 0.5 ml/min. 10% acetonitrile (v/v) was added to maintain good chromatographic efficiency and recovery from the SCX column. Separated peptides were loaded onto a Hypersil Gold PFP column (50 mm × 2.1 mm, 5 μm) by a column-switching valve for the second dimension reversed-phase chromatography with an increasing organic gradient elution to further separate the surrogate peptide from background noise, and therefore, to provide a clean chromatogram (Figure 42.6). The overall run time was 4.5 min per sample.

Key Analytical Method Features

A protease inhibitor was used to stabilize the peptide in human plasma. 1% BSA was used to address nonspecific binding for standard preparation. Full-length stable isotope-labeled peptide PEG conjugate was used as ISs to track each of the analytical steps for robust bioanalytical performance. Two-dimensional HPLC with contrasting chromatographic mechanisms was used to reduce coeluting matrix components for ESI-MS/MS detection, leading to a net increase in S/N. LC-MS/MS detection of the N-terminal surrogate peptide combined with immunoaffinity extraction on the PEG carrier not only provides high specificity and sensitivity but also overcomes a rate-limiting step of ELISA, which is the production of peptide-specific antibody.

42.4 CONCLUSIONS

LC-MS and LC-MS/MS for quantitative peptide analysis in biomatrix, using either intact peptides or peptide surrogates, have increasingly become the preferred analytical technique as they allow for very selective and sensitive measurements. However, despite the inherent selectivity and sensitivity of LC–MS and LC-MS/MS, peptide bioanalysis still presents significant challenges as described in Section 42.2. These challenges manifest themselves differently on a peptide-by-peptide basis, which typically requires a fit-for-purpose strategy to make the appropriate choices for sample preparation procedures, LC separations, and mass spectrometric detection platforms as exemplified in the sample protocol section. With advances in mass spectrometry, chromatography, and biological sample preparation procedures, peptide bioanalysis by LC-MS is clearly becoming more sensitive, reliable, robust, and cost-effective.

ACKNOWLEDGMENTS

We would like to thank Dr. Sarah Wilson, Dr. Mark Rose, and Dr. Yang Xu for the permission to use their publications in the example experimental protocols.

REFERENCES

Berna M, Schmalz C, Duffin K, Mitchell P, Chambers M, Ackermann B. Online immunoaffinity liquid chromatography/tandem mass spectrometry determination of a type II collagen peptide biomarker in rat urine: investigation of the impact of collision-induced dissociation fluctuation on peptide quantitation. Anal Biochem 2006;356:235–243.

Berna M, Ott L, Engle S, Watson D, Solter P, Ackermann B. Quantification of NTproBNP in rat serum using immunoprecipitation and LC/MS/MS: a biomarker of drug-induced cardiac hypertrophy. Anal Chem 2008;80:561–566.

Chang D, Kolis SJ, Linderholm KH, et al. Bioanalytical method development and validation for a large peptide HIV fusion inhibitor (Enfuvirtide, T-20) and its metabolite in human plasma using LC-MS/MS. J Pharm Biomed Anal 2005;38:487–496.

Churchwell MI, Twaddle NC, Meeker LR, Doerge DR. Improving LC-MS sensitivity through increases in chromatographic performance: comparisons of UPLC-ES/MS/MS to HPLC-ES/MS/MS. J Chromatogr B Analyt Technol Biomed Life Sci 2005;825:134–143.

Cutillas PR. Analysis of peptides in biological fluids by LC-MS/MS. Methods Mol Biol 2010;658:311–321.

Dalluge JJ. Mass spectrometry: an emerging alternative to traditional methods for measurement of diagnostic proteins, peptides and amino acids. Curr Protein Pept Sci 2002;3:181–190.

De Kock SS, Rodgers JP, Swanepoel BC. Growth hormone abuse in the horse: preliminary assessment of a mass spectrometric procedure for IGF-1 identification and quantitation. Rapid Commun Mass Spectrom 2001;15:1191–1197.

Ewles M, Goodwin L. Bioanalytical approaches to analyzing peptides and proteins by LC–MS/MS. Bioanalysis 2011;3:1379–1397.

Geng X, Regnier FE. Retention model for proteins in reversed-phase liquid chromatography. J Chromatogr 1984;296:15–30.

Hager JW, Le Blanc JC. High-performance liquid chromatography–tandem mass spectrometry with a new quadrupole/linear ion trap instrument. J Chromatogr A 2003;1020:3–9.

Halquist MS, Karnes TH. Quantitative liquid chromatography tandem mass spectrometry analysis of macromolecules using signature peptides in biological fluids. Biomed Chromatogr 2010;25:47–58.

Ji QC, Rodila R, Gage EM, El-Shourbagy TA. A strategy of plasma protein quantitation by selective reaction monitoring of an intact protein. Anal Chem 2003;75:7008–7014.

John H, Walden M, Schafer S, Genz S, Forssmann WG. Analytical procedures for quantification of peptides in pharmaceutical research by liquid chromatography–mass spectrometry. Anal Bioanal Chem 2004;378:883–897.

Johnson RS, Martin SA, Biemann K, Stults JT, Watson JT. Novel fragmentation process of peptides by collision-induced decomposition in a tandem mass spectrometer: differentiation of leucine and isoleucine. Anal Chem 1987;59:2621–2625.

King R, Bonfiglio R, Fernandez-Metzler C, Miller-Stein C, Olah T. Mechanistic investigation of ionization suppression in electrospray ionization. J Am Soc Mass Spectrom 2000;11:942–950.

Li WW, Nemirovskiy O, Fountain S, Rodney Mathews W, Szekely-Klepser G. Clinical validation of an immunoaffinity LC-MS/MS assay for the quantification of a collagen type II neoepitope peptide: a biomarker of matrix metalloproteinase activity and osteoarthritis in human urine. Anal Biochem 2007;369: 41–53.

Li H, Rose MJ, Tran L, et al. Development of a method for the sensitive and quantitative determination of hepcidin in human serum using LC-MS/MS. J Pharmacol Toxicol Methods 2009;59:171–180.

Li H, Rose MJ, Holder JR, Wright M, Miranda LP, James CA. Direct quantitative analysis of a 20 kDa PEGylated human calcitonin gene peptide antagonist in cynomolgus monkey serum using in-source CID and UPLC-MS/MS. J Am Soc Mass Spectrom 2011;22:1660–1667.

Lien S, Lowman HB. Therapeutic peptides. Trends Biotechnol 2003;21:556–562.

Londry FA, Hager JW. Mass selective axial ion ejection from a linear quadrupole ion trap. J Am Soc Mass Spectrom 2003;14:1130–1147.

Mesmin C, Dubois M, Becher F, Fenaille F, Ezan E. Liquid chromatography/tandem mass spectrometry assay for the absolute quantification of the expected circulating apelin peptides in human plasma. Rapid Commun Mass Spectrom 2010;24:2875–2884.

Michalski A, Damoc E, Hauschild JP, et al. Mass spectrometry-based proteomics using Q Exactive, a high-performance benchtop quadrupole Orbitrap mass spectrometer. Mol Cell Proteomics 2011;10:M111.011015.

Miranda LP, Holder JR, Shi L, et al. Identification of potent, selective, and metabolically stable peptide antagonists to the calcitonin gene-related peptide (CGRP) receptor. J Med Chem 2008a;51:7889–7897.

Miranda LP, Winters KA, Gegg CV, et al. Design and synthesis of conformationally constrained glucagon-like peptide-1 derivatives with increased plasma stability and prolonged in vivo activity. J Med Chem 2008b;51:2758–2765.

Motoyama A, Venable JD, Ruse CI, Yates JR 3rd. Automated ultra-high-pressure multidimensional protein identification technology (UHP-MudPIT) for improved peptide identification of proteomic samples. Anal Chem 2006;78:5109–5118.

Motoyama A, Xu T, Ruse CI, Wohlschlegel JA, Yates JR 3rd. Anion and cation mixed-bed ion exchange for enhanced multidimensional separations of peptides and phosphopeptides. Anal Chem 2007;79:3623–3634.

Roepstorff P, Fohlman J. Proposal for a common nomenclature for sequence ions in mass spectra of peptides. Biomed Mass Spectrom 1984;11:601.

Sato AK, Viswanathan M, Kent RB, Wood CR. Therapeutic peptides: technological advances driving peptides into development. Curr Opin Biotechnol 2006;17:638–642.

Schwartz JC, Senko MW, Syka JE. A two-dimensional quadrupole ion trap mass spectrometer. J Am Soc Mass Spectrom 2002;13:659–669.

Shen JX, Motyka RJ, Roach JP, Hayes RN. Minimization of ion suppression in LC-MS/MS analysis through the application of strong cation exchange solid-phase extraction (SCX-SPE). J Pharm Biomed Anal 2005;37:359–367.

Shipkova P, Drexler DM, Langish R, Smalley J, Salyan ME, Sanders M. Application of ion trap technology to liquid chromatography/mass spectrometry quantitation of large peptides. Rapid Commun Mass Spectrom 2008;22:1359–1366.

Shushan B. A review of clinical diagnostic applications of liquid chromatography-tandem mass spectrometry. Mass Spectrom Rev 2010;29:930–944.

Song KH, An HM, Kim HJ, Ahn SH, Chung SJ, Shim CK. Simple liquid chromatography-electrospray ionization mass spectrometry method for the routine determination of salmon calcitonin in serum. J Chromatogr B Analyt Technol Biomed Life Sci 2002;775:247–255.

Thevis M, Thomas A, Delahaut P, Bosseloir A, Schanzer W. Doping control analysis of intact rapid-acting insulin analogues in human urine by liquid chromatography–tandem mass spectrometry. Anal Chem 2006;78:1897–1903.

Van Den Broek I, Sparidans RW, Huitema AD, Schellens JH, Beijnen JH. Development and validation of a quantitative assay for the measurement of two HIV-fusion inhibitors, enfuvirtide and tifuvirtide, and one metabolite of enfuvirtide (M-20) in human plasma by liquid chromatography-tandem mass spectrometry. J Chromatogr B Analyt Technol Biomed Life Sci 2006;837:49–58.

Van Den Broek I, Sparidans RW, Schellens JH, Beijnen JH. Enzymatic digestion as a tool for the LC-MS/MS quantification of large peptides in biological matrices: measurement of chymotryptic fragments from the HIV-1 fusion inhibitor enfuvirtide and its metabolite M-20 in human plasma. J Chromatogr B Analyt Technol Biomed Life Sci 2007;854:245–259.

Van Den Broek I, Sparidans RW, Schellens JH, Beijnen JH. Quantitative bioanalysis of peptides by liquid chromatography coupled to (tandem) mass spectrometry. J Chromatogr B Analyt Technol Biomed Life Sci 2008;872:1–22.

Van Den Broek I, Sparidans RW, Schellens JH, Beijnen JH. Quantitative assay for six potential breast cancer biomarker peptides in human serum by liquid chromatography coupled to tandem mass spectrometry. J Chromatogr B Analyt Technol Biomed Life Sci 2010;878:590–602.

Wilson SF, Li H, Rose MJ, Xiao J, Holder JR, James CA. Development and validation of a method for the determination of a therapeutic peptide with affinity for the human B1 receptor in human plasma using HPLC-MS/MS. J Chromatogr B Analyt Technol Biomed Life Sci 2010;878:749–757.

Wolf R, Hoffmann T, Rosche F, Demuth HU. Simultaneous determination of incretin hormones and their truncated forms from human plasma by immunoprecipitation and liquid chromatography–mass spectrometry. J Chromatogr B Analyt Technol Biomed Life Sci 2004;803:91–99.

Wolf R, Rosche F, Hoffmann T, Demuth HU. Immunoprecipitation and liquid chromatographic–mass spectrometric determination of the peptide glucose-dependent insulinotropic polypeptides GIP1-42 and GIP3-42 from human plasma samples. New sensitive method to analyze physiological concentrations of peptide hormones. J Chromatogr A 2001;926:21–27.

Wu ST, Ouyang Z, Olah TV, Jemal M. A strategy for liquid chromatography/tandem mass spectrometry based quantitation of pegylated protein drugs in plasma using plasma protein precipitation with water-miscible organic solvents and subsequent trypsin digestion to generate surrogate peptides for detection. Rapid Commun Mass Spectrom 2011;25:281–290.

Xu Y, Mehl JT, Bakhtiar R, Woolf EJ. Immunoaffinity purification using anti-PEG antibody followed by two-dimensional liquid chromatography/tandem mass spectrometry for the quantification of a PEGylated therapeutic peptide in human plasma. Anal Chem 2010;82:6877–6886.

Xue YJ, Akinsanya JB, Liu J, Unger SE. A simplified protein precipitation/mixed-mode cation-exchange solid-phase extraction, followed by high-speed liquid chromatography/mass spectrometry, for the determination of a basic drug in human plasma. Rapid Commun Mass Spectrom 2006;20:2660–2668.

Yamaguchi K, Takashima M, Uchimura T, Kobayashi S. Development of a sensitive liquid chromatography–electrospray ionization mass spectrometry method for the measurement of KW-5139 in rat plasma. Biomed Chromatogr 2000;14:77–81.

43

LC-MS BIOANALYSIS OF NUCLEOSIDES

LAIXIN WANG AND MIN MENG

43.1 INTRODUCTION

Nucleosides are the basic building blocks of the nucleic acids ribonucleic acid (RNA) and deoxyribonucleic acid (DNA). The chemical structure of nucleosides is composed of a base moiety and a ribose (for RNA) or deoxyribose (for DNA) sugar. Both RNA and DNA contain four natural nucleosides: adenosine (A), uridine (U)/thymidine (T), guanosine (G), and cytidine (C) (Figure 43.1).

Nucleoside analogs are widely used in anticancer, antiviral, and immunosuppressive therapy (Jansen et al., 2011). A vast majority of FDA-approved drugs to treat viral infections are nucleoside analogs, including but not limited to zidovudine (AZT or azidothymidine), didanosine (ddI or 2′,3′-dideoxyinosine), zalcitabine (ddC), emtricitabine (FTC), and lamivudine (3TC) for HIV; entecavir, lamivudine, and telbivudine for HBV; acyclovir for herpes simplex virus (HSV). With regard to chemotherapy, nucleoside or nucleobase analogs were among the first to be introduced for the treatment of cancer. Some examples include cytarabine (Ara-C) for acute myeloid leukemia, fluorouracil (5-FU) for skin cancer, and gemcitabine (dFdC) for breast, pancreatic, lung, and ovarian cancers.

Modified nucleosides have also been studied as potential biomarkers for the presence, development, and remission of many cancers (Dudley 2010; Hsu et al., 2011). Elevated levels of pseudouridine, 2-pyridone-5-carboxamide-N1-ribofuranoside, N^2,N^2-dimethyl-guanine, 1-methylguano sine, 2-methylguanosine, and 1-methyladenosine have been found in the urine from a variety of cancer patients (Seidel et al., 2006). 8-Hydroxy-2′-deoxyguanosine (8OHdG) is used as biomarkers for oxidative stress, either in cellular DNA or as an elimination product in urine (Bolin and Cardozo-Pelaez, 2007). Therefore, quantitative analysis of nucleoside analogs in biological matrices is critically important for both clinical diagnosis and drug developments. For this purpose, capillary electrophoresis (Yang et al., 2009), reverse-phase (RP) high-performance liquid chromatography (HPLC), ion-pairing (IP) HPLC, porous graphitic carbon (PGC) chromatography (Jansen et al., 2009a; Pabst et al., 2010; Vainchtein et al., 2010), and more recently, hydrophilic interaction chromatography (hydrophilic interaction liquid chromatography, HILIC) (Marrubini et al., 2010) have been used to separate nucleoside analogs as well as their metabolites. The use of mass spectrometry detector has significantly improved method specificity, sensitivity, and productivity (Jansen et al., 2011).

43.2 METHODS AND APPROACHES

43.2.1 Principles and Methodology

Liquid chromatography–mass spectrometry (LC-MS/MS) is a well-established technique in bioanalysis due to its inherent specificity, sensitivity, and high-throughput (Jansen et al., 2011). Reverse-phase HPLC has been commonly used for chromatography separations in bioanalytical assays. Unlike most of the nucleotide analytes that need appropriate ion-pair reagents in mobile phases to be adequately retained on reverse-phase columns (Cohen et al., 2010; Jensen et al., 2011), nucleoside analogs are usually able to be separated sufficiently without the assistance of ion-pair reagents (Renner et al., 2000; Zhao et al., 2004; Carli et al., 2009; Heudi et al., 2009; Hsu et al., 2011). However, since most of the nucleoside analogs are very polar compounds, the mobile

Handbook of LC-MS Bioanalysis: Best Practices, Experimental Protocols, and Regulations, First Edition. Edited by Wenkui Li, Jie Zhang, and Francis L.S. Tse.

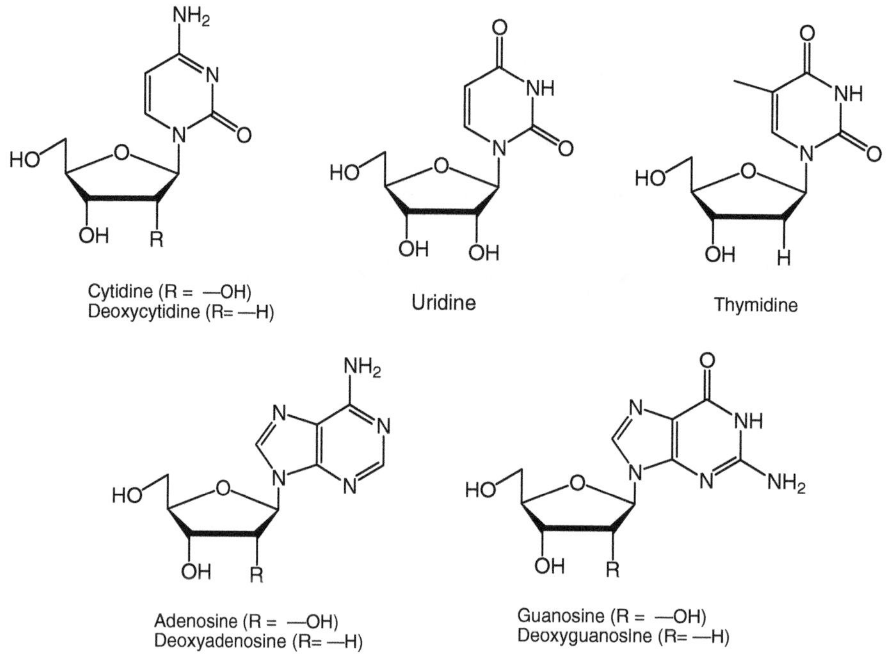

Cytidine (R = —OH)
Deoxycytidine (R= —H)

Uridine

Thymidine

Adenosine (R = —OH)
Deoxyadenosine (R= —H)

Guanosine (R = —OH)
Deoxyguanosine (R= —H)

FIGURE 43.1 Chemical structures of natural nucleosides and deoxyribonucleosides.

phases usually contain very high aqueous contents (up to 100% of aqueous). Therefore, the preferred reverse-phase columns are endcapped or polar embedded columns, such as Waters Atlantis dC18 (or T3), Varian Polaris C18, Phenomenex Hydro-RP, YMC hydrosphere, and Agilent Zorbax columns.

In addition to conventional reverse-phase columns, the PGC column has been used for LC-MS/MS analysis of nucleoside analogs. The retention mechanism on PGC column is complexing (Hanai, 2003). The stationary phase is involved in π–π interactions and dispersive interactions with aromatic analytes. The graphitic carbon surface acts as a Lewis base toward polar solutes. Therefore, nucleoside analogs usually have good retentions and separations on PGC columns (Pabst et al., 2010; Vainchtein et al., 2010). However, PGC columns have been found to be affected by redox reagents such as hydrogen peroxide or sulfite, which leads to retention time shifts (Jansen et al., 2009b; Melmer et al., 2010). The unpredictable and variable retention properties make it an important drawback for it to be used for GLP studies.

Restek Ultra IBD column is another unique column that can readily separate many nucleoside analogs under regular reverse-phase conditions. The Ultra IBD stationary phase is composed of a polar group within an alkyl chain. The polar group gives extra retention for many polar analytes as well as unique selectivity. It also provides a high level of base deactivation and compatibility with highly aqueous mobile phases. Therefore, the Ultra IBD column is a good alternative for LC-MS/MS analysis of nucleoside analogs.

Recently, HILIC is also increasingly being used for LC-MS/MS analysis of nucleoside analogs. HILIC is a version of normal phase chromatography. It is commonly believed that in HILIC, the mobile phase forms a water-rich layer on the surface of the polar stationary phase versus the water-deficient mobile phase, creating a liquid–liquid extraction system. The analyte is distributed between these two layers. However, HILIC is more than just simple partitioning and includes hydrogen donor interactions between neutral polar species as well as weak electrostatic mechanisms under the high organic solvent conditions. This distinguishes HILIC as a mechanism distinct from ion exchange chromatography. The more polar compounds will have a stronger interaction with the stationary aqueous layer than the less polar compounds. Thus, a separation based on a compound's polarity and degree of solvation takes place. Therefore, the peak shapes and separations of nucleoside analogs on HILIC columns are very sensitive to the compositions of the reconstitute solvents of the extracts and the mobile phases.

Nucleoside analogs need to be extracted out of the biological matrices prior to LC-MS/MS analysis. There are three major sample extraction methods: protein precipitation extraction (PPE), liquid–liquid extraction (LLE), and solid phase extraction (SPE). Since nucleoside analogs are very polar compounds, LLE usually does not provide adequate extraction recoveries. SPE (Renner et al., 2000; Hsu et al., 2011) and PPE (Carli et al., 2009; Heudi et al., 2009; Vainchtein et al., 2010) have been commonly used to purify nucleoside analogs from biological samples.

The quantitative LC-MS/MS analysis is usually conducted on a triple quadrupole-like instrument in selected reaction monitoring (SRM) mode, in which a particular fragment ion of a selected precursor ion was detected. The specific pair of *m/z* values associated to the precursor and fragment ions is referred to as a "transition." Multiple SRM transitions can be measured within the same experiment on the chromatographic timescale by rapidly toggling between the different precursor/fragment pairs (SRM transitions). Both positive-mode and negative-mode electronspray ion source (ESI) are commonly used for detection of nucleoside analogs during LC-MS/MS analysis (Jansen et al., 2011). Most recently, atmospheric pressure chemical ionization (APCI) was also used to detect adenosine (Van Dycke et al., 2010) and cytosine analogs (Honeywell et al., 2007; Montange et al., 2010).

43.2.2 Troubleshooting

Matrix stability is a potential problem for some nucleosides and their analogs. In plasma, adenosine is rapidly deaminated into hypoxanthinosine by deaminase. Cordycepin is deaminated by adenosine deaminase and quickly metabolized to 3′-deoxy-hypoxanthinosine (Tsai et al., 2010). Cytidine deaminase is another enzyme that converts cytidine, capecitabine, cytarabine, gemcitabine, and nucleoside analogs into corresponding metabolites in biological matrices (Figure 43.2). Therefore, low temperature and keeping the extraction procedure simple are critical for the success of the bioanalysis of these compounds. Adding appropriate inhibitor(s) to samples sometimes is necessary to stabilize the analytes during sample storage and/or extraction. Erythro-9-(2-hydroxy-3-nonyl) adenine (EHNA) is a potent adenosine deaminase inhibitor (Tsai et al., 2010, Van Dycke et al., 2010). It has been used to stabilize adenosine and cordycepin in biological matrices for quantitative analysis. Tetrahydrouridine (THU) is a common cytidine deaminase inhibitor. It has been successfully used for quantitative analysis of cytidine and its analogs in pharmacokinetic samples (Bowen et al., 2009; Li et al., 2011). Usually THU is added to the blood collection tubes prior to blood sample collections. The recommended final concentrations of THU are in the range of 10–100 μg/ml.

If the samples are not stabilized with appropriate inhibitor(s) at the time of collections, the samples need to be kept at reduced temperatures all the times as possible to minimize the analyte degradations. It is important to thaw and keep the plasma samples in a wet-ice bath. The inhibitor(s) should be added into the samples as early as possible during the sample preparations. In the example protocol, internal standard and ammonium hydroxide were added into the wells of the extraction plate before the plasma samples were aliquoted. As such the enzymes in the plasma samples were deactivated immediately in the extraction plate. In addition, ammonium hydroxide disrupts the analyte–matrix bindings to increase the extraction recovery and assay precision. Therefore, it is critical to keep the ammonium hydroxide solution at the appropriate concentration. Since ammonium hydroxide is very volatile, it is recommended to prepare this reagent fresh daily.

PPE is the simplest and most cost-effective method for nucleoside analog extraction. PPE involves denaturation (loss of tertiary and secondary structures) of proteins present in biological matrix by external stress, such as the use of a strong acid/base or appropriate amount of organic solvent. Most of the bioanalytical methods employ the addition of a minimum of three parts of organic solvent to one part of aqueous sample, followed by mixing and centrifugation or filtration. The extraction can be easily automated by using standard 96-well deep well plates and appropriate liquid handling systems. If the PPE filter plate is used, the extraction recovery of the analyte can be compromised due to nonspecific bindings (Kocan et al., 2006). Sometimes the filters of the PPE plates can also be clogged or leak precipitative particles. Therefore, centrifugation is a preferred method to remove the precipitates in the samples. To improve the assay accuracy and precision, it is critical to mix the internal standard

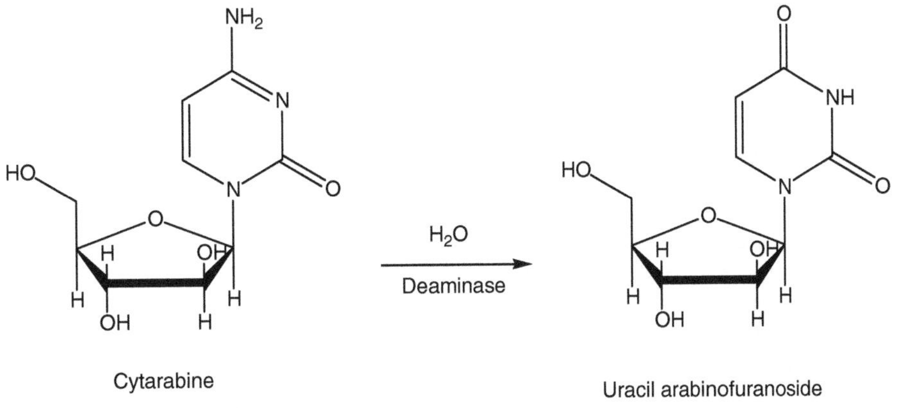

Cytarabine

H_2O

Deaminase

Uracil arabinofuranoside

FIGURE 43.2 Chemical structure of cytarabine and its conversion to uracil arabinofuranoside.

and the samples adequately in aqueous buffers before adding organic solvents. This step is particularly important for analytes that tend to form matrix-binding complexes.

Matrix effect is a significant source to impact the assay accuracy and precision of a LC-MS/MS method. Matrix effect results from coeluting matrix components that affect the ionization of the target analyte, either in ion suppression or ion enhancement. Matrix effect can be highly variable and can be difficult to control or predict. They are caused by numerous factors, including, but not limited to endogenous phospholipids, dose vehicles, formulation agents, and mobile phase modifiers. Typical "matrix effect" can manifest themselves by drifting or fluctuating analyte or internal standard response during analysis which can cause inaccuracy and imprecision of quantitative results. This may indicate that certain matrix components are being inadvertently retained on the LC column and subsequently eluted off the column unpredictably, which prevents consistent ionization of the analytes and/or internal standards as they enter the MS system. Some researchers have focused on optimizing sample preparation to reduce matrix effect, while others have focused on manipulating chromatographic parameters. It has been demonstrated that matrix effect is a greater concern for ESI than for APCI, and that the degree of the effect varies by instrument platform and manufacturer (Mei et al., 2003). A simple, yet effective measure to reduce this "matrix effect" is to incorporate either a forward flush or a back flush (flush the column in a reverse direction) with a high proportion of organic solvent (acetone, acetonitrile, methanol) into the LC program between sample injections to remove these unwanted matrix components from the LC column. Since PPE primarily removes the proteins from the plasma samples, it is important to optimize chromatographic condition to minimize the matrix effect. Tandem Labs had demonstrated that the residual plasma phospholipids in the extracts are a major source of ion suppression and identified specific phospholipids that cause matrix effect (Bennett and Liang, 2004). It has been found that none of the endogenous phospholipids were able to be eluted out of the reverse-phase columns until the organic components in the mobile phase reach 35% or above. Since most of the nucleoside analogs are eluted out of the column with less than 20% of organic solvent, endogenous phospholipids will likely stay on the column under these conditions. To prevent the phospholipids from accumulating on the LC column, the most effective way is to back flush the column with high organic solvent (such as 90/10 MeCN/water) after each injection.

Broad peak shape or low retention on LC column is another common problem for LC-MS/MS analysis of nucleoside analogs. Trifluoroacetic acid (TFA) has a great propensity to protonate nucleoside bases (Xing et al., 2004). Addition of TFA to mobile phases usually can improve the peak shape and separation of nucleoside compounds through ion-pair reverse-phase mechanism. Unfortunately, for MS analysis, TFA also causes signal suppression and must be dealt with if it is used in the mobile phases (Annesley, 2003; Li et al., 2007). There are several options that can be used to minimize or correct for ion suppression during LC-MS/MS analysis. Addition of acetic acid into mobile phases can alleviate the sensitivity loss due to the use of TFA (Shou and Naidong, 2005; Heudi et al., 2009). The use of surface-tension-lowering modifiers in the ESI source has also been reported to be successful when TFA is used as a mobile phase modifier. The postcolumn addition of acids and solvent carriers to displace TFA from compounds and aid ionization has also been used successfully. This is a process commonly referred as "TFA Fix" (Annesley, 2003).

Protocol: Simultaneous quantitation of cytarabine and uracil arabinofuranoside in human plasma using LC-MS/MS.

Test Material and Matrix

- Cytarabine (Toronto Research Chemicals)
- Uracil 1-β-D-arabinofuranoside (Sigma-Aldrich)
- Cytarabine-^{13}C3 (Toronto Research Chemicals)
- Human plasma (K_2EDTA) (BioChemEd)

Equipment and Reagents

- *Liquid handling system*: Microlab Nimbus 96, Hamilton Robotics, Inc.
- *HPLC system*: LC10AD HPLC Pumps and SCL-10AVP System Controller, Shimadzu Scientific Instruments
- *Autosampler*: CTC PAL Workstation, LEAP Technologies
- *Standard 6-port switching valve*: VICI Cheminert 10U-0263H
- *Mass spectrometer*: Sciex API4000 Triple Quadrupole MS/MS, Applied Biosystems, Inc.
- *HPLC column*: Varian Polaris C18-A 3μ, 2 × 50 mm
- *N,N*-dimethylformamide (DMF), HPLC grade, Burdick & Jackson
- Ammonium acetate, ACS grade, Sigma-Aldrich
- Ammonium hydroxide (NH_4OH) (~28% to 30% purity), ACS grade, Sigma-Aldrich
- Acetonitrile (MeCN), HPLC grade, EMD
- Acetone, HPLC grade, EMD
- Formic acid, ACS grade, EMD
- Methanol (MeOH), HPLC grade, EMD
- Water, deionized, Type 1 (typically 18.2 MΩ cm) or equivalent
- *Needle wash 1*: 50/50 water/DMF
- *Needle wash 2*: 0.1% FA in water
- *Mobile phase A*: 10 mM ammonium acetate, pH unadjusted

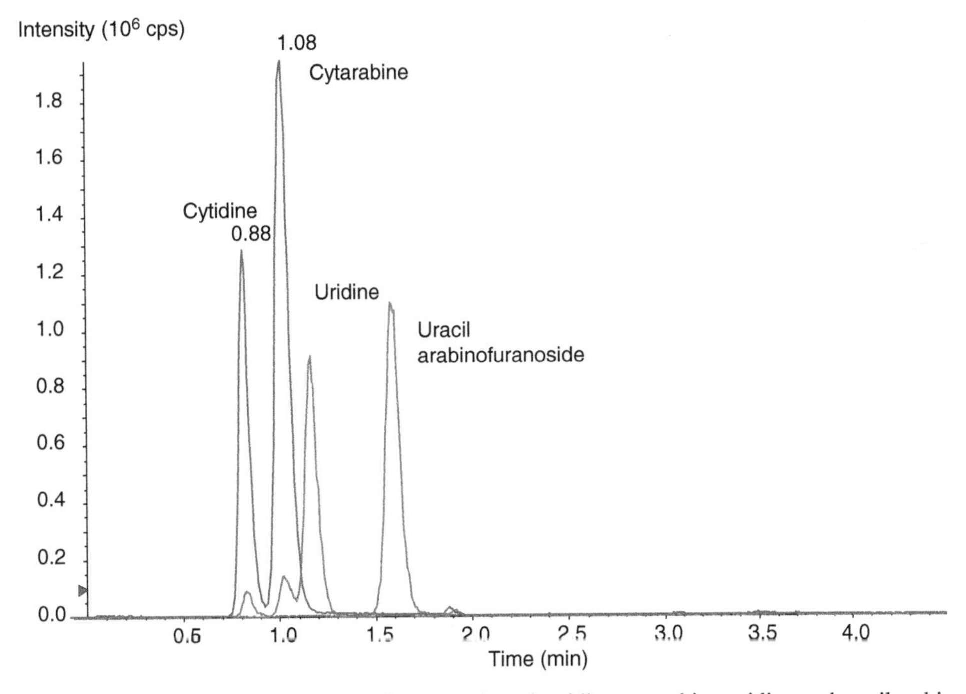

FIGURE 43.3 Typical chromatography separation of cytidine, cytarabine, uridine, and uracil arabinofuranoside in the described experimental LC-MS/MS conditions (cytidine and uridine are endogenous compounds that have same molecule weights with corresponding analytes, and therefore need to be adequately separated chromatographically).

- *Mobile phase B*: MeCN
- *Mobile phase C*: 90/10 MeCN/water (to be used for back-flushing column at 0.8 ml/min flow rate)

Method

Solution and Sample Preparation

1. Weigh an appropriate amount of cytarabine and Uracil 1-β-D-arabinofuranoside and dissolve in the required volume of DMF for 16 mg/ml solutions. All stock solutions are stored in polypropylene tubes at 1–8°C without foil protection.
2. Prepare a combo solution in DMF containing 8 mg/ml of cytarabine and Uracil 1-β-D-arabinofuranoside from the stock solution.
3. Prepare the standard curves and quality control (QC) samples in ice-cold human plasma in a wet-ice bath from the combo solution or other appropriate intermediate solution or samples. The calibration curves contain eight concentrations: 0.2, 0.5, 1, 2, 5, 20, 40, and 50 μg/ml. The QCs contain five levels: 0.2, 0.6, 20, 38, and 200 μg/ml.
4. Weigh an appropriate amount of cytarabine-^{13}C3 and dissolve in the required volume of 50/50 water/MeOH to give a 50 μg/ml internal standard (IS) working solution.

Sample Extraction

1. Add 50 μg/ml of cold IS working solution (50 μg/ml of cytarabine-^{13}C3 in 50:50 water:MeOH) into the corresponding wells of a clean 96-well plate sitting on a wet ice.
2. To all wells add 100 μl of freshly prepared 50/50 (28–30% ammonium hydroxide)/water (v/v).
3. Manually aliquot 50 μl of each sample into the corresponding wells of 96-well plate sitting on a wet-ice bath.
4. Vortex the samples for ~2 min.
5. To all samples add 700 μl of freshly prepared 5% ammonium hydroxide in MeCN.
6. Vortex-mix the samples at high speed on the multitube vortexer for ~10 min.
7. Centrifuge the samples at ~3000 rpm for ~5 min.
8. Transfer 100 μl of the supernatant into a clean 96-well plate.
9. Dry down the samples completely in a Turbo-vap at ~40°C.
10. Reconstitute in 500 μl of 10 mM ammonium acetate in water (pH unadjusted).
11. Vortex-mix the samples using the multitube vortexer for ~2 min on medium setting.

12. Centrifuge the samples at ~3000 rpm for 10 min.
13. Store the samples at room temperature until being analyzed.

LC-MS/MS Analysis

1. The LC system was equipped with a standard six-port switch valve (VICI Cheminert 10U-0263H) to facilitate the back flush of the LC column after the last peak of each sample was eluted out of the column. The LC conditions are as follows (Figure 43.3):
   a. *Flow rate*: 0.4 ml/min
   b. *Column temperature*: 30°C
   c. *Gradient*:

Time	0.0′	1.0′	2.0′	2.5′	3.0′	3.5′	4.0′
B%	0.0	0.0	E1[a]	30.0	E2[a]	0.0	0.0

[a]The column was back-flushed from minute 2 (E1) to minute 3 (E2) with mobile phase C at 0.8 ml/min.

2. The mass spectrometer was operated in positive ionization mode with ESI source. The MRM of *m/z* for cytarabine, uracil 1-β-D-arabinofuranoside, and cytarabine-$^{13}C3$ are 244.1 → 112.1, 245.1 → 113.1, and 247.1 → 115.1, respectively. The typical MS conditions are as follows: dwell time, 100 ms; source temperature, 500°C; IS voltage, 4000 V; declustering potential (DP), 25 V; curtain gas, 30 psi, nebulizer gas (GS1), 50 psi; turbo gas (GS2), 70 psi; collision energy, 35 eV.

REFERENCES

Annesley TM. Ion suppression in mass spectrometry. Clin Chem 2003;49(7):1041–1044.

Bennett P, Liang H. 2004. Overcoming matrix effects resulting from biological phospholipids through selective extractions in quantitative LC/MS/MS. Presented at the 52nd ASMS Conference. Available at http://www.tandemlabs.com/documents/PatrickASMSPaper.pdf. Accessed Apr 4, 2013

Bolin C, Cardozo-Pelaez F. Assessing biomarkers of oxidative stress: Analysis of guanosine and oxidized guanosine nucleotide triphosphates by high performance liquid chromatography with electrochemical detection. J Chromatogr B Analyt Technol Biomed Life Sci 2007;856(1–2):121–130.

Bowen C, Wang S, Licea-Perez H. Development of a sensitive and selective LC-MS/MS method for simultaneous determination of gemcitabine and 2,2-difluoro-2-deoxyuridine in human plasma. J Chromatogr B Analyt Technol Biomed Life Sci 2009;877(22):2123–2129.

Carli D, Honorat M, Cohen S, et al. Simultaneous quantification of 5-FU, 5-FUrd, 5-FdUrd, 5-FdUMP, dUMP and TMP in cultured cell models by LC-MS/MS. J Chromatogr B Analyt Technol Biomed Life Sci 2009;877(27):2937–2944.

Cohen S, Jordheim LP, Megherbi M, Dumontet C, Guitton J. Liquid chromatographic methods for the determination of endogenous nucleotides and nucleotide analogs used in cancer therapy: a review. J Chromatogr B Analyt Technol Biomed Life Sci 2010;878(22):1912–1928.

Dudley E. Analysis of urinary modified nucleosides by mass spectrometry. In: Banoub JH, Limbach PA, editors. Mass Spectrometry of Nucleosides and Nucleic Acids. New York: CRC Press; 2010. p 163–194.

Hanai T. Separation of polar compounds using carbon columns. J Chromatogr A 2003;989(2):183–196.

Heudi O, Barteau S, Picard F, Kretz O. A sensitive LC-MS/MS method for quantification of a nucleoside analog in plasma: application to in vivo rat pharmacokinetic studies. J Chromatogr B Analyt Technol Biomed Life Sci 2009;877(20–21):1887–1893.

Honeywell R, Laan AC, van Groeningen CJ, et al. The determination of gemcitabine and 2′-deoxycytidine in human plasma and tissue by APCI tandem mass spectrometry. J Chromatogr B Analyt Technol Biomed Life Sci 2007;847(2):142–152.

Hsu WY, Lin WD, Tsai Y, et al. Analysis of urinary nucleosides as potential tumor markers in human breast cancer by high performance liquid chromatography/electrospray ionization tandem mass spectrometry. Clin Chim Acta 2011;412(19–20):1861–1866.

Jansen RS, Rosing H, Schellens JH, Beijnen JH. Simultaneous quantification of 2′,2′-difluorodeoxycytidine and 2′,2′-difluorodeoxyuridine nucleosides and nucleotides in white blood cells using porous graphitic carbon chromatography coupled with tandem mass spectrometry. Rapid Commun Mass Spectrom 2009a;23(19):3040–3050.

Jansen RS, Rosing H, Schellens JH, Beijnen JH. Retention studies of 2′-2′-difluorodeoxycytidine and 2′-2′-difluorodeoxyuridine nucleosides and nucleotides on porous graphitic carbon: development of a liquid chromatography-tandem mass spectrometry method. J Chromatogr A 2009b;1216(15):3168–3174.

Jansen RS, Rosing H, Schellens JH, Beijnen JH. Mass spectrometry in the quantitative analysis of therapeutic intracellular nucleotide analogs. Mass Spectrom Rev 2011;30(2):321–343.

Kocan G, Quang C, Tang D. Evaluation of protein precipitation filter plates for high-throughput LC-MS/MS bioanalytical sample preparation. Am Drug Discov 2006;1(3):21–24.

Li W, Luo S, Li S, et al. Simultaneous determination of ribavirin and ribavirin base in monkey plasma by high performance liquid chromatography with tandem mass spectrometry. J Chromatogr B Analyt Technol Biomed Life Sci 2007;846(1–2):57–68.

Li W, Zhang J, Tse FL. Strategies in quantitative LC-MS/MS analysis of unstable small molecules in biological matrices. Biomed Chromatogr 2011;25(1–2):258–277.

Marrubini G, Mendoza BE, Massolini G. Separation of purine and pyrimidine bases and nucleosides by hydrophilic interaction chromatography. J Sep Sci 2010;33(6–7):803–816.

Mei H, Hsieh Y, Nardo C, et al. Investigation of matrix effects in bioanalytical high-performance liquid chromatography/tandem mass spectrometric assays: application to drug discovery. Rapid Commun Mass Spectrom 2003;17(1):97–103.

Melmer M, Stangler T, Premstaller A, Lindner W. Solvent effects on the retention of oligosaccharides in porous graphitic carbon liquid chromatography. J Chromatogr A 2010;1217(39):6092–6096.

Montange D, Bérard M, Demarchi M, et al. An APCI LC-MS/MS method for routine determination of capecitabine and its metabolites in human plasma. J Mass Spectrom 2010;45(6):670–677.

Pabst M, Grass J, Fischl R, et al. Nucleotide and nucleotide sugar analysis by liquid chromatography–electrospray ionization–mass spectrometry on surface-conditioned porous graphitic carbon. Anal Chem 2010;82(23):9782–9788.

Renner T, Fechner T, Scherer G. Fast quantification of the urinary marker of oxidative stress 8-hydroxy-2′-deoxyguanosine using solid-phase extraction and high-performance liquid chromatography with triple-stage quadrupole mass detection. J Chromatogr B Biomed Sci Appl 2000;738(2):311–317.

Seidel A, Brunner S, Seidel P, Fritz G, Herbarth O. Modified nucleosides: an accurate tumour marker for clinical diagnosis of cancer, early detection and therapy control. Br J Cancer 2006;94(11):1726–1733.

Shou WZ, Naidong W. Simple means to alleviate sensitivity loss by trifluoroacetic acid (TFA) mobile phases in the hydrophilic interaction chromatography–electrospray tandem mass spectrometric (HILIC-ESI/MS/MS) bioanalysis of basic compounds. J Chromatogr B Analyt Technol Biomed Life Sci 2005;825(2):186–192.

Tsai YJ, Lin LC, Tsai TH. Pharmacokinetics of adenosine and cordycepin, a bioactive constituent of Cordyceps sinensis in rat. J Agric Food Chem 2010;58(8):4638–4643.

Vainchtein LD, Rosing H, Schellens JH, Beijnen JH. A new, validated HPLC-MS/MS method for the simultaneous determination of the anti-cancer agent capecitabine and its metabolites: 5′-deoxy-5-fluorocytidine, 5′-deoxy-5-fluorouridine, 5-fluorouracil and 5-fluorodihydrouracil, in human plasma. Biomed Chromatogr 2010;24(4):374–386.

Van Dycke A, Verstraete A, Pil K, et al. Quantitative analysis of adenosine using liquid chromatography/atmospheric pressure chemical ionization–tandem mass spectrometry (LC/APCI-MS/MS). J Chromatogr B Analyt Technol Biomed Life Sci 2010;878(19):1493–1498.

Xing J, Apedo A, Tymiak A, Zhao N. Liquid chromatographic analysis of nucleosides and their mono-, di- and triphosphates using porous graphitic carbon stationary phase coupled with electrospray mass spectrometry. Rapid Commun Mass Spectrom 2004;18(14):1599–1606.

Yang FQ, Ge L, Yong JW, Tan SN, Li SP. Determination of nucleosides and nucleobases in different species of Cordyceps by capillary electrophoresis–mass spectrometry. J Pharm Biomed Anal 2009;50(3):307–314.

Zhao M, Rudek MA, He P, et al. Quantification of 5-azacytidine in plasma by electrospray tandem mass spectrometry coupled with high-performance liquid chromatography. J Chromatogr B Analyt Technol Biomed Life Sci 2004;813(1–2):81–88.

44

LC-MS BIOANALYSIS OF NUCLEOTIDES

Sabine Cohen, Marie-Claude Gagnieu, Isabelle Lefebvre, and Jérôme Guitton

44.1 INTRODUCTION

A nucleotide is composed of a nucleobase (nitrogenous base), a five-carbon sugar (either ribose or 2-deoxyribose), and one, two, or three phosphate groups. The phosphate groups form bonds with the 2, 3, or 5 carbon of the sugar, with the 5-carbon site most common (Figure 44.1). Cyclic nucleotides form when the phosphate group is bound to two of the sugar's hydroxyl groups. Nucleotides contain either a purine or a pyrimidine base. Ribonucleotides are nucleotides in which the sugar is ribose, while deoxyribonucleotides are nucleotides in which the sugar is deoxyribose (Figure 44.1) (Koolman and Roehm, 2005). Nucleotides are molecules that, when joined, make up the individual structural units of the nucleic acids RNA and DNA. In addition, nucleotides participate in cellular signaling (cGMP and cAMP) and are incorporated into important cofactors of enzymatic reactions (coenzyme A, FAD, FMN, and NADP+). Endogenous ribonucleotides and deoxyribonucleotides, especially the triphosphates (ATP and GTP), have a fundamental role in cell function and play central roles in metabolism, in which capacity they serve as sources of chemical energy (Rodwell, 2003).

Nucleoside analogs are commonly used in immunosuppressive, anticancer, and antiviral therapies, and their mechanism of action is based on their structural similarity leading to compete with natural nucleotides into the nucleotide metabolic pathway.

The highly active antiretroviral therapy used for the treatment of patients infected by HIV involves different classes of drugs: nucleoside reverse transcriptase inhibitors (NRTIs), non-nucleoside reverse transcriptase inhibitors, integrase inhibitors, protease inhibitors, fusion inhibitors, and coreceptor antagonists (Back et al., 2005). NRTIs are administered as prodrugs, which require host cell entry and phosphorylation by cellular kinases. Several kinases are involved in these sequential enzymatic phosphorylation steps, such as thymidine kinase, adenylate kinase, deoxycytidine kinase, or 5′-nucleoside diphosphate kinase. Triphosphate metabolites are the active components of all NRTIs (Figure 44.2). Tenofovir and adefovir are nucleotide reverse transcriptase inhibitors whose active form is a diphosphate. All NRTIs compete with endogenous analogs and stop DNA elongation. When a virus infects a cell, reverse transcriptase copies the viral single-stranded RNA genome into a double-stranded viral DNA. The viral DNA is then integrated into the host chromosomal DNA, which then allows host cellular processes, such as transcription and translation to reproduce the virus. Since the 3′-hydroxyl group is lacking into the pentose structure of the NRTIs, the formation of a 3′,5′-phosphodiester bond between the NRTIs and 5′-nucleoside triphosphates does not occur (Arts and Hazuda, 2012). Thus, NRTIs block reverse transcriptase's enzymatic function and prevent completion of synthesis of the double-stranded viral DNA, thus preventing HIV from multiplying.

The nucleotide metabolism is also one of the cellular targets in cancer treatment. Deoxyadenosine analogs such as fludarabine or cladribine, deoxycytidine analogs such as gemcitabine or cytarabine, and fluoropyrimidine such as 5-fluorouracil are used in the treatment of cancer. As described earlier for NRTIs, nucleoside analogs are activated by phosphorylation by intracellular kinases. Several mechanisms of action are described for these anticancer drugs such as incorporation into DNA leading to the production of non-natural nucleic acids, chain termination, and cell cycle arrest.

Handbook of LC-MS Bioanalysis: Best Practices, Experimental Protocols, and Regulations, First Edition. Edited by Wenkui Li, Jie Zhang, and Francis L.S. Tse.

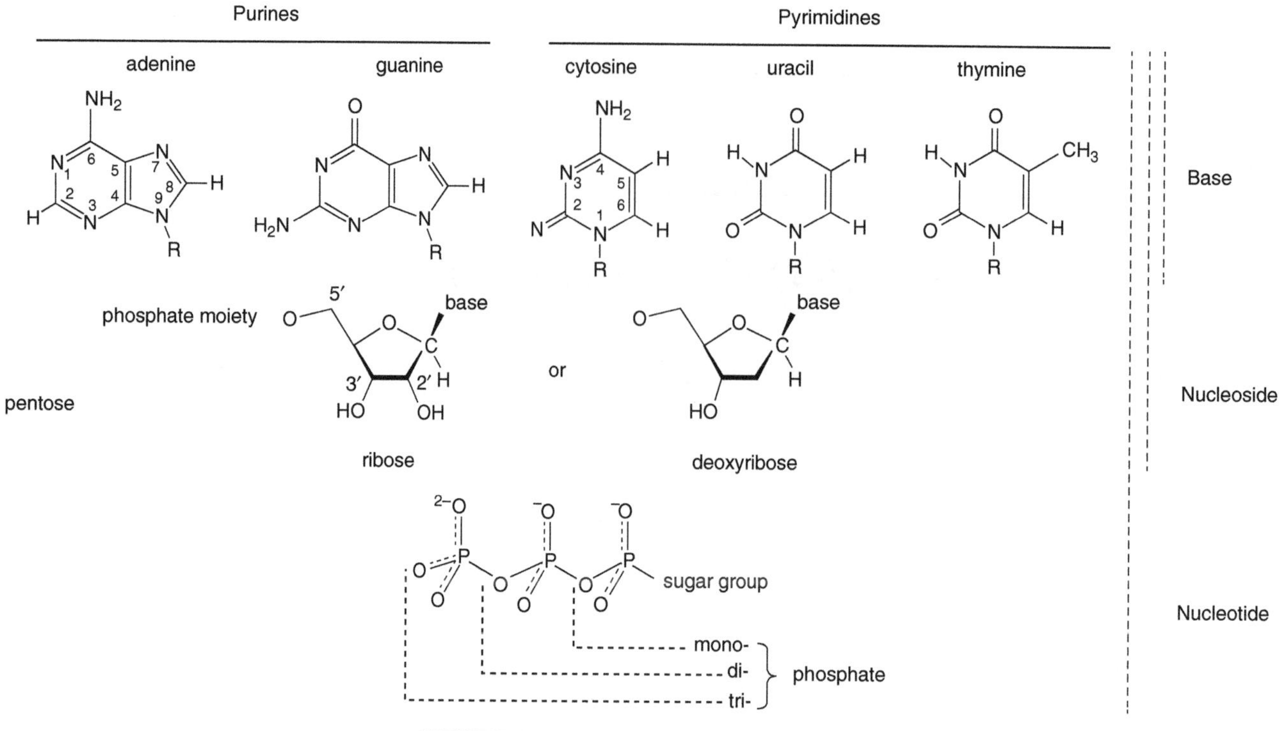

FIGURE 44.1 Basic structure of nucleotides.

Gemcitabine diphosphate and the triphosphorylated forms of deoxyadenosine analogs are inhibitors of ribonucleotide reductase (RNR or ribonucleoside diphosphate reductase) (Xie and Plunkett, 1995). RNR catalyzes the formation of deoxyribonucleotides from ribonucleotides. Thus, the substrates for RNR are ADP, GDP, CDP, and UDP, and RNR plays a critical role in regulating the total rate of DNA synthesis. 5-FU induces cell death by RNA and DNA incorporation of triphosphorylated forms and inhibits the enzyme thymidylate synthase by the fluorodeoxyuridine monophosphate leading to the depletion in thymidine triphosphate (Longley et al., 2003).

From a bioanalysis point of view, the technical approach is the same for both the natural and the analog nucleotides. Numerous LC-MS/MS assays have been published for the quantification of natural as well as analog nucleotides. This abundant production of articles reflects not only the scientific relevance to quantify nucleotides but also probably the analytical complexity of their determination. The LC-MS/MS quantification of nucleotides is challenging for several aspects: (1) highly variable concentration between ribonucleotides and deoxyribonucleotides, (2) chemical and biological instabilities and/or reactivity, (3) the high polarity with the presence of a number of phosphate groups, (4) the potential interferences from endogenous nucleotides as well as between endogenous and analog nucleotides, (5) ion suppression occurring into the source of the mass detector due to the composition of mobile phase selected, and (6) at last but not the least the matrix effect due to the biological complexity of cells.

In this chapter, we present one protocol for the preparation of peripheral blood mononuclear cells (PBMCs), and two for the extraction procedure. The different possibilities to build a standard curve according to the determination of natural or analog nucleotides are proposed. Then, the expression of nucleotides concentration is mentioned. Most chromatographic separations are based on graphite or C18 stationary phase. Both columns are usually used with ion pairing. Few assays described weak anion exchange (WAX) or hydrophilic liquid interaction chromatography (HILIC). We selected six protocols indicative of the approaches described for the separation of nucleotides.

44.2 PREANALYTICAL METHODS AND APPROACHES

44.2.1 Matrix Used for the Determination of Nucleotides

The sample preparation is probably the major critical step in the analytical process for the quantification of nucleotides. The initial phase consists of cell isolation, followed by the extraction of nucleotides from cells. As previously described, nucleoside analogs undergo intracellular anabolic phosphorylation to be converted into the triphosphorylated

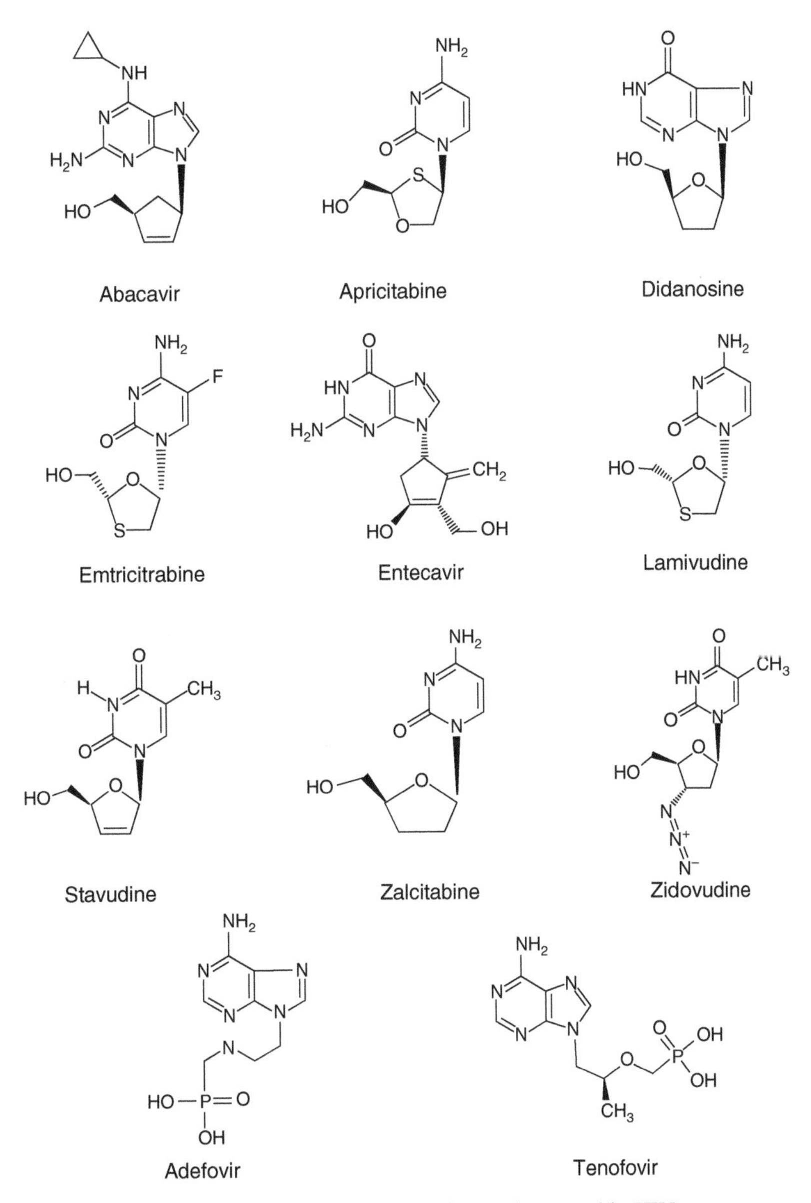

FIGURE 44.2 Structure of nucleoside and nucleotide analogs used in HIV-treatment as reverse-transcriptase inhibitors. Triphosphate metabolites are the active form of all these analogs.

form corresponding to the active compound. Since the nucleoside triphosphate is trapped into the cell, the pharmacokinetic profile is different from the nucleoside form in the plasma. For example, the half-life of the triphosphorylated form of the NRTIs is longer than the plasma half-life of the parent compound (Back et al., 2005). Thus, determination of intracellular concentration, that is, at the target site (PBMCs or leukemic cells) of the active compound, should be a more relevant data according to pharmacokinetics, clinical efficacy, and toxicity point of view (Jansen et al., 2012). PBMCs can be isolated using either Ficoll gradient centrifugation or a commercial tube for cell preparation. Blood samples should be brought to the laboratory as quickly as possible and procedures of cell isolation should be performed rapidly and at 4°C to inhibit enzymatic activity (cells containing enzymes involved in the nucleoside metabolism), to inhibit potential instability of certain compounds, and to limit the active drug efflux. Erythrocyte contamination should be avoided for two reasons. On the one hand, endogenous nucleotides from erythrocytes may enhance the matrix effect. On the other hand, phosphorylated nucleoside analogs may also be present in the erythrocyte and contaminate the PBMC fraction (Durand-Gasselin et al., 2007).

Numerous assays described the LC-MS/MS quantification of nucleotides in PBMCs. The protocol proposed later is used in our laboratory. It is a combination of several protocols described in the literature.

Equipment and Reagents

- Cell Preparation Tubes® (CPTs®) (Becton Dickinson).
- NaCl 0.9%.
- Hypotonic ammonium chloride solution containing NH_4Cl (3.5 g) and $NaHCO_3$ (0.036 g) in 500 ml of distilled water.

Method

1. 7 ml of peripheral blood is drawn in CPTs®.[a]
2. PBMCs are separated according to manufacturer protocol in order to obtain a diffuse layer containing cells just above the gel (centrifugation at room temperature (18–25°C) for 25 min at 1500–1800*g*).[b]
3. After removing the plasma, PBMCs are poured into a tube and washed using NaCl 0.9% stored at 4°C with centrifugation at 4°C (for 10 min at 300*g*).
4. Remaining cells present on the CPTs® are poured again with NaCl 0.9%.
5. After centrifugation at 4°C (for 10 min at 300*g*) and removing the washing solution, cells are counted using an appropriate method (classical cell recovery: 5–20 × 10^6 cells for 7 ml of peripheral blood).
6. If necessary, in order to eliminate red blood cell (RBCs) lysis, 2 ml of a hypotonic ammonium chloride solution at 4°C is added to the cell pellet.
7. After 2 min, centrifugation is performed to eliminate RBCs lysis solution, the supernatant is discarded, and the remaining cells are washed in cold NaCl 0.9% solution.
8. Dry pellets are stored at −80°C.

Notes

[a]The use of CPTs® is less time-consuming than the conventional Ficoll gradient centrifugation method. Recovery of PBMCs is equivalent with both usual methods of preparation: Ficoll Histopaque or CPTs®.

[b]It is necessary to collect cell ring as soon as possible since some nucleotides are particularly unstable (such as d4T-TP with a degradation rate about 40% in 40 min).

44.2.2 Extraction from Cell Matrix

Several sample extraction procedures of nucleotides from cells have been proposed (Cohen et al., 2010; Jansen et al., 2011). The point common to all methods is a precipitation step. Strong acid and particularly perchloric acid was used. Neutralization of the supernatant is then performed using a basic solution. Although extraction recovery is reported as satisfactory, attention must be given to the eventual impact of strong acid on the stability of the compounds of interest, the chromatographic separation, and the pollution of mass detector. According to these potential limitations, protein precipitation is currently performed with a mixture of acetonitrile or methanol (proportion from 60% to 80%) and water. Under this condition, the extraction yield is mostly between 70% and 100%. In several assays is added a treatment such as freezing–thawing cycle, sonication in an ice bath, or an additional step with solid phase extraction (SPE). At the end of the procedure the supernatant containing organic solvent can be evaporated in order to concentrate the nucleotides.

Here, we focused on the protocols used for the extraction of nucleotides (mono-, di-, and tri-) from cells obtained after preparation of PBMCs or cell culture.

Protocol 1: Extraction of the nucleotides by a simple precipitation. This cellular lysis technique is most often described in the literature.

Equipment and Reagents

- Ice-cold methanol
- Tris–HCl buffer (50 mM, pH 5)

Method

1. 1–5 ml of a mixture containing methanol/Tris–HCl buffer 70/30 (v/v) is added to each cell pellet.[a,b,c]
2. Cells are disrupted by vortexing and scraping the tube (for total extraction of nucleotides).
3. Centrifugation of 18,000*g* for 15 min at 4°C to remove cellular debris.
4. Supernatant is evaporated to dryness at 40°C under a nitrogen stream.
5. Residue is dissolved in a volume of water (or mobile phase) according to the needed sensitivity of the method.

Notes

[a]Appropriate amounts of working calibration solutions for calibration standards and internal standard should be added just before the addition of the mixture.

[b]The proportion of methanol in the mixture must not exceed 70% since a precipitation of nucleotide triphosphate may occur with higher proportion.

[c]For the cellular pellets treatment, other protocols replace methanol/Tris–HCl buffer by perchloric acid or a mixture of methanol/water (60/40 or 70/30, v/v).

Protocol 2: A combination of a methanol precipitation followed by SPE. From our laboratory experience, the signal-to-noise ratio is improved under this combined extraction.

Equipment and Reagents

- SPE cartridge OASIS® WAX (60 mg) (Waters).
- Ammonium acetate solution (50 mM), pH adjusted to 4.5 with acetic acid.
- *Solution A*: methanol/water/ammonium acetate solution (80/15/5, v/v/v) prepared extemporaneously.

Method

1. Conditioning of the SPE cartridge with 2 ml of methanol, and then with 2 ml of ammonium acetate solution.[a]
2. Loading onto the SPE column of the sample preparation containing cell pellet previously precipitated with a mixture of methanol/water (60/40, v/v).
3. Wash the SPE cartridge with 2 ml of ammonium acetate solution.
4. Elution of the analytes from the SPE column with 2 ml of the solution A.[b]
5. Eluate is evaporated to dryness at 40°C under nitrogen.
6. Residue is reconstituted in a volume of water (or mobile phase) according to the sensibility of the method.

Notes

[a]Conditioning of the SPE cartridge with ammonium acetate solution improves nucleotides retention. At pH 4.5, WAX sorbent (pK_a ~6) is under ionized form (R_4–N^+), which strongly improved ionic interactions with phosphate groups of the nucleotides.

[b]Elution of the nucleotides is obtained by using a basic pH since WAX sorbent became not ionized.

44.2.3 Preparation of Calibration Standard Standards and QCs and Use of Internal Standard

The challenge for quantitative analysis of nucleotide analogs and endogenous nucleotides is clearly different. For nucleotide analogs, as these compounds are xenobiotics, they are naturally absent from cells. Blank cell pellet can be spiked with the compounds of interest for preparation of the standard standards and quality controls (interference of endogenous compounds and/or endogenous nucleotides are excluded here, for this point see Section 44.3). However, few nucleotide analogs are commercially available. Thus, the laboratories must obtain them from drug companies or require performing an elaborate and tedious synthesis. Recently, Jansen et al. proposed a method to phosphorylate nucleosides into their mono-, di-, and triphosphate derivatives. From stable isotope labeled nucleosides, a mixture of labeled nucleotides can be synthesized, which can be used as internal standard (Jansen et al., 2010).

For endogenous nucleotides, the compounds are naturally present in cells. The first approach is to base the quantification on the addition calibration method. Stable isotope labeled nucleotides may be used as internal standard. The second approach is to build the standard curve with the stable isotope labeled analogs commercially available. For example, stable isotope labeled dATP is used as a surrogate for the quantification of dATP. It is considered that the response factor on LC-MS/MS system is identical for both compounds. Moreover, special attention should be given concerning the purity of the molecule and the percentage of labeling. With this approach, the unlabeled internal standard is used such as bromo- or chloro-nucleotide (Cohen et al., 2009). From our experience, the last approach is more accurate and precise than the addition calibration method (unpublished data).

44.2.4 Expression of Nucleotide Concentrations

The concentration of nucleotides is related to the number of cells present in the sample. The number of cells must be accurately measured and different parameters can be used. The most common approach involves counting the number of cells using microscope and Malassez cell, for example, hemocytometer or flow cytometry. Manual cell counting is time-consuming and requires some experience. Sources of error may also be observed with automated cell counting device in the case of, for example, too many or too few cells into the samples, contamination of the apparatus. The determination of the protein concentration into the cell pellet is also proposed. Several protein assays using bovine serum albumin as standard are available. Another approach is to quantify DNA since a relationship between DNA content in the cell pellet and the PBMC count has been established. DNA detection is performed using the fluorescent dye, SYBR green (Benech et al., 2004).

44.2.5 LC-MS/MS

44.2.5.1 Chromatography Based on Alkyl Silica Analytical Column Analysis of nucleotides with C18 analytical column is based on the addition of ion-pairing agent in the mobile phase. The principle of this separation mode involves the dynamic modification of the surface of a reversed-phase chromatographic packing by the adsorption of hydrophobic ions. The ion-pair reagent attaches to the stationary phase and creates a charged surface thus increasing the interaction with the column resulting in greater retention of the sample. The ion-pair chromatography reagents are large ionic molecules having a charge opposite to the analyte of interest, as well as a hydrophobic region to interact with the stationary phase. The majority of separation are performed on conventional alkyl silica stationary phases (C18 and C8). The eluent pH is a key factor to control analyte charge status. The selectivity can be controlled on both ion-pairing reagents concentration and organic modifier percentage in the eluent. Ion-pair reagents commonly used are either anionic or cationic surfactants such as alkyl sulfonates or tetraalkyl ammonium salts, which are typically nonvolatile and likely to cause ion suppression when used with LC-MS. However, this type of liquid chromatography can be coupled with mass spectrometry detection if low amounts of regular ion-pairing agents are used (Claire, 2000; Vela et al., 2007) or if volatile ion-pairing agents are present in the mobile phase (Fung et al., 2001; Becher et al., 2002a). Another alternative is to prefer column miniaturization in order to reduce the amount of ion-pairing introduced into the ESI source (Cichna et al., 2003).

Protocol 1: The simultaneous quantification of intracellular natural (dATP, dCTP, dGTP, TTP) and antiretroviral nucleosides (abacavir (ABC), emtricitabine (FTC), tenofovir disoproxil fumarate (TDF), amdoxovir (DAPD), zidovudine (ZDV)), and their intracellular mono- and triphosphorylated metabolites.

This assay is based on ion-pairing chromatography coupled with tandem mass spectrometry in positive or negative electrospray ionization. Compounds were measured in human macrophages obtained from PBMC after Histopaque separation technique, cultured in the presence of monocyte colony stimulating factor and exposed to antiretroviral nucleosides (Fromentin et al., 2010).

Equipment and Reagents

- Hexylamine, ammonium phosphate, acetonitrile.
- *Chromatography system*: Dionex Packing Ultimate 3000 modular LC system.
- *Mass spectrometer*: TSQ Quantum Ultra triple quadrupole (Thermo Electron).
- *Analytical column*: Hypersil Gold-C18 column 100 mm × 1 mm, 3 μm particle size (Thermo Electron).

Method

1. Chromatographic separation was achieved using a linear gradient:
   - *Mobile phase* A: 2 mM ammonium phosphate buffer[a] containing 3 mM hexylamine[b] (pH 9.2).[c]
   - *Mobile phase* B: acetonitrile.
   - Gradient starts with 9% acetonitrile and reaches 60% in 15 min.
   - Then, immediate return to initial conditions for an equilibration time (14 min).
   - Flow rate at 50 μl/min, oven temperature: 30°C.[d]
   - *Injection volume*: 45 μl with autosampler at 4°C.
   - Cleaning solution sprayed on the ion source before 2 min and after 13 min[e]
2. Transitions and detection mode:
   - All nucleosides and nucleotides in positive mode from 2 to 11 min (3TC: 230 → 112; 3TC-MP: 310 → 112; 3TC-TP: 470 → 112; DXG: 254 → 152; DXG-MP: 334 → 152; DGX-TP: 494 → 152; TFV: 288 → 176; TFV-DP: 448 → 270; ZDV-MP: 348 → 81; ZDV-TP: 506 → 380; FTC-TP: 488 → 130; CBV-TP: 488 → 52; dATP: 492 → 136; dGTP: 508 → 152; dCTP: 468 → 112; TTP: 483 → 81; [$^{13}C^{15}N$]dATP: 507 → 146; [$^{13}C^{15}N$]dGTP: 523 → 162; [$^{13}C^{15}N$]dCTP: 480 → 119; [$^{13}C^{15}N$]TTP: 495 → 134)
   - ZDV-TP between 11 and 13 min in negative mode.[f]

Notes

[a]Four compositions of buffer A were tested: ammonium phosphate at 1 and 2 mM and ammonium hydrogen carbonate at 10 and 20 mM. Ammonium hydrogen carbonate was associated with peak tailing, decrease in effective plate number, and decomposition resulting in higher variability of chromatography. Ammonium phosphate at 2 mM allowed reproducible chromatographic conditions with a better peak shape for all nucleotides (except for DXG-TP).

[b]Hexylamine at 12 mM was tested and gave the lowest peak tailing, the highest effective plate number, and the greatest sensitivity, but the ion source was clogged following 10 injections and the sensitivity was decreased. The chosen concentration of 3 mM was sufficient to obtain a symmetric peak without clogging the MS source.

[c]Different pH (7, 8, 9.2, and 10) of buffer A were tested in order to obtain the best quality of the chromatography (peak tailing and capacity and effective plate number), and pH 9.2 was optimal for all nucleotides.

[d]Oven temperature of 30°C was the best value between 25°C and 35°C to obtain a significant improvement of the chromatography.

[e]Cleaning solution is a mixture of acetonitrile and water (80/20, v/v) sprayed on the ion source in order to avoid loss of signal.

[f]Optimization of the chromatographic method is achieved in order to obtain an efficient separation between the last two eluted nucleotides CBV-TP and ZDV-TP to enable a switch in polarity (from positive to negative) at 11 min.

Protocol 2: A combination of two assays proposed by the same team for the quantification of intracellular antiretroviral nucleosides (stavudine (d4T), lamivudine (3TC), and didanosine (ddi)).

This assay is also based on ion-pairing chromatography coupled with tandem mass spectrometry in negative electrospray ionization. However, the ion-pairing agent used in these assays is different from the one used in Protocol 1. Since numerous methods published are based on ion-pairing chromatography, we propose two distinct protocols. Antiretroviral nucleotides were measured in human PBMC obtained from whole blood after Ficoll Histopaque or CPTs separation techniques (Pruvost et al., 2001; Becher et al., 2002).

Equipment and Reagents

- *N′*,*N′*-dimethylhexylamine (DMH), acetonitrile, methanol, formic acid ammonium salt.
- *Internal standard*: 2-chloroadenosine 5′-triphosphate (Cl-ATP).[a]
- *Chromatography system*: HP1100 Series liquid chromatograph (Agilent Technologies).
- *Mass spectrometer*: API 3000 (Sciex, Applied Biosystems).
- *Analytical column*: Supelcogel ODP-50, 5 μm, 150 × 2.1 mm (Sigma-Aldrich-Supelco)[b].

Method

1. Chromatographic separation was achieved using a linear gradient:
   - *Mobile phase A*: DMH (10 mM) and ammonium phosphate (3 mM), pH 11.5.
   - *Mobile phase B*: DMH (20 mM) and ammonium formate (6 mM)/acetonitrile (1/1, v/v).
   - *Gradient*:
     - From 0 to 2 min: A/B = 70/30 (v/v).
     - From 2 to 12 min: linear gradient from A/B = 70/30 to A/B = 35/65 (v/v).
     - From 12 to 13 min: linear gradient from A/B = 35/65 to A/B = 0/100 (v/v).
     - From 13 to 16 min: B = 100.
   - Then, immediate return to initial conditions.
   - Equilibration time 10 min, with A/B = 70/30 (v/v).
   - Flow rate at 300 μl/min, oven temperature: 30°C.
   - Injection volume: 40 μl with autosampler at 4°C.
2. Transitions and detection mode:
   - All molecules are detected using negative electrospray ionization.
   - d4T-TP: 463 → 159; dT-TP: 481 → 159; ddA-TP: 474 → 159; dA-TP: 490 → 159; 3TC-TP: 468 → 159; dC-TP: 466 → 159; Cl-ATP: 540 → 159 and 542 → 159.

Notes

[a]Cl-ATP is selected as internal standard rather than 8-bromoadenosine 5′-triphosphate, adenosine-5′-*O*-(3-thiotriphosphate) or guanosine-5′-*O*-(3-thiotriphosphate) for the following reasons: no degradation in PBMC extracts, baseline regular on each side of the Cl-ATP peak, no interfering peak at the retention time on blank extract.

[b]Several analytical columns were tested: SMT-C18, 5 μm, 150 × 2.1 mm (Separation Methods Technologies, Interchim); PLRP-S, 5 μm, 150 × 2.1 mm (Polymer Laboratories); X-Terra MS C18, 5 μm, 150 × 2.1 mm (Waters); Supelcogel ODP-50, 5 μm, 150 × 2.1 mm (Sigma-Aldrich-Supelco). The better peak efficiency was obtained with Supelcogel ODP column.

44.2.5.2 Chromatography Based on Porous Carbon Analytical Column Another approach to analyze nucleotides is the use of carbon column (Xing et al., 2004; Wang et al., 2009). The retention mechanisms on porous graphitic carbon are complex. The strength of interaction for individual analytes is largely determined by the molecular area of the analyte in contact with the surface, and the strength of interaction is dependent on the nature of the functional groups present at the point of interaction with the flat surface. Porous carbon material excels at the separation of highly polar compounds with closely related structures. The retention mechanism is a mixture of hydrophobic and electrostatic interactions. This support has the disadvantage of being very retentive and so some compounds may not be eluted from porous carbon material. This stationary phase is totally stable across the entire pH range of 1–14. Controlling the pH of the mobile phase, and hence the degree of ionization of the analyte, can be a strong tool in the development of a chromatographic method on a porous carbon column.

Protocol 3: The simultaneous determination of 2′,2′-difluorodeoxycytidine (gemcitabine, dFdC), 2′,2′-difluoro deoxyuridine (dFdU), and their mono-, di-, and triphosphates (dFdCMP, dFdCDP, dFdCTP, dFdUMP, dFdUDP, and dFdUTP) using a porous graphitic carbon column without an ion-pairing agent.

The method was applied for the quantification of these eight compounds in PBMCs from a patient receiving dFdC (Jansen et al., 2009).

Equipment and Reagents

- Tetrahydrofuran, formic acid, acetonitrile, hydrogen peroxide, ammonium acetate, ammonium bicarbonate.
- *Chromatography system*: Shimadzu LC 20 series HPLC system.

- *Mass spectrometer*: TSQ Quantum Ultra triple quadrupole (Thermo Electron).
- *Analytical column*: Hypercarb 100 mm × 2.1 mm, 5 μm particle size (Thermo Electron).
- *Guard column*: Hypercarb 10 × 2.1 mm ID, 5 μm.

Method

1. Chromatographic separation was achieved using a linear gradient[a]:
   - *Mobile phase A*: 1 mM ammonium acetate in acetonitrile/water (15/85, v/v), pH adjusted to 5 with glacial acetic acid.[b]
   - *Mobile phase B*: 25 mM ammonium bicarbonate in acetonitrile/water (15/85, v/v)[b].
   - *Gradient elution*: 0–0.1 min: 0% B; 0.1–2 min: from 0% to 100% B; 2–12 min: B maintained at 100%; 12–15 min: 0% B.
   - Flow rate at 250 μl/min, oven temperature: 45°C.
   - *Injection volume*: 25 μl with autosampler at 4°C.
   - During the first 4.5 min, the eluate was directed to waste.
   - Between each sample a reconditioning phase was programed by injecting 100 μl diluted formic acid (10% in water) in a flow of 100% A with the column outlet directed to the waste for 5 min.[c]
   - After each analytical batch, the column was backflushed with about 20-column volumes tetrahydrofuran before storage on mobile phase A.
2. Transitions and detection mode:
   - All compounds were analyzed in positive ionization mode except dFdU, which exhibited a most intense signal in negative mode.
   - Reaction monitoring transitions: dFdC: 264 → 112; [^{13}C,^{15}N]-dFdC: 267 → 115; dFdU: 263 → 111; [^{13}C,^{15}N]-dFdU: 266 → 114; dFdCMP: 344 → 246; [^{13}C,^{15}N]-dFdCMP: 347 → 249; dFdUMP: 345 → 247; [^{13}C,^{15}N]-dFdUMP: 248 → 250; dFdCDP: 424 → 326; [^{13}C,^{15}N]-dFdCDP: 427 → 329; dFdUDP: 425 → 327; [^{13}C,^{15}N]-dFdUDP: 428 → 330; dFdCTP: 504 → 326; [^{13}C,^{15}N]-dFdCTP: 507 → 329; dFdUTP: 505 → 247; [^{13}C,^{15}N]-dFdUTP: 508 → 250.

Notes

[a]Before each analytical run (about 60 samples), the analytical column was treated (30 min, 250 μl/min) with preconditioning buffer constituted by 1 mM ammonium acetate in acetonitrile/water (15/85, v/v) with 0.05% hydrogen peroxide, pH adjusted to 4 with glacial acetic acid. The redox state of PGC and pH has an extreme importance on the stability of the retention of the nucleotides. They are controlled using a pH-buffered hydrogen peroxide solution (preconditioning buffer).

[b]Analytes are completely separated within 15 min. The retention of the nucleosides is governed by the amount of organic modifier (acetonitrile), whereas the retention of the nucleotides is controlled by the amount of eluting ion (bicarbonate).

[c]Direct analysis of a second sample after running a gradient is not possible because of the loss of separation capabilities of the PGC column. The injection of 100 μl diluted formic acid restores the separation properties of the column and avoids a long reequilibration (30–60 min).

Protocol 4: The simultaneous analysis in cells of eight endogenous ribonucleoside triphosphates and deoxyribonucleoside triphosphates.

This assay is based on a liquid chromatography ion-pair using a PGC column coupled to a tandem mass spectrometer operating in both modes of negative and positive electrospray ionization. The quantification of these analytes is described in the murine leukemia cell line (L1210). The sample pretreatment is based on a protein precipitation coupled to an SPE, which is described in Section 44.2.2 (see Protocol 2). The concentration of endogenous compounds is calculated using calibration curves of their stable [^{13}C,^{15}N]-labeled isotope analogs (Cohen et al., 2009).

Equipment and Reagents

- Acetonitrile, methanol, acetic acid, ammonium acetate, ammonia hydroxide, diethylamine (DEA), hexylamine (HA)
- Chromatography system: Surveyor AS autosampler injector, a column thermostater, a Surveyor MS quaternary pump (Thermo Electron)
- Mass spectrometer: TSQ Quantum Ultra triple quadrupole equipped with an electrospray ionization source (Thermo Electron)
- Analytical column: Hypercarb 100 mm × 2.1 mm, 5 μm particle size (Thermo Electron)

Method

1. Chromatographic separation was achieved using a multistep gradient:
   - *Mobile phase A*: 5 mM HA–0.5% DEA in water, pH adjusted to 10 with acetic acid.
   - *Mobile phase B*: 50% acetonitrile in water.
   - *Gradient elution*: 0–13 min: from 0% to 24% B; 13–35 min: B maintained at 24%; 35–45 min: from 24% to 50% B; 45–46 min: from 50% to 100% B; 46–48 min: B maintained at 100%; 48–49 min: from 100%

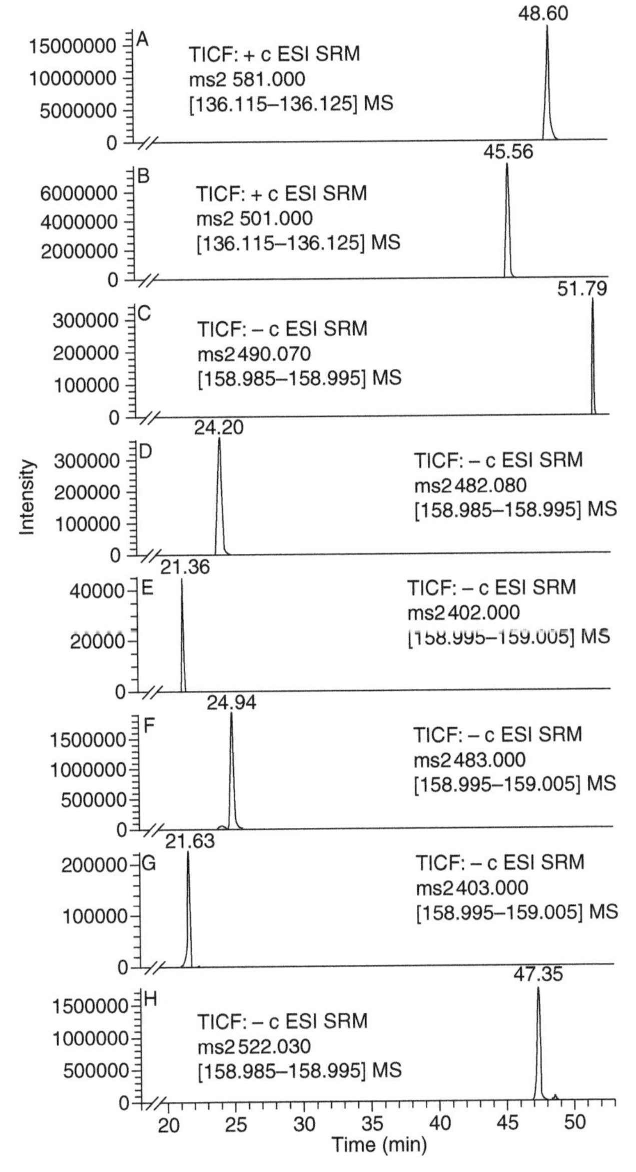

FIGURE 44.3 Example of MRM ion chromatograms obtained with the analytical conditions described in Protocol 4. (a) ATP ([ATP + DEA + H]$^+$), (b) ADP ([ADP + DEA + H]$^+$), (c) dATP.

to 0% B; 49–68 min: B maintained at 0% (Figure 44.3).

- Flow rate at 250 μl/min, oven temperature: 30°C.
- *Injection volume*: 10 μl with autosampler at 5°C.

2. Transitions and detection mode:
   - All compounds were analyzed in negative ionization mode except ATP and dGTP, which are analyzed in positive mode.[a] The internal standard Br-ATP is monitored both in negative and positive ionization modes
   - *Reaction monitoring transitions*: ATP[b,c]: 581.0 → 136.1; [$^{13}C_{10}$,$^{15}N_5$]-ATP[b,c]: 596.2 → 146.1; dATP: 490.1 → 159.0; [$^{13}C_{10}$,$^{15}N_5$]-dATP: 505.0 → 159.0; GTP: 522.1 → 159.0; [$^{13}C_{10}$,$^{15}N_5$]-GTP: 537.0 → 159.0; dGTP[b]: 581.0 → 152.1; [$^{13}C_{10}$,$^{15}N_5$]-dGTP[b]: 596.2 → 162.1; CTP[d]: 482.1 → 159.0; [$^{13}C_9$,$^{15}N_3$]-CTP: 494.0 → 159.0; dCTP: 466.1 → 159.0; [$^{13}C_9$,$^{15}N_3$]-dCTP: 478.0 → 159.0; TTP[d]: 481.1 → 159.0; [$^{13}C_{10}$,$^{15}N_2$]-TTP: 493.0 → 159.0; UTP[d]: 483.0 → 159.0; [$^{13}C_9$,$^{15}N_2$]-UTP: 494.0 → 159.0; Br-ATP[b](+): 585.9 → 159.0; Br-ATP(−): 660.9 → 215.9

Notes

[a]ATP and dGTP have the same molecular weight and the same transition in the negative mode. Using DEA as the counterion and the positive mode both compounds may be detected simultaneously (see Section 44.3).

[b]Adduct forms: The presence in the mobile phase of a counterion such as DEA masks negative charges on the phosphate group and enhances the protonation of the nucleobase. The adduct ions [NTP–DEA–H]$^+$ have a better sensitivity than the [NTP + H]$^+$.

[c]In our own experience, we observed greater stability of the signal when ATP is monitored with the negative ionization mode (*m/z* 506 → 159.0). This is possible because the proportion of intracellular dGTP relative to ATP is very low (about 1/1000) and does not interfere with the quantification of ATP.

[d]It is necessary to obtain a good chromatographic separation between CTP and UTP since these two compounds have a molecular weight which differs only by 1 amu and they have the same fragmentation pathway. This is also observed between TTP and CTP.

44.2.5.3 Chromatography Based on Other Analytical Columns Analysis of nucleotides is mostly based on the chromatography approach described earlier. However, although few assays are proposed at this time, WAX or HILIC is a relevant mode for nucleotide analysis and should be presented. The retention model for WAX liquid chromatography is dependent on the ion capacity of the stationary phase and the competing anion in the mobile phase. Elution of the analytes can occur by an increase of the pH (within a limited range) or by an increase of the amount of the competing anion. It should be noted that ion strength of the matrix can have an impact on chromatographic behavior (retention time, peak resolution) since WAX is not as robust as reverse-phase chromatography (Shi et al., 2002). HILIC is a version of normal-phase chromatography. In HILIC conditions, retention increases with the polarity of the analyte and decreases with the polarity of the mobile phase. It was suggested that the retention mechanism for HILIC was a partitioning between the bulk eluent and a water-rich layer, partially immobilized on the stationary phase, whereas retention in conventional normal phase chromatography is predominantly governed by surface adsorption phenomena.

Protocol 5: The quantification of intracellular triphosphate metabolite of D-D4FC, an investigational HIV nucleoside reverse transcriptase inhibitor, in human PBMC samples (Shi et al., 2002).

This assay is based on WAX liquid chromatography coupled with tandem mass spectrometry in positive electrospray ionization. Experiments conducted in this study were performed from blank human PBMCs spiked with D-D4FC-TP or from *in vitro* human PBMCs incubated with D-D4FC. The cells were prepared with Histopaque system.

Equipment and Reagents

- Methanol, glacial acid acetic, acetonitrile, ammonium acetate, 28% ammonia hydroxide aqueous solution.
- *Chromatography system*: Shimadzu LC-10ADVP binary pumps.
- *Mass spectrometer*: Sciex API 4000 triple quadrupole.
- *Analytical column*: Keystone BioBasic column 20 mm × 1 mm, 5 μm particle size (Thermo Fisher).

Method

1. Chromatographic separation is achieved using a step gradient:
   - *Mobile phase A*: 10 mM ammonium acetate in 30/70 (v/v) acetonitrile/water (pH adjusted at 6 by the addition of glacial acetic acid)
   - *Mobile phase B*: 1 mM ammonium acetate in 30/70 (v/v) acetonitrile/water (pH adjusted at 10.5 by the addition of ammonia hydroxide)
   - Step gradient with phase B such as 0% from 0 to 0.2 min, 20% from 0.21 to 0.7 min, 60% from 0.71 to 1.1 min, 100% from 1.11 to 1.5 min, and 0% from 1.51 to 2 min[a]
   - Flow rate at 500 μl/min
   - *Injection volume*: 40 μl with autosampler at 5°C
2. Transitions and detection mode:
   - Positive ionization mode[b] (all compounds are detected between 1 and 2 min).
   - Reaction-monitoring transitions (D-D4FC-TP: 468 → 130, D-D4FC-DP: 388 → 130, D-D4FC-MP: 308 → 130, [$^{13}C_5$]-D-D4FC-TP: 473 → 130) used as internal standard.

Notes

[a]The LC conditions as described were not adequate for the separation of different NTPs. For this purpose, the authors proposed different possibilities such as to use another mobile phase with buffer capacity around 8.

[b]Negative mode is about four times more sensitive but dCTP is isobaric to D-D4FC-TP.

Protocol 6: An LC-MS/MS method for the quantification of 2′C-methyl-cytidine-triphosphate (2′-Me-Cy-TP) in liver tissue samples (Pucci et al., 2009).

This method is based on a HILIC using an aminopropyl column. Experiments conducted in this study are performed from blank liver sample homogenates spiked with 2′-Me-Cy-TP and from liver homogenates from rats dosed orally with a 2′-C-methyl cytosine prodrug. An automated solid phase procedure is implemented to extract 2′-Me-Cy-TP from liver samples using 96-well SPE plate (WAX sorbent, Strata-X-AW, 30 mg, Phenomenex).

Equipment and Reagents

- Acetonitrile, ammonium acetate.
- *Chromatography system*: Agilent 1100 Series liquid chromatograph (Agilent Technologies).
- *Mass spectrometer*: API4000 triple quadrupole equipped with an electrospray ionization source (Applied Biosystems).
- *Analytical column*: Luna® aminopropyl column 100 mm × 2.0 mm, 3 μm particle size (Phenomenex).[a]

Method

1. Chromatographic separation was achieved using a multistep gradient:
   - *Mobile phase A*: 20 mM ammonium acetate in acetonitrile/water (5/95, v/v) with a pH adjusted at 9.45.
   - *Mobile phase B*: acetonitrile.
   - *Gradient elution*: 0–10 min: from 30% to 100% A; 10–18 min: A maintained at 100%; then the gradient starts to decrease to 30/70% (A/B) remaining constant to this ratio until 30 min.
   - Flow rate at 300 μl/min, the column being maintained at room temperature.
   - *Injection volume*: 50 μl with autosampler at 4°C.
2. Transitions and detection mode:
   - Negative ionization mode.
   - Reaction monitoring transitions: 2′-Me-Cy-TP: 495.9 → 158.8; 7-deaza-2′-C-methyl-adenosine triphosphate (internal standard): 519.2 → 158.8.

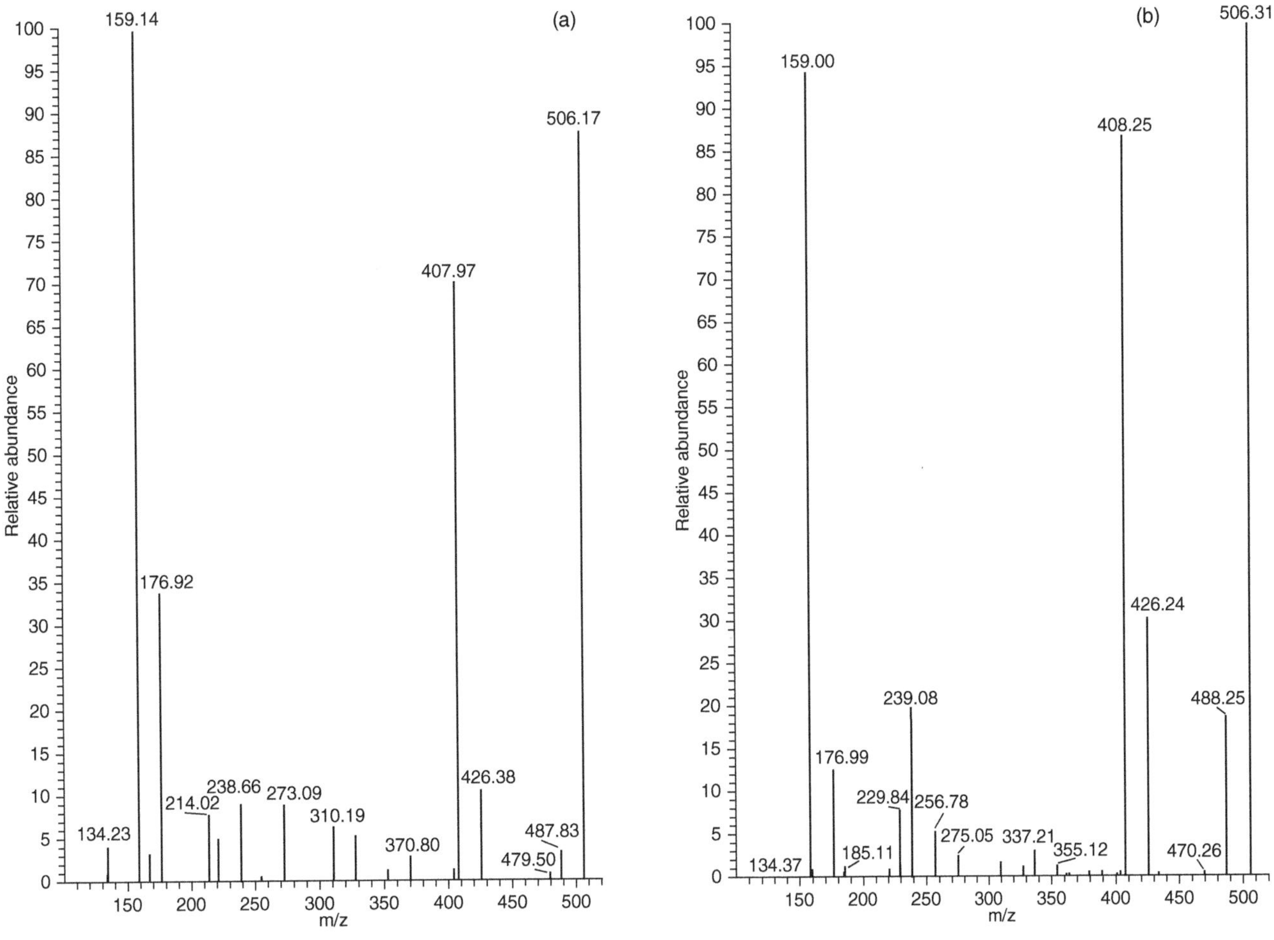

FIGURE 44.4 Product ion scans of ATP (a) and dGTP (b) in negative ionization mode, the parent ions are $[ATP - H]^-$ and $[dGTP - H]^-$, with a *m/z* 506. Ions produced are the same in both spectra except an ion at *m/z* 257 for dGTP. This specific ion can distinguish dGTP to ATP, but the sensitivity is very poor. (*Continued*).

Notes

[a]Another aminopropyl column has been tested Polaris-NH_2 (Varian); however, with this column a poor extraction column efficiency was observed.

44.2.5.4 MS/MS Detection From a theoretical point of view, anionic compounds such as nucleotides should be analyzed in negative ionization mode. However, most natural nucleotides and nucleotide analogs can be detected in negative and positive ionization modes (Cohen et al., 2009; Jansen et al., 2012). Some exceptions exist such as UDP, which do not produce a signal in negative ionization mode (Cordell et al., 2008). In the same way, nucleotides of dFdU were detectable in negative ionization mode, whereas dFdU was not (Jansen et al., 2009). The intensity of the protonation of nucleotides derived from uracil and thymine is lower than those observed for nucleotides derived from adenine, guanine, and cytosine. This is due to the presence of the amino group into the structure of the base of these last compounds (Pruvost et al., 2008). According to the assay, when the negative ionization mode is used, the most abundant fragment analyzed is mostly the base fragment ion (cleavage of the glycosidic bond), or the pyrophosphate ion (*m/z* 159). When the positive ionization mode is selected, the monitored ion fragment was mostly the nucleic base ion (Cohen et al., 2009; Fromentin et al., 2010; Jansen et al., 2012).

44.3 TROUBLESHOOTING

- Although tandem mass spectrometry allows obtaining high selectivity, LC-MS/MS bioanalytical challenges remain as several nucleotide analogs as well as endogenous nucleotides exhibit the same molecular weight and similar MS/MS fragmentation pattern. For example, $[^{13}C,^{15}N_2]$-dFdC has the same parent mass ion

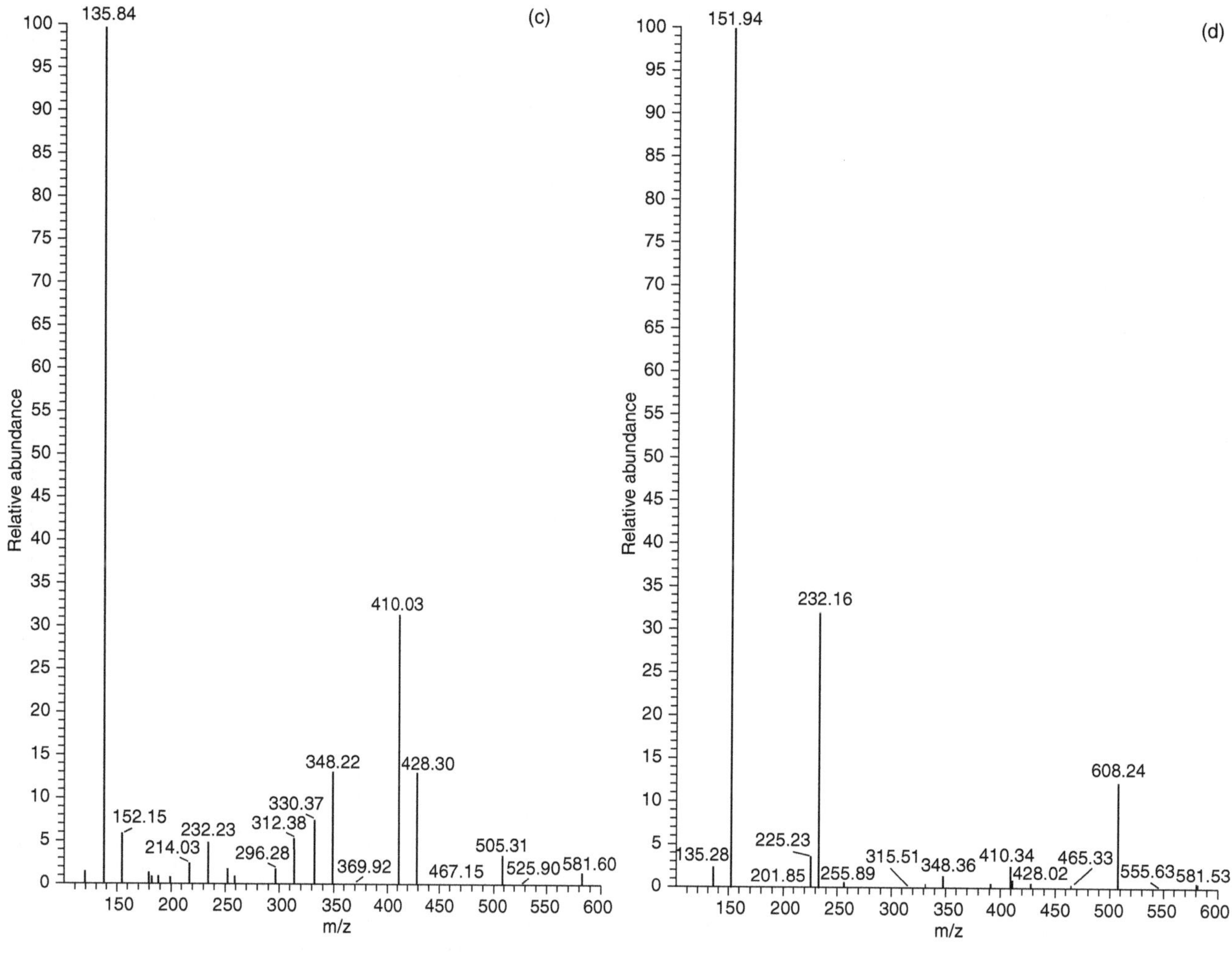

FIGURE 44.4 (*Continued*) Product ion scans of ATP (c) and dGTP (d) in positive ionization mode with a DEA as counterion in a mobile phase, parent ion are $[ATP + DEA + H]^+$ and $[dGTP + DEA + H]^+$, with a *m/z* 581. The presence of DEA in the mobile phase masked negative charges on the phosphate group and enhanced the protonation of the nucleobase and the fragmentation pathway is modified. The main ion corresponds to adenine (*m/z* 136) for ATP and guanine (*m/z* 152) for dGTP. From our experience, the adduct ion $[NTP + DEA + H]^+$ presents a better sensitivity than the $[NTP + H]^+$.

as dFdU and produces a signal in the dFdU transition. It is the same for the phosphorylate derivatives. The masses of the $[^{13}C,^{15}N_2]$-dFdU nucleotides are the same as the adenosine, deoxyguanosine, and zidovudine nucleotides (Jansen et al., 2009). ATP, dGTP, and ZDV-TP (zidovudine-triphosphate) have the same molecular weight and similar fragmentation pattern in negative ionization mode (Henneré et al., 2003). It is the same for ZDV-MP, AMP, and dGMP (Fromentin et al., 2010). The Na adduct of 3TC-TP interferes with the analysis of dATP (Pruvost et al., 2008). CTP interferes with the analysis of araCTP (Crauste et al., 2009). In the same way, CTP, UTP, and TTP shift only of 1 amu (respectively, mw 483, 484, 482) and they produce the same daughter ion (*m/z* 159.0 pyrophosphate ion) using the negative ionization mode. Thus, the interference due to the isotopic abundance in MS/MS detection can be significant. It is the same problem for the di- and monophosphate derivates of these three nucleosides (Tuytten et al., 2002). Thus, in all cases, it is essential to obtain a chromatographic separation of these compounds for their simultaneous determination. Recently, the quantification of decitabine triphosphate (aza-dCTP), a deoxycytidine analog used in myelodysplastic syndrome, was published (Jansen et al., 2012). Aza-dCTP and dCTP have a mass of 467 and 468 Da, respectively. However, with the natural isotopic abundance and the chromatographic coelution of both compounds, dCTP induces interference for the aza-dCTP quantification. Thus, authors calculated the interference caused by dCTP from dCTP signal by additional monitoring of the dCTP mass transition in each sample (Jansen et al., 2012). Sometimes, there is an alternative as proposed for ATP and dGTP. Both compounds have the same molecular weight (mw 507) and the same MS/MS transition (*m/z* 506 → 159.0) in negative ionization mode (Figure 44.4). However, different adducts were formed when these two compounds were analyzed in the presence of a counterion such as DEA and monitored using positive ionization mode. Under these conditions, the fragmentation pathway was modified. The daughter ion corresponding to adenine (*m/z* 136) for ATP and guanine (*m/z* 152) for dGTP allowed distinguishing between the two compounds (Cohen et al., 2009).

- It was reported that the matrix effect varies with different batches of PBMCs. It was suggested that contamination by lysed red blood cells may be the cause (Shi et al., 2002). The matrix effect shifts from around 0.8 with no colored PBMCs to around 0.34 with reddish PBMCs. The anionic compounds coeluted with the molecule of interest are involved in the matrix effect (see Section 44.2.1 for the preparation).
- Below pH 3, nucleoside diphosphates are unstable and are likely to lose one phosphate group (Cordell et al., 2008).
- From several works, it was suggested that the phosphate group of nucleotides interacts with stainless steel causing a peak tailing on the chromatogram. The peak tailing is reduced by replacing all stainless steel devices by PEEK materials in the instrument (Tuytten et al., 2006). It was also reported that the use of ammonium salts (ammonium hydrogen carbonate, ammonium hydroxide, or ammonium phosphate in the mobile phase prevents peak tailing (Asakawa et al., 2008).

REFERENCES

Arts EJ, Hazuda DJ. HIV-1 antiretroviral drug therapy. Cold Spring Harbor Perspectives in Medicine 2012;2:a007161.

Asakawa Y, Tokida N, Ozawa C, Ishiba M, Tagaya O, Asakawa N. Suppression effects of carbonate on the interaction between stainless steel and phosphate groups of phosphate compounds in high-performance liquid chromatography and electrospray ionization mass spectrometry. Journal of Chromatography A 2008;1198–1199:80–86.

Back DJ, Burger DM, Charles W, Flexner CW, Gerber JG. The pharmacology of antiretroviral nucleoside and nucleotide reverse transcriptase inhibitors. Implications for once-daily dosing. Journal of Acquired Immune Deficiency Syndromes 2005;39:S1–S23.

Becher F, Schlemmer D, Pruvost A, et al. Development of a direct assay for measuring intracellular AZT triphosphate in humans peripheral blood mononuclear cells. Analytical Chemistry 2002a;74(16):4220–4227.

Becher F, Pruvost A, Goujard C, et al. Improved method for the simultaneous determination of d4T, 3TC and ddl intracellular phosphorylated anabolites in human peripheral-blood mononuclear cells using high-performance liquid chromatography/tandem mass spectrometry. Rapid Communications in Mass Spectrometry 2002b;16(6):555–565.

Benech H, Théodoro F, Herbet A, et al. Peripheral blood mononuclear cell counting using a DNA-detection-based method. Analytical Biochemistry 2004;330:172–174.

Cichna M, Raab M, Daxecker H, Griesmacher A, Müller MM, Markl P. Determination of fifteen nucleotides in cultured human mononuclear blood and umbilical vein endothelial cells by solvent generated ion-pair chromatography. Journal of Chromatography B 2003;787(2):381–391.

Claire RL 3rd. Positive ion electrospray ionization tandem mass spectrometry coupled to ion-pairing high-performance liquid chromatography with a phosphate buffer for the quantitative analysis of intracellular nucleotides. Rapid Communications in Mass Spectrometry 2000;14(17):1625–1634.

Cohen S, Megherbi M, Jordheim LP, et al. Simultaneous analysis of eight nucleoside triphosphates in cell lines by liquid chromatography coupled with tandem mass spectrometry. Journal of Chromatography B 2009;877(30):3831–3840.

Cohen S, Jordheim LP, Megherbi M, Dumontet C, Guitton J. Liquid chromatographic methods for the determination of endogenous nucleotides and nucleotide analogs used in cancer therapy: a review. Journal of Chromatography B 2010;878(22):1912–1928.

Cordell RL, Hill SJ, Ortori CA, Barrett DA. Quantitative profiling of nucleotides and related phosphate-containing metabolites in cultured mammalian cells by liquid chromatography tandem electrospray mass spectrometry. Journal of Chromatography B 2008;871(1):115–124.

Crauste C, Lefebvre I, Hovaneissian M, et al. Development of a sensitive and selective LC/MS/MS method for the simultaneous determination of intracellular 1-beta-D-arabinofuranosylcytosine triphosphate (araCTP), cytidine triphosphate (CTP) and deoxycytidine triphosphate (dCTP) in a human follicular lymphoma cell line. Journal of Chromatography B 2009;877(14–15):1417–1425.

Durand-Gasselin L, Da Silva D, Benech H, Pruvost A, Grassi J. Evidence and possible consequences of the phosphorylation of nucleoside reverse transcriptase inhibitors in human red blood cells. Antimicrobial Agents and Chemotherapy 2007;51(6):2105–2111.

Fromentin E, Gavegnano C, Obikhod A, Schinazi RF. Simultaneous quantification of intracellular natural and antiretroviral nucleosides and nucleotides by liquid chromatography-tandem mass spectrometry. Analytical Chemistry 2010;82(5):1982–1989.

Fung EN, Cai Z, Burnette TC, Sinhababu AK. Simultaneous determination of Ziagen and its phosphorylated metabolites by ion-pairing high-performance liquid chromatography-tandem mass spectrometry. Journal of Chromatography B 2001;754(2):285–295.

Henneré G, Becher F, Pruvost A, Goujard C, Grassi J, Benech H. Liquid chromatography-tandem mass spectrometry assays for intracellular deoxyribonucleotide triphosphate competitors of nucleoside antiretrovirals. Journal of Chromatography B 2003;789(2):273–281.

Jansen RS, Rosing H, Schellens JH, Beijnen JH. Retention studies of 2′-2′-difluorodeoxycytidine and 2′-2′-difluorodeoxyuridine nucleosides and nucleotides on porous graphitic carbon: development of a liquid chromatography-tandem mass spectrometry method. Journal of Chromatography A 2009;1216(15):3168–3174.

Jansen RS, Rosing H, Schellens JHM, Beijnen JH. Facile small scale synthesis of nucleoside 5′-phosphate mixtures. Nucleosides, Nucleotides and Nucleic Acids 2010;(29):14–26.

Jansen RS, Rosing H, Schellens JH, Beijnen JH. Mass spectrometry in the quantitative analysis of therapeutic intracellular nucleotide analogs. Mass Spectrometry Review 2011;30(2):321–343.

Jansen RS, Rosing H, Wijermans PW, Keizer RJ, Schellens JH, Beijnen JH. Decitabine triphosphate levels in peripheral blood mononuclear cells from patients receiving prolonged low-dose decitabine administration: a pilot study. Cancer Chemotherapy and Pharmacology 2012;69(6):1457–1466.

Koolman J, Roehm K-H. Color Atlas of Biochemistry. 2nd ed. Stuttgart, Germany: Thieme; 2005.

Longley DB, Harkin DP, Johnston PG. 5-fluorouracil: mechanisms of action and clinical strategies. Nature Reviews Cancer 2003;3(5):330–338.

Pruvost A, Becher F, Bardouille P, et al. Direct determination of phosphorylated intracellular anabolites of stavudine (d4T) by liquid chromatography/tandem mass spectrometry. Rapid Communications in Mass Spectrometry 2001;15(16):1401–1408.

Pruvost A, Théodoro F, Agrofoglio L, Negredo E, Bénech H. Specificity enhancement with LC-positive ESI-MS/MS for the measurement of nucleotides: application to the quantitative determination of carbovir triphosphate, lamivudine triphosphate and tenofovir diphosphate in human peripheral blood mononuclear cells. Journal of Mass Spectrometry 2008;43(2):224–233.

Pucci V, Giuliano C, Zhang R, et al. HILIC LC-MS for the determination of 2′-C-methyl-cytidine-triphosphate in rat liver. Journal of Separation Science 2009;32(9):1275–1283.

Rodwell VW. Nucleotides. In: Murray RK, Granner DK, Mayes PA, Rodwell VW, editors. Harper's Illustrated Biochemistry. 26th ed. New York: McGraw-Hill Companies; 2003. p 286–292.

Shi G, Wu JT, Li Y, et al. Novel direct detection method for quantitative determination of intracellular nucleoside triphosphates using weak anion exchange liquid chromatography/tandem mass spectrometry. Rapid Communications in Mass Spectrometry 2002;16(11):1092–1099.

Tuytten R, Lemière F, Dongen WV, Esmans EL, Slegers H. Short capillary ion-pair high-performance liquid chromatography coupled to electrospray (tandem) mass spectrometry for the simultaneous analysis of nucleoside mono-, di- and triphosphates. Rapid Communications in Mass Spectrometry 2002;16(12):1205–1215.

Tuytten R, Lemière F, Witters E, et al. Stainless steel electrospray probe: a dead end for phosphorylated organic compounds? Journal of Chromatography A 2006;1104(1–2):209–221.

Vela JE, Olson LY, Huang A, Fridland A, Ray AS. Simultaneous quantitation of the nucleotide analog adefovir, its phosphorylated anabolites and 2′-deoxyadenosine triphosphate by ion-pairing LC/MS/MS. Journal of Chromatography B 2007;848(2):335–343.

Wang J, Lin T, Lai J, Cai Z, Yang MS. Analysis of adenosine phosphates in HepG-2 cell by a HPLC-ESI-MS system with porous graphitic carbon as stationary phase. Journal of Chromatography B 2009;877(22):2019–2024.

Xie C, Plunkett W. Metabolism and actions of 2-chloro-9-(2-deoxy-2-fluoro-beta-D-arabinofuranosyl)-adenine in human lymphoblastoid cells. Cancer Research 1995;55(13):2847–2852.

Xing J, Apedo A, Tymiak A, Zhao N. Liquid chromatographic analysis of nucleosides and their mono-, di- and triphosphates using porous graphitic carbon stationary phase coupled with electrospray mass spectrometry. Rapid Communications in Mass Spectrometry 2004;18(14):1599–1606.

45

LC-MS BIOANALYSIS OF STEROIDS

JIE ZHANG AND FRANK Z. STANCZYK

45.1 INTRODUCTION

A variety of steroid hormones are produced in women and men. As shown in Table 45.1, they can be classified into several categories, based on the number of carbon atoms and structural characteristics. The cholanes and cholestanes will not be discussed in this chapter. Early assays for the measurement of steroids were enzyme-based bioassays and chemical-based assays, with colorimetric and fluorometric detection. Other methods employed gas chromatography separations combined with a quantitative detection such as electron capture. These assays were restricted to the analysis of urine samples and lacked adequate sensitivity (in mg and μg ranges). Subsequently, development of the gas chromatography–mass spectrometry (GC-MS) assay and radioimmunoassay (RIA) made it possible to quantify steroid hormones routinely in serum and plasma with relatively high sensitivity and specificity. Development of a GC-MS interface in the 1960s set the standard for steroid hormone analysis by GC-MS. By 1965, measurement of androsterone sulfate and dehydroepiandrosterone (DHEA) sulfate in serum was reported (Sjövall and Vihko, 1965), and in 1966 the first comprehensive urinary steroid profile was published (Horning et al., 1966). An improved derivatization technique resulted in a complete profile of all human-excreted steroid hormones of interest (Horning et al., 1968).

A few years after steroids began to be measured by GC-MS, the RIA method for steroid hormones was developed. In 1969, Abraham reported the measurement of estradiol (E_2) in serum (Abraham, 1969), and subsequent RIA methods were developed for other steroid hormones. Due to the relative ease in technological performance and low cost of RIAs compared to GC-MS assays, RIAs have become more widely used in research and diagnostic laboratories. The immediate impact of the introduction of RIA methods was that it allowed the measurement of an immensely wide range of compounds of clinical and biological importance, and opened up new horizons in the field of endocrinology. RIAs used in numerous studies began to enrich this field with new knowledge. In addition, their use in diagnostic testing provided physicians with valuable information for diagnosing and treating countless number of patients.

In the meantime, advances were made in GC-MS assay methodology. A key development in circa 1970 was the computerization of GC-MS assay data processing and the adoption of glass capillary columns for high-resolution steroid separations. During the 1970s, GC-MS was used for clinical steroid analysis, employing repetitive scanning techniques for urinary profile analysis of steroid hormone metabolites and related ion monitoring for targeted steroid hormones in serum. By 1979, a flexible fused silica column replaced the fragile glass columns that were used, leading to more robust analyses. Finally in 1982, the introduction of a computer-controlled bench-top instrument having a mass selective detector, marketed by Hewlett-Packard, firmly established GC-MS as a routine technique.

In the past two decades, high-performance liquid chromatography (HPLC) coupled with tandem mass spectrometry (LC-MS/MS) has revolutionized the measurement of steroids. In addition, the invention of an electrospray ion source and subsequent development of atmospheric chemical ionization allowed for the simple coupling of the liquid chromatograph to the mass spectrometer, often negating the need for steroid derivatization required for typical GC-MS analysis. These advances reduced the complexity of the mass spectrometry (MS) assay and shortened the assay run time

Handbook of LC-MS Bioanalysis: Best Practices, Experimental Protocols, and Regulations, First Edition. Edited by Wenkui Li, Jie Zhang, and Francis L.S. Tse.

TABLE 45.1 Classification of Steroid Hormones and Representative LC-MS Methods

Representative steroids and structures	Key physiological information	Representative LC-MS method and related key information
Estranes (C_{18} steroids)		
Estradiol (E_2) $C_{18}H_{24}O_2$, 272.38 g/mol	• Has the highest estrogenic activity in the body • Secreted predominantly by the ovaries in premenopausal women • Formed in peripheral tissues (primarily fat) in postmenopausal women and men • About 37% is bound to SHBG in premenopausal women and 20% in men • Serum levels in the follicular, periovulatory, and luteal phases of the menstrual cycle are 20–150 pg/ml, 150–750 pg/ml, and 75–300 pg/ml; in postmenopausal women and men they are <25 pg/ml and 10–50 pg//ml, respectively	• 0.2 ml serum, 1 pg/ml LOD • PPT plus online extraction • C_8 column and 10-min gradient elution with CH_3OH/H_2O • Sciex API5000, ESI-, 271/145 • Guo et al. (2008)
Estrone (E_1) $C_{18}H_{22}O_2$, 270.37 g/mol	• Has substantial estrogenic activity • Secreted partly by the ovaries and also formed in peripheral tissues, primarily fat • Circulating levels follow profile of estradiol during the menstrual cycle, but the levels are lower	• 0.25 ml serum, 1 pg/ml LOQ • picolinoyl derivatization • SPE • C_{18} column and 10-min gradient elution with $CH_3CN/CH_3OH/CH_3COOH$ • Sciex API5000, ESI +, 376/157 • Yamashita et al. (2007)
Estriol (E_3) $C_{18}H_{24}O_3$, 288.38 g/mol	• Weak estrogen, which is present in serum at very low levels during the menstrual cycle but very high levels during pregnancy • Secreted predominantly by the placenta	• Refer to E_2
Androstanes (C_{19} steroids)		
Testosterone (T) 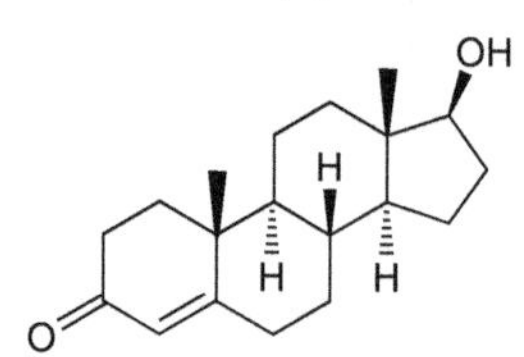 $C_{19}H_{28}O_2$, 288.42 g/mol	• In men, predominantly secreted by the testes, and in premenopausal women about 65% is derived from the ovaries and 35% from adrenals • About 44% is bound to SHBG in men and 66% in premenopausal women • Serum levels in men are 3.5–12 ng/ml and 0.20–0.70 ng/ml in premenopausal women	• 0.15 ml serum, 3 pg/ml LOQ • Online with HTLC • C_{12} column and 4.5-min gradient elution with $CH_3CN/H_2O/CH_3COOH$ • Finnigan TSQ Quantum, APCI +, 289/109 • Salameh et al. (2010)
Dihydrotestosterone (DHT) 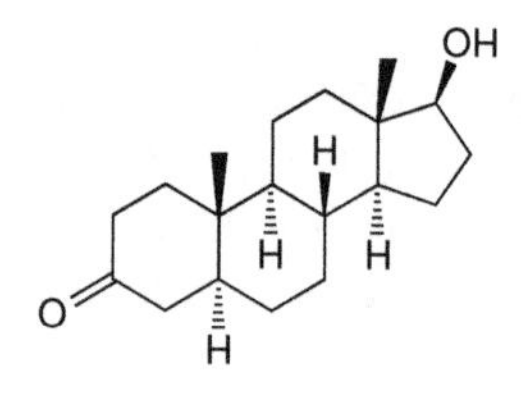 $C_{19}H_{30}O_2$, 290.44 g/mol	• Most potent androgen in the body • Formed in peripheral tissues • About 60% is bound to SHBG in men and 78% in premenopausal women • Serum levels in men are 0.25–0.75 ng/ml and 0.05–0.3 ng/ml in women	• 0.3 ml serum, 10 pg/ml LOQ • LLE and SPE in 96-well plate • Derivatization with 2,3-pyridinedicarboxylic anhydride • UPLC C_{18} column and 5-min gradient elution with $CH_3CN/H_2O/CH_3COONH_4$ • Sciex API5000, ESI +, 291/111 • Licea-Perez et al. (2008)

TABLE 45.1 *(Continued)*

Representative steroids and structures	Key physiological information	Representative LC-MS method and related key information
Androstenedione 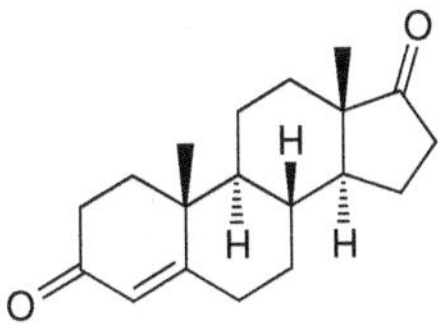 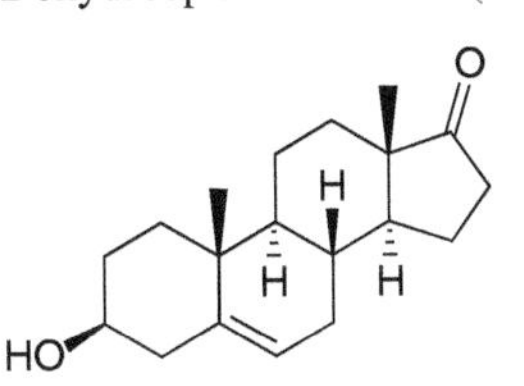 $C_{19}H_{26}O_2$, 286.40 g/mol	• In premenopausal women, secreted by adrenal and ovaries in approximately equal amounts • Important as a precursor of testosterone and estrone	• 15 μl serum or 14 μl blood (dry blood spot), 400 pg/ml LOQ • PPT • C18 column and 6-min gradient elution with $CH_3OH/H_2O/HCOOH$ • Waters Ultima, ESI +, 287/97 • Janzen et al. (2008)
Dehydroepiandrosterone (DHEA) 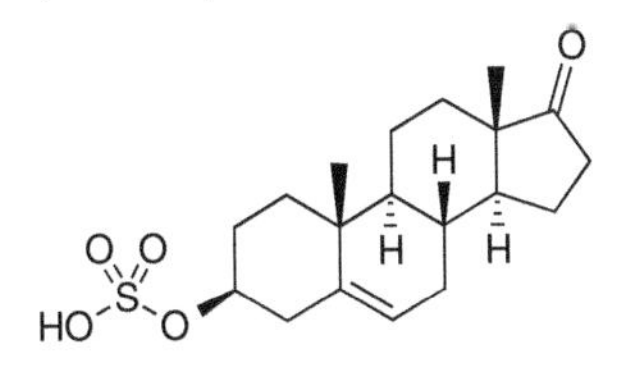 $C_{19}H_{28}O_2$, 288.42 g/mol	• Produced predominantly in the adrenals and to a lesser extent in the gonads • Serves as a precursor of androgens and estrogens • Serum levels decline progressively from early to late adulthood and are 1.8–12.5 ng/ml in males and 1.3–9.8 ng/ml in females	• 0.2 ml serum, 10 pg/ml LOD • PPT plus online extraction • C_8 column and 11-min gradient with CH_3OH/H2O • Sciex API5000, APPI +, 271/213 • Guo et al. (2006)
Dehydroepiandrosterone sulfate (DHEAS) 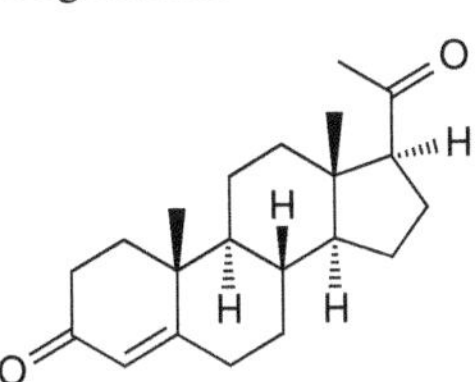 $C_{19}H_{28}O_5S$, 368.49 g/mol	• Produced predominantly by the adrenals • Serves as a good marker of adrenal androgen production • Serum levels decline with age, similar to DHEA and are 0.8–5.6 μg/ml in men and 0.15–2.8 μg/ml in women	• Refer to DHEA
Pregnanes (C_{21} steroids)		
Progesterone 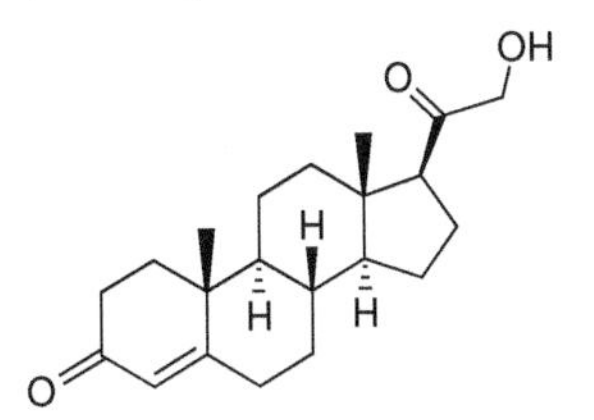 $C_{21}H_{30}O_2$, 314.46 g/mol	• Formed in the corpus luteum during the luteal phase of the menstrual cycle • Present in very high concentrations in pregnancy • Serum levels are $<$0.5 ng/ml and 4–20 ng/ml in the follicular and luteal phases of the cycle, respectively, and as high as 200 ng/ml or greater in late pregnancy	• Refer to DHEA
Deoxycorticosterone (11-deoxycorticosterone, DOC) $C_{21}H_{30}O_3$, 330.46 g/mol	• Formed in the adrenals from progesterone • Precursor of corticosterone and aldosterone • Elevated levels may reflect the hypertension observed in patients with 11-hydroxylase deficiency, the second leading cause of congenital adrenal hyperplasia	• 0.6 ml serum, 100 pg/ml LOQ • PPT plus online extraction with monolithic column • C_{18} column and 15-min gradient elution with $CH_3OH/H_2O/HCOONH_4$ (pH = 3.0) • Waters Quattro Primer, APCI +, 331/97 • Carvalho et al. (2008)
Corticosterone 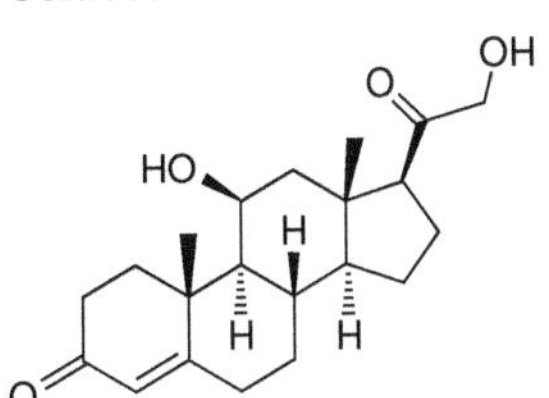 $C_{21}H_{30}O_4$, 346.46 g/mol	• Weak mineralocorticoid • Important intermediate in the formation of aldosterone in the adrenals	• Refer to DHEA

TABLE 45.1 (*Continued*)

Representative steroids and structures	Key physiological information	Representative LC-MS method and related key information
Aldosterone $C_{21}H_{28}O_5$, 360.44 g/mol	• The primary mineralocorticoid • Formed in the adrenals from corticosterone via the intermediate, 18-hydroxycorticosterone • Its secretion is mediated by the renin–angiotensin system • Useful diagnostic marker for evaluation of blood pressure disorders and abnormal serum sodium or potassium disorders	• 0.5 ml serum, 10 pg/ml LOQ • LLE with dichloromethane/ethyl ether (60/40) • C_{18} column and 10-min gradient elution with CH_3OH/H_2O • Sciex API3000, ESI-, 359/189 • Turpeinen et al. (2008)
Cortisol (hydrocortisone) $C_{21}H_{30}O_5$, 362.46 g/mol	• The major glucocorticoid • Formed in the adrenals from 17α-hydroxyprogesterone via the intermediate 11-deoxycortisol • Approximately 90% is bound to CBG in men and women • Serum levels are highest in the morning (50–250 ng/ml) and lowest in the evening (25–125 ng/ml) in men and women • Serum levels are decreased in Addison's disease and increased in Cushing's disease	• Human saliva • Online SPE • C_{18} column, and 5 min gradient elution with $CH_3OH/CH_3COONH_4/HCOOH$ • Waters Quattro Primer, ESI +, 363/121 • Perogamvros et al. (2009)
Cholanes (C_{24} steroids)		
Cholic acid $C_{24}H_{40}O_5$, 408.57 g/mol	• A key bile acid • Serum levels are elevated in intrahepatic cholestasis of pregnancy	• Not discussed in this chapter
Chenodeoxycholic acid 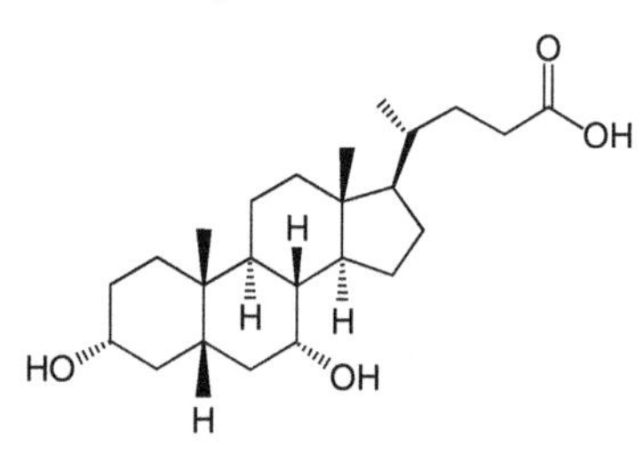 $C_{24}H_{40}O_4$, 392.57 g/mol	• A key bile acid • Serum levels are elevated in intrahepatic cholestasis of pregnancy	• Not discussed in this chapter
Cholestanes (C_{27} steroids)		
Cholesterol 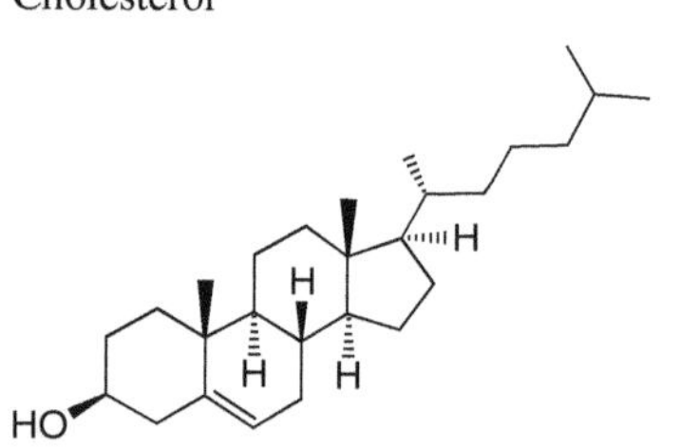 $C_{27}H_{46}O$, 386.65 g/mol	• Synthesized *de novo* from acetate in steroid-forming cells • Measurement of total cholesterol, LDL-cholesterol and HDL-cholesterol provides valuable information for the risk of coronary artery disease	• Not discussed in this chapter

dramatically, greatly increasing the throughput of patient samples while still providing highly accurate and precise results.

Although there seems to be a general consensus that MS assays will become the gold standard for steroid hormone measurements, there are many challenges to overcome before this occurs. Major challenges include the following: (a) the affordability by smaller laboratories to purchase MS instruments and pay for operating costs; (b) improving assay sensitivity, especially for measuring low E_2 levels in postmenopausal women and patients treated with aromatase inhibitors; (c) developing assays for profiles of steroid hormone metabolites in serum and tissues; (d) obtaining reliable reference ranges; and (e) standardizing steroid MS analyses.

The most difficult challenge of MS assays is generally agreed to be the standardization of their measurements. It is important to realize that MS methodology has variability issues similar to those of conventional RIAs and direct immunoassays. The results of one study (Vesper et al., 2009) showed that eight MS assay methods were being used in different laboratories with the same measurement principles, but with major differences in assay performance. Differences in accuracy among MS assay methods appear to be due to calibrators, whereas differences in assay precision may be explained in part by variation in sample preparation.

The Centers for Disease Control and Prevention (CDC) has recently initiated a project to overcome differences in testing. Differences such as the large variability in steroid hormone measurements among different assay methods and the lack of valid reference intervals for clinical diagnosis and treatment particularly for epidemiologic studies were being examined. The goal of the CDC is to assure that measurement results from a sample are the same, independent of the methodology or technology used in a laboratory. Although it is evident from previous efforts by the CDC that vast improvements in measurement performance can be achieved through assay standardization, it could take years to accomplish such an achievement for the substantial number of principal steroid hormones that require quantitation in women and men.

45.2 METHODS AND APPROACHES

45.2.1 Calibration Standards and Quality Control Samples

In general, steroids are hydrophobic and can be dissolved well in many organic solvents. Typically, methanol and acetonitrile are the initial solvent choices used in the preparation of stock and/or spiking solutions of steroids for quantitative analysis of samples. Calibration standards and QC samples are prepared in biological matrices either by a serial dilution of stock solutions with blank matrices or by spiking prediluted analyte solutions at various concentrations into biological matrices. It is not recommended to use neat organic or aqueous-solution-derived calibration curves to quantitate steroids in biological samples, as this cannot compensate for sample extraction efficiency from biological matrices or ionization differences caused by matrix effects that occur during LC-MS analysis and detection.

Analyte-free biological matrices are generally required in LC-MS quantitative analysis of steroids to avoid or minimize issues of interference. Due to the presence of endogenous steroids in biological systems, it has been very challenging to find suitable matrices to use in the preparation of calibration standards and QC samples for steroid bioanalysis. A great amount of effort on the evaluation and pretreatment of biological matrices is required in LC-MS method development for steroid analysis of biological samples. Multiple lots of matrices should be screened for the presence of steroids of interest prior to the preparation of calibration standards and QC samples. Typically, pretreatment of matrices to remove endogenous steroids from the blank matrix is necessary. A widely accepted approach has been to use activated charcoal to extract steroids from biological matrices. For example, for quantitative analysis of prednisolone and cortisol in human plasma and urine (AbuRuz et al., 2003), 4 g of activated charcoal was added to 100 ml of human plasma or urine. After mixing with a magnetic stirrer for 2 h, the samples were centrifuged at 1610*g* for 3 h and then the supernatant was passed through a sintered glass filter. Since this steroid removal approach is labor intensive, users can purchase steroid-free matrices directly from vendors, for example, Bioreclamation (Westbury, NY) and Golden West Biologicals (Temecula, CA). Some laboratories use albumin solutions as surrogates for plasma and serum. For example, in a method for quantitative analysis of estrogens and their metabolites in human serum, calibration standards were prepared in a 4% albumin solution (Guo et al., 2008). A more complicated approach was described recently by Ming et al. (2011) in their work on the preparation of an analyte-free matrix. In this approach, 200 ml of human urine was loaded on a 60-ml SPE column (Sigma-Aldrich Discovery DSC C_{18}). The first 10 ml of eluate from the column was discarded, and the subsequent eluates were collected into 20-ml scintillation vials. All urine fractions collected were then checked by LC-MS/MS for the steroids of interest, and the fractions with nondetectable steroids were combined and used as steroid-free urine matrix.

The concerns about using artificial matrices in LC-MS bioanalysis comes from the fact that the surrogate matrix does not contain all matrix components, anticoagulants, etc. presented in the study samples. This could potentially cause differences in ionization efficiency in the ion source leading to a bias of the mass spectrometer's response between calibration and study samples due to matrix effects. In addition, steroid analyte recoveries of extraction from nonidentical matrices could be different, resulting in quantitation bias.

Caution should be taken in processing quantitation data to make sure that the presence of endogenous steroids in the samples does not jeopardize accurate determination of analyte concentration. If necessary, the lower limit of quantitation (LLOQ) of an assay may be elevated to a level where the interference peak areas in the blank samples are below 20% of the analyte peak areas in the LLOQ samples.

45.2.2 Internal Standards

Internal standards play a very important role in the quantitative analysis of steroids from biological samples using LC-MS/MS. The purpose of using internal standards is not only to correct for the variation of analytes during sample processing and chromatography, but also to compensate for variable ionization of analytes in the ion source of the MS detector. Whenever possible, stable isotope-labeled steroids should be used as internal standards for the analytes to be analyzed. Most stable isotope-labeled steroid internal standards are labeled with either deuterium or ^{13}C. Between the two, the deuterium-labeled steroids are more commonly available due to lower cost and easier synthetic preparation steps. Stable isotope-labeled steroids are commercially available from many vendors such as Toronto Research Chemicals Inc. (North York, Ontario, Canada), CDN Isotopes (Pointe-Claire, Quebec, Canada), and Cambridge Isotope Laboratories (Andover, MA). Two considerations should be given, in general, when selecting a stable isotope-labeled steroid internal standard. First, a stable isotope-labeled compound should not contain any unlabeled target steroids to be analyzed. The contribution of a trace amount of steroid analyte from the internal standard to the target analyte may cause disqualification of an analytical run due to the presence of the interference. On the other hand, the presence of natural isotopes of any target analyte can contribute to the intensity of the internal standard ions, if the mass between an analyte and its stable isotope-labeled internal standard is too close. This type of interference is concentration-dependent, and therefore, may impact on the linear dynamic range of an assay. To avoid this type of isotope effect on quantitation, a stable isotope-labeled internal standard should be greater than or equal to three mass units. If stable isotope-labeled internal standards are not available during method development, structural analogues (i.e., compounds with similar structures and slight differences in functional groups) can also be used alternatively. In this case, careful evaluation of the extraction recoveries and matrix effects of the analyte and the internal standard is warranted during method development and validation. If significant variation of instrument response is observed for an analyte and its internal standard, efforts should be given to explore alternative internal standards by using other chemical analogues. Quantitative analysis without using an internal standard(s) should be avoided. For detailed information about the application of internal standards in bioanalysis, refer to Chapter 17.

45.2.3 Sensitivity Enhancement via Derivatization

Sample treatment with chemical derivatization has been extensively used in the past for steroid analysis using LC-MS and GC-MS, as observed in the literature where a large number of publications give detailed chemical derivatization procedures as part of sample preparation. These studies demonstrated that derivatization was useful to increase the molecular weight of a precursor ion, improve ionization efficiency, change the polarity of the detection, improve the fragmentation patterns, eliminate the interferences, and reduce the matrix effects. Overall, sample derivatization can lead to profound improvements in assay performance, and consequently increases detection sensitivity by several orders of magnitude. In a practical sense, this is very important during steroid analysis with MS since most steroids are generally not efficiently ionized in ESI or APCI compared to many other types of compounds. Derivatization also serves to optimize chromatography by improving separation. In recent years, with the availability of modern LC-MS technology and instrumentation, derivatization is no longer absolutely required in the analysis of steroids in biological samples using LC-MS. Regardless, it is always an option for those methods performed on older analytical instruments and dealing with some unusual situations during method development.

Derivatization of steroid analytes is achieved by coupling a polar moiety to either a keto or hydroxyl group of the steroid molecule. As reviewed by Santa in Chapter 19 and his previous publication (Santa, 2011), typical derivatization reagents used for reacting with a keto group of steroids include hydroxylamine, 1-(carboxymethyl)pyridinium chloride hydrazide (Girard's reagent P), 2-hydrazino-1-methylpyridine (HMP), 2-hydrazinopyridine (HP). For reacting with the hydroxyl group of a steroid, derivatization reagents are 5-dimethylamino-1-naphthalene sulfonyl chloride (dansyl chloride), picolinic acid, 2-fluoro-1-methylpyridinium *p*-toluenesulfonate (FPMTS), 1-(2,4-dinitro-5-fluorophenyl)-4-methylpiperazine (PPZ), 4-(4-methyl-1-piperazyl)-3-nitrobenzoyl azide (APZ). For steroids containing more than one keto or hydroxyl group, reagents such as HP or isonicotinoyl azide could be used for producing multiple derivatives. One must consider that these types of steroid derivatives with multiple charged groups may end up losing the sensitivity gained from MS when a single derivative is used.

During method development, various derivatization reagents need to be carefully considered and selected. The key to success is to identify a derivatization reagent that is specific for the analyte of interest. Typically, derivatization reagents can generate signature product ions on tandem MS due to the loss of the moiety added by the derivatization. For example, dansyl chloride (5-dimethylamino-1-naphthalene sulfonyl chloride) has a dimethylamino group as an ionization moiety and a reactive sulfonylchloride group. The generated derivatives provided a predominately single product ion at m/z 171 by collision-induced dissociation (CID), which was assigned to the protonated dimethylaminonaphtyl moiety. This derivatization reagent has been successfully used in the analysis of E_1 and E_2 in human plasma (Nelson et al., 2004) for increasing the detection sensitivity by 6- to 10-fold. Using the same derivatization agent, Xu et al. (2007) were able to simultaneously analyze 15 estrogens in human serum at a high sensitivity level (8 pg/ml LLOQ). In cases when derivatives lack analyte-specific distinct product ions, it is necessary to rely on more extensive chromatographic separations to achieve the required specificity of these types of analyses.

Biological sample treatment with derivatization is typically performed after liquid–liquid extraction (LLE) or solid phase extraction (SPE). It is always a major concern whether an added derivatization step to the overall sample processing could compromise the quality of a bioanalytical method. Obviously, the assay precision is highly dependent on the recovery of a derivatization reaction. Factors that can significantly impact the recovery of a derivatization reaction are the derivatization agent, pH of solution, incubation temperature, and incubation time. Therefore, optimization and confirmation of the derivatization conditions must be clearly defined during method development to obtain consistent batch-to-batch recovery of the steroid of interest. A good derivatization reaction should be simple, easy to perform, and have a high reaction yield. One example in this regard was given by Xu et al. (2005), who demonstrated a well-defined dansyl derivatization procedure optimized for the reaction heating time and temperature, dansyl chloride concentration, pH, and the presence of L-ascorbic acid. A 0.1 ml aliquot of 0.1 M sodium bicarbonate buffer (pH 9.0) and 0.1 ml of 1 mg/ml dansyl chloride solution in acetone were added to the dried extracts of urine samples from LLE using dichloromethane. The reaction time was only 5 min at 60°C, and the reaction solution was directly injected into an LC-MS/MS system for the analysis of the dansyl chloride derivatives of estrogens.

Method development for derivatization of steroid analytes is a tedious process, and the establishment of some systematic and efficient approaches is necessary for success. Licea-Perez et al. (2008) exemplified a very useful approach in their work for developing a derivatization method in the sample preparation for the simultaneous analysis of testosterone (T) and 5α-dihydrotestosterone (DHT). The structures of the two analytes were almost identical, both containing a hydroxyl and a carbonyl group, except that the double bond in ring A of DHT was reduced. The authors realized that the unique 3-oxo-Δ4 structure allowed the achievement of the sensitivity required, even without derivatization. Therefore, the focus of method development was on derivatization conditions for DHT only. Initial efforts included derivatizing the carbonyl moiety with hydroxylamine, nitrobenzyl hydroxylamine, and methyl hydroxylamine, which failed to meet the sensitivity needs. Therefore, derivatization conditions for reacting with the hydroxyl moiety of DHT were evaluated. A number of derivatization reagents such as 2-fluoro-1-methylpyridine, nitrobenzoyl chloride, *N*-methyl isotonic anhydride, 3,4-pyridine dicarboxylic anhydride, 2,3-pyridine dicarboxylic anhydride, phthalic anhydride, 3-nitrophthalic anhydride, 2,3-pyrazinedicarboxylic anhydride were tested. These reagents were all ruled out due to either the presence of a large amount of precipitate after hydrolysis or the lack of sensitivity of the DHT derivatives on the mass spectrometer. While the derivatization of T and DHT with 3,4-pyridinedicarboxylic anhydride was found to allow the quantitation of DHT at the desired LOQ in serum, chromatographic separation between the DHT derivative and other endogenous steroid derivatives (e.g., androsterone) presented in the samples was not achieved. Finally, the derivatization of DHT and T with 2,3-pyridinedicarboxylic anhydride was found to generate two positional isomeric products with one domination of each in both positive and negative ionization LC-MS. After reaching a 0.01 ng/ml LOQ of DHT using negative ion mode, resulting in a high-efficiency derivatization reaction, sufficient separation of T and DHT derivatives from other endogenous steroid derivatives was achieved. The derivatization method was then finalized followed by a successful validation in this laboratory.

45.2.4 Extraction for LC-MS Analysis

Biological sample cleanup through extraction is an important step in the bioanalytical process for steroid analysis using LC-MS. The purpose of extraction is to remove the proteins and interfering substances from samples to assist the chromatography of an analyte of interest on HPLC columns, to minimize matrix effects during ionization on MS, as well as to increase assay sensitivity. Sometimes, extraction is also used to concentrate an analyte in samples in order to improve the sensitivity of an analytical assay. Selection of extraction techniques depends on the compound, matrix, sensitivity target, and throughput. Historically, LLE and SPE are the two major sample cleanup methods in steroid bioanalysis. The use of protein precipitation (PPT) in sample extraction has been increased in recent years along with the significant advancement of modern LC-MS instrumentation. Steroids in a biological complex such as plasma, serum, or whole blood may not be directly analyzed using LC-MS without extraction, unless special techniques such as online extraction and high-turbulence liquid chromatography (HTLC) are employed. Sample analysis using a "dilution and shoot" approach may be limited to analysis of urinary steroids only.

45.2.4.1 Liquid–Liquid Extraction LLE is based on the differential solubility, acid–base equilibrium, and partitioning of analyte molecules between aqueous and organic phases. Prior to extraction, a biological sample is typically treated with the addition of a chemical reagent, for example, formic acid, to release steroids from the proteins to which they are bound. Internal standard solution containing a minimum amount of organic solvent is added to the sample prior to extraction with a water immiscible organic solvent. Prior to the availability of the modern high-sensitivity mass spectrometers and related sample preparation techniques, large volumes of biological samples and extraction solvents were required to perform extraction in 10 ml or larger tubes. Currently, in a typical bioanalytical laboratory in the pharmaceutical industry, the sample volume taken for analysis and organic solvent volume used for LLE extraction are <0.1 and 1 ml, respectively. Thus, high-throughput LLE based on a 96-well plate format has become a popular practice in advanced bioanalytical laboratories.

An organic solvent for LLE is selected on the basis of the polarity or type of the steroid of interest. While the polarity of the steroid molecule can incrementally increase by substituting a polar moiety such as keto, hydroxyl, or carbolic acid into the molecule, significant changes in the polarity of steroid molecules occur after they are conjugated with glycine, taurine, sulfuric acid, and glucuronic acid. Typical immiscible organic solvents used for steroid extraction from biological samples are heptane, hexane, pentane, dichloromethane, diethyl ether, ethyl acetate, and methyl *tertiary*-butyl ether (MTBE) (ranked according to polarity from low to high). A standard LLE procedure used in steroid analysis was exemplified in the study on quantitative determination of norethindrone (NE) and ethinyl estradiol (EE) by LC-MS/MS in human plasma (Li et al., 2005). The LLE was performed using a 0.5 ml aliquot of plasma and 4 ml of *n*-butyl chloride in 16 × 125 mm glass tubes with screw caps. After adding a 25 μl aliquot of the stable isotope-labeled internal standard in methanol to each sample, the tubes were capped and vortex mixed at high speed for 3 min. The samples were then centrifuged at 4000 rpm for 10 min. The aqueous layer was frozen in a dry ice–acetone bath which allowed efficient decantation of the organic layer into prelabeled 13 × 100 mm glass tubes. The organic layer was evaporated to dryness under a stream of nitrogen at 40°C. The sample residue was reconstituted and injected into an LC-MS/MS instrument for analysis. A similar method was also established by a different laboratory using a 96-well plate format with automation (Licea-Perez et al., 2007).

As a modified version of conventional LLE, supported liquid extraction (SLE) has been implemented in LC-MS bioanalysis of steroids in recent years. SLE is a flow-through-based technique utilizing specially cleaned and sized porous adsorbents, including diatomaceous earth, as the supporting media where an efficient LLE takes place between the very thin aqueous stationary film formed after sample loading and a water immiscible organic solvent. SLE is commercially available in plate and cartridge format as SLE+ (Biotage, Charlotte, NC) and Aquamatrix (Orochem Ontario, CA). Compared to the traditional LLE, SLE is more suitable for high throughput of analysis. In one application by Wu et al. (2010), Biotage's 96-well SLE plate containing 200 mg diatomaceous earth was used for sample preparation for quantitative analysis of hydrocortisone in mouse serum using LC-MS/MS. A 100 μl aliquot of sample premixed with 25 μl of the internal standard working solution and 100 μl aliquot of deionized water was loaded onto a 96-well SLE plate, allowing the samples to pass through the sorbent by gravity. After several minutes, 600 μl of MTBE (three volumes of the extraction solvent) were added to each well of the plate. At the end, a vacuum of ~10 Hg was applied to the plate for complete elution of the analyte from the SLE plate. The MTBE eluate was then further processed for analysis of the steroid analyte by LC-MS/MS.

45.2.4.2 Solid Phase Extraction Solid phase extraction has been widely implemented in sample extraction for steroid analysis since it was introduced to the steroid field about 30 years ago, and it is considered to be the most important tool of sample extraction in quantitative analysis of steroids. Compared to other extraction techniques, SPE generally provides a cleaner sample extract that benefits assay specificity.

SPE cartridges and disks are available from many vendors with a variety of stationary phases (10–200 mg amount), each of which can separate analytes according to different chemical properties. Most stationary phases are based on silica that has been bonded to a specific functional group. Some of these functional groups include hydrocarbon chains of variable length (reverse-phase SPE), quaternary ammonium or amino groups (anion exchange), and aliphatic sulfonic acid or carboxyl groups (cation exchange). The reverse-phase SPEs are mainly used for the extraction of unconjugated steroids and the ion-exchange SPEs for conjugated steroids. SPE cartridges are also available in a 96-well format and can be handled using liquid handling automation instruments such as the Tomtec Quadra 96, Packard Multiprobe II, Hamilton STAR, etc. Compared to individual sample processing, the 96-well format with a parallel sample processing can dramatically improve assay efficiency and reduce the turnaround time.

In a typical reverse SPE procedure, the steroid analyte of interest in biological samples is mixed with an aliquot of internal standard solution. Steroid–protein binding should be disrupted by adding an aliquot of a chemical reagent to samples. The choice of the reagent includes formic acid, disodium EDTA, trichloroacetic acid, acetic acid, and phosphoric acid. The samples are then applied to an SPE cartridge. As the sample goes through the cartridge, the steroid analyte(s) is retained, and the unwanted components (such as salts and proteins) from the biological samples are washed out of the cartridge with water and aqueous organic solvent mixture. The analyte(s) retained in the stationary phase is eluted with organic solvents such as methanol or acetonitrile and collected for further treatment prior to LC-MS analysis. A manifold designed for SPE allows application of vacuum (negative pressure), which can be used to speed up the extraction process by pulling the liquid sample through the stationary phase completely. Based on the same principle, an SPE extraction using a 96-well plate can also be processed using a centrifugation approach (positive mode). The speed of centrifugation is typically no more than 1000*g*. The advantage of this approach is that effective gravitational force is equally applied to all samples on the plate.

The key factors in SPE method development to be considered are SPE sorbent selection (i.e., type and size), sample volume determination, composition and volume of the washing solution, and the elution solution. In addition, a sample pretreatment procedure for steroid–protein binding disruption needs to be defined. A pH condition in SPE may not be relevant for most steroid analytes but necessary for steroid conjugates. Qin et al. (2008) demonstrated a well-defined SPE procedure for achieving greater than 90% recovery of seven estrogen conjugates in human urine. Waters Oasis HLB cartridges were preconditioned with 0.3% aqueous phosphoric acid, and the urine sample was acidified before loading. After the cartridge was washed with water and 2% acetic acid in methanol/water (60/40, v/v), the analytes were eluted with methanol containing 2% ammonium hydroxide.

45.2.4.3 Protein Precipitation The advances in modern MS sensitivity and specificity in recent years have provided opportunities for implementing the PPT extraction procedure, the simplest and the most straightforward sample preparation technique, for steroid bioanalysis. The PPT procedure for sample preparation is achieved by adding an aliquot of organic solvent to a sample followed by vortex mixing and centrifugation, resulting in the release of analytes from the matrix while allowing them to remain dissolved in the supernatant liquid. The most commonly used organic solvents are methanol and acetonitrile, which can effectively remove proteins from plasma or serum samples when used in a ratio of 3:1 or greater. The supernatant from PPT can be directly injected into an LC-MS system for analysis if the PPT solvent composition is similar to the HPLC mobile phase. It is also very common to evaporate a supernatant to dryness for reconstitution with different solvents to achieve better chromatography of an analyte in an LC-MS system. The disadvantage of the PPT technique is that it is ineffective in removing phospholipids that can cause significant matrix effects during MS analyses. To address the issue, it is highly recommended to use stable isotope-labeled internal standards for quantitative analysis of steroids in biological matrices. One example for steroid bioanalysis using a PPT approach in sample preparation is the work presented by Janzen et al. (2008) who developed and validated a fast (6 min run time) LC-MS/MS method for analysis of 10 pregnane steroids. In the sample preparation, a 15-μl aliquot of serum was mixed with 200 μl of acetone/acetonitrile 50:50 (v/v) and 20 μl of the internal standard solution in a microtiter plate well. The samples were injected into the LC-MS/MS system for analysis after gentle vortex mixing and centrifugation.

45.2.4.4 Online Extraction As a fundamental part of LC-MS/MS analysis, manual sample preparation which includes extraction is not only tedious and time consuming but also considered to be one of the major sources of analytical method development failure. One way to deal with this challenge is to implement online extraction technologies, which have become more commonly used over the last few years with the introduction of highly automated two-dimensional HPLC systems (Carvalho et al., 2008; Guo et al., 2008; Singh, 2008). In these experiments the sample is injected into the initial loading or trap column (first dimension) in the system, which is normally a variant of SPE or anion-exchange packing materials in a column format that is almost identical to a traditional HPLC column. The sample is fractionated in a similar fashion to off-line versions of the same technique, but this is done in an automated fashion using a programmed experiment and different mobile phase compared to HPLC. The application of these eluting phases is often applied in steps, with each step stripping (i.e., heart cutting) a portion of the analyte peak eluting from the stationary phase. These analytes are then diverted directly onto an analytical HPLC column (the second dimension) and separated by a mobile phase gradient.

A variation of the two-dimensional LC approach utilizes HTLC. In HTLC-MS, the sample is prepared by first acidifying it with 1.5- to 2-fold volumes of 10–20% formic acid in water (exact specifications need to be determined experimentally and vary from compound to compound) to release the analyte from any binding proteins. Internal standard is then added and the sample vigorously mixed prior to incubation at room temperature for 30 min. The sample is placed onto an Aria TLX-4 HTLC system (Cohesive Technologies Inc.; Franklin, MA; part of Thermo Fisher Scientific) and the prepared sample injected into an extraction column at high flow rate. The turbulence created inside the column allows analytes to bind to the large particles of the extraction column, while protein and other debris flow freely through to waste. The flow is then reversed, slowed, and the analytes are eluted and transferred to a reverse-phase analytical column. The analytes are then quantitated using a tandem mass spectrometer. The whole process of extraction, separation, and detection is automated through the use of the Aria TX-4 system that controls the entire chromatography process as well as controlling the collection of the data on the mass spectrometer. Using this technique, Quest Diagnostics

Nichols Institute validated a highly sensitive HTLC-MS/MS assay for the analysis of testosterone in human serum, and the LLOQ was 3 pg/ml (Salameh et al., 2010).

45.2.5 Hydrolysis of Steroid Phase II Conjugates

Conjugation with glucuronic and/or sulfonic acid is one of the major metabolic pathways for steroids in the human. Therefore, steroids are present in biological matrices in both conjugated and unconjugated forms. Due to lack of reference standards for steroid conjugates, the quantitative determination of steroid conjugates has been typically performed using an indirect approach, in which the steroid conjugates are analyzed after enzymatic and/or chemical hydrolysis reactions. The concentration of a conjugated steroid in a sample is then calculated by the determination of the difference of free steroid concentration before and after hydrolysis.

Hydrolysis of glucuronide or sulfate conjugates is performed by the incubation of samples with β-glucuronidase or sulfatase in aqueous solutions. These enzymes are obtained from various sources, such as the bacteria *Escherichia coli*, mammalian bovine liver, or the molluscan snail Helix pomatia. A β-glucuronidase or sulfatase product with absolute pure enzyme activity may not be easily found from a commercial source as both types of enzymatic activities are most likely coexisting in the products obtained, which make it hard to determine concentrations of glucuronide and sulfate conjugates individually from samples using the enzyme hydrolysis approach. Incubation conditions need to be optimized for amounts of enzyme, temperature, duration, and pH to reach the highest reaction yield of the hydrolysis reaction of conjugates in samples. Typically, between 1 and 20 units of the enzyme preparation is used per microliter of the biological sample. The pH of the hydrolysis incubation is generally maintained at acidic conditions. The hydrolysis reaction duration may vary from 30 min to 22 h, depending on the compound and source of enzyme (Gomes et al., 2009). A typical hydrolysis experiment procedure in steroid analysis was exemplified in the study for the analysis of serum E_2 and E_1 and their metabolites by Xu et al. (2007). In their work, to a 0.5 ml aliquot of serum sample, a 0.5 ml volume of freshly prepared enzymatic hydrolysis buffer was added, which contained 2 mg of L-ascorbic acid, 5 μl of β-glucuronidase/sulfatase (Helix pomatia, type HP-2 from Sigma Chemical Co.), and 0.5 ml of 0.15 M sodium acetate buffer (pH 4.1). After incubation for 20 h at 37°C, each sample underwent slow inverse extraction at 8 rpm with 8 ml of dichloromethane for 30 min. After extraction, the aqueous layer was discarded and the organic phase was transferred into a clean glass tube followed by further treatment prior to LC-MS/MS analysis.

To deal with situations where conjugates are resistant to enzyme hydrolysis, cleavage of steroid glucuronides and sulfates can also be achieved using a chemical hydrolysis-based approach. This approach employs strong acids, high temperature, and long reaction duration. However, this approach should proceed with caution, as the treatment may generate unwanted side products and damage structurally fragile steroids. In one successful study, Tang and Crone (1989) incubated steroid conjugate samples with anhydrous methanolic hydrogen chloride for 10 min at 60°C, which resulted in efficient and effective hydrolysis of steroid conjugates.

The significant improvement of LC-MS technologies over the last two decades has provided opportunities for the measurement of intact steroid conjugates and related compounds to avoid the hydrolysis step prior to analysis. Qin et al. (2008) reported a hydrophilic interaction liquid chromatography (HILIC) separation with tandem MS detection method for the analysis of seven urinary estrogen conjugates without sample pretreatment with hydrolysis. The assay had an LLOQ of 2 pg/ml and was validated using 1 ml of urine sample. Since the mobile phase used in the gradient elution on HILIC typically starts with a high organic solvent content, it also provides opportunities for analyzing intact steroid conjugates in biological samples using a PPT extraction-HILIC-MS platform.

45.2.6 LC-MS

LC-MS is now well established as a pivotal tool for the bioanalysis of small molecules in biological matrices in the pharmaceutical industry and clinical laboratories. LC-MS has been widely used in steroid profile analysis and quantitation, although GC-MS as a traditional standard tool still has its advantage in determining new metabolic pathways and identification of unknown steroids.

Prior to quantitation by MS, chromatographic separation of steroids was performed on HPLC, based on the selective distribution of steroids between a liquid mobile phase and a stationary phase. Reverse-phase HPLC columns (typically 5–10 cm × 2–4.6 mm) packed with nonpolar types of stationary phases (C_{18}, C_8, and phenyl-bonded silica) are the main choices of columns for steroid chromatographic separation, along with methanol, acetonitrile, and water as mobile phases. The use of certain mobile phase additives such as buffers and pH modifiers may improve chromatography and ionization efficiency. For example, in the development of a highly sensitive and selective LC-MS/MS method for the determination of testosterone and DHT in human serum (Licea-Perez et al., 2008), using a Waters Acquity UPLC BEH C_{18} stationary-phase column and water–acetonitrile as mobile phase, it was observed that the higher the concentration of ammonium acetate present in the mobile phase, the better the peak resolution and the longer the peak retention for the two steroid analytes. The study also demonstrated that UPLC, as a more advanced liquid chromatography technology, has been successfully implemented into the pharmaceutical laboratories in steroid analysis by offering not only high productivity but also significant increased analytical resolution, speed, and sensitivity.

A mass spectrometer analyzes compounds according to their m/z ratios. The two key components in this process of MS analysis are the ion generation in the ion source and ion sorting in the mass analyzer. The introduction of atmospheric pressure ionization (API) techniques allows the analysis of various compounds, including steroids, by LC-MS. In API, the analyte molecules are ionized first at atmospheric pressure. The analyte ions are then mechanically and electrostatically separated from neutral molecules. Several different types of ion sources are commonly used in LC-MS, and each is suitable for different classes of compounds. Electrospray ionization (ESI) relies in part on chemistry to generate analyte ions in solution before the analyte reaches the mass spectrometer. Atmospheric pressure chemical ionization (APCI) is applicable to a wide range of polar and nonpolar molecules. Atmospheric pressure photoionization (APPI) is a relatively new technique that is similar to APCI but the difference is that the nebulizer gas (N_2) in APCI is ionized via the corona discharge needle that in turn ionizes gas-phase solvent molecules, which then transfer the charge to the analyte molecules through chemical reactions (chemical ionization). In contrast, in APPI the molecules in the gas phase are ionized by a discharge lamp that generates photons in a narrow range of ionization energies. In general, ionization of steroid compounds can be performed in either positive or negative ion modes, and the ESI source is typically more sensitive than the APCI source for polar compounds, and therefore, it is an ideal device for the direct analysis of steroid conjugates without hydrolysis. For nonpolar steroid molecules, the sensitivity on the ESI source is less satisfactory than the APCI source. Derivatization of steroids with polar groups has been historically used as an effective means to increase assay sensitivity by improving ionization efficiency of steroids on an ESI source.

The triple quadrupole mass spectrometer provides far better sensitivity, specificity, and dynamic range in quantitation application compared to the single quadrupole mass spectrometer. In MS/MS, the first mass discriminating quadrupole analyzer (Q1) is used for the selection of the molecular ions of analyte(s); these preselected ions are passed into the nonmass discriminating quadrupole (Q2) known as the collision cell, where they undergo collisions with a wall of inert gas molecules (such as N_2 or Ar) leading to energy transfer into the analyte and fragmentation of the molecule in a predictable manner. These fragment ions are then selected by a second mass discriminating quadrupole analyzer (Q3) according to their m/z ratios. The mode of acquisition for monitoring the transition from one parent ion to one or more daughter ions is called multiple reaction monitoring (MRM). During method development, mass transitions observed for the analytes should be carefully evaluated and selected as they are directly related to assay sensitivity and selectivity. A mass transition that has the highest intensity is not necessarily the best one, and this needs to be determined based on the selectivity and specificity of an assay. A mass transition from a unique fragment usually has a better assay specificity. It is a good approach to examine multiple mass fragments of target steroids and internal standards obtained from spiked biological samples. Another parameter related to mass transition is the dwell time, which is the time set for data collection from each MRM transition cycle during the analysis. Typically, it is set to allow acquisition of 12–15 scans over each peak of interest.

45.2.7 Major Challenges of LC-MS Assays for Steroid Hormones

Although LC-MS is rapidly becoming a gold standard in the pharmaceutical industry and clinical diagnostic laboratories as the key analytical tool for quantitative analysis of steroids in biological matrices, there are many challenges still remaining. From a method development perspective, the major challenges are related to assay sensitivity and specificity.

Steroids that are present at the mid to high pg/ml and ng/ml concentration levels in the biological samples can routinely be measured by scientists in pharmaceutical and clinical laboratories. However, many steroid hormones circulate at very low concentrations in healthy subject or patients. Ultrasensitive assays that can be used to measure steroids at a low pg/ml or even at fg/ml levels are required to address some clinical diagnostic issues. For example, measurements of low levels of E_2 may be important for the prediction of risk of fractures, and for monitoring the extent of E_2 suppression in women receiving aromatase inhibitor therapy and the response to antiestrogens for the prevention of breast cancer (Stanczyk and Clarke, 2010). Another example is that there is presently a great interest in the role of catechol and 16α-hydroxylated estrogens in the etiology of breast cancer. However, these estrogen metabolites (EMs) are present in very low concentrations in serum and breast tissue. There is a major need to obtain valid ultrasensitive assays for the unconjugated forms of these estrogens. In addition, drug–drug interaction studies of new drug candidates such as low-dose oral contraceptives (OCs) containing EE are common in drug development. These studies require highly sensitive and selective bioanalytical methods for accurate measurement of low levels in human blood.

Improvements in MS instrumentation have had a significant impact on our ability to achieve better sensitivity in steroid quantitative analysis. The direct and effective way to deal with the sensitivity challenge is obviously to use a state-of-the-art tandem mass spectrometer such as the AB Sciex API4000 or API5000, Thermo Scientific TSQ Quantum Ultra, Waters Xevo TQ, etc. (also refer to the methods highlighted in Table 45.1). The impact of biological sample matrix on detection sensitivity is a key factor to be considered during method development. Endogenous biological components can elevate the nonspecific background signals, interfere with analyte peaks, or suppress analyte ionization. Phospholipids and coextracted or coeluted interfering compounds are known to contribute to matrix effects. Due to their hydrophobic nature, phospholipids show strong retention on reverse-phase columns. When gradient conditions are not sufficient to remove these interfering components, they may build up on the column or elute in subsequent injections, resulting in unpredictable instrument responses for the analyte(s). A common procedure for the evaluation of matrix effects is postcolumn infusion of the target analyte in the effluent of the chromatographic column and the introduction of matrix blank samples. Negative peaks eluting at the retention time of the analyte of interest are signs of ion suppression. Another approach for the evaluation of matrix effects is based on direct comparison of the signal intensity of the analyte measured with and without the biological sample matrix. Several approaches can be taken to reduce and minimize matrix effects: (1) modification of sample extraction procedures by, for example, replacing a PPT with an LLE or SPE, and by optimizing organic solvents used in the extraction; (2) optimization of chromatography by making changes to gradient conditions, mobile phase strength, and pH to shift the retention of the analyte away from chromatographic regions affected by ion suppression. Optimized chromatography also allows resolving issues with interfering peaks; (3) optimization of ionization conditions. Selection of the LC-MS ionization source and ionization polarity can impact the extent of matrix effects, and in general, switching from ESI to APCI can be an effective strategy for dealing with matrix effects. ESI is generally utilized for assays requiring high sensitivity; however, it is known to be more susceptible to matrix effects compared to APCI. On the other hand, many neutral steroids are poorly ionized on ESI. Therefore, the selection of an ionization source for steroid analysis should be made based on the evaluation of instrument responses and matrix effects obtained from using both sources; (4) compensation of matrix effects through use of stable isotope-labeled internal standards.

One reason that LC-MS has become a popular tool in bioanalytical and clinical laboratories is due to its high specificity and the ability of simultaneous analysis of multiple drugs and endogenous analytes in a biological sample. Assessment of the specificity of an LC–MS method should be performed during method development. Three groups of compounds are considered to have an impact on assay specificity: (1) isobars of the target steroid analyte (i.e., either structural isomers that share an elemental formula or steroids having the same nominal molecular mass, for example, cortisone/prednisolone, cortisol/tetrahydroprednisolone, T/DHEA, cortisol/fenofibrate); (2) steroid M + 2 isotopes (e.g., T and DHT, E_1 and E_2, cortisone and cortisol, prednisone and prednisolone, prednisolone and cortisol); and (3) coeluted substances that have the same characteristic precursor and product ions. These structurally related steroids typically have very similar mass fragmentation patterns or even identical product ions.

Chromatographic separation is the most effective way in dealing with the specificity issues in steroid bioanalysis. Great efforts must be given to developing HPLC conditions to achieve the best separation of steroid analytes in biological samples.

The commonly used approach for the assessment of the specificity of analysis is to start with the evaluation of full product ion mass spectra of an analyte and its related MRM transitions. The relative intensities of the MRM mass transitions derived from different fragments of the targeted molecule and the internal standard in the matrix spiked with analytes and in the blank are compared to rule out potential interferences from endogenous compounds. Before wrapping up a method development activity, it is important to verify the specificity of an assay by analysis of incurred samples collected from subjects and especially the patients, to which the method will be used for the analysis of steroids. This is to avoid redundant work on method development if significantly higher concentrations of an interfering compound(s) are observed unexpectedly at the onset of sample analysis. Establishment of a specific mass transition to avoid interference issues may rely on the acquisition of several mass transitions of a target analyte and internal standard from the analysis of incurred samples. It is very helpful to have knowledge about all potential interferences to steroid analytes that are present in the samples to be analyzed. An excellent publication by Kushnir et al. (2005) described several approaches used in the assessment of assay specificity. One can read the paper for very comprehensive discussions on the topic.

Isotopic crosstalk among steroids on LC-MS analysis can be exaggerated when their concentration ranges in biological matrices are significantly different, which is not uncommon in healthy volunteers or patients with steroid hormone disorders. In these situations, a steroid with an average molecular weight of 1 or 2 Da less than another steroid can cause very significant isotopic crosstalk in Q1 in the tandem MS. Unless the two steroids have different collision-induced fragmentation patterns, the crosstalk can contribute to overestimation of the steroid concentration. Therefore, using a well-defined MRM transition in quantitative LC-MS analysis of steroids is particularly important to achieve high assay specificity.

45.3 REPRESENTATIVE PROTOCOLS OF LC-MS BIOANALYSIS OF STEROID HORMONES

Numerous LC-MS methods for the analysis of steroids in biological matrices have been reported, and several review papers are also available to readers (Kushnir et al., 2010,

2011; Shackleton, 2010; Stanczyk and Clarke, 2010). In order to provide some key technical elements to guide LC-MS/MS method development and validation, several examples of established methods for the analysis of steroid hormones are highlighted in this section. Although the specific LC-MS equipment employed in the examples may not be applicable to some laboratories, compatible instruments from different vendors may work equally well. Most importantly, the principles of the methodologies and the strategies employed in the method development will apply to all laboratories in general for the development of new bioanalytical methods.

45.3.1 Estrogens

Estrogens are responsible for the development and maintenance of female secondary sex characteristics, reproductive function, regulation of the menstrual cycle, and maintenance of pregnancy. An accurate measurement of low concentrations of estrogens is important for the diagnosis of sex-hormone-related disorders. Challenges in the measurement of estrogens using LC-MS are low physiologic concentrations and the presence of endogenous interference compounds. The approaches used for enhancing the sensitivity were either through the use of a more sensitive instrument, or through the use of the derivatization. Implementation of the UPLC technology and 96-well plate format has demonstrated the advantage of increasing the speed of sample analysis.

Protocol 1: Simultaneous determination of estradiol (E_2), estrone (E_1), estriol (E_3), and 16α-hydroxyestrone (16α-OHE$_1$) in human serum without derivatization by Guo et al. (2008).

Method Highlights

Calibration standards and QC samples were prepared in 4% albumin solution. A 200 μl aliquot of serum sample was mixed with 300 μl of acetonitrile containing 1 ng/ml of deuterium-labeled internal standards. The supernatant obtained from centrifugation was diluted with water and then injected to a SCIEX API 5000 with an ESI in negative ion mode. Chromatographic separation was carried out on a C_8 column from Supelco (33 $\times$ 3 mm, 3 μm) over 8 min using methanol and water as mobile phase. A switching valve was used to allow changing the mobile phase between sample loading at a flow rate of 1 ml/min and analyte elution at a flow rate of 0.6 ml/min. The MRMs monitored in quantitation for E_2, E_1 and E_3, and 16α-OHE$_1$ were m/z 287/145(183), 269/145(143), 287/171(145), and 285/145 (143), respectively. High sensitivity was achieved with limits of detection of 1 pg/ml for E_1 and 16α-OHE$_1$ and 2 pg/ml for E_2 and E_3 in serum.

Technical Notes

- For a confirmation purpose, the authors also monitored an additional MRM transition (shown in parenthesis above) for each analyte.
- The authors' concern was that overestimation of estrogens may occur as a result of the hydrolysis of conjugated estrogens in serum samples during derivatization or LLE and SPE extraction.
- Analyte peak baseline separation was achieved for E_3, 16α-OHE$_1$, and E_2, but not for E_1 and E_2; however, the interference and crosstalk between E_1 and E_2 are considered to be negligible.
- Two approaches were applied to increase sensitivity: (1) two segments on MRM transition with increased dwell times for all the analytes measured; (2) maximum injection of the sample extract (about 1/3 of total volume).
- Stable isotope-labeled internal standards were used for all analytes except for 16α-OHE$_1$
- A 10-fold increase in sensitivity and much less interference were realized by using ESI negative ion mode as opposed to the APPI positive ion mode.

Protocol 2: Simultaneously quantitative measurement of 15 endogenous estrogens in human urine and serum with derivatization by Xu et al. (2005, 2007).

Method Highlights

Calibration standards and quality control samples were prepared in charcoal-stripped human urine or serum in the presence of 0.1% of L-ascorbic acid, and the lower limits of quantitation were 40 pg/mL for urine and 8 pg/ml for serum. A 500 μl aliquot of serum sample was pretreated for 20 h at 37°C with enzymatic hydrolysis buffer containing 2 mg of L-ascorbic acid, 5 μl of β-glucuronidase/sulfatase, and 0.5 ml of 0.15 M sodium acetate buffer (pH 4.1). To the dried sample after hydrolysis, 100 μl of 0.1 M sodium bicarbonate buffer (pH at 9.0) and 100 μl of dansyl chloride solution (1 mg/ml in acetone) were added. After 5 min at 60°C, the reaction solution was directly injected into a Finnigan TSQ Quantum MS/MS system coupled with ESI operated in the positive ion mode. HPLC was carried out on a Phenomenex Synergi Hydro-RP 150 $\times$ 2 mm column maintained at 40°C. The mobile phase, operating at a flow rate of 0.2 ml/min, consisted of methanol and 0.1% (v/v) formic acid in water with a linear gradient elution. A unique daughter ion, m/z 171 or 170, produced from the dansyl derivatives of estrogens and their metabolites was monitored for each of the analytes.

Technical Notes

- The advantages of analyzing estrogens using dansyl derivatization were significant gain in the sensitivity (through the introduction of a tertiary amine in the structure), mild reaction conditions, and quantitative yield of the products.

- For the measurement of conjugated EMs, the samples were treated with enzymatic hydrolysis using β-glucuronidase and sulfatase. For the measurement of unconjugated EM, the identical sample preparation procedures were used with the exclusion of the hydrolysis step.
- To obtain full baseline separation for all 15 EMs (except 16α-OHE_1 and 16-ketoE_2) by reverse phase C_{18} chromatography, a 70-min, long-gradient program was applied.
- It was found that ESI was superior over APCI and APPI in detecting and quantitating the EM-dansyl species. No matrix-induced MS ion suppression was observed in EM-dansyl and D-EM-dansyl eluted regions. The unique product ion is observed at m/z 170 for catechol estrogens (i.e., 1,2-dihydroxybenzene) and m/z 171 for the remaining estrogens.
- Optimization for estrogen metabolite derivatization with dansyl chloride was made for reaction heating time and temperature, dansyl chloride concentration, pH, and the yield of dansylation in the presence of L-ascorbic acid. The best yield of dansylation for all EMs was achieved at 60°C for 5 min containing 0.1% (w/v) ascorbic acid. Increasing dansyl chloride concentration from 1 to 3 mg/ml did not improve the yield of dansylation under the same conditions. No significant change in the extent of dansylation for noncatechol estrogens at pH 8.5–11.5 in the presence or absence of 0.1% (w/v) L-ascorbic acid was observed. The absence of 0.1% (w/v) ascorbic acid did result in a significant decrease in the dansylation efficiency of catechol estrogens.

Protocol 3: A semi-automated high-throughput UPLC-MS/MS method for simultaneous determination of OCs containing ethinyl estradiol (EE), norethindrone (NE), or levonorgestrel (LN) in human plasma by Licea-Perez et al. (2007).

Method Highlights

Assisted with a Hamilton STAR liquid handler, 0.3 ml of each plasma sample was transferred to an Arctic White 96-well polypropylene plate followed by the transfer of internal standard solution and 1 ml of MTBE. After vortex mixing and centrifugation, the organic phase was transferred to 1.2-ml polypropylene 96-well tubes using a TomTec liquid handler. After evaporation under a stream of nitrogen at 45°C, 0.1 ml of 100 mM sodium bicarbonate (pH 11.0) and 0.1 ml of 1 mg/ml dansyl chloride in acetone were added for derivtization at 60°C for 5 min. The steroid derivatives were extracted on an LLE hydromatrix plate and eluted with hexane prior to LC-MS/MS analysis. The chromatographic separation was achieved on either a conventional HPLC (Genesis C_{18}, 50 × 2 mm, 3 μm) or UPLC (Waters BEH C_{18}, 50 × 2 mm, 1.7 μM) column, respectively, using acetonitrile/water/formic acid as the mobile phase under gradient elution. The MRM transitions on a Sciex API 4000 equipped with a TurboIonSpray interface and operated in the positive ionization mode were m/z 530/171, 534/171, 299/231, 305/237, 313/245, and 319/251 for dansyl-EE, dansyl-EE-d4, NE, NE-d6, LN, and LNd6, respectively.

Technical Notes

- When compared to HPLC-MS/MS, the LC run time using UPLC was reduced by more than half while the analytes of interest were better resolved from the interfering endogenous plasma components. The signal-to-noise ratio was also increased by a factor of two for the dansyl-EE but remained unchanged for NE and LN. Overall the signal-to-noise ratios at LOQ were >10 for all the analytes.
- PPT was not considered because of its inadequacy to remove endogenous interferences. The SPE method using Waters Oasis HLB 96-well plates was discarded due to the presence of an unexpected interference peak that coeluted with the analyte.
- Two semi-automated LLE methods in 96-well plate format were developed to increase sample throughput: (1) a Varian Combilute 96-well SLE (260 mg/well) was used to assist in automating the LLE. However, due to slightly lower recovery, the method required at least 500 μl of sample to achieve the desired LLQ for the OCs; (2) ArcticWhite 2 ml 96-well plates with ArctiSeal mats were used for extraction of 300 μl plasma with 1 ml MTBE. The plate and seal mat needed to be washed with MTBE prior to extraction to clean the plastic residue and remove a minor interfering peak.
- Sodium bicarbonate (pH = 11.0) used in the derivatization strongly suppressed the ionization of NE and LN, so it was removed with a Varian Combilute 96-well SLE (260 mg/well) using hexane as an elution solvent.
- The use of deactivated (silanized) glass inserts was crucial due to the complete loss of dansyl-EE and dansyl-EE-d_4 because of tube absorption when the extracts were stored in polypropylene or other nondeactivated glass vials.
- A good separation of the contraceptive steroids from the interferences was critical for the accurate determination of analyte concentrations in the low concentration level samples. The LLOQ of 0.01 and 0.1 ng/ml were for EE and NE (or LN), respectively.

45.3.2 Androgen

Androgen is a group of steroid hormones that stimulates or controls the development and maintenance of male characteristics. Testosterone is the major androgen in males and predominant bioactive androgen in women. Testosterone plays an important role not only in the development and maintenance of the male phenotype but also in physiological functions and diseases of women and children. Quantitative determination of T and DHT in human serum with a sensitive and accurate bioanalytical method is required for the diagnosis and treatment of androgen deficiency in men.

Protocol 4: Simultaneous measurement of serum testosterone and dihydrotestosterone using LC-MS by liquid chromatography–tandem MS without derivatization by Shiraishi et al. (2008).

Method Highlights

Charcoal-treated human serum was used as the blank matrix of the assay. An aliquot (100 μl) of serum sample was mixed with a 25 μl aliquot of the internal standard solution in a glass tube followed with a LLE procedure using a 2 $\times$ 2 ml volume of ethyl acetate/hexane (3:2 by volume). To the pooled extracts, a 0.35 ml volume of 0.1 M sodium hydroxide was added to remove acidic contaminants in the extract, which may interfere in the analysis. The LC-MS/MS analysis was performed with a Sciex API 5000 LC-MS/MS instrument operated with ESI in the positive mode. A Thermo Hypersil GOLD column (100 $\times$ 1 mm, 3 μm) with a gradient profile at a slow flow rate of 0.045 ml/min and mobile phase of methanol and water containing 2% methanol and 26 mM formic acid was used. The 17.5-min gradient profile allowed baseline separation of the two analytes. The MRM transitions were m/z 289.2/109.0 for T and 291.2/110.9 for D2-T, 291.2/255.2 for DHT and 294.2/258.2 for D3-DHT, respectively.

Technical Notes

- Compared to other published methods, this method employed a simple LLE method without derivatization and only required 0.1 ml volume of sample to achieve the desired specificity and sensitivity (20 pg/ml), allowing the method to be easily adopted for routine use by clinical chemistry laboratories.
- Due to a low extraction recovery, SPE was not considered.
- Specificity was evaluated by running dilutions of potential cross-reacting steroids (androgens, progestins, and estrogens) at concentrations up to 100-fold greater than the concentration of the highest calibrator of the assay.
- Potential for chemical interference by over-the-counter drugs (ascorbic acid, salicylic acid, or acetaminophen) and by additives used in blood-collection tubes (EDTA, citrate, or heparin anticoagulant) were evaluated. There was no interference identified.
- Matrix effects were evaluated by measuring the changes in the baseline response of the respective MRM transition for the analytes and internal standards to the injection of solvent and steroid-free serum extract into the LC-MS/MS.

45.3.3 Pregnane Derivatives

Pregnane derivatives contain a basic 21-carbon pregnane skeleton and include progesterone (P), pregnenolone (Preg), and their 17α-hydroxylated derivatives, namely, 17α-hydroxyprogesterone (17OHP) and 17α-hydroxypregnenolone (17OHPreg), as well as the mineralocorticoids and glucocorticoids. As the majority of the steroids are normally present at relatively low concentrations in blood, the bioanalytical methods for measurement of these steroids in a complex sample matrix require high sensitivity and specificity. The difficulties in the analysis of 11-deoxycortisol (11DC), Pregn, and 17OHPregn with LC-MS/MS methods in the past were related to poor ionization of these molecules, and their nonspecific MRM mass transitions in the tandem MS. The protocols collected below are to show (1) some efforts that have been made over the years by enhancing the sensitivity of the mass spectrometric detection and improving the fragmentation patterns of pregnane derivatives via incorporation in the structure of a functional group that promotes ionization, and (2) application of the LC-MS method in analysis of pregnane derivatives in dried blood samples.

Protocol 5: Simultaneous analysis of 11-deoxycortisol, 17OHP, 17OHPreg, and pregnenolone in human serum by Kushnir et al. (2006).

Method Highlights

Calibration standards were prepared in 10 mg/ml bovine serum albumin. Aliquots (0.2 ml) of samples were transferred into glass tubes containing 2 ml of water followed by the addition of 20 μl of deuterated internal standard solution (11DC-d2, 17OHP-d8, 17OHPr-d3, and Pr-d4). The samples were applied to Phenomenex Strata X SPE columns prewashed with water and 20% acetonitrile in water. After a 15-min dry period, steroids were eluted with 3 ml of MTBE. The organic phase was treated to dryness, and the sample residues were redissolved in 0.3 ml of hydroxylamine solution (1.5 M, pH 10.0) for a derivatization reaction at 90°C for 30 min. The derivatives were extracted with 2 ml of MTBE before performing LC-MS/MS analysis. The chromatographic separation was carried out on a 50 $\times$ 2.0 mm Phenomenex Synergy Fusion RP (C_{18}) column using a gradient elution with mobile phase of methanol, water, and formic acid at a flow rate of 0.5 ml/min. The MRM transitions on a Sciex API4000 LC-MS/MS equipped with an ESI source and operated in positive ion mode were m/z 377/124 (112), 361/124(112), 348/330 (312), and 332/86 (300) for 11DC, 17OHP, 17OHPreg, and Preg, respectively.

Technical Notes

- To determine potential interference from cross-talk, 43 steroids and steroid metabolites were checked during method development. The limits of detection were 0.025 ng/ml for 11DC, 17OHP, and Preg, and 0.10 ng/ml for 17OHPreg.
- The poor ionization efficiency and CID fragmentation of 11DC, 17OHPreg, and Preg on a Sciex API3000 resulted in using a derivatization approach. The most promising derivatization of those evaluated was the oximation reaction, with hydroxylamine reacting with the keto group of the steroids to form oxime derivatives. ESI in positive ion mode showed the best sensitivity for the oxime derivatives.
- The authors found that ionization was more efficient with the addition of formic acid to the mobile phase compared with the format buffer.
- The Preg concentrations in the samples collected in sodium EDTA were found to be 280% higher than that in serum and sodium heparin plasma. Therefore, collection tubes containing EDTA were not used.
- A comprehensive analysis of the relationship between the efficiencies of ionization methods and the steroid structures with and without oxime derivatization was made by the authors.
- Multiple mass transitions for the analytes of interest were monitored to assess selectivity of the analysis in MS/MS.

Protocol 6: Simultaneous determination of 17α-hydroxyp regnenolone and 17α-hydroxyprogesterone in dried blood spots by Higashi et al. (2008).

Method Highlights

A 3-mm disk (equivalent to 2.65 μl of whole blood) was cut from a 10-mm diameter dried blood spot. Three disks of the dried blood filter paper were used for the extraction of each sample in a test tube. Methanol (200 μl) containing two internal standards (100 pg each) was added to the tube followed by ultrasonic extraction at ambient temperature for 30 min. The methanol extract was diluted with water (1 ml) and purified using a Phenomenex Strata-X 60 mg cartridge. After successive washing with water (2 ml), methanol/water (1:1, v/v, 2 ml) and hexane (1 ml), the steroids were eluted with ethyl acetate (1 ml). The dried residue was dissolved in ethanol (50 μl) and then subjected to derivatization with a freshly prepared solution of 10 μg of HP in 50 μl of ethanol containing 25 μg of trifluoroacetic acid under ultrasonication at ambient temperature for 15 min. The HP derivatives were further analyzed on a Sciex API2000 LC-MS/MS system. The separation was achieved with reverse-phase chromatography. The MRM transitions on ESI in positive-ion mode were m/z 424.3/253.0 for 17OHPreg-HP, m/z 513.4/364.1 for 17OHP-bisHP; 427.3/253.0 for IS1-HP and 520.4/368.1 for IS2-bisHP. The limits of quantitation were 1.0 and 0.5 ng/ml for 17OHPreg and 17OHP, respectively.

Technical Notes

- This method combined with derivatization of the oxo-group of the steroids was developed to deal with different ionization efficiencies on ESI or APCI for the structural related steroids 17OHP (3-oxo-4-ene-steroid) and 17OHPreg (3β-hydroxy-5-ene-steroid).
- Recoveries of about 80% were achieved for extraction of both analytes from the dried blood spot samples.
- On ESI in positive mode, $[MH]^+$ of the 17OHPreg and 17OHP HP derivatives had characteristic product ions at m/z 253.0 and 364.1, respectively.
- Derivatization of the oxo-steroids formed E- and Z-isomers, resulting in twin peaks on the chromatograms. With the specified HPLC condition (i.e., YMC-Pack Pro C_{18} RS column and the mobile phase of acetonitrile–methanol–10 mM ammonium acetate, 5:3:1 by volume), the derivatized 17OHPreg and 17OHP appeared as two single peaks, although the peak shape of 17OHP-HP was not symmetrical.
- Assay specificity was evaluated by the analysis of more than 10 endogenous progestogens, corticoids, and androgens under the same LC-MS/MS conditions, which confirmed that the endogenous steroids did not interfere with the quantification of 17OHPreg and 17OHP.
- The limits of quantitation were 1.0 and 0.5 ng/ml for 17OHPreg and 17OHP, respectively. It is assumed that a 100-fold increased sensitivity can be further easily achieved if the Sciex API 2000 is replaced with an advanced tandem MS instrument.

ACKNOWLEDGMENTS

We would like to thank Dr J. Flarakos, Novartis Institutes for BioMedical Research, for his valuable comments on this manuscript. We gratefully acknowledge Dr Nigel Clarke, Quest Diagnostics Nichols Institute, for valuable discussions and suggestions during the preparation of this chapter.

REFERENCES

Abraham GE. Solid-phase radioimmunoassay of estradiol-17 beta. J Clin Endocrinol Metab 1969;29(6):866–870.

AbuRuz S, Millership J, Heaney L, McElnay J. Simple liquid chromatography method for the rapid simultaneous determination of prednisolone and cortisol in plasma and urine using hydrophilic lipophilic balanced solid phase extraction cartridges. J Chromatogr B Analyt Technol Biomed Life Sci 2003;798(2):193–201.

Carvalho VM, Nakamura OH, Vieira JG. Simultaneous quantitation of seven endogenous C-21 adrenal steroids by liquid chromatography tandem mass spectrometry in human serum. J Chromatogr B Analyt Technol Biomed Life Sci 2008;872(1–2):154–161

Gomes RL, Meredith W, Snape CE, Sephton MA. Analysis of conjugated steroid androgens: deconjugation, derivatisation and associated issues. J Pharm Biomed Anal 2009;49(5):1133–1140.

Guo T, Gu J, Soldin OP, Singh RJ, Soldin SJ. Rapid measurement of estrogens and their metabolites in human serum by liquid chromatography-tandem mass spectrometry without derivatization. Clin Biochem 2008;41(9):736–741.

Guo T, Taylor RL, Singh RJ, Soldin SJ. Simultaneous determination of 12 steroids by isotope dilution liquid chromatography-photospray ionization tandem mass spectrometry. Clin Chim Acta 2006;372(1–2):76–82.

Higashi T, Nishio T, Uchida S, Shimada K, Fukushi M, Maeda M. Simultaneous determination of 17alpha-hydroxypregnenolone and 17alpha-hydroxyprogesterone in dried blood spots from low birth weight infants using LC-MS/MS. J Pharm Biomed Anal 2008;48(1):177–182.

Horning EC, Brooks CJ, Vanden Heuvel WJ. Gas phase analytical methods for the study of steroids. Adv Lipid Res 1968;6:273–392.

Horning MG, Knox KL, Dalgliesh CE, Horning EC. Gas-liquid chromatographic study and estimation of several urinary aromatic acids. Anal Biochem 1966;17(2):244–257.

Janzen N, Sander S, Terhardt M, Peter M, Sander J. Fast and direct quantification of adrenal steroids by tandem mass spectrometry in serum and dried blood spots. J Chromatogr B Analyt Technol Biomed Life Sci 2008;861(1):117–122.

Kushnir MM, Rockwood AL, Bergquist J. Liquid chromatography-tandem mass spectrometry applications in endocrinology. Mass Spectrom Rev 2010;29(3):480–502.

Kushnir MM, Rockwood AL, Nelson GJ, Yue B, Urry FM. Assessing analytical specificity in quantitative analysis using tandem mass spectrometry. Clin Biochem 2005;38(4):319–327.

Kushnir MM, Rockwood AL, Roberts WL, et al. Development and performance evaluation of a tandem mass spectrometry assay for 4 adrenal steroids. Clin Chem 2006;52(8):1559–1567.

Kushnir MM, Rockwood AL, Roberts WL, Yue B, Bergquist J, Meikle AW. Liquid chromatography tandem mass spectrometry for analysis of steroids in clinical laboratories. Clin Biochem 2011;44(1):77–88.

Li W, Li YH, Li AC, Zhou S, Naidong W. Simultaneous determination of norethindrone and ethinyl estradiol in human plasma by high performance liquid chromatography with tandem mass spectrometry—experiences on developing a highly selective method using derivatization reagent for enhancing sensitivity. J Chromatogr B Analyt Technol Biomed Life Sci 2005;825(2):223–232.

Licea-Perez H, Wang S, Bowen CL, Yang E. A semi-automated 96-well plate method for the simultaneous determination of oral contraceptives concentrations in human plasma using ultra performance liquid chromatography coupled with tandem mass spectrometry. J Chromatogr B Analyt Technol Biomed Life Sci 2007;852(1–2):69–76.

Licea-Perez H, Wang S, Szapacs ME, Yang E. Development of a highly sensitive and selective UPLC/MS/MS method for the simultaneous determination of testosterone and 5-alpha-dihydrotestosterone in human serum to support testosterone replacement therapy for hypogonadism. Steroids 2008;73(6):601–610. Epub 2008 Feb 2.

Ming DS, Heathcote J, Garg A, Darbyshire J, Eagleston E, Bajkov TL. The determination of urinary free and conjugated cortisol using UPLC-MS/MS. Bioanalysis 2011;3(3):301–312.

Nelson RE, Grebe SK, OKane DJ, Singh RJ. Liquid chromatography–tandem mass spectrometry assay for simultaneous measurement of estradiol and estrone in human plasma. Clin Chem 2004;50(2):373–384.

Perogamvros I, Owen LJ, Newell-Price J, Ray DW, Trainer PJ, Keevil BG. Simultaneous measurement of cortisol and cortisone in human saliva using liquid chromatography-tandem mass spectrometry: application in basal and stimulated conditions. J Chromatogr B Analyt Technol Biomed Life Sci 2009;877(29):3771–3775.

Qin F, Zhao YY, Sawyer MB, Li XF. Hydrophilic interaction liquid chromatography-tandem mass spectrometry determination of estrogen conjugates in human urine. Anal Chem 2008;80(9):3404–3411.

Salameh WA, Redor-Goldman MM, Clarke NJ, Reitz RE, Caulfield MP. Validation of a total testosterone assay using high-turbulence liquid chromatography tandem mass spectrometry: total and free testosterone reference ranges. Steroids 2010;75(2):169–175.

Santa T. Derivatization reagents in liquid chromatography/electrospray ionization tandem mass spectrometry. Biomed Chromatogr 2011;25(12):1–10.

Shackleton C. Clinical steroid mass spectrometry: a 45-year history culminating in HPLC-MS/MS becoming an essential tool for patient diagnosis. J Steroid Biochem Mol Biol 2010;121(3–5):481–490. Epub Feb 25, 2010.

Shiraishi S, Lee PW, Leung A, Goh VH, Swerdloff RS, Wang C. Simultaneous measurement of serum testosterone and dihydrotestosterone by liquid chromatography–tandem mass spectrometry. Clin Chem 2008;54(11):1855–1863.

Singh RJ. Validation of a high throughput method for serum/plasma testosterone using liquid chromatography tandem mass spectrometry (LC-MS/MS). Steroids 2008;73(13):1339–1344.

Sjövall J, Vihko R. Determination of androsterone and dehydroepiandrosterone sulfates in human serum by gas-liquid chromatography. Steroids 1965;6(5):597–604.

Stanczyk FZ, Clarke NJ. Advantages and challenges of mass spectrometry assays for steroid hormones. J Steroid Biochem Mol Biol 2010;121(3–5):491–495.

Tang PW, Crone DL. A new method for hydrolyzing sulfate and glucuronyl conjugates of steroids. Anal Biochem 1989;182:289–294.

Turpeinen U, Hämäläinen E, Stenman UH. Determination of aldosterone in serum by liquid chromatography-tandem mass spectrometry. J Chromatogr B Analyt Technol Biomed Life Sci 2008;862(1–2):113–118.

Vesper HW, Bhasin S, Wang C, et al. Interlaboratory comparison study of serum total testosterone [corrected] measurements performed by mass spectrometry methods. Steroids 2009;74(6):498–503.

Wu S, Li W, Mujamdar T, Smith T, Bryant M, Tse FL. Supported liquid extraction in combination with LC-MS/MS for high-throughput quantitative analysis of hydrocortisone in mouse serum. Biomed Chromatogr 2010;24(6):632–638. PubMed

Xu X, Roman JM, Issaq HJ, Keefer LK, Veenstra TD, Ziegler RG. Quantitative measurement of endogenous estrogens and estrogen metabolites in human serum by liquid chromatography-tandem mass spectrometry. Anal Chem 2007;79(20):7813–7821.

Xu X, Veenstra TD, Fox SD, et al. Measuring fifteen endogenous estrogens simultaneously in human urine by high-performance liquid chromatography-mass spectrometry. Anal Chem 2005;77(20):6646–6654.

Yamashita K, Okuyama M, Watanabe Y, Honma S, Kobayashi S, Numazawa M. Highly sensitive determination of estrone and estradiol in human serum by liquid chromatography-electrospray ionization tandem mass spectrometry. Steroids 2007;72(11–12):819–827.

46

LC-MS BIOANALYSIS OF LIPOSOMAL DRUGS AND LIPIDS

TROY VOELKER AND ROGER DEMERS

46.1 INTRODUCTION

A liposome is an artificially prepared spherical vesicle composed of natural phospholipids and other lipid chains with surfactant properties (Edwards, 2006). The major types of liposomes are the multilamellar vesicle (MLV), the small unilamellar vesicle (SUV), and the large unilamellar vesicle (LUV). Per design, liposome encapsulates a region of aqueous solution inside a hydrophobic membrane; therefore, dissolved hydrophilic solutes cannot readily pass through the lipids. However, hydrophobic chemicals can be dissolved into the membrane. On the other hand, liposomes can fuse with other bilayers such as the cell membrane. These features allow liposomes to carry both hydrophobic molecules and hydrophilic drug molecules in formulation for an improved delivery and efficacy (Torchilin, 2006). Liposomal formulations have introduced a new way of increasing bioavailability of some active therapeutic agents. One particular example is the selective drug delivery of the anticancer agent doxorubicin in polyethylene glycol liposomes for the treatment of solid tumors in patients with breast-carcinoma metastases with improved patient survival (Perez, 2002).

The components of liposomal drug products are the drug substance, the lipids, and other formulation ingredients. The physicochemical properties of the liposome drug products are critical to ensuring drug product quality. Therefore, regulatory agent requires the evaluation of not only liposomal drugs but also each constituent such as the cationic and pegylated lipids. In order to assess the impact of the formulation and, in particular, the specialized lipids on the pharmacokinetic and pharmacodynamic profile of the study the cationic and pegylated lipids may also need to be analyzed.

46.2 METHODS AND APPROACHES

46.2.1 Principles

One key component to consider in the bioanalysis of preclinical or clinical samples collected after administration of liposomal drugs is the measurement of total (free, protein-bound and encapsulated) and unencapsulated (including unbound) active pharmaceutical ingredients. Measuring only the free portion (unbound and unencapsulated) of the active drug molecule(s) in the study samples obtained after administration of liposomal drug can be a challenge due to the possible instability of the liposome during the bioanalysis process. The stability of liposome is dependent on many environmental factors such as pH, temperature, and coexisting chemicals. Among the many factors, storage temperature and effects of freeze/thawing are the important ones that need to be evaluated in developing a bioanalytical assay for liposome drug, especially with the formulations that are thermally sensitive. Once the possible instability of the liposome has been thoroughly evaluated and the instability issues have been addressed, the rest of the bioanalytical assay development and validation can be expected to be straightforward.

The most popular strategy to analyzing free (unbound and unencapsulated) and total (unbound + unencapsulated + encapsulated) drugs in biological samples is based on measuring the former portion and subtracting it from the total amount to calculate the bound amount of the active ingredient. A common approach that has been published to achieve this is separating the liposomal component from the free active ingredient by solid phase extraction (SPE) or size exclusion (Druckmann, 1989; Thies, 1990).

Handbook of LC-MS Bioanalysis: Best Practices, Experimental Protocols, and Regulations, First Edition. Edited by Wenkui Li, Jie Zhang, and Francis L.S. Tse.

As constitutional components of liposomal drug products, lipids possess both hydrophobic and hydrophilic characteristics. This feature presents a challenge for liquid chromatography in LC-MS bioanalysis. We have observed that cationic lipids have better retention and peak shape with normal phase chromatography than other means, while pegylated lipids can be chromatographed under reverse phase conditions. Another important aspect to consider in the pegylated lipid analysis is the fluctuation of mass (*m/z*) of the same lipids detected from one product to the other due to the variation in polymerization, a process used to make lipids for liposomal drug formulation.

46.2.2 Methodology

46.2.2.1 Analysis of Liposome Drugs The measurement of unencapsulated drug molecules, for example, doxorubicin, in plasma samples was carried out by using an SPE method, where the liposomal plasma samples were passing through the sorbent and the free drug (e.g., doxorubicin) was retained on SPE. The drug was then eluted from the SPE cartridge and analyzed with LC-MS/MS. In order to obtain the total drug concentrations, the plasma samples were treated with MeOH to deconstruct the liposome and then taken through the same SPE extraction.

46.2.2.2 Analysis of Pegylated and Cationic Lipids The analysis of the lipids can be accomplished by protein precipitation and then the resulting extracts split into two plates for subsequent analysis of the pegylated lipid and the cationic lipid. The reason for the separate analysis is that pegylated lipids elute with better peak shape under reverse phase conditions while cationic lipids have better peak shape with normal phase conditions. Due to the nature of the pegylated lipid having strong hydrophobic and hydrophilic chemical properties, the use of SPE proved to be difficult for low lower limit of quantitation (LLOQ) since the recovery was poor. Liquid/liquid extractions also had very low recovery due to these chemical properties. It was found that protein precipitation with isoprapanol (IPA)/acetonitrile yielded a better recovery than the SPE or LLE methods.The use of IPA as an extraction solvent improved recovery due to the limited solubility of the pegylated lipid.

46.3 PROTOCOLS

46.3.1 Quantitative Determination of Free Doxorubicin and Doxorubicinol in Human Plasma Using LC-MS/MS (Figures 46.1, 46.2, and 46.3)

Test Material and Matrix

- Doxorubicin
- Doxorubicinol
- Daunorubicinol (internal standard)
- Human Plasma (K_2EDTA) [BioChemEd]

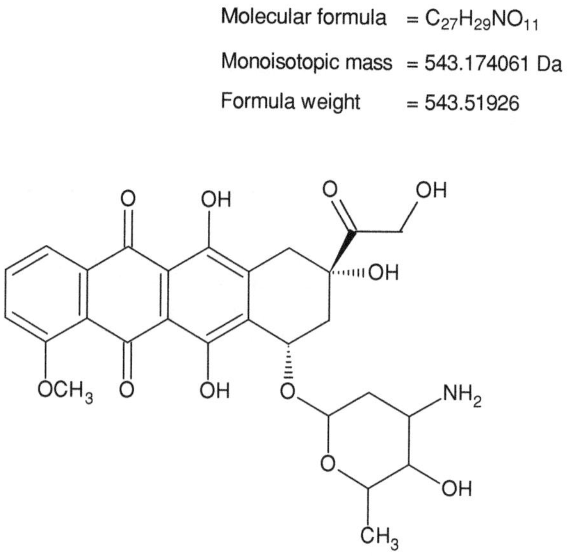

Molecular formula = $C_{27}H_{29}NO_{11}$
Monoisotopic mass = 543.174061 Da
Formula weight = 543.51926

FIGURE 46.1 Molecular structure and weight for Doxorubicin.

Equipment and Reagents

- *Liquid handling system*: Tomec Quadra 96 or Hamilton MICROLAB® NIMBUS 96
- *HPLC system*: Shimadzu LC-10D
- *HPLC column*: Atlantis dC18 50 × 2.1, 5μm, Waters Corp., Milford, MA
- *Autosampler*: SIL-5000, Shimadzu, Columbia, MD
- *Mass spectrometer*: API5000
- Acetone, HPLC grade (EMD Chemicals Inc., Gibbstown, NJ)
- Formic acid, min 96% purity, HPLC grade (EMD)
- Acetonitrile (MeCN), HPLC grade (EMD)

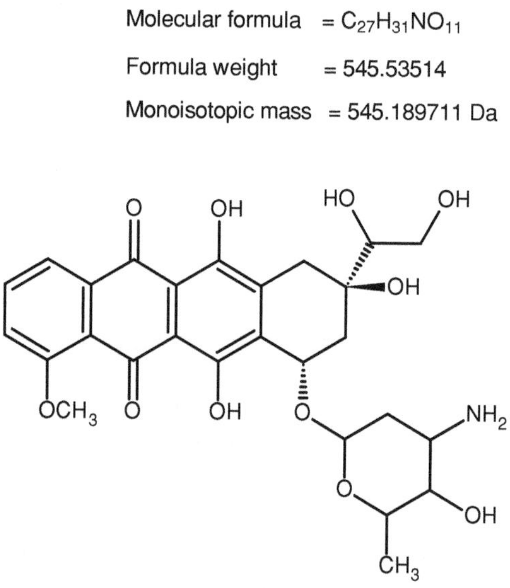

Molecular formula = $C_{27}H_{31}NO_{11}$
Formula weight = 545.53514
Monoisotopic mass = 545.189711 Da

FIGURE 46.2 Molecular structure and weight for Doxorubicinol.

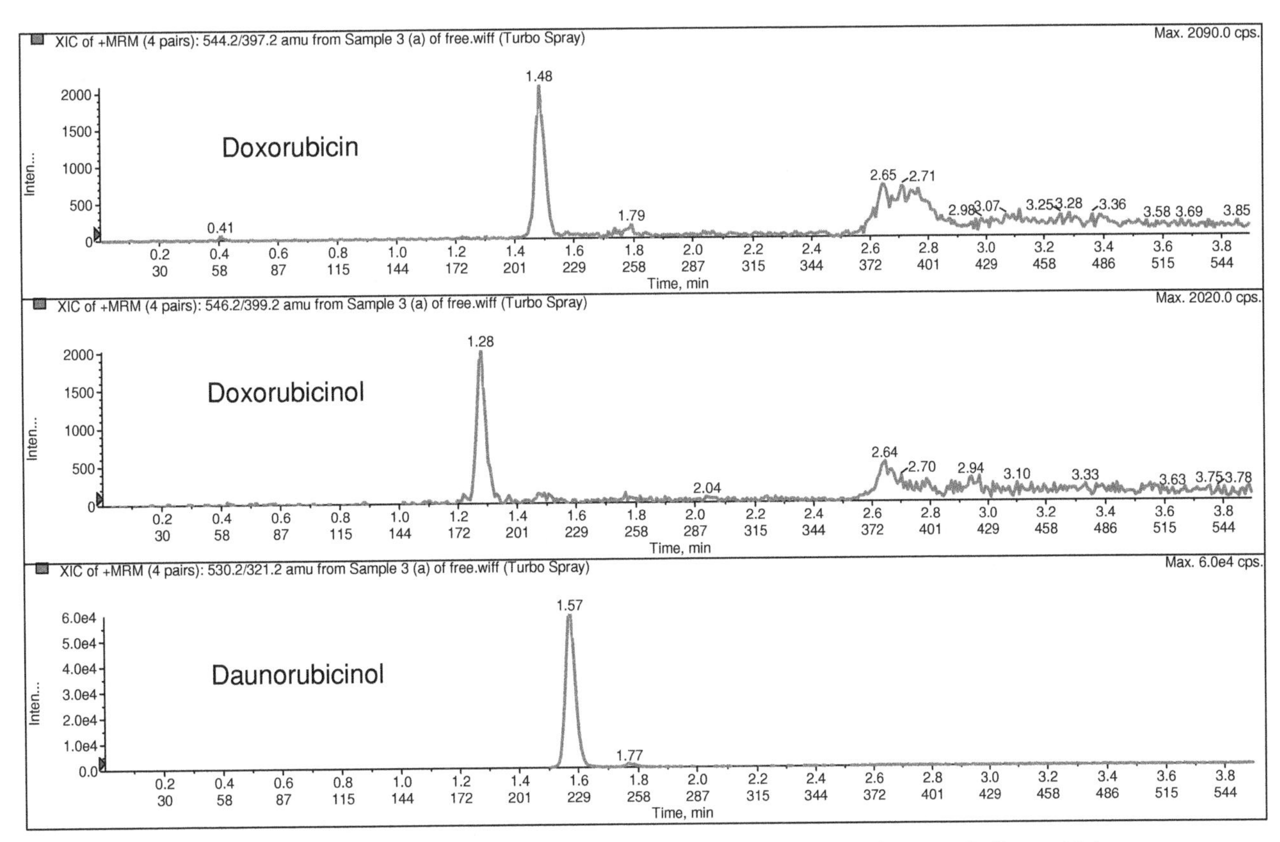

FIGURE 46.3 Representative LC-MS/MS chromatogram of LLOQ sample extract for Doxorubicin (5.00 ng/ml, top panel), Doxorubicinol (5.00 ng/ml, middle panel), and internal standard (bottom panel).

- Ammonium acetate (NH_4OAc), ACS grade (Sigma-Aldrich)
- Ammonium hydroxide (NH_4OH), ACS grade, 28–30% (EMD)
- Methanol (MeOH), HPLC grade (EMD)
- Phosphate buffered saline 1 × (PBS), pH 7.4, Gibco, Invitrogen
- Water, deionized, Type 1 (typically 18.2 MΩ cm) or equivalent (HPLC grade)
- *Needle wash 1*: 5% NH_4OH in methanol
- *Needle wash 2*: MeCN:water:formic acid, 20:80:0.1 (v,v,v)
- *Mobile phase A*: 0.1% formic acid in water
- *Mobile phase B*: 0.1% formic acid in MeCN
- *Mobile phase C*: Acetone

Method

Solution and Sample Preparation

1. Weigh an appropriate amount of Doxorubicin, Doxorubicinol, and Daunorubicinol (IS) and dissolve in the required volume of methanol:water (50:50, v:v) to give a 1.00 mg/ml, 1.00 mg/ml, and 0.500 mg/ml stock solution. Prepare duplicate stock solutions. One is designated to prepare standard spiking solution and another one is used to prepare quality control (QC) spiking solution.
2. Prepare working spiking solutions in methanol:water (50:50, v:v). There is one standard spiking solution at the corresponding concentration of 100,000 ng/ml, and then, the curve is prepared fresh in human plasma at the following concentrations: 5, 10, 50, 100, 500, 1000, 4500, and 5000 ng/ml. The QCs are prepared in human plasma from the 1 mg/ml stock solution and their corresponding concentrations are 5, 15, 245, 2000, 4000, and 25,000 ng/ml.

Sample Extraction

1. Aliquot 100 μl of the standard or QC in the corresponding well of a 96-well plate.
2. Add 100 μl of internal standard working solution 200 ng/ml in PBS (Daunorubicinol) to the samples except matrix blanks.
3. To the matrix blanks, add 100 μl of PBS (Daunorubicinol) to the matrix blanks.
4. To all samples add 300 μl of PBS.
5. Vortex-mix the samples at the high setting on the multitube vortexer for approximately 3 min.

6. Condition Cerex CBA SPE Cartridges (20 mg) with 500 μl of methanol at approximately 1 drop/s (gravity feed or positive pressure).
7. Condition Cerex CBA SPE Cartridges (20 mg) with 500 μl of water at approximately 1 drop/s (gravity feed or positive pressure).
8. Use liquid handling robot to transfer 250 μl of samples to cartridges. Load samples onto respective cartridges and push through slowly using positive pressure at approximately 1 drop/s (gravity feed or positive pressure).
9. Wash cartridges with 500 μl of water. Use positive pressure to elute at approximately 1 drop/s (gravity feed or positive pressure).
10. Wash cartridges with 500 μl/ml of methanol:water (40:60, v:v). Use positive pressure to elute at approximately 1 drop/s (gravity feed or positive pressure).
11. Elute into a 96-well plate with 300 μl of 200 mM NH_4OAc in methanol:water (80:20, v:v). Use positive pressure to elute at approximately 1 drop/s (gravity feed or positive pressure).
12. To all samples, add 700 μl of water.
13. Vortex-mix the samples at the low setting on the multitube vortexer for approximately 1 min.
14. Centrifuge the samples at approximately 3000 rpm for approximately 1 min.
15. Store samples at room temperature while waiting for analysis.

LC-MS/MS Analysis

1. The LC system was equipped with two standard six-port switch valves (VICI Cheminert 10U-0263H) facilitate the flow diversion of the LC eluent after the last peak of each sample was eluted out of the column and a column backflush. The LC conditions are as follows:
   a. *Flow rate*: 0.300 ml/min
   b. *Column temperature*: ambient
   c. Isocratic:

Time (min)	0	2	2.03	3.3	3.33	4
B%	20	50	95	95	20	Stop

Note: The column was backflushed from 2.03 to 3.3 min to elute any interference that remained on the column before the next injection.

The mass spectrometer was operated in positive ionization mode with an electrospray ion source (ESI). The SRM transition for Doxorubicin, Doxorubicinol and Daunorubicinol are 544.2 → 397.2, 546.2 → 399.2, and 530.2 → 321.2, respectively. The typical MS conditions are: dwell time, 100 ms; source temperature, 400°C; IS voltage, 5000 V; declustering potential (DP), 50; curtain gas, 35 psi; nebulizer gas (GS1), 75 psi; turbo gas (GS2), 70 psi; collision energy, 17 eV.

46.3.2 Quantitative Determination of Total Doxorubicin in Human Plasma Using LC-MS/MS

Test Material and Matrix

- Doxorubicin
- Doxorubicinol
- Daunorubicinol (internal standard)
- Human Plasma (K_2EDTA) [BioChemEd]

Equipment and Reagents

- *Liquid handling system:* Tomec Quadra 96 or Hamilton MICROLAB® NIMBUS 96
- *HPLC system*: Shimadzu LC-10D
- *HPLC column*: Atlantis dC18 50 × 2.1, 5 μm, Waters Corp., Milford, MA
- *Autosampler*: SIL-5000, Shimadzu, Columbia, MD
- *Mass spectrometer*: API5000
- Acetone, HPLC grade (EMD)
- Formic acid, min 96% purity, HPLC grade (EMD)
- Acetonitrile (MeCN), HPLC grade (EMD)
- Ammonium acetate (NH_4OAc), ACS grade (Sigma-Aldrich)
- Ammonium hydroxide (NH_4OH), ACS grade, 28–30% (EMD)
- Methanol (MeOH), HPLC grade (EMD)
- Phosphate buffered saline 1 × (PBS), pH 7.4, Gibco, Invitrogen
- Water, deionized, Type 1 (typically 18.2 MΩ cm) or equivalent (HPLC grade)
- *Needle wash 1*: 5% NH_4OH in methanol
- *Needle wash 2*: MeCN:water:formic acid, 20:80:0.1 (v,v,v)
- *Mobile phase A*: 0.1% formic acid in water
- *Mobile phase B*: 0.1% formic acid in MeCN
- *Mobile phase C*: Acetone

Method

Solution and Sample Preparation

1. Weigh an appropriate amount of Doxorubicin, Doxorubicinol and Daunorubicinol (IS) and dissolve in the required volume of methanol:water (50:50, v:v) to give a 1 mg/ml, 1 mg/ml, and 0.500 mg/ml stock solution. Prepare duplicate stock solutions. One is designated

to prepare standard spiking solution and another one is used to prepare QC spiking solution.

2. Prepare working spiking solutions in methanol:water (50:50, v:v). There is one standard spiking solution at the corresponding concentration of: 100,000 ng/ml, and then, the curve is prepared fresh in human plasma at the following concentrations: 5, 10, 50, 100, 500, 1000, 4500, and 5000 ng/ml. The QCs are prepared in human plasma from the 1 mg/ml stock solution and their corresponding concentrations are 5, 15, 245, 2000, 4000 and 25,000 ng/ml.

Sample Extraction

1. Aliquot 100 μl of the standard or QC in the corresponding well of a 96-well plate.
2. Add 100 μl of internal standard working solution 200 ng/ml in PBS (Daunorubicinol) to the samples except matrix blanks.
3. To the matrix blanks, add 100 μl of PBS (Daunorubicinol) to the matrix blanks.
4. To all samples, add 300 μl of MeOH.
5. Cover block with sealing mat and vortex for approximately 5 min at speed (speed setting 2–4 on multitube vortexer).
6. Centrifuge at 3000 rpm for 5 min at room temperature.
7. Use liquid handling robot to transfer 100 μl of samples to a clean 96-well plate.
8. To all samples, add 400 μl of PBS.
9. Vortex-mix the samples at the high setting on the multitube vortexer for approximately 3 min.
10. Condition Cerex CBA SPE Cartridges (20 mg) with 500 μl of methanol at approximately 1 drop/s (gravity feed or positive pressure).
11. Condition Cerex CBA SPE Cartridges (20 mg) with 500 μl of water at approximately 1 drop/s (gravity feed or positive pressure).
12. Use liquid handling robot to transfer 250 μl of samples to cartridges. Load samples onto respective cartridges and push through slowly using positive pressure at approximately 1 drop/s (gravity feed or positive pressure).
13. Wash cartridges with 500 μl of water. Use positive pressure to elute at approximately 1 drop/s (gravity feed or positive pressure).
14. Wash cartridges with 500 μl/ml of methanol: water (40:60, v:v). Use positive pressure to elute at approximately 1 drop/s (gravity feed or positive pressure).
15. Elute into a 96-well plate with 300 μl of 200 mM NH_4OAc in methanol:water (80:20, v:v). Use positive pressure to elute at approximately 1 drop/s (gravity feed or positive pressure).
16. To all samples add 100 μl of water.
17. Vortex-mix the samples at the low setting on the multitube vortexer for approximately 1 min.
18. Centrifuge the samples at approximately 3000 rpm for approximately 1 min.
19. Store samples at room temperature while waiting for analysis.

LC-MS/MS Analysis

1. The LC system was equipped with two standard six-port switch valves (VICI Cheminert 10U-0263H) facilitate the flow diversion of the LC eluent after the last peak of each sample was eluted out of the column and a column backflush. The LC conditions are as follows:
   a. *Flow rate*: 0.300 ml/min
   b. *Column temperature*: Ambient
   c. *Isocratic*:

Time (min)	0	2	2.03	3.3	3.33	4
B%	20	50	95	95	20	Stop

Note: The column was backflushed from 2.03 to 3.3 min to elute any interference that remained on the column before the next injection.

The mass spectrometer was operated in positive ionization mode with an ESI. The SRM transition for Doxorubicin, Doxorubicinol and Daunorubicinol are 544.2 → 397.2, 546.2 → 399.2 and 530.2 → 321.2, respectively. The typical MS conditions are dwell time, 100 ms; source temperature, 400°C; IS voltage, 5000 V; DP, 50; curtain gas, 35 psi; nebulizer gas (GS1), 75 psi; turbo gas (GS2), 70 psi; collision energy, 17 eV.

46.3.3 Quantitative Determination of Cationic Lipid in Human Plasma Using LC-MS/MS

Test Material and Matrix (Figures 46.4 and 46.5)

- Cationic lipid
- Human Plasma (K_2EDTA) [BioChemEd]

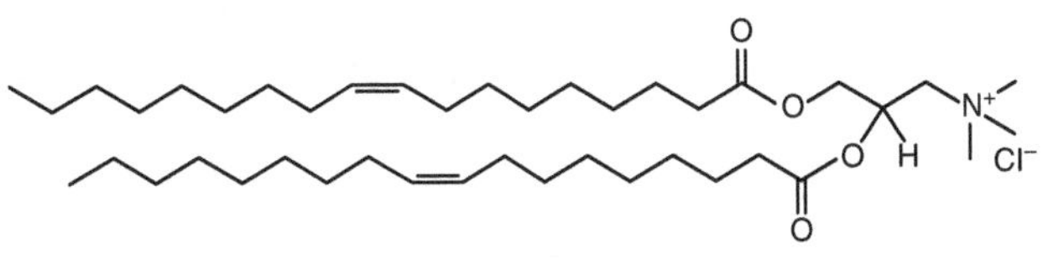

FIGURE 46.4 Structure of representative cationic lipid.

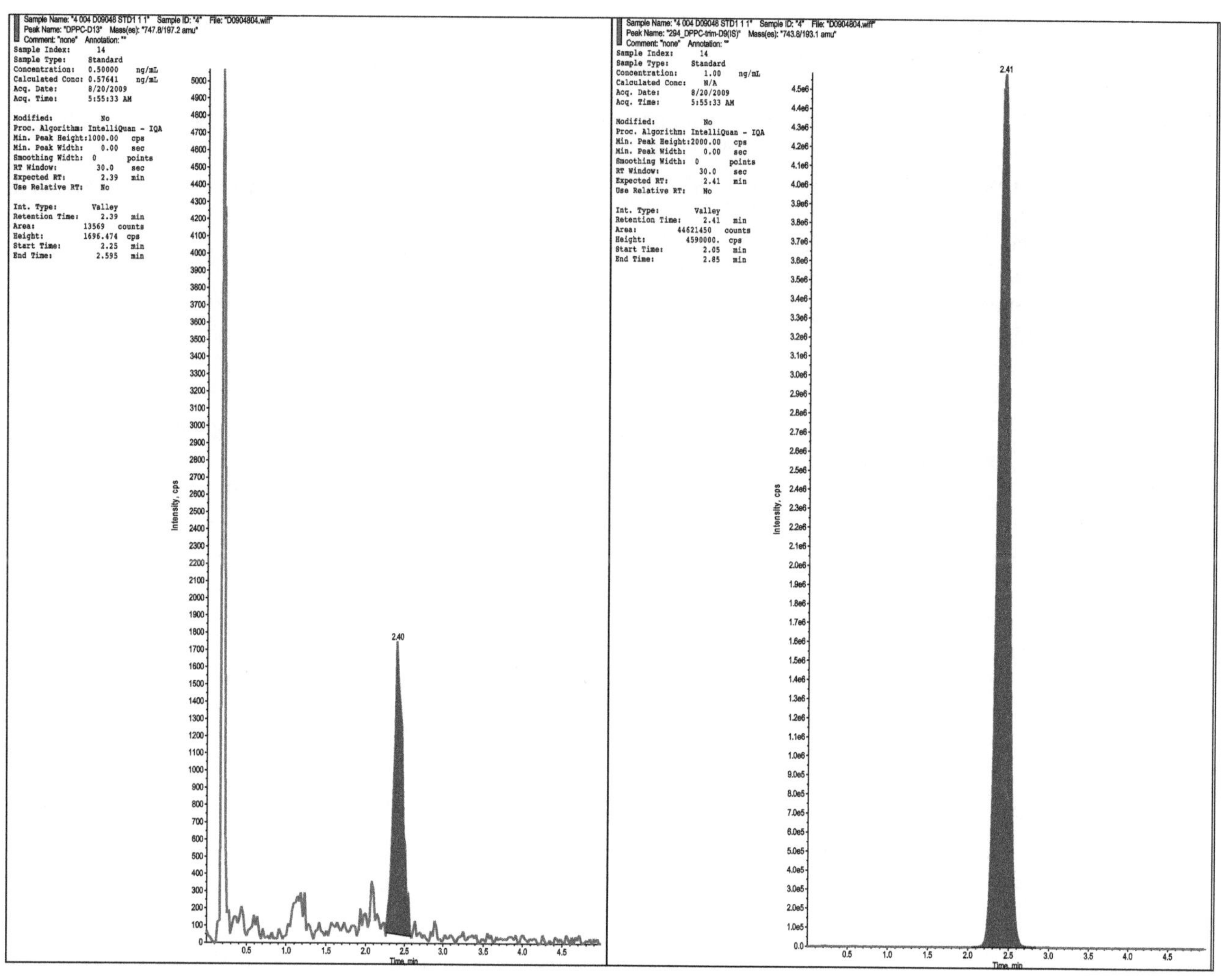

FIGURE 46.5 Representative LC-MS/MS chromatogram of a cationic lipid at LLOQ (0.100 ng/ml, left panel) and internal standard (right panel).

Equipment and Reagents

- *Liquid handling system*: Tomec Quadra 96 or Hamilton MICROLAB® NIMBUS 96
- *HPLC system*: Shimadzu LC-10D
- *HPLC column*: Phenomenex Luna Silica 2.0 × 50 mm 5 μm
- *Autosampler*: LEAP PAL LC Autosampler
- *Mass spectrometer*: API5000
- Acetonitrile (MeCN), HPLC grade (EMD)
- Ammonium formate, ACS grade (Sigma-Aldrich)
- Isopropyl alcohol (IPA), HPLC grade (EMD)
- *N,N*-dimethylformamide (DMF), HPLC grade (Burdick & Jackson)
- Water, deionized, Type 1 (typically 18.2 MΩ cm) or equivalent (HPLC grade)
- *Needle wash 1*: DMF
- *Needle wash 2*: 90/10 MeCN/water (v/v)
- *Mobile phase A*: 10 mM ammonium formate, pH unadjusted
- *Mobile phase B*: 50/50 IPA/MeCN (v/v)
- *Mobile phase C*: 70/30 MeCN/water (v/v) to be used for diversion and backflush

Method

Solution and Sample Preparation

1. Weigh an appropriate amount of cationic lipid and internal standard and dissolve in the required volume of DMF for 1.20 mg/ml (cationic lipid) 1.63 mg/ml (internal standard) stock solutions. Prepare duplicate stock solutions. One is designated to prepare standard spiking solution and another one is used to prepare QC spiking solution.

2. Prepare working spiking solutions in DMF. There are eight standard spiking solutions and their corresponding concentrations are 2, 4, 20, 100, 400, 1200, 1800, and 2000 ng/ml. Similarly, there are five QC spiking solutions and their corresponding concentrations are 2, 6, 400, 1600, and 8000 ng/ml.
3. Prepare the standard curves and QC samples in human plasma from the spiking solution. The standard calibrators contain eight concentrations: 0.1, 0.2, 1, 5, 20, 60, 90, and 100 ng/ml. The QCs contain five levels: 0.1 (LLOQ), 0.3 (low), 40 (medium), 80 (high), and 400 (dilution) ng/ml.

Sample Extraction

1. Add 400 μl of internal standard working solution [500 ng/ml of IS in IPA:MeCN (50:50 v/v)] to the samples except matrix blanks.
2. To the matrix blanks, add 400 μl of IPA:MeCN (50:50 v/v).
3. Vortex-mix the samples at the high setting on the multitube vortexer for approximately 5 min.
4. Centrifuge the samples at approximately 3000 rpm for approximately 5 min to spin down the precipitate.
5. Manually transfer approximately 250 μl of supernatant into a clean 96-well plate.
6. Evaporate the samples to dryness at approximately 45°C under nitrogen in the Turbovap for approximately 40 min.
7. Reconstitute the samples with 300 μl of MeCN.
8. Vortex-mix the samples at the low setting on the multitube vortexer for approximately 1 min.
9. Centrifuge the samples at approximately 3000 rpm for approximately 1 min.
10. Store samples in refrigerator while waiting for analysis.

LC-MS/MS Analysis

1. The LC system was equipped with two standard six-port switch valves (VICI Cheminert 10U-0263H) facilitate the flow diversion of the LC eluent after the last peak of each sample was eluted out of the column and a column backflush. The LC conditions are as follows:
   a. *Flow rate*: 0.8 ml/min
   b. *Colum temperature*: 30°C
   c. *Isocratic*:

Time (min)	0.0′	3.5′
B%	90	Stop

Note: The column was diverted and only the LC eluent from 0.5 to 2 min is introduced to MS. The column was back flushed from 1.3 to 2 min to elute any interference that remained on the column before the next injection.

The mass spectrometer was operated in positive ionization mode with an ESI. The SRM transition for cationic lipid and internal standard are 642.5 → 132 and 655 → 83, respectively. The typical MS conditions are dwell time, 100 ms; source temperature, 500°C; IS voltage, 5000 V; DP, 25; curtain gas, 30 psi; nebulizer gas (GS1), 50 psi; turbo gas (GS2), 70 psi; collision energy, 40 eV.

46.3.4 Quantitative Determination of Pegylated Lipids in Human Plasma Using LC-MS/MS

Test Material and Matrix (Figures 46.6 and 46.7)

- Pegylated lipid
- Human Plasma (K_2EDTA) [BioChemEd]

Equipment and Reagents

- *Liquid handling system*: Tomec Quadra 96 or Hamilton MICROLAB® NIMBUS 96
- *HPLC system*: Shimadzu LC-10D
- *HPLC column*: Waters X-Bridge, C8 2.1 × 50 mm 3.5 μm
- *Autosampler*: LEAP PAL LC Autosampler
- *Mass spectrometer*: API5000
- *N,N*-dimethylformamide (DMF), HPLC grade (Burdick & Jackson)
- Ammonium acetate, ACS grade (Sigma-Aldrich)
- Acetonitrile (MeCN), HPLC grade (EMD)
- Formic acid (FA), ACS grade (EMD)

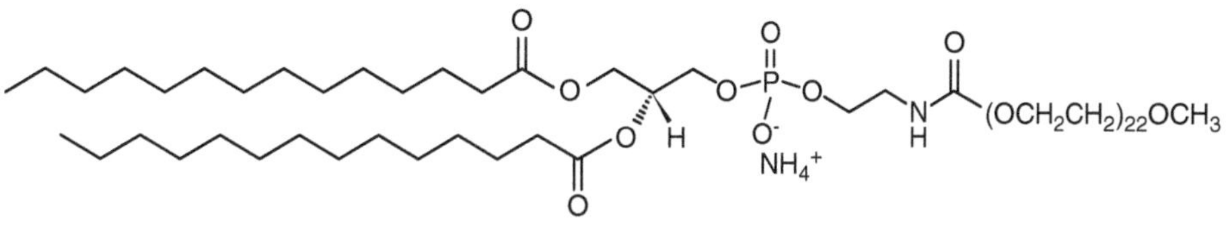

$(OCH_2CH_2)_{22}OCH_3$ NH_4^+

FIGURE 46.6 Representative structure of pegylated lipid.

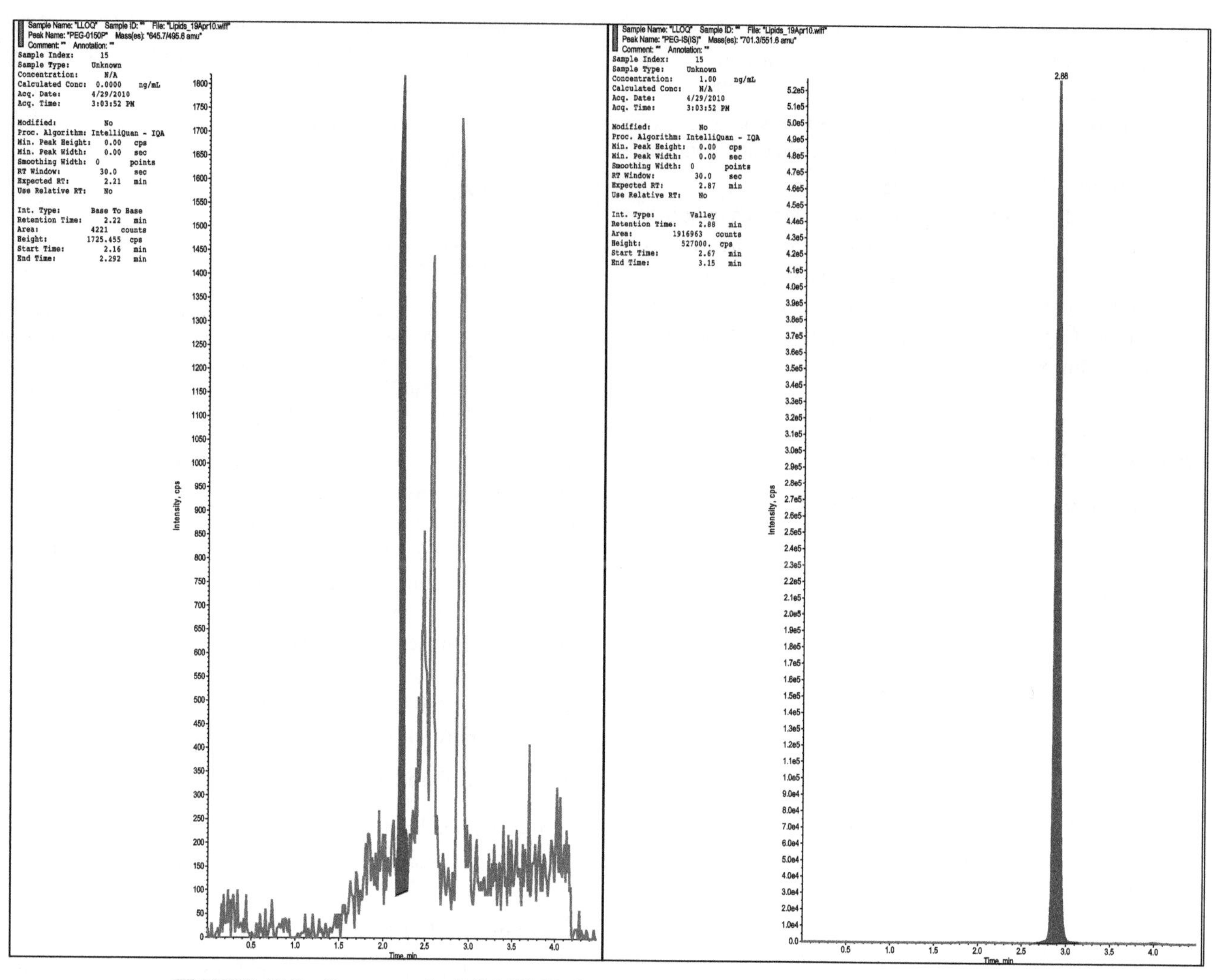

FIGURE 46.7 Representative LC-MS/MS chromatogram of a pegylated lipid at LLOQ (5 ng/ml, left panel) and internal standard (right panel).

- Methanol (MeOH), HPLC grade (EMD)
- Water, deionized, Type 1 (typically 18.2 MΩ cm) or equivalent (HPLC grade)
- Isopropyl alcohol (IPA), HPLC grade (EMD)
- Ammonium hydroxide (NH_4OH), ACS grade (Sigma-Aldrich)
- Ammonium bicarbonate (AmBicarb), ≥99.0% (Sigma-Aldrich)
- *Needle wash 1*: 0.1% formic acid in [water:MeCN (10/90 v/v)]
- *Needle wash 2*: 0.1% formic acid in water
- *Mobile phase A*: 0.2% ammonium hydroxide in 5mM ammonium bicarbonate, pH unadjusted
- *Mobile phase B*: 50/50 MeOH/MeCN (v/v)
- *Mobile phase C*: 90/10 MeCN/water (v/v) to be used for diversion and backflush

Method

Solution and Sample Preparation

1. Weigh an appropriate amount of pegylated lipid and internal standard and dissolve in the required volume of DMF for 0.541 mg/ml (cationic lipid) 2.09 mg/ml (internal standard) stock solutions. Prepare duplicate stock solutions. One is designated to prepare standard spiking solution and another one is used to prepare QC spiking solution.
2. Prepare working spiking solutions in DMF. There are four standard spiking solutions and their corresponding concentrations are 250, 1000, 5000, and 20,000 ng/ml. Similarly, there are four QC spiking solutions and their corresponding concentrations are 500, 4000, 2000, and 574,000 ng/ml.
3. Prepare the standard curves and QC samples in human plasma from the spiking solution. The standard

calibrators contain eight concentrations: 5, 10, 50, 250, 1000, 3000, 4500, and 5000 ng/ml. The QCs contain five levels: 5 (LLOQ), 15 (low), 2000 (medium), 4000 (high), and 20,000 (dilution) ng/ml.

Sample Extraction

1. Add 100 μl of 100 mM ammonium acetate pH unadjusted to all samples.
2. Vortex-mix the samples at the low setting on the multitube vortexer for approximately 1 min.
3. Add 50 μl of internal standard working solution [6000 ng/ml of IS in (1% FA in water):IPA (50:50 v/v)] to the samples except matrix blanks.
4. To the matrix blanks, add 50 μl of [1% FA in water:IPA (50:50 v/v)].
5. Centrifuge the samples at approximately 3000 rpm for approximately 1 min.
6. Vortex-mix the samples at the high setting on the multitube vortexer for approximately 1 min.
7. Add 800 μl of MeCN to the samples.
8. Vortex-mix the samples at the low setting first on the multitube vortexer, then increase the speed to the medium setting slowly to avoid splash for approximately 3 min.
9. Centrifuge the samples at approximately 3000 rpm for approximately 5 min to spin down the precipitate.
10. Decant the supernatant into a clean polypropylene tube.
11. Evaporate the samples to dryness at approximately 50°C under nitrogen in the Turbovap for approximately 40 min.
12. Reconstitute the samples with 200 μl of [1% FA in water:IPA (50:50 v/v)].
13. Vortex-mix the samples at the high setting on the multitube vortexer for approximately 1 min.
14. Centrifuge the samples at approximately 3000 rpm for approximately 1 min.
15. Transfer the samples into a clean tapered well-plate and centrifuge the plate at approximately 3000 rpm for approximately 1 min.
16. Store samples in either a chilled autosampler or a refrigerator.

LC-MS/MS Analysis

1. The LC system was equipped with two standard 6-port switch valves (VICI Cheminert 10U-0263H) facilitate the flow diversion of the LC eluent after the last peak of each sample was eluted out of the column and a column back flush. The LC conditions are as follows:
   a. *Flow rate*: 0.800 ml/min
   b. *Colum temperature*: 30°C
   c. *Isocratic*:

Time (min)	0.0′	3.5′
B%	90	Stop

Note: The column was diverted and only the LC eluent from 1 to 2 min is introduced to MS. The column was back flushed from 2 to 3 min to elute any interference that remained on the column before the next injection.

The mass spectrometer was operated in positive ionization mode with an ESI. The SRM transition for cationic lipid and internal standard are 893.2 → 726.1 and 903.5 → 728.1, respectively. The typical MS conditions are: dwell time, 100 ms; source temperature, 500°C; IS voltage, 4500 V; DP, 40; curtain gas, 25 psi; nebulizer gas (GS1), 50 psi; turbo gas (GS2), 60 psi; collision energy, 45 eV.

REFERENCES

Druckmann S, Gabizon A, Barenholz Y. Separation of liposome-associated doxorubicin from non-liposome-associated doxorubicin in human plasma: implications for pharmacokinetic studies. Biochim Biophys Acta 1989;980:381–384.

Edwards KA, Baeumner AJ. Analysis of liposomes. Talanta 2006;68:1432–1441.

Perez AT, Domenech GH, Frankel C, Vogel CL. Pegylated liposomal doxorubicin for metastatic breast cancer: the Cancer Research Network, Inc., Experience. Cancer Investigation 2002;20:22–29.

Thies RL, Cowens DW, Cullins PR, Bally MB, Mayer LD. Method for rapid separation of liposome-associated doxorubicin from free doxorubicin in plasma. Anal Biochem 1990;188: 65–71.

Torchilin VP. Recent approaches to intracellular delivery of drugs and DNA and organelle targeting. Annu Rev Biomed Eng 2006;8:343–375.

47

LC-MS BIOANALYSIS OF PROTEINS

ZIPING YANG, WENKUI LI, HAROLD T. SMITH, AND FRANCIS L.S. TSE

47.1 INTRODUCTION

The biopharmaceutical industry has shown a growing trend in developing biologics such as recombinant proteins and monoclonal antibodies (mAbs) that continue to present promises in addressing unmet medical needs. Therefore, the timely development of bioanalytical assays for these therapeutic proteins in biological fluids such as plasma is essential in order to elucidate their pharmacokinetic and toxicokinetic properties in animals and humans. The conventional choice of assay for proteins is immunoassay, which can be highly sensitive and quite specific to the target protein. However, immunoassays have a number of commonly encountered difficulties including a relatively narrow dynamic range, assay imprecision, matrix interference, and a relatively long assay development time.

Whereas liquid chromatography–mass spectrometry (LC-MS) is commonly used for small molecule analysis, this technique has only recently been applied to the bioanalysis of therapeutic proteins (Yang et al., 2007; Heudi et al., 2008; Ji et al., 2009). In contrast to immunoassays, LC-MS assays have the advantages of fast assay development, good specificity, relatively wide dynamic range, and assay precision and accuracy. LC-MS methodologies typically contain several common steps including enzyme digestion to break down proteins into peptides, monitoring one or more peptides selected from the target protein in LC-MS, and use of the readout of the peptides to quantify the protein. As plasma has a very complex proteome that could affect the sensitivity of the assay, the target protein usually needs to be enriched from the specimen prior to the enzyme digestion step. To control the variability due to multiple sample processing steps and potential LC-MS instrument fluctuations, a reliable internal standard is required to ensure a good assay performance.

This chapter describes the procedures that have been generally used in the bioanalysis of therapeutic proteins in plasma (Figure 47.1). Different sample processing approaches that have been used to improve assay sensitivity, accuracy, and throughput are reviewed.

47.2 METHODS

47.2.1 Selection of Signal Peptide(s)

In general, the LC-MS approach for the bioanalysis of a protein is to quantify a unique peptide (signal peptide) selected from the target protein, and use the LC-MS/MS readout of the signal peptide to assay the protein. The selection of a signal peptide from the target protein can start from an *in silico* list of the potential peptides cleaved by a protease, such as trypsin. The list is then narrowed by excluding the peptides containing amino acid residues which could undergo post-translational modifications, for example, potential glycosylation sites including asparagines, serine, and threonine. Peptides containing cysteines also should be avoided as they can be derivatized during enzyme digestion. Another selection criterion is the size of the peptide. A signal peptide containing <20 amino acid residues would yield good ionization in the mass spectrometer. The initially selected peptides should then be screened against a protein database, such as Swissprot, to ensure the uniqueness of the peptides from the matrix proteomes. Specifically for antibody proteins, *in silico* peptide selection can be performed only at the sequence of the variable regions of the light or heavy chains.

Handbook of LC-MS Bioanalysis: Best Practices, Experimental Protocols, and Regulations, First Edition. Edited by Wenkui Li, Jie Zhang, and Francis L.S. Tse.

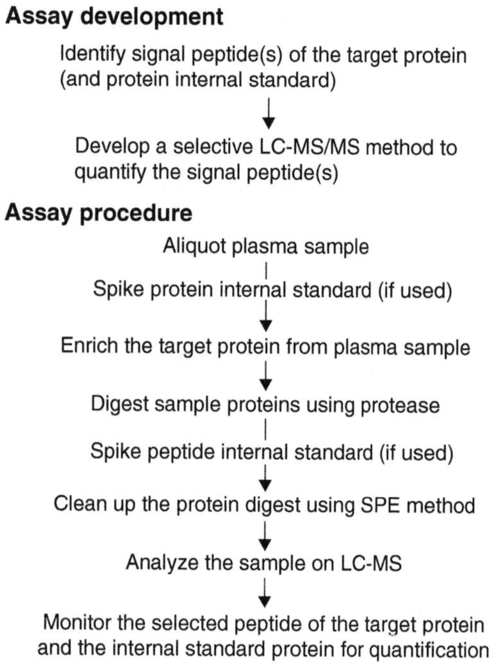

Assay development

Identify signal peptide(s) of the target protein
(and protein internal standard)

↓

Develop a selective LC-MS/MS method to
quantify the signal peptide(s)

Assay procedure

Aliquot plasma sample

Spike protein internal standard (if used)

↓

Enrich the target protein from plasma sample

↓

Digest sample proteins using protease

Spike peptide internal standard (if used)

↓

Clean up the protein digest using SPE method

↓

Analyze the sample on LC-MS

↓

Monitor the selected peptide of the target protein
and the internal standard protein for quantification

FIGURE 47.1 General scheme of the bioassay development and procedure for therapeutic proteins.

The selection of the signal peptides is further optimized based on the intensity of their observed signals in LC-MS/MS. In addition, the uniqueness of the signal peptide needs to be confirmed by the absence of major interference from the matrix to the peptide's LC-MS/MS signal, which also helps to reduce background noise.

47.2.2 Internal Standard

47.2.2.1 Stable Isotope-Labeled Peptide Many LC-MS/MS assays of protein quantification have employed stable isotope-labeled signal peptides as internal standards. The synthesis of stable-labeled peptides is often costly and quite time-consuming. It can be a rate-limiting step in the development of the assay.

A recent study applied a differential dimethyl-labeling approach to the synthesis of the isotope-labeled peptide.In this approach, the peptides are derivatized at the N-terminus and ε-amino group of lysine through reductive amination by a simple reagent, formaldehyde. The derivatization adds two methyl groups to an amino group, so the molecular weight of the peptide after the reaction with a single derivatization site shifts by 28 Da, or 32 Da if deuterium-labeled formaldehyde is used. This shift of molecular weight can be in multiples of 28 or 32 Da if the peptide has multiple dimethyl derivatization sites.

In the bioanalysis of a target protein, the reference standard of the target protein is digested, followed by derivatization using D-labeled formaldehyde, while the digested mixture of the biological samples reacts with unlabeled formaldehyde. The derivatized protein standard digest is then spiked into the sample digest to be used as the internal standard. The molecular weight of the peptide originating from the internal standard will have 4, 8, or 4X Da more than that originating from the biological sample, if the peptide has 1, 2, or X derivatization sites. The reductive amination reaction condition is mild. At room temperature, a high yield can be obtained in about 2 h or less for some peptides. The reagents are commercially available and inexpensive. More importantly for the LC-MS analysis, the charge state of the dimethyl-labeled peptides is preserved at ESI-MS. One drawback of this internal standard approach is that the mass difference between the D-labeled and H-labeled peptides can be as small as 4 Da. With multiple charged ions, the pair of peptide ions may be barely resolved. In this case, cross-interference between LC-MS/MS signals of the analyte and the internal standard may occur (Hsu et al., 2003). Therefore, it would be wise to select a signal peptide with more than one derivatization sites to minimize the cross-interference. It is also known that formaldehyde is a toxic chemical and is allergenic and carcinogenic; thus, the reaction should be performed in a fume hood.

In general, using a stable isotope-labeled peptide as an internal standard facilitates monitoring of the sample following enzyme digestion and identifying any fluctuation of the analysis on the LC-MS system. However, this approach does not address variations that might occur during protein digestion or any sample logistics prior to this step.

47.2.2.2 Analog Protein Using a suitable analog protein as the internal standard is a convenient and efficient approach to developing an LC-MS/MS bioassay for therapeutic proteins. For instance, bovine fetuin was used as the internal standard for the bioassays of an mAb and somatropin (recombinant human growth hormone) (Yang et al., 2007). Bovine fetuin produces many tryptic peptides that are different from the usual sample matrixes as bovine is not a common species used for *in vivo* testing of drug candidates. These peptides have different physicochemical properties and cover a wide range of elution times in LC-MS analysis (Figure 47.2). Therefore, it is relatively easy to find a peptide that has a LC retention time close to the signal peptide of the target protein to allow good monitoring of some common effects on LC-MS analysis, such as ion suppression and solvent effects. Furthermore, bovine fetuin is added at the beginning of the assay and goes through the complete sample processing steps along with the analyte; therefore, the variability introduced by the multistep sample preparation can be normalized.

47.2.2.3 Stable Isotope-Labeled Protein The ideal internal standard for LC-MS/MS bioassay of protein is the molecule labeled with a stable isotope. This is exemplified in the recent work by Heudi et al. (2008) for the quantification

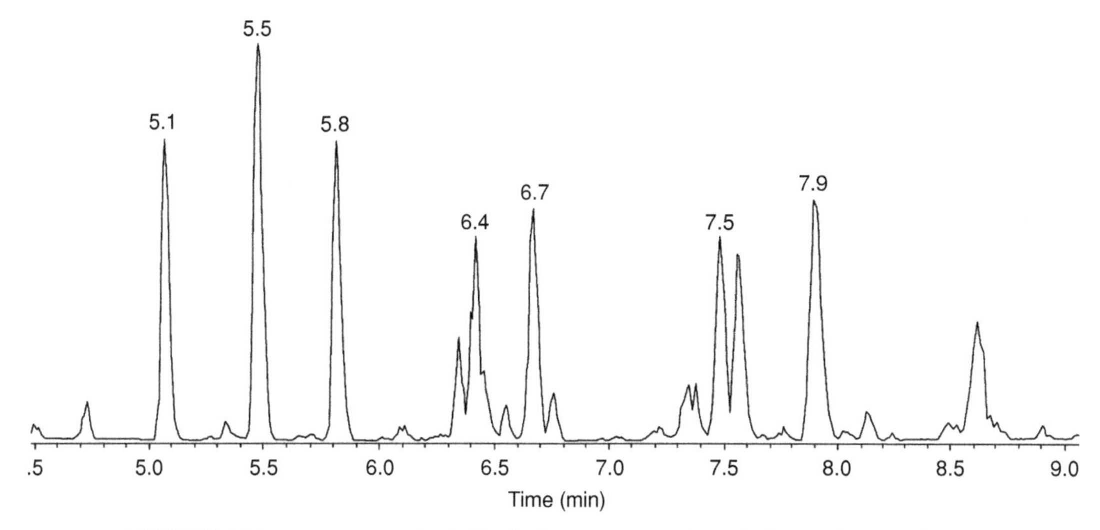

FIGURE 47.2 A representative LC-MS chromatogram of tryptic digest of bovine fetuin.

of a therapeutic mAb. The stable isotope-labeled mAb was produced by adding stable isotope-labeled threonine amino acid to the nutrition of the medium where the cells were cultivated. All the threonine residues of the resulting protein contained four ^{13}C and one ^{15}N. In another case reported by Liu et al. (2011), the stable isotope-labeled protein contained both ^{13}C-arginnie and ^{13}C-lysine residues.

The stable isotope-labeled protein internal standard has the same physical and biochemical properties as the analyte protein. When introduced to the biological samples at the very beginning of the assay, the stable isotope-labeled protein is useful for monitoring the variations throughout the entire assay procedure from enzyme digestion to the LC-MS/MS analysis of the signal peptide. One additional benefit of using a labeled protein as an internal standard is that the selection of the signal peptide can be more flexible and multiple signal peptides can be monitored without multiple isotope-labeled peptides. Although the labeled protein is considered an ideal internal standard for protein assay, the approach has not been broadly applied, primarily due to the time and cost associated with its production.

47.2.3 Sample Enrichment

47.2.3.1 Enrichment of Antibody Drugs Using Protein A Protein A was originally found in the cell wall of the bacterium *Staphylococcus*. It was found to have the ability to bind the heavy chain with the Fc region of immunoglobulins, especially IgGs of mammalian species. This special ability affords its use in the extraction and purification of IgGs. The commercially available protein A products include protein A magnetic beads and protein A agorose gel resins. In the antibody extraction procedure, protein A is mixed and incubated with the biological sample to allow its binding with antibodies in the sample. The antibodies are then extracted from the remaining components in the sample by centrifugation or attraction using a magnet. The antibodies are separated from protein A for further digestion and LC-MS analysis.

The enrichment procedure is normally performed under mild conditions and easy to handle. In general, immunoglobulins account for about 10% of the total plasma proteins. Therefore, with the protein A enrichment, about 90% of the plasma proteins are removed, which greatly reduces the burden on the follow-up enzyme digestion, LC separation, and MS/MS analysis. Interestingly, the reported sensitivity for antibody drug assays using protein A enrichment was about 1–2 μg/ml (Lu et al., 2009; Liu et al., 2011), which is similar to the results obtained using the assays without the enrichment procedure. One possible explanation could be the loss of proteins during the multiple steps of sample treatment. Furthermore, the binding capacity of the protein A product may limit the dynamic range of the assays.

47.2.3.2 Immunoprecipitation Protein A gel or beads are also widely used for immunoprecipitation. In this technique, protein A first binds to the antibody specific to the target antigen, and then is mixed and incubated with the biological sample to allow affinitive binding between the antibody and the antigen. Following the protein A extraction procedure, the target protein is precipitated from the sample solution. Via immunoprecipitaiton enrichment, assay sensitivity can be improved to ~0.1 μg/ml or better (Lu et al., 2009). However, the procedure requires the development of a specific antibody, which can be costly and time-consuming.

47.2.3.3 SPE Enrichment A protein enrichment approach using a C18 solid phase extraction (SPE) plate with a monolithic sorbent bed was described to enrich pegylated interferon from human serum (Yang et al., 2009). Using this approach, the assay sensitivity for pegylated interferon in human serum was about 3.6 ng/ml. The method may be applicable to other low molecular weight proteins, including some therapeutic proteins and biomarkers.

The general function of the SPE is to remove large molecules and retain small molecules. The mechanism of protein retention in the monolithic SPE sorbent is not clear, but it is possible that the monolithic SPE has a more active surface area, which allows for the retention of molecules with slightly higher molecular weights.

47.2.4 Sample Protein Digest

To produce the signal peptide from the target protein in the biological samples, the proteins in the samples need to be broken down into peptides using protease enzymes, such as trypsin. A general procedure for trypsin digestion starts with denaturing of the protein, and reduction and alkylation of the cysteine residues, followed by incubation with the enzyme. The success of the digestion process is critical for the subsequent LC-MS/MS analysis, as the yield of the signal peptide directly impacts the sensitivity of the assay. The following example is a procedure to digest 10 μl plasma, which is a reasonable sample size for the assay of a target therapeutic protein in biological samples obtained from animal or clinical studies.

To denature and reduce (break disulfate bonds) the 10 μl plasma sample, add 90 μl of 6 M guanidine chloride in 0.1 M ammonium bicarbonate buffer (pH 8) and 2 μl of 1 M dithiothreitol (DTT) to reach a final DTT concentration of 5 mM. Incubate at 75°C for 1 h with gentle shaking. Freshly prepare 1 M iodoacetamide (IAA) in 0.1 M ammonium bicarbonate, and transfer 10 μl to the sample to reach a final IAA concentration of 0.02 M. Incubate in the dark at room temperature for 2 h (IAA is sensitive to light) to alkylate and thereby protect the –SH group of cysteine, the reduction product. Filter the samples using 10 kDa Molecular Weight Cut-Off (MWCO) Amicon filters (Millipore, Billerica, MA) to remove the high concentration of salt. A larger MWCO filter can be used in the assay for a larger size target protein. The remaining portion of the sample is then reconstituted with 0.5 ml of 0.1 M ammonium bicarbonate (pH 8). After addition of 7.5 μg of trypsin, the sample is incubated at 37°C for overnight.

This procedure can be easily scaled up or down according to the starting sample volume. Some reagents are interchangeable with other chemicals with similar chemical properties. For example, the guanidine chloride can be substituted with urea without impact on the results. In addition, studies have shown that the digestion time can be reduced considerably if this is performed under microwaves.

47.2.5 Sample Digest Cleanup

The biological sample becomes a very complex peptide mixture after protein digestion and has a high concentration of salts that would result in high matrix effect and noise in LC-MS analysis. A sample cleanup step is therefore required to remove these excess compounds and improve the signal-to-noise ratio, or the sensitivity, of the assay.

SPE is commonly used for high throughput sample cleanup. Reversed-phase (RP)-SPE removes salts and highly hydrophobic components from the samples. By adjusting the strength of the wash and elution solvents, the sample remaining after RP–SPE should contain the target signal peptide and components within a narrow range of hydrophobicity. This step therefore reduces the burden on the RP–LC separation and allows for a slower and shorter LC gradient of organic solvent within the retention time period when the analyte elutes. Ultimately, the analyte can be better resolved from the noise peaks. In addition, with the use of a shorter separation gradient, the analytical LC-MS/MS run time for each sample is reduced, improving the overall throughput.

Ion exchange SPE, including strong cation ion exchange (SCX)-SPE and strong anion ion exchange (SAX)-SPE, uses another dimension in the sample cleanup. Complementing with RP–SPE, ion exchange SPE removes the unwanted matrix components according to their basicity/acidity. Through the optimization of the SPE wash and elution conditions, most of the peptides in the sample digest are either washed away or trapped on the SPE sorbent. Only the peptides of interest and a minimum of other components remain in the final sample extract. This ion exchange SPE sample cleanup can eliminate some peptides that may coelute with the peptides of interest on RP–LC, therefore reducing the matrix effects as well as improving the assay sensitivity.

Studies have shown the benefits of combining RP-SPE and ion exchange SPE sequentially (two-dimensional SPE) in sample digest cleanup (Yang et al., 2007), although this approach adds an extra step to sample processing.

47.2.6 LC-MS/MS

The common LC columns used to resolve the complex peptide mixture in the biological sample digest are typically analytical size columns with an inner diameter ranging from 1 to 4.6 mm. Conventional high-performance LC systems can deliver suitable flow rate for the separation solvents used for these types of columns without further modification on the system setup. If a reversed-phase LC column is used, peptides are normally eluted with 5–40% of the organic solvent (e.g., acetonitrile) at a suitable flow rate. Therefore, a slow organic solvent gradient within this range is most effective in resolving the peptides.

Capillary LC coupled with nano- or micro-ESI-MS provides a greater sensitivity in the detection of peptides compared to conventional analytical LC. However, it often requires very long LC separation times that eventually limit the assay throughput. On the other hand, an ultra-performance LC system has recently been applied for bioanalysis of proteins and showed an improved assay throughput due to high flow rate and high efficiency of the separation.

Triple quadrupole mass spectrometers operating in multiple-reaction mode (MS/MS) are often used for the assay of a target peptide. After positive electrospray ionization on the mass spectrometer, tryptic peptides normally become doubly charged ions, as protonation occurs at the two $-NH_2$ groups, one at the N-terminal and the other at the side chain of the C-terminal lysine or arginine. The double-charge reduces the mass/charge ratio (*m/z*) of the peptide parent ion to half of the peptide molecular weight. As a result the *m/z* can fall into the typical mass range of a mass spectrometer even for a large peptide. Peptide ions can be easily fragmented to multiple product ions using medium collision energy. The selection of the MS/MS transition can be optimized by the signal intensity as well as the level of interference or noise from the matrix.

47.3 SUMMARY

This chapter describes a general procedure for developing a bioanalytical assay for therapeutic proteins using LC-MS/MS that has been shown to provide better assay accuracy and precision than immunoassays. Several different formats for each of the assay steps have been discussed, including the choice of internal standard and the approach for sample enrichment and digestion. When looking for an appropriate protein bioassay format for a particular project, the project situation, such as the budget, allowed time, availability of materials, and targeted sensitivity (lower limit of quantification) have to be considered. Furthermore, the LC-MS/MS methodology described in this chapter is focused on the analysis of the total target protein concentration in the biological samples. It does not differentiate the protein in free form from that bound to its receptors.

The LC-MS approach for protein bioassay in support of biologics development has only been investigated recently. There are challenges ahead. For example, assay sensitivity must be improved to support clinical studies using low doses, and assay throughput needs to be increased in order to handle the large number of samples typically generated during drug development Nevertheless, the faster LC-MS/MS assay development compared to immunoassays should render this technology a preferred option for the bioanalysis of protein drugs, especially during discovery and the early stages of development.

REFERENCES

Heudi O, Barteau S, Zimmer D, et al. Towards absolute quantification of therapeutic monoclonal antibody in serum by LC-MS/MS using isotope-labeled antibody standard and protein cleavage isotope dilution mass spectrometry. Anal Chem 2008;80:4200–4207.

Hsu JL, Huang SY, Chow NH, Chen SH. Stable-isotope dimethyl labeling for quantitative proteomics. Anal Chem 2003;75:6843–6852.

Ji C, Sadagopan N, Zhang Y, Lepsy C. A universal strategy for development of a method for absolute quantification of therapeutic monoclonal antibodies in biological matrices using differential dimethyl labeling coupled with ultra performance liquid chromatography–tandem mass spectrometry. Anal Chem 2009;81:9321–9328.

Liu H, Manuilov AV, Chumsae C, Babineau ML, Tarcsa E. Quantitation of a recombinant monoclonal antibody in monkey serum by liquid chromatography–mass spectrometry. Anal Biochem 2011;414:147–153.

Lu Q, Zheng X, McIntosh T, et al. Development of different analysis platforms with LC-MS for pharmacokinetic studies of protein drugs. Anal Chem 2009;81:8715–8723.

Yang Z, Hayes M, Fang X, Daley M, Ettenberg S, Tse FLS. LC-MS/MS approach for quantification of therapeutic proteins in plasma using a protein internal standard and 2D-solid-phase extraction cleanup. Anal Chem 2007;79:9294–9301.

Yang Z, Ke J, Hayes M, Bryant M, Tse FLS. A sensitive and high-throughput LC-MS/MS method for the quantification of pegylated-interferon-α2a in human serum using monolithic C18 solid phase extraction for enrichment. J Chromatogr B Analyt Technol Biomed Life Sci 2009;877:1737–1742.

48

LC-MS BIOANALYSIS OF OLIGONUCLEOTIDES

Michael G. Bartlett, Buyun Chen, and A. Cary McGinnis

48.1 INTRODUCTION

Oligonucleotides contain the basic information for all cellular processes and therefore represent some of the most important information needed by biomedical researchers. Oligonucleotides are routinely used to modulate the expression of genes and are increasingly being targeted as biomarkers for many disease states. Recent renewed interest in therapeutic oligonucleotides has caused a dramatic increase in the number of these molecules undergoing clinical trials. Therefore, methods to selectively and sensitively measure oligonucleotides are increasingly needed.

Current methods to determine oligonucleotide targets are primarily based on molecular biological approaches. These include real-time reverse transcriptase polymerase chain reaction (RT-qPCR) methods or more indirect methods such as measuring the loss of expression of a target protein using western blots (Overhoff et al., 2004; Yu et al., 2004; Chen et al., 2005; Ro et al., 2006; Wei et al., 2006; Liu et al., 2008; Stratford et al., 2008; Cheng et al., 2009; Kim et al., 2009; Kroh et al., 2010; Tijsen et al., 2010). While these approaches are excellent for tracking the effectiveness of oligonucleotides on their gene targets, they do not directly measure the levels of oligonucleotides, nor do they have the ability to interrogate metabolic and chemical degradation products. Current methods that directly measure oligonucleotides at low concentration levels include stem loop RT-qPCR and hybridization immunoassays. However, both of these methods have the drawback of requiring primer design and validation for each oligonucleotide, and these methods also cannot track unknown metabolites or degradation products. In addition, these methods cannot differentiate modified oligonucleotides from unmodified oligonucleotides, and they cannot replicate these modifications within their amplification processes (Chen et al., 2005; Cheng et al., 2009).

Alternatives to these biological approaches that directly measure oligonucleotides would include techniques like liquid chromatography (especially ion exchange and ion pair), mass spectrometry, and capillary electrophoresis. These methods all have the added benefit of recognizing unknowns, and with the aid of mass spectrometry, determining the identity of the unknown. There are several methods reported in the literature; however, many of these methods suffer from limited or unreported accuracy and precision and also have relatively limited sensitivity (Leeds et al., 1996; Chen et al., 1997; Srivatsa et al., 1997; Beverly et al., 2005; Dai et al., 2005; Lin et al., 2007; Zhang et al., 2007; Zou et al., 2008; Issaq and Blonder, 2009; McCarthy et al., 2009; Deng et al., 2010). For methods such as LC-MS to become routine for the determination of oligonucleotides, it is imperative that they be able to measure ng/ml levels and lower of these molecules. This would open up the study of many basic intracellular mechanisms and also the study of the pharmacokinetics/pharmacodynamics (PK/PD) and metabolism for oligonucleotide-based therapeutics. A summary of the ranges covered by the major techniques for oligonucleotide analysis is shown in Table 48.1.

In this chapter, we highlight both the challenges and successes of using LC-MS for the determination of oligonucleotides. While there have been some past successes using continuous-flow fast-atom bombardment (van Breemen et al., 1991; Vollmer and Gross, 1995) for the determination of oligonucleotides, all of the recent success in the literature has been accomplished using electrospray ionization. Therefore, that is the major focus of this chapter with a few other technologies highlighted. There are two major research fields

Handbook of LC-MS Bioanalysis: Best Practices, Experimental Protocols, and Regulations, First Edition. Edited by Wenkui Li, Jie Zhang, and Francis L.S. Tse.

TABLE 48.1 Concentration Ranges Covered by Various Techniques Used for the Quantitative Determination of Oligonucleotides

Techniques	Concentration range
LC-UV	Mid ng/ml – High μg/ml
LC-MS	Low ng/ml – Low μg/ml
Hybridization	Mid pg/ml – Low μg/ml
q-PCR	Low pg/ml – high ng/ml

within the area of oligonucleotides: molecular biology and clinical therapeutics. While the goals of each of these fields are often quite different, the majority of the challenges when using LC-MS are similar.

48.2 PHYSIOCHEMICAL PROPERTIES AND MODIFICATIONS OF OLIGONUCLEOTIDES

Before any discussion of the LC-MS determination of oligonucleotides can begin, we first need to understand the physical and chemical characteristics that make these molecules so challenging to analyze.

48.2.1 Physiochemical Properties of Oligonucleotides

Oligonucleotides are the most polar of the major classes of biomolecules. They consist of three major features: (1) nucleic acid bases, (2) a furanose sugar (typically ribose), and (3) a phosphate linkage (Figure 48.1). The most common nucleic acid bases are adenine, guanine, cytosine, thymine, and uracil. While many modifications to the bases occur in nature to improve their biological function, modifications to nucleobases are not widely observed in therapeutic oligonucleotides, but there are instances where they have been shown to increase potency (Chiu and Rana, 2003). The sugar is the main determinant in whether an oligonucleotide is classified as a ribonucleic acid (RNA) or a deoxyribonucleic acid (DNA). While the presence or absence of the 2′ hydroxyl group appears to be a modest modification, its effect on the global structure and stability of oligonucleotides is profound. For example, RNA is far less stable than DNA in large part due to the presence of this 2′-hydroxyl group. This hydroxyl group is easily deprotonated, both chemically and enzymatically, initiating nucleophilic attack on the adjacent phosphorus atom, resulting in cleavage of the phosphate backbone.

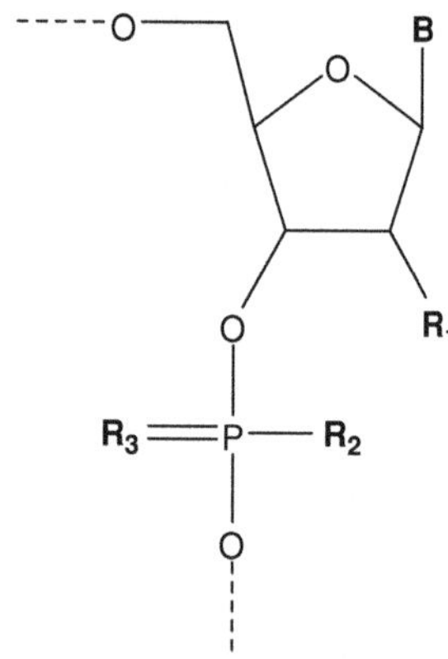

FIGURE 48.1 The structures of various oligonucleotides where B represents various nucleobases, R_1 is typically –H (DNA), –OH (RNA), –OMe, –OEt, or –F, R_2 is typically –OH (DNA or RNA), –Me (methylphosphonate), and R_3 is typically =O (DNA or RNA) or =S (phosphorothioate).

The phosphate group that links from the 5′ hydroxyl group of one ribose unit to the 3′ hydroxyl group of the next ribose in the biopolymer is a key structural feature of oligonucleotides. The angles between the ribose ring and the phosphate group are quite restricted and provide the rigidity associated with this class of compounds. At physiological pH, the phosphate group is highly charged and is also often associated with cations such as sodium and magnesium. This leads to another significant challenge associated with oligonucleotide determination—the need for efficient means of desalting samples that by their very nature have high affinities for these cations.

Oligonucleotides can be either single stranded or double stranded. Electrospray ionization mass spectrometry can be used to detect noncovalently associated oligonucleotide duplexes. However, it is difficult to preserve these hydrogen-bonding interactions through an LC column. Of the major types of chromatography, ion-exchange methods have shown the ability to consistently maintain these duplexes intact although there are instances of nondenaturing ion-pair chromatography (Beverly et al., 2005; Seiffert et al., 2011). While there have been some successes interfacing ion-exchange chromatography with mass spectrometry, the elution buffers (sodium chloride or sodium perchlorate) needed for oligonucleotides are not easily compatible with electrospray ionization. Therefore, the major focus of LC-MS methods for oligonucleotides has been on the single-stranded (ss) forms.

48.2.2 Modifications

Many modifications in oligonucleotides are made in an effort to improve the resistance of these biopolymers to hydrolysis, especially RNA. One of the most common in nature is the 2′-*O*-methyl modification. This blocks the deprotonation event that initiates base-catalyzed hydrolysis of RNA, which is the main mechanism of most ribonucleases (Spahr and Hollingworth, 1961). Some therapeutic oligonucleotides have emulated this strategy, in addition to trying larger alkyl groups such as O-ethyl, *O*-methoxyethyl and even total replacement of the 2′ hydroxyl group with a fluoride group (Chiu and Rana, 2003). Substitutions on the 2′ position have also been shown to increase the binding affinity of various RNAs for their complementary mRNA strands and to protect against off-target effects (Braasch et al., 2003; Chiu and Rana, 2003; Crooke, 2004; Watts et al., 2008).

Besides decreasing degradation, modifications also can decrease toxicity and increase potency by increasing pharmacokinetic properties such as the volume of distribution (De Fougerolles et al., 2007). They also can enhance binding specificity. 2′-*O*-methoxyethyl modifications have been shown to dramatically increase the elimination half-life of oligonucleotides up to 30 days in plasma and tissues in a wide variety of species (Geary et al., 2001; Geary et al., 2001). 2′-F modifications reduce the hydrophobicity of the oligonucleotide, which aids with entry into the cell (Watts et al., 2008). In general, modifications of the 2′ position of ribose decrease an oligonucleotides affinity for the enzyme RNase H, thus favoring the RNA interference pathway over the antisense pathway.

Methyl ether and allyl ethers have been used to end-cap the 3′-overhangs of small interfering RNA to protect them against exonucleases (Amarzguioui et al., 2003; Prakash et al., 2005). Interestingly, although these bulkier allyl groups do not work well for internal modifications in RNA, they retain excellent activity when used in anti-microRNAs that can be used to down-regulate endogenous microRNA (Davis et al., 2006).

Backbone modifications including phosphorothioate and methylphosphonate have been used both to increase hydrophobicity and to increase resistance to exonucleases (De Clerq et al., 1969; Miller et al., 1981). These modifications also increase plasma protein binding, which improves tissue distribution and prevents renal excretion (De Paula et al., 2007). While methylphosphonates have not had much success clinically, phosphorothioate modifications have been widely used to maintain the DNA-like properties in oligonucleotides (Eckstein, 2000). Phosphorothioate containing oligonucleotides are also excellent substrates for the enzyme RNase H and they possess low protein binding to serum albumin (Stein et al., 1988). Phosphorothioates are rapidly absorbed and distributed following intravenous administration, and they have a long elimination half-life (greater than 60 h in humans) (Crooke, 2004). Their clearance is primarily due to metabolic degradation, including desulfuration and hydrolysis via several nucleases.

48.3 METHODS AND APPROACHES

The majority of LC-MS approaches for oligonucleotides involve the use of electrospray ionization. The high degree of charging along the phosphate backbone makes this a particularly successful approach. However, the heavy reliance on electrospray severely limits the chromatographic approaches that can be used to analyze oligonucleotides. The most common approach taken for LC-MS of oligonucleotides involves using ion-pair chromatography. This approach involves the addition of a cationic ion-pairing agent to the mobile phase in order to provide an additional weak anion exchange mechanism to the separation.

48.3.1 Sample Preparation

48.3.1.1 Conventional Protein Precipitation, Liquid–Liquid Extraction, and Solid Phase Extraction Bioanalytical methods begin with sample extraction from a biological matrix. The goal of sample preparation is to isolate the targeted oligonucleotide(s) from interfering compounds. The goal is to maximize the recovery of the analyte while removing unwanted components. Among the many approaches, deproteinization is one of the most desirable due to its simplicity and speed. There are two basic approaches that have been tried with oligonucleotides to remove proteins from the biological sample: (1) methanol precipitation or (2) proteinase K digestion (Chen et al., 1997; Raynaud et al., 1997; Arora et al., 2002). However, it should be noted that none of these methods involved mass spectrometric detection. Therefore, to date these approaches have not met with any significant success for LC-MS determinations of oligonucleotides.

Liquid–liquid extractions (LLE) have been widely employed for the isolation of oligonucleotides for many years. The most common of type of LLE involves phenol/chloroform extractions or modifications thereof (Griffey et al., 1997; Beverly et al., 2005; Beverly et al., 2006; Waters et al., 2000). During LLE of phosphorothioate DNA, the addition of 5% ammonium hydroxide to the phenol–chloroform has been found to aid in the distribution of oligonucleotides to the aqueous phase (Zhang et al., 2007). For the determination of RNA, a mixture of phenol/ chloroform/isoamyl alcohol (25:24:1) and 1 mM Tris/EDTA at pH 8.0 has been found to provide recovery between 14% and 40% and was directly adaptable to LC-MS (Beverly et al., 2005).

The most widely used sample preparation method for oligonucleotides involves solid phase extraction (SPE) (Bellon et al., 2000; Gilar et al., 1997; Dai et al., 2005; Deng et al., 2010). Most of the SPE methods employ similar buffer systems as those used in chromatographic separations. Dai et al. (2005) have proposed a straightforward sample cleanup procedure for a phosphorothioate DNAs. The biological samples were mixed with a loading buffer containing the ion-pairing agent, triethylammonium bicarbonate (TEAB) and extracted using a C18 SPE cartridge. The proteins and salts were easily eluted with TEAB and water. The oligonucleotides were retained on the cartridge due to their increased hydrophobicity and pseudoneutral properties following ion pairing with TEAB. The ammonium ion also shields the oligonucleotide from binding with sodium and potassium ions. The recovery of this procedure is reported to be 43–64% depending on the oligonucleotide concentration. The breakthrough from the SPE cartridge was measured by UV and found to be negligible, indicating that irreversible binding to the SPE support may be the cause for the reduced recovery. A similar method using triethylammonium acetate (TEAA) as the ion-pair agent was employed for the quantitation of liposome-entrapped antisense DNA (Johnson et al., 2005).

Finally, combinations of LLE and SPE are also quite common approaches for the determination of oligonucleotides. For example, Zhang et al. (2007) employed LLE followed by SPE for the determination of a phosphorothioate-containing DNA from plasma. Even for RNA the combination of LLE and SPE has been found to be beneficial.

In a method to determine a therapeutic RNA from ocular tissue, the sample was lysed and homogenized and most of the cellular components and DNA were removed by a mixture of phenol:chloroform (5:1) isoamyl alcohol at pH 4.7. The aqueous layer was then mixed with ethanol to increase affinity toward the glass support and finally, the mixture was twice passed through a glass fiber filter. The small RNAs were immobilized on the filter and larger RNAs were washed off. The analyte was recovered by washing the filter with deionized water.

With increasing demands being made on bioanalytical methods, the need for high sample throughput has become a significant criterion in method development. Many kits are now available for DNA and RNA extraction. However, these are mostly geared toward polymerase chain reaction (PCR) applications, but several are also being marketed for small-length oligonucleotide applications, especially miRNAs. A substantial need continues to exist for simple, high-throughput methods for extracting small oligos. The methods must be amenable to a variety of biological matrices, including cells and tissues. The primary need is for the robust recovery of small amounts of RNA and DNA for PK/PD analysis, as well as, for the study of cellular mechanisms of oligonucleotides.

48.3.1.2 Enzymatic Digestion of Oligonucleotides The conventional approaches described above are excellent for small oligonucleotides. However, as oligonucleotide sequences become larger, there is a greater difficulty in their sample preparation. In these cases, it is often beneficial to use enzymes to reduce the size of the oligonucleotides to something more manageable (typically less than 25 bases) (Bellon et al., 2000; Hossain and Limbach, 2007). These approaches are most often used for larger RNAs, such as those involved in the ribosome. However, with increasing interest in stem-loop RNA, micro RNAs approaches involving digestion following by quantitation of smaller pieces that are representative of the larger molecules would appear to be reasonable approaches.

It is even possible to use this approach for sequencing smaller oligonucleotides or even obtaining partial sequencing information for larger oligonucleotides. There are two basic approaches used depending on the size of the oligonucleotides and the information that is desired. For smaller oligonucleotides, exonucleases in combination with LC-MS can be effective in aiding in determining either partial or full sequences. In this case, aliquots can be taken from the digestion reaction at fixed time intervals and the molecular weight of the remaining oligonucleotide measured. The difference in the measured molecular weights then determines the terminal nucleotide that was lost. This approach can be employed using either a 3′- or a 5′-specific exonuclease, allowing approximately 3–4 bases from each end of the oligonucleotides to be determined (Hossain and Limbach, 2007). Farand and colleagues have extended this approach by combining the use of exonucleases with chemical degradation to fully sequence siRNAs containing modifications, such as 2′-F, 2′-*O*-Me, and some abasic residues. Their methods even allow for the de novo sequencing of both strands of highly modified siRNAs (Farand and Beverly, 2008; Farand and Gosselin, 2009).

48.3.2 Chromatography

48.3.2.1 Ion-Pair Liquid Chromatography The first chromatographic separations of oligonucleotides used TEAA because of its good separation efficiency (Fritz et al., 1978; Huber et al., 1992; Kamel and Brown, 1996). Initially used to purify PCR products and primers, TEAA is an attractive ion-pairing agent for LC-MS because it is volatile and therefore, can be evaporated during the desolvation process in the mass spectrometer source (Huber et al., 1993; Gilar, 2001).

Huber et al. (1993) have used the TEAA mobile phase with a 14-mer single-stranded DNA. Their method had a wide linear dynamic range and a limit of detection of 8 ng/ml. Another paper from this group also showed separation of mixtures of homopolymer oligonucleotides ranging from 12 to 30 nucleotides demonstrating the ability to chromatographically differentiate between oligonucleotides differing by additional bases (Huber and Krajete, 1999) (Figure 48.2). Ion-pair LC using TEAA has also been used to separate synthetic primers up to 30-mers, as well as long DNA sequences in excess of 450 bases (Gilar et al., 2002). Dickman et al. used ion-pair LC with TEAA to analyze curved and bent duplex DNAs containing large poly A tracks (Dickman, 2005). Using ion-exchange LC methods, curved DNAs have longer-than-expected retention times, which cause them to coelute with higher molecular weight fragments. However, when using ion-pairing LC with TEAA a 378-base-pair DNA either curved or bent coeluted demonstrating that the higher order structure of the oligonucleotides did not appreciably contribute to the retention mechanism.

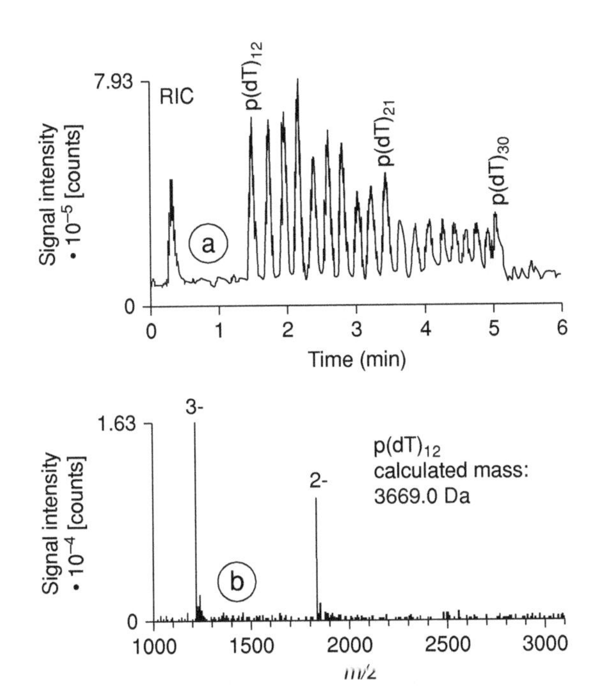

FIGURE 48.2 (a): Separation of polyT DNA by LC-MS using triethylammonium acetate ion-pairing mobile phase. (b) Mass spectrum of dT_{12} from LC-MS run shown above (adapted from Huber and Krajete (1999)).

Gilar et al. (2002) have extensively investigated the retention of single-stranded DNA when using ion-pair LC with TEAA. They analyzed a wide range of mixed base oligonucleotides and used the data to create a mathematical model that could be used to predict retention behavior. They applied their model and successfully predicted the elution of 39 different mixed base single-stranded DNA products. Their data showed that when determining the hydrophobicity of a sequence, the contributions from the bases C and G are less than from the bases A and T, which corresponded to earlier findings (Fritz et al., 1978; Huber et al., 1993; Gilar, 2001; Gilar et al., 2002).

In a minor change of the ion-pairing buffer system, TEAB has also been used for the separation of DNA. The first report of TEAB for the separation of oligonucleotides was in 1999 by Huber and Krajete (Huber and Krajete, 1999). This mobile phase also uses a volatile counter ion (bicarbonate) and thus is compatible with mass spectrometry. They demonstrated that TEAB enhances chromatographic resolution relative to TEAA for smaller oligonucleotides, but was similar when used with larger oligonucleotides (>30-mer). A thorough overview of this work is available in a review by Huber and Oberacher (2001).

Diisopropylammonium acetate is another ion-pairing agent that has demonstrated the ability to provide excellent separation of oligonucleotides. Bothner et al. (1995) have used this ion-pairing agent on several types of modified oligonucleotides including phosphorothioate and methylphosphonate containing oligonucleotides. In one of the applications, an oligonucleotide with half of its backbone converted to methylphosphonates was analyzed along with its $n - 1$ and $n - 2$ metabolites. While their products were only partially resolved chromatographically, MS detection with selected ion monitoring allowed for the unambiguous determination of each metabolite. The results were similar for phosphorothioate-containing oligonucleotides.

More recently, hexylammonium acetate (HAA) has been proposed as an ion-pairing agent for the determination of oligonucleotides with reasonable comparable mass spectral sensitivity to triethylamine (TEA) (McCarthy et al., 2009). McCarthy et al. used HAA for semipreparative purification of synthetic RNA to separate unmodified duplex RNA from an excess of the single strands as well as duplex impurities (Figure 48.3).

48.3.2.2 Capillary Liquid Chromatography Most LC-MS applications involving oligonucleotides have been accomplished using columns with 1–2 mm diameter columns. However, the improved chromatographic and ionization efficiencies offered by capillary LC offer advantages for the determination of oligonucleotides. Dickman and Hornby (2006) used capillary LC to analyze miRNA from total RNA extracted from HeLa cells. They were specifically targeting *Let-7* miRNA, which they found could be separated from the total RNA fraction. This study is an excellent proof of concept for research applications involving miRNA research and one of the few oligonucleotide applications using a commercially available capillary LC column.

Oberacher et al. (2001) used capillary columns to successfully separate a pool of 21-mer ssDNA primers of the same length with different compositions, as well as, double-stranded (ds) DNA primer products. They proposed that electrostatic interaction of the duplex was the retention mechanism for dsDNA, whereas the hydrophobic interactions of the nucleic acid bases were primarily responsible for the separation of the ssDNA. The duplexes could be analyzed for base substitutions by raising the column temperature above the melting temperature (Tm) of the oligonucleotide. At this temperature, the single strands were easily resolved and characterized.

Holzl et al. (2005) also used capillary LC for RNA separations. A synthetic unmodified 55-mer RNA was prepared in a 900-fold molar excess concentration of ethylenediaminetetraacetic acid (EDTA). The EDTA was added to the sample preparation to reduce the observed cationic adduction in the electrospray ionization mass spectra. The chromatography easily separated the EDTA and allowed the detection of the desalted 55-mer. This same approach was applied to the analysis of an intact 5S rRNA subunit (Figure 48.4). The authors also showed separation of a synthetic 21-mer RNA from its

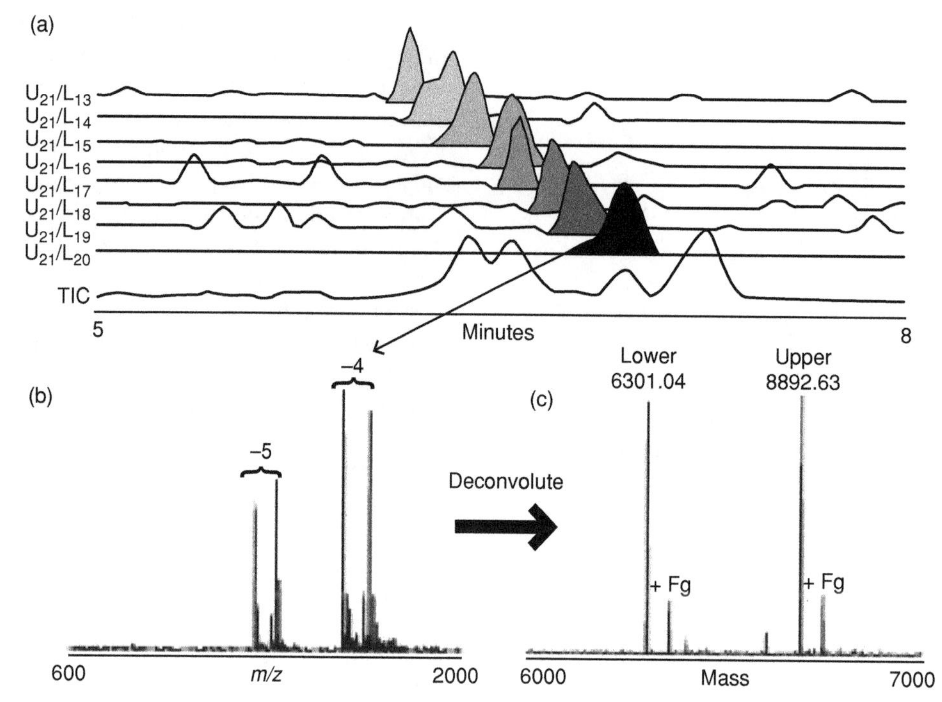

FIGURE 48.3 (a) Total ion chromatogram and reconstructed ion chromatograms from select impurities of a duplex siRNA. (b) Mass spectrum for truncated duplex with one nucleotide missing from the lower strand. (c) Deconvoluted mass spectrum of two strands of siRNA (adapted from McCarthy et al. (2009)).

failure sequences down to a 7-mer, although coelution of some of the $n - 1$ sequences was observed. This paper also provides an example of on-line MS sequencing of a particularly large oligonucleotide (32-mer).

48.3.2.3 Other Chromatographic Approaches Hydrophilic interaction liquid chromatography (HILIC) for LC-MS has gained significant interest recently due to the greater analyte response that can be obtained from the higher percentage of organic solvent used in the mobile phase. One application of HILIC for the determination of oligonucleotides also introduced the use of inductively coupled plasma mass spectrometry (ICP-MS) for detection (Easter et al., 2010). The chromatography was only capable of baseline separating the full-length unmodified oligonucleotide from the $n - 5$ sequences (Figure 48.5). So while there is need for improved chromatography, the novel use of ICP-MS to measure oligonucleotides via detection of the phosphate backbone as phosphorous oxide ion (*m/z* 47) is noteworthy. Detection limits for a dT_{30} were 0.336 ng/ml. This method could also potentially be used for selective determination of phosphorothioate-containing oligonucleotides by measuring the response from sulfur oxide. Further work on improving separations could make this technique an excellent choice for the quantitative analysis of oligonucleotides.

48.3.3 Tandem Mass Spectrometry of Oligonucleotides

Structural characterization of oligonucleotides using tandem mass spectrometry has been used for many years. However, there are few examples of its use online with liquid chromatography. This is due to many factors including (1) the more limited sensitivity observed in general for negative ion mass spectrometry and (2) the high molecular weight and thus the large number of degrees of freedom found in oligonucleotides. However, the need to characterize endogenous and therapeutic oligonucleotides is greatly facilitated by tandem mass spectrometry.

The full characterization of therapeutic oligonucleotides will be required for regulatory approval. Identification of process impurities, as well as of the metabolites of these therapeutic oligonucleotides, is needed to provide a more comprehensive set of data to answer questions concerning both the potency and potential side effects. Characterization of the metabolite profiles can also reveal details on the metabolic processing of the oligonucleotides and highlight areas of the molecule that may require modification to alter PK/PD properties. The combination of liquid chromatography with mass spectrometric analysis can provide extensive structural information. However, this will occur only when the concentrations of the oligonucleotides are sufficiently

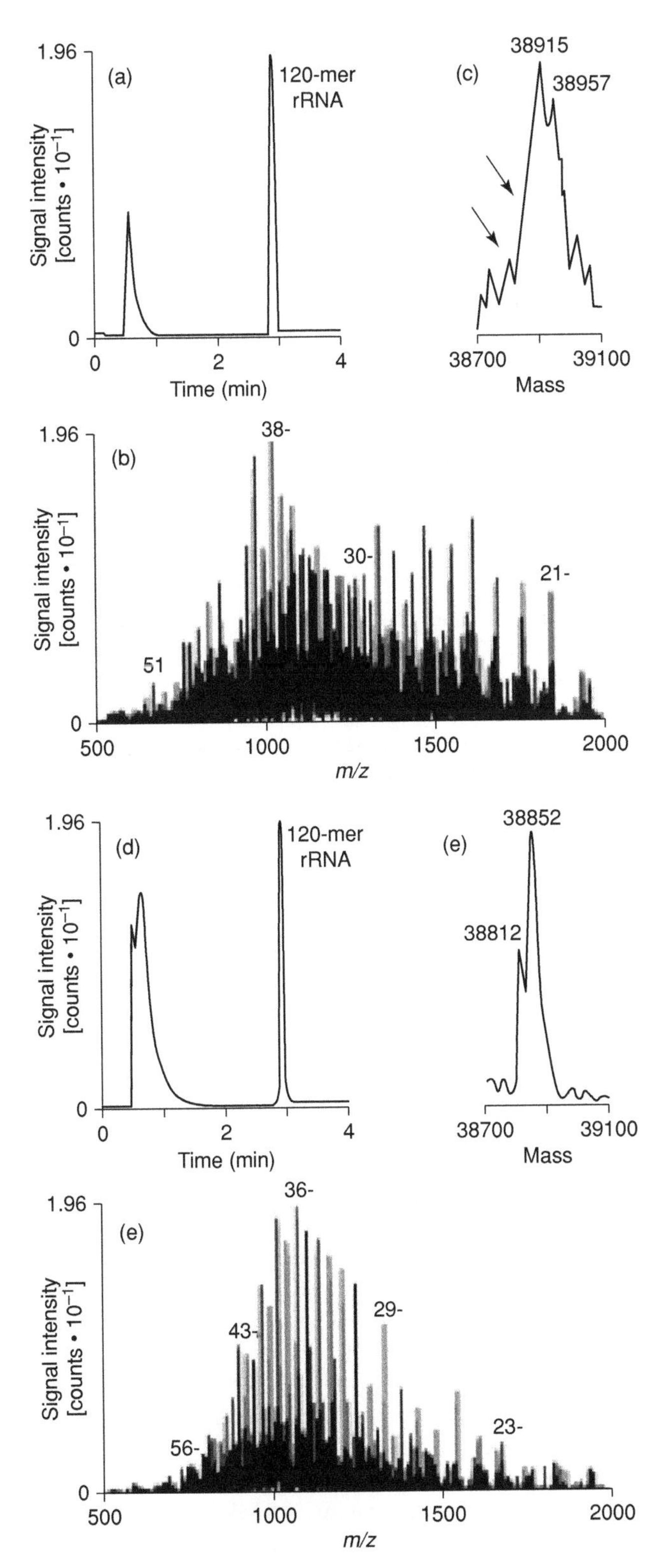

FIGURE 48.4 Electrospray ionization mass spectrum of 5S ribosomal subunit from *E. coli* (a–c) from water (d–f) from 25 mM EDTA solution (adapted from Holzl et al. (2005)).

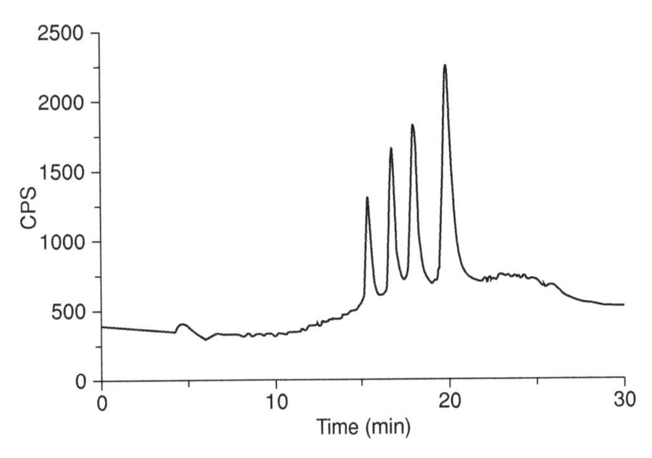

FIGURE 48.5 Hydrophilic interaction liquid chromatography inductively coupled plasma mass spectrometry chromatogram of dT_{10}, dT_{15}, dT_{20}, and dT_{30} (adapted from Easter et al. (2010)).

high and the peak widths are broad enough to allow high-quality tandem mass spectra to be obtained.

Most small ss oligonucleotides can be directly sequenced by collision-induced dissociation (McLuckey, 1992; McLuckey and Habibi-Goudarzi, 1993). Sequencing oligonucleotides with more than 25 bases is complicated due to secondary fragmentation patterns that make interpretation of the resulting mass spectrum difficult. Additional enzyme digestion may therefore be necessary before MS analysis (Crain and McCloskey, 1998). An excellent review of the tandem mass spectrometry of oligonucleotides is available from Beverly (Beverly, 2011). In general, a-B and w fragmentation ions are the most abundant in the MS/MS spectra of DNA oligonucleotides, while c and y ions are the most abundant in the MS/MS spectra of RNA oligonucleotides (McLuckey, 1992). However, different modifications of the oligonucleotide structure have varying impacts on the fragmentation pattern. For example, Ni et al. (1996) observed that b ions and y ions are significantly more abundant in phosphorothioate-containing DNA oligonucleotides relative to those with phosphodiester backbones. This suggests that these ions may be more useful for sequencing of these molecules.

Computational algorithms greatly enhance the use of mass spectral data to either identify or confirm the sequence of a given oligonucleotide (Watson and Sparkman, 2008). Initially, oligonucleotide sequencing involved laborious manual interpretation of the data by comparing tables of calculated *m/z* values corresponding to expected peaks present in a mass spectrum (Gaus et al., 1997). One of the first algorithms was developed by Pomerantz et al. (1993), which would generate lists of DNA and RNA base compositions for a given molecular mass. With the wider application of oligonucleotides in both laboratory and clinical settings, internet-based computer programs have become available to assist with mass spectral interpretation. Among the most useful of these is the MONGO algorithm developed by McCloskey and coworkers which is available from a website at the University of Utah (Rozenski and McCloskey, 1999; Rozenski and McCloskey, 2002). This algorithm is capable of calculating the molecular mass, *m/z* values of an electrospray series and the masses of potential fragment ions arising from a precursor with a defined charge state. However, this program supports few modifications to the oligonucleotide structure. The more advanced SOS algorithm, later developed by the same group, allows users to define any combination of base, sugar, or backbone modifications in the structure of the oligonucleotide and generate fragment ions (Rozenski and McCloskey, 2002).

48.3.4 LC-MS/MS

48.3.4.1 LC-MS/MS of Therapeutic Oligonucleotides

The metabolism and degradation patterns of oligonucleotides have been studied for more than 20 years using LC-MS. Among the many oligonucleotides, antisense phosphorothioate DNAs and transfer RNA have been the most extensively characterized. LC-MS of phosphorothioate-containing oligonucleotides can be conducted with greater success than tRNA mainly due to their increased hydrophobicity and thus the greater organic solvent content of the mobile phase during their elution relative to RNAs.

The *in vitro* metabolism of a 21-mer phosphorothioate oligonucleotide was studied following incubation with liver homogenate (Crooke et al., 2000). Exonuclease activity at the 3′-end was found to predominate; however, 5′-exonuclease activity and endonuclease activity were also observed. The most significant *in vivo* metabolic study of a phosphorothioate DNA was conducted by Isis Pharmaceuticals on a 20-mer clinical candidate (Isis 2302) (Gaus et al., 1997). The metabolites were fraction collected using anion-exchange chromatography and desalted prior to LC-MS/MS analysis. For most metabolites, a-B ions and w ions were observed from fragmentation. Only 3′ chain shortened metabolites were observed ($n + 1$ to $n - 3$). A small amount of depurination from the 3′-end was also observed. Interestingly, $n + 1$ metabolites were also observed although no definitive structural information was obtained for these.

Similar findings were also seen from *in vivo* metabolic studies of other phosphorothioate-containing oligonucleotides (Gaus et al., 1997). The most abundant metabolites were from losses of nucleotides from the 3′-end. Depurination of the adenine base was also observed in liver metabolism. The metabolic profiles from kidney and liver tissues displayed distinctive patterns. There were greater concentrations of 5′ metabolism observed in the kidney relative to the liver. Somewhat surprisingly, greater overall metabolism was observed in the kidney versus the liver.

Dai et al. (2005) characterized the metabolites of a Bcl-2 antisense phosphorothioate DNA G3139 in human and rat plasma and urine. Six major metabolites were generated from 3′ metabolism. It is worth noting that since the base on both the 3′ and 5′ ends were thymines, the deconvoluted mass spectra of the precursor ions did not provide enough information to rule out either one of the metabolites. The researchers obtained tandem mass spectra of the standard of both metabolites and compared the w series and a-B ions with the unknown metabolites and were able to assign the structure to as the 3′ $n - 1$ chain shortened sequence.

Metabolite characterization of ds oligonucleotides can be more complicated due to the presence of both ss and ds metabolites. In addition, the ds ion species can dissociate in the electrospray source, which will further complicate the identification process unless sufficient chromatographic separation is achieved prior to ionization. Beverly and colleagues studied the *in vitro* and *in vivo* metabolism of an siRNA in urine and ocular samples (Beverly et al., 2005; Beverly et al., 2006). The siRNA was modified with abasic protecting groups and two thymines on the 5′ end while the 3′ end was modified with a phosphorothioate backbone. The metabolism of both the single strands (urine) and the duplex (urine and vitreous samples) showed exonuclease activity on the 3′ end while the 5′ end exhibited endonuclease activity (Figure 48.6). The author proposed a mechanism by which the endonucleases jumped the protected abasic site to start the cleavage and the exonucleases took over the degradation from there. Different cleavage sites on the phosphate backbone were also observed among different metabolites. The degradation of the antisense strand started later in the time course because of the nuclease protection from the *O*-methyl group located at the 2′ position of ribose on each purine base on this strand. This slower onset of degradation was important for therapeutic reasons as the antisense strand would directly interact with the mRNA. Due to the denaturing effects of the ESI process, the signal of both sense and antisense strands were observed in the mass spectrum of the duplex. The author pointed out that single strands formed from the denatured duplex could reanneal, and give false information on duplex metabolites. This issue was addressed by spiking ss metabolites (5′ $n - 3$ sense strand) at twice the concentration of the duplex and analyzing the result. The mass spectrum did not contain a mass corresponding to a duplex containing the $n - 3$ metabolite, indicating that annealing between the full length and metabolite strands had not occurred. In the *in vivo* metabolism study of the duplex, siRNA sequences shorter than $n - 2$ on one strand are all unambiguously identified, including 3′ $n - 3$ and 3′ $n - 5$ antisense strands binding to a full-length sense strand, respectively, and a 3′ $n - 5$ antisense strand binding to a 5′ $n - 5$ sense strand both containing terminal phosphates. However, a sequence of $n - 2$ antisense strands binding to a full-length sense strand could not be assigned as a 3′ chain shortened metabolite or a 5′ end one

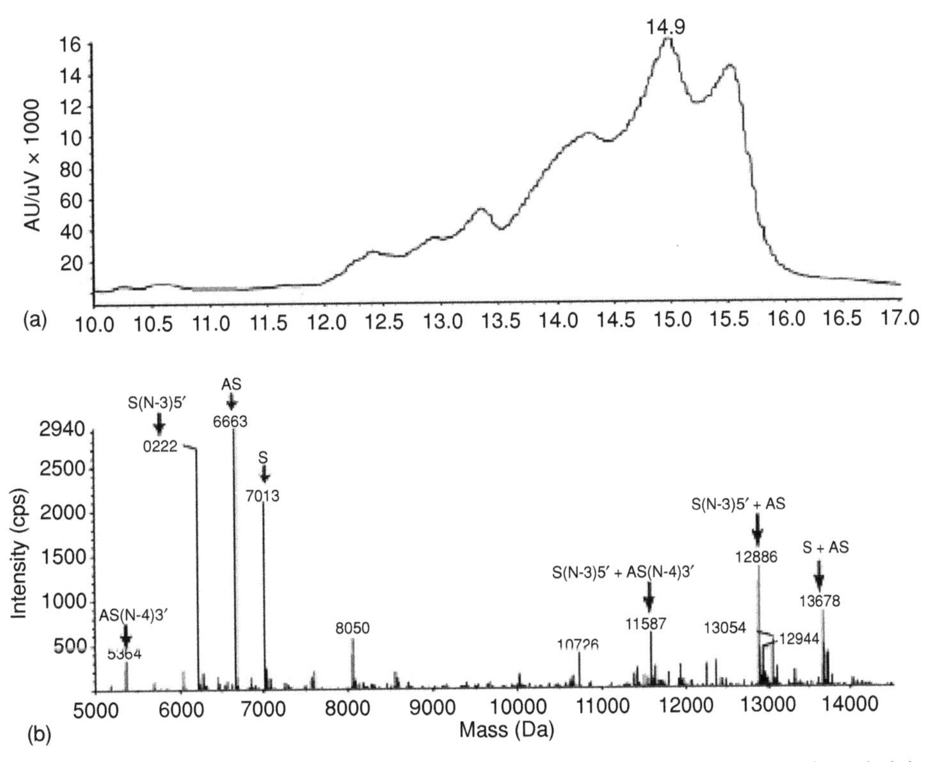

FIGURE 48.6 (a) LC-UV (A260 nm) of a vitreous sample taken four days following administration of a therapeutic siRNA. (b) Deconvoluted mass spectrum taken from 14 to 16 min from the chromatogram shown in (a). The full-length duplex is shown at 13,678 Da, as are two duplexes made up of chain-shortened metabolites observed at 12,886 and 11,587 Da. The single strands are visible at 7013 and 6663 Da and the two metabolites at 6222 and 5364 Da (adapted from Beverly et al. (2006)).

with a terminal phosphate with enough resolution (0.05 Da difference).

48.3.4.2 LC-MS of Transfer RNA (tRNA) and Ribosomal RNA (rRNA) Naturally occurring RNAs have significant numbers of modifications to enhance their biological function. Transfer RNAs have the greatest number of modifications and are a reasonable size (60–80 bases) and therefore have received the most attention. Recently, intact tRNAs have been fully sequenced using similar strategies employed in top-down proteomics (Huang et al., 2010; Taucher and Breuker, 2010). Notably, these are among the largest ever sequenced in a single mass spectrometry experiment. While the sequencing of such large oligonucleotides is a tremendous accomplishment, both of these studies used infusion of the samples. Therefore, it is not known if oligonucleotides of this size can be sequenced during LC-MS runs.

Typically, larger RNAs are digested into smaller sequences (2–10 bases). This approach was pioneered by Pomerantz and McCloskey (1990) more than 20 years ago. These smaller oligonucleotides can then be analyzed using LC-MS and if necessary sequenced using tandem mass spectrometry. More recently this approach has been labeled RNase mapping and has been successfully combined with MS/MS and LC-MS/MS (Matthiesen and Kirpekar, 2009). The modifications found in tRNA have been summarized and can be found in frequent updates from McCloskey and coworkers (Cantara et al., 2011).

Of the many modifications that are found in RNA, one of the most interesting is pseudouridine. Psuedouridine is found in the loop that bears its name in tRNA, as well as in other places in RNA (Durairaj and Limbach, 2008). Pseudouridine replaces the normal N-linked glycosidic bond between the base and ribose with a carbon-carbon bond. Since loss of the nucleobase is the first step in initiating many of the major fragment ions in oligonucleotides, it is not surprising that the presence of pseudouridine dramatically alters the appearance of fragment ions at these positions in the sequence (Figure 48.7). However these alterations, as well as the appearance of a diagnostic ion at *m/z* 207, draw attention to this position, allowing for follow-up experiments to confirm the presence of this unusual base (Pomerantz and McCloskey, 2005; Addepalli and Limbach, 2011).

The approaches used to place modifications in tRNAs have also been applied to larger rRNAs. McCloskey and coworkers were able to place the modifications in the 5S, 16S, and 23S rRNA of the archaeal bacteria *Sulfolobus solfataricus* (Bruenger et al., 1993; Noon et al., 1998). They have also

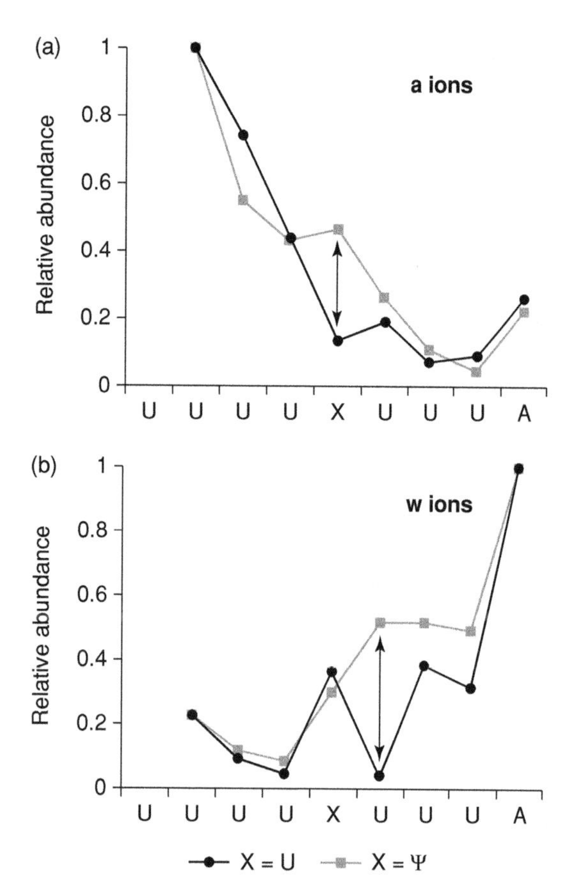

FIGURE 48.7 Normalized sum, observed over all charge states of product ion abundances for two oligonucleotide isomers. (a) The a-ion series normalized to the a_2 ion. (b) The w-ion series normalized to the w_1 ion. Arrows represent differences observed at the site of pseudouridylation (adapted from Pomerantz et al. (2005)).

placed the modifications into various portions of the rRNA from *Haloferax volcanii*, *Thermatoga maritima*, *Thermus thermophilus*, *Pyrodictium occultum*, and *Escherichia coli* (Bruenger et al., 1993; Kowalak et al., 1995, 1996, 2000; Guymon et al., 2006, 2007).

More recently, LC-MS approaches have been used to monitor relative changes in tRNA pools in cells (Castleberry and Limbach, 2010; Chan et al., 2011). This experiment measures small oligonucleotides from the tRNA following T1 nuclease digestion. Among the many small oligonucleotides, there are unique sequences that represent each of the more than 70 tRNAs used for protein synthesis. This approach has the ability to quantify changes in the tRNA pool arising from alterations in the cellular environment.

48.3.4.3 Balancing the Strength of Ion-Pairing Agents in Chromatography and Electrospray Ionization Efficiency in Mass Spectrometric Detection for LC-MS Quantification of Oligonucleotides A thorough understanding of the role of ion-pairing agents in both the separation and desorption process during LC-MS is necessary in order to evaluate the many options available. The mobile phases generally use a weak cationic ion-pairing agent and use hexafluoroisopropanol (HFIP) as the counterion. This mobile phase system was first employed by Hancock and coworkers and although there have been some changes to the identity of the ion-pairing agent; the approach has been relatively unchanged for almost two decades (Apffel et al., 1997a, 1997b).

A complete list of ion-pairing agents used in LC-MS of oligonucleotides along with their physical properties is shown in Table 48.2. It is worthwhile to note that all of the weak cationic agents that have been successfully used are amines and therefore, the extent of their charging is governed by the pH of the mobile phase. In the LC column the ion-pairing agent is needed to modify the stationary phase and allow for a dual mechanism of separation to occur. There is still a typical hydrophobic separation from the alkyl (C-8 or C-18) stationary phase but there is now an ion-exchange type of an interaction from the amines that have been introduced into the mobile phase. However, as opposed to most LC methods employing ion-pairing agents, in electrospray ionization, the desorption process is highly dependent on the presence of the ion-pairing agents to assist with efficient transfer of the oligonucleotides from the solution phase into the gas phase. The mechanism of this desorption process has been the subject of several studies (Mack et al., 1970; Whitehouse et al., 1985; Greig and Griffey, 1995; Muddiman et al., 1996; Griffey et al., 1997; de la Mora, 2000; Kebarle and Peschke, 2000; Wang and Cole, 2000; Nguyen and Fenn, 2007).

Before discussing the role of ion-pairing agents in the LC-MS of oligonucleotides, it is important to remember that electrospray ionization is a surface desorption process. Therefore, solution conditions that alter the equilibrium between bulk solution and the surface of droplets will be reflected by changes in the appearance of the mass spectra. Analyte hydrophobicity has long been recognized as a factor influencing the rate of migration of a molecule to the surface of an electrospray droplet. In an excellent demonstration of this phenomenon, Muddiman et al. compared the signal intensity from an equimolar solution of complimentary strands of DNA (Null et al., 2003; Shuford and Muddiman, 2011). The more hydrophobic single strand had an approximately 30% greater signal intensity. Interestingly, this effect can be eliminated by the heating of the electrospray capillary (Figure 48.8) (Frahm et al., 2005). The exact mechanism for this effect has not been determined but is likely a combination of reduced surface tension from the formation of smaller electrospray droplets and improved desorption from the droplet surface due to changes in the activation energy barrier for ion release. Ion-pairing agents have been proposed to operate through a similar mechanism. The association of the ion-pairing agent with the oligonucleotides masks the charge site on the phosphate backbone and increases the rate of transport from the bulk solution to the surface of the droplet.

TABLE 48.2 A Complete List of Ion-Pairing Agents Used in LC-MS of Oligonucleotides Along with Their Physical Properties

Ion-pairing agents	pKa	Boiling point (°C)	Log *P*	Gas phase basicity (kJ/mol)
Triethylamine	10.75	88	1.47	951
Hexylamine	10.56	131—132	1.565	893.5
Dipropylamine	11.0	105—110	1.504	929.3
Diisopropylamine	11.05	84	1.444	NA
Piperidine	11.22	106	0.655	921
Dimethylbutylamine	10.2	93	1.78	938.2
Diisopropylethylamine	10.5	127	2.064	963.5
Tripropylamine	NA	156	3.175	991
Tributylamine	10.89	216	4.656	967.6
Imidazole	14.5	256	−0.08	909.2

NA, values not available.

While a significant amount of the earliest work involving electrospray ionization of oligonucleotides employed TEAA, more recently other ion-pairing systems have become more widespread in their use. The HFIP/TEA mobile phase developed by Apffel et al. is currently the most widely applied in LC-MS applications measuring oligonucleotides (Apffel et al., 1997a, 1997b; Beverly et al., 2005; Dai et al., 2005; Beverly et al., 2006; Lin et al., 2007; Zhang et al., 2007; Kenski et al., 2009; Deng et al., 2010). This mobile phase buffering system was optimized by Gilar et al. (2003) to 400 mM of HFIP and 16.3 mM of TEA to maximize the separation efficiency and mass spectral signal intensity for ssDNA.

Apffel et al. proposed that HFIP/TEA buffer assists the transfer of oligonucleotides into the gas phase by raising the pH of the droplet toward 10 as the HFIP is selectively evaporated from the droplet (Apffel et al., 1997a, 1997b). As

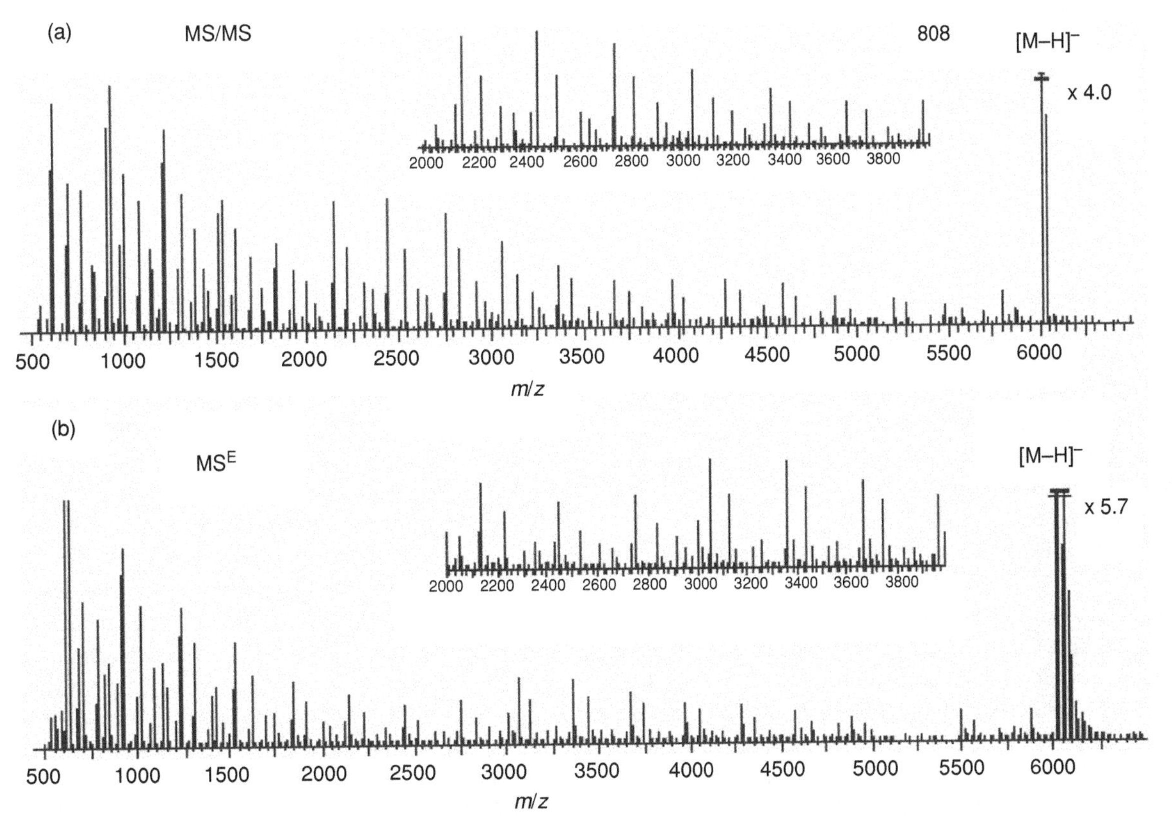

FIGURE 48.8 Comparison of (a) collision-induced dissociation mass spectrum and (b) MS^E mass spectrum of a 20-nucleotide polydeoxythymidine (adapted from Ivleva et al. (2010)).

the pH increases, TEA dissociates from the phosphodiester-backbone enhancing oligonucleotide desorption in to the gas phase. The lower pKa of HFIP (9.3) also allows for rapid evaporation because it is not significantly dissociated at the typical pH used with these mobile phase buffers.

There have been extensive studies of the impact of the acidic counter ion on electrospray signal intensity. When considering the desolvation of an electrospray droplet, the boiling points of both parts of the buffer are important to consider. Experimental evidence has shown that TEAA gives lower signal intensities than HFIP/TEA or TEAB (Apffel et al., 1997a, 1997b; Huber and Krajete, 1999; Huber and Krajete 2000a; Gilar et al., 2003). To explain this observation it should be noted that TEA has a lower boiling point than acetic acid (89°C vs. 118°C). Therefore, during droplet desolvation in electrospray ionization, the TEA evaporates more rapidly than the acetic acid. The loss of TEA alters the pH in the droplet leading to lower ionization efficiencies for the sample, likely due to competition for ionization between the acetic acid and the oligonucleotide. Another factor is that TEAA is less hydrophobic than other ion-pairing buffer systems. Therefore, it requires less organic modifier to elute the oligonucleotides from the LC system, so the surface tension of the droplet is higher causing decreased ionization efficiency.

In addition to acetate, bicarbonate has been used as the counter ion in different ion-pairing systems for the determination of oligonucleotides. One advantage of bicarbonate is it is far more volatile than amines such as TEA and butyldimethylamine (BDMA). Huber et al. showed that, with the addition of an acetonitrile sheath liquid, TEAB gave up to sevenfold higher signal intensity than TEAA with equivalent chromatographic performance (Huber and Krajete, 1999). In a separate paper, Huber et al. found that a butyldimethylammonium bicarbonate (BDMAB) mobile phase increased sensitivity even more than TEAB (Oberacher et al., 2001). This effect is likely due to the fact that BDMAB is significantly more hydrophobic than either TEAA or TEAB. This means that chromatographically it will require a higher concentration of acetonitrile to elute the oligonucleotide. This higher organic concentration will reduce the surface tension of the electrospray droplet and provide greater signal intensity.

The conductivity of the counter ion can also play a role in the ionization. Huber and Krajete investigated the role of various counter ions with TEA in the electrospray ionization process. They used acetate, bicarbonate, formate, and chloride in solutions containing 20% acetonitrile at a pH of 8.9. They found that the volatility of the counter ion did not correlate with signal intensity that was somewhat unexpected (Huber and Krajete, 1999). The acetate counter ion gave the highest signal intensity under these conditions. This led them to conclude that the volatility of the counterion does not completely explain the effects of these ion-pairing agents on the electrospray desorption of oligonucleotides. They proposed that the increased conductivity of the counter ion is involved in ion suppression of the oligonucleotide signal through direct competition for ionization. This is supported by the work of Gilar et al. who showed that lowering the concentration of HFIP increased the signal from the oligonucleotides but eventually the desorption of the oligonucleotides is adversely impacted (Gilar et al., 2003). Therefore, the ion-pairing agents improve ion desorption but also cause ion suppression and these two phenomena will need to be balanced for optimal LC-MS of oligonucleotides.

Ionization efficiency must also be balanced with alterations in the charge state distribution of the oligonucleotide. Charge state reduction can allow for greater signal intensity over fewer *m/z* values, and potentially increase the method detection limits. In general, the charge state is reduced when protons remain associated with the phosphodiester backbone, leaving fewer negative charges. This process occurs during the transfer of ions to the gas phase, as well as within the gas phase. This process was studied by Muddiman et al. who proposed a mechanism for charge state reduction in which the hydrogen bound proton was shared as a dimer between the phosphodiester bond and triethlyamine (Muddiman et al., 1996). Because of the higher proton affinity of the oligonucleotides backbone, the TEA is lost as a neutral during the electrospray desorption process. The addition of HFIP also shifts the charge state envelope to higher *m/z* values and does not produce the charge state reduction observed with other ion-pairing buffer systems such as TEAB and BDMAB (Oberacher et al., 2001). This indicates that the proton affinity of HFIP is higher than bicarbonate (Huber and Krajete, 2000b; Beverly et al., 2005).

Recently, Ivleva et al. offered additional work using LC-MS for the characterization of both phosphorothioate and locked nucleic acid based siRNAs (Ivleva et al., 2010). The authors used MS^E a technique developed by Waters that employs alternating scans as either high or low collision energies. The low collision energy data provided molecular weight information about the oligonucleotides. The elevated collision energy data provided ion fragmentation for sequence information for all analytes. This provides an advantage over more targeted MRM detection methods because all of the ions are fragmented and detected, not just predetermined ions. Therefore, if new peaks arose in a chromatogram, MS^E would be able to analyze these without needing to modify the MS method. The authors used this method to characterize a modified synthetic siRNA and its failure sequences and by-products as well as metabolites from an *in vitro* hydrolysis procedure. A comparison of MS/MS and MS^E sequencing of a 20-mer is shown in Figure 48.9.

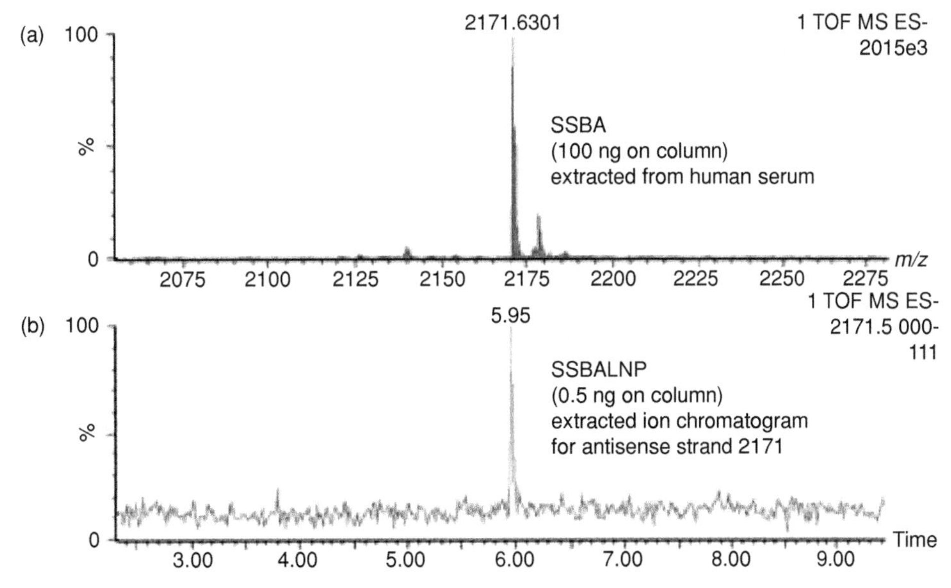

FIGURE 48.9 Mass chromatograms from ion-pair LC-MS determination of a modified siRNA from plasma. (a) Concentration of 10 μg/ml. (b) Concentration of 50 ng/ml (adapted from Beverly (2011)).

48.4 PROTOCOLS

From the preceding sections, it is clear that there are many different applications for LC-MS to the study of oligonucleotides. The following section contains procedures for the most common types of analyses for the determination of oligonucleotides from various *in vitro* or *in vivo* samples. These protocols have been created to facilitate the use of LC-MS by others entering this challenging field.

48.4.1 Antisense DNA

Among the therapeutic oligonucleotides the most well studied are the antisense DNAs. The greatest successes in the class have come from ss phosphorothioate-modified oligonucleotides in the range of 18–24-mers. However, oligonucleotides of this type have several unique challenges, most notably their increased hydrophobicity and their susceptibility to oxidation. Therefore, sample preparations and chromatographic conditions optimized for this class of compounds may not be particularly successful when applied to other more highly polar oligonucleotides. In addition, careful attention needs to be paid to evaporation steps and storage conditions that may promote conversion of the phosphorothioate backbone back to the normally occurring phosphodiester linkage. The addition of antioxidants is common in many of these methods but there is need to carefully evaluate the impact of using high concentrations of these agents on method recovery and analyte ion suppression.

Protocol 1: Protocol for antisense DNA from plasma (adapted from Zhang et al. (2007)).

Part 1: Liquid–Liquid Extraction

1. To a 200 μL sample of plasma, add 500 μL of a 5% solution of ammonium hydroxide and 100 μL of chloroform/phenol, 1:2 v/w.
2. Vortex sample for 2 min.
3. Centrifuge sample for 10 min at 3000*g*.
4. Transfer aqueous portion of sample to a clean tube.
5. Add 800 μL of buffer containing 17.2 mM triethylamine, 200 mM hexafluoroisopropanol at pH 8.5.
6. Vortex sample for 2 min.
7. Sample is now prepped for solid phase extraction.

Part 2: Solid Phase Extraction

1. Condition a 1 ml solid phase extraction cartridge containing 10 mg of a hydrophilic–lipophilic balance stationary phase with 1 ml of acetonitrile, followed by 2 ml of a buffer containing 8.6 mM triethylamine, 100 mM hexafluoroisopropanol at pH 8.5.
2. Load sample from liquid–liquid extraction onto the cartridge.
3. Wash the cartridge with 300 μL of a buffer containing 8.6 mM triethylamine, 100 mM hexafluoroisopropanol at pH 8.5.

4. Next wash the cartridge with 500 μL of 100 mM triethylammonium bicarbonate.
5. Elute the analyte using 500 μL of acetonitrile: 100 mM triethylamine, 60:40 (w/v).

Part 3: Prepare Sample for LC-MS Analysis

1. Evaporate eluent from solid phase extraction to near dryness under nitrogen (approximately 2–3 μL).
2. Reconstitute sample in 100 μL of 7 mM methanol/triethylamine in water, 1:9 (v/v).
3. Vortex sample for 2 min.
4. Inject 40 μL into the LC-MS system.

Part 4: Liquid Chromatography Conditions

1. The LC mobile phase consists of A: 1.7 mM triethylamine and 100 mM hexafluoroisopropanol (pH 7.5) in water and B: methanol.
2. Liquid chromatography gradient follows the program (time (min), % mobile phase B): (0,8); (0.5,8); (10,28); (17,40); (17.5,8); (22.5:8).
3. Separation is conducted using a 50 mm × 2.1 mm Hypersil Gold C18 column packed with 1.9 μm particles. The column is maintained at 35 °C and the flow rate was 0.15 ml/min.
4. LC flow is directed into the mass spectrometer between 5 and 10 min and diverted to waste at all other times.
5. The autosampler needle is washed twice with water and methanol postinjection to reduce sample carry over.

Part 5: Mass Spectrometry Conditions

1. The mass spectrometer is operated in the negative ion mode using multiple-reaction monitoring (MRM) on a triple quadrupole mass analyzer.
2. The most abundant charge state of the oligonucleotide is selected using the first quadrupole and fragmented using Argon as the collision gas.
3. The MRM transition monitored for the analysis is from the molecular signal to the singly charged $[W_1 - H_2O]^-$ ion.
4. The general instrument conditions are as follows: declustering potential (DP) −10 V, focusing potential (FP) −80 V, entrance potential (EP) −10 V, collision energy (CE) −30 V, collision exit potential (CXP) −15 V, nebulizing gas (NEB) 13, curtain gas (CUR) 10, collision gas (CAD) 4, ionspray needle voltage (IS) −3000 V, temperature (TEM) 550 °C.

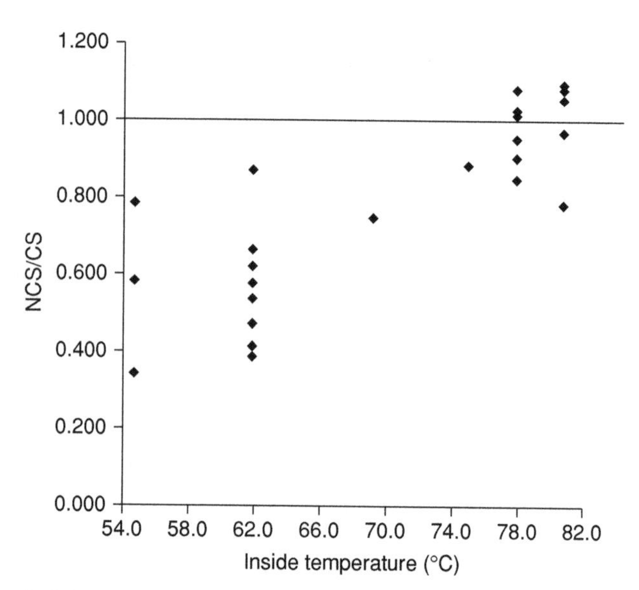

FIGURE 48.10 Plot of the ion abundance ratios from noncoding strand (NCS) and coding strand (CS) versus the temperature of the heated electrospray ionization source. The line indicates equal response from the NCS and the CS (adapted from Frahm et al. (2005)).

48.4.2 Small Interfering RNA and Micro RNA

The addition of the 2′-hydroxyl group in RNA causes it to be more hydrophilic than DNA. Therefore, chromatographic conditions that are used for DNA often need to be modified for RNA to maintain similar separation efficiencies. The need for more aqueous mobile phases for RNA analysis also leads to decreased ionization efficiency in electrospray and thus lower LC-MS response (Figure 48.10). In addition, RNA samples may be either single or double stranded. Chromatographic methods that will maintain noncovalently bound ds RNA are often quite different than those required for the determination of single-stranded RNA.

Protocol 2: Protocol for determination of single-stranded small interfering RNA and micro RNA from plasma (adapted from Ye and Beverly (2011)).

Part 1: Preparation of Magnetic Beads

1. 20 μL of strong anion exchange beads at a concentration of 50 mg/ml were equilibrated with 180 μL of water (pH 4.0, 0.1% Tween 20) to obtain a homogeneous suspension.
2. Bead solution was loaded onto a 96-well plate. The plate was placed into a magnetic-ring magnet stand for 2 min to bring the beads to the walls of the plate and allow the loading solution to be removed.

Part 2: Isolation of siRNA from Serum Sample

1. A 20 μL serum sample has 2 μL of internal standard added and is brought up to a total volume of 200 μL with pH 4.0 water.
2. Add a 200 μL diluted serum sample to the 96-well plate.
3. Mix sample by pipetting up and down 8 $\times$.
4. Allow sample to extract for 2 min.
5. Place sample in magnetic-ring magnet for 2 min to separate beads.
6. Remove supernatant from sample well.
7. Add 200 μL of pH 4.0, 0.01% Tween solution to wash beads.
8. Remove wash solution.
9. Elute siRNA from beads by adding 200 μL of 1 M Ammonium Chloride.
10. Samples were filtered using a 0.2 μm PVDF filter to ensure removal of magnetic beads prior to injection into the LC-MS system.

Part 3: LC-MS Determination of siRNA

1. Separation of the siRNA sample was conducted using a 2.1 $\times$ 50 mm Waters BEH C18 column with a 1.7 μm particle size.
2. The LC mobile phase consisted of A: 200 mM hexafluoroisopropanol/8.5 mM triethylamine in water and B: Methanol.
3. Liquid chromatography gradient used the following program (time (min), % mobile phase B): (0,1); (10,1); (20,8.5).
4. The column was heated to 75°C to denature the double stranded siRNA allowing each strand to be chromatographically resolved.
5. The flow rate was 300 μL per min and the injection volume was 10 μL.
6. The time-of-flight mass spectrometer was operated in negative ion mode using electrospray ionization.
7. The capillary voltage was set to 4000 V, the cone voltage was 30 V, the extraction voltage was 3 V, the source temperature was 125 °C and the desolvation temperature was 350 °C. The cone gas flow was 40 L/hr and the desolvation gas flow was set to 600 l/hr.

48.4.3 Transfer and Ribosomal RNA

These groups of RNAs have been widely studied due to their involvement in protein biosynthesis and also because the nucleoside modifications have been proposed as a way to organize the phylogenetic connection between species (McCloskey et al., 2001; Cole et al., 2009). These oligonucleotides are larger than those described in the previous protocols and are all single stranded, although there is significant secondary and tertiary structure associated with these molecules. The pioneering work in this field was done, over the course of several decades, by McCloskey and coworkers (Crain and McCloskey, 1998). However, most of the recent work on RNA modifications has involved the use of LC-MS. These RNAs come in a variety of sizes from roughly 75-mers (tRNA) to 2900-mers (23S rRNA). For species with a known genome, the sequences of these RNAs are known with the exception of any modifications. However, it is the identity and location of these modifications that is needed to increase knowledge of the molecular mechanisms of protein biosynthesis.

Protocol 3: Protocol for the determination of transfer and ribosomal RNA (adapted from Castleberry and Limbach (2010)).

Part 1: Enzymatic Digestion of RNA

1. RNase T1 was precipitated from the original solution with acetone, resuspended, and eluted in 1 ml of 75% aqueous acetonitrile from a C18 solid phase extraction cartridge. Solutions were then taken to dryness and resuspended in sterile water.
2. For digestion, 500 units of RNase T1 was added to 10 μg of RNA and 5 μL of 220 mM ammonium acetate. The reaction was incubated for 2 hr at physiological temperature.
3. Following digestion, the samples were lyophilized and reconstituted in 16.3 mM triethylamine/400 mM hexafluoroisopropanol at pH 7.0 at a concentration of 0.5 μg/ml.

Part 2: LC-MS Determination of RNA Digestion Products

LC-MS Determination of siRNA

1. Separation of the RNA sample was conducted using a 1 $\times$ 150 mm Waters XTerra MS C18 column with a 3.5 μm particle size and a 50 μm pore size.
2. The LC mobile phase consisted of A: 400 mM hexafluoroisopropanol/16.3 mM triethylamine in water (pH 7.0) and B: Methanol.
3. Liquid chromatography gradient used the following program (time (min), % mobile phase B): (0,5); (5,20); (7,30); (50,95); (55,95); (55.1,5); (70,5).
4. The flow rate was 40 μL/min and the injection volume was 10 μL.
5. The FT-ICR mass spectrometer was operated in negative ion mode using electrospray ionization.

6. The capillary voltage was set to 4250 V, the source temperature was 325°C. The mass range of the instrument was restricted to *m/z* 360–2000 to avoid interference from the dimer of hexafluoroisopropanol.

48.5 TROUBLESHOOTING OLIGONUCLEOTIDE LC-MS BIOANALYSIS

When developing an LC-MS method there are always problems that can arise. In this section, we discuss key issues that uniquely impact LC-MS methods for oligonucleotides. Oligonucleotide separations are quite challenging and the interactions between the analyte and the stationary phase are complex. In addition, the need to interface the separation to a mass spectrometer restricts the choices of mobile phase composition in order to maintain the necessary volatility needed to support electrospray ionization. Liquid chromatographic separations of oligonucleotides have been primarily conducted using ion-exchange or ion-pairing mechanisms. While a few ion-exchange methods have been successfully interfaced with mass spectrometry, the number of salts used for displacing analytes that are compatible with mass spectrometry is limited. To date, none of the mass spectrometry compatible salts has been successfully used in an ion-exchange separation of oligonucleotides. Therefore, all current LC-MS methods for oligonucleotides have employed ion-pair chromatography.

48.5.1 Ion-Pair Selection

It is important to recognize that ion suppressing effects from the ion-pair agents are still possible with oligonucleotides. Gilar et al. addressed this and demonstrated that when moving from LC-UV methods to LC-MS methods, lower concentrations of both TEA and HFIP did improve signal intensity but there was a threshold where these concentrations became too low and signal intensity decreased due to inefficient analyte desorption (Fountain et al., 2003). While the optimized values for the HFIP/TEA buffer system have been proposed, the same rigor has not been applied to other combinations with HFIP. It is important to recognize this as several studies have demonstrated advantages for other amines over TEA for the LC-MS of oligonucleotides. Since oligonucleotides differ widely in their properties, it is likely that different cationic amines will be appropriate for different modifications and classes (DNA vs. RNA or ss vs. ds). From the limited number of studies using different cationic amines for the LC-MS of oligonucleotides, it is still difficult to determine general guidelines for the selection of cationic amines, however, 10–20 mM is likely a good starting point.

Over the past several years there has been a general trend away from using ion-pairing agents in the mobile phase of electrospray ionization methods due to ion suppression effects, which often lower the overall analyte response. Therefore to minimize the ion suppressing effects of ion-pair agent, LC columns have been covalently modified to incorporate the ion-pairing agent into the stationary phase of the column (Bicker et al., 2008). In the case of oligonucleotides though, the ion-pairing agent has an important role in enhancing analyte desorption and therefore, removing it from the mobile phase will significantly decrease analyte signal.

48.5.2 Solvent and Buffer Compatibility

The selection of solvents for LC-MS of oligonucleotides is primarily determined by the buffer system. Most electrospray ionization methods use acetonitrile for the organic portion of the mobile phase due to the increased sensitivity that is generally observed when using this solvent. However, almost all LC-MS methods for oligonucleotides use methanol since HFIP has limited solubility in acetonitrile (Apffel et al., 1997a, 1997b). The concentration of HFIP that is used is typically 100 mM or higher, so it is critical to keep this fact in mind when establishing your mobile phase. There are no solubility limitations on the choice of cationic amines when choosing the composition of the mobile phase.

While almost all LC-MS methods have stayed with combinations of water and methanol, it is worth noting that blended mixtures of methanol and acetonitrile can be used with HFIP. This would allow one to take advantage of the increased volatility of acetonitrile but maintain the solvating properties of methanol.

48.5.3 Stability and Storage

There are significant differences in the stability of various oligonucleotides. Overall, DNA molecules are relatively stable and can be stored for extended periods of time under either refrigerated (−4°C) or frozen (−20°C or −80°C) conditions. High salt content can destabilize ds oligonucleotides (as well as potentially contribute unwanted adduction in electrospray ionization). When dealing with RNA, there are many precautions that need to be taken to minimize the exposure of these oligonucleotides to ubiquitous RNases. It is important to clean bench spaces with a bleach solution and then ethanol and heat glassware that will be in contact with RNA containing solutions to at least 200°C for 4 h. RNase-free consumable products can also be purchased.

Like many different types of molecules, oligonucleotides are also subject to nonspecific binding to surfaces. Most of these losses are due to electrostatic interactions with the surfaces of containers, extraction devices or instrumentation. Glass surfaces can be particularly problematic, but this can often be reduced by silanizing the surface as was reported by Zhang et al. (2007). However, removing this concern is not as simple as switching from glass to plastic containers, as many plastic containers also have high nonspecific adsorption. In addition to containers, sample preparation processes with high surface areas, such as SPE may also be problematic. However, the magnitude of this effect will depend significantly on the surface chemistry of the SPE materials.

As is true with most bioanalytical methods, the use of inert surfaces on the interior of the liquid chromatography system, such as polyether ether ketone (PEEK), will minimize nonspecific adsorptive losses. Because of the significant amount of nonspecific losses in general, the use of higher concentrations of the internal standard or the introduction of additional oligonucleotides into the sample as sacrificial materials to cover these reactive surfaces can be beneficial.

48.5.4 Ion Suppression

A ubiquitous problem with LC-MS assays employing electrospray ionization is ion suppression. There are several sources of molecules that can cause ion suppression that need to be considered. One of the first is the ion-pairing agents that are needed for the chromatographic separations. Both HFIP and TEA or other amines can significantly suppress the signals observed. This effect was studied in detail by Fountain et al. (2003) who recommended significantly reducing the amount of HFIP from 400 mM to 100 mM and TEA from 16.3 mM to 8.6 mM.

In addition to the mobile phase, the choice of sample preparation may cause differences in analyte response due to other coeluting compounds that remain in the sample. The use of liquid–liquid extraction techniques like chloroform–phenol extractions while excellent at extracting oligonucleotides, are equally effective at extracting phospholipids. Therefore, the combination of SPE with liquid–liquid extraction may be beneficial in reducing the phospholipids from the sample (Zhang et al., 2007). In addition to the endogenous biomolecules, ion suppression can also result from sources such as the excipients used in drug formulations (Murugaiah et al., 2010). As mentioned in the previous section, the practice of using increased concentrations of internal standards or other sacrificial oligonucleotides to mask active surfaces may also be a source of ion suppression.

Finally, the polyanionic nature of oligonucleotides leads to adduction with a number of cations like sodium and potassium. If the sample preparation techniques are not able to reduce the concentration of these salts in the sample they will divide the signal over a greater number of ion channels, and significantly reduce method sensitivity. While this is not classically considered ion suppression, it has a similar effect. The presence of high concentrations of amines in the mobile phase such as TEA, piperidine, and imidazole, have been shown to dramatically reduce the presence of these alkali metal adducts and therefore increase method sensitivity (Greig and Griffey, 1995).

48.6 SUMMARY

When developing methods for the determination of oligonucleotides using electrospray ionization LC-MS there are a number of key factors. Many of these factors are unique to this class of biomolecules and failure to adequately control these factors can have adverse consequences on methods. The following represent the most significant features of these LC-MS methods.

- Ion-pair chromatography provides excellent chromatographic separation and mass spectrometric signal intensity. However, it is important to carefully evaluate both the type of ion pair and its concentration early in method development.
- Generally, the electrospray ionization response for DNA is greater than RNA due to its greater hydrophobicity.
- Sample preparation will likely involve a combination of liquid–liquid extraction and SPE to obtain both high recoveries and low ion suppression.
- Be aware of the limited solubility of many of the common mobile phase additives in acetonitrile. This is the reason that most LC-MS methods use methanol as the organic solvent.
- Significantly greater care is needed when handling RNA due to its susceptibility to degradation via RNases. Solutions and materials need to be sterile and free from RNases to ensure the robustness of methods.

REFERENCES

Addepalli B, Limbach PA. Mass spectrometry-based quantification of pseudouridine in RNA. J Am Soc Mass Spectrom 2011;22(8):1363–1372.

Amarzguioui M, Holen T, Babaie E, Prydz H. Tolerance for mutations and chemical modifications in a siRNA. Nucl Acids Res 2003;31(2):589–595.

Apffel A, Chakel J, Fischer S, Lichtenwalter K, Hancock WS. Analysis of oligonucleotides by HPLC–electrospray ionization mass spectrometry. Anal Chem 1997a;69(7):1320–1325.

Apffel A, Chakel J, Fischer S, Lichtenwalter K, Hancock WS. New procedure for the use of high-performance liquid chromatography–electrospray ionization mass spectrometry for the analysis of nucleotides and oligonucleotides. J Chromatogr A 1997b;777(1):3–21.

Arora V, Knapp DC, Reddy MT, Weller DD, Iversen PL. Bioavailability and efficacy of antisense morpholino oligomers targeted to c-myc and cytochrome P-450 3A2 following oral administration in rats. J Pharm Sci 2002;91(4):1009–1018.

Bellon L, Maloney L, Zinnen SP, Sandberg JA, Johnson KE. Quantitative determination of a chemically modified hammerhead ribozyme in blood plasma using 96-well solid-phase extraction coupled with high-performance liquid chromatography or capillary gel electrophoresis. Anal Biochem 2000;283(2):228–240.

Beverly M, Hartsough K, Machemer L. Liquid chromatography/electrospray mass spectrometric analysis of metabolites from an inhibitory RNA duplex. Rapid Commun Mass Spectrom 2005;19(12):1675–1682.

Beverly M, Hartsough K, Machemer L, Pavco P, Lockridge J. Liquid chromatography electrospray ionization mass spectrometry analysis of the ocular metabolites from a short interfering RNA duplex. J Chromatogr B Analyt Technol Biomed Life Sci 2006;835(1–2):62–70.

Beverly MB. Applications of mass spectrometry to the study of siRNA. Mass Spectrom Rev 2011;30(6):979–998.

Bicker W, Lammerhofer M, Lindner W. Mixed-mode stationary phases as a complementary selectivity concept in liquid chromatography-tandem mass spectrometry-based bioanalytical assays. Anal Bioanal Chem 2008;390:263–266.

Bothner B, Chatman K, Sarkisian M, Siuzdak G. Liquid chromatography mass spectrometry of antisense oligonucleotides. Bioorg Med Chem Lett 1995;5(23):2863–2868.

Braasch DA, Jensen S, Liu Y, et al. RNA interference in mammalian cells by chemically-modified RNA. Biochemistry 2003;42(26):7967–7975.

Bruenger E, Kowalak JA, Kuchino Y, et al. 5S rRNA modification in the hyperthermophilic archaea Sulfolobus solfataricus and Pyrodictium occultum. FASEB J 1993;7(1):196–200.

Cantara WA, Crain PF, Rozenski J, et al. The RNA modification database, RNAMDB: 2011 update. Nucleic Acids Res 2011;39(Suppl 1):D195–D201.

Castleberry C, Limbach P. Relative quantitation of transfer RNAs using liquid chromatography mass spectrometry and signature digestion products. Nucleic Acids Res 2010;38(16): e162.

Chan CT, Dyavaiah M, DeMott MS, Taghizadeh K, Dedon PC, Begley TJ. Correction: a quantitative systems approach reveals dynamic control of tRNA modifications during cellular stress. PLoS Genet 2011;7(2). DOI: 10.1371/annotation/6549d0b1-efde-4aa4-9cda-1cef43f66b30.

Chen C, Ridzon D, Broomer AJ, et al. Real-time quantification of microRNAs by stem-loop RT-PCR. Nucleic Acids Res 2005;33(20):e179.

Chen SH, Qian M, Brennan JM, Gallo JM. Determination of antisense phosphorothioate oligonucleotides and catabolites in biological fluids and tissue extracts using anion-exchange high-performance liquid chromatography and capillary gel electrophoresis. J Chromatogr B Biomed Sci Appl 1997;692(1):43–51.

Cheng A, Li M, Liang Y, et al. Stem-loop RT-PCR quantification of siRNAs in vitro and in vivo. Oligonucleotides 2009;19(2):203–208.

Chiu Y, Rana T. siRNA function in RNAi: a chemical modification analysis. RNA 2003;9(9):1034–1048.

De Clerq E, Eckstein F, Merigan TC. Interferon induction increased through chemical modification of a synthetic polyribonucleotide. Science 1969;165(3898):1137–1139.

Cole J, Wang Q, Cardenas E, et al. The Ribosomal Database Project: improved alignments and new tools for rRNA analysis. Nucleic Acids Res 2009;37:D141–D145.

Crain PF, McCloskey JA. Applications of mass spectrometry to the characterization of oligonucleotides and nucleic acids. Curr Opin Biotech 1998;9(1):25–34.

Crooke RM, Graham MJ, Martin MJ, Lemonidis KM, Wyrzykiewiecz T, Cummins LL. Metabolism of antisense oligonucleotides in rat liver homogenates. J Pharmacol Exp Ther 2000;292(1):140–149.

Crooke ST. Antisense strategies. Curr Mol Med 2004;4(5):465–487.

Dai G, Wei X, Liu Z, Liu S, Marcucci G, Chan KK. Characterization and quantification of Bcl-2 antisense G3139 and metabolites in plasma and urine by ion-pair reversed phase HPLC coupled with electrospray ion-trap mass spectrometry. J Chromatogr B Analyt Technol Biomed Life Sci 2005;825(2):201–213.

Davis S, Lollo B, Freier S, Esau C. Improved targeting of miRNA with antisense oligonucleotides. Nucleic Acids Res 2006;34(8):2294–2304.

De Fougerolles A, Vornlocher HP, Maraganorc J, Lieberman J. Interfering with disease: a progress report on siRNA-based therapeutics. Nat Rev Drug Discov 2007;6(6):443–453.

de la Mora F. Electrospray ionization of large multiply charged species proceeds via Dole's charged residue mechanism. Anal Chim Acta 2000;406(1):93–104.

De Paula D, Bentley MV, Mahato RI. Hydrophobization and bioconjugation for enhanced siRNA delivery and targeting. RNA 2007;13(4):431–456.

Deng P, Chen X, Zhang G, Zhong D. Bioanalysis of an oligonucleotide and its metabolites by liquid chromatography-tandem mass spectrometry. J Pharmaceut Biomed Anal 2010;52(4):571–579.

Dickman M. Effects of sequence and structure in the separation of nucleic acids using ion pair reverse phase liquid chromatography. J Chromatogr A 2005;1076(1–2):83–89.

Dickman M, Hornby D. Enrichment and analysis of RNA centered on ion pair reverse phase methodology. RNA 2006;12(4):691–696.

Durairaj A, Limbach PA. Mass spectrometry of the fifth nucleoside: a review of the identification of pseudouridine in nucleic acids. Anal Chim Acta 2008;623(2):117–125.

Easter RN, Kröning KK, Caruso JA, Limbach PA. Separation and identification of oligonucleotides by hydrophilic interaction liquid chromatography (HILIC) inductively coupled plasma mass spectrometry (ICPMS). Analyst 2010;135:2560–2565.

Eckstein F. Phosphorothioate oligodeoxynucleotides: What is their origin and what is unique about them? Antisense Nucleic Acid Drug Dev 2000;10(2):117–121.

Farand J, Beverly M. Sequence confirmation of modified oligonucleotides using chemical degradation, electrospray ionization, time-of-flight, and tandem mass spectrometry. Anal Chem 2008;80(19):7414–7421.

Farand J, Gosselin F. De novo sequence determination of modified oligonucleotides. Anal Chem 2009;81(10):3723–3730.

Fountain K, Gilar M, Gilar M. Analysis of native and chemically modified oligonucleotides by tandem ion-pair reversed-phase high-performance liquid chromatography/electrospray ionization mass spectrometry. Rapid Commun Mass Spectrom 2003;17(7):646–653.

Frahm JL, Muddiman DC, Burke MJ. Leveling response factors in the electrospray ionization process using a heated capillary interface. J Am Soc Mass Spectrom 2005;16(5):772–778.

Fritz H, Belagaje R, Brown EL, et al. High-pressure liquid chromatography in polynucleotide synthesis. Biochemistry 1978;17(7):1257–1267.

Gaus HJ, Owens SR, Winniman M, Cooper S, Cummins LL. On-line HPLC electrospray mass spectrometry of phosphorothioate oligonucleotide metabolites. Anal Chem 1997;69(3):313–319.

Geary RS, Khatsenko O, Bunker K, et al. Absolute bioavailability of 2′-O-(2-methoxyethyl)-modified antisense oligonucleotides following intraduodenal instillation in rats. J Pharmacol Exp Ther 2001;296(3):898–904.

Geary RS, Watanabe TA, Truong L, et al. Pharmacokinetic properties of 2′-O-(2-methoxyethyl)-modified oligonucleotide analogs in rats. J Pharmacol Exp Ther 2001;296(3):890–897.

Gilar M. Analysis and purification of synthetic oligonucleotides by reversed-phase high-performance liquid chromatography with photodiode array and mass spectrometry detection. Anal Biochem 2001;298(2):196–206.

Gilar M, Belenky A, Smisek DL, Borque A, Cohen AS. Kinetics of phosphorothioate oligonucleotide metabolism in biological fluids. Nucleic Acids Res 1997;25(18):3615–3620.

Gilar M, Fountain K, Budman Y, Holyoke JL, Davoudi H, Gebler JC. Characterization of therapeutic oligonucleotides using liquid chromatography with on-line mass spectrometry detection. Oligonucleotides 2003;13(4):229–243.

Gilar M, Fountain KJ, Budman Y, et al. Ion-pair reversed-phase high-performance liquid chromatography analysis of oligonucleotides: retention prediction. J Chromatogr A 2002;958(1–2):167–182.

Greig M, Griffey R. Utility of organic bases for improved electrospray mass spectrometry of oligonucleotides. Rapid Commun Mass Spectrom 1995;9(1):97–102.

Griffey R, Greig M, Gaus HJ, et al. Characterization of oligonucleotide metabolism in vivo via liquid chromatography/electrospray tandem mass spectrometry with a quadrupole ion trap mass spectrometer. J Mass Spectrom 1997;32(3):305–313.

Griffey R, Sasmor H, Greig MJ. Oligonucleotide charge states in negative ionization electrospray-mass spectrometry are a function of solution ammonium ion concentration. J Am Soc Mass Spectr 1997;8(2):155–160.

Guymon R, Pomerantz SC, et al. Influence of phylogeny on post-transcriptional modification of rRNA in thermophilic prokaryotes: the complete modification map of 16S rRNA of Thermus thermophilus. Biochemistry 2006;45(15):4888–4899.

Guymon R, Pomerantz SC, Ison JN, Crain PF, McCloskey JA. Post-transcriptional modifications in the small subunit ribosomal RNA from Thermotoga maritima, including presence of a novel modified cytidine. RNA 2007;13(3):396–403.

Hölzl G, Oberacher H, Pitsch S, Stutz A, Huber CG. Analysis of biological and synthetic ribonucleic acids by liquid chromatography- mass spectrometry using monolithic capillary columns. Anal Chem 2005;77(2):673–680.

Hossain M, Limbach P. Mass spectrometry-based detection of transfer RNAs by their signature endonuclease digestion products. RNA 2007;13(2):295–303.

Huang TY, Liu J, McLuckey SA. Top-down tandem mass spectrometry of tRNA via ion trap collision-induced dissociation. J Am Soc Mass Spectrom 2010;21(6):890–898.

Huber C, Krajete A. Analysis of nucleic acids by capillary ion-pair reversed-phase HPLC coupled to negative-ion electrospray ionization mass spectrometry. Anal Chem 1999;71(17):3730–3739.

Huber C, Krajete A. Comparison of direct infusion and on line liquid chromatography/electrospray ionization mass spectrometry for the analysis of nucleic acids. J Mass Spectrom 2000a;35(7):870–877.

Huber C, Krajete, A. Sheath liquid effects in capillary high-performance liquid chromatography-electrospray mass spectrometry of oligonucleotides. J Chromatogr A 2000b;870(1–2):413–424.

Huber C, Oberacher H. Analysis of nucleic acids by on-line liquid chromatography–mass spectrometry. Mass Spectrom Rev 2001;20(5):310–343.

Huber C, Oefner P, Bonn GK. High-performance liquid chromatographic separation of detritylated oligonucleotides on highly cross-linked poly-(styrene-divinylbenzene) particles. J Chromatogr A 1992;599(1–2):113–118.

Huber C, Oefner P, et al. High-resolution liquid chromatography of oligonucleotides on nonporous alkylated styrene-divinylbenzene copolymers. Anal Biochem 1993;212(2):351–358.

Issaq HJ, Blonder, J. Electrophoresis and liquid chromatography/tandem mass spectrometry in disease biomarker discovery. J Chromatogr B Analyt Technol Biomed Life Sci 2009;877(13):1222–1228.

Ivleva V, Yu Y, Gilar M. Ultra performance liquid chromatography/tandem mass spectrometry (UPLC/MS/MS) and UPLC/MSE analysis of RNA oligonucleotides. Rapid Commun Mass Spectrom 2010;24(17):2631–2640.

Johnson JL, Guo W, Zang J, et al. Quantification of raf antisense oligonucleotide (rafAON) in biological matrices by LC MS/MS to support pharmacokinetics of a liposome entrapped rafAON formulation. Biomed Chromatogr 2005;19(4):272–278.

Kamel A, Brown P. High-performance liquid chromatography versus capillary electrophoresis for the separation and analysis of oligonucleotides. Am Lab 1996;28:40–51.

Kebarle P, Peschke M. On the mechanisms by which the charged droplets produced by electrospray lead to gas phase ions. Anal Chim Acta 2000;406(1):11–35.

Kenski DM, Cooper AJ, Li JJ, et al. Analysis of acyclic nucleoside modifications in siRNAs finds sensitivity at position 1 that is restored by 5′-terminal phosphorylation both in vitro and in vivo. Nucl Acids Res 2009;38(2):660–671.

Kim E-J, Park TG, Oh Y-K, Shim C-K. Assessment of siRNA pharmacokinetics using ELISA-based quantification. J Control Release 2009;143(2):660–671.

Kowalak JA, Bruenger E, Crain PF, McCloskey JA. Identities and phylogenetic comparisons of posttranscriptional modifications in 16 S ribosomal RNA from *Haloferax volcanii*. J Biol Chem 2000;275(32):24484–24489.

Kowalak JA, Bruenger E, et al. Structural characterization of U*-1915 in domain IV from Escherichia coli 23S ribosomal RNA as 3-methylpseudouridine. Nucleic Acids Res 1996;24(4):688–693.

Kowalak JA, Bruenger E, Hashizume T, Peltier JM, Ofengand J, McCloskey JA. Posttranscriptional modification of the central loop of domain V in Escherichia coli 23 S ribosomal RNA. J Biol Chem 1995;270(30):17758–17764.

Kroh E, Parkin R, Mitchell PS, Tewari M. Analysis of circulating microRNA biomarkers in plasma and serum using quantitative reverse transcription-PCR (qRT-PCR). Methods 2010;50(4):298–301.

Leeds JM, Graham MJ, Truong L, Cummins LL. Quantitation of phosphorothioate oligonucleotides in human plasma. Anal Biochem 1996;235(1):36–43.

Lin Z, Li W, Dai G. Application of LC-MS for quantitative analysis and metabolite identification of therapeutic oligonucleotides. J Pharmaceut Biomed Anal 2007;44(2):330–341.

Liu W, Stevenson M, Seymour LW, Fisher KD. Quantification of siRNA using competitive qPCR. Nucleic Acids Res 2008;37(1):e4.

Mack LL, Kralik P, Rheude A, Dole M. Molecular beams of macroions. II. J Chem Phys 1970;52:4977.

Matthiesen R, Kirpekar F. Identification of RNA molecules by specific enzyme digestion and mass spectrometry: software for and implementation of RNA mass mapping. Nucleic Acids Res 2009;37(6):e48.

McCarthy SM, Gilar M, Gebler. Reversed-phase ion-pair liquid chromatography analysis and purification of small interfering RNA. Anal Biochem 2009;390(2):181–188.

McCloskey JA, Graham DE, Zhou S, et al. Post-transcriptional modification in archaeal tRNAs: identities and phylogenetic relations of nucleotides from mesophilic and hyperthermophilic Methanococcales. Nucleic Acids Res 2001;29(22):4699–4706.

McLuckey S. Principles of collisional activation in analytical mass spectrometry. J Am Soc Mass Spectrom 1992;3(6):599–614.

McLuckey S, Habibi-Goudarzi S. Decompositions of multiply charged oligonucleotide anions. J Am Chem Soc 1993;115(25):12085–12095.

Miller PS, McParland KB, Jayaraman K, Ts'o PO. Biochemical and biological effects of nonionic nucleic acid methylphosphonates. Biochemistry 1981;20(7):1874–1880.

Muddiman D, Cheng X, Udseth HR, Smith RD. Charge-state reduction with improved signal intensity of oligonucleotides in electrospray ionization mass spectrometry. J Am Soc Mass Spectrom 1996;7(8):697–706.

Murugaiah V, Zedalis W, Lavine G, Charisse K, Manoharan M. Reversed-phase high-performance liquid chromatography method for simultaneous analysis of two liposome-formulated short interfering RNA duplexes. Anal Biochem 2010;401(1):61–67.

Nguyen S, Fenn J. Gas-phase ions of solute species from charged droplets of solutions. Proc Natl Acad Sci USA 2007;104(4):1111.

Ni J, Pomerantz SC, Rozenski J, Zhang Y, McCloskey JA. Interpretation of oligonucleotide mass spectra for determination of sequence using electrospray ionization and tandem mass spectrometry. Anal Chem 1996;68(13):1989–1999.

Noon KR, Bruenger E, McCloskey JA. Posttranscriptional modifications in 16S and 23S rRNAs of the archaeal hyperthermophile Sulfolobus solfataricus. J Bacteriol 1998;180(11):2883–2888.

Null AP, Nepomuceno AI, Muddiman C. Implications of hydrophobicity and free energy of solvation for characterization of nucleic acids by electrospray ionization mass spectrometry. Anal Chem 2003;75(6):1331–1339.

Oberacher H, Oefner P, Parson W, Huber CG. On line liquid chromatography mass spectrometry: a useful tool for the detection of DNA sequence variation. Angew Chem Int Edit 2001;40(20):3828–3830.

Oberacher H, Parson W, Muhlmann R, Huber CG. Analysis of polymerase chain reaction products by on-line liquid chromatography- mass spectrometry for genotyping of polymorphic short tandem repeat loci. Anal Chem 2001;73(21):5109–5115.

Overhoff M, Wunsche W, Sczkiel G. Quantitative detection of siRNA and single-stranded oligonucleotides: relationship between uptake and biological activity of siRNA. Nucleic Acids Res 2004;32(21):e170.

Pomerantz SC, Kowalak J, McCloskey JA. Determination of oligonucleotide composition from mass spectrometrically measured molecular weight. J Am Soc Mass Spectrom 1993;4(3):204–209.

Pomerantz SC, McCloskey J. Analysis of RNA hydrolyzates by liquid chromatography–mass spectrometry. Method Enzymol 1990;193:796–824.

Pomerantz SC, McCloskey J. Detection of the common RNA nucleoside pseudouridine in mixtures of oligonucleotides by mass spectrometry. Anal Chem 2005;77(15):4687–4697.

Prakash TP, Allerson CR, Dande P, et al. Positional effect of chemical modifications on short interference RNA activity in mammalian cells. J Med Chem 2005;48(13):4247–4253.

Raynaud F, Orr R, Goddard PM, et al. Pharmacokinetics of G3139, a phosphorothioate oligodeoxynucleotide antisense to bcl-2, after intravenous administration or continuous subcutaneous infusion to mice. J Pharmacol Exp Ther 1997;281(1):420–427.

Ro S, Park C, Jin J, Sanders KM, Yan W. A PCR-based method for detection and quantification of small RNAs. Biochem Biophys Res Commun 2006;351(3):756.

Rozenski J, McCloskey JA. 1999. From http://library.med.utah.edu/masspec/mongo.htm.

Rozenski J, McCloskey JA. SOS: a simple interactive program for ab initio oligonucleotide sequencing by mass spectrometry. J Am Soc Mass Spectrom 2002;13(3):200–203.

Seiffert S, Debelak H, Hadwiger P, et al. Characterization of side reactions during the annealing of small interfering RNAs. Anal Biochem 2011;414(1):47–57.

Shuford CM, Muddiman DC. Capitalizing on the hydrophobic bias of electrospray ionization through chemical modification in mass spectrometry-based proteomics. Expert Rev Proteomics 2011;8(3):317–323.

Spahr P, Hollingworth B. Purification and mechanism of action of ribonuclease from Escherichia coli ribosomes. J Biol Chem 1961;236(3):823–831.

Srivatsa GS, Klopchin P, Batt M, Feldman M, Carlson RH, Cole DL. Selectivity of anion exchange chromatography and capillary gel electrophoresis for the analysis of phosphorothioate oligonucleotides. J Pharm Biomed Anal 1997;16(4):619–630.

Stein CA, Subasinghe C, Shinozuka K, Cohen JS. Physicochemical properties of phospborothioate oligodeoxynucleotides. Nucleic Acids Res 1988;16(8):3209–3221.

Stratford S, Stec S, Jadhav V, Seitzer J, Abrams M, Beverly M. Examination of real-time polymerase chain reaction methods for the detection and quantification of modified siRNA. Anal Biochem 2008;379(1):96–104.

Taucher M, Breuker K. Top-down mass spectrometry for sequencing of larger (up to 61 nt) RNA by CAD and EDD. J Am Soc Mass Spectrom 2010;21(6):918–929.

Tijsen A, Creemers E, Moerland PD, et al. MiR423–5p as a circulating biomarker for heart failure. Circ Res 2010;106(6):1035–1039.

van Breemen RB, Martin LRB, Le JC. Continuous-flow fast atom bombardment mass spectrometry of oligonucleotides. J Am Soc Mass Spectrom 1991;2(2):157–163.

Vollmer DL, Gross ML. Cation-exchange resins for removal of alkali metal cations from oligonucleotide samples for fast atom bombardment mass spectrometry. J Mass Spectrom 1995;30(1):113–118.

Wang G, Cole RB. Charged residue versus ion evaporation for formation of alkali metal halide cluster ions in ESI. Anal Chim Acta 2000;406(1):53–65.

Waters J, Webb A, Cunningham D, et al. Phase I clinical and pharmacokinetic study of bcl-2 antisense oligonucleotide therapy in patients with non-Hodgkin's lymphoma. J Clin Oncol 2000;18(9):1812.

Watson JT, Sparkman OD. *Introduction to Mass Spectrometry: Instrumentation, Applications, and Strategies for Data Interpretation.* 4th ed. Chichester, UK: John Wiley & Sons, Ltd.; 2008.

Watts J, Deleavey G, Damha MJ. Chemically modified siRNA: tools and applications. Drug Discov Today 2008;13(19–20):842–855.

Wei X, Dai G, Marcucci G, et al. A specific picomolar hybridization-based ELISA assay for the determination of phosphorothioate oligonucleotides in plasma and cellular matrices. Pharmaceut Res 2006;23(6):1251–1264.

Whitehouse C, Dreyer R, Yamashita M, Fenn JB. Electrospray interface for liquid chromatographs and mass spectrometers. Anal Chem 1985;57(3):675–679.

Ye G, Beverly M. The use of strong anion exchange (SAX) magnetic particles for the extraction of therapeutic siRNA and their analysis by liquid chromatography/mass spectrometry. Rapid Commun Mass Spectrom 2011;25(21):3207–3215.

Yu RZ, Geary RS, Levin AA. Application of novel quantitative bioanalytical methods for pharmacokinetic and pharmacokinetic/pharmacodynamic assessments of antisense oligonucleotides. Curr Opin Drug Discov Devel 2004;7(2):195–203.

Zhang G, Lin J, Srinivasan K, Kavetskaia O, Duncan JN. Strategies for bioanalysis of an oligonucleotide class macromolecule from rat plasma using liquid chromatography–tandem mass spectrometry. Anal Chem 2007;79(9):3416–3424.

Zou Y, Tiller P, Chen IW, Beverly M, Hochman J. Metabolite identification of small interfering RNA duplex by high resolution accurate mass spectrometry. Rapid Commun Mass Spectrom 2008;22(12):1871–1881.

49

LC-MS BIOANALYSIS OF PLATINUM DRUGS

TROY VOELKER AND MIN MENG

49.1 INTRODUCTION

Platinum (Pt) drugs are used to treat various types of cancers, including sarcomas, some carcinomas (e.g., small-cell lung cancer and ovarian cancer), lymphomas, and germ cell tumors. Platinum drugs react *in vivo* with DNA via inter- and intrastrand crosslink, interfere with gene transcription and DNA replication, ultimately trigger apoptosis (programmed cell-death) (Rosenberg et al., 1969). Although platinum drugs are among the most active chemotherapeutic agents available in clinic and they are effective against various cancers, their clinical usage is hindered by their toxic side effects and drug resistance. Short-term exposure to platinum drugs could trigger abdominal pain, diarrhea, and emesis. Prolonged exposure of platinum drugs could result in nephrotoxicity in which the platinum drugs accumulate in the renal tubular fluid. Currently, there are two platinum drugs that are clinically used worldwide as antitumor drugs: cisplatin and carboplatin. Early research showed that, compared to cisplatin, carboplatin offers similar efficacy but less nonhematological toxicity and significantly better patient quality of life. Currently, the research for new platinum drugs has focused on improving tolerability and resistance. New platinum drugs under development that show promising results in ovarian cancer are oxaliplatin, nedaplatin, lobaplatin, satraplatin (JM216), BBR3464, and ZD0473 (Manzotti et al., 2000; Barefoot, 2001; Piccart et al., 2001; Boulikas and Vougiouka, 2003).

As shown in Figure 49.1, all platinum drugs are metal complexes composed of a central platinum moiety and multiple inorganic or organic ligands. Cisplatin is the simplest platinum drug in which the platinum atom is bound to two amine groups and two chloride leaving groups. Carboplatin, oxaliplatin, nedaplatin, lobaplatin, satraplatin (JM216), and ZD0473 are derivatives of cisplatin in which the amine ligands are replaced and modified by various organic groups. In contrast, BBR3464 is a novel cationic trinuclear platinum which is structurally distinctive from other platinum drugs.

49.2 METHODS AND APPROACHES

49.2.1 Principles

Quantitative analysis of platinum drugs in biological matrices is critically important for both clinical diagnosis and new drug developments. Currently, there are three approaches to measure platinum drugs in biological matrix: (1) to measure the amount of platinum-DNA adducts; (2) to measure the amount of metal platinum; and (3) to measure the amount of intact platinum drugs.

All platinum drugs form inter- and intrastrand crosslinking DNA adducts to a certain extent. For example, approximately 65% of cisplatin binds to 1,2-guanine-guanine to form CP-d(GpG) adduct *in vivo*. Pt-DNA adducts are excellent biomarkers for the measurement of platinum drugs. The common methods used to measure Pt-DNA adducts are antibody probe and ^{32}P postlabeling (Iijima et al., 2004; Baskerville-Abraham et al., 2009). Another alternative approach is to measure the content of the metal platinum using atomic absorption spectroscopy (AAS) and inductively coupled plasma–mass spectrometry (ICP-MS) (Bosch et al., 2008). The measurement of metal platinum represents the total amount of platinum *in vivo* regardless of the biological activity of certain forms of platinum format (free or bound or adduct). Between these two

Handbook of LC-MS Bioanalysis: Best Practices, Experimental Protocols, and Regulations, First Edition. Edited by Wenkui Li, Jie Zhang, and Francis L.S. Tse.

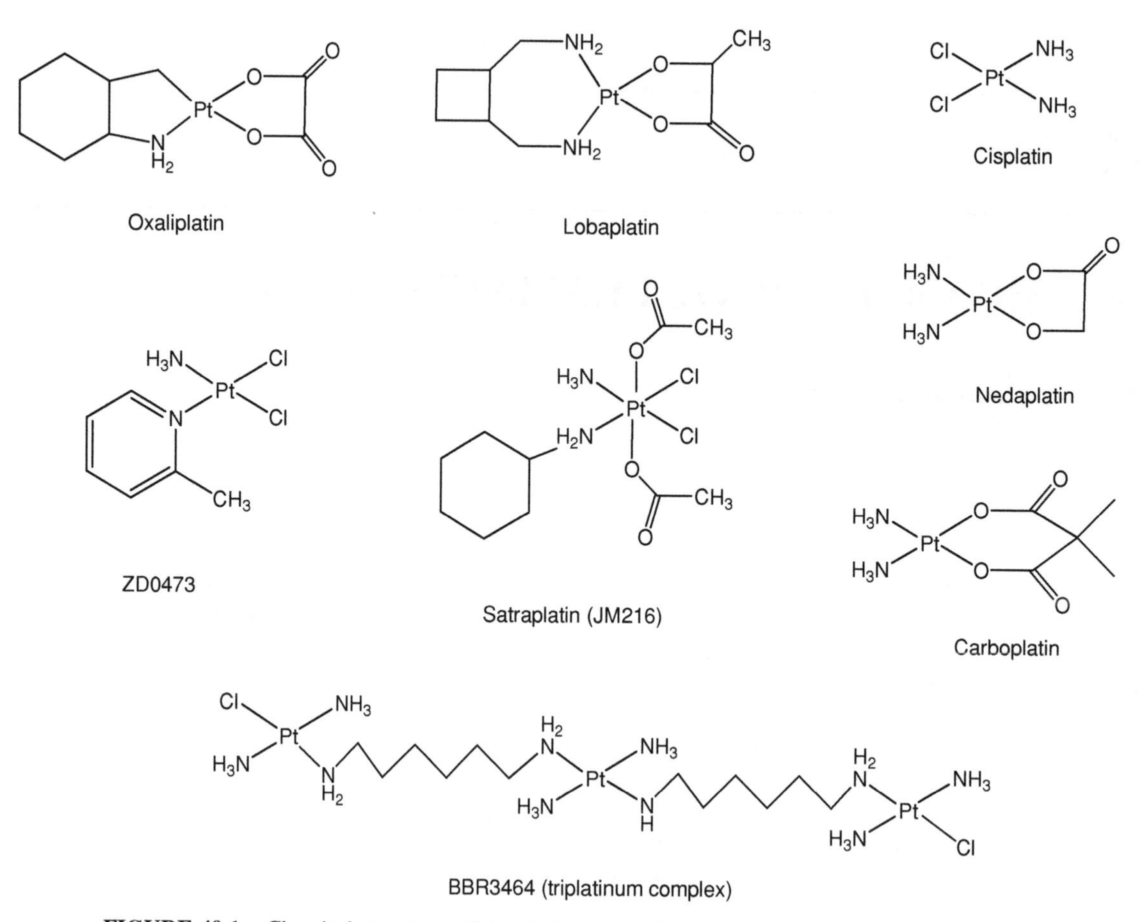

FIGURE 49.1 Chemical structures of the platinum complexes cisplatin, carboplatin, oxaliplatin, nedaplatin, lobaplatin, satraplatin (JM216), BBR3464, and ZD0473.

techniques, ICP–MS has higher sensitivity than AAS; thus it has been used to analyze clinical samples. Because bound platinum has no cytotoxicity, the measurement of intact platinum complex is important for clinical diagnosis and new drug development. Due to its inherent advantage of specificity, sensitivity, and high throughput, the most commonly used technique for the quantitation of intact platinum complex is liquid chromatography–tandem mass spectrometry (LC-MS/MS) (Guetins et al., 2002; Guo et al., 2003; Stokvis et al., 2005). This protocol focuses on the quantitation of intact platinum complex in biological matrix using LC-MS/MS.

49.2.2 Methodology

Platinum drugs are very polar compounds. It cannot be extracted directly via liquid–liquid extraction (LLE) unless derivatization is involved. Because the therapeutical level of platinum drugs is relatively high, solid phase extraction (SPE) which typically yields cleaner extract is unnecessary, thus seldom implemented. The most commonly used extraction procedures for platinum drugs are dilution extraction and protein precipitation extraction (PPE).

Over the years, several LC-MS/MS methodologies have been developed in our laboratory using PPE or dilution extraction (Meng et al., 2003; Meng et al., 2006; Liu et al., 2011; Protocol 1 and Protocol 2) to quantify carboplatin, oxaliplatin, and cisplatin in which the matrix was either human plasma or human plasma ultrafiltrate. Plasma ultrafiltrate provides true unbound measurement of platinum drugs. It can be generated from plasma by transferring blank human plasma into Amicon Microcon centrifugal filter devices (Microcon® YM-30) and followed by centrifuging at approximately 6000*g* for 10 min. Because plasma ultrafiltrate is virtually 100% free of major and macro endogenous molecules, the most efficient and convenient approach to analyze plasma ultrafiltrate sample is to dilute sample with buffer or water and inject directly onto a reverse-phase LC column (Protocol 2). In contrast, plasma samples need further sample cleanup due to the presence of proteins and other macromolecules. The most efficient and convenient approach to analyze plasma samples is to precipitate them with organic solvent such as acetonitrile (ACN) and inject

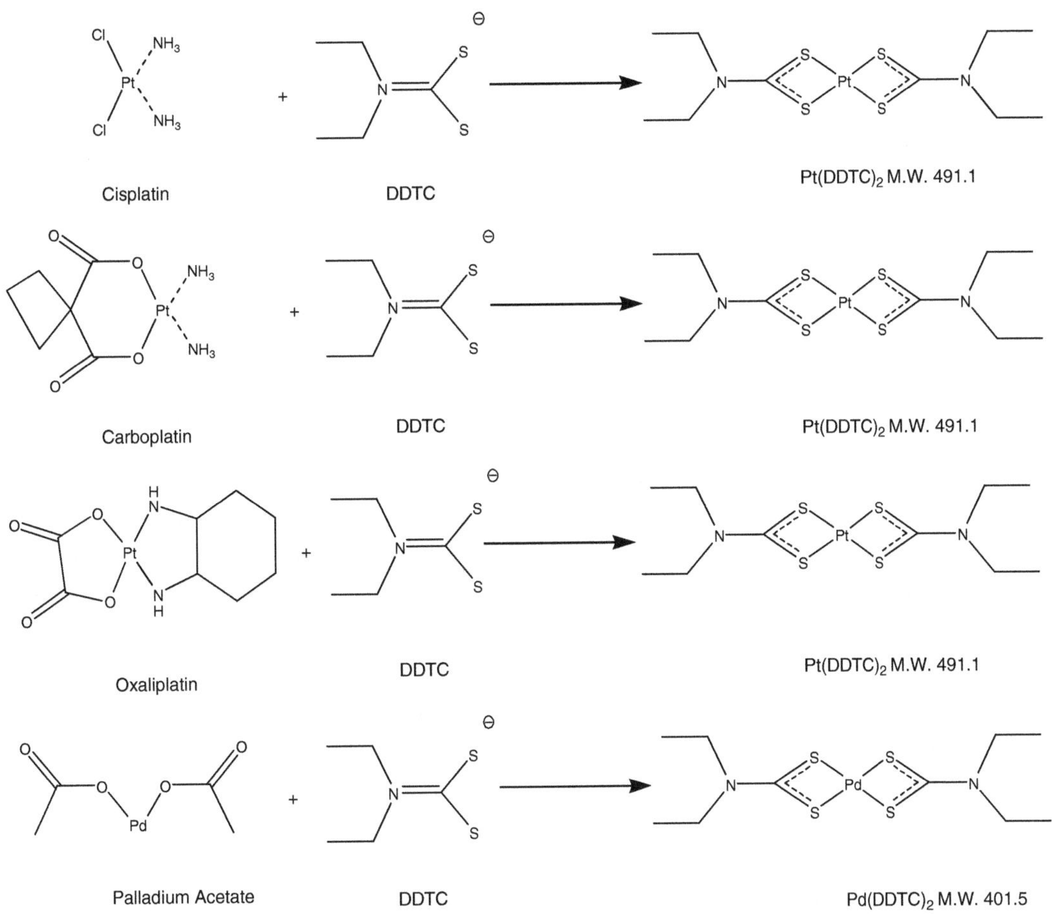

Cisplatin + DDTC → $Pt(DDTC)_2$ M.W. 491.1

Carboplatin + DDTC → $Pt(DDTC)_2$ M.W. 491.1

Oxaliplatin + DDTC → $Pt(DDTC)_2$ M.W. 491.1

Palladium Acetate + DDTC → $Pd(DDTC)_2$ M.W. 401.5

FIGURE 49.2 Chemical structures of cisplatin, carboplatin, oxaliplatin, palladium (II) acetate, and diethyldithiocarbamate (DDTC).

directly on a normal phase LC column (Protocol 1). Both the approaches omit an evaporation step and thus improve the efficiency of sample preparation.

Although similar methodologies can be applied to carboplatin and oxaliplatin, it was found that the quantitation of cisplatin is very challenging using LC-MS/MS directly. Infusion of three platinum complexes into LC-MS/MS directly showed that only carboplatin and oxaliplatin can be directly ionized. There is no detectable ion found for cisplatin regardless of the types of solvents and acid modifiers utilized necessitating the need to make a derivative of cisplatin for LC-MS/MS quantitation. A novel methodology has been developed using diethyldithiocarbamate (DDTC) derivative reagent followed by LLE to quantify platinum from cisplatin (Andrews et al., 1984; Meng et al., 2006). Because DDTC is a strong chelating reagent, it replaces all inorganic or organic ligands and form Pt-$(DDTC)_4$ exclusively (Figure 49.2). Thus, the same methodology can be used to quantify other platinum drug complexes such as carboplatin and oxaliplatin.

49.3 TROUBLESHOOTING

Oxaliplatin is not stable in biological matrix and water at room temperature. For the quantitation of oxaliplatin, the sample preparation has to be conducted on a wet-ice bath. The evaporation or concentration of the extract under elevated temperature was prohibited. The final extract should be kept refrigerated while waiting for analysis with a chilled autosampler.

While dilution extraction is the fastest extraction procedure, it also yields the dirtiest extract. Therefore, it is very critical to achieve sufficient separation and retention on the LC column. It is recommended that the minimal K′ of the platinum compounds should be >2.0. Additionally, in order to keep the MS interface clean, it may be necessary to incorporate divert switching valve to the LC program so that the early unretentive eluent is diverted to waste and not to the instrument. This setup could avoid decrease of the signal and divergent curves due to buildup on the face plate.

Because platinum complexes are extremely polar, their retention on reverse-phase columns is very challenging. The mobile phase usually contains very high aqueous contents, even up to 100% aqueous. Therefore, the preferred reverse-phase columns are endcapped or polar embedded which can tolerate high amount of aqueous mobile phase. In both the applications (Meng et al., 2003; Protocol 2), either Metasil Inertsil® ODS-2 column or Inertsil C4 column was utilized to obtain sufficient retention and K'.

Another approach to achieve retention is to use normal-phase LC column as shown in two applications (Liu et al., 2011; Protocol 1). Normal phase chromatography is very practical for MeCN extracts from PPE for plasma assays. However, it was found that phospholipids, one of the major endogenous matrix effect components in biological matrix, could cause potential problems (Bennett and Van Horne, 2003; Liu et al., 2011). In Liu's application, phospholipids coelute with carboplatin under initial LC condition. New LC conditions can separate carboplatin from phospholipids but it requires more than 10 min of cycle time. To resolve matrix effect by phospholipids and develop a rapid and robust method, a two-dimensional chromatography was developed in which a trapping guard column (Luna C18(2)®, 30 × 2 mm, 5 μm), an analytical column (Alltima™ HILIC column 50 × 2 mm, 5 μm), and a switching valve were utilized. Using this strategy, phospholipids are trapped on the guard column and the phospholipid-free eluent was introduced to the analytical column and MS. Despite the complications, this design keeps the MS clean and shortens the cycle time to just 3.5 min.

Protocol 1: Quantitative determination of total carboplatin in human plasma using LC-MS/MS.

Test Material and Matrix

- Carboplatin (Sigma)
- Oxaliplatin (Sigma)
- Human plasma (K_2EDTA) (BioChemEd)

Equipment and Reagents

- *Liquid handling system*: Tomec Quadra 96 or Hamilton MICROLAB® NIMBUS 96
- *HPLC system*: Shimadzu LC-10D
- *HPLC column*: Phenomenex Luna Silica 3 × 50 mm 3 μm
- *Autosampler*: LEAP PAL LC Autosampler
- *Mass spectrometer*: API5000
- Ammonium acetate, ACS grade, Sigma-Aldrich
- Acetonitrile (MeCN), HPLC grade, EMD
- Formic acid (FA), ACS grade, EMD
- Water, deionized, type 1, (typically 18.2 MΩ cm) or equivalent
- *Needle wash 1*: 0.1% FA in (water/MeCN 90:10 v/v)
- *Needle wash 2*: 0.1% FA in water
- *Mobile phase A*: 0.1% FA in 5 mM ammonium acetate, pH unadjusted
- *Mobile phase B*: MeCN
- *Mobile phase C*: 90/10 MeCN/water (to be used for diversion)

Method

Solution and Sample Preparation

1. Weigh an appropriate amount of carboplatin and oxaliplatin (internal standard) and dissolve in the required volume of water for 10 mg/ml (carboplatin) and 0.5 mg/ml (oxaliplatin) stock solutions. Prepare duplicate stock solutions. One is designated to prepare standard spiking solution and another one is used to prepare quality control (QC) spiking solution.
2. Prepare working spiking solutions in water. There are eight standard spiking solutions and their corresponding concentrations are 2, 4, 20, 100, 300, 600, 900, and 1000 μg/ml. Similarly, there are five QC spiking solutions and their corresponding concentrations are 2, 6, 400, 800, and 4000 μg/ml.
3. Prepare the standard curves and QC samples in human plasma from the spiking solution. The standard calibrators contain eight concentrations 0.1, 0.2, 1, 5, 15, 30, 45, and 50 μg/ml. The QCs contain five levels 0.1 (LLOQ), 0.3 (low), 20 (medium), 40 (high), and 200 (dilution) μg/ml.

Sample Extraction

1. Manually aliquot 50 μl of each sample into the corresponding wells of 96-well plate (including calibrators, QCs, unknown samples, QC0, and blank human plasma).
2. Add 50 μl of internal standard working solution (10,000 ng/ml of oxaliplatin in water) to the samples except matrix blanks.
3. To the matrix blanks, add 50 μl of water.
4. Centrifuge the samples at approximately 1500*g* for approximately 1 min.
5. Add 400 μl of MeCN to the samples.
6. Vortex-mix the samples at the low setting on the multitube vortexer for approximately 2 min.
7. Centrifuge the samples at approximately 1500*g* for approximately 5 min to spin down the precipitate.
8. Add 200 μl of MeCN into a clean 96-well plate.

9. Transfer 200 μl of the supernatant into the 99-well plate using Tomec Quadra 96 or Hamilton Nimbus.
10. Vortex-mix the samples using the multitube vortexer for approximately 2 min on medium setting.
11. Centrifuge the samples at approximately 1500*g* for 1 min.
12. Store the samples at 1–8°C until being analyzed.

LC-MS/MS Analysis

1. The LC system was equipped with a standard six-port switch valve (VICI Cheminert 10U-0263H) facilitate the flow diversion of the LC eluent after the last peak of each sample was eluted out of the column. The LC conditions are as follows:
   a. *Flow rate*: 1 ml/min
   b. *Column temperature*: 25°C
   c. *Gradient*:

Time	0.0′	1.60′	1.62′	2.60′	2.61′	3.5′
B%	90	85	50	50	90	Stop

Note: The column was diverted and only the LC eluent from 0.5 to 1.6 min is introduced to MS.

2. The mass spectrometer was operated in positive ionization mode with an electrospray ion source (ESI). The SRM transition for carboplatin and oxaliplatin are 372.2 → 294.2 and 398.2 → 306.3, respectively. The typical MS conditions are as follows: dwell time, 100 ms; source temperature, 500°C; IS voltage, 4000 V; declustering potential (DP), 25; curtain gas, 30 psi; nebulizer gas (GS1), 50 psi; turbo gas (GS2), 70 psi; collision energy, 35 eV (Figure 49.3).

Protocol 2: Quantitative determination of free carboplatin in human plasma ultrafiltrate using LC-MS/MS.

Test Material and Matrix

- Carboplatin (Sigma)
- Oxaliplatin (Sigma)
- Human plasma (K_2EDTA) (BioChemEd)
- Human plasma (K_2EDTA) Ultrafiltrate (generated in house)

Equipment and Reagents

- *HPLC system*: Shimadzu LC-10D
- *HPLC column*: Varian Inertsil C4, 3 μm, 3 × 150 mm
- *Autosampler*: LEAP PAL LC Autosampler
- *Mass spectrometer*: API5000
- Ammonium acetate, ACS grade, Sigma-Aldrich
- Acetonitrile (MeCN), HPLC grade, EMD
- Formic acid (FA), ACS grade, EMD
- Methanol (MeOH), HPLC grade, EMD
- Water, deionized, type 1, (typically 18.2 MΩ cm) or equivalent
- *Needle wash 1*: 0.1% FA in (water/MeCN 90:10 v/v)
- *Needle wash 2*: 0.1% FA in water
- *Mobile phase A*: 0.1% FA in water
- *Mobile phase B*: MeCN

Method

Solution and Sample Preparation

1. Weigh an appropriate amount of carboplatin and oxaliplatin (internal standard) and dissolve in the required volume of water for 10 mg/ml (carboplatin) and 0.5 mg/ml (oxaliplatin) stock solutions. Prepare duplicate stock solutions. One is designated to prepare standard spiking solution and another one is used to prepare QC spiking solution.
2. Prepare working spiking solutions in water. There are eight standard spiking solutions and their corresponding concentrations are 2, 4, 20, 100, 300, 600, 900, and 1000 μg/ml. Similarly, there are five QC spiking solutions and their corresponding concentrations are 2, 6, 400, 800, and 4000 μg/ml.
3. Prepare the standard curves and QC samples in human plasma ultrafiltrate from the spiking solution. The standard calibrators contain eight concentrations: 0.1, 0.2, 1, 5, 15, 30, 45, and 50 μg/ml. The QCs contain five levels: 0.1 (LLOQ), 0.3 (low), 20 (medium), 40 (high), and 200 (dilution) μg/ml.

Sample Extraction

1. Manually aliquot 20 μl of each sample into the corresponding wells of 96-well plate (including calibrators, QCs, unknown samples, QC0, and blank human plasma ultrafiltrate).
2. Add 500 μl of internal standard working solution (1000 ng/ml of oxaliplatin in water) to the samples except matrix blanks.
3. To the matrix blanks, add 500 μl of water.
4. Vortex-mix the samples at the low setting on the multitube vortexer for approximately 2 min.
5. Centrifuge the samples at approximately 1500*g* for approximately 5 min to spin down the precipitate.
6. Store the samples at 1–8°C until being analyzed.

LC-MS/MS Analysis

1. The LC conditions are as follows:
   a. *Flow rate*: 0.6 ml/min

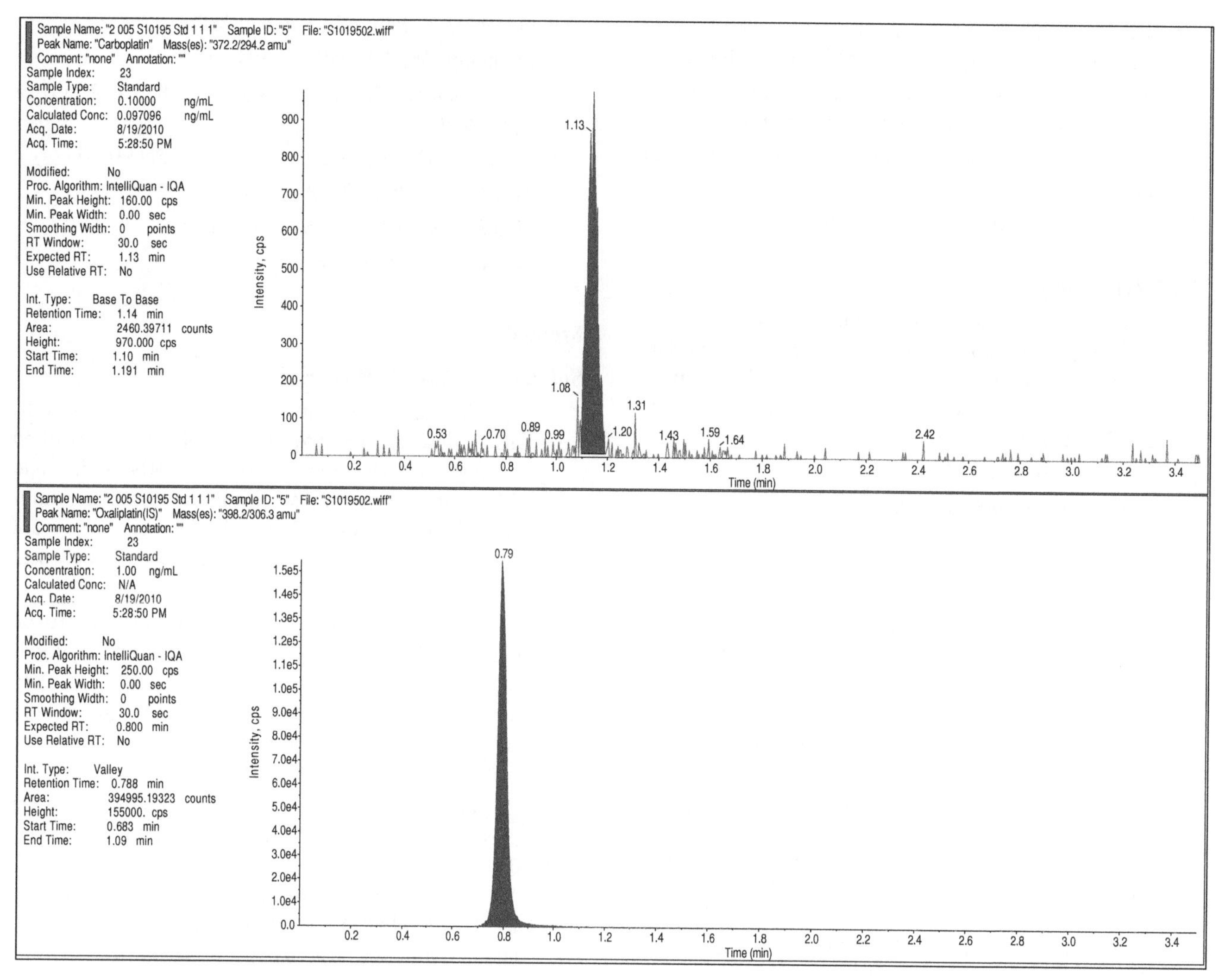

FIGURE 49.3 Typical chromatography of low standard (0.1 μg/ml) from human plasma using PPE coupled with normal phase LC-MS/MS. Top panel: carboplatin. Bottom panel: oxaliplatin.

b. *Column temperature*: 30°C

c. *LC program*:

Time	0′	6′
B%	5	Stop

Note: The column was forward flushed from 3 to 5 min at 0.8 ml/min.

2. The mass spectrometer was operated in positive ionization mode with an ESI. The SRM transition for carboplatin and oxaliplatin are 372.2 → 294.2 and 398.2 → 306.3, respectively. The typical MS conditions are dwell time, 100 ms; source temperature, 500°C; IS voltage, 4000 V; DP, 25; curtain gas, 30 psi; nebulizer gas (GS1), 50 psi; turbo gas (GS2), 70 psi; collision energy, 35 eV (Figure 49.4).

Protocol 3: Quantitative determination of unbound cisplatin, carboplatin, and oxaliplatin in human plasma ultrafiltrate by measuring platinum–DDTC complex with LC–MS/MS.

Test Material and Matrix

- Carboplatin (Sigma)
- Oxaliplatin (Sigma)
- Cisplatin (Sigma)
- DDTC (Sigma)
- Human plasma (K_2EDTA) (BioChemEd)

Equipment and Reagents

- *Liquid handling system*: Tomec Quadra 96 or Hamilton MICROLAB® NIMBUS 96

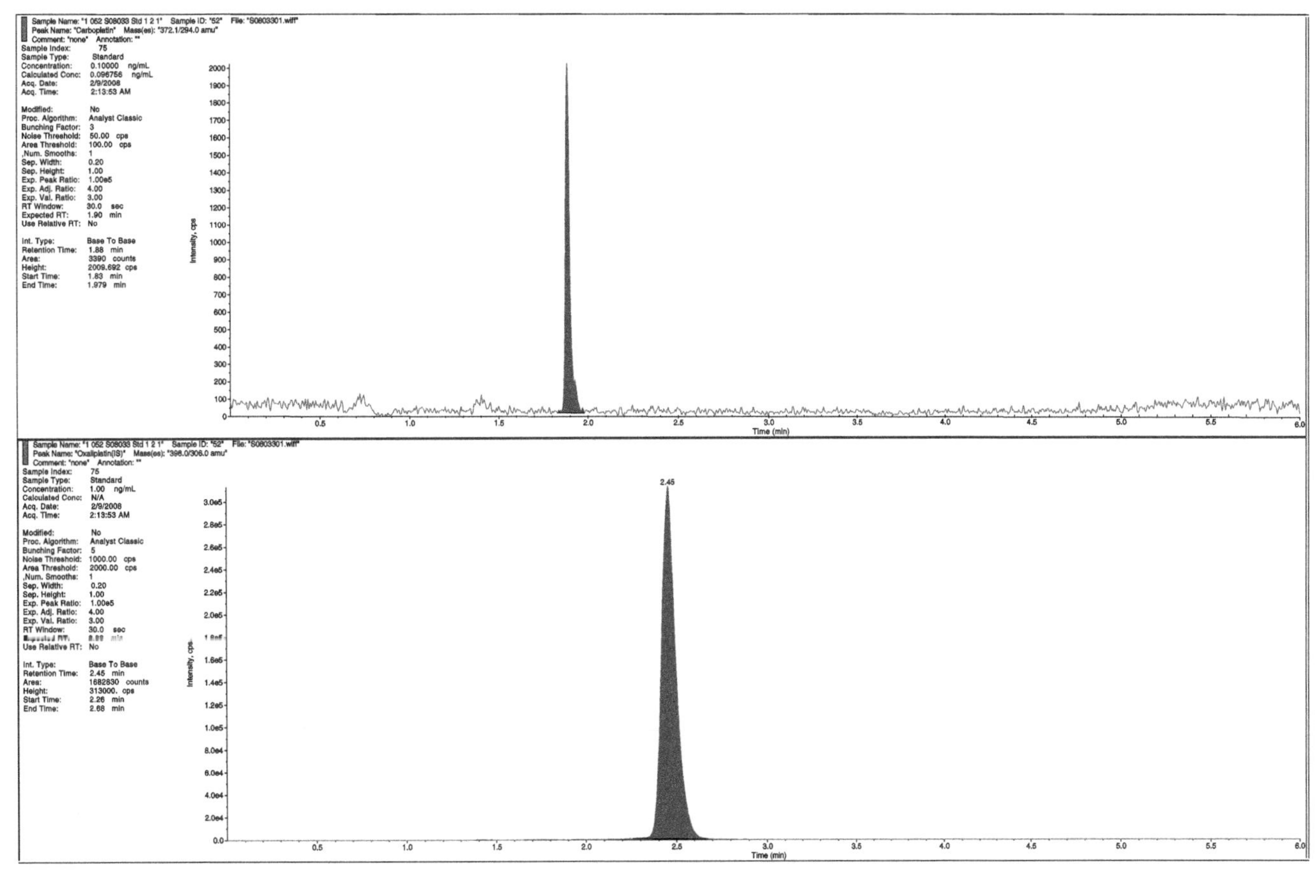

FIGURE 49.4 Typical chromatography of low standard (0.1 μg/ml). Top panel: carboplatin. Bottom panel: oxaliplatin.

- *HPLC system*: Shimadzu LC-10D
- *HPLC column*: Phenomenex Luna Gemini C18, 5 μm, 2 × 50 mm
- *Autosampler:* LEAP PAL LC Autosampler
- *Mass spectrometer*: API4000
- Ammonium acetate, ACS grade, Sigma-Aldrich
- Acetonitrile (MeCN), HPLC grade, EMD
- Formic acid (FA), ACS grade, EMD
- Water, deionized, type 1, (typically 18.2 MΩ cm) or equivalent
- *Needle wash 1*: 0.1% FA in (water/MeCN 90:10 v/v)
- *Needle wash 2*: 0.1% FA in water
- *Mobile phase A*: 0.1% formic acid in water
- *Mobile phase B*: MeCN
- *Mobile phase C*: 90/10 Acetone/water (for diversion and backflush)

Method

Solution and Sample Preparation

1. Weigh an appropriate amount of carboplatin and oxaliplatin (internal standard) and dissolve in the required volume of water for 10.0 mg/ml (carboplatin) and 0.5 mg/ml (oxaliplatin) stock solutions. Prepare duplicate stock solutions. One is designated to prepare standard spiking solution and another one is used to prepare QC spiking solution.
2. Prepare working spiking solutions in water. There are eight standard spiking solutions and their corresponding concentrations are 20, 100, 200, 1000, 5000, 10,000, 15,000and 20,000 ng/ml. Similarly, there are five QC spiking solutions and their corresponding concentrations are 20, 60, 8000, and 16,000 ng/ml.
3. Prepare the standard curves and QC samples in human plasma from the spiking solution. The standard calibrators contain eight concentrations: 1, 5, 10, 50, 250, 500, 750, and 1000 ng/ml. The QCs contain five levels: 1 (LLOQ), 3 (low), 400 (medium), 800 (high) ng/ml.

Sample Extraction

1. Manually aliquot 50 μl of each sample into the corresponding wells of 96-well plate (including calibrators, QCs, unknown samples, QC0, and blank human plasma).

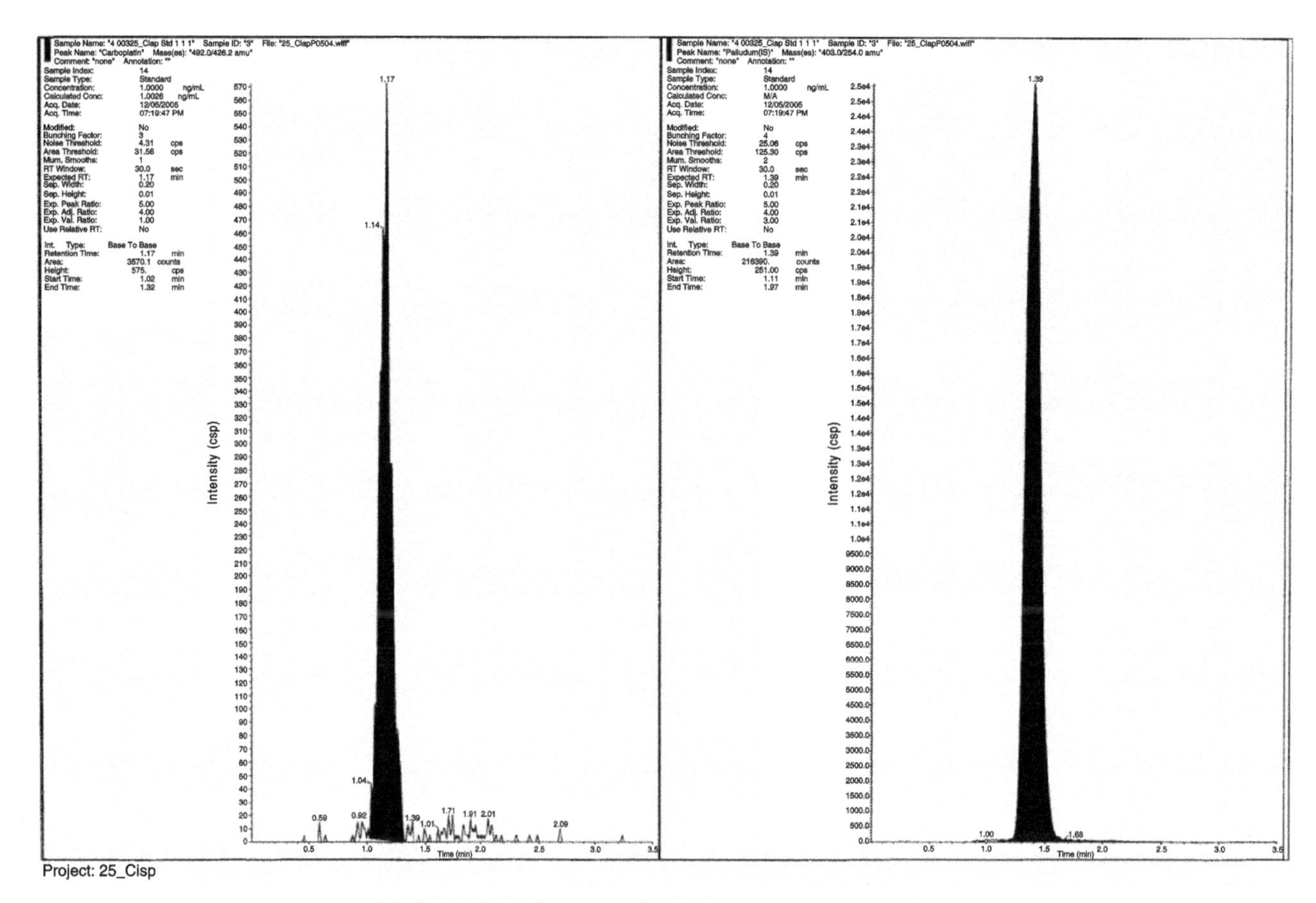

FIGURE 49.5 Representative SRM chromatogram (LLOQ).

2. Add 100 μl of internal standard Pd-acetate (1000 ng/ml in Water) into 100 μl platinum-fortified human plasma ultrafiltrate.
3. To the matrix blanks, add 50 μl of water.
4. Add freshly prepared 5% diethyldithiocarbamate (DDTC) in 0.2 N NaOH into all samples.
5. Incubate all samples at 45°C for 30 min.
6. Centrifuge the samples at approximately 1500*g* for approximately 1 min.
7. Add 2 ml of methyl *t*-butyl ether (MtBE).
8. Shake samples for 15 min.
9. Centrifuge the samples at approximately 1500*g* for approximately 5 min to spin down the water layer.
10. Freeze aqueous lay in acetone–dry ice bath.
11. Transfer top organic layer to clean tubes.
12. Dry samples completely in Turbovap at 45°C for approximately 20 min.
13. Reconstitute sample with 200 μl MeCN.
14. Vortex-mix the samples at the low setting on the multitube vortexer for approximately 2 min.
15. Centrifuge the samples at approximately 1500 rpm for 1 min.
16. Transfer samples to an autosampler vial.

LC-MS/MS Analysis

1. The LC system was equipped with a standard six-port switch valve (VICI Cheminert 10U-0263H) facilitate the flow diversion of the LC eluent after the last peak of each sample was eluted out of the column. The LC conditions are as follows:
   a. *Flow rate*: 0.3 ml/min
   b. *Column temperature*: 30°C
   c. *Isocratic*:

Time	0.0′	3.5′
B%	80	Stop

Note: The column was diverted and only the LC eluent from 0.8 to 2.2 min is introduced to MS. The column is backflushed with mobile phase C from 2.3 to 3.5 min at 0.5 ml/min.

The mass spectrometer was operated in positive ionization mode with an ESI. The SRM transition for all platinum compounds [Pt-$(DDTC)_2$] and palladium [Pd-$(DDTC)_2$] are 492 → 426 and 403 → 254, respectively. The typical MS conditions are dwell time, 100 ms; source temperature, 400°C; IS voltage, 4000 V; declustering potential (DP), 25; curtain gas, 30 psi; nebulizer gas (GS1), 50 psi; turbo gas (GS2), 70 psi; collision energy, 35 eV (Figure 49.5).

REFERENCES

Andrews PA, Wung WE, Howell SB. A high-performance liquid chromatographic assay with improved selectivity for cisplatin and active platinum (II) complexes in plasma ultrafiltrate. Anal Biochem 1984;143(1):46–56.

Barefoot RR. Speciation of platinum compounds: a review of recent applications in studies of platinum anticancer drugs. J Chromatogr B Biomed Sci Appl 2001;751(2):205–211.

Baskerville-Abraham IM, Boysen G, Troutman JM, et al. Development of an ultraperformance liquid chromatography/mass spectrometry method to quantify cisplatin 1,2 intrastrand guanine–guanine adducts. Chem Res Toxicol 2009;22(5):905–912.

Bennett PK, Van Horne KC. Identification of the major endogenous and persistent compounds in plasma, serum, and tissue that cause matrix effects with electrospray LC/MS techniques. Presented at the 2003 AAPS Annual Meeting and Exposition, Salt Lake City, Utah, October 2003.

Bosch ME, Sánchez AJ, Rojas FS, Ojeda CB. Analytical methodologies for the determination of cisplatin. J Pharm Biomed Anal 2008;47(3):451–459.

Boulikas T, Vougiouka M. Cisplatin and platinum drugs at the molecular level [Review]. Oncol Rep 2003;10(6):1663–1682.

Guetins G, Boeck GD, Highley MS, et al. Hyphenated techniques in anticancer drug monitoring: II. Liquid chromatography-mass spectrometry and capillary electrophoresis-mass spectrometry. J Chromat A 2002;976:239–247.

Guo P, Li S, Gallo JM. Determination of carboplatin in plasma and tumor by high-performance liquid chromatography–mass spectrometry. J Chromatogr B Analyt Technol Biomed Life Sci 2003;783(1):43–52.

Iijima H, Patrzyc HB, Dawidzik JB, et al. Measurement of DNA adducts in cells exposed to cisplatin. Anal Biochem 2004;333(1):65–71.

Liu Y, Demers R, Wentzel D, Hess E, Cojocaru L. Novel assay with phospholipids column trapping and switching valve system for the determination of carboplatin in human plasma by LC/MS/MS. Presented at the 2011 ASMS Conference, Denver, Colorado, June 2011.

Manzotti C, Pratesi G, Manta E, et al. BBR3464: a novel triplatinum complex, exhibiting a preclinical profile of antitumor efficacy different from cisplatin. Clin Cancer Res 2000;6:2626–2634.

Meng M, Liu S, Bennett P. Simple and rapid determination of carboplatin in human plasma ultrafiltrate using LC/MS/MS by direct injection coupled with post column addition. Presented at the 2003 AAPS Annual Meeting and Exposition, Salt Lake City, Utah, October 2003.

Meng M, Kuntz R, Fontanet A, Bennett P. A novel approach to quantify unbound cisplatin, carboplatin, and oxaliplatin in human plasma ultrafiltrate by measuring platinum-DDTC complex using LC/MS/MS. Presented at the 2006 ASMS Conference, Seattle, Washington, May 2006.

Piccart MJ, Lamb H, Vermorken JB. Current and future potential roles of the platinum drugs in the treatment of ovarian cancer. Ann Oncol 2001;12(9):1195–1203.

Rosenberg B, VanCamp L, Trosko JE, Mansour VH. Platinum compounds: a new class of potent antitumour agents. Nature 1969;222(5191):385–386.

Stokvis E, Rosing H, Beijnen JH. Liquid chromatography–mass spectrometry for the quantitative bioanalysis of anticancer drugs. Mass Spectrom Rev 2005;24(6):887–917.

50

MICROFLOW LC-MS FOR QUANTITATIVE ANALYSIS OF DRUGS IN SUPPORT OF MICROSAMPLING

HEATHER SKOR AND SADAYAPPAN V. RAHAVENDRAN

50.1 INTRODUCTION

There is an increased emphasis on validation of pharmacological targets and development of preclinical efficacy models in drug discovery in an attempt to reduce the failure of drug candidates during phase 2 clinical development due to lack of efficacy (Wehling, 2009). The mouse is an important animal model used for assessing the efficacy of targeted therapies in oncology (Caponigro and Sellers, 2011). Mice are used due to their small size, relatively low cost for husbandry, ability to reproduce rapidly with large litters, and suitability for genetic modifications (Cheon and Orsulic, 2011). Initial pharmacokinetic (PK) studies are typically conducted in mice by drug metabolism groups supporting small-molecule discovery to ensure adequate drug exposure before conducting mouse efficacy (PK/PD) studies by oncology biologists.

Typically, a single PK profile of a discovery compound is obtained through dosing two to three mice and collecting a total of 7–9 time points of blood over a 24-h period (3 time points of blood collected/mouse between 75 and 150 μl of blood per time point). This method is known as a composite blood sampling technique, resulting in increased compound and animal use as well as increased PK variability. Composite blood sampling is a type of sparse sampling technique in which blood samples are collected at a subset of designed time points from each animal. With the utilization of high sensitivity LC-MS/MS methods, microsampling now enables the collection of a complete PK profile of a compound from a single mouse through serial sampling of smaller volumes of blood per time point (i.e., 10–50 μl of blood per time point). The advantages of using serial microsampling include a reduction in number of mice per study resulting in lower cost, reduced compound quantity for dosing, less discomfort to animals due to less blood removed, the ability to obtain more results from a single animal (measurement of analyte, metabolites, and biomarker levels in plasma providing both PK and PD endpoints), and collection of a complete PK profile from the same mouse eliminating interanimal variability (Balani et al., 2004). Figure 50.1 illustrates both serial and composite sampling of blood time points in mice. Blood collected via serial microsampling are typically processed by one of four methods: liquid blood, liquid plasma, dried blood spot (DBS), or dried plasma spot (DPS). In this protocol only liquid plasma and DBS approaches are addressed.

With the collection of small volume blood samples from serially bled mice, the ability to accurately quantitate low levels of analytes has resulted in the resurgence of interest in microbore LC-MS/MS (Rainville, 2011; Rahavendran et al., 2012). The development and utilization of microbore columns (defined as columns with internal diameter (ID) of $\leq$1 mm, also known as capillary columns) was initially investigated by Horvath et al. in the late 1960s (Horvath et al., 1967), and more extensively in the 1970s by Novotny and other researchers (Ishii et al., 1977; Novotny, 1980). These early innovators reported the following benefits using microbore columns: reduction in mobile phase solvent consumption, use of smaller sample volumes, and increased sensitivity. The increased sensitivity obtained with microbore columns, when compared to conventional columns (i.e., 2.1 or 4.6 mm ID), is due to reduced chromatographic dilution of the analyte, which results in a higher response in a concentration-dependent detector such as an electrospray ionization mass spectrometer (ESI-MS) (Vissers et al., 1997). Qu and colleagues also described the application of low level

Handbook of LC-MS Bioanalysis: Best Practices, Experimental Protocols, and Regulations, First Edition. Edited by Wenkui Li, Jie Zhang, and Francis L.S. Tse.

Composite sampling: 125 μl blood/sampling

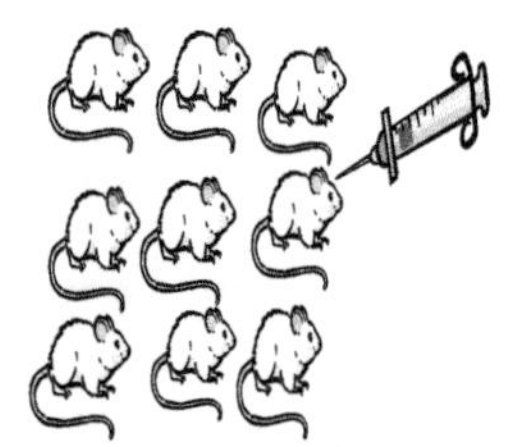

Serial sampling: 10 – <50 μl blood/sampling

Animal #	5 min	15 min	30 min	1 h	2 h	4 h	6 h	8 h	24 h
1	X			X			X		
2	X			X			X		
3	X			X			X		
4		X			X			X	
5		X			X			X	
6		X			X			X	
7			X			X			X
8			X			X			X
9			X			X			X

Animal #	15 min	30 min	1 h	2 h	4 h	6 h	8 h	24 h
1	X	X	X	X	X	X	X	X
2	X	X	X	X	X	X	X	X
3	X	X	X	X	X	X	X	X

FIGURE 50.1 Study design illustrating composite and serial samplings of mice.

quantitation of corticosteroids in plasma samples and paclitaxel at low pg/ml levels using microbore LC–MS/MS (Qu et al., 2007; Yu et al., 2008). Sensitivity enhancement at lower flow rates is observed with electrospray ionization, where the efficiency of ion creation and transfer into the vacuum system of a mass spectrometer is optimal because small droplets are easier to evaporate than large droplets. It is thus easier to create and charge small droplets at low flow rates (microbore chromatography) than from high flow rates (conventional chromatography) (Covey et al., 2009). It has been reported that microbore LC-MS/MS enables shorter run times, smaller sample volumes, less matrix effects and reduced solvent usage, a cleaner ionization source of the mass spectrometer for longer duration of time, and comparable sensitivity to conventional LC-MS/MS (Lim and Lord, 2002; Neyer and Hobbs, 2009; Smith et al., 2011).

50.2 METHODS AND APPROACHES

50.2.1 Principles of Microsampling in Mice

The acceptable maximum volume of blood that may be drawn from a mouse weighing 25 g (70 ml/kg blood volume) for an acute study (<24 h) is approximately 350 μl which represents 20% of its circulating blood volume (Diehl et al., 2001; Burnett, 2011). In many laboratories, an upper limit of 10–15% of circulating blood volume is specified in the Animal Care and Use Protocol (ACUP) during the course of an acute PK study. For a chronic study extending up to 4 weeks, 20% of circulating blood may be removed, depending on the institution's IACUC (Institutional Animal Care and Use Committee) regulations. The calculation of the maximum blood sample volume from a mouse is shown in Equation 50.1.

$$\begin{aligned}\text{Maximum blood volume} &= \text{mean blood volume (70 ml/kg)}\\ &\times \text{mouse weight (0.025 kg)}\\ &\times \text{\% of blood volume}\end{aligned} \tag{50.1}$$

Serial blood sampling from tail vein, saphenous venepuncture, submandibular and retro-orbital sites in mice has been reported previously to support PK studies (Avni et al., 2009; Peng et al., 2009; Kurawattimath et al., 2012). Tail vein and saphenous venepuncture techniques do not require the use of anesthesia; however, retro-orbital bleeds require administration of anesthesia. Retro-orbital bleeds also require the services of a skilled phlebotomist. It is a technique used for composite blood sampling and also as a serial sampling technique in combination with saphenous venepuncture when speed and larger volumes of blood are required (Avni et al., 2009). Saphenous venepuncture is a fast, reliable, route of blood sampling that is less stressful to the mouse (Hem et al., 1998; Oruganti and Gaidhani, 2011). Blood collection by a combination of sites such as saphenous venepuncture, retro-orbital, tail vein, and terminal bleed via cardiac puncture offers more than a single blood sampling route in mice and allows for micro- (up to 40 μl) and larger volume (>50 μl) sampling from a single mouse (Rahavendran et al., 2012).For serial sampling using plasma as the analytical matrix, approximately 30–50 μl blood is collected from the bleed site using an EDTA-coated capillary tube, then transferred into a 100-μl, EDTA-coated microcentrifuge tube and centrifuged to yield approximately 15–25 μl volume of plasma for analyses. Jonsson et al.

reported the collection of approximately 32 μl of blood via tail vein sampling in mice directly into a hematocrit tube coated with EDTA (Jonsson et al., 2012). One end of the hematocrit tube was plugged with wax and placed in a 4.5-ml tube and centrifuged. After centrifugation, the hematocrit tube was cut above the blood cell phase using a ceramic cutter and an exact volume of 8 μl plasma was collected with a capillary micropipette, from the end of the plasma part of the hematocrit tube. Finally, the capillary containing the plasma sample was placed in a 1.1-ml polypropylene tube with screw cap and stored at -2°C pending sample analyses. When conducting serial sampling by DBS, approximately 10–20 μl of blood is collected from the bleeding site at each time point using an EDTA-coated capillary tube. The blood is then spotted directly onto a DBS paper and allowed to dry for 2–4 h at room temperature. The liquid plasma or DBS samples are then processed and analytes assayed by LC-MS/MS.

Before investigating the exposure–efficacy relationship (PK/PD) of a compound in tumor-bearing mice, the oral exposure of the compound is determined in naive mice of the same strain as the tumor-bearing mice at a number of different doses. These experiments serve two purposes: (1) to ensure that the compound plasma exposures are at or above the efficacious concentration over the duration of the dosing interval and (2) to assess the maximum tolerated dose. The plasma efficacious concentration (C_{eff}) is determined from an *in vitro* IC_{50} cell-based assay. Once the maximum tolerated dose at which the exposure of the compound has been established to provide C_{eff} coverage over the duration of the dosing interval, the compound is dosed into tumor-bearing mice over the duration of days/weeks to demonstrate tumor growth inhibition. Both plasma and tumor concentrations of compound and biomarker are measured to develop the PK/PD model for the compound.

50.2.2 Principles of Microbore LC-MS/MS

Microbore liquid chromatography is used to describe columns with IDs of 1 mm or less. A precise nomenclature (Vissers et al., 1997) is that microbore chromatography employs columns with ID of 0.5–1.0 mm; capillary chromatography uses columns with ID of 0.1–0.5 mm and nanoscale chromatography uses columns with ID of 0.01–0.1 mm. Microbore columns are available with stationary phases used in conventional 2.1–4.6 mm ID columns, providing the same separation efficiency (plate counts) and chromatographic selectivity. Stationary phases have a lower resistance factor when they are packed in microbore columns. Therefore, operation of a microbore HPLC system at the same high pressure can provide faster linear velocities and faster analyses than conventional HPLC systems. The significance of moving to microbore HPLC includes the reduced column volumes and reduced volumetric flow rates needed to achieve linear flow velocities when compared to conventional HPLC. Decreased volumetric flow rates provide benefits for conducting faster analyses. Gradient mixing can be achieved in the absence of active mixers and in substantially smaller volumes compared to conventional HPLC. In order to ensure optimal performance for microbore columns at low flow rates (5–150 μl/min), postmixing volumes and, in particular, postcolumn dead volumes must be minimized. Most commercially available microbore LC systems today are ultrahigh-pressure liquid chromatography (UHPLC) systems with the ability to provide rapid separations at higher pressures (10,000 psi) compared to conventional HPLC systems (approximately 5000 psi). Microbore systems using variants of dual reciprocating piston pump designs and syringe pump designs are currently marketed (Dionex; Michrom; ABsciex websites). The Eksigent Express HT UHPLC system, for example, employs microfluidic flow control (MFC) and pneumatically actuated syringe pumps to provide accurate high-pressure delivery of each mobile phase. Each pressure source is coupled to a fast response flow meter that monitors the fluid delivery of each channel. A feedback loop makes real-time adjustments to the flow rate. By continuously monitoring the flow from each of the pumps, the flow rate can be accurately adjusted many times per second that results in reproducible retention times and less background noise due to pulsation-free mobile phase delivery (Neyer and Hobbs, 2009).

Compared to conventional HPLC, microbore UHPLC increases the signal-to-noise (S/N) ratio considerably when ESI–MS/MS is employed as the detector and because microbore UHPLC results in a much smaller dilution of peak concentration. This is shown by Equation 50.2 (Qu et al., 2007).

$$D = C_{\mathrm{end}}/C_{\mathrm{inj}} = \varepsilon\pi r^2(1+k)(2\pi LH)^{1/2}/V_{\mathrm{inj}}, \quad (50.2)$$

where ε is the column porosity; R is the column radius; K is the retention factor; L is the column length; H is the plate height; V_{inj} is the injection volume.

During chromatographic separation, the dilution D of an injected sample $= C_{\mathrm{end}}/C_{\mathrm{inj}}$, where C_{end} is the concentration after chromatography and C_{inj} is the concentration injected. D is in direct proportion to the square of column radius.

A drawback when using microbore chromatography is the reduced column loading capacity when compared to conventional chromatography, and therefore, counteracts the gain in sensitivity that can be achieved through lower flow rate. Therefore, to increase the concentration sensivity of microbore LC-MS/MS relative to conventional LC-MS/MS, the volume of sample injected onto the column needs to be increased by initially incorporating a trapping column as a concentration step and then an analytical column for chromatographic separation (Yu et al., 2008).

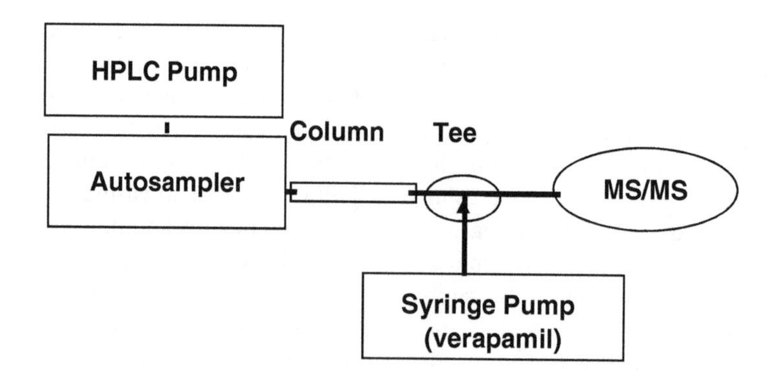

Conventional LC–MS/MS

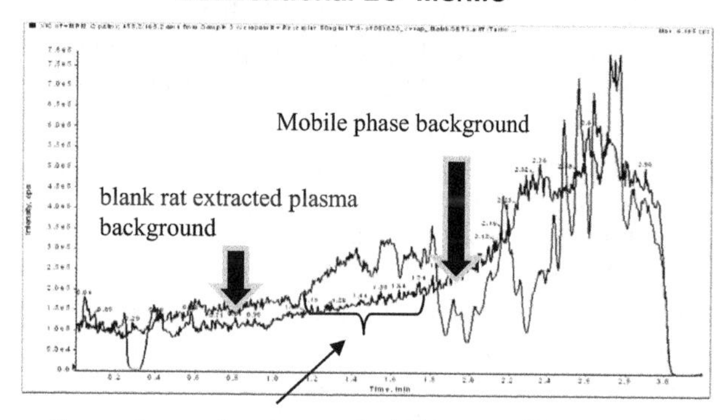

- Some ion enhancement at Verapamil t_r
- Significant matrix effect early and late in the run

Microbore LC–MS/MS

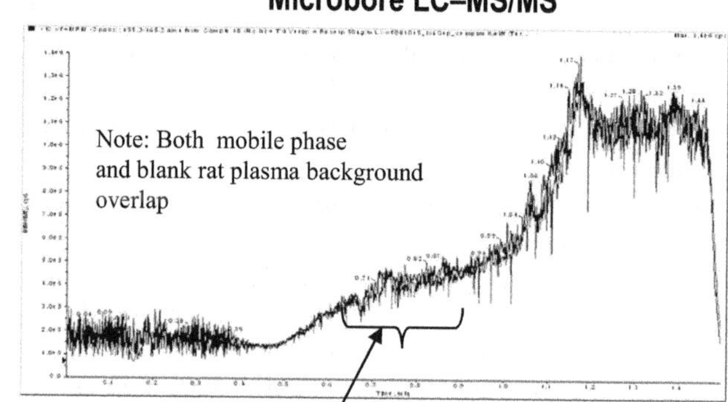

- No significant matrix effect at Verapamil t_r
- Very little matrix effect

FIGURE 50.2 Matrix effects in rat plasma using the T-infusion method with conventional and microbore LC-MS/MS.

Postcolumn band broadening should be addressed when optimizing microflow LC-MS/MS systems in order to achieve optimal performance. This can be achieved by modifying the standard turbo V ion source of a Sciex mass spectrometer (e.g., API4000, API5000, or API5500) with a smaller ID (25–50 μm) electrode needle. One of the important advantages observed with microbore LC-MS/MS, compared to conventional LC-MS/MS, is the reduced matrix effect. The postcolumn infusion technique can be used to qualitatively assess matrix effect (Chambers et al., 2007) by identifying chromatographic regions most likely to experience ion suppression and ion enhancement. In this technique, extracted blank matrix is injected into the LC-MS/MS system using the standard method program (injection volume, column, and gradient) while a solution of the compound of interest is continuously infused into the system postcolumn using a T-union and an infusion pump. The multiple reaction monitoring (MRM) transition of the compound of interest is monitored across the chromatogram. The chromatogram is then compared to a blank chromatogram, one acquired under the same conditions but with a null injection (injection of mobile phase solvent). Differences in chromatographic response between the injected blank matrix and the null injection can then be used to assess chromatographic regions that may be potentially impacted by ion suppression or enhancement. In this example, the chromatographic systems, conventional and microbore, were operated using their respective columns and gradient methods. Verapamil at 50 ng/ml concentration was teed in postcolumn at a flow rate of 10 μl/min using an infusion pump. Details of the postcolumn infusion setup, LC-MS/MS system conditions and resultant chromatograms (ion suppressograms) are shown in Figure 50.2. The chromatograms monitor the MRM transition of verapamil and show that ion enhancement exists at the retention time of verapamil in the conventional system, but is minimal using the microbore LC–MS/MS system.

50.3 PROTOCOLS

50.3.1 Study Design, Plasma Sample Collection, and Microbore LC-MS/MS Analysis After Oral Dosing of PF-A in Solution and Suspension in Mice

50.3.1.1 Dosing and Sampling In this study design, mice are orally dosed with PF-A either in a solution or in a suspension formulation. Serial blood samples are collected from each mouse at 30 min, 2 h, 4 h, 7 h, and 24 h, respectively,

TABLE 50.1 Treatment Information

Compound	Group	Dose	Dose units	Route of administration	Formulation	Regimen	Fasted/fed	Vehicle
PF-A	1	25	mg/kg	Oral	Solution	Single	Fed	PEG200:DMSO:saline = 70:10:20
PF-A	2	25	mg/kg	Oral	Suspension	Single	Fed	0.5% Methyl cellulose

and centrifuged to yield plasma. Tables 50.1, 50.2, and 50.3 provide the details for treatment, subject (animal), and blood sampling schedules, followed by sample preparation and LC-MS/MS analysis.

- Bleed techniques used include retro-orbital, saphenous vein, which are employed serially, and cardiac puncture as the terminal bleed procedure.
- The blood volume collected/time point was approximately 0.05 ml using 75-mm, EDTA-coated capillary tubes (Drummond Scientific, PA, USA). The collected blood in the capillary tube is transferred into a 100-μl microcentrifuge tube (Iris Sample Processing, MA, USA), and centrifuged for 15 min at 1500g at 4°C to yield approximately 20–25 μl of plasma in each sample. The plasma samples are stored at −80°C in a 96-well shallow conical plate until sample analysis.

50.3.1.2 Sample Preparation

- Stock solutions of compound (1 mg/ml) and the subsequent standard solutions are prepared in methanol.
- Structural analog is used as internal standard (IS) and prepared at 500 ng/ml in methanol. All stock solutions are stored at −20°C.
- Calibration standards were prepared in nu/nu mouse plasma and consisted of seven concentrations at 1, 5, 10, 50, 200, 1000, and 2500 ng/ml. Standards were prepared by addition of 2 μl of stock standards into 20 μl of blank mouse plasma. QC standards were prepared at levels of 15, 75, 250, and 750 ng/ml in mouse plasma.
- A generic plasma protein precipitation method was employed. Approximately 10 μl aliquots of plasma standards, QCs, and study samples are transferred into a 96-well plate (0.65-ml microtubes, National Scientific Supply Co., Inc., CA, USA). A 10 μl aliquot of IS working solution (500 ng/ml) was added to all samples except double blanks, followed by addition of 40 μl of acetonitrile/methanol (1:1). Following vortex-mixing and centrifugation at 4°C (3000g for 5 min), 25 μl of supernatant was transferred to 0.65 μl tubes containing 50 μl, 10% acetonitrile in water and 1.5 μl aliquot was injected into microbore LC–MS/MS system.

50.3.1.3 Microbore LC-MS/MS Analysis

- The microbore LC-MS/MS system consisted of a CTC-HTS-PAL autosampler equipped with a cool stack (LEAP Technologies, Carrboro, NC), an Eksigent Express HT Ultra UHPLC system (Applied Biosystems, Foster City, CA, USA), and an API5500 QTRAP quadrupole mass spectrometer (AB Sciex, Foster City, CA, USA). Mobile phase A consisted of water containing 0.1% formic acid, and mobile phase B consisted of acetonitrile containing 0.1% formic acid. A Halo RP C18, 1 × 50 mm, 2.7-μm column (Eksigent, CA, USA) at a flow rate of 0.150 ml/min. A gradient elution program was utilized with the initial solvent composition held at 10% B for 0.15 min and then increased linearly to 90% B over 0.75 min and held at 90% B for an additional 0.4 min. The column was then reequilibrated at initial conditions (10% B) for 0.2 min. The total run time is 1.5 min. The mass spectrometry was operated in turbo ion spray mode (either positive or negative) using MRM. The turbo-ion voltage was set to 4.5–5.5 kV and the auxiliary gas temperature was maintained at 450–500°C. High purity nitrogen is used for GAS 1, GAS 2, curtain, and collision gases. Declustering potential, collision energy, entrance potential, and collision cell exit potential conditions are optimized for respective compound and IS.
- Analyst® Software (AB Sciex, Foster City, CA, USA), version 1.5.1 is used for data acquisition and chromatographic peak integration. The standard calibration curve

TABLE 50.2 Subject Information

Group	Mouse	Treatment	Strain and gender	Weight (g)	Concentration in vehicle (mg/ml)	Injection volume (ml/kg)	Time dosed
1	01	PO-sol	Female nu/nu	20–25	2.5	10	
1	02	PO-sol	Female nu/nu	20–25	2.5	10	
2	03	PO-susp	Female nu/nu	20–25	2.5	10	
2	04	PO-susp	Female nu/nu	20–25	2.5	10	

TABLE 50.3 Blood Sampling Time—the Blood Sampling Schedule after Oral Dosing of PF-A in Both Solution and Suspension Formulations Are Shown in the Table

Mouse no.	Weight (g)	Volume injected (μl)	Injection time	$T = 30$ min Blood sampling time	$T = 2$ h Blood sampling time	$T = 4$ h Blood sampling time	$T = 7$ h Blood sampling time	$T = 24$ h Blood sampling time
01								
02								
03								
04								

was constructed using a weighted ($1/x^2$) linear regression. Pharmacokinetic parameters were calculated using Watson Bioanalytical LIMS software version 7.2.0.0.3 (Thermo Fisher Scientific, Waltham, MA, USA).

- Figure 50.3 shows the unbound plasma ($f_u = 0.02$) PK profile after a single 25 mg/kg oral dose of PF-A in both solution and suspension formulations.

50.3.2 Study Design, DBS Collection, and Microbore LC-MS/MS Analysis After QID and TID Oral Dosing of PF-A in Mice

50.3.2.1 Dosing and Sampling Protocol In this study design, PF-A are dosed orally to mice, four times on day 1 and thrice on day 2 to assess duration of PF-A blood levels above C_{eff}. Tables 50.4, 50.5, and 50.6 provide the details for treatment, subject (animal), and serial blood sampling schedules, followed by DBS collection, sample processing, and LC-MS/MS analysis.

- Tables 50.6a and 50.6b show the dosing and blood sampling schedules for PF-A. PF-A was dosed orally to $n = 3$ female nu/nu mice at 25 mg/kg four times on day 1 and three times on day 2. Nine serial blood sampling time points were collected from each of the three mice over the 2 days of the study. Saphenous vein for serial blood sampling and cardiac puncture as terminal bleed were the bleed techniques employed.

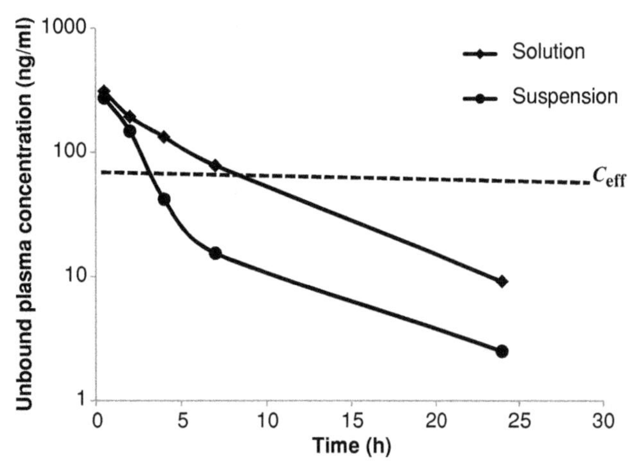

FIGURE 50.3 Oral unbound plasma exposure of PF-A (25 mg/kg dose) in mice.

- The volume of blood collected/time point directly from the saphenous vein was approximately 15 μl aliquot using a 75-mm, EDTA-coated capillary tubes (Drummond Scientific, PA, USA) and blood was carefully spotted onto the DBS card (DBS untreated paper was cut from 31 ET CHR sheets (47 cm × 57 cm) and purchased from GE-Whatman (catalog# 3031–915, NJ, USA). The spotted DBS paper was dried for approximately 2–4 h at room temperature and was stored in a desiccator at room temperature until sample analysis.

50.3.2.2 Sample Preparation

- Stock solutions of compound (1 mg/ml) were prepared in methanol and subsequent standard solutions are prepared in methanol.
- Structural analog was used as IS and prepared at 500 ng/ml in methanol and all stock solutions were stored at −20°C.
- Calibration standards were prepared in nu/nu mouse blood and consisted of seven concentrations at 1, 5, 10, 50, 200, 1000, and 2500 ng/ml. Standards were prepared by addition of 2 μl of stock standards into 20 μl of blank mouse blood. QC standards were prepared at levels of 15, 250, and 750 ng/ml in mouse blood.
- Approximately 15 μl aliquots of blood calibration standards and QC standards were spotted onto the DBS paper by holding the pipette tip or capillary tube a few millimeters above the card. The DBS paper containing calibration standards and QCs were dried thoroughly (approximately 2–4 h) at room temperature and QCs were stored in the desiccator at ambient temperature until sample analysis.
- In house, results have shown that the DMPK-C cards and the untreated 31 ET CHR sheets provided by GE-Whatman provide similar results. The 31 ET CHR

TABLE 50.4 Treatment Information

Compound	Dose	Dose units	Route of administration	Formulation	Regimen	Fasted/fed	Vehicle
PF-A	25	mg/kg	Oral	Suspension	Multiple	Fed	0.5% Methyl cellulose

TABLE 50.5 Subject Information

Group	Mouse	Treatment	Strain and gender	Weight (g)	Concentration in vehicle (mg/ml)	Injection volume (ml/kg)	Time dosed
1	01	PO-suspension	Female nu/nu	20–25	2.5	10	
1	02	PO-suspension	Female nu/nu	20–25	2.5	10	
1	03	PO-suspension	Female nu/nu	20–25	2.5	10	

sheets were cut to size and used for all subsequent DBS studies both for cost-savings and for flexibility.

- 3.2-mm discs of DBS samples, calibration standards, and QC standards were punched using an automated Wallac DBS puncher model 1296-071 (Perkin Elmer Life and Analytical Sciences, CT, USA). A 10 μl aliquot of IS working solution (500 ng/ml) was added to each well containing a punched DBS (except double blanks) and was vortex-mixed for 1 min, followed by a 50 μl aliquot of 60% aqueous methanol and sonication for 20 min. Following centrifugation at 4°C (3000*g* for 10 min), the supernatant was transferred into a clean 96-well plate. A 1.5 μl aliquot of the supernatant was injected onto the microbore LC column for mass spectral analysis.
- Microbore LC–MS/MS analysis conditions are described in Section 3.1.3.
- Figure 50.4 shows the unbound plasma exposure profile after multiple 25 mg/kg oral doses of PF-A and the duration of PF-A levels above C_{eff}. The blood concentrations obtained from LC–MS/MS analyses were converted to total plasma concentrations using an *in vitro* blood-to-plasma partition ratio value (BPR = 0.83). Unbound plasma concentrations were calculated by correction with fraction unbound in plasma ($f_u = 0.02$).

50.4 TROUBLESHOOTING

In Life Issues

- Saphenous vein bleeds are appropriate for microsampling (50 μl or less) and less suitable for large-volume sampling due to the slow bleed rate. Saphenous vein bleeds work well in some mouse strains such as CD-1, nu/nu, C57 BL6; however, are more difficult in strains with low blood pressure such as SCID/beige mice. An alternate route to use would be the tail vein.
- When harvesting plasma from small blood volumes (i.e., 30 μl bleed or less), extra caution is required to determine if the plasma is hemolyzed.

TABLE 50.6 Dosing and Blood Sampling Schedule

(a) Day 1 Dosing and Blood Sampling

Mouse no.	Weight (g)	Volume injected (μl)	Injection time	$T = 0$ h Dose 1	$T = 0.5$ h Sampling time	$T = 4$ h Dose 2	$T = 7.5$ h Sampling time	$T = 8$ h Dose 3	$T = 12$ h Dose 4	$T = 12.5$ h Sampling time	$T = 23.5$ h Sampling time
1											
2											
3											

(b) Day 2 Dosing and Blood Sampling

Mouse no.	Weight (g)	Volume injected (μl)	Injection time	$T = 24$ h Dose 5	$T = 24.5$ h Sampling time	$T = 27.5$ h Sampling time	$T = 28$ h Dose 6	$T = 31.5$ h Sampling time	$T = 32$ h Dose 7	$T = 32.5$ h Sampling time	$T = 35.5$ h Sampling time
1											
2											
3											

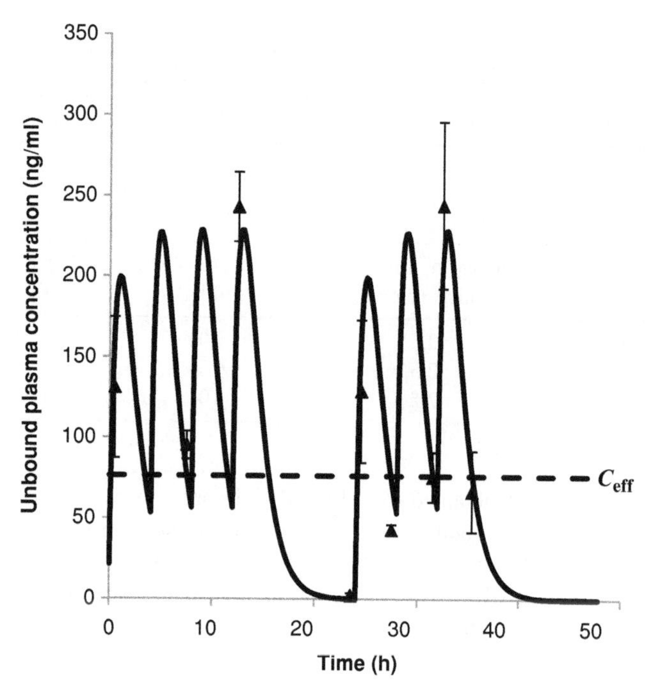

FIGURE 50.4 Oral concentration—time profile of PF-A (25 mg/kg doses) after QID (day 1) and TID (day 2) in mice.

DBS

- Do not touch the pipette to the DBS card when spotting the blood. Allow the blood to drop from the pipette tip onto the card.
- Extra drying time may be required based on laboratory temperature and humidity.
- DBS cards should be dried first and then stored in a desiccator.
- When collecting multiple blood spots onto the same card, allow adequate space between blood spots to prevent cross-contamination.
- Reduce possible cross-contamination when punching DBS by punching from a blank card between sample punches.

Microbore LC-MS/MS Analysis

- Leaks can be difficult to visualize in microbore LC–MS/MS. Use a laboratory wipe to check for leaks along the flow path. Monitor changes in pressure to determine if leaks or clogs are present in the flow path.
- When purchasing and installing microbore columns, check vendor's requirements for suitable end fittings. Incompatible fittings and ferrules can contribute to poor peak shape.
- Be careful not to over tighten microbore fittings. Follow the vendor's recommendations for force required to tighten fittings. Replace fittings that leak after excessive tightening, they have been damaged by over compression and continued tighten is unlikely to solve the leak.

ACKNOWLEDGMENTS

The authors would like to thank Robert Hunter and David Paterson for the in-life work and Paolo Vicini for discussion of the PK dosing regimen for PF-A.

REFERENCES

Avni D, Goldsmith M, Ernst O, et al. Modulation of TNFα, IL-10 and IL-12p40 levels by a ceramide-1-phosphate analog, PCERA-1, in vivo and ex vivo in primary macrophages. Immun Lett 2009;123:1–8.

Balani SK, Li P, Nguyen J, et al. Effective dosing regimen of 1-aminobenzotriazole for inhibition of antipyrine clearance in guinea pigs and mice using serial sampling. Drug Metab Dispos 2004;32:1092–1095.

Burnett JEC. Dried blood spot sampling: practical considerations and recommendation for use with preclinical studies. Bioanalysis 2011;3(10):1099–1107.

Caponigro G, Sellers WR. Advances in the preclinical testing of cancer therapeutic hypotheses. Nat Rev Drug Discov 2011;10:179–187.

Chambers E, Wagrowski-Diehl DM, Lu X, Mazzeo JR. Systematic and comprehensive strategy for reducing matrix effects in LC/MS/MS analyses. J Chromatogr B 2007;852:22–34.

Cheon DJ, Orsulic S. Mouse models of cancer. Ann Rev Pathol 2011;6:95–119.

Covey TR, Thomson BA, Schneider BB. Atmospheric pressure ion sources. Mass Spectrom Rev 2009;28:870–897.

Diehl K-H, Hull R, Morton D, et.al. A good practice guide to the administration of substances and removal of blood including routes and volumes. J Appl Tox 2001;21:15–23.

Hem A, Smith AJ, Solberg P. Saphenous vein puncture for blood sampling of the mouse, rat, hamster, gerbil, guinea pig, ferret and mink. Lab Anim 1998;32:364–368.

Horvath CG, Preiss BA, Lipsky SR. Fast liquid chromatography—an investigation of operating parameters and the separation of nucleotides on pellicular ion exchangers. Anal Chem 1967;39(12):1422–1428.

Ishii D, Asai K, Hibi K, Jonokuchi T, Nagaya M. A study of micro high performance liquid chromatography. I. Development of technique for miniaturization of high performance liquid chromatography. J Chromatogr A 1977;144(2):157–168.

Jonsson O, Villar RP, Nilsson LB, Eriksson M, Konigsson K. Validation of a bioanalytical method using capillary microsampling of 8 μL plasma samples: application to a toxicokinetic study in mice. Bioanalysis 2012;4(16):1989–1998.

Kurawattimath V, Pocha K, Mariappan TT, Trivedi RK, Mandlekar S. A modified serial blood sampling technique and utility of dried-blood spot technique in estimation of blood concentration: application in mouse pharmacokinetics. Eur J Drug Metab Pharmacokinet 2012;37:23–30.

Lim CK, Lord G. Current developments in LC-MS for pharmaceutical analysis. Biol Pharm Bull 2002;25(5):547–557.

Neyer D, Hobbs S. Application of microbore UHPLC-MS-MS to the quantitation of in vivo pharmacokinetic study samples. Curr Trends Mass Spectrom 2009;7:40–44.

Novotny M. Capillary HPLC: columns and related instrumentation. J Chromatogr Sci 1980;18:473–478.

Oruganti M, Gaidhani S. Routine bleeding techniques in laboratory rodents. Intl J Pharm Sci Res 2011;2(3):516–524.

Peng SX, Rockafellow BA, Skedzielewski TM, Huebert ND, Hageman W. Improved pharmacokinetic and bioavailability support of drug discovery using serial blood sampling in mice. J Pharm Sci 2009;98(5):1877–1884.

Qu J, Qu Y, Straubinger RM. Ultra-sensitive quantification of corticosteroids in plasma samples using selective solid-phase extraction and reversed-phase capillary high-performance liquid chromatography/tandem mass spectrometry. Anal Chem 2007;79:3786–3793.

Rahavendran SV, Vekich S, Skor H, et al. Discovery pharmacokinetic studies in mice using serial microsampling, dried blood spots, and microbore LC/MS/MS. Bioanalysis 2012;4(9):1077–1095.

Rainville P. Microfluidic LC-MS for analysis of small volume biofluid samples: where we have been and where we need to go. Bioanalysis 2011;3(1):1–3.

Smith D, Tella M, Rahavendran SV, Shen Z. Quantitative analysis of PD 332991 in mouse plasma using automated micro-sample processing and microbore liquid chromatography coupled with tandem mass spectrometry. J Chromatogr B Analyt Technol Biomed Life Sci. 2011;879:2860–2865. doi:10.1016/j.jchromb.2011.08.009

Vissers JPC, Claessens HA, Cramers CA. Microcolumn liquid chromatography: instrumentation, detection and applications. J Chromatogr A 1997;779:1–28.

Wehling M. Assessing the translatability of drug projects: what needs to be scored to predict success? Nat Rev Drug Discov 2009;8:541–546.

Yu H, Straubinger RM, Cao J, Qu J. Ultra-sensitive quantification of paclitaxel using selective solid-phase extraction in conjunction with reversed-phase capillary liquid chromatography/tandem mass spectrometry. J Chromatogr A 2008;1210: 160–167.

Thermoscientific. Ultimate 3000RSLC nano system. Available at http://www.dionex.com/en-us/products/liquid-chromatography/lc-systems/rslc/rslcnano/lp-81238.html. Accessed Dec 13, 2012.

Bruker-Michrom. Advance splitless nano-capillary LC. Available at https://www.michrom.com /Products/LCInstruments/AdvanceSplitlessnanoLC/tabid/116/Default.aspx. Accessed Apr 8, 2013.

AB SCIEX. Eksigent HT ultra UHPLC. Available at http://www.absciex.com/company/news-room/ab-sciex-expands-eksigent-microflow-uhplc-capabilities-for-lcms-workflows. Accessed Apr 8, 2013.

51

QUANTIFICATION OF ENDOGENOUS ANALYTES IN BIOFLUIDS BY A COMBINATION OF LC-MS AND CONSTRUCTION OF CALIBRATION CURVES USING STABLE-ISOTOPES AS SURROGATE ANALYTES WITH TRUE BIOLOGICAL CONTROL MATRICES

WENLIN LI, LUCINDA COHEN, AND ERICK KINDT

51.1 INTRODUCTION

The selectivity of a method is the measure of its ability to determine a particular analyte in the presence of other interfering components in the matrix. A method that is perfectly selective for an analyte is specific (USFDA, 2001; Lee et al., 2006). The three main contributors to the selectivity of a liquid chromatographic–mass spectrometry (LC–MS) method are as follows: (1) sample preparation, including directed extraction of the analyte from interfering components, (2) LC separation, and (3) selected reaction monitoring (SRM) detection based on the precursor and product ion mass-to-charge (m/z) values. Unfortunately, none of these contributors can successfully solve the problem of background interference when the analyte itself is present in the matrix used to prepare calibration standards. Therefore, a fundamental challenge in developing and validating robust quantitative biomarker assays, in contrast to assays developed for drug measurements, is the lack of relevant biological matrix that is devoid of the analytes of interest (Figure 51.1). Not only does this challenge require an assessment and optimization of assay selectivity, but also necessitates a strategy for preparing assay calibrators and quality control (QC) samples. Often, a substitute calibrator matrix cannot be identified because of the lack of availability, interfering components, or an inadequate analyte recovery compared to the biological matrix of interest. Another possible solution to the problem, the classic method of standard additions, could be conducted by successively adding known amounts of analyte to individual aliquots of the standard, and then back-extrapolating the original unknown concentration (Skoog et al., 1998). However, this is generally impractical due to the small volume of sample typically obtained in animal studies (~10 to 50 μl of plasma, ~10 to 100 mg tissue, etc.) and the tedious nature of this type of analysis. Table 51.1 describes some strategies, advantages, and disadvantages that scientists have used to prepare calibrators for biomarker quantification using LC-MS/MS. Similarly, QC samples prepared in the biological matrix of interest are required to assess assay robustness over the calibration range. However, their preparation can be troublesome depending upon the biomarker endogenous level in the matrix, as well as the degree of modulation during study sample analysis.

Isotope dilution has been a widely practiced and accepted method for quantitative mass spectrometry analysis, which involves spiking a known amount of isotope-labeled analyte into each standard, QC, and experimental sample (De Leenheer et al., 1985; Giovannini et al., 1991). The incorporation of ^{13}C, ^{15}N, ^{2}H, or other heavy elements into the synthetic biomarker allows it to maintain similar recovery,

Handbook of LC-MS Bioanalysis: Best Practices, Experimental Protocols, and Regulations, First Edition. Edited by Wenkui Li, Jie Zhang, and Francis L.S. Tse.

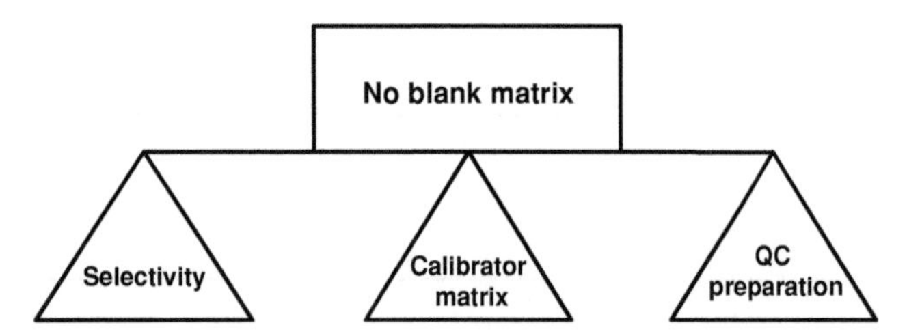

FIGURE 51.1 A fundamental issue in developing quantitative biomarker methods using LC-MS/MS is the lack calibrator matrix that does not contain measurable amounts of the analyte of interest. This challenge requires a strategy to address assay selectivity and the preparation of quality control samples and calibrators.

chromatographic and ionization properties as the endogenous form, but also provides the shift in mass to enable resolution from the endogenous form on the mass spectrometer. In this traditional approach, the stable isotope-labeled analyte functions as an internal standard to compensate for variability due to loss of analyte or matrix effects in sample preparation and LC-MS/MS analysis. Linear least-squares regression analysis is then conducted to generate a calibration curve that plots the standard concentration versus the ratio of analyte peak area to stable isotope-labeled internal standard peak area.

By applying the concept of isotope dilution, an alternative "surrogate standard" approach to remedy the issue of identifying an appropriate assay calibrator matrix for mass spectrometry-based methods has been successfully applied and reported by Cohen and others (Li and Cohen, 2003; Penner et al., 2010). This technique involves the use of a stable-labeled isoform of the biomarker compound as the calibrator.

51.2 METHODS AND APPROACHES

51.2.1 Principles of "Surrogate Analyte" Approach

The "surrogate analyte" approach for mass spectrometric quantitation of endogenous analytes involves a combination of mass spectrometry and a surrogate analyte, the stable isotope-labeled analog as reference standard. It exploits the unique selectivity of mass spectrometry in differentiating isotopes of the same element. Unlike isotope dilution, in which stable isotope-labeled standard is used as an internal standard, in this procedure, the stable labeled material is used as a surrogate analyte. Due to the absence of background signal from the stable labeled compound in biological matrices, calibration standards can be generated in any control biofluid. The intrinsic chemical and physical properties of the labeled and natural analyte are almost identical; hence, the recovery of endogenous analyte from biofluid after sample extraction may be represented by that of the stable isotope-labeled standard. Furthermore, their chromatographic and mass spectrometric ionization behaviors should be essentially identical. Therefore, a calibration curve from the stable labeled compound can be used directly for the quantitation of the endogenous analyte. From this perspective, the interference from existing endogenous analytes in control biofluid is removed, simplifying method development and validation and providing enhanced precision and accuracy over conventional approaches.

The fundamental concept is very simple, which is to generate a standard calibration curve using the peak area response from a surrogate analyte and calculate the sample concentrations of endogenous analyte based on the regression equation from the stable labeled standard. Unlike isotope dilution, the stable isotope-labeled analyte is not added to every standard

TABLE 51.1 Advantages and Disadvantages of Some Analytical Techniques for Preparing Calibrators for Use in Quantitative LC-MS/MS Biomarker Methods

	Calibrator technique	Advantage	Disadvantage
1	Use of aqueous or organic neat solution	Simplicity	May not account for sample extraction efficiency from a biological matrix or matrix effects that occur during analysis and detection
2	Pretreatment of biological matrix to remove endogenous analyte	May be devoid of detectable endogenous analyte levels; may account for extraction efficiency from biological matrix	May be labor and cost intensive and may be only partially successful
3	Use biological matrix; perform endogenous background subtraction during data processing	Matrix is similar to assay experimental samples	Impractical if the background levels of analyte are significantly greater than the expected experimentally measured levels
4	Serial addition of standards to individual aliquots of the sample (Bader, 1980)	Matrix is the same as assay experimental samples	May require an extensive amount of sample, may be extremely time-consuming and the linear range of the analysis can be limited; so only sample concentrations greater than the baseline matrix concentration are quantifiable

and sample, but instead is spiked and quantitated only in the calibration standards. Thus, the stable isotope-labeled analyte functions not as an internal standard but as a surrogate analyte. Method development of the analytical procedure involves the following steps.

1. Obtain authentic reference material of the natural (endogenous) analyte, a surrogate analyte (the stable isotope-labeled standard, which has at least 2-Da mass difference from the endogenous analyte), and, if desired, an internal standard (either a structural analog or a secondary stable isotope-labeled analog). If neither a structural analog nor a secondary stable isotope-labeled standard is available, quantitation can be conducted using an external standard method.
2. Identify the prominent precursor ion peaks in the mass spectra of each analyte and examine and eliminate potential interferences among these analytes.
3. Measure by LC-MS/MS the peak areas of the endogenous and surrogate analyte using a neat solution containing the two components. Determine the response factor (RF) of the surrogate analyte to the endogenous analyte to eliminate any isotope effects or differences in ionization efficiency.
4. Prepare calibration standards of the surrogate analyte over a desired concentration range in control biofluid.
5. Extract the calibration standards, QCs, and samples via an appropriate sample preparation method.
6. Analyze the extracted samples and standards by LC-MS/MS. The peak area responses of the endogenous analyte, surrogate analyte, and internal standard are measured. The calibration curve is constructed based on the peak area ratio of the surrogate analyte to that of an internal standard (IS).
7. The concentrations of the endogenous analyte in biofluid are calculated based on the regression equation of the calibration curve and the peak area ratio of the endogenous analyte to internal standard.

51.2.2 General Applications

For endogenous LC-MS/MS assays, matrix effects can be more complex than for xenobiotic assays. When an analyte-free matrix is not available, the calibration curve has to be generated in an alternative substitute matrix, such as stripped plasma/urine, bovine serum albumin, pure solvents or combinations thereof. Hence, the difference in matrix suppression in standards and incurred samples is inherent and can strongly influence reproducibility and assay accuracy. The use of a stable isotope-labeled analog as an internal standard can be considered to compensate for this problem. However, there have been examples in the literature to the contrary. Jemal et al. (2003) reported that the use of a stable isotope-labeled analog does not always compensate for the difference between the incurred samples and alternative substitute standard matrix, which raises a big hurdle for developing a rugged biomarker method. They investigated the feasibility of using a surrogate analyte for the quantitation of mevalonic acid (MVA) and demonstrated that MVA can be successfully analyzed with adequate accuracy and precision, which was not achievable by isotope dilution. There are several other recent examples (Ahmadkhaniha et al., 2010; MacNeill et al., 2010; Sharma et al., 2011; Shi et al., 2011) showing that the surrogate approach offers improved accuracy, precision, speed, and simplicity for biomarker analysis on a routine basis compared to other currently available methodology. In general, the surrogate approach can be applied with any mass spectrometric instrumentation for the quantification of biomarkers in complex matrices such as food, soil, or tissue that exhibit high background levels of the analyte of interest.

51.2.3 Methodology

Protocol 1: Surrogate analyte approach—setting up an LC-MS/MS system.

Equipment and Reagents

- Endogenous analyte reference standard
- Surrogate analyte reference material: (^{13}C, ^{15}N, or ^{2}H) analog of the endogenous analyte
- Internal standard reference material
- HPLC grade reagents and pH modifiers for the preparation of mobile phase and stock solution
- Suitable reverse phase (RP) analytical columns
- Liquid or gas chromatographic separation system of user's choice (HPLC, UPLC, Micro LC, or Nano LC)
- Mass spectrometer (single or triple quadrupole, hybrid)

Procedure

1. Prepare standard at 5 μg/ml in neat solution for each of the analytes.
2. Acquire precursor and product ions via flow injection or infusion.
3. Develop and optimize the liquid chromatographic and mass spectrometry conditions by injection of a mixture of the analytes in neat solution to achieve desired retention, separation, and peak shape. Figure 51.2 is an example of setting up the LC separation and mass detection for the quantification of endogenous α-ketoisocaproic acid (KIC) in rat plasma using $^{2}H_3$-KIC as surrogate analyte and ketocaproic acid as internal standard (Li and Cohen, 2003).

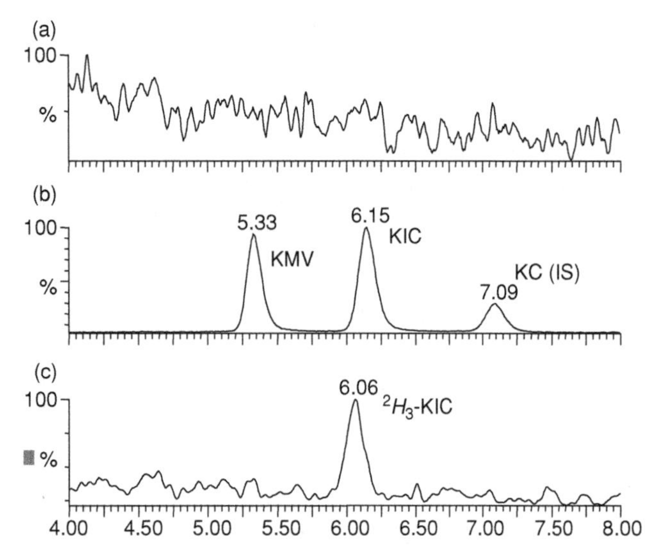

FIGURE 51.2 LC-MS chromatograms of rat plasma extracts. (a) A control rat plasma sample showing $^2H_3$-KIC (*m/z* 132) is absent in the sample matrix. (b) Endogenous KIC, *m/z* 129 and KMV (ketomethylvalerate, an isobaric analog of KIC) as well as spiked internal standard KC (250 ng/ml). They all bare the same *m/z* and were well separated. (c) 10 ng/ml spiked $^2H_3$-KIC. Reproduced with permission from ACS Publications.

4. Access instrument response for the labeled and unlabeled analytes by preparing a mixture of standard containing equal moles of the labeled and unlabeled reference standards at the upper limit of quantification level, and then serially diluting in injection solution over the concentration range of the calibration curve. At least three replicates at each concentration level are prepared and the peak areas of labeled and unlabeled analytes in each sample are measured on the LC-MS/MS system. The RF of the labeled to the unlabeled analyte is determined using the following equation:

$$\text{RF} = \frac{\text{Peak area}_{(\text{surrogate analyte})}}{\text{Peak area}_{(\text{analyte})}}$$

(at equivalent concentration)

5. Conduct statistical analysis of the RF value by one-way analysis of variance using commercially available software (e.g., SigmaStat, JMP) to determine the difference of the intrinsic ionization efficiency between the labeled and unlabeled isotopes. If there is no significant difference among the RF values from each concentration level, the mean RF can be used in data analysis. Table 51.2 provides an example for the determination of RF.

The RF was determined by the analysis of four replicate samples of KIC and $^2H_3$-KIC at each of the calibration standard concentration levels (10, 25, 50, 100, 250, 500, 1000, 2500, 5000, 10,000 ng/ml) prepared as a mixture in reconstitution solvent. The RF value was calculated by comparing the response of $^2H_3$-KIC to the response of KIC at equimolar concentrations. One-way ANOVA was conducted using SigmaStat. Results indicated that there was not a statistically significant difference ($P = 0.134$) among the mean RF values determined at each concentration level. The data suggested the RF values were concentration independent in the tested range and it was close to the theoretical value 1, which was used for quantitation of KIC (Li and Cohen, 2003).

Note The first stage in developing a reliable method is the selection of a suitable surrogate analyte. Considerations should include the chemical stability, purity, interference

TABLE 51.2 Assessing the Response Factor ($\text{Area}_{2\text{H3-KIC}}/\text{Area}_{\text{KIC}}$) at Different Concentrations

	Replicates						
Conc. (ng/ml)	1	2	3	4	Average	SD	%RSD
10	0.984	0.982	0.950	NA	0.972	0.0194	2.00
25	0.855	1.04	0.936	0.984	0.955	0.0795	8.33
50	1.09	0.977	1.04	1.08	1.05	0.0535	5.10
100	0.912	1.08	1.08	1.00	1.02	0.0799	7.85
250	1.02	1.04	0.992	1.03	1.02	0.0213	2.09
500	1.04	0.976	1.03	1.04	1.02	0.0301	2.95
1000	1.03	1.01	1.01	1.03	1.02	0.0108	1.06
2500	1.02	1.04	1.02	1.06	1.03	0.0181	1.75
5000	0.994	1.05	1.00	1.05	1.03	0.0320	3.212
10000	1.05	1.05	1.06	1.00	1.04	0.0290	2.78
Pooled average	1.02						
Pooled SD	0.0482						
Pooled %RSD	4.74						

with the natural analyte, and cost. Among the commonly utilized isotopes are ^{2}H (deuterium), ^{13}C, ^{15}N, ^{18}O, and ^{3}H (tritium) analogs, labeled at stable positions within the molecule to prevent back exchange. These are the most desirable and commonly employed in terms of precision, accuracy, reproducibility, ease of synthesis, and low cost (Pickup and McPherson, 1976). However, the chromatographic retention time will increase as the number of deuteriums increases in the analog, which may negatively affect the accuracy of LC-MS/MS analysis. Another consideration should be that the analyte and the stable isotope-labeled standard yield unique product ions, rather than a fragment resulting from the loss of the portion of the molecule tagged with the stable isotopes in order to prevent crosstalk between the SRM channels.

When a stable labeled compound is used as an internal standard (isotope dilution), the purity is critical. A minor amount of unlabeled material present in the labeled analog will interfere with the endogenous analyte, which will affect the detection limit and calibration linearity. The procedure described in this work does not suffer from this problem. First, the internal standard need not be a stable labeled analog because nonendogenous analogs with different mass transition values can play this role. Second, only the response from labeled analyte is taken into account when the calibration curve is generated. Using this procedure, the analyte purity and the percentage of parent are included when calculating the calibration curve results. Ultimately, the nominal concentrations are reflected by the detected response. Additionally, the contribution to the actual analyte signal from the incompletely labeled standard is generally orders of magnitude less than the interfering background signal present in the biological matrix.

The value of the described RF is defined as the instrument response of the stable isotope-labeled standard to that of the endogenous analyte. Theoretically, the ratio should be close to 1, which means equal amounts of stable labeled and natural compound produce the same MS/MS response intensity. Practically, matrix effects introduced through the mobile phase and the extracts from biofluid samples will usually cause ionization suppression. The RFs determined by comparison of signal (peak areas) from neat solutions of labeled and unlabeled analytes do not take into account ionization suppression (or enhancement) caused by biological matrix effects. The most definitive method for determining matrix effects, as described by King et al. (2000) is unfortunately inappropriate when the analyte itself occurs in the blank matrix biofluid. Evaluation of matrix effects via removal of the endogenous analyte from blank plasma is both laborious and costly. Nevertheless, the matrix effect for the endogenous analytes can be accessed by recovery of QC samples.

Protocol 2: Surrogate analyte approach—preparation of calibration standards, quality control, and study samples.

Equipment and Reagents

- Control biological matrix that resembles incurred sample matrix, collected using the same procedure where possible
- Stock solution of the labeled and unlabeled reference standard and IS
- 96-well formatted plates or tubes to accommodate high-throughput sample processing
- Liquid handling devices such as multichannel pipettes and robotics, (e.g., Tomtech, Hamilton)
- 96-well plate compatible centrifuge, evaporator, and plate sealer

Procedure

1. Prepare "pooled matrix" by pooling a minimum of two or more independent sources of control biological matrix containing the analyte. The matrix used to prepare calibration standard and QCs should be the same as that intended for study samples. Prepare calibration standards of labeled analyte in the "pooled matrix" by one of the following schemes:
   a. Serial dilution of working standard in control biological matrix over a predefined concentration range.
   b. Spiking prepared intermediate working standards (in neat solution) into adequate volumes of control matrix, with the final biological content in the samples at least 95%.
2. Prepare QC samples in the pooled matrix by one of the following schemes:
   a. Quality controls are prepared by spiking a known concentration of labeled analyte in the 'pooled matrix' in the same way as calibration standards. Concentrations of the QC samples should be distributed across the calibration range, such as one concentration at approximately two to five times the lower limit of quantification (LQC), one concentration at approximately the mid point of the calibration curve (MQC), and one concentration toward the high end of the calibration curve (HQC).
   b. Quality controls are prepared by spiking known concentrations of the unlabeled analyte in the "pooled matrix". The endogenous baseline (EB) level of the analyte in the pooled matrix needs to be estimated prior to the sample analysis. The nominal concentration of each QC sample is calculated as EB plus spiked concentration of unlabeled analyte standard. The levels of QCs should cover the anticipated concentrations of the biomarker in the samples.

TABLE 51.3 Quality Controls in Dry Bloodspot Assay Using LC-MS/MS for Quantification of Serotonin in Rat Whole Blood

	QC1 100 ng/ml $^2H_4$-serotonin	QC2 500 ng/ml $^2H_4$-serotonin	QC3 EB	QC4 EB + 250 ng/ml serotonin	QC5 EB + 500 ng/ml serotonin	QC6 EB + 1500 ng/ml serotonin
Nominal conc. (ng/ml)	100	500	2195	2445	2695	3695
Intrarun mean	89.5	448	2140	2540	3040	3810
Intrarun SD	4.55	25.5	135	252	140	85.4
Intrarun %CV	5.1	5.7	6.3	9.9	4.6	2.2
Intrarun %RE	−10.5	−10.4	−4.5	1.6	10.9	1.9
n	3	3	6	3	3	3
Intrarun mean	93.1	501	2220	2850	3000	4430
Intrarun SD	6.9	30.1	190	64.3	50.3	223
Intrarun %CV	7.4	6	8.6	2.3	1.7	5
Intrarun %RE	−6.9	0.2	−0.9	14	9.5	18.4
n	3	3	6	3	3	3
Intrarun mean	95.1	438	2224	2420	2800	4140
Intrarun SD	4.3	21.8	963	75.1	137	150
Intrarun %CV	4.5	5	36.9	3.1	4.9	3.6
Intrarun %RE	−4.9	−12.4	16.5	−3.2	2.2	10.7
n	3	3	5	3	3	3
Interrun mean	92.6	462.3	2195	2604	2944	4127
Interrun SD	5.27	37.0	171	236	149	303
Interrun %CV	5.69	8.00	7.79	9.05	5.07	7.35
Interrun %RE	−7.44	−7.53	0.0	6.52	9.26	11.7
n	9.00	9	15	9	9	9

   c. Quality controls are prepared by spiking known concentrations of both labeled and unlabeled analytes in the "pooled matrix". The spiked QCs consist of either individual labeled analog and unlabeled analytes or a mixture of the two analytes.

3. Add an adequate amount of internal standard to all the samples including standards, QCs, and study samples. Labeled and unlabeled analytes can then be isolated from the biological matrix by commonly used processes such as protein precipitation via methanol or acetonitrile, liquid–liquid extraction, or solid phase extraction.
4. Analyze samples on LC-MS/MS systems.

Note Selection of a suitable format for preparing control samples is critical to demonstrate the assay accuracy and precision. For QC preparation scheme (2a), a necessary assumption has to apply that the matrix effect for the unlabeled analyte and its labeled analogs is comparable. However, this may not always be true practically. Scheme (2b) is considered the standard operating procedure of QC preparation for the quantification of endogenous analytes in biological matrix. Recovery of the endogenous analyte from the pooled matrix calculated against the calibration curve of labeled analog takes into account matrix effects and best reflects the accuracy for sample analysis.

Under most circumstances, the dilution of the study specimens does not allow, or is not necessary, for the surrogate-analyte approach. In the case of relatively high EB of the analyte in biological matrix and lower concentration (down modulation) of the analytes is expected in study samples due to treatments, QC sample with a concentration lower than the EB is necessary in order to be relevant to the concentration in study samples. Scheme (2c) is a practical approach to accommodate such issues. Table 51.3 provides an example of QC preparation by scheme (2c) in a dry bloodspot assay using HPLC tandem mass spectrometry for quantification of serotonin.

The calibration curve was generated with $^2H_4$-serotonin prepared in rat whole blood by spiking an intermediate standard prepared in water via serial dilution and using methyl-serotonin as internal standard. Six levels of QC samples including two levels of $^2H_4$-serotonin (100, 500 ng/ml) and four levels of serotonin (EB, EB + 250, EB + 500, EB + 1500 ng/ml) were prepared in pooled whole blood to cover the anticipated concentration of serotonin in study samples. Rat whole blood was collected in heparin-coated plasmatic tubes and dispensed (20 μl) on a filter paper card (Schleicher & Schuell BioScience). A disc of 1/8 sample spot was collected into a 1.2-ml plastic tube. To each vial, add 300 μl of methanol with 2% formic acid including internal standard. Vortex for 15 min and centrifuge for 10 min at 4000 rpm.

Transfer 50 μl of supernatant to a 1-ml 96-well plate. Dry down and reconstitute with 150 μl water. Analytes were separated on an RP HPLC column (Metasil Thermo Heparin C18 AQ 2.0 $\times$ 100 mm, 5 μm) and detected on an API 4000 triple quadruple mass spectrometer.

Protocol 3: Surrogate analyte approach—processing unknown samples.

After a surrogate calibration curve has been generated, the concentrations of the endogenous analyte in biofluid can be calculated by an Excel™ spreadsheet using the regression parameters and the peak area ratio of endogenous analyte to internal standard. The RF of the stable isotope-labeled standard to that of the endogenous analyte should be incorporated in the calculation if the value does not equal to one.

1. Processing unknown samples using linear regression algorithm.

Equations for the linear least-squares regression of the surrogate analyte are as follows:

$$y = \boldsymbol{a}_0 + \boldsymbol{a}_1\boldsymbol{x} \quad \text{or} \quad x = (y - \boldsymbol{a}_0)/\boldsymbol{a}_1$$

where a_1 is the slope of the regression line, a_0 is the intercept, y is the ratio of the peak area of the surrogate analyte to the peak area of internal standard, and x is the concentration of the surrogate analyte. Therefore, the concentration of the surrogate analyte is calculated by

$$\text{Conc.}_{(\text{surrogate analyte})} = \frac{\left(\dfrac{\text{Peak area}_{(\text{surrogate analyte})}}{\text{Peak area}_{(\text{internal standard})}}\right) - \boldsymbol{a}_0}{\boldsymbol{a}_1}$$

Using the regression parameters (slope and intercept) derived for the surrogate analyte, along with the

	A	B	C	D
1	Regression Parameters Generated by Calibration Curve of Surrogate Analyte			
2	Peak Name: 13C5-15N2 Glutamine			
3	Internal Standard: D5 Glutamine			
4	Q1/Q3 Masses: 154.00/89.00 Da			
5	Fit	Linear ($Y = a_0 + a_1x$)	Weighting	1 / x
6	a0	0.013		
7	a1	0.00666		
8	Correlation coefficient	0.9999		
9	Use Area			
10				
11		Entry of calibration Parameters		
12	a0	a1	RF	
13	0.013	0.00666	1	
14				
15	Processing Unknown Samples (Glutamine)			
16	Sample ID	$\text{Area}_{\text{analyte}}/\text{Area}_{\text{IS}}$	$\text{Conc}_{\text{Analyte}}$ (ng/mL)	Formula for calutating the C
17	QC1 (EB, endogenous basal conc.)	0.08	10.4	(B17*$C$13-$A$13)/$B$13
18	QC1 (EB, endogenous basal conc.)	0.06	6.5	(B18*$C$13-$A$13)/$B$13
19	QC1 (EB, endogenous basal conc.)	0.06	6.9	(B19*$C$13-$A$13)/$B$13
20	QC2 (EB+20 ng/ml Glutamine)	0.19	26	(B20*$C$13-$A$13)/$B$13
21	QC2 (EB+20 ng/ml Glutamine)	0.21	29	(B21*$C$13-$A$13)/$B$13
22	QC2 (EB+20 ng/ml Glutamine)	0.17	23	(B22*$C$13-$A$13)/$B$13
23	QC3 (EB+200 ng/ml Glutamine)	1.40	208	(B23*$C$13-$A$13)/$B$13
24	QC3 (EB+200 ng/ml Glutamine)	1.44	214	(B24*$C$13-$A$13)/$B$13
25	QC3 (EB+200 ng/ml Glutamine)	1.38	206	(B25*$C$13-$A$13)/$B$13
26	QC4 (EB+2000 ng/ml Glutamine)	13.4	2005	(B26*$C$13-$A$13)/$B$13
27	QC4 (EB+2000 ng/ml Glutamine)	14.1	2119	(B27*$C$13-$A$13)/$B$13
28	QC4 (EB+2000 ng/ml Glutamine)	12.8	1927	(B28*$C$13-$A$13)/$B$13
29	Unknown Sample 1	1.29	192	(B29*$C$13-$A$13)/$B$13
30	Unknown Sample 2	1.95	291	(B30*$C$13-$A$13)/$B$13
31	Unknown Sample 3	1.52	226	(B31*$C$13-$A$13)/$B$13
32	Unknown Sample 4	1.39	206	(B32*$C$13-$A$13)/$B$13
33	Unknown Sample 5	1.36	202	(B33*$C$13-$A$13)/$B$13
34	Unknown Sample 6	1.60	238	(B34*$C$13-$A$13)/$B$13
35	Unknown Sample 7	1.45	216	(B35*$C$13-$A$13)/$B$13
36	Unknown Sample 8	1.27	189	(B36*$C$13-$A$13)/$B$13
37	Unknown Sample 9	1.31	194	(B37*$C$13-$A$13)/$B$13
38	Unknown Sample 10	1.27	188	(B38*$C$13-$A$13)/$B$13
39	Unknown Sample 11	1.23	183	(B39*$C$13-$A$13)/$B$13

FIGURE 51.3 Layout of an Excel™ spreadsheet for processing unknown samples by linear regression algorithm. The concentrations of endogenous glutamine in cell lysate samples were interpolated based on the linear regression equation of a calibration curve generated by a stable labeled $^{13}C_5,^{15}N_2$-glutamine.

RF, the concentration of the endogenous analyte in unknown samples can be calculated by

$$\text{Conc.}_{(\text{analyte})} = \frac{\left(\dfrac{\text{Peak area}_{(\text{analyte})}}{\text{Peak area}_{(\text{internal standard})}}\right) \times \text{RF} - \boldsymbol{a_0}}{\boldsymbol{a_1}}$$

To create a results table for the endogenous analyte in unknown samples on an Excel™ spreadsheet, you need to define specific sections of cells for the regression parameters and columns designated for sample ID, peak ratio, and computed concentrations in the unknown samples. Figure 51.3 shows a typical layout of a spreadsheet for processing unknown samples by linear regression algorithm. In this case, the concentrations of endogenous glutamine in cell lysate samples were interpolated based on the linear regression equation of the calibration curve generated by stable labeled $^{13}C_5$, $^{15}N_2$-glutamine. Simply fill the a_0, a_1, and RF values in the entry of the calibration parameter section, and copy the sample ID and the corresponding peak ratio to the designated columns. Then, set the correct formula for computing the concentration of endogenous analyte in the first unknown sample and copy/fill down to populate the concentrations for the rest of the samples

2. Processing unknown samples using quadratic regression algorithm.

   The equations for quadratic fitting of the surrogate analyte are as follows:

$$\boldsymbol{y = a_0 + a_1x + a_2x^2}$$

where a_0, a_1, and a_2 are three constants, y is the ratio of the peak area of the surrogate analyte to the peak area of internal standard, and x is the concentration of the surrogate analyte.

$$y = \frac{\text{Peak area}_{(\text{surrogate analyte})}}{\text{Peak area}_{(\text{internal standard})}}$$

$$x = \text{Conc.}_{(\text{surrogate analyte})}$$

	A	B	C	D	E	F	G	H
1	Regression Parameters Generated by Calibration Curve of Surrogate Analyte							
2	Peak Name: 13C5-15N1 Glutamic Acid							
3	Internal Standard: D5 Glutamine							
4	Q1/Q3 Masses: 154.00/89.00 Da							
5	Fit	Quadratic ($Y = a_0 + a_1x + a_2x^2$)	Weighting	1 / x				
6	a0	0.00832						
7	a1	0.00341						
8	a2	-1.93E-07						
9	Correlation coefficient	0.9983						
10								
11		Entry of calibration Parameters						
12	a_0	a_1	a_2	RF				
13	0.00832	0.00341	-1.93E-07	1				
15	Processing Unknown Samples (Glutamic Acid)							
16	Sample ID	Area$_{analyte}$/Area$_{IS}$	Conc$_{Analyte}$ (ng/mL)	Formula for calculating the Conc. of the Analyte				
17	Unknown Sample 1	0.03	6.36	(SQRT($B$13*$B$13-4*$C$13*($A$13-B17*$D$13))-$B$13)/(2*$C$13)				
18	Unknown Sample 2	0.063	16.05	(SQRT($B$13*$B$13-4*$C$13*($A$13-B18*$D$13))-$B$13)/(2*$C$13)				
19	Unknown Sample 3	0.085	22.52	(SQRT($B$13*$B$13-4*$C$13*($A$13-B19*$D$13))-$B$13)/(2*$C$13)				
20	Unknown Sample 4	0.093	24.87	(SQRT($B$13*$B$13-4*$C$13*($A$13-B20*$D$13))-$B$13)/(2*$C$13)				
21	Unknown Sample 5	0.131	36.05	(SQRT($B$13*$B$13-4*$C$13*($A$13-B21*$D$13))-$B$13)/(2*$C$13)				
22	Unknown Sample 6	0.177	49.61	(SQRT($B$13*$B$13-4*$C$13*($A$13-B22*$D$13))-$B$13)/(2*$C$13)				
23	Unknown Sample 7	0.241	68.50	(SQRT($B$13*$B$13-4*$C$13*($A$13-B23*$D$13))-$B$13)/(2*$C$13)				
24	Unknown Sample 8	0.249	70.86	(SQRT($B$13*$B$13-4*$C$13*($A$13-B24*$D$13))-$B$13)/(2*$C$13)				
25	QC1 (EB, endogenous basal conc.)	0.145	40.17	(SQRT($B$13*$B$13-4*$C$13*($A$13-B25*$D$13))-$B$13)/(2*$C$13)				
26	QC1 (EB, endogenous basal conc.)	0.138	38.11	(SQRT($B$13*$B$13-4*$C$13*($A$13-B26*$D$13))-$B$13)/(2*$C$13)				
27	QC1 (EB, endogenous basal conc.)	0.13	35.76	(SQRT($B$13*$B$13-4*$C$13*($A$13-B27*$D$13))-$B$13)/(2*$C$13)				
28	QC2 (EB+10 ng/ml Glutamic Acid)	0.162	45.18	(SQRT($B$13*$B$13-4*$C$13*($A$13-B28*$D$13))-$B$13)/(2*$C$13)				
29	QC2 (EB+10 ng/ml Glutamic Acid)	0.163	45.48	(SQRT($B$13*$B$13-4*$C$13*($A$13-B29*$D$13))-$B$13)/(2*$C$13)				
30	QC2 (EB+10 ng/ml Glutamic Acid)	0.164	45.77	(SQRT($B$13*$B$13-4*$C$13*($A$13-B30*$D$13))-$B$13)/(2*$C$13)				
31	QC3 (EB+100 ng/ml Glutamic Acid)	0.454	131.68	(SQRT($B$13*$B$13-4*$C$13*($A$13-B31*$D$13))-$B$13)/(2*$C$13)				
32	QC3 (EB+100 ng/ml Glutamic Acid)	0.471	136.74	(SQRT($B$13*$B$13-4*$C$13*($A$13-B32*$D$13))-$B$13)/(2*$C$13)				
33	QC3 (EB+100 ng/ml Glutamic Acid)	0.465	134.95	(SQRT($B$13*$B$13-4*$C$13*($A$13-B33*$D$13))-$B$13)/(2*$C$13)				
34	QC4 (EB+1000 ng/ml Glutamic Acid)	3.002	926.50	(SQRT($B$13*$B$13-4*$C$13*($A$13-B34*$D$13))-$B$13)/(2*$C$13)				
35	QC4 (EB+1000 ng/ml Glutamic Acid)	3.365	1046.33	(SQRT($B$13*$B$13-4*$C$13*($A$13-B35*$D$13))-$B$13)/(2*$C$13)				
36	QC4 (EB+1000 ng/ml Glutamic Acid)	3.039	938.63	(SQRT($B$13*$B$13-4*$C$13*($A$13-B36*$D$13))-$B$13)/(2*$C$13)				

FIGURE 51.4 Layout of an Excel™ spreadsheet for processing unknown samples by quadratic regression algorithm. The concentrations of endogenous glutamic acid in cell lysate samples were interpolated based on the nonlinear regression equation of a calibration curve generated by a stable labeled $^{13}C_5$,$^{15}N_1$-glutamic acid.

Hence, the concentration of the surrogate analyte is calculated by

$$\text{Conc.}_{(\text{surrogate analyte})} = \frac{(-a_1) + \sqrt{a_1^2 - 4(a_0 - y) \times a_2}}{2a_2}$$

Using the regression parameters derived for the surrogate analyte, along with the RF, the concentration of the endogenous analyte in unknown samples can be calculated by

$$\text{Conc.}_{(\text{analyte})} = \frac{(-a_1) + \sqrt{a_1^2 - 4(a_0 - y') \times a_2}}{2a_2}$$

$$y' = \frac{\text{Peak area}_{(\text{analyte})}}{\text{Peak area}_{(\text{internal standard})}} \times \text{RF}$$

Figure 51.4 shows how to use Excel™ spreadsheet to quantitatively process unknown samples by quadratic regression algorithm. In this example, the concentrations of endogenous glutamic acid in cell lysate samples were measured based on the regression parameters generated by a stable labeled$^{13}C_5$, $^{15}N_1$-glutamic acid. The steps involved in creating a result table by nonlinear regression are similar as described in the section for linear regression.

Note A few recent versions of software for liquid chromatographic mass spectrometry data integration provide versatile calibration options that allow the quantification of one MRM transition against another directly from the data integration processes. For example, a ^{13}C-labeled analyte can be quantified against a nonlabeled standard curve on QuanLynx (Waters) software. Similarly, the endogenous species can be quantified against a calibration curve constructed from an isotope-labeled analog right from within MultiQuant (AB Sciex) software. For detailed information about the features and functionalities of these software, refer to the vendor's user manual or software reference guide. In addition, data can be reduced using the regression analysis program in Watson LIMS. Before importing the raw data in Watson, you need to replace the peak area of the endogenous analyte in each standard sample with the peak area of correspondent labeled standard. Data reduction on instrument controlling software or Watson is efficient, fast, and more accurate than manual approaches.

REFERENCES

Ahmadkhaniha R, Shafiee A, Rastkari N, Khoshayand MR, Kobarfard F. Quantification of endogenous steroids in human urine by gas chromatography mass spectrometry using a surrogate analyte approach. J Chromatogr B Analyt Technol Biomed Life Sci 2010;878(11–12):845–852.

Bader M. A systematic approach to standard addition methods in instrumental analysis. J Chem Educ 1980;57(10):703.

De Leenheer AP, Lefevere MF, Lambert WE, Colinet ES. Isotope-dilution mass spectrometry in clinical chemistry. Adv Clin Chem 1985;24:111–161.

USFDA. Guidance for Industry: Bioanalytical Method Validation. 2001.

Giovannini MG, Pieraccini G, Moneti G. Isotope dilution mass spectrometry: Definitive methods and reference materials in clinical chemistry. Ann Ist Super Sanita 1991;27(3):401–410.

Jemal M, Schuster A, Whigan DB. Liquid chromatography/tandem mass spectrometry methods for quantitation of mevalonic acid in human plasma and urine: Method validation, demonstration of using a surrogate analyte, and demonstration of unacceptable matrix effect in spite of use of a stable isotope analog internal standard. Rapid Commun Mass Spectrom 2003;17(15):1723–1734.

King R, Bonfiglio R, Fernandez-Metzler C, Miller-Stein C, Olah T. Mechanistic investigation of ionization suppression in electrospray ionization. J Am Soc Mass Spectrom 2000;11(11):942–950.

Lee JW, Devanarayan V, Barrett YC, et al. Fit-for-purpose method development and validation for successful biomarker measurement. Pharm Res 2006;23(2):312–328.

Li W, Cohen LH. Quantitation of endogenous analytes in biofluid without a true blank matrix. Anal Chem 2003;75(21):5854–5859.

MacNeill R, Sangster T, Moussallie M, Trinh V, Stromeyer R, Daley E. Stable-labeled analogues and reliable quantification of nonprotein biomarkers by LC-MS/MS. Bioanalysis 2010;2(1):69–80.

Penner N, Ramanathan R, Zgoda-Pols J, Chowdhury S. Quantitative determination of hippuric and benzoic acids in urine by LC-MS/MS using surrogate standards. J Pharm Biomed Anal 2010;52(4):534–543.

Pickup J, McPherson K. Theoretical considerations in stable isotope dilution mass spectrometry for organic analysis. Anal Chem 1976;48(13):1885–1890.

Sharma K, Singh RR, Kandaswamy M, et al. LC-MS/MS-ESI method for simultaneous quantitation of three endocannabinoids and its application to rat pharmacokinetic studies. Bioanalysis 2011;3(2):181–196.

Shi J, Liu HF, Wong JM, Huang RN, Jones E, Carlson TJ. Development of a robust and sensitive LC-MS/MS method for the determination of adenine in plasma of different species and its application to in vivo studies. J Pharm Biomed Anal 2011;56(4):778–784.

Skoog DA, Holler FJ, Nieman TA. Principles of Instrumental Analysis. Philadelphia, PA: Saunders College Publishing; 1998. p 15–18.

APPENDIX 1

BODY AND ORGAN WEIGHTS AND PHYSIOLOGICAL PARAMETERS IN LABORATORY ANIMALS AND HUMANS

		Mouse	Rat	Rabbit	Monkey	Dog	Human	Reference
Body weight (kg)		0.02	0.25	2.5	5	10	70	1
Organ weight (g)	Brain	0.36	1.8	14	90	80	1400	1
	Liver	1.75	10.0	77	150	320	1800	1
	Kidneys	0.32	2.0	13	25	50	310	1
	Heart	0.08	1.0	5	18.5	80	330	1
	Spleen	0.1	0.75	1	8	25	180	1
	Adrenals	0.004	0.05	0.5	1.2	1	14	1
	Lung	0.12	1.5	18	33	100	1000	1
Blood volume (ml)		1.7	13.5	165	367	900	5200	1
Blood pH		–	7.38	7.35	–	7.36	7.39	2
Blood hematocrit (%)		45	46	36	41	42	44	2
Urine pH		–	–	–	–	–	4.5–8.0	2

Reference 1. Davies B, Morris T. *Physiological parameters in laboratory animals and humans*. Pharm Res 1993;10(7):1093–1095.
Reference 2. Kwon Y. *Handbook of Essential Pharmacokinetics, Pharmacodynamics, and Drug Metabolism for Industrial Scientist*. Springer; 2002.

Handbook of LC-MS Bioanalysis: Best Practices, Experimental Protocols, and Regulations, First Edition. Edited by Wenkui Li, Jie Zhang, and Francis L.S. Tse.

APPENDIX 2

ANTICOAGULANTS COMMONLY USED IN BLOOD SAMPLE COLLECTION

Anticoagulant	Structure	Mechanism	Counter ion	Concentration
Heparin	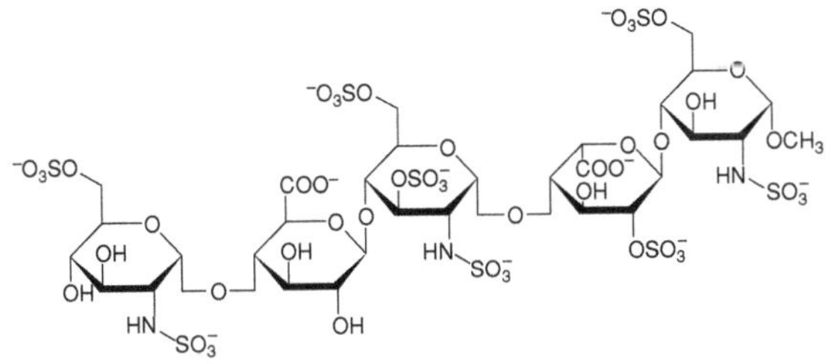	Inhibition of coagulation proteins	Li, Na, K, NH_4	~20 units/ml (~0.2 mg/ml)
Ethylenediamine-tetraacetic acid (EDTA)	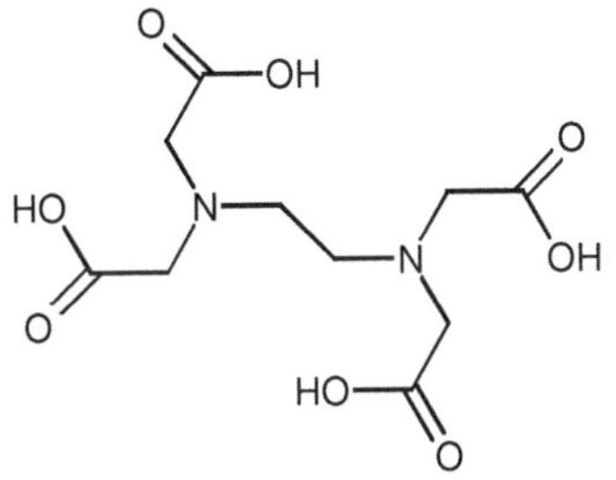	Calcium chelation	Na, K_2 (solid form), K_3 (liquid form)	~1.5 mg/ml
Citrate	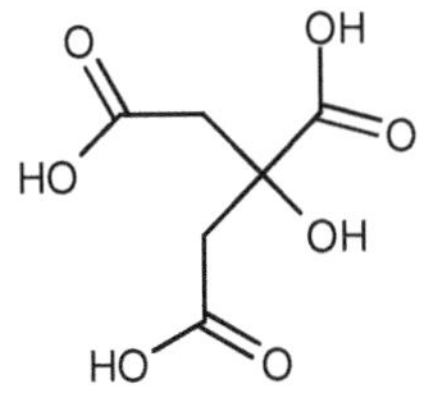	Calcium chelation	Na	3.2%
Oxalate	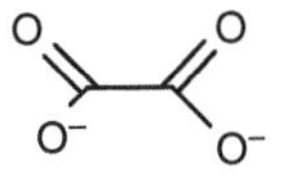	Calcium chelation	Li, Na, K, NH_4	1–2 mg/ml

Handbook of LC-MS Bioanalysis: Best Practices, Experimental Protocols, and Regulations, First Edition. Edited by Wenkui Li, Jie Zhang, and Francis L.S. Tse.

APPENDIX 3

SOLVENTS AND REAGENTS COMMONLY USED IN LC-MS BIOANALYSIS

Solvent/reagent	Molecular Formula	MW (Da)	pK_a
Acetic acid	CH_3CO_2H	60.05	4.8
Acetone	C_3H_6O	58.08	-
Acetonitrile	CH_3CN	41.05	-
Ammonium acetate	$NH_4CH_3CO_2$	77.08	4.8/9.2
Ammonium bicarbonate	NH_4HCO_3	79.06	6.4/9.2/10.3
Ammonium formate	NH_4HCO_2	63.06	3.8/9.2
Ammonium hydroxide	NH_4OH	35.04	9.2
Chloroform	$CHCl_3$	119.38	-
Dichloromethane	CH_2Cl_2	84.93	-
Diethyl ether	$(CH_3CH_2)_2O$	74.12	-
Diethylamine	$C_4H_{11}N$	73.14	11.0
Dimethyl sulfoxide	C_2H_6OS	78.13	-
Ethanol	CH_3CH_2OH	46.08	-
Ethyl acetate	$CH_3CO_2CH_2CH_3$	88.12	-
Formic acid	HCO_2H	46.03	3.8
Heptane	$CH_3(CH_2)_5CH_3$	100.21	-
Hexane	$CH_3(CH_2)_4CH_3$	86.18	-
Isopropanol	$CH_3CH(OH)CH_3$	60.11	-
Methanol	CH_3OH	32.04	-
Methyl *tert*-butyl ether	$(CH_3)_3COCH_3$	88.15	-
n-Propanol	$CH_3CH_2CH_2OH$	60.11	-
Tetrahydrofuran	C_4H_8O	72.12	-
Triethylammonium acetate	$(CH_3CH_2)_3NHOCOCH_3$	161.24	4.8/11.0
Triethylammonium formate	$(CH_3CH_2)_3NHOCOH$	147.22	3.8/11.0
Trifluoroacetic acid	CF_3CO_2H	114.02	0.2
Tris(hydroxymethyl) aminomethane	$(HOCH_2)_3CNH_2$	121.14	8.3
Water	H_2O	18.02	-

Handbook of LC-MS Bioanalysis: Best Practices, Experimental Protocols, and Regulations, First Edition. Edited by Wenkui Li, Jie Zhang, and Francis L.S. Tse.

APPENDIX 4

GLOSSARY OF TERMS USED IN LC-MS BIOANALYSIS

Absolute Bioavailability The fraction of a dose reaching the systemic circulation intact.

Anticoagulant A substance added to the blood during collection to prevents blood coagulation (clotting).

Accelerator Mass Spectrometry (AMS) Highly sensitive isotope ratio analytical method that separates and measures ^{12}C, ^{13}C and ^{14}C. The utilization of AMS allows for very low levels of ^{14}C to be administered to human volunteers in a mass-balance study.

Accuracy The closeness of the determined value to the value which is accepted either as a conventional true value or an accepted reference value. Accuracy is defined as (determined value/true value) x100%.

Additive A chemical, typically in a liquid form, added to the HPLC mobile phase to improve the LC-MS analysis.

Adsorbent A material having capacity or tendency to adsorb another substance.

Anion Exchange Chromatography A chromatographic process that is used to separate anion analyte(s) from other neutral and cation components by using an ionized positively charged stationary phase.

Analyte The chemical entity to be analyzed, which can be intact drug, biomolecule, metabolite and/or degradation product in a biological matrix.

Analytical Method A technique and procedure for the qualitative and quantitative analysis of a drug and its related components in a given sample.

Analytical Run A complete set of analytical and study samples with appropriate number of calibration standards and QC samples.

Analytical Procedure The detailed experiment steps of the analysis performed.

Antibody–Drug Conjugates (ADCs) Monoclonal antibodies (mAbs) attached to biologically active drugs by chemical linkers with labile bonds.

Antisense Oligonucleotide A short string of nucleotides that can bond to mRNA and block the process of gene expression.

Area Under the Curve (AUC) The area under the plot of plasma concentration of drug against time after drug administration.

Archive Collection and storage of records (raw data, protocols, reports, etc.) for a defined period of time.

Atmospheric Pressure Chemical Ionization (APCI) A supplementary technique to electrospray ionization for analyzing smaller, thermally stable polar and nonpolar compounds. In APCI, the analyte solution is introduced into a pneumatic nebulizer and desolvated in a heated field before interacting with the corona discharge to produce ions of analytes.

Atmospheric Pressure Photo Ionization (APPI) An ionization technique in which compounds are ionized by UV light.

Automation The use of robotic devices to complete analytical tasks.

Autosampler Stability Stability of analyte of interest in a processed sample that is stored in an autosampler set at a predefined temperature.

Auxiliary Gas Nitrogen used in a mass spectrometer ion source to aid in solvent removal.

Baseline The chromatographic signal of a LC-MS system in the absence of an analyte.

Batch A bioanalytical run consisting of groups of calibration standards, quality control samples, blanks, zero samples and study samples.

Bias The difference between a measured concentration value and a nominal value or a reference value.

Bioanalysis The process of analyzing drugs, drug metabolites, chemicals and/or endogenous biomarkers present

Handbook of LC-MS Bioanalysis: Best Practices, Experimental Protocols, and Regulations, First Edition. Edited by Wenkui Li, Jie Zhang, and Francis L.S. Tse.

in biological matrices to provide concentration values of analytes of interest.

Bioanalytical Assay A method for quantitative determination of drug and/or its metabolite(s) or endogenous molecule(s) of interest in biological matrices.

Bioanalytical Method Transfer Transferring a validated bioanalytical method from one laboratory to another.

Bioavailability The fraction of an administered dose of drug that reaches the systemic circulation unchanged.

Bioequivalence Equivalence in the rate and extent of absorption of the active moiety from two pharmaceutical formulations having identical active ingredients.

Biofluids See **Biological Fluids**.

Biological Fluids Liquids originating from inside the bodies of humans or animals. They include fluids that are excreted or secreted from the body. Common biological fluids for bioanalysis are blood, plasma, serum, cerebrospinal fluid (CSF), bile, saliva, semen, tears, feces and urine.

Biological Matrix Fluid or tissue obtained from a biological system. Examples are blood, serum, plasma, urine, feces, saliva, sputum, and various tissues.

Biological Specimens See **Biological Matrix**.

Biomarker A characteristic that is objectively measured and evaluated as an indicator of normal biologic processes, pathogenic processes or pharmacologic responses to a therapeutic intervention.

Biopharmaceuticals Medicines that are derived from living organisms. They include proteins (including antibodies) and nucleic acids (DNA, RNA or antisense oligonucleotides) used for therapeutic or *in vivo* diagnostic purposes.

Biotransformation The chemical alteration of a drug substance in a biological system. The alteration may inactivate the drug or result in the production of an active molecule of an inactive parent compound.

Blank Matrix A biological matrix that is free of analytes to be analyzed.

Blood to Plasma Ratio Concentration of the test compound in whole blood compared with that in plasma.

Bonded Phase A stationary phase which is covalently bonded to the support particles or the inside wall of a tube.

Calibration An operation that, under specified conditions, establishes the relationship between the values determined by a measuring instrument and the corresponding known values.

Calibration Curve A set of several calibration standards at various known concentrations analyzed in an analytical run for determining the concentration of an analyte in an unknown sample.

Calibration Range The interval between the upper and lower concentration of analyte in the sample for which it has been demonstrated that the analytical procedure meets the predefined acceptance criteria for assay precision, accuracy and linearity.

Calibration Standard A biological sample prepared by adding a known amount of analyte to the biological matrix. Calibration standards are used to construct calibration curves from which the concentrations of analytes in QCs and in study samples are determined.

Capillary LC Liquid chromatography performed by using a capillary column. The column consists of a tube and packing contained by frits in each end of the tube.

Cation Exchange Chromatography A chromatographic process that is used to separate cations by using an ionized negatively charged stationary phase.

Carry-over The detection of an analyte in blank sample after the analysis of samples with a high concentration of the analyte.

Capillary Electrophoresis (CE) An analytical technique that separates charged analytes based on their electrophoretic mobility with the use of an applied voltage.

Certificate of Analysis (CoA) The reporting of the analytical characterization results released for intended use in a GMP or GLP laboratory.

Chiral Column Chromatography An analytical technique that separates chiral compounds using a HPLC column having a single enantiomer of a chiral compound built in the stationary phase. The separation of two enantiomers is achieved based on their different affinities to the stationary phase.

Chiral Compound A compound that contains an asymmetric center or chiral center.

Chiral Separation See **Chiral Column Chromatography**.

Chiral Stationary Phases A material that discriminates between stereoisomers allowing their chromatographic separation.

Chromatography A physical method of separation in which the components in a sample are separated based on differential partitioning between the stationary and mobile phases.

Clinical Samples Biological specimen taken from human subjects in clinical studies.

Clinical Trial A research study to evaluate a new drug or treatment in humans subjects.

Collision Activated Dissociation (CAD) A mechanism by which molecular ions are fragmented in the gas phase in mass spectrometry. The molecular ions are usually accelerated by an electrical potential to high kinetic energy and then allowed to collide with neutral molecules (often helium, nitrogen or argon). In the collision some of the kinetic energy is converted into internal energy which results in bond breakage and the fragmentation of the molecular ion into smaller fragments, which then can be analyzed by a mass spectrometer.

Collision Cell A chamber located between the quadrupole mass filters referred to as Q1 and Q3 in a tandem mass spectrometer. The collision cell may be composed of a quadrupole or superposed lens and is filled with an inert gas such as nitrogen, argon or helium. Adequate voltage is applied to make the ions collide with the molecules of gas.

Collision Induced Dissociation (CID) See **Collision Activated Dissociation**.

Column Switching An arrangement with switching valve(s) and tubing that enables the separation of a complex mixture on multiple columns, whereby fractions eluting from the first column are directed to a second column for further separation, preferably with a different separation mechanism.

Column Dead Volume The sum of the interstitial volume (intra-particle volume and inter-particle volume) plus extra volume contributed by injector, connecting tubing, and end-fittings.

Conjugate A product derived from combining two or more substances together.

Contamination The presence of unwanted analyte(s) in a sample.

Contract Research Organization (CRO) A service organization that provides support to the pharmaceutical and biotechnology industries in the form of outsourced pharmaceutical research services.

Control Matrix See **Blank Matrix**.

Counter-ion An ion having a charge opposite to that of the substance with which it is associated.

Covalent Binding Two molecules form one intact molecule through the formation of a shared bond.

Cross-validation Assessment of method performance when the same method is used in more than one laboratory or by more than one analyst, in order to ensure that the analytical results from the different sources can be directly compared.

Curtain Gas A flow of countercurrent drying gas emanating from between two closely spaced plates, each with a central orifice, along the axis of a spray ionization source.

Cytochrome P450 A superfamily of heme-containing enzymes that are widely distributed throughout mammalian species, which can metabolize a variety of endogenous and exogenous compounds via multiple reaction pathways.

Daughter Ion See **Tandem Mass Spectrometry**.

Derivatization An analytical technique in which a compound is transformed into a different chemical form by adding a specific functional group in order to improve separation and detection.

Dried Blood Spot (DBS) A technique for the collection of blood on cellulose-based paper for bioanalysis.

Dried Matrix Spot (DMS) A DBS technique extended to nonblood matrices such as plasma, serum, tears, saliva, urine and synovial fluid.

Drug Degradation The change of a drug into a byproduct(s) which is typically less complex.

Drug–Drug Interaction (DDI) The effect of one drug on the absorption, transport, metabolism and excretion of a coadministrated drug, and vice versa.

Drug Metabolism See **Biotransformation**.

Drug Metabolite(s) Altered form(s) of an administered drug formed generally by biotransformation.

Electrospray Ionization (ESI) A soft ionization technique used in mass spectrometry by which a sample solution is sprayed across a high potential difference from a needle into an orifice to produce ionized species in the gas phase.

Elution The passing of mobile phase through a chromatographic bed/stationary phase or the passing of elution solvent through a solid phase extraction (SPE) bed to remove compounds.

Endogenous Compound A substance that originates within an organism, tissue or cell.

Endogenous Matrix See **Blank Matrix**.

Enzymes Large biological molecules (mainly proteins) with specific three-dimensional structures. Enzymes are highly selective catalysts that metabolize drugs *in vivo* and might also cause instability of analyte(s) of interest in the intended biological matrix under *ex vivo* conditions.

Esterase An enzyme that converts an ester into an acid and an alcohol.

Fit-for-Purpose Method Validation A process of defining study intent and establishing with experimental data whether the assay performance characteristics are reliable for the intended application.

Flow Rate The volume of the mobile phase that passes through the column per unit time.

FDA Form 483 (Form 483, or 483) A form used by the US FDA investigators to document and communicate their findings and concerns during inspections.

Full Validation Establishment of all validation parameters to apply to sample analysis for the bioanalytical method for each analyte according to SOP.

Good Laboratory Practice (GLP) A quality system of management controls for research laboratories and organizations to ensure the uniformity, consistency, reliability, reproducibility, quality, and integrity of nonclinical safety tests.

Gradient Elution A chromatographic technique by which a mobile phase gradient is used to modulate the retention times of the analyte(s) of interest.

Guard Column A small column that is placed between the injector and the analytical column to protect the analytical column.

High Performance Liquid Chromatography (HPLC) A form of liquid chromatography to separate compounds that are dissolved in solution. HPLC instruments consist of reservoirs of mobile phases, pumps, an injector, a separation column, a column oven, and a detector.

Hematocrit The proportion (typically as the volume percentage) of the red blood cells in blood.

Hemolysis Lysis of red blood cells resulting in rupture of erythrocytes (red blood cells) and release of hemoglobin into the plasma.

Heparin A highly sulfated glycosaminoglycan used as an anticoagulant for blood sample collection.

High-Field Asymmetric Waveform Ion Mobility Spectrometry (FAIMS) A mass spectrometry technique in which ions at atmospheric pressure are separated by the application of a high-voltage asymmetric waveform at radio frequency combined with a static waveform applied between two electrodes.

In vitro A procedure performed in a controlled environment outside of the body.

Incurred Samples Study samples from dosed subjects or animals.

Incurred Sample Reanalysis (ISR) Reanalysis of selected study samples taken in a clinical or preclinical study to determine assay reproducibility by comparing the results of the repeat and initial analyses.

Interference A compound (endogenous component, co-medication or contaminant) that produces a signal at the same mass-to-charge ratio (*m/z*)and retention time as the analyte of interest.

Internal Standard A chemical substance added in a constant amount to the study samples, calibration standards and quality control samples in a quantitative analysis in order to determine the concentration of an analyte. This substance is structurally similar but not identical to the analyte of interest.

Ion Mobility Spectrometry A technology used to separate ionized molecules in the gaseous phase via differences in their mobility through a carrier gas.

Ion Pairing The addition of a reagent to mobile phases to shield charged groups and increase hydrophobicity of the analyte(s) of interest for an improved reversed-phase chromatographic performance.

Ion Source An electro-magnetic device in a mass spectrometer that ionizes the analyte molecules and separate the resulting ions from the HPLC eluents. Two common types of ion sources in LC-MS bioanalysis are ESI and APCI.

Ion Suppression/Enhancement Reduction/enhancement of the response of analyte and/or internal standard in mass spectrometer due to coelution of endogenous substances from the matrix.

Ion Trap Mass Spectrometer A type of mass spectrometer with a device that functions both as an ion store in which gaseous ions can be confined for a period of time and as a mass spectrometer of considerable mass range and variable mass resolution.

Isobaric Compounds Structural isomers that share the same elemental formula or have the same nominal molecular mass.

Isocratic Elution Use of a constant-composition of mobile phase in liquid chromatography.

Isotope Dilution An analytical technique in which a known quantity of a stable isotope-labeled analogue of the analyte is added to the sample with unknown analyte concentration. The ratio of the two signal intensities is used to determine their relative proportion and, consequently, from this signal intensity ratio the unknown analyte concentration can be determined.

Laboratory Information Management System (LIMS) A software based laboratory and information management system that offers workflow and data tracking support.

Least-squares A statistical method of determining a regression equation which best describes the relationship between two variables.

Linear Regression A method of modeling the relationship between two variables. In bioanalysis, the two variables are concentration and instrument response (typically peak area ratio of analyte and internal standard). In a simple linear regression model, the relationship between variables is established by a straight line, mathematically expressed as $y = ax + b$, where y is instrument response, x is concentration, a is the slope of the regression equation, and b is the y-intercept of the regression equation.

Liquid Chromatography (LC) See **HPLC**.

Liquid Chromatography–Tandem Mass Spectrometry (LC-MS, LC-MS/MS) An analytical technique that combines the physical separation capabilities of liquid chromatography (or HPLC) with the mass analysis capabilities of mass spectrometry or tandem mass spectrometry (MS or MS/MS). LC-MS or LC-MS/MS is a powerful technique with very high sensitivity and selectivity for qualitative and quantitative applications.

Liquid–Liquid Extraction (LLE) A sample extraction technique by which two immiscible solvents are used (usually water and an organic solvent) to partition a target analyte(s) from one liquid phase into another liquid phase.

Lower Limit of Quantification (LLOQ) The lowest amount of analyte in a sample which can be quantitatively determined with pre-defined precision and accuracy.

Mass Spectrometry (MS) An analytical technique for the qualitative and quantitative detection of analytes based on measuring the mass to charge ratio (*m/z*) of ionized molecules.

Mass-to-Charge Ratio (*m/z*) A physical quantity of mass spectrometry formed by dividing the mass number of an ion by its charge number.

Matrix Effects Suppression or enhancement of ionization of analytes by the presence of unintended analytes (for analysis) or other interfering substances in the sample.

Matrix Factor The quantitative measurement of matrix effects during bioanalytical method validation, calculated as the ratio of the peak response of the analyte in the presence of matrix to the corresponding peak response in the absence of matrix.

Metabolism See **Biotransformation**.

Metabolite A compound produced by a metabolic reaction or biotransformation.

Metabolite Profiling Detection, characterization and quantitation of metabolites present in biological samples.

Method Validation The process of performing tests designed to verify that a bioanalytical method is suitable for its intended purpose and is capable of providing useful and valid analytical data.

Mobile Phase The aqueous and organic solvents used in chromatography.

Monoclonal Antibodies (MAbs) Antibodies produced by a single clone of b lymphocytes and that are therefore identical in structure and antigen specificity.

Monolithic Column A HPLC column that is filled with a skeleton of porous material rather than particulate beads to bind with analytes. It is comprised of macro- and mesoporous channels that generate low column back pressures, even at relatively high flow rates.

Nominal Concentration Theoretical or expected concentration.

Multiple Reaction Monitoring (MRM) A quantitative analytical method used in tandem mass spectrometry in which an ion of a particular mass is selected in the first stage (Q1) of a tandem mass spectrometer and a product ion of a collision associated dissociation of the precursor ion (Q2) is selected in the third stage (Q3) for detection. The precursor/product pair (also referred to as a transition) is used in the quantitative determination of an analyte in biological matrix.

New Drug Application (NDA) A regulatory submission process in the USA through which an applicant (or sponsor) formally propose that the Food and Drug Administration (FDA) approves a new pharmaceutical for sale and marketing in the USA.

Nonspecific Binding (NSB) Interaction between molecules of interest in the intended matrix and the surface of the container for the matix without specificity of the connection sites.

Normal Phase Chromatography A type of HPLC separation technique using a polar stationary phase and a nonpolar and nonaqueous mobile phase.

Oligonucleotide A short nucleic acid polymer, typically with twenty or fewer bases.

Orbitrap A Fourier transform-based mass analyzer consisting of an outer barrel-like electrode and a coaxial inner spindle-like electrode that form an electrostatic field with quadro-logarithmic potential distribution.

Parent Ion See **Tandem Mass Spectrometry**.

Partial Validation Series of analytical experiments where only relevant parts of the validation are repeated after modifications are made to the validated bioanalytical method.

Perforated Dried Blood Spot A microsampling technique where a fixed volume of whole blood is deposited on a perforated filter paper.

Phararmacokinetics (PK) The study of the rate and extent of absorption, distribution, metabolism and excretion of an administered drug in the body.

Pharmacodynamics (PD) The study of the biochemical and physiological effects of drugs on the body.

Phospholipids A class of lipids that are a major component of all cell membranes as they can form lipid bilayers. Most phospholipids contain a diglyceride, a phosphate group, and a simple organic molecule such as choline; one exception to this rule is sphingomyelin, which is derived from sphingosine instead of glycerol.

pKa The negative base-10 logarithm of the acid dissociation constant of a solution.

Plasma See **Biological Matrix**.

Precision The closeness of agreement (degree of scatter) between a series of measurements obtained under the prescribed conditions. Precision is defined as the ratio of standard deviation/mean (%).

Prodrug A drug molecule that is intended to undergo *in vivo* chemical and/or enzymatic transformation to produce the active moiety.

Protein Precipitation Addition of miscible organic solvents to a sample resulting in the precipitation of proteins in the matrix.

Qualification The action of proving that an equipment or process works correctly and consistently and produces the expected results. Qualification for analytical equipment includes installation qualification (IQ), operation qualification (OQ), and performance qualification (PQ).

Quality Control Operational techniques and activities that are used to fulfill the requirements for quality in compliance with specification.

Quality Control Sample A spiked sample used to monitor the performance of a bioanalytical method and to assess the integrity and validity of the results of the unknown samples analysed in an individual batch.

Quantification Determination of amounts or concentrations of a compound present in a biological matrix.

Quantification Range see **Calibration Range**.

Reactive Metabolite Molecule that results from the metabolism of a drug into an electrophilic (reactive) species, which could then be attacked by a nucleophile in the cell, such as glutathione or a nucleophilic protein amino acid.

Recovery The extraction efficiency of an analytical process, reported as a percentage of the known amount of an analyte carried through the sample extraction and processing steps of the method.

Reference Material A compound characterized for one or more chemical properties, accompanied by a certificate of analysis (CoA) that provides the value of the specified property.

Regulated Bioanalysis Bioanalytical practice in compliance with GLP.

Reproducibility The ability of an experiment result to be accurately reproduced or replicated independently.

Reversed Phase Chromatography A type of HPLC separation technique using a nonpolar stationary phase and a polar aqueous mobile phase.

Selected Reaction Monitoring (MRM) See Multiple Reaction Monitoring.

Selectivity The ability of the bioanalytical method to measure and differentiate the analyte(s) of interest and internal standard in the presence of other components in the sample.

Serum See **Biological Matrix**.

Signal to Noise Ratio (S/N) A measure used in analytical chemistry that compares the level of a desired analytical signal (S) to the level of background instrument noise (N).

Solid-Phase Extraction (SPE) A separation process using packed solid sorbent (silica or polymer based) and appropriate elution solvent(s), by which compound(s) of interest in a liquid mixture are separated from other compounds in the same mixture according to their physical and chemical properties.

Specificity The ability of a bioanalytical method to measure the analyte accurately in the presence of other compounds, either exogenous or endogenous, in the matrix.

Stable Isotope-Labeled Internal Standard (SILIS) The isotope analogy of an analyte in which a few atoms of the analyte are exchanged with heavy isotopes ^{2}H, ^{15}N, or ^{13}C. It is commonly used as an internal standard in LC-MS bioanalysis.

Standard Curve See **Calibration Curve**.

Standard Deviation Measurement of variability in statistics, showing how much the individual value varied from the average.

Standard Operating Procedure (SOP) A written document describing the established steps and processes for completing a specific task.

Statistical Outlier An observation or data point that appears to deviate markedly from others in the comparison group.

Stock Solution A solution prepared from a powder form of analyte which is then used in preparation of working solutions of the analyte as part of the analytical procedures.

Supported Liquid Extraction (SLE) An extraction technique, which is different from traditional liquid–liquid extraction (LLE) and uses a cartridge packed with a modified form of diatomaceous earth and water-immiscible organic solvent.

Tandem Mass Spectrometry A mass spectrometry technique using two or more mass analyzers. With two in tandem, the precursor ions (or parent ions) are mass-selected by a first mass analyzer and focused into a collision region where they are fragmented into product ions (daughter ions) which are then characterized by a second mass analyzer.

Tissue See **Biological Matrix**.

Total Ion Chromatogram A plot of the total ion signal in each of a series of mass spectra that are recorded as a function of chromatographic retention time.

Turbulent Flow A LC separation technique using narrow-diameter columns packed with large particles to create high linear velocity (turbulence) to separate larger compounds from smaller ones which diffuse into the particle pores before eluting.

Two Dimensional (2D) Chromatography A combined chromatographic mode which is based on different separation mechanisms to achieve better separation for analytes in complex samples.

Ultra Performance Liquid Chromatography (UPLC) A variant of HPLC using columns with particle size smaller than 2 μm, providing significantly better and faster separation than the traditional columns.

Upper Limit of Quantification (ULOQ) The highest amount of analyte in a sample which can be quantitatively determined with pre-defined precision and accuracy.

Urine See Biological Matrix.

Validation The process used to establish documented evidence that provides assurance that a specific process will consistently produce a product meeting its predetermined specifications and quality attributes.

Validation Protocol A written plan stating how validation will be conducted, including test parameters, product characteristics, production equipment, and decision points on what constitutes acceptable test results.

Van Deemter Equation An equation used to explain band broadening in chromatography. The equation, $h = A + B/v + Cv$, represents the height equivalent of a theoretical plate and has three terms. The A-term is used to describe eddy diffusion, which allows for the different paths a solute may follow when passing over particles of different sizes. The B-term is for the contribution caused by molecular diffusion or longitudinal diffusion of the solute while passing through the column. The C term is the contribution of mass transfer and allows for the finite rate of transfer of the solute between the stationary phase and mobile phase.

Whole Blood Blood drawn from the body from which no component, such as plasma or platelets, has been removed.

INDEX

Handbook of LC-MS Bioanalysis: Best Practices, Experimental Protocols, and Regulations, First Edition. Edited by Wenkui Li, Jie Zhang, and Francis L.S. Tse.